Environmental Division

Core Programming Topic at the 2011 AIChE Annual Meeting

Minneapolis, Minnesota, USA
16-21 October 2011

Volume 1 of 2

ISBN: 978-1-61839-730-0

Printed from e-media with permission by:

Curran Associates, Inc.
57 Morehouse Lane
Red Hook, NY 12571

Some format issues inherent in the e-media version may also appear in this print version.

Printed by Curran Associates, Inc. (2012)

AIChE
3 Park Avenue
New York, NY 10016-5991

Phone: (203) 702-7660
Fax: (203) 775-5177

www.aiche.org

Additional copies of this publication are available from:

Curran Associates, Inc.
57 Morehouse Lane
Red Hook, NY 12571 USA
Phone: 845-758-0400
Fax: 845-758-2634
Email: curran@proceedings.com
Web: www.proceedings.com

TABLE OF CONTENTS

Volume 1

Volume 2

(8a) Dispersion of Contaminants In Urban Areas: Coupling Dispersion and CFD Models

Jos-Francisco Perales, Eva Gallego and Xavier Roca, Chemical Engineering, Polytechnic University of Catalonia, Barcelona, Spain

Title: DISPERSIONOF CONTAMINANTS IN URBAN AREAS: COUPLING DISPERSION AND CFD MODELS.

Authors: [1*]Perales, J.F.; [1]Gallego,E.; [1]Roca F.X ; [2]Guardino X.

[1]Laboratoridel Centre de Medi Ambient.Universitat Polit°Cnica deCatalunya

(LCMA-UPC). Avda.Diagonal, 647. E 08028 Barcelona. Phone: 34934016683,Fax: 34934017150, e-mail: Lcma.info@upc.edu

[2]CentroNacional de Condiciones de Trabajo. INSHT. C/Dulcet,2-10. E 08034

Barcelona. Phone: 34932800102, Fax: 34932803642,e-mail: cnctinsht@mtas.es

* Author towhom correspondence should be addressed

e-mails: jose.francisco.perales@upc.edu

eva.gallego@upc.edu

fco.javier.roca@upc.edu

cnctinsht@mtin.es

Air dispersion models are usedto obtain the inmission concentration values of atmosphericcontaminants emitted by well defined focuses. There are normative models asAERMOD and CALPUFF in USA for example that must be applied before new industrialactivities begin to operate in order to asses that the maximum legislated inmission levels are not surpassed. These models give theconcentration values of contaminants for hexahedral nodes over the territory. Theygive only one averaged concentration value for each of the nodes. These nodes over the terrain have a size ofhundreds or thousands of meters, so the information they give is scarce whenreferred to an urban zone. For these dispersion models the characterization ofthe terrain is made through the 'land use' file data which is then introducedin the calculation process with values of terrain roughness, that are used topredict the wind speed profile over this terrain and these wind profile is usedlater to calculate the transport of atmospheric contaminants and then theirsconcentrations.

In order to obtain moreaccurate results when urban zones are studied a new approach is used in thiswork: the coupling of tridimensional wind fields calculated for the areas ofinterests using the meteorological part of atmospheric dispersion models andthe use of Computational Fluid Dynamic (CFD) software to take into account theelements of the urban configuration: streets, buildings, parks, etc. It usesthe meteorological results from The Atmospheric Pollution Model (TAPM, SCIRO, Australia) which is an Eulerianmodel that solves approximations of the fundamental fluid dynamics and scalartransport equations to predict meteorology and pollutant concentrations from mesoscalar meteorological data. Nesting is used to achievethe detailed wind field for the area of interest.

Examples of wind fieldscalculated with TAPM at three different territorial scales are shown in figure 1.As can be observed, the orography of the zone istaken into account and influences the wind trajectories calculated.

Thesecalculated wind field is then used as initial and boundary conditionsfor the CFD software, COMSOL Multiphysics, whereurban elements could be very well defined, as seen in figure 2, that shows theeffect of the emission of a chimney over a regular distribution of buildings.

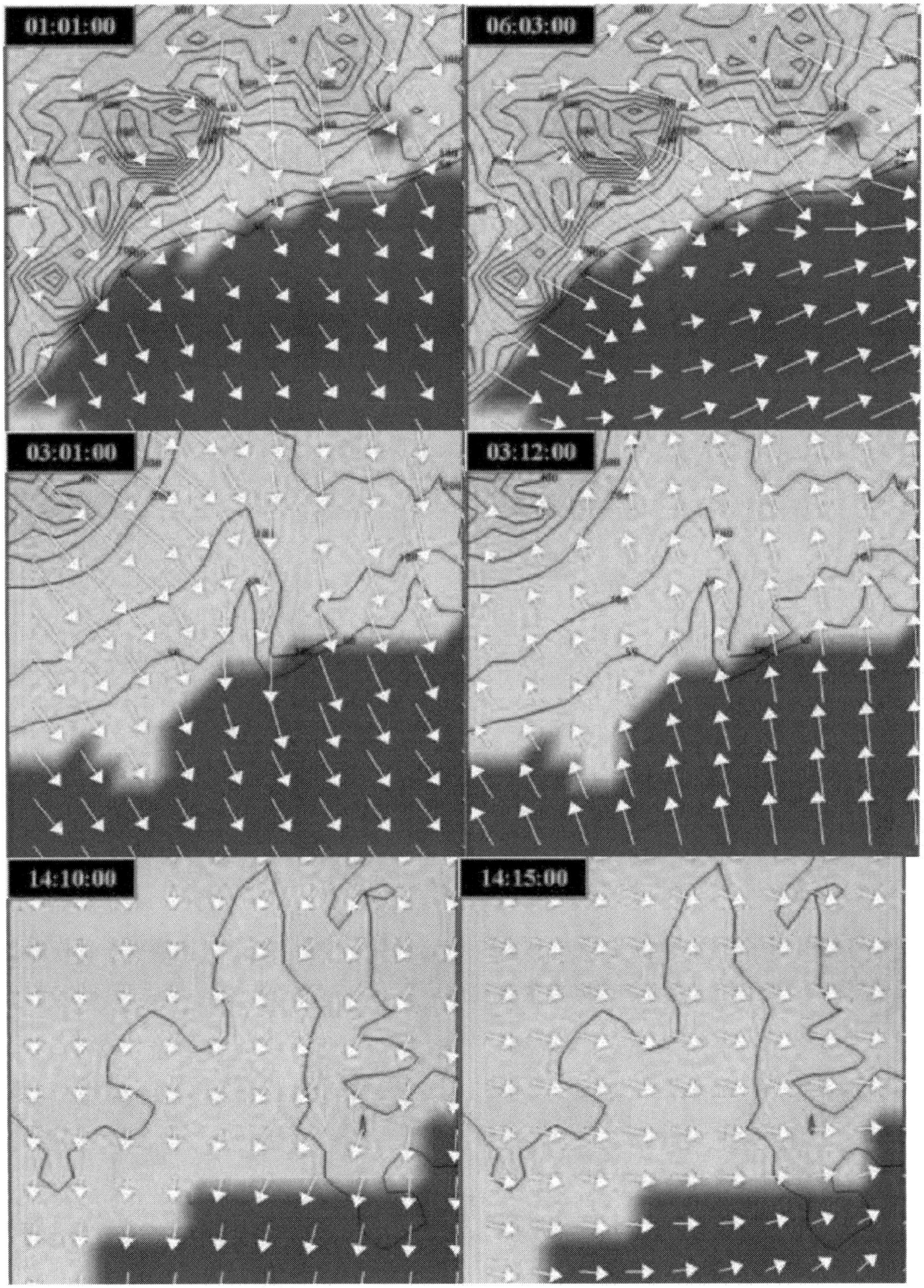

Figure1. Wind field examples obtained with TAPM model with thethree inner grids of the four nested grid used. Size of terrain in top twopictures: 100x100 Km (nodes 5x5 Km). Middle pictures: 30x30 Km (nodes400x400m). Bottom pictures: 8x8 Km (nodes 400x400 m).

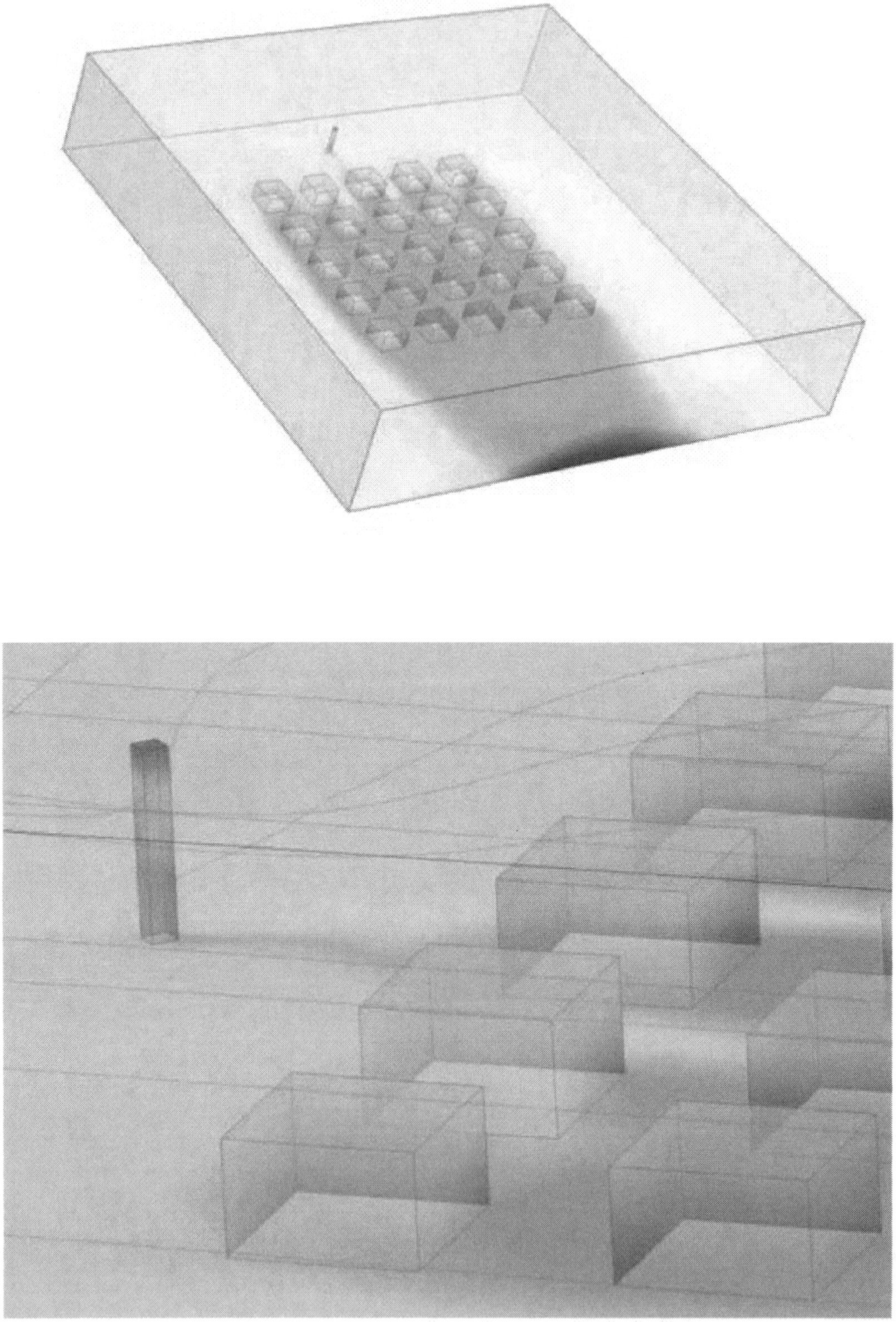

Figure2. CFD results for the study of the impact of a chimneyemission on an array of buildings in a hexahedral node whose size is similar tothe nodes of the bottom of figure 1.

Atmospheric Dispersion Using Commercially Available CFD Software

Eric Peterson, Morgan Reed and Chaitanya Sheth
Scandpower, Inc., a member of the Lloyd's Register Group, Houston, Texas, USA

Abstract

Atmospheric dispersion is often performed on effluents from point sources within the Oil, Gas, Petrochemical and Chemical industries. Results may be required to: comply with mandated regulations; demonstrate due diligence; assess the boundaries of effluent plumes. Two-dimensional modeling software uses the economy of reduced dimensionality to efficiently solve the requisite equations of mathematics and physics. Inclusion of a third dimension in modeling of interconnected relationships, offer more realistic results with confidence that generally allow smaller contingencies in process operations. Results from two commercially available software packages are presented and evaluated for identical scenarios, to compare and contrast their efficacy in modeling atmospheric dispersion.

Introduction

Computational Fluid Dynamics (CFD) is often used to obtain numerical solutions for dispersion studies on offshore platforms, drill ships and floating production, storage and offloading (FPSO) units. Most of the commercially available software is based on finite volume method which allows studies to be performed comparing such software packages. This paper compares two commercially available software programs, Kameleon FireEx (KFX) and Flame Acceleration Simulator (FLACS), in their ability to analyze identical exhaust dispersion scenarios on an FPSO unit. Specific comparisons to be made are simulation run time and difference between results obtained.

Modeling Approach

The modeling approach implemented in this study is a basic vapor cloud dispersion emanating from a main engine exhaust stack. Meteorological data and exhaust release characteristics are seen in Table 1 and 2, respectively. Information associated to exhaust release and environmental conditions were kept as strict as possible to ensure comparison details are identical. Each simulation was set to run for 90 seconds of simulation time. The courant number was set to 20 for both FLACS and KFX.

Table 1: Meteorological Data used for Exhaust Dispersion Simulations

Ambient temperature	32 C
Relative humidity	75 %
Wind speed (m/s)	2
Solar radiation	None
Wind direction from bow	0 degrees

Table 2: Exhaust Release Characteristics

Release Point	# of Exhausts	Exhaust Diameter (m)	Total Mass Flow Rate (kg/s)	T (°C)	Release Direction
Main Engine	1	0.24	17.0	291	Vertical

FLACS

Typically, CFD based models are not the simplest to use. A mock FPSO unit was created with engine exhausts jettisoned upwards from an exhaust stack fin. A helideck, with helicopter, has been positioned aft of the exhausts. Results will be presented as the vapor cloud impinges upon the availability of the helideck. The FLACS geometry was then converted to a form readily usable by KFX.

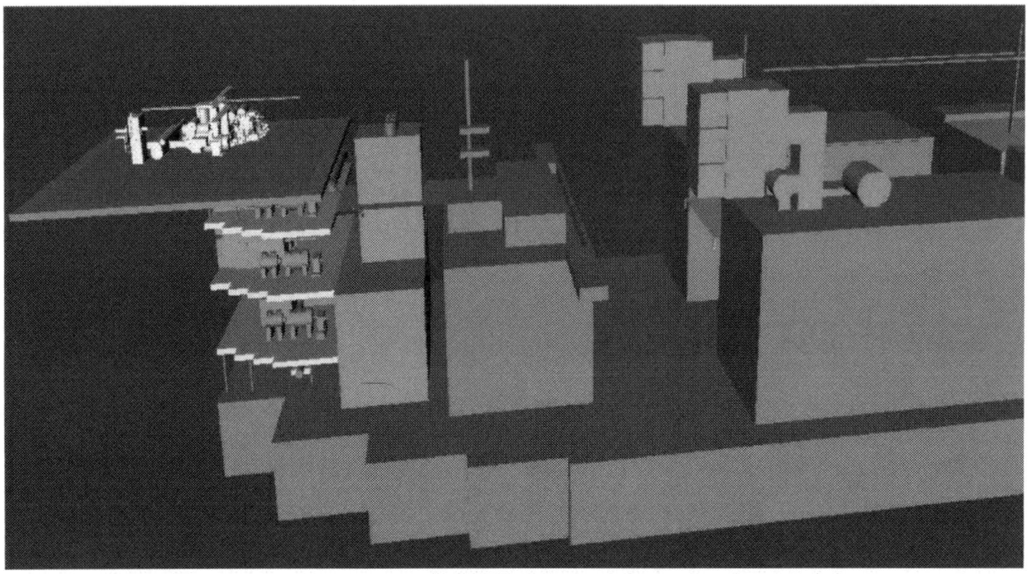

Figure 1: Geometry of FPSO Viewed from Starboard Aft

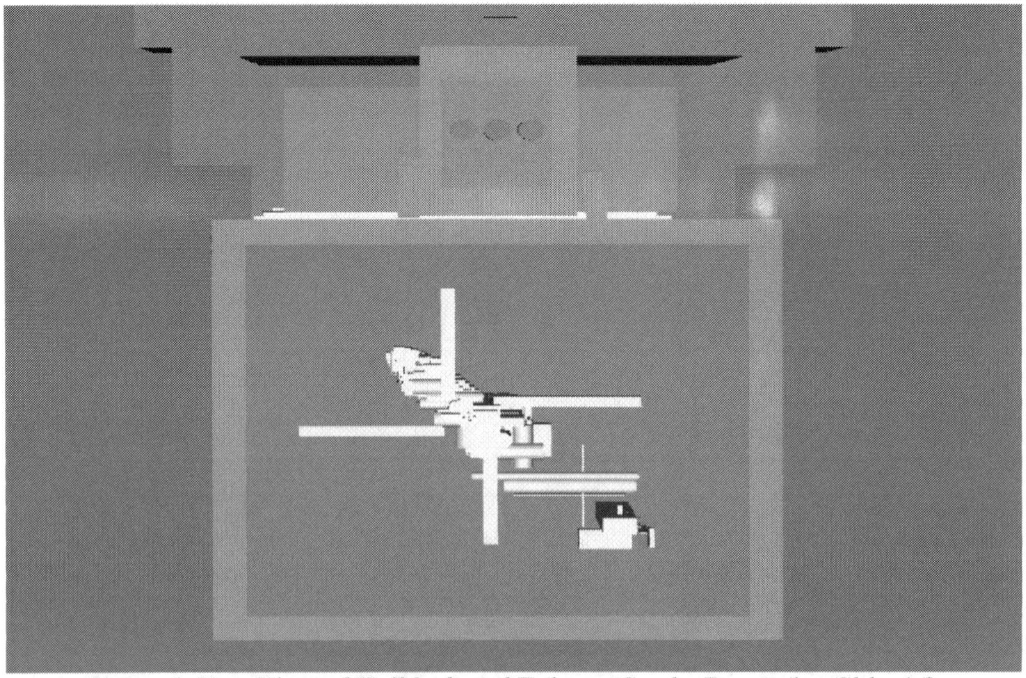

Figure 2: Top View of Helideck and Exhaust Stacks Located at Ship Aft

The release position for FLACS is located at the center of the main exhaust (center cylinder) and located at elevation 54 meters. This release location is identical to that set in the KFX coordinate system (e.g. same x-y-x coordinates).

KFX

The geometry images show in Figures 3 and 4 are taken from the KFX software. The 3D geometrical layouts of the two software packages are observed. The purpose of creating the geometry in FLACS and converting to a format usable by KFX is to ensure they are nearly completely identical geometries.

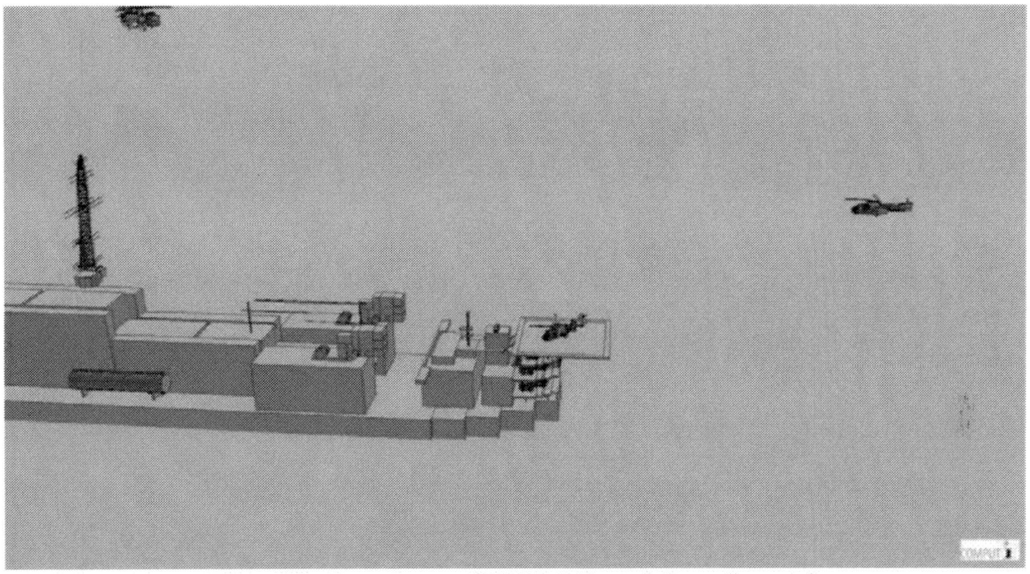

Figure 3: Geometry of FPSO Viewed from Port Aft

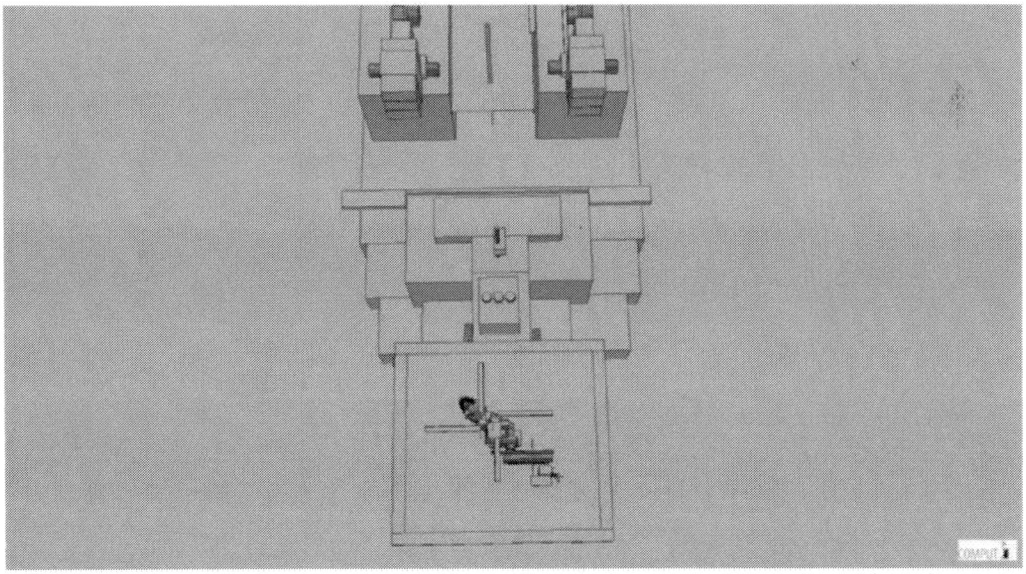

Figure 4: Top View of Helideck and Exhaust Stacks Located at Ship Aft

Discussion

FLACS

The results from the FLACS simulations were obtained after 72 hours of computational time and seen in Figures 5 and 6.

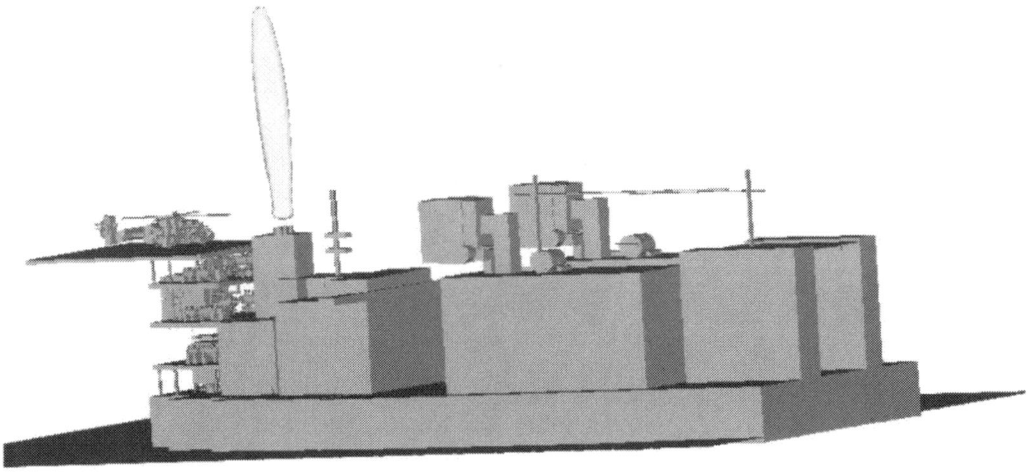

Figure 5: Results from Post Processor FlowVis after 90 Seconds of Simulation Time

Figure 6: Results from Post Processor FlowVis after 90 Seconds of Simulation Time

The results are extracted from flowvis (the post processing software utilized by FLACS) and presented as a gas cloud that contains the LFL of methane (4.5%/vol). The wind speed used during the simulation is low, 2 m/s, and does not cause the exhaust plume to affect the Helideck availability.

KFX

The results obtained using KFX are able to be extracted after 8 hours of computational time, many hours before those available in FLACS. The user tool to examine results in KFX is the post processor tool kfxview. KFXview offers complete 3D manipulation and 360 degree freedom of rotation and results are seen in Figures 7 and 8.

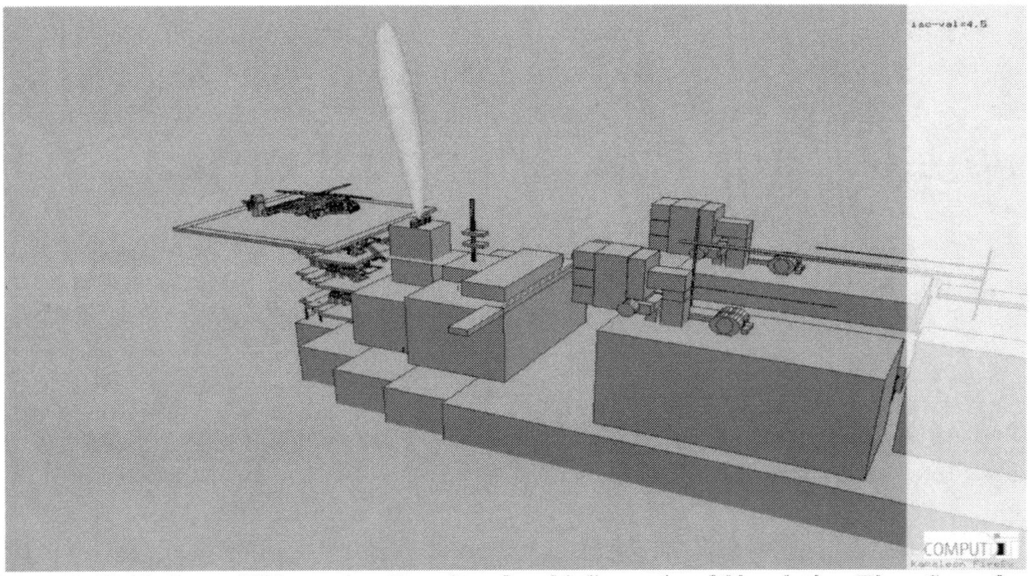

Figure 7: KFXview of Dispersion Results after 90 Seconds of Simulation Time Seen from Starboard Bow

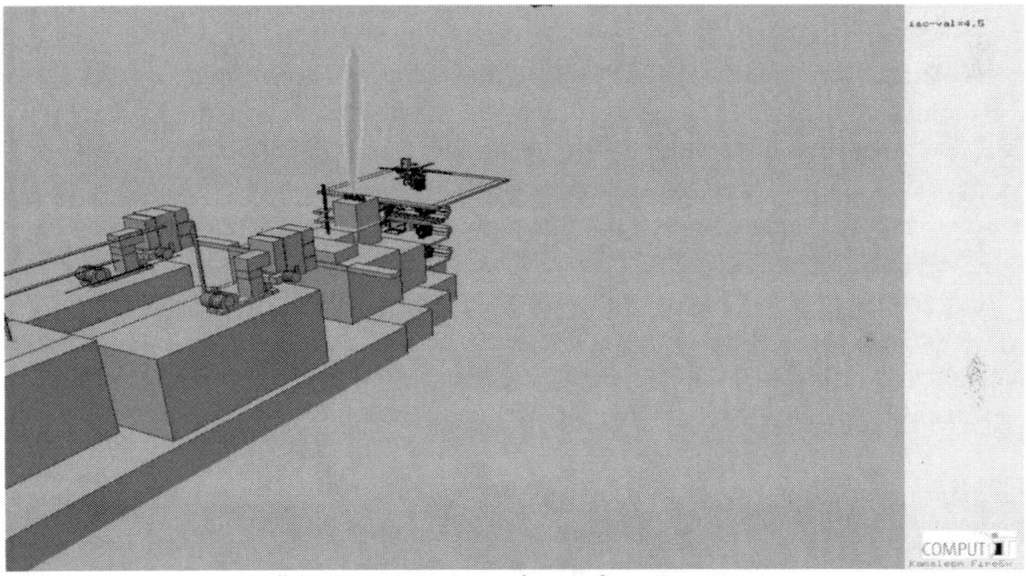

Figure 8: KFXview of seen from Port Bow

Conclusion

Both 3D software packages provide users with state of the art simulation capabilities. KFXview and FlowVis both offer a wide variety of post processing views from 2D through 3D displays. In the simulation scenario examined in this paper, KFX gave nearly identical results to those produced in FLACS but at drastically reduced computational time.

(8c) Investigating Atmospheric Inorganic Aerosol Sensitivity Relationships Through Thermodynamics and Population Dynamics

Shannon L. Capps[1], Shunliu Zhao[2], Amir Hakami[2], Matthew D. Turner[3], Daven K. Henze[3], Armistead G. Russell[4] and Athanasios Nenes[5], (1)School of Chemical & Biomolecular Engineering, Georgia Institute of Technology, Atlanta, GA, (2)Department of Civil and Environmental Engineering, Carleton University, Ottawa, ON, Canada, (3)Department of Mechanical Engineering, University of Colorado-Boulder, Boulder, CO, (4)School of Civil & Environmental Engineering, Georgia Institute of Technology, Atlanta, GA, (5)Schools of Earth & Atmospheric Sciences and Chemical & Biomolecular Engineering, Georgia Institute of Technology, Atlanta, GA

Atmospheric aerosols alter climate and can harm human health. Characterizing the chemical state and sizes of aerosols as these effects occur is challenging due to complex atmospheric processing. Quantitative estimates of the impact of aerosols are often sought through mathematical representations of the chemical and physical transformations occurring throughout the atmospheric residence time (i.e., chemical transport modeling). These models must account for both directly emitted aerosol and contributions from precursor gases through nonlinear pathways. The complexity of the system obscures relationships between the system input and output; thus, extensive modeling efforts are required to provide predictive modeling capabilities. An alternative approach employs sensitivity analysis, which provides quantitative estimates of the effect of a selected condition (i.e., concentration of a precursor gas) on a final value (i.e., aerosol component concentration). The adjoint of a model efficiently elucidates the sensitivity of a selected model output to a number of input values.

The regional Community Multiscale Air Quality model is used to investigate regional-scale scientific questions and to quantify emissions controls necessary to attain air quality standards. Within it, condensation and evaporation of inorganic gaseous species are modeled by thermodynamic equilibrium with ISORROPIA (Nenes et al, 1998; Fountoukis and Nenes, 2007), and aerosol population dynamics are quantified by a three-moment scheme (Whitby et al., 1991; Whitby and McMurry, 1997; Binkowski and Roselle, 2003). In this work, the recently developed adjoint of these aerosol processes is employed to elucidate sensitivity relationships of inorganic aerosol populations to concentrations of aerosol species and precursor gases. In particular, this work focuses on the ammonium-nitrate-sulfate-sodium-chloride aerosol system. Finally, sensitivities are determined for atmospheric conditions commensurate with summer in the northeastern United States.

See more of this Session: Atmospheric Chemistry and Physics I
See more of this Group/Topical: Environmental Division

(8d) Aqueous-Phase Organic-Inorganic Aerosol Chemistry: Model Simulations

V. Faye McNeill, Department of Chemical Engineering, Columbia University, New York City, NY

We have developed GAMMA (Gas-Aerosol Model for Mechanism Analysis), a photochemical box model with coupled gas-phase chemistry and detailed aqueous aerosol chemistry. The aqueous-phase mechanism includes organosulfate formation via recently proposed epoxide and sulfate radical pathways. The model also simulates changes in aerosol absorption caused by the dark formation of light-absorbing organics from methylglyoxal, glyoxal, and acetaldehyde (e.g. Schwier et al. 2010, Noziere et al. 2007).

We find that under low-NOx conditions, organosulfate and SOA formation from isoprene is dominated by the isoprene-derived epoxydiol pathways proposed by Paulot et al. (2009) and Surratt et al. (2009). >80% of SOA mass after 12 hours is composed of 2-methyltetrol, and 13% is 2-methyltetrol sulfate ester. 2-methyltetrol sulfate ester comprises over 99% of organosulfate mass under these conditions. Under high-NOx conditions, organosulfate species make a negligible contribution to isoprene-derived SOA mass.

Overnight, the formation of light-absorbing compounds from methylglyoxal and glyoxal is responsible for <10% of SOA mass. However, sufficient concentrations of light-absorbing products are achieved to significantly affect aerosol optical properties. GAMMA output is used to predict aerosol single scattering albedo (SSA) and absorption Angstrom exponent based on laboratory data, Kramers-Kronig, and Mie theory.

See more of this Session: Atmospheric Chemistry and Physics I
See more of this Group/Topical: Environmental Division

(8e) Exploring Atmospheric Organic Aerosol Source Apportionment In the United States for An Entire Simulation Year

Benjamin N. Murphy, Chemical Engineering, Carnegie Mellon University, Pittsburgh, PA, Kristina M. Wagstrom, Department of Civil Engineering, University of Minnesota - Twin Cities, Minneapolis, MN and Spyros N. Pandis, Chemical Engineering, University of Patras, Patras, Greece

The contribution of specific sources, both anthropogenic and biogenic, to atmospheric pollution varies widely throughout space and time. The relationship between these sources and ambient receptors is complicated by variability in the sources themselves, meteorological conditions, and chemistry that occurs during the pollutants' atmospheric lifetime. These relationships must be well-characterized for relevant pollutant species in order to inform policy decisions that seek to improve air quality with the most effective emissions reduction strategies. The Particulate Matter Source Apportionment Technology (PSAT) algorithm [Wagstrom et al., 2008] was developed and applied to a regional-scale, three-dimensional chemical transport model (PMCAMx) to address this need from several perspectives for inorganic constituents of atmospheric particulate matter (PM) (e.g. sulfate, ammonium, nitrate, elemental carbon, etc).

For this study, we focus on organic aerosol (OA), a major contributor to the PM mass composition. Understanding the OA system is more complicated than understanding the inorganic aerosol constituents (ammonium, sulfate, nitrate, chloride, etc) due to the large number of distinct organic species present, widely varying structure and properties, and their complex oxidation pathways. Previously, we have modeled OA formation and continued atmospheric oxidation with the regional-scale CTM, PMCAMx-2008 [Murphy and Pandis, 2009]. Here we combine PSAT and PMCAMx-2008 to simulate organic aerosol formation with source resolution for the US throughout an entire year, 2008.

A notable conceptual improvement in PMCAMx-2008 is the treatment of POA evaporation upon dilution in the atmosphere. In previous work, a uniform representative volatility distribution was applied to OA mass from all sources. In this work, we expand that analysis to treat source-specific volatility distributions from diesel, gasoline, and biomass burning constituents to quantify the effect of that assumption on the dominance of particular sources to the ambient OA loading. The continued gas-phase oxidation of organic vapors can change their volatility over time and this has an effect on the long-range transport and properties of these species. We investigate this relationship by distinguishing between local, short range, mid range and long range pollutants at several receptor sites in the US. We finally assess the average chronological age of OA mass throughout the US domain while taking into account the continued oxidation and semivolatile partitioning of the aerosol and vapors. The analysis from this study confirms the highly regional, dispersed nature of organic aerosol pollution and suggests that local emissions reduction efforts alone may be insufficient for effective mitigation of this pollutant.

Wagstrom, K. M., et al. (2008), Development and application of a computationally efficient particulate matter apportionment algorithm in a three-dimensional chemical transport model, *Atmos. Environ.*, 42, 5650– 5659, doi:10.1016/j.atmosenv.2008.03.012.

Murphy, B. N., and S. N. Pandis (2009), Simulating the formation of semivolatile primary and secondary organic aerosol in a regional chemical transport model, *Environ. Sci. Technol.*, 43, 4722–4728.

See more of this Session: Atmospheric Chemistry and Physics I
See more of this Group/Topical: Environmental Division

(8f) Discrete Phase Modeling of Atmospheric Particulate Transport From Mine Tailings

Andrea C. Landzuri[1], Jacqueline Brauneis[1], Eduardo Sez[1] and Eric Betterton[2], (1)Chemical and Environmental Engineering, University of Arizona, Tucson, AZ, (2)Atmospheric Sciences, University of Arizona, Tucson, AZ

Arid regions such as Arizona are susceptible to dust storms. Metal and metalloids contained in the tailings dust can travel long distances and deposit in vegetated areas, soil and water. An approximate geometry of a tailing in Dewey-Humboldt has been constructed and meshed using ANSYS mesh generator. This tailing is massive in size and it is close in proximity to near-by residents which can be affected by coarse particles that can deposit in the lung airways.

Particulate transport was investigated using the discrete phase model (DPM) available in the computational fluid dynamics (CFD) software ANSYS FLUENT. The wind flow field was simulated using the k-epsilon turbulent model and appropriate equations and constants that solve a simple logarithmic and unidirectional wind profile that resembles the atmospheric boundary layer. North-west winds at a reference height of 10 m ranged from 1 to 15 m/s. Qualitative and quantitative information was obtained from the trajectories of particles with densities comparable to that of dust (about 1000-2500 kg/m^3). Particles 3-100 mm in diameter were released from various locations of the tailing with the objective of investigating how particles dispersed and how they compare to ideal cases where topography is not included.

Results suggest that the vertical drag exerted on the particle can have less or no effect on particles greater than 15 mm during the initial stage of injection, therefore gravitational force is dominant causing these particles to reach the same maximum heights. As particle motion progresses, horizontal drag becomes more important for smaller particles causing them to settle first. Variables such as angle of injection, velocity of injection, and height of injection were also investigated.

See more of this Session: Atmospheric Chemistry and Physics I
See more of this Group/Topical: Environmental Division

(8g) Optimizing Data Assimilation Parameters for Improved CMAQ PM2.5 Estimates Over the United States to Inform Epidemiological Studies

Sinan Sousan[1], Jaemeen Baek[2], Scott Spak[2], Naresh Kumar[3], Jacob Oleson[4], Gregory Carmichael[1] and Charles Stanier[1], (1)Chemical and Biochemical Engineering Department/Center for Global and Regional Environmental Research, University of Iowa, Iowa City, IA, (2)Center for Global and Regional Environmental Research, University of Iowa, Iowa City, IA, (3)Department of Geography, University of Iowa, Iowa City, IA, (4)Department of Biostatistics, University of Iowa, Iowa City, IA

$PM_{2.5}$ exposure estimates for United States at high spatial and temporal resolution are highly desirable to conduct further epidemiological studies focused on identifying more and less harmful types of particulate matter. Data assimilation via optimal interpolation (OI) was used to improve the Models-3 Community Multiscale Air Quality Model (CMAQ) modeling system estimates of aerosol pollution. Model predicted concentrations over North America for 2002 with and without assimilation of MODIS satellite-based aerosol optical depth are compared to $PM_{2.5}$ measurements for performance evaluation.

Our results show that OI can be done in three successful optimization approaches. Each approach is characterized by slightly different sets of relative error statistics on the model and the satellite observations, and also the different choices of temporal and spatial aggregation in modeled AODs and OI parameters. The first successful optimization scheme uses a spatially- and temporally-invariant settings and simple temporal (monthly) averaging of model and satellite data. The second approach uses either customized OI settings for specific regions or time periods. The third approach uses customized settings specific to time periods and geographical locations. In all cases, we find it is critical to exclude from optimization MODIS data from months and regions where it tends to degrade rather than improve model skill. Results from the first approach can improve average fractional bias from 1.31 to 0.59 (relative to IMPROVE $PM_{2.5}$ measurements) and from 0.34 to 0.22 (relative to STN $PM_{2.5}$ measurements). The other two schemes have better performance, but at the expense of requiring more month- and region- specific settings. The robustness of the technique has been applied by training the OI system on odd months of 2002 and testing settings on even months, and vice versa.

See more of this Session: Atmospheric Chemistry and Physics I
See more of this Group/Topical: Environmental Division

(8h) A Hybrid Approach for Developing Time-Space Resolved Estimates of Ambient Air Pollution Exposure

Naresh Kumar, Epidemiology and Public Health, University of Miami, Miami, FL

Integrating satellite remote sensing, data assimilation techniques and spatiotemporal Kriging, we demonstrate how to develop time-space resolved estimates of exposure to ambient particulate of different sizes (PM) for the health effect studies. The proposed methodology is implemented in three sequential steps. First, using the empirical relationship between aerosol optical depth (AOD) from satellites and *in situ* PM measurements at EPA sites, predict PM for all valid AOD data points. Second, employ data assimilation techniques to validate and improve the predicted PM. Third, employ spatiotemporal Kriging to develop exposure estimates (using the predicted PM) at the required spatiotemporal scales, such as daily PM estimate at any point location.

To demonstrate the application of the Hybrid approach, daily estimates of PM ($PM_{2.5}$, PM_{10} and $PM_{10-2.5}$) will be developed from 2000 to 2009 at 2.5km spatial resolution for Cleveland Metropolitan Statistical Area (MSA). The 2km AOD (~2.3 million data points at 2km spatial resolution from 2000 to 2009) were extracted using the data from Moderate Resolution Imaging Spectroradiometer (MODIS), aboard Terra and Aqua satellites. *In situ* measurements of PM were acquired from EPA and meteorological data from the National Climatic Data center. Using these data a systematic grid of daily PM was developed at 2.5km spatial resolution for the Cleveland MSA from 2000 to 2005; 90% of the predicted values were robust. The proposed methodology is likely advance the exposure modeling to the next level by providing robust and reliable PM estimate at the (desired) spatiotemporal scales of health data.

See more of this Session: Atmospheric Chemistry and Physics I
See more of this Group/Topical: Environmental Division

(10a) Fermentation Cost Improvement Through Propagation Advances for a Commercially Relevant Biomass-to-Ethanol Process

Kate Brandon Sutton, B. Emme, L. Jones, S. McHatton, C. Kang, C. Corbett, L. Putnam and M. Torry-Smith, Novozymes North America Inc., Franklinton, NC

Compared to their starch-to-ethanol counterparts, pentose-utilizing organisms for biomass-to-ethanol processes are less robust while having to ferment mixed sugars in a significantly more inhibitory substrate. This inhibition can generally be overcome with a high yeast pitch to maximize xylose uptake rates in fermentation. The cost driven trade-off of a high yeast pitch during fermentation is the necessity of dedicated onsite propagation, as opposed to purchase of all yeast. Novozymes' Advance Biofuel cost model suggest the majority of the fermentation cost when using high yeast pitches can be attributed to the propagation variable costs, most significantly from the carbohydrate source. Therefore, maximizing the cost performance of C5 yeast propagation is important to ensure an economically feasible biomass-to-ethanol fermentation step.

Novozymes has identified two approaches to reduce the fermentation cost: first, increase propagation cell yields from sugars to reduce the realized organism cost, preferably on cheap carbohydrate sources, or second, produce a more robust organism from propagation with a higher realized xylose uptake rate in fermentation which would enable fermentation at a lower yeast pitch in a reasonable processing time. This presentation will discuss cost model results in conjunction with reactor data to pinpoint potential savings and opportunities to lower overall fermentation cost for second generation ethanol. Sensitivities on cost from propagation and fermentation process variables targets will be presented. Propagation optimization with respect to higher realized xylose uptake rate in fermentation will be addressed as well. Lastly, the holistic minimum ethanol selling price implications for process optimizations will be given.

See more of this Session: Biological Conversions and Processes for Renewable Feedstocks I
See more of this Group/Topical: Sustainable Engineering Forum

(10b) Combined Computational and Experimental Investigations of Processive Cellulases for Engineering Activity Improvements

Gregg T. Beckham[1], **Larry Taylor**[2], Michael F. Crowley[3], William S. Adney[3] and Michael E. Himmel[3], (1)National Bioenergy Center, National Renewable Energy Laboratory, Golden, CO, (2) Biosciences Center, NREL, Golden, CO, (3)Biosciences Center, National Renewable Energy Laboratory, Golden, CO

The Family 7 cellobiohydrolase (Cel7A) from *Hypocrea jecorina* (*Trichoderma reesei*) is one of the most well-characterized, active cellulases known for deconstruction of cellulose to soluble sugars. With the aim to engineer enhanced cellulase specific activity, we have investigated the Cel7A mechanism on cellulose with molecular dynamics simulations and biochemical experiments. We have elucidated new functions for the carbohydrate-binding module, determined the effect of glycosylation on the stability of the linker and carbohydrate-binding module, examined the role of aromatic and polar residues in the catalytic domain tunnel, and examined the structure of the catalytically-active complex of Cel7A on crystalline cellulose up to the microsecond time scale. These new insights combined with experimental activity data have led to a large-scale experimental screening of new enzymes to examine both natural diversity and engineered enzymes for activity and stability improvements. Our combined computational and experimental approach has demonstrated that unprecedented improvements in specific activities of processive cellulases are possible, which when combined in industrial cocktails, will result in lower enzyme loadings for cell wall deconstruction and lower overall production costs for renewable biofuels.

See more of this Session: Biological Conversions and Processes for Renewable Feedstocks I
See more of this Group/Topical: Sustainable Engineering Forum

(10c) Exploration of Significant Enzyme Activities Required for the Deconstruction of Ionic-Liquid Pretreated Biomass

Christopher J. Barr[1], Jeffery Mertens[2] and Constance Schall[1], (1)Chemical and Environmental Engineering, University of Toledo, Toledo, OH, (2)NCAUR, USDA-ARS, Peoria, IL

Terrestrial biomass is a second generation renewable feedstock, comprised of polysaccharides derived from cellulose and hemicellulose, which can be used to produce carbon-based chemicals and fuels. The recalcitrance to deconstruction of biomass is one of the main impediments in the conversion of biomass to monomeric sugars and the production of sugar platform fuels and chemicals. Ionic Liquid (IL) pretreatment creates unique biomass substrates composed of amorphous cellulose with largely intact hemicellulose and lignin. IL pretreated substrates differ from many other pretreatments which remove hemicellulose and retain cellulose crystallinity. This substrate poses an interesting challenge for enzymatic deconstruction using traditional commercial mixtures.

Mono-component cellulase and hemicellulase enzymes from *Aspergillus nidulans* have been expressed, purified, and characterized. Key enzyme activities and/or interactions that significantly contribute to the conversion of a variety of lignocellulosic biomass substrates to monomeric sugars were examined by enzyme combinations in series of hydrolysis experiments. Auxiliary hemicellulases seem to be essential to the mono-component enzyme mixtures. Investigation of these accessory enzymes also yields insight into an understanding of the differences in hemicellulose structure between feedstocks. Some auxiliary hemicellulases include acetyl xylan esterases (AXE), arabinofuranosidases, and α-glucuronidases. In addition to mono-component enzyme mixtures, next generation commercially-available enzyme mixtures were characterized and formulated using key activities to optimize hydrolysis at low enzyme loadings. These simplified mixtures will be dependent on the substrate. As part of these studies, an analysis of variance (ANOVA) was used to statistically model the main and interaction effects of key enzyme activities for each substrate.

See more of this Session: Biological Conversions and Processes for Renewable Feedstocks I
See more of this Group/Topical: Sustainable Engineering Forum

(10d) Inhibition of ENZYMATIC Hydrolysis by Soluble Sugars ON MODEL CELLULOSE Thin Film USING QUARTZ CRYSTAL Microbalance

Hsin-Fen Li[1], Ravinder Garlapalli[1], Michael D. Flythe[2], Sue E. Nokes[3], Stephen E. Rankin[1] and Barbara L. Knutson[1], (1)Chemical and Materials Engineering, University of Kentucky, Lexington, KY, (2)Department of Animal and Food Sciences, University of Kentucky, Lexington, KY, (3)Department of Biosystems and Agricultural Engineering, University of Kentucky, Lexington, KY

INHIBITION OF ENZYMATIC HYDROLYSIS BY SOLUBLE SUGARS ON MODEL CELLULOSE THIN FILM USING QUARTZ CRYSTAL MICROBALANCE

Hsin-Fen Li [1], Ravinder Garlapalli[1]. Michael D. Flythe [2,3], Sue E. Nokes [4], Stephen E. Rankin[1], Barbara L. Knutson[1]

[1]University of Kentucky, Department of Chemical and Materials Engineering, Lexington, Kentucky 40506

[2]USDA, Agricultural Research Service, Forage-Animal Production Research Unit; Lexington, Kentucky 40546

[3]University of Kentucky, Department of Animal and Food Sciences, Lexington, Kentucky 40546

[4]University of Kentucky, Department of Biosystems and Agricultural Engineering, Lexington, Kentucky 40506

Soluble sugars produced from the hydrolysis of cellulose are one of the most promising feedstocks for production of fuels and commodity chemicals from renewable plant-based resources. The recalcitrance of cellulose to degradation is well documented and attributed to multiple sources, including high cellulose crystallinity and the hydrogen binding framework that is embedded in the matrix substances (lignin and hemicellulose). Cellulase is a cellulytic enzyme that possess a catalytic domain capable of hydrolyzing the beta-1, 4-glycosidic bonds in cellulose. However, during biomass conversion to soluble sugars, its hydrolytic efficiency is limited by the presence of degradation products including cellobiose. To study the inhibitory effect of cellobiose and the interaction of the inhibited enzyme and the cellulose substrate, the change in mass and surface properties of cellulose thin films were measured using Quartz Crystal Microbalance with Dissipation (QCM-D). The cellulose thin film was synthesized from thermochemically dissolved cellulose and was coated on a quartz crystal resonator. The mass of the spin-coated cellulose film was detected by the resonance frequency, and the cellulose surface coverage was observed by atomic force microscopy (AFM). Cellulase adsorption, activity, and inhibition on model cellulose thin films as were measured directly as changes to the piezoelectric properties of the cellulose thin films by QCM-D. At 5 g/L, cellobiose completely inhibits cellulase (0.5 %v/v) activity. The inhibited cellulases bind reversibly to the cellulose and reduce subsequent enzymatic hydrolysis rates by more than 50%. Significantly higher activity is observed when cellulase concentration is decreased from 1 %v/v to 0.1 %v/v, suggesting that cellulase acts progressively on the cellulose chain, and the hydrolysis is restricted by saturated enzyme conditions.

See more of this Session: Biological Conversions and Processes for Renewable Feedstocks I
See more of this Group/Topical: Sustainable Engineering Forum

(10e) Selection and Characterization of Acid-Stable Cellulases Against Pretreated Oil Palm Empty Fruit Bunch From a Colombian Extreme Environment

Luis Miguel Medina[1], Luisa Cabezas[2], Isabella Bahamon[1], Laura Palma[1], Andrs Gonzlez[1] and Silvia Restrepo[2], (1)Chemical Engineering, Centro de Diseo de Productos y Procesos, Universidad de los Andes, Bogot, Colombia, (2)Biological Sciences, Laboratorio de Micologa y Fitopatologa (LAMFU), Universidad de los Andes, Bogot, Colombia

The rapid growth of the oil palm industry around the world generates environmental concerns, with a production of more than 28.0 million tons per year of oil palm around the world, generating a great amount of by-products and wastes, such as liquid effluent, gaseous emission, and solid waste materials. Nowadays, solid waste is used commonly as manure, nevertheless the composting process is reported to be complex[1]. Therefore this agro-industry should take into consideration the use of sustainable technology. A material balance of an oil palm plant yields, in terms of mass, for every Ton of fresh fruit processed 220 kg of oil palm empty fruit bunch (OPEFB), 125-130 kg of fiber, around 50 kg of shell and 195-200 kg of palm oil [2]. Studies on the lignocellulosic content of the OPEFB have shown that display a high cellulose and hemicellulose and low lignin content [2]. This feature allows to use this waste as a substrate to obtain simple sugars, which can be used as carbon source in different fermentation processes to enable the production of alcohols that can be used as fuels. The biodegradation of cellulose is carried out by three classes of enzymes: endoglucanase (carboxymethylcellulase), exogluconase (cellobiohydrolase) and β-glucosidase, which together with a preliminary treatment of cellulose, manage to break the glycosidic bonds to produce simple sugars [3]. However, these enzymes hydrolyze in different parts of the polymer so the relation between the substrate and the enzyme should not be extrapolated due to different lignocellulosic structure [4]. Taking all this into account the objective of this study was to explore the cellulolytic potential of endophytes from a Colombian extreme environment for the saccharification of oil palm empty fruit bunch. It was determined the cellulolytic capacity for the saccharification *in vitro* of the pretreated oil palm empty fruit bunch of one hundred Colombian native endophytes isolated from young leaves attacked by insects (with herbivory) as well with no symptom of attack (without herbivory) from *Espelieta grandiflora* and *E. corymbosa* plants from the Páramo Crúz Verde. The microorganisms were screened to asses which of them have cellulolytic capacity on solid agar with CMC as substrate followed by Congo red staining method [5]. The microorganisms that showed cellulolytic activity were grown on cellulase inducting medium, using pretreated OPEFB as substrate, the activity of the three main components of cellulase (CMCase, β-glucosidase and Cellobiohydrolase) was measured, and total cellulolytic activity (Filter paper activity. FPA), using the procedures described by Zhang *et. al (2009)*. Candidates obtained, (two entophytes), were partially purified through partial precipitation with ammonium sulfate (90%) [6] and chromatography using DAE-shepadex A-50 column previously equilibrated with sodium acetate buffer (pH 5, 0.05 M) and the fractions were withdrawn with a linear NaCl gradient, the fractions with cellulolytic activity were pooled and then concentrated by ultrafiltration with a centriprep (Amicon) having a 10 kDa molecular mass cut-off. The screenings allowed us to identify 28 endophytes with cellulolytic activity, and the best obtained were a *Penicilium sp.* and an *Aspergillus sp.* with maximum enzyme activities, CMCase, FPase, exogluconase and β-glucosidase, of 0.50-0.16, 0.035-0.032, 0.87-0.99 and 0.068-0.01 U/ml respectively. CMCase and the FPA activities of the partially purified cellulase was assessed at pHs ranging from 1 to 10, and temperatures ranging from 30 to 80 °C in order to find the optimum conditions for the cellulases. The cellulases from *Penicillium sp. and Aspergillus sp.* were stable at low pH which could be a valuable tool for hydrolysis of cellulose under acidic conditions.

References

1. Yusoff, S., *Renewable energy from palm oil - innovation on effective utilization of waste.* Journal of Cleaner Production, 2006. **14**(1): p. 87-93.

2. Umikalsom, M.S., et al., *The treatment of oil palm empty fruit bunch fibre for subsequent use as substrate for cellulase production by Chaetomium globosum Kunze.* Bioresource Technology, 1997. **62**(1-2): p. 1-9.

3. Sun, Y. and J. Cheng, *Hydrolysis of lignocellulosic materials for ethanol production, a review.* Bioresource Technology, 2002. **83**(1): p. 1-11.

4. Zhang, Y.H.P., M.E. Himmel, and J.R. Mielenz, *Outlook for cellulase improvement: Screening and selection strategies.* Biotechnology Advances, 2006. **24**: p. 452-481.

5. Kasana, R., et al., *A Rapid and Easy Method for the Detection of Microbial Cellulases on Agar Plates Using Gram's Iodine.* Current Microbiology, 2008. **57**(5): p. 503-507.

6. Scopes, R.K., *Protein purification principles and practice* Second ed. Spinger Advanced texts in chemistry, ed. C.R. Cantor. 1982: Springer-Verlag.

See more of this Session: Biological Conversions and Processes for Renewable Feedstocks I
See more of this Group/Topical: Sustainable Engineering Forum

(10f) Effect of AFEX Pretreatment Degradation Products On Enzymatic Hydrolysis and Microbial Fermentation by Saccharomyces Cerevisiae 424A (LNH-ST)

Xiaoyu Tang[1], Leonardo da Costa Sousa[1], Shishir Chundawat[1], Mingjie Jin[1], Lau Ming[1], James Humpula[1], Nirmal Uppugundla[1], Kevin Chambliss[2], Ramin Vismeh[3], A. Daniel Jones[3], Zeyi Xiao[4], Bruce Dale[1] and **Venkatesh Balan**[1], (1)Chemical Engineering and Material Science, Michigan State Univeristy, Lansing, MI, (2)Chemistry and Biochemistry, Baylor University, Waco, TX, (3) Biochemistry and Molecular Biology, Michigan State University, East Lansing, MI, (4)School of Chemical Engineering, Sichuan University, Chengdu, CA, China

Several degradation products are formed or released during AFEX pretreatment process, which can be divided into five groups: 1) carbohydrates, 2) aliphatic acids, 3) nitrogenous compounds, 4) furans, 5) aromatic compounds. Some of these degradation products are expected to be inhibitory to *Saccharomyces cerevisiae* 424A(LNH-ST), thus affecting the xylose fermentation, as well as various saccharolytic enzymes. In this work, the effect of various AFEX degradation products on co-fermentation by *Saccharomyces cerevisiae* 424A(LNH-ST) were tested individually and in combinations. A synthetic media composed of similar nutrient composition as AFEX corn stover hydrolyzate was designed and successfully used as a control fermentation media in this study. Nitrogenous compounds showed the inhibitory effect on xylose fermentation, followed by carbohydrates, aliphatic acid and aromatic compounds. The inhibitory effect caused by nitrogenous compounds and carbohydrates were attributed to the presence of amides and oligomeric xylan, respectively. To our knowledge, this is the first report of inhibitory effect of oligomeric xylan and amides on xylose fermentation by *Saccharomyces cerevisiae*. When testing the amides (feruloyl amide, coumaroyl amide and acetamide) and their corresponding acids (ferulic acid, coumaric acid and acetic acid) individually, the amides were less inhibitory than their acids with respect to cell growth and xylose fermentation. From this point, the shift of degradation products from organic acids to amides may be responsible for the high fermentability of AFEX pretreated lignocellulosic biomass. We also evaluated the extent of enzymatic inhibition by various molecular fractions of AFEX treated corn stover degradation products. The pretreated biomass extracted products were fractionated using ultracentrifugation, gel filtration, and C18 resin solid phase extraction and further analyzed by mass spectrometry. The results indicate that both large and small molecule compounds play a role in enzyme/ microbial inhibition.

See more of this Session: Biological Conversions and Processes for Renewable Feedstocks I
See more of this Group/Topical: Sustainable Engineering Forum

MODELLING SIMULTANEOUS SACCHARIFICATION AND FERMENTATION OF MICROCRYSTALLINE CELLULOSE USING COMPUTATIONAL FLUID DYNAMICS

Josebus Maree van Zyl, Stellenbosch University, Stellenbosch, RSA
Thomas Michael Harms, Stellenbosch University, Stellenbosch, RSA
Eugéne van Rensburg, Stellenbosch University, Stellenbosch, RSA
Willem Heber van Zyl, Stellenbosch University, Stellenbosch, RSA
Lee Rybeck Lynd, Dartmouth College, Hanover, NH

Presentation ID#: 233146

Biography

J.M. van Zyl
email: jmvz@sun.ac.za
Tel: +27 (0)21 808 4272

Extended Abstract

Cellulose is a polysaccharide and one of the most abundant organic compounds found on the Earth surface [1]. It holds the potential to reduce human reliance on fossil derived fuels by producing ethanol through the hydrolysis and fermentation of cellulose. Ethanol is a highly combustible liquid fuel which has been shown to be a viable alternative to petrol. Producing ethanol from cellulose however has proved most challenging, particularly overcoming the recalcitrance of cellulose to enzymatic hydrolysis.

Cellulose to ethanol conversion has successfully been demonstrated at laboratory scale for a variety of feedstock and production methods and in a few cases at demonstration scale. Commercialization of cellulosic ethanol is however limited by the lack of large capital funds required to construct such facilities. A compounding effect is that investors still consider the technology as immature and high risk and as such are hesitant to commit. Obtaining the necessary capital funding requires de-risking the technology as well as policy frameworks and incentives from government.

Engineers have recently begun the development of numerical models for implementation with computational fluid dynamic (CFD) software that will allow investigation of secondary effects of cellulosic ethanol production scale-up. The advantage of CFD is the modelling of the flow patterns and conditions that could exist in a commercial scale fermenter. This includes shear-rates, power requirements, degree of particle suspension and through user-coded modules the reaction-rates of biological or chemical processes. This methodology could allow a significant reduction in the risk associated for investors at a reasonably low financial expense.

Presented in this paper are results of the reaction kinetic modelling (Figure 1) of microcrystalline cellulose to ethanol conversion using cellulases from the fungus *Trichoderma reesei* and an industrial grade yeast *Saccharomyces cerevisiae*, commonly used in the food industry [2]. Observations from the results indicated that the number of available bonding sites on the particle surface appeared to control the rate at which hydrolysis occurred. Enzymes appear to catalyse a reaction at a constant rate thus more sites available to attach should result in higher reaction rates. However due to the fibril structure of microcrystalline cellulose a large number of bonding sites are available resulting in a large number of exoglucanase and endoglucanase enzymes initially attaching to sites on the cellulose chains. Once the initial available sites diminish, a large percentage of the exoglucanase enzymes detach from the particle surface and enters the fermentation broth, effectively reducing the rate of hydrolysis. It thus suggests that if the work rate of the endoglucanase enzyme could be significantly increased, high hydrolysis rates could be achieved as more reduction ends become available for the exoglucanase enzymes.

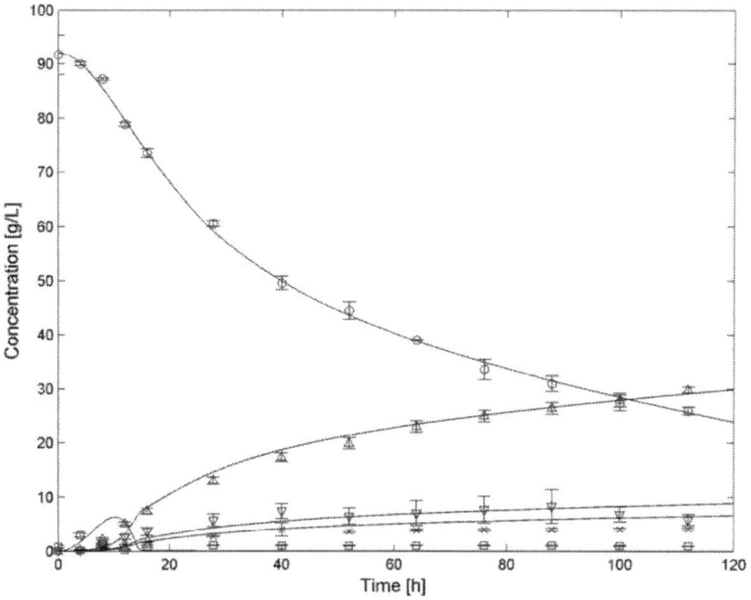

Figure 1: Kinetics model for hydrolysis and fermentation of Avicel PH-101
SSF of Avicel (○) forming glucose (□) fermented to ethanol (Δ) and glycerol (x). Additional yeast cells grown (∇). Simulated results are presented with solid lines.

Microcrystalline cellulose is an insoluble compound which needs to be modelled as particles. It is further required to determine the effects of each substance in the fermentation broth on the viscosity of the system. Results indicated microcrystalline particles occur in various shapes and sizes and exhibit significantly higher drag resistance when the average particle sized is in the order of 50 µm and smaller. This effect is possibly due to the combination of particle topology, structured water [3] and near surface effects [4]. Furthermore the presence of particles has the greatest influence on the apparent dynamic viscosity of the fermentation broth (Figure 2). A numerical model was developed for use in CFD codes which could predict the effects of the particles along with ethanol and glycerol on the apparent dynamic viscosity of the fermentation broth with an average error of 11.1 %.

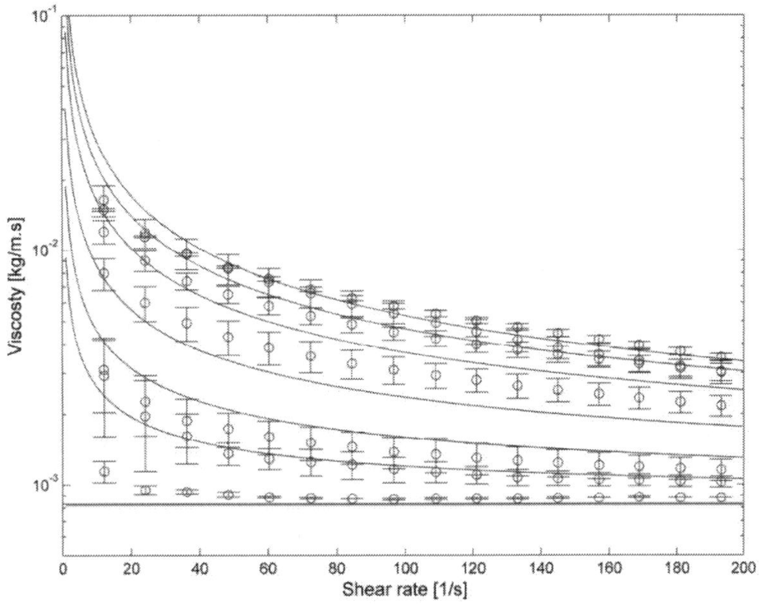

Figure 2: Dynamic viscosity influence of Avicel PH-101 particles

Experimental results (○) with error-bars indicating the standard deviation compared to the numerical predictions (–)

The above mentioned models were incorporated into the commercial CFD package STAR-CCM+ [5] to evaluate the dominant flow patterns and particle distribution within a typical batch mixing vessel. The kinetic models were used to calculate the efficiency of the mixing vessel in-terms of ethanol production rate based on the substrate and reactants concentrations. Results indicated that as cellulose conversion occurs the dynamic viscosity of the fermentation medium increases while the particle sizes decrease. This leads to improved particle suspension and homogeneity which in-turn approaches the optimal conditions for biological hydrolysis and thus ethanol production.

References

1. Lynd LR, Weimer PJ, van Zyl WH, and Pretorius IS, 2002, Microbial cellulose utilization: Fundamentals and Biotechnology, Microbiology and Molecular Biology Reviews, 66 (3): 506-577.
2. van Zyl JM, van Rensburg E, van Zyl WH, Lynd LR, and Harms TM, 2010, A kinetic model for simultaneous saccharification and fermentation of avicel with saccharomyces cerevisiae, Biotechnology and Bioengineering. 108:924-933.
3. J.F. Matthews, C.E. Skopec, P.E. Mason, P. Zuccato, R.W. Torget, J. Sugiyama, M.E. Himmel, J.W. Brady, Computer simulation studies of microcrystalline cellulose Iβ, Carbohyd. Res. 341 (2006) 138-152.
4. J. Zheng, G.H. Pollack, Long-range forces extending from polymer-gel surfaces. Phys. Rev. E, 68 (2003) 031408.
5. STAR-CCM+, version 6.02.007, CD-Adapco, London, UK

(43a) Modelling Acid Gas Reactions In Air Products Sour Compression Process

Frank Petrocelli[1], Vince White[2] and Jeremy Immer[1], (1)Tonnage Gases, Equipment and Energy - Technology, Air Products and Chemicals, Inc., Allentown, PA, (2)Tonnage Gases, Equipment and Energy - Technology, Air Products PLC, Walton-on-Thames, United Kingdom

Oxyfuel combustion in a coal-fired power station produces a raw CO_2 product containing contaminants such as water vapor plus oxygen, nitrogen and argon derived from the excess oxygen for combustion, impurities in the oxygen used, and any air leakage into the system. There are also acid gases present, such as SO_3, SO_2, HCl and NO_X produced as byproducts of combustion. The overall process for purifying the CO_2 takes the impure CO_2 from the power boiler and cools it to condense water vapor, remove traces of ash and dissolve soluble gases such as SO_3 and HCl. The cooled, raw CO_2 is then compressed to about 30 bar and the CO_2 is purified by partial liquefaction and phase separation.

In 2006, Air Products first presented reactions that gave a pathway for SO_2 to be removed as H_2SO_4 and NO and NO_2 to be removed as HNO_3 during the compression of the raw CO_2 from the Oxyfuel combustion process. This proprietary process is now widely known as Air Products' sour compression process. Then, in 2008, we presented initial results from the Oxycoal-UK project in which these reactions were studied experimentally to provide the important reaction kinetic information at elevated pressure that was so far missing from the literature. This experimental work was carried out at lab scale in Imperial College London using synthetic flue gas, and then using actual flue gas via a sidestream from Doosan Babcock's 160kW coal-fired oxyfuel rig. From these experimental results, it was clear that the main reaction pathways are viable and the rates of reaction sufficient to produce the desired results: SO_x and NO_x removal by compression and contact with water. And, in 2010, we presented further results on experimental work on improving our understanding of the sour compression process.

Simultaneous with the above efforts, Air Products has pursued two opportunities to demonstrate the sour compression process at larger scale: 1) a U.S. Department of Energy (DOE) project to design and construct a proprietary CO_2 purification system for slip stream pilot-scale tests at the site of a 15 MW_{th} tangentially-fired (T-fired) oxycombustion test facility; and 2) a slip stream pilot plant to demonstrate proprietary Oxyfuel purification technology at Vattenfall's 30 MWth wall-fired Oxyfuel boiler research and development facility in Schwarze Pumpe, Germany. Both projects allow us to better understand the sour compression reactions, using real Oxyfuel-derived raw CO_2, with a vapor-liquid contacting device equivalent to that forseen for full scale plants.

At present, our pursuit of larger demonstration plant projects for the Oxyfuel CO_2 purification technology relies, in part, on further enhancing our understanding of the sour compression process. To that end, in late 2010 Air Products commissioned a purpose built lab that is being used for studying the sour compression reactions in more detail than previous experimental work allowed. To date, the experimental program has explored the effect of key process variables, including SO_2 and NO_x concentrations, pressure, residence time, and humidity; through experiments run under several different gas-liquid contacting configurations. Analysis of these data has allowed us to better elucidate the fundamental reaction network and to develop a robust kinetic model that is being utilized for our design work on several commercial scale demo plant projects.

In this paper, we will present improvements we have made in our understanding of the reaction network and in our process model, as a result of the recent lab work.

See more of this Session: Oxycombustion of Coal and Other Fuels I
See more of this Group/Topical: Energy and Transport Processes

(43b) The Effect of Compression and Intercooling On the Composition of a Pulverized Coal Oxycombution Flue Gas

Stephen J. Gerdemann, Casey Carney, John Clark, Sivaram Harendra, Thomas Ochs and Danylo Oryshchyn, National Energy Technology Laboratory, U.S. Department of Energy, Albany, OR

Oxy-fuel combustion uses conventional technology to combust pulverized coal in denitrified air resulting in a flue gas which is primarily CO_2 and water. Integrated Pollutant Removal (IPR®) is an NETL patented process that compresses the oxycombustion flue gas to pipeline pressure. The IPR process as it is realized at the Jupiter Oxygen Burner Test Facility is a spray tower (direct-contact heat exchanger) followed by 4 stages of compression with intercooling. The spray tower can be run either with water or with water plus a reagent used to control the pH and precipitate the reactive components of the gas. Latent and sensible heat given up by the processed gas at the cooling steps are partially recovered and in a commercial process could be used to preheat the boiler feed water. A FTIR and a gas chromatograph were used to measure the entering gas composition, the exiting composition as well as the composition at several intermediate points within the IPR system. The experimental results are compared to an ASPEN model of the system. The entering flue gas is approximately 30% water vapor with the remainder mostly CO_2. The gas also contains SO_2, various oxides of nitrogen (mostly from nitrogen in the coal), O_2, CO, HCl and small amounts of other gasses derived from these. The initial spray tower removes much of the water and soluble gasses. More water and soluble gases are removed in each intercooling step until the exit gas is primarily CO_2. The increasing pressure also changes the relative amounts of the different oxides of nitrogen.

See more of this Session: Oxycombustion of Coal and Other Fuels I
See more of this Group/Topical: Energy and Transport Processes

2011 AIChE Annual Meeting; Abstract # 225838; Energy and Transport Processes

A novel high pressure oxyfuel process with staged combustion

Author(s): **Krish R. Krishnamurthy**, Clean Energy and Innovation Management, Linde LLC, Murray Hill, NJ and Hanno Tautz, Research and Development, Linde AG, Linde Engineering Division, Pullach, Germany

Introduction

The application of pure oxygen for the combustion of coal provides the advantage to reduce the flue gas flow significantly and to operate the combustion and steam production at higher pressure. This results in higher electrical net efficiency and a reduction of investment costs.

High pressure oxyfuel process concept

There are already several oxyfuel processes in development. One approach is to keep the technology as close as possible to the existing atmospheric boiler concept. In this case, the combustion temperature of coal with oxygen has to be limited by dilution of the pure oxygen with a recycled CO_2 flow.

Another approach is the application of elevated pressure to the combustion process. In this case as well, a recycle flow of CO_2 or condensed water or a mixture of this is necessary to limit the combustion temperature.

To overcome the problem of intensive cooling of coal combustion with oxygen in the new process covered by this presentation, the oxidation is performed in two steps. In the first step, the coal is gasified with pure oxygen and, in the second step, the syngas produced by the gasifier is burnt in several stages in a unique heat exchanger equipment and process for generating steam. Between the two steps above, a cyclone and hot gas filtration system are employed for gas cleaning from solids.

The above process avoids the need for recycle CO_2, reduces cooling water requirement and enables increased preheating of boiler feed water. As the combustion is mainly in reducing atmosphere, the NOx formation is strongly reduced. At the pressure of 80 bars, the water can be condensed at about 50°C. With a further reduction of the temperature to 30°C CO_2 is condensed in a relatively pure condition and needs none or minimum purification requirement prior to pumping to 120 bars. The desulphurization can be done in the water phase by neutralization of H_2SO_4 with caustic, see figure 1.

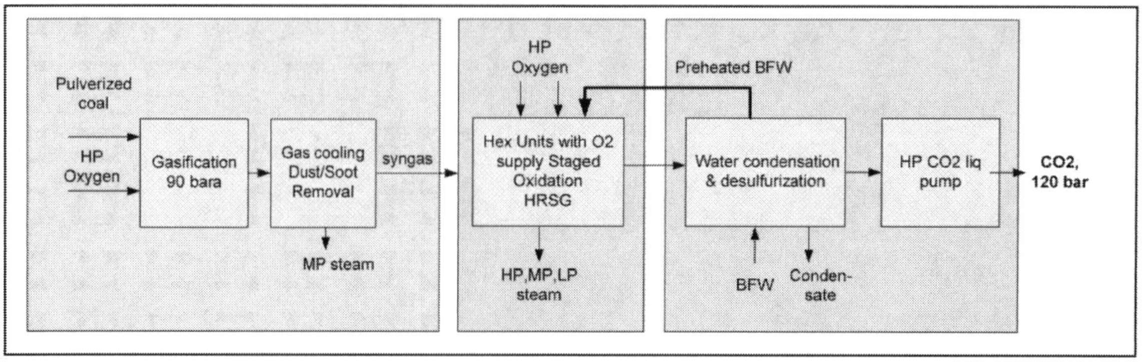

Figure 1: Process for high pressure oxyfuel process with staged combustion

Thermodynamic consideration

(i) Increase of thermal energy input by condensation energy

From the thermodynamic viewpoint, the combustion of coal at elevated pressure has the advantage in that the thermal energy of water condensation can be used at reasonable temperatures for the preheating of boiler feed water. The difference of this thermal energy is shown in a QT diagram in figure 2. The values are calculated based on a power plant with 2200 MW thermal energy input, for the combustion of pre-dried lignite with an approximately 22,000 kJ/kg ash free lower heating value. At 80 bars, the condensation starts at a temperature of about 145°C compared with about 80°C at atmospheric pressure. With this higher condensation temperature the boiler feed water can be preheated to about 160°C instead of 100°C and about 110 MW more thermal power can be used for the process.

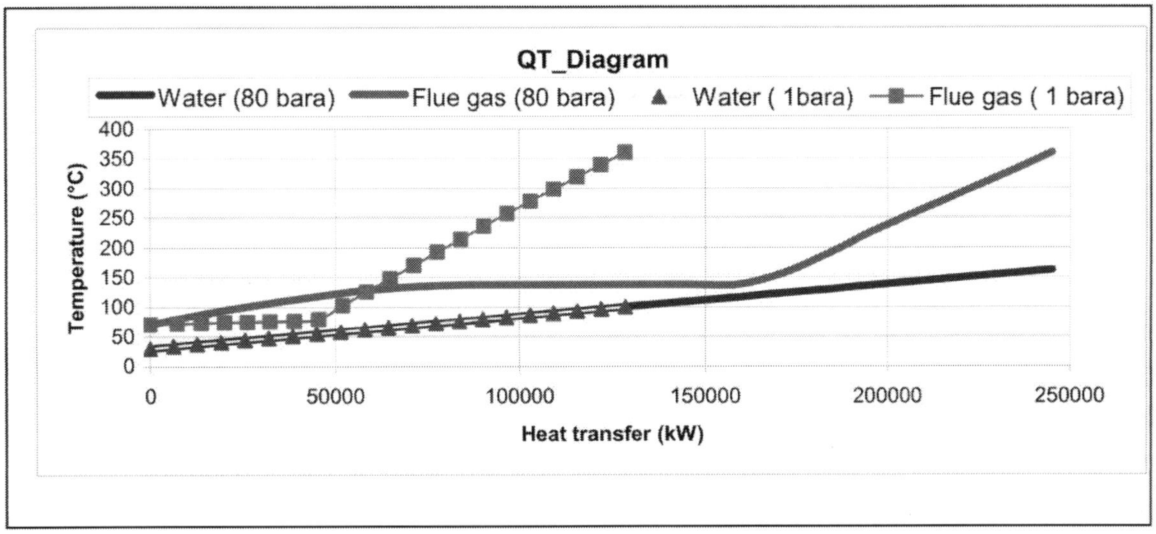

Figure 2: QT-diagram for atmospheric and high pressure flue gas case.

(ii) Increase of turbine efficiency with higher steam temperature

There are several groups investigating high temperature materials development, which can be applied in steam boilers for steam generation up to 700°C at 350 bars. With the currently available materials, the introduction of these operating conditions is not expected within the next several years in atmospheric boiler plants.

The high pressure oxyfuel process gives the opportunity to apply these process parameters already with existing materials. The reasons allowing this possibility of higher temperature and higher pressure steam generation are highlighted below:

- Reduction of differential pressure between steam and flue gas by 80 bars
- Increased heat transfer because of elevated pressure
- Reduction of heat and fouling resistance because of gas cleaning in front of heat exchanger section

Therefore the flue gas and material temperature and the tube material wall thickness in the heat exchangers can be lower than in atmospheric boilers for the same heat transfer power. The expected increase of electrical efficiency with the 700°C technology and equipment optimization is 4% points.

Key equipment for steam production

The production of high pressure steam at temperatures above 500°C needs specially constructed heat exchanger equipment. Typically used tube sheets in a shell and tube arrangement would become too thick at the high temperatures and pressures. A suitable construction for high pressure flue gas and staged combustion can be a spiral wound heat exchanger, see figure 3.

The steam can be overheated in tube bundles, which are connected over different tube baffles to the high pressure pipes. With this, the diameter and the mechanical load of the tube sheets can be reduced. Additional burning or oxidizing gases can be introduced in the heat exchanger.

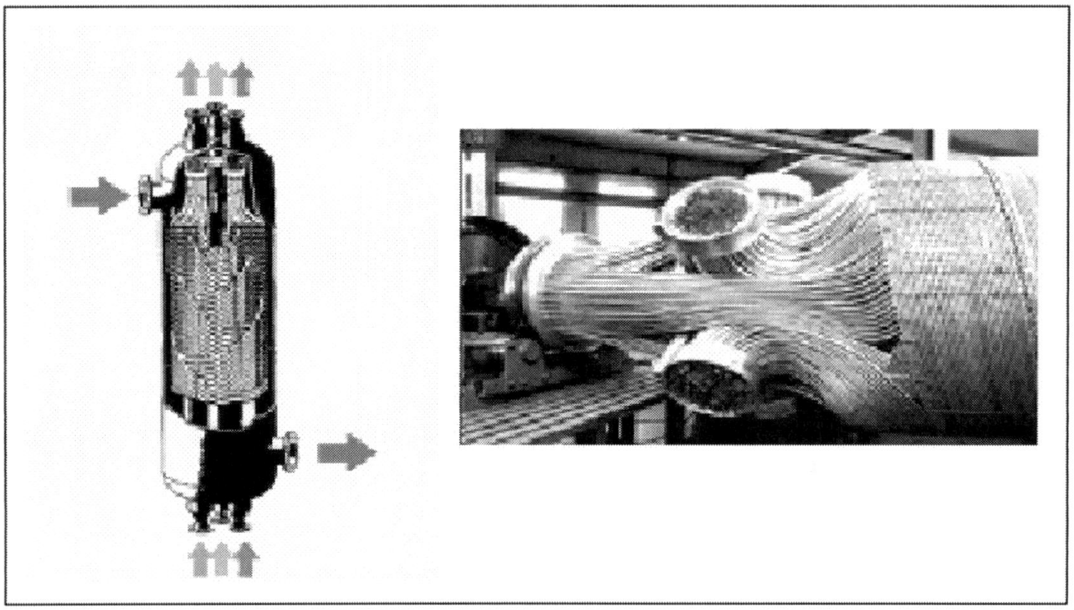

Figure 3: Spiral wound heat exchanger for high temperature steam production.

Results of process simulation

A detailed process simulation was conducted for the calculation of the electrical efficiency of the high pressure oxyfuel process and is compared with published data for the atmospheric oxyfuel process and an IGCC process with CO_2 capture. The results can be seen in table 1:

Table 1: Efficiencies of oxyfuel processes with CCS.

	Atmospheric Oxyfuel	IGCC & CCS	HP Oxyfuel 600 °C steam	HP Oxyfuel 700 °C steam
Net efficiency (%)	40 /3/	35 /2/	42	42+4=46
CO_2 capture (%)	90	90	99	99

Process concept and construction

The high pressure oxyfuel process results in a significant reduction of effective gas flow. This leads to a compact plant layout, which can be seen in figs 4 and 5.

A preliminary estimation of equipment weight shows a strong reduction of material for the boiler construction. The main equipment can be pre fabricated in a workshop, transported and directly installed at the construction site. With this the erection time and costs can be reduced significantly.

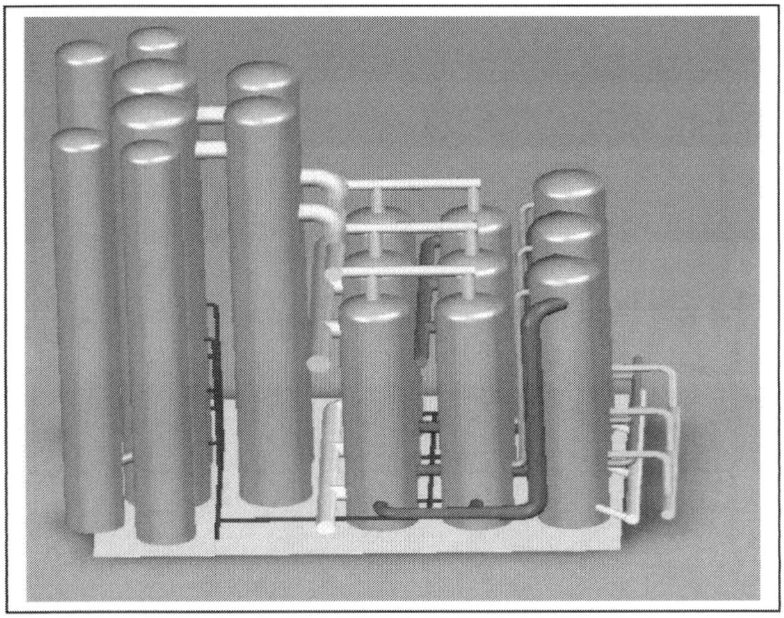

Figure 4: Sketch of high pressure oxyfuel process proportion Boiler and HP oxyfuel

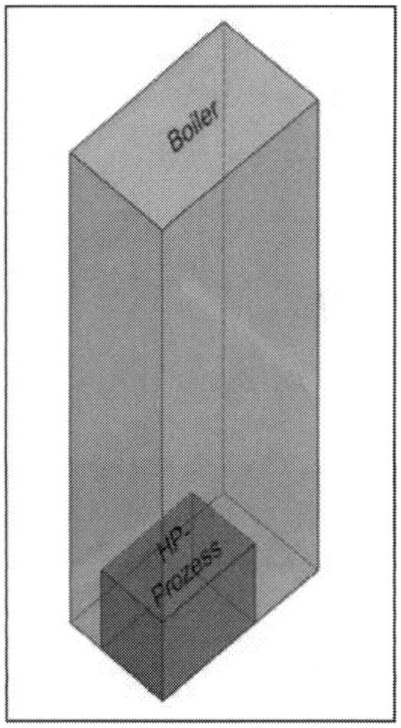

Figure 5: Size plant

Summary

The high pressure oxyfuel process involving staged combustion of gasified syngas to generate steam offers a number of advantages summarized below:

- Higher electrical efficiency by > 2-6 % points compared to atmospheric oxyfuel process.
- Strongly reduced cost by unit standardisation, prefabrication and material reduction.
- Strong reduction of boiler size and material demand.
- Simplified gas cleaning and CO_2 capture
- Option for 700°C steam technology possible
- The production of syngas gives the possibility to use the gasification for poly-generation processes in times of lower electricity demand.

(43d) High Temperature Oxygen Production

Michael K. Opoku, Chemical and Biological Engineering, South Dakota School of Mines and Technology, Rapid City, SD and Jan A. Puszynski, Chemical and Biological Engineering Department, South Dakota School of Mines and Technology, Rapid City, SD

A potentialalternative to produce clean coal technologies would be to combust fossil fuelsin pure oxygen instead of air, which contains approximately 78% nitrogen byvolume. If nitrogen was removed from the process, flue gas streams would have amuch higher concentration of CO_2, reducing or eliminating the needfor costly CO_2 capture. Moreover, NO_X emissions and thesubsequent need for processing would be reduced significantly. The main problemwith this method is separating oxygen from the air. This is usually completedcryogenically which requires a lot of energy (for a typical 500MW coal-firedpower station supplying pure oxygen requires at least 15% of the electricitythe plant generates annually). Cryogenic air separation technology forproducing high purity oxygen is energy intensive and requires air to be cooledto about - 300 degree Fahrenheit for separation. The cold oxygen must be heatedto high temperatures in order to couple this air separation process with coalpower plants. The high cost of producing oxygen, however, has made oxy-combustiona cost-prohibitive option for commercial use in most power plants. Developingtechnologies in oxygen and ion transport membranes have the potential to reducethe cost of oxygen production and increase oxy-combustion's cost-effectiveness.

In this research, multicomponentoxides containing Mn in different oxidation states were explored as reactantsfor two-step high temperature high purity oxygen production. High temperaturereduction and oxidation of these compounds were carried out in an inert gas andair atmosphere, respectively, using TGA-DSC. For synthesized manganese-based oxidematerials to separate air, they should be able to incorporate oxygen atoms into their crystal structure and release oxygen at the same or elevatedtemperatures while varying the oxygen partial pressure of the system.

$$Air + MO_{red(s)} \rightarrow MO_{oxid(s)} + gases(mostly\ N_2)$$

$$MO_{oxid(s)} \rightarrow MO_{red(s)} + O_{2(g)}$$

The results obtainedfrom TGA-DSC showed all investigated compounds can produce high purity oxygenat elevated temperatures. Details of multicomponent oxides preparation, characterization,and high temperature conversion of the oxides under reduction and oxidationconditions will be discussed.

(43e) Design, Construction and Operation of a Pilot Scale Integrated Pollutant Removal (IPR) Process for Oxy-Combustion

Stephen J. Gerdemann and Danylo Oryshchyn, National Energy Technology Laboratory, U.S. Department of Energy, Albany, OR

NETL has built an Integrated Pollutant Removal (IPR®) system at Jupiter Oxygen's 15MWth oxy-fuel test facility in Hammond, Indiana. An ASPENPLUS model of this installation is also in use to inform the development of the next stage of the IPR system-development. The installed system is designed to process a slip stream of flue gas from an oxy-coal boiler. This flue is composed primarily of CO_2 (~55%) and water (~30%) with smaller amounts O_2, N_2, Ar and other combustion products such as NO_X and SO_X. The IPR process is designed to output a relatively pure (>90%) CO_2 stream at pipeline pressure while recovering as much energy as possible and integrating that recovery into the overall combustion process. Water condensed from the flue-gas is also recovered. Water-treatment process design is being refined through experimentation.

IPR consists of a spray tower (direct-contact heat exchanger) followed by 4 stages of compression (3, 14, 40 & 140 atm) with intercooling between each stage. The process has been completely instrumented to flow rate, temperature and pressure at each stage. Material samples include gas and liquid throughout IPR, and solids (where they fall out in the first IPR operation – a direct contact heat exchange spray tower). In the field, IPR has shown that the achieved the desired output and has further defined challenges for IPR design and the design of its instrumentation, for example accurate flow measurement of a corrosive/erosive fluid stream, minimizing corrosion in components, and in-stream condensation which also efficiently removes soluble gasses. This paper discusses these challenges, presents results from IPR operation, and identifies opportunities for optimization of an oxy-combustion IPR system, also visiting the design of the system model in ASPEN plus.

See more of this Session: Oxycombustion of Coal and Other Fuels I
See more of this Group/Topical: Energy and Transport Processes

() From Risk Management to Sustainability: in Theory and in Practice

Subhas Sikdar, National Risk Management Research Lab, US EPA, Cincinnati, OH

Environmental risk management for years has allowed interventions, which depended on understanding and quantification of risk. Sustainability now is superseding various earlier risk management approaches such as waste treatment, waste minimization, pollution prevention, and design for environment. Industrial, and government entities at state, city, and municipal levels, have espoused the principles of sustainability in implementing development and growth agenda. Approaches vary but all use criteria, indicators and metrics to determine favorable sustainability outcomes. There is still much subjective judgment in these determinations, despite attempts to quantify such assessments. Of the three sustainability domains, industry struggles with quantifying societal impacts or benefits, especially at the product and process levels. Lack of transparency from one to another company is a significant barrier to researchers for building a uniform framework and acceptable set of metrics for quantifying the sustainability state of products and processes. At the level of the corporation, much public reporting is now available. But that sort of reporting, while it is important for public relations, has little research value. In practice, the technical actions to improve the environmental stewardship for products and processes have moved not far from traditional yet locally effective approaches of risk management. Sustainability being an inherently multidisciplinary idea, only integrative approaches that use life cycle thinking in technical enquiry and data gathering will move us to a better position to judge sustainability of products and processes.

See more of this Session: Plenary Session: 1st Annual World Congress On Sustainable Engineering
See more of this Group/Topical: 1st Annual World Congress on Sustainable Engineering

() Sustainable Forest Resource Biorefineries Vision for the Future

Shri Ramaswamy, Bioproducts and Biosystems Engineering, University of Minnesota, St Paul, MN

The forest products industry has been a globally competitive industry in the US and North America for much of the last century. Abundance of forest resources, high quality infrastructure for its collection and processing at large scale, availability of human and financial capital and accessibility to the markets are the key competitive advantages. This industry has been primarily focused on producing low cost cellulose fiber and fiber products including building materials. The industry's markets are mature and the products have become commodities and as a result, profit margins have dwindled. High investment costs, increasing costs of raw materials, and leveling of operational efficiency and the technology playing field have thrown the industry's future economic viability into question. The industry is responding by looking to expand the the product portfolio to increase revenue streams and profitability. A diverse portfolio of bioenergy and biomaterials products can transform conventional pulp and paper mills into full fledged biorefineries. With significant expertise in the understanding, logistics, handling and processing of lignocellulosic biomass and sustainable forestry practices, the forest industry has much to offer in contributing to the future development and engineering of sustainable biorefineries. In addition to high value cellulose fiber and fiber based products including nano-cellulose composites, the individual constituents of biomass namely hemi-cellulose, lignin, extractives and such can be sustainably extracted and converted to value-added products. In addition to second and third generation biofuels, bioenergy and other bio-based products offer a tremendous renewable products opportunity for the US and much of the world. Thermochemical or biochemical approaches or a combination of the two hold promise for the future. There are still unsolved scientific and engineering challenges in achieving future sustainable biorefineries including recalcitrance of biomass, density of biomass feedstock, low cost biomass pretreatment approaches, separation and purification of sugar and non-sugar streams, bioengineering of fermentative organisms, removal of fermentation inhibitors and purification of product streams, novel bioprocessing approaches suitable for mixture of biomass feedstocks and value added co-product development. The approaches and solutions need to include the local and regional natures of biomass species and sustainability considerations and they can be in combination with existing forest and other biomass based industries.

See more of this Session: Plenary Session: 1st Annual World Congress On Sustainable Engineering
See more of this Group/Topical: 1st Annual World Congress on Sustainable Engineering

() Sustainable Hydrogen, Gasoline, and Diesel Biofuels From Grasses and Crop Residues

Ram B. Gupta, Chemical Engineering, Auburn University, Auburn, AL

Today world is facing two critical problems: (1) high fuel prices, and (2) climatic changes. Both of these are linked to the overdependence on the fossil fuels: petroleum, natural gas, and coal. The transport is almost totally dependent on fossil particularly petroleum based fuels such as gasoline, diesel fuel, liquefied petroleum gas, and natural gas. Of the special concerns are the liquid fuels for use in the automobiles. Hence, there is a widespread recent interest in obtaining liquid fuels from non-fossil sources. The combination of rising oil prices, issues of security, climate instability and pollution, and deepening poverty in rural and agricultural areas, is propelling governments to enact powerful incentives for the use of biomass-based fuels, which is in turn sparking investment.

Production of grain-based ethanol and vegetable-oil-based bio-diesel are being practiced today with difficulties due to the competition with the food supply. Now the industry is looking toward using biomass which does not compete with the food or fiber supplies; however it is facing unsolved technological/scientific difficulties. Cellulosic ethanol involves the use of expensive cellulase enzymes. Biomass can be gasified to produce synthesis gas which can be converted into liquid fuels by Fischer-Tropsch technology. Another exciting technology is hydrothermal process in which biomass components (cellulose, hemi-celluloses, and lignin) are first depolymerized to their respective monomers, followed by hydrogenation to produce liquid fuels. The process is carried out in sub- or supercritical water, which can utilize wet biomass without any need for drying. Since the carbon dioxide released upon burning of the biomass-based fuels is captured-back in producing new biomass, there is no net emission of CO_2. The technology can significantly address both fuel and climatic challenges as today's world has a large surplus of biomass or biomass-producing capacity.

Biomass can also be used to produce hydrogen, which can be directly used as a clean fuel or combined with heavy crude in the upgradation step. The former use can produce petroleum which has part of the energy from biomass, suitable for near term sustainability.

See more of this Session: Plenary Session: 1st Annual World Congress On Sustainable Engineering
See more of this Group/Topical: 1st Annual World Congress on Sustainable Engineering

(70a) Biogenic Organic Aerosol: Aging Under Various Light Conditions and Oxidant Concentrations

Kaytlin M. Henry and Neil M. Donahue, Chemical Engineering, Carnegie Mellon University, Pittsburgh, PA

Fine particulate matter (PM) affects visibility and cloud formation, scattering and/or absorbing light, and ultimately affecting climate (IPCC, 2007). These small particles also impact human health, causing damage to respiratory and cardiovascular systems (Pope et al., 2002). Atmospheric PM consists of many inorganic species (such as sulfate, nitrate, sea salt, and dust), which have been relatively well constrained. However, between 20-90% of the total PM is organic, consisting of thousands of compounds, and this complex mixture is not well understood (Kanakidou et al., 2005; Zhang et al., 2007). In addition, most studies have focused on only the first few hours of organic aerosol (OA) formation. PM can remain in the atmosphere for ~1 week, therefore we need to understand further chemical processing of these compounds, which is termed "aging".

An additional layer of complexity is added when we consider the volatile organic compounds (VOCs) in the atmosphere. This gas phase is in equilibrium with the OA; therefore, gas-phase reactions in turn affect the particle phase for both initial OA formation and aging. This work ages the OA system formed from alpha-pinene ozonolysis. alpha-Pinene is a biogenic compound, emitted from pine trees, and a model for many biogenic systems with ring compounds containing an endocyclic double bond. Because of the equilibrium between all gas and aerosol species, the presence of biogenic aerosol may influence the formation of anthropogenic aerosol and vice versa. Biogenic VOC emissions can be up to 10 times higher than anthropogenic VOCs; therefore it is essential that we isolate the biogenic system to understand its atmospheric processing. This aging is done by gas-phase reactions with the OH radical, the most powerful atmospheric oxidant. We then look for an affect on the particle phase, as our concerns for human health and climate lie with PM concentrations.

Experiments were performed at the Carnegie Mellon Smog Chamber, a 12 m^3 Teflon reactor. OH aging is performed via photolysis of either HOOH or HONO using UV lights or in the dark via tetra-methyl ethylene ozonolysis. In this way we can isolate the features of both light and dark aging scenarios. In addition, the HONO source produces higher concentrations of OH, allowing us to probe the OH concentration variable. Results show that this alpha-pinene OA system is very photochemically rich and dynamic. In the dark, OA concentrations grow, while under the UV lights the aerosol evaporates. This evaporation can be combated by an order of magnitude increase in OH concentrations with the HONO source. This complex mechanism and equilibrium partitioning must continue to be explored as it may impact other, chemically similar systems. Finally this aging can be included in chemical transport models, which can aid policy makers in decisions regarding emissions reductions.

References:

IPCC: Forth Assessment Report: Climate Change, 2007.

Kanakidou, M., Seinfeld, J. H., Pandis, S. N., Barnes, I., Dentener, F. J., Facchini, M. C., Van Dingenen, R., Ervens, B., Nenes, A., Nielsen, C. J., Swietlicki, E., Putaud, J. P., Balkanski, Y., Fuzzi, S., Horth, J., Moortgat, G. K., Winterhalter, R., Myhre, C. E. L., Tsigaridis, K., Vignati, E., Stephanou, E. G., and Wilson, J.: Organic aerosol and global climate modelling: a review, Atmospheric Chemistry and Physics, 5, 1053-1123, 2005.

Pope, C. A., Burnett, R. T., Thun, M. J., Calle, E. E., Krewski, D., Ito, K., and Thurston, G. D.: Lung cancer, cardiopulmonary mortality, and long-term exposure to fine particulate air pollution, JAMA-J. Am. Med. Assoc., 287, 1132-1141, 2002.

Zhang, Q., Jimenez, J. L., Canagaratna, M. R., Allan, J. D., Coe, H., Ulbrich, I., Alfarra, M. R., Takami, A., Middlebrook, A. M., Sun, Y. L., Dzepina, K., Dunlea, E., Docherty, K., DeCarlo, P. F., Salcedo, D., Onasch, T., Jayne, J. T., Miyoshi, T., Shimono, A., Hatakeyama, S., Takegawa, N., Kondo, Y., Schneider, J., Drewnick, F., Borrmann, S., Weimer, S., Demerjian, K., Williams, P., Bower, K., Bahreini, R., Cottrell, L., Griffin, R. J., Rautiainen, J., Sun, J. Y., Zhang, Y. M., and Worsnop, D. R.: Ubiquity and dominance of oxygenated species in organic aerosols in anthropogenically-influenced Northern Hemisphere midlatitudes, Geophys. Res. Lett., 34, 6, L1380110.1029/2007gl029979, 2007.

See more of this Session: Atmospheric Chemistry and Physics II
See more of this Group/Topical: Environmental Division

(70b) Size Dependent Growth Rates of Freshly Nucleated Particles

Chongai Kuang[1], Modi Chen[2], Peter H. McMurry[2] and Jian Wang[1], (1)Environmental Science, Brookhaven National Laboratory, Upton, NY, (2)Mechanical Engineering, University of Minnesota, Minneapolis, MN

Atmospheric aerosols influence climate and climate change on local to global scales by affecting the atmospheric radiation balance directly through scattering and absorbing incoming solar radiation and indirectly as cloud condensation nuclei (CCN). New particle formation (NPF) by photochemical reactions of gas-phase precursors greatly increases the number concentrations of atmospheric aerosols, and is often followed by rapid growth to a CCN-active size, leading to a significant enhancement of the global CCN population. This rapid growth, often two to fifty times that of the growth assuming the condensation of sulfuric acid alone, is not well understood or represented in global models, limiting the ability to accurately assess the impact of NPF on the global surface CCN population and the aerosol indirect effect. Understanding this rapid particle growth and representing it in models require direct measurements of the particle growth rate, especially immediately following nucleation, at which stage freshly nucleated particles are most vulnerable to scavenging by existing particles. Recently, size distribution measurements of freshly nucleated 1 – 6 nm particles were acquired during an intensive field campaign in Boulder, CO using a diethylene glycol based ultrafine condensation particle counter optimized for detection of 1 nm condensation nuclei as a detector in a SMPS instrument. These measurements provide one of the first direct measurements of size and time-dependent particle growth rates of freshly nucleated particles. Observed growth rates for 1 – 6 nm particles were found to range from 1 – 30X that assuming only sulfuric acid condensation, with these growth rate enhancements increasing approximately linearly with diameter below 3 nm.

See more of this Session: Atmospheric Chemistry and Physics II
See more of this Group/Topical: Environmental Division

(70c) On the Nature of Water-Soluble Organic Aerosols In the Southern California Region

Armin Sorooshian[1], Hanh Duong[1], Anna Wonaschuetz[2], Scott Hersey[3], Jill Craven[3], Andrew Metcalf[3], Athanasios Nenes[4], Haflidi Jonsson[5], Richard C. Flagan[6] and John H. Seinfeld[7], (1)Chemical and Environmental Engineering, University of Arizona, Tucson, AZ, (2)Atmospheric Sciences, University of Arizona, Tucson, AZ, (3)Caltech, Pasadena, CA, (4)Georgia Institute of Technology, Atlanta, GA, (5) Center for Interdisciplinary Remotely Piloted Aircraft Studies, Marina, CA, (6)Chemical Engineering, California Institute of Technology, Pasadena, CA, (7)California Institute of Technology, Pasadena, CA

Water soluble organic carbon (WSOC) species are important owing to their ability to influence aerosol water-uptake properties and radiative forcing. The chemical composition, sources, and atmospheric evolution of WSOC are not well established. A particle-into-liquid sampler quantified WSOC during the airborne deployment of the Center for Interdisciplinary Remotely-Piloted Aircraft Studies (CIRPAS) Twin Otter as part of the CalNex campaign (May 2010). Eighteen flights were conducted to study the nature of aerosol particles in the San Joaquin Valley, the Los Angeles Basin and outflow regions. The measurements indicate that WSOC accounts for a significant portion of the organic aerosol mass, with likely contributions from both primary emissions and secondary production processes. The highest levels of WSOC were associated with smoke plumes in the Los Angeles Basin. On 11 of 16 flights, WSOC exhibited its highest concentration above the first 500 meters of the boundary layer, usually near the Banning and Cajon Passes, and also near the hilly terrain over corridor between San Bernadino and Riverside on the eastern edge of the Los Angeles basin. Aging and transport of polluted aerosol from the western edge of the Los Angeles basin often led to peak WSOC levels near the eastern edge of the basin near Riverside and San Bernadino during the afternoons. This work will discuss the extent to which concentrations and the spatiotemporal distribution of WSOC are dependent on secondary production, transport, meteorology, and the influence of biomass burning.

See more of this Session: Atmospheric Chemistry and Physics II
See more of this Group/Topical: Environmental Division

(70d) Inferring New Particle Formation and Growth From Measurements of Aerosol Properties

Robert Bullard, Chemical and Biochemical Engineering, University of Iowa, Iowa City, IA and Charles O. Stanier, Chemical & Biochemical Engineering, University of Iowa, Iowa City, IA

Aerosols play an important role in Earth's radiation balance and their climate forcing effects represent the largest uncertainties in the Intergovernmental Panel on Climate Change's 2007 fourth assessment report. The nucleation and growth of gas phase molecules into liquid or solid phase particles is the main source of ultrafine aerosol particles in the atmosphere. The complicated underlying physical properties and conditions for nucleation are not well understood. While data on nucleation is limited both temporally and spatially, data on atmospheric particle number concentration is readily available in many locations. Our current goal is to improve methods of data analysis in order to infer information about new particle formation and growth from long term measurements of particle number and other aerosol properties. As a test case, we investigate a five year (2006 – 2010) record of 1 minute particle number concentration data from Bondville, IL to infer nucleation rates in rural continental environments. The probability of nucleation can be quantified by optimization of agreement between possible size distributions (constrained in total number by observations) and collocated optical and physical measurements. The suggesting of an aerosol size distribution will be aided by data from a 2005 field campaign in which the aerosols size distribution was measured. Parallel efforts at modeling the nucleation activity will also be discussed.

See more of this Session: Atmospheric Chemistry and Physics II
See more of this Group/Topical: Environmental Division

(70e) Compositional and Mixing State Impacts On Size-Resolved CCN Concentrations In Atlanta, Georgia, During 2008-2009

Richard H. Moore, Chemical and Biomolecular Engineering, Georgia Institute of Technology, Atlanta, GA, Luz T. Padro, Civil & Environmental Engineering, Tufts University, Molly McLaughlin, School of Chemical & Biomolecular Engineering, Georgia Institute of Technology, Atlanta, GA and Athanasios Nenes, Georgia Institute of Technology, Atlanta, GA

The effect of organic species on the cloud droplet formation process constitutes a large source of uncertainty in aerosol-cloud-climate interactions studies. In-situ size-resolved measurements of the Cloud Condensation Nucleus (CCN) activity of ambient aerosol can unravel such complex chemical effects. The city of Atlanta provides a unique environment for exploring the interplay between the anthropogenic emissions of a major metropolitan area (believed to be predominantly SO_2 and $(NH_4)_2SO_4$ with some organic contribution) with biogenic emissions from surrounding forest and farmland (believed to be comprised of mostly organics), and their relative contribution to the aerosol size distribution and chemical mixing state, and, thus, cloud condensation nuclei (CCN).

We present a comprehensive characterization of aerosol and CCN sampled in Midtown Atlanta during the period of August 2008 to August 2009, which includes continuous observations over a nine-month period on the Georgia Tech Rooftop Measurement Platform (in midtown Atlanta) and also during the month-long 2008 AMIGAS and 2009 NCCN intensive campaigns at the Jefferson Street monitoring site. CCN were measured with a Droplet Measurement Technologies Continuous-Flow Streamwise Thermal-Gradient Chamber[1,2], operated in spectrometer mode using Scanning Flow CCN Analysis[3] (SFCA), or size-resolved counter mode using Scanning Mobility CCN Analysis[4] (SMCA). SFCA and SMCA were used to provide size-resolved CCN concentrations over a variety of particle sizes (40-120 nm) and supersaturations (0.1-1%) with high temporal resolution.

The daily trend of the mixing state, CCN activity, and droplet growth kinetics of the size-resolved CCN are presented and compared to chemical composition measurements. Inferences about the impact of photochemistry and mixing on CCN activity, hygroscopic uptake and droplet activation kinetics are carried out. We also quantify the predictive uncertainty associated with simplifying compositional assumptions (e.g., size-invariant composition, organic insoluble/soluble with constant hygroscopicity) used for predicting CCN concentrations in global climate models.

[1]Roberts, G. and A. Nenes. *Aerosol Sci. Technol.*, 2005

[2]Lance et al., *Aerosol Sci. Technol.*, 2006

[3]Moore, R.H. and A. Nenes. *Aerosol Sci. Technol.*, 2009

[4]Moore, R.H., A. Nenes, and J. Medina. *Aerosol Sci. Technol.*, 2010

See more of this Session: Atmospheric Chemistry and Physics II
See more of this Group/Topical: Environmental Division

(70f) Dynamical Equilibrium States In Low Temperature Cirrus

Athanasios Nenes, Georgia Institute of Technology, Atlanta, GA and Donifan Barahona, Chemical Engineering, Georgia Institute of Technology, Atlanta, GA

Low ice crystal concentration and sustained in-cloud supersaturation, commonly found in cloud observations at low temperature, challenge our understanding of cirrus formation. Heterogeneous freezing from effloresced ammonium sulfate, glassy aerosol, dust and black carbon are proposed to cause these phenomena; this requires low updrafts for cirrus characteristics to agree with observations and is at odds with the gravity wave spectrum in the upper troposphere. Background temperature fluctuations however can establish a "dynamical equilibrium" between ice production and sedimentation loss (as opposed to ice crystal formation during the first stages of cloud evolution and subsequent slow cloud decay) that explains low temperature cirrus properties. This newly-discovered state is favored at low temperatures and does not require heterogeneous nucleation to occur (the presence of ice nuclei can however facilitate its onset). Our understanding of cirrus clouds and their role in anthropogenic climate change is reshaped, as the type of dynamical forcing will set these clouds in one of two "preferred" microphysical regimes with very different susceptibility to aerosol.

See more of this Session: Atmospheric Chemistry and Physics II
See more of this Group/Topical: Environmental Division

(70g) CCN and Morphology Imaging of Numerous Oxidized VOCs In a Smog Chamber

William L. Madry[1], Dabrina Dutcher[2] and **Timothy Raymond**[1], (1)Chemical Engineering, Bucknell University, Lewisburg, PA, (2)Bucknell University, Lewisburg, PA

Biogenic sources emit approximately 1150 Tg C/yr in the form of thousands of volatile organic compounds (VOCs) [1]. VOCs nucleate to form secondary organic aerosol (SOA), which participate in atmospheric processes such as the absorption and scattering of radiation, activation as cloud droplets, and other heterogeneous reactions. We have constructed a small, indoor smog chamber for use in studying SOA formation and properties. We generated SOA through the oxidation of VOC directly with injected ozone and indirectly via the photochemical production of ozone from NOx precursors. We studied the time-dependent size distribution and cloud condensation nucleus (CCN) activation potential of aerosol generated from both biogenic and anthropogenic VOCs, including common terpenes such as alpha- and beta-pinene and d-limonene and industrial byproducts such as benzene and toluene. We find that the VOCs nucleate rapidly and the resultant aerosol activates at low supersaturations of water vapor, though the generated SOA is found to be less CCN active than inorganic salts such as sodium chloride and ammonium sulfate. CCN activity is observed to increase with aerosol age, and may influence cloud properties through enhancement of CCN number concentration. The evolution of the SOA size distribution has been found dependent on the initial reactant species, with mixtures of VOC displaying an average of the individual species' distributions. Aerosol particles produced from different reactants are in the process of being characterized by Atomic Force Microscopy. Early microscopy results indicate that the SOA particle shape is not consistent across all reactant species. Our research suggests that models of cloud formation should not neglect contributions and enhancements to total CCN activity from the oxidation of VOCs.

[1] Goldstein and Galbally, ES&T, March 2007.

See more of this Session: Atmospheric Chemistry and Physics II
See more of this Group/Topical: Environmental Division

(70h) Investigation of Hygroscopic and Morphological Properties of Atmospheric Aerosols

Laura Cook and **Timothy Raymond**, Chemical Engineering, Bucknell University, Lewisburg, PA

The first objective of this research was to examine the hygroscopic and morphological nature of various substances through the use of an Environmental Scanning Electron Microscope (ESEM). The hygroscopic growth and changes in morphology for pure-component aerosols were studied for particles greater than two microns in size. Hygroscopic growth was observed through changes in relative humidity (RH) and hygroscopic growth curves were created for ammonium sulfate, ammonium chloride, sodium chloride, calcium chloride, glutaric acid, and malonic acid. The ESEM system is advantageous over the Hygroscopic Tandem Differential Mobility Analysis (HTDMA) system becuase it offers a in-situ view of changes in morphology as particles experience water uptake.
The second objective of this research, the hygroscopic growth of multi-component aerosol mixtures, was studied using Hygroscopic Tandem Differential Mobility Analysis (HTDMA). The size distribution for an aerosol stream was determined before and after the stream was subjected to an increase in relative humidity. The complex mixtures examined contained combinations in equal mass amounts of adipic acid, glutaric acid, maleic acid, aspartic acid, tartaric acid, succinic acid, malonic acid, phthalic acid, and azelaic acid. Pure solutions of sodium chloride, ammonium sulfate, ammonium chloride, ammonium nitrate, calcium chloride, maleic acid, tartaric acid, malonic acid, and phthalic acids were also run through the HTMDA system. Understanding the hygroscopic growth and morphology of aerosols are essential for understanding cloud formation and climate change. This research will provide insight into the physical morphological changes that atmospheric particles exhibit when experiencing changes in humidity.

See more of this Session: Atmospheric Chemistry and Physics II
See more of this Group/Topical: Environmental Division

(78a) Climate Change and Greenhouse Gas Emissions: Current Law, Voluntary Initiatives and Future Implications

Mary Ellen Ternes, McAfee& Taft, Oklahoma City, OK

This presentation will provide a review and update of legal and policy developments regarding greenhouse gas (GHG) emissions and mitigation. Specifically, this presentation will review the current and potential future regulatory, legislative and judical landscape impacting climate change drivers, including powerplant and other energy sector regulation under the Clean Air Act and energy conservation programs. This presentation will also summarize California programs, This session will be of particular interest to those interested in research opportunities resulting from legislatively mandated technology forcing as well as regulatory implementation

See more of this Session: Climate Change Legislation and Regulation
See more of this Group/Topical: Environmental Division

(78b) Total Global Warming Potential: An Expanded Definition of GWP

Paul Blowers, Chemical and Environmental Engineering, The University of Arizona, Tucson, AZ, Christina Canter, Dept. of Chem. and Env.Engr., The University of Arizona, Tucson, AZ and Daniel David Galvan, Chem. and Env. Engr., Univ. of Arizona, Tucson, AZ

Total Global Warming Potential: An Expanded Definition of GWP

Paul Blowers,Christina Canter, Daniel David Galvan

Global warming potential was created as a referenced valuefor comparing chemicals that may contribute to global climate change on aconvenient basis. The formal generallyaccepted definition is:

$$GWP = \frac{\int_0^{TH} a_i[x_i(t)]dt}{GWP_{CO_2}}$$

where TH is an arbitrary time horizon,normally chosen to be 20, 100, or 500 years, a_i is the radiativeforcing of a compound, i, defined as the energy absorbed per square meter ofatmosphere per ppb of concentration of compound i, and $x_i(t)$ is thetime dependent concentration time profile for chemical i. This definition inherently assumes that theparent compound i is the only species that will contribute to capturing energyand it is removed from the atmosphere by hydroxyl radical attack, the dominantremoval mechanisms. However, largercompounds may fracture apart into daughter species that may form stablecompounds which go on to have their own atmospheric lifetimes and radiativeforcing values that may additionally contribute to global climate change. This would change the above equation to:

$$GWP_{total} = \sum_j \frac{\int_0^{TH} a_{i,j}[x_{i,j}(t)]dt}{GWP_{CO_2}}$$

where i is the parent species and j representsthe various daughter species which may temporally form and dissipate. The expanded definition opens up the need forother data that may be experimentally inaccessible due to the difficulty ofexamining total degradation mechanisms and in identifying other removalmechanisms like dissolution into raindrops and then rain-out of species. This work applies quantum chemicalcalculations to evaluate several candidate species to highlight when it isimportant to use an expanded GWP definition to capture potential environmentalliabilities of compounds. Whereexperimental data are unavailable, quantum chemical calculations are used topredict intermediate values for radiative forcing and kinetic degradation rateconstants.

See more of this Session: Climate Change Legislation and Regulation
See more of this Group/Topical: Environmental Division

(78c) Regional Carbon Footprint of Dairy Feed Rations for Milk Production In the United States

Felix Adom[1], Ashley Maes[1], Charles Workman[1], Zara Clayton-Nierderman[2], Greg Thoma[2] and David Shonnard[1], (1)Chemical Engineering, Michigan Technological University, Houghton, MI, (2) Department of Chemical Engineering, University of Arkansas, Fayetteville, AR

A carbon footprint analysis was conducted for cultivation and harvesting of feed rations used for the production of dairy milk in the U.S. The goal is to determine the greenhouse (GHG) emissions on a basis of 1 kg of ration harvested or produced (kg CO_2eq. / kg of feed rations) in the U.S on a regional basis and to make recommendations to reduce GHG emissions from the crop production chain.

Commonly used dairy feedstuffs in the U.S. such as soybeans, corn grain/silage alfalfa, etc were identified based on a recent literature review and information from dairy farm surveys. By dividing the U.S into five dairy regions the following input data for the cultivation and harvesting of dairy rations were collected for each region; crop production data, energy input, soil amendments, and crop protection chemicals. Life cycle inventory (LCI) data for these inputs were mainly collected from the USDA National Agricultural Statistical Service on a state-by-state basis as well as from state extension services forage crop budget estimates. Apart from consulting other LCA studies and published articles and reports, this cradle-to-farm gate GHG impact analysis was conducted utilizing the EcoInvent unit processes in SimaPro version 7.1©.

The final GHG results (kg CO_2 eq. / kg of feed ration) varied regionally depending on a number of factors such as the lime and fertilizer application rates. The average national U.S. GHG results of the main crops are as follows: corn grain (0. 331), corn silage (0.071), dry distiller grains (0.818), wet distiller grains (0.268), oats (0.736), soybeans (0.344), soybean meal (0.406), winter wheat (0.395), alfalfa hay (0.147) and forage mix (0.137).

Dairy production region 2 (the southeast) generally showed a relatively high level of carbon footprint for most crops grown locally and this can be attributed mostly to the higher application rates of both synthetic fertilizers and lime. The highest contributor to carbon footprint for most regions (apart from soybean) was identified to be the application of inorganic N fertilizer. Finally, efficient transfer of knowledge to farmers with regards to Fertilizer Best Management Practices such as precision application of farm nutrients as well as efficient management of crop residues can contribute significantly to reduce the regional carbon footprints.

See more of this Session: Climate Change Legislation and Regulation
See more of this Group/Topical: Environmental Division

(78d) An Energetic Comparison Between Heat Integrated Pressure-Temperature-Swing Adsorption Using Alkaline-Earth Oxide Sorbents and Amines for the Capture of CO2 From Coal Fired Plants

Paul Ghougassian, Chemical Engineering, UCLA, Los Angeles, CA

In this work, we present a comparison on an energy basis between a heat integrated process using Alkaline-Earth Oxide sorbents and traditional amines for the capture of CO_2 from the flue gas of coal-fired plants. State of the art amines require 1550 Btu/lb of CO_2 captured. This drastically reduces the output and efficiency of the plant since this energy is drawn from low-pressure steam used to heat the reboiler of the amine process instead of producing useful work. An alternative design relies on using Alkaline-Earth Oxide sorbents in a multi-bed arrangements and cycle of desorption/adsorption steps to adsorb the CO_2 from the flue gas and desorb it in a water stream at lower pressure (higher temperature) before sending the mixture for separation, liquefaction and sequestration. The heat integration part involves using a heat-pump cycle around the adsorber/desorber unit to channel the heat released by the lower temperature exothermic adsorption reaction in the adsorber to the higher temperature endothermic desorption reaction occurring in the desorber. Current studies based on modeling the adsorber/desorber beds as equilibrium reactors show a 10% increase in output at the same level of CO_2 capture when compared with amines.

See more of this Session: Climate Change Legislation and Regulation
See more of this Group/Topical: Environmental Division

(78e) Overview and Perspectives of CO2 Emission In Chinese Cement Industry

De-Long Xu[1], Hui Li[2], Jian Zhang[3], Yongxiang Ren[2], Yong-Hua Duan[3] and Yong Min[2], (1)Xi'an University of Architecture and Technology, Xi'an, China, (2)Xi'an University of Architecture &Technology, Institute of Powder Engineering, Xi'an, China, (3)Material Institute, Xi'an University of Architecture and Technology,, Xi'an, China

Since 1985, China became the world's largest cement-producing countries in the world-BThe total output of China's cement amount and its market share in the world has increased year after year. Due to the technology improvement, the unit energy consumption of Chinese cement industry shows a declining trend. As a example, the heat consumption declined by 11% in 2005 compared with 1995; the total energy consumption in Chinese cement industry declined from 11.81% in 2001 dropped to 7.44% in 2007.

CO2 emissions from the cement industry comes mainly from limestone and other carbonate material contained in the decomposition and combustion of fossil fuels. With the gradual improvement of production technology, the direct emission of CO2 per unit of product shows a declining trend. The direct CO2 emissions from 0.787t CO2 / t in 1995 reduced to 0.592t CO2 / t in 2005.

The presentation will introduce following areas:
(1) for the next 10-30 years, China's cement industry CO2 emission reduction targets;
(2) to develop low-carbon economy of China's cement industry to achieve the overall strategic thinking;
(3) analysis of various emission reduction technology skills, technology maturity and the reduction potential of technology promotion in each period to determine the focus of the key technical issues to be carried out;
(4) the formation of China's cement industry to achieve a low carbon economy, technology roadmaps;
(5) specific suggestions and assurance measures.

See more of this Session: Climate Change Legislation and Regulation
See more of this Group/Topical: Environmental Division

(78f) Control of Hg Emissions From Coal-Fired Power Plants In China

Fan Zhang[1], **Shuang Deng**[1], Yu Liu[1], Hongmei Wang[1] and Zi Liu[2], (1)Chinese Research Academy of Environmental Science, Beijing, China, (2)Environmental Protection of the People's Republic of China, Beijing, China

Mercury emissions are regarded as one of the most problematic environmental issues in the world because of their bio-accumulation within the aquatic food chain. Through bio-accumulation, mercury may result in neuron damage to humans. Coal combustion processes are found to be the main source of anthropogenic mercury emission to the atmosphere. These processes account for about 45% of the total global anthropogenic mercury emissions. Coal consumption was about 3 billion tons in China in 2009. Mercury contamination is a serious problem in China. Chinese government attaches great importance to prevention and control of mercury pollution. The Ministry of Environmental Protection of the People's Republic of China initiated the trial of mercury pollution control in coal-fired power plants on September 2010.

In this paper, the mercury emissions from 20 coal-fired power plants in china are investigated. Simultaneously sampling of coal, bottom ash, fly ash, FGD slurry, flue gases, and particles associated with the gas phase has been performed. Ontario-Hydro method and 30B method were used to analyze the mercury in flue gas, and the effects of the existing air pollutant control devices on mercury emission in coal-fired power plants were investigated. According to these results, some suggestions of controlling the mercury emission from the coal-fired power plants are made and submitted to the Ministry of Environmental Protection, in order to provide technical support for the Chinese government formulating strategies for the control of Hg emission in China.

See more of this Session: Climate Change Legislation and Regulation
See more of this Group/Topical: Environmental Division

(90a) Sustainability of Biomass Crop Production

Shijie Liu, Paper and Bioprocess Engineering, SUNY College of Environmental Science and Forestry, Syracuse, NY

Woody biomass is renewable only if sustainable production is imposed. How to evaluate the benefit of different biomass crop species on a sustainability scale is a difficult task. Site dependence and regional diversity are the most important factors, while biomass productivity determines the viability and selection of a biorefinery facility. An optimum and sustainable biomass stand production rate is found to be one with the incremental growth rate at harvest equal to the average overall growth rate. Forest biomass is found to fevorable over algae and other higher cost crops.

See more of this Session: Feedstock, Supply Chain, and Logistics of Forest Biorefineries
See more of this Group/Topical: International Congress on Energy 2011

(90b) Determining and Addressing Feedstock Variability for Thermochemical Biomass Conversion

Jessica L. Olstad[1], Kim Magrini[1], Richard Boardman[2] and Garold L. Gresham[3], (1)National Renewable Energy Laboratory, Golden, CO, (2)Chemical Engineering, Idaho National Laboratory, Idaho Falls, ID, (3)Interfacial Chemistry, Idaho National Laboratory, Idaho Falls, ID

A common and significant problem in quantitative analysis is working with a small sample, taken from tons of material, for characterization and generating a variety of values for a single property. The question that arises from the situation is whether the variability comes from the analysis or the sample itself. In many instances, it is assumed that the analysis holds inherent errors and no thought is given to the inconsistency or variability of the given sample.

The quality of feedstock for any production process is extremely important for consistent performance, and the same can be said about biomass feedstock for green energy processes. Having a uniform biomass feedstock can ensure good quality of the resulting fuel and other chemical products. Biomass feedstocks are naturally different, starting with the fact that feedstocks cover a range of lignin based materials including corn stover, switch grass, wheat straw, as well as hard and soft woods; in addition, the material properties for each of these feedstock categories differ depending on the location of the feedstock, time of year for growth, water availability and soil properties, as well as how it was harvested, processed, and stored. The Idaho National Laboratory (INL) has been tasked with developing uniform feedstocks for use in biomass thermochemical processes. The National Renewable Energy Laboratory (NREL) is developing thermochemical biomass conversion processes to fuels and chemicals. Consequently, INL and NREL have formed a collaboration to address and minimize these issues in the future. INL and NREL will jointly characterize an assortment of biomass feedstocks in an effort to define sample heterogeneity and determine the most effective sampling proceeses that address these variabilites. Analytical techniques to be used include ultimate and proximate analysis, surface area, and particle size analyses. These results, to be discussed, will be tabulated in an online database for thermochemical research.

See more of this Session: Feedstock, Supply Chain, and Logistics of Forest Biorefineries
See more of this Group/Topical: International Congress on Energy 2011

(90c) Variability of Switchgrass, a Dedicated Bioenergy Crop

Paul B. Filson[1], Nicole Labb[1], Samuel W. Jackson[2], Elena Doukhanina[3], Parthiban Radhakrishnan[3] and Joon-Hyun Park[3], (1)Center for Renewable Carbon (Forestry Wildlife and Fisheries), University of Tennessee, Knoxville, TN, (2)Office of Bioenergy Programs, University of Tennessee, Knoxville, TN, (3)Ceres Inc., Thousand Oaks, CA

The utilization of switchgrass as a dedicated energy crop is well documented in the scientific community. However, there has not been any large-scale production of switchgrass as a feedstock for the production of biofuels. In this work, three different varieties of switchgrass were grown on 600 acres of land as a feedstock in the production of bioethanol in Tennessee. Samples of each variety were harvested from the various fields at an interval of eight weeks for three consecutive periods. The variability, quality and suitability of the three different varieties of switchgrass as a feedstock for the production of bioethanol were assessed using near infrared spectroscopy, standard chemical analyses, and other structural properties in a process analytical technology.

See more of this Session: Feedstock, Supply Chain, and Logistics of Forest Biorefineries
See more of this Group/Topical: International Congress on Energy 2011

(90d) Anaerobic Bacterial Treatment of Corn Stover to Prepare a Biorefining Feedstock

James MacLellan, Biosystems and Agricultural Engineering, Michigan State University, East Lansing, MI

Anaerobic Bacterial Treatment of Corn Stover to Prepare a Biorefining Feedstock

James MacLellan, Zhengbo Yue, Robert Kraemer, Wei Liao*

Department of Biosystems and Agricultural Engineering, Michigan State University

East Lansing, Michigan 48823

Abstract

The objective of this study is to apply an anaerobic bacterial consortium to uptake the hemicellulose in corn stover to prepare a relatively cellulose-enriched feedstock for biofuel production. Previous research demonstrated that submerged anaerobic bacterial culture is capable of converting hemicellulose in raw lignocelluloses to methane gas. The in-depth study is needed to optimize the culture, enhance the hemicellulose utilization efficiency, and improve the quality of treated cellulose as a biorefining feedstock. An anaerobic bacterial consortium from animal manure will be screened and enriched as the seed for this study. Two feedstocks of raw corn stover and steam-treated corn stover will be used to compare the effects of structure on the anaerobic culture of hemicellulose uptake. A completely Randomized Design (CRD) on seed-to-feedstock ratio, culture temperature, and culture time will be carried out. Fiber composition, methane production, and microbial consortia will also be analyzed. An optimal condition can then be established based on the quality and quantity of cellulose in the treated corn stover. The ethanol production on the treated corn stover will be fulfilled in order to elucidate the effects of anaerobic bacterial treatment on biofuels production.

*: Corresponding author

Dr. Wei Liao

Department of Biosystems & Agricultural Engineering

Michigan State University

202 A.W. Farrall Hall

East Lansing, MI 48824-1323

Tel: 517-432-7205

Fax: 517-432-2892

E-mail: liaow@msu.edu

See more of this Session: Feedstock, Supply Chain, and Logistics of Forest Biorefineries
See more of this Group/Topical: International Congress on Energy 2011

(90e) Pellets From Pretreated Biomass

M. Toufiq Reza[1], Charles J Coronella[2], Joan G Lynam[2] and Victor R. Vasquez[2], (1)Chemical and Materials Engineering, University of Nevada, Reno, Reno, NV, (2)Chemical Engineering, University of Nevada, Reno, Reno, NV

Biomass supply is complicated by diversity of feedstocks, seasonal availability, and widely distributed feedstock. The costs of biomass supply and logistics hinder commercialization for advanced fuel and power production. Hydrothermal carbonization (HTC, or wet torrefaction) is a pretreatment process for making a homogenized, carbon rich, and energy dense solid fuel, called HTC biochar, from lignocellulosic biomass. Compared to raw biomass, HTC biochar is both more hydrophobic for better storage and more friable for better processing. In this pretreatment method, the biomass is treated with hot compressed water in an inert environment in the temperature range of 200-260°C. The range of mass yield of biochar is 60-90% and its energy yield is 75-95%, indicating that the process causes an energy densification of 6-25%. Higher reaction temperature decreases mass yield and increases energy densification.

Kinetic data show that hemicelluloses and water extractives react very quickly compared to cellulose and that lignin is relatively inert, resulting in a biochar with increased lignin content. Lignin extracted from raw wood by Van-Soest method exhibits a glass transition temperature at about 135-165 ° C. Similarly, lignin extracted from HTC biochar exhibits a glass transition temperature in the same range. Lignin acts as a natural binder for pelletization of HTC biochar, if pelletizing is done in the temperature range of glass transition. A hydraulic press with a controllable heated die was used for producing uniform pellets. The materials used to make pellets were raw loblolly pine and pine pretreated at 200°C, 230°C, and 260°C. The sample was compressed into a 13 mm die to make cylindrical pellets 8-10 mm long. The pressure applied was 1000 MPa while a temperature of 140 ° C was maintained for 30 seconds to allow the lignin showing the glass transition behavior.

Pellets made from HTC biochar exhibit favorable properties resulting from high levels of lignin, including increased abrasion resistance, increased durability, increased modulus of elasticity, increased energy density, and increased mass density. On the other hand, ultimate breaking strength of pellets is decreased with increasing pretreatment temperature. Pellets produced from biomass pretreated at 260°C, have volumetric fuel value 70% greater than pellets produced from untreated pine and 142% greater than the raw biomass. Equilibrium moisture content (EMC) of pellets of HTC biochar pretreated at 260 ° C is about one fourth that of pellets produced from raw biomass. Pellets produced from HTC biochar maintain structural integrity significantly longer than do pellets produced from raw woody biomass.

See more of this Session: Feedstock, Supply Chain, and Logistics of Forest Biorefineries
See more of this Group/Topical: International Congress on Energy 2011

CHANGE OF STRUCTURE OF SWEET SORGHUM BAGASSE AFTER SOLID-STATE FERMENTATION

Ji-hong Li[1], *Shi-Zhong Li*[1], *Zhi-pei Yan*[1] , *Qing Li*[2], *Liang-cai Peng*[3]

[1]*Institute of New Energy Technology, Tsinghua University, Beijing, China*

[2]*Colledge of Science, Huazhong Agricultural University. Wuhan, China*

[3]*Colledge of Plant Science & Technology, Huazhong Agricultural University, Wuhan, China*

Abstract

Solid–state fermentation technology of sweet sorghum is an efficient, simple and low cost process to produce ethanol. In this study, the structure of sweet sorghum bagasse before and after solid-fermentation was determined by analysis of composition, IR, SEM and XRD. The results indicated that the sweet sorghum bagasse after solid-fermentation (SSBA) has lower crystallized cellulose (15.87%), higher amorphous cellulose (3.69%) and significantly higher xylan (23.32%) than the sweet sorghum bagasse before solid-fermentation (SSBB). And the corresponding composition of SSBB was 22.41%, 3.39% and 14.44% respectively. Moreover, the crysrallinity index and degree of polymerization (DP) of SSBA was also slightly lower than that of SSBB. The image obtain from SEM indicated the surface of SSBA was rougher than that of SSBB. This means yeast destruct the cell wall of sweet sorghum stem to some extent. It is interesting that the composition of lignin of sweet sorghum also changed significantly after solid-fermentation. Total lignin increased after solid-fermentation, while the ratio of S and H unit of lignin decreased. All results revealed that yeast not only utilized the soluble sugars in sweet sorghum stem, but also metabolized some composition of cell wall. So this sweet sorghum bagasse with relative higher xylan could be a potential feedstock for xylose or furfural. It is also a potential feedstock for lignocellulosic ethanol due to its relative feasibly accessible structure. b

(93a) Sustainable Business Operation and Environmental Preservation - the Evolving 21st Century Thinking On the Achievability of Both

Jacqueline Sibblies, Environmental Compliance Consulting, Somerset, NJ

The governmental implementation of environmental protection measures of the 1970's generated passionate debate among stakeholders on the results that these effected procedures could or would make. A large section of public and several public advocacy groups felt that requirements were strict enough on the industry to bring about the kind and degree of operational changes needed to appreciably mitigate the state of environmental deterioration. Members of the industrial sector viewed the environmental protection measures as very restrictive, undermining business profitability and possibly threatening to business survival. The regulatory agencies considered the requirements reasonable and adequate in addressing the then prevailing environmental concerns.

The arguments and diverse perspectives on environmental protection requirements have persisted over the decades, intensifying at pivotal junctures. As scientific research and human health evaluation data revealed increasing and/or new negative environmental effects from industrial operations, governmental agencies have sought to reduce these factors through further regulations, either by amendments of existing regulations or promulgation of new ones. Such moves have left industry with a heightened sense of being subjected to overly burdensome requirements that could not be complied with economically. Some thought that there were requirements that were simply not necessary and that industry was in the best position to devise procedures for controlling practices that negatively impact the environment, should be allowed to develop appropriate countermeasures.

Indeed by the 1980's, the chemical industry embarked on a set of self derived standards of operation known as Responsible Care, developed in Canada. Participation is voluntary and requires that members follow recommended practices of safe handling and disposal of chemicals, monitor these practices and keep records of specified metrics. Later, the European non-governmental organization (NGO), International Organization for Standardization (ISO), developed standards for an environmental management system which included requirements for regulatory compliance.

Despite these efforts, the 1990's brought even more stringent environmental regulations due to demands for a cleaner environment and growing concerns regarding climate change by the public, public advocacy groups and environmental groups. Internationally, consumers began to indicate their preferences for goods and services that incorporate environmental preservation. Members of the industrial sector started to look for ways in which they could fulfill consumer expectations and still maintain acceptable business returns.

During the first decade of the 21st century, various sectors of industry led by prominent leader in those sectors developed operational/business models that have succeeded in sustaining high environmental standards and economic performance. This paper will present discuss the development of the new thinking that generated this industrial paradigm shift and offer a projection into an era of fuller stake holder collaboration towards sustainability.

See more of this Session: Green Engineering and Sustainability In Collegiate Curriculum
See more of this Group/Topical: Environmental Division

(93b) Preparing Students for Careers In Which Sustainability Will Grow In Importance

Larry Erickson, Keith L. Hohn and John R. Schlup, Chemical Engineering, Kansas State University, Manhattan, KS

There is a growing interest in sustainability among students because of the growing importance of renewable energy, sustainable ecosystems, life cycle analysis, and biobased products. Recent experience with a Research Experiences for Undergraduates program on sustainable energy, a sustainability seminar program, new courses for a Biobased Products and Bioenergy Certificate Program, and other sustainability efforts will be described. Distance education processes enable universities to work cooperatively to offer more new courses. Recent progress of the Great Plains Interactive Distance Education Alliance will be presented.

See more of this Session: Green Engineering and Sustainability In Collegiate Curriculum
See more of this Group/Topical: Environmental Division

(93d) Teaching the Concepts of Life Cycle Assessment Through the Evaluation of Pharmaceutical Syntheses

Mariano J. Savelski, C. Stewart Slater and David Hitchcock, Chemical Engineering, Rowan University, Glassboro, NJ

The concepts of green engineering and life cycle assessment can be integrated into the curriculum using a drug manufacturing example. The pharmaceutical industry has one of the highest waste generation rates per pound of product produced and the highest amount of organic solvents used per pound of product produced for any commercial sector. The solvents result in wastes released into the environment through the drug manufacturing process and throughout the life cycle of the solvent. Students can easily evaluate chemical use and waste reduction that result from new drug synthesis routes, but few understand how these new processing routes impact the overall life cycle emissions from cradle to grave. This paper describes our efforts to educate students in life cycle thinking in making green design decisions in chemical syntheses.

A case study in drug manufacture is used to illustrate how a sustainability analysis is performed to select the design with the lower overall carbon footprint. A tutorial was developed to show how a life cycle assessment is performed in a process and the key issues such as defining the boundaries of the system and calculating the appropriate life cycle inventory data for all chemicals involved in the analysis. The case study can be used in a condensed version for lower level instruction or expanded for upper level process design classes. Students see how changes made in manufacturing lead to environmental impacts well beyond the production plant and define the true carbon footprint of a manufacturing operation.

This project is sponsored by the NSF Engineering Research Center on Structured Organic Particulate Systems.

See more of this Session: Green Engineering and Sustainability In Collegiate Curriculum
See more of this Group/Topical: Environmental Division

Sustainability Initiatives at Several U.S. Institutions of Higher Education

M.P. Sharma, Ph.D.
Chemical and Petroleum Engineering Department
University of Wyoming
Laramie, Wyoming

Matthew Winslett
Engineering Manager, Energy Management
Facilities Management Department
University of Alabama at Birmingham
1530 3rd Avenue South
Birmingham, Alabama 35294-4564

Robert W. Peters, Ph.D., P.E.
Department of Civil, Construction, and Environmental Engineering
University of Alabama at Birmingham
1075 13th Street South
Birmingham, Alabama 35294-4440

Abstract

Recently the American College and University Presidents' Climate Commitment (ACUPCC) released its 2009 Annual Report showing building momentum on comprehensive action towards clean energy and greenhouse gas pollution reductions in higher education. The following samples of observations made by some of the leaders of institutions of higher education point to a trend (perhaps and a new phenomenon) at campuses of institutions of higher education in the nation (and the world): "Colleges and universities must exercise leadership in their communities and throughout society by modeling ways to eliminate global warming emissions, and providing the knowledge and educated graduates to achieve climate neutrality campuses that address the climate challenge by eliminating global warming emissions..... will better serve their students and meet social mandate..." – The signatories of the ACUPCC.

"As the national debate on climate change moves in a positive direction after years of neglect, leaders of higher education are inspiring the rest of society to address the critical challenge of global climate disruption by reducing emissions on campus and by starting by starting to reorient the education of the nation's future professionals." -- Dr. Anthony D. Cortese, one of the ACUPCC's supporting organizations). "More than ever, universities must take leadership roles to address the grand challenges of the twenty-first century, and climate change is paramount amongst these". – Michael Crow, President, Arizona State University.

The authors of this paper have been involved with the Campus Sustainability Programs at their own universities in conducting and studying projects. This paper will present the results and discussions of design, implementation and outcome of campus sustainability projects initiated, completed and underway at several colleges and institutions of higher education including the University of Wyoming (UW) and the University of Alabama at Birmingham (UAB). The vision, the lessons learnt, and impacts obtained from the implementation of these campus sustainability projects will be discussed in the paper. A part of this paper is a work-in-progress. The goal is to compare, contrast, and draw conclusions to learn from different models of "campus sustainability initiatives and projects" that are emerging at different institutions of higher education.

The University of Wyoming (UW) is positioning itself to become more sustainable [1]. Over the last five years, it has made great advances. President of UW created the Campus Sustainability Committee (CSC) in 2005 "to serve in an advisory capacity to all campus departments, colleges and other entities in an effort to advance environmental and economic sustainability at the University of Wyoming." In 2007, the President signed the American College and University Presidents Climate Commitment (ACUPCC), committing UW to become carbon neutral by 2050. In 2009, UW submitted its Climate Action Plan (CAP), UW's guiding document to reducing carbon emissions and becoming more sustainable. This is a review of UW's advances in sustainability in 2010 and serves as a marker for where it currently stands on its path to more sustainable operations.

Since signing the ACUPCC, the University of Wyoming has made considerable progress in its goal. Each fiscal year since the signing of the ACUPCC, the University has conducted a greenhouse gas emissions inventory (GGEI) [2]. The GGEIs show that the University has remained relatively steady in the amount of greenhouse gases (GHG) it emits since the baseline year of 2005. As more efficient campus buildings begin to come online – including the Kendall House, the Berry Center, Visual Arts, and the College of Business – and as the development and use of alternative energy sources become more feasible at UW, emissions relative to the total square footage are expected to decrease more significantly. Creating sustainability awareness among the campus community and beyond, and encouraging behavior that leads to less resource consumption, is also critical to reducing carbon emissions. UW has made great strides to that end, as well. Now on to presenting and discussing selected accomplishments toward sustainability commitments and GHG reductions at UW [3].

Sustainability Initiatives Underway at UW

Accomplishments

The University of Wyoming has taken steps toward reducing institutional greenhouse gas emissions, particularly within the following key areas: energy, facilities, grounds, water, transportation, purchasing and procurement, recycling and waste management, curriculum and research, and behavioral change.

Energy

UW has begun to reduce its energy consumption through a phased approach. Phase I (2009 – 2015) calls for ramping up energy efficiency across campus. Phase II (2015 – 2030) will include the implementation and use of alternative energy. Phase III (2030 – 2050) will continue building on the first two phases and include purchase of carbon offsets where increased efficiency and alternative energy are inappropriate. The following list highlights progress made in 2010 toward energy efficiency.

- Student interns completed the third annual Greenhouse Gas Emissions Inventory in the spring 2010 semester.
- A team has been created as a sub-committee of the CSC to focus on energy efficiency in utilities and facilities.
- Interior lighting systems in several core campus buildings have been upgraded.
- Windows in five core campus buildings were replaced in 2009.
- One biomass test burn in the Central Energy Plant was conducted in 2009 and a second in 2010.
- Phase one of a steam trap upgrade project was completed in summer 2010.

- Phase I of a campus metering program is complete with installation of condensate, water and electric meters for core campus buildings.
- Interior lighting upgrades continue to replace T-12 lamps with T-8 lamps, add lighting controls where appropriate and replace incandescent bulbs with compact fluorescent bulbs.
- Twenty-five vending misers (mechanisms that turn vending machines off during non-peak hours or when outdoor beverage machines do not need to cool due to weather) have been installed with funding from Auxiliary Enterprises. These machines are in the lowest traffic areas on campus so the savings will be the greatest. Also, after analysis of the usage of vending machines, 17 were deemed underperforming and were removed. This program will be expanded to the cardswipe type of vending machines if a demonstration test of the vending miser is successful.
- All light bulbs/ tubes in Washakie Center have been switched from T-12 to energy saving T-8 bulbs as they have needed to be replaced.
- The Residence Hall Association and the Residence Life & Dining Services Sustainability GA piloted a program to reduce electrical consumption by 6,379 kilowatt hours against the previous three year average in the month of February.

Facilities

Through the Facilities Planning office, all new buildings on campus are now constructed to Leadership in Energy & Environmental Design (LEED) standards. More efficient buildings directly affect UW's greenhouse gas emissions and result in a more comfortable and healthy indoor environment for students, faculty and staff. The following highlights UW's commitment to building toward LEED principles.

- All new buildings are being constructed to at least LEED Silvers standards.
- Two buildings will be LEED certified. The College of Business is certified LEED Gold, and the Kendall House is expected to be certified LEED Silver by spring 2011.
- ACRES Farm has a new passive solar shed, which is expected to be LEED certified in 2011.

Grounds

University Grounds Services continues to operate and develop in ways that reduce impact on the campus and community ecosystem. Tree-planting is at the heart of this goal. Trees beautify campus and sequester carbon. In the last two years, the UW Physical Plant has added nearly 200 trees.

- Over 100 trees were planted in several key campus areas in 2009.
- An additional 60 trees were planted in several targeted campus locations in spring 2010.
- More efficient maintenance equipment is being acquired as replacement purchases are made.
- Environmentally friendlier fertilizer and pesticides are being used to maintain campus grounds.

Water

Water conservation involves educating the campus community, reducing water use, and upgrading plumbing fixtures. The first step in water conservation at UW is to better understand use so that education programs can be developed and upgrades can be made in key areas. A metering program is now in place and will guide conservation measures in the coming years.

- Installation of water meters for each major facility on core campus will provide necessary data to track water usage, identify problem areas and recognize usage improvements.
- An irrigation system upgrade is being planned to replace individual controls with a centrally controlled automated system.

- Use of ground water and rain water is being evaluated for feasible applications.

Transportation

Transportation practices and services impact UW's institutional carbon emissions significantly. Facilities Planning, TransPark and Transit Services aim to help reduce UW's emissions through expanded shuttle services, bike and pedestrian routes, and adding hybrid and other alternative fuel vehicles to the fleet.

- The campus shuttle system has been expanded to include express shuttles from a remote south lot and long-term east lot.
- As part of the Long Range Development Plan (LRDP), defined transportation routes for shuttles, vehicles, bicycles and pedestrians have been developed.
- The number of electric fleet vehicles was expanded in 2009 from three to eight.
- A number of fleet vehicles can run on flex fuel including a 20% bio fuel mix.
- Six efficient diesel buses are on order and should be received by mid-2011.

Purchasing and Procurement

From paper and cleaning products to food and food service containers, UW is working to purchase products with post-consumer recycled content and products that are compostable.

- A contract was negotiated and executed to purchase 100% recycled content paper products for restrooms in campus facilities.
- Through the Procurement Office, most departments contract for 30% recycled content paper.
- Dining Services uses green cleaning products for every application where there is a certified green product on the market including: Geo Fusion 9000 Machine Detergent, Geo Fusion Silver Fusion Presoak, Geo Mag Fusion (Pot and Pan detergent), Geo Crystal Fusion Rinse Additive.
- Wherever possible, dining locations use compostable boxes, flatware, cups, plates, bowls, to-go boxes, sushi containers, salad and sandwich containers, and other packaging. The number of compostable/biodegradable products issued in February 2010 was 8,038,800, which approximates an average month.
- Dining Services sourced a supplier and product to be able to serve only cage-free liquid egg mix.
- Dining Services has worked to increase procurement of locally-grown foods when possible. Sources within 200 miles have been found for beef patties, eggs, lamb, and lettuce when in-season. A source for apples has also been found in Western Colorado when in-season. Milk is purchased through a local dairy in Denver, CO.
- Dining Services continues to work with ACRES Farm by purchasing locally grown produce when available.
- Dining Services serves organically-grown, fair-trade coffee in Washakie Dining Center, University Catering, Book and Bean, Elements and Ross Hall.

Recycling and Waste Management

Recycling efforts at UW have advanced significantly over the last year. The improvements can be generally categorized as campus-wide collection efforts and residence hall specific efforts. The following list details the landfill diversion achievements UW has enjoyed in 2010.

- UW participated in Recyclemania in 2009 and 2010 – coordinated by UW Recycling, the Campus Sustainability class and support by the Wyoming Union. In 2010, UW placed 51/267 in the Grand Champion category; 120/346 in the Per Capita Classic category; and 62/199 in the Waste Minimization category.

- The campus recycling operation has been expanded to include additional pickup locations, one additional truck, one additional employee and a drive-through Recycling Drop-Off Center.
- Recycling pickups have been increased from bi-monthly to weekly resulting in increased tonnage collected.
- Twenty cardboard-specific dumpsters have been purchased and placed resulting in significantly increased cardboard collections.
- Recycling efforts in the Athletic venues have been expanded by adding collection points, coordinating pick-ups after events and expanding the Recycling Cyclers Program in the tailgating lots.
- Recycling and composting efforts have been expanded in Washakie Dining Hall.
- Through the collection efforts in the residence halls, a total of 11.5 tons of material collected over the academic year was redirected away from the landfill.
- Three tons of cardboard were collected from the Residence Halls on move-in weekend.
- CSC and Students for Sustainability purchased two pedal pickups (modified three-wheel bicycles) for recycling collection at UW events.
- In May 2010, 2.383 tons of reusable items were collected and donated to local charities during the "When You Move Out, Don't Throw It Out" campaign.
- One project of the ENR 4600/5600 Campus Sustainability Class (spring 2010, 20 students) was a Furniture Round-Up to gather, store, and redistribute unwanted furniture after a semester ends and before a semester starts.
- Summit '10 Welcome Picnic on Prexy's Pasture (August) was a zero waste event, where all waste from the event was either recycled or composted.
- Washakie Dining Center conducted surveys/research on going tray-less within the dining hall and held two tray-less days, one each semester. These tray-less days netted a savings in pulp waste of approximately 300 pounds each day in addition to raising awareness about food waste.
- Dining Services removed drink straws from the Washakie Dining Center.
- Insulated drink containers are given to all students living in the residence halls, in an effort to promote the use of reusable, rather than disposable containers.
- Approximately 70-105 gallons of compostable waste are sent to ACRES farm everyday from Washakie Dining Center, University Catering, and the Wyoming Union Food Court.
- Dining Services worked with the Laramie landfill to test the composting of food waste from the Washakie Dining Center.
- Dining Services worked with the College of Engineering to test a machine that would convert used fryer oil to Bio-Diesel.
- Students from the Center for Volunteer Service assisted in the going tray-less project at Washakie Dining Center.
- The Good Mule Project Conference used compostable dining utensils and compostable materials.
- Summit View Apartments deconstruction project estimates 85% of materials from deconstruction are being recycled (approximately 15,000 yards of materials).

Behavioral Change

Increasing awareness and encouraging changes in behavior among the students, faculty and staff can result in a significant decrease in GHG emissions. The following list highlights some of the lifestyle and educational programming sponsored by the CSC and others last year.

Behavioral Change – lifestyle

- The University of Wyoming ranked among the nation's 286 most environmentally responsible colleges, according to the Green Guide, published by the Princeton Review and the U.S. Green Building Council.
- University of Wyoming received a B on the Sustainable Endowment Institute's green report card. This is an improvement from the 2009 score of B-.
- A Green Cleaning program was implemented in 2010 with a first year reduction of 25% in routine cleaning chemicals.
- Three students were hired as interns to work on CSC efforts - two on the FY 2010 GHG emissions inventory and one on planning for the Wyoming Sustainability Summit.
- The CSC wrote a successful proposal for an AmeriCorps intern position for 2010-2011. A UW student was hired into the position to create and conduct online surveys for students, faculty and staff in 2011.
- Over 100 students helped weatherize homes around Laramie for the 2010 Days of Dialogue Service Day.
- Alternative Spring Break trips were led to the Grand Canyon and to the Everglades where students spent spring break helping preserve the native ecology and wildlife.
- Insulated drink containers were given to all students living in the residence halls, in an effort to promote the use of reusable, rather than disposable containers.
- RHA and the GA of Sustainability awarded the winning residents of the Recycling and Electrical Consumption Challenge with recycled grocery bags.
- RHA created a new executive position, the Sustainability and Community Service Officer, committed to sustainability issues within Residence Life and Dining Services.
- The RHA menu committee worked with Dining Services on a tray-less campaign in Washakie. A trayless day was held November 2 in the Washakie Dining Hall.

Behavioral Change - education and outreach

- A team has been created as a sub-committee of the CSC, which focuses on education and outreach to the campus community.
- A website has been created, and is continually maintained for the CSC, at http://www.uwyo.edu.sustainability/ to make the sustainability vision, goals, reports and information readily available.
- A Campus Sustainability group has been created on WyoWeb, the UW intranet portal, to share files, collaborate on action steps and provide work areas for subcommittees and teams.
- A media relations program was implemented in cooperation with UW Media to publish sustainability-related articles twice per month.
- Development of a CSC logo was unveiled at Earth Day 2010 events. The logo was chosen through a logo contest held in the spring of 2010. The winner was a UW student, chosen from 58 entries and awarded a $300 prize.
- The first Wyoming Sustainability Summit was held April 12-13, 2010 at the UW Conference Center in Laramie. Over 100 people attended the two-day event. The Summit was co-sponsored by the Campus Sustainability Committee, the Haub School of Environment and

Natural Resources, and Residence Life and Dining Services with support from the Argosy Foundation.

- UW's collegiate chapter of the American Marketing Association and Alpha Kappa Psi co-sponsored the 3rd annual Sustainable Business Practices forum in October 2010.
- The Environment and Natural Resources program organized a "Green for Green" 5k on April 17th, 2010. Volunteers held signs along the race route to inform runners of various sustainability tips such as eating locally, properly insulating homes, and using CFL bulbs.
- Academic Affairs and the Office of Research and Economic Development sponsored a public lecture by world-renowned ecologist, E.O. Wilson in September 2010.
- Over 10 members of the CSC attended the Association for the Advancement of Sustainability in Higher Education (AASHE) annual conference in Denver in October. Several committee members gave presentations and participated in panel discussions on topics ranging from implementing a CAP with a campus-wide committee to institutional recycling to teaching a project-oriented sustainability class.
- The CSC participated in the UW Homecoming Parade in October. The CSC entry featured the Physical Plant's electric truck and the Recycling Cylers pedal pickups. Recyclables were collected along the parade route.
- The CSC hosted an informational booth in the Union Breezeway on Campus Sustainability Day (October 20).
- University of Wyoming was presented with the first Channel 5 Sustainability Award by Channel 5 News in Cheyenne. A 30-minute program featured some of UW's achievements in green building and recycling.
- RHA hosted information booths on campus on Campus Sustainability Day (October 20) and Earth Day (April 22nd).
- The GA of Sustainability, with support from RHA, hosted many educational events throughout the school year that educate residents on various topics such as drinking water, electrical usage and real benefits of recycling.

Sustainability Initiatives Underway at UAB

The University of Alabama (UAB) Energy Management Department's function is to ensure the most energy efficient operation of UAB's facilities. UAB's Energy Management Department is responsible for managing all of the energy usage, billing, metering, and various conservation projects for the more than 150 UAB campus and hospital buildings. We are constantly striving to save energy and promote sustainability without sacrificing a comfortable environment throughout the campus. Some of the things we do include customer support regarding utilities, building space surveys and analysis, lighting surveys and analysis, utility billing, usage and rate analysis, monthly utility billing, utility sub-metering, project reviews, track usage and rebated funds from sewer credits, maintain all of UAB's utility records with the use of Energy-CAP database, provide customer requested utility usage reports, and energy conservation and sustainability projects.

<u>Examples of Sustainability Projects at UAB</u>

With the increasing cost of utilities, energy conservation and sustainability projects have been instrumental in achieving ongoing costs reductions campus-wide. Some of the recent projects UAB has accomplished include building equipment setbacks (during unoccupied times), reduced air exchanges of research areas (during unoccupied times), carbon dioxide (CO_2) monitoring of occupied space to control introduced outside air quantities, efficient upgrades of air handling equipment – controllable by building automation systems, installation of variable frequency drives, replacement of water cooled medical/laboratory gas pumping systems, ice makers, and freezers, demand controlled domestic water pumping systems, T-12 to T-8/T-5/LED fluorescent lighting upgrades, installation of occupancy sensed lighting controls, steam trap surveys and system upgrades, utility rate renegotiations, fin water condensation recovery reclaimed to cooling tower, evaporative cooling tower upgrades, vegetative (green) roof installations, hot water temperature setbacks based on outside air temperatures, and boiler combustion control upgrades.

Equipment Setbacks and Reduced Air Exchanges

Unoccupied space conditioning equipment setbacks and reduced air exchanges have by far been the greatest return on investment. Starting in 2007 many of UAB's building were surveyed at intervals around the clock to determine occupancy frequency habits of students and employees. Occupancy data was gathered and analyzed with help from students from the UAB Civil, Construction, and Environmental Engineering Department and resulting analysis determined that most buildings are unoccupied nightly from hours between 10:00 p.m. and 6:00 a.m. as well as most weekends and holidays. Each building's automation system was programmed to shut-off the air handling motors and decentralized chillers during unoccupied time saving btu's and electrical horsepower. If an internal space temperature reaches above 85F in the summer or below 55F in the winter the respective system starts up insuring safe operations. At present, 41 UAB buildings are being setback during unoccupied times generally from 10:00 p.m. until 6:00 a.m. nightly with savings realized at over $1,000,000 per year.

Other unoccupied setbacks include reduced air exchanges of research laboratories. Normally used in lighting control applications, ceiling mounted occupancy sensors have been utilized to signal the building's automation system (BAS) unit to decrease air flows. Normal air exchanges for laboratories are 8–12 per hour depending on the building during occupied times. After 30 minutes of vacancy, the occupancy sensor directs the BAS to lower the number of air exchanges to three per hour. A group effort between UAB's Departments including School of Medicine, Campus Maintenance, Occupational Health and Safety, and Energy Management has set a cost saving precedence with this endeavor saving utility costs in all areas of utilities including electricity, steam, natural gas, and chilled water. At present, two research buildings are utilizing this control strategy with four others planned for the future.

Domestic Water Conservation and Recovery Projects

Domestic water conservation and recovery projects have also been very beneficial in reducing costs at UAB. Compressed air and vacuum pumping systems are found throughout the UAB campus laboratories and hospital settings. These traditionally inefficient pumps require water for compression as well as a cooling source. With escalating water and sewer costs this method of cooling was deemed unfeasible. These projects replace liquid-ring pumps with ones that do not require water for cooling including claw-type vacuum systems and scroll-type compressed air systems. In most applications less electricity is required to do the same work due to increased pump efficiencies. Depending on building type, which determines electrical rates ($0.08 – $0.10/kW-hr), for every horsepower that is saved by a more efficient pump, electrical costs are reduced approximately $400 - 500 per year for continuous duty operations. A constant single-pass through water flow for cooling is especially expensive if sewer costs are also required. The local cost for each constant year-round one gallon per minute cooling water source is almost $7,500 per year including sewer charges. Therefore system paybacks are very short – usually within one to two years. See Table 1 below for details and Figures 1 and 2 for an example typical installation.

Table 1. UAB Air/Vacuum Pump Replacement Savings.

Pumping System Building Location	System Type	Required GPM Cooling Water	Annual Savings Gallons	Annual Savings Water/Sewer	Horse Power Reduction	Annual Savings Electrical
Tinsley-Harrison Research	Vacuum	4.5	2,366,820	$33,750	0	$0
School of Dentistry	Medical Air	4.5	2,366,820	$33,750	50	$20,000
Bevill Biomedical Sciences	Lab Air	3.5	1,840,860	$26,250	5	$2,000
McCallum Health Science	Vacuum	2.5	1,314,900	$18,750	0	$0
Spain-Wallace Hospital	Medical Air	3.5	1,840,860	$26,250	15	$6,000
Kaul Human Genetics	Vacuum	2.5	1,314,900	$18,750	15	$6,000
Biomedical Research II	Vacuum	2.5	1,314,900	$18,750	15	$6,000
TOTALS		23.5	12,360,060	$176,250	100	$40,000

Figure 1. Spain-Wallace Hospital Medical Vacuum Pumping System (before).

Figure 2. Spain-Wallace Hospital Medical Vacuum Pumping System (after).

UAB has been a leader in recovering and utilizing condensate water from air handler coils. Many of UAB buildings incorporate 100% outside air into conditioned space. The harsh and humid Alabama summer weather causes huge amounts of condensation to form from the air hander chilled water coils. Collection and recovery systems were designed and installed to utilize this source of water. As condensate water is gravity drained from air handler chilled water coils, it is intercepted and collected in small basins positioned locally at each air handler. Each local basin incorporates a small level-controlled sump pump sized to transfer collected water to a larger (500 gallon) collection tank. The 500 gallon collection tank employs level controls that pump high-pressure, filtered, 55°F condensate back into the centralized chilled-water return line before it leaves the building. This rise in pressure is sensed blocks away at the central plant and water there is locally relieved from the chilled water return line into the evaporative cooling tower. This process reduces tower makeup water required during the evaporative cooling process. If the small basin sump pump, high pressure return pump, or Central Plant tower relief valve fails, the water safely exits down the local drain found in the building as it has for many years. While this process may have been accomplished before we may indeed be the first to incorporate this manner of utilizing pressure relieved water in our cooling towers from long distances. We presently have two sites collecting fin water condensate since May 2011 including McCallum/Kaul and Zeigler/Tinsley-Harrison Tower (see Table 2 below) with future sites underway including Bevill Biomedical Research, Biomedical Research II, Research Support Building, North Pavilion Hospital, and Shelby Biomedical Research Building yielding a hopeful 3,000,000 gallons per month during the cooling season.

Table 2. UAB Fin Water Recovery Savings.

Fin Water Recovery Location	Month	Gallons	GPM	Value
McCallum/Kaul Genetics	MAY	166,200	3.72	$699.81
	JUN	430,300	9.96	$1,811.85
	JUL	615,604	13.79	$2,592.10
Zeigler/Tinsley-Harrison	MAY	183,700	4.12	$773.50
	JUN	285,100	6.60	$1,200.46
	JUL	410,497	9.20	$1,728.47
TOTALS		2,091,401		$8,806.19

Figure 3. Zeigler/Tinsley-Harrison Fin Water Collection System.

UAB utility usage and costs are on the decline as a result from these many projects. Operating on an October to September fiscal year (FY), FY0910 realized close to a $3 million reduction over the previous fiscal year. It is also worth noting that during this time period Alabama experienced the harshest winter and summer weather on record – all within the same FY. FY1011 looks to be an exceptional year of savings as well. At the half-way mark of the FY1011 fiscal year (October through March) savings compared to the previous year are down $5 million.

Pilot Vegetated Roof on Top of Hulsey Center

UAB implemented a pilot vegetated roof system installed on top of Hulsey Center in July 2008. The roof occupies approximately 14,940 ft^2 (i.e., occupying ~81.2% of the roofing area, or approximately 0.34 acres). Figure 4 shows a photograph of the pilot green roof, and Figure 5 shows several of the sedum plants on the roof. [Prior to installing the pilot green roof, the civil engineering senior design class investigated the loading of a wet vegetative roof on top of Hulsey Center, and found that the current building infrastructure could withstand the loading associated with the vegetative roof. Hulsey Center was originally designed to have two additional floors in the building]. This pilot green roof contains more than 20,000 sedum plants. In the construction of this vegetated roof, the existing roof was removed down to the structural concrete deck. A waterproof roofing membrane was installed directly to the concrete deck. A layer of 3-inches of extruded polystyrene roofing insulation was then applied, to which an 8-oz non-woven geo-textile scrim sheet was applied. Then, 2-ft x 2-ft x 2-in prestressed pavers or brick pavers were installed for design, decoration, and access to the vegetative roof. Then, 3½-inches of light weight engineered soil were applied. Sedum plants were planted at the rate of 150 plants per 100-ft^2. The cost of retrofitting the roof and installing a pilot green roof on top of Hulsey Center was ~$150,000. The usage of natural gas, water, and electricity are lower than that compared to the utility usage prior to installation of the green roof on top of Hulsey Center. Implementation of the vegetated roof system has reduced the utility bills of Hulsey Center by 20–25% compared with the utility bills prior to implementation of the vegetated roof system.

Figure 4. Vegetative roof on top of Hulsey Center (June 2009).

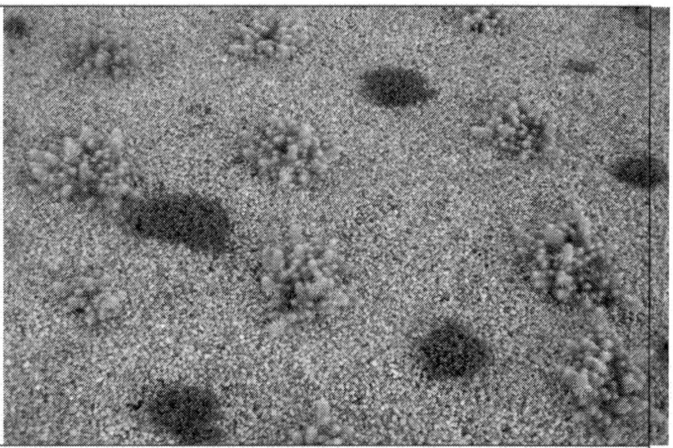

Figure 5. Close-up of vegetative roof on top of Hulsey Center (June 2009).

The UAB Energy Management team is passionate and dedicated to reducing utility usage and costs for the University. Consisting of four engineers and one clerical personnel, everyone is focused on achieving our common goal – *Think Green*, *Save Gold*!

Summary and Conclusions

Over the past eighteen months, the University of Wyoming and the University of Alabama at Birmingham have been successful in advancing campus sustainability in several major areas. The University of Wyoming's particular focus in the areas of energy, facilities, grounds, water, transportation, purchasing and procurement, recycling and waste management, policy, and behavioral change suggest that the UW campus is on its way to achieving its goal of significantly reducing its carbon emissions. In fact, even in light of tremendous growth in square footage (about 400,000 SF since 2006), campus stationary and purchased electricity have remained fairly steady. This is a testament to ramping up energy efficiency in existing buildings and adding more efficient buildings through capital construction projects.

In addition to the University of Wyoming's progress in reducing GHG emissions, the education and outreach efforts show a commitment to encouraging behavior change among the campus community. Approaches that involve students directly and provide them with hands-on experience, such as the greenhouse gas emissions inventory internships and the sustainability GA in Residence Life and Dining Services, conserve staff resources and provide valuable learning opportunities for students.

As illustrated by the activities and initiatives underway at the University of Wyoming and the University of Alabama at Birmingham, campuses are becoming more proactive in the area of sustainability to make the campuses more energy efficient and more environmentally friendly.

Acknowledgements

Sections in this document correspond to different sections in the three references cited below in the List of References. Reference [1] is the main source for use of direct citations of the sections. The authors also acknowledge the technical assistance provided by UAB's Facilities Management Department.

Reference Citations

[1] Campus Sustainability Committee (CSC) Progress Report, University of Wyoming, February 2011.

[2] www.uwyo.edu/sustainability.

[3] Climate Action Plan, University of Wyoming, September 15, 2009.

(93f) Teaching Life Cycle Assessment and Related Concepts

Arunprakash T. Karunanithi, Department of Civil Engineering, University of Colorado Denver, Denver, CO

This presentation shares the authors experience in teaching a graduate course on Life Cycle Assessment based concepts. The presentation will cover various concepts discussed in the course including Life Cycle Assessment, Exergy analysis, Emergy analysis, and Ecological foot printing. Impact assessment tools used for teaching will be discussed. Effective case studies developed for this course will be shared.

See more of this Session: Green Engineering and Sustainability In Collegiate Curriculum
See more of this Group/Topical: Environmental Division

(97a) Global Climate Change-the Sustainability Challenge

Frank Princiotta, Office of Research of Development, USEPA, Research Triangle Park, NC

Population growth and developmental pressures, spawned by an increasing demand for resource intensive goods, foods and services, are altering the planet in ways that threaten the long-term well being of humans and other species. The presentation discusses the role that climate change plays in challenging long-term sustainability. It also identifies the factors that are responsible for potentially unsustainable global impacts, including climate change. Transparent modeling tools and the most recent literature are used, to quantify the challenge posed by climate change and potential technological remedies. This presentation aims to provide a succinct integration of the projected warming the earth is likely to experience in the decades ahead, the emission reductions that may be needed to constrain this warming, and the technologies needed to help achieve these emission reduction. Factors driving CO2 emissions, the importance of key countries especially China, and how different emission trajectories could affect warming this century are analyzed. It is concluded that it is too late too avoid substantial warming; the best result that appears achievable, would be to constrain warming to about 2.0 °C (range of 1.3 to 2.7 °C) above pre-industrial levels by 2100. In order to constrain warming to such a level, the current annual 3% CO2 emission growth rate needs to transform rapidly to an annual decrease rate of from 2 to 3% for decades as soon as possible. Also presented is a sector-by-sector summary of mitigation options, and R&D priorities. Summary information will be provided for the critical power generation and transportation sectors. The importance and challenges associated with carbon capture and storage (CCS) for coal-fired power generation will be discussed. It is concluded that the current generation of energy generation and end use technologies are capable of achieving less than half of the emission reduction needed for such a major mitigation program. New technologies will have to be developed and deployed at a rapid rate, especially for the key power generation and transportation sectors. Current energy technology research, development, demonstration and deployment programs fall far short of what is required, although there has been recent funding increases which are significant. Of particular significance is the Advanced Research Projects Agency-Energy (ARPA-E). It is a new and promising Department of Energy program, focusing on "transformational' advances in energy technology.

See more of this Session: International Forum on Energy Sustainability (plenary)
See more of this Group/Topical: International Congress on Energy 2011

(97b) Converting Carbon Dioxide and Water Into Hydrocarbons with Solar Engergy and Nanostructured Metal Surfaces

Mengyan Shen, Cong Wang, Haibin Huo, Haizhou Ren and Michael Johnson, Department of Physics and Applied Physics, and Nanomanufacturing Center, University of Massachusetts Lowell, Lowell, MA

We report femtosecond laser induced self-assembled metal nanostructure arrays can directly synthesize liquid and solid hydrocarbon compounds from carbon dioxide, water, and sunlight at a high production rate which is much higher than those in previous studies. Moreover, different from previous artificial photosynthesis works, this phenomenon presents a new mechanism that, through a surface-enhanced photodissociation process, nature-like photosynthesis can be performed artificially.

See more of this Session: International Forum on Energy Sustainability (plenary)
See more of this Group/Topical: International Congress on Energy 2011

(97c) Electrochemical Conversion of CO2 with Ionic Liquids In a Microfluidic Reactor

Michael R. Thorson[1], Devin T. Whipple[2] and Paul J.A. Kenis[1], (1)Department of Chemical & Biomolecular Engineering, University of Illinois at Urbana Champaign, Urbana, IL, (2)Chemical and Biomolecular Engineering, University of Illinois at Urbana-Champaign, Urbana, IL

Energy storage solutions compatible with the current energy grid need to be developed for renewable and carbon neutral energy sources to replace fossil fuels. Electrochemical conversion of CO_2 into fuels can provide a means for the storage of electrical energy in chemical form, which is important considering that most carbon neutral energy sources such as wind and solar produce electricity. Storage of this electrical energy in a chemical form would enable the use of renewable energy in the transportation sector, which makes up a large portion of our energy use but has only 3% supplied by renewable sources [1]. Additionally, this technology could provide a means of storage for renewable energy sources such as wind and solar that are intermittent and, as a result, require output leveling.

Conversion of CO_2 into CO [2,3] is attractive due to the versatility of CO (with hydrogen) as a feedstock for the Fischer-Tropsch process, which enables the synthesis of a variety of products including liquid hydrocarbon fuels. Presently, key challenges reducing large overpotentials and improving selectivity, thereby resulting in a search for electrolytes and catalysts that stabilize CO_2- intermediate. Work by Rosen *et al.* has shown that the ionic liquids may stabilize the intermediate and lower the necessary overpotentials [4].

Here, we use a microfluidic reactor that employs a flowing electrolyte to investigate the influence of ionic liquids in a flowing cell [5]. In the flow cell, we found ionic liquids improve the selectivity for CO as opposed to H_2 and increase the current densities, both at low currents. However, ionic liquids were found to be mass transport limited, thereby limiting the selectivity at high currents. To address this limitation, we study several key parameters including, temperature, and pressure with regards to current efficiency, faradaic efficiency, current densities and overpotentials. Additives are used to improve the selectivity for CO and minimize anodic poisoning. massand make design changes to overcome these mass transport limitations. The mass transport limitations can be overcome via the use of temperature as well as design and operational changes.

References

[1] Whipple, D. T., and Kenis, P. J. A., *J. Phys. Chem. Lett.*, **2010**, 1, 3451.

[2] Hori, Y., Wakebe, H., Tsukamoto, T., Koga, O., *Electrochim. Acta,* **1994**, 39, 1833.

[3] Delacourt, C., Ridgway, P. L., Kerr, J. B., Newman, J., *J. Electrochem. Soc,* **2008**, 155, B42.

[4] Rosen, B. A., Salehi-Khojin, A., Masel, R., Artificial Photosynthesis: Conversion of CO_2 at Low Temperature and Potential Using Ionic Liquids, *AIChE Annual Meeting* , 2010.

[5] Whipple, D. T., Finke E. C., and Kenis, P. J. A., *Electr. & Solid-State Lett.*, 2010, 13, B109-B111.

See more of this Session: International Forum on Energy Sustainability (plenary)
See more of this Group/Topical: International Congress on Energy 2011

(97d) Status and Challenges to Implement Large-Scale Energy Storage Into the Grid

John Lemmon, Pacific Northwest National Laboratory, Richland, WA

Growing world energy consumption, environmental concerns over the use of fossil fuels and the resource constraints have spurred great interests in renewable energy from wind and solar. To effectively dispatch renewable energy into the national grid requires electrical energy storage (EES). EES is also useful to improve reliability, stability and efficiency of a future grid that is expected to be able to provide the electrical requirements to plug-in hybrid vehicles. Among the most promising ones are electrochemical storage technologies or batteries that are capable of storing and releasing electrical energy in accordance with demand. While a number of technologies have been developed or even demonstrated, there remain significant challenges in performance and cost for broad market penetration. As such PNNL has carried out wide research into advanced materials and EES technologies. This paper will report the efforts and discuss the needs, status and remaining challenges to implement large scale ESS.

See more of this Session: International Forum on Energy Sustainability (plenary)
See more of this Group/Topical: International Congress on Energy 2011

(97e) Investigations On Geothermal Heat Pump Systems Using Treated Waste Water As Energy Source

Huajun Yuan, Maghana Cherukuri and Sohail Murad, Department of Chemical Engineering, University of Illinois at Chicago, Chicago, IL

Geothermal energy can be used for generating electricity and providing direct heat for applications such as space heating and cooling, water heating, aquaculture and industrial processes. Geothermal heat pump (GHP) is a central heating/cooling system that pumps heat to or from the ground to provide heating and air condition. There has been a significant recent interest in geothermal energy as an efficient alternative to conventional heat, ventilation and air conditioning (HVAC) systems. This interest primarily lies in the almost 50% reductions in energy cost historically achieved with such systems, and a relative rapid rate of return of additional capital associated with geothermal systems. Extreme weather conditions in Illinois make it an ideal state for heating/cooling using geothermal energy. On the other hand, water harvesting and reuse as well as renewable energy play an important role in greater sustainability of ecosystems. For systems which have access to water, another attractive alternative is to use open or closed loop geothermal pond systems. Energy exchange with water is possible at a much higher rate than with soil and installation costs are significantly lower. We have designed and installed a geothermal HVAC system at the Kirie Waste Water Treatment Facility in Des Plaines, IL. The effluent water in the water treatment plant is discharged at temperatures between 55-65 ^{o}F and at rather high flow rates (50 - 100 MGD) which is ideal for use in geothermal systems. In this work we designed two different geothermal systems using treated waste water as energy source, i.e., open loop and close loop geothermal systems. The main difference between the open and close loop pond geothermal system lies in whether the water from the pond is circulated in the geothermal system for more efficient heat exchange, or if the exchange in energy takes between water in a closed system exchanging energy with the water in the pond, without actually removing any water from the pond. The two processes were simulated using Aspen Plus simulation package, the efficiencies were compared and coefficients of performance (COP) of these two systems at various conditions were calculated. The possibility of further increasing the efficiency and applicability of geothermal systems was also explored. The work can contribute to the increased use of geothermal energy in Illinois and developing more sustainable energy systems.

See more of this Session: International Forum on Energy Sustainability (plenary)
See more of this Group/Topical: International Congress on Energy 2011

OPTIMIZING ENERGY CONSUMPTION IN GAS OIL SEPARATION PLANT USING LEAN SIX SIGMA METHODOLOGY

BANDAR JUBRAN ALQAHTANI & DR. MAHMOUD BAHY NOUR ELDIN, SAUDI ARAMCO (SAUDI ARABIAN OIL COMPANY), DHAHRAN, SAUDI ARABIA

Energy resources conservation is essential if population continue to rise and if responsible and caring industrial community and the rest of the developing world is to retain a life style equal to or even better than the one that enjoys today. The economics of industrial production, the limitations of global energy supply, and the realities of environmental conservation are an enduring concern for all industries. Wherever you turn, there's another entreaty to save energy, reduce carbon emissions and protect the environment for posterity.

An energy assessment study was conducted to technically and economically identify energy conservation opportunities at one of the largest Gas Oil Separation Plants (GOSPs) in Saudi Aramco. The energy assessment study used Lean Six Sigma (LSS) methodology to identify, quantify, and rank potential energy conservation initiatives. It is instructive to mention here that such approach is the first LSS application for energy conservation in Saudi Aramco.

Six Sigma uses statistical process analysis and control tools to find and address root causes of energy wastes and variation. Lean Six Sigma project follow a methodology composed of five phases each; bear the acronyms DMAIC which stands for Define, Measure, Analyze, Improve and Control phases. Although Lean Six Sigma originated as a set of practices originally designed to improve manufacturing processes and eliminate defects, its application nowadays have been extended to other types of business processes such as energy performance improvement.

In this approach, the group first defines the project and the plant's processes to be improved, then measures process performance. Measured data is analyzed and process bottlenecks and problems are identified. After analysis, the group develops an improvement program and determines how to remove defects and minimize the energy consumption. At the end, the team develops mechanisms to track the implementation and the energy savings out of the study's recommendations.

The assessment team initially identified 24 potential areas of energy saving. Using a process map and cause-and-effect matrix, followed by a traditional cost-benefit analysis, and finally by a detailed engineering analysis, the list was reduced to eight major areas of improvement. This paper will present the subject energy assessment methodology along with the major recommendations to reduce the power consumption index in the studied Gas Oil Separation Plant.

A Global Challenge to Adsorption Processes in Public Water Supply Production: Mechanisms, Effects, Design Considerations, and Strategic Aversion.

Walter J. Weber, Jr., The Gordon M. Fair and Earnest Boyce Distinguished University Professor, Emeritus, Department of Chemical Engineering, The University of Michigan, Ann Arbor, MI 48109-2125

Abstract

Fixed-bed reactors (FBRs) containing granular microporous activated carbons or polymeric resins comprise integral parts of many public potable water supply treatment systems. These adsorbents and reactor configurations are highly effective in removing potentially harmful synthetic organic contaminants found increasingly in raw source waters. The existence in most raw source waters of complex and highly variable natural organic materials (NOMs) is essentially a concomitant reality, While NOMs themselves pose no direct contamination problems, various molecular fractions thereof are in fact also actively adsorbed, but in preloading" mode" by activated carbons and synthetic polymer adsorbents. This adsorbent "preloading" by NOMs has profound negative effects on the subsequent ability of the adsorbents to remove the potentially hazardous organic contaminants for which they were intended. Unfortunately, these effects are seldom factored adequately into FBR initial designs. This paper presents a rational approach by which the impacts of NOM preloading can be incorporated into the design of FBR adsorber systems on a source-water specific basis. As suggested by its title, the scope of the paper is broad. It begins with an examination of phenomena and mechanisms associated with NOM preloading of FBR adsorbers. The results of a series of investigations from which we have evolved a quantitative approach to the integration of preloading effects in the design of FBR adsorbers are then detailed and a specific case study described. By way of an enticing brief conclusion, a clever means for Strategic Aversion of the global NOM challenge described is referenced.

Introduction

The principal purpose of adsorption systems in water treatment operations is to remove specific types of undesirable organic compounds such as halogenated solvents, pesticides, and fuel residues that pose threats to human health). Activated carbons and several synthetic organic polymers serve this function by providing large surface areas populated by different types of functional groups generated in various "activation processes".. Nuisance and potentially harmful organic compounds are generally present at relatively small concentrations (i.e., ppb levels) in most potable water sources. Conversely, a heterogeneous array of natural organic compounds (e.g., humic and fulvic acids) are commonly present in the same source waters at much higher concentrations (i.e., ppm levels). While compounds comprising this heterogeneous natural organic matter (NOM) are not necessarily harmful to public health on their own accord, some of their constituents interact with halogen-based disinfectants to form potentially harmful "disinfection by-products" (e.g., chlorine and bromine substituted aliphatic and aromatic compounds).

By design, activated carbon is a highly effective adsorbent for removing a broad range of different types of undesirable organic compounds targeted for removal during water treatment (i.e., "target" compounds). Given these inherent properties, GAC is thus effective also in removing many compounds comprising NOM in sources waters. This is desirable from one perspective; i.e., GAC reduces the number of NOM components available to color water, cause odors or tastes, and function as precursors for formation of disinfection by-products. From a different perspective however it is undesirable; i.e., the diverse array of chemical moieties comprising NOM compete effectively with target compounds for the surfaces and reactive sites of activated carbons and polymeric adsorbents, thus interfering with and diminishing its adsorption of target compounds.

Preloading Phenomena

On a one-to-one "molecular" basis, NOM moieties are generally not adsorbed as effectively as target contaminants. However, their presence in order-of-magnitude higher concentrations makes them effective competitors for adsorption sites on GAC. Another less obvious, and somewhat ironic, reason NOM moieties compete effectively for GAC sites is that most of them are in fact not as effectively adsorbed as most target contaminants. This behavior allows them more rapid access than target compounds to the fresh activated carbon regions of a typical fixed bed adsorber (FBR), as depicted by the wavefront behaviors shown in Figure 1. In accessing the fresh GAC more rapidly, the NOM moieties "preload" and "precondition" the fresh GAC in ways that potentially change its properties and reduce its effectiveness for adsorption of target contaminants that come along later. In other words, the fresh GAC intended for target compound removal rapidly becomes a "used" GAC of unknown, but generally lesser, effectiveness for subsequently removing target compounds.

Figure 1. Relative Wavefront Behaviors Leading to Preloading

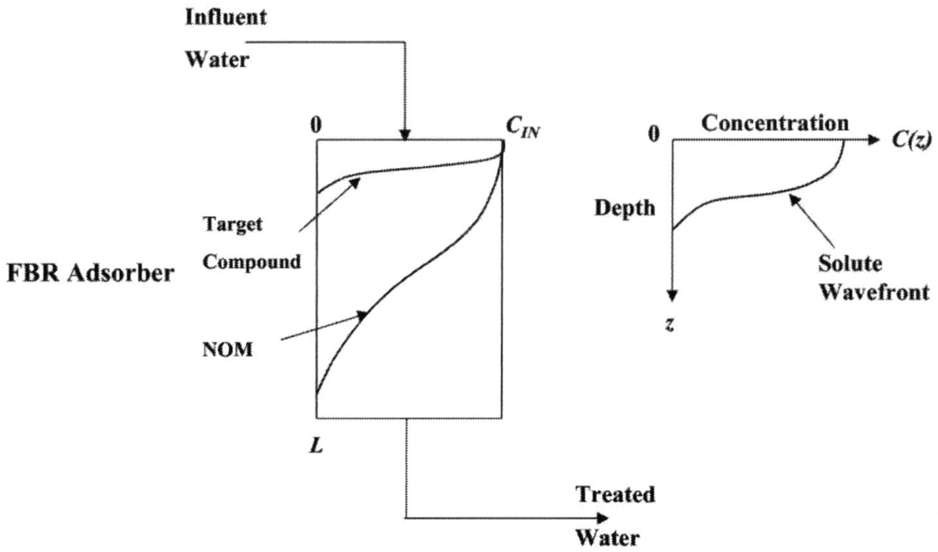

It would seem at first consideration a relatively simple, albeit costly, matter to add more carbon to account for the decreases in its capacity caused by adsorption of the NOM compounds. However, it is impossible to do this in a rigorous engineering manner without knowing the nature and magnitude of the effects of the NOM on the abilities and capacities of the GAC for target compound removal. Unfortunately, these effects can neither be predicted from any available adsorption theories nor can they be assessed by simple lumped-parameter NOM concentration measurements (e.g., total organic carbon). This is the case because: i) the water containing the NOM and the target compounds is markedly changed in character once it moves a short distance in the GAC bed; and ii) the character of the GAC once so contacted is changed as well from what it was initially. To make the situation even more complicated, the changes in each of these two media relate to the specific changes that occur in the other as functions of time and position in the

FBR. Such changes in turn relate to the chemical composition of the water being treated and to the flow characteristics of the FBR.

Adsorption of NOM

Figure 2 presents a schematic representation of the effects of various parameters on the adsorption of NOMs by GAC, plotted as functions of an important background water control variable; i.e., pH. This qualitative representation illustrates the variety of factors that can influence NOM adsorption behavior, which in turn determine how a given GAC will be affected by preloading.

Figure 2. Parameters Affecting NOM Sorption (Karanfil et al, 1998)

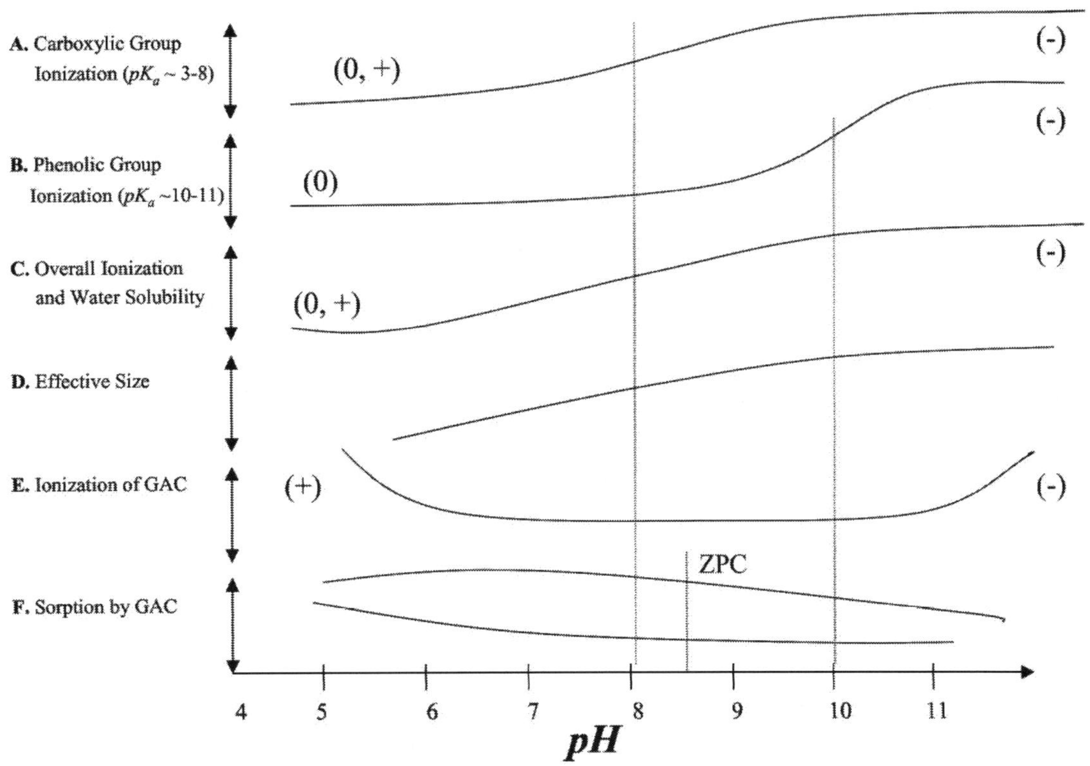

To provide a quantitative example of some of the effects shown in Figure 2, consider the equilibrium adsorption behaviors of the seven NOMs identified and characterized in

Table 1. To characterize equilibrium conditions for this example, we use the following modified Freundlich equation to represent the adsorption isotherm

$$q_e = K_{F,M} \, (C_e/D_o)^n \qquad (1)$$

where q_e is the solid phase concentration of NOM at equilibrium expressed as milligrams of total organic carbon (TOC) per gram of GAC, C_e is the corresponding solution phase concentration (mg TOC/L), D_o is the initial GAC dosage (g/L), $K_{F,M}$ is a Freundlich "affinity" parameter for the chemically heterogeneous NOM, and n is related to the magnitude of the adsorption driving force and to the distribution of energy sites on the GAC (Weber and DiGiano, 1996). The ratio C_e/D_o is the GAC-dosage normalized residual NOM concentration; i.e., the amount of unadsorbed NOM per unit of carbon mass present in the reactors used to collect individual q_e and C_e data points for isotherm plotting and fitting with Equation 1. This modification of the conventional Freundlich equation, which is identical to Equation (1) but does not normalize C_e for D_o, has been found suitable for characterizing the isotherms of heterodisperse polymers. Both the modified and conventional Freundlich equations have been used in different research projects described in this paper, the modified form given in Equation 1 has been used for the most part in the NOM adsorption equilibrium work described in the first half of this paper. The conventional form was employed in the case study example of preloading effects on target compound removal.

Table 1. NOM Characteristics (Karanfil et al, 1996)

NOM	Carbon/Oxygen Contents (%) C	O	O/C	Aromaticity (%)	Phenolic Acidity (meq/g TOC)	Carboxylic Acidity (meq/g TOC)	Weight-Average MW
Aldrich HA	50.7	31.4	0.47	57.7	3.6	7.9	4006
Soil HA (IHSS)	58.0	33.7	0.44	50.0	2.8	5.9	6202
Peat HA(IHSS)	56.6	35.4	0.47	47.0	3.5	7.0	5584
Leonardite HA (IHSS)	63.1	31.0	0.37	58.0	4.4	5.5	5867
Suwannee River FA (IHSS)	53.5	41.3	0.58	24.8	3.3	8.1	1920
Laurentian HA	51.9	39.9	0.58	33.4	5.7	8.9	3982
Laurentian FA	45.1	49.7	0.83	28.2	9.3	11.7	2402

Data obtained by Karanfil, *et al* (1996) for the individual isotherms for the seven NOMs identified in Table 1 are plotted in Figure 3, along with the best fit of the modified Freundlich model to each set of data. Values of $K_{F,M}$ and n for these different NOMs range from 6.6 to 1.7 (mg TOC/g GAC)$^{1-n}$ and from 0.43 to 0.22, respectively, the relative adsorption capacities of the GAC for the NOMs decline in the order: SRFA(1920)>AHA(4006)>LaHA(3982)>LaFa(2402)>LeHa(5867)~SHA(6202), where the quantities in parentheses are the number-average molecular weights of the NOMs determined by size exclusion chromatography.

Figure 3. Humic and Fulvic Acid Sorption Isotherms (Karanfil, 1996)

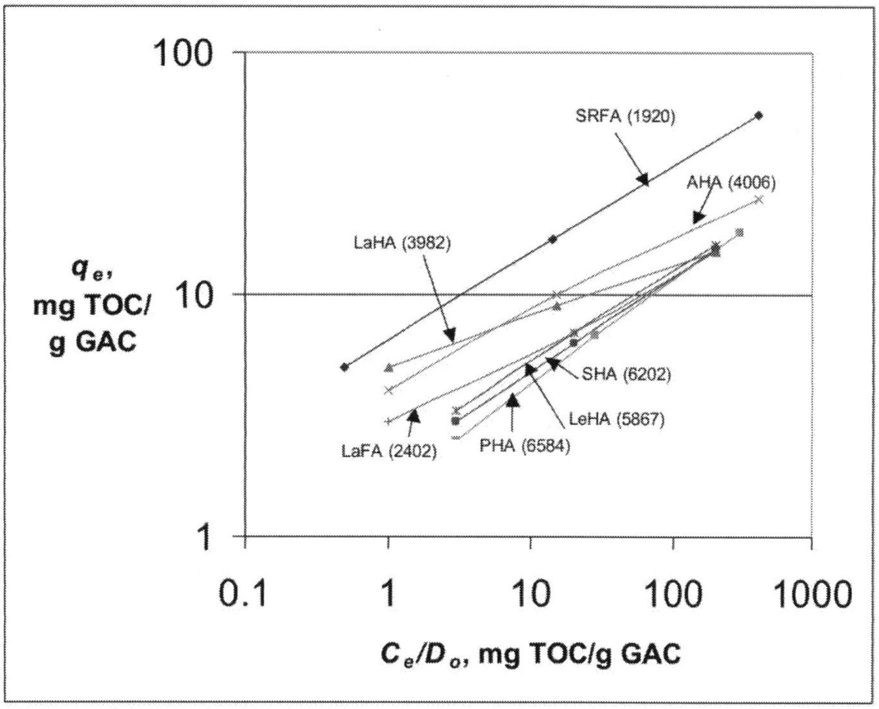

Figure 4 depicts the different adsorption behaviors of various size fractions of one of the NOMs identified in Table 1, i.e., the Laurentian Humic Acid (LaHA). Values of $K_{F,M}$ and n for the different size fractions of this LaHA were found to range from 18.9 to 3.5 (mg TOC/g GAC)$^{1-n}$ and from 0.12 to 0.25, respectively, compared to values of 5.3 (mg TOC/gGAC)$^{1-n}$ and 0.22 for the whole LaHA. It is clear from Figure 4 that, for the constituents of a given NOM mixture, the relative adsorption capacity of the GAC decreases with increasing molecular size. Figure 3, however, just as clearly indicates that other properties of NOM mixtures are also important determinants of adsorption, as suggested earlier in Figure 2.

Figure 4. Sorption Isotherms for Components of Laurentian HA

(Karanfil et al, 1996)

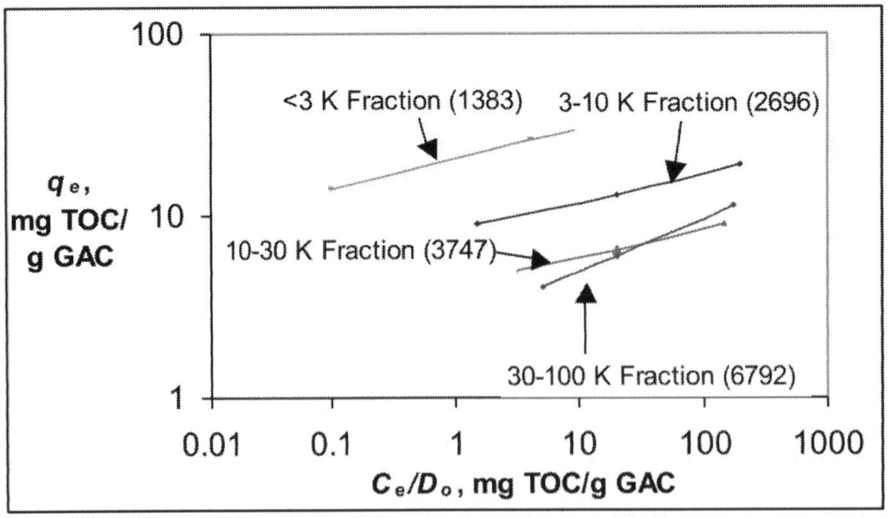

To bring the discussion relating to factors affecting NOM adsorption behavior to a close, Figures 5 and 6 after Karanfil *et al* (1998) show the relative effects of three general solution properties on the adsorption of the Laurentian Fulvic Acid (LaFA) identified in

Table 1. Parts A and B of Figure 5 show isotherms for adsorption of the LaFA from oxygenated and deoxygenated water for three different solution pH values and three different ionic strengths, respectively. Parts A and B of Figure 6 present corresponding parametric trends for the ratio $(K_{F,M})_{oxic}/(K_{F,M})_{anox}$ as functions of solution pH and ionic strength, respectively.

Figures 5. Laurentian HA Isotherms as Functions of pH (A) and Ionic Strength (B) (Karanfil et al, 1998)

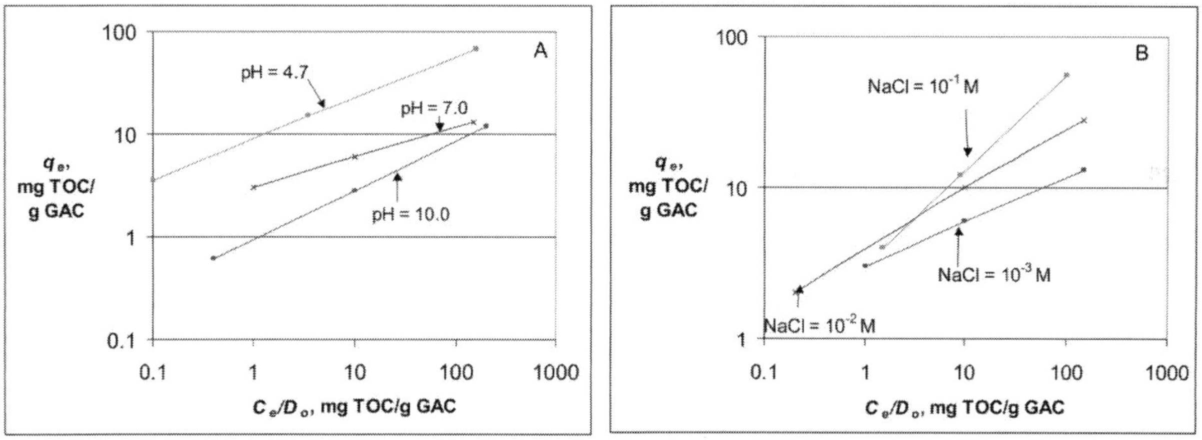

Figure 6. *pH* (A) and Ionic Strength (B) Effects on Oxygen Sensitivity (Karanfil et al, 1998)

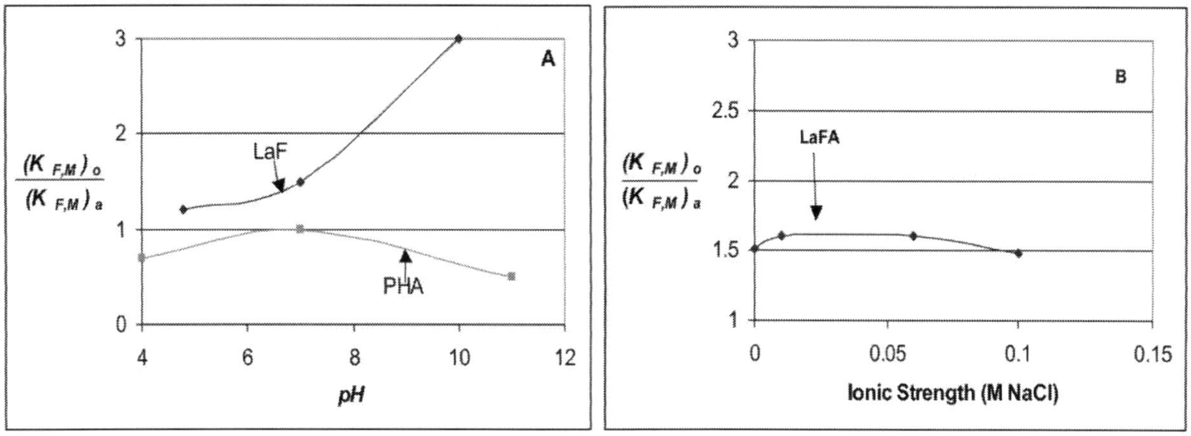

Preloading Mechanisms

The dual mechanism concept advanced by Carter *et al* (1992) involving sequential occupation of high energy surfaces and subsequent pore blocking as an explanation for preloading effects has more or less stood the test of time. This concept derives from observations of changes in the exponent n of the Freundlich isotherm model (modified or otherwise) as a function of increasing time of preloading of GAC in CMBRs and short-bed adsorbers (SBAs). As noted earlier in discussion of Equation 1, the parameter n is representative of the distribution of energy sites on solid adsorbents. The concept is depicted schematically in the series of four panels shown in Figure 7 with the n values given in the panels being actual experimental values measured for adsorption of NOM from Huron River water, a raw water source for the city of Ann Arbor, MI.

Figure 7. Schematic Representation of Preloading Mechanisms (Carter et al, 1992)

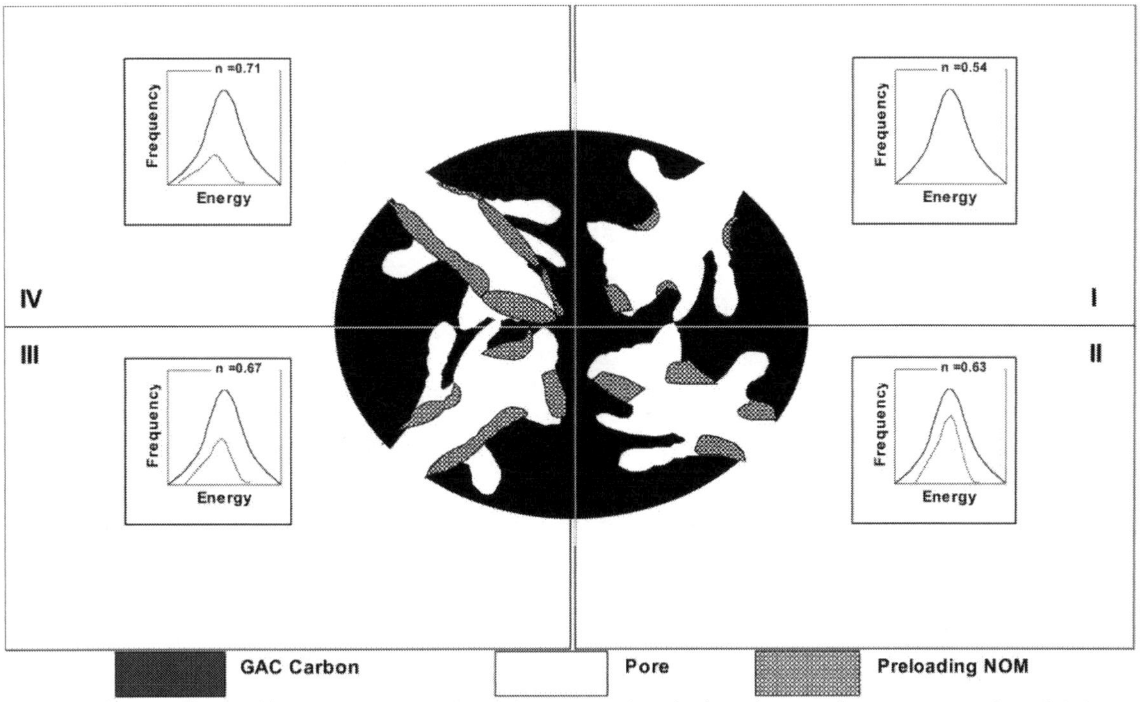

Panel I in Figure 7 represents an initial non-preloaded carbon reference state in which the surfaces and pores are free from NOM. In this state the GAC has highest energy site heterogeneity and lowest *n* value along with the corresponding hypothetical site energy distribution shown in Figure 7. Panel II depicts early preloading effects, with certain NOM species rapidly moving through the larger pores and occupying the most readily accessible high-energy sites on the GAC. The resulting site energy distribution is thereby narrowed from that of non-preloaded carbon, with losses occurring especially in high-energy regions. This is reflected in a rapid initial increase in *n* for early preloading. Panel III represents an intermediate stage in the preloading process as the smaller pore openings begin to be blocked and the remaining readily accessible high-energy sites are lost.

Losses in numbers of sites then begin to occur more evenly across all energy levels, resulting in a gradual leveling off of K_F (or $K_{F,M}$) and *n* values at extended preloading times. This plateauing or leveling of parameter values suggests that most of the readily accessible high energy sites have been exhausted, and that further decreases in capacity occur more slowly via ongoing pore blockage. Panel IV in Figure 7 represents the state of an extensively preloaded carbon. Primary capacity reduction at and beyond this state is mechanistically attributable principally to pore blockage, and all isotherms tend toward parallel slopes (i.e., equal *n* values).

The above mechanism is consistent with a distributed reactivity model (DRM) concept advanced to explain the observed sorption behaviors of heterogeneous soils and sediments in subsurface systems (Weber, *et al*, 1991). In the context of the present discussion, this model projects a sequential compression of an initially wide distribution of sites having different adsorption energies to a more narrowly distributed range by virtue of an initial and uncontested occupation of higher-energy sites on the GAC by selected NOM moieties, followed by a gradual shifting of the dominant capacity reduction mechanism to one of pore blockage by different NOM constituent species. The concept visualizes complex sorption phenomena as comprising one or more typically linear absorption processes, and one or more typically non-linear adsorption processes. Each absorption and adsorption contribution to the overall process manifests a different level of reactivity. For such situations, the sum of absorptive processes increases with increasing saturation of surfaces and pore volumes by organic matrices, and the overall sorption phenomenon thus trends to more linear overall behavior.

An initially fresh (non-preloaded) activated carbon accordingly may act to remove organic compounds primarily by highly non-linear adsorptive mechanisms that exhibit *n* values much lower than unity. This is shown in panel I of Figure 7. As the carbon is preloaded, however, NOM molecules begin to coat the surfaces of the pores, and perhaps even begin to fill some pore spaces completely, essentially forming an organic phase into which absorption now occurs. Panels II through IV of Figure 7 depict this progression of organic accumulation within the carbon pores. As the organic saturation of the carbon increases, the sorption balance shifts away from surface adsorption mechanisms and more toward absorption or "partitioning" mechanisms, with corresponding increases in *n* values. At infinite preloading time, all pore volumes would presumably be filled with NOM organic matrix, and the removal of the target compounds would occur primarily by a nearly linear absorption-like process.

Removal Effects on Target Compounds: A Case Study

As noted earlier and depicted in Figure 1, the most common reactor configuration for GAC applications in water and wastewater treatment is that of a fixed-bed reactor (FBR).

To illustrate the effects of preloading on FBR adsorption behavior with respect to target organic compounds we will examine a specific case study based on work by Carter and Weber (1994) involving a water treatment grade carbon (Filtrasorb 400, Calgon Corp., Pittsburgh, PA), a common chlorinated solvent target contaminant (trichloroethylene, TCE) and a public surface water supply source (Huron River water,

Ann Arbor, MI) having a background NOM as any other might be reasonably representative.

FBR Model

A pore diffusion model was used for calibration of rate data and simulation of all FBR breakthrough data. The model equations and solutions are based upon earlier work by Crittenden *et al.* (1986). The model was reduced to a set of simultaneous ordinary differential equations (ODEs) in time using the method of orthogonal collocation. The coefficients of the ODEs incorporate the isotherm and rate information provided by the equilibrium parameters K_F (note use of conventional Freundlich model form), n and the rate parameters kf (external film mass transfer coefficient), and Dp (internal pore diffusion coefficient). The breakthrough simulation was then obtained by calling a general ODE solver for stiff equations (LSODE) to solve the ODEs through a series of time steps until the total run time was met. The validity of this approach and the ability of the pore diffusion model to describe adsorption in fixed beds have been widely demonstrated. The conventional constant parameter version of the pore diffusion model was used for calibration of SBA rate data and for simulation of DBA adsorber data. A second version of the pore diffusion model referred to here as the time-variable parameter model was developed in which each of the isotherm and rate parameters was input as a function of model run time, e.g., $K_F = \phi(t)$. The isotherm and rate parameters were then updated at each time step, resulting in ODE coefficients that changed over the model run time according to the parametric input expressions.

GAC Preloading

The general chemical characteristics of the Huron River water (HRW) used for preloading are given in Table 2. This water was collected in one batch, dosed to a level of 100 mg/L with sodium azide, and stored under refrigerated conditions. Immediately prior to its use in both isotherm experiments and column studies the HRW was filtered through a 1.0-μm filter. A GAC of US Standard Sieve Size 30/40 (0.512mm arithmetic mean diameter) was preloaded with HRW in FBR adsorbers at a nominal surface loading of approximately 11 cm/sec. The four parallel beds of GAC were preloaded for one week, for two weeks, for three weeks, and four weeks, respectively. Breakthrough of TOC was measured as a function of time in each preloading bed. The mass of background NOM loaded onto the four-week bed was 27 mg TOC/g GAC.

Parameter	Value
Total Organic Carbon (mg/L)	7.4
pH	8.3
Alkalinity (mg/L as CaO_3)	248
Calcium (mg/L as CaO_3)	178
Hardness (mg/L as CaO_3)	260
Ionic Strength (mM as NaCl)	6.9
Mol wt of Organic Matter (amu) (peak weight)	1000-5000 (1860)

Table 2. Huron River Water (Carter and Weber, 1994)

Equilibrium Studies

Conventional Freundlich Isotherm constants for adsorption of TCE from HRW in CMBR systems for the cases of no preloading and preloading for two, three, and four weeks are presented in Table 3. The data indicate significant loss of equilibrium capacity with increased preloading time with the biggest drop occurring by the end of the second week. K_F values dropped from 1.39 to 0.287 over four weeks of preloading. Correspondingly, Freundlich n values increased from 0.542 to 0.713.

Table 3. Isotherm and Rate Parameters (Carter and Weber, 1994)

Preloading Time	K_F (1)	n (2)	k_f (3)	T (2)
None	1.4	0.54	3.5	1.2
One Week	-	-	-	-
Two Weeks	0.49	0.63	1.4	5.0
Three Weeks	0.41	0.67	1.2	8.0
Four Weeks	0.29	0.71	1.0	6.0

(1) Based on $q_e = K_F\, C_e^n$ (with q_e = (µg/mg) and C_e = (µg/L))
(2) Dimensionless
(3) cm/sec * 10^3

Rate Studies

TCE rate parameters were gathered in the background waters for various levels of preloading using the short-bed adsorber (SBA) technique developed by Weber and Liu (1980). In the pore diffusion model described above the relevant rate parameters are k_f, the film transfer coefficient which characterize mass-transfer resistance across a hydrodynamic boundary layer surrounding a carbon particle, and D_p, the intraparticle pore diffusion coefficient which quantifies TCE diffusion through the liquid contained within internal particle pores. The SBA is designed to be sufficiently short so that virtually immediate solute breakthrough occurs. Thus, for early time data, (i.e., before significant intraparticle mass transport occurs) simulations using the pore diffusion model are sensitive only to changes in the film transfer coefficient, k_f. By calibrating only early time data, an explicit determination of k_f is made. Once k_f is determined, it is held constant and the entire SBA breakthrough data set then simulated by varying D_p until a best fit of the model output to the experimental data is obtained. The pore diffusivity, D_p, is incorporated here in terms of an impedance factor, τ, using the relationship $D_p = D_l/\tau$, where D_l is the free liquid diffusivity of the solute.

External Film Transfer

The parameter values resulting from model calibrations of SBA data are also presented in Table 3. A statistically significant decrease in the external film transfer coefficient with increasing preloading was observed with k_f values dropping from 3.5×10^{-3} to 1.0×10^{-3} cm/s over four weeks of preloading. These findings suggest that preloading may impact diffusion processes on the surfaces of GAC particles. One possible explanation may be that the local fluid viscosity of the film surrounding a particle is increased by the presence of the relatively large molecular weight background organic matter preloaded on the particle surface, resulting in decreased rates of diffusion of TCE across the hydro- dynamic boundary layer. Another possibility is that some effective area for TCE mass-transfer flux into the carbon particle is lost due to a partial blockage of external pores by preloaded background organic matter.

Intraparticle Diffusion

The values measured for the pore diffusion impedance factor, τ, can be related to changes in resistance to TCE diffusion with preloading. Impedances near unity imply that the total mass diffusion is on the order of that given by the free liquid diffusivity of the solute, i.e., the carbon pore spaces are relatively open and free of obstruction. Impedances several factors greater than unity indicate that intraparticle diffusion of TCE is greatly restricted by the presence of preloaded background organic matter in the carbon pores. The impedance factors presented in Table 3 can be interpreted within the above framework. For unpreloaded carbon, the τ value was 1.2. This indicates that significant intraparticle transport of TCE occurred in the unpreloaded carbon. For all of the Huron River water preloaded carbons, however, the impedance factors were found to be significantly greater than unity. Values of τ between five and ten indicate that intraparticle diffusion was 5-10 times slower than that for TCE in pure water. This suggests that preloaded background organic matter posed a significant resistance to TCE diffusion within the carbon pores. The high upper confidence limits of τ values greater than three was caused by an increasing insensitivity of the model to this parameter as intraparticle impedance became large.

Time-Variable Parameter Model Sensitivity

To assess the relative contribution of each isotherm and rate parameter to the overall shape of the breakthrough profile in the time-variable parameter modeling method, each parameter was varied individually from a constant baseline condition. The baseline case was produced by simulating breakthrough in the bench-scale fixed-bed adsorber using unpreloaded carbon isotherm and rate parameters based on the CMBR isotherms and corresponding SBA rate parameters for Huron River water over a period typical of the run times used in this research. For the time-variable parameter simulations, a nonlinear regression software package (SYSTAT, Systat, Inc., Evanston, IL) was used to obtain continuous functions of the isotherm and rate parameter values with preloading time that were incorporated into the model. Exponential and linear expressions, which give parameter values as functions of pre- loading time, were fitted to the isotherm and rate data given in Table 3. The resulting equations are presented in Table 4. In the sensitivity analysis, parameter expressions for the Huron River water CMBR-based system were varied one at a time from the baseline condition, and the resulting simulations are plotted in Figure 8.

Figure 8. Time-Variable Modeling Sensitivity to Input Parameters (Carter and Weber, 1994)

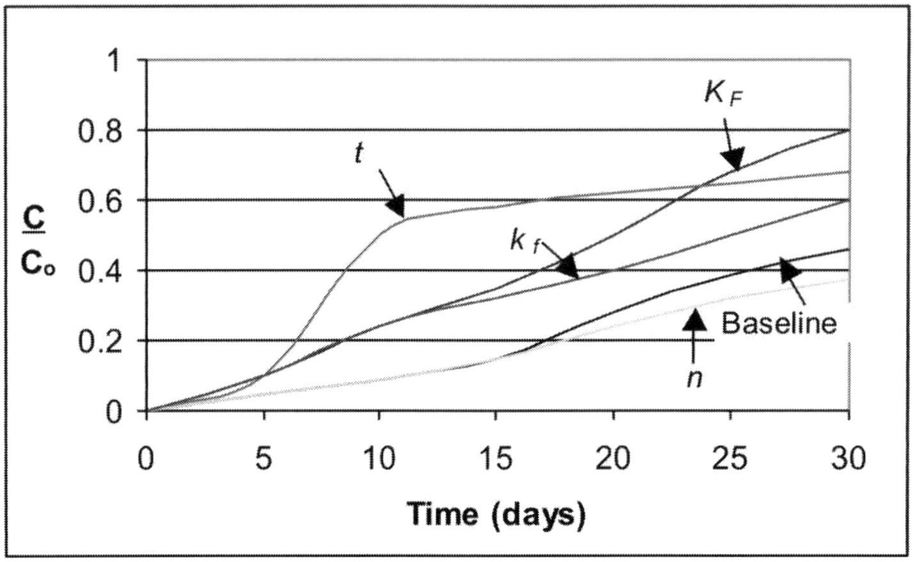

Examination of the breakthrough curves presented in Figure 8 reveals that the greatest impact on this profile was caused by changes in Freundlich K_F and impedance factor values. Decreases in the Freundlich K_F value caused a sharp rise in the latter portion of the breakthrough profile. This rapid rise was probably due to the coupled effect of an accelerated approach to equilibrium as the carbon became loaded with solute and a simultaneous lowering of capacity as K_F values declined. The effect produced by increasing impedance largely impacted intermediate time data. The decreasing rate of uptake of solute by the carbon as τ increased promoted faster breakthrough of solute. Smaller impacts were noted for variations in Freundlich n and film transfer coefficient (k_f) values. Increases in n actually produced a decrease in the breakthrough profile in the sensitivity analysis. This result is an artifact arising from changing τ while holding K_F constant which, in effect, results in an apparent increase in isotherm capacity. In actuality, increases in n are always accompanied by decreases in K_F. The notable feature of the curve resulting from the n variation is that the effect is relatively small, although it still would provide some offset to the impacts of declining K_F values. The range of impact of decreasing film-transfer coefficient (k_f) values is surprising; changes in this parameter are significant over the entire period simulated. The mechanistic nature of this parameter would suggest that k_f variations would primarily impact early time data and diminish thereafter (Weber and Liu, 1980). This behavior may be due to the relatively large drop in k_f (~70%) that was seen with these data. It may also relate to the portion of the breakthrough curve simulated, which reflected only about 50 % breakthrough. As breakthrough approaches completion, the impact of changes in k_f may very well decrease.

Fixed-Bed Adsorber Experimental Data and Modeling Results

Breakthrough data were obtained for TCE in the background water over a period of four weeks using the bench-scale fixed-bed adsorber system. For each set of fixed-bed breakthrough data, both constant parameter and time-variable parameter simulations were made using the pore diffusion model. For the constant parameter simulations, parameter sets obtained at each of the individual preloading levels were used in the pore diffusion model to simulate bench-scale fixed-bed adsorber TCE breakthrough behavior. The time-variable parameter pore diffusion model was used to ascertain whether better predictive capability could be achieved if equilibrium and rate parameters are made to vary with time of operation.

The data and the constant parameter predictions for TCE breakthrough in the Huron River system are presented in Figure 9. The data are bracketed by predictions from the unpreloaded carbon and two-week preloaded parameters, with the unpreloaded carbon set over predicting removal and the two-week set under predicting removal. The close proximity of the four-week CMBR set to the two-week set prediction in Figure 9 suggests that preloading by the Huron River water was beginning to level off. Predictions resulting from use of the time-variable parameter expressions are shown in Figure 10. The predicted curve exhibited a sharper breakthrough profile, with more removal of TCE at early times.

Figures 9. Constant Parameter Predictions and Data for TCE Breakthrough (Carter and Weber, 1994)

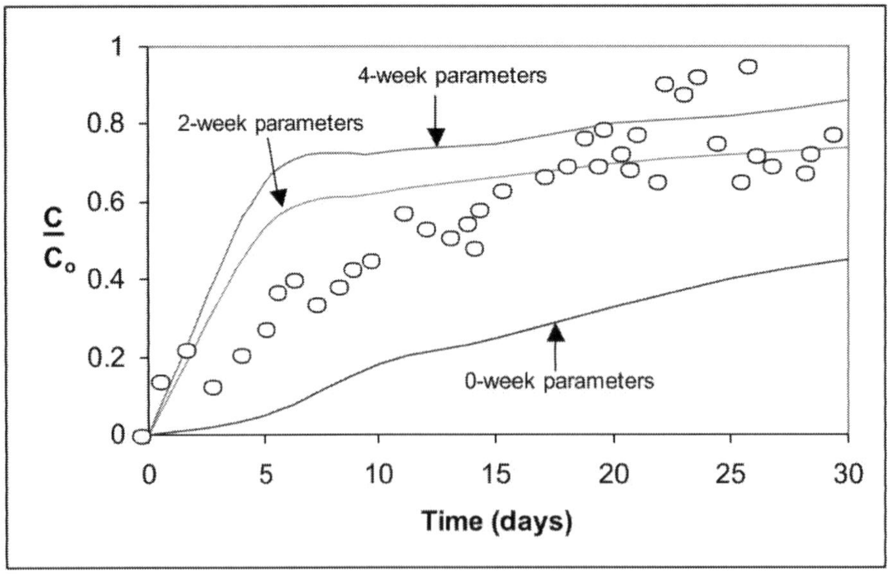

Figure 10. Time-Variable Model Predictions and Data for TCE Breakthrough (Carter and Weber, 1994)

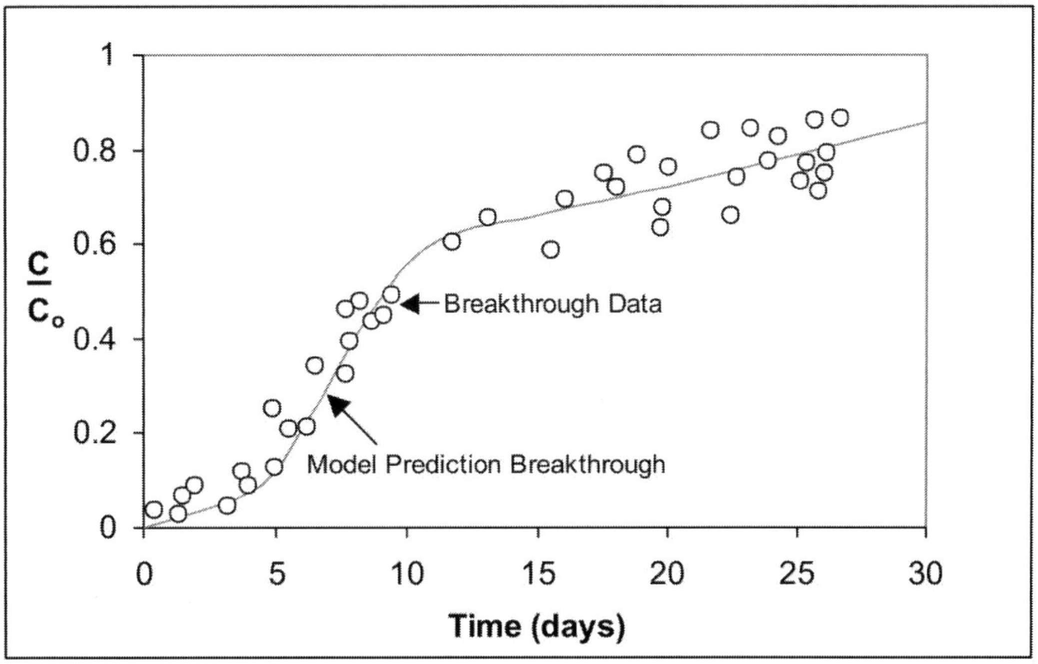

The sensitivity analysis presented earlier in Figure 8 can provide insight into the reasons why a time-variable parameter approach better fits the data. Intraparticle pore diffusion resistance was shown to primarily influence intermediate time data, causing a rapid breakthrough over this region, while decreasing capacity (as measured by K_F) significantly impacted later time data. Individually, the unpreloaded carbon parameters estimated too much capacity and intraparticle mass transport, while the two-week parameters gave levels that were too much reduced. With the time- variable parameter model, the impacts due to preloading appear to be apportioned over the run time in a manner representative of the changes in equilibrium capacity and rates of adsorption that occurred in the bed. Thus, the transition from low to high values for τ caused an acceleration of breakthrough due to higher mass-transfer resistances when capacity was high over intermediate time data. At later times, when intraparticle resistance reached a limiting value, the enhanced upward trend of the breakthrough data was continued as the K_F capacity effects became more significant.

For systems in which adsorption rate and equilibrium parameters do not change markedly over run time, fixed-bed adsorber breakthrough data may be adequately simulated using a constant-parameter modeling approach and some representative "average" parameter set.

Indeed, fixed-bed adsorber removal of target organic compounds in a background organic matter matrix has been successfully modeled using a single set of isotherm and rate parameters obtained on unpreloaded carbons (Smith and Weber 1989, Summers et al 1989, and Speth and Miltner 1989). These studies either used concentrations of target compounds at least an order of magnitude higher than those used in this work or were conducted over significantly shorter periods of time (a few days). Enhanced target compound adsorption rates from higher concentrations or a reduced time for preloading from the shorter run times would reduce the scope of any parameter changes which might occur, resulting in reasonably accurate fixed- bed adsorber modeling with a single parameter set. Conversely, in the time-variable-parameter modeling approach necessary for accurate predictions of TCE breakthrough data for the Huron River water, parameters varied a great deal over the period of study. Thus, a time-variable parameter modeling approach may be beat suited when target compound concentrations are low relative to background organic matter and fixed-bed adsorber run times are long enough for significant preloading to occur.

Conclusions and Practical Implications

The HRW case study breakthrough data summarized in the preceding section observed to lie between model simulations based on parameters derived from unpreloaded carbon data and those based on preloaded carbons. This suggests that, in practice, competitive sorption posed by background organic matter to a target solute will not be strictly simultaneous or strictly preloading in nature. For background waters that do not cause extensive changes in parameter values with preloading, the use of parameters obtained on unpreloaded or an intermediately preloaded carbon may reasonably simulate fixed-bed adsorber behavior. One indication of this state may be that intraparticle diffusion in a preloaded carbon is not yet substantially diminished from the free liquid diffusion of the target solute. In cases where isotherm and rate parameters are observed to change with pre- loading over the time period of interest, the time-variable parameter approach is the likely candidate for accurate predictions. An intraparticle diffusion coefficient several factors less than free liquid diffusion of the target solute may be a sign for this condition.

The tendency observed in the Huron River water data for preloading effects to approach limiting values suggests a third modeling regime; i.e., use of one set of constant parameters representative of a highly preloaded state. This approach was investigated by simulating breakthrough from a one-meter long column over a period of one year by both time-variable and constant parameter methods, using parameters representative of the preloading effects found in the Huron River treatment system. In the time-variable simulation, parameters were allowed to vary over the first four weeks of preloading in the same fashion as for the shorter time predictions discussed earlier. After four weeks of simulation, the values were fixed at those levels. In the constant parameter approach, the parameters were set at the four-week preloading level over the entire simulation time. A breakpoint of 10% of influent concentration was chosen for comparison purposes. The constant parameter modeling approach gave a time to breakpoint of 107 days while the time-variable parameter simulation predicted 143 days. The "extra" capacity afforded by the time-variable parameter simulation over the first four weeks of preloading (i.e., before the model parameters became equivalent) is surprisingly large given the one-year simulated run time. Nonetheless, use of a constant parameter approach with parameters representative of an extensively preloaded state would, in practice, provide a conservative estimate of overall adsorber capacity that would be acceptable for preliminary design calculations.

While three possible modeling regimes have been proposed, the uniqueness of each adsorption system precludes determination of which predictive approach is most appropriate without performing some isotherm and rate parameter measurements. Because of the heterogeneous characteristics of background organic matter, the amount of TOC preloaded onto a given mass of carbon is not a reliable indicator of the resulting preloading effect on adsorption of a target compound. Moreover, characteristics of the solute itself will influence the impact of preloading on its adsorption equilibria and rates. The impact of preloading may be more severe for a weakly hydrophobic compound such as TCE than for a highly hydrophobic solute that would tend to sorb to preloaded organic matter. Exact identification of the mechanisms of the preloading process, assessment of key properties that can be used to quantify the preloading potential of a background organic material, and an understanding of how solutes interact with preloaded organic matter in the preloaded state are areas requiring further research.

Acknowledgements

This paper draws on an invited presentation by the author at an International Water Association (IWA) sponsored Workshop on Biological Activated Carbon Filtration held at Delft University of Technology in the Netherlands nine years ago. Much more significantly, the material described in the paper is drawn from an extensive series of investigations conducted by former students of the author, as evident in the many literature citations thereto. All credits for any such intrinsic value this "capstone" paper may offer is due them. Accountability for flaws and/or shortcomings associated with this effort to compile and describe the overall results of their work accrues on the other hand solely to the author.

References

Carter, M.C., Weber, W.J. Jr., and Olmstead, K.P. (1992) "Effects of Background Dissolved Organic Matter on TCE Adsorption by GAC" *Journal of the American Water Works Association, 84*, 8, 81-91

Carter, M.C., Weber, W.J. Jr., (1994) "Modeling Adsorption of TCE by Activated Carbon Preloaded by Background Organic Matter" *Environmental Science and Technology*, 28, 4, 614-623

Crittenden, J.C., Hutzler, N.J., Geyer, D.G., Oravitz, J.L., Friedman, G. (1986) "Transport of Organic Compounds With Saturated Groundwater Flow: Model Development and Parameter Sensitivity" *Water Resources Research*, 22, 271-284

Karanfil, T, Kilduff, J.E., Schlautman, M.A., and Weber, W.J.Jr (1996) "Adsorption of Organic Macromolecules by Granular Activated Carbon. 1. Influence of Molecular Properties under Anoxic Solution Conditions" *Environmental Science and Technology* 30, 7, 2187-2194

Karanfil, T, Kilduff, J.E., Schlautman, M.A., and Weber, W.J.Jr (1998) "The Oxygen Sensitivity of Organic Macromolecule Sorption by Activated Carbon: Effects of Solution Chemistry" *Water Research*, 32, 1, 154-164

Smith, E.H. and Weber, W.J. Jr. (1989) "Evaluation of Mass Transfer Parameters for GAC Adsorption from Complex Organic Matrices" *Environmental Science and Technology*, 23, 713-722

Speth, T.F. and Mitner, R.J. (1989) "Effect of Preloading on the Scale-up of GAC Microcolumns" *Journal of the American Water Works Association*, 81, 4, 141-148

Summers, R.S., Haist B., Koehler, J., Ritz, J., Zimmer, G. and Santheimer, H. (1989) "The Influence of Background Organic Matter on GAC Adsorption" *Journal of the American Water Works Association*, 81, 5, 66-74

Weber, W.J.Jr. and DiGiano, F.A. (1996) *Process Dynamics in Environmental Systems*, Wiley-Interscience, New York

Weber, W.J.Jr., and Liu, K.T. (1980) "Determination of Mass Transport Parameters for Fixed-Bed Adsorbers" *Chemical Engineering Communications*, 6, 49-60Weber, W.J. Jr., Katz, L.A., and McGinley, P.A. (1992) "A Distributed Reactivity Model for Sorption by

Soils and Sediments: 1. Conceptual Basis and Equilibrium Assessments" *Environmental Science and Technology*, 26, 10, 1955-1962 (1992)

Nomenclature

n	=Freundlich isotherm exponent	[-]
K_F	=Freundlich isotherm capacity factor	[*]
$K_{F,M}$	=Modified Freundlich isotherm capacity factor	[**]
k_f	=film mass transfer coefficient	[cm/s]
C_e	=equilibrium solution-phase concentration	[mg or μg/L]
D_o	=initial sorbent dosage	[g/L]
D_p	=pore diffusion coefficient	[cm^2/s]
D_l	=free liquid diffusion coefficient	[cm^2/s]
q_e	=equilibrium solid-phase concentration	[mg or μg/g]
$\phi(x)$	=function of x	
τ	=diffusion impedance	

*Units dependent upon units of q_e and C_e
**Units dependent upon units of q_e, C_e and D_o

Abbreviations

CMBR	=completely mixed batch reactor
FBR	=fixed bed reactor
GAC	=granular activated carbon
HRW	=Huron River water
NOM	=natural organic matter
ppb	=parts per billion
ppm	=parts per million
SBA	=short bed adsorber
TCE	= trichloroethylene
TOC	=total organic carbon

(115b) Water Challenges and Opportunities In the Emerging World

Adil Dhalla, GE Singapore Water Technology Centre at NUS, Singapore, Singapore

The human population has grown more in the last 50 years than it did in the previous 4 million years. We can see this as both a challenge and an opportunity. Two of our key resources – power and water, sustain the quality of our life, and the renewing of these is economically sound. Less than 1% of the world's freshwater is readily available for human use. The amount of water that's safe for us to use is declining sharply due to pollution and contamination. We need water not only to survive, but also to prosper and grow. With water demand tripling by 2030, we have to be three times as smart. Our business, GE Water & Process Technologies brings together experienced professionals and advanced technologies to solve the world's most complex challenges related to water availability and quality, increased productivity and cost reduction, and environmental regulations.

See more of this Session: Plenary Session on Water Technology Challenges and Advances (invited speakers)
See more of this Group/Topical: Topical 1: Water Technology for Developed and Developing Countries (see also Separations Division)

(115c) Epa's Research and Development Program for Innovative Water Technologies

Thomas Speth, National Risk Management Research Laboratory, USEPA, Office of Research and Development, Cincinnati, OH

EPA's Research and Development Program for Innovative Water Technologies

Thomas F. Speth

The EPA's Administrator Lisa Jackson has released a drinking water strategy to protect public health from contaminants in drinking water. An element of this strategy is to foster development of new drinking water technologies to address health risks posed by a broad array of contaminants. Specifically, EPA wishes to 1) develop, identify, and evaluate drinking water systems of the future that have minimal carbon and water footprints and are more sustainable than current practices; 2) conduct strategic demonstration projects with small communities to address a broad suite of contaminants while maintaining safe drinking water through their distribution systems at reasonable and predictable costs; 3) develop infrastructure management that accommodates the new generation of technologies and monitoring strategies. As part of this effort, the EPA along with the Small Business Administration are catalyzing the formation of a water technology innovation cluster that involves industry, government, academia, and non-government organization partners working together to bring novel technologies to the drinking water market. Emphasis is being placed on systems-based approaches that will utilize and benefit from integrated trans-disciplinary research. This presentation will address the rationale for the program and the efforts that are currently underway.

See more of this Session: Plenary Session on Water Technology Challenges and Advances (invited speakers)
See more of this Group/Topical: Topical 1: Water Technology for Developed and Developing Countries (see also Separations Division)

(115d) Water Treatment Opportunities Utilizing the Dramatic Nano-Fluidic Properties of Carbon Nanotube Membranes

Bruce J. Hinds, Chemical and Materials Eng., University of Kentucky, Lexington KY, KY

Carbon nanotubes have three key attributes that make them of great interest for novel membrane applications 1) atomically flat graphite surface allows for ideal fluid slip boundary conditions 2) the plasma cutting process to open CNTs inherently places functional chemistry at CNT core entrance and 3) CNT are electrically conductive allowing for electrochemical reactions and application of electric fields gradients at CNT tips. In general, the transport mechanisms through CNT membrane are a) ionic diffusion is near bulk expectation with no enhancement from CNT b) gas flow is enhanced by ~1-2 order of magnitude due to specular reflection off of flat graphitic surface c) and pressure driven flux of a variety of solvents are 4-5 ORDERS OF MAGNITUDE FASTER than conventional Newtonian flow due to atomically flat graphite planes inducing nearly ideal slip conditions. Facilitated transport, electrocatalytic defouling and electroosmotic pumping through CNTs are observed. For water treatment applications such, as RO desalination, a 2 wt% CNT polymer has the potential to result in a 100 fold reduction in membrane area. However one of the key difficulties is that to impart membrane chemical selectivity, chemical functionalization of CNT surfaces ruin the slip boundary condition and dramatically reduce flow enhancements. Efforts in the field generally rely on placing high charge density at CNT tips to reject ions or to precisely control CNT diameter to allow only single file columns of water.

See more of this Session: Plenary Session on Water Technology Challenges and Advances (invited speakers)
See more of this Group/Topical: Topical 1: Water Technology for Developed and Developing Countries (see also Separations Division)

Rules of Thumb for Reactor Design and Process Conditions for Thermochemical Treatment of Chromated Copper Arsenate (CCA) Treated Wood Waste

Lieve Helsen, University of Leuven (K.U.Leuven), Heverlee, Belgium
Frederic Cuypers, University of Leuven (K.U.Leuven), Heverlee, Belgium

Extended abstract

Lab-scale pyrolysis experiments have been performed to gain insight in the mechanisms influencing the metal behavior during carbonization of chromated copper arsenate (CCA) treated wood chips. To this end a fixed bed low temperature pyrolysis reactor and a thermogravimetric (TG) analyser have been used. Besides CCA treated wood chips, mixtures of metal and organic model compounds (As_2O_5, As_2O_3, $CrAsO_4$, $CrAsO_4$ + $Ca(OH)_2$, As_2O_5 + glucose, As_2O_5 + lignin, As_2O_5 + activated carbon, As_2O_3 adsorbed on activated carbon, $CrAsO_4$ + glucose, $CrAsO_4$ + lignin, $CrAsO_4$ + $Ca(OH)_2$ + glucose, $CrAsO_4$ + $Ca(OH)_2$ + lignin) have been pyrolyzed in a controlled way. The aim of these model compound experiments is not to imitate the real CCA wood pyrolysis process, but to isolate some individual mechanisms (without being influenced by a variety of complex interactions) and to study the influence of different process parameters on these mechanisms. The parameters controlled during pyrolysis (and their corresponding range) are: temperature (330°C - 430°C), heating rate (5°C/min – 20°C/min), residence time (10 min – 40 min), reactor pressure (0 bar gage – 5 bar gage), heater gas flow rate (1000 Nl/h – 1800 Nl/h), oxygen concentration (1 vol% - 3 vol%) and steam concentration (5 wt% - 15 wt%).

Based on the experimental observations, reactor design guidelines are derived in order to identify an operating window that minimizes the volatilization of metals during thermochemical conversion of CCA treated wood, while ensuring a good quality of charcoal product.

The main observations related to metal and organic model compound experiments are:

1. The interactions between inorganic compounds (metals and minerals) and organic compounds play an important role during thermal decomposition. On the one hand the presence of organic compounds (wood, carbon, glucose, ...) promotes the reduction of As_2O_5 and subsequent volatilization of As_2O_3. On the other hand formation of complexes or agglomerates and adsorption effects may help to prevent arsenic loss below 400°C and may facilitate metal recuperation.
2. Arsenic adsorption on the solid residue may prevent the volatilization of As_2O_3, and may facilitate the formation of agglomerates of As_2O_3 with wood minerals, giving rise to thermally more stable micro-structures. Furthermore, in a fixed bed or moving bed reactor, small fractions of As_2O_3 that volatilize at the bottom of the reactor can be re-adsorbed by above lying wood or partially pyrolyzed wood layers, giving rise to thermally more stable compounds.
3. Mixtures of metal model compounds and respectively glucose and lignin behave in a different way. Glucose, which is a representative organic model compound for tar formation and secondary char formation, shows foam formation during pyrolysis which hampers the release of arsenic trioxide. Elevated pressure has a pronounced effect on the foam density and the generation of pyrolysis gas, resulting in almost 5 wt% higher arsenic retentions at 5 bar gage compared to 0 bar gage in glucose + As_2O_3 and glucose + As_2O_5 samples. Lignin, which is a representative organic model compound for

char production, shows no foam formation. The pyrolysis residue of lignin helps to stabilize arsenic trioxide.

4. Chromium arsenate forms thermally stable compounds with wood minerals consisting of Cr(III), As(V) and Ca(II), at temperatures below 310°C. Elevated pressure seems to have no significant influence on this process. Chromium arsenate, which had not formed stable compounds, decomposes to Cr_2O_3 (which may form volatile organo-chromium compounds) and As_2O_5, at temperatures below 310°C. Lignin helps to stabilize $CrAsO_4$, and thus counteracts this decomposition.
5. Once As_2O_5 is formed, it is readily reduced to As_2O_3 (following a radicallary mechanism which is favored by elevated pressure) in the presence of reductive pyrolysis vapors. Once As_2O_3 is formed, mass transfer limitations and adsorption on the pyrolysis residue (which are both enhanced by elevated pressure) play an important role in retaining arsenic in the pyrolysis residue.

The CCA wood experiments show that:

1. Percentages of metals volatilized depend strongly on the operating conditions of the process. Therefore, arsenic release can be limited by choosing the operating conditions very carefully.
2. The effect of residence time on arsenic and chromium retentions is limited for the ranges considered.
3. Heating rate mainly affects chromium retentions, primarily through a higher thermal lag of the inner particle temperature at higher heating rates.
4. Reactor temperatures lower than 400°C result in a carbon-like product, having the same matrix as the original wood, only porosity has increased. This allows the minerals and metals to migrate during the thermal process, and to be liberated afterwards since the solid structure does not collapse (in contrast to conventional carbonization processes).
5. Arsenic retentions decrease slightly with increasing temperature till 390°C, with a sharp decrease at higher temperatures, which may be attributed to accelerated desorption of arsenic trioxide at temperatures above 390°C. Chromium retentions show a less pronounced effect of temperature.
6. Arsenic and chromium retentions are affected in a different way by elevated pressure. Arsenic retentions increase about 5% at 5 bar gage compared to 0 bar gage, which is attributed to a higher mass transfer resistance. In the case of chromium, the effect is less clear; the increased mass transfer resistance might be counteracted by enhanced heat transfer at the early stages of pyrolysis.
7. Since pressure increase does not only influence the mode of occurrence of arsenic, but also the char structure and extent of tar reactions (elevated pressure enhances secondary reactions and heat transport), the metal behavior and ease of metal recuperation depends on the reactor pressure.
8. Oxygen concentration mainly affects the reactor temperature (a change of inlet oxygen concentration from 1 vol% to 3 vol% increases the peak temperature from 420°C to 550°C at the bottom of the reactor, and from 390°C to 420°C at the top of the reactor), which results in a decrease in arsenic retention with increasing oxygen concentration.
9. A higher gas flow rate and the addition of steam mainly result in a more homogeneous reactor temperature profile. The net impact on metal retention is limited.
10. Scanning electron microscopy coupled to energy dispersive X-ray analysis (SEM-EDXA) shows that during pyrolysis the metal and mineral compounds form agglomerates, which suggests a relatively easy (mechanical) separation between

metals/minerals and carbon. The mineral agglomerates (already existing in the initial wood before pyrolysis) may assist in the metal agglomeration process.

Although further research is needed to investigate the cross correlation between the various parameters, the experimental observations can be translated in following recommendations for reactor design and process conditions:

1. To ensure a good quality and brittle pyrolysis residue a residence time of at least 20 min and a process temperature of at least 370°C is necessary. However, the process temperature has to be limited to 390°C to limit metal release.
2. A more homogeneous wood input, with less fine particles, will result in a more homogeneous pyrolysis residue and reduced arsenic and chromium volatilization.
3. A longer residence time and lower temperature result in a similar wood conversion while reducing chromium and arsenic volatilization, however a lower plant capacity is obtained.
4. An operating pressure of 5 bar (or even higher) results in 5% higher arsenic retentions and mass retentions. Furthermore, elevated pressures allow to use lower heater gas temperatures to obtain the same wood conversion, resulting in even higher metal retentions.
5. Small gas flow rates are important with respect to:
    a. localizing the release of free and adsorbed water (drying) on the one hand and tar formation on the other hand in distinct phases
    b. minimizing the dilution of combustible vapors
6. Tar problems can be avoided by:
    a. preventing mixing of tar and water compounds (which would result in sticky tar compounds),
    b. facilitating secondary reactions (such as cracking of tar); repeated cracking of organic compounds (stimulated by stratification of pressure and temperature in the reactor column) eliminates the liquid phase.
7. The inlet oxygen concentration has to be strictly controlled below 1 vol% or even 0.5 vol% (at a heater gas flow rate of 1400 Nl/h and without steam addition), which increases both the arsenic retention and mass retention with 5% compared to an oxygen concentration in the range 1-3 vol%.
8. If oxygen control below 1 vol% is not feasible, the heater gas flow rate should be limited to 500 Nl/h to limit arsenic release. However, this could hamper the heating of the reactor bed. Nevertheless, heater gas flow rates should be lower than 1800 Nl/h to reach acceptable charcoal yields.
9. Steam addition results in better reactor temperature control, and as a consequence increased metal and mass retentions. Moreover, thermal lag is reduced which is beneficial with respect to process time. However, significant amounts of steam are needed (40 wt% steam at an initial oxygen concentration of 2 vol% and 70 wt% steam at an initial oxygen concentration of 3 vol% to limit the reactor bed temperature to 390°C).

These recommendations still need to be tested in an industrial scale reactor.

(118b) Biomass Pyrolysis and Gasification: Engineering Rules of Thumb

Michael J. Antal, Hawaii Natural Energy Institute, University of Hawaii, Honolulu, HI

In this presentation I review important findings recorded in the archival literature concerning the thermodynamics, kinetics, mechanisms, and catalysis of biomass pyrolysis and gasification that serve as rules of thumb for reactor design.

See more of this Session: Reactor Engineering for Biomass Feedstocks
See more of this Group/Topical: Sustainable Engineering Forum

(118c) Rule of Thumb for Design of Catalytic Pyrolysis of Biomass In Bubbling Fluid Bed Reactor

Foster Agblevor, Biological Engineering, Utah State University, Logan, UT and Ofei Mante, Biological Systsems Engineering, Virginia Polytechnic Institute and State University, Blacksburg, VA

The production of stable biomass pyrolysis depends strongly on both reactor design and the effective design of a condensation train to collect the pyrolysis products. There are several challenges in the design of the pyrolysis system including solids feeding, char separation from the pyrolysis vapors, condensation of the pyrolysis vapors and compression of the non-condensible gases. In the feeding of solid biomass into the reactor, the feedstock properties such is softening point are critical in the feeding. Bridging properties are also very important and that is dictated by the method of biomass preparation, either knife milling or hammer milling. Hammer milled and disc-milled feedstocks tend to bridge easily whereas knife milled material bridges less. The reactor operating temperature is strongly influenced by the composition of the feedstocks. Biomass feedstocks such as poultry litter which are high in ash tend to form clinker in the reactor and therefore reactor temperature control is very critical for successful operation. The condensation train; especially with recycling of the gases tend to create major problems especially if an electrostatic precipitator is used to condense the aerosols. For systems employing electrostatic precipitator, the cooling temperature in the condensers must be very low because any water vapor escaping the condensers tend to cause electrical short-circuiting in the unit and reduces efficiency of the aerosol capture. Catalytic pyrolysis tends to produce more water and therefore the cooling system must be very efficient. The compressor for the non-condensible gases must be able to handle dirty gases (large clearances) otherwise it is easily clogged with the pyrolysis oils that tend to degrade the performance of the compressor and requires frequent replacements.

See more of this Session: Reactor Engineering for Biomass Feedstocks
See more of this Group/Topical: Sustainable Engineering Forum

(118d) Rule of Thumb for Reactor Design and Process Conditions for Supercritical Water Gasification of Wet Biomass

Yukihiko Matsumura, Division of Energy and Environmental Engineering, Hiroshima University, Higashi-Hiroshima, Japan

The Rule of thumb for reactor design and process conditions for supercritical water gasification of wet biomass is introduced. The important aspects includes heat transfer and heating rate, reaction rate, plugging, gas composition, inorganic element behavior, and heat of reaction. The know-hows were developed by the laboratory scale, and pilot scale reactors.

See more of this Session: Reactor Engineering for Biomass Feedstocks
See more of this Group/Topical: Sustainable Engineering Forum

Rule of Thumb for Reactor Design and Process Condition for Biodiesel Production

Masato Kouzu, Doshisha University, Kizugawa, Kyoto, Japan

Introduction

For the last decade, the planet earth clearly tends toward warming due to the greenhouse gas emissions. One of solutions for the globally environmental problem is to utilize biomass resources, which are generated from CO_2 and H_2O with the aid of photosynthesis, as alternative fuel and chemicals. Plant oil is one of biomass resources, and is used as feedstock of alternative to petroleum diesel oil. The alternative fuel is extensively termed as "Biodiesel". Biodiesel has advantages of not only renewable nature mentioned above but also good fuel properties such as high flush point and excellent lubricity [1-2]. Moreover, it was reported that emissions of both carbon monoxide and particulate were reduced by fueling the engine with biodiesel [3-4]. In 2003, European Community decided to replace at least 5.75 % of the yearly consumed fossil fuels with biofuels, by the year 2010. This decision accelerated the use of biodiesel and its production has been constantly growth. The total of biodiesel yearly produced in the world was 7.75 milion metric tons in 2008 [5].

Biodiesel consists of fatty acids methyl ester (FAME), which was produced by transesterifying plant oil with methanol. For the industrial biodiesel production, plant oil is transesterified with the help of alkali catalyst dissolved in methanol. The alkali catalyst such as sodium hydroxide is so active that the yield of FAME is above 90 % for 0.1 h at 333 K under atmospheric pressure [6]. Glycerol that is by-produced from plant oil is separated from biodiesel by decantation. Then, purification to eliminate impurity such as soap from biodiesel is carried out.

In this presentation, the rule of thumb for reactor design and process condition for biodiesel production is introduced. The important aspect includes type of plant oil used as feedstock, an amount of methanol mixed with plant oil and quality of the produced biodiesel. Recently, much interest has been taken in transesterifying plant oil with the novel manners: enzymatic and heterogeneous catalytic reactions. Also, it is very important to consider the way for recycling glycerol of the by-product.

Type of plant oil

Plant oil consists of triacylglycerides having fatty acid moieties, and contains some minor components. Among the minor components, water and free fatty acids (FFA) inhibit the alkali-catalyzed transesterification that is applied to the industrial biodiesel production. Water brings about the undesirable hydrolysis of triglycerides. It is easy to limit the water content to the trace level by drying, because of the hydrophobic property of plant oil. FFA neutralizes the alkali catalyst. Moreover, soap produced by the neutralization causes formation of stable emulsion with the result that it is difficult to decant biodiesel from glycerol. An industrial way to eliminate FFA is the pre-neutralization with alkali hydroxide. For the pre-neutralization, aqueous solution of alkali hydroxide is stirred into the plant oil. Then, the soap sediment and the FFA-free oil separate out by decantation. Recently, much interest has been taken in pre-esterifying FFA into FAME with the help of acid catalyst such as sulfonated cation-exchange resin [7].

Waste cooking oil is the cost-advantageous feedstock for biodiesel production. In addition to FFA and water, polar compounds are impurities of the waste cooking oil. Heat for food preparation such as frying leads to thermolytic and oxidative reactions of triglycerides having unsaturated fatty acid chains, which produces polar compounds. Polar compounds seem to be of no negative effect on the base-catalyzed transesterification [8].

Furthermore, it should be noted that fatty acids composition of plant oil affects quality of biodiesel. As palm oil is abundant in triacylglycerides consisting of saturated fatty acid chains, the resultant biodiesel is inferior in the low-temperature properties such as cold filter plugging point (CFPP).

Methanol/oil molar ratio

An amount of methanol mixed with plant oil is the most important variable affecting the yield of FAME, because transesterification of plant oil with methanol is a reversible reaction. Due to the chemical equilibrium, excess of methanol is required for achieving the conversion of plant oil into biodiesel. The required methanol/oil molar ratio seems to be 6, while the stoichiometric ratio is 3 [6]. Although an increase in the methanol/oil molar ratio is effective in not only preventing the reverse reaction but also enhancing the conversion rate, a large amount of methanol mixed with the transesterified products makes it difficult to decant biodiesel from glycerol.

The efficient way to achieve the perfect conversion with the minimized amount of methanol is the two step transesterification. In this way, glycerol produced by the primary transesterification is removed in advance of the secondary transesterification. Due to the removal of glycerol, for the secondary transesterification, the chemical equilibrium shifts towards the forward reaction that is useful in maximizing the FAME yield.

Quality of biodiesel

As properties of biodiesel must meet the quality standard such as EN 14214 and ASTM D6751, purification is a very important operation for biodiesel production. Biodiesel collected after the decantation which is carried out after the transesterification contains soap deriving from alkali catalyst. Moreover, it is difficult to achieve the perfect removal of glycerol by only the decantation. Since these impurities are soluble in water, biodiesel is commonly purified by washing with water. As a large amount of soap leads to formation of stable emulsion, an amount of alkali catalyst employed for the transesterification should be optimized. Use of hot water is effective in preventing the emulsion formation. Water mixed with biodiesel for the purification can be eliminated perfectly by decantation followed by drying.

On the other hand, there are some water-free procedures to purify biodiesel. Cation-exchange resin serves adsorbent for eliminating soap and glycerol from biodiesel [9]. Also, it was reported that impurities of biodiesel was eliminated by utilizing amorphous magnesium silicate as the absorbent [10]. Some researchers studied vacuum distillation to collect highly-pure biodiesel [11].

Recycle of glycerol

It is essential for biodiesel production to recycle glycerol of the by-product. Glycerol collected after the transesterification followed by the decantation is contaminated with methanol, water and soap. Soap is decomposed with sulfuric acid, and the resultant alkali-sulfate is removed from stream of the glycerol purification. Fatty acids from soap are decanted from a mixture of glycerol, methanol and water. Finally, methanol and water are topped by heating in order to obtain the crude glycerol used as raw material of chemicals.

Also, it is important to study novel conversion routes of glycerol into chemicals, because intensive production of biodiesel has caused overproduction of glycerol that makes its market price plummeting. Corma et al. showed a variety of the novel catalytic reactions to convert glycerol into chemicals in their research paper

[12]. Hydrogenolysis into propanediol is one of the potential routes for promoting the recycle of glycerol as chemicals.

Catalyst selection

Sodium and potassium hydroxides are the alkali catalyst of commercial availability. Although sodium hydroxide is cheaper than potassium hydroxide, easy handling of glycerol that contains soap deriving from potassium hydroxide is advantageous to operation of biodiesel production process.

For the purpose of improving biodiesel production process, many researchers have focused their attentions on enzymatic and heterogeneous catalytic traansesterifications [13-14]. Both of them bring about not only the water-free purification of biodiesel but also easy recycle of glycerol. For the enzymatic transesterification, plant oil is converted into biodiesel with the help of lipase acting as bio-catalyst. Major characteristic of the enzymatic transesterification is mild reaction conditions: the transesterification plant oil is catalyzed at room temperature. Furthermore, the enzymatic transesterification is insensitive to FFA and water contained in the low-quality feedstock. For the practical use of the enzymatic process, the effect of immobilization on the catalytic activity of lipase has been investigated [15].

Zinc-aluminum mixed oxide was utilized as the heterogeneous catalyst for a biodiesel production process, which was termed as "Esterfif-H™" [16]. The heterogeneous catalyst seemed to be active at the temperature of 483-523 K under the pressure of 3-5 MPa. The catalytic activities of calcium oxide and sulfonated amorphous carbon are close to that of alkali catalyst [8, 17].

Reference

1. Dorado, M. P., E. Ballesteros, J. M. Arnal, J. Gomez and F. J. Lopez (2003), "Exhaust emissions from a diesel engine fueled with transesterified waste olive oil", *Fuel*, 82, pp. 1311-1315.

2. Lin, C. Y., H. A. Lin and L. B. Hung (2006), "Structure and properties of biodiesel produced by peroxidation process" *Fuel* ,86, pp. 1743-1749.

3. Karmee, S. K. and A. Chadha (2005), "Preparation of biodiesel from crude oil of *Pongamia pinaanta", Bioresour. Technol.*, 96, pp. 1425-1429.

4. Szybist, J. P., J. Song, M. Alam and A. L. Boehman (2007), "Biodiesel combustion, emissions and emission control", *Fuel Process. Technol.*, 88, pp. 679-691.

5. Knothe, G (2010), "Biodiesel: Current trends and properties", *Top. Catal.* 53, pp. 714-720.
6. Freedman, B. ,E. H. Pryde and T. L. Mounts (1984), "Variables affecting the yield of fatty esters from transesterified vegetable oils", *J. Am. Oil Chem. Soc.*, 61, 1638-1643.
7. Russbueldt, B. M. E. and W. F. Hoelderich (2009), "New sulfonic acid ion-exchange resins for the preesterification of different oils and fats with high content of free fatty acids", *Appl. Catal. A: Gen.*, 362, pp. 47-57
8. Kouzu, M., T. Kasuno, M. Tajika, Y. Sugimoto, S. Yamanaka and J. Hidaka (2008), "Calcium oxide as a solid base catalyst for transesterification of soybean oil and its application to biodiesel", *Fuel*, 87, pp. 2798-2806.
9. Berrios, M. and R. L. Skelton (2008), "Comparison of purification methods for biodiesel", *Chem. Eng. J.*, 144, pp. 459-465.
10. Kucek, K. T., M. A. F. Cesar-Oliveira, H. M. Wilhelm and L. P. Ramos (2007), "Ethanolysis of refined soybean oil assisted by sodium and potassium hydroxides", *J. Am. Oil Chem. Soc.*, 84, pp. 385-392.
11. Sakai, T., A. Kawashima and T. Koshikawa (2009), "Economic assessment of batch production processes using homogeneous and heterogeneous alkali catalysts", *Bioresour. Technol.*, 100, pp. 3268-3276
12. Corma, A., S. Iborra and A. Velty (2007), "Chemical routes for the transformation of biomass into chemicals", *Chem. Rev.*, 107, pp. 2411-2502.
13. Kulkarni, M. G. and Dalai A. K. (2006), "Waste cooking oil -An economical source for biodiesel: A review", *Ind. Eng. Chem.*, 45, pp. 2901-2913.
14. Yan, S., C. DiMaggio, S. Mohan, M. Kim, S. O. Salley and K. Y. Simon Ng (2010), "Advancements in heterogeneous catalysis for biodiesel synthesis", *Top. Catal.*, 53, pp. 721-736.
15. Oda, M., M. Kaieda, S. Hama, H. Yamaji, A. Kondo, E. Izumoto and H. Fukuda (2005), "Facilitatory effect of immobilized lipase-producing *Rhizopus oryzae* cells on acyl migration in biodiesel-fuel production", *Biochem. Eng. J.*, 23, pp. 45-51.
16. Bournay, L., D. Casanave, B. Delfort, G. Hillion and J. A. Chodorge (2005), "New heterogeneous process for biodiesel production: A way to improve the quality and the value of the crude glycerin produced by biodiesel plants", *Catal. Today*, 106, pp. 190-192.
17. Hara, M. (2010), "Biodiesel production by amorphous carbon bearing SO_3H, COOH and phenolic OH groups, a solid Bronsted acid catalyst", *Top. Catal.*, 53, pp. 805-810.

(118f) Rule of Thumb for Design and Analysis of Biomass Utilization System

Kazuhiro Mochidzuki, IIS, Univ. of Tokyo, Tokyo, Japan and Akiyoshi Sakoda, Institute of Industrial Science, University of Tokyo, Tokyo, Japan

This paper deals with the rule of thumb for design and analysis of overall biomass utilization system, including, the matching of available biomass and conversion technologies, the market and demand of biomass-based energy/products, the evaluation of material and energy balances throughout the system, the environmental impacts, and the sustainability. When laying a plot of biomass utilization, it is important to develop comprehensive strategy based on a reasonable scenario. Nevertheless every plot has different circumstances and viewpoints on it. There are endless varieties of factors regarding to design and analysis of the biomass systems. The problems spread across wide-ranged issues from agriculture/forestry to industry, from political/social sectors to scientific/engineering sectors, from local to worldwide, and so on. To establish a reasonable biomass system, it is necessary to make clear the whole picture of it, organizing the key factors and problems.

We have developed concepts and methods to design biomass utilization systems. Our approach includes the followings: 1) screening the key factors, 2) drawing up a scenario and its modeling, 3) matching the available biomass sources and utilization technologies, 4) analysis/optimization of material and energy balances throughout the whole system, and 5) evaluation of economical and environmental issues. In this paper our recent progress in this field is shown with some results of case studies.

See more of this Session: Reactor Engineering for Biomass Feedstocks
See more of this Group/Topical: Sustainable Engineering Forum

(124a) GREENSCOPE: A Method for Modeling Chemical Process Sustainability

Raymond L. Smith, Gerardo J. Ruiz-Mercado and Michael A. Gonzalez, US EPA, Cincinnati, OH

Current work within the U.S. Environmental Protection Agency's National Risk Management Research Laboratory is focused on the development of a method for modeling chemical process sustainability. The GREENSCOPE methodology, defined for the four bases of Environment, Economics, Efficiency, and Energy, can evaluate processes with over a hundred different indicators. These indicators provide a means for realizing the principles of green chemistry and green engineering in the context of sustainability.

Development of the methodology has centered around three focal points. One is a taxonomy of impacts that describe the indicators and provide absolute scales for their evaluation. The setting of best and worst limits for the indicators allows the user to know the status of the process under study in relation to understood values. Thus, existing or imagined processes can be evaluated according to their relative indicator scores, and process modifications can strive towards realizable targets.

A second area of focus is in advancing definitions of data needs for the many indicators of the taxonomy. Each of the indicators has specific data that is necessary for their calculation. Values needed and data sources have been identified. These needs can be mapped according to the information source (e.g., input stream, output stream, external data, etc.) for each of the bases. The user can visualize data-indicator relationships on the way to choosing selected ones for evaluation.

Finally, the methodology is applied to an example, considering the production of biodiesel with various process parameters. Evaluations and process improvements are shown using the taxonomy and data developments described above. Analyses show where biodiesel processes can be improved and made more sustainable in terms of environmental, economic, efficiency, and energy measures.

See more of this Session: Sustainability Metrics At the Process and Product Level
See more of this Group/Topical: Environmental Division

(124b) Life Cycle Comparison of Lithium Ion Battery Manufacturing Methods

Qiang Dai and Christian M. Lastoskie, Civil & Environmental Engineering, University of Michigan, Ann Arbor, MI

Abstract

Although lithium ion batteries(LIB) are considered to be a promising power storage option for batteryelectric vehicles (BEVs), little work has been done to improve and optimize themanufacturing of LIBs from a life-cycle perspective. In this study, theenvironmental impacts of five commercially available LIBs based on differentcathode chemistries, including $LiCoO_2$, $LiMn_2O_4$,$LiCoMnO_4$, $LiNiMnO_4$ and $LiFePO_4$, are examined toaddress the differences in material flows in the production phase of LIBs. In addition, solid state and sol-gelsynthesis pathways are investigated for each cathode materials to account forthe differences between wet chemistry and dry chemistry, with a particularinterest in process heat input and organic solvent use. The results indicatethat using solid state pathway, $LiMn_2O_4$appears to be the most environment-benign cathode material, followed by $LiFePO_4$. For all of the cathode materials, the environmental burdens of sol-gel pathwayoutweigh that of the solid state synthesis method,with organic solvent as the largest contributor. In both cases, startingmaterials containing nickel and cobalt are responsible for a considerable shareof the environmental impacts pertinent to the manufacturing of LIBs. Acomparison of the environmental impacts of anode active materials and theoperational phase impacts of BEVs relative to internal combustion enginevehicles will also be presented in this talk.

See more of this Session: Sustainability Metrics At the Process and Product Level
See more of this Group/Topical: Environmental Division

(124c) Use of Life Cycle Assessment to Evaluate the Sustainable Manufacture of the Active Pharmaceutical Ingredient Pregabalin

Mariano J. Savelski[1], **C. Stewart Slater**[1], Peter J. Dunn[2], Donald J. Knoechel[3] and Christine M. Visnic[4], (1)Department of Chemical Engineering, Rowan University, Glassboro, NJ, (2)Global Research and Development, Pfizer, Inc., Sandwich, United Kingdom, (3)Global Supply, Pfizer, Inc., Kalamazoo, MI, (4)Global Environment, Health and Safety, Pfizer, Inc., Peapack, NJ

Three different routes to synthesize pregabalin, the active ingredient in the pain relieving medicine Lyrica®, have been analyzed. A life cycle assessment has been conducted to evaluate emissions generated from cradle to gate for the three syntheses. The first process studied was a classical chemical synthesis resolution route. The second manufacturing path involves the use of an enzyme to perform a biocatalytic resolution. The third production path, also a biocatalytic route, integrates the recycle of a wrong enantiomer. The analysis shows how transitioning from an entirely chemical synthesis to an enzyme-based biosynthesis greatly reduces organic solvent use and also the usage of the sponge Nickel catalyst. A life cycle analysis associated with the raw materials used, energy consumed, and waste generated was performed. Life cycle emissions resulting from the raw materials manufacture represent the majority of the carbon footprint for the three processes. Upon using the biocatalytic process, the resulting amount of raw materials, waste and energy, and their respective carbon footprint are significantly reduced. These changes are quantified and their contribution to the life cycle evaluated. For example, the sponge nickel catalyst, is one of the largest contributors on a per mass basis and three times that of the lipolase enzyme, but small when the LCA is performed. Organic solvents are a major contributor to the environmental footprint for the classical route. Since the biocatalytic route is performed mainly in water, the majority of organic solvents are eliminated and their associated life cycle emissions are significantly reduced.

See more of this Session: Sustainability Metrics At the Process and Product Level
See more of this Group/Topical: Environmental Division

(124d) Sustainable Supply Chain Design by the P-Graph Framework

Istvan Heckl, Department Computer Science and Systems Technology, University of Pannonia, Veszprem, Hungary, Heriberto Cabezas, Office of Research and Development, U.S. Environmental Protection Agency, Cincinnati, OH, Ferenc Friedler, Department of Computer Science and Systems Technology, University of Pannonia, Veszprm, Hungary and L. T. Fan, Department of Chemical Engineering, Kansas State University, Manhattan, KS

The present work proposes a computer-aided methodology for designing sustainable supply chains in terms of sustainability metrics by resorting to the P-graph framework. The methodology is an outcome of the collaboration between the Office of Research and Development (ORD) of the U.S. EPA and the research group led by the founders of the P-graph framework. U.S. EPA/ORD has substantial creditable experience with the development of indicators and metrics for sustainability, while the P-graph framework is deemed to be highly effective for algorithmic design of process networks including supply chains. The integration of supply chain design and sustainability is the main focus of the collaboration.

The P-graph framework provides a mathematically rigorous procedure for synthesizing optimal and alternative suboptimal networks subject to multiple objectives and constraints, which include profitability and sustainability in the proposed methodology. Specifically, to evaluate the sustainability of a given process under construction including its supply chain, sustainability metrics are incorporated into the design procedure.

The proposed methodology is demonstrated with the optimal design of a supply chain for the infrastructure providing heat and electric power to an agricultural region with relatively limited area where agricultural wastes can potentially be recovered as renewable resources. The objective functions for optimization comprise the profit, ecological footprint, exergy dissipation, green net regional product, and ratio between renewable and total emergies.

See more of this Session: Sustainability Metrics At the Process and Product Level
See more of this Group/Topical: Environmental Division

(124e) Risk Assessment Tools Use Through Process/Product Life-Cycle: Start with Questions, Not Data !

Joao Machado, Tiago Robalo and **Jose Menezes**, Technology Services & Products, 4Tune Engineering Ltd, Lisbon, Portugal

Risk Assessment is done today in the Pharma and Bio-pharma industries with a wide range of tools (FMEA, Ishikawa diagrams, Process Layouts, etc.) that are scattered across the organization and addressed in different meetings and sometimes by different teams. This arises to many problems on the workflow because the communication between each tool is almost non-existing and in order to keep consistency, all tools must be revisited each time a change is made in one of them. Also, most of these tools are not suitable for brainstorm meetings but only to aggregate the key-points. All of these problems lead to redundant workflow and interruptions in the middle of very productive brainstorm sessions on complex and multidisciplinary subjects.

Therefore, it is crucial to tackle this problem with a solution that holds in itself all the necessary tools for risk analysis and, most important, all data is connected between the tools so there is no wasted time revisiting the project.

Moreover, Risk Assessment is many times done in a brainstorming and intense environment where the ability to transport all the ideas in a quick but, at the same time, accurate and professional way is crucial.

For this, an interactive and user-friendly platform where anyone can contribute is crucial not only to hasten the Risk Assessment process, but actually improve it. The use of such tool will be demonstrated with the implementation of a case study.

See more of this Session: Sustainability Metrics At the Process and Product Level
See more of this Group/Topical: Environmental Division

(124f) Sustainable Energy Bio-Sources, An Environmental Performance Approach: Life Cycle Assessment From WWTP to BESs

Gabriel Blejman[1], Roberta Pacciani[2], Sonia Guri[2] and **Lourdes F. Vega**[2], (1)MATGAS Research Center, Barcelona, Spain, (2)Carburos Metlicos, Air Products Group, Barcelona, Spain

Sustainableenergy bio-sources, an environmental performance approach: life cycleassessment from WWTP to BESs

G. Blejman[1], R.Pacciani[1,2], S. Guri[1,2], and L.F. Vega[1,2]

[1] MATGAS Research Center, Campus UAB, 08193 Bellaterra,Barcelona, Spain

[2] Carburos Metlicos,Air Products Group, C/ Aragn300, 08009 Barcelona, Spain

There is a huge impetus for developing newtechnologies for the production of both energy and bioproductsin a sustainable way. For example, Bio-Electrochemical Systems (BESs), such as MicrobialFuel Cells (MFCs) and Microbial Electrolysis Cells (MECs) are noveltechnologies which promise to replace energy intensive wastewater treatmentprocesses (WWTPs) by converting the organic waste fraction present inwastewater into hydrogen or electricity. Their potential as sustainabletechnologies for combined energy and biofuelproduction is very high, considering the huge amount of sludge generated(average 30 kg dry matter/inhabitant-year). Moreover,defining the characteristics and the environmental burdens of the wastewater cominginto a WWTP plays a key role in understanding the possible applications of thesludge for generating energy and hydrogen with MFCs and MECs, respectively.

Inthis work we analyse the environmental impact of (i)a Wastewater Treatment Plant and (ii) the production of Hydrogen and biogas fromthe sludge generated in the WWTP itself.

Firstlywe assess the environmental impact of 1 m3 of water entering the Waste Water Treatment (WWTP) process using Life Cycle Assessment methodology. We then identify theenvironmental hot spots in the process chain, focusing on the most important environmentalburdens of a WWTP: energy use, main pollutants from personal care products,heavy metals and final utilization of sludge as a bioproduct.

Theenvironmental impact is assessed with Simapro,Version 7.2, with CML2000 method. The impact categories considered are: AP(acidification potential, global, kg SO2 eq.), GWP100a (global warmingpotential, kgCO2 eq.), EP (eutrophication potential,global, kg PO_4 eq.), PHO (photochemical oxidation, kg formed ozone),DAR (depletion of abiotic resources, kg antimonyeq.), ODP (ozone depletion potential, kg CFC-11 eq.), ETP (ecotoxicitypotential, kg 1,4-DCB eq.) as well as CED (Cumulative EnergyDemand). Bearing in mind that water reuse and reclamation is one of the mainissues of water sustainability, we add to the standard CML impact categories theFreshwater Ecosystem Indicator, (direct effect on biodiversity of freshwaterecosystem) and the Freshwater Depletion (FD), to properly estimate the impactof the process in the water cycle.

This project is partially financedby Carburos Metlicos andthe Spanish Government, CDTI, in the CENIT BIOSOS, belonging to the Program Ingenio 2010.

See more of this Session: Sustainability Metrics At the Process and Product Level
See more of this Group/Topical: Environmental Division

(136a) HRTEM & XPS Applied to Particulate Emissions

Randy L. Vander Wal[1], Chung-Hsuan Huang[2] and Chethan Kumar Gaddam[2], (1)Energy and Mineral Engineering & The EMS Energy Institute, Pennsylvania State University, University Park, PA, (2) Energy and Mineral Eng. and The EMS Energy Institute, Pennsylvania State University, University Park, PA

Characterization of source (particulate) emission profiles is integral to formulation of source profiles towards development of receptor-based models and extraction of source contributions to monitored sites. Exposure and health effects can then be accurately evaluated, atmospheric impacts accurately assessed and regulatory statutes specifically targeted.

Electron microscopic and spectroscopic methods have great potential for carbonaceous aerosol characterization. High-resolution transmission electron microscopy (HRTEM) provides information of the carbon nanostructure (or relative lack thereof). Image quantification allows extraction of various statistical parameters including lamella length, separation distance and tortuosity. These may be summarized in the form of histograms. X-ray photoelectron spectroscopy (XPS) provides a means for probing the surface chemistry of materials. It provides not only information about the atomic composition of a sample but also information about the structure and oxidation state of the constituent elements.

Combustion produced soot is a highly variable material. Physically the nanostructure can range from amorphous to graphitic to fullerenic. Chemically nearly any element could be included, though the surface functional groups are predominantly oxygen-based. Results will be presented for analysis of HRTEM images of the physical nanostructure and XPS analysis of the chemical composition of soots collected from plant, institutional and residential scale oil-fired boilers, diesel engine, gas-turbine combustor (jet engines) and a wildfire. Physically soots from these emission source classes may be differentiated based on carbon lamella length, mean separation and tortuosity. Chemically these soots may also be distinguished by elemental composition, surface (oxygen) functional groups and carbon bonding.

Characterization of source (particulate) emission profiles is integral to formulation of source profiles towards development of receptor-based models and extraction of source contributions to monitored sites. Exposure and health effects can then be accurately evaluated, atmospheric impacts accurately assessed and regulatory statutes specifically targeted.

See more of this Session: Aerosol Science and Technology Enabling Environmental and Energy Studies I
See more of this Group/Topical: Environmental Division

(136b) Formation of Secondary Organic Aerosol From the Direct Photolytic Generation of Organic Radicals

Sean H. Kessler[1], Theodora Nah[2], Anthony J. Carrasquillo[3], John T. Jayne[4], Douglas A. Worsnop[4], Kevin R. Wilson[5] and Jesse H. Kroll[6], (1)Chemical Engineering, Massachusetts Institute of Technology, Cambridge, MA, (2)Chemistry, University of California, Berkeley, Berkeley, CA, (3)Civil and Environmental Engineering, Massachusetts Institute of Technology, Cambridge, MA, (4)Aerodyne Research, Inc., Billerica, MA, (5)Chemical Sciences Division, Lawrence Berkeley National Laboratory, Berkeley, CA, (6)Chemical Engineering, Civil and Environmental Engineering, Massachusetts Institute of Technology, Cambridge, MA

The immense chemical complexity of organic aerosol, which inhibits predictions of the explicit kinetic behavior of the formation and destruction of these aerosols, is caused primarily by the nearly uncountable number of reactive pathways by which constituent molecules can be oxidized. In this talk, we present a novel method for generating secondary organic aerosol (SOA) by using a photolabile precursor to generate important radical intermediate species directly, for example by photolyzing an iodoalkane to produce alkylperoxy radicals. These radical species, which are the same as those that participate in atmospheric oxidation, react further to form lower-volatility products that subsequently condense onto particles. This method presents two main advantages over standard oxidant-initiated studies: first that subsequent oxidative steps are suppressed in the absence of a gas-phase oxidant, and second that by tailoring the precursor species to have a specified structure, we can examine the effect that the location of the radical site within a molecule has on its ability to form new SOA. Measurements of yields and chemical composition (by aerosol mass spectrometry with either electron impact or vacuum ultraviolet ionization) allow for the comparison of the aerosol generated using this technique with those from chamber experiments, wherein oxidation is initiated by the hydroxyl radical.

See more of this Session: Aerosol Science and Technology Enabling Environmental and Energy Studies I
See more of this Group/Topical: Environmental Division

GeoEngineering - How and How Much Will It Cost

Terry A. Ring, Chemical Engineering, University of Utah, Salt Lake City, UT 84112

Abstract

GeoEngineering has been suggested as a way to mitigate the effects global warming due to the CO_2 buildup in the atmosphere caused by the combustion of fossil fuels. One avenue available for global warming mitigation is the scattering of light back into space from the upper atmosphere. This work looks at the engineering of aerosol particles that can be dispersed into the upper atmosphere, that resist settling out and that have a large scattering efficiency; answering the questions: 1) What is the particle loading required to mitigate the effects of the buildup of CO_2 in the atmosphere?, 2) How would these particles be loaded into the upper atmosphere? and 3) What would this all cost? The costs to mitigate the effects of the last 10 years of global warming are compared to the cost to move a city of 6 million people that might be displaced by global warming induced raises in sea level or by desertification.

Introduction

As politics in the US struggles to do something significant about global warming, the political dilemma internationally is even worse. This lack of serious and concerted global approach is made very obvious by those that signed the Kyoto Protocal themselves have not delivered on their promises which call for signatory nations to reduce their emissions to below 1990 levels by 2012 – on average a 5% decrease. The recent recession is the only thing that has had any serious impact on the reduction of CO_2 emissions. Given this international political logjam, it is becoming increasing clear that various means of GeoEngineering have to be evaluated and one or more methods readied for a global warming crisis – abrupt climate change caused by ocean conveyor-belt shutdowns – should it be needed in a hurry. There are various ideas floated for GeoEngineeing which are shown in Figure 1, one of the most promising being solar radiation management. A 0.5% albedo increase would roughly halve the effect of CO_2 doubling[1]. It should be pointed out that none of these solar radiation management techniques addresses the problem of ocean acidification. As a result, there are political factions worldwide that are against any GeoEngineering experimentation. So how do we go proceed?

It is suggested that we first look at the risks to the planet of the various methods of GeoEngineering, then look at the economics and finally look in detail at the science and evaluate the technical feasibility. Only then should some small scale experiments be performed to ready some technology for deployment should the need arises.

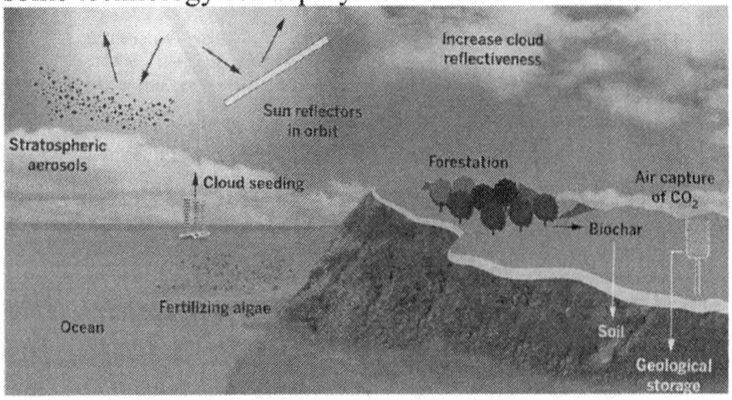

Figure 1 Methods for the control of Global Warming.

Some work has been done in this regard already. One of the first serious evaluations in he 1970s was that of the Russian climatologist Mikhail Budyko who calculated the amount of SO_2, 35 million tons per year, to be placed into the stratosphere to generated droplets of sulfuric acid aerosols in the upper atmosphere. With this additional stratospheric aerosol, the earth's increased reflectivity could compensate for the global warming effects of 150 ppm CO_2. In 1984, John Nuckolls and Wallace Broecker found that 35 million tons per year of SO_2 would cost \$20 billion and the fleet of Boeing 747s needed to deliver the gas to the stratosphere would cost another \$20 billion both in 1984 \$s. In 1995 Paul Crutzen[2], who won the Nobel Prize in 1995 for his work on the ozone depletion in the upper atmosphere revived this idea pointing out that 55 million tons per year of SO_2 is released into the lower atmosphere due to combustion, where SO_2's cooling effect is presently offsetting global warming. In addition, Crutzen suggested that balloons be used in place of Boeing 747s with a cost of \$40 to \$50 billion in 1995 \$s. The stratospheric SO_2 needs to be replenished each year because it will fall to earth as rain and snow. Certainly the additional SO_2 would further increase acid rain and further add to the acidification of the ocean which is a significant environmental problem. So are there other means that have lower environmental costs?

A preliminary study of solar radiation management by Edward Teller and others in 1997 presented the pros and cons of various relatively "low-tech" proposals to mitigate global warming through scattering/reflecting sunlight away from the Earth via insertion of various materials in the upper stratosphere, low earth orbit, and L_1 locations[3]. Of these the use of dust is particularly attractive. In 1992, a report by the US National Academy of Sciences (NAS)[4] on GeoEngineering noted that dust is a better choice compared to sulfur, because dust is from natural soil and so should have no noticeable effect on the ground as it gradually falls into the troposphere and rains out. It estimated that about $\sim 10^{10}$ kg dust would be required to mitigate the warming from a doubling of atmospheric CO_2 or about 1 kg dust per 100 t of carbon emissions. Examples of the effects of the imposition of aerosol particles in the atmosphere can be found in history. Comets have been blamed for the dramatic but brief cooling period which commenced in 1159 BC, and resulted in widespread disruption to civilisation at the time. Other examples of climate change events linked to comets include the famines around 536 BC. The majority of recent atmospheric aerosol additions have been in the troposphere due to pollution and forest fires, rarer examples are due to the eruption of volcanoes, which affect mainly the stratosphere. For example, average global temperatures dropped about a degree Fahrenheit for about two years after the eruption of Mount Pinatubo in 1991, and very cold temperatures caused crop failures and famine in North America and Europe for two years following the eruption of Tambora in 1815[5]. For stratospheric aerosols the experiment has been done. And it was successful!

This paper considers the use of fractal aggregates of TiO_2 a material with a high refractive index used ubiquitously as a white pigment in paint. TiO_2 has one of the lowest hydroxide solubility products[6] known to mankind so its environmental impact would thereby be minimized. Fractal aggregates especially those with a coating of water or ice do not have a light scattering and light absorption theory. This work fills in this gap using these results to identify the quantity of TiO_2 particles required to mitigate global warming. Fractal aggregates due to their structure scatter more light and absorb much less light making them particularly attractive for solar radiation management. Less TiO_2 in this fractal form is required to mitigate global warming. In addition, TiO_2 is rather dense and one would think that TiO_2 particles would settle to earth quickly but due to the fractal aggregate structure their terminal velocity is substantially lowered compared to an equivalent sphere. Finally the costs associated with the use of TiO_2

particles to mitigate the effects of global warming are calculated using refurbished supersonic Concord aircraft to disperse an aerosol into the stratosphere at 60,000 feet (18,000 m).

Light Scattering of Coated Fractal Aggregates

Light scattering and light absorption for fractal aggregates is a complicated problem where particle are so close to one another that multiple scattering must be considered. Berry and Percival[7] developed a theory for the scattering of fractal clusters of N Rayleigh spheres with a radius of gyration given by R_F. They use the near field solutions to the light scattering equations for the neighboring spherules in the fractal and account for incident light and light scattered from each of the other individual spherules in the cluster, a process called infinite order multiple scattering or equivalently a mean field approximation. Clustering statistics are accounted for using the radial probability distribution function of the fractal that depends upon the fractal dimension of the cluster, D_F. Berry and Percival give the absorbance and scattering enhancement factors averaged over all angles as:

$$\frac{Q_{scat\_N}(\lambda)}{NQ_{scat}(\lambda)} = 1 + \frac{2(N-1)}{(D_F-1)(2-D_F)\left(\frac{2\pi R_F}{\lambda\sqrt{(D_F+1)D_F}}\right)^2}\left\{[1+\left(\frac{2\pi R_F}{\lambda\sqrt{(D_F+1)D_F}}\right)^2]^{1-\frac{D_F}{2}}\cos\left[(2-D_F)\tan^{-1}\left(\frac{2\pi R_F}{\lambda\sqrt{(D_F+1)D_F}}\right)\right]-1\right\} \quad [1]$$

and

$$\frac{Q_{abs\_N}(\lambda)}{NQ_{abs}(\lambda)} = 1. \quad [2]$$

where Q is the scattering (scat subscript) or absorption (abs subscript) efficiencies of an individual primary particle, R_F is the radius of gyration of the fractal particle and D_F is the fractal dimension. The lack of absorption enhancement greater than 1.0 given in Equation 2 neglects the fact that the scattered light from one primary particle might fall upon another primary particle in the fractal structure and become absorbed or go through multiple scattering events before being absorbed. As a result, Equation 2 should be considered simply a first order term in a series of terms. We can account for scattered light that is then absorbed by another primary particle in the fractal structure, a second order term, by using the probability that a light beam will interact with the fractal structure, $[R_F/a]^{(D_F-2)}$ given by Witten and Cates[8], and multiplying by the fractal scattering efficiency which would give an absorbance efficiency on a per-primary-particle basis:

$$\frac{Q_{abs\_N\,Fractal}(\lambda)}{NQ_{abs}(\lambda)} = 1 + [(\frac{R}{a})^{D_F-2}]\,Q_{scat\_N}(\lambda). \quad [3]$$

Equation 3 gives an absorption enhancement that is related to the fractal scattering efficiency, which is very large compared to the scattering efficiency of a single particle.

To account for the certainty that any oxide particle in the stratosphere will be coated with water or more likely ice, the scattering and absorption efficiencies are determined by from a theory developed by Aden and Kerker[9] of a coated sphere follows a similar mathematical formulation as that of the Mie theory for uncoated primary particles with dimensionless radius,

$\alpha=2\pi a/\lambda$, and with the radius of the coating of thickness δ, made dimension less by the wavelength of light λ, i.e. $\beta = 2\pi(a+\delta)/\lambda$, and the relative refractive indexes for the sphere core to that of the medium is given by $m_1(\lambda)$ and that of the coating to that of the medium is given by $m_2(\lambda)$. This theory uses the rigorous development of the Mie theory where α is not limited to values less than one required by the simpler Rayleigh light scattering theory. Applying this to fractal TiO_2 using the wavelength dependence of the refractive index[10] of TiO_2, water coating the TiO_2 particles and air as a medium was obtained from so that $m_1(\lambda)$ and $m_2(\lambda)$ could be determined to be used in

$$Q_{scat}(\lambda) = \frac{2}{\alpha^2}\sum_{n=1}^{\infty}(2n+1)\left\{\left|A_n(\alpha,\beta,m_1(\lambda),m_2(\lambda))\right|^2 + \left|B_n(\alpha,\beta,m_1(\lambda),m_2(\lambda))\right|^2\right\} \quad [4]$$

$$Q_{abs}(\lambda) = \frac{2}{\alpha^2}\sum_{n=1}^{\infty}(2n+1)\{\mathrm{Re}[A_n(\alpha,\beta,m_1(\lambda),m_2(\lambda)) + B_n(\alpha,\beta,m_1(\lambda),m_2(\lambda))]\} - Q_{scat}(\lambda) \quad [5]$$

where Re(z) is the real part of the imaginary number z with A_n and B_n are determined from a complex formulation of half-integer order Bessel, Neumann and Hankel functions and are wavelength dependent. Note that refractive indexes and their ratios, $m_1(\lambda)$ and $m_2(\lambda)$, are complex numbers. The scattering and absorption efficiency is used to determining the attenuation of sun light with elevation, z, according to the Beer-Lambert law.

$$-\frac{d\, I(z,\lambda)}{dz} = N\,\pi\, a^2\,(Q_{abs}(\lambda) + Q_{scat}(\lambda))\, I(z,\lambda) \quad [6]$$

The boundary condition

$$I(z=\infty,\lambda) = I_{sun}(\lambda) \quad [7]$$

corresponds to the sun's light intensity[11] as a function of wavelength. With Equation 6 the scattering and absorption of light as the sunlight penetrates the atmosphere can be determined.

The results of this analysis are that there is a scattering enhancement factor of 200 to 10,000 as the fractal dimension increases from 1.8 to 3.0 for large fractal objects with a large number of primary particles in them. As a result of these calculations $1\text{x}10^8$ particles/m^3 corresponding to result $3.6\text{x}10^{10}$kg of TiO_2 with 0.5 micron (radius of gyration) fractal aggregates of 10 nm TiO_2 primary particles in a fractal structure with a fractal dimension of 1.8 are required to totally eliminate the global warming due to all anthropogenic sources[12] (2.4 W/m^2 max including error estimate) if placed in a 1 km section of the stratosphere.

Settling of Fractal Aggregates

The sedimentation of a 1 micron diameter fractal aggregate of 10 nm primary particles can be calculated by a balance of the drag and buoyancy forces.

$$F_G = F_D$$

$$N\frac{4\pi (R_o)^3}{3}(\rho_p - \rho_f)g = N\,6\pi\mu_f u_t R_o$$

where N is the number of primary particles of radius, R_o, in the fractal aggregate ($N=[R_F/R_o]^{D_F}$), ρ is the density of the particle (p subscript) and fluid (f subscript) and μ_f is the viscosity of the fluid, air. This force balance gives the terminal settling velocity, u_t, for the fractal aggregate particle which is equivalent to that for an individual primary particle. The terminal settling velocity of a 10 nm (radius) TiO_2 particle is $1.7x10^{-8}$ m/s suggesting that it will take 33,300 years for the particle to settle to the sea from 18,000 m where the airplane will place it not considering any convection. Should the TiO_2 particle be the size of the radius of gyration and not be a fractal structure, the sedimentation velocity would be $1.2x10^{-4}$ m/s with a settling time of 4.8 yrs. This shows the significance of using fractal aggregates on the terminal settling velocity of stratospheric aerosols used for solar radiation management. Using fractal particles the particles will stay aloft for a very long time and as such may be considered a nearly permanent global warming fix.

Cost of Aerosol Manipulation of Solar Radiation

The costs associated with putting $3.6x10^{10}$kg of TiO_2 into a 1 km section of the stratosphere are significant. The TiO_2 costs alone are $360 Billion. The approach that was used to cost this scenario involves refurbishing 10 Concord airplanes at a cost of $19 million each. Ten Concords correspond to 1/3 of all the Concords in airplane graveyards around the world. An additional $10 million is estimated as the cost to equip each Concord with an aerosol generator, pneumatic conveying system and TiO_2 storage tanks. The storage capacity of a Concord is limited so that 327,000 flights are required to disperse the required TiO2 aerosol. A Concord can fly at 18,000 m altitude with a range of 7,250 km and uses jet fuel at a rate of 13.2 kg/km. Using the cost of jet fuel each flight will cost $15,000 for a flight across the Atlantic. As a result the total cost to mitigate the effects of all of global warming due to all anthropogenic sources some 2.4 W/m^2 of the earth's surface is $412.9 Billion in 2011 $s. Please keep in mind that these are one time costs and that at any time along the progression of this work the experiment can be stopped giving a limited reduction of global warming. This mitigation cost is on the order of what it would cost to relocate a US city of 6 million people, $200 Billion in 2011 $s, assuming that each family's home would need to be rebuilt at a cost of $100,000 each. One final point to be reiterated, solar radiation management can lower and even mitigate the effects of global warming but it will not prevent the acidification of the oceans as CO_2 build up in the atmosphere is not affected by the addition of aerosols to the stratosphere.

References

[1] Committee on Science, Engineering, and Public Policy (1992). *Policy Implications of Greenhouse Warming: Mitigation, Adaptation, and the Science Base*. National Academy Press. p. 447. ISBN 0-309-04386-7. http://fermat.nap.edu/openbook/0309043867/html/447.html.

[2] Crutzen, Pau J. "Albedo Enhancement by Stratospheric Sulfur Injections: A Contribution to Resolve a Policy Dilemma?" Climatic Change 77(200)6, 211-19.

[3] Edward Teller; Roderick Hyde, and Lowell Wood (1997) (PDF). *Global Warming and Ice Ages: Prospects for Physics-Based Modulation of Global Change*. Lawrence Livermore National Laboratory. https://e-reports-ext.llnl.gov/pdf/231636.pdf. Retrieved 2018-10-30. See pages 10-14 in particular.

[4] National Academy of Sciences, Policy implications of greenhouse warming: Mitigation, adaptation and the science base. National Academy Press, Washington DC, 1992, pp. 433–464.

[5] http://www.cotf.edu/ete/modules/volcanoes/vclimate.html

[6] K_{sp}=3.1x10^-40 http://www.google.com/url?sa=t&source=web&cd=20&ved=0CGIQFjAJOAo&url=http%3A%2F%2F203.158.253.140%2Fmedia%2Fe-

Book%2FEngineer%2FEnvironmental%2FWater%2520Quality%2520and%2520Treatment%2520Handbook%2F16593_10.pdf&rct=j&q=solubility%20product%20ti(OH)4&ei=ptdbTp3gOabniAL58fGrCQ&usg=AFQjCNGFTvvkOKsSv2uFlQarQHPBCMI0Rg&sig2=fMNrU04UjqO9Du_-VLT3SA&cad=rja

[7] Berry M.V. and Percival, I.C. , "Optics of fractal clusters such as smoke," Optica Acta, 33, 5, 577-591,(1986).

[8] Witten, T.A., and Cates, M.E., Science 323,1607-11(1986).

9 Aden, A.L. and Kerker, M. J.,,, Scattering of electromagnetic waves from two concentric spheres,,, Appl. Phys., 22, 1242, (1951).

[10] http://refractiveindex.info/

[11] http://lasp.colorado.edu/sorce/data/tsi_data.htm or http://www.ngdc.noaa.gov/stp/solar/solarirrad.html

[12] Intergovernmental Panel on Climate Change Report of Working Group I

(136d) Studies of Atmospheric Aerosol Chemistry Using Aerosol Chemical Ionization Mass Spectrometry

V. Faye McNeill, Department of Chemical Engineering, Columbia University, New York City, NY

This talk will highlight the use of Aerosol Chemical Ionization Mass Spectrometry (Aerosol-CIMS) as enabling technology for the study of atmospheric aerosol chemistry. Aerosol-CIMS is particularly well-suited for laboratory studies of aerosol organics; chemical ionization leads to low fragmentation, enabling speciated measurements of organic species with the high sensitivity and fast time response of CIMS. Recent work in which Aerosol-CIMS was used to study the formation of organosulfates, oligomers, and nitrogen-containing species in secondary organic aerosols will be discussed, as well as the application of Aerosol-CIMS to study the oxidative aging of organic aerosol material. Application of Aerosol CIMS to the analysis of ambient aerosols will be discussed.

See more of this Session: Aerosol Science and Technology Enabling Environmental and Energy Studies I
See more of this Group/Topical: Environmental Division

(136e) Optimization of Laminar Flow Ultrafine Condensation Particle Counters for the Enhanced Detection of 1 Nanometer Condensation Nuclei

Chongai Kuang[1], Modi Chen[2], Peter H. McMurry[2] and Jian Wang[1], (1)Environmental Science, Brookhaven National Laboratory, Upton, NY, (2)Mechanical Engineering, University of Minnesota, Minneapolis, MN

Atmospheric aerosols influence climate and climate change on local to global scales by affecting the atmospheric radiation balance directly through scattering and absorbing incoming solar radiation and indirectly as cloud condensation nuclei. New particle formation (NPF) by photochemical reactions of gas-phase precursors greatly increases the number concentrations of atmospheric aerosols, and therefore their impact on climate. Although methods for measuring sizes and concentrations of newly formed particles of diameter greater than 3 nm are well established, measurements of nanoparticles smaller than this are needed to constrain nucleation rates and to better understand nucleation mechanisms. A diethylene glycol-based laminar flow ultrafine condensation particle counter (DEG-UCPC) has recently been developed for sub-2 nm detection, enabling the detection of negatively charged sodium chloride with an efficiency of 2% at a mobility diameter of 1.16 nm. By increasing the aerosol flow rate and operating temperature difference in the DEG-UCPC, this detection efficiency has been increased by a factor of 10 at the same particle size without nucleation of the working fluid. Similar non-destructive operating modifications to a commercial butanol-based CPC (TSI 3025A) have increased the detection efficiency of 1.68 nm mobility diameter particles from less than 0.5% to over 35% for negatively charged sodium chloride. Laboratory characterization of CPC detection efficiency as a function of particle size, charge, and composition will be presented for both instruments. Ambient size distributions acquired with this modified DEG-UCPC will also be presented.

See more of this Session: Aerosol Science and Technology Enabling Environmental and Energy Studies I
See more of this Group/Topical: Environmental Division

(136f) Online Measurements of Structure and Mass Concentration for Airborne Nanoparticle Agglomerates

Zhun Liu[1], Jacob Swanson[1], Jing Wang[2], Heinz Fissan[3] and David Y. H. Pui[1], (1)Mechanical Engineering, University of Minnesota, Minneapolis, MN, (2)Institute of Environmental Engineering, ETH Zurich, Zurich, Switzerland, (3)IUTA, Duisburg, Germany

A variety of engineered nanomaterials, such as carbon black, fumed silica and tinania, are produced in flame aerosol synthesis as nanoparticle agglomerates. Other important sources of nanoparticle agglomerates include the soot particles generated from the burning of biomass and diesel fuel. Particles formed in these high-temperature processes are commonly found to grow from nearly-spherical primary particles that are usually tens of nanometers or smaller into fractal-like agglomerates that can be microns in size or larger. This change in particle morphology has profound implications on particle properties, including light scattering, surface chemistry, transport, and potential health effects. The time and labor expenses of offline electron microscopy inhibit its application in situations, such as quality monitoring for nanomaterial manufacturing, where fast measurement of agglomerate structure is required. Therefore, an online characterization tool for the structure of nanoparticle agglomerates is highly desirable. Moreover, real-time information of airborne agglomerate mass concentration is of interest to both the nanotoxicologists and the diesel engine industry. It was found that lower mass dose of nanoparticles is needed for them to have the same toxicity as particles of micron size or larger due to the increased active surface area of the former. In addition, the increasingly stringent environmental regulations call for cleaner diesel engines and it is difficult to use the gravimetric filter method to measure very low mass concentrations of diesel soot emissions. Thus, advances in nano environmental health and safety (nanoEHS) study require a sensitive instrument that can measure low levels of agglomerate mass concentration.

With the aid of a newly developed instrument, Universal Nanoparticle Analyzer (UNPA), we conducted online structural characterization and real-time mass measurements for metallic nanoparticle agglomerates generated by spark discharge and diesel soot aggregates. The instrument uses mobility classification in conjunction with measurement of unipolar charging properties to extract information about primary particle size and agglomerate compactness. Reasonable agreement was found between the primary particle sizes measured by UNPA and transmission electron microscope (TEM). Additionally, the instrument was able to detect the change in morphology of metallic nanoparticle agglomerates due to sintering. Based on the agglomerate number distribution from a mobility scan, the UNPA calculates the agglomerate mass distribution using mobility and charging models for loose ideal agglomerates. Parallel measurements of the mass concentration were performed for the loose agglomerates in our study (i.e. metallic agglomerates and diesel soot aggregates without treatment after production) using filter collection and UNPA. Agreement between the instrument-based and the filter-based mass concentrations is good in the case of loose silver agglomerates and reasonable for diesel soot aggregates.

See more of this Session: Aerosol Science and Technology Enabling Environmental and Energy Studies I
See more of this Group/Topical: Environmental Division

(136g) Simultaneous Reduction of NOx, Solid, and Semi-Volatile Particles Using 4-Way Catalyzed Filtration Systems

Jacob J. Swanson[1], Joe R. Sweeney[1], Chris Larson[1], David B. Kittelson[1], Robert A. Newman[2] and Robin P. Ziebarth[3], (1)Mechanical Engineering, University of Minnesota, Minneapolis, MN, (2)Core R&D - Materials Science and Engineering, The Dow Chemical Company, Midland, MN, (3)Core R&D - Chemical Sciences, The Dow Chemical Company, Midland, MN

Diesel engines are widely used inautomotive, transportation, construction equipment, stand-by power generators,and marine engines and off road applications. In 2007, the U.S. EnvironmentalProtection Agency (EPA) reduced the diesel particulate matter emission (DPM)standard for heavy-duty diesel engines used on-road to 0.0134 g/kWh and in2010, the emission standard for NOx (NO + NO_2) was reduced to 0.27g/kWh, roughly a 10-fold reduction over previous standards. To meet the on- andoff-road DPM standards most manufacturers rely on exhaust emission controldevices such as diesel particulate filters (DPF), a wall-flow filter in whichalternate channels are blocked forcing filtration to take place as the exhaustgases pass through the channel walls while the DPM is retained in the filter[1, 2].

The next generation of emissioncontrol devices includes 4-way catalyzed filtration systems (4WCFS) thatinclude both NOx and DPM removal technology. Toyota Motor Corporation introducedthe first successful 4WCFS concept in the early 2000s that they referred to asa "diesel particulate - NOx reduction" system or DPNR [3]. In brief, thisversion of a 4WCFS consists of a bare DPF substrate that has been washcoatedwith a NOx storage catalyst. Thus, DPM and NOx are reduced simultaneously by asingle control device. The bare DPF used in this study is made from an advancedceramic material (ACM). The ACM manufacturing process is controlled so that themicrostructure, total porosity, and pore size distribution are tailored to meetrequirements for DPM emission control. Additionally, ACM is suitable forcatalyzed applications. By increasing the porosity of the ACM filters, it washypothesized that larger amounts of NOx storage catalyst could be loaded ontothe filters. Previous studies have shown that ACM DPFs demonstrate highfiltration efficiency, low-pressure drop, high-temperature handling capability,and excellent mechanical integrity at a porosity of 60% or higher [4, 5, 6, 7].

The focus of this work is twofold.The first objective was to develop a methodology tosimultaneously evaluate the NOx and DPM control performance of mini4WCFS that are challenged with diesel exhaust from a 2005 John Deere off-roaddiesel engine. The experimental mini filters evaluated were 1.9 x 1.9 x 7.5 cmsquare prisms. Thus, the method differs from traditional tests where full-sizedaftertreatment exhaust systems are evaluated. It is a cost effective method toevaluate prototype filters rapidly and consistently with control oftemperature, flow rate, and face velocity. The second objective was to evaluatethe impact of catalyst loading and substrate porosity on (1) back pressure, (2)catalytic performance, (3) the particle filtration performance of catalyst-coatedstandard (STD) and high porosity

(HP) test filters as stand-alone 4-waycatalyzed filtration systems. The catalyst coatings used in this study werebasic research coatings and were used due to availabilitywithin the project timeframe to test the catalyst loading capacity of the STDand HP test filters and are not intended for production applications. Further work would be needed with commercial automotive catalyst companies tooptimize for their commercial coatings.

Experimental measurements includedsimultaneous and time resolved total and solid particle filtration efficiency(FE), size resolved FE, pressure drop (ΔP), and NOx removal performance.Preliminary results indicated that the use of HP ACM 4WCFS results in 98%reductions in NOx and 95% reductions in solid and semi-volatile particulatematter averaged over 10 regeneration cycles. Similarly, STD ACM 4WCFS exhibitedhigh filtration efficiencies but slightly reduced NOx control performance. Therich / lean cycling that is used to regenerate the filter has almost no impacton solid particle filtration efficiency, but impacts NOx removal efficiencies.Shorter lean times (more frequent regeneration) lead to higher removalefficiencies but more reductant is consumed. Overall, the new methodology forsimultaneously evaluating the FE and NOx removal performance of small scale emission control devices has proven to be avaluable developmental tool. Future work includes optimization and evaluationof additional filters and catalysts and scaling up the system to evaluatelarger prototypes.

References:

[1] T.V. Johnson, *PlatinumMetals Rev.* 52(1):23-37, 2008.

[2] T.V. Johnson, 2010.*PlatinumMetals Rev.* 54(1):37-41, 2010.

[3] K. Nakatani, S. Hirota, S. Takeshima, K. Itoh, *SAE TechnicalPaper* 2002-01-0957, 2002.

[4] A.J. Pyzik, C.G. Li, *Int.J. Appl. Ceram. Technol.* 2(6):440-451, 2005.

[5] A.J. Pyzik, C.S. Todd, C. Han, *J. European Ceramic Society* 28:383-391, 2008.

[6] C.G. Li, H. Koelman, R. Ramanathan, U. Baretzky, G. Forbriger, T. Meunier, *SAE Int. J. Fuels Lubr.*1(1): 1307-1312, 2008.

[7] J. Swanson, M.Schumacher, W. Watts. D. Kittelson, R. Newman, R. Ziebarth, *29th AAAR Conference*,Portland, OR, October 25 - 29, 2010.

See more of this Session: Aerosol Science and Technology Enabling Environmental and Energy Studies I
See more of this Group/Topical: Environmental Division

(136h) The Importance of Pt Dispersion In DeNOx Catalysts

Robert B°Chel, ETH Zurich, Zurich, Switzerland, Sotiris E. Pratsinis, Department of Mechanical and Process Engineering, ETH Zrich, Zrich, Switzerland and Alfons Baiker, Inst. Chemical and Bioengineering, ETH Zurich, Zurich, Switzerland

NO_x storage-reduction (NSR) is successfully applied for exhaust gas treatment of lean fuel engines. Under fuel lean conditions, exhaust NO_x is oxidized over a noble metal (Pt) and then trapped on an alkali- or alkaline earth metal (Ba) in the form of metal-nitrates. During the subsequent short fuel rich period the NO_x originating from decomposition of the metal-nitrates is reduced on the metal component to nitrogen and the cycle restarts. [1]

The noble metal (Pt) was found to play an important role in NSR systems [2] in order to oxidize NO under fuel lean conditions and to reduce NO_x species to N_2 under fuel rich condition. To use the Pt in the most efficient way, the available Pt is finely dispersed over the support, increasing the NO oxidation reaction [3]. But higher Pt dispersions, and therefore smaller Pt clusters, do not always lead to the highest NSR conversion rates, as the catalytic activity of Pt is also a function of the coordination and geometry of the Pt and smaller Pt clusters tend to form PtO_x [4].

With a flame spray pyrolysis (FSP) setup [5, 6] the production to cooling rate was varied changing the catalyst specific surface areas from 80 to 220 m^2/g. For these materials a Pt dispersion of 5 to 25% was measured: With higher SSA the Pt dispersion decreased, what is unexpected but was confirmed by counting Pt clusters on high resolution scanning transmission electron microscopy (STEM) pictures. The Pt was preferentially located on Ba and for small $BaCO_3$ particles a bimodal size distribution of the Pt was seen, whereas for bigger $BaCO_3$ particles the size distribution of the Pt clusters was becoming narrower. The measured NSR activity changes with particle size and can be correlated with the resistance of Pt to be oxidation under fuel lean conditions.

References:

[1] W.S. Epling, L.E. Campbell, A. Yezerets, N.W. Currier, J.E. Parks, Cat. Rev. - Sci. Eng. 46 (2004) 163-245.

[2] M.O. Symalla, A. Drochner, H. Vogel, R. Büchel, S.E. Pratsinis, A. Baiker, Appl. Catal., B 89 (2009) 41-48.

[3] J.H. Lee, H.H. Kung, Catal. Lett. 51 (1998) 1-4.

[4] L. Olsson, E. Fridell, J. Catal. 210 (2002) 340-353.

[5] R. Strobel, L. Madler, M. Piacentini, M. Maciejewski, A. Baiker, S.E. Pratsinis, Chem. Mater. 18 (2006) 2532-2537.

[6] R. Büchel, R. Strobel, F. Krumeich, A. Baiker, S.E. Pratsinis, J. Catal. 261 (2009) 201-207.

See more of this Session: Aerosol Science and Technology Enabling Environmental and Energy Studies I
See more of this Group/Topical: Environmental Division

(153a) Towards Sustainable Manufacturing: An Approach Evaluating Energy Reduction Techniques for System Improvement

Tamer O. Girgis, Department of Chemical Engineering and Materials Science, Wayne State University, Detroit, MI and **Yinlun Huang**, Dept. of Chemical Engineering & Material Science, Wayne State University, Detroit, MI

Manufacturing industries are facing major challenges due to economic globalization, increased energy costs, etc. Capital investments are highly dependent on the economy and driven by customer demands. Hitherto, a variety of technologies have been developed for energy efficiency improvement, such as those developed by the Department of Energy, which have been used for industrial system assessment. By these technologies, system performance improvement opportunities may be identified, such as those through equipment/process modification, control adjustment, workforce training, system management improvement, etc. However, how to quantify the sustainability performance of the technologies and how to use them in a systematic way is still a challenge.

In this paper, we introduce an energy technology integration approach for sustainable manufacturing. By this approach, the known energy reduction technologies will be assessed quantitatively by a set of suitable sustainability metrics. The process system to be improved will be assessed by the same set of sustainability metrics as well. The assessment results will be analyzed together and the most suitable set of energy reduction technologies will be selected to improve process system's sustainability to the expected level, for which the investment will be still under budget limit. The introduced approach can provide a scientific guidance to industry sustainability enhancement. The efficacy of the approach will be demonstrated through case studies on electroplating systems.

See more of this Session: Energy Sustainability, Challenges and Solutions
See more of this Group/Topical: Sustainable Engineering Forum

(153b) Optimal Synthesis of Fuel Gas Networks for Energy Sustainability

M. M. Faruque Hasan and **Iftekar A. Karimi**, Dept of Chemical & Biomolecular Engineering, National University of Singapore, Singapore, Singapore

The world is facing two critical and inseparable issues in energy and climate change. The fossil fuels (oil, coal, and natural gas) that supply over 85% of the current primary worldwide demand will continue to be the major source for the near future. However, coal, oil, and natural gas emit 208, 164, and 117 pounds of CO_2 per one million Btu of energy respectively. Since the cleanest energy is the one that we never use, immediate measures to reduce their usage by conserving energy are extremely critical in arresting the dreadful impact on the environment due to the GHG (Green House Gas) emissions.

In the USA, the chemical industry consumes nearly 20% of the total industrial energy consumption (Worrell et al., 2000).[1] Energy contributes as much as 40% to the total operating cost of a chemical plant (Zargarzadeh et al., 2007).[2] Most plants buy electricity or fuel externally to satisfy their energy demands. Some even consume a portion of their raw materials, products, or byproducts as fuel. For instance, refineries use products such as vaporized LPG and fuel oil as fuel supplements. This is called FFP or Fuel From Product. On the other hand, base-load liquefied natural gas (LNG) plants use the feed (natural gas) as a fuel to run their gas turbine drivers and generators. This is called FFF or Fuel From Feed. Obviously, FFF and FFP directly reduce plant capacity and revenue, and measures to reduce their consumptions are certainly desirable.

Many chemical plants produce a variety of hydrocarbon gases with fuel value. To this end, an attractive option is to use the usually many impure, waste, surplus, byproducts, purge, or side streams with some heating value as fuel, instead of sending them to flare. In general, most plants have units that can consume such fuels. A fuel gas network (FGN) integrates and uses these fuel gases appropriately to make their best use. The FGNs are critical components of most chemical plants such as LNG and refinery. With the increasing demand of chemical and petrochemical products, decreasing natural resources, and stringent restrictions on GHG emissions, integrating Energy, Environment and Economics (IEEE) is the key to sustainability. The use of a well-designed and well-operated FGN is an excellent example of IEEE. It can yield "greener and leaner" operation through reduced emissions arising from reduced flaring and substantial savings arising from greater fuel efficiency. In addition to reduced fuel usage, waste, and pollution, the benefits of an FGN may also include higher profit margin, increased plant energy security, augmented product quality, and even stable operation.

However, a systematic approach to design and operate realistic FGNs has been missing in the literature. The mixing and re-distribution of fuels is essentially a generalization of the well known pooling problem. While the work on the pooling problem and its extensions and variations is extensive, limited work exists on the integration of "fuel streams" (versus refinery fuels) per se to reduce the energy cost as an issue. De Carli et al. (2002)[3] addressed the intelligent management and control of a fuel network. Wicaksono et al. (2006)[4] proposed a mixed-integer nonlinear programming (MINLP) model for integrating various fuel sources in an LNG plant. Wicaksono et al. (2007)[5] extended this to integrate JBOG (Jetty BOG) as an additional source. Hasan et al. (2008)[6] addressed the optimal synthesis of FGN and presented two models based on two superstructures, one with 1-stage and the other with 2-stage mixing. However, all these works focused primarily on mass and property integration to meet customer needs and demands. They assumed isothermal and isobaric systems and isothermal mixing. These are limiting assumptions, as pressure and temperature variations are significant and often necessary in most plants. Lastly, even those works that addressed FGNs neither accommodated various fuel quality requirements of the consumers rigorously nor provided a comprehensive model to optimize FGN synthesis.

In this work, we address the optimal synthesis of an FGN with many practical features such as auxiliary equipment (valves, pipelines, compressors, heaters/coolers, etc.), non-isobaric and non-isothermal operation, non-isothermal mixing, nonlinear fuel quality specifications, fuel/utility costs, disposal and treatment costs, and emission standards. We develop a nonlinear program (NLP) based on a superstructure that includes essential equipment such as compressors, valves, and utility heaters/coolers in addition to sources and sinks. Thus, it embeds plausible alternatives for heating/cooling, moving, mixing, and splitting. We successfully apply our model to three real-life case studies from LNG and refinery industries to demonstrate that an FGN can save 40-50% of total energy cost of a plant and reduce fuel-from-feed or fuel-from-product consumptions by similar amounts. We also address the issue of retrofitting new sources of fuel (e.g. integrating JBOG in an LNG plant) into an existing FGN. Finally, we highlight the utility of optimal FGN synthesis in reducing the energy cost of fuel-using and energy-intensive chemical plants. This work represents an important contribution towards conserving energy, preserving environment, and improving plant economics using advanced techniques of process systems optimization.

References

1. Worrell, E.; Phylipsen, D.; Einstein, D.; Martin, N. Energy Use and Energy Intensity of the U.S. Chemical Industry. *Environmental Energy Technologies division*. 2000. http://ies.lbl.gov/iespubs/44314.pdf.
2. Zargarzadeh, M.; Karimi, I. A.; Alfadala, H. E. Olexan: a tool for online exergy analysis. In Proceedings: *ESCAPE17*, Bucharest, Romania, May 27–30, 2007.
3. De Carli, A.; Falzini, S.; Liberatore, R.; Tomei, D. Intelligent Management and Control of Fuel Gas Network. In: Proceedings of the 28th Annual Conference of the *IEEE Industrial Electronics Society*, 2002, p. 2921–2926.
4. Wicaksono, D. S.; Karimi, I. A.; Alfadala, H. E.; Al-Hatou, O. I. Optimization of Fuel Gas Network in an LNG Plant. Presented in *AIChE Annual Meeting*, San Francisco, CA, USA, Nov. 12–17, 2006.
5. Wicaksono, D. S.; Karimi, I. A.; Alfadala, H. E.; Al-Hatou, O. I. Integrating Recovered Jetty

Boil-Off Gas as a Fuel in an LNG Plant. In Proceedings: *ESCAPE17*, Bucharest, Romania, May 27–30, 2007.
6. Hasan, M. M. F.; Karimi, I. A. Optimal Synthesis of Fuel Gas Networks. *INFORMS Annual Meeting*, Washington D. C., USA, Oct. 12-15, 2008.

See more of this Session: Energy Sustainability, Challenges and Solutions
See more of this Group/Topical: Sustainable Engineering Forum

(153c) Nationwide Energy Supply Chain Analysis for Transportation Fuels

Josephine A. Elia[1], Richard C. Baliban[2] and Christodoulos A. Floudas[2], (1)Chemical Engineering, Princeton University, Princeton, NJ, (2)Chemical and Biological Engineering, Princeton University, Princeton, NJ

Hybrid energy processes are an attractive avenue towards a more sustainable energy future due to their flexibility in converting multiple types of inputs into a consistent range of products. For the transportation sector, Fischer-Tropsch synfuels can be generated from a diverse of feedstock inputs to replace petroleum-based products [1]. Using the novel hybrid coal, biomass, and natural gas to liquids (CBGTL) process, the United States can utilize domestically available resources to produce gasoline, diesel, and kerosene to fulfill the fuel demands for the country [2-6]. Further, the incorporation of different types of biomass as feedstocks helps mitigate greenhouse gas emissions and further the efforts for renewable energy production [7]. Such hybrid processes, however, require the cooperation between multiple economic domains to achieve efficient resource allocation. Thus, an integrated and systematic approach that evaluates the interconnection between key stakeholders is crucial to attain a cost-competitive energy supply chain.

We developed an optimization framework to solve a large-scale, nationwide energy supply chain problem that takes into account the varying degrees of resource availabilities and the demand profile for the United States [4]. The supply chain begins at the feedstock source locations, ends at the demand locations, consists of optimized CBGTL facilities using the process synthesis with simultaneous heat, power, and water integration method [6]. These facilities are the 108 optimal plant topologies analyzed in [6] with distinct combinations of six coal feedstock types, three biomass feedstock types, one natural gas composition, three plant capacities, and two carbon management options (i.e., with or without carbon sequestration). A superset of candidate facility locations across the country is derived from a filtered list of United States counties in the lower 48 states based on the availability and transportation cost of feedstocks to each county.

The CBGTL facilities and the candidate locations, along with information on the nationwide configuration for the feedstock, fuel demands, and modes of transportation obtained from published government-based databases [4], serve as parameter inputs to the formulated large-scale mixed-integer linear optimization (MILP) model that represents discrete and continuous decisions in the energy supply chain. The model is solved using CPLEX to obtain the optimal supply chain topology with the minimum overall cost of fuel production for the entire network, the facility selections and locations, and the interconnections between feedstock sources, facility locations, and demand locations.

Ten case studies are presented to illustrate the performance of the optimal supply chain network under three major scenarios, namely the supply chain network with i) unrestricted topology, where the selections of the facility locations are not restricted to any geographical constraints, ii) imposed degrees of plant distribution, where a minimum number of selected facilities in the network is introduced, and iii) restricted topology, where the selection of the plant locations are restricted to be in proximity to current energy infrastructures. Facets of the model solution provide a quantitative basis to evaluate and examine trade-offs in investment decisions under these different scenarios.

[1] Takeshita, T.; Yamaji, K. Important roles of Fischer-Tropsch synfuels in the global energy future. *Energy Policy* 2008, *36*, 2773-2784.

[2] Baliban, R. C.; Elia, J. A.; Floudas, C. A. Toward novel hybrid biomass, coal, and natural gas processes for satisfying current transportation fuel demands, 1: Process alternatives, gasification modeling, process simulation, and economic analysis. *Industrial & Engineering Chemistry Research* 2010, *49*, 7343-7370.

[3] Elia, J. A.; Baliban, R, C.; Floudas, C. A. Toward novel hybrid biomass, coal, and natural gas processes for satisfying current transportation fuel demands, 2: Simultaneous heat and power integration. *Industrial & Engineering Chemistry Research* 2010, *49*, 7371-7388.

[4] Elia, J. A.; Baliban, R. C.; Xiao, X.; Floudas, C. A. Optimal energy supply network determination and life cycle analysis for hybrid coal, biomass and natural gas to liquid (CBGTL) plants using carbon-based hydrogen production. *Computers and Chemical Engineering* 2011, In press.

[5] Baliban, R. C.; Elia, J. A.; Floudas, C. A. Optimization Framework for the Simultaneous Process Synthesis, Heat and Power Integration of a Thermochemical Hybrid Biomass, Coal, and Natural Gas Facility. *Computers and Chemical Engineering* 2011, In press.

[6] Baliban, R. C.; Elia, J. A.; Floudas, C. A. Simultaneous Process Synthesis, Heat, Power, and Water Integration of Thermochemical Hybrid Biomass, Coal, and Natural Gas Facilities, 2011, In preparation.

[7] Somerville, C.; Youngs, H.; Taylor, C.; Davis, S. C.; Long, S. P. Feedstocks for Lignocellulosic Biofuels. *Science* 2010, *329*, 790-792

See more of this Session: Energy Sustainability, Challenges and Solutions
See more of this Group/Topical: Sustainable Engineering Forum

(153d) Development of An Integrated CO2 Capture, Transportation and Storage Infrastructure for the UK and North Sea Using An Optimisation Framework

Paolo Prada[1], **N.V.S.N. Murthy Konda**[2] and Nilay Shah[2], (1)E&P Division, Eni S.p.A., Milan, Italy, (2) Centre for Process Systems Engineering, Department of Chemical Engineering, Imperial College London, London, United Kingdom

Carbon Capture and Storage (CCS) is widely regarded as an effective medium-term CO_2 abatement technology. However, high cost of the capture technologies, at their current state of the art, and the huge upfront costs of the CO_2 transport pipelines and storage facilities are hindering the large-scale development and full-chain development at commercial scale. Thus, it is of utmost importance to minimize the associated costs, as much as possible, to facilitate optimal design of CCS networks – this emphasizes the need for rigorous optimization based approach such as the one we employed in this study.

Despite the aforementioned challenges, especially within the context of the UK, given the presence of large number of CO_2 emitters and the vicinity of large storage sites in the Southern North Sea, CCS potential for the UK is significant. Thus, it is worthwhile to investigate the economic feasibility of the development of an integrated CCS infrastructure (which may benefit from shared transport and storage components, and economies of scale) together with the possible physical layout and evolution of such a large-scale infrastructure.

In this study, we investigated the optimal deployment of such an infrastructure based on overall cost (including capital and operating costs) minimisation. This is formulated as a Mixed Integer Linear Programming (MILP) problem. The scope of this work includes 33 large UK CO_2 emitters and 6 Southern North Sea sinks. Relevant cost estimates and integrated infrastructure layouts are generated through 'snapshot' model for several scenarios with varying CO_2 reduction targets from 5 to 105 $MtCO_2$/year.

Our results show that, for all the scenarios, practical integrated infrastructures can be developed. An initial 'seed' infrastructure is required in the Southern Yorkshire and Humber region and captured CO_2 is routed towards Southern North Sea depleted gas fields via the Theddlethorpe terminal. With increasing reduction targets, the infrastructure evolves and incorporates Thames estuary area emitters whose CO_2 is separately routed offshore via the Bacton terminal. In the first stage, the CCS system is dominated by coal power stations, primarily Drax, while CCGTs' contribution becomes significant above 75 $MtCO_2$/year reduction targets. An optimal overall cost of 81.5 $/$tCO_{2,Total}$ is achieved at 30 $MtCO_2$/year after an initial decline primarily due to increasingly shared transport and storage infrastructures. As the reduction target increased beyond 30 Mt/year, more expensive CO_2 sources are included in the network – subsequently, the overall cost continually increases up to a maximum of 99.8 $/$tCO_{2,Total}$ when the capture target increased to 105 $MtCO_2$/year.

Sensitivity analysis performed on key parameters – i.e., scaling factors and gas price – revealed that not only the overall costs and but also the timing of retrofit of gas-based power generation are affected. On the other hand, oil refineries, cement works and smaller capacity CCGTs are discarded as cost-optimal emitters to capture in all cases.

The MILP model is thus shown to be an effective framework for the optimal planning and, thereby facilitating, large-scale deployment of CCS infrastructure. (Disclaimer: This work is done while the 1st and 2nd authors were at Imperial College London and the content/opinions/results discussed do not necessarily represent those of the associated institutions)

See more of this Session: Energy Sustainability, Challenges and Solutions
See more of this Group/Topical: Sustainable Engineering Forum

(153e) Energy Balances for the Production of Cellulosic Ethanol

Kyriacos Zygourakis and Venetia Rigou, Chemical and Biomolecular Engineering, Rice University, Houston, TX

Cellulosic ethanol has the potential to meet a substantial fraction of the US needs for liquid transportation fuels. To remove the perceived major barriers to the large-scale commercialization of this technology, a lot of attention has been focused lately on overcoming the recalcitrance of plant cell walls to deconstruction, on the development of more efficient fermentation methods and on genetically engineering innovative energy crops that maximize fuel production. However, the commercial success of cellulosic ethanol will depend to a large extent on our ability to design energy efficient plants.

We report here some recent results from a systematic analysis of the energy requirements of cellulosic ethanol production. A comprehensive process model for the production of ethanol from cellulosic biomass (corn stover, switchgrass etc.) was developed and implemented using the SuperPro simulator package. We paid particular attention to the operation of the utilities section of the plant, a combined heat and power (CHP) unit that generated the steam and electricity needed for the purification of ethanol and the other plant units.

A detailed parametric analysis was carried out to study how the energy balances of each plant section and the overall energy efficiency of the plant were influenced by changes in (a) the composition of the biomass feedstock, and (b) the conversion levels of the pretreatment, hydrolysis and fermentation stages. We carefully analyzed the requirements of the utilities section of the ethanol plant where byproduct streams of the production process were utilized for energy generation. One of the key questions asked was whether these streams were a sufficient fuel source for meeting the energy requirements of the plant or whether we needed to provide extra fuel in the form of additional biomass or natural gas.

Our calculations indicate a significant trade-off between ethanol production and external energy inputs. Increases in the cellulose and hemicelulose contents of the feedstock or increases in the conversion of the fermentation stage would result (as expected) to higher ethanol production rates. This increase, however, would be offset by an increase in the requirement for extra energy inputs in the form on natural gas or biomass fed directly to the utilities section. Therefore, any changes designed to increase ethanol production may be accompanied by lower overall efficiencies and net energy ratios for the biorefinery. These results cast some doubt on the ultimate effectiveness of efforts aimed at engineering innovative energy crops with high cellulose/hemicellulose content in order to maximize fuel production.

See more of this Session: Energy Sustainability, Challenges and Solutions
See more of this Group/Topical: Sustainable Engineering Forum

(153f) High Photon Flux Photoelectrochemical Reactor: Modelling and Operation

Klaus Hellgardt[1], C.-K. Ong[2], G. Kelsall[2] and S. Dennison[2], (1)Chemical Engineering, Imperial College London, London, United Kingdom, (2)Imperial College London, London, United Kingdom

Production of hydrogen via sustainable technologies using renewable resources is an important goal for future global energy generation and storage. Photoelectrolysis is an attractive approach in this context, both, for its simplicity and the opportunity of employing relatively cheap materials. Considerable effort has been expended on the development of materials for photoelectrolysis - less so on the development of photoelectrochemical reactors. Here we model and demonstrate the operation of a high photon flux photoelectrochemical reactor and identify some of the key criteria for the rational design of an efficient, practical device. Mathematical models have already been developed to account for fluid flow and current distribution. An additional model rationalises the minority carrier distribution within the photoelectrochemical film itself. Process intensification through solar concentration will be demonstrated with a flat plate device, leading to a 5-fold increase in hydrogen production rate per unit active semiconductor area, or, conversely a significant reduction in device cost.

See more of this Session: Energy Sustainability, Challenges and Solutions
See more of this Group/Topical: Sustainable Engineering Forum

(155a) The Kinetics of Redox Reactions of Hexavalent Chromium and Trichloroethylene In Wastes Containing Both Contaminants

Jacob Mlusu and Ramesh Chawla, Chemical Engineering, Howard University, Washington, DC

This paper presents the feasibility of using oxidizing and reducing agents for the simultaneous remediation of heavy metals and chlorinated hydrocarbons in aqueous and soil systems. Chelsea soil was selected as a model soil while trichloroethylene (TCE) and hexavalent chromium (Cr(VI)) were selected as model contaminants. Ferrous ion, Fe(II) was used as a reducing agent while permanganate, (MnO_4^-), was used as an oxidizing agent. Bench-scale experiments were conducted to determine the reaction kinetics and extent of redox reactions of Cr(VI) and TCE in separate and combined systems in both aqueous and soil systems. In the redox reaction, Cr(VI) is reduced to the nonhazardous Cr(III), and Fe (II) is oxidized to ferric ion, Fe(III); while TCE is oxidized to chloride, hydrogen, and carbon dioxide and permanganate is reduced to manganese dioxide (MnO_2.). In the experimental analysis, the Cr(VI)-Fe (II) reaction yielded 100% and 99% Cr(VI) reduction when Fe(II) was used in water and Chelsea soil, respectively. The overall reaction was determined to be second order, first order with respect to both Cr (VI) and Fe(II). The TCE-$KMnO_4$ reaction yielded 100%, and 82% TCE oxidation by excess MnO_4^- in water and Chelsea soil, respectively. The overall reaction was determined to be second order reaction, first order with respect to both TCE and MnO_4^- in both the aqueous and soil systems. The Cr(VI)-TCE reaction yielded 92% Cr(VI) reduction in water when excess TCE was used in the absence of Fe(II) and $KMnO_4$. The overall reaction was determined to be second-order overall, first order with respect to both Cr(VI) and TCE. TCE was found to have a significant effect on the reduction of Cr(VI) and vice versa. When TCE was present in the aqueous system, the Cr(VI)-Fe(II) reaction with the same experimental conditions as before yielded 94% Cr(VI) reduction. Presence of TCE was advantageous to the reduction of Cr(VI), as it acted as an additional reductant. When Cr(VI) was present in the aqueous system, TCE-KMnO4 reaction with the same conditions as before yielded 97% TCE oxidation.

See more of this Session: Fundamentals of Environmental Process and Reaction Engineering
See more of this Group/Topical: Environmental Division

(155b) Kinetic Modeling of Gas-Phase Mercury Oxidation by Chlorine and Bromine In Combustion Effluents During Oxy-Combustion

Itsaso Auzmendi-Murua, Chemical, Biological, and Pharmaceutical Engineering, New Jersey Institute of Technology, Newark, NJ and Joseph W. Bozzelli, New Jersey Institute of Technology, Newark, NJ

Around 29 Gt of CO_2 are emitted annually worldwide. The main sources of these emissions are transportation and electricity-generating plants, which use coal and fuel combustion. Since CO_2 capturing techniques are still expensive, in order to reduce CO_2 emissions from fuel combustion, oxy-combustion has been proposed as a solution, with the aim of obtaining combustion flue gases with higher concentrations of CO_2 that are easier to remove. The implementation of oxy-combustion affects the speciation of the rest of the combustion products, including the speciation of mercury. Different techniques have been proposed for the control of mercury emissions, due to its high toxicity. In this work, the addition of halogens (chlorine and bromine) to the flue gases have been studied in order to oxidize mercury from its elemental form to oxidized form – HgX_2. As these mercury halides are water soluble, it is feasible to capture the HgX using conventional air pollution control devices. An elementary reaction mechanism has been developed for modeling the oxidation of mercury by the addition of halides, in both regular combustion and oxy-combustion conditions. Thermochemical properties and kinetic parameters have been calculated for Hg and Halogen – NO_x species using B3LYP and CBS-QB3 methods. The elementary reaction mechanism has been constructed with use of chemical activation analysis for association and addition reactions with quantum RRK analysis for k(E) and Master Equation analysis for fall-off. Results indicate that the presence of NO_x inhibits Hg oxidation by chlorine addition, through several catalytic cycles for loss of Cl. As for bromine addition, similar catalytic cycles are observed, but in this case, the overall thermochemistry favours more mercury oxidation. Therefore, results show that during oxy-combustion conditions, Hg oxidation is enhanced, since flue gases contain lower concentrations of NO_x.

See more of this Session: Fundamentals of Environmental Process and Reaction Engineering
See more of this Group/Topical: Environmental Division

A Computational Model to Evaluate Solar Offset and Water Generation of Various Cooling Systems Using Location-Specific Annual Climate Data

Dia Milani[a], Ali Abbas[a], Matteo Chiesa[b]

[a]Chemical and Biomolecular Engineering School, The University of Sydney, Sydney, Australia

[b]Masdar, Abu Dhabi Future Energy Company, Abu Dhabi, UAE

Abstract

An algorithm has been developed to address conventional vs. renewable dilemma. The debate of using renewable energy to replace conventional sources in various cooling systems to partially compensate this energy staggering operation and improve cooling efficiency is still everlasting. Because of numerous uncertainties associated with climatic data and the effect of climate change, the accuracy of predicting the trend of solar cooling's performance is not sufficient; and therefore, the confidence over capital vs. operational expenditure and payback time calculation was diluted. In this project, annual statistical weather data of any location can be analyzed to predict the possible offset that renewable solar can offer to slash energy requirement of a particular cooling or dehumidification system. Matching that analysis against total cost of installation and operation will provide solid ground for investment decision making criteria. The project takes into account location specific boundary conditions such as the cooling demand time series, solar resource availability, climatic conditions, component cost and performance parameters. This methodology evaluates the techno-economic performance of the solar collector systems include solar thermal and solar electric cooling options and account for water generation from ambient moisture as another significant product to be offset from total cost. Testing this algorithm for hot spot locations such as Abu Dhabi revealed an interesting potential for solar cooling investment, although more validation of these parameters are needed.

Keywords: Solar Cooling, Cooling Generation Cost, Photovoltaic, Adsorption Chillers, Thermoelectric Cooler, Absorption Chillers

(155d) Toxicity of Flue Gas In Microalgae CO2 Mitigation Systems

Adriana Pacheco, Mario M. Alvarez and Javier Lara-Gil, Biotechnology, Tecnologico de Monterrey, Monterrey, Mexico

Microalgal-based CO_2 capture from flue gas is an attractive mitigation strategy for industry. However, other gases are also present that may be harmful to microalgae growth, such as NO_x and SO_x. In this study, we simulated flue gas by adding to the culture medium salts of nitrite, sulfite, or bisulfite. Two high CO_2 tolerant species were evaluated: an environmental isolate (Rio) and *Scenedesmus* sp. (UTEX 1589). Toxicity tests were performed in 96-well microplates under an atmosphere of 2% (v/v) CO_2 to determine the concentration range of trace gases for microalgae growth. Both algae were able to tolerate concentrations of nitrite (>1600 ppm) and sulfite (400 ppm) higher than those reported in effluent gas. However, bisulfite inhibited growth at 50 ppm. This effect was associated with a drastic decrease in medium pH. As a result, we explored the use of cement dust and $CaCO_3$ to lessen the decline in pH. We used concentrations of 300 ppm and 240 ppm of cement dust and $CaCO_3$, respectively, to increased pH from 2.6 to 5.0. This increment in pH resulted beneficial to the growth of both species, tolerating concentrations of 400 ppm bisulfite (2X upper limit concentration found in flue gas). In all tests, environmental isolate (Rio) generated more biomass than the collection strain. Therefore, our microalgae isolate shows a better tolerance to the flue gas constituents, where cement dust might be a beneficial component of the mitigation system.

See more of this Session: Fundamentals of Environmental Process and Reaction Engineering
See more of this Group/Topical: Environmental Division

(155e) Reducing Air Quality Impact From Plant Start-up Emissions by Integrating Air Quality Modeling and Plant Start-up Simulation

Jian Zhang, Qiang Xu and Thomas C. Ho, Dan F. Smith Department of Chemical Engineering, Lamar University, Beaumont, TX

Chemical industry concentrated regions experience highly localized and transient air pollution events. For instance, chemical plant flaring activities, especially intensive flaring during plant startup process, emit huge amounts of volatile organic compounds (VOC) and nitrogen oxides (NO_X), causing elevated ground-level ozone concentrations that violate the National Ambient Air Quality Standards. Emissions are inevitable during plant startups, their air quality impacts, however, are different under different meteorological conditions. This provides opportunities to identify cost-effective control strategies for reducing air quality impacts from plant start-up emissions.

In this paper, detailed process system knowledge of plant startup emission and speciation is integrated with air quality modeling and simulation for the first time to quantitatively study regional air quality impact from ethylene plant startup emissions. The general process has three major steps: emission source characterization, flaring emission speciation and quantification, and air quality simulation and decision support. In the first step, plant-wide dynamic simulation and optimization are employed to minimize start-up emissions and identify benchmark emission dynamic profiles during a plant start-up. In the second step, simplified flaring emission speciation and quantification are conducted to identify dynamic profiles of ozone precursors, such as various VOC and NOx. Then, the flaring precursor information is input to the air quality model, CAMx. In the third step, multiple air quality simulations are performed for start-up emission impact analysis and decision support. This is the most complex step. Firstly, a background air quality simulation is performed. The simulation needs atmospheric reaction mechanism, geological domain information, emission inventory data, meteorology information, initial conditions and boundary conditions. The simulation gives temporal and spatial background ozone and other air pollutants distributions. It also gives different ozone formation mechanisms for different regions. Secondly, the startup procedures are arranged into different time windows, and start-up emissions are integrated into the background air quality model to generate different cases. Through multiple air quality simulations, regional air pollutant increments can be obtained. Through comparison and analysis, the best start-up time window will be identified.

A typical ethylene plant startup procedure has been studied. Results show that local maximum ozone increment can be significant due to an ethylene start-up emission. However, if appropriate start-up time window is selected, the maximum ozone increment can also be substantially reduced comparing with the worst case.

(155f) Modeling and Optimization of a Modified Claus Process As Part of An Integrated Gasification Combined Cycle (IGCC) Plant with CO2 Capture

Debangsu Bhattacharyya[1], **Dustin D. Jones**[2], Richard Turton[1] and Stephen E. Zitney[3], (1)Department of Chemical Engg., West Virginia University, Morgantown, WV, (2)Chemical Engineering, West Virginia University, Morgantown, WV, (3)Collaboratory for Process and Dynamic Systems Research, National Energy Technology Laboratory, U.S. Department of Energy, Morgantown, WV

The modified Claus process is one of the most common technologies for sulfur recovery from acid gas streams. In the case of an Integrated Gasification Combined Cycle (IGCC), the feeds to the Claus process are the off-gas from the sour water stripper (SWS) and the acid gas from the stripper of the acid gas removal (AGR) unit. Because of the stringent environmental regulations considered in this work, the tailgas from the Claus process is further treated in a hydrogenation unit, compressed from about 1 atm to 50 atm, and recycled back to the AGR unit for recapture. Important design criteria for the Claus unit, when part of an IGCC power plant, are the ability to destroy ammonia completely and deep sulfur recovery from a relatively low purity acid gas stream without sacrificing flame stability. Due to these criteria, modifications are often required to the process, resulting in a modified Claus process. For studies discussed here, these modifications include the use of a 95% pure oxygen stream as the oxidant, a split flow configuration, and preheating of the feeds with the intermediate pressure steam generated in the waste heat boiler (WHB).

In the future, for IGCC plants with CO_2 capture, the Claus unit must satisfy the emission standards without sacrificing the plant efficiency in the face of typical disturbances of an IGCC plant such as rapid change in the feed flowrates due to load-following and wide changes in the feed composition because of changes in the coal feed to the gasifier. The Claus unit should be adequately designed and its control design should be appropriately implemented to satisfy these objectives. A representative dynamic model is needed for these studies. Even though the Claus process has been commercialized for decades, most papers treat the key reactions in a Claus process as equilibrium reactions. Such models may be validated by manipulating the temperature approach to equilibrium for a set of steady-state operating data, but are of limited use for dynamic studies.

In a Claus process, especially in the furnace and the WHB, a number of reactions take place along with the widely-considered Claus reaction. In this work, a set of linearly independent reactions has been identified and kinetic models of the flame zone, anoxic zone, WHB, and catalytic reactors have been developed. Various approaches are taken to derive the kinetic rate expressions which are either missing in the open literature or found to be inconsistent. A set of plant data is used for optimal estimation of the kinetic parameters while a set of plant data is used for the model validation. The final model agrees well with the published plant data.

Using the developed kinetics models of the Claus reaction furnace, WHB, and catalytic stages, sensitivity studies are completed to examine whether important process variables could be controlled in the face of the typical process disturbances from the IGCC plant. Additionally, a study is carried out to examine the optimal operation of a Claus plant considering three, highly interactive process variables in the context of an IGCC power plant with CO_2 capture. The first variable is the single pass conversion of H_2S in the Claus unit. Because of a very high operating pressure of the AGR process in this IGCC plant, considerable compression cost is associated with recycling the unconverted tail gas. The second variable is the flow of hydrogen to the AGR unit from the Claus unit. Hydrogen sent to the AGR unit can be converted to electric power more efficiently by the gas turbine than burning it in the Claus furnace and raising steam. The third variable is the consumption of oxygen in the Claus unit. Considerable operating cost is incurred in the air separation unit for producing oxygen. An optimization study examines the tradeoffs of these process variables along with the net power requirement of the Claus unit.

See more of this Session: Fundamentals of Environmental Process and Reaction Engineering
See more of this Group/Topical: Environmental Division

(187a) Modeling a Net Zero Future with Historic Buildings

Phillip Saieg, Alliance for Sustainable Colorado, Denver, CO

This presentation will be an in-depth look at how to retrofit historic buildings toward the goals of LEED EBOM Platinum, and a net zero energy balance. First, we will make the case for why we should care about our historic buildings, including the trends and future of the green building industry from a climate perspective and a financial perspective. Second, there will be an assessment of the technology that exists today for retrofitting, the financial models that make sense for this type of work, and how LEED can best be best utilized for these projects – including how to approach a LEED EBOM platinum certification with a historic building. Third, we get into the nuts and bolts of how to retrofit: walking through the process, players, and unique steps to successfully create a green, net zero historic building – including an extensive discussion integrating preservation and sustainability. Finally, we will look at this process in a case study of the Alliance Center, a model building for energy-efficient historic retrofits and hub for sustainability, currently re-certifying its LEED EBOM Gold status to Platinum status.

Though this presentation gives special credence toward historic buildings, the vast majority of concepts, processes, and action steps described in this presentation can be applied to existing buildings at-large via the following learning points:

- Green building industry trends
- Achieving LEED EBOM Platinum
- Steps for greening historic buildings
- Combination passive/active design
- Financial models to make green retrofits
- Defining and achieving Net Zero in historic buildings

See more of this Session: Sustainable Building Materials
See more of this Group/Topical: Environmental Division

(187b) Flame Retardant Siloxane Polymers Produced by Eco-Friendly Synthetic Methods

Romy Kirby, Ravi Mosurkal and Jason W. Soares, U.S. Army Natick Soldier Research Development and Engineering Center, Natick, MA

There is atremendous interest in the area of flame-retardant (FR) materials due to theirnumerous applications in clothing, construction, aviation andtelecommunications. Current flame retardant (FR) materials on the market arebased on halogenated polymers and additives that are either classified asharmful/toxic or give off such products upon combustion. While these materialshave offered cost-effective solutions for FR applications, there is now anincreased global emphasis on discontinuing the use of these materials andreplacing them with environmentally and human safe alternatives. Thus, thereexists the need to develop a new generation of environmentally-safe, economicalpolymeric FR materials with efficient FR performance. Our work centers on thecreation of novel polysiloxane copolymers (PS) by incorporating additives (e.g.boron, organoclays and fillers) through a facile, environmentally benign,solvent-free synthetic approach. It is well-known in the literature that boronis inherently thermally stable, environmentally safe and has extreme durabilityto air oxidation. Furthermore, boroxo silanol derivatives, used asceramic precursors, exhibit efficient FR properties.[1] Wehave synthesized boron containing polysiloxane copolymers (polyborosiloxanes -PBS) without the use of solvents. Several different terpolymers,consisting of a boronic acid (PBA) and an aromatic dianhydride (Oxy or DAH) atvarious PBA/dianhydride molar feed ratios, have also been synthesized. PBSterpolymers have exhibited efficient flame and thermal properties with adecomposition temperature (T_d) of 400-430°C and a heat releasecapacity (HRC) of 170-250 J/g-K. Concurrently, PS-organoclay nanocompositeshave been synthesized. When polymers are combined with small amounts oforganoclays, they can exhibit improved thermal, mechanical and FR properties.[2]Here, polysiloxane copolymers containing Oxy or DAH have been blendedwith three different organoclays and two different fillers. PS-organoclaynanocomposites exhibit a T_d of 400°C and an HRC of 160-200J/g-K. A synergistic approach is being employed to determine if furtherenhancement of the FR properties can be achieved by creating PBS terpolymer organoclaynanocomposites. PS-filler blends created to determine the influence offillers such as cellulose and chitosan on thermal and flame retardant propertieswill also be discussed. The eco-friendly synthesis of PS-based nanocompositesand blends and resulting flame retardant and thermal properties represent anopportunity for a new generation of environmentally and human safe FR polymersfor military and civilian applications.

[1] (a) Armitage, P.; Ebdon, P.R.; Hunt, B.J.; JonesM.S.; Thorpet, F.G. Polymer Degradation and Stability, 1996, 54, 387. (b)Martn, C.; Ronda, J.C.; Cdiz, V. PolymerDegradation and Stability, 2006, 91,747.

[2] A.B. Morgan and C.A.Wilkie, (Eds.) Flame RetardantPolymer Nanocomposites, Wiley-Interscience,

New York, 2007.

See more of this Session: Sustainable Building Materials
See more of this Group/Topical: Environmental Division

(187c) The Healthy Energy and Envelope Balance

Margaret Pauls, Hutton Architecture Studio, Denver, CO and Phillip Saieg, Alliance for Sustainable Colorado, Denver, CO

Choosing energy efficient healthy indoor environments is linked to exterior envelopes. Understanding this linkage involves many facets that include; 1. Understanding the impacts of building envelope materials on energy performance and the indoor environment. 2. Identifying available tools for analyzing envelope material choices. 3. Discussing current building industry trends and their effect on energy performance and occupancy health. 4. Identifying current barriers to improving how designers synthesize design choices relating to energy, systems, and material selection. The advent of modern mechanical, heating, air conditioning, and ventilation, energy efficient buildings have become increasingly dependent on tight and highly insulated envelopes. This talk will focus on building materials and developing the details of how energy and envelopes both effect building energy use, indoor air quality and the overall occupied built environment. Some of the the potential risks and benefits of current building materials will be covered and the gaps and areas for improvement identified. Finally, some of the tools for making more informed choices will be reviewed.

The relationships between energy performance and healthy air and indoor environmental quality is critical to the future of sustainable architecture and engineering. This presentation will seek to answer these vital questions and chart a path to a healthy and sustainable built environment.

See more of this Session: Sustainable Building Materials
See more of this Group/Topical: Environmental Division

deTerra® Biobased Polymer and It's Use in Durable Applications

Dr. Jeffrey J. Cernohous, Brandon J. Cernohous, and Dr. Adam R. Pawloski
Interfacial Solutions, LLC. River Falls, WI

Abstract

In the past decade, the market for biobased polymers has grown dramatically. Much of this growth has been in the packaging market, where biobased polymer films (i.e., PLA and PHA) have displaced conventional LDPE and LLDPE. Packaging is a natural application for biobased polymers given the short life cycle and consumable nature of this product. Recently, consumers have begun to demand green alternatives in applications where durable plastics have historically been utilized. For example, PVC has been under intense scrutiny in recent years. PVC's inherent chemical nature and the additives it is often formulated with have been identified by several organizations targeting chemicals that can harm the environment. In certain markets, like the building and construction market, there is a salient need for a biobased material that can be used as a PVC alternative. LEED certification and other local codes and regulations are creating a new market for durable biobased polymers. One may consider durable biobased polymers a misnomer. However, the inherent nature of biopolymers like PLA, is such that they can function in one environment (e.g., interior building applications), but degrade in another environment (i.e., compost). This work describes several grades of polymeric compounds that been recently developed by Interfacial Solutions for durable applications. These products are currently marketed under out deTerra® Biobased Polymer tradename.

Introduction

Over the past several years, Interfacial Solutions has been actively developing durable biopolymer compounds that have very high biobased content. Biopolymers, like PLA, typically have excellent stiffness and good clarity. However, biopolymers are often have marginal impact strength and poor temperature resistance. Biopolymers also present processing issues because they are susceptible to hydrolytic cleavage and depolymerization during melt processing. This can lead to a catastrophic decrease in ultimate molecular weight and poor mechanical properties in the final article. To combat these issues, Interfacial Solutions developed a process to hyperbranch biopolymers. This reactive extrusion technique provides a handle to control melt rheology of biopolymers. The resulting materials have improved processibility and mechanical properties when compared to unmodified biopolymers (e.g., PLA). We have also engaged in extensive research and development efforts to develop flame retardant versions of biopolymers. We have developed biopolymer grades that meet both E84 and UL94 requirements. However, there still exists opportunities for biopolymer compounds that have the following attributes:

1) Improved impact resistance
2) Improved temperature resistance (i.e., HDT)
3) Improved hydrolytic stability
4) Improved processing characteristics
5) Lower cost

Herein, we describe some of our recent efforts to improve the impact resistance for flame retardant and non-flame retardant biopolymers, that we are currently marketing under our deTerra® Biobased Polymer tradename.

Materials

The deTerra® biobased polymer grades that are described herein are given in Table 1.

Table 1. deTerra® Biobased Polymer Grades

deTerra® Biobased Polymer Grade	*DESCRIPTION*
XP696	Flame retardant extrusion grade
XP443	Extrusion grade
XP345	Flame retardant injection molding grade
XP698	Injection molding grade

Results and Discussion

The test methods used to characterize deTerra® biobased polymer grades are given in Table 2. The key mechanical and physical properties of deTerra® biobased polymer grades given in Table 1 are summarized in Tables 3-6. As can be seen from the tables, our new non-FR grades have an excellent combination of flexural and impact properties. The flame retardant grades have excellent flexural properties and impact properties that are markedly improved of conventional PLA based formulations. It is also notable that our hyperbranching technology allows us to control the melt rheology and melt index of these biopolymer formulations. We can dial in melt indexes between 0.3 g/10 min and 50 g/10 min using this process. Finally, it is notable that all of our deTerra Biobased Polymer grades have exceptionally high biobased content, in excess of 70%.

Table 2. List of Test Methods

Property	Test Method
Specific Gravity	Archimedes method
Melt Flow Index	ASTM D1238
Flame Resistance	UL94 or E84
Flexural	ASTM D790
Impact	ASTM D256

Table 3. Selected Properties deTerra® Biobased Polymer Grades

deTerra® Biobased Polymer Grade	*Specific Gravity (g/cc)*	*Melt Flow Index* (g/10 min)*	*Flame Rating (Test)*
XP696	1.26	1.0	Class 1A (E84) V2 (UL94)
IP9974	1.24	1.0	Not Rated
XP345	1.30	18.0	V0 (UL94)
XP698	1.25	27.0	Not Rated

*Test performed at 190 °C using 2.12 Kg load

Table 4. Impact Properties deTerra® Biobased Polymer Grades

deTerra® Biobased Polymer Grade	*Izod Impact-Notched (ft-lbs/in)*	*Izod Impact-Unnotched (ft-lbs/in)*
XP696	0.65	10.5
IP9974	1.45	15.0
XP345	0.95	11.0
XP698	16.5	43.0

Table 5. Flexural Properties deTerra® Biobased Polymer Grades

deTerra® Biobased Polymer Grade	*Flexural Strength (psi)*	*Flexural Modulus (psi)*
XP696	12,600	458,000
IP9974	11,500	360,000
XP345	11,200	498,000
XP698	8,600	330,000

Table 6. Estimated Biobased Content of deTerra® Biobased Polymer Grades

deTerra® Biobased Polymer Grade	*Estimated Biobased Content*
XP696	93
IP9914	87
XP345	76
XP698	85

Figure 1 shows a photograph of articles that have been produced using various extrusion and injection molding grades of deTerra® Biobased Polymer.

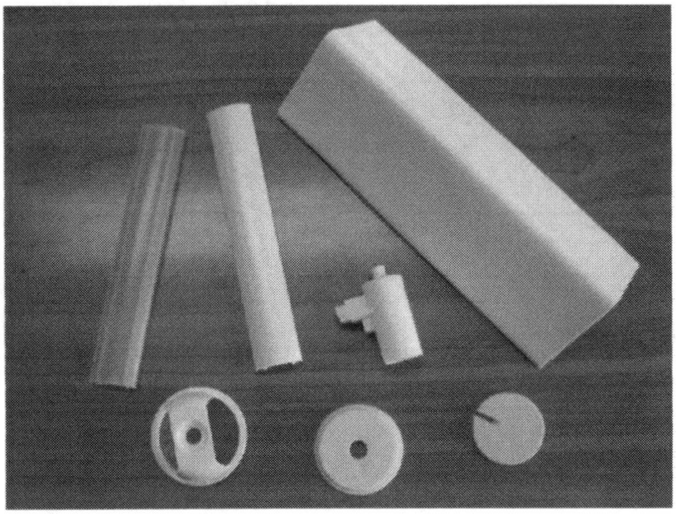

Figure 1. Articles produced using deTerra® Biobased Polymers

Conclusions

In conclusion, Interfacial Solutions has developed several new biobased polymer compounds that have improved mechanical properties and processing characteristics. We are continuing these efforts in the future to provide the plastics industry with durable biopolymer compounds that have differentiated performance while maintaining very high biobased content.

Roofing Materials as Effective Means to Decrease Heat Loads on Buildings

Sandra Nunez
Graduate Student
Department of Civil, Construction and Environmental Engineering
University of Alabama at Birmingham
Birmingham, Alabama

Ronald Sherrod
Undergraduate Student
Department of Chemical Engineering
Worchester Polytechnic Institute
Worchester, Massachusetts

Matthew Winslett
Engineering Manager, Energy Management
Facilities Management Department
University of Alabama at Birmingham
1530 3rd Avenue South
Birmingham, Alabama 35294-4564

Robert W. Peters, Ph.D., P.E.
Professor of Environmental Engineering
Civil, Construction, and Environmental Engineering
University of Alabama at Birmingham
Birmingham, Alabama

Abstract

The overall goal of the research involving the green roofs on the University of Alabama at Birmingham (UAB) campus is to identify a suitable roofing material combination which reduces energy cooling costs, by measuring surface and internal temperatures and comparing the data. In addition to finding a suitable roofing structure it is hoped that the long term effects of the retrofits produce a decrease in the urban heat island effect, on the UAB campus, and an increase in air quality.

During the past 3+ years, UAB has been investigating external and internal temperatures to determine a way to lower utility costs, but more recently the investigation of internal temperatures has been more pertinent. Due to automated data capture, we have been able to plot data at a very high resolution. This high resolution prompted us to model the temperature changes inside of the roof to better understand how the roof heats and cools on a yearly basis. Given this chance to model the internal temperatures of the roofing structures, it is our hope to better understand the thermal properties at work on a larger scale to reduce utility costs and reduce the urban heat island effect.

Introduction

When examining how to lower air-conditioning costs effectively and mitigate the urban heat island, the materials of the roofing structure come into play. Different roofing materials transfer heat into a structure in different ways, but by the same mechanisms. Heat that is being transferred into the roofing structure is primarily transferred through conduction and convection. The internal temperatures that have been investigated are affected by convection, conduction, and radiation; this is caused due to an energy transfer from the exterior surfaces of the structure to the internal surfaces [Clements and Sherif, 1998]. These are the key factors that affect the performance of the air-conditioning system. This transfer of heat is what the air-conditioning is trying to compensate for. Therefore, if a roof allows more heat to be transferred through the various mechanisms, the more work the air-conditioning unit must perform.

The University of Alabama at Birmingham is located in the Southeastern United States. This region is characterized by its hot summers and relatively mild winters. Due to these hot summers during which the temperatures can approach 100 °F, along with conditions of high humidity, it is crucial to have a roof which minimizes cooling costs. These hot summers also create an urban heat island in the Birmingham area due to the current roofing structures and facility usage. An urban heat island is defined on the basis of a temperature difference between urban and rural area, caused by human modifications to the natural geographic landscape [Cermark *et al.*, 1995]. The human modifications observed here are the buildings and roads added to the natural habitat. These structures retain the heat transferred from the sunlight increasing the temperature. This property results in an increased temperature in the highly modified areas (urban downtown areas) versus a less modified area (rural areas).

Heat transfer of absorptive structures in the city (UAB) and long wave radiation by buildings are forces which help form urban heat islands [Bretz *et al.*, 1997]. Bretz *et al.* [1997] primarily addresses mitigating urban heat islands, by the reflection of radiation, integrated over the solar spectrum (0.3–2.5 mm). This includes specular and diffuse reflection [Bretz *et al.*, 1997]. By increasing the albedo of a structure we increase how reflective the object is to the solar radiation. This property will result in a lower surface temperature and less heat being transferred into the building. Therefore, UAB can mitigate the urban heat island effect and lower cooling costs through an albedo increase. Bretz *et al.* [1997] continue to address the long term efficiencies of high and low albedo structures by saying, dark roofing materials tend to increase in albedo over time, because of dirt collection and oxidation. The light roofs, on the other hand, have a decrease in albedo as a response to the same weathering.

In addition to lowering the air-conditioning costs of UAB, the mitigation of the urban heat island effect also would increase urban air quality. The increase in air quality is an indirect result of the lower air-conditioning load itself. The higher air-quality is derived from the air-conditioning units putting out less greenhouse gases. These greenhouse gases cause retention of heat in the area along with smog producing incidents. Therefore, by reducing air-condition unit operation loads, smog formation can be cut down, providing an increase in air quality.

Historically, studies on green roofs have explored their energy performance compared with traditional roofs. Thermal performance indicated a significant reduction (~40%) of a building cooling load during the summer period [Spala *et al.*, 2008]. Similar results were achieved for a nursery school, with reductions ranging from 6% to 49%, and reduction ranging from 12% to 87% on the last floor of the nursery school [Santamouris *et al.*, 2007]. Wong *et al.* [2007] note that green roofs tend to experience lower surface temperatures than the original exposed roof, especially in areas well covered by vegetation. When green roofs are well covered by vegetation, the resulting substrate moisture will tend to keep substrate temperature lower than the original exposed bare roof. These studies determined that over 60% of the heat gain was mitigated by vegetated roof systems. Summertime data have indicated significant lower peak roof surface temperature and higher nighttime surface temperature for green roofs as compared to conventional roofs [Sonne, 2006]. The maximum average daily temperature seen for the conventional roof surface was 54.4°C in his study, while the maximum average day green roof surface temperature was 32.8°C (~21.7°C lower than the conventional roof). Green roofs offer cooling potential (~3.02 kWh/day) to maintain an average room air temperature of 25.7°C [Kumar, 2005]. Green roofs help minimize environmental burdens, conserve energy, and extend the life span of the roofing system in overall sustainability [Cavanaugh, 2008]. Up to 30% of total rooftop cooling is due to plant transpiration [Takakura *et al.*, 2000]. Bell and Spolek [2009] compared different types of plants for use in increasing the thermal resistance (*R*-value) of green roofs, and found that ryegrass delivered the highest effective *R*-value compared with bare soil, *Vinca major, Trifolium repens,* and *Sedum hispanicum.* Also, though increasing the depth of bare soil from 5 to 14 cm increased the *R*-value, no difference was found for different depths of planted soil. This implies that the bulk of benefit toward *R*-value is from evapotranspiration and leaf shading, rather than the moist soil [Bell and Spolek, 2009].

<u>Mini-Roofs</u>

During this study, 15 mini-roof combinations were observed for trends in internal temperatures. The various 15 mini-roof combinations are summarized in Table 1. To investigate the thermal properties of the roofing structures an ambient temperature probe was placed inside of each roof to record data every 10 minutes. This data was then automatically sent to a data logger and placed into an Excel file for review later. The temperature probe reports the temperature to the nearest hundredth of a degree Fahrenheit.

The roofing materials used are all standard commercial flat roof materials. Flat roof materials were only looked at during the study, since the primary application for the roofing combinations will be on a commercial flat roof top, and not a slanted roof structure. Each mini-roof is 8-ft long x 4-ft wide x 4-ft deep (see Figure 1). A number of different roofing systems are being examined for their energy performance. All roofs are insulated with 2.0-in. of extruded polystyrene . Then the particular roofing combination being investigated is applied over the insulation and sealed. The roofs also include a proper drainage spout, to ensure correct water evacuation, such as on a real roof. The roofing systems being studied using the 15 mini-roofs are listed in Table 1 below.

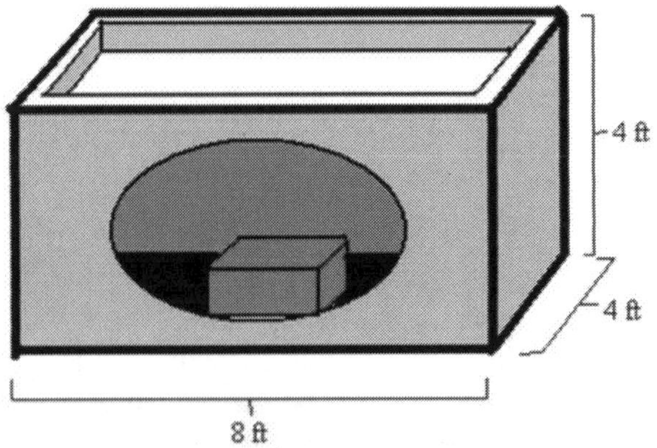

Figure 1. Depiction of a Typical Mini-Roof System.

Table 1. Mini-roof descriptions.

Mini-Roof No.	Mini-Roof Description
1 (Sensor B)	White TPO/PVC/Elvaloy fully adhered, FiberTite Membrane.
2 (Sensor C)	Black 60-mil EPDM fully adhered/coated/white urethane, Mule Hide Membrane.
3 (Sensor D)	Black 60-mil EPDM fully adhered, Mule Hide Membrane.
4 (Sensor E)	Beige TPO/PVC/Elvaloy fully adhered, FiberTite Membrane.
5 (Sensor F)	White granular modified, Firestone SBS Modified Membrane.
6 (Sensor G)	Black granular modified, Firestone SBS Modified Membrane.
7 (Sensor H)	Black granular modified coated/white urethane, Firestone SBS Modified Membrane.
8 (Sensor I)	Bituthene IRMA with lightweight "T Clear" pavers.
9 (Sensor J)	Bituthene IRMA with river rock ballast.
10 (Sensor K)	Bituthene IRMA with vegetative green roof, ½-in. drain mat, 350-lbs dry soil.
11 (Sensor L)	Bituthene IRMA with vegetative green roof, 1-in. drain mat, 350-lbs dry soil.
12 (Sensor M)	Black 60-mil EPDM loose, ballasted with river rock, Mule Hide Membrane.
13 (Sensor N)	Black 60-mil EPDM loose, ballasted with #300 marble chips, Mule Hide Membrane.
14 (Sensor O)	White TPO/PVC/Elvaroy loose laid, ballasted with river rock, FiberTite Membrane.
15 (Sensor P)	Bituthene IRMA with vegetative green roof, ½-in. drain mat, 350-lbs dry soil.
Sensor S	Sensor inside mini-roof No.10 (inside the soil of the green roof).
Sensor T	Sensor under the white TPO/PVC/Elvaloy loose laid, ballasted with river rock.

We have also collected surface temperature measurements during the month of June on the various roofing materials used with our mini-roof systems. During this time period, we have observed roofing surface temperatures ranging from 20.6°C to 82.2°C. The lower surface temperature values, 20.6°C to 48.3°C, are found in a loose rock or stone roof combination. The higher temperature values occur on the roofing combinations primarily made of a coating or membrane; these roofs exhibit temperatures roughly between 25.6°C to 82.2°C. The vegetative mini-roofs exhibit surface temperatures ranging between 21.7°C to 52.2°C, while their large counterpart, the pilot roof on top of Hulsey Center, exhibits higher temperatures ranging from 31.7°C to 61.1°C. When looking at the specific roofs contained in each subset of Table 2, the lighter the roof, generally the cooler it is. For example, the white Firestone SBS is cooler than the black Firestone SBS on any given day. This thermal property is observed due to the reflectivity of the roof. The darker colored roofs absorb more incoming light radiation than the light colored roofs causing the dark roofs to become hotter. (Since the roofs temperatures observed are taken during the late spring and early summer, it can be inferred that overall roof temperature will increase during late summer).

Table 2. Surface temperature ranges* observed on the mini-roofs.

General Roof Type	Surface Temperature Range
Coating/Membrane (e.g., PVC or Urethane or Firestone SBS) [Colors: Black and White]	25.6°C–82.2°C: Black Firestone SBS: 40.0°C–82.2°C; White Firestone SBS: 38.9°C–70.0°C)
Vegetative (1–2 plant species/dark and light small rock substrate)	21.7°C–52.2°C
Loose Rock (e.g., river rock or white	20.6°C–48.3°C

marble chips)	
White Pavers Stone	20.6°C–48.3°C
Pilot roof (~20 plant species/dark and light small rock substrate/gray pavers stone)	31.6°C–61.1°C

*Temperature ranges are based on the average values per reading between June 1st, 2009 and July 2nd, 2009.

Observations of the Internal Temperature of the Mini-Roofs

Trends seen in statistical comparisons of internal temperatures of similar roofs (using the null hypotheses: $\mu_1 \geq \mu_2$) are summarized below:

- Between the two river rock roofs, Roof 9 will most likely always be hotter or equal to roof 12.
- Between the 3 vegetative roofs, all the roofs are statistically equal to each other in thermal properties.
- Both SBS Firestone roofs are statistically the same, but roof 6 is usually hotter.
- The TPO/PVC/Elvaloy Roofs are statistically the same.
- The 60-mil EPDM roofs were also statistically the same.
- Statistically speaking, all roofs behave similarly, this comes from the work done by insulation on creating an adequate thermal barrier.

Sinusoidal Regression Method

After acquiring the necessary raw data from the mini-roof sensors, Microsoft Excel (version 2010) was used to plot the data to discover trends in the temperature readings. After discovering the trend, we attempted to fit a mathematical model to the data. We noticed that the temperatures cycled in a sinusoidal fashion on yearly and daily time frames, so using the general form sine function was an appropriate model to describe the data. Taking the raw data, we used a program called FFT.exe (version 3.2) to determine the frequency. [ω_0 from the general form $x = A \sin(\omega_0 t + \phi) + C$]. This program calculates the Fast Fourier Transform of a time and then outputs the results as a text file which we then plotted in Excel to determine the frequency of the yearly oscillation. Observing the graph, two major peaks were displayed – one representing the yearly frequency and the other representing the daily oscillations. The yearly peak was very broad so the frequencies of this peak were averaged to determine the ω_0 in radians. After trial and error, we used an ω_0 of 3.18×10^{-8} rad. After calculating this frequency, another program called sinefind.exe (version 1.7) was used to determine the phase angle (ϕ) and the amplitude ("A") of the general form. Sinefind.exe is a program which performs a sine curve fit of input data. The amplitude resulting from sinfind.exe was too large to be reasonable; after trial and error, we determined that this was also the case with the phase angle. We contemplated the issue and we arrived at the conclusion to remove the mean of the data to determine if the problem was solved. After mean removal, the sinfind.exe output yielded a more reasonable amplitude which we used, but it resulted in the phase angle being too low. The conclusion was drawn that the phase angle was between the two values, so each one was weighted in appropriate ratios and the true phase angle was determined. Using the general form sine function, an Excel spreadsheet was developed to obtain an r^2 value to observe the change in the coefficient to determine the optimum weighting between the two phase angles output by analyzed data with and without the mean. We used the already determined frequency and amplitude in addition to the mean of the raw data for "C" in the general form. The optimum weighting was determined to be approximately 70% the phase angle output from the raw data and 30% the phase angle output from mean removal data. Using this methodology an optimum format for the function was determined to be:

$$T = (A_{\text{from sinfind.exe of data with mean removed}}) \times (\sin(2\pi\omega_{\text{from FFT.exe of raw data}} + ((0.7\,\phi_{\text{output of raw data}}) + 0.3\,\phi_{\text{output of mean removed data}}))) + \mu_{\text{of raw data}}$$

Using this general formula for determining the model, we applied this methodology for all 15 roofs in the study.

Figure 2 shows an example of the internal temperature variation within the mini-roof structure over a several year period. The data is observed to follow a sine-wave type functionality.

Results

Using the discussed method, we determined 15 different formulas to describe each roof internal thermal trends. The determined functions supported that most roofs are statistically different from one another from an amplitude aspect but the phase angles are statistically the same. We also have discovered that most roofs are significantly different from the standpoint of the average mean roof temperature, but the significance is most prevalent in the summer months and during the other times of the year the roofs behave in a more similar fashion to one another. (See Tables 2 and 3 for statistics and means of Roofs). Table 4 summarizes the best sine-wave fits to the internal temperature data for the various mini-roofs. Looking at the data, we see that the roofs have amplitudes ranging from 19.5 to 20.7 degrees and phases angles roughly 4.2 to 4.3 radians. The r^2 values ranged from 0.69 to 0.71showing the data works to help understand the trends. The best model developed was for roof 3 (Figure 3), having the highest correlation coefficient of all 15 mini-roofs.

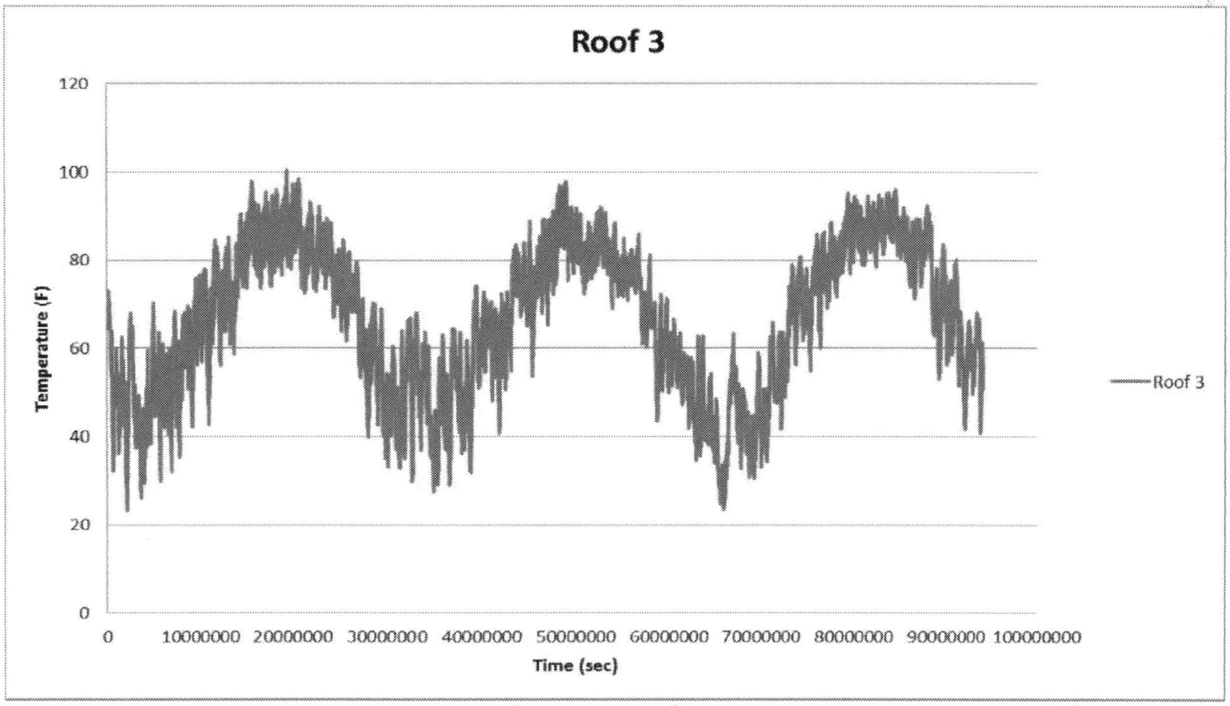

Figure 2. Temperature variation for mini-roof No.3 over a several year period.

Table 2. Summary of statistics for the phase means angles from curve fitting.

Statistics of Phase Angles	
Mean	4.0303
Standard Error	0.00343
Median	4.0276
Mode	4.0242
Standard Deviation	0.01329
Sample Variance	0.000177
Kurtosis	12.0174
Skewness	3.3228
Range	0.05512
Minimum	4.0211
Maximum	4.0762
Sum	60.4541
Count	15

Table 3. Summary of statistics for the from curve fitting.

Statistics of Means	
Mean	66.67
Standard Error	0.18
Median	66.62
Mode	#N/A
Standard Deviation	0.68
Sample Variance	0.46
Kurtosis	3.11
Skewness	1.11
Range	3.05
Minimum	65.42
Maximum	68.47
Sum	1000.12
Count	15

Table 4. Fitted sine-wave functionalities describing the internal temperatures in the mini-roofs.

Roof	Final Equation	Amplitude from sinefind.exe	Phase Angle	r^2	Mean
1	T = 20.1632*sin ((2*π*3.17894e(-8)*t) +4.021975)+66.68283	20.163	4.022	0.696	66.683
2	T = 20.5499*sin ((2*π*3.18e(-8)*t) +4.02272)+66.63564	20.550	4.023	0.701	66.636
3	T = 20.392*sin ((2*π*3.18e(-8)*t) +4.07624)+66.47947	20.392	4.076	0.712	66.479
4	T = 20.482*sin ((2*π*3.18e(-8)*t) +4.021125)+66.35505	20.482	4.021	0.698	66.355
5	T = 20.4891*sin ((2*π*3.18e(-8)*t) +4.03214)+67.54093	20.349	4.032	0.701	67.541
6	T = 20.7812*sin ((2*π*3.18e(-8)*t) +4.03142)+68.46883	20.781	4.031	0.699	68.469
7	T = 20.2886*sin ((2*π*3.18e(-8)*t) +4.02415)+67.08382	20.289	4.024	0.701	67.084
8	T = 20.7812*sin ((2*π*3.18e(-8)*t) +4.02415)+66.18969	20.781	4.024	0.695	66.190
9	T = 20.3285*sin ((2*π*3.18e(-8)*t) +4.024975)+66.91944	20.329	4.025	0.701	66.919
10	T = 19.8608*sin ((2*π*3.18e(-8)*t) +4.02758)+66.6981	19.861	4.028	0.697	66.698
11	T = 20.0128*sin ((2*π*3.18e(-8)*t) +4.026215)+66.62328	20.013	4.026	0.697	66.623
12	T = 20.0479*sin ((2*π*3.18e(-8)*t) +4.027853)+66.27383	20.048	4.028	0.694	66.274

13	T = 19.9914*sin ((2*π*3.18e(-8)*t) +4.0325)+66.47272	19.991	4.033	0.696	66.473
14	T = 20.0661*sin ((2*π*3.18e(-8)*t) +4.032494)+65.42343	20.066	4.032	0.689	65.423
15	T = 19.4628*sin ((2*π*3.18e(-8)*t) +4.028672)+66.27177	19.463	4.029	0.692	66.272

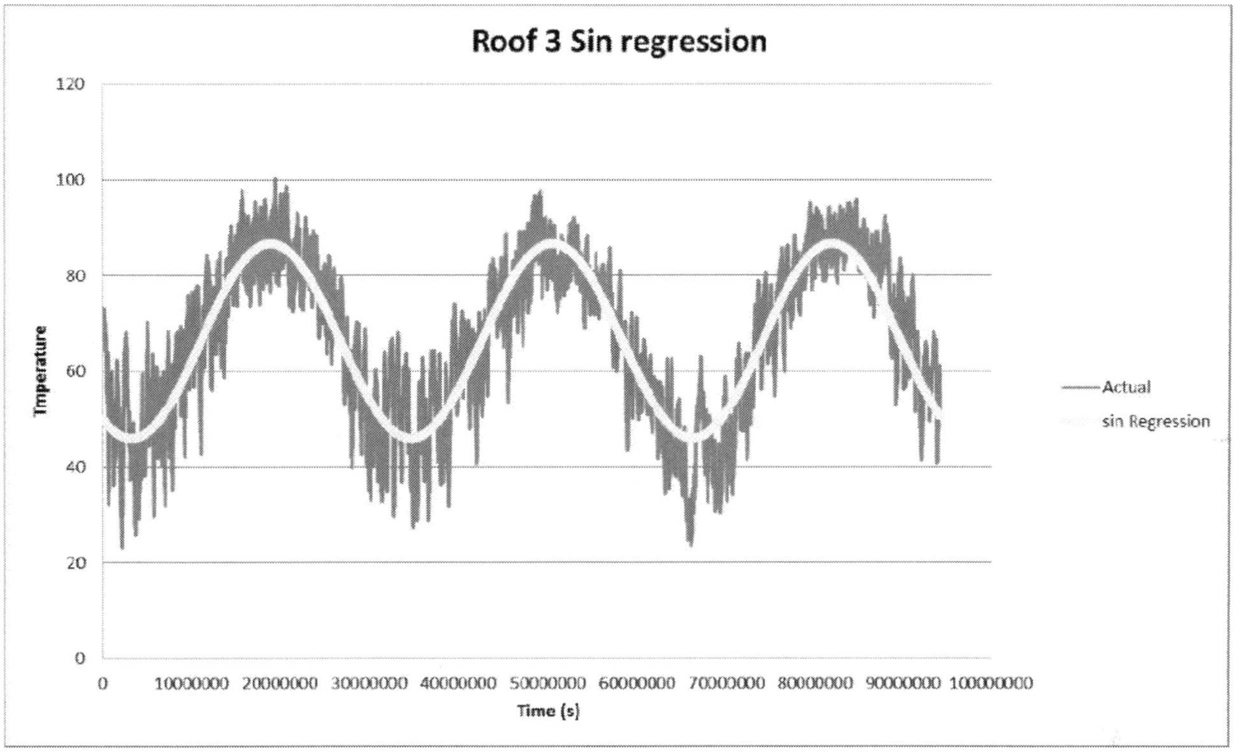

Figure 3. Fitted sine wave regression for a mini-roof system.

Summary and Conclusions

In summary, the general method to develop a mathematical model works well to understand the average temperatures observed during the course of a year. While the model does represent yearly cycles quite well, it lacks the detail to better represent the daily cycles which could become a problem if a prediction of more specific daily temperatures were needed for research. Overall, it is our position that the model does well to describe the average daily temperature during a particular time of the year. This model will work best in a yearly aspect for its use here at the University of Alabama at Birmingham, but should be refined to be used on a daily resolution.

Acknowledgments

The authors acknowledge the technical assistance provided by UAB's Facilities Management Department in the conduct of this research.

References Cited:

Bell, H. and G. Spolek, 2009. "Measured Energy Performance of Greenroofs", in ***Proc. Seventh Annual Greening Rooftops for Sustainable Communities Conference***, The Cardinal Group, Toronto.

Bretz, S., H. Akbari, and A. Rosenfeld, 1997. "Practical Issues for Using Solar-Reflective Materials to Mitigate Urban Heat Islands", ***Atmospheric Environment***. ***1***: 95–101.

Cavanaugh, L.M., 2008. "Redefining the Green Roof", ***J. Archit. Eng.***, ***14***(1): 4–6.

Cermark, J., A. Davenport, E. Plate, and D. Viegas, 1995. ***Wind Climate in Cities***. Kluwer Academic Publishers, The Netherlands.

Clements, J.A. and S.A. Sherif, 1998. "Thermal Analysis of Roof-Spray Cooling", ***International Journal of Energy Research***, ***22***: 1337 – 1350.

Kumar, R., 2005. "Performance Evaluation of Green Roof and Shading for Thermal Protection of Buildings", ***Build. Environ.***, ***40***(11): 1505–1511.

Santamouris, M., C. Pavlou, P. Doukas, G. Mihalakakou, A. Synneta, A. Hatzibiros, and P. Patargias, 2007. "Investigating and Analysing the Energy and Environmental Performance of an Experimental Green Roof System Installed in a Nursery School Building in Athens, Greece", ***Energy***, ***32***(9): 178–1788.

Sonne, J., 2006. "Evaluating Green Roof Energy Performance", ***ASHRAE J.***, ***48***(2): 59–61.

Spala, A., H.S. Bagiorgas, M.N. Assimakopoulos, J. Kalavrouziotis, D. Matthopoulos, and G. Mihalakakou, 2008. "On the Green Roof System. Selection, State of the Art and Energy Potential Investigation of a System Installed in an Office Building in Athens, Greece", ***Renew. Energy***, ***33***(1): 173–177.

Takakura, T., S. Kitade, and E. Goto, 2000. "Cooling Effect of Greenery Cover over a Building", ***Energy Build.***, ***31***: 1-6.

Wong, N.H., P.Y. Tan, and C. Yu, 2007. "Study of Thermal Performance of Extensive Rooftop Greenery Systems in the Tropical Climate", ***Build. Environ.***, ***42***(1): 25–54.

Comparison of miscibility between Paraffin and three types of Polyethylene in Polyethylene/Paraffin blends as form-stable phase change materials

Fang Chen, Michael Wolcott

Washington State University

Background

What are PCMs?

Phase change materials

1. High heats of fusion

absorb or release a lot of energy before melting or solidifying

2. Temperature remains constant during the phase change

keep the object at a uniform temperature

Example: paraffin, lauric acid and ......

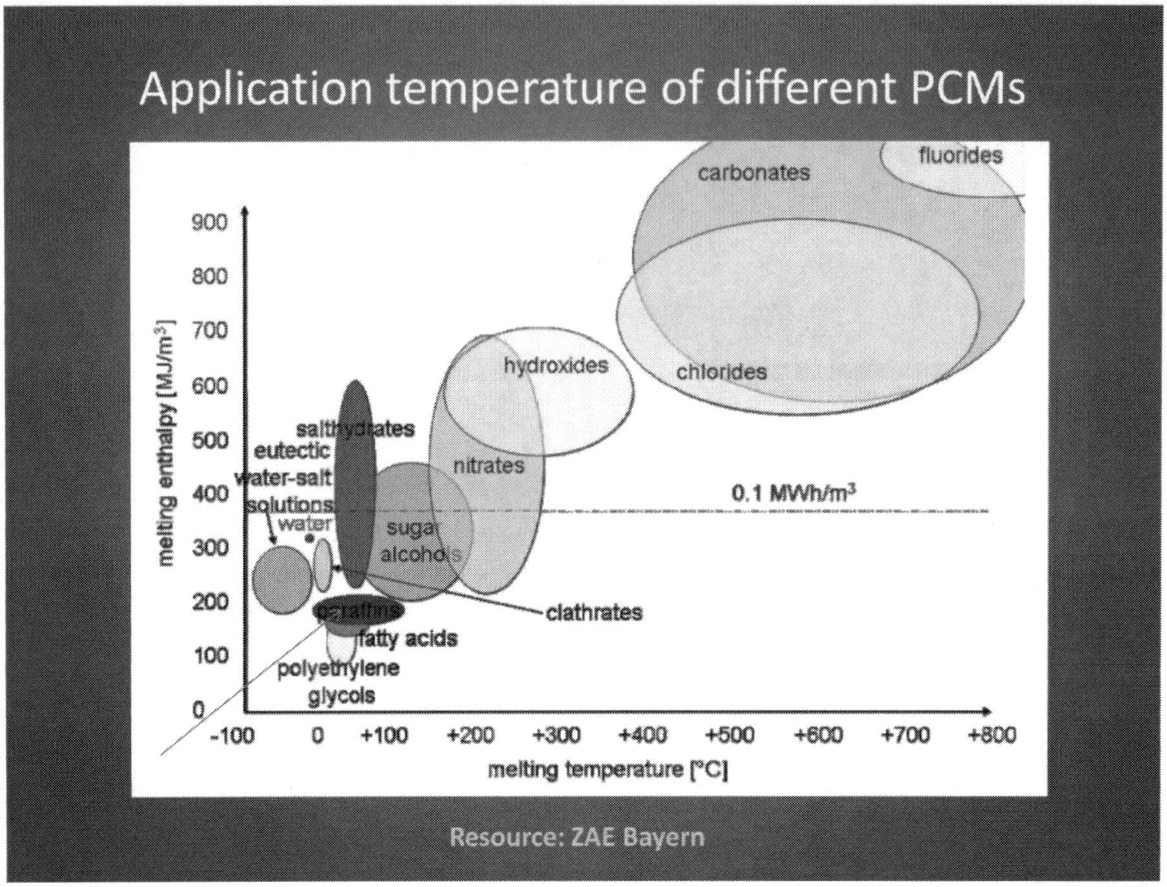
Application temperature of different PCMs
melting enthalpy [MJ/m³]
melting temperature [°C]
900
800
700
600
500
400
300
200
100
0
-100
0
+100
+200
+300
+400
+500
+600
+700
+800
fluorides
carbonates
chlorides
hydroxides
nitrates
salthydrates
eutectic water-salt solutions
water
sugar alcohols
0.1 MWh/m³
clathrates
paraffins
fatty acids
polyethylene glycols
Resource: ZAE Bayern

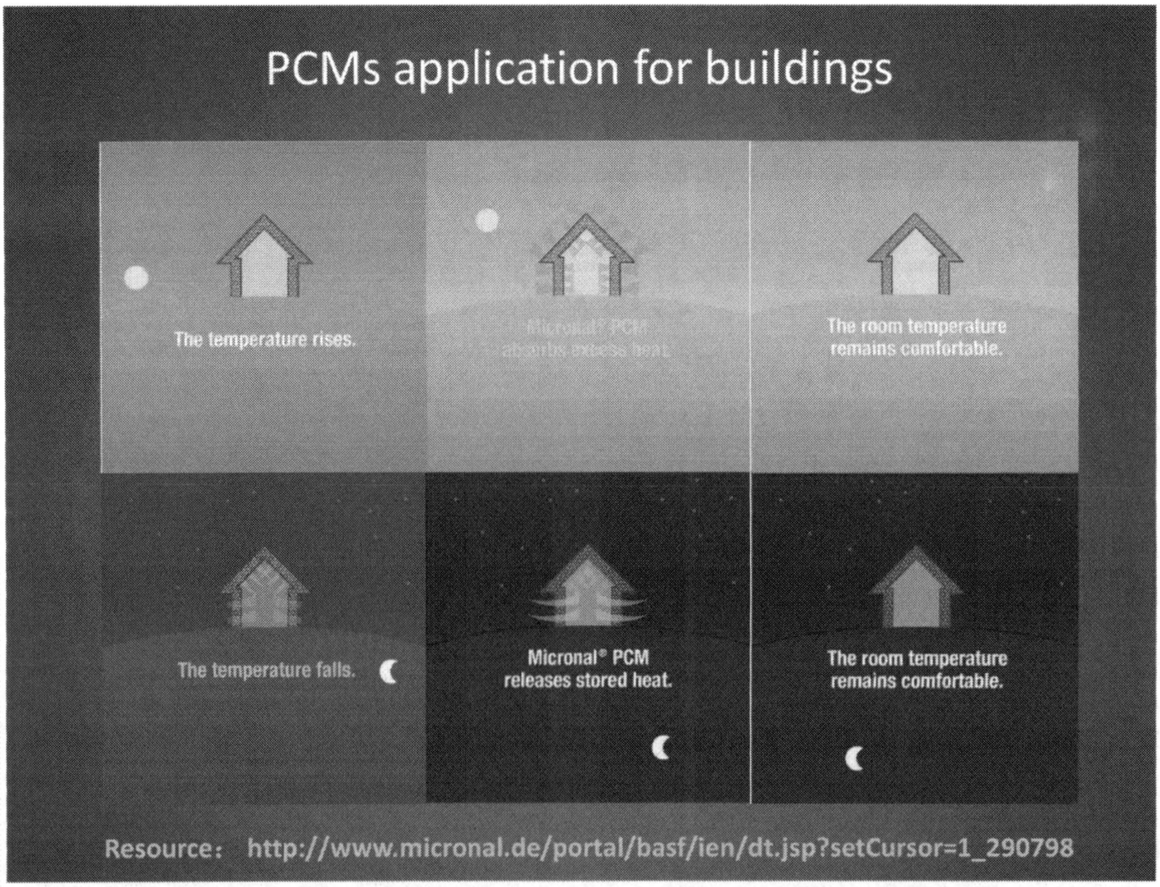
PCMs application for buildings
The temperature rises.
Micronal® PCM absorbs excess heat.
The room temperature remains comfortable.
The temperature falls.
Micronal® PCM releases stored heat.
The room temperature remains comfortable.
Resource： http://www.micronal.de/portal/basf/ien/dt.jsp?setCursor=1_290798

PCM Problem statement

Leakage solution-----------------Polymer/paraffin blends

Polymer/paraffin blend leakage---------------Miscibility

Widely used polymer-------------------------------------PE

Study----------------------miscibility of PE/paraffin blend

Topics and Objectives

1. Morphology (AFM Tapping mode)
Characterize dispersion difference of paraffin in 3 types of PE.
2. Crystallization
a. Paraffin crystallization separation
Distinguish changes on paraffin crystallization separation in 3 types of PE.
b. Crystallinity of PE and paraffin
Investigate intensity difference of interaction between paraffin and 3 types of PE on crystallinity.
3. Paraffin equilibrium melting temperature (include thickening ration)
a. Equilibrium melting temperature and melting point depression
Discriminate paraffin equilibrium melting temperature depression in 3 types of PE.
b. interaction parameter X12
Estimate miscibility of 3 types of PE/Paraffin quantitatively using Flory-Huggins equation

Single objective number: 5

Objective1

Morphology (AFM Tapping mode)

Characterize dispersion difference of paraffin in 3 types of PE.

Method:

Hot pressing extruded blends PE70/Paraffin30 at 180°C
for 5min to make a flat film and cooling the film down at room
Temperature. Then AFM will be carried on to take images.

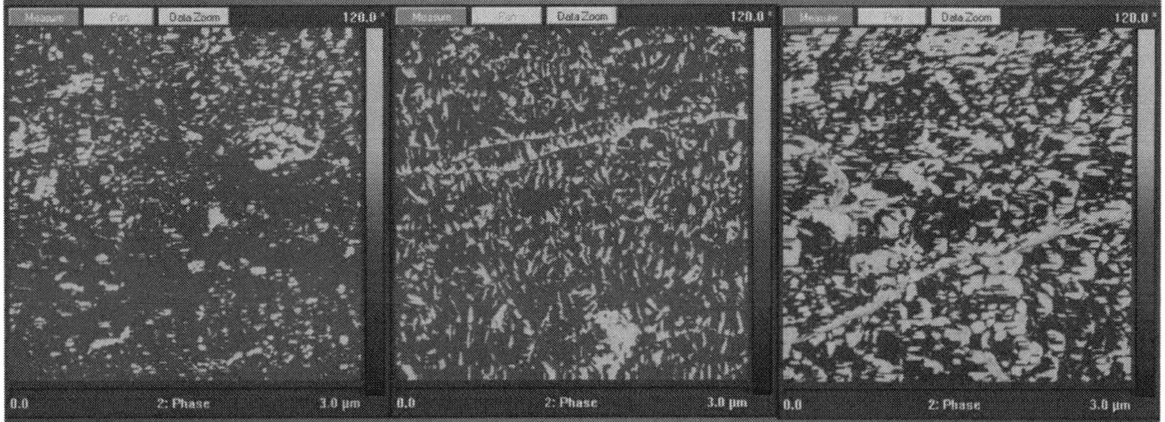

H70 L70 LL70

Conclusion 1

PE and paraffin are not completely miscible because there are at least 2 obvious phases.

HDPE is less miscible with paraffin than LDPE and LLDPE because of less intermedia color in AFM picture.

Objective 2

Paraffin crystallization separation

Distinguish changes on paraffin crystallization separation in 3 types of PE.

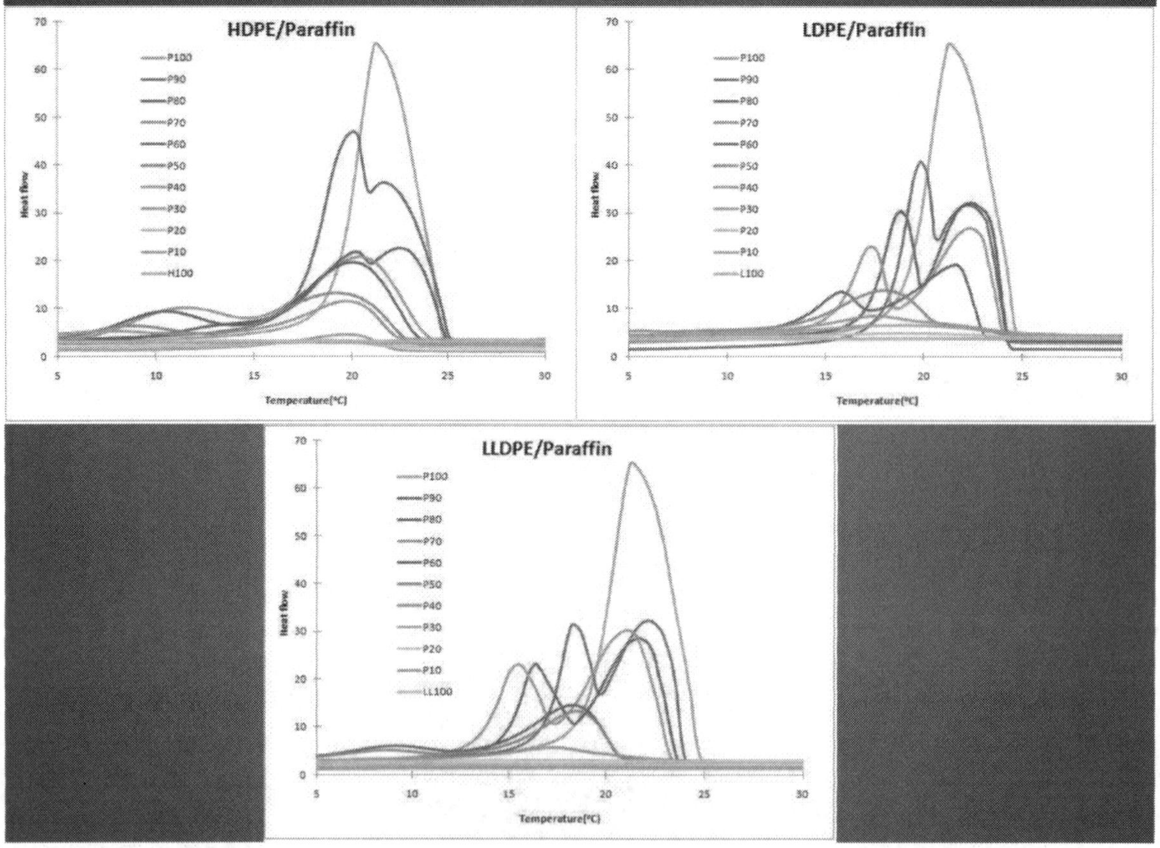

Conclusion 2

There are two crystallization peaks for PE/Paraffin blends, and enthalpy growth on one is at the expense of the other.

Effect of LDPE on paraffin crystallization is more than HDPE and less than LLDPE based on extent of crystallization peak shifting.

Objective 3

Paraffin equilibrium melting temperature (include thickening ratio)

Discriminate paraffin equilibrium melting temperature depression in 3 types of PE.

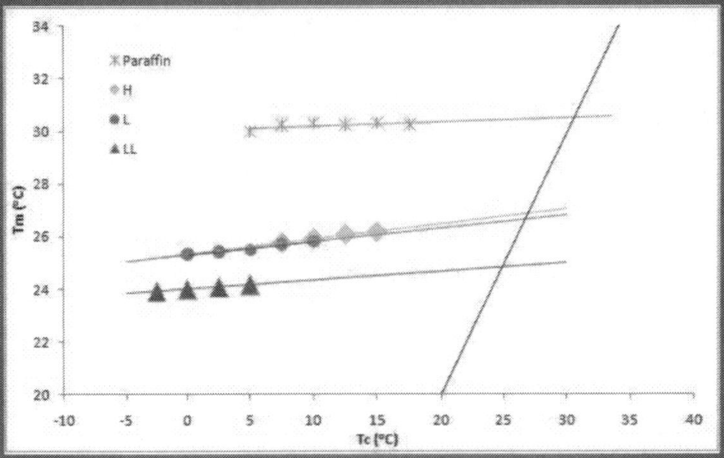

wax content	type	slope	intercept	thickening ratio	equilibrium Tm
100	Paraffin	0.0161	30.054	**31.06**	**30.55**
24	H	0.0572	25.334	**8.74**	**26.87**
20	L	0.0512	25.3	**9.77**	**26.67**
23	LL	0.0332	24.001	**15.06**	**24.83**

Thickening ratio of paraffin

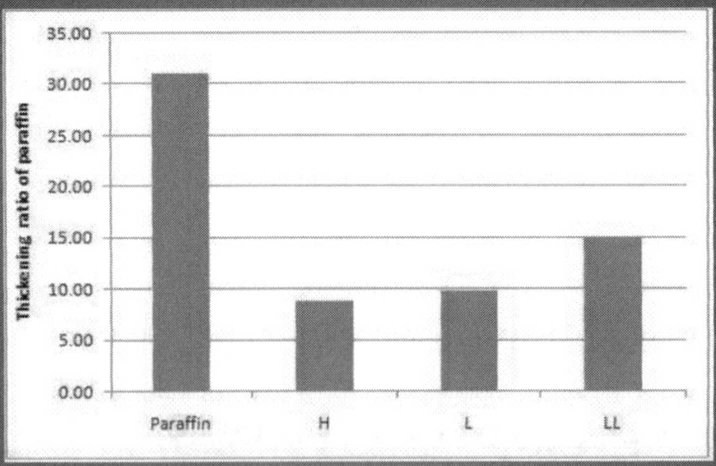

Conclusion 3

Effect of LDPE on paraffin melting is more than HDPE and less than LLDPE based on extent of equilibrium melting temperature depression and thickening ratio dropping.

Objective 4

Crystallinity of PE and paraffin

Investigate intensity difference of interaction between paraffin and 3 types of PE on crystallinity.

DSC Method:

50.0-150.0C	50C/min
150.0C	2.0 min
150.0- -10C	-10C/min
-10-150C	10C/min

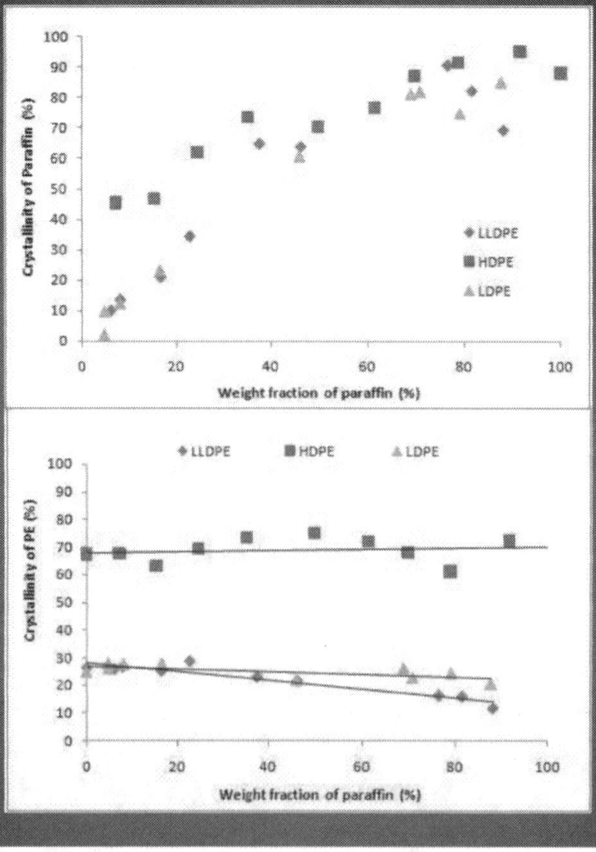

Conclusion 4

Paraffin and PE crystallinities drop with increasing of the other component dramatically and slightly.

Based on intensity of crystallinity drop,

Intensity of PE crystallinity drop: LLDPE>LDPE>HDPE
Intensity of paraffin crystallinity drop: (LLDPE and LDPE)>HDPE

Objective 5

Interaction parameter X12

Estimate miscibility of PE/Paraffin quantitatively using Flory-Huggins Equation.

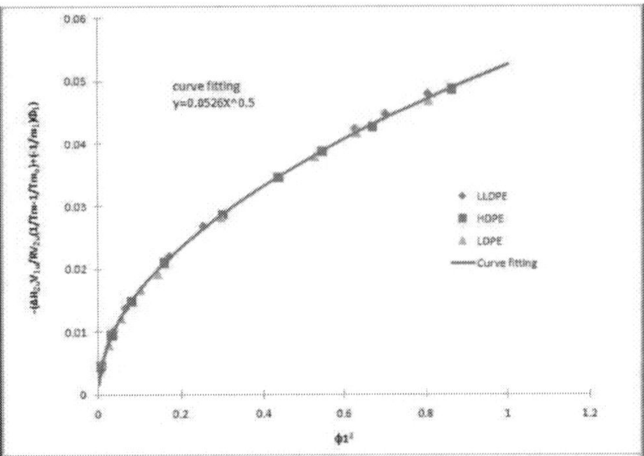

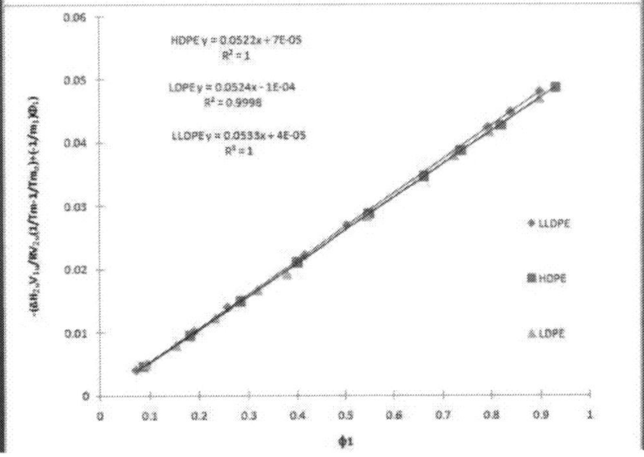

Interaction parameter X12

Model	Allometric		
Equation	y = a*x^b		
Reduced Chi-Sqr	1.15094E-7		
Adj. R-Squa	0.99952		
		Value	Standard Er
B	a	0.0526	1.65588E-4
B	b	0.5008	0.0032

Y=0.0526X^0.5

X12=Slope= first derivative of y

y'=0.0263x^(-0.5)

Composition dependent. Increasing of x will lead to a smaller X12 value.

Trend line did not go through zero. It's due to residual entropic effect which is neglected in the derivation of the equation.

Conclusion 5

All the interaction parameters of 3 types of PE/paraffin blends were estimated to be formulation dependent.

We can not distinguish 3 types of PE/Paraffin based on interaction parameter because of accuracy and amount of duplication due to high similarity of 3 types of PE.

Final Conclusion

Degree of Miscibility

in an order of

HDPE/paraffin<LDPE/Paraffin<LLDPE/Paraffin

(187h) Recent Advances In Selected Types of Sustainable Building Materials

Tom R. Marrero, chemical Engineering, University of Missouri, Columbia, MO

The purpose of this presentation is to review developments in masonry and polymeric building materials that

are identified as sustainable. Materials reported in reviewed journals, government reports and commercial

literature published since 2007 are reviewed. The anticipated results are trends in sustainable building material

usage in different locations in the United States and a few oversea countries, changes in the use of masonry, and usage of ploymers. Polymers includes synthetic and natural (biopolymers) materials.

The anticipated results should indicate future consumption of sustainable building materials and research needs.

a little sustainability

See more of this Session: Sustainable Building Materials
See more of this Group/Topical: Environmental Division

(188a) Removal of Cr(VI) In Aqueous Solution by Fe3O4 Nanoparticles In the Presence and Absence of Mn(II)

Yong Liu[1], Shaoyi Jia[1], Songhai Wu[1] and Yongli Sun[2], (1)School of chemical engineering and technology, Tianjin University, Tianjin, China, (2)Tianjin University, Tianjin

Abstract: Chromium is widely used in various industrial process such as metal electroplating, leather tanning and manufacturing of products for corrosion protection. At the same time, it is one of the major pollutants around the world, which is found in soil, waste sites, and water. Chromium exists in the environment primarily in two valence states: Cr(VI) and Cr(III). Chromium (VI) is highly toxic compared with Cr(III) to biological activities. In the past decades, Cr(VI)-containing waste water was treated by various adsorbent via adsorption process, including activated carbon, chitosan, functionalized polymers and ferric oxides. Among these adsorbents, magnetite (Fe_3O_4) nanoparticles have been studied extensively. However, little is known regarding the effect of Mn(II) on the adsorption process. In this study, removal of cr(VI) in the presence and absence of Mn(II) by Fe_3O_4 nanoparticles was investigated. Fe_3O_4 nanoparticles were firstly synthesized by modified hydrolysis method, morphology and physical properties was measured by TEM, SEM, FTIR, XRD and VSM. Then effect of Mn(II) on Cr(VI) removal by obtained Fe_3O_4 nanoparticles was investigated under different pH value and ratio of Cr(VI) and Mn(II). The experiment was conducted in two ways: pre-treated method and simultaneous method. In the pre-treated method, at pH=4, the final removal percentage of Cr(VI) was 114.3%, 116.1% and 102.5% compared with the control when Mn(II):Cr(VI) was 1:2, 1:1 and 2:1, respectively, meanwhile at pH=8, the final removal percentage of Cr(VI) was 56.8%, 87.6% and 88.5%. In the simultaneous method, at pH=4, the final removal percentage of Cr(VI) was 93.3%, 94.5% and 92.8% with the control when Mn(II):Cr(VI) was 1:2, 1:1 and 2:1, respectively. However, an interesting phenomenon was that at pH=8, the final removal percentage of Cr(VI) was 128.8%, 204.9% and 349.4% with the control. The results indicated that Mn(II) was not an inhibitor but an accelerant in the process of Cr(VI) removal by Fe_3O_4 nanoparticles.

See more of this Session: Sustainable Water Use and Management
See more of this Group/Topical: Sustainable Engineering Forum

(188b) Effect of Airborne Deposition On Water Acidity In the Rio Grande River Basin

Qin Qian[1], Badri Parajuli[1], Luke Prinsloo[1], Jerry Lin[1], Jack Hopper[2] and **T. C. Ho**[2], (1)Department of Civil Engineering, Lamar University, Beaumont, TX, (2)Department of Chemical Engineering, Lamar University, Beaumont, TX

The Upper Rio Grande River extends from the Texas-New Mexico state line downstream to the International Amistad Dam, a length of 650 miles (1,045 km). With increasing population and urbanization, the water in this segment is becoming high in salts and bacteria. The water quality problem compounded with the concern over water shortage is a serious issue for this Rio Grande River Basin. Airborne pollutants such as SO_4^{2-} and NO_3^- can reach earth surface via wet deposition or dry deposition, which can pollute the earth's valuable water resources. The objectives of this study were to collect and analyze data to understand the effect of airborne particulate pollution on the water quality in the Rio Grande Basin and to simulate the acidity of the water in the Amistad Reservoir using the CE-QUAL-W2 under different flow and precipitation conditions. An analysis of the collected data on wet deposition has indicated that the pH values and the concentration of ions in the rainwater vary with the amount of rainfall in each rain event. A higher concentration of SO_4^{2-} and NO_3^- increases the acidity of the rain water in the ranges between pH=4.9 and pH=5.4 while a higher concentration of Ca^{2+} and Na^+ decreases the acidity of the rainwater in the ranges between pH=5.5 and pH=5.8. A higher amount of rainfall in each rain event tends to decrease the concentration of ions which may increase or decrease the pH of the rainwater depending on the dominating ion species, i.e., SO_4^{2-}/NO_3^- or Ca^{2+}/Na^+. An analysis of the collected river water data show that the pH in the Rio Grande River water is much higher than the rainwater pH and is in the ranges between pH=7.8 and pH=8.3 due to the buffering effect of Ca^{2+} in the river water. The pH in the river water has been observed to vary with the water temperature, which decreases with an increase in water temperature. In addition, the pH decreases when the river flow rate increases due to rainfall, which mainly is because of the decrease in Ca^{2+} concentration due to the rainwater dilution and the delayed dissolution of $CaCO_3$ into Ca^{2+} in the river water. The results from pH simulations of the Amistad Reservoir water in 2004 based on the two-dimensional water quality and hydrodynamics model, CE-QUAL-W2, have indicated that the precipitation decreases the pH by about 0.1 to 0.15 as compared to the no rain condition. The water pH in the reservoir is observed to vary inversely with water temperature. However, the pH variation is insignificant when the rates of river flow are significantly reduced from the original flow rates.

See more of this Session: Sustainable Water Use and Management
See more of this Group/Topical: Sustainable Engineering Forum

(188c) Removal of Boron From Waste Water by Electro-Coagulation Process

Krishan K. Garg Jr., Chemical Engg., IIT ROORKEE, Hindaun CITY, India and **Dr. B. Prasad Sr.**, Chemical Engg., IIT ROORKEE, Roorkee, India

EC application needs no chemical reagents so this is very important process for waste water treatment. Initial boron concentration in the reactor changes from lower to higher. According to this paper we can see properly the effect of different type of parameters on boron removal efficiency such as p^H , time, temp., current density, distance b/w Al electrodes, changing electrolyte concentration, initial boron concentration. Effect of changing concentration of boron & electrolyte dosages on energy consumption. Analytical treatment of boron had been done by using of Cur-cumin method after that absorbance measured at 428 nm by spectrophotometer. Characterization of boron samples has been done by SEM, EDAX, and XRD. Concentration of boron taken at 50 mg/l to 250 mg/l. boron standard solution prepared synthetically of 500 mg/l. In the result of the boron removal efficiency reaches in optimum condition up to 94% and energy consumption varies inversely proportional to the initial boron concentration and concentration of electrolyte.

See more of this Session: Sustainable Water Use and Management
See more of this Group/Topical: Sustainable Engineering Forum

(188d) Effect of Makeup Water Properties On the Condenser Fouling In Power Plant Cooling System

Iman Safari[1], Michael E. Walker[1], Javad Abbasian[1], Hamid Arastoopour[2], Ming-Kai Hsieh[3], David A. Dzombak[3] and David C. Miller[4], (1)Chemical and Biological Engineering, Illinois Institute of Technology, Chicago, IL, (2)Department of Chemical and Biological Engineering, Wanger Institute for Sustainable Energy Research (WISER), Illinois Institute of Technology, Chicago, IL, (3)Department of Civil and Environmental Engineering, Carnegie Mellon University, Pittsburgh, PA, (4)U.S. Department Of Energy, National Energy Technology Laboratory, Morgantown, WV

The thermoelectric power industry in the U.S. uses a large amount of fresh water. As available freshwater for use in thermoelectric power production becomes increasingly limited, use of nontraditional water sources is of growing interest. Utilization of nontraditional water, in cooling systems increases the potential for mineral precipitation on heat exchanger surfaces. In that regard, predicting the accelerated rate of scaling and fouling in condenser is crucial to evaluate the condenser performance. To achieve this goal, water chemistry should be incorporated in cooling system modeling and simulation.

This paper addresses the effects of various makeup water properties on the cooling system, namely pH and aqueous speciation, both of which are important factors affecting the fouling rate in the main condenser. Detailed modeling of the volatile species desorption (i.e. CO_2 and NH_3), the formation of scale in the recirculating system , and the relationship between water quality and the corresponding fouling rates is presented.

See more of this Session: Sustainable Water Use and Management
See more of this Group/Topical: Sustainable Engineering Forum

(188e) Complete Mineralization of Sulphonated Azo Dyes Using Bacterial Consortia

Arti Tiwari and **Sanjeev Garg**, Chemical Engineering, IIT Kanpur, Kanpur, Uttar Pradesh, India

Azo dyes are widely used in a number of industries such as textile, food, cosmetics, paper, etc. About 60–70% of more than 10, 000 dyes used in the textile industry are azo dyes. Azo dyes production is around 50% of total dye manufactured annually. These compounds have an azo bond (R1–N=N–R2), where R1 and R2 are aromatic groups. During their synthesis and use, large quantities of dyes are lost in waste discharges and domestic wastewaters, thus, leading to surface and ground water pollution.

The aim of the current study was to elucidate the genetic factors responsible for azo dyes degradation and complete mineralization by microbial consortia. These consortia were isolated and characterized (TJ-1 and TJ-2) by 16 S rRNA technique. Azo dyes cannot be easily degraded under aerobic conditions. Nonetheless, it has been shown that the azo bond was reduced anaerobically, which caused color removal. Aromatic amines were obtained from the cleavage of the azo bond. These amines are carcinogenic compounds and are quite stable in anaerobic environments and further mineralization of these compounds was achieved mainly under aerobic conditions. Plasmid curing experiments were performed and it was observed that the genes encoding the enzyme for azo bond reduction were plasmid encoded while the genes encoding for enzymes necessary for mineralization of aromatic amines were chromosomally encoded.

See more of this Session: Sustainable Water Use and Management
See more of this Group/Topical: Sustainable Engineering Forum

(188f) Field Sampling of Water/Sediments for Atrazine and Amoxicillin Contaminations In the Rio Grande Basin

Manohar Palla[1], **Daniel H. Chen**[2], Jewel Andrew Gomes[2], Kevin Urbanczyk[3] and Guy Falzarano[3], (1) Dan F. Smith Dept. of Chemical Engineering, Lamar University, Beaumont, TX, (2)Dan F. Smith Department of Chemical Engineering, Lamar University, Beaumont, TX, (3)Dept. of Geology, Sul Ross State University, Alpine, TX

Atrazine ($C_8H_{14}ClN_5$) is the most extensively used Herbicide (Weed Killer for corn, sugarcane and other crops) with an annual usage > 60 million pounds and much makes its way into streams, rivers, lakes and drinking water supplies. Atrazine adsorbed in sediments is determined by hot solvent extraction followed by HPLC (Agilent 1100 HPLC with a Zerobax Eclipse C-8 Column and a variable wavelength detector at 222 nm). Field Sampling efforts included 11 expeditions and 114 sites visited in CO, NM, and TX. The project provides a valuable profile on the extent of atrazine and amoxicillin pollutions in term of time and location and identifies the environmental risks of polluted water and sediments along the Rio Grande Basin due to herbicide/antibiotic uses. We are currently analyzing both atrazine and its main degradation byproduct: hydroxyatrazine to determine whether many of the sites are in compliance with the EPA intermediate-term level of concern: 37.5 ppb over a 90-day period. Soxhlet Hot Solvent extraction was done to 6 sediment samples from the Summer 2010 expedition and the results are compared with the prediction from the Soil/Sediment Sorption Coefficient of atrazine and the soil organic carbon content. Atrazine entrained in sediment sometimes exceeds this level. Further sampling and analysis are thus warranted to ensure the water quality in the Rio-Grande Basin. We are currently analyzing amoxicillin ($C_{16}H_{19}N_3O_5S$) contaminants in the water samples collected from the field with new HPLC columns (Zorbax SB-C18 3.5 μm, 4.6 X 75 mm) and a Variable Wavelength UV detector. The Amoxicillin concentrations range from 1021- 53871ppb from 19 field water samples.

See more of this Session: Sustainable Water Use and Management
See more of this Group/Topical: Sustainable Engineering Forum

(188g) Aerobic SBR Processes In Treating PTA Manufacturing Wastewater

Zheng Yanmei[1], Yufeng Han[2], Jiale Huang[1] and Qingbiao Li[1], (1)Department of Chemical and Biochemical Engineering, College of Chemistry and Chemical Engineering, Xiamen Universiy, Xiamen, China, (2)Assets Management Co.. Ltd, Xiamen Universiy, Xiamen, China

Purified terephthalic acid(PTA) is among the top 50 chemicals manufactured in the world. As one of the most important raw material, PTA was used to make many synthetic products, especially polyester fiber. PTA manufacturing wastewater is a kind of wastewater difficult to treat due to the high level of organic contamination and the complex components in it. The aerobic sequencing batch reactor activated sludge process(SBR) was studied for the technological feasibilities in treating PTA manufacturing wastewater.

It was showed that the simulated PTA manufacturing wastewater could be rapidly treated by the aerobic SBR process, when initial TA concentration was lower than 1500mg L^{-1}, the aerobic SBR system removed more than 90% of COD after only 4 hours treatment. Aerobic SBR system displayed a higher degradation activity when aeration rate, settling time and feeding way were all in the proper conditions. COD removal efficiency could reach 96.0% after just aerating for 10 hours when initial TA concentration was 4000mg L^{-1}. However, COD in the effluent reached a limitation of 235 mg L^{-1} which was not allowed to discharge directly, because COD in the effluent had to be lower than 150 mg L^{-1} to prevent the pollution of environment. In the presence of higher initial TA concentration, COD in the effluent and surplus biomass production each cycle were both higher, indicating that the aerobic SBR process should be improved because of the high organic concentration of PTA manufacturing wastewater.

The hydrolytic-aerobic recycling SBR process was more efficiently in treating PTA manufacturing wastewater than the other intermittent biological processes, so this process may have broad applications in the future.

See more of this Session: Sustainable Water Use and Management
See more of this Group/Topical: Sustainable Engineering Forum

GASIFICATION OF ORANGE RESIDUE CHARS IN CO_2

Astrid Altamar[1], Suélen Amorim[1], Valquiria Balvedi[1], Regina F.P.M. Moreira[1] and Humberto J. José[1]

[1]Laboratory of Energy and the Environment. Federal University of Santa Catarina. Florianópolis, SC. Brazil.

Gasification of chars from orange residues and wood sawdust were studied in order to compare their reactivities using thermogravimetric analysis (TGA). Char samples were obtained by pyrolysis in a tubular reactor of quartz at a temperature of 900 °C and heating rate of 250 °C/min. The gasification of the chars using CO_2 was done on a thermogravimetric analyzer isothermally at temperatures up to 950 °C, and non-isothermally at temperatures up to 980 °C. Several morphological changes due to vaporization during pyrolysis were identified by comparing the SEM images of the chars and the corresponding biomass.

The biomass particle size was below 150 µm, and the sample size low enough to reduce heat and mass transfer side effects, thus ensuring that the reaction kinetics is the limiting step. The reactivity in CO_2 of both biomasses increased with conversion. However, differences in reactivity were observed between the two types of biomass. The char of orange residue exhibited a higher reactivity than wood sawdust. This could be due to the catalytic effect of the inorganic content, specifically potassium, which was found in the orange residue ash but not in the sawdust ash. The activation energies found for the wood sawdust char gasification were 197.4 and 187.7 kJ/mol for the isothermal and non-isothermal methods respectively, while for the orange residue char were 126.5 and 129.3 kJ/mol.

1. INTRODUCTION

Biomass waste from forest and industrial residues are an important source of energy, in particular orange residues can be considered in regions with high orange production like Brazil, where orange production was 2.02×10^7 Ton in 2010, which is 37.2% of the world production [1].

Biomass residues must be conditioned in order to be used in energy conversion devices, which require high energy content and homogeneity in their feed streams. Several techniques for biomass conditioning have been developed, being gasification one of the most promising because of its relative simplicity and high performance.

Gasification is a process where the carbon present in pyrolyzed samples reacts with a stream containing carbon dioxide, producing carbon monoxide and hydrogen. The kinetic study of the gasification is performed by thermogravimetric analysis, with the advantage of doing both isothermal and non-isothermal kinetic evaluation.

Isothermal kinetic analysis has been the traditional method used to find the reaction order and parameters such as activation energies and preexponential factor in chemical reactions. However, recently the benefits of doing this determination by non-isothermal experiments has been shown by measuring the weight change of solid samples while they react in an environment where the temperature is controlled with a dynamic gradient temperature called the heating rate (HR). [2-5].

Nomenclature			
c_{CO2}	concentration of carbon dioxide (mol/m^3)	R_i	reactivity of char (s^{-1})
Ea	activation energy (kJ/mol)	R_{50}	reactivity at 50% conversion (s^{-1})
HR	heating rate (K/s)	R	ideal gas law constant (0,008314) (kJ/(mol K))
km	reaction rate constant (m^3 kg^{-1} s^{-1})	t	time (s)
km,0	pre-exponential factor (m^3 kg^{-1} s^{-1})	T	temperature (K)
m_C	mass of carbon (kg)	T_0	initial temperature (K)
$m_{C,0}$	initial mass of carbon (kg)	TG	thermogravimetry
r	reaction rate (kg s^{-1})	X_c	Conversion of carbon
Mc	molecular weight of carbon (kg/mol)		

Results from Jess et al. [6] show that this method gives information not only about chemical reactions, but also about the mass transfer effect in a reaction by performing a single experiment.

2. THEORETICAL BACKGROUND

2.1. Isothermal Analysis

The gasification of lignocellulosic biomasses is considered according to the kinetic expression:

$$r = -\frac{dm_C}{dt} = k_m c_{CO_2} m_C Mc \tag{1}$$

Where r is the reaction rate, expressed as the weight loss in time, essentially from the carbon remaining in the pyrolyzed sample.

The reaction rate constant is expressed as:

$$k_m = k_{m,0} e^{(-E_A/RT)} \tag{2}$$

The procedure used for the data analysis is the same as the one used by other researchers [7]. Based on the homogeneus reaction model the reactivity of the char is defined as:

$$R_i = -\frac{1}{m_C}\frac{dm_C}{dt} \tag{3}$$

A representative reactivity value at 50% conversion R_{50} for each test was calculated. For isothermal analysis, R_{50} can be expressed as:

$$\ln(R_{50}) = \ln(k_{m,0} c_{CO_2}) - \frac{E_A}{RT} \tag{4}$$

Subsequently, values of preexponential factor and activation energy may be estimated from the intercept and slope, respectively, of a plot ln R_{50} versus 1/T.

2.2. Non-Isothermal Analysis

For the non-isothermal experiments the Eq. (1) can be re-arranged to a better form for the data provided by the thermogravimetric analyzer. Introducing the heating rate,

$$HR = \frac{dT}{dt} \tag{5}$$

The expression becomes:

$$-\frac{dm_C}{dT}\frac{HR}{m_C} = k_{m,0}c_{CO_2}M_C e^{(-E_A/RT)} \tag{6}$$

And in terms of the conversion X_C, calculated as

$$X_C = 1 - \frac{m_C}{m_{C,0}} \tag{7}$$

The following expression is obtained:

$$\ln\left(\frac{dX_C}{dT}\frac{HR}{(1-X_C)}\right) = \ln\left(k_{m,0}c_{CO_2}M_C\right) - \frac{E_A}{RT} \tag{8}$$

So, the plot of the term in the left side versus 1/T must be a straight line where the slope and intercept are used to calculate the activation energy and the preexponential factor respectively. The reason why it is possible to calculate these terms in a single experiment is that being the time dependent variable (HR) a constant, it is possible to directly tie the conversion to the temperature change, which leads to the direct kinetic parameters determination.

Eq. (6) can be integrated to obtain an expression for the weight of the sample as a function of temperature when a constant heating rate is applied,

$$-\int_{m_{C,0}}^{m_C}\frac{dm_C}{m_C} = ln\frac{m_{C,0}}{m_C} = \frac{k_{m,0}c_{CO_2}M_C}{HR}\int_{T_0}^{T} e^{(-E_A/RT)}dT \tag{9}$$

There is no analytical solution to the integral in the right side, but it can be solved through series expansion, according to:

$$\int_{T_0}^{T} e^{(-E_A/RT)}dT \approx \frac{RT^2}{E_A}e^{-\frac{E_A}{RT}}\left[1 + \frac{2!}{-E_A/RT} + \frac{3!}{(-E_A/RT)^2} + \cdots\right]\Bigg|_{T_0}^{T} \tag{10}$$

For activation energies greater than 100 kJ/mol, the terms corresponding to T_0 are negligible when the temperature difference $T\text{-}T_0$ is more than 200 K, and the series expansion can be reduced only to the first two terms as a good approximation, so the integral is calculated as:

$$\int_{T_0}^{T} e^{(-E_A/RT)}dT \approx \frac{RT^2}{E_A}e^{-\frac{E_A}{RT}}\left(1 - \frac{2RT}{E_A}\right) \tag{11}$$

So the expression for the sample weight leads to:

$$TG = \frac{m_C}{m_{C,0}} = \exp\left[\frac{k_{m,0} c_{CO_2} M_C}{HR} \frac{RT^2}{E_A}\left(1 - \frac{2RT}{E_A}\right) e^{-\frac{E_A}{RT}}\right] \quad (12)$$

This expression is useful to determine the accuracy of the prediction of the model when compared to the experimental data obtained from the TGA.

3. EXPERIMENTAL

Wood sawdust and orange residues from the southern Brazilian region of Santa Catarina state were used in this study. The proximate and elemental analysis and the chemical composition of their ashes are shown in Table 1. The biomass samples were pyrolyzed in a tubular reactor of quartz. Approximately 1 g of biomass was loaded into the reactor which was then the reactor was inserted into the furnace. The biomass samples were placed at the center of the furnace and heated from ambient temperature to 900 °C using a heating rate of 250 °C/min under nitrogen flow (400 ml/min). The samples were kept at 900 °C for 15 min to complete the pyrolysis. An inert atmosphere (N_2) was maintained until the reactor had cooled down to the room temperature to prevent reaction with air. The char residue was stored under inert conditions in a desiccator for later use. The diameter of the char of biomass was smaller than 150 μm. Table 2 present its chemical composition.

Table 1
Analysis biomass samples (wt%, db)

	Wood sawdust	Orange residue		Wood sawdust	Orange residue
Proximate analyses			**Ash analyses**		
Fixed Carbon	20,13	29,26	Fe_2O_3	4,44	6,17
Volatile matter	79,44	66,38	CaO	31,27	16,11
Ash	0,43	4,36	MgO	11,64	4,18
Elemental analyses			Na_2O	1,67	0,16
C	55,06	46,40	K_2O	10,44	31,48
H	7,11	10,09	SiO_2	15,69	10,17
N	0,21	1,70	Al_2O_3	12,30	3,84
O (by difference)	37,14	37,39	TiO_2	3,94	0,25
			P_2O_5	2,74	8,46
			MnO	2,06	0,07
			SO_4	3,02	2,89
			Others	0,79	14,88

The kinetic study of the gasification was done by both, isothermal and non-isothermal experiments. The isothermal results were used as a reference to validate the non-isothermal ones. The isothermal and non-isothermal gasification experiments were carried out by means of a DTG60/60H Shimadzu thermogravimetric analyser (TGA) operating at atmospheric pressure. A constant amount (3-4 mg for wood sawdust and 5-6 mg for orange residue) was placed in the alumina crucible and the temperature was measured by the thermocouple located close to the sample. Carbon dioxide was used as the gasifying agent with a flow rate of 100 ml/min. Non isothermal runs proceeded at heating rates of 10, 20, 30°C/min from 35 to 980°C. Isothermal runs, were carried out at 800, 850, 900, 950°C, a char sample was kept at gasifying temperature to ensure complete carbon conversion. All weight measurement experiments were corrected with blank runs under the same heating conditions.

Table 2
Analyses of the char

	Char wood sawdust	Char orange residue
Elemental analysis (wt%, db)		
C	75,40	67,79
H	1,77	1,82
N	0,34	1,60
O (by difference)	22,22	28,55

Scanning electron microscopy (SEM) analysis of the raw material and char particles was employed to determine the effect of the pyrolysis on the microstructure. SEM analysis was performed using an JEOL JSM-6390LV.

4. RESULTS AND DISCUSSION

4.1. SEM ANALYSIS

Samples were analyzed by SEM, as may be seen in Figures 1 and 2, where the formation of cavities along the fibers is evident. This is the direct result of the materials volatilization.

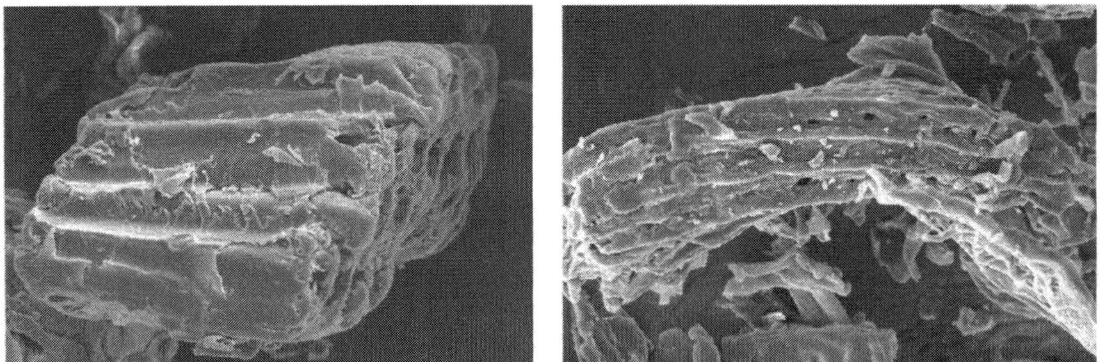

(a) (b)

Figure 1. SEM images of wood sawdust (a) raw material, (b) pyrolyzed at 900°C

(a) **(b)**

Figure 2. SEM images of orange residue (a) raw material, (b) pyrolyzed at 900°C.

4.2. KINETIC ANALYSIS

4.2.1. Isothermal Experiments

Behavior of the gasification reaction of wood sawdust and orange residue chars in CO_2 are shown in Figure 3. It can be seen that the orange residue exhibits more reactivity than the wood sawdust char.

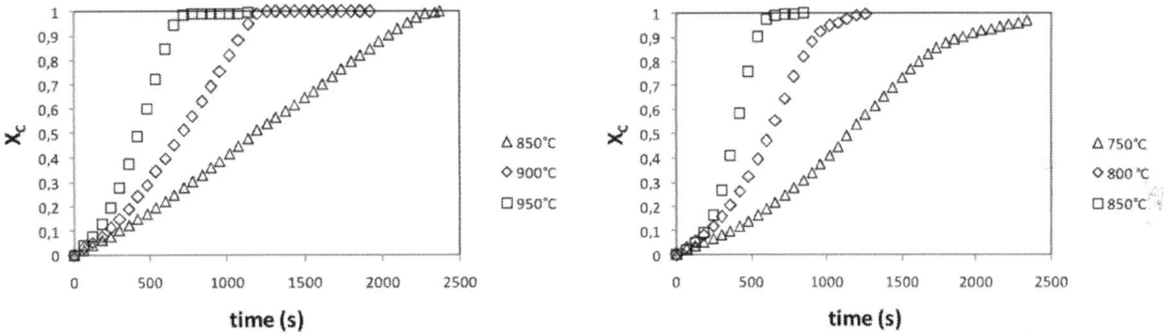

(a) **(b)**

Figure 3. Conversion of (a) wood sawdust and (b) orange residue chars in CO_2 at different temperatures.

To determine the activation energy, the methodology proposed by Ollero et al. [7], as described in section 2.1 was applied. Figure 4 shows the reactivity of wood sawdust in carbon dioxide as a function of conversion. The activation energy was determined by regression analysis and it was found to be 197.4 kJ/mol at 850-950 °C. This value is in agreement with literature. [3, 8-10].

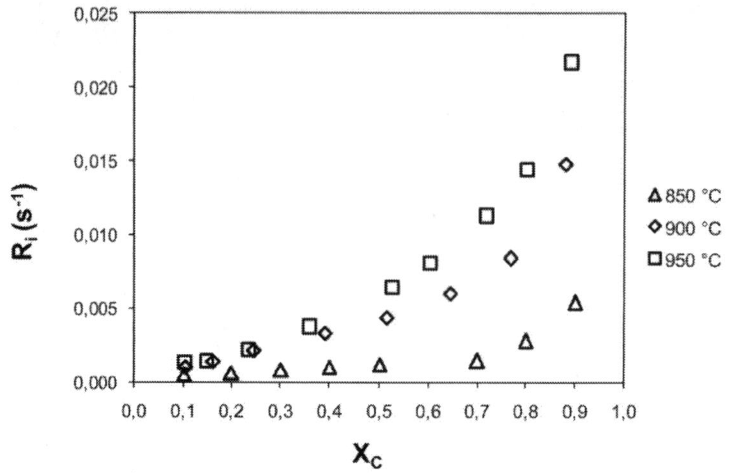

Figure 4. Reactivity as a function of conversion for wood sawdust char gasification in CO_2.

Figure 5 shows the reactivity as a function of conversion for the orange residue char. Using the same methodology as the one used with wood sawdust, the activation energy found to be 126.5 kJ/mol in the range 750-850 °C. [2, 7, 11, 12].

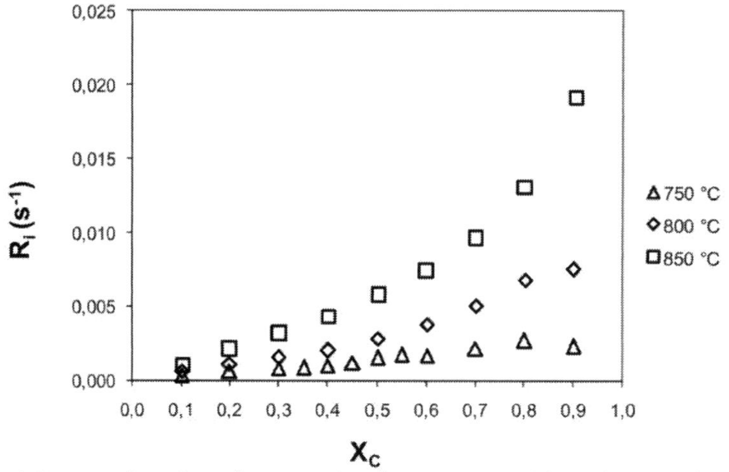

Figure 5. Reactivity as a function of conversion for orange residue char gasification in CO_2.

4.2.2. Non-Isothermal Experiments

Figure 6 shows the experimental results for the conversion of the chars in CO_2 at a heating rate of 20 K/min. It can be seen that the gasification for the orange residue char takes place at lower temperature than the one for the wood sawdust char.

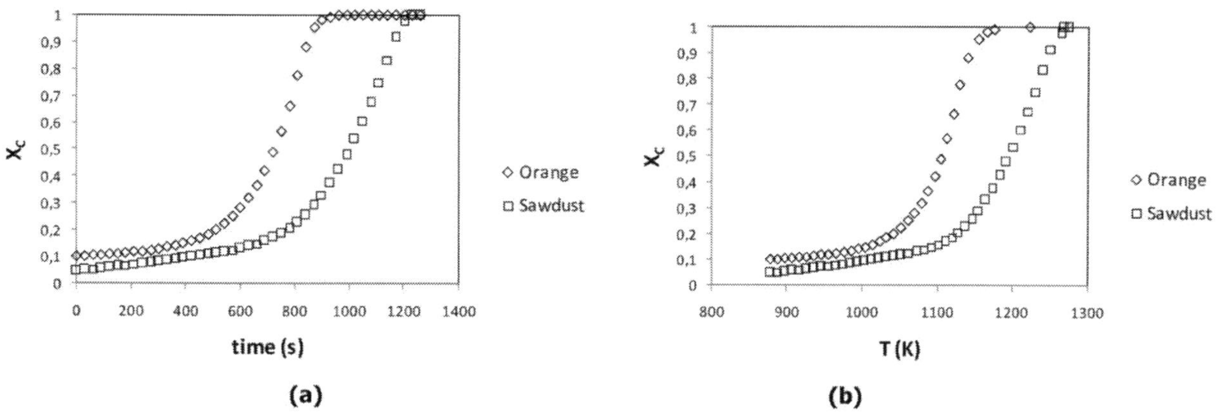

Figure 6. Conversion versus time (a) and temperature (b) of chars from orange residue and wood sawdust in CO_2 at a heating rate of 20 K/min.

As described in section 2.2, the plot of $\ln\left(\frac{dX_C}{dT}\frac{HR}{1-X_C}\right)$ vs 1/T (Figure 7), for the wood sawdust char, allows the calculation of the activation energy for the gasification. In this case the value obtained is 187.7 kJ/mol, which is close to the obtained for the isothermal method.

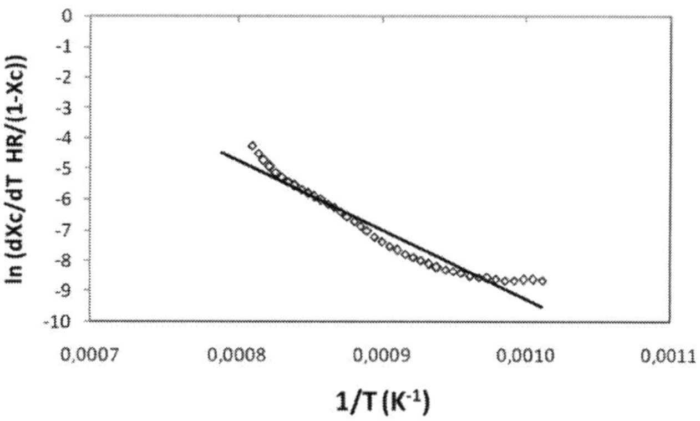

Figure 7. Plot to determine the activation energy of wood sawdust char gasification in CO_2 at heating rate of 20 K/min.

Figure 8 shows the values of the experimental and simulated thermogravimetric analysis as a function of temperature, that is, the fraction of the initial mass of the sample obtained from the TGA data, compared to the calculated values using Eq. (12). For conversions between 7.3% and 79.3%, the mean error associated in this range is 2.9%, calculated as the root mean square error (RMS).

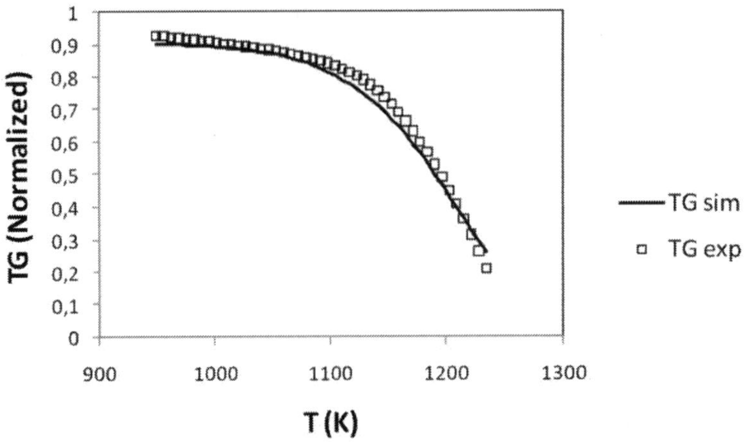

Figure 8. Comparison of experimental and simulated TG values for wood sawdust at heating rate of 20 K/min.

Figure 9 shows the plot used to estimate of the activation energy for the orange residue char, the obtained value was 129.3 kJ/mol. The tendency of the slope to the increase at higher temperatures indicates that there are several chemical reactions with different activation energies. The reactions with lower energies occur at lower temperatures, as a result, these low energy reactants disappear, leaving behind a sample with higher energy reactants, which becomes evident in the non-isothermal analysis with the increase of the slope. On the other hand, this tendency also confirms that there is no mass transfer limitation on the reactions.

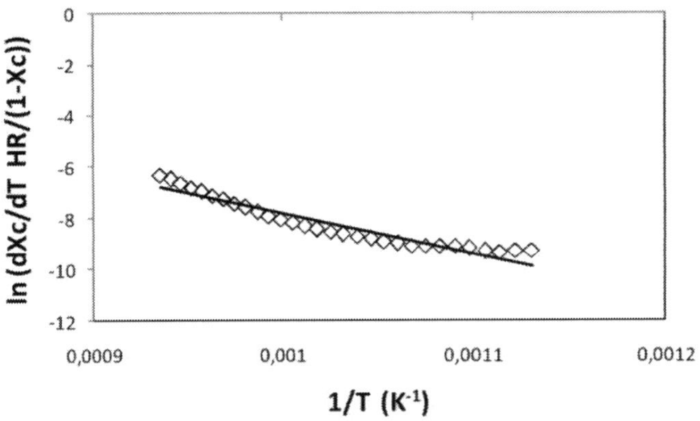

Figure 9. Plot to determine the activation energy of orange residue char gasification in CO_2 at heating rate of 20 K/min.

The comparison of the experimental and simulated values for the thermogravimetric analysis is shown in Figure 10. For the range of conversion between 10.0 and 54.0%, the associated error for the predicted values of the thermogravimetric analysis is 2.8%.

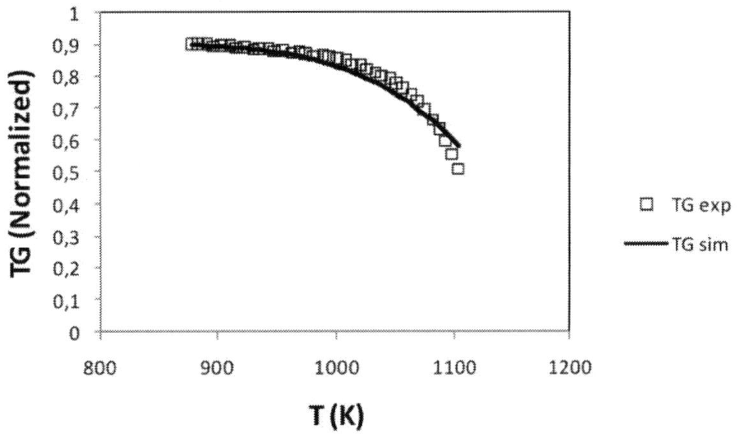

Figure 10. Comparison of experimental and simulated TG values for orange residue at heating rate of 20 K/min.

The activation energies calculated for the wood sawdust and orange residue chars are shown in Table 3. It can be seen that there is good agreement between the values obtained for the isothermal and non isothermal methods in each case, and there is consistency with literature.

Table 3
Activation energy calculated for wood sawdust char and orange residue char gasification

		Wood Sawdust Char		
Method	**Ea, kJ/mol**	**T (K)**	**X_C**	**Error (%)**
Isothermal	197.4	1123-1173		
Non Isothermal	187.7	905-1235	0.07-0.79	2.9
		Orange Residue Char		
Method	**Ea, kJ/mol**	**T (K)**	**X_C**	**Error (%)**
Isothermal	126.5	1023-1123		
Non Isothermal	129.3	880-1110	0.10-0.54	2.8

5. CONCLUSIONS

The orange residues char exhibit higher reactivity than wood sawdust due to its higher mineral content, which acts as catalyst for the gasification reactions. The use of these orange residues as biomass for gasification is an interesting alternative either directly or mixed with wood sawdust in order to improve the efficiency of the process.

The results obtained in this study show that the non-isothermal kinetic analysis is a powerful and reliable tool for the analysis of multiple chemical reactions, allowing the determination of the kinetic parameters, as well as the identification of possible interference from mass transfer diffusion.

6. ACKNOWLEDGEMENTS

The authors acknowledge CAPES and CNPq, Brazil, for financial support.

7. REFERENCES

(1) http://www.fas.usda.gov/psdonline/psdHome.aspx. Visited on 15 august 2011.

(2) Huang, Y., Yin, X., Wu C., Wang, C., Xie, J., Zhou, Z., Ma, L., Li, H. (2009), "Effects of metal catalysts on CO2 gasification reactivity of biomass char", *Biotechnology Advances*, Vol 27, pp. 568–572.

(3) Khalil, R., Várhegyi, G., Jäschke, S., Grønli, M. and Hustad, J (2009). "CO2 Gasification of Biomass Chars: A Kinetic Study", *Energy & Fuels*, Vol. 23, pp. 94-100.

(4) Sima-Ella, E., Yuan, G., Mays, T. (2005), "A simple kinetic analysis to determine the intrinsic reactivity of coal chars", *Fuel*, Vol 84, pp. 1920–1925.

(5) Vamvuka, D., Karouki, E. and Sfakiotakis, S. (2011), "Gasification of waste biomass chars by carbon dioxide via thermogravimetry. Part I: Effect of mineral matter", *Fuel*, Vol. 90. pp. 1120-1127.

(6) Jess, A., Andresen, A.-K., (2010), "Influence of mass transfer on thermogravimetric analysis of combustion and gasification reactivity of coke", *Fuel*, Vol 89, pp. 1541–1548.

(7) Ollero P, Serrera A, Arjona R, Alcantarilla S. (2003), "The CO2 gasification kinetics of olive residue", *Biomass Bioenergy*, Vol 24, pp. 151–61.

(8) Barrio, M., Hustand, J.E (2001), "CO2 gasification of birch char and the effect of CO inhibition on the calculation of chemical kinetics", in: Bridgwater AV, editor. *Progress in thermochemical biomass conversion*. Oxford: Blackwell Science Ltd., pp 47-60.

(9) Moilanen A, Saviharju K., (1997), "Gasification reactivities of biomass fuels in pressurized conditions and product gas mixtures", in: Bridgwater AV, Boocock DGB, editors. *Developments in thermochemical biomass conversion*. London: Blackie A & P, pp. 828–37.

(10) Risnes H, Sorensen LH, Hustad JE., (2001), "CO2 reactivity of chars from wheat, spruce and coal" in: Bridgwater AV, editor, *Progress in themochemical biomass conversion*. Oxford: Blackwell Science Ltd., pp. 61–72.

(11) Gómez-Barea, A., Ollero, P. and Fernández-Baco, C. (2006), "Diffusional Effects in CO_2 Gasification Experiments with Single Biomass Char Particles. 1. Experimental Investigation", *Energy & Fuels*, Vol. 20, pp. 2202-2210.

(12) Marquez-Montesinos F, Cordero T, Rodriguez-Mirasol J, Rodriguez JJ. (2002), "CO2 and steam gasification of a grapefruit skin char", *Fuel*, Vol 81, pp.423–429.

(199b) A Numerical Study On Bio-Oil Gasification Using A Multicomponent Approach

Lei Zhang and **Song-Charng Kong**, Mechanical Engineering, Iowa State University, Ames, IA

A Numerical Study on Bio-Oil Gasification Using A Multicomponent Approach

Bio-oil,also called pyrolysis oil, is a complex biorenewable fuel produced from fastpyrolysis of biomass. Since bio-oil is carbon dioxide neutral, it can be anideal replacement for petroleum fuel to be used in many kinds of combustion devices.It can also be gasified to produce synthesis gas, which in turn can be used forpower generation or synthesized to produce liquid fuels. The objective of thisstudy is to investigate the gasification/vaporization characteristics ofbio-oil drops under gasfier and combustor conditions. In this study, numerical methodsconsidering the actual composition of bio-oil are mainly used in the simulationof gasification/vaporization processes. Literature using experimentalapproaches such as Gas Chromatography/Mass Spectrometer has shown that bio-oil componentscan be categorized into several groups, including carbohydrates, furans,phenols, guaiacols, syringols, water, etc. By analyzing the composition ofbio-oil produced from different sources, a group of ten dominant components areidentified from the above groups. Using group contribution methods, thecritical and physical properties of these dominant components are estimated. Simplifyingbio-oil as a mixture of these dominant components, a multi-component dropvaporization model using a discrete component approach is developed in thesimulation of gasification/vaporization. Using the vaporization model, thevaporization histories of single bio-oil drops with different initialcompositions, which are the mass fractions of the components determined by thefeedstock and pyrolysis process, are calculated. In addition, the dynamics ofbio-oil spray is also studied numerically, and the distributions of the vaporsof various bio-oil components in the reactor are also predicted.

See more of this Session: Advances In Gasification Research
See more of this Group/Topical: Energy and Transport Processes

(199c) Realistic Impurity Generation Via Coal Gasification

Robert R. Mota, Chemical Engineering, University of North Dakota, Grand Forks, ND

Coal gasification is more efficient and environmentally friendly way to utilize coal than conventional coal combustion systems. However, there are some disadvantages to using coal as an energy source. Gasification comprises of many reactions, of which the two primary reactions; pyrolysis and steam gasification are endothermic. During the gasification process, impurities are transformed into gas, liquid and solid phases. The degrees to which these materials are transformed are dependent upon the impurity type and association in the fuel. During gas cooling the vapor phases condense and react with other impurities in the gas phase and with heat transfer surfaces such as gas cooling systems. Likewise, the liquid droplets will solidify into particles. These interactions are important in studying their effect on syngas cooling systems. Previous research[1] has shown that certain impurities have a high potential of condensing in the syn-gas cooler. The impurities found were lead, germanium, tin, antimony, bismuth, titanium, cadmium, zinc and arsenic. The focus of this research will be to find ways to mitigate the negative effects these impurities have on syn-gas coolers and hot gas cleanup by designing a small scale, fluidized bed, gasifier and characterizing the product gas through a syn-gas cooling system.

See more of this Session: Advances In Gasification Research
See more of this Group/Topical: Energy and Transport Processes

(199d) Comparison of 1D Modeling Methods for Entrained Coal Gasification

Moon-Kyeong Hwang, Mechanical Engineering, Pusan national university, Busan, South Korea

Comparison of 1D modeling methods for entrained coal gasification

Moonkyeong Hwang, Minsung Kim

School of Mechanical Engineering, Pusan National University, 30 Jangjeon-dong, Gumjeong-ku, Busan, 609-735, South Korea

Chunghwan Jeon, Youngjune Chang, Juhun Song*

Pusan Clean Center, Pusan National University, 30 Jangjeon-dong,

Gumjeong-ku, Busan, 609-735, South Korea

*Corresponding Author: jxs704@pusan.ac.kr

Telephone: +82-51-510-7330

FAX: +82-51-582-9818

Abstract

In the past, a number of one-dimensional modeling work have been made to predict the local gasification behavior of pulverized coals entrained at the normal operating conditions of tubular gasifier. In this work, three different modeling methods to compute the coal gasification at the gasifier exit in simpler manner are developed and compared. These include thermodynamic method, reaction kinetics method and well stirred reactor (WSR) method. With these models, important variables such as effect of oxygen addition and coal type on the gasification are evaluated in terms of coal conversion, the compositions and temperature of synthesis gas products at the exit.

Keywords

Coal gasification, 1D modeling, thermodynamic method, reaction kinetics method, WSR

See more of this Session: Advances In Gasification Research
See more of this Group/Topical: Energy and Transport Processes

(199e) Rigourous Modeling of An Entrained Flow Gasifier Considering a Heuristic Recirculation Model and a Detailed Energy Balance

Job Kasule[1], Richard Turton[1], Debangsu Bhattacharyya[1] and Stephen E. Zitney[2], (1)Department of Chemical Engg., West Virginia University, Morgantown, WV, (2)Collaboratory for Process and Dynamic Systems Research, National Energy Technology Laboratory, U.S. Department of Energy, Morgantown, WV

Coal-fired gasifiers are the centerpiece of integrated gasification combined cycle (IGCC) power plants. The gasifier produces synthesis gas that is cleaned and sent for combustion in a gas turbine. Several mathematical models have been developed to study the physical and chemical processes taking place inside the gasifier. Such models range from simple one-dimensional (1D) steady-state models to sophisticated dynamic 3D computational fluid dynamics (CFD) models. In this work, a 1D model of a single-stage, entrained-flow, General Electric Energy (GEE)-type gasifier has been developed.

In our model, mass, momentum and energy balance equations for solid and gas phases are considered. The model includes a number of heterogeneous and homogeneous chemical reactions along with devolatilization and drying of the slurry feed. The unreacted-core shrinking model is used for calculating the rates of the heterogeneous reactions. A detailed model of the radiative heat transfer has been developed considering interactions between all internal surfaces (side wall, top, and bottom surfaces) of the gasifier and the solids along with the interactions between the surfaces themselves. Unlike models developed in many previous studies, no arbitrary wall temperature profile is assumed but an energy balance across the wall including the heat loss from the gasifier wall to the environment is carried out. In the slurry-fed gasifiers, the recirculation near the inlet of the gasifier is promoted for rapid mixing of the slurry feed with a part of the hot reaction products. This vigorous mixing results in a significant rise in temperature that helps in evaporating the water and devolatilizing the coal. The recirculation is achieved by appropriately designing the feed burner, and feeding the oxygen through a swirling annular injector. In the current model, a heuristic recirculation model has been developed and the conservation equations have been appropriately modified.

The above formulation results in a highly nonlinear system of differential algebraic equations (DAE) that is solved using a Newton-type method in Aspen Custom Modeler® (ACM). The model is validated with the data from experimental studies and pilot plants. The paper includes species concentration profiles and temperature profiles of the gas and solid phases. A number of parametric studies are carried out to gain more insight into the gasifier performance as the inlet and operating conditions change. The results include the effects of the recirculation ratio on the initial heat up of the slurry feed as the type of coal and its quality are changed. For various inlet conditions, the model can be utilized to predict the profiles of a number of key variables which are very difficult, if not impossible, to measure in an actual gasifier.

See more of this Session: Advances In Gasification Research
See more of this Group/Topical: Energy and Transport Processes

Process Design and Validation of Operation in a Coal-to-Methanol Gasification Plant using Probabilistic Methodologies

Raul E. Ayala and Judeth B. Corry, GE Energy, 1333 W. Loop South, Houston, TX 77072

After a chemical process has been developed, designed, and launched at commercial scale, validation of the large scale results against design assumptions is essential. Typically, the engineer encounters factors that affect the quality of the data collected at a plant, both from a variability standpoint as well as from a data availability perspective, or lack thereof. This paper describes Six Sigma and probabilistic methodologies for uncertainty reduction used to ensure successful validation plans, data collection strategies, and methods for data analysis in order to validate a large-scale commercial system. The strategies will be illustrated using the GE designed and validated operation of a coal-to-methanol gasification facility in China using the aforementioned error-reducing strategies.

In a GE coal-to-methanol plant, coal is first ground as slurry and used as a feedstock to the entrained gasifier, where the slurry is converted to a gas mixture primarily of carbon monoxide, hydrogen, carbon dioxide and water to produce the raw syngas stream. After conditioning, the syngas is used as a reactant for the production of methanol in the downstream portion of the plant. A slag stream is the byproduct of the gasification process that contains the inert ash originally in the raw coal.

The GE validation process is used to ensure the test is executed in a manner to ensure high quality results. This process includes preparing for the test with reviews of the test plans, checklists, sample points, safety reviews, testing equipment required, and execution details as well as specifics required to define readiness such as "steady state" which can often be challenging in the field. During a field test, the test preparation and checklists provide the needed guidance at this time. After the test is completed, the team reviews the test execution to determine if it delivered the desired information along with the location of the data for future use. Lessons learned are also captured so that future tests benefit.

As it is challenging to simultaneously obtain discrete physical samples from all existing streams and vessels in a commercial plant, one requires a variety of statistical tools to make predictions about the non-measured point locations in various sub-systems or (black box) control volumes. This set of statistical tools makes up the overall probabilistic (or stochastic) approach to determine the overall plant performance analysis, which is the goal of the validation process.

Data from three major black box subsystems in the plant were used to carry out our probabilistic analysis: slurry preparation area, gasification island area, and slag handling area. Six Sigma techniques were used to achieve the data reconciliation across the subsystems by measuring time averages (means), local variability (standard deviations), and certainty in the instrumentation calibration and estimate of measurement drift (MSA), before doing a minimization of residuals across key parameters. The five-day validation test period was called the Baseline Test, and digital data from the distributed control system (DCS) was combined with daily chemical analysis of discrete physical samples (solid, liquid, and gases) to obtain snapshots of the overall plant operation. An in-house method of minimization of residuals in the mass and energy balances provided the means to data reconciliation across subsystem tie streams.

Once mass and heat balance reconciliation was achieved by minimizing residuals, three critical parameters were used to estimate the overall efficiency of the gasification and syngas production plant: total syngas production rate (SG flow rate), specific oxygen consumption (SOC, vol O_2/ vol ($CO+H_2$)), and specific coal consumption (SCC, mass dry coal/ vol ($CO+H_2$)). In addition, percent carbon conversion, another of the most difficult plant performance parameters to accurately quantify in a plant, was also estimated with improved precision using our probabilistic (stochastic) methodology than by standard deterministic methodologies.

Overall, results showed that our validation models enabled us to help existing customers improve the performance of their plants and enhance our designs for new projects. The expected error in the validated averages for carbon conversion was reduced over the test period because of the higher confidence in the measured and calculated values in each of the black-box subsystem control volumes. The results and recommendations provided to the customer defined a clear path to systematic improvements of plant operation in the future.

A NEW PROCESS FOR IMPROVED DUST REMOVAL BY CYCLONE

Sanaz Akhbari Far, Parrto Institute, Tehran, Iran
Sorood Zahedi, Dept. Chem. Eng. Iran Univ. of Sci. & Tech., Narmak, Tehran, Iran
Mansour Shirvani, Dept. Chem. Eng. Iran Univ. of Sci. & Tech., Narmak, Tehran, Iran
Sepideh Akhbari Far, Dept. Chem. Eng. Iran Univ. of Sci. & Tech., Narmak, Tehran, Iran

Abstract

In this paper a process is presented for improving the efficiency of cyclone de-dusting systems. The de-dusting process includes cyclones connected with a specially designed cylindrical chamber which includes a rotating tube inside it with air impinging nozzles drilled on the peripheral surface of the tube. The impingement clean gas from the drilled nozzles on the peripheries of the rotating internal tube help in pushing and throwing the particles to the high radius distances during downward flow of dust included air stream inside the chamber. Thereby, the dust particles will become accumulated near the surface of the cylindrical chamber which is collected then by cyclone(s) from a recycling stream, taken from the higher radius zone of the chamber in the lower part of the chamber. The exhaust dusty flow of the cyclone is recycled to the jet-impingement chamber for further de-dusting. Experiments conducted on a setup of the process shows that both of parameters; the length of the chamber as well as the recycle rate impart a considerable effect on the overall efficiency of the de-dusting process. However, the effect of jet-impingement is much limited and lower.

Keyword
Cyclone, Efficiency Improvement, Dust Removal Process, Jet-Impingement, Recycle Stream.

Introduction

Many researchers have contributed to the large volume of work on improving the efficiency of cyclones by introducing either improved design and operating variables or by new modification in the design of the equipment [1]. In [2, 3] an auxiliary device called post cyclone (PoC) was introduced and tested for its ability to reduce the emissions of fines from industrial cyclones. In order to increase the efficiency of conventional cyclones, a double cyclone has been proposed and tested [4]. Another recent innovation for improving the collection efficiency is the use of a centrifugal impeller inside the cyclone in place of the immersion tube for the purpose of improved repulsion of the dust escaping from the cyclone [5]. Also the idea of (PoC) had been tested as a recycle stream from the outlet stream to the feed stream of the cyclone in order to give another chance to the particles to become separated by the cyclone. In this method an especial design chamber were used for dividing the outlet dust included gas stream from the cyclone into two streams [6]. In the design of the chamber it was tried to take the benefits of increasing the concentration of particle dusts in the recycle stream compared to the cleaned out stream which are both taken from the dividing chamber. In this way significant improvements in removal efficiency of the cyclone were obtained. Since that the concept of this paper is an extension and improvement of the idea of [6], for better description of the subject the schematic diagram of the recycle cyclone in [6] is shown in Figure 1.

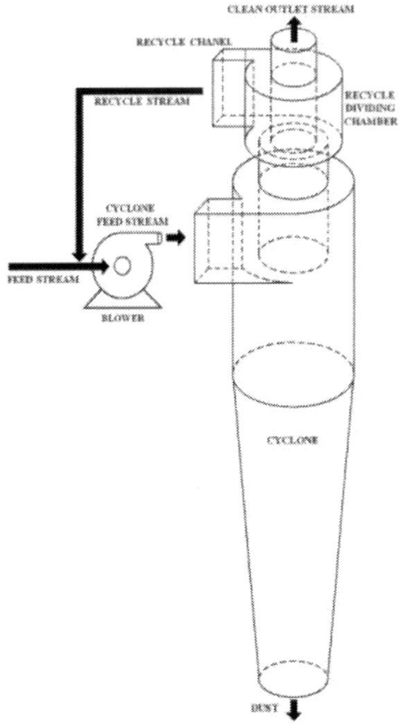

Figure 1. Recycling cyclone [6]

In this paper the impact of modification is put on the design of the dividing chamber of the idea in [6]. The dividing chamber is used for dividing a dust entrained gas stream into two streams; one with higher particle concentration and the other with lower concentration. The low concentration stream is considered as cleaned-out stream from the overall dust separating process. The other one is sent to a cyclone for dust removal and the cleaned out stream from the cyclone is recycled back and mixed with the feed to the dividing chamber. A flow diagram is shown in Figure 2 for the proposed dust separator tested in this paper. The main attempt in the design of the cylindrical chamber is to achieve to more particle concentration in the recycle flow which is taken from the chamber to be sent into the cyclone. This flow is taken from the near wall streams from an annular cross section at the bottom of the chamber. However, the cleaned out stream is taken from circular cross section at the bottom of the chamber. Increasing the concentration of Particles in the output recycle stream from the chamber is achieved by two mechanisms: one is the centrifugal force imposed on the particles due to the tangential input of feed stream to the chamber and the other is the jet-impingement radial flow which is applied by a rotating tube installed inside the dividing chamber for providing radial jet-impingement streams from its peripheral nozzles for throwing and concentrating the particles in zones near to the wall of the cylindrical chamber. This process is called here jet-impingement-cyclone-recycling dust separator. The required flow meters and blowers are also shown in this figure. The outlet stream from the impingement nozzles is entered from the lower part of the nozzle tube. The more concentrated stream flows from the annular section in the lower section of the impingement chamber and the less concentrated stream enters the internal cylinder duct. The impinging jets around the rotating inner tube which are fed by clean air should be designed such that it works more effective in throwing particles and putting them much far from the axis of the tube. The jet-impingement-recycling-cyclone dust separating process introduced in this paper can be considered to be much effective than a cyclone de-duster alone provided that it is designed in a cyclone feed input process instead of the chamber feed input design which is applied in this paper. However, the design of chamber feed input is used in this work due to the fact that the objective is to determine the performances of the chamber itself.

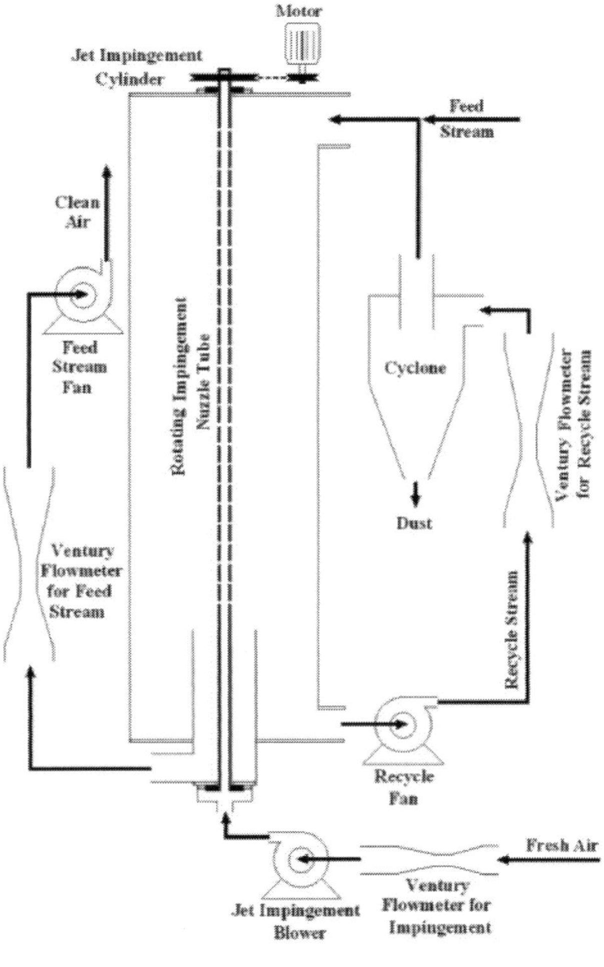

Figure 2. Jet impingement-recycling-cyclone dust separating Process with dividing chamber feed input design

Experiments

A pilot scale apparatus were designed and built for obtaining data. Two cyclones of stairmand design with cylindrical diameter equal to 15 cm was used with a jet-impingement diameter of 30 cm and height equal to 165 cm. 300 nozzles of 6 mm diameter were used around the nozzle tube. The apparatus was tested according to CCD (Central Composite Design) experimental design method. For this purpose 20 experiments for variation of three variables including: feed flow rate, recycle flow rate and jet-impingement flow rate were conducted. By CCD experimental method optimal operating conditions can also be explored from the data. The following three dimension model is used for multi regression technique for obtaining the optimum values of the three variables.

$$Y=B_0+B_1X_1+B_2X_2+B_3X_3+B_{11}X_1^2+B_{22}X_2^2+B_{33}X_3^2+B_{12}X_1X_2+B_{13}X_1X_3+B_{23}X_3 \quad (1)$$

The coefficients, B_i, must be evaluated according to the set of data. The factors X_i are the design factors that are introduced in Table 1. As it is shown, the Coded level of the design factors are chosen in the range of 1.6818 to -1.6818 which is refer to the special experimental space defined in the CCD (Central Composite Design) methodology in which it has one centre point which is introduced by 0 in the coded level and two star point at the corner which is introduced by α that can be calculated by the following relation:

$$\alpha=[2^k]^{\frac{1}{4}} \quad (2)$$

Where k is the number of variable and it is equal to three in this article. The calculation shows -1.6818 and 1.6818 for the star points for this research. Depending on the nature of experiments all of the coded values have equivalent un-coded level which are introduced the second, third and fourth rows.

Table 1. The coded and un-coded design factors for the three variables

Design Factors	1.6818	1	0.0	-1	-1.6818
X1 =Feed Ratio	0.84	0.7	0.5	0.3	0.16
X2 =Recycle Ratio	0.84	0.7	0.5	0.3	0.16
X3 =Jet-Impingement Ratio	0.88	0.7	0.44	0.178	0.0

The SQP(Sequential quadratic Programming) method is used to minimize the maximum error of model from the experiment. The results of this method and the related statistical tests revealed that the regression of the model is equal to 0.978 which shows that the model can predict the experimental results very well. The maximum values of the variables mentioned in Table 2 are as: 14, 30 and 15 cubic meters per hour, respectively.

Table 2. The results of experiments for feed from jet-impingement chamber

Number of Experiments	Feed Ratio	Recycle Ratio	Jet Ratio	Efficiency
1	-1	-1	-1	93.64
2	0	0	0	94.86
3	-1	-1	1	94.63
4	-1	1	-1	94.68
5	0	0	0	95.18
6	-1	1	1	94.99
7	1	-1	-1	95.52
8	0	0	0	94.00
9	1	1	1	90.23
10	1	1	-1	95.72
11	0	0	0	94.68
12	1	1	1	96.43
13	-1.6818	0	0	95.10
14	1.6818	0	0	94.25
15	0	-1.6818	0	96.17
16	0	1.6818	0	92.03
17	0	0	0	96.68
18	0	0	-1.6818	94.18
19	0	0	1.6818	96.39
20	0	0	0	94.69

For more precision the experiments have been done four times for each data.

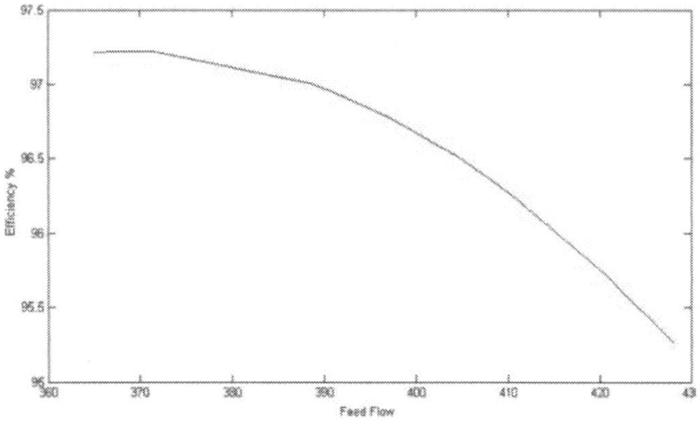

Figure 3. Effect of feed flow rate on efficiency at constant jet-impingement and recycle flow rates recycle flow rate=159.9 m^3/hr jet flow rate=76.9 m^3/hr

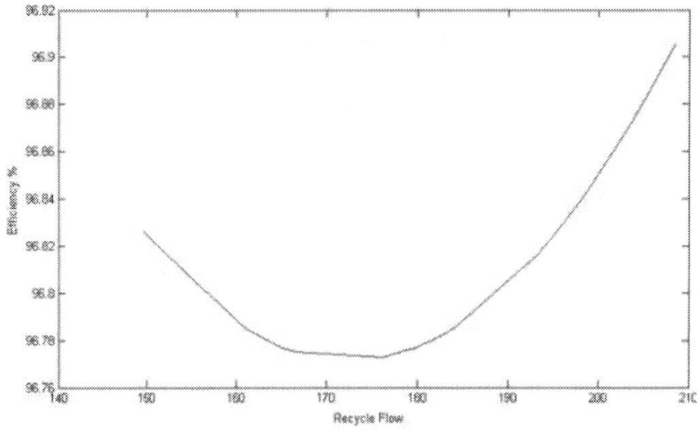

Figure 4. Effect of recycle flow rate on efficiency at constant jet-impingement and feed flow rates feed flow rate=358.0 m^3/hr jet flow rate=76.9 m^3/hr

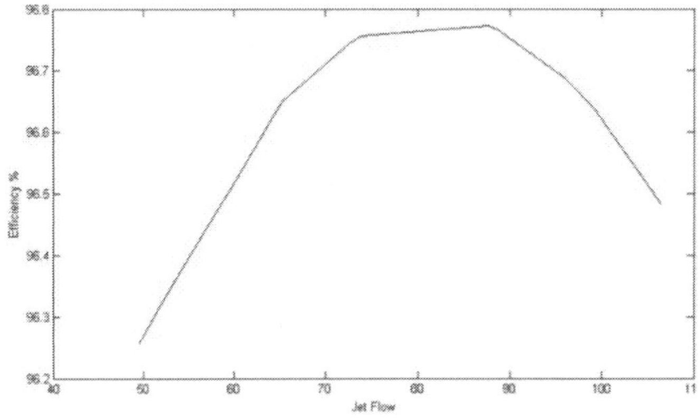

Figure 5. Effect of jet-Impingement flow rate on efficiency at constant feed and recycle flow rates recycle flow rate=159.9 m^3/hr feed flow rate=358.0 m^3/hr

Conclusion

A new technique is proposed in this paper for improvement cyclone efficiency. It makes use of jet-impingement flow effects on particle displacement in radial direction when the dusty flow is passing through a cylindrical chamber. Several experiments, based on the central composite design were designed and carried out to determine the effects of recycle flow rate, Jet Impingement and Feed flow rate on the efficiency of the system. The results of experiments reveal that increasing the Jet flow increases the efficiency to a maximum value and then decreases it. On the other hand increasing the recycle flow rate decreases the efficiency to a minimum point and then raise it gradually. Regarding to the obtained results to reach to the optimum efficiency the Jet flow ratio, recycle flow ratio and feed flow ratio should be set at 0.44, 0.5 and 0.5 respectively. The optimum efficiency at this operating condition is 96.68.

References

1. Molerus O., Gluckler M. (1996), "Powder Tech.," 86, p. 37.
2. Ray M. B., Luning P. E., Hoffmann A. (1997), "Ind. Eng. Chem. Res., " 36, p. 2766.
3. Jo Y., Tien C., Ray M. B. (2000), "Powder Tech.," 113, p. 97.
4. Lim K. S., SKim H., Lee K. W. (2004), "Aerosol Sci.," 35, p. 103.
5. Jiao J., Zheng Y., Sun G., Wang J., (Sep. 2006), "Purif. Technol.," 49, p. 157.
6. Sadighi S., Shirvani M., Esmaeli M. and Farzami R., 2006, "Chem. Eng. & Tech.," 29, p. 1242.

METAL NANOPARTICLE SYNTHESIS IN AEROSOL REACTORS

Jun Liu and Sean C. Garrick, University of Minnesota, Minneapolis, MN

Introduction

Aerosol Technology holds a lot of promise for the invention & engineering of new nanomaterials of environmental and energy applications[1-4]. Many of the envisioned products and processes of clean energy resources involve production of metal nanoparticles. One important application is motivated by the use of solar energy to produce hydrogen via hydrolysis, which is sustainable, carbon-neutral and makes use of renewable resources[5-7]. The process is one where the surface of metal particles is used to perform hydrolysis. However the particles must be very high quality and optimized quantity. The synthesis and dynamics of such small particles aren't clearly known and producing particles at those sizes is done in a somewhat ad hoc manner. This work contributes to improvements in both areas.

A practical method to produce metal nanoparticle is via the cooling of hot metal vapor in aerosol reactors[8-10]. As the vapor cools, molecules nucleates in the absence of external medium, produce nanoparticles. Homogeneous nucleation process is not entirely clear. The most common physical/mathematical description is referred to as the classical nucleation theory (CNT), which is very attractive because it is conceptually and mathematically simple[11]. Unfortunately, recent research has shown that it's not necessarily appropriate for many types of problems. One major assumption in the CNT is that the properties of the material are the same at every length-scale[12,13]. Correcting these "size dependent" properties yields very different results[14,15]. On the other hand, the interrelation of flow, thermal transport with the nucleation processes in practical complex flows has not been fully studied before[16-18]. Specifically, a more accurate model of nanoparticle nucleation is urgently needed. Nowadays, with the fast development in supercomputers and computational algorithms, the cost of numerical simulations has been greatly reduced and the accuracy has been significantly increased. With considerable knowledge accumulated via advanced diagnostics in fluid and particle dynamics, as well as the wealth of information provided by molecular dynamics, studying the interrelated and interdisciplinary processes in a more accurate manner shows great potential for fruitful future investigations.

In this work I will combine numerical methods with the physical models to probe the formation and growth of particles in shear flows. Specifically, this work will (1) examine the effects of size-dependent surface tension on the formation metal nanoparticles in laminar flows; (2) consider the effects of small-scale and large-scale mixing on nanoparticle formation; and (3) examine the interactions between the fluid, thermal and particle fields in high-speed flows.

Formulation

Flow and Scalar Transport

All simulations are obtained by solving the fluid, thermal and chemical fields as functions of time and space. The flows under consideration are governed by the Navier-Stokes equations. The primary transport variables are the velocity vector , the fluid pressure , the fluid density , and the enthalpy , and they are solved by the conservation equations as following,

$$\frac{\partial \rho}{\partial t} + \frac{\partial \rho u_j}{\partial x_j} = 0$$

$$\frac{\partial \rho u_i}{\partial t} + \frac{\partial \rho u_i u_j}{\partial x_j} = -\frac{\partial p}{\partial x_i} + \frac{\partial \tau_{ij}}{\partial x_j}$$

$$\frac{\partial \rho h}{\partial t} + \frac{\partial \rho h u_j}{\partial x_j} = \frac{\partial}{\partial x_j}\left(\frac{k}{c_p}\frac{\partial h}{\partial x_j}\right)$$

$$\frac{\partial \rho Y_i}{\partial t} + \frac{\partial \rho u_j Y_i}{\partial x_j} = \frac{\partial}{\partial x_j}(\rho D_i \frac{\partial Y_i}{\partial x_j}) + \dot{\omega}_i$$

where is the viscous stress tensor for a Newtonian fluid, k is the coefficient of thermal conduction, and is the specific heat at constant pressure. The thermal conduction coefficient and the specific heat are related to the analytical component of the mixture. The change of enthalpy is associated with temperature by The system is closed with the ideal gas equation.

Nucleation

The expression for classical nucleation rate derived by Frenkel is given by

$$J_{Fr} = \frac{(\rho Y_m)^2}{\rho_m m_m}\sqrt{\frac{2\sigma}{\pi m_m}}\exp\left(-\frac{\pi \sigma {d_p}^2}{3k_B T}\right)$$

where is the bulk density of condensing species, is the mass of a molecular and is the density of the fluid. The critical size of the nucleus is given by the Kelvin Equation:

$$d_p = \frac{4\sigma v_m}{k_B T \ln(S)}$$

where is the surface tension of the particle, is the volume of a monomer, S is the saturation ratio of the condensing vapor.

Girshick and Chiu (1990) derived the nucleation rate expression with self-consistent correction, which considered the kinetic effect and the surface energy of the monomers. The calibrated expression is given by

$$J_{ss} = \frac{1}{S}\exp\left(\frac{s_1 \sigma}{k_B T}\right) J_{Fr}$$

where s1 is the total surface of the monomers which form a critical cluster, the surface tension of cluster is still assumed a function of temperature, as consistent with the classical theory. Since the nucleation rate expressions are in the unit of number per unit volume and time. The mass conversion rate from vapor phase is given by

$$\dot{\omega}_m = \frac{\pi}{6}\rho_m {d_p}^3 J_{ss}$$

the above expression can be integrated into the conservation equations for the chemical species and particle fields.

Size-dependent surface tension

Classical nucleation theory implies that the surface tension changes with temperature. However recent developments gleaned from several experimental studies show size-dependent surface tension is necessary in better describing the formation of small nanoparticles A. A. Onischuk *it al*{Onischuk:2006ba}. derived a homogeneous nucleation rate expression for various liquid metals that account for the effect of size-dependent surface tension on the critical, or nucleating, diameter. This is an important factor as metal nanoparticles find more applications and interest in industrials.

The nucleation rate expression with size-dependent surface tension can now be derived. However, in the original article there is a mathematical omission when deriving the final result. The physical significance of the omission depends on the critical, or nucleating radius, and the metal under consideration. In the Eq. (44) of the original paper by A. A. Onischuk it al., the authors reach the expression for homogeneous nucleation by

$$J \approx \frac{n_1}{2\pi R_S}\sqrt{\frac{3\ln S}{\rho\kappa}}\sqrt{\frac{R_S}{R_e}}\exp\left(-\frac{16\pi m_m^2}{3\rho^2(\ln S)^2}\left(\frac{\sigma}{k_B T}\right)^3\right)$$

The size-dependent surface tension is discovered to be a universal behavior for various metal materials (Fig. 1)[15]. It is reasonable since for superfine nanoparticles, the molecular configuration could affect the surface free energy.

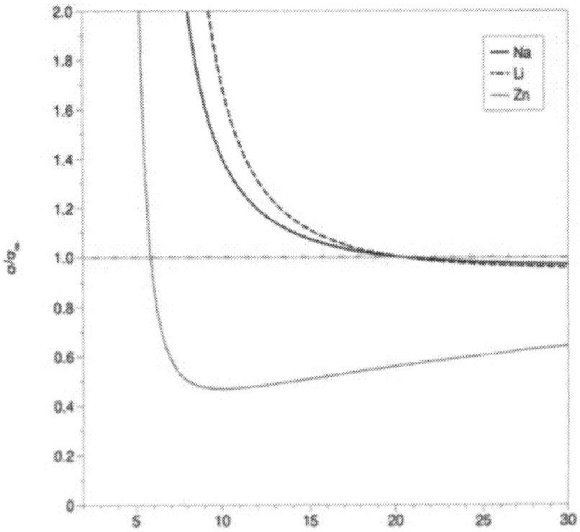

Figure 1 Size-dependent surface tension for metals

Simulation configuration

The configuration for the simulations is shown in Fig. 2. Laminar flows are simple and the steady velocity, thermal and chemical fields make them relatively straightforward to analyze. Also, laminar aerosol reactors in the shape of round jets have practical significance in controlling the physical conditions for particle synthesis. To further validate the surface-dependent surface tension model, and study the effect of fluid mixing on nanoparticle nucleation, the first step is to examine the metal nanoparticle nucleation in round jet laminar flows.

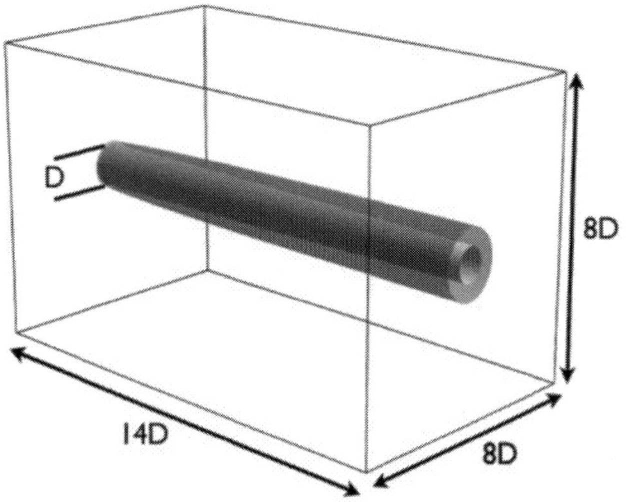

Numerical simulations of lithium nanoparticle nucleation in three-dimensional round jet laminar flows are performed. The flows consist of jet of lithium vapor (diluted in argon) issuing into a co-flowing stream of argon. The simulation parameters are presented in Table. 1.

Species	Mass fraction	T_c (K)	T_∞ (K)	Re	Pr	Sc	Le
Zn	0.02	1000	450	224	0.77	0.57	1.35
Li	0.0016	1615	726	130	0.84	0.65	1.22
Mg	0.02	1400	630	167	1.06	1.62	0.65

Table 1 Thermal, chemical and transport properties

Initially, the molecules collide with each other but hardly stick together because the energy barrier is not sufficient to balance the kinetic energy of the molecules. As the fluid travels downstream, and molecular/thermal diffusion occurs, the vapor cools and the vapor becomes super-saturated and vapor to particle conversion begins to occur. Simulations are performed for zinc, lithium and magnesium respectively under the conditions of laminar round jet flows primarily.

Selected Results

Figure 3 shows that the simulation results correlate much better correspondence with the experimental data. As compared to the experimental data, at relatively lower temperature (T<1000K), the classical nucleation theories (CNT) fails to provide the accurate relation of lithium nucleation saturation ratio S and temperature T under specific nucleation rate Js. The underestimation of saturation ratio under the same temperature and nucleation means actually much less lithium nanoparticles are formed under the same physical condition, since the requirement for super-saturation is stricter in experiments than in classical theory. At the same temperature range, the scaled nucleation theory (SNT) yields a much better correlation with the experimental data. At higher temperature (T > 1000K), the CNT results becomes closer to the experiments predicted, and the SNT has significant systematic drift from the experimental result. While the CNT and SNT both fail to provide the accurate model for metal nanoparticle nucleation, size-dependent surface tension model still shows its advantages. The colored scalar points represent samples taken from the simulation, and the trend correlate well with the experimental data. The results show that at higher temperature, the accuracy of classical theory increases since the corresponding physical assumptions in classical theories are satisfied: larger particles are formed at higher temperature so that the size-dependent surface tension is non-significant; The collision frequency is higher at higher temperature for the molecules so that the assumption of molecular "flux" onto the surface of a nuclei is more valid, and stable nuclei is formed by single molecule attaching and leaving the surface subsequently. Similarly, the SNT could predict the relation well around the chosen "critical" point, which is useful in short temperature ranges but failed to provide a more universal solution. Among them, the size-dependent surface tension theory provides a universal expression, which does not need to be changed for specific materials but the critical physical properties, and it correlates well with the experimental results.

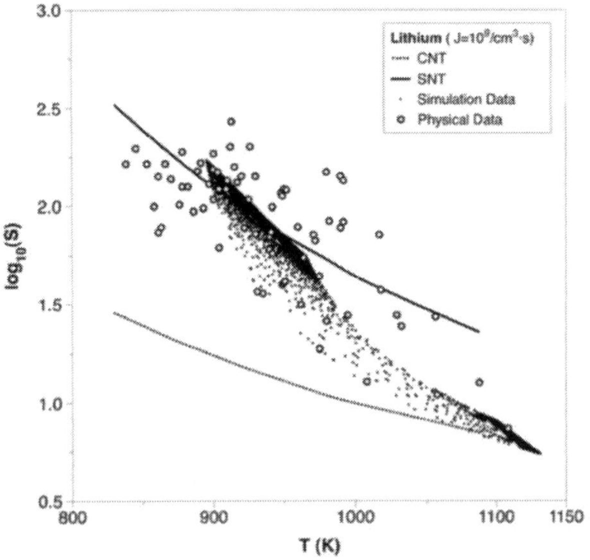

Figure 3 Comparison of simulation and physical data (Li)

Similar trends are observed in the zinc nucleation (Fig. 4) and the magnesium nucleation (Fig. 5) In Fig. 4, the CNT under estimated the zinc nanoparticle nucleation by a much higher requirement of saturation ratio (6<ln(S)<10) to result the nucleation rate Js~ $10^{10}/(cm^3s)$ at temperature T, where the physical data indicates that the same nucleation rate could occur with lower saturation ratio (2<\ln(S)<4), which is 0.24% to 1.8% of the CNT predicted. The simulation data with size dependent surface tension theory, represented by the colored scalar points, correlates very well with the physical data in the variation trend of saturation ratio S with temperature T, and the majority of samples are taken in the temperature range 580K<T<730K. By comparing Fig. 4 with lithium nucleation in Fig. 3, the difference is that the CNT underestimate the zinc nucleation rate while overestimated the lithium nucleation rate, the reason is that the CNT over predicted zinc surface and underestimated lithium surface tension, both at small particle sizes (dp~ 1nm). Figure 5 shows the comparison of simulation data, scaled nucleation theory (SNT) and the experimental data for magnesium nucleation. The CNT cannot be located within the same figure, however SNT and size dependent theory both predict satisfied relation with the experiment. Due to the stricter limitations of SNT, where a critical reference point of temperature Tcr and saturation ratio Scr need to be priory indicated, it limited the application of certain method since it is highly material dependent. Meanwhile the size dependent surface tension theory makes use of the classical expression with a few modifications, which is computational economic and efficient in applying to a broader range of materials being considered. By comparing magnesium nucleation in Fig. 5 with lithium and zinc nucleation in Fig. 3 and Fig. 4 respectively, size dependent surface tension theory shows its advantages in better describing different metal nanoparticle nucleation via the same formulation, and it is computational economic and efficient to use, easier to port for different materials. Although in both cases the SNT could give satisfied correlation given that critical reference properties are carefully chosen, the size dependent surface tension theory gives a better correlation with the experimental data, and it is much more straightforward to implement into computational fluid dynamics.

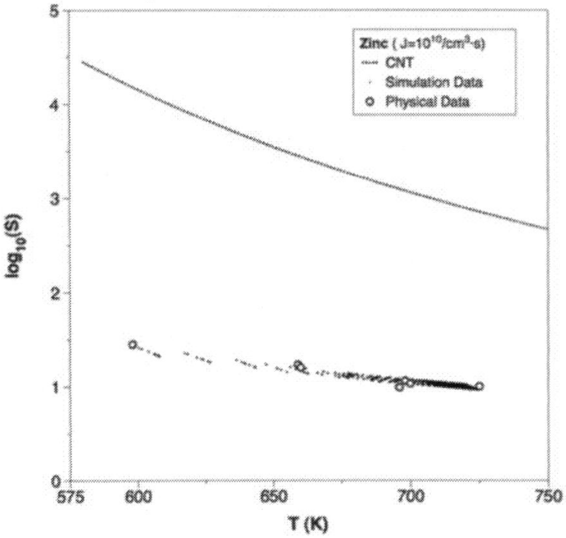

Figure 4 Comparison of simulation and physical data (Zn)

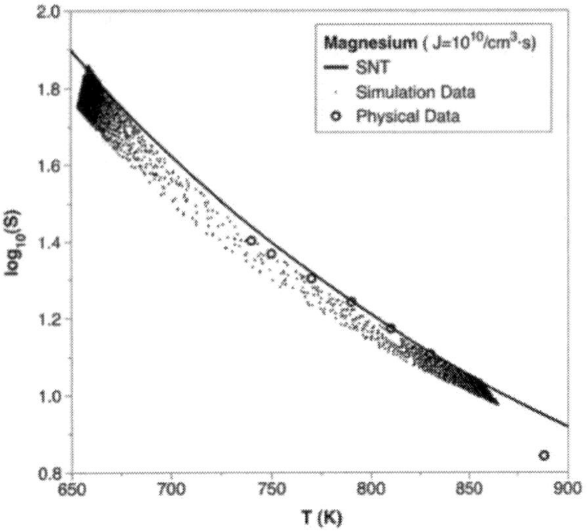

Figure 5 Comparison of simulation and physical data (Mg)

Conclusions

Direct numerical simulations of metal nanoparticle nucleation in laminar jets are performed. The flows consist of metal vapor diluted in argon issuing into a cooler argon stream. The metal vapor becomes supersaturated and metal nanoparticles are formed as a result of phase conversion process. The formation of metal nanoparticles is treated as an irreversible isothermal process and is modeled using a homogeneous nucleation theory with size-dependent surface tension. This approach promises to be more accurate than classical nucleation theory (CNT) while maintaining the ease with which CNT can be implemented in computational fluid dynamics. Simulations are performed for different types of metal materials: lithium, zinc and magnesium. The simulations elucidate the spatial and temporal evolution of the combined fluid, chemical, and particle fields in a Eulerian manner. Results are compared with current available nucleation theories and recent physical data. The effect of fluid mixing, dilution, and cooling on the resultant particles' size distribution is also accessed for the effect of fluid evolution on the resultant particles' physical properties.

The results show that in the context of model optimization, a better degree of accuracy in correlating the particles' production rate with physical conditions is established by taking size-dependent surface tension into account. While the CNT is advantageous in its simple form of expression and easy implantation into computational fluid dynamics, its accuracy relative to physical data is often unacceptable especially for small particles, due to the negligence of physical property variation at small scales. Some other nucleation theories, such as the SNT, density functional theory (DFT) and molecular dynamics techniques (MD), enjoy a better degree of accuracy, however sacrifice the computational expense and implantation complexity, or their usages are limited for only certain materials. The nucleation model with size-dependent surface tension not only attains high accuracy with respect to physical data, but also committed to compatibility, computational efficiency and interoperability by minimum amelioration based on the CNT for various materials.

In the context of fluid mixing, dilution, and cooling interactions with particle nucleation: In laminar flows, the nucleation occurs in the region of contacting layer, and is dominated by mixing caused by the molecular diffusion. The trend of decreasing temperature along the radial direction from the core of the jet leads to higher super-saturation of metal vapor, and smaller critical size. The counter effect of diluted vapor restrains its production rate. Both effects contribute to the preference on specifically sized nuclei with optimal production rate. The majority of the resultant metal particles' sizes range around or less than $1.0 nm$, which is a common phenomena for metal particle nucleation. The Lewis number Le, which is material dependent and also varies with physical conditions, is proven to be controlling the two-way balancing effects, and as a result the resultant particles' qualities such as size distribution. Practically in aerosol reactors it is easier to control the temperature and vapor concentration distributions, which act as adjustment levers to the optimizations of the resultant physical and chemical properties.

The limitation of the laminar flows is that the fluid fluctuation mixing effect is not significant, although they have significant applications in certain aerosol reactor's design. For long, DNS of particle formation processes in real flows are not feasible due to either poor model performance or high computational expense. The optimized nucleation model allows large scale DNSs of practical flow systems, which enables more complex phenomena to be analyzed under this context. For example the geometry analyzed in this work only requires roughly 200 CPU-hours. This numerical approach has proven to be promising and economic to predict metal nucleation in spatially and temporally developing flows. Moreover, the use of such model will facilitate the inclusion of more physics and chemistry and enable the simulation of complex flows.

References

1. Williams, J. et al. Nanoparticle drug delivery system for intravenous delivery of topoisomerase inhibitors. *Journal of Controlled Release* **91**, 167–172 (2003).
2. Hrkach, J.S., Peracchia, M.T., Bomb, A., Lotan, N. & Langer, R. Nanotechnology for biomaterials engineering: structural characterization of amphiphilic polymeric nanoparticles by 1H NMR spectroscopy. *Biomaterials* **18**, 27–30 (1997).
3. Jong, W.H.D. & Borm, P.J. Drug delivery and nanoparticles: applications and hazards. *International Journal of Nanomedicine* **3**, 133–149 (2008).
4. *NanoStructured Materials. NanoStructured Materials*
5. Weidenkaff, A., Reller, A.W., Wokaun, A. & STEINFELD, A. Thermogravimetric analysis of the ZnO/Zn water splitting cycle. *Thermochim Acta* **359**, 69–75 (2000).
6. Abu Hamed, T., Davidson, J.H. & Stolzenburg, M. Hydrolysis of Evaporated Zn in a Hot Wall Flow Reactor. *J. Sol. Energy Eng.* **130**, 041010 (2008).
7. Fletcher, E. & Moen, R. Hydrogen and oxygen from water. *Science* **197**, 1050–1056 (1977).
8. Wegner, K., Walker, B., Tsantilis, S. & Pratsinis, S.E. Design of metal nanoparticle synthesis by vapor flow condensation. *Chemical Engineering Science* **57**, 1753–1762 (2002).
9. Panda, S. & Pratsinis, S.E. Modeling the Synthesis of Aluminum Particles by Evaporation-Condensation in an Aerosol Flow Reactor. *NanoStructured Materials* **5**, 755–767 (1995).
10. Wang, G. & Garrick, S. Modeling and simulation of titania formation and growth in

temporal mixing layers. *J Aerosol Sci* **37**, 431–451 (2006).
11. Friedlander, S.K. *Smoke, Dust and Haze: Fundamentals of Aerosol Behavior.* (John Wiley and Sons: New York, NY, 1977).
12. Friedlander, K.S. *Smoke, dust and haze: fundamentals of aerosol dynamics. Book* (2000).
13. Tolman, R.C. The effect of droplet size on surface tension. *J Chem Phys* **17**, (1949).
14. Bartell, L.S. Tolman's [delta], surface curvature, compressibility effect and the free energy of drops. *J Phys Chem B* **105**, 11615–11618 (2001).
15. Onischuk, A., Purtov, P., Baklanov, A., Karasev, V. & Vosel, S. Evaluation of surface tension and Tolman length as a function of droplet radius from experimental nucleation rate and supersaturation ratio: Metal vapor homogeneous nucleation. *J Chem Phys* **124**, – (2006).
16. Kennedy, I.M. Flow field effects on nucleation in a reacting mixing layer. *Phys. Fluids A* **28**, 3515–3524 (1985).
17. Nguyen, H.V. et al. Homogeneous and Heterogeneous Nucleation in a Laminar Flow Aerosol Generator. *J Colloid Interf Sci* **119**, 491–504 (1987).
18. Liu, J., Murfield, N. & Garrick, S.C. The Effects of Fluid Turbulence on Nanoparticle Nucleation in Shear Flows. *Proceedings of the 2010 International Aerosol Conference* (2010).

(201c) Plasma-Assisted Aerosol Synthesis of Silicon Nanocrystals for Solar Cells

David J. Rowe and Uwe R. Kortshagen, Mechanical Engineering, University of Minnesota, Minneapolis, MN

Theoretical efficiency limits and high production costs imposed by single crystal semiconductors may ultimately restrict the use of photovoltaics for widespread energy generation. Therefore, decreasing cell fabrication cost and increasing cell efficiency should be targeted simultaneously to make photovoltaics more commercially viable. One way to achieve such reductions is to use silicon nanocrystals (SiNCs) synthesized from a nonthermal plasma process to replace one or more layers in a p-n (or p-i-n) junction solar cell. Thin films of SiNCs can be deposited on flexible or inexpensive substrates, and quantum confinement effects associated with nanoscale silicon theoretically hold the potential to surpassing the Shockley-Queisser limit for photovoltaic energy conversion. Furthermore, by depositing the SiNCs directly from the aerosol phase, extra processing steps associated with liquid-phase ink preparation and film deposition can be circumvented.

In this study, a capacitively-coupled argon-silane nonthermal rf plasma doped with phosphine is employed in a flow-through reactor to generate phosphorus-doped silicon nanocrystals. In a simple approach to deposition, the doped SiNCs are accelerated towards a substrate via a pressure drop caused by inserting a rectangular nozzle downstream of the plasma synthesis region. By moving a substrate through the sheet of particles created by the nozzle, a continuous film of doped nanocrystals is formed. To investigate the potential of this strategy to form SiNC thin films, such films were used as the n-type layer of a p-n junction solar cell. The SiNC layer is deposited onto a p-type single crystalline wafer and annealed to modify the SiNC surface states. RF sputtered ITO or doped zinc oxide from atomic layer deposition was used as a top contact, and 200 nm of evaporated aluminum is used for the back contact.

In this study we characterized the effect of in-situ and post-deposition processing conditions. We studied the effect that the surface chemistry has on the solar cell performance by varying the plasma synthesis conditions. Increased SiNC diameter is observed to significantly increase the performance of the solar cells. By adjusting the plasma input power and the precursor gas mixtures, we were able to identify the effect of nanocrystal hydrogen passivation on the solar cell properties. Furthermore, by increasing the standoff distance of the substrate from the nozzle, we can control the porosity of the SiNC films. Then, post-deposition surface modification was studied by annealing films in a variety of environments, such as forming gas, argon, nitrogen, and 10% oxygen.

This work was supported partially by the MRSEC Program of the National Science Foundation under Award Number DMR-0819885 and in part by the 3M Science and Technology Fellowship. Part of this work was carried out in the College of Science and Engineering Nanofabrication Center, University of Minnesota, which receives partial support from NSF through the NNIN program.

(201d) Fundamental Design of Microfibrous Materials As Pleated Filter Media

Yanli Chen, Guomin Xu and Bruce J. Tatarchuk, Chemical Engineering, Auburn University, Auburn, AL

Microfibrous-supported sorbent materials (MSSM) invented by the Center for Microfibrous Materials Manufacturing at Auburn University have shown applicability in removing particulate matters and harmful airborne contaminants such as volatile organic compounds (VOC's). However, these dual-functioning materials have substantially higher flow resistances than traditional filtration media due to decreased fiber diameters, tighter packing densities, and the presence of adsorbent/catalyst particles. The utilization of these filters results in a significantly higher energy consumption; thus, these materials are limited in their use as media for air filters. A pleated filter design is a common approach in the air filtration industry to increase the available media area and reduce the pressure drop. Multi-element structured arrays (MESA's) developed by our group represent a new approach that can further increase the available filtration area and reduce the pressure drop across the system by integrating multiple pleated filter elements into a single filtration unit. In this study, we experimentally investigated the effect of design parameters (filter depth, pleat number, media thickness, element count) on filtration performance, such as pressure drop, dirt loading capacity, and particle removal efficiency for single pleated filters and MESA units, respectively. In addition, semi-empirical pressure drop models for pleated filters and MESA units were applied to optimize those parameters from energy efficiency and lower life-cycle cost standpoints.

See more of this Session: Aerosol Science and Technology Enabling Environmental and Energy Studies II
See more of this Group/Topical: Environmental Division

(201e) Synthesis of TiO_2 Nanoparticles Modified by Transition Metals for Solar Fuel Production From CO_2 and Water

Ying Li and Fei Gao, University of Wisconsin-Milwaukee, Milwaukee, WI

Titanium dioxide (TiO_2) nanoparticles modified by transition metals (Ag, Cu) have been synthesized through a one-step, low-cost spray pyrolysis method in this study. A BGI Colison Nebulizer was used to atomize the precursor solution consisting of TTIP, ethanol, silver nitrate/copper nitrate, and acetic acid. The precursor aerosols were delivered to a quartz reactor inside a tubular furnace where the Ag/Cu-TiO_2 composite particles were produced and collected. The materials were characterized by XRD, UV-vis diffuse reflectance, SEM, TEM, etc. The particle size, crystal phase, optical behavior, and silver dispersion were correlated with precursor composition, furnace temperature, and residence time. The Ag/Cu-TiO_2 nanoparticles were tested as photocatalysts for two renewable fuel applications: (1) hydrogen production from photocatalytic water splitting and (2) CO_2 photoreduction with water for production of fuels (CO, methane, or methanol, etc), both under solar irradiation. A 450 W Xe lamp and a 150 W Solar Simulator with AM 1.5G filter were used as the light sources. A Gas Chromatograph (GC) equipped with TCD/FID/MS detectors was used to identify the products and measure the yields of solar fuels. The activities of catalysts with different Ag/Cu concentrations and TiO_2 particle sizes were investigated and compared with benchmark Ag/Cu-P25 catalysts prepared by incipient wet-impregnation method.

See more of this Session: Aerosol Science and Technology Enabling Environmental and Energy Studies II
See more of this Group/Topical: Environmental Division

(201f) Low Temperature Synthesis of N-TiO2 Particles by Hydrazine-Assisted Aerosol Pyrolysis of Titanium Tetraisopropoxide Nitric Solution

Wei-Ning Wang, Pornpan Namkhang, Koyar S. Rane and Pratim Biswas, Department of Energy, Environmental and Chemical Engineering, Washington University in St. Louis, St. Louis, MO

A single step, low temperature synthesis of N-TiO_2 particles by a hydrazine-assisted aerosol pyrolysis method was developed. Titanium tetraisopropoxide (TTIP) nitric solution and hydrazine hydrate were used as precursors. The size, crystallinity, bandgap, and surface area/pore size of the particle properties were tailored by manipulating the precursor concentration, N_2H_4/Ti molar ratio, and temperature, which were characterized by means of SEM, TEM, XRD, XPS, and UV-VIS. The preliminary results showed that hydrazine played a key role in incorporating nitrogen into TiO_2 lattices and promoting crystal growth by supplying additional energy due to exothermic reactions of hydrazinated precursor with oxygen. A mechanistic explanation of the process will be described.

See more of this Session: Aerosol Science and Technology Enabling Environmental and Energy Studies II
See more of this Group/Topical: Environmental Division

(201g) Modeling Air-Borne Nanoparticle Diffusion Through Filters

Tsz Yan Ling[1], Jing Wang[2] and David Y. H. Pui[1], (1)Mechanical Engineering, University of Minnesota, Minneapolis, MN, (2)Institute of Environmental Engineering, ETH Zurich, Switzerland

Nanoparticles are being used in many commercial products and this increases the chance of release of nanoparticles to the environment. Air filtration can be an effective way to remove nanoparticles from the air stream. Due to the small size and mass, diffusion is the dominant filtration mechanism for nanoparticles, and the total collection efficiency will be close to collection efficiency due to diffusion only. The objective of this study is to perform numerical simulation of air-borne nanoparticle diffusion across filters using commercial computational Fluid dynamics model (CFX, ANSYS).

The model methodology was first verified with the case of nanoparticle penetration across standard screen filter, which has a highly regular and repeating cross-shaped configuration. The modeling results agree well with the experimental findings reported in the literature. The method is then extended to real filters. The 3-dimensional geometries of filters are constructed based on electron microscopy observations. Calculations show that filters are very efficient in trapping particles if only concentration-gradient-driven diffusion (zero air flow) is considered. In the case of convective diffusion, when a finite air flow is allowed to pass through the medium, particle penetration increases. The change in penetration is a function of face velocity of the air flow. Details of the simulation results and their comparison with experiments will be discussed.

See more of this Session: Aerosol Science and Technology Enabling Environmental and Energy Studies II
See more of this Group/Topical: Environmental Division

(201h) Efficient Luminescence From Silicon Nanocrystals: Role of the Plasma Afterglow

Rebecca Anthony[1], David J. Rowe[2], Matthias Stein[3], Jihua Yang[1] and **Uwe R. Kortshagen**[4], (1) Department of Mechanical Engineering, University of Minnesota, Twin-Cities, Minneapolis, MN, (2) Mechanical Engineering, University of Minnesota, Minneapolis, MN, (3)Department of Electronic Materials and Nanostructures, University of Duisburg - Essen, (4)Department of Mechanical Engineering, University of Minnesota, Minneapolis, MN

The pursuit of efficient and sustainable lighting technologies has led to great interest in the optical characteristics of nanoscale silicon. Silicon is abundant and largely nontoxic, and silicon nanocrystals (SiNCs) exhibit efficient and tunable light emission that can be used in the fabrication of visible and IR light-emitting devices (LEDs)[1], as well as in other applications. In order to maximize the performance of SiNCs in LED structures, it is important to study and understand the parameters of SiNC synthesis and processing that lead to optimal light-emission.

For synthesis of high-quality SiNCs, the nonthermal plasma reactor provides excellent reliability and control. Following a surface functionalization step, these SiNCs have been shown to exhibit photoluminescence (PL) quantum yields (QYs) of greater than 60%[2]. The current study examines the effects of the plasma afterglow on SiNC PL efficiency. We held constant the flowrates of argon and silane (5% bal. He), the reactor pressure, and rf power. The only parameter changed was the injection of gas through a downstream sidearm of the reactor into the plasma afterglow. We chose hydrogen, helium, and argon as sidearm injection gases, keeping the partial pressure of injected gas constant.

Using FTIR analysis, we saw a change in the Si-H_x stretch between the H_2, He, and Ar samples, with decreasing contributions from Si-H_3, respectively. This indicates a change in surface based on gas injection. For PL measurements, we surface-functionalized the SiNCs in a thermal hydrosilylation reaction, exchanging some of the surface hydrogen with dodecyl ligands. This improves PL QY and allows clear colloidal suspsensions of SiNCs in nonpolar solvents to be formed. When comparing the PL QYs of these samples, we find a surprisingly strong dependence of the PL QY on the type of the injected gas. Argon injection leads to the lowest QY of only ~12%, while hydrogen leads to the highest QY with almost 55%. Helium injection yields QYs between these two extremes.

After synthesis, when a SiNC passes through the plasma afterglow, interactions with energetic gas species could cause desorption of H from the SiNC surface. One possibility is that sidearm gas has a quenching effect on this H desorption. H_2 and He have similar thermal conductivities, while Ar has a lower thermal conductivity and also easily absorbs energy through production of excited states. This hypothesis may explain why argon sidearm gas leads to the poorest PL performance.

However, hydrogen also has the special capability to further passivate the SiNC surface in the plasma effluent. To test the possibility of injected hydrogen reacting with the SiNC surface, we injected deuterium into the afterglow. Chemically, deuterium and hydrogen should be have alike—however, if the deuterium binds to the SiNC surface, we should be able to distinguish Si-H_x peaks from Si-D_x peaks in FTIR. We saw that the hydride-covered surface we see for other injection schemes is almost fully exchanged for a deuterated surface. Quantum yields for SiNCs produced with deuterium and hydrogen injection are comparable. This result demonstrates that hydrogen injection into the plasma afterglow leads to an exchange or replacement of surface species, and may create a more defect-free surface, resulting in PL enhancement. We confirmed the reduced defect density of hydrogen-injection SiNCs compared to argon-injection SiNCs using EPR analysis.

In conclusion, we have shown that the injection of gas into the afterglow of the plasma leads to measurable changes in PL performance of the nanocrystals. We hypothesize that hydrogen injection yields SiNCs with the most efficient photoluminescence by a combination of quenching of surface hydrogen desorption and additional hydrogen passivation of the SiNC surfaces. The results of this work will hopefully enable more groups to achieve high-QY SiNCs in aerosol or plasma reactions, and possibly allow better control of SiNC luminescence efficiency, leading to higher-performance light-emission technologies such as LEDs.

Acknowledgements: This work was primarily supported by NSF through the MRSEC grant DMR-0819885.

References:

1. Cheng, Anthony, Kortshagen, and Holmes. *Nano Letters* **11**(articles ASAP) 2011.

2. Jurbergs, Rogojina, Mangolini, and Kortshagen. *Applied Physics Letters,* **88** (233116) 2006.

See more of this Session: Aerosol Science and Technology Enabling Environmental and Energy Studies II
See more of this Group/Topical: Environmental Division

(213a) Competing Forces On the Extraction of Heavy Metals In Soil Remediation

Cameron A. Grey, Mario Oyanader and Pedro E. Arce, Chemical Engineering, Tennessee Technological University, Cookeville, TN

Heavy metal contaminants found in soil locations pose a tremendous problem to their removal and proper disposal. As to their removal, the extraction of these metals from the ground has been tested theoretically with the application of an electrical field in a soil matrix. This method has been reported to extract specific contaminants with little to no excavation through low permeability soils. In the presence of low permeability soils, these electric fields also theoretical prove to be very promising; however, there is still the need for furthering this technology in modeling as well as in its physical application. A major key in determining the driving forces of this system are the different components of the force (electroosmosis, electromigration, buoyancy, and advection) and how they contribute to the bulk flow of the fluid in the ground. Optimizing this force is one such discovery that will be unveiled as well as the different geometries and the analysis of their benefits. The Debye-Hückel approximation will be employed in order to help solve the differential equations of the electrostatic potential. As a result of this research, these contaminant sites will be readily cleaned with an efficient technology that is easily applied to the area; as well as further the advancement of the use of electric fields in cleaning applications.

See more of this Session: Contaminant Transport and Site Remediation
See more of this Group/Topical: Environmental Division

(213b) Behavior of Mercury During Chemical Processing In the Defense Waste Processing Facility

John R. Zamecnik, David C Koopman and Michael E Stone, Savannah River National Laboratory, Aiken, SC

The Defense Waste Processing Facility (DWPF) at the Savannah River Site vitrifies High Level Waste (HLW) for disposal. DWPF has been processing radioactive waste since March 1996 to produce over 3,000 canisters filled with HLW glass. Sludge waste consists of insoluble metal hydroxides (primarily iron, aluminum, calcium, magnesium, manganese, and uranium) and soluble sodium salts (carbonate, hydroxide, nitrite, nitrate, and sulfate).

The pretreatment process in the Chemical Processing Cell (CPC) Sludge Receipt and Adjustment Tank (SRAT) adds nitric and formic acids to the sludge to lower pH, destroy nitrite and carbonate, and chemically reduce mercury and manganese. Acidification of the sludge also allows higher solids concentrations to be processed by lowering the yield stress of the sludge. During these processes, hydrogen can be produced by noble metal catalyzed decomposition of excess formic acid. The ratio of nitric to formic acid is adjusted to target a glass pool with 20% of the iron having a valence of Fe^{2+}.

The feed to the SRAT contains mercury as mercuric oxide (HgO). The reduction by formic acid converts the HgO to elemental mercury which is subsequently stripped off by boiling. The maximum concentration of mercury in allowed in the product is about 0.8 wt% of the total solids to reduce deposition of mercury compounds in the melter offgas system. The theoretical stripping ratio, based on vapor pressures, is 1g mercury evaporated per 250g water evaporated; early studies showed that the actual ratio was about 1:750. Recent experience in DWPF has shown that the ratio is often much greater than 1:750 resulting in extremely long stripping times. Also, recovery of mercury in the offgas system in a decanter has been poor.

Laboratory studies were performed to improve the understanding of the mechanisms of mercury reduction, stripping, and recovery. The effects of other redox species, chlorides, mercury drop size, and slurry concentration on reduction and stripping were examined. Recovery of mercury was studied to determine the effect of non-condensable gases, and other species condensed with mercury that may chemically react or inhibit condensation and coalescence.

See more of this Session: Contaminant Transport and Site Remediation
See more of this Group/Topical: Environmental Division

(213c) Attenuation of Tetrachloroethylene In Groundwater by Bioremediation

Sathishkumar Santharam[1], Larry Davis[2], Larry Erickson[1] and **Jwan Ibbini**[2], (1)Chemical Engineering, Kansas State University, Manhattan, KS, (2)Biochemistry, Kansas State University, Manhattan, KS

A pilot scale feasibility study was conducted for attenuation of tetrachloroethylene (PCE) contaminated site. PCE concentration in groundwater at the pilot study area was about 15 mg/L (ppm) in the deep zone of the aquifer. Nutrient solutions prepared with different combinations of soy oil methyl esters, cheese whey, yeast extract, lactate and glucose were injected for biostimulation and preconditioning of ground water prior to KB-1 (a consortium of *Dehalococcoides*) injection. Nutrient solution was injected four times during the first year of the pilot study to supplement microbial growth. During biostimulation, no appreciable degradation beyond dichloroethylene (DCE) was evident. The addition of KB-1, during second nutrient injection, reduced PCE and DCE concentrations in the pilot study area to less than 10 percent of the original levels. Total chlorinated ethene concentrations did not reach the original levels two years after the last nutrient addition. The combination of substrates injected provided both short term and long term supply of nutrients for the microbes. Tracer study results demonstrated that microbial growth decreased ground water velocity during the study, but returned to normal conditions one year after the last nutrient addition.

See more of this Session: Contaminant Transport and Site Remediation
See more of this Group/Topical: Environmental Division

(213d) Isolation and Characterization of a TNT-Degrading Bacterium

Liangdong Zou, Diannan Lu and Zheng Liu, Department of chemical engineering, Tsinghua University, Beijing, China

2,4,6-Trinitrotolene(TNT) is one extensively used explosive for various purposes and is harmful for soil and water. In the present study, four strains named as Thu-A, Thu-B, Thu-C and Thu-Z, were isolated from contaminated sites. Thu-Z had the highest degradation efficiency compare with other strains. It was identified by 16S rDNA assay that Thu-Z was a ***Pantone*** spp. The similarities in both biochemical properties and morphology to the representatives of ***Pantoea agglomerans*** suggested Thu-Z was ***Pantoea agglomerans*** strain. Thu-Z could use TNT as sole nitrogen source while cleavage NO_2 from nitroaromatic ring via direct aromatic ring reduction. Under nitrogen-limited condition, 96.6%N of TNT was used by Thu-Z for growth and with $NaNO_2$ as sole nitrogen source were done as compare. Also trace nitro reduction metabolites were identified. This meant that a dual pathway was found in Thu-Z, and direct aromatic ring reduction was predominant. Addition of other nitrogen source ($(NH_4)_2SO_4$) stimulated the growth of Thu-Z and accelerated TNT degradation.

See more of this Session: Contaminant Transport and Site Remediation
See more of this Group/Topical: Environmental Division

(213e) Chemical Release During Dredging of Contaminated Sediment

R. Ravikrishna, Prathamesh Ekawde and Kanithi Mahesh, Chemical Engineering, Indian Institute of Technology-Madras, Chennai, India

One of the options forthe remediation of contaminated sediments in rivers, lakes or coastal bay areasis dredging and subsequent relocation of the dredged material for storageand/or disposal. Dredging induces resuspension of bedsediment, the extent of which depends on the type of dredging as well as theefficiency of execution. One of the concerns from a risk assessment perspectiveis the release of chemicals into water and possibly to air as a result of desorption of the hazardous chemicals from the contaminatedsolids, which are resuspended [1]. From a riskassessment perspective, it is necessary to evaluate the chemical release duringdredging since this affects water quality and potentially the air quality aswell.

The primary point ofconcern is the rate of chemical release from the resuspendedsolids during dredging by desorption to the waterlayer, where it can become 'bioavailable' orevaporate. Traditionally, the prediction of the magnitude of chemical releaseis based on mass balance measurements of suspended solids concentrations andthe bed sediment chemical concentration assuming .uniform chemical loading inthe undisturbed bed sediment with respect to particle sizes. Previous experiments in laboratorysimulations suggested that the chemical concentration associated with thesolids may be particle size dependent. Since at different times, the suspendedsolids concentration and particle size distribution vary, a model that predictsthe aqueous phase concentration must incorporate the dynamic behaviour of the solids. Correspondingly, the measurementof chemical concentration as a function of particle size must be made for themodel to be useful.

The model to describethe unsteady state change in aqueous phase chemical concentration in the absenceof evaporation, assuming the aqueous phase to be well mixed is shown inequations 1a and 1b(1) - 1b(n). Equation 1a represents the change in aqueousphase concentration, which is the sum of contributions of desorptionof chemical from each discrete particle size. Equations 1b(1-n) representchemical balance on each of the 'n' particle fractions of size dp_i .The desorption kinetics are dependent on thesolid-water mass transfer coefficient, k^2_{A3}. The driving force is the concentrationdifference between the aqueous concentration and the concentration inequilibrium with the chemical present on the solids, where K_{A32} isthe solid-water partition constant. The second term in equation 1a representsthe rate of evaporation.

$$\frac{d}{dt}(\rho_{A2}) = \sum_{i=1}^{n} \frac{6\, k^2_{A3}\rho_{33,i}}{\rho_p d_{p,i}} \cdot \left[\frac{(\rho_{33,i} W_{3,1})}{\rho_{33,i} K_{A32}} - \rho_{A2}\right] - K_{A}\left(\rho_{A2} - \rho^{*}_{A2}\right) \quad (1a)$$

$$\frac{d}{dt}(\rho_{ss}W_s)_i = -\frac{6\,k_{B}^{2}\rho_{ss,i}}{\rho_p d_{p,i}} \cdot \left[\frac{(\rho_{ss,i}W_{s,i})}{\rho_{ss,i}K_{B}} - \rho_{B}\right] \quad \text{(1b-1to 1b-n)}$$

The quantity $\rho_{ss,i}$ is the suspended solidconcentration that is dependent on the particle size and changes with time. Inthe model, it is assumed that the suspended solids concentration changes as aresult of the settling. The equationswere solved using Matlab along with the estimate of $\rho_{ss,i}$ at each time step. Thismodel is based on an earlier version of the model which did not consider thetime-dependency of suspended solids concentrations or the particle sizedependent loading [2].

One of the primary inputs to the model described above isthe particle size dependent distribution of contaminant concentration (loading)in the solids. An experimental technique based on the separation of differentfractions based on settling is also described here. Tests were conducted toevaluate the efficiency of separation and the different fractions. In thesetests, the solids, chemical and organic carbon balance was measured to validatethe separation process. The separation of solids using the settling method isreasonably effective with a solids balance that is satisfactory. Relativeloading of different size fractions were collected and were different. Themodel was tested for conditions and the performance was evaluated withreference to the previous model as well as experimental data obtained forchemical desorption.

See more of this Session: Contaminant Transport and Site Remediation
See more of this Group/Topical: Environmental Division

(213f) BP Oil Ganglia Transport In Transparent Porous Media At Low Capillary Number

Peixi Zhu and Kyriakos Papadopoulos, Chemical and Biomolecular Engineering Department, Tulane University, New Orleans, LA

Usingvideo-capillary-microscopy, microscopic packed beds were constructed insidecapillaries, using cryolite as the porousmedium. When such packed bed was filled with an aqueous surfactantsolution, the cryolite was renderedtransparent. Oil droplets were injected inside or next to theporous media and their flow/mobilization was monitored. Pressure-drivenflow was induced at varying pressure gradients and at different concentrationsof key surfactants. At low capillary number, e.g., superficial flowrate on the order of 10^{-2}ml/min, the motion of oil phase is mainlydriven by capillary force. The change of the oil phase from ganglia tosmall droplets of a size comparable to that of the pores,was captured in film. Subsequent analysis of the multiphase flow revealedthe evolution of the flow patterns along the axis of the capillary. Thetalk will show videos of the oil mobilization phenomena and will present theconclusions of this on-going research to date.

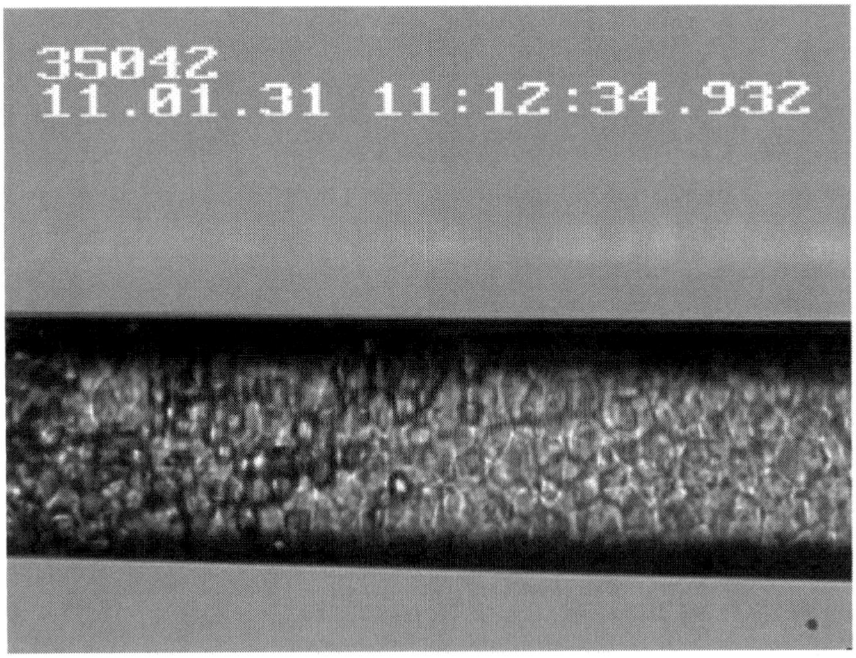

See more of this Session: Contaminant Transport and Site Remediation
See more of this Group/Topical: Environmental Division

(217a) Assessment of Potential Exposure to Carbon Nanotubes In the Production of Polymer Nanocomposites

Drew J. Thompson[1], Lin Li[1], Jing Wang[2] and David Y.H. Pui[1], (1)Mechanical Engineering, University of Minnesota, Minneapolis, MN, (2)Institute of Environmental Engineering, ETH Zurich, Switzerland

Recent animal toxicology studies have shown that carbon nanotubes (CNTs) may pose a significant health risk to those exposed to this nanomaterial in the workplace. However, many questions regarding the toxicology of CNTs still remain unanswered. There has yet to be agreement on which dose metric would not only correlate with the negative health effects but also be able to be successfully implemented in field measurements. The roles of the inclusion of metal catalyst particles in and the morphology of CNTs and/or CNT structures in the toxicology of CNTs are still being investigated. While these matters are being researched further by toxicologists there remains a dearth of information on the actual exposure to CNTs. Even if CNTs are found to be hazardous, there will be no risk without exposure. To fully understand this potential risk effort must be taken to measure the occupational exposure to CNTs.

In this study the potential occupational exposure to CNTs in the production of polymer nanocomposites, where CNTs were used as a filler material, was assessed. The potential exposure to CNTs was quantified by analyzing respirable dust samples collected by personal sampling for elemental carbon. The mass concentration of elemental carbon, of which CNTs mainly consist, was determined by thermal-optical analysis of the sample. Potential exposure sources were identified and characterized by both real-time measurements and area sampling. The potential sources investigated were comprised of both production equipment and production related tasks, which included the handling of unbound CNTs. The real-time measurements were performed by instrumentation which provided number concentration, surface area concentration, respirable mass concentration, and size distributions for micron-sized particles. Area samples were analyzed by transmission electron microscopy to characterize the morphology and to calculate the number concentration of CNT structures. Other samples were taken for analysis by inductively coupled plasma mass spectrometry to measure the concentration of trace metals. The concentration of these metals was then used to estimate the CNT mass concentration based on known impurities. Thermal-optical analysis was performed on additional area samples to measure the elemental carbon mass concentration.

See more of this Session: Environmental Health and Safety (EHS) Concerns of Nanomaterials
See more of this Group/Topical: Topical C: Environmental Aspects, Applications, and Implications of Nanomaterials and Nanotechnology

(217b) A Novel Technique for In-Vivo Toxicological Characterization of Engineered Nanomaterials

Georgios A. Sotiriou[1], Edgar Diaz[2], Mark Long[2], John Godleski[2], Joseph Brain[2], Sotiris E. Pratsinis[1] and Philip Demokritou[2], (1)Department of Mechanical and Process Engineering, Particle Technology Laboratory, ETH Zurich, Zurich, Switzerland, (2)Department of Environmental Health, Harvard School of Public Health, Boston, MA

How do we balance the potential hazards from new and inadequately characterized engineered nanomaterials with the potential and power of nanotechnology? It has become apparent, that potential environmental health effects are critical factors in the likely success or failure of new nanotechnologies. The current nanotoxicology approach of evaluating the toxic properties of the large universe of existing and emerging ENMs at one material at the time, places a significant financial burden to an emerging industry and may slow down significantly the development of this emerging industry. Alternatively, a promising approach will be to examine families of engineered and rigorously characterized particles, and to study the role of such factors as particle size, composition, shape, charge and surface chemistry and their link to specific biological outcomes.

Here, a novel method is explored which is suitable for ENM *in-vivo* inhalation and *in-vitro* toxicological characterization studies. ENMs are produced continuously in the gas phase using industry relevant flame spray pyrolysis reactors, allowing their continuous transfer to inhalation chambers, without altering their state of agglomeration. Important properties of the generated aerosols (i.e. particle size, concentration, shape, state of agglomeration, surface chemistry) can be modified allowing for both *in-vitro* and *in-vivo* investigations of toxicity. The ability of the developed technique to generate a variety of industry relevant, property controlled exposure atmospheres for inhalation studies was systematically investigated. The suitability of the technique to characterize the toxicity of inhaled ENMs in intact animal models was also shown in an *in-vivo* study involving Sprague-Dawley rats, using freshly generated nano-iron oxide (Fe_2O_3) as a test aerosol. We demonstrated both pulmonary and systemic toxicity using a variety of toxicological assays under acute exposure conditions. The future use of this novel method will improve our understanding in terms of the health and safety of ENMs and help us to investigate further our central hypothesis that physical and chemical characteristics of ENMs determine their bioavailability, redistribution, and toxicity in the lungs.

See more of this Session: Environmental Health and Safety (EHS) Concerns of Nanomaterials
See more of this Group/Topical: Topical C: Environmental Aspects, Applications, and Implications of Nanomaterials and Nanotechnology

(217c) Evaluation of the Toxicity of Nanomaterials Based On Knowledge Extraction From High Throughput Screening of Biological Toxicity Data

Rong Liu[1], Saji George[1], Bryan France[1], Robert Damoiseaux[1], Robert Rallo[2], Kenneth Bradley[1], Andre E. Nel[3] and Yoram Cohen[4], (1)California NanoSystems Institute, Center for the Environmental Implications of Nanotechnology (CEIN), Los Angeles, CA, (2)Enginyeria Informatica i Matematiques, Universitat Rovira Virgili, Tarragona, Spain, (3)Division of NanoMedicine, UCLA, Los Angeles, (4) Chemical & Biomolecular Engineering, University of California, Los Angeles, Los Angeles, CA

Nano-sized materials are increasingly utilized in many industrial products and processes, with over 1000 commercial products utilizing engineered nanomaterials (eNMs), primarily due to their unique nano-scale properties. Despite the many beneficial uses of nanomaterials, there is a growing public concern regarding the potential release of eNMs to the environment and the potential adverse impacts of exposures to eNMs. Although there are mounting studies on the toxicity of eNMs, understanding of the general principles governing eNMs toxicity potential and the long-term environmental health and safety associated with eNM containing products is in its infancy. In this regard, high throughput toxicity screening is critical for characterization of the potential hazard of eNMs, which is in turn indispensable information for subsequent risk assessment and development of environmental and health regulatory policies. In the present work an approach to knowledge extraction from high throughput screening (HTS) of nanoparticle toxicity is presented, based on RAW and BEAS-2B mammalian cell lines, focusing on identification of toxicity outcomes (i.e., "hit identification"), development of predictive nano-quantitative-structure-activity relations (nano-SARs), identification of pathway linkages and correlation of cell signaling pathways and cytotoxicity. Based on newly developed feature selection approach, a classification based nano-SAR was developed for cytotoxicity of metal and metal oxide nanoparticles. The approach was demonstrated for a library of metal and metal oxide nanoparticles to develop a classification based nano-SAR that can be used to identify decision boundaries with respect to hazard ranking of nanoparticles. In addition, self-organizing maps (SOM) analysis was performed to identify clusters corresponding to sub-lethal pro-inflammatory responses associated with ROS generation, lethal genotoxic responses, and cytotoxicity. Finally, complex network theory methods were applied to identify relationships between cell responses (pathway activation and cytotoxicity measures) as well as physicochemical properties of nanoparticles. For example, the analysis based on the above cell lines revealed a hierarchical activity-activity pattern where sub-lethal effects such as ROS generation and the intracellular Ca^{2+} flux are strongly related to lethal effects (e.g. cell membrane damage and cell death) and that eNM primary size, aggregation and its dissolution tendency were critical parameters affecting the observed toxicity behavior.

(217d) Human Blood Biocompatibility of Silver Nanoparticles

Jonghoon Choi[1], Vytas Reipa[2], Victoria M. Hitchins[3], Peter L. Goering[3] and Richard A. Malinauskas[3], (1)Department of Chemical Engineering, The David H. Koch Institute for Integrative Cancer Research, Massachusetts Institute of Technology, Cambridge, MA, (2)Biochemical Science, National Institute of Standards and Technology, Gaithersburg, MD, (3)Center for Devices and Radiological Health, Food and Drug Administration, Silver Spring, MD

Silver nanoparticles are increasingly being used in medical devices for their antimicrobial properties. Present toxicological techniques may not be sufficient to fully determine their safety for use in medical devices. In this talk, *in vitro* assessment of the hemolytic properties of silver particles in human blood will be discussed. In addition to the observed hemolysis, the possible mechanisms of red blood cell damage were explored in regard to the physical and chemical properties of silver particles. Four silver particle (two nano-sized, two micro-sized) preparations were dispersed in water and characterized using multiple analytical methods. Particle size and agglomeration were dependent on the suspension media. Under similar conditions to the hemolysis assay, i.e., with particles added to phosphate buffered saline and plasma, the size of the nanoparticles increased compared to particles suspended in water. Nano-sized particles exhibited higher hemolysis than micro-sized particles at mass concentrations greater than 220 ug/ml. While significant particle size changes occurred due to interaction with media components, the higher level of hemolysis observed with nanoparticles compared to micro-sized particles may be related to greater surface area, increased silver ion release, and direct nanoparticle interactions with red blood cells. These *in vitro* hemolysis results suggest that blood-contacting medical devices containing silver nanoparticles should be carefully evaluated for their hemolytic potential.

See more of this Session: Environmental Health and Safety (EHS) Concerns of Nanomaterials
See more of this Group/Topical: Topical C: Environmental Aspects, Applications, and Implications of Nanomaterials and Nanotechnology

(217e) Nanotechnology - What's So Big about the Small Stuff?

William C. Looney, Environment, AECOM, Milwaukee, WI

Nanotechnology - What's So Big about the Small Stuff?

Nanotechnology is "the Next Big Thing." It is a truly internationalphenomenon that "will have a major impact on the health, wealth and security ofthe world's people that will be at least as significant in this century asantibiotics, the integrated circuit, and manmade polymers." [1]

The market for nanotechnology-based products is expected to reach $2.5trillion by 2015, up from $147 billion in 2007.[2]

Nanotechnologypresents the promise of a diverse array of manufactured goods and products thatincorporate improved and innovative properties. Studies have determined that, due to their unique properties, nanomaterials (NM) have unpredictableproperties in biological systems and that the health implications from NMexposures are not yet known. Although numerous productsincorporating nanotechnology are here today, specific regulatory drivers for NM are still emerging. Companies producing or using NM must addressthe potential hazards of NM in the workplace to comply with existing workplace standards under the Occupational Healthand Safety Act. For example, the GeneralDuty Clause requires that employers provide each employee a place of employmentthat is free from recognized hazards that are causing or are likely to causedeath or serious physical harm and the Hazard Communication Standard (29 CFR1900.1200) provides workers with the right to know the hazards that they areexposed to in the workplace. Uncertaintyabout adverse health effects from exposures to NM and which work practices toadopt today should not impede efforts by companies to develop worker safetyprograms that minimize potential health effects. Such an approach can reduce the likelihood ofworker exposures to NM and the financial costs of subsequent liability claimsand litigation.

This presentation will provide an overview ofnanotechnology basics, an inventory of potentially applicable regulations andguidelines and the benefits of a proactive approach.

[1]National Science and Technology CouncilCommittee on Technology

[2]*Lux Populi Newsletter*, February 22, 2010, LuxResearch

Contact Information:

William C. Looney

Senior Program Manager

Director AECOM Nanotechnology Initiative

AECOM Environment

D 414.577.1362   C 262.893.0658

bill.looney@aecom.com

AECOM

11425 W. Lake Park Drive, Suite 100

Milwaukee, WI 53224

T 414.359.3030 F 414.359.0822

www.aecom.com

See more of this Session: Environmental Health and Safety (EHS) Concerns of Nanomaterials
See more of this Group/Topical: Topical C: Environmental Aspects, Applications, and Implications of Nanomaterials and Nanotechnology

(217f) Forensic Analysis of Nano-Material Environmental Regulation

Coby A. Scher and Dennis L. Caputo, Quest Consulting, Inc., Bellaire, TX

Forensic Analysis of Nanomaterial Environmental Regulation

Coby A. Scher – Quest Consulting, Inc. (Presenter)

& Dennis L. Caputo – Quest Consulting, Inc.

Some considerable effort has recently been displayed by manufacturers, researchers, regulatory bodies, professional organizations, NGOs, and other interested parties addressing how nanomaterials and their manufacture should be environmentally regulated. In the USA, everyone including EPA, FDA, ABA, universities and a host of other entities (including companies) are trying to voice an opinion. The executive branch of the federal government with congress's approval and funding has created an investigatory group examining the regulatory potential. Internationally, the EU and ISO are investigating the subject of "nano-regulation". There is a body of thought within science and engineering termed forensics. In fact, the pure definition of forensic according to Webster's Collegiate Dictionary, 10th edition, is "1; belonging to, used in, or suitable to courts of judicature or to public discussion and debate 2: Argumentative, rhetorical 3: Relating to or dealing with the application of scientific knowledge to legal problems." All three parts of the definition are applicable to the issue at hand – the application of forensics to the environmental regulation of nanomaterials. It is the authors' contention that use of environmental forensics to encompass scientific investigative methods will provide the answer to what level of regulation is required. Forensic environmental science and engineering is a state of mind; an analytical approach to solving this problem. And it need not be complicated, just complete and scientifically defensible. This paper will update the current governmental movement toward environmental regulation of nanomaterials and apply a forensic approach to solving the problem of nanomaterial environmental regulation.

See more of this Session: Environmental Health and Safety (EHS) Concerns of Nanomaterials
See more of this Group/Topical: Topical C: Environmental Aspects, Applications, and Implications of Nanomaterials and Nanotechnology

(259a) Solar Thermal Production of Hydrogen Using Alumina Supported Ferrites: Thermodynamic Evaluation and On-Sun Validation

Paul Lichty, Department of Chemical and Biological Engineering, University of Colorado, Boulder, CO, Xinhua Liang, Chemical and Biological Engineering, University of Colorado at Boulder, Boulder, CO, Carl Bingham, National Renewable Energy Laboratory, Golden, CO and Alan W. Weimer, Department of Chemical and Biological Engineering, University of Colorado at Boulder, Boulder, CO

Alumina supports were coated with $CoFe_2O_4$ using an alternating atomic layer deposition process. Nano-powder and hollow skeletal shells were used as supports. Large, high surface area polymer particles (~600μm, 43.5 m^2/g) have been coated via ALD with several layers of alumina. The alumina layers were applied using TMA/water ALD chemistry. The particles were then heated to 800°C in air to burn out the polymer substrate. Thermodynamic modeling of the particles shows the creation of a hercynite/spinel phase that is capable of being cycled with water to produce hydrogen. The process requires reduction temperatures several hundred degrees below those of conventional ferrites supported on inert materials. Cycling has been completed on-sun at the National Renewable Energy Laboratory in Golden, CO. Powders were housed in a prototype concentrated solar reactor and subjected to cycling. Hydrogen production was quantified using a mass spectrometer. Results were used to estimate solar to hydrogen efficiency of the prototype reactor.

See more of this Session: Renewable Hydrogen Production I
See more of this Group/Topical: International Congress on Energy 2011

(259b) Comparison of On-Board Hydrogen Production From Several Non-Fossil Fuel Feedstocks

Wenhua H. Zhu, Center for Microfibrous Materials, Department of Chemical Engineering, Auburn University, Auburn, AL, Ying Zhu, Department of Chemical Engineering, Auburn University, Auburn, AL and Bruce J. Tatarchuk, Center for Microfibrous Materials manufacturing , Department of Chemical Engineering, Auburn University, Auburn, AL

In comparison with other types of fuel cells, the PEFC is compact and lightweight because of its better volumic-power and power density characteristics. It usually works at a temperature range from 40°C to 90°C. On the other hand, the high-temperature PEFC operating from 120-180°C is suitable for H_2 fuel produced by fuel processors, because of its CO tolerance up to 3%. These features make it suitable for automotive applications due to requirement of fast dynamic response to the speed and power-need change. But liquid hydrogen fuel faces many challenges such as transportation and on-board H_2 energy storage. This work compares and discusses in details for potential methods on-board H_2 production from non-fossil fuel feedstocks, such as ammonia and methanol fuels, which is useful for low or high-temperature PEFC on-board applications. Hydrogen fuel from fuel processors is one of the solutions to on-board energy storage because of the high cost and risk of liquid hydrogen or metal hydride storage. Hydrogen fuels from non-fossil fuel feedstocks are normally from water /ammonia /methanol electrolysis, methanol steam reforming, or reforming of bio-fuels. Due to the difficulties of hydrogen storage and transportation, ammonia and methanol are two of preferred choices of hydrogen fuel sources for low-temperature or high-temperature PEFC applications because of their existing production, storage, and transport technologies.

See more of this Session: Renewable Hydrogen Production I
See more of this Group/Topical: International Congress on Energy 2011

(259c) Energy Optimization of Hydrogen Production From Biomass

Mariano Martn, Chemical Engineering, University of Salamanca, Salamanca, Spain and Ignacio E. Grossmann, Department of Chemical Engineering, Carnegie Mellon University, Pittsburgh, PA

Energyoptimization of Hydrogen production from biomass

Mariano Martn[a,b],Ignacio E. Grossmann[b][1]

mariano.m3@usal.es; grossmann@cmu.edu

[a] Departamentode Ingeniera Qumica yTextil. Universidad de Salamanca. Plz. Caidos 1-5 37008, Salamanca (Spain)

[b]Department of ChemicalEngineering. Carnegie Mellon University 5000 Forbes Avd. Pittsburgh PA 15213

Hydrogen is a promising source of energy for thetransportation sector. It is clean and the technology is becoming available asmany car manufactures plan to offer it in the future (Cole 2007). However, itsdeployment in the market is not ready due to the strength and convenience ofthe liquid fuels industry (the energy density of liquid fuels is much higher),as well as the lack of infrastructure for hydrogen distribution, handling andstorage, resulting in high production costs using current technologies such asreforming of natural gas ($2.6/kg) or water electrolysis ($2.6-$4.2/kg)(Levene, Kroposki, & Sverdrup, 2006). The key issue to make hydrogen anattractive alternative fuel for the transportation sector is to optimize theproduction process from renewable raw materials instead of the more commonprocesses such as natural gas reforming or electrolysis of water (Rand &Dell, 2008).

Although biomass has been traditionally regarded asa source of liquid biofuels (i.e. bioethanol, green diesel), it can in factalso be considered as a source of hydrogen as has been reported recently. Gasificationof biomass generates a fair amount of H_2 which has to be purifiedand/or separated from the other gases resulting from the gasification(Tanksale, Beltramini, & Lu, 2010). Therefore, recent studies haveevaluated the production of hydrogen from gasification of biomass using asimulation-based approach selecting the technologies involved with good potentialin terms of yield and energy consumption (Feng, Wang,&Min, 2009; Gao,Aimin,&Quan, 2009; Hamelinck & Faaij, 2002; Ji, Feng, & Chen,2009a, 2009b; Lau et al., 2002; Li, Zeng, & Fan, 2010; Mueller-langer,Tzimas, Kalschmitt, &

Peteves, 2007).

In this paper we address the conceptual design forthe production of hydrogen from lignocellulosic raw materials. The optimizationof the process is formulated as an MINLP using short-cut models, design andempirical correlations for modeling a superstructure embedding two differentgasification technologies, direct and indirect gasification, and two reformingmodes, partial oxidation or steam reforming. Finally, following recent trends,a water gas shift reactor (WGSR) with membrane separation is used to obtainpure hydrogen.

The problem is solved by constraining the binaryvariables of the MINLP so as to select each gasifier and reforming modeyielding four NLP's. The operating conditions in the gasifiers and at the WGSRare optimized for hydrogen production in the solution of each NLP. Next, theenergy is integrated, and finally, an economic evaluation is performed todetermine the production cost of each of the four alternatives. It is shownthat indirect gasification with steam reforming is the preferred technologyproviding higher production yields than the ones reported in the literature forhydrogen from natural gas and at a potentially lower and promising productioncost 0.67 $/kg with a yield of 0.11 kg/kg wetbiomass.

Finally, an important issue topoint out is the fact that this process generates CO_2, 1.2 kg/ $kg_{Biomass}$together with 0.11kgH_2 /$kg_{Biomass}$.It has recently been reported that the production of biodiesel from algae is apromising technology with a yield to biodiesel 10-100 times higher than usingthe current raw materials like soy bean or rapeedseed (Yusuf, 2007). However,it needs plenty of CO_2. This process could be a source for growingthe algae. Biodiesel production requires in the range of 3.6 kg of CO_2per kg of biodiesel (Pate, 2008; Sazdanoff, 2006). Thus, on a kg of biomassbasis, it would be possible to obtain 0.33 $kg_{Biodiesel}$/$kg_{Biomass}$together with the hydrogen produced which provides full use of the biomasswhile producing short-term and long term biofuels with good integration ofdifferent processes. According to the DOE, by using algae for the productionof biodiesel it would be possible to meet the demand for biodiesel with 2-5% ofthe current cropland and with no consumption of water, since salt water canalso be used.

Keywords: Energy, Biofuels,Alternative fuels, Fuel cells, Water

Cole, D. E.(2007). Issues facing the auto industry: Alternative fuels, technologies, and policiesACP meeting eagle crest conference center

Levene, J.,Kroposki, B., & Sverdrup, G. (2006).Wind energy and production of hydrogenand electricity=opportunities for renewable hydrogen. In Conference paperNREL/CP-560-39534, March.

Feng, X., Wang,L., & Min, S. (2009). Industrial energy evaluation for hydrogen productionsystem from biomass and natural gas. Applied Energy, 86, 1767-1773.

Gao, N., Aimin,L. I., & Quan, C. (2009). A novel reforming method for hydrogen productionfrom biomass steam gasification. Bioresource Technology, 100, 4271-4277.

Hamelinck, C.N., & Faaij, A. P. C. (2002). Future prospects for productionofmethanol and hydrogen from biomass. Journal of Power Sources, 111, 1-22.

Ji, P., Feng,W., & Chen, B. (2009a). Production of ultrapure hydrogen from biomass gasificationwith air. Chemical Engineering Science, 64, 582-592.

Ji, P., Feng,W., & Chen, B. (2009b). Comprehensive simulation of an intensified processfor H2 production from steam gasification of biomass. Industrial and EngineeringChemistry Research, 48, 3909-3920.

Lau, F. S.,Bowen, D. A., Dihu, R., Doong, S., Hughes, E. E., Remick, R., et al. (2002). Techno-economicanalysis of hydrogen production by gasification of biomass. Final technicalreport work performed under DOE. Contract number: DE-FC36- 01GO11089

Li, F., Zeng,L., & Fan, L. S. (2010). Techno-economic analysis of coal-based hydrogen andelectricity cogeneration processes withCO2 capture. Industrial and EngineeringChemistry Research, 49(21), 11018-11028

Mueller-langer,F., Tzimas, E., Kalschmitt, M., & Peteves, S. (2007). Techno-economicassessment of hydrogen production process for the hydrogen economy for theshort andmediumterm. International Journal of Hydrogen

Tanksale, A.,Beltramini, J. N., & Lu, G. M. (2010). A review of catalytic hydrogenproduction processes from biomass. Renewable and Sustainable Energy Reviews,14, 166-182.

[1]Corresponding author. Tel.:+1-412-268-3642; Fax: +1-412-268-7139.

Email address:grossmann@cmu.edu (I.E. Grossmann)

See more of this Session: Renewable Hydrogen Production I
See more of this Group/Topical: International Congress on Energy 2011

(259d) Steam Reforming of Ethanol-Phenol Mixture As a Model for Biomass Tar

Gabriella Garbarino and **Guido Busca**, Department of Chemical Engineering, University of Genova, Genova, Italy

Introduction

Hydrogen is produced by biomasses through a variety of pyrolysis and gasification processes. Most of these processes are hampered by the coproduction of tars, a mixture of heavy compounds that may result in fouling of heat exchangers and reactors. A steam reforming step is needed to remove such noxious compounds. Studies report that phenol is the main component of tars after low temperature gasification. However, tars are complex mixtures of heavier and lighter compounds. In previous paper we have reported the results on steam reforming of bioethanol. To model tar feed we decided to mix phenol with ethanol. The presence of ethanol in ethanol-phenol-water solutions allows to increase phenol-to-water mole ratio in a single liquid phase.

Preparation of catalysts

Differet families of catalysts have been used. 5%Ni/Al_2O_3, 2.5%Ni-2.5%Co/Al_2O_3, 5%Co/Al_2O_3 and Ni50%/Al_2O_3 have been prepared by conventional wet impregnation of Siralox 5/170 support (Alumina with 5% SiO_2 from Sasol, 170 m^2/g) using Ni and Co hexahydrate nitrate water solutions. After impregnation, drying at 363K for 8 hours and calcination at 973K for 5 hours have been performed.

A 1 to 1 Ni:Al atomic ratio coprecipitate has been prepared as reported by Busca et al. 1992 . The amorphous precipitate filtered and the cake has been dried at 363K for 8 hours and calcined at 973K for 5 hours. The final catalyst is actually composed of a NiO-$NiAl_2O_4$ mixed phase. The preparation of Ni-YSZ and Ni-Co-YSZ catalysts has been described by Resini et al. (2008). The preparation and characterization of NiZnAl, NiMgAl, CoZnAl and CoNiZnAl coprecipitated catalysts has been described previously (Resini et al., 2009; Barattini et al., 2009; Busca et al., 2010).

Catalytic tests

The catalytic experiments were carried out in a fixed-bed tubular quartz flow reactor, operating isothermally, loaded with $30*10^{-6}$ kg-$44*10^{-6}$ kg of catalyst mixed with $270*10^{-6}$ kg-$440*10^{-6}$ kg of quartz particles (both of them 60-70 mesh sieved). The catalytic tests were performed by feeding stoichiometric water to substrate ratios sufficient to realize steam reforming reactions, or excess water. The results reported are based on reactant conversions, selectivity to carbon-containing products and hydrogen yield vs. temperature, as measured at each temperature when apparent steady state was reached, generally in 1 h.

Products analysis was performed with a gas-chromatograph Agilent 4890 equipped with a Varian capillary column "Molsieve 5A/Porabond Q Tandem" and TCD and FID detectors in series. Between them a Nickel Catalyst Tube was employed to reduce CO to CH_4. A six-port valve with a $0.5*10^{-6}$ m^3 loop was used for the gas sampling of the outlet gases. The sampling of the outlet gases was also made by injection, using a gas-tight with a nominal volume of $0.25*10^{-6}$ m^3. Furthermore, it is also available a sampling injection point at the end of the preheating zone to analyse the reagents and determine the possible decomposition or change in the feed. Products analysis was also performed on GC/MS (HP GCD plus), in order to have a precise identification of the compounds.

Steam Reforming of ethanol

Ethanol steam reforming experiments have been performed in dilute conditions over several nickel and cobalt containing catalysts. Experiments have been performed by varying catalyst surface area, reactants flow rate, contact time, reactants feed composition and temperature. The data suggest that adsorbed acetaldehyde and acetic acid play an important role as intermediates of ethanol steam reforming, while also acetone may have a role in the ethanol steam reforming reaction. The key step for high hydrogen yield during ethanol steam reforming is represented by the evolution of acetate species, either towards decomposition giving rise to methane and CO_x, or to steam reforming to CO_2 and H_2. At high temperature hydrogen production depends on approaching methane steam reforming and reverse water gas shift equilibria. Ethylene end dimethyl ether are parallel products found at low conversion. With excess water acetaldehyde is not found among the products, and hydrogen yields as high as 95% have been obtained at 853 K.

Co based or Co-Ni alloy containing catalysts appear to be even more active and selective than Ni based catalysts in particular at intermediate temperature.

The most relevant selectivity determining step is supposed to be the evolution of acetate ions, assumed to be the key intermediate. The high selectivity to hydrogen and CO_2 obtained in the 770-870 K range is attributed to the less reduced state of cobalt in reaction conditions, allowing the evolution of acetate species to CO_2 and hydrogen. On more or completely reduced surfaces (with Ni and/or less water in the feed) acetates decompose preferentially producing CH_4 and CO_x.

Hydrogen rich gases can be produced by steam reforming (bio)ethanol at moderate temperature, i.e. near 750 K, which is definitely lower than that needed to steam reform liquid hydrocarbons. Co-based catalysts seem to be better than NiO-based catalyst to work in the moderate temperature range.

Steam reforming of ethanol/phenol mixtures.

The catalytic test were performed feeding 80*10^{-6} Nm^3/min containing 70% He(carrier gas), 27% water, 2% ethanol and 1% phenol. Experiments have been performed using Co/Al_2O_3, Ni/Al_2O_3 and Co-Ni/ Al_2O_3 catalysts (5% metal). In this case Co–containing catalysts appear to be far less active than pure Ni catalyst, in fact at low temperaure we have different alkylates and at high temperature we have different lighter byproducts as methane and ethylene.

Over the 5% Ni/Al_2O_3 catalyst we find a very high conversion of both ethanol and phenol already at 721 K. However, most of phenol is actually converted into alkylate C_8 and C_{10} compounds. It is in fact well known that both acid and basic catalysts can catalyse the phenol alkylation reaction by ethanol. Several different alkylated phenol compounds are observed. Additionally, ethanol is also converted into ethylene and acetaldehyde and, to a small extent, to CO and CO_2. By increasing temperature to 810 K the phenol alkylation reaction disappears and conversion of phenol drops to near zero. Also ethanol conversion decreases, still producing ethylene, acetaldehyde and carbon oxides. By further increasing reaction temperature, also steam reforming of phenol occurs. At 951 K both ethanol and phenol are almost totally steam reformed to give a mixture of CO and CO_2. The hydrogen yield obtained at this temperature is very high being really only limited by position of the water gas shift equilibrium at high temperature. Experiments performed with impregnated catalysts richer in Ni i.e. 20% and 50% Ni/Al_2O_3 showed that alkylation activity at low temperature is much decreased by increasing Ni content. This allowed us to attribute the alkylation catalytic behaviour to alumina not covered by Ni. On the other hand, by increasing Ni loading the steam reforming activity is strongly increased. The catalyst richer in Ni at these temperatures converts almost all the feed, ethanol and phenol, to $CO+CO_2$, with very small amounts of methane, ethylene, ethane and benzene. At 870 K CO and CO_2 with traces of methane are the only detectable products over 50% Ni/Al_2O_3 catalyst.

Over $Ni-NiAl_2O_4$ catalyst prepared by coprecipitation (Ni:Al a.r.= 1) the alkylation of phenol is negligible with the only detection of vinylphenol, while also some dehydroxylation to benzene is observed. However this catalyst has similar activity in steam reforming as the 50%Ni/Al_2O_3 catalyst prepared by impregnation, allowing complete steam reforming of both reactants already at 860 K. Hydrogen yield ranges 90% at high temperature (850-950 K). Preliminary duration tests showed that the catalytic activity of the coprecipitated Ni-Al catalyst did not change after 6 h on stream at 863 K.

The ex hydrotalcite NiZnAl catalyst (Ni:Zn:Al 50:21.29) is also very active allowing the almost complete conversion of both reactants already at 824 K. The only relevant difference is the production of significant amounts of benzene by conversion of phenol at 738 K. Also in this case, yields to hydrogen are very high at high temperature (90%).

References.

L. Barattini, G. Ramis, C. Resini, G. Busca, M. Sisani, U. Costantino. 2009. Reaction path of ethanol and acetic acid steam reforming over Ni–Zn–Al catalysts. Flow reactor studies *Chemical Engineering Journal*, 153:43-49

G. Busca, V. Lorenzelli and V. Sanchez Escribano.1992.Preparation, solid-state characterization and surface chemistry of high-surface area $Ni_xAl_{2-2x}O_{3-2x}$ mixed oxides. *Chemistry of Materials*, 4, 595-605

G. Busca, U. Costantino, T. Montanari, G. Ramis, C. Resini, M. Sisani. 2010. Nickel versus cobalt catalysts for hydrogen production by ethanol steam reforming: Ni–Co–Zn–Al catalysts from hydrotalcite-like precursors. *International Journal of Hydrogen Energy*, 35:5356-5366

C.Resini, M.Delgado, S. Presto, L. J. Alemany, P. Riani, R. Marazza, G. Ramis, G. Busca. 2008.Yttria-stabilized zirconia (YSZ) supported Ni–Co alloys (precursor of SOFC anodes) as catalysts for the steam reforming of ethanol *International Journal of Hydrogen Energy*, 33:3728

C. Resini, T. Montanari, L. Barattini, G. Ramis, G. Busca, S. Presto, P. Riani, R. Marazza, M. Sisani, F. Marmottini, U. Costantino. 2009. Hydrogen production by ethanol steam reforming over Ni catalysts derived from hydrotalcite-like precursors: Catalyst characterization, catalytic activity and reaction path. *Applied Catalysis A: General*, 355:83-93

See more of this Session: Renewable Hydrogen Production I
See more of this Group/Topical: International Congress on Energy 2011

(259e) Synthesis of Ru Nanoparticles Supported Over Carbon Film by Pulsed Laser Deposition for Hydrogen Production by Catalytic Hydrolysis of NaBH4

Rohan Fernandes[1], Nainesh Patel[1], Antonio Miotello[1] and **Dushyant Kothari**[2], (1)Physics, Universita degli Studi di Trento, Povo (Trento), Italy, (2)Centre for Nanosciences & Nanotechnology, University of Mumbai,, Mumbai, India

Sodium borohydride ($NaBH_4$), with H_2 storage capacity of 10.8 wt. %, produces pure hydrogen at room temperature in alkaline solution: this makes it a good source of hydrogen for fuel cell used for the portable and cellular technologies. However, catalysts play a vital role to control the hydrogen production during the hydrolysis reaction of $NaBH_4$. In the present work, Ru nanoparticles supported over carbon porous film (Ru-C) are synthesized using Pulsed Laser Deposition (PLD) and used as a catalyst in the hydrolysis of $NaBH_4$ to produce molecular hydrogen. Various Ar pressures (10 to 50 Pa) were used during deposition of carbon films to obtain extremely irregular and porous-carbon support with high surface area prior to Ru nanoparticles deposition which is done in-situ. The hydrogen generation rate produced by Ru-C catalyst film is significantly higher as compared to unsupported Ru film and commercial Ru-C powder. The catalyst films were characterized by SEM, TEM, XPS, EDS, XRD, Raman spectroscopy and BET surface area measurement to understand the enhanced activity of the catalyst. Morphological analysis showed that Ru nanoparticles with average size below 50 nm are well dispersed in the carbon matrix. Carbon films deposited under Ar atmosphere contain high amount of sp^2 coordinated graphitic carbon as observed by Raman spectra. Nanoparticles produced by laser ablation process act as active catalytic centers for hydrolysis while the carbon support provides high initial surface area for the Ru nanoparticles with better dispersion and tolerance against aggregation. The catalytic activity of the film was found to be dependent on the amount of Ru loading. Kinetic studies on the hydrolysis reaction of $NaBH_4$ with Ru-C catalyst film reveal that the concentrations of both $NaBH_4$ and NaOH have essentially no effects on hydrogen generation rate.

See more of this Session: Renewable Hydrogen Production I
See more of this Group/Topical: International Congress on Energy 2011

(259f) Hydrogen Release During Experimental Solid Sodium Borohydride Tablet Hydrolysis Reactions

Yih-Hang Chen Sr., Department of Chemical and materials engineering, Tamkang University, Taipei, Taiwan

The objective of this research is to generate a step shape curve of the hydrogen release rate for PEMFCs used. The experiments use solid sodium borohydride and Ru/Al_2O_3 catalyst powder to compress and form the tablet. And using the fuel tablet as a reactant immerses it into the water and generates hydrogen. In order to achieve to step shape of hydrogen release rate, the operating variables of the system include the ratio of water/$NaBH_4$ by weight, the ratio of catalyst/$NaBH_4$ by weight and Surrounding Temperature(T_s). The result shows that hydrogen release rate is quite stable at 25ml/min when operating conditions are w/s=6, cat/s=0.015, and T_s=25°C, respectively. Up to here, we realize that heat removal rate is a key domain variable for sodium borohydride hydrolysis reaction. In order to obtain higher hydrogen release rate, the adiabatic operation is used to operate the system around 80°C. Finally, we propose a method to overcome the problems of the low hydrogen release rate and how to boost it, which is to initially apply heat to the hydrolysis system. The hydrogen release rate of the solid $NaBH_4$ with an initial temperature of 70°C can reach 275ml/min without boost-up behavior.

See more of this Session: Renewable Hydrogen Production I
See more of this Group/Topical: International Congress on Energy 2011

(264a) Enhanced Food Waste Hydrolysis for Integrated Biofuel Production

Jingwei Ma[1], Liang Yu[2], Yubin Zheng[1] and Shulin Chen[3], (1)Department of Biological Systems Engineering, Washington State University, Pullman, WA, (2)Washington State University, Pullman, WA, (3)Biological Systems Engineering, Washington State University, Pullman, WA

In recent years, there has been an increasing interest in using anaerobic digestion as a means of reducing the volume of municipal solid waste (MSW) and generating methane biogas for energy recovery. In order to increase process efficiency of food waste treatment and obtain valuable product, hydrolysis of food waste were enhanced by optimizing pH and total solids (TS), then two strategies of volatile fatty acid (VFA) utilization for biofuel production were proposed in this study. One of the strategies was methanization of VFA abundant liquid in high rate anaerobic digester, such as up-flow anaerobic sludge bed reactor (UASB) as the methanogenic reactor. Another strategy was production of single cell oil (SCO) or microbial lipid by oleaginous yeast *Cryptococcus curvatus* to be used as biodiesel feedstocks. High concentration of VFA production was the optimizing objective during the hydrolysis of food waste. After 20 d incubation time at pH 6.0 and 12% TS, VFA concentration, VFA yield and volatile solids (VS) reduction were reached to 53.9g/L, 0.38gVFA/gVS and 59.2%, respectively. Biofuel production of two strategies was compared and the mass balance of two systems was calculated. At steady-state, methane production rate of 2.28L/L/d and lipid production of 1.1g/L/d were achieved in two systems, respectively. Ammonia and phosphor recovery were added into these two systems to produce valuable products and alleviate inhibition.

See more of this Session: Sustainable Fuel From Renewable Resources I
See more of this Group/Topical: Environmental Division

(264b) Biodiesel Production by Utilizing by Products From Fish Oil Concentration

Martin Ernst, R&D BioFuels, BDI BioEnergy International AG, Grambach/Graz, Austria

Due to the health-promoting properties of polyunsaturated fatty acids (PUFA) shown in numerous medical studies, omega 3 concentrates based on fish oil are enjoying increasing popularity in the food industry. According their occurrence in natural fish oils, concentrates with high concentrations of PUFAs of Eicosapentaenoic acid (EPA) and Docosahexaenoic acid (DHA) are used. The applications range from nutritional supplements and food supplements up to pharmaceutical preparations. For large scale industrial production, high-quality fish oils with concentrations about 30% in total in EPA/DHA are commonly used for the production of EPA/DHA ethyl ester and triglyceride concentrates. For concentration to desired EPA/DHA amounts the fish oil is transesterified to ethyl esters. Adjacent to the transesterification the EPA and DHA esters are concentrated by molecular distillation. During the concentration step more than 60% of the starting ethyl ester amount forms the so called ethyl ester light fraction which is low in content EPA and DHA. Due to the low content of poly unsaturated fatty acids, this light fraction is usually thermal utilized. However, since the quantities of high-quality fish oils are limited, there is consequently an increasing trend also using fish oils with considerably lower concentrations of these omega-3-fatty-acids. Naturally by converting fish oils which are low in EPA/ DHA content also the amount of ethyl ester light fraction will increase. Therefore a further upgrade of this light fraction to higher value products is important to improve the overall economics.

Target of the ongoing research is the evaluation and necessary treatment of ethyl ester light fraction for utilization as alternative fuel. Therefore fish oil with lower content of polyunsaturated fatty acids was transesterified to ethyl ester. By molecular distillation this ester was used to produce marketable ethyl ester concentrates. The resulting ethyl ester light fraction with low contents of PUFAs was evaluated for its ability to produce biodiesel according ASTM standard. Compared to conventional methyl ester several parameter of the light fraction has to be improved to fulfil the ASTM standard. Especially the quality parameters of free glycerine, carbon residue and atmospheric residue are standard parameters out of specification. By utilizing this side stream, a renewable fuel could be generated, offering also the opportunity to recover remaining PUFAs in the light fraction.

See more of this Session: Sustainable Fuel From Renewable Resources I
See more of this Group/Topical: Environmental Division

(264c) A Sustainable Energy Campus

Earl R. Beaver, Practical Sustainability, LLC, Chesterfield, MO

The Sustainable Energy Campus combines multiple technologies from different industries to create a truly sustainable "waste-to-profit" campus where the waste from one technology being used becomes the input feedstock to another technology that utilizes that waste to create a full spectrum of highly profitable fuel products. All the while, greenhouse gases are reduced, materials are recovered, energy is optimized and virtually no emissions are released to the environment. Further, the reality is highly profitable immediately and for the long term. Thus, genuine waste-to-profit.

Technologies from the rubber, chemical and petroleum industries are at play in the Sustainable Energy Campus. A Carbon Black recovery production line takes scrap automobile and truck tires and uses a vacuum pyrolysis process to convert those tires into a product called Carbon Black Alternative and a stream of pyrolysis oil. Carbon Black is present in 75-85% of all new tires produced worldwide. A new cracking and reaction technology combined with a sophisticated separation process converts the pyrolysis oil into raw materials for gasoline, kerosene, diesel, naphtha, jet A fuel base and liquified low molecular weight hydrocarbons (propane, butane and pentane). While each step can be profitable independently, combining them in a single site with cogeneration allows for energy and materials optimization as well as maximizing revenue. Further, minor components are captured and isolated which have extremely high value. The key materials are cyclic terpenes and high been recovered at high yields and high purities.

Since a single scrap tire weighs approximately 21 pounds and one tire is produced per year per person in industrialized countries, the Sustainable Energy Campus provides a means of producing massive amounts of valuable products which otherwise would be produced from incremental fossil fuels and raw materials at significant cost to society and the environment. A full scale tire pyrolysis unit utilizing this teechnology has been in operation for more than one year and has produced more than ten million pounds of carbon black along with the co-products cited above. The data related to mass of products, energy savings and net environmental impact will be presented and discussed.

See more of this Session: Sustainable Fuel From Renewable Resources I
See more of this Group/Topical: Environmental Division

(264d) Air Quality Impacts of Achieving U.S. Renewable Fuel Mandates

Kristina M. Wagstrom[1], Christopher Tessum[1], Jason Hill[2] and Julian Marshall[1], (1)Department of Civil Engineering, University of Minnesota - Twin Cities, Minneapolis, MN, (2)Department of Bioproducts and Biosystems Engineering, University of Minnesota - Twin Cities, St. Paul, MN

The current US Renewable Fuel Standard (RFS2) requires increasing amounts of biofuel production through 2022. In this analysis we use lifecycle assessment, air dispersion modeling, and health risk assessment to estimate air pollution and related health impacts of the production and use of an additional 7.5 billion gallons of corn grain ethanol and 5 billion gallons of corn stover cellulosic ethanol.

Specifically, we considered three potential increases in ethanol volumes associated with moving from the Renewable Fuel Standard – Phase 1 (RFS1) to Phase 2 (RFS2): (i) a 5 billion gallon increase in cellulosic (corn stover-derived) ethanol, (ii) a 7.5 billion gallon increase in corn-derived ethanol, and (iii) a combination of the two (5 billion gallons of cellulosic ethanol plus 7.5 billion gallons of corn ethanol).

We use the Greenhouse Gases, Regulated Emissions and Energy Use in Transportation Model (GREET) to estimate the life cycle emissions associated with the production and use of both the ethanol fuels and an energy equivalent amount of gasoline. Temporal and spatial information was also utilized in order to allocate the lifecycle emissions as necessary for photochemical modeling. The Sparse Matrix Operator Kernel Emissions (SMOKE) model was used to calculate the base case emissions. We use the CAMx reactive photochemical air quality model to estimate the resulting changes in ozone and $PM_{2.5}$ concentrations, relative to a non-RFS scenario (i.e., relative to consuming an energy-equivalent amount of gasoline), for the entire United States using 2005 as a base year. Finally, we use BenMAP to evaluate the potential health impacts from the anticipated air quality changes.

Preliminary results for February and August indicate that population-weighted ambient $PM_{2.5}$ concentration increases are up to 189% higher for RFS than for the non-RFS scenario; analogous values for nitrate $PM_{2.5}$ are 15%-150% higher for RFS. Concentration impacts vary in space and time; predicted concentrations exhibit greater spatial variability for RFS than for the non-RFS scenario. Preliminary findings from the $PM_{2.5}$ health risk assessment suggest greater impacts from RFS2 scenarios than from an energy equivalent amount of gasoline.

See more of this Session: Sustainable Fuel From Renewable Resources I
See more of this Group/Topical: Environmental Division

(264e) Multi-Objective Footprints-Based Synthesis of Integrated Bioethanol Production Systems

Lidija •u•ek and Zdravko Kravanja, University of Maribor, Faculty of Chemistry and Chemical Engineering, Maribor, Slovenia

Abstract

Energy consumption in the transport sector depends almost exclusively on fossil fuels – oil. The forecasts show a growing tendency of this sector in the following years. The price of oil on global markets continues to remain high. Peak oil production is predicted to start shortly or even that it has been already started (Narodoslawsky, 2007; Robelius, 2007; The Oil Drum). At the same time, a significant rise in carbon dioxide emissions from the burning of fossil fuels is raising the earth's temperature and is causing climate change (Hamilton, 2008). Therefore it is important to take actions towards energy efficiency, emissions reductions, fuel supply security, and low-carbon economy.

Biomass is renewable only if production is sustainable. Also, different research studies have showed that lower carbon footprints are achieved by utilizing biomass, when compared to energy generated from fossil sources. On the other hand higher amounts of nitrogen are emitted into air and water (Cherubini and Strømman, 2011; Bauer, 2008; •u•ek et al., 2011).

This contribution focuses on the simultaneous integration of different technologies for converting starchy and lignocellulosic raw materials to bioethanol. Different paths are selected: biochemical, thermo-chemical and thermo-biochemical and integrated in one superstructure so that operating costs, energy and agricultural land are minimized and profit is maximized. The ethanol production is assumed to be centralized. Raw materials are assumed to grow on the agricultural area of the maximum of 50 000 hectares. The fossil energy used in the supply chain of bioethanol production is also calculated and analyzed. The degree of (un)sustainability of the bioethanol production system is evaluated by the carbon and nitrogen footprint and compared to gasoline supply chain. Multi-objective optimization is performed on bioethanol supply chain where net footprints are minimized. The optimization of the system is formulated as a mixed-integer non-linear programming (MINLP) problem. The problem is solved by the use of the Mixed-Integer Process Synthesizer MIPSYN using mass and energy balances and other constraints with profit maximization as the optimization criterion.

References

Bauer C., 2008, Life Cycle Assessment of Fossil and Biomass Power Generation Chains. An analysis carried out for ALSTOM Power services, PSI-report No. 08-05, Paul Scherrer Institute, Villigen, PSI,

Switzerland

Cherubini F., and Strømman A. H., 2011, Life cycle assessment of bioenergy systems: State of the art and future challenges, Bioresource technology, 102, 437-451

•u•ek L., Klemeš J. J., and Kravanja Z., 2011, Overview of footprints and relations between carbon and nitrogen footprints, Chemical Engineering Transactions, 25, 923-928, DOI: 10.3303/CET1125154

Hamilton L., 2008, Firewood and Woody Biomass and their Role in Greenhouse Gas Reduction, Agriculture Notes < www.homeheat.com.au> Accessed: April 2011

Narodoslawsky M., 2007, Sustainable Processes – The Challenge of the 21st Century for Chemical Engineering, Proceedings of European Congress of Chemical Engineering (ECCE-6)

Robelius F., 2007, Giant Oil Fields -The Highway to Oil, Giant Oil Fields and Their Importance for Future Oil Production, Acta Universitatis Upsaliensis, Uppsala, Sweden

The Oil Drum <www.theoildrum.com> Accessed: April 2011

See more of this Session: Sustainable Fuel From Renewable Resources I
See more of this Group/Topical: Environmental Division

(264f) Dilute Acid Pretreatment of Farm-Waste Cultured Algae and Its Effect On Enzymatic Hydrolysis of Cellulose

Rui Chen, Zhengbo Yue, Lauren A. Deitz, Jin Wang, Yan Liu and Wei Liao, Biosystems and Agricultural Engineering, Michigan State University, East Lansing, MI

The objectives of this study are: (1) to determine the optimal conditions of hydrolysis of farm-waste cultured algal biomass using dilute acid method, and (2) to investigate the effects and correlation of algal hydrolysate and enzyme loadings on enzymatic hydrolysis of pretreated lingocellulose. We statistically studied the effect of reaction conditions of acid concentration, temperature, and reaction time on the accumulation of total sugar released in dilute sulfuric acid treatment of AD effluent cultured algal biomass. Based on the regression model, the optimal conditions for sugar accumulation were 4% H_2SO_4 (w/w), 116°C reaction temperature and 30 min reaction time. Under these conditions, sugar concentration reached 28.18% of total biomass (DW), which is 99% of total sugar in algal biomass. Furthermore, the algal hydrolysate was applied as medium in the enzymatic hydrolysis of alkaline pretreated AD fiber, it demonstrated a significant increase ($P < 0.001$) in glucose yield compared to the control of hydrolyses using citrate buffer (conventionally employed in enzymatic hydrolysis) and deionized water. This study presented a potential pathway of integrated utilization of both liquid (algal cultivation) and solid (enzymatic hydrolysis) wastes from aerobically digested dairy manure for biofuels production. Implementation of such process will provide a mutually beneficial solution for agricultural operations, bioenergy production, and environmental protection.

See more of this Session: Sustainable Fuel From Renewable Resources I
See more of this Group/Topical: Environmental Division

(264g) High Quality Biodiesel Fuel Production From Crude Jatropha Oil without Upstream and Downstream Processing

Naomi Shibasaki-Kitakawa, Kei-ichi Kanagawa, Kazunori Nakashima and **Toshikuni Yonemoto**, Department of Chemical Engineering, Tohoku University, Sendai, Japan

1. Introduction

Jatropha curcas oil has been received much attention as a renewable source of biodiesel fuel with no competing food uses. Crude *Jatropha* oil hashigh contents of free fatty acid (FFA) and water. FFA and water in the raw oil significantly decrease the yield and quality of the product biodiesel in the industrial production process using the homogeneous alkali catalyst. Thus, the present production process is usually accompanied by a refining process to remove FFA in the raw oil to less than 0.5 wt% and a dewatering process of the raw oil. In addition, the downstream purification process to remove the by-products, such as soap and glycerin, from the biodieselis required to satisfy the fuel's quality requirements. These upstream and downstream processes raise the production cost of the biodiesel fuel.

In this study, the bench-scale production system connected the expanded-bed reactor packed with the cation-exchange resin and that packed with an anion-exchange resin in series was constructed. The cation-exchangeres in has a catalytic ability to convert FFA to biodiesel by esterification reaction without inhibition by water[1]. The anion-exchange resin has acatalytic ability to convert glycerides to biodiesel by transesterification reaction without soapformation[2] as well as an absorbent ability to remove the impurities such as glycerin, water and pigment contained in the product[3]. Thus, high quality biodiesel fuel would be produced from the crude *Jatropha* oil without any pretreatment of the raw oil and the purification of the products.

2. Experimental methods

Awater-jacketed column packed with the cation-exchange resin, Diaion PK208LH, and two columns packed with the anion-exchange resin, Diaion PA306S, were connected in series in the bench-scale production system (Fig.1). The temperature of each column was kept constant at 50 °C using a hot-water recirculating system. The mixed-solution of the crude *Jatropha*oil (FFA content of 2 wt%, water content of 3300 mg/kg) and methanol at the stoichiometric molar ratio of methanol to the total fatty acid residue in the oils was supplied to the bottom of the first column at the constant flow rate. The effluent solutions from the top of the columns were collected, and the concentrations of the reactants and products in the solution were determined.

3. Results and discussion

Figure 2 shows the HPLC chromatograms of the feed (a) and the effluents from the first column (b), the second column (c), and the third column (d). Several peaks of the FFA, diglyceride (DG), and triglyceride (TG) were observed in the feed. In the effluent from the first column, the peaks of the FFA completely disappeared and the peaks of the fatty acid methyl ester (FAME) appeared. The peaks of diglyceride and triglyceride became smaller in the effluent from the second column and completely disappeared in the effluent from the third column. Thus, the FFA and glycerides in the crude *Jatropha* oil were completely converted to FAME.

Table 1 shows the analytical results of the product biodiesel compared with standards of USA (ASTM D6751)and Europe (EN14214). The acid value and the contents of free glycerin and total glycerin in the product fully satisfied with both standards without downstream processing. In addition, the contents of remaining reactants (monoglyceride, diglyceride,triglyceride), water and FAME also satisfied with the standards of Europe. The glycerinand water as well as FFA and pigment were confirmed to be removed by the adsorption on the resin. Therefore, the proposed system permitted the production of high quality biodiesel fuel without upstream and downstream processing.

4. Reference

1) N.Shibasaki-Kitakawa*et al., Energy Fuels,* **24**, 3634 (2010)

2) N.Shibasaki-Kitakawa *et al., Bioresour. Technol.,* **98**, 416 (2007)

3) T.Tsuji *et al., Energy Fuels,* **23**, 6163 (2009)

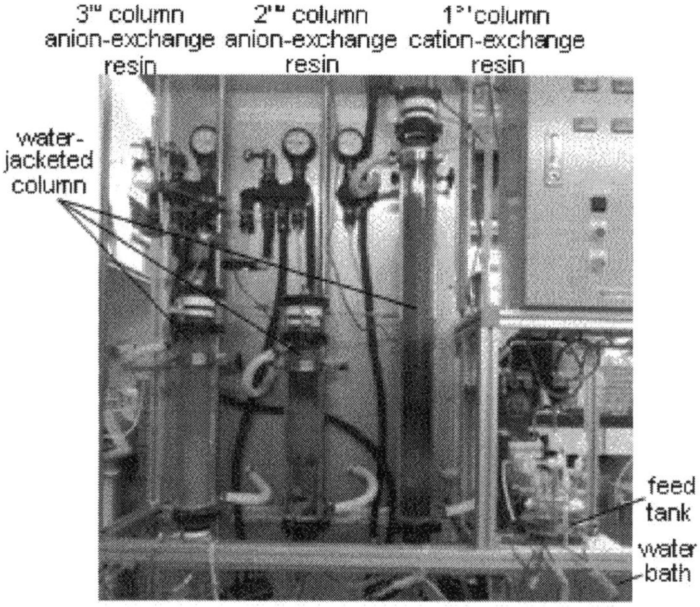

Fig. 1 Photograph of bench-scaleproduction system with cation- and anion-exchangeresins

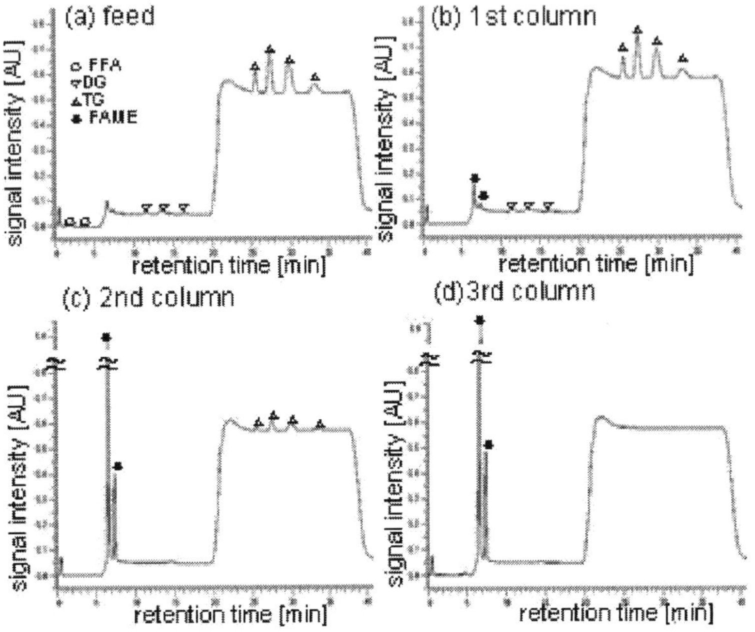

Fig. 2 HPLC chromatograms of feed andeffluents from each column

Table 1 Comparison ofproduct properties with standards ofbiodiesel fuel

property	units	product	standards	
			ASTM D6751	EN14214
acid value (FFA)	[mg-KOH/g]	0.03	≤0.8	≤0.5
free glycerin	[wt%]	0.01	≤0.02	≤0.02
total glycerin	[wt%]	0.03	≤0.24	≤0.25
monoglyceride	[wt%]	0.06	-	≤0.80
diglyceride	[wt%]	0.01	-	≤0.20
triglyceride	[wt%]	0.10	-	≤0.20
water	[mg/kg]	148	-	≤500
FAME	[wt%]	99.1	-	≥96.5

See more of this Session: Sustainable Fuel From Renewable Resources I
See more of this Group/Topical: Environmental Division

(267a) Pyrolysis Vs. Gasification: Effect of Oxygen In Reaction Atmosphere On Biochar Properties

Catherine E. Brewer, Center for Sustainable Environmental Technologies, Iowa State University, Ames, IA and Robert C. Brown, Mechanical Engineeering, Iowa State University, Ames, IA

Since biochar properties can vary widely depending on feedstock and processing conditions, making meaningful comparisons between biochar produced by different processes (pyrolysis, gasification, kiln carbonization) and at different locations can be difficult. Biochar characterization methods can provide some useful metrics such as van Krevelen diagrams, fixed carbon to volatiles ratios, and aromatic ring cluster size. One of the key parameters shown to influence biochar properties is the highest heating temperature (HHT) reached during the reaction; clear trends can be observed in the characteristics of slow pyrolysis biochars over the 200-800°C HHT range. These trends do not hold up as well, however, when biochars made at the same HHT by fast pyrolysis or gasification are considered. For low and high temperature gasification, the presence of oxygen in the reaction atmosphere creates biochars with higher O/C molar ratios and higher apparent HHTs than analogous slow pyrolysis reactions. This study used a large sample size thermogravimetric analyzer (TGA) to produce corn stover biochars reaching different HHTs under slow pyrolysis conditions with varying levels of oxygen present in the flow gas. The biochars were characterized by proximate analysis, CHNS elemental analysis and ^{13}C solid-state nuclear magnetic resonance spectroscopy (NMR) to understand the combined effects of HHT and oxygen on biochar properties. The goal of the study is to determine if the presence of oxygen in the reaction atmosphere at a given HHT would be beneficial for the creation of oxygenated functional groups on biochar surfaces similar to biochars that have "aged" in the soil environment.

See more of this Session: Thermochemical Conversion
See more of this Group/Topical: International Congress on Energy 2011

(267b) Pretreatment of Loblolly Pine for Fast Pyrolysis Via Torrefaction

Jiajia Meng, Forest Biomaterials, North Carolina State University, Raleigh, NC

Topical 6: Lignocelluosics: Biorefineries and Sustainable Energy

T6007 Thermalchemical Conversion I

AIChE 2011, October 16~21, 2011

Minneapolis, MN

Pretreatment of Loblolly Pine for Fast Pyrolysis via Torrefaction

Jiajia Meng, David Tilotta, and Sunkyu Park*

Department of Forest Biomaterials, North Carolina State University

*Corresponding author: sunkyu_park@ncsu.edu

Fluidized-bed fast pyrolysis was performed on torrefied loblolly pine with differing degrees of torrefaction severity. Compared to the pyrolysis of non-torrefied loblolly pine, which has a bio-oil yield of 65%, the bio-oil produced from thermally processed loblolly pine has a 10-30% lower yield depending on torrefaction severity. In order to increase the liquid recovery efficiency from the pyrolysis of intensely treated wood, optimal biomass particle size and pyrolysis temperature were investigated in this study. GC/MS analysis was applied to quantify several important chemical markers in the bio-oils, such as levoglucosan, glycolaldehyde and acetol. It was found that the decomposition of torrefied wood with different thermal treatments leads to varied content of these chemical components in the bio-oils. This result suggests that torrefaction may exert power on the pyrolysis pathway of the major components in the woody biomass. For instance, the opportunity for ring scission and the formation of levoglucosan from cellulose may be altered. Also, the decreasing amount of guaiacols and guaiacyl derivatives in the bio-oil, with respect to the torrefaction temperature, implies the reduction of methoxy groups either in the torrefaction process or in the condensation process occurring in fast pyrolysis. To understand the effect of torrefaction pretreatment on the variation of bio-oil composition, py-GC/MS and MBMS analysis of both torrefied wood sample and the residues from acid hydrolysis of torrefied wood was utilized. In addition, the stability and the acidity of the bio-oils produced from torrefied wood were examined to identify the bio-oil quality improvement gained from torrefaction. These results indicate that fast pyrolysis of the torrefied wood may be an effective way to produce high quality bio-oil.

See more of this Session: Thermochemical Conversion
See more of this Group/Topical: International Congress on Energy 2011

(267c) Characterization of Biochar, the by-Product of Energy Production From Biomass Via Thermochemical Processes

Ellen M. Knapp, Chemical Engineering, Villanova University, Villanova, PA and Justinus A. Satrio, Department of Chemical Engineering, Villanova University, Villanova, PA

Characterization of Biochar, the By-Product of Energy Production from Biomass via Thermochemical Processes

Ellen M. Knapp and Justinus A. Satrio

Chemical Engineering Department, Villanova University, PA 19085

Biochar is the solid product obtained through thermochemically processing biomass, namely through fast pyrolysis, slow pyrolysis, and gasification. The use of biochar has been shown to hold great potentials to aid in the sequestration of carbon dioxide and to act as a soil restoring agent. The thermochemical processes for energy production from biomass that stores carbon as biochar can be considered carbon negative due to high recalcitrant property of biochar. As a soil restoring agent, biochar has been shown to improve the quality of soils by increasing microbial activity, lowering soil acidity and density, and improving nutrient and water retention. It has also been shown that during thermochemical processing of biomass, a significant portion of mineral nutrients contained in the original biomass are recovered in biochar.

It is believed that all biochars are not created equal and will have different physical and chemical properties depending on their biomass origins and the thermochemical processing used for producing the biochars. The purpose of this study is to evaluate the effects of biomass precursors and process conditions of thermochemical processes on the chemical and physical properties of the produced biochars. For this study, biochars are synthesized by fast pyrolysis and slow pyrolysis processing of selected biomass materials using a small tubular reactor furnace at selected process conditions. A variety of biomass materials, commonly available in the state of Pennsylvania, particularly in the region where Villanova University is situated, are used for the study, which include wheat straw, switch grass, woodchips and spent mushroom substrate (SMS). The last material, SMS, is the waste product from mushroom growing industries and is produced in very large quantities in Chester County, which is known as the "mushroom growing capital of the world." Currently, Chester County accounts for approximately 37% of the mushroom production in the U.S.A. Biochar produced from various biomass materials by various process conditions will be analyzed for their physical and chemical characteristics/properties. Analysis will include proximate analysis, elemental analysis, heating value, surface area/porosity, and composition of mineral matters. Focus will be placed on the comparison of the chars formed from each biomass precursor at different process conditions. In particular, emphasis will be placed on comparing the characteristics of SMS char versus wheat straw char. SMS is a derivative of wheat straw that has been combined with various manures and grains. The comparison of chars produced from these two biomass precursors in particular will provide insight into the effects of additional materials added to the straws in the SMS on the properties of chars.

See more of this Session: Thermochemical Conversion
See more of this Group/Topical: International Congress on Energy 2011

(267d) Pyrolysis of Cellulose with the Use of Ionic Liquid for Efficient Production of Levoglucosenone

Shinji Kudo[1], Zhenwei Zhou[2], Koyo Norinaga[3] and Jun-ichiro Hayashi[3], (1)Research and Education Center of Carbon Resources, Kyushu University, Fukuoka, Japan, (2)Department of Applied Science for Electronics and Materials, Kyushu University, Fukuoka, Japan, (3)Institute for Materials Chemistry and Engineering, Kyushu University, Fukuoka, Japan

Pyrolysis is a most practical method to directly convert cellulose into condensable organic product, which is in general termed tar, with high yield. The tar consists of a variety of organic compounds, such as anhydrosugars and furan derivatives, and this is due to progress of various reactions in parallel and in series. For selective production of valuable chemicals, thus, the pyrolysis is currently less suitable compared to indirect methods via saccharification. A compound, levoglucosenone, can be produced by means of pyrolysis more easily than by general chemical methods, but it is difficult for any methods reported so far to convert cellulose to this particular compounds with either high selectivity or yield. This study proposes the pyrolysis of cellulose with the use of a type of ionic liquid (IL). ILs have widely been used as solvents for catalytic conversion of cellulose due to their characteristic properties such as solvent powers for cellulose, low vapor pressures and thermal stabilities. The proposed pyrolysis employs a type of IL, 1-butyl-2,3-dimethyl imidazolium trifluoromethanesulfonate ([BMMIM][OTf]), with an expectation of its high thermal stability occurring from alkyl substituent to 2-carbon. TG/DTA analyses in fact demonstrated its perfect stability at temperature up to 350 °C, where the most of the ILs underwent self-decomposition. The anion of [BMMIM][OTf], i.e., [OTf], is the conjugated base of a superacid, trifluoromethanesulfonic acid ($pK_a < -14$), and it was therefore expected catalysis of this IL in the pyrolysis of cellulose. Mixed with [BMMIM][OTf] at a mass ratio of 1 and temperature of 60 °C, cellulose was swollen. The swelling was maintained even after the mixture was cooled down to temperature below the melting point of [BMMIM][OTf], 41 °C. A TG analysis showed that the pyrolysis of cellulose in the mixture started at 160 °C, caused rapid and slow mass releases at 200–275 °C and at higher temperature up to 350 °C. Neither decomposition nor evaporation of [BMMIM][OTf], if any, was detected in the course of heating. The products from the pyrolysis at 300 °C were 32% char, 30% tar, 33% water, and 3–5% gas on a cellulose mass basis. Most importantly, levoglucosenone accounted for more than 70% and 27 mol-% on tar mass and cellulose monomer bases, respectively. Formation of levoglucosenone during the pyrolysis of a phosphoric-acid-treated cellulose was reported previously, but its yield was less than half of that from the cellulose-IL mixture. Such a high yield of levoglucosenone seems to be attributed to a catalytic role of catalysis of the [OTf] anion. Repeated use of [BMMIM][OTf] was also examined. The IL was separated from the char by washing the mixture with ethanol, and then recovered by means of evaporative removal of ethanol. The IL recovery was complete within the margin of experimental error. TG analysis of the used IL and also its repeated use in the subsequent pyrolysis of cellulose confirmed little or no change in the property of the IL in the processes.

Modeling of the formation and reaction properties of biomass

Karsten Müller, Olga Lobanova, Liudmila Mokrushina, Wolfgang Arlt
Friedrich-Alexander-University Erlangen-Nuremberg,
Chair of separation science and technology

Biomass as an alternative source of energy to replace fossil fuels has drawn much attention in recent years. Many of the biomass utilization processes developed in the past later turned out to be ineffective in terms of their energy balance. For example, some publications indicate that the amount of energy needed for the production of biodiesel and bioethanol exceeds the energy derived from the respective fuels.

A systematic development of new processes, avoiding the development of energetically inefficient processes, requires knowledge of the thermodynamic data of the biomass used. Especially for complex ingredients like polysaccharides and lignin, these data are not available from the literature, yet.

This research project aims at the determination of the formation and reaction properties of such biomass compounds as well as other thermodynamic properties like heat capacity.

In a succeeding step thermodynamics of reactions utilizing biomass as a feedstock have been analyzed. A group of model reactions, such as the conversion of saccharides or lignin to biofules like methane or hydrogen have been evaluated. For an appropriate thermodynamic modeling valid substance data are necessary. In the course of the project it could be shown that the free energy of formation is linear function of the degree of polymerization. For details see [1].

By balancing enthalpy and Gibbs' energy of the respective reactions, the equilibrium yields and conversions have been estimated. The first reaction has already been implemented as a high temperature process. Since in terms of energy, it is desired to carry out reactions at mild conditions, the feasibility of doing this was researched from the thermodynamic viewpoint.

It could be shown that e.g. the production of gaseous products such as hydrogen or methane from saccharides at low temperatures is thermodynamically possible. The equilibrium yield can be further increased by lowering the system pressure. This allows the gaseous products to leave the system and therefore shifts the equilibrium position towards product formation. The addition of inert gas can also be used to increase the formation of hydrogen.

Many biomass compounds are not, or only limitedly, soluble in water. Hence, solvents have to be used. Calculations showed that the usage of an appropriate solvent does not lower equilibrium conversion and can even help increasing the conversion.

This project has been supported by the Deutsche Forschungsgemeinschaft (Projektnummer: AR 236/34-1).

[1] O. Lobanova, K. Müller, L. Mokrushina, W. Arlt, „Estimation of Thermodynamic Properties of Polysaccharides", Chemical Engineering & Technology, 2011, 32, No. 6, 867–876

(267f) Effect of Pressure and Heating Rate On Biomass Devolatilization and Gasification

Pradeep K. Agrawal, School of Chemical & Biomolecular Engineering, Georgia Institute of Technology, Atlanta, GA and Kristiina Iisa, National Renewable Energy Laboratory, Boulder, CO

The goal of present study is to characterize the effect of pressure and heating rate on the devolatilization behavior of different biomass species. Three biomass species (southern Pine, corn stover, and switchgrass) were studied. A combination of two complementary gasification devices are utilized in this study: (i) pressurized entrained flow reactor (PEFR) provides high heating rates ($\geq 10^3$ $^oC/sec$) and short residence times (1-20 sec), and (ii) pressurized thermo-gravimetric analyzer (PTGA) provides lower heating rates (0.1-1 $^oC/sec$), but larger residence times (5-100 min). The pressurized thermogravimetric analyzer connected to a mass spectrometer and a Fourier-transform infrared analyzer provides information on the evolution of gaseous species during pyrolysis.

The char yield appeared to increase with an increase in pyrolysis pressure. The yields of the lighter gas components (CO, CO_2, H_2, CH_4, H_2O) increased as the pressure was increased. The increase in the amount of char and lighter gases was accompanied by a decrease in the formation of tar species. This was supported by an observed decrease in furan and furfural and higher hydrocarbon fragments with increasing pressure. There were clear differences in the evolution of gaseous species between the woody and herbaceous and agricultural residue feedstocks. The evolution of char morphology is strongly influenced by pyrolysis pressure and heating rates. The results suggest a complex role of pyrolysis conditions and biomass ash content on the char reactivity.

See more of this Session: Thermochemical Conversion
See more of this Group/Topical: International Congress on Energy 2011

(267g) Technoeconomic Analysis of Biomass Gasification System Utilizing Geothermal Energy for Processing

Sudhanya Banerjee[1], Yanan Zhang[1], **Song-Charng Kong**[1], Robert C. Brown[1] and Brian Anderson[2], (1) Mechanical Engineering, Iowa State University, Ames, IA, (2)Chemical Engineering, West Virginia University, Morgantown, WV

The use of biomass as a source of energy has been studied extensively inrecent years, and special attention has been paid to biomass gasification. In this study, ASPEN Plus is used to conduct technoeconomic analysis of a large biomass gasification system. A biorefinery scenario consisting of 2,000 dry metric tonnes per day of corn stover was developed using the model. In the original setup, steam is purchased and used in various processes. In this study, we explore the use of geothermal energy by substituting the purchased steam with steam derived from geothermal resources. Geothermal energy is in in theform of steam with temperature of 150°C. The flow rate of the geothermalsteam is in the range of 40 to 60 kg/sec and the cost is estimated at \$3 – \$7/MMBTU. This technoeconomic analysis assesses the use of geothermal steam instead of the purchased steam for various processes. Results are reported including the feasibility of the process, product characteristics, and the technical and economic aspects of using geothermal steam for biomass gasification system.

See more of this Session: Thermochemical Conversion
See more of this Group/Topical: International Congress on Energy 2011

(274b) Biodegradation of Hardwood Lignocellulosics by the Western Poplar Clearwing Borer, Paranthrene Robiniae (Hy. Edwards)

Jing Ke, Dhrubojyoti D. Laskar and Shulin Chen, Washington State University, Pullman, WA

Plant lignin is a source of useful chemicals, and also the major barrier for saccharification of lignocellulosic biomass for ethanol conversion. The enzymatic lignin degradation/modification could bypass the need for chemical pretreatment and thereby facilitate bioprocess consolidation. Herein, we reveal our new discovery on the elucidation of the prototype of hardwood lignin modification/ degradation in clearwing borer system, *Paranthrene robiniae*. The wood-boring clearwing borer, *P. robiniae*, effectively tunnels hardwood structures during the larval stage, and its digestion products from wood components has not yet been investigated. Subsequent solid state ^{13}C cross polarization magic angle spinning (CP/MAS) nuclear magnetic resonance (NMR) spectroscopy, attenuated total reflectance fourier transform infrared (ATR-FTIR), pyrolysis-gas chromatography/mass spectrometry (Py-GC/MS), and thermogravimetric (TG) analysis on tunnel walls and frass produced, provided evidence for structural alterations and lignin degradation during such hardwood digestion process. Our results strongly suggest that the structural alteration of lignin primarily involved a preferential degradation of syringyl units accompanied with oxidation on the side chains of lignin guaiacyl moieties. This study also further indicated that, unlike the wood-feeding termite, the clearwing borer does not target cellulose as an energy source, and thus its lignin degradation ability should provide potential information on how to disassemble and utilize hardwood lignin. Overall, this new biological model with an efficient lignin disruption system will provide the new insights into novel enzyme system required for effective plant cell wall disintegration for enhanced cellulose accessibility and production of value-added lignin derived products.

See more of this Session: Biological and Chemical Conversion Processes In Forest Biorefineries I: Enzymatic and Other Biological Processes
See more of this Group/Topical: International Congress on Energy 2011

Improved Ethanol Production through Anaerobic Thermophilic Co-Culture From Various Carbohydrates

Lei Xu*, Ulrike Tschirner

Bioproducts and Biosystems Engineering University of Minnesota, St. Paul, MN 55108, USA

1. Introduction

Affordable bio-fuel has become popular due to an imbalance between supply and demand of existing energy sources as well as the problems associated with recent fuel refinery processes (1). Lignocellulosic biomass offers a great potential as a bio-fuel resource due to their abundance and high conversion potential.

Consolidate Bioprocess Process (CBP) fermentation with proper microbes is a promising solution offering the potential for higher efficiency and lower production cost compared to other current conversion processes (2). The major attempts of this study were to evaluate the advantages of integrating both *C.themocellum* and *C.thermolacticum* into the CBP fermentation on the ethanol and acetate formation under extreme pH (5-10) on various saccharides (3). In order to optimize the co-culture, by overcoming the inhibitor effect of high pH and high ethanol concentrations, allowing for an effective merging of pretreatment and fermentation processes to bring the technology closer to economically feasible biofuels from non-food based biomass.

2. Materials and Methods

Both the *C. thermocellum* (ATCC 27405) and *C. thermolacticum* (ATCC 43739) were cultured with media described by He et al. (2009) under controlled conditions (80%N_2/20%CO_2). The conversions of selected materials including cellulose, hemicellulose and their mono- or di-saccharide intermediates are investigated. The major saccharides studied include glucose, cellobiose, and xylose. A low initial substrate/medium ratio (1% w/v) was used and a 5 mL homogeneous microbe culture with a cell weight of 5 g/L was transferred into 100 mL anaerobic bottles containing 50 mL of medium prepared under anaerobic conditions. Cultures incubated at 57 °C without shaking at pH 5 to pH 10 were analyzed for the end-products (ethanol and acetate). The impact of the ethanol production and accumulation within the medium on total solvent production was studied by adding various levels of ethanol at the beginning of the fermentation. The addition of 0 g/L of ethanol represents the beginning stage of the fermentation when no ethanol was produced. By adding ascending amounts of ethanol from 0.5 g/L to 4 g/L at the start, this study simulated the ethanol accumulation progress as a sequence of the continuous fermentation. The influences of pH, ethanol concentration and substrate types were further compared between *C. thermocellum/C. thermolacticum* mono-cultures and co-culture fermentation by analyzing the ethanol and acetate production using HPLC.

3. Results and Discussion

The innovative co-culture is extremely efficiently in fermenting separate components of lignocellulosics, especially MCC, for ethanol production due to enzymatic characteristics of the strains (4, 5). Almost all the ethanol yields obtained from co-culture fermentation are above any of its mono-culture fermentation yields, often almost tripled as compared to the mono-cultures under identical conditions on all the substrate. In addition, this co-culture exhibited a shorter lag period (48 h) than any of their mono-culture before the initiation of the selected sugar fermentations. However, the increase of ethanol yield caused by applying a co-culture was observed not to correlate to ethanol/acetate ratio improvement.

Based on ethanol and acetate yield after 10 d fermentation, statistical analysis indicated that the solvent yields were significantly related to the growth conditions, including the interactions among application of co-culture, initial ethanol level, pH value and substrate type. Fermentation yields for both strains were sensitive to pH changes. The sensitivity towards pH change was enhanced with the increase of initial ethanol concentration. When the starting ethanol concentration reaches 4 g/L, the pH influence has the maximum input on ethanol production for all substrate conversions and for all strains. Alkaline conditions, especially pH equal to 9, is preferred for *C. thermocellum/C. thermolacticum* co-culture fermentation if the goal is high solvents production. Almost all of the ethanol yields obtained from co-culture fermentation at this pH were higher than the sum of their mono-culture fermentations. Also, the co-culture produced the largest amount of acetate during fermentation under high initial ethanol levels (4 g/L), and the optimum pH of 9. Under this condition, the ethanol yield from xylose was increased from 0.1 g/L for *C. thermolacticum* and 0 g/L for *C. thermocellum* to 4.5 g/L using a co-culture, which almost approached 90% of the theoretical xylose fermentation efficiency. For MCC fermentation, improvement was detected to be from 1.4 g/L for *C. thermocellum* and 0.1 g/L for *C. thermolacticum* to a maximum of 3.8 g/L for co-culture. This represents 1.5 mol of ethanol per mol of sugar utilized for co-culture fermentation (around 75% of the theoretical efficiency) (6), almost three times the ethanol yield of the combined mono-cultures. Hemicellulose fermentation appeared to be slower and less effective than conversion of other fermentable sugars due to the lack of effective enzymes (4, 5). In this case, by applying the co-culture of *C. thermolacticum* and *C. thermocellum* to lignocellulosic fermentation, optimization of the conditions should focus on the degradation conditions for cellulosic material and the hemicellulose fragments, mainly xylose. Furthermore, in order to archive higher solvent productions, initial mild alkaline condition, prompt removal of acetate from the system, or pH-control is essential.

4. Conclusion

In the present work, a detailed analysis of the co-culture fermentation with various carbohydrates under extreme pH with existing of ethanol was carried out. The results demonstrated the technical feasibility of using our innovative simultaneous *C. thermocellum/C. thermolacticum* co-culture to converting components of lignocellulosics directly into ethanol and acetate because of shorter lag time and higher solvent production. An interaction among pH, substrates and initial ethanol level to the solvent production could be shown. To obtain maximum ethanol yield, providing a comparatively higher starting pH and higher initial ethanol level are preferred in this co-culture anaerobic fermentation.

Acknowledgements

The authors acknowledge the support of the University of Minnesota and the Buckman Fellowship fund for supporting this research.

References

1. Kumar P, Barrett DM, Delwiche MJ, Stroeve P. Pretreatment of lignocellulosic biomass for efficient hydrolysis and biofuel production. Industrial & Engineering Chemistry Research. 2009;48(8):3713-3729.

2. Perlack RD, Wright LL, Turhollow AF, Graham RL, Stokes BJ, Erbach DC. Biomass as feedstock for a bioenergy and bioproducts industry: The technical feasibility of a billion-ton annual supply. [Internet]. 2005:May 20th, 2011. Available from: http://feedstockreview.ornl.gov/pdf/billion_ton_vision.pdf.

3. Willaert RG, Baron GV. Gel entrapment and micro-encapsulation: Methods, applications and engineering principles. Rev Chem Eng. 1996;12(1-2):1-205.

4. Raman B, Pan C, Hurst BG, Rodriguez M, McKeown KC, Lankford KP, et al. Impact of pretreated switchgrass and biomass carbohydrates on *clostridium thermocellum* ATCC 27405 cellulosome composition: A quantitative proteomic analysis. PLoS ONE 4. 2009;4(4):e5271.

5. Pellerina P, Gosselina M, Lepoutrea J, Samaina E, Debeire P. Enzymic production of oligosaccharides from corncob xylan. Enzyme Microb Technol. 1991;13(8):617-21.

6. Kim YJ, Weigand AW. Experimental analysis of a product inhibited fermentation in an aqueous two-phased system. Appl Biochem Biotechnol. 1992;34-35(1):419-30.

(274d) Thermostable Cellulases and Xylanases

Aditya Bhalla[1], Venkat Kunreddy[1], Sudhir Kumar[1], Kenneth Bischoff[2], Lew P. Christopher[3], Stephen R. Hughes[4] and Rajesh K. Sani[5], (1)Chemical and Biological Engineering, South Dakota School of Mines & Technology, Rapid City, SD, (2)Bioproducts and Biocatalysis Research Unit, USDA-ARS, NCAUR, Peoria, IL, (3)Center for Bioprocessing Research & Development, South Dakota School of Mines & Technology, Rapid City, SD, (4)USDA-ARS, NCAUR, Peoria, IL, (5)Chemical & Biological Engineering, South Dakota School of Mines & Technology, Rapid City, SD

Thermostable cellulases and xylanases offer several advantages including greater rates of hydrolysis for lignocellulosic substrates, lowered risk of contamination, and increased flexibility with respect to process design. We have isolated several thermophilic cellulose- and xylan-degrading cultures belonging to the genera *Bacillus*, *Clostridium*, and *Geobacillus* using soil samples collected from the Homestake Gold Mine, SD and a local compost facility. One of the isolates, *Bacillus* sp. (DUSELR13 strain) produced xylanase which had optimum pH and temperature of 7 and 75^{0}C, respectively. Other isolate *Geobacillus* sp. (WSUCF1 strain) secreted endoglucanase having maximum activity at pH 5 and temperature 70^{0}C. The thermal stability of enzymes secreted by DUSELR13 and WSUCF1 was tested by determining the enzyme activity remaining after incubation of the enzymes at different temperatures (50 -100^{0}C) for a period of 40 days. WSUCF1 endoglucanase showed remarkable residual activity (>50%) after prolonged incubation for 25 days at 60^{0}C compared to others reported in literature. DUSELR13 xylanase retained >50% residual activity after incubation for 35 days at 60^{0}C. The effect of various lignocellulosic feedstocks on the endoglucanase and xylanase secretion by these isolates was also examined. K_m values for DUSEL R13 xylanase and WSUCF1 endoglucanase were calculated using Lineweaver–Burk plots. In case of DUSELR13 xylanase, the K_m value with xylan from birchwood was 1.4 mg/ml while for WSUCF1 endoglucanase with carboxymethylcellulose this value was 1.08 mg/ml. These thermostable enzymes were overexpressed extracellularly using *E.coli* as a heterologous host. Comparison studies were carried out between recombinant enzymes and native enzymes. These thermostable cellulases and xylanases are currently being used to develop a novel cost-effective and more efficient form of the simultaneous saccharification and fermentation process for biorefineries.

See more of this Session: Biological and Chemical Conversion Processes In Forest Biorefineries I: Enzymatic and Other Biological Processes
See more of this Group/Topical: International Congress on Energy 2011

(274e) Ionic Liquid Pretreatment of Wood Biomass Leading to Enhanced Enzymatic Delignification

Muhammad Moniruzzaman, Department of Material & Energy Science, Okayama University, Japan, Tsutomu Ono, Department of Material & Energy Science, Okayama University, Hiroshi Kanzaki, Department of Chemistry and Bioactive Compounds, Okayama University and Kenji Okada, Department of Life Science, Kurashiki University of Science and the Arts, Japan

The biological pretreatment of wood biomass for degrading lignin, a complex aromatic polymer, has received much attention as an environmentally safe or "green" process. Generally, enzymes isolated from naturally occurring fungi, or with enzymes produced by genetically engineered fungi have been used for wood biodegradation. However, this approach for lignin degradation in aqueous solution has been found to be very slow mainly due to the difficulties in enzyme accessibility to the hemicelluloses and the poor solubility of the lignin. To address these issues, this presentation will describe a new approach for enzymatic delignification of wood biomass using room temperature ionic liquids (RTILs) as pretreated agents or/and (co)solvents. It is well known that ionic liquids (ILs), a potentially attractive "green" recyclable alternative to environmentally harmful organic solvents, have been increasingly exploited as solvents and/or (co)solvents and/or reagents in a wide range of applications including pretreatment of lignocellulosic biomass. The pretreatment of wood biomass comprises using the hydrophilic IL [emim] [OAc] (1-ethyl-3-methylimidazolium acetate) that possesses very high solubility of wood biomas and lignin, which makes biomass very suitable for enzymatic degradation. It has been shown that the cellulose rich materials obtained from combination effects of IL and biological pretreatment contained significantly lower amounts of lignin as compared to the amounts found when each method applied alone. The produced cellulose rich materials were characterized by acid hydrolysis, Fourier-transform infrared spectroscopy (FTIR), Scanning electron microscopy (SEM), Thermalgravimetric analysis (TGA) and X-ray diffractometry (XRD). SEM and XRD revealed considerable microstructural and crystallinity index changes in the pretreated cellulose rich materials. We believe that this newly developed environmentally friendly process will play a great role in converting cellulosic biomass, the most abundant renewable biomaterials in the world, to biomaterials, biopolymers, biofuels, bioplastics and so on.

See more of this Session: Biological and Chemical Conversion Processes In Forest Biorefineries I: Enzymatic and Other Biological Processes
See more of this Group/Topical: International Congress on Energy 2011

(274f) The Evaluation of Inhibition of Enzymatic Hydrolysis Following Dilute Acid Pretreatment of Herbaceous, Softwood and Hardwood

Mahdieh Aghazadeh, Chemical and Biochemical Engineering, University of Maine, Orono, ME and G. Peter van Walsum, Chemical and Biological Engineering, University of Maine, Orono, ME

Presenter: Mahdieh Aghazadeh

The Evaluation of Inhibition of Enzymatic Hydrolysis following Dilute Acid Pretreatment of herbaceous, softwood and hardwood

Peter Van Walsum* and Mahdieh Aghazadeh

Department of Chemical and Biological Engineering

University of Maine, Orono, Maine 04496

Phone: 207-581-2226

Fax: 207-581-2323

Email: pvanwalsum@umche.maine.edu

ABSTRACT

The concept of severity combining residence time and temperature is often used in the pulp and paper industry. The influence of the severity of dilute acid hydrolysis of lignocellulosic biomass on sugar yield and on the subsequent enzymatic hydrolysis of the hydrolysate by commercial enzyme preparation (from Novozymes) was investigated. A dilute acid pretreatment was performed at 0.7 % (w/v) solid concentration with 100 mL Accelerated Solvent Extraction (ASE Dionex Corp.). The pretreatment conditions (residence time, temperature) were treated as a single parameter, s*everity factor* (log R_0), with constant acid concentration. The decrease in the yield of oligosaccharides coincided with the maximum concentrations of furfural and 5-hydroxymethylfurfural (5-HMF). With the further increase in s*everity factor*, the concentrations of furfural and 5-HMF decreased while the formation of formic acid and levulinic acid increased. The effect of acetic acid, formic acid, levulinic acid, furfural, and 5-HMF on enzymatic hydrolysis was investigated. In former studies the effect of the sugar produced from the hydrolysate itself had not been studied - in this study the different enzyme inhibition effect on three major types of biomass, herbaceous (corn stover), soft wood (pine) and hardwood (maple/oak) were observed.

See more of this Session: Biological and Chemical Conversion Processes In Forest Biorefineries I: Enzymatic and Other Biological Processes
See more of this Group/Topical: International Congress on Energy 2011

(274g) Role of Enzyme Blend Composition, Biomass Particle Size and Solids Loading On Pretreated Biomass Saccharification Efficiency

Shishir PS Chundawat[1], Nirmal Uppugundla[1], Mary Lipton[2], Margaret Magyar[1], Dahai Gao[1], Venkatesh Balan[1] and Bruce Dale[1], (1)Chemical Engineering and Material Science, Michigan State University, Lansing, MI, (2)Biological Sciences, Pacific Northwest National Laboratory, Richland, WA

One of the limitations of screening enzyme combinations using microplate based assays (Gao et al. 2010, Bioresour Technol 101, 8:2770) is that these are typically carried out using milled biomass (<500 mms) at low solids loading (<1%, w/v). Biomass particle size and solids loading can both influence the enzyme set necessary for efficient hydrolysis. In this study, three different harvest varieties of AFEX pretreated corn stover were hydrolyzed at 0.6-18% solids loading using multiple combinations of cellulases, hemicellulases and other accessory enzymes for extended hydrolysis time periods. Three particle size ranges were tested (<100 mm, <500 mm, >1 mm). The crude glycosyl hydrolase mixtures were also analyzed using a high-throughput proteomics approach to identify enzyme activities necessary for efficient hydrolysis. We show that proteomics based spectral counting approach is a reasonably accurate and rapid analytical technique that can be used to determine protein composition of complex glycosyl hydrolase mixtures that also correlates with the specific activity of individual enzymes present within the mixture. We also find that both particle size and solids loading can significantly impact the optimal combination of hydrolytic enzymes necessary for efficient hydrolysis. In summary, we found that optimal cellulase-hemicellulase mixtures can achieve near-theoretical glucan and xylan yields at greater than fourfold lower total protein loading than using individual commercial enzyme blends alone.

See more of this Session: Biological and Chemical Conversion Processes In Forest Biorefineries I: Enzymatic and Other Biological Processes
See more of this Group/Topical: International Congress on Energy 2011

(280a) Investigation of CuO-Based Oxygen Carriers As CLOU Materials for Solid Fuels

Mehdi Arjmand[1], Henrik Leion[1], Abdul-Majeed Azad[2], Tobias Mattisson[2] and Anders Lyngfelt[2], (1)Department of Chemical & Biological Engineering, Chalmers University of Technology, Gothenburg, Sweden, (2)Department of Energy and Environment, Chalmers University of Technology, Gothenburg, Sweden

Chemical-loopingcombustion (CLC) is a promising combustion technique with efficient CO_2-captureat low cost due to inherent gas separation. In a CLC process, a metal oxide iscirculated between two reactors, wherein they undergo changes from an oxidizedform in the air reactor to a reduced one in the fuel reactor. In the case ofchemical-looping with oxygen uncoupling (CLOU), the oxygen carrier releases oxygenin the gas phase in the fuel reactor as shown in the figure. This enables a highrate of conversion of char from solid fuels, as it eliminates the need for thegasification step needed in normal CLC with solid fuels.

Only certain puremetal oxides of Co, Cu and Mn have an equilibriumpartial pressure of oxygen suitable for CLOU. Due to its favorablethermodynamics (facile redox properties) and agile kinetics (high reactivity),CuO is a promising CLOU material, in addition to being an excellent oxygencarrier in the combustion of gaseous fuels.

In this work, theperformance of CuO-based oxygen carriers supported on ZrO_2stabilized by calcium oxide, magnesia and ceria will be presented to assesstheir suitability as oxygen carriers for solid fuel combustion. The samples wereprepared by freeze-granulation and the reactivity was examined in a laboratoryfluidized bed reactor under cyclic oxidizing and reducing conditions,simulating the CLOU process. Owing to the fast release of oxygen by CLOUparticles, the reaction rates were investigated using coke as the fuel ofchoice. The influence of reaction temperature and oxygen concentration on therate of reaction was also studied.

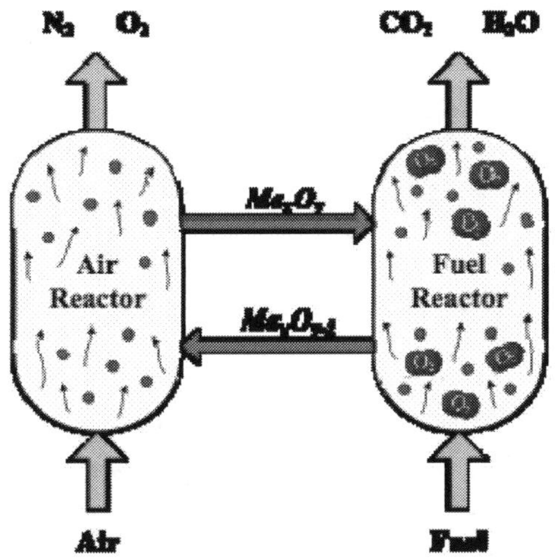

Schematicof the chemical-looping combustion via oxygen uncoupling (CLOU) process

See more of this Session: Chemical Looping Processes I
See more of this Group/Topical: Energy and Transport Processes

(280b) Oxygen Uptake and Release Kinetics of Cu/ Cu_2O/CuO On Al_2O_3 and ZrO_2-SiO_2 As Materials for Chemical Looping Combustion

Björn Brandt[1], Philipp Gruene[2], Frank Rosowski[2], Bob Farrauto[3] and Marco J. Castaldi[1], (1) Department of Earth and Environmental Engineering (HKSM), Columbia University in the City of New York, New York City, NY, (2)BASF SE, Ludwigshafen, Germany, (3)R & D, BASF Catalysts LLC, Iselin, NJ

One key parameter in ChemicalLooping Combustion (CLC) is the oxygen transport capacity of the CLC materialsbetween the two process reactors; in practice, it may be limited by acombination of many factors, including the metal content of the inert binder,the oxygen uptake and release kinetics and the thermodynamic equilibrium underthe process conditions in the two reactors, the respective residence times, therate of contaminant deposition and removal (such as coke, sulfur, or ashdeposits), and deterioration of the oxygen carrier. Such factors can have a significantimpact on the economics and viability of a CLC power plant, because it isexpected that many tons of CLC materials would be required in a practical CLCpower plant (the required inventory of CLC material can be estimated to be inthe range of 0.5 to 6 tons/MW power for typical CLC materials and processoperation conditions).

With our ongoing program, weaim for a systematic study of those interconnected aspects, and for obtaining amore quantitative description of CLC from the perspective of gas-surfacekinetics. As CLC materials, we have chosen copper oxide supported on two inertbinders: 1. ZrO_2-SiO_2(a high-performance material used in the field of catalysis), and 2. γ-$Al_2O_3$for comparison (a more economic, generic binder previously described inliterature); those materials are prepared by incipient wetness impregnationfollowed by a quantification of the prepared formulation by calibrated X-Ray fluorescence. The study is centered on TGA/DTA/DSC experiments, which allow for testing theoxygen uptake and release kinetics as a function of important processparameters; they will be amended with morphological characterizations (BET, Hg-porosimetry, electron microscopy, EDX elementaldispersion).

We will present experimentsobtained in the temperature range from 450 to 1000 C, using CLC materialswith Cu loadings ranging from ~ 5 to 50 wt.-%. Our first resultsindicate that the ZrO_2-SiO_2 material (17 wt.-% Cu)can be cycled many times (more than 30 reduction/oxidation cycles have been studied) with lessthan 2 % change of its oxygen uptake and release capacity. The experimentsshow that the material can be looped quantitatively between the fully oxidizedstate (CuO), and the fully reduced (metallic) stateusing methane as fuel at 750 C; at higher temperatures (starting around800 C), the material begins to spontaneously decompose to $Cu_2O$when the air is removed, likely due to thermodynamic driving forces; thus, thistemperature constitutes a significant upper boundary in order to maximize theoxygen transport capacity of the material in CLC. The observed oxygen uptakerates are $2.5x10^{-8}$ mole O_2 sec^{-1} (for 1mg of this CLC material), and varied by less than 4 % over atemperature rate from 550 to 750 C; this apparent lack of variationof the rate is probably due to extrinsic limitations of the oxygen uptake step– *e.g.* due to bulk or porediffusion processes. On the other hand, oxygen release rates under methaneexposure displays a strong temperature dependence; the reduction rate variesbetween $1.8x10^{-8}$ mole O_2 sec^{-1} at 750 C,and $3.1x10^{-9}$ mole O_2 sec^{-1} at 550 C (for1 mg of the material, respectively). Moreover, reduction rates are lower thanthe oxygen uptake rates at the same temperatures (26 % slower at750 C, 79 % slower at 550 C). These temperature findingsindicate that oxygen diffusion in the lattice is not the limiting factor forthe reduction rates. It will be discussed whether the observation of differentreduction rates is linked to the kinetics of the fuel activation on the CLCmaterials. Finally, to begin to understand the impact of a more realistic fuellooping cycle, tests done with simulated gasified coal including steam andppm-levels of H_2S will be shown.

See more of this Session: Chemical Looping Processes I
See more of this Group/Topical: Energy and Transport Processes

(280c) Investigation of Cobalt Ferrite and Nickel Ferrite for Chemical Looping Hydrogen Production

Victoria J. Aston[1], Anthony H. McDaniel[2], Brian W. Evanko[1], Jonathan R. Scheffe[3], Mark D. Allendorf[2] and Alan W. Weimer[4], (1)Chemical and Biological Engineering, University of Colorado at Boulder, Boulder, CO, (2)Sandia National Laboratories, Livermore, CA, (3)Department of Mechanical and Process Engineering, ETH Zurich, Zurich, CO, Switzerland, (4)Department of Chemical and Biological Engineering, University of Colorado at Boulder, Boulder, CO

The chemical looping hydrogen production process uses synthesis gas or methane to reduce a metal oxide material, forming a waste stream of CO_2 and H_2O from which CO_2 can be easily sequestered. The reduced metal oxide is then oxidized with steam to produce pure H_2 and regenerate the original metal oxide. This system requires a metal oxide material that achieves high H_2 and CO conversions to H_2O and CO_2 during the reduction step while maintaining fast oxidation and reduction kinetics through many cycles. In this study, the use of nano-scale cobalt- and nickel-ferrite ($CoFe_2O_4$ and $NiFe_2O_4$) based materials was investigated for chemical looping hydrogen production. Nano-thin films of ferrite were deposited on high surface area ZrO_2 support material using the atomic layer deposition (ALD) process to minimize sintering and diffusion limitations on reaction kinetics. The kinetics and cyclability performance of these thin film materials were analyzed and compared with the performance of conventionally made impregnation ferrite samples on the same high surface area substrates.

See more of this Session: Chemical Looping Processes I
See more of this Group/Topical: Energy and Transport Processes

(280d) Evaluation of Iron Based Oxygen Carriers for Chemical Looping Dry Reforming

Saurabh Bhavsar, Department of Chemical Engineering, University of Pittsburgh, US DOE-National Energy Technology Laboratory, Pittsburgh, Pittsburgh, PA, Michelle Najera, Department of Chemical Engineering, University of Pittsburgh, Mascaro Center for Sustainable Innovation, University of Pittsburgh, Pittsburgh, PA and Gtz Veser, Department of Chemical Engineering, University of Pittsburgh, US DOE-National Energy Technology Laboratory, Pittsburgh; Mascaro Center for Sustainable Innovation, University of Pittsburgh, Pittsburgh, PA

'Chemical Looping Combustion' (CLC) is an emerging clean combustion technology which offers an efficient and elegant route for fossil fuel combustion with inherent CO_2 capture. In CLC, an oxygen carrier (typically a metal) undergoes cyclic reduction and oxidation with fuel and air, respectively. By changing the oxidant, the process can be applied for purposes beyond combustion: Replacing air with CO_2 as oxidant results in dry reforming of the fuel, producing CO as result of the carrier oxidation. This process, Chemical Looping Dry Reforming (CLDR), hence results in utilization of CO_2, providing a potential alternate pathway to carbon mitigation via sequestration. However, since CO_2 is a weaker oxidant than air, significant challenges lie in the slow oxidation kinetics. While nano-sizing the oxygen carrier can be expected to increase carrier activity and result in higher carrier utilization, it also exacerbates the carrier stability issues at the demanding high-temperature cyclic operation conditions.

In the present study, iron was chosen as oxygen carrier for CLDR based on a prior thermodynamic evaluation of various metals. Different nanocomposite iron carriers, including Fe-BHA (barium hexaaluminate) and Fe@SiO_2 core-shell materials, were synthesized with the aim to combine high reactivity and high carrier utilization with good thermal stability. The carriers were characterized by TEM, SEM, XRD, and BET, and then evaluated in fixed bed reactor and TGA (thermogravimetric analysis) CLDR studies. Our results indicate that sufficiently small Fe nanoparticles show high CO_2 conversion activity down to temperatures well below 400°C, but that the oxide matrix required to render the carriers thermally stable can result in significant mass transfer limitations at high-temperature conditions (~800°C). Overall, our results indicate that CLDR could offer a novel way for efficient CO_2 utilization over a broad temperature range.

See more of this Session: Chemical Looping Processes I
See more of this Group/Topical: Energy and Transport Processes

The Effect of Fly Ash from Different Coal Sources on the Reactivity of an Iron-based Oxygen Carrier for Chemical Looping Combustion

Aurora Rubel*, Yi Zhang, James K. Neathery, and Kunlei Liu
University of Kentucky, Center for Applied Energy
2540 Research Park Drive
Lexington, KY 40511

Introduction

The University of Kentucky, Center for Applied Energy Research is investigating oxygen carriers (OC) for chemical looping combustion (CLC) with the goal of developing a pressurized chemical looping combustion/gasification process for generating electricity from coal. An important issue in the application of CLC to a solid fuel is the effect of ash on the performance of the OC. A previous study using a bituminous coal ash obtained from an electric utility suggested that the ash was not detrimental to the capacity and reactivity of two iron based OCs and iron oxide already present in the ash may have contributed additional OC activity[1]. With the wide variety of coals used to generate electricity world-wide, it is unlikely that all ashes will have the same effect on OCs. Recently four ashes obtained from bituminous, sub-bituminous, and lignite coals were used in a comparative study of their effect on the reactivity of an iron based OC to combust a high carbon beneficiated coal char under dry inert gas (argon) conditions in a TGA[2]. As expected the ash effect on the OC varied with the different ashes. The lignite ash did not affect the performance of the OC up to the highest concentration (75%) tested. The Bulgarian subbituminous which had a Fe_2O_3 concentration of nearly 14% caused a 20% reduction in the performance of the OC which was contrary to anticipated results. The Powder River Basin (PRB) ash produced a slight reduction in carbon removal ability as did the China bituminous ash. However, the results for the China ash which had 8.5% residual carbon indicated that the OC at the right carbon to OC ratio could also combust the residual carbon in the ash as well as the carbon in the fuel char. This study was undertaken to further investigate the impact on an iron based OC of these four different fly ashes in the presence of water vapor (WV) consistent with conditions which would exist during CLC in comparison to dry gas results.

Experimental

Four ashes were collected from power plants in Bulgaria (subbituminous), China (bituminous) and the US (Texas lignite and Powder River Basin subbituminous). The ashes were analyzed for CHNS and major and minor elements. The ashes were used as-received The four ashes were mixed with a 1:9 by weight mixture of char and OC so that the final mixture of char:OC:ash was 5:45:50. A 2 gram sample was heated in a TGA to 950°C in a 200 ml/min (STP) flow of Argon with and without 10% WV added. The sample was heated at 950°C for 5 hours. The OC was generated at the CAER by freeze granulation (FG) by procedures published elsewhere[3]. The FG OC used in this study was 50% Fe_2O_3 on an Al_2O_3 support. Control runs included the OC, char, and each ash run alone. A (1:9) char:OC mixture with no ash added was considered the base case. Aliquots of the before and after heating of the sample were submitted for carbon determinations to determine the percent carbon removed from the starting mixtures. Results were normalized to the base case of no ash added samples. Reaction rates were determined from the differential weight loss curves obtained by TGA.

Results

The ashes varied in chemical composition as determined from elemental and inorganic analyses (Table 1). Residual carbon and iron oxide contents varied from 0.07 to 8.52% and 4.22 to 13.71% respectively. These were considered the most important parameters but other oxides which might contribute oxygen (CaO, MgO, SO_3) were also present in high concentrations in some of the ashes. The residual carbon in all ashes except the China bituminous ash was less than 1%.

Table 1. Fly ash type and composition.

Ash	%C	%Fe_2O_3	CaO	MgO	SO_3
Bul Sub	0.78	13.71	7.48	2.52	2.78
Chi Bit	8.52	4.99	8.08	1.16	1.00
PRB	0.07	5.73	21.13	4.38	1.72
TX Lig	0.81	4.22	14.68	2.02	1.46

Heating the ashes under wet gases removed the 99, 97, 99, 58% of that carbon for the lignite, China , Bulgarian, and PRB ashes respectively (Figure 1). Under a flow of dry argon, comparable carbon removal from the ashes were 89, 15, 78, and 58% indicating the advantage of wet gases providing some gasification of the ash carbon. The effect of gasification on carbon removal was very evident by the increase in carbon removal from the char and char:OC controls from dry to wet gases of 1.2 to 62.3% and 56.7 to 86.8% respectively.

The benefits of the OC were improved carbon removal of 27.5% under wet gases and 55.5% under dry gases in comparison to the char alone control (Figure 2). Under wet gas conditions, the char:OC:ash mixtures all had carbon removals equal to or exceeding the zero ash base case indicating that none of the ashes affected the capacity of the OC to oxidize carbon. This was not true for the dry gas runs where the China, Bulgarian, and PRB ash all produced some decrease in the OC capacity.

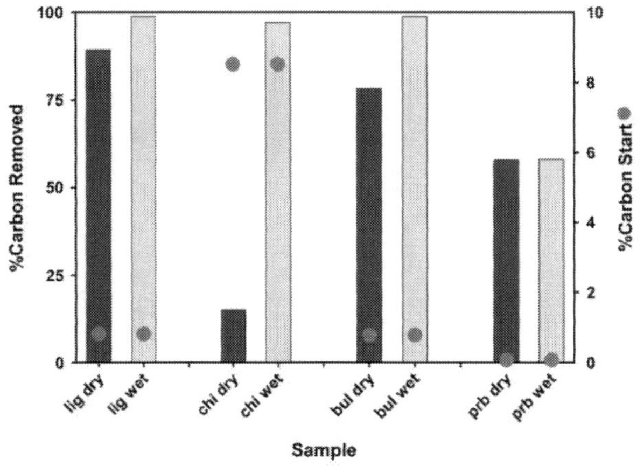

Figure 1. Carbon removed from ashes alone under experimental conditions.

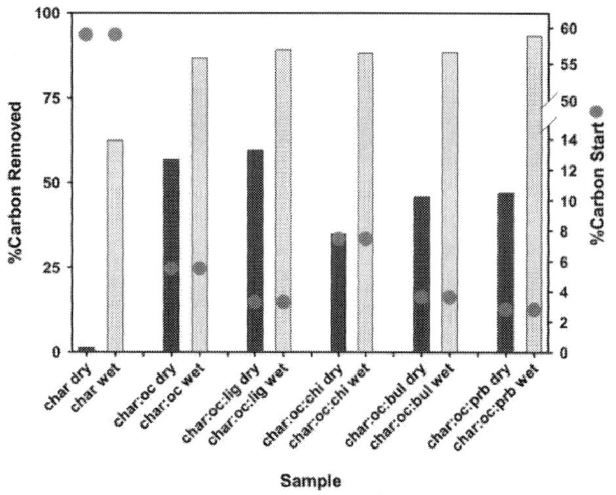

Figure 2. Carbon removed from char alone and char:FGOC:ash mixtures under experimental conditions.

The influence of the different ashes on the reaction rate of carbon removal from the char was determined from the TGA results. Rates under wet gas conditions were significantly slower than under dry gases. For instance, the maximum char gasification rate (no OC or ash) was -3.5 mg/g/min wet and -8.5 mg/g/min dry. The addition of OC improved the maximum char gasification rate for wet gases to -7.0 mg/g/min and the dry gas rate remained unchanged at -8.4 mg/g/min. Maximum carbon removal rates for the ash mixtures varied from a low of -6.5 mg/g/min for the lignite to a high of -9.9 mg/g/min for the PRB ash under dry condition and from a low of -3.9 mg/g/min for lignite to a high of -6.3 mg/g/min for the PRB ash mixtures. All ashes did reduce the maximum char removal rate under wet gases. Reductions were 45, 13, 16, 10% for lignite, China, Bulgarian, and PRB ashes respectively. Whereas under dry gases, all carbon removal rate except for the lignite were equal to or more than the base case of zero ash. Another outcome from this work was the fact that 50% or more of the carbon was removed during the first 10 minutes of dry gases and 20 minutes for wet gases. Since a CLC process will have short residence time in comparison to the 5 hours used in this study, faster initial reaction rates are important.

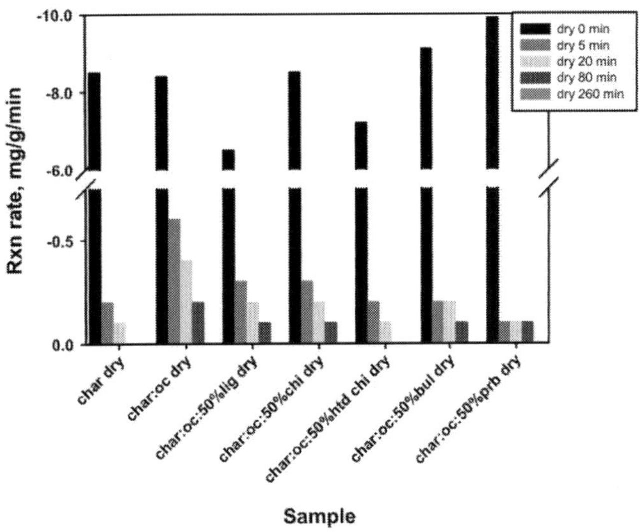

Figure 3. Carbon removal from char, char:FGOC, and char:FGOC: ash mixtures under dry gas experimental condition.

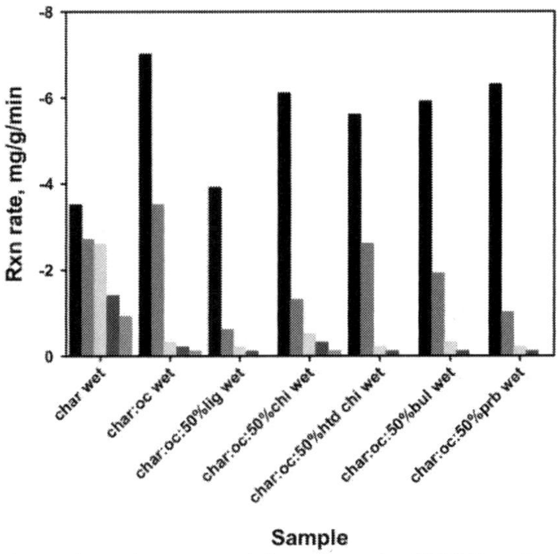

Figure 4. Carbon removal from char, char:FGOC, and char:FGOC: ash mixtures under wet gas experimental condition.

Conclusions

Both positive and negative results were obtained for the effect of ash with WV on the performance of the OC. The effect of each ash was dependent on the source coal. WV had a beneficial effect eliminating the detrimental effect of ash seen under dry conditions for the China, Bulgarian, and PRB ashes on the OC capacity to oxidize carbon. The OC had beneficial impact on carbon removal under both wet and dry gas. Carbon oxidation rates were significantly increased by the OC in wet gases over the rate of carbon gasification only. WV negatively impacted the OC's overall carbon removal rate over comparable dry runs with 50% removal of the carbon requiring 10 minutes under dry conditions and 20 under wet gas conditions. The conclusion was that in the presence of WV, ashes have no effect on the capacity of the OC to oxidize the carbon in the fuel but there was a reduction in the rate of oxidation.

References

1. Rubel, A.; Zhang, Y.; Liu, K.; Neathery, J. Effect of Ash on Oxygen Carriers for Chemical Looping Combustion of a High CarbonChar *Oil & Gas Science and Technology* [Online], 2011.

2. Rubel, A.; Zhang, Y.; Neathery, J.; Liu, K. In *Comparative Study of the Effect of Different Coal Fly Ashes on the Performance of Oxygen Carriers for CLC*, 7th International Symposium on Coal Combustion, Harbin, China, July 17-18, 2011; Harbin, China, 2011.

3. Zhang, Y.; Rubel, A.; Lui, K.; Neathery, J. In *Development of Freeze Granulated Oxygen Carriers for Chemical Looping Combustion of Solid Fuels*, 2010 Spring AIChE Conference, San Antonio, TX, March 21-25; San Antonio, TX, 2010.

(280f) Investigation of Oxygen Diffusivity within Oxygen Carrier In Chemical Looping Process

Siwei Luo[1], Zhenchao Sun[2], Fanxing Li[3], Deepak Sridhar[4], Liang Zeng[1] and Liang-Shih Fan[5], (1) William G. Lowrie Department of Chemical and Biomolecular Engineering, Ohio State University, Columbus, OH, (2)Chemical and Biomolecular Engineering, Ohio State University, Columbus, OH, (3) Chemical and Biomolecular Engineering, North Carolina State University, Raleigh, NC, (4)Chemical and Biomolecular Engineering, The Ohio State University, Columbus, OH, (5)William G. Lowrie Department of Chemical and Biomolecular Engineering, The Ohio State University, Columbus, OH

Chemical looping process is effective in converting carbonaceous fuels into hydrogen and electricity while generating a sequestration-ready CO_2 stream. The process scheme is enabled by the cyclic reduction and oxidation reactions of an oxygen carrier (OC). The properties of OCs are of critically importance for the performance of chemical looping process. Our previous experimental studies show that composite iron oxide synthesized by adding support materials to the pure iron oxide can serve as good OCs. Unlike the unsupported iron oxide which deactivates within the first few redox cycles, the composite iron oxide maintains good reactivity over a long time although pore volume decreases significantly in the first few cycles. This phenomenon indicates that traditional understanding on support effect, mainly based on morphological analysis, may not be accurate. In this study, we investigated the role of support through both marker experiments and periodic Density Functional Theory (DFT) calculations. Ilmenite ($FeTiO_3$) was used to model the simplified TiO_2 supported iron (II) oxide material. Both the oxygen diffusion barrier and vacancy formation energy in ilmenite ($FeTiO_3$) and wustite ($Fe_{1-x}O$) were calculated. The diffusivities of oxygen in both materials were estimated based on the calculation results. It was found that the energy barrier for oxygen anion diffusion in ilmenite was significantly lower than that in wustite. The calculated oxygen diffusivity in ilmenite was also higher. This finding was corroborated by the marker experiment which indicates notably enhanced oxygen anion conductivity in the supported iron oxide.

See more of this Session: Chemical Looping Processes I
See more of this Group/Topical: Energy and Transport Processes

THE STATUS OF CHEMICAL LOOPING PROCESSES DEVELOPED AT THE OHIO STATE UNIVERSITY

Presented by

Deepak Sridhar
William G. Lowrie Department of Chemical and Biomolecular Engineering
The Ohio State University
Columbus, OH, USA

CHEMICAL LOOPING AT OSU

OSU CCR Process Demonstration

OSU 25KW$_{th}$ Calcium Looping System for Hydrogen Production

OSU 25KW$_{th}$ Syngas Chemical Looping System for Hydrogen Production

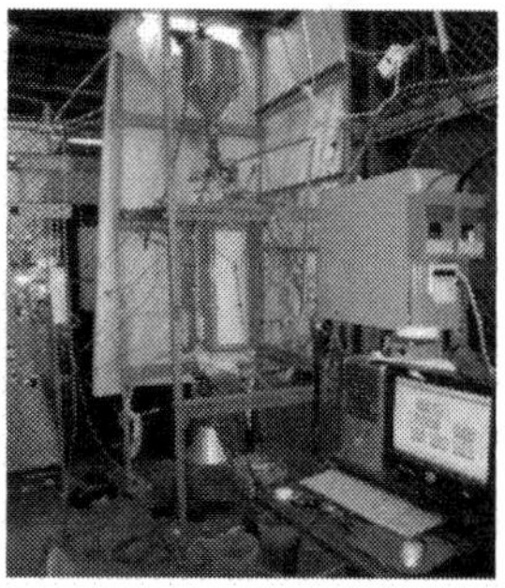

OSU 2.5KW$_{th}$ Coal Direct Chemical Looping System – Bench Unit

Syngas Chemical Looping (SCL) Process

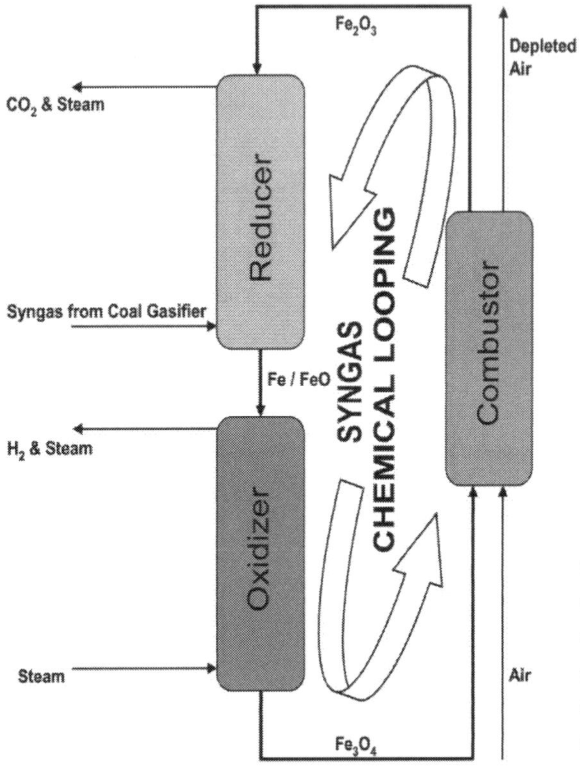

Main reactions:

Reducer $C_xH_yO_z + Fe_2O_3 \rightarrow CO_2 + H_2O + Fe$

Oxidizer $Fe + H_2O \rightarrow Fe_3O_4 + H_2$

Combustion Train $Fe_3O_4 + O_2 \rightarrow Fe_2O_3 + Heat$

Overall reaction

$C_xH_yO_z + H_2O/O_2 \rightarrow CO_2 + H_2/Heat$

Possible Oxygen Carriers:

Fe_2O_3 – Fe / FeO

General Observations:

- 2 Moving Bed + 1 Entrained bed reactors
- Very High Fuel Conversion
- Near 100% in-situ CO_2 capture
- High Purity H_2 generation
- High Solid Conversion
- Low Solid Circulation Rate

Syngas Chemical Looping – Applications

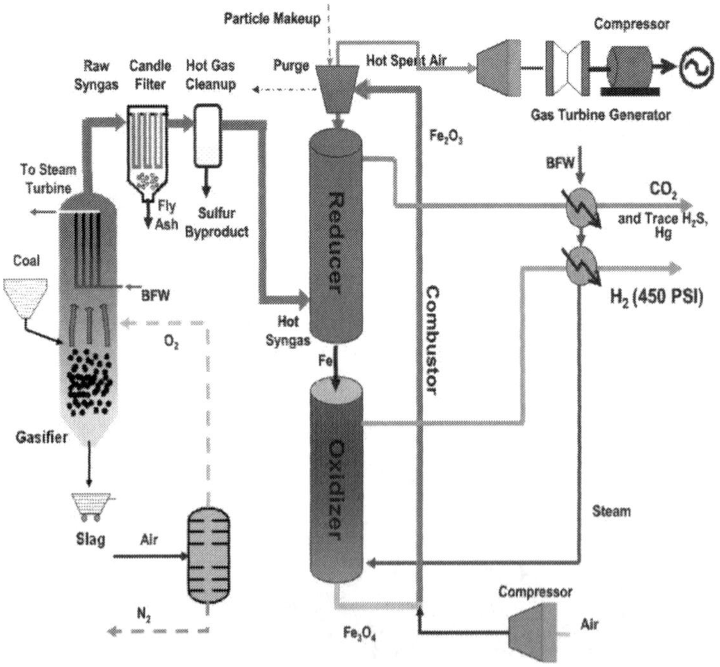

General Process Flow scheme

Product Flexibility

- Can co-generate any ratio of H_2 and Electricity from gaseous fuels
- Coupled with F-T process to generate liquid fuels efficiently

THE OHIO STATE UNIVERSITY

Chemical Looping Reactor Design

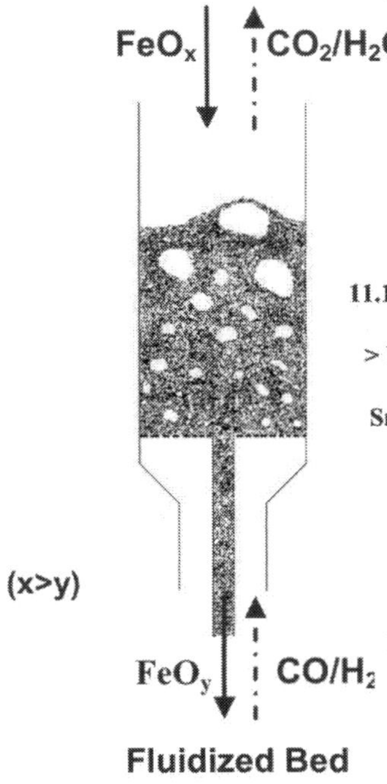

Fluidized Bed *v.s.* Moving Bed

11.11%	← Maximum Solid Conversion →	**50.00%**
> U_{mfv}	← Gas Velocity →	**< U_{mfv}**
Small	← Particle Size →	**Large**

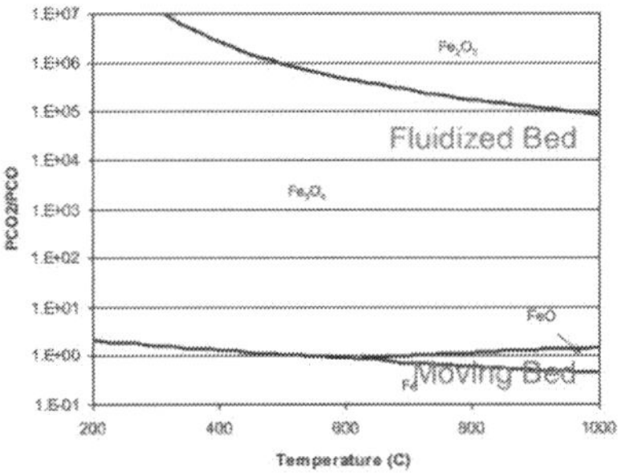

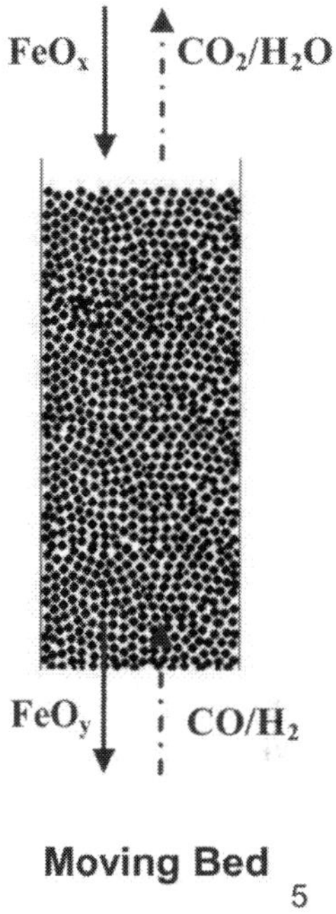

5

THE OHIO STATE UNIVERSITY

Syngas Chemical Looping Process Development

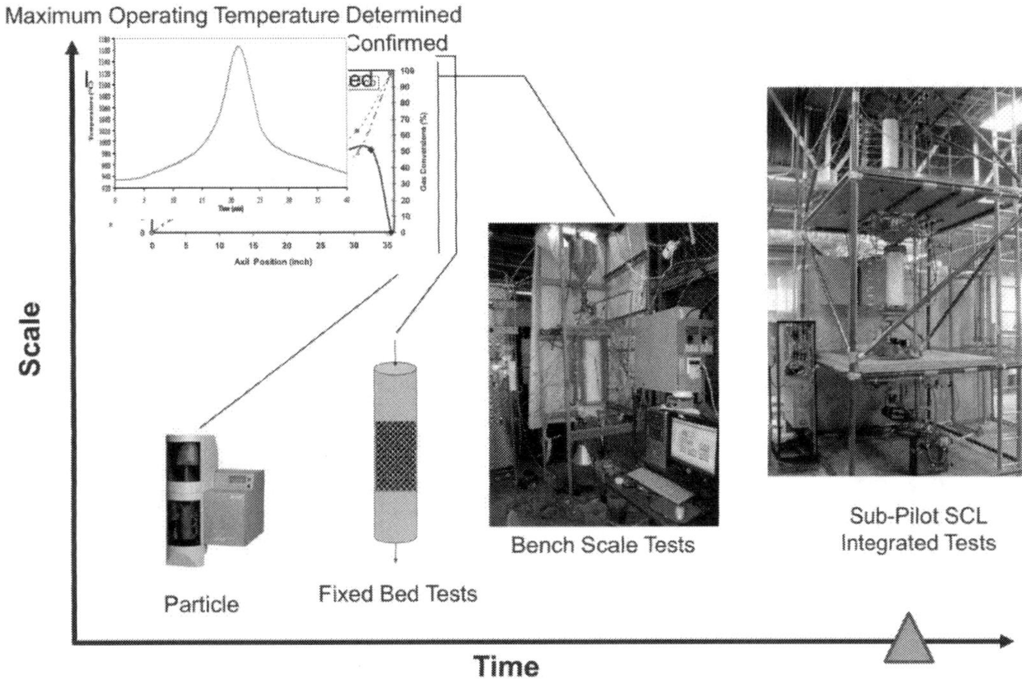

Techno-Economic Evaluation of the SCL process

	IGCC Process	SCL Process Electricity	CDCL Process Electricity	Conventional to Hydrogen Process	SCL Process Hydrogen	CDCL Process Hydrogen
Coal Feed (ton/hr)	132.9	132.9	132.9	132.9	132.9	132.9
Carbon Capture (%)	90	100	100	90	100	100
Hydrogen (ton/hr)	0	0	0	14.4	15.6	20.0
Net Power (MW)	321	365	435	2.1	26	3
Efficiency (%HHV)	32.1	36.5	43.5	57.8	64.1	79.0

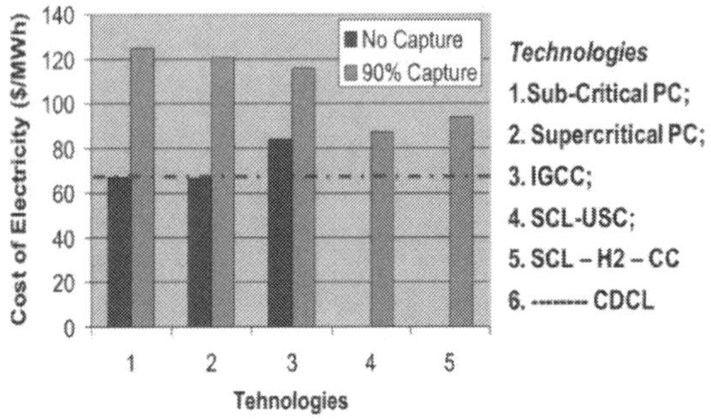

Process unit performance conditions and Assumptions used are similar to those adopted by the DOE baseline studies given in "Cost and Performance Baseline for Fossil Energy Plants", DOE/NETL-2007/1281.

- Improvement in Efficiency over conventional processes
- Lowest cost of electricity under a CO_2 regulated scenario
- SCL can pave the way to CDCL demonstrations in the future
- H_2 Production: SCL is 7 – 10% More Efficient Than Conventional Process (Analyzed by SAIC)
- Electricity Generation: SCL is 4.5 – 7% More Efficient Than IGCC with 100% CO_2 Capture (Analyzed by Noblis)
- When Integrated CTL, SCL can Increase the Liquid Fuel Yield by 10% with 19% Reduction in CO_2 Emission (Analyzed by Noblis)

SCL – Sub-Pilot Scale Unit Design

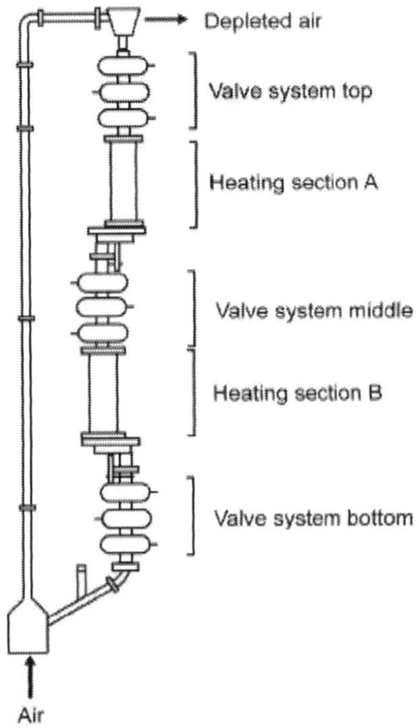

SCL – Sub-Pilot Scale Unit Features

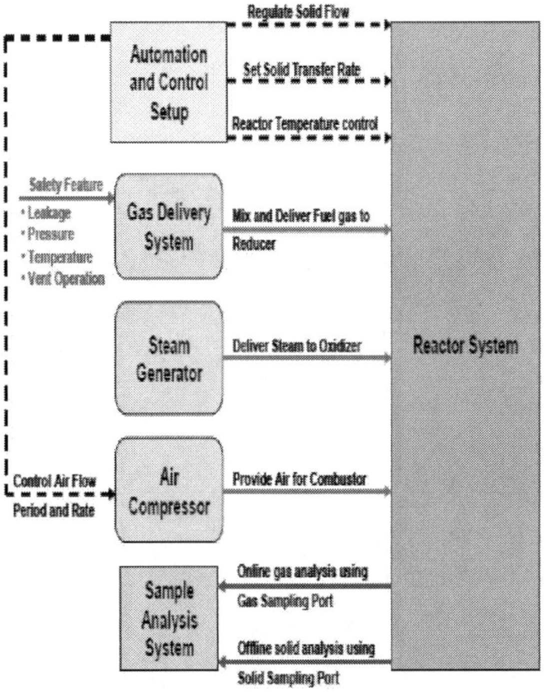

Salient Features:

- Completely automated process
  - Reactor temperature regulation
  - Valve operation
  - Air entrainment
- Regulated Gas flow
  - gas mixing panel could deliver any fuel combination
  - accurate steam and air flow
- Multiple Analyzers to obtain reliable data
  - Varian Micro GC
  - Sick Maihak Gas Analyser
- Safety features
  - Pressure Control
  - Temperature Control
  - Gas Leak Control
  - Vent Failure Control

Initial Sub-Pilot Scale Unit Operational Results

- 40+ hours of operational experience gained
  - 4 test runs
  - Final run was for 24 hours continuous
  - Average Syngas conversion – 98.9%
  - Average H_2 purity – 93.5%

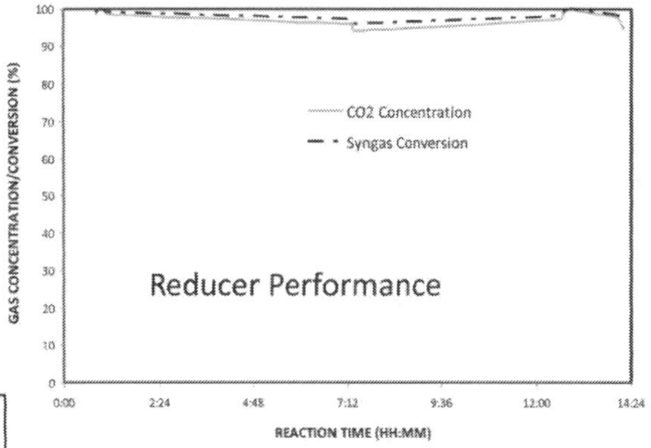

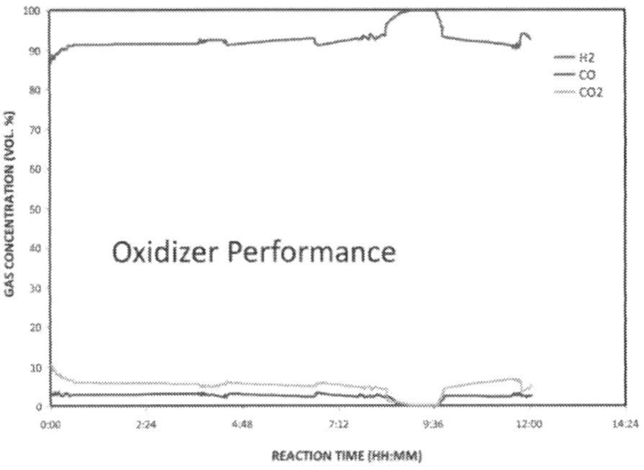

- Longer tests were not performed due to failure of the mechanical valves and solid feeder system
- Carbon deposition detected due to lack of gas pre-heating
- Attrition of 1% per cycle observed (more than double the inert condition attrition test results)

Sub-Pilot Scale Unit Performance Improvement

- **Longer Continuous Duration Testing required**
  - Changing from two point solid flow control to one point control
  - Reducing the number of mechanical valves used
- **Reduce Particle Attrition**
  - Change Combustor bend design
  - Use the improved oxygen carrier
- **Improved Reactor Performance**
  - Add pre-heating of inlet gases for all the three reactors

2008

2009

2010

Performance Results Modified Sub-pilot Unit

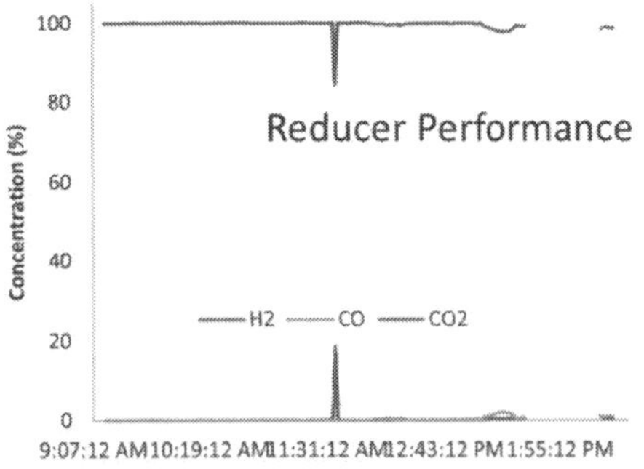

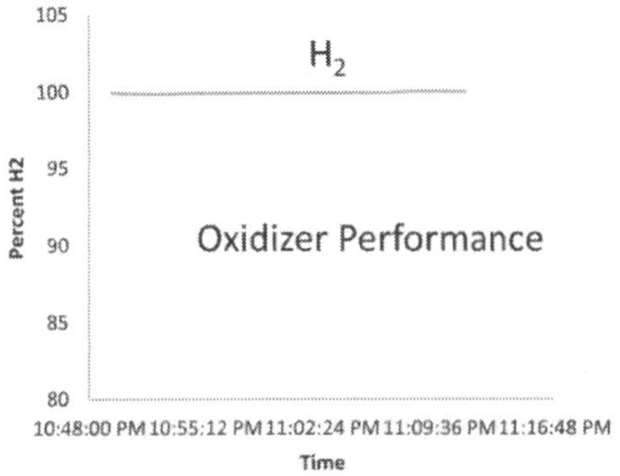

- Longer tests were not performed spanning more than 48 hours
- Carbon deposition was not observed
- Attrition of 0.53% per cycle observed (similar to the inert condition attrition test results)
- Syngas conversion of ~100% was observed.
- H2 purity of >99.9% was observed.
- Particle performance post the sub-pilot unit operation was similar to fresh particles

Accomplishments of the SCL Demonstrations

- 2.5 kW_{th} Demonstration Completed, > 99.5% Pure CO_2 and >99.95% Pure Hydrogen Obtained
- H_2 Production: SCL is 7 – 10% More Efficient Than Conventional Process (Analyzed by SAIC)
- Electricity Generation: SCL is 4.5 – 7% More Efficient Than IGCC with 100% CO_2 Capture (Analyzed by Noblis)
- When Integrated CTL, SCL can Increase the Liquid Fuel Yield by 10% with 19% Reduction in CO_2 Emission (Analyzed by Noblis)
- A 25 kW_{th} Sub-Pilot unit constructed and successfully operated, >99% in syngas conversion and >93.5% in H_2 purity obtained
- ***DOE/ARPA-E grant awarded, a 250 kW_{th} plant is being designed***

Successful Bench and Sub-pilot scale tests + Encouraging Techno-economic analysis

↓

Pilot Scale Design, Construction and Testing

Coal Direct Chemical Looping – Process Flow

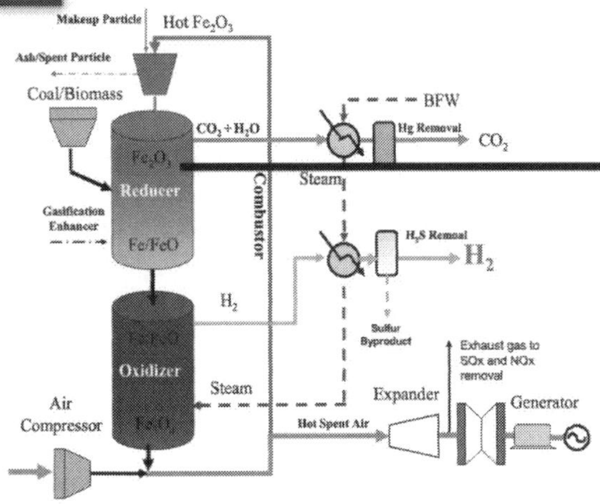

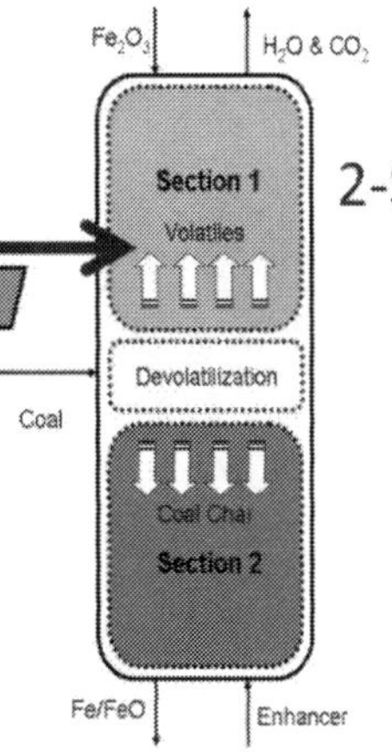

2-Stage Moving Bed

- Counter-current flows
- Gaseous volatiles
- Coal char

More challenging reducer operation less developed compared to SCL due to handling of solids at different sizes : Pulverized coal (~µm) vs. OC (2-5mm)

- **No need for Capital/Energy Intensive Gasifier and Air Separation Unit**
- **Great Scheme for the Retrofit to PC**

Reaction Enhancer

- Initiates the solid-solid reaction
- Direct enhancer : CO_2, H_2O, O_2
  - $CO_2 + C \rightarrow 2CO$
  - $H_2O + C \rightarrow H_2 + CO$
- Indirect enhancer : H_2
  - $3H_2 + Fe_2O_3 \rightarrow 2Fe + 3H_2O$

CDCL Current Status and Future Work

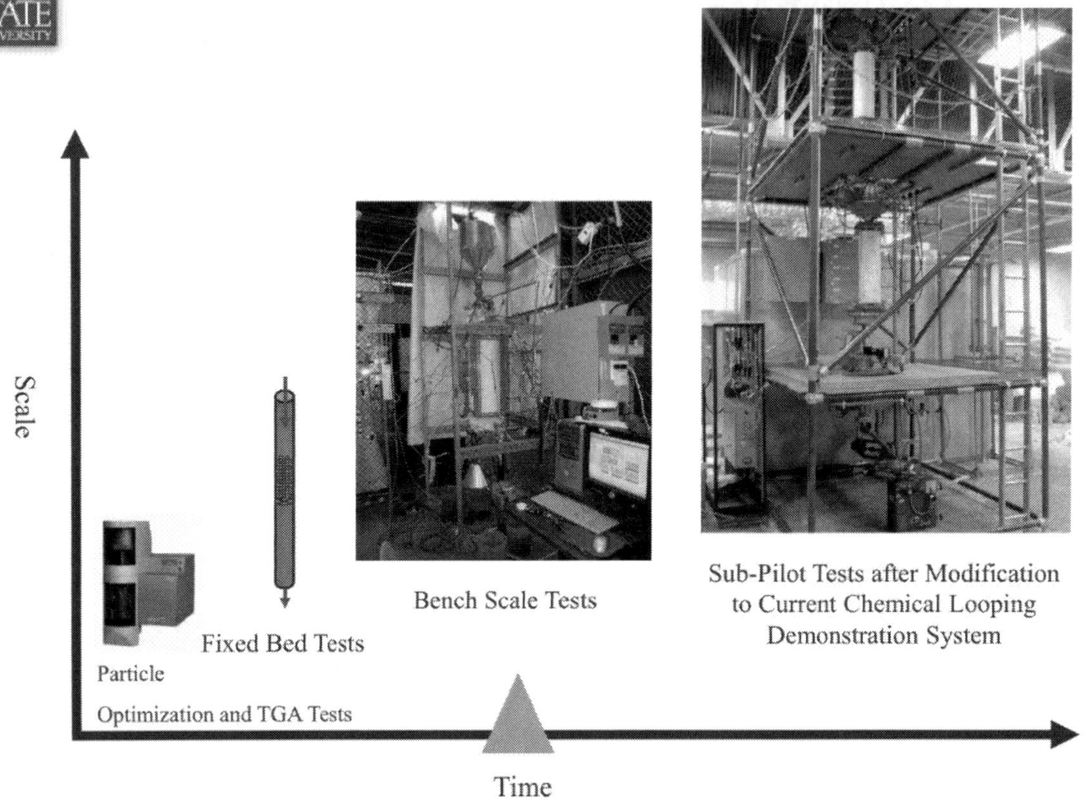

Coal Direct Chemical Looping – Key Results

Type of Fuel	Coal Volatile (CH_4)	Lignite Char	Ohio Coal Char	Anthracite Coal
Fuel Conversion (%)	**99.8**	**94.9**	**90.5**	**95.5**
CO_2 Concentration in Exhaust (% Dry Basis)	**98.8**	**99.23**	**99.8**	**97.3**

Particle Reactivity After Multiple Bench Tests

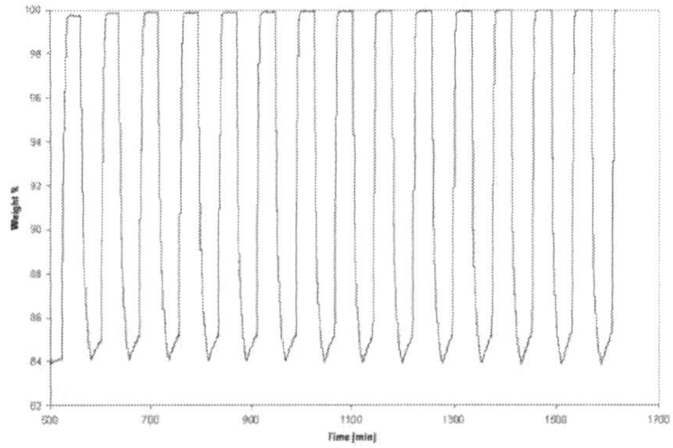

Comparison Among SCL, CDCL and Traditional Coal to Hydrogen/Electricity Processes

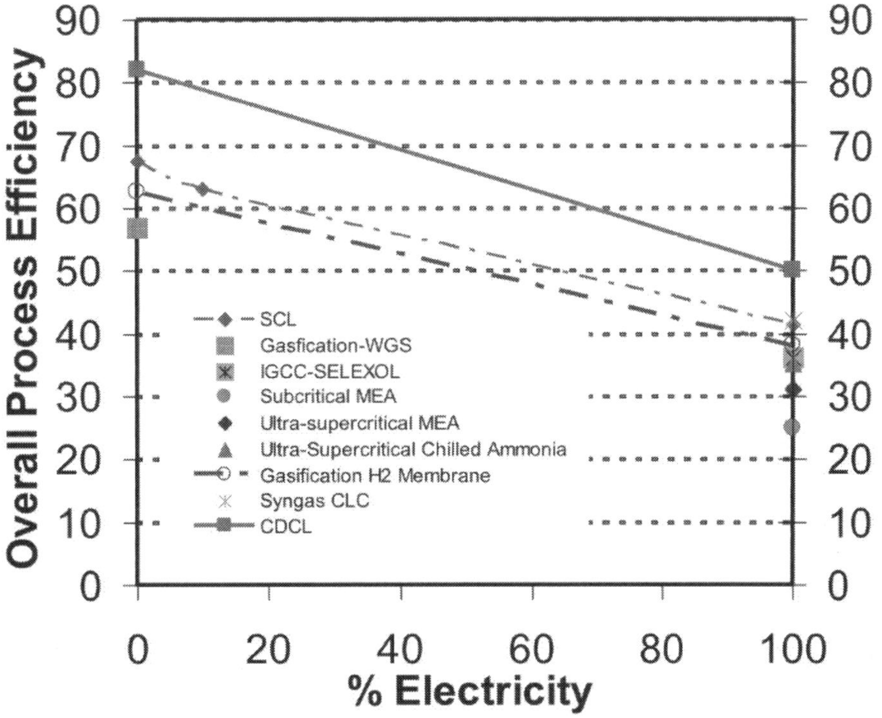

Team Work is Key

Dr. Liang-Shih Fan

Deepak Sridhar	Dr. Dawei Wang	Nihar Phalak
Ray Kim	Dr. Qiang Zhou	Niranjini Deshpande
Andrew Tong	Liang Zeng	William Wang
Elena Chung	Siwei Luo	Yao Wang
Mandar Kathe	Aining Wang	Zhenchao Sun
Samuel Bayham	Hui Yang	

Acknowledgements

Government Agencies and Industrial Corporations

- Department of Energy
- Ohio Coal Development Office
- Ohio Department of Development

Thank You

- Clear Skies
- Noblis/Metritek
- CONSOL Energy
- PSRI
- AEP
- Duke Energy
- Babcock & Wilcox
- Air Products
- Shell
- CRI/Criterion
- First Energy
- Carmeuse Lime and Stone
- LittleFord Day
- Specialty Minerals Inc.

(287a) Multi-Period Optimisation Design of Post-Combustion CO_2 Capture Processes

Niall Mac Dowell, Chemical Engineering, Imperial College London, London, United Kingdom and Nilay Shah, Centre for Process Systems Engineering, Department of Chemical Engineering, Imperial College London, London, United Kingdom

There have been several design and optimisation studies in the context of amine-based post-combustion CO_2 capture (PC-CC) from coal-fired power stations, and it is well accepted that the principal contributor to the capital-cost (CAPEX) of amine-based PC-CC is the absorber, with the majority of operating cost (OPEX) being accounted for by solvent regeneration[1]. However, the majority of these studies consider only the maximal stable generation (MSG) point of the power station. In reality, many power-plants do not operate consistently at the MSG. Thus, the cost optimal design of a PC-CC process must account for the dynamic operation of many fossil-fuel power-plants. The economic operation of a decarbonised fossil-fired power-plant is a complex trade-off between the time varying cost-of-electricity and cost of fuel. These variables will in turn affect the cost-optimal degree of CO_2 capture, and thus the design and operation of the CO_2 capture plant.

In this work, we present a new approach for the multi-period design of an integrated coal-fired power-plant and an alkanolamine-based CO_2 capture process. A validated[2] rate-based model of an absorber/stripper system for the chemisorption of acid gas in aqueous solvent solutions is used to represent the CO_2 capture process. To account for the non-idealities that are typical of amines and water, the statistical associating fluid theory for potentials of variable range (SAFT-VR)[3],[4] is used. This is a molecular approach, specifically suited to associating fluids. The SAFT formalism is used to represent some of the equilibrium reactions characterising the system, thereby simplifying the description of the chemical reactions[5]. The dynamic operation of the power-plant is described using data obtained from a real power station describing the variation in fuel burn rates in response to changing power demand. As we lack a thermodynamically consistent method in describing the degradation effects of SO_x on the solvents, we consider a model flue gas comprising N_2, CO_2 and H_2O. All models are implemented in the gPROMS[6] software package.

Cost-optimal process design and operation are identified via a constrained, multi-parametric optimisation approach. The objective function is the minimisation of total process cost (both CAPEX and OPEX) per unit of CO_2 captured. The design variables are the flowrate and state of the lean solvent stream as well as column geometry. Key performance indicators such as CO_2 emissions, solvent losses to the environment as well as the minimisation of energy required for solvent regeneration and also minimisation of CO_2 emissions to the atmosphere are included in the optimality criteria.

[1] Mac Dowell, N. et al, Energy. Environ. Sci., 3, 2010, 1645-1669

[2] Mac Dowell, N., Galindo, A., Jackson, G and Adjiman, C. S., Comp. Aided Chem. Eng., 28, 2010, 1231-1236

[3] Chapman, W.G., Gubbins, K.E., Jackson, G. & Radosz, M., Ind. Eng. Chem. Res. 29, 1709-1721., 1990

[4] Gil-Villegas, A., Galindo, A., Whitehead, P. J., Mills, S. J. & Jackson, G., J. Chem. Phys. 106 (10), 1997

[5] Mac Dowell, N., Llovell, F., Adjiman, C. S., Jackson, G and Galindo, A., Ind. Eng. Chem. Res., 49 (4), 1883-1899, 2010

[6] Process Systems Enterprise (PSE) Ltd. http://www.psenterprise.com/index.html

See more of this Session: CO2 Capture, Control and Sequestration
See more of this Group/Topical: Sustainable Engineering Forum

(287b) Optimal Design of Integrated Power Generation and CO2 Capture Systems

Yingru Zhao, Department of Chemical Engineering, Imperial College London, London, United Kingdom, Niall Mac Dowell, Chemical Engineering, Imperial College London, London, United Kingdom, Nouri J. Samsatli, Department of Chemical Engineering, Imperial College London, London and Nilay Shah, Centre for Process Systems Engineering, Department of Chemical Engineering, Imperial College London, London, United Kingdom

There have been extensive efforts to date related to the design and optimisation of CO_2 capture processes in isolation. However, very little work has been performed on the evaluation of CCS from integrated power plants and CO_2 capture processes. To improve the overall efficiency, increased process integration of the power generation process and capture process is required. Furthermore, it is well known that CCS from gas-fired power-plants presents a very different challenge to CCS from coal-fired power-plants.

Insights regarding the integrated plant operation can be provided through dynamic modelling and simulation. In this study, we present novel dynamic models of both decarbonised coal- and gas-fired power plants which account for the various transients associated in the power generation process. Sub-critical plants are considered as the dynamics of their operation are of more interest than those of the super-critical plants. Owing to its technological maturity, the decarbonisation technology of choice is post-combustion CO_2 capture using amine-based chemisorption. Detailed mass and energy balances were performed for each of the sub-models. All sub-models and their integration are implemented in gPROMS, and finally a generic system modelling framework is developed to study the performance and cost implications.

Based on a rigorous whole system analysis, the interaction of CO_2 capture level and power plant output control are investigated. The cost-optimal degree of CO_2 capture from both kinds of power station are also determined. In solving this optimisation problem, we account for the time-varying value of electricity as well as market prices associated with CO_2 emissions. Moreover, an objective function is proposed which maximises the revenue stream associated with energy production whilst simultaneously minimising the CO_2 emitted to atmosphere.

See more of this Session: CO2 Capture, Control and Sequestration
See more of this Group/Topical: Sustainable Engineering Forum

(287c) Synthesis of Optimal Adsorptive Carbon Capture Processes

YoungJung Chang[1], Alison Cozad[1], Hosoo Kim[2], Andrew Lee[2], Panagiotis Vouzis[1], N.V.S.N. Murthy Konda[3], A.J. Simon[4], Nick Sahinidis[1] and **David C. Miller**[3], (1)Chemical Engineering, Carnegie Mellon University, Pittsburgh, PA, (2)Computational Science Division, National Energy Technology Laboratory, Morgantown, WV, (3)U.S. Department Of Energy, National Energy Technology Laboratory, Morgantown, WV, (4)Lawrence Livermore National Laboratory, Livermore, CA

Solid sorbent carbon capture systems have the potential to require significantly lower regeneration energy compared to aqueous monoethanol amine (MEA) systems. To date, the majority of work on solid sorbents has focused on developing the sorbent materials themselves. In order to advance these technologies, it is necessary to design systems that can exploit the full potential and unique characteristics of these materials.

The Department of Energy (DOE) recently initiated the Carbon Capture Simulation Initiative (CCSI) to develop computational tools to accelerate the commercialization of carbon capture technology. Solid sorbents is the first Industry Challenge Problem considered under this initiative. An early goal of the initiative is to demonstrate a superstructure-based framework to synthesize an optimal solid sorbent carbon capture process. For a given solid sorbent, there are a number of potential reactors and reactor configurations consisting of various fluidized bed reactors, moving bed reactors, and fixed bed reactors. Detailed process models for these reactors have been modeled using Aspen Custom Modeler; however, such models are computationally intractable for large optimization-based process synthesis. Thus, in order to facilitate the use of these models for process synthesis, we have developed an approach for generating simple algebraic surrogate models that can be used in an optimization formulation. This presentation will describe the superstructure formulation which uses these surrogate models to choose among various process alternatives and will describe the resulting optimal process configuration.

See more of this Session: CO2 Capture, Control and Sequestration
See more of this Group/Topical: Sustainable Engineering Forum

INTEGRATED DESIGN AND DYNAMIC OPERATION OF SOLVENT-BASED PCC PROCESSES

Rajab Khalilpour, Ali Abbas

School of Chemical and Biomolecular Engineering,
The University of Sydney, Sydney, Australia
Tel: +61 4100 91600; Fax: +61 2 9351 2854; E-mail: r.kh@usyd.edu.au

2011 AIChE Annual Meeting: Minneapolis, MN, October 16-21

Abstract

Despite notable developments in solvent-based Post-combustion Carbon Capture (PCC) processes, the implementation of this technology in a power plant still burdens notable energy penalty mainly due to solvent regeneration. The energy penalty though varies by the type of power plant and by different techno-economic studies, is estimated to be above 20% (Khalilpour and Abbas, 2011). This results in serious reduction in power plant load.

One of the ways for power plants to reduce the profit loss, due to reduction of power load, is to schedule their carbon capture process. Considering the fact that the legal carbon capture requirements will be annual-based, a power plant would be able o choose to operate its PCC process in full capacity when the market electricity price is not high (e.g. off-peak). Contrarily, at a time with high electricity demand, the PCC process can be ramped down (or maybe shut-down) to allow maximum power generation. This strategy is very important for societies with liberated electricity markets.

In a monopoly electricity market, the government is usually responsible for investment in power plants and for satisfying the society electricity demands. In such condition, the load plan of power plants is defined by regional demand. Contrarily, in a liberated electricity market, such as in Australia, power plant operators plan their loading based on market price. This means that a combination of various factors such as weather temperature, working-day/holiday and others results in different demand and defines market electricity price (usually at every 30-minute interval). Consequently, the wholesale electricity price shows high volatility. Therefore, it is vital and economically attractive to study the dynamic behavior of power plants integrated with PCC process. This can allow power operators to analyze the complicacy of decision-making and load scheduling in tomorrow's power plants.

In this study we present our recent simulation study of a coal-fired power plant with solvent-based PCC. The study presents results that examine the behavior of the power plant with changes in load, flue gas flowrate and/or composition, electricity demand, etc. Interesting insight is shed on the effects of dynamic behaviors of the power plant on the PCC process sizing. Overall performance of the plant is also assessed in terms of various parameters such as solvent type and key operating conditions.

References

KHALILPOUR, R. & ABBAS, A. 2011. HEN optimization for efficient retrofitting of coal-fired power plants with post-combustion carbon capture. *International Journal of Greenhouse Gas Control,* 5, 189-199.

(287e) A Methodology for Risk Assessment of CO2 Sequestration Based On Surrogate Models of Detailed Simulations

Yan Zhang and Nick Sahinidis, Chemical Engineering, Carnegie Mellon University, Pittsburgh, PA

Large scale implementation of CO2 geologic sequestration requires detailed uncertainty analysis and risk assessment for leakage potential. Probabilistic risk assessment based on Monte Carlo simulation is one option for quantifying the risk. It involves repeated random realizations of a simulator to generate probability distributions of interesting variables such as leakage amount. However, in the context of modeling CO2 plume migration underground after injection, numerical solution of non-linear equation systems for heat and mass balances is too time-intensive for Monte Carlo simulation. To cope with this computational expense, investigators are often forced to make simplifying assumptions and resort to approximate models. To address this challenge, instead of solving a large system of non-linear equations, we first develop a surrogate model for the multi-component multi-phase flow processes involved in the injection of CO2 into deep saline aquifers. In particular, we use TOUGH2 in conjunction with adaptive sampling and cross validation to obtain a surrogate algebraic model. Then, probabilistic risk assessment for CO2 leakage potential is performed based on Monte Carlo simulation using the surrogate model. We demonstrate that our approach not only reduces significantly the computational requirements of risk assessment but also provides accurate results. The accuracy of the model was validated by comparisons to TOUGH2 itself.

See more of this Session: CO2 Capture, Control and Sequestration
See more of this Group/Topical: Sustainable Engineering Forum

(287f) The Application of a Model Predictive Control-Based Approach to CO2 Capture Processes: Towards Operational Cost Minimisation

Alicia Arce, MATGAS Research Center, Barcelona, Spain, Niall Mac Dowell, Chemical Engineering, Imperial College London, London, United Kingdom, Nilay Shah, Centre for Process Systems Engineering, Department of Chemical Engineering, Imperial College London, London, United Kingdom and Lourdes F. Vega, MATGAS Research Center and Carburos Metlicos, Air Products Group, Barcelona, Spain

Amine-based chemisorption of CO_2 is a promising near term option for the decarbonisation of large, fixed-point CO_2 emission sources. However, the operational expenditure (OPEX) associated with this technology imposes a significant energy penalty on the power plant. As it is the solvent which determines the thermodynamic and kinetic efficiency of the process, the design of advanced solvents provides a real opportunity to reduce the OPEX. Typically, when one refers to the dynamic operation of CCS-type processes, it is the transient behaviour of the start-up and shut-down of this process which is considered. However, the so-called steady-state operation of power-plants is itself dynamic, i.e., the flowrate, temperature and composition of the inlet flue-gas can vary in real time. Thus, a given solvent and mode of process operation which may be optimal for a particular operating regime, may be sub-optimal for another operating regime. Consequently, the implementation of advanced control strategies present an important opportunity for the intensification of these processes resulting in a significant reduction in the lifetime operational expenditure associated with CCS systems.

This work focuses on the design of an explicit model predictive controller (MPC) for the real-time control of solvent composition and process operation as successfully implemented in other applications such as .To achieve this, we integrate molecular-based fluid theories with high-fidelity process models. Based on this model, an approximate model is developed and a model predictive controller is formulated for the reduced model where our multi-parametric algorithms are applied to derive a suitable and robust explicit MPC controller. By incorporating the explicit controller expressions (control laws) in the original model, a validation step is then carried out.

In this way, we present a unified-systems based methodology for the OPEX reduction of solvent-based post-combustion CO_2 capture processes.

IPCC, 2005: IPCC Special Report on Carbon Dioxide Capture and Storage. Prepared by Working Group III of the IPCC, Cambridge University Press, Cambridge, United Kingdom and New York, USA

Pistikopoulos, E.N., Bozinis, N., Dua, V., Perkins, J. & Sakizlis, V. (2004). Improved Process Control, European Patent EP1399784

Arce A., del Real A. J., Bordons C. and Ramírez D. R. IEEE Transactions on Industrial Electronics, 57 (6), 1892-1905, 2010.

Mac Dowell, N. et al., Ind. Eng. Chem. Res., 49(4), 1883-1899, 2010

Mac Dowell, N. et al., ESCAPE 20, 2010

See more of this Session: CO2 Capture, Control and Sequestration
See more of this Group/Topical: Sustainable Engineering Forum

Effect of Shear Stress on Cytotoxicity of Silica Nanoparticles

Donghyuk Kim, Yu-Shen Lin, and Christy L. Haynes

In this work, nanotoxicity in the bloodstream was modeled as the cytotoxicity of sub-50 nm mesoporous silica nanoparticles to human endothelial cells was investigated under microfluidic flow conditions. Compared to traditional *in vitro* cytotoxicity assays performed under static conditions, unmodified mesoporous silica nanoparticles show higher and shear stress-dependent toxicity to endothelial cells under flow conditions. Interestingly, even under flow conditions, highly organo-modified mesoporous silica nanoparticles show no significant toxicity to endothelial cells. This paper clearly demonstrates that shear stress is an important factor to be considered in *in vitro* nanotoxicology assessments and provides a simple device for pursuing this consideration.

(292b) Quantifying the Origin of Nanosilver Ions and Their Antibacterial Activity

Georgios A. Sotiriou[1], Andreas Meyer[2], Jesper Knijnenburg[1], Sven Panke[2] and Sotiris E. Pratsinis[1], (1) Department of Mechanical and Process Engineering, Particle Technology Laboratory, ETH Zurich, Zurich, Switzerland, (2)Department of Biosystems Science and Engineering, ETH Zurich

Nanosilver is one of the most prominent components in several nanotechnology products, primarily for its bactericidal and plasmonic properties [1,2]. Its use, however, is questioned as U.S. EPA has been petitioned to label it as pesticide. To better understand its antibacterial activity, nanosilver with closely controlled size immobilized on nanostructured silica is generated by flame aerosol technology [3]. The antibacterial activity of small (<10 nm) nanosilver particles is attributed mostly to released Ag^+ ions from their surface upon contact with water [1]. The origin of these ions is explored systematically now by conditioning the surface of various nanosilver particles (4-9 nm) immobilized on nanostructured silica by wet-impregnation and dry flame-spray pyrolysis. Silica prevents nanosilver flocculation without the need of any surface functionalization (e.g. citrate) that would compromise the Ag^+ ion release kinetics. Washing or reducing the Ag/SiO_2 nanoparticles removes the oxide layer from the nanosilver surface minimizing their ion leaching. The amount of leached Ag^+ ions corresponds to the dissolution of one or two silver oxide monolayers. Finally, the antibacterial activity of small (< 10 nm) nanosilver particles that have their surface oxide layer removed or reduced is significantly lower than the freshly prepared ones by either wet or dry methods.

1. Sotiriou, G. A.; Pratsinis, S. E., Antibacterial activity of nanosilver ions and particles. *Environ. Sci. Technol.* **2010,** *44*, 5649.

2. Sotiriou, G. A.; Sannomiya, T.; Teleki, A.; Krumeich, F.; Vörös, J.; Pratsinis, S. E., Non-Toxic Dry-Coated Nanosilver for Plasmonic Biosensors. *Adv. Funct. Mater.* **2010,** *20*, 4250.

3. Sotiriou, G. A.; Teleki, A.; Camenzind, A.; Krumeich, F.; Meyer, A.; Panke, S.; Pratsinis, S. E. Nanosilver on nanostructured silica: Antibacterial activity and Ag surface area. *Chem. Eng. J.* **2011,** *170,* 547.

See more of this Session: Environmental Implications of Nanomaterials: Biological Interactions
See more of this Group/Topical: Topical C: Environmental Aspects, Applications, and Implications of Nanomaterials and Nanotechnology

(292c) Oxidation of Biomolecules by Emerging Inorganic Nanoparticles

Antonia Luna-Velasco[1], Reyes Sierra-Alvarez[2] and Jim A. Field[2], (1)Chemical and Environmental Engineering, University of Arizona, Tucson, AZ, (2)Chemical & Environmental Engineering, University of Arizona, Tucson, AZ

Oxidationof Biomolecules by Emerging Inorganic Nanoparticles

Antonia Luna-Velasco, Reyes Sierra-Alvarez, Jim A.Field

Departmentof Chemical and Environmental Engineering

Universityof Arizona, P.O. Box 210011, Tucson, Arizona 85721, USA

In recent years, inorganicnanoparticles (NPs) are finding increasingapplication in manufacturing and consumer products. The unusual properties ofthese emerging pollutants (e.g., very small size, high surface area andreactivity) has lead to growing public and regulatory concern about theirpotential toxicity. An important mechanism suggested for nanotoxicityis oxidative stress caused by reactive oxygen species (ROS). Testing thechemical generation of ROS by the interaction of NPs with biomoleculescould be a helpful tool for nanotoxicity screening. Inthis study, a rapid screening assay was developed to determine the chemicalgeneration of ROS by NPs through their interaction with biomoleculessusceptible to oxidation such as, L-3,4-dihydroxyphemylalanine(L-dopa) and 1,2-dihydroxybenzene (catechol). The assay utilized the dye dichlorofluorescin(DCFH), which is oxidized to its highly fluorescent product dichlorofluorescein(DCF) by ROS. Additionally, oxidation of proteins by NPs was evaluated using anELISA assay (based on detecting carbonyl groups in amino acid side chains).

Theinteraction of nanosized CeO_2 with L-dopa orcatechol significantly enhanced the oxidation of theROS indicator dye. Nanoscale CeO_2, $Fe_2O_3$and Fe^0 enhanced ROS production during the autooxidationof L-dopa by more than four-fold in reactions that were dependent on O_2. Mn_2O_3 NPs oxidized DCFHin the absence of O_2 or L-dopa, suggesting a direct oxidation of thedye. Mn_2O_3 NPs also significantly oxidized the protein bovineserum albumin indicating their potential reactivity with cell components. ROSgeneration by NPs (CeO_2, Mn_2O_3 and Fe^0)in aerobic aqueous suspensions was confirmed by electron paramagnetic resonance(EPR) signature for hydroxyl radicals with spin traps. The results taken as awhole indicate that NPs can generate ROS via chemical reactions with mediumcomponents and biomolecules susceptible to oxidation,such as L-dopa. The rapid assay with L-dopa and DCFH is a method proposed toscreen for chemical ROS production by NPs.

(292d) Demonstration of Nanoparticle-Bound Polymer Biodegradation and Resulting Nanoparticle Destabilization

Teresa L. Kirschling[1], Patricia L. Golas[2], Krzysztof Matyjaszewski[2], Kelvin B. Gregory[3], Gregory V. Lowry[3] and **Robert D. Tilton**[4], (1)Department of Chemical Engineering, Carnegie Mellon University, Pittsburgh, PA, (2)Department of Chemistry, Carnegie Mellon University, Pittsburgh, PA, (3) Department of Civil and Environmental Engineering, Carnegie Mellon University, Pittsburgh, PA, (4) Department of Chemical Engineering and Department of Biomedical Engineering, Carnegie Mellon University, Pittsburgh, PA

Many engineered nanoparticles are modified by adsorption or grafting of polymers onto their surfaces in order to impart colloidal stability in suspension, to promote adhesion with matrix materials, to increase biocompatibility, or to impart some particular biomolecular targeting capability. By controlling nanoparticle flocculation and surface deposition, surface-bound polymer layers strongly influence nanoparticle fate, transport, and subsequent biological impact in environmental systems. Biodegradation is one potential route to the breakdown of surface-bound polymers, but it is unknown whether such nanoparticle-bound polymers are bioavailable or if the surface confinement restricts microbial access to the chains so that they become recalcitrant to biodegradation. Most engineered nanoparticles are in a size range that cannot be internalized by bacteria, so the bound chains must be accessed extracellularly. Here we demonstrate, for the first time, that a polymer brush bound to nanomaterials can be bioavailable to microbes. Model core/shell nanomaterials with covalently attached polyethylene oxide (PEO) chains were prepared for this study. These were synthesized to obtain a nanomaterial in which biodegradation was the only available coating breakdown mechanism. Taking the form of a densely crosslinked bottlebrush copolymer with a hydrophobic core, these nanomaterials consisted of poly(ethylene oxide) methacrylate macromonomers crosslinked by hydrophobic divinyl benzene. PEO is a common constituent of consumer products, and PEO-based polymers are common, effective steric stabilizers for nanoparticles and other colloids. Accordingly, this an appropriate model system to represent a class of nanoparticle coatings that is likely to be introduced to the environment. The nanoparticles were approximately 30 nm in hydrodynamic radius and were colloidally stable in the culture media used in the biodegradation experiments. Microbial samples isolated from river water were enriched by supplying PEO homopolymer as the sole carbon source. The PEO-homopolymer degrading microbial enrichment cultures were transferred to fresh media, then supplied with PEO brush-bound nanoparticles as the sole carbon source. Biomass and CO_2 production measurements indicated that the particle-bound PEO brush was indeed bioavailable to these microbes. DNA sequencing indicated that the mixed culture responsible for degradation included *Novosphingobium sp.*, *Pseudomonas sp.*, and *Hydrogenophaga sp.* Brush biodegradation resulted in nanoparticle flocculation. Control experiments demonstrated that byproducts of microbial metabolism or cell debris did not cause the observed nanoparticle flocculation. This points to the need to understand biologically mediated transformations of nanoparticle coatings in order to understand the fate and transport of nanoparticles in the environment.

See more of this Session: Environmental Implications of Nanomaterials: Biological Interactions
See more of this Group/Topical: Topical C: Environmental Aspects, Applications, and Implications of Nanomaterials and Nanotechnology

Partitioning of Fullerene between Water and Synthetic Membrane Materials: Effects of Temperature and Membrane Compositions

Yeonjeong Ha, Howard M. Liljestrand, Lynn E. Katz

Department of Civil, Architectural and Environmental Engineering

The University of Texas at Austin, 1 University Station C1786, Austin, Texas, 78712

1. Introduction

As the usage of engineered nanomaterials increases, concerns on the unknown risks of these nanoparticles towards humans and the environment have grown. Despite these growing concerns about nanomaterials in aquatic environments, many unknowns remain. Investigating bio-uptake and bio-accumulation of nanomaterials is of critical importance not only to determine bioavailability but also to evaluate fate and transport in the environment of these compounds.

For hydrophobic contaminants and ionized organic chemicals, many models have been developed to determine their bio-concentration and toxicity. For example, octanol-water partitioning (K_{ow}) has been successfully used to evaluate the bioavailability of chemicals. More recently, liposome-water partitioning (K_{lipw}) has been proposed as being more appropriate than octanol-water partitioning (K_{ow}) because synthetic membrane vesicles are better surrogates and K_{lipw} can reduce thermodynamic differences from actual biota. For engineered nanomaterials, K_{ow} values of C_{60} (Jafvert et al., 2008) and carbonnanotubes (Petersen et al., 2010) have been determined. However, few attempts have been made to measure K_{lipw} of nanoparticles to investigate bioavailability of engineered nanomaterials.

Carbon fullerene (C_{60}), which was discovered in 1985 (Kroto et al., 1985), is a major material of nanoscale science and engineering. The production and application of fullerene have steadily increased due to increasing commercial interest in its unique properties, such as cage like structure, strong hydrophobicity, and high reactivity. It is estimated that more than 1500 tons per year of carbon nanoparticles including fullerenes and nanotubes are produced (Ringwood et al., 2009). However, there is limited information concerning the impact of fullerene on the natural environment and human health.

Bioavailability of fullerene in aquatic environment can be investigated by measuring partition coefficient between liposome and water (K_{lipw}) similar to that for other chemicals. Many researchers (Yamamoto et al., 2004, Kwon et al., 2006) have determined K_{lipw} using an equilibrium dialysis technique in which water and liposome solutions are separated by a dialysis membrane through which the chemical can diffuse but the liposome cannot. However this approach is less useful for fullerene, because the dialysis membrane can prevent the diffusion of engineered nanoparticles after their aggregation in water to a particle size much bigger than the molecular scale of hydrophobic or ionizing organic chemicals.

In this study, solid supported lipid membrane (SSLM) was chosen as a model biological membranes. Nonporous silica beads were used as the solids, and these were coated with the lipids. K_{lipw} of fullerene was investigated with various types of lipid membrane materials and temperatures. In addition, ternary lipid (liquid ordered phase lipid, liquid disordered phase lipid and cholesterol) mixture membrane vesicles were prepared at two different temperatures (above and below the transition temperature) to investigate whether K_{lipw} of nanoparticles can be influenced by the phase height difference between liquid disordered and liquid ordered/cholesterol. Lipids used in this study are 1,2 dimyristoyl-sn-glycero-3-phosphochloline (DMPC), 1,2 dioleoyl-sn-glycero-3 phosphocholine (DOPC), 1,2 dierucoyl-sn-glycero-3-phosphochloline (DEruPC), DOPC/ sphinogosylphosphochloline (SM) /cholesterol, DMPC/SM/cholesterol, DEruPC/SM/cholesterol.

2. Materials and Methods

2.1 Chemicals

Fullerene (C_{60}, 99.5+ %) was purchased from SES Research (Houston, TX). Unsaturated lipids such as 1,2 dimyristoyl-sn-glycero-3-phosphochloline (DMPC, C14:1, C14:1), dioleoyl-sn-glycero-3 phosphocholine (DOPC, C18:1, C18:1), and 1,2 dierucoyl-sn-glycero-3-phosphochloline (DEruPC, C22:1, C22:1), and sphinogosylphosphochloline (SM) were obtained from Avanti Polar Lipids. Cholesterol was purchased from Aldrich Chemical Co., and non-functionalized silica microspheres (mean diameter of 5.2 μm, 100 % solids content) were obtained from Bang's Laboratories.

2.2 Aqueous C_{60} solution

Fullerene has negligible solubility in water due to its strong hydrophobicity. However, many studies have shown that fullerene can form highly stable, negatively charged colloidal aggregates (nC_{60}) when it is exposed to water. Stable aggregated forms of fullerene have more than 8-11 orders of magnitude greater solubility than its hydrophobic form (Ruoff et al., 1993). Therefore, it is evident that the transition from its extreme hydrophobic form to water stable forms have critical implications for fullerene transport, transformation and biological effects (Isaacson et al., 2009).

The water stable suspensions of fullerene can be formed by various preparation methods in the laboratory. In this study, a modified SON/nC_{60} method was applied to make the aqueous C_{60} solution. Briefly, the fullerenes nanoparticles were formed by first adding pure fullerene to toluene. The solution was then mixed with a large volume of the deionized water and the toluene was removed by evaporation under ultrasonic treatment. The fullerene suspension in water was filtered through a 0.8 μm membrane filter. The mean size of C_{60} particles in water is in the range between 50 and 100 nm as measured by Dynamic Light Scattering (Model Zetasizer nano ZS).

2.3 Preparation of solid supported lipid bilayer

The synthetic membrane vesicles of each lipid compositions can be prepared using the thin film hydration technique followed by the rapid extrusion process. The detailed procedure is described elsewhere (Escher et al., 2000, Kwon et al., 2006). Briefly, the organic solution of lipid was evaporated using a nitrogen purging and the thin film residue was dissolved in water. After storage and equilibration at 4°C for a day, the lipid suspension was extruded several times through a polycarbonate membrane to reduce polydiversity of vesicles.

Solid supported lipid bilayers were formed using methods described in Bayerl et al., (1990) and Baksh et al., (2004). Silica beads were washed three times with methanol. The lipid suspensions were incubated at 30 °C for 30 min, and the lipid dispersions were sonicated (Branson 2000) for 15 min. The resulting vesicle dispersions were incubated at 60 °C for 5 min and then poured onto the dry glass beads with 60 sec rigorous vortex mixing. After vortexing, the silica beads and lipid dispersions mixture were centrifuged for 10 min at 2,000 g and then supernatants were discarded. Excess vesicles not adsorbed at the solid surface were removed by redispersing in water, vortexing, and the supernatants were discarded. The mass of lipids which were adsorbed on the silica beads was determined by measuring the difference between initial lipid concentration and supernatant concentration. Each lipid concentration was measured by TOC Analyzer (Tekmar Dohrmann Apollo 9000).

For solid supported ternary lipid mixture membrane vesicles, weight ratio of unsaturated lipids: SM: cholesterol was 2:1:1. Two replicate samples were prepared. One sample was stored at below the transition temperature (T_m), and the other one was first stored at above T_m overnight and then placed at below T_m to create phase height difference between liquid disordered lipid and liquid ordered lipid/cholesterol.

2.4 Determination of K_{lipw}

The solid supported lipid bilayer and aqueous fullerene dispersion were placed into 1.8 mL amber vial with PTFE/silicon septa. The reactors were incubated for 80 hr using a custom-made tumbling device (4 °C, 11 °C, and 25 °C) or shaker (30 °C and 50 °C, at 180 rpm). For ternary lipid mixture, vials were incubated at 25 °C with a tumbling device. The preliminary experiment showed that equilibrium was obtained after 80 hrs using both tumbler and shaker. After 80 hrs, each vial was placed for 1 hr to settle down the solid supported lipid bilayer. Then supernatants contained only free fullerene without lipid vesicles, and concentration of fullerene in supernatants (C_w) were measured using a Waters 2690 high-performance liquid chromatography system equipped with a Waters 996 photodiode array detector (Milford, MA, USA) after liquid extraction using toluene and 0.1 M of $Mg(ClO_4)_2$. Double or Triplicate samples were prepared for each test. The partition coefficient (K_{lipw}) was calculated using

$$K_{lipw}(L/kg-liposome)=\frac{C_{lip}}{C_w m}=\frac{C_0-C_w}{C_w m}$$

Where C_{lip} is the concentration of fullerene in the liposome side, which is same as the difference between the initial or control C_{60} concentration (C_0) and fullerene concentration in supernatants (C_w) and m is the concentration of liposome (kg-liposome/L).

3. Results and Discussion

Table 1 shows K_{lipw} values between water and DMPC ($K_{lipw,DMPC}$), and between water and DEruPC ($K_{lipw,DEruPC}$) at five different temperatures (4 °C, 11 °C, 25 °C, 30 °C, and 50 °C). The log K_{lipw} (L/kg) values range from 3.3 to 5.3. As far as we know, the K_{lipw} value of fullerene has not been reported. Some studies have concerned about Bioconcentration factor (BCF) of fullerene: Jafvert et al., (2008) estimated log BCF values of fullerene as 4.51, 5.37, 5.71 for cod, earthworms and salmon, respectively, and Tervonen et al., (2010) measured log BCF values of fullerene for *Daphnia magna* ranged from 3.30 and 3.88. Those reported values are generally consistent with the results of this study.

Table 1. Liposome – Water Partition Coefficients (K_{lipw}, L/kg-liposome) of fullerene obtained at five different temperatures using 1,2 dimyristoyl-sn-glycero-3-phosphochloline (DMPC) and 1,2 dierucoyl-sn-glycero-3-phosphochloline (DEruPC)*

	4 °C	11 °C	25 °C	30 °C	50 °C
$K_{lipw,DMPC}$	1.96×10^3	4.34×10^3	7.87×10^3	1.52×10^4	2.18×10^4
$K_{lipw,DEruPC}$	9.92×10^3	1.07×10^4	2.64×10^4	5.62×10^4	1.89×10^5

*Average values of duplicates analysis are used

Both $K_{lipw,DOPC}$ and $K_{lipw,DEruPC}$ values of fullerene increase with increasing temperatures (Figure 1). It has been reported that for most of hydrophobic chemicals such as EDCs (Kwon et al., 2007), K_{lipw} values with liquid crystalline phase liposomes decreased with increasing temperature. Therefore, partitioning processes of fullerene between water and liposome can be different from those of hydrophobic chemicals.

At all different temperatures, log $K_{lipw,DEruPC}$ values are higher than log $K_{lipw,DMPC}$ by more than 0.4 log units (Figure 1). Based on this result, it can be concluded that K_{lipw} values of fullerenes are larger for lipid bilayers composed of longer acyl chains. Wong-Ekkabut et al., (2008) indicated that when aggregated fullerenes are located close to the lipid head group, only small clusters form and rapidly penetrate into the bilayer and then disaggregate after entering the membrane. This result shows that, despite of large size of fullerene aggregates, which are from ten to a few hundreds of nanometers, they can penetrate through lipid bilayers. Therefore, the length of acyl chain of lipids can affect the K_{lipw} of fullerenes. However, more work is required to confirm this hypothesis.

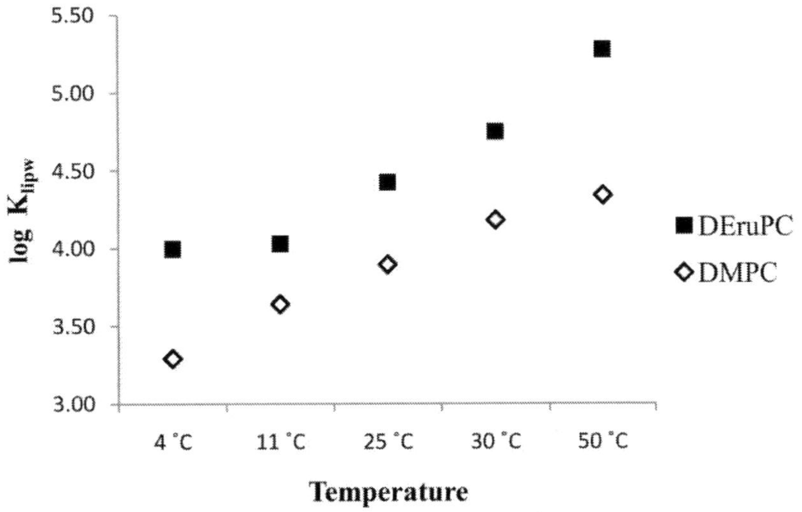

Figure 1. Effects of temperatures on lipid – water partitioning coefficients (K_{lipw}, L/kg-liposome) of fullerene using DMPC and DEruPC (average values of duplicates analysis are used).

4. References

1. Jafvert, C.T., Kulkarni, P.P. (2008), "Buckminsterfullerene's (C_{60}) octanol-water partition coefficient (Kow) and aqueous solubility," *Environ. Sci. Technol.* 42. pp. 5945-5950.
2. Petersen, E.J., Huang, Q., Weber, W.J.(2010), "Relevance of octanol-water distribution measurements to the potential ecological uptake of multi-walled carbon nanotubes," *Environ. Toxicol. Chem.* 29. pp. 1106-1112.
3. Kroto, H.W.,Heath, J.R., O'Brien, S.C., Curl, R.F., Smalley, R. E.(1985), "C60-Buckminsterfullerene," *Nature.* 318. pp. 162-163.
4. Ringwood, A.H., Levi-Polyachenko,N., Carroll, D.L. (2009), " Fullerene Exposures with Oysters: Embryonic, Adult, and Cellular Responses," *Environ. Sci. Technol.* 43. pp. 7136-7141.
5. Yamamoto, H., Liljestrand, H.M., Shimizu, Y. (2004), "Effects of dissolved organic matter surrogates on the partitioning of 17ß-Estradiol and p-Nonylphenol between synthetic membrane vesicles and water," *Environ. Sci. Technol.* 38(8). 2351-2358.
6. Kwon, JH., Liljestrand, H.M., Katz, L.E. (2006), "Partitioning of moderately hydrophobic endocrine discruptors between water and synthetic membrane vesicles," *Environ. Toxicol. Chem.* 25(8). pp.1984-1992.
7. Ruoff, R. S., Tse, D. S., Malhotra,R., Lorents, D.C. (1993), "Solubillity of C_{60} in a variety of solvents, " *J. Phys. Chem.* 97. pp. 3379-3383.
8. Isaacson, C.W., Kleber, M., Field, J.A. (2009), "Quantitative analysis of fullerene nanomaterials in environmental systems: A critical review," *Environ. Sci. Technol.* 43. pp. 6463-6474.

9. Escher, B.I., Schwarzenbach, R.P. (2000), "Evaluation of liposome – Water partitioning of organic acids and bases. 1. Development of a sorption model," *Environ. Sci. Technol.* 34(18). pp.3954-3961.
10. Bayerl, T.M., Bloom, M. (1990), "Physical properties of single phospholipid bilayers adsorbed to micro glass beads," *Biophys. J.* 58. pp. 357-362.
11. Baksh, M.M., Jaros, M., Groves, J.T. (2004), "Detection of molecular interactions at membrane surfaces through colloid phase transitions," *Nature.* 427. pp. 139-141.
12. Tervonen, K., Waissi, G., Petersen, E.J., akkanen, J., Kukkonen, J.V. (2010), "Analysis of fullerene-C60 and kinetic measurements for its accumulation and depuration in Daphnia magna," *Environ. Toxicol. Chem.* 29. pp. 1072-1078.
13. Kwon, JH., Liljestrand, H.M., Katz, L.E., Hiroshi, Y. (2007), "Partitioning thermodynamics of selected endocrine disruptors between water and synthetic membrane vesicles: Effect of membrane compositions", *Environ. Sci. Technol.* 41. pp. 4011-4018.
14. Wong-ekkabut, J., Baoukina, S., Triampo, W., Tang, I.M., Tieleman, D.P., Monticelli, L. (2008), "Computer simulation study of fullerene translocation through lipid membrane", *Nature Nanotechnology.* 3. pp. 363-368.

(292f) Single Wall Carbon Nanotubes Enter Cells by Endocytosis and Not Membrane Penetration

Peter N. Yaron[1], Brian D. Holt[2], Phillip A. Short[2], Mathias Lsche[3], Mohammad F. Islam[4] and Kris N. Dahl[5], (1)Department of Chemical Engineering, Carnegie Mellon University, Pittsburgh, PA, (2) Biomedical Engineering, Carnegie Mellon University, Pittsburgh, PA, (3)Physics, Carnegie Mellon University, Pittsburgh, PA, (4)Material Science, Carnegie Mellon University, Pittsburgh, PA, (5) Chemical Engineering, Carnegie Mellon University, Pittsburgh, PA

Carbon nanotubes are increasingly being tested for use in cellular applications. Determining the mode of entry is essential to control and regulate specific interactions with cells, to understand toxicological effects of nanotubes, and to develop nanotube-based cellular technologies. We investigated cellular uptake of Pluronic copolymer–stabilized, purified ~145 nm long single wall carbon nanotubes (SWCNTs) through a series of complementary cellular, cell-mimetic, and *in vitro* model membrane experiments. SWCNTs localized within fluorescently labeled endosomes as measured by fluorescence lifetime imaging microscopy. Confocal Raman spectroscopy showed a dramatic reduction in SWCNT uptake into cells at 4 °C compared with 37 °C. Electrochemical impedance spectroscopy and Langmuir monolayer film balance measurements showed that Pluronic-stabilized SWCNTs associated with membranes but did not possess sufficient insertion energy to penetrate through the membrane. SWCNTs associated with vesicles made from plasma membranes but did not rupture the vesicles. These measurements, combined, demonstrated that Pluronic-stabilized SWCNTs did not penetrate the membrane via non-specific physical interactions but entered cells via endocytosis, and association of SWCNTs to membrane likely increased uptake. These results suggest that specific targeting of SWCNTs for use in therapeutics should focus on endocytosis as the preferential method for the cellular uptake.

See more of this Session: Environmental Implications of Nanomaterials: Biological Interactions
See more of this Group/Topical: Topical C: Environmental Aspects, Applications, and Implications of Nanomaterials and Nanotechnology

(301a) Opportunities for Government to Influence Technology Innovation In the Water Industry

Jeffrey Q. Adams, National Risk Management Research Laboratory, US EPA, Cincinnati, OH

Opportunities for Government to Influence Technology Innovation in the Water Industry

Jeffrey Adams, USEPA NRMRL, WSWRD, WQMB

This presentation will discuss recent activities initiated by EPA's National Risk Management Research Laboratory to collaborate with numerous organizations including Federal and State regulatory and permitting agencies, water utilities & communities, water treatment manufacturers & vendors, engineering/science community, and business/venture capital sectors to promote technology innovation and markets focusing on the drinking water industry.

Ongoing environmental science research provides guidance for the development Federal and State water supply policies and regulations which become major industry drivers for what role certain environmental services and technology applications will play in the water supply business and markets. The Federal government often sets the initial framework for defining water policy, issues, and needs. This presentation will explore some scenarios on what government can do with limited resources to influence and promote technology innovation and help streamline the complex technological and institutional hurdles that must be confronted as new emerging tools and water treatment applications move across the research continuum from bench concept to market-ready penetration and sustainable deployment.

On April 27, 2011, the Obama Administration released a national "Clean Water Framework" which affirms a comprehensive commitment to protecting the nation's waters. The framework emphasizes the use of partnerships among States, tribes, local governments, and diverse stakeholders to promote innovative approaches for achieving sustainable water supplies. It directs the use of the latest science and research to improve water regulations/policies and identify and address emerging pollution challenges. This complements EPA's new set of strategies for managing drinking water policy issues including: addressing water contaminants as groups as much as feasible rather than one at a time; fostering development on new drinking water technologies by utilizing private innovators and small business; using the authority of multiple statutes to protect water supplies; partner with States to develop shared access to all public water systems monitoring data.

On the forefront of consideration for regulatory groups of contaminants to manage are numerous selected carcinogenic Volatile Organic Chemicals (VOCs), several Nitrosamines, and numerous chlorination Disinfection Byproducts (DBPs). Other groups that may emerge include: Perfluorinated compounds, Organophosphate Pesticides, Carbamate Pesticides, Triazine Pesticides, Chloroacetanilides, and Cyanotoxins. Various individual chemicals receiving renewed interest include Fluoride, Chromium-VI, and Perchlorate, and various microbials and chemicals from the Contaminant Candidate List-3 (CCL3). In addition, initiatives to shift from defining drinking water MCLs from micrograms/L to nanograms/L will have an impact on regulatory policy consideration as the industry grapples with new analytical methods, how to determine health impacts at such low levels of exposure, and determining cost-effective treatment applications.

In conjunction with regulatory drivers, business market dynamics play a role in stimulating the development of innovative technology applications. In 2010, it was estimated that global venture capital investment in the water sector represented only 2% of all cleantech industry investments with the U.S. share of global investments under 30%. The estimated total U.S. water equipment market (excluding engineering/operator services and chemical consumables) was about $28 billion across drinking water, wastewater, and industrial water. The drinking water equipment market segment alone was estimated at about $12 billion, and an analysis indicated the additional available market innovation opportunity in this sector was only $1.2 billion. Factors contributing to the weak investment in water technology include the low profit potential for an industry that does not utilize full cost pricing policies for water production and distribution and the difficulties encountered with federal, state, local government regulatory, permitting, and procurement policies. The analysis indicated that facing competition from well-financed global competitors, U.S. innovators would benefit from technology test-beds and incubators, particularly those sponsored by government, providing demonstration of technology performance within the framework of current regulatory policies or that stimulate reformation of policies.

EPA's Office of Water develops policies and regulatory guidance that influences State decision-making for technology review/permitting/acceptance. Federal guidance often lists technology applications and associated treatment credits for various applications. However, if an innovative technology is developed that does not fit within the current guidance framework, it may be difficult to get States approval to use it. Though this may be an unintended consequence, State regulatory agencies are reluctant to permit technologies that do not have a credible record of performance, or technologies that are designed/configured/operated in a manner that falls outside of the expertise of a State's review staff or cannot be assessed under current federal guidance procedures. These technologies may very well be innovative and prove to be sustainable solutions down-the-road but are viewed as black-box devices that are difficult to evaluate under a current framework.

Several efforts initiated by EPA's National Risk Management Research Laboratory to promote and support technology innovation in the water industry will be discussed. These include providing technology development test-beds at EPA facilities thru equipment vendor partnerships thru CRADA mechanisms, field and lab performance testing of various types of market-ready water treatment technologies under the Environmental Technology Verification (ETV) Program, fifty full-scale demonstration studies of new and innovative arsenic removal technologies deployed at operating water utilities, early-phase testing of new technologies under the SBIR Program, collaborations with the Water

Research Foundation, and Water Environment Research Foundation, and EPA's collaboration with a Water Technology Innovation Cluster (WTIC).

These programs allow the government to act as a catalyst to support independent credible evaluation / verification of various technologies thru third-party partnerships and collaborators. Funding for testing studies varies across different programs ranging from complete government funding, partial vendor subsidy with other collaborator funding, to complete vendor cost coverage of the testing. Testing protocols have been developed thru a stakeholder process that allows feedback from regulatory, science/ engineering, water utility/user community, and technology manufacturer/vendor viewpoints. The independent performance testing results can offer many contributions to technology innovation. Results may inform future regulatory policy reform. Testing results provide manufacturers with identification of strengths and weaknesses of their specific products to foster improvements in their design and operation based on lessons learned. These programs have provided State and local government permitting and procurement staff with credible data in standardized reports to inform decision-making. Efforts have provided protocols for the basis of standardized testing in the scientific/engineering community and stimulation of NSF-type Certification programs to help streamline State approval confidence of innovative technologies. Test-bed studies on market-ready equipment have in some cases provided positive market outcomes for vendors in achieving market penetration and States permitting advantages. Under these EPA programs, study results and lessons learned have been communicated to federal and State regulatory Agency staff and various stakeholder groups thru numerous EPA sponsored Workshops and Webinars in recent years.

Under the WTIC effort, EPA's role is a catalyst to collaborate with other organizations within a regional technology innovation cluster framework in the Ohio, Kentucky, Indiana tri-state region. The Cluster is a geographic concentration of interconnected firms - businesses, suppliers, and service providers - and supporting institutions - local government, business chambers, universities, investors, and others - that work together in an organized manner to promote economic growth and technological innovation, in this case associated with the water technology sector. The mission of WTIC is to develop and commercialize innovative technologies to solve environmental challenges and spur sustainable economic development and job creation.

See more of this Session: Industry and Government Activities In Water Purification
See more of this Group/Topical: Topical 1: Water Technology for Developed and Developing Countries (see also Separations Division)

Navy Shipboard Desalination

Ian Peek, Code 985, Naval Sea Systems Command - Philadelphia, Philadelphia, PA
David Nordham, Code 923, Naval Sea Systems Command - Philadelphia, Philadelphia, PA
Paul Armistead, Office of Naval Research, Arlington, VA

Shipboard desalination is required to produce potable water on U.S. Navy ships, for personnel as well as machinery uses. The purpose of this talk is to provide an overview of Navy shipboard desalination, from distillation through legacy reverse osmosis systems to the development and deployment of future systems. A discussion of the driving forces behind the design process are included, as well as the targeted performance metrics.

Shipboard Desalination using Distillation

Prior to the introduction of reverse osmosis, shipboard desalination was accomplished using distillation. Distillation is very energy intensive, requiring approximately 424 kWh/1000gal. A significant amount of maintenance is required, shown by an operational availability of approximately 60%. Distillation units also require a large amount of chemical additions, in the form of antiscalant.

Existing Shipboard Reverse Osmosis Systems

The first RO unit on a U.S. Navy ship was installed on the *USS Fletcher* (DD 992) in 1988. Successful evaluation showed an operational availability greater than 99%. Since then, the ships have a standardized design, the Navy Standard Reverse Osmosis (NSRO), which has a 12,000 gal/day capacity. The process schematic may be seen in Figure (1).

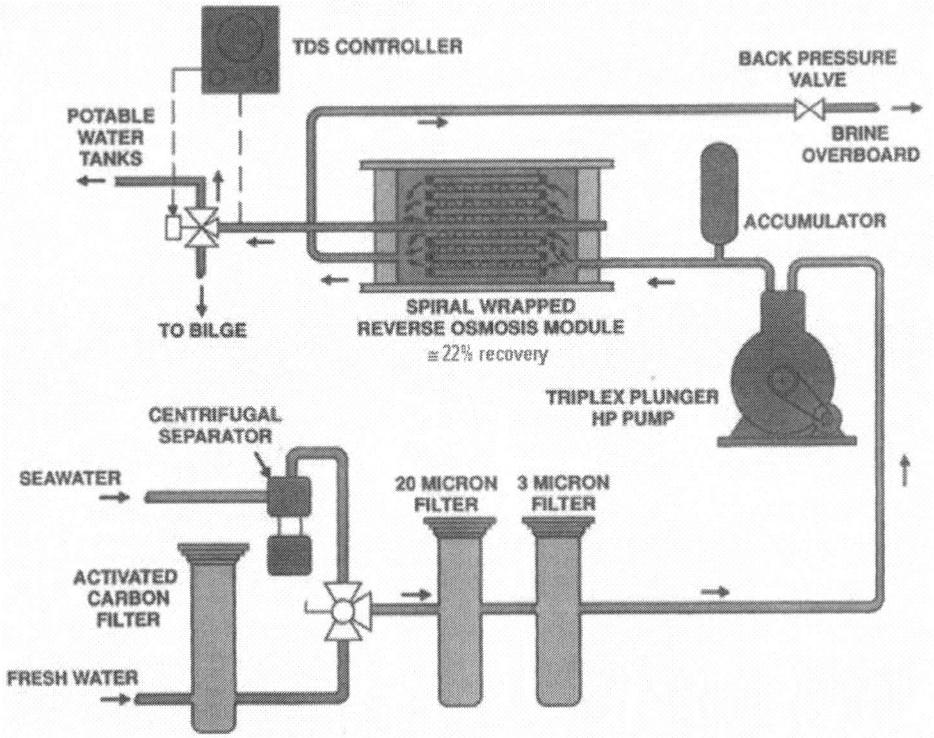

Figure (1) – Navy Standard Reverse Osmosis Process Flow Schematic

Depending on an individual ship's water requirement, the design is scaled up or down. Prefiltration is accomplished with a centrifugal strainer followed by cartridge filtration, at the 20 and 3 micron sizes. The RO section consists of a high pressure plunger pump into commercial spiral-wound membranes, after which the permeate goes to the shipboard potable water storage and the brine is discharged overboard. The system operates at 22% recovery (ratio of permeate to feed flows). Ships requiring higher purity water utilize multiple pass RO to achieve the desired level of purification.

While this is a robust and proven design, having been installed on Navy ships since the 1980s, there are several areas for improvement. The focus of the design was for deep blue sea, i.e., the open ocean. Given recent conflicts, littoral operations are more frequent. In littoral operations, the cartridge filters clog frequently, requiring dedicated manpower to replace the filter elements. This leads to additional storage requirements for new filters as well as used cartridges prior to disposal. The lifetime of cartridge filters decreased from four to six weeks in the open oceans to as little as seven hours in turbid littoral waters. The added strain on the cartridge filters increases the fouling on the RO elements, often requiring their replacement as well. Chlorination of feed seawater piping for biofouling and MIC protection raise operational concerns for RO membranes and require additional process steps for dechlorination. Lastly, the desire to increase potable water generation for improved "quality of life" for sailors cannot be met in the existing volumetric and gravimetric footprints with the NSRO design.

Advancing Technology – the EUWP Program

In 2003, the Office of Naval Research initiated a program to advance the technology used in shipboard desalination systems, the EUWP program (Expeditionary Unit Water Purification). The objectives of this program were to reduce energy requirements, minimize system size and weight, increase ship operational capabilities in the littorals, improve reliability and reduce maintenance, lower logistics requirements and lower operating and total ownership costs. The products of the EUWP program were two large scale units, with capacities of 100,000 gal/day and 300,000 gal/day, respectively.

In addition to the construction of the units, the Navy learned several critical pieces of information as a result of endurance testing. Innovative membrane prefiltration could successfully filter highly turbid coastal seawater with little maintenance requirements in 2400 hours. The use of energy recovery devices allowed for reduced energy, size, noise, and maintenance requirements of high pressure pumps. High flux RO membrane elements and hybrid configurations allowed for reduced space, weight, and energy requirements. An advanced electrodeionization device allowed for the production of superior high quality water with no operator intervention or ion exchange resin replacements in 1700 hours. A comparison between a 100,000 gal/day distillation unit and the 100,000 gal/day EUWP unit may be seen in Table (1).

Table (1) – Comparison of Flash Distillation and Advanced RO Systems

	Distiller	**Advanced RO (EUWP)**
Dimensions (L x W x H)	22 x 10.7 x 9.25	20 x 8 x 8
Weight (lbs)	51,716	33,000
Volumetric Production (gal/day/cu ft)	46	50
Gravimetric Production (gal/day/lb)	1.9	3.7
Operational Availability	< 65%	> 95%
Energy (kWh/1000 gal)	424	65
Cost ($/1000 gal)	$24.75	$3.78

Goals for the Next Generation of Shipboard Desalination Systems

Based on the success of the EUWP program, the Navy selected Shipboard Desalination as a Future Naval Capability Program from FY2010 to FY2014. The primary objective of the program is to develop specifications for advanced shipboard desalination systems that cost the same or less than traditional NSRO systems with 40% less volume, 40% less weight, 65% less energy, greater than 95% operational availability, and 50% lower total ownership costs. The products of the FNC effort will be demonstrator units that will be capable of being installed on a ship. The first two years of the program are dedicated to maturing component level technologies including advanced membrane filtration, antifouling and chlorine resistant RO membranes, energy recovery pumps, and system controls for optimizing shipboard operation. The latter two years of the program are dedicated to designing complete desalination systems (pretreatment and desalination). The completed systems will range in size from 4,000 gal/day to 18,000 gal/day, and 100,000 gal/day.

Design Considerations for Shipboard Desalination Systems

Land based municipal water desalination systems are designed to minimize capital and operational costs and thus energy efficiency is a major concern. For Navy shipboard systems, energy efficiency is an important but not overriding concern since the power to propel a ship through the water dominates overall energy use. Capital costs are important to the Navy, however the Navy will design for a long system lifetime and use highly corrosion resistant materials. Space on a ship is limited and so the desalination system must be compact. High operational availability and minimal maintenance are also required.

There are additional engineering requirements for a successful shipboard design. For combatant craft (e.g., destroyers, cruisers, carriers, submarines), all systems must demonstrate the capability to withstand and continue operating after shock events, to ensure operation after potential explosions near the ship's hull. The systems must also be capable of withstanding vibrations caused by other shipboard equipment. The control systems must meet military requirements for electromagnetic interference, and materials of construction cannot cause issues if a fire exists, limiting plastic components. If the systems are to be installed on existing ships, the construction of the system must be accounted for, i.e., whether it has to be lifted into the ship through the hatchways.

(301c) Regulatory Compliance and Water Treatment Advances: From Research to Application

John Albert, Water Research Foundation, Denver, CO

Water utilities must comply with a myriad of regulations from stage two disinfection by-product rule to the revised total coliform rule. Many of these rules are rigorous and at times conflicting. Precisely navigating the regulatory landscape and meeting water quality objectives while being vigilant about unintended consequences has become a major focus for utilities now more than ever.

Evaluating appropriate technology choices to comply with multiple and/or conflicting water quality goals is a priority for utilities. Currently, a number of decision support tools have been developed to aid utilities in this effort. This talk will highlight one such tool, the Simultaneous Compliance Tool that was recently developed by the Water Research Foundation. The tool is designed for both large and small water systems.

As regulations become more prescriptive, customer expectations increase, and the quality of current source waters is expected to degrade over time many utilities need to consider treatment options above and beyond conventional treatment. Extensive research has been conducted on advanced treatment options such as biological treatment, advanced oxidation processes, membranes, UV, and ozone as a means to meet these challenges.

Moving these technologies from research studies to field application involves extensive benefits analyses including; cost, operations trade-offs, and long-term forecasting for sustainable use. Recent applications of these technologies and case-studies will be highlighted.

See more of this Session: Industry and Government Activities In Water Purification
See more of this Group/Topical: Topical 1: Water Technology for Developed and Developing Countries (see also Separations Division)

Inland Brackish Groundwater as a Resource: Federal Research, Issues, and Challenges

Michelle Chapman
Bureau of Reclamation, Denver, CO
mchapman@usbr.gov

Abstract

The mission of the Bureau of Reclamation is to manage, develop, and protect water and related resources in an environmentally and economically sound manner in the interest of the American public. In the past, that meant building dams and irrigation systems for farmers. Toward the last quarter of the 20st century, Reclamation was called to assist with municipal and industrial water supply through the Title XVI Water recycling and Reuse program for recovering impaired water sources. Now that program has grown into the WaterSMART program (Sustain and Manage America's Resources for Tomorrow) for improving water utility infrastructure, and the Rural Water Supply program to assess potable water supply needs and identify options for fulfilling those needs. Advanced water treatment research has been funded for the past 15 years through the Desalination Act of 1996 and congressional support to the WateReuse Foundation. Through these programs and other federally funded research, we have come to the point where technology is available to treat almost any source of water. The challenge now is to determine the most efficient treatment solution for a particular situation. It is necessary to consider the aquifer water chemistry, productivity, and replenishment, concentrate management, and power options in designing a sustainable customized technical solution to a treatment problem.

Introduction

Why should the Federal Government be concerned with brackish groundwater sources? Congress established the Bureau of Reclamation in 1902 to reclaim the desert of the western United States, providing family farms for our growing country. This was important to solidify our claim on the west. Starting with small irrigation projects, Reclamation negotiated the complex Colorado River Compact between states bordering the Colorado River to allocate its flow. The Compact cleared the way for construction of the Hoover Dam and the All-American Canal starting in 1928. During the Depression, Congress authorized 40 large projects to expand infrastructure and put people to work. Since then Reclamation has evolved to meet new challenges as they arose. It is one thing to design and build a huge dam, but quite another to predict what will happen to fisheries, nesting areas, dam foundations once they are filled with water, irrigated soils, and drainage patterns after the dam and irrigation system are built. Reclamation developed new areas of expertise to solve problems as necessary.

Toward the last quarter of the 20st century, Reclamation was called to assist with municipal and industrial water supply through the Title XVI Water Recycling and Reuse program for recovering impaired water sources. That program has now grown into the WaterSMART program (Sustain and Manage America's Resources for Tomorrow) for improving water utility infrastructure. Congressional support drove several rural water projects in the 1980's and 90's such as the $500 million Mni Wiconi water project. The project is composed of a pumping and filtration plant on the Missouri River and 4,200 miles of pipeline to bring water to central South Dakota for around 10,000 families. Now, however, the Rural Water Supply Act of 2005 gives Reclamation authority to assess, plan, and prioritize these projects rather than having Congress authorize projects individually. As Reclamation delves into the backlog of rural water supply project needs assessments and appraisal studies, the question of treating local brackish ground water arises regularly.

Six of the past ten years have brought significant drought to different parts of the western US. Traditional water supplies are not as reliable as they were historically. Whether one calls it climate change or normal variance in weather patterns, rain and snowfall are not coming when and where they are wanted. Water utility managers realize that they need to have a more diverse portfolio of water sources to draw from. Non-traditional sources, such as municipal wastewater, brackish groundwater, and the sea, are becoming the back-up supply, or in some areas, the primary supply of water.

Fortunately, advanced water treatment research funded for the past 15 years through the Desalination Act of 1996 and through congressional support for the WateReuse Foundation has progressed rapidly. Through these programs and other federally funded research, we have come to the point where technology is available and affordable to treat most any source of water wherever it may be found.

Reclamation's Desalination Research Facilities

Reclamation has two desalination research facilities. The first was the Water Quality Improvement Center (WQIC) in Yuma, Arizona associated with the Yuma Desalting Plant. The original purpose of the WQIC was to evaluate improvements to the desalting process for the larger treatment plant and provide drinking water to the administrative offices at the plant. Changes in pretreatment were tested out in the WQIC. As the treatment process settled, Reclamation offered the WQIC to the public as a place to test treatment ideas for Colorado River water. Clients could choose from a variety of water sources: irrigation return flow from the canal or river water; raw or pretreated; concentrate or permeate; chlorinated or not. There are a variety of reverse osmosis skids, micro-ultra filtration, chemical injection, and conventional pretreatment systems available for clients to use in their studies.

The second facility is the Brackish Groundwater National Desalination Research Facility (BGNDRF) in Alamogordo, NM. Built in 2005 through the Desalination and Water Purification Research Program, the facility is designed to test and evaluate treatment processes and concentrate management solutions. There are four wells with total dissolved solids levels ranging from 550 to 4200 mg/L. There are four outside test pads for medium (60 gpm) to large (130 gpm) pilot studies and six bays inside for test systems treating 30 gpm or less. The facility has two evaporation ponds for concentrate management studies, an agricultural research area and space for wind and solar powered process development.

Projects tested at the facility so far address many aspects of brackish aquifer utilization. There are projects for testing alternatives for desalination concentrate minimization or elimination, new membrane development, processes for produced water treatment, a new method of electrodialysis, and a variety of brackish groundwater pretreatment processes. The number of projects is increasing each year. There are at least eight scheduled for 2012.

Issues to Resolve

We really have a lot to learn about brackish groundwater aquifers. Traditionally, the USGS and EPA have gathered data on fresh water wells. Since people generally do not use saline wells, data is not collected. The oil and gas industry have information on deep saline aquifers near oil and gas wells, but their databases are not available to the public. What happens to brackish aquifers as water is withdrawn? Can brackish aquifers be used in a sustainable manner? There are many sides to the issue – or many issues depending on ones perspective.

- Engineering perspective
    - How can we make BW desalination more robust?

  - How can we integrate alternative energy sources into the desalination process?
  - Can we optimize the mineral recovery process to be as reliable as food processing?
  - What is the best design point – lowest capital cost or most reliable?
- Hydro-geologist perspective
  - How is saline groundwater connected to fresh groundwater and surface water?
  - How are these other water sources affected by extracting saline groundwater?
  - Will the quality of the saline groundwater change over time?
  - Can we sustainably manage the concentrate?
- Legal perspective
  - Who has rights to brackish water?
  - Are those who desalt brackish groundwater responsible for the concentrate until it reaches the ocean?
  - Are they responsible to ensure that it is properly mixed when it enters the ocean?
- Socio-economist perspective
  - Is desalted brackish water more precious than fresh because of the added cost?
  - Should it be used to wash the car and water cattle?
  - Is it for industry or only for drinking?
  - If it is used for industry, should the industry pay back the development cost?
  - Should it be reused often or returned to the hydrologic cycle?
  - Should the government support desalination projects in the desert or encourage migration to water rich areas.

Conclusion

As we go forward in our efforts to make rural Western US a kinder place by providing water for all no matter where they live, we need to remember our past mistakes in reclaiming the west, study potential consequences, and find answers to these questions to aid us in our plans.

References

Brief History Bureau of Reclamation - http://www.usbr.gov/history/2011NEWBRIEFHISTORYV1.pdf

2008, Brokaw, C. "Mni Wiconi water reaching Pine Ridge reservation" News From Indian Country, accessed from http://indiancountrynews.net/index.php?option=com_content&task=view&id=4387

Mni Wiconi Rural Water Supply Project (15.522) Agency: Department of the Interior Office: Bureau of Reclamation. Accessed from: http://www.federalgrantswire.com/mni-wiconi-rural-water-supply-project.html

Yuma Water Quality Improvement Center: http://www.usbr.gov/lc/yuma/facilities/wqic/yao_facilities_wqic.html

Brackish Groundwater national Desalination Research Facility Fact Sheet: http://www.usbr.gov/pmts/water/research/BGNDRF%20FACT%20SHEET%20Final.pdf

(301e) Electrodialysis Reversal Pilot Studies Performed At the Brackish Ground Water National Desalination Research Facility (BGNDRF) In Alamogordo, NM

Neil E. Moe[1], John Barber[1], Ali Sharbat[2] and Abbas Ghassemi[2], (1)GE Water & Process Technogies, Las Cruces, NM, (2)Institute for Energy and the Environment, New Mexico State University, Las Cruces, NM

Electrodialysis Reversal (EDR) technology uses voltage driving force and ion selective membranes to remove salt from water. The "Reversal" in EDR refers to the process of periodically reversing the polarity of the electrodes in order to inhibit scale formation. EDR has been an established desalination technology for almost 40 years, and yet the understanding of EDR is less developed than that of other technologies such as distillation and reverse osmosis. For example, all the major suppliers of reverse osmosis membranes also provide advanced simulation tools that enable engineers to reliably design systems. In contrast, there is no such design software publicly available for EDR and very few literature references that provide operating data for commercial sized systems. EDR is cloaked in mystery.

Extensive studies were performed using full sized pilot EDR equipment, where the electrodes, spacers, and membranes were sized identical to commercial systems, while the number of cell pairs was typically 40, compared to 600 at most commercial installations. The experiments were performed at BGNDRF utilizing the various sources of brackish water available on site. Experimental variables included water chemistry, temperature, recovery, velocity, volts/amps, stack configuration, and membranes. Thousands of individual experiments were performed and over 900 water samples were collected for analysis.

Results were analyzed in a number of ways. First, the relationship between volts, amps, and cut rate of dissolved salts is defined as function of operating conditions. Second, the selectivity for various ions was determined, indicating that selectivity was highest at high temperatures and low cut rates. Third, limiting current was determined and found to have unexpected functional dependencies. Temperature had very little effect over the range 7 – 40 C, the nature of the ionic species had no discernible effect, and limiting current was nearly linearly dependent on fluid velocity. The quantitative coupling of operating conditions (temperature and flow velocity) with performance (volts, amps, conductivity) and water chemistry led to greatly increased understanding of EDR process and provided the foundation for an EDR process model.

See more of this Session: Industry and Government Activities In Water Purification
See more of this Group/Topical: Topical 1: Water Technology for Developed and Developing Countries (see also Separations Division)

Feasibility Study of High Cut Rate of Dissolved Mineral Salts in Electrodialysis Reversal (EDR) Process

Connor Hanrahan[1], Ali Sharbat[1], Neil E. Moe[2], John Barber[2] and Abbas Ghassemi[1]
(1) Institute for Energy and the Environment, New Mexico State University, Las Cruces, NM; (2) GE Water and Process Technologies, Las Cruces, NM

Abstract

Relatively high capital expenditure (CapEx) of desalination technologies has limited the practical employment of saline and brackish water desalination. Hence, any attempt to reduce CapEx will result in greater utilization of desalination technologies. Conventional separation percentages of total dissolved salts in EDR technologies is between 40-60% per stage for current industrial-scale systems [1,2]. Thus, depending on the saline/brackish water quality, most EDR systems are designed to desalinate water in multiple stages, which results in higher CapEx costs. In this research, the feasibility of increasing the cut rate of dissolved mineral salts in an EDR process has been studied in an effort to lower CapEx costs by reducing the number of required stages.

Full pilot-scale experimentation was performed on a 41 cell pair electrodialyser with 12 gpm influent flow rate at the Brackish Groundwater National Desalination Research Facility (BGNDRF) in Alamogordo, New Mexico. Brackish feed-water was provided from a deep aquifer in the Tularosa Basin with an average conductivity of 1794 μS/cm, which is within the spectrum for water quality in the southwestern region of the United States[3,4]. Using this feed water, short term limiting current experiments were performed that demonstrated an entitlement of 88-99% cut rate of dissolved solids per stage, directly related to the limiting current of the system. Next, long-term experiments were performed to demonstrate the feasibility of continuous high-removal operation. More than 150 hours of continuous operation were successfully completed at $\geq$ 90% removal of dissolved solids.

Introduction

According to the 2003 "World Water Development Report" by the United Nations, more than 50 percent of nations will face water stress or water shortages by 2025, and as much as 75 percent of the world's population could face water scarcity by 2050. Fueled by rapid population growth, much of the Southwest United States is already witnessing an increased demand on their limited fresh water supply, and is looking towards its extensive brackish water resources to address this issue. Where inland brackish water desalination has the potential to create "new" water resources in these arid regions, implementation is hindered by the relatively high cost of treating brackish waters.

Multiple processes exist for treating saline waters such as reverse osmosis (RO), electrodialysis reversal (EDR), evaporation, membrane distillation and ion exchange. Where each technology supports a different role in treating saline waters, electrodialysis reversal (EDR) has proven particularly useful for treating hard waters. EDR utilizes an electric field to remove charged particles from a saline solution through permselective ion-exchange membranes. Where sparingly soluble salts accumulate on the surface of membranes during standard desalination procedures, EDR has the advantage of a periodic polarity reversal in which the salts are cleared from the membrane surfaces before becoming permanently attached. This makes EDR particularly well-suited for treating brackish waters where the concentrations of sparing soluble salts, such as calcium sulfate and calcium carbonate, are high.

However, EDR suffers from relatively high CapEx costs, when compared with an RO process, in which the stack components are the primary cost. In fact, EDR is most competitive when a one or two stage system is used to desalt raw water sources [5]. Though extensive laboratory testing has shown an entitlement of up to 99% removal per stage for EDR systems, current EDR plants are operated at 40-60 % removal per stage [1,2]. Therefore, an opportunity exists to reduce the CapEx costs of EDR by increasing the single-stage removal and decreasing the number of required stages.

Full pilot-scale experimentation at BGNDRF has successfully completed more than 150 hours of continuous operation at >90% single-stage removal, with 12 gpm influent flow from a brackish water aquifer in the Tularosa Basin near Alamogordo, New Mexico.

Theory

Significant modeling work has been done to show that as dissolved salts are passed across permselective membranes, the electrolyte concentration decreases in the boundary layer, which is formed on the desalting surface of the membrane [6]. At the limiting current density, the concentration of dissolved salts on the surface of the ion-exchange membranes approaches zero, where the transmembrane flux does not increase with increasing driving force [7]. Above the limiting current density, water molecules will dissociate and lead to the precipitation of scale-forming inorganics or cause membrane damage. Care must therefore be taken to understand the limiting current density of a particular EDR assembly with its specified feed water.

A current-voltage curve is typically used to determine the membrane resistance, the limiting current and the boundary layer thickness of a system containing ion exchange membranes [8]. The curve is used to relate the current passed through a membrane with the voltage drop across the membrane and its adjacent boundary layer. By adjusting the electrical potential between the electrodes and monitoring the change in current, a resistance profile was used to determine the limiting current (I_{lim}) from the inflection of the V/I vs 1/I plot (Fig. 1).

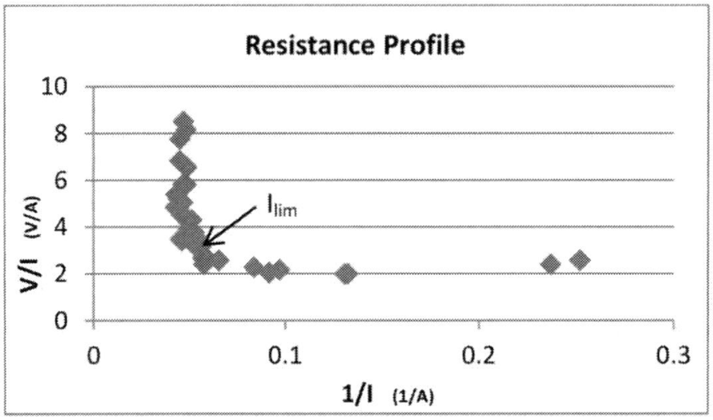

Fig. 1 V/I vs. 1/I plot for determining limiting current.

The current-voltage curve is the most common graphical technique for determining the limiting current of an EDR system. The current-voltage curve from our data determined a limiting current density of 18.72 A, which stands as the theoretical limit of operation in order to avoid water splitting and the subsequent membrane damage/scaling. However, experimental data on '% Removal vs. Current' reveals that the stack is capable of operating above 20 amps without witnessing the drastic pH changes caused by water splitting.

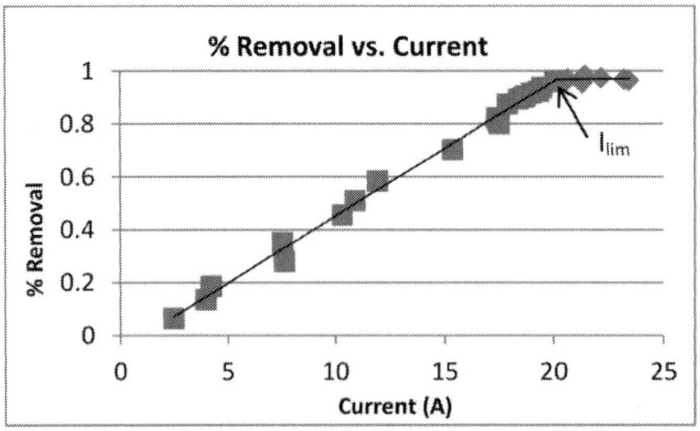

Fig. 2 '% Removal vs. Current' plot for determining limiting current.

This discrepancy of results in determining the limiting current density is most likely due to the effects of very high, single-stage removal. Where most researchers performing limiting current experiments are working with higher salinity feed-water and laboratory-scale systems, by the time our stack removes 90% of the dissolved minerals, the diluate stream begins to add to the total stack resistance, due to a lack of ions. The current-voltage curve proposed by other researchers [6-8] addresses the resistance across the membrane surfaces and adjacent boundary layers, but not the influence of resistance caused by a nearly demineralized diluate stream. Based on this observation, our experimental design was based upon the '% Removal vs. Current' plot, targeting 19.72 amps (97% of I_{lim}).

Materials and Methods

Full-scale pilot studies were carried out at BGNDRF, using brackish water from a deep aquifer in the Tularosa Basin near Alamogordo, New Mexico. The system includes a multi-media filtration tank, a cartridge filter, an EDR stack, pumps, control valves and a main control panel. The raw feed water was passed through a multi-media filter and a GE cartridge filter before reaching the hydraulic system, where it was separated into concentrate and diluate feed to the EDR stack.

The experiments were performed using an Ionics Aquamite I EDR stack (23x25.5x50 cm^3) with two electrical stages of 41 cell pairs each. GE Cation-CR67-HMR and Anion-AR204-SZRA membranes were used, separated by Mark IV spacers (0.76 mm thick), with an effective membrane surface area of 3,450 cm^2. Due to experimental design considerations, only the first electrical stage was used in the removal of dissolved solids, while the voltage of the second stage was set only to deter reverse electrodialysis (no current flow). Under these design parameters, the system had a total effective membrane area of 14.15 m^2 and an average current density of 59.4 A/m^2 (aprox. 97% of limiting current density). The voltage between the electrodes was controlled by a power supply that provided a constant electrical potential difference. An oscilloscope was used to measure the voltage and current between the electrodes (stage V & stage A), and a multimeter was used to monitor the stack voltage for irregularities. The polarity reversal period was set for 15 min.

Due to peculiarities of the system, the positive and negative cycles operate with slightly different product flow rates – 7.5 and 7.8 gpm respectively. As the system provides a constant voltage, this difference in flow rate causes a variance in amperage, which was remedied by targeting 90% removal based on separate drive setting for the positive and negative cycles. The system was therefore operated in positive polarity at 66.2 V and around 19.2 A, and in negative polarity at 71.3 V and around 21.3 A, which led to minor discrepancies in product and concentrate water conductivities.

Due to the nature of the experiment, and the high potential for $CaCO_3$ scaling from the feed water, a continuous acid feed of 15% HCl was fed to the concentrate stream to keep the LSI below 2.1, and an ECIP solution of pH 2.5 was run continuously to the electrode chambers.

A limiting current experiment was performed at the beginning of the test to ensure that the stack performance was optimal and to provide a target range for 90% single-stage removal.

Results and Discussion

Table 1 –
Run-Time Data at 90% Removal

Polarity	Conductivity (μS/cm)			pH			Voltage (V)	Current (A)	Stack Voltage (V)	% Removal
	Feed	Product	Concentrate	Feed	Product	Concentrate				
Positive	1794	159.5	5616	8.3	8.2	8.1	66.2	19.1	62.1	91.1
Negative	1794	197	5930	8.3	8.1	8.1	71.3	21.3	67.8	89.0

Table 2 –
Water Quality Data at 90% Removal

	Ca^{2+}	Mg^{2+}	Na^{+}	K^{+}	Sr^{2+}	Cl^{-}	SO_4^{2-}	HCO_3^{-}	Fl^{2+}	Si	PO_4^{3-}
FEED	58.5	13.5	326.6	5.95	1.9	38.6	632.3	196	2.61	26	0.6
Product	6.5	1.7	28	<0.5	.22	1.9	50	34.5	0.5	29	< 0.4
Concentrate	278.2	56.862	842	22	8.7	123	2105	251.7	4.5	26.5	0.7

* Concentrations, mg/L

The average run-time data and water quality analyses are presented in tables 1 and 2. The feedwater conductivity varied between 1753 and 1849 μS/cm, and the product and concentrate conductivities varied according to polarity reversal period as discussed previously. The electrical potential and current remained constant throughout operation. The stack voltage was monitored to ensure that the stack was operating properly, and it also remained constant throughout the experiment. The product blowdown was 7.5 gpm in positive polarity and 7.8 gpm in negative polarity, while the concentrate blowdown remained constant at 2.2 gpm.

Conclusion

In an effort to reduce the required CapEx for EDR processes, high single-stage removal was demonstrated using brackish groundwater from a deep well in the Tularosa Basin, near Alamogordo, New Mexico. Based on previous limiting current experiments, more than 150 hours of continuous operation were successfully completed targeting 90% single-stage removal.

Acknowledgements

We wish to thank Dr. Abbas Ghassemi, Reclamation, General Electric, and entire staff at BGNDRF for their continued support.

References

1. Veza J.M. (2001) "Desalination in Canary Islands: An update." Desalination 133
2. Valero, F., Arbós, R. (2010) "Desalination of brackish river water using Electrodialysis Reversal (EDR): Control of the THMs formation in the Barcelona (NE Spain) area." Desalination, 253.
3. Gleick, P. (2010) "Roadmap for sustainable water resources in southwestern North America." PNAS, 14 Dec
4. Lansford, R., Hernandez, J., Enis, P.J., Truby, D. and Mapel, C. (1987) "Evaluation of available saline water resources in New Mexico for the production of microalgae." FY 1986 Aquatic Species Program Annual Report; Solar Energy Research Institute, Golden, Colorado, SERI/SP-231-3071.
5. Reahl, E. R. (2008). "Half A Century of Desalination with Electrodialysis." from http://www.gewater.com/pdf/Technical Papers_Cust/Americas/English/TP1038EN.pdf.
6. Tanaka, Y., Iwahashi, M., Kogure, M. (1994) "Distribution of electrodialytic condition in an ion-exchange membrane electrodialyzer and limiting current density." Journal of Membrane Science. 92
7. Ibanez, R., Stamatialis, D.F., Wessling, M. (2004) "Role of membrane surface in concentration polarization at cation exchange membranes." Journal of Membrane Science, 239.
8. Długołęcki, P., Anet, B., Metz, S., Nijmeijer, K., Wessling, M. (2010) "Transport limitations in ion exchange membranes at low salt concentrations." Journal of Membrane Science, Vol. 346, No. 1

() Green Engineering and Sustainability: A Systems Analysis Perspective

Urmila Diwekar, Vishwamitra Research Institute, Center for Uncertain Systems: Tools for Optimization and Management, Clarendon Hills, IL

This talk presents a systems analysis perspective that extends the traditional plant design framework to green engineering, green energy and industrial ecology leading to sustainability. For green engineering this involves starting the design decisions as early as chemical and material selection stage on one end, and managing and planning decisions at the other end. However, uncertainties and multiple and conflicting objectives are inherent in such a design process. Green engineering principles are illustrated here using a green energy sector case study. Uncertainties increase further in industrial ecology. The concept of overall sustainability goes beyond industrial ecology and brings in time dependent nature of the ecosystem and multi-disciplinary decision making. Optimal control methods and theories from financial literature can be useful in handling the time dependent uncertainties in this problem. Decision making at various stages starting from green plant design, green energy, to industrial ecology, and sustainability is illustrated for the mercury cycling. Power plant sector is a major source of mercury pollution. In order to circumvent the persistent, bioaccumulative effect of mercury, one has to take decisions at various levels of the cycle starting with greener power systems, industrial symbiosis through trading, and controlling the toxic methyl mercury formation in water bodies and accumulation in aquatic biota.

See more of this Session: Lawrence K. Cecil Award Lecture
See more of this Group/Topical: Environmental Division

() WISE Presentation: Addressing Regulatory Barriers to Grid-Scale Energy Storage Deployment

Anita Luong, Johns Hopkins University, Baltimore, MD

Renewable energies are a fast-growing source of electricity in America, bringing with it many benefits of clean, sustainable energy. However, along with these benefits comes several challenges, most notably intermittency and supply-demand mismatch, which limit the growth of these new energy sources. Electric energy storage is one solution that can address these challenges. This presentation will investigate regulatory and market barriers that hinder the deployment of electric energy storage onto the power grid.

See more of this Session: Lawrence K. Cecil Award Lecture
See more of this Group/Topical: Environmental Division

() WISE Presentation: Creating a Federal Mandate for State Renewable Electricity Standards (RES)

Mikkel VandeBergh, Oregon State University, Corvallis, OR

A federal renewable electricity standard (RES) has been proposed in Congress for the last ten years with no success. Difficulties with varying states' ability to generate renewable energy and with making state and federal standards compatible lead to a new solution. Instead of a federal RES, a federal mandate for state renewable electricity standards should be implemented to gain the benefits of a RES program, while garnering acceptance from a broader audience.

See more of this Session: Lawrence K. Cecil Award Lecture
See more of this Group/Topical: Environmental Division

(305a) Life Cycle Assessment of Advanced Regional Biomass Processing Depots Integrated with Sustainable Cellulosic Feedstock Landscapes

Pragnya L. Eranki, Bryan Bals, Seungdo Kim and Bruce E. Dale, Department of Chemical Engineering, Michigan State University, Lansing, MI

Lignocellulosic biomass can help fulfill escalating demands for liquid fuels and mitigate the environmental impacts of petroleum-derived fuels. Cellulosic biofuels also offer potential solutions to many socio-economic and sustainability problems caused by petroleum dependence. However, the large biorefineries which process cellulosic biomass to fuels face many logistical problems because they are fully integrated, centralized facilities in which all units of the conversion process are present in a single location. Biomass logistical problems are further exacerbated by physical characteristics of feedstocks such as low bulk density, compositional variability, seasonality and their tendency to decompose.

A potential solution to these problems might be a network of distributed processing facilities called "Regional Biomass Processing Depots" (RBPDs). In their simplest configuration these depots procure, preprocess /pretreat, densify and deliver feedstock to the biorefinery while providing a single co-product animal feed to end-users. However, the RBPDs can be configured into other, more complex setups by employing multiple technologies such as the ammonia fiber expansion pretreatment (AFEX), Leaf Protein Concentrate (LPC) extraction, leaf/stem separation, pyrolysis and anaerobic digestion, as dictated by feedstock characteristics and local conditions, to produce a variety of valuable co-products and for energy generation via synergies among these technologies.

In a previous analysis we considered the differences between distributed processing and centralized processing combined with a variety of landscape scenarios. This study established that combining high yielding perennial grasses with distributed processing that includes densification of AFEX treated biomass prior to transportation to the biorefinery may generate greater energy yields and lower greenhouse gas emissions compared to centralized processing. In this study we aim to extend this analysis to watershed- scales in various landscape configuration scenarios using more sophisticated geospatial information system tools as well as by including a wider range of feedstocks and sustainable management practices, including the use of marginal lands. On the processing side, we evaluate the energy balances and environmental impacts of advanced RBPDs and their associated transport systems. We then intend to integrate the various elements within this overall system and conduct a life cycle analysis (LCA) using the Gabi software.

Through an integrated systems-wide analysis and by developing flexible models we seek the most sustainable landscape-processing-transportation arrangements for cellulosic biofuel production. We believe that greater energy and environmental benefits can be added to biofuel production by not merely aiming at improved logistics but by reimagining agricultural systems and landscapes, minimizing the challenges associated with biomass supply and its processing and by investigating multiple, potentially synergistic processing technologies and related complexities.

See more of this Session: Life Cycle Analysis of Renewable Feedstock-Based Processes and Products I
See more of this Group/Topical: Sustainable Engineering Forum

(305b) Lifecycle Assessment and Policy: Implications of the Renewable Fuel Standard for Upper-Midwest Energy Supply

Michael J. Brodeur-Campbell[1], Jordan Klinger[1], Kathleen Halvorsen[2] and David Shonnard[1], (1)Chemical Engineering, Michigan Technological University, Houghton, MI, (2)School of Forest Resourses and Environmental Science and Department of Social Sciences, Michigan Technological University, Houghton, MI

LifecycleAssessment and Policy: Implications of the Renewable Fuel Standard for

Upper-MidwestEnergy Supply

Michael Brodeur-Campbell[1], Jordan Klinger[1],Kathleen Halvorsen[2,3], and David Shonnard[1,4]

[1] Department of ChemicalEngineering, Michigan Technological University

[2] Department of SocialSciences, Michigan Technological University

[3] School of Forest Resourcesand Environmental Science, Michigan Technological University

[4] Director Sustainable FuturesInstitute, Michigan Technological University

The use of biofuels derived from lignocellulosic biomass has receivedsignificant public attention with Congressional policies such as the EnergyPolicy Act of 2005 (EPAct), updated in the EnergyIndependence and Security Act of 2007 (EISA) which included the Renewable FuelStandard (RFS). The RFS mandates theblending of ethanol and advanced / cellulosic biofuels into the U.S.transportation fuel supply for each year until 2022. "The Renewable Fuel Standard has a lifecycleperspective and contains lifecycle greenhouse gas emissions reductionrequirements (relative to petroleum gasoline) for the production ofbiofuels." Conventional ethanol derivedfrom corn starch must meet a 20% reduction level and is capped at 15 billiongallons/yr. "°Cellulosic" ethanol mustmeet a 60% reduction standard and is expected to make up 16 billion gallons/yr.by 2022. "Advanced" biofuel that meet a50% reduction must make up the remaining5 billion gallons/yr mandated by 2022 (36 billion gallons/yr total ethanol). In addition to demanding an accounting for bothdirect and indirect land use change CO_2 emissions, the policy alsospecifies several other environmental and social metrics to consider inevaluating biofuel production.Environmental metrics specified include airquality, water quantity and quality, wetlands and ecosystem health, andwildlife habitat. Other indirect effectson society include energy security, commercial fuel production andinfrastructure, consumer fuel prices, job creation, agriculture impacts, ruraleconomic development, and future food prices.

While many studies have been performed on agricultural residues such ascorn stover (Sheehan, Aden et al. 2004, Hsu 2010) and wheat straw (Kabel, Bos et al. 2007), muchless attention has been given to woody feedstocks. For this LCA, several regionally importantpotential woody feedstocks are analyzed (hybrid poplar, hybrid willow, andmixed hardwood logging residues) as well as two herbaceous feedstocks (switchgrassmonoculture, and a diverse prairie grass ecosystem). This LCA is directly driven by therequirements of the RFS and includes metrics for water and air quality, as wellas global warming potential. Eutrophicationpotential was chosen as the most important measure for water quality, as theeffects of fertilizer use in biomass cultivation generally dominate allcategories of water quality.

Particulatematter emissions were chosen as the most important measure for air quality, asthis category has the greatest implication for human health.

SimaPro 7.2 and the EcoInventdatabase were used to model the process and assign environmental burdens. A co-product credit is assigned to therenewable electricity generated from the non-fermentable portion of thebiomass. This electricity is modeled todisplace grid electricity at the regional mix for the states of Michigan,Wisconsin, Minnesota, Iowa, and Illinois, which is derived approximately 70%from coal-fired power plants. Inputs andoutputs for the conversion were based on the National Renewable EnergyLaboratory Technical Report NREL/TP-510-32438 (Aden, Ruth, et al. 2002). The ratio of fuel to electricity producedfrom each feedstock was modified based on typical values for cellulose,hemicellulose, lignin, ash, and extractives for each feedstock. Table 1 below summarizes the results forGlobal Warming Potential, Eutrophication Potential, and Particulate MatterEmissions for the feedstocks analyzed.

Table1: Summary of LCA Results for Cellulosic Ethanol

	Logging Residue	**Prairie**	**Switchgrass**	**Willow**	**Poplar**
Global Warming Potential (g CO_2 eq. / MJ)	9.0	-11.0	0.9	-2.4	-7.9
Eutrophication Potential (g N eq. / MJ)	0.0	0.0	0.3	0.1	0.1
Particulate Matter <2.5 m (mg / MJ)	0.4	-1.4	-1.1	-2.8	-3.1

All fuel pathways analyzed meet 60% greenhousegas reduction requirements for a cellulosic biofuel, even before any co-productcredit is applied. With co-productcredit some pathways result in net negative carbon emissions due to coaldisplacement. Eutrophication potentialis increased in all cases due to feedstock production activities; the highestis for switchgrass which is the most fertilizer-intensive feedstock, while thelowest is for multispecies prairie. Particulateemissions are reduced for all cases except logging residues, largely due toavoided coal power. Particulate matteremissions for logging residues are high because of the low productivity peracre and long transportation distances required to supply enough material forconversion.

Future work will focus on refining ourfeedstock conversion emission estimates using the ASPEN Plus process simulationsoftware, and on including land use change estimates for carbon flows.

References

Aden, A., M. Ruth, K. Ibsen, J. Jechura, K. Neeves, J. Sheehan,B. Wallace, L. Montague, A. Slayton, and J. Lukas (2002). *Lignocellulosic Biomass to Ethanol ProcessDesign and Economics Utilizing Co-Current Dilute Acid Prehydrolysisand Enzymatic Hydrolysis for Corn Stover*. National Renewable EnergyLaboratory, Golden, Colorado.

Hsu,D., D. Inman, G.A. Heath, E.J. Wolfrum, M.K. Mann,and A. Aden (2010). *Life CycleEnvironmental Impacts of Selected U.S. Ethanol Production and Use Pathways in2022*. Environmental Science &Technology; 44, 5289-5297.

Kabel, M.A., Bos,G., Zeevalking, J., Voragen,A.G.J., Schols, H.A., (2007). *Effect of pretreatment severity on xylansolubility and enzymatic breakdown of the remaining cellulose from wheat straw*.Bioresource Technology; 98, 20342042.

Sheehan J, Aden A, Paustian K, Killian K,Brenner J, Walsh M, et al. (2004) *Energyand environmental aspects of using corn stover for fuel ethanol*. Journal of Industrial Ecology;7(4):117e46.

See more of this Session: Life Cycle Analysis of Renewable Feedstock-Based Processes and Products I
See more of this Group/Topical: Sustainable Engineering Forum

(305c) Optimal Design and Operations of Cellulosic Biofuel Supply Chains Under Uncertainty

Fengqi You, Dept of Chemical Engineering, Carnegie Mellon University, Pittsburgh, PA and Belinda Wang, Department of Chemical and Biological Engineering, Northwestern University, Evanston, IL

Concerns about climate change, energy security, and the diminishing supply of fossil fuels are causing our society to search for new renewable sources of transportation fuels. Domestically available biomass has been proposed as part of the solution to our dependence on fossil fuels. Biofuels, especially liquid transportation fuels produced from cellulosic materials, have the benefits of significantly reducing greenhouse gas (GHG) emissions and leading to new jobs and greater economic vitality in rural areas. Renewable Fuels Standard (RFS), part of the Energy Independence and Security Act of 2007, establishes a target of 16 billion gallons of cellulosic biofuel annual production by 2022 [1, 2], although the United States produced less than 1 billion gallons of liquid fuels from cellulosic materials in 2010 [3, 4]. In observance of this mandatory production target, many new cellulosic biomass-to-biofuels supply chains will be designed and developed in the coming decade for better economic, environmental and social performances. However, uncertainty resulting from supply and demand variations may have significant impact on the biofuel supply chain. Therefore, an efficient optimization strategy is urgently needed to for the design and operations of sustainable and robust cellulosic biofuel supply chains.

In this work, we address the optimal design and planning of cellulosic biofuel supply chains under supply and demand uncertainty. A two-stage stochastic mixed-integer linear programming (SMILP) model combined with Monte Carlo sampling and the associated statistical analysis [4, 5] is proposed to deal with different types of uncertainty, and it is incorporated into a multi-period planning model that takes into account the main characteristics of the biofuel supply chains, such as seasonality of feedstock supply, biomass deterioration with time, geographical diversity and availability of biomass resources, feedstock density, diverse conversion technologies and byproducts, infrastructure compatibility, demand distribution, regional economic structure, and government incentives. In the two-stage framework, the supply chain network design and capacity planning decisions are made "here-and-now" prior to the resolution of uncertainty, while the production, transportation and storage decisions for each time period are postponed in a "wait-and-see" mode. The SMILP model integrates decision making across multiple temporal and spatial scales and simultaneously predicts the optimal network design, facility location, technology selection, capital investment, production operations, inventory control, and logistics management decisions. In order to solve the resulting large scale SMILP problems effectively, a decomposition algorithm based on sampling average approximation [5] and multi-cut L-shaped method [6, 7] is proposed by taking advantage of the problem structure.

In addition to the economic objective of minimizing the annualized net present cost, the SMILP model is also extended to integrate with life cycle assessment (LCA) and regional economic input-output (EIO) analysis through a multiobjective optimization scheme to include two other objectives: the environmental objective measured by life-cycle greenhouse gas emissions and the social objective measured by the number of accrued local jobs resulting from the construction and operation of the cellulosic biofuel supply chain. The multiobjective optimization framework allows the model to establish tradeoffs among the economic, environmental, and social performances of the cellulosic biofuel supply chains in a systematic way. The multiobjective optimization problem is solved with an ε-constraint method and produces Pareto-optimal curves that reveal how the optimal annualized cost and the supply chain network structure change with different environmental and social performance of the entire supply chain [10, 11].

The proposed optimization model and solution method is illustrated through county-level case study for the state of Illinois. Three major types of biomass, including crop residues, energy crops, and wood residues, and three major conversion pathways, including biochemical conversion, gasification followed by Fischer-Tropsch synthesis and fast pyrolysis followed by hydroprocessing are considered. Uncertainty information is generated from the time series analysis [8] based on the historical data of biomass feedstock supply [9] and liquid fuel demand [1]. County-level results will be presented that provide regionally-based insight into transition pathways of biomass production and conversion. Computational results also demonstrate the effectiveness of the proposed decomposition algorithm for the solution of large-scale SMILP problems.

References:

[1] Biomass Program Multi-Year Program Plan 2010; EERE, U.S. DOE, March 2010.

[2] National Biofuels Action Plan; Biomass Research and Development Board: U.S. U.S. Energy Information Administration.

[3] National Renewable Fuel Standard Program for 2010 and Beyond; U.S. EPA, February 2010

[4] Department of Agriculture and U.S. Department of Energy: 2008.

[5] Shapiro, A.; Homem-de-Mello, T., A simulation-based approach to two-stage stochastic programming with recourse, Mathematical Programming, 1998, 81, 301-325

[6] Birge, J.R.; Louveaux, F., Introduction to Stochastic Programming, Springer Verlag, New York, 1997

[7] You, F.; Wassick, J. M.; Grossmann, I. E., Risk management for global supply chain planning under uncertainty: models and algorithms. AIChE Journal 2009, 55, 931-946.

[8] Enders, W., Applied Econometric Time Series. Wiley: Hoboken, NJ, 2004.

[9] National Agricultural Statistics Service.

[10] You, F.; Tao, L.; Graziano, D. J.; Snyder, S. W., Optimal Design of Sustainable Cellulosic Biofuel Supply Chains: Multi-objective Optimization Coupled with Life Cycle Assessment and Input-Output Analysis. AIChE Journal 2011, In press, DOI: 10.1002/aic.12637.

[11] You, F.; Wang, B., Life Cycle Optimization of Biomass-to-Liquids Supply Chains with Distributed-Centralized Processing Networks. Industrial & Engineering Chemistry Research 2011, Submitted.

See more of this Session: Life Cycle Analysis of Renewable Feedstock-Based Processes and Products I
See more of this Group/Topical: Sustainable Engineering Forum

(305d) Pennycress-Derived Jet Fuel and Diesel LCA: Sustainable Advanced Biofuels

Jiqing Fan, Chemical Engineering, Michigan Technological University, Houghton, MI, David Shonnard, Michigan Technological University, Houghton, MI, Tom N. Kalnes, Research and Development, UOP LLC, Des Plaines, IL and Peter Johnsen, Arvens Technology Inc, Peoria, IL

Pennycress-derived jet fuel and diesel show promise as emerging biofuels that can help reduce dependence on petroleum imports, lower the trade deficit, and most importantly, lower greenhouse gas (GHG) emissions. Pennycress is a prolific seed producer with yields of more than 2000 pounds of per acre. The seeds contain 36% oil by weight, nearly twice the amount as soybeans. In the envisaged biofuel scenario, Pennycress would be grown between segments of a typical corn/soybean rotation, thus avoiding the concerns of a fuel crop competing with food crop production. As useful applications for pennycress meal develop, the oil is expected to become a sustainable alternative for advanced biofuels production.

The goal of this study is to determine the life cycle GHG emissions, cumulative energy demand, and fossil energy demand of pennycress-derived Hydroprocessed Renewable Jet (HRJ) fuel and Renewable Diesel (RD) as produced by the UOP Ecofining™ process, and compute the GHG savings per MJ of fuel compared to petroleum-based jet fuels and diesel. The scope of this study encompasses the entire life cycle from pennycress cultivation and raw materials acquisition through the production and use of the fuels in vehicles and aircraft operations. The pennycress is grown in the Midwestern United States as a winter annual. After harvesting, the pennycress seeds are transported to a centralized processing facility where the oil is recovered and a de-oiled meal co-produced for use as a secondary energy source or animal feed. The oil is then transported to a hydroprocessing plant as a source of renewable feedstock for HRJ and RD production. Inventory data for pennycress cultivation, transportation, and oil recovery were provided by Arvens Technology Inc. Data for conversion of the pennycress oil to HRJ and RD were obtained from engineering design data, and was supplied by UOP. All the inventory data were assembled based on energy content of 1 MJ of final fuel product, which was the functional unit of this life cycle assessment (LCA) study. In the base case, energy allocation was applied to distribute environmental impacts among various products and co-products along the life cycle of each biofuel. The life cycle GHG emissions (excluding land use change) for RD and HRJ are 26.5 and 28.5 g CO_2 eq./MJ fuel, which represent savings relative to petroleum fuel counterparts of 72% and 69% respectively. Total energy consumption for each pennycress biofuels product is comparable to petroleum fuels with the exception that most of the energy is derived from renewable biomass as opposed to non renewable fossil. Scenario analyses were also conducted to determine response to model assumptions and data uncertainty, including allocation methodology, fertilizer application rate, source of H_2, and land use change. The results show that pennycress derived biofuels could qualify as advanced biofuels as defined by the Renewable Fuels Standard (RFS2).

(305e) Effect of Biomass Feedstock On the Sustainability Metrics of Fischer-Tropsch Fuels

Ignasi Palou-Rivera and M.Q. Wang, Argonne National Laboratory, Argonne, IL

The use of biomass as feedstock for the manufacture of Fischer-Tropsch fuels has significant effects on their sustainability metrics. A number of studies have looked at this biomass use [1,2], with an emphasis on techno-economic results. Some of the studies incorporate limited life-cycle analyses [1,2,3], especially considering green-house gas emissions. Our study is a complete Well-To-Wheels Life Cycle Analysis of biomass-fed Fischer-Tropsch liquid fuels (gasoline, diesel, and jet-fuel) performed using our GREET(Greenhouse gases, Regulated Emissions, and Energy use in Transportation) Model developed at Argonne national laboratory under DOE sponsorship. The presented LCA has a particular focus on the effects of several design and operational parameters:

- ***Biomass characteristics***: forestry and agricultural residues, humidity level, relative amount (when not used as only feedstock), etc.
- ***Gasification characteristics***: biomass co-gasification with coal, mix of biomass-derived syngas with other (e.g. coal- and natural gas-derived syngas), gasifier technologies, etc.
- ***Process characteristics***: CO_2 capture and sequestration (CCS) degree, process integration, FT synthesis characteristics and catalyst type, etc.

The LCA includes an encompassing sensitivity analysis of the effects of all these parameters on major sustainability metrics. The considered metrics include energy use (total, fossil, etc), green-house gas emissions (CO_2e GHG), and emissions of other criteria pollutants (NO_x, PM10, PM2.5, SO_x, VOC, CO)

[1] Liu, G.; Larson, E.D.; Williams, R.H.; Kreutz, T.G.; Guo, X. *Making Fischer-Tropsch Fuels and Electricity from Coal and Biomass: Performance and Cost Analysis*. Energy Fuels, 2011, 25, 415-437

[2] NETL/DOE (National Energy Technology Laboratory/U.S. Department of Energy), *Affordable, Low-Carbon Diesel Fuel from Domestic Coal and Biomass*, January 2009.

[3] Xie, X.; Wang, M.; Han, J. *Assessment of Fuel-Cycle Energy Use and Greenhouse Gas Emissions for Fischer-Tropsch Diesel from Coal and Cellulosic Biomass*. Environ. Sci. Technol., 2011, 45, 3047-3053.

See more of this Session: Life Cycle Analysis of Renewable Feedstock-Based Processes and Products I
See more of this Group/Topical: Sustainable Engineering Forum

(305f) An Integrated Framework for Design of Sustainable Chemical Process Based On Life Cycle Assessment

Masahiko Hirao and Yasunori Kikuchi, Department of Chemical System Engineering, The University of Tokyo, Tokyo, Japan

In this work, we propose a supporting framework for the design of chemical process integrating computer-aided process engineering (CAPE) and life cycle assessment (LCA) tools. LCA has been a useful tool for quantifying the environmental impacts from existing product life cycle and its application to process development enables us to design environmentally conscious product and process. During process development, however, implementation of LCA is difficult due to lack of inventory data. CAPE tools are used for simulation and evaluation of chemical processes, but have potential for the estimation of process inventories used in LCA. Thus, we modeled activities and information flows using integration definition for function modeling (IDEF0) method for the sustainable chemical process development in which CAPE and LCA tools are integrated. Applicability of the proposed activity model is demonstrated and verified by case studies on the process development of bioethanol-derived chemicals.

See more of this Session: Life Cycle Analysis of Renewable Feedstock-Based Processes and Products I
See more of this Group/Topical: Sustainable Engineering Forum

(334a) Application of Ultrasound with Various Pretreatments to Improve Biological Hydrogen Production From Food Waste

ElSayed ElBeshbishy, Civil and Environmental Engineering, The University of Western Ontario, London, ON, Canada, Bipro Ranjan Dhar, Chemical & Biochemical Engineering, The University of Western Ontario, London, ON, Canada, Hisham Hafez, GreenField Ethanol Inc. and George Nakhla, Chemical and Biochemical Engineering, University of Western Ontario, London, ON, Canada

Biological hydrogen production from waste has a great potential to meet the growing energy demand as well as to mitigate environmental pollution. However, one of the major limitations of biological hydrogen production technology is lower hydrogen yield. This study has been done to investigate the impact of ultrasound with thermal, acid and base pretreatment on biological hydrogen production via dark fermentation of food waste. Four individual pretreatment methods: ultrasound, thermal, acid, and base and three combined pretreatments combining ultrasound with heat (UH), ultrasound with acid (UA), and ultrasound with base (UB) were conducted. The effect of pretreatments has been evaluated in terms of solubilization of food waste and the improvement in biological hydrogen production without using extra seed. Compared to the control, ultrasound with base UB pretreatment provided the highest increase in soluble chemical oxygen demand (SCOD) 33%. However, the relative ranking of pretreatment based on the hydrogen production is: UA > Ultrasound> UH > Heat or Thermal > UB > Acid > Base.

See more of this Session: Renewable Hydrogen Production II
See more of this Group/Topical: International Congress on Energy 2011

(334b) Hydrogen Production Via Fermentation of Avicel and Corn Stover In a Sequencing Batch Reactor

Lauren R. Magnusson, Mohagheghi Ali, Thammannagowda Shivegowda, Melvin Tucker and Maness Pin-Ching, National Renewable Energy, Golden, CO

When produced from waste lignocellulosic materials, hydrogen is one of the most promising options for a long-term, net-zero emission energy carrier. A major barrier to industrial scale production of bio-H_2 is its techno-economic feasibility. Advances in consolidated bioprocessing are required to improve the rates of cellulose hydrolysis and its subsequent fermentation to H_2. *Clostridium thermocellum* is a thermophilic anaerobic microbe with one of the highest known rates of cellulose degradation. The goal of this study is to improve substrate utilization by employing a sequencing batch reactor, which would increase substrate hydrolysis leading to higher rate and yield of hydrogen. Reactor parameters and substrate loading were first optimized in batch mode, using *C. thermocellum* grown on avicel and pretreated cornstover at 1, 2.5 and 5 g/L. We observed that an increase in carbon loading led to higher rates of H_2 production yet with a lower H_2 molar yield (mol H_2/mol hexose) in both substrates. These data guide the fermentation in sequencing batch mode, which was initiated by running a 2.5 g/L fermentation for 24 hours followed by settling the unfermented solid substrate along with bacteria for one hour. This was then proceeded by removing 500 mL of the clear supernatant using a peristaltic pump, adding 500 mL of fresh media (gassed out with nitrogen gas) and substrate and turning the agitation back on. Four cycles were completed for each carbon loading condition. The hydraulic residence time (HRT) tested was 48 hrs, with a carbon loading of 2.5 and 5.0 g/L/d.

Initial results indicate that at a carbon loading of 2.5 g/L/day, all of the substrate was consumed and the bacteria did not settle when agitation was turned off. The average concentration of protein in the supernatant after settling for an hour was 164 mg/mL. At 5.0 g/L/d loading, we observed an excess of substrate present in the reactor and the avicel turned orange, which is an indicative of presence of cellulosome hence bacterial attachment. Protein concentrations were measured before settling (208 mg/mL), and in the supernatant after settling (38 mg/mL). The data suggest that by operating sequencing-batch fermentation with excess cellulose, approximately 80% of the bacterial biomass can be retained, presumably fully adapted to degrade cellulose. Additionally, the average rate of hydrogen production was increased from 0.60 to 0.92 mmol/L/hr during the 5 g/L/d carbon loading condition. One of the main benefits of sequencing-batch operation is the removal of the lag phase at the beginning of conventional batch-fermentations, once adapted. The time to peak hydrogen production in batch one is almost 19 hours. By the fourth fed-batch cycle, this number drops to 4 hrs. Collectively our findings form the basis for further optimization to improve the techno-economic feasibility of hydrogen production via fermentation.

(334c) Biological Hydrogen Production From Municipal Solid Waste

Suvarna Talluri[1], Mohan Raj Subramanian[2] and **Lew P. Christopher**[1], (1)Center for Bioprocessing Research and Development and Department of Chemical and Biological Engineering, South Dakota School of Mines & Technology, Rapid City, SD, (2)Center for Bioprocessing Research and Development, South Dakota School of Mines & Technology, Rapid City, SD

The handling and disposal of municipal solid waste (MSW) is of growing concern in the U.S. as this waste continues to increase. MSW contains ~60% of biodegradable resources which can be utilized to produce biorenewable energy and value-added bioproducts. This study focuses on the isolation of new and efficient bacterial strains for production of biohydrogen from MSW via dark fermentation. The H_2-producing bacteria were isolated from MSW and granular sludge after several enrichment steps on various substrates, including MSW. Out of 65 isolates, two bacterial isolates were chosen as the most promising H_2 producers when grown on MSW as sole carbon source in a mineral salt medium. Analysis of the substrate utilization revealed that both isolates could utilize complex polymeric substrates such as starch, cellulose and hemicelluloses, in addition to C5 and C6 monosugars. The major byproducts from both isolates were ethanol, propionate, lactate and acetate.

See more of this Session: Renewable Hydrogen Production II
See more of this Group/Topical: International Congress on Energy 2011

(334d) Optimization of Photosynthetic Hydrogen Yield From Platinized Photosystem I Complexes

Rosemary K. Le[1], Ifeyinwa J. Iwuchukwu[2], Ernest A. Iwuchukwu[3], Barry D. Bruce[4], Rupy S. Sawhney[3] and Paul D. Frymier[1], (1)Chemical and Biomolecular Engineering, University of Tennessee - Knoxville, Knoxville, TN, (2)Biochemistry and Molecular Biology, University of Georgia, Athens, GA, (3)Industrial Engineering, University of Tennessee - Knoxville, Knoxville, TN, (4)Biochemistry and Molecular Biology, University of Tennessee - Knoxville, Knoxville, TN

Fractional factorial design was utilized to optimize light-dependenthydrogen production by platinized Photosystem I isolated from thecyanobacterium *Thermosynechoccocuselongatus* BP-1 using response surface methodology (RSM). Using JMP, statistical data analysissoftware, a model equation was developed to predict an optimum for hydrogenevolved, after platinization, under the optimal conditions for temperature,light intensity and platinum salt concentration. The total H_2 yield had asignificant dependence on platinum salt concentration and temperature duringplatinization. Light intensity duringplatinization had a minimal effect on the total H_2 yield. The surface and contour plots yielding themodel equation indicated that an optimal hydrogen yield would be maximized at alight intensity of 236 µ E m^{-2} s^{-1},platinum salt concentration of 636 µ M and temperature of 30.8^{0}C.The predicted optimum was 8.17 mol H_2 h^{-1} mg chl^{-1}. To validate the model, an experiment wasperformed with observed optimal conditions during platinization, yielding 8.02 molH_2 h^{-1} mg chl^{-1} during hydrogen evolution, thehighest value observed in the series of experiments.

See more of this Session: Renewable Hydrogen Production II
See more of this Group/Topical: International Congress on Energy 2011

(334e) Synthesis of Bifunctional Catalyst for Aqueous Phase Reforming of Biomass

Richa Tungal[1], Rajesh Shende[2], Carlos Landaverde Alvarado[2] and Kenneth M. Benjamin[2], (1) Chemical and Biological Engineering, South Dakota School of Mines & Technology, Rapid City, SD, (2) Chemical and Biological Engineering, South Dakota School of Mines and Technology, Rapid City, SD

This study reports the synthesis of novel bifunctional catalytic material and its utilization in catalytic aqueous phase reforming of biomass to enhance selectivity and conversion into hydrogen and higher hydrocarbons. The synthesis involved the preparation of mesoporous aluminosilicate catalyst by using the surfactant assisted sol-gel method followed by immobilization of Ni/Cu/Ru. The alkoxides precursors for aluminoslicate were chosen in a desired weight ratio and mixed with the Pluronic 123 surfactant. After hydrolysis and condensation steps, the gel obtained was aged for 24 h, treated in ethanol at 60°C for 2 hrs and finally calcined at 400-600°C at a heating rate of 1°C/min. The mesoporous structure was analyzed using TEM and the phase composition was determined using powder X-ray diffraction. The mesoporous aluminosilicate material was contacted with ethanol/water solution of Ni and Cu salts for 8 h at 40-50°C. The powder obtained was dried at 105°C for 2 h and reduced under H_2 environment for 2 h and further characterized by SEM/EDX to determine the loading of metal ions on the mesoporous aluminosilicate material. As received biomass (pine wood and cellulose) were subjected to aqueous phase liquefaction process at 250°C in a high pressure reactor in presence of homogeneous Ni or Cu salts as catalyst and H_2 generation was monitored. The slurry obtained after liquefaction process was processed further to recover the residue and bio-oil. The Ni/Cu loaded meosporous catalyst was contacted with the processed slurry obtained after liquefaction and the H_2 and higher hydrocarbon formation was investigated. The results obtained on H_2 generation during catalytic liquid phase liquefaction processing of a biomass, SEM/EDX analysis of Ni/Cu loaded aluminosilicate, and selective conversion into higher hydrocarbons will be presented. In addition the mechanistic model based on catalytic free-radical chemistry will also be presented, and results discussed.

See more of this Session: Renewable Hydrogen Production II
See more of this Group/Topical: International Congress on Energy 2011

(334f) Catalytic Hydrolysis of Ammonia Borane Using Co-B Nanoparticles Embedded In Mesoporous Silica Particles: a Very Efficient Catalyst for Hydrogen Production

Nainesh Patel[1], Rupali Dholam[2], Renuka Jaiswal[2], Namdeo B. Lihitkar[2], Rohan Fernandes[1], Antonio Miotello[1] and **Dushyant C. Kothari**[2], (1)Physics, Universita degli Studi di Trento, Povo (Trento), Italy, (2)Centre for Nanosciences & Nanotechnology, University of Mumbai,, Mumabi, India

Ammonia Borane (AB), with high hydrogen storage capacity (19.5 wt.%), is predicted to be possibly a future fuel for providing pure H_2 to fuel cell at room temperature for on-board and portable applications. However, a heterogeneous catalyst plays a vital role for first initiating and then controlling H_2 production during hydrolysis reactions. Thus, the present work is focused to develop nano-catalyst to be used for dehydrogenation reactions of AB. Firstly, mesoporous silica nanoparticles (~ 100 nm) with pore size of ~ 3 nm were synthesis by hydrolysis and condensation of alkoxy silane using CTAB surfactant as a template. Later, cobalt salt was embedded inside the pores of the mesoporous silica by chemical impregnation method followed by reduction using $NaBH_4$ to form Co-B nanoparticles. It was observed that Co-B nanoparticles act as highly active catalytic centers to produce significantly higher rate (about 6 times) of H_2 by hydrolysis of AB than the same amount of the corresponding Co-B powders (produced by similar chemical reduction method but without mesoporous silica support). Almost complete conversion (95%) of AB was obtained during hydrolysis, as confirmed by ^{11}B NMR, by using nano-catalyst at room temperature. The nano-catalyst were characterized by TEM, SEM, AFM, XPS, EDS, XRD and BET surface area measurement to understand the enhanced activity of the catalyst. Availability of a large number of under-coordinated Co active atoms owing to the size and shape of nanoparticles, better dispersion & polycrystalline nature of nanoparticles, and optimum interaction with reactant provided by electron enrichment on Co sites from B, are the main features acquired by the Co-B nanoparticles that exhibit high catalytic efficiency. The efficient nature of Co-B nanoparticles is well supported by the very low activation energy and high H_2 generation rate obtained from the hydrolysis of AB.

See more of this Session: Renewable Hydrogen Production II
See more of this Group/Topical: International Congress on Energy 2011

(338a) Comparing Resource Consumption and Sustainability of Biobased Materials Using Eco-LCA

Erin F. Landers and Bhavik R. Bakshi, William G. Lowrie Department of Chemical and Biomolecular Engineering, The Ohio State University, Columbus, OH

Ecosystem goods and services are essential to all human activity. In spite of this importance, the Millennium Ecosystem Assessment has found that many ecosystem services are highly degraded. In many engineering analyses, these necessary goods and services are undervalued, or even ignored completely. It is therefore crucial to develop and modify existing methods to allow these services to be valued and accounted for in decision making. Conventional life cycle assessment (LCA) tends to only look at energy consumption and emissions. However, Ecologically Based LCA (Eco-LCA) is a thermodynamically based LCA that allows for many ecosystem goods and services used throughout our economy to be quantified, in addition to the traditional LCA results of energy consumption and emissions.

To create the Eco-LCA model, a nationally based economic model must first be obtained, to which the resources will then be added. Eco-LCA uses the national input-output economic model that the Bureau of Economic Analysis provides every five years. This current model, based on the 2002 national economy, shows the interactions between different industries and is used to track the use of ecosystem goods and services through the economy, which will be compared with previous research done to create an Eco-LCA inventory for the 1997 economic model. Resource use must then be added to the model; this is done by finding values for resource use from government agencies and various research groups. The thermodynamic properties of exergy and emergy allow for the many different units of ecosystem goods and services to be converted to the common basis of thermodynamic work. This common basis means the results of Eco-LCA analyses can be aggregated and directly compared in metrics such as Cumulative Exergy Consumed and Percent Renewable. After adding the ecosystem goods and services to the national economic model, resource intensities are calculated for every resource and economic industry by taking into account the resources entering the economy, the interactions between industries, and the total monetary activity of every industry. This then provides a measure of individual and total resource use per dollar of economic activity in each industry. These new resource intensities can be directly compared to the resource intensities from the 1997 Eco-LCA model and provide insight as to changes in both resource use and interactions between industries from 1997 to 2002.

A case study will be presented using Eco-LCA to evaluate biobased materials such as biofuels and biopolymers. The sustainability of biofuels will be discussed by investigating the resources required and emissions produced for various feedstocks, including changes since 1997. This will allow for a more dynamic view of the progress towards the most viable, sustainable biofuels. Resource consumption to make biopolymers will be compared to traditional polymers, and conclusions will be drawn as to the most sustainable polymer materials.

See more of this Session: Sustainability and Biobased Materials
See more of this Group/Topical: Forest and Plant Bioproducts Division

(338c) Integrated Biorefinery Operations: Biomass Pretreatment As It Relates to Biochemical and Thermochemical Products

Brian L. Cooper, Jeff London and Robert Mellon, Hazen Research, Inc., Golden, CO

Biomass based chemical refining represents many new opportunities in engineering and process design. The population of companies seeking to develop their technology is large and diverse, from pure liquid fuel producers to integrated multi-product refineries. A common flow sheet among many of these organizations shows a hybrid biochemical/thermochemical process, whereupon the carbohydrate values in a biomass are removed and the resultant lignin portion is utilized for co-products or process heat. As a company with capabilities in both areas, Hazen Research is conducting a series of experimental programs that generate data for a hypothetical process for the biorefining of Lodgepole Pine (Pinus contorta) and Ponderosa Pine (Pinus ponderosa) into chemical values and thermal chemical products. Initial studies demonstrated a clear correlation between biomass pretreatment conditions and the resultant pyrolysis/gasification products generated from the hydrolysis residue. The presentation will focus on new data generated in a second round of experiments, which include optimizing carbohydrate yields in hydrolysis as well as running a continuous thermal processing in place of batch experiments.

See more of this Session: Sustainability and Biobased Materials
See more of this Group/Topical: Forest and Plant Bioproducts Division

Green Poly (lactic acid) Composites Reinforced with Sugarcane Bagasse Residues

Letian Wang, Zhaohui Tong

Agricultural and Biological Engineering Department, IFAS, University of Florida,

Rogers Hall Museum Road, PO Box110570, Gainesville, FL, 32611-0570

Email: ztong@ufl.edu

Poly (lactic acid) (PLA) composites reinforced by sugarcane bagasse residues were prepared using twin-extrusion. By using near zero-value bio-based residues, these composites are not only 100% renewable and biodegradable, but also have the competitive cost in comparison with petroleum-based composites. The sugarcane bagasse residues (10-30 wt.%), containing different content of fibers and all the remaining lignin, were used to prepare lignocelluloses reinforced PLA composite with desirable physical and chemical properties. In this study, the sugarcane bagasse residues include the residues originally from the sugar extraction process, after pretreatment, and after fermentation of biorefinery processes. The varieties of different bagasse residues include the composition, particle size and its surface reactivity during the chemical and biological (enzymes, microbes) treatments that the biomass is subjected to. The effect of these varieties on the morphologies, thermal properties, mechanical properties and the molecular weight of PLA composites was studied. Furthermore, a coupling agent (Desmodur® VKS 20) was used to increase the interfacial bonding of PLA with lignocelluloses. Desmodur® VKS 20 is a mixture of diphenylmethane-4, 4'-diisocyanate (MDI) with isomers and higher functional homologues (PMDI). The microstructure and the properties of the composites in the presence of couple agents were also investigated in this study. It was found that this coupling agent could extend the polymer chain and therefore improved the mechanical properties of the composites.

ALTERNATIVE USES OF SUGARCANE MOLASSES AND JAGERRY FOR PRODUCTION OF VALUE ADDED FATTY ACIDS, ESPECIALLY OXALIC ACID

P.Sanyal*, S. Gupta, S. Chattopadhyay and A.K. Ray
Department of Paper Technology
Indian institute of technology roorkee
Saharanpur campus, Saharanpur-247001, India
*National Sugar Institute, Ministry of Food and Civil Supplies, Govt. of India, India

ABSTRACT

Sugarcane molasses and jaggery provide a unique source for large number of downstream products which include livestock and human food sources, potable alcohol and absolute alcohol, alcohol based chemicals like oxalic acid, citric acid, tartaric acid, lactic acids, acetic acid, diethyl oxalate, ethyl acetate, acetic anhydride etc., which are raw materials for manufacture of pharmaceuticals, plastics, synthetic rubber, base for yeast, lysine, protein and innumerable other products. In this present study, status and economics of Indian industries who utilize molasses to produce organic acids through both chemical and biochemical routes like lactic, citric, tartaric, aconitric and oxalic acid etc. are reviewed with particular reference to oxalic acid production and updated research findings. The oxalic acid or its derivatives have widespread industrial applications such as textiles, leather, oil refinery, metallurgy ,nuclear power plant, dyes, catalysts, explosives, straw bleaching, printing, marble polishing, metal and cloth cleaning, pharmaceuticals and antibiotics. It is also a very important chemical in petroleum, rare-earth, ink, rust, corrosion inhibitor, and dental adhesive processing.

Oxalic acid can be produced from a number of sources from sugar itself, from sugarcane or sugar industry wastes, sugarcane trash, wastes from paper industry by oxidation of their main components of carbohydrates e.g. sucrose, glucose, and other sugar bearing raw materials, from dry powder of parthenium hysterophorus L., and other cellulosic wastes, wood dust, wood cellulose etc. Sugar cane or sugar beet molasses, sugarcane jaggery, deionized cane juice can be a potential source from sugar industry wastes. Both chemical as well as biochemical process may be followed to produce this important chemical. In chemical route generally catalytic oxidation of the above ingredients by nitric acid or by formed nitrogen oxides in batch or continuous stirred tank reactors produces oxalic acid, acid sachharate, and tartaric acid. Production of oxalic acid by fermentations of carbohydrates using *Aspergillus niger* and *Pseudomonas fluorescens* and radiation processing of carbonate solutions and molasses are reported. Production of tartaric acid and citric acid under certain conditions can also be produced. However, the majority of work involves oxidation of sugar bearing materials by nitric acid to produce oxalic acid. In this present study, an attempt has also been made to optimize the processes of producing oxalic acid from sugarcane molasses and sugar cane jaggery by oxidation with nitric acid in presence of dehydrant and catalyst with or without the presence of catalyst promoters. Parameters affecting the reaction were: air flow rate, temperature, the amount of V_2O_5, catalyst promoters and the concentrations of reactants and dehydrant. The maximum yields obtained by different processes were compared. Treatments of NO_x vapour by absorbing in a suitable medium (in soda ash or sulphuric acid), to avoid air pollution (and also to produce value added by-products, glauber salt and sodium nitrate), mother liquor and catalyst recycling were attempted to enhance the economy of the plant.

INTRODUCTION:

India is the largest producer of sugar in the world [1-4]. In terms of sugarcane production, India and Brazil are almost equally placed now a days. Two to three years' before it was the second largest producer of sugar in the world next to Brazil. At present, total world production is of the order of 110-116 million tons from both beet and cane sugar.India is expected to produce round 27.5 million metric tonnes (MMt)and 30.5MMt in 2011-12 and 2016-17 respectively.

It is wellknown that one of the important by- product of the Sugar Industry is the cane molasses, which is a strong building block for many chemical industries. In spite of producing cane molasses in large quantity in the country, several molasses based products are being imported in substantial quantity[4]. Department of Science and technology in India submitted a report which detailed the demand supply scenario and status of the Indian molasses based industry concerning the five products namely Lactic Acid, L-Lysine, Gluconic Acid, Glutamic Acid/Monosodium Glutamate and Itaconic Acid. It is necessary that India should set up facility for the production of the above molasses based products for the reasons [4]such as to utilize the molasses availability, to avoid the import of the molasses based downstream products, to export the molasses based downstream products in the global market. To achieve breakthrough in the fermentation technology for the molasses based products, as molasses is the byproduct of the sugarcane industry it is important to highlight regarding the status of this industry in this country.

STATUS OF SUGAR CANE INDUSTRY IN INDIA:

Sugar industry, worth Rs 60,0000 millions is also one of the largest agro-based industries in India-a country that happens to be also the largest consumer of white plantation sugar.[1-2] It has manufactured 26 million MT in 2007-08 which has declined over the following years due to varieties of reasons, 14.7 million tons 2008-2009 and again increased to 16.0 million tons 2009-2010. As already indicated in the coming years, the production of sugar is likely to go up steadily in the country [1-3].With approximately 25 million tons of sugar production in 2010-11 and 24 million tons of consumption, both sugar production and consumption are bound to increase lineally. The growth rate of sugar production in India is shown in Fig.1 and the same for molasses is given in Fig.2.

Indian sugar consumption is growing at a CAGR of 3.75% .Sugar production in India has occupied a place of pride in the Indian economy being one of the five largest industries comprising of more than 571 units spread all over the country.

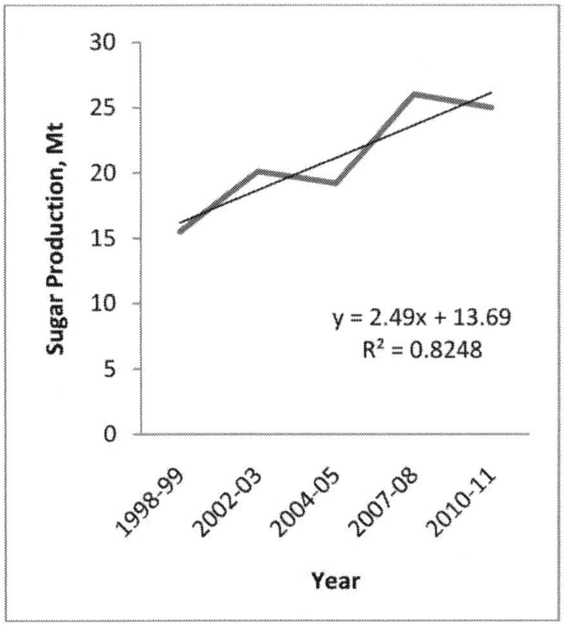

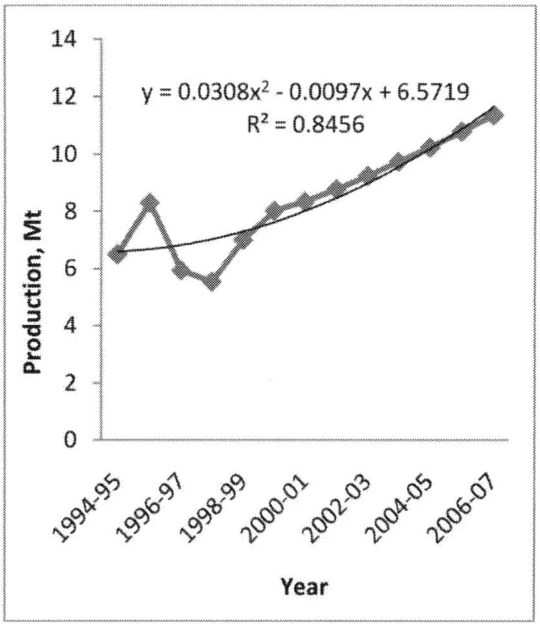

Fig.1 Production of white plantation sugar. Fig. 2 Production of Molasses

The average yield of cane in India is around 70 -77 t/ha in tropical states to about 52t/ha in subtropical stases[1] though in UP(which produces 40% sugar) the yield has decreased from 55 MT/ha to 45 MT/ha in the recent past. High yielding new varieties, namely Co 238, Co 239 and Co 118 identified can be used to increase the productivity of sugar cane and sugar. The area under sugarcane is presently less than 2% of total cultivable area in the country and about 3% of the irrigated area. There is considerable scope for increasing the area under sugarcane considering the fact that it is more profitable compared to other crops.

INDIAN MOLASSES CHARACTERISTICS AND ITS UTILIZATION:

Molasses is the uncrystallizable syrup which remains as final residue in process of manufacture of sugar. Its composition varies due to a number of factors, including cane maturity, variety, agricultural conditions, industrial processes etc. A typical composition of final molasses(all in %) in Indian Sugar Mill is as under:

H_2O	-	17-25	
Sucrose	-	30-40	} 50% Saccharine matter
Glucose	-	4-9	
Fructose	-	5-12	
Other reducing sugars	-	1-4	
Total reducing sugar	-	10-25	
Carbohydrates, Gums, Starch	-	2-5	
Water soluble carbonates	-	7-5	
Nitrogen	-	2.5-4.5	
Others	-	0.5-1.5	

Waxes, sterols phosphorus - 0.1-1.0
Vitamins - varying amounts

As molasses contains substantial quantity of sucrose as well as reducing sugars, it must produce many chemicals acids as pure sucrose does as shown in the following diagram[1]:

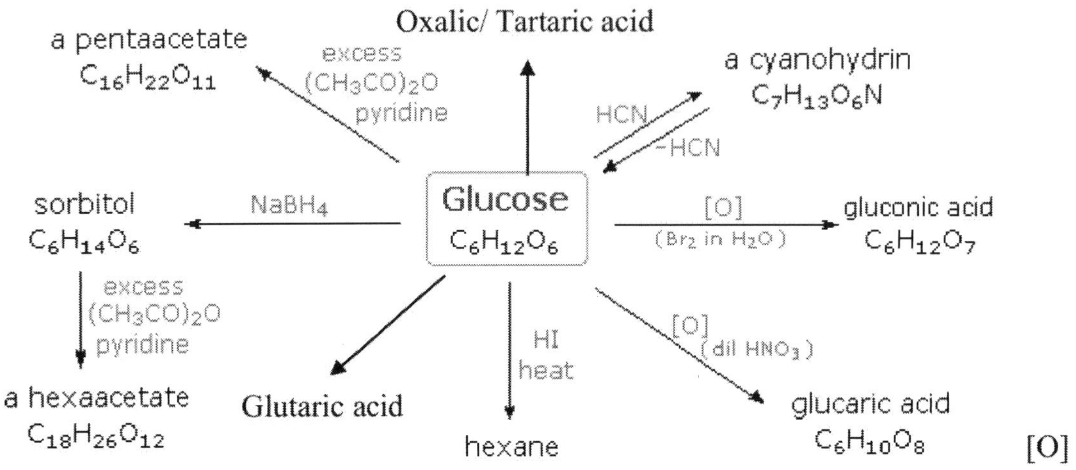

Most of the chemicals are however imported in India.

Use of Molasses

(a) Direct utilization: Fertilizer, animal feed
(b) Distilling industry: Ethyl alcohol, rectified spirit, power alcohol, alcohol derivatives, potable liquor, such as beverages
(c) Other fermentation industries: Vinegar and acetic acids, acetone and butanol, citric acid, lactic acid, glycerine, yeast,
(d) Other chemical industries: Oxalic acid, and tartaric acid
(e) Miscellaneous industries: Aconitic acid, monosodium glucamate, dextran,L-Lysine, Xanthenes gum, 2-keto gluconic, 5-keto gluconic , a-ketoglutaric, fumaric acids , glycolic acid,kojic acid and glucaric acid etc.

Besides the above, molasses can be used for many other value added products.Some of the important value added chemicals (mostly imported) are briefly described below:

Molasses is the only raw material available in large quantities in India for an economic production of alcohol until petrochemical source develops. Almost all the commercial molasses in the world is cane molasses because the beet-sugar producing countries usually consume it domestically. The most significant use of final molasses in many industrialized, countries is in animal feed. Other uses of molasses are as the raw material for alcohol production and as the base for yeast, lysine and citric acid production and may others. The chemicals derived from molasses could be produced cost effectively and be profitably exported [3,5-40].

Molasses contains small amount of Lysine, or L-lysine, an essential amino acid, 2,6-diaminohexanoic acid, building blocks of protein, required for proper growth but the body can't manufacture it; lysine has to be obtained from food. Lysine it plays an essential role in the production of carnitine, a nutrient responsible for converting fatty acids into energy and helping to lower cholesterol. Lysine appears to help the body absorb calcium, and plays an important role in the formation of collagen, a substance important for bones and connective tissues including skin, tendon, and cartilage. Glucaric acid is a member of a larger group of compounds known as sugar acids – and more specifically, aldaric acids. Glucaric acid has gained attention because it was identified as one of the top 12 renewable building block chemicals.

Fumaric acid

The principal industrial use of fumaric acid is as a mordant in textile dyeing. It appears that it could find application as a replacement for tartaric acid in leavening agents, as well as in the manufacture of resins and plastics for special purposes. Itaconic and fumaric acids, if sufficiently cheap, could flow into plastics. The same is true of citraconic acids which may be derived from citric acid. In addition, this group is outstanding as acidulating agents in foods. Citric, lactic, and gluconic acids are known to be safe ingredients of the diet and are extensively employed in foods, soft drinks, confectionary and pharmaceuticals.

Itaconic acid

Itaconic acid, 2-Ketogluconic acid is produced by the action of various species of bacteria of the genus *Pseudomonas* on glucose in aerated cultures in the presence of calcium carbonate. This is formed by *Aspergillus itaconicus* and is now available commercially. It has the possibilities in the preparation of resins and plastics because it is a substituted acrylic acid. It is now possible to get 37-percent weight yields in 12 days at about pH 2.2. With submerged fermentations, yields as high as 45% have been obtained in 5 days at a pH of 1.8.Weight yields of 90 percent can be obtained in a 30-hour submerged fermentation. The keto acids can be converted to d-araboascorbic acid which is a good antioxidant for food use. In this capacity, it can serve in the preservation of flavors in the canning industry and in preventing the development of oxidation flavors.

Kojic acid

It is very chemically reactive compound and many derivatives have been prepared from it. It can be prepared readily in yields of more than 50 percent by molds growing on a variety of carbon sources like glucose and xylose, with the use of members of the *Aspergillus flavus, A.oryzae* group.

5 keto gluconic acid is produced by various species of *Acetobacter* in weight yields exceeding 90 percent. When oxidized by air in the presence of vanadium pentoxide, it is converted to d-tartaric acid with good yield. A ketoglutaric acid can be produced in a 25 percent yield when fermentation of glucose by *Pseudomonas fluorescens* is allowed to proceed beyond the 2-ketogluconic acid stage to the point where no more reducing sugar is observed. Good yields of bionic acids such as maltobionic and lactobionic acids could be obtained by the action of *Pseumonas graveolens* on maltose and lactose. The pentonic acids such as d-arabonic, and d-

ribonic acids are formed with good yields during oxidation of d-arabinose , d-xylose and d-ribose by the genus *Pseudomonas*.

Lactic Acid:

Five tons of molasses would be needed to produce one ton of lactic acid.

Lactic acid is used primarily as an acidulant and preservative agent in the pharmaceutical, chemical, and food industries like beverages, beer, jelly, etc. Described as ``the chemical of the future" because of its use in the manufacture of polylactic acid, it is reckoned that polylactic acid could substantially replace plastic bags, since it is bio-degradable [9-10]. The current interest lies in making biodegradable plastics used for general and medical applications as a delivery vehicle for vaccinations.

Lactic acid production parameter optimization[9] was performed using cane molasses by the design of experiment(DOE).Eight factors with three levels studied were yeast extract, $CaCO_3$, $MnSO_4$, pH, temperature, molasses, urea, and Tween 80. Lactic acid production was significantly affected by interactions of two factors such as temperature-urea, pH-$CaCO_3$, temperature- Tween 80, but individually they have minimum impact on lactic acid production. Individually pH, yeast extract, molasses, and urea are the most significant factors in lactic acid production. Lower amount of $CaCO_3$ and $MnSO_4$ and Tween 80 enhanced lactic acid production. The expected yield of lactic acid under optimal condition was 98.66% and the actual yield was 96.8%. Overall 6% improvement in yield has been observed when compared with same (mutant) species grown on cane molasses.Lourdes R. Fragante [5] has conducted fifteen experiments using *Lactobacillus delbreukii* in aqueous molasses and molasses/coco water solutions containing 9, 10, 11 and 12% sugar as invert. The additives used were 2% nitrogen, 1% phosphate and 8.3% invertase per 35g. molasses. The incubation temperature was 45°C. Highest yields were obtained at 10% sugar level for all substrates. Coconut water was found to support lactic acid production. Substrate-Molasses in coco water solution with invertase further increased the yields but only up to 10% sugar level. The presence of higher glucose/fructose levels seemed to inhibit lactic acid production. All substrates gave decreasing yields at 11 and 12% sugar.

Citric Acid

Seven tons of molasses would be required to produce one ton of citric acid.

This chemical is mainly used as a preservative in food processing, pharmaceutical and cosmetic industries. It is also used to manufacture sodium citrate, used in the production of detergents. Industrial production of citric acid by fermentation is one of the most efficient examples of fungal fermentation technology[7-8].The food industry is thought to comprise as much as 70% of this market; the remainder being largely pharmaceutical produce, though the metal cleaning, agriculture and detergent industries would also be of notable importance.

Tartaric Acid:

Tartaric acid is used above all in many applications in the food and textile industries. For example, tartaric acid is used as a food additive, especially as an antioxidant and acidification agent. In addition, tartaric acid is used as a building material additive for modification of the setting characteristics of cement and serves as a starting substance for numerous chemical reactions, especially for chiral synthesis.

A method of preparing tartaric acid in which 5-ketogluconate is oxidized in a carbonate buffer at an alkaline pH preferably in the range of 8 to 10 using vanadate has been described [11-13]. One way of producing tartaric acid is to recover it from tartar, a substance which is produced during the production of wine.

However, the crude tartar which results from the production of wine is as a rule contaminated by organic materials so that expensive preparation is necessary to obtain the tartaric acid from the tartar.

It has been found surprisingly that under these reaction conditions, i.e. also without the addition of vanadate or another catalyst, a conversion of 5-ketogluconate to tartaric acid results whereby a conversion of 5-ketogluconic acid to tartaric acid of up to 55% is achieved. If vanadate is additionally supplied to the reaction mixture, conversions up to 75% of the 5-ketogluconate used to tartaric acid are obtained[11-12].

A further process for producing tartaric acid resides in the conversion of glucose to tartaric acid through microorganisms like *Acetobacter* or *Gluconobacter* in the presence of vanadate as trace element. It can be demonstrated that in these processes the tartaric acid production is carried out in two steps. In the first step 5-ketogluconate is excreted into the medium of microorganisms cultured in a glucose-containing medium, while in a second step the conversion of 5-ketogluconate to tartaric acid is carried out purely chemically with the vanadium salt as trace element, i.e. independently of the presence of microorganisms[13].

PRODUCTION STATISTICS AND ECONOMICS:

Although India is a major producer of molasses (fig.2). Sugar mills produce around 8 million tons (mt) of molasses, which is projected to increase to 10.5 mt over the next ten years. About 14 per cent of the molasses produced today is exported, without any value addition. There is a growing demand world over for molasses-based chemicals, which are presently imported [6]. The huge export opportunity is being lost, because, principally, the Indian industry is not fully aware of the business potential in this area. There is a possibility of manufacturing these chemicals with indigenous resources. Here again, a huge export opportunity exists manufacturing this chemical at competitive prices, further harnessing of overseas business opportunities. On the other hand, the country imports molasses-derived chemicals in very large amounts

Estimated demand for lactic acid (partly imported) is 2500 tons in the year 2015 A.D. Lactochem Ltd of the Malladi group has a 200 tpa plant in operation. Karnataka-based Munirabad Chemical Co, which has a 180 tpa plant, is another example of a successful plant. Orchem Ltd, an associate unit of Atul Drug House, attempted to put up a 300-tpa plant, but did not succeed. The company first attempted to manufacture the chemical synthetically on a laboratory scale, but ran into difficulties when getting into commercial scale operations. Another company called Sangita Bio-Chem of Orissa also attempted a venture, but is said to be held up for want of funds. Two new units are coming up; one is Prathista Industries in Andhra Pradesh, with 10000 tpa capacity and Interspice Biotech near Chennai. The latter is coming up with technology from GaLactic S.A. of Belgium. The study recommends setting up of a plant of 3,000 tons capacity, which would cost around Rs 27 crore.

In 2002, citric acid production by fermentation was estimated to be in the region of $7.0x10^5$ ton/year [8]. Such production levels are necessary to meet the high demands of the now massive market. From the design of the original fermentation process right up to the present day, citrate production has largely been overseen by Pfizer. Deviation from the market profile of the 60s and 70s has seen the once dominant pharmaceutical market demand overshadowed by that of the ever-growing food market, particularly the carbonated drinks industry. This shift has seen the price of citrate fall significantly, thus encouraging further improvements in the rate and efficiency of its production. Unlike oxalic acid, the technology for producing citric acid from molasses is established and available. The Indian industries worth mentioning are Bharat Starch Industries of the Thapar group, Citric India Ltd of Mumbai and Andhra Citrates of Hyderabad, Citurgia Biochemicals of Mumbai.

For the first time, a detailed study (commissioned by the Centre and conducted by a consultancy company called Nandini Chemical Consultancy[3] has identified some molasses-based products and the business potential for each of them. The study, which advocates setting up of plants for the identified molasses-based chemicals, makes two postulates for its recommendations. First, the price fluctuation of molasses in the country is immaterial because the value addition in the manufacture of these chemicals is so high that volatility in prices is inconsequential. Second, the target market is the world market and any entrepreneur who wants to first establish a ``good domestic base" might as well look for opportunities elsewhere.The following table indicates the approximate business statistics for three principal molasses derived acids.

Table Major Molasses based acids, their demand-Supply, growth rate, export potential

Acids	Global Demand, tons(year)/ Growth Rate,%	Demand/Installed and Production Capacity, tons	Export, tons
Oxalic acid	250,000; 3.0	30,000 /22,000	9,000
Citric acid	703,200(2001), $7.0x10^5$ (2002),1,040,900(2011)	12,000(2001)/ 38,475(2011)	4,610(0.4% of world market)
Lactic acid	70,000(2001)/100,000(2002) 181,560(2005),500,000 (2011)	560(2001)/ 1,980 (2011)	5,000(3% of world share)

Techno-Economic feasibility studies:

Department of Science and technology (DST), Government of India, has estimated the likely gap in supply by 2015 AD in India and recommended the Installed Capacity of the following chemicals. The detailed project report has been made based on number of 300 working days per annum and 24 numbers of working hours.

Lactic acid, is presently consumed 200 Mt, partially met by imports, likely to increase to 600 M t. Only M/s Lactochem India Ltd. is manufacturing it. L-Lysine, itaconic acid, gluconic acid and glutamic/ MSG acid all are imported, India will need 18,000, 1800, 1500 and 10,000 Mt respectively. Adequate electrochemical process technology has been developed for the production of Gluconates from Glucose by Central Electrochemical Research Institute, Karaikudi to produce the later two chemicals.

The Nandini Chemicals study has also said that huge export possibilities exist for other molasses-based products such as lysine, glutamic acid, itaconic acid, gluconic acid, ephedrine hydrochloride and yeast.

Oxalic Acid:

To produce one ton of oxalic acid, 2.5 tons of molasses would be required.
Oxalic acid and its derivatives have widespread industrial applications like textile, leather or tanning, oil refinery, catalysts, dyes, metallurgy, pharmaceutical, marble polishing, petroleum, nuclear engineering, explosives, detergent, plastic, film industries and antibiotics.
Other uses include dental adhesive processing, corrosion inhibition, straw bleaching, printing, precipitating agent in rare-earth mineral processing, bleaching agent in the textile activities, wood pulp bleaching, wood bleaching agent(used to lighten and brighten dingy gray wood), remover of cement residues, rust remover for metal treatment, commercial rust removers to remove rust stains from tubs and sinks, food and rust stains from kitchen countertops, grinding agent, such as marble polishing ,waste water treatment, removing calcium in water, cleaning and sterilizing home brewing equipment, as a reducing agent for photography and ink removal ,as purifying agent in pharmaceutical industry, clean rock collectors, clean mineral specimens, , plumbing fixtures and fabric.

Indian Statistics for oxalic acid manufacture:

Nandini Chemicals recommends a minimum economic size of 15,000 tons, which will call for an investment of about Rs 17 crore. In India, oxalic acid is produced from sugar, by a dozen-odd companies, most of them Mumbai-based. Two companies have attempted to produce oxalic acid from molasses instead of sugar - but both attempts failed. One was a Chennai-based company called Krebs India, which put up a 300 tpa plant at Pondicherry. The other was Punjab Chemicals and Pharmaceuticals Ltd, the largest producer of oxalic acid through the sugar route. The cost advantage can be gauged by the fact that while it takes 600 kg of sugar to produce one ton of oxalic acid, it takes 2.5 tons of molasses to produce the same.

Present cost of molasses based acids:

The present market cost of the following acids from molasses in Indian currency is as under:

Citric acid:Rs. 51.74,Oxalic acid, Rs.52,Fumaric acid(Food grade):60.72,Lactic acid:Rs.68.33,L-Tartaric acid:122.75,D-Tartaric acid :251.09,. These are in the ratio of 1:1.005:1.174:1.321:2.372 :4.853. It is reported that the ratio of prices of some important acids including the mentioned one are as under: Glycolic: Formic acid: Acetic Acid: Oxalic acid : Tartaric acid: Butyric Acid=1:1.44:1.5:1.61:3.56:3.78. From the current market study it is usual noticed that the imported chemicals are found to be cheaper than those indigenously manufactured.

R& D efforts for Oxalic acid production:
Literature Survey:

There are various methods for manufacturing oxalic acid [14-40]. Process depends upon the raw material selected. The conventional method of production generally consists of glucose or sucrose oxidation in presence of nitric acid and vanadium pentoxide is used as a catalyst. Attempts have been made to prepare oxalic acid from ethylene, ethylene glycol, propylene, lignin, jaggery, cane and beet molasses, sugarcane, sugars, flour, plant wastes, formate, carbonate and bicarbonate salts, cellulose and wood wastes by using different methods. These methods are: oxidation of pure sucrose or impure commercial sugars, cellulose and other carbohydrates by nitric acid, or oxides of nitrogen, fermentation of carbohydrates, fusion of sawdust with caustic soda, oxidation of olefins(ethylene , propylene) and glycols, radiation processing of carbonate solutions and molasses, and decomposition of formates.

Of the above the first two received much attention.Other methods such as thermal decomposition of sodium formate (Radulescu et al., 1993)have little bearing on economic production of oxalic acid.

Fermentations of carbohydrates to oxalic acid have been reported for *Aspergillus niger* (Rujiter et al., 1999) and *Pseudomonas fluorescens* (Hamel *et al*., 1999).

Number of attempts have made for the production of oxalic acid by the nitric acid oxidation of carbohydrates such as glucose, sucrose, cellulose and commercial waste materials (Metin Gürü,*et al*.,2001,Guo *et al*., 1997, Sattar *et al*., 1988, Mane *et al.,* 1986; Mane *et al*., 1988, 1986, Nowak,1986 , Sullivan *et al*.,1983,Dube *et al*., 1982, Shukla *et al*., 1982 ,Sanyal *et al.* 1981,82,Gupta *et al*., 1980, Despande and Vyas ,1979,. Agrawal *et al*., 1979, Raha *et al*., 1976, Bose *et al*.,1971, Preen,1942)

Bretschneider *et. al,* [15] reported oxidation of sucrose by nitric acid and discussed the kinetics of the process for tartaric acid as the main product., not oxalic acid.
Berntsen and Degering(1949) oxidized glucose to produce potassium acid sacchharate with oxides of nitrogen and Chia and Lin(1962) oxidized sucrose to produce gluco-saccharic acid. Tanz and Kao (1944) reported work on the selection of catalyst and promoters for this oxidation reaction using oxides of Vanadium, Molybdenum and Tungsten. The oxidation of ethylene glycol to oxalic acid has been discussed by Vishwalingam *et al*., (1980) and kinetics was studied. The production of oxalic acid from other sugar bearing commercial by-product materials such as jaggery (gur) , beet molasses , cane molasses and wood cellulose have also been tried by a very few investigators [14].
Raha *et al*., [30] have described especially the method for producing oxalic acid from Gur. In most of these cases after pretreatment the raw material was oxidized by nitric acid in presence of vanadium pentoxide. Agarwal and Rao [14] oxidized cellulosic materials by nitric acid and found ammonium-meta-vanadate which is a cheaper material to be equally effective as the oxide of vanadium.
The attempts for producing oxalic acid using waste materials like final molasses have not been very economical. Bose *et al.,* [17] attempted the preparation of oxalic acid from sugarcane

molasses using nitric acid and vanadium pentoxide and obtained about 55 percent of the theoretical yield.

Deshpande and Vyas [19] have produced oxalic acid from sugar and the nitrogen oxides evolved were absorbed to produce sodium nitrite as a by- product . Gupta *et al.*, [22] reported the production of oxalic acid from deionized sugar cane juice.

Optimum conditions of reaction as developed earlier were employed in the present investigation and the raw materials (sugarcane molasses) of three different compositions were utilized.

Metin Gürü,*et al.*, [23]conducted exhaustive investigation for the production of oxalic acid from sugar beet molasses using a series of three reactors with formed nitrogen oxides in the following two reactors Parameters, air flow rate, temperature(45 °C to 58°C), the amount of V_2O_5 catalyst and the concentrations of molasses and H_2SO_4 were optimized. The process is however a noble one in respect of recovery of NOx, to prevent air pollution ,to reduce the unit cost of reactant for oxalic acid production and the equipment arrangement. The yield was reported of the order of a maximum 77.5% and 74.1% of theoretical yield obtained by using the absorption solutions with NOx.

Though the process is claimed to be Novel, definitely improvement over the earlier one, in India firstly the raw material used by Guru *et al.*, beet molasses is scarcely available(almost negligible) whereas cane molasses are abundantly available. Besides, beet molasses from refinery goes for cattle feed, hence not available. It however contains more sucrose than that of molasses but lower than jaggery. The sugar beet molasses used in the experiment contained 48.1% sucrose, 3.2% invert sugar and 19.4% water. These are compared in the following table.

Investigators	**Yield of theoretical,%**	**Raw Materials employed**
Metin Guru *et al.*,[23]	78.9/ 77.5 /74.1	Sugar beet Molasses
Sanyal *et.al.*, [33]	74	Jaggery (Gur)
Raha *et al.*, [30]	74	Jaggery(Gur)
Shukla *et al.*,[35]	48.5	Cane Molasses
Agarwal *et al.*[14]	< 50	Cellulosic materials
Sattar *et al.*,[34]	48.5	Molasses
Mane *et al.*,[27]	51.6/60	Sugar cane trash/Saw dust
Sullivan [36]	80.2	Saw dust
Bose *et al.*,[17]	55	Sugar cane molasses

Therefore it can be ascertained if final cane molasses is at all allowed to use , manufacture of oxalic acid, and tartaric acid with the three CSTR as proposed by Metin Gürü,*et al.*, with optimum process conditions of temperature, catalyst concentration etc can give the best yield.

Therefore the work done previously by Sanyal *et al.*, [33] are reproduced here under for effect of two different raw materials , namely jaggery and final cane molasses for further critical analysis;

Raw Material and Chemicals Used:

1. Sugarcane Molasses: Khandsari and final sugar cane factory molasses were used and their analyses are as follows:

Specifications on Sugarcane molasses

Sl. No.	Item	Molasses Sample		
		A	B	C
(i)	Brixs (Bx^0)	70.30	80.30	84.00
(ii)	Pol.	33.00	34.90	-
(iii)	Purity	46.94	43.46	-
(iv)	Sucrose %	29.90	28.00	39.98
(v)	Sucrose Purity	42.53	34.80	41.80
(vi	Reducing Sugar %	21.42	18.36	20.80

2. Gur(Sugar cane Jaggery):

Large quantities of gur are produced in India mostly from small scale units and the properties are likely to change depending upon the sugar cane used. Sudies were made using a single batch product with the following composition:

Specifications of Sugarcane Jaggery (Gur)

Parameters	Percentage
Sucrose	76.3
Reducing sugars	9.5
Purity	84.65
Ash	3-6
Colour	Yellow Brown
pH	6.3

1. Acids: Commercial grade Nitric Acid (60 % HNO_3) and A.R. grade Dehydrant(H_2SO_4)
2. Catalyst: LR grade Vanadium Pentoxide (V_2O_5) in powder and pellet forms.
3. Promoters: LR grade oxides of Tungsten(WO_3) Molybdenuim(MoO_3) and Iron(Fe_2O_3)
4. Soda Ash: commercial grade Soda Ash (40% Na_2CO_3).

Experimental:

The experimental set-up for the batch reactor and other accessories used has been described elsewhere [33, 35]. The reactor consisted of long glass tube vessel of 1 liter capacity fitted with internal glass cooling coil through which constant temperature water was circulated. Three side tubes were attached for liquid feed, for air bubbling and for continuous removal of the reaction gases.

Procedure:

Predetermined amounts of gur/ molasses, distilled water and catalyst 0.08 % were charged in the reactor. The air compressor was started to introduce at a rate of 9-10 l/min and the acid mixture containing measured amount of nitric acid and sulfuric acid was added slowly. Reaction was allowed to proceed for a particular batch time and the temperature was maintained at 54.0±1.0°C.

After completion of the batch whole mass was transferred to a flask and was left overnight at a room temperature of 25.0±4.0°C. Needle shaped crystals of oxalic acid dehydrate were formed which were filtered, washed and dried. This product gave the yield for first crop. The filtrate was again cooled in the freezer at 2.0±1.0°C for about 12 hrs. which gave further crystals and the product gave the yield for the second crop. The two crystal yields were combined to get the total yield.

The repeated experiments were carried out for both jaggery as well as molasses at desired air flow rates, reaction time, temperature and catalyst concentration. Product yield was defined as the ratio of oxalic acid dehydrate produced to the amount of oxalic acid dehydrate that was expected by reaction stoichiometry from a particular amount of sucrose present in the jaggery sample. For further experiments on molasses, the filtered air flow rate was kept constant and the reactant to acid ratio was varied. Oxalic acid yield was computed after separating the crystals from mother liquor.

Results and Discussions:

Experimental results on sugar cane molasses were shown in Fig.3 indicating how reactant ratio(molasses : dehydrant: HNO_3).The molasses however, has been kept in the ration as 1.

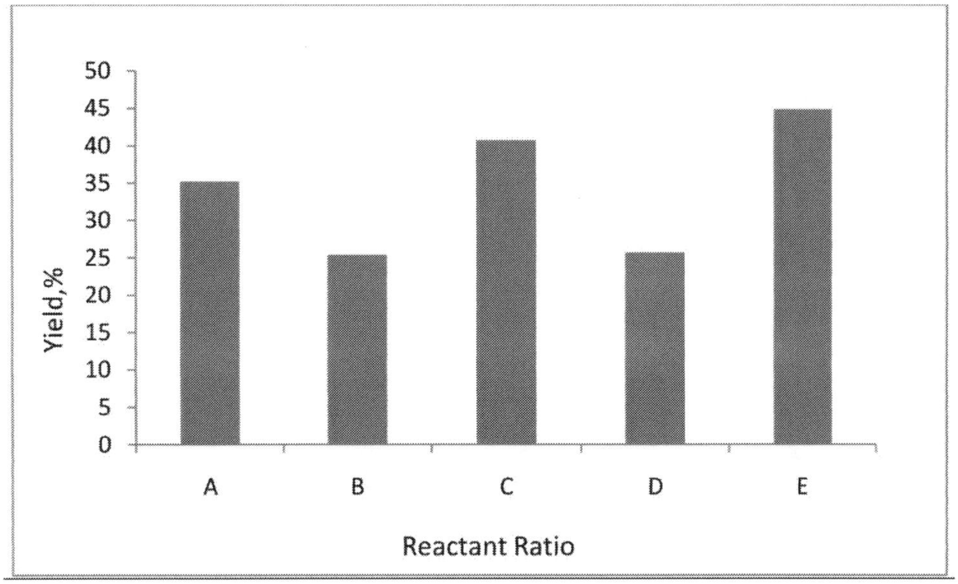

A=1:0.92:4.20, B=1:1.38:4.06, C=1:1.84:4.06, D=1: 1.895:3.60, E=1:2.134:5.41

Fig.3 Yield of oxalic acid as a function of reactant ratio

It is clear from the plot that slightly higher amount of dehydrant and nitric acid gave maximum yield of oxalic acid about 48% of theoretical yield. There was not much difference observed between C and E. considering the cost of dehydrant and nitric acid C might be chosen as the optimum yield.

Vapor Treatment: Reaction vapor products consisting of moisture and nitrogen oxides were allowed to pass through soda ash solution. Thus the gas absorption with chemical reaction took place in the absorber. After the batch completion, the solution was concentrated and cooled for crystal formation. The crystalline product was analyzed for carbonate and nitrate fractions. It is evident from the results that product ratio of $NaNO_3$ and Na_2CO_3 ranges between 1.69 -1.90 with the percentage conversion of soda ash between 60-65%.

Mother Liquor Treatment and recycling: After complete possible extraction of oxalic acid, the mother liquor was treated with soda ash, under constant stirring, till neutral solution was obtained. This neutralized solution was concentrated and cooled to get glauber salt (Na_2SO_4) mixed with unreacted soda ash (Na_2CO_3). The yields and analysis of results indicated that $CO_3^{//}$: $SO_4^{//}$ and yield of Na_2SO_4 (in percentage of theoretical) were between 1:2.70- 1:3.35 and 67.0 - 78.2 respectively. Analytical procedure given in the literature was employed [37- 38].Mother liquor recycling results gave yield (in % of theoretical yield) 18.01% -22.60%.

Product Purity: Oxalic acid crystals were tested chemically and physically using Bausch a Lomb Spectrophotometer and Erma Colorimeter. A comparison was made with I.S.I. grade sample and the product was found to be 99.5% pure.

Parametric relationship for jiggery utilization:

Effect of Reaction Time:

The effect of reaction time on yield was evaluated by varying the reaction time from 1 to 12 hours at constant 0.08% catalyst concentration as shown in Fig.4. The experimental results indicated that the yield of about 74.0% of oxalic acid was attained at a reaction time of 10.0 hrs, beyond which the yield started decreasing. This

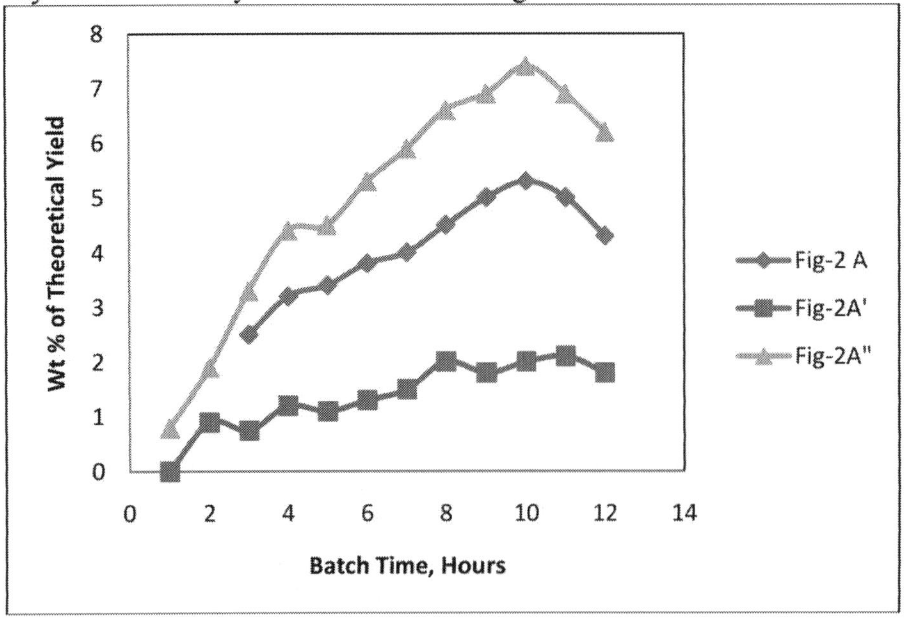

Fig.4 Percent of theoretical yield as a function of reaction time

Effect of Catalyst Concentration:

Vanadium pentoxide powder was used as catalyst. The effect of catalyst concentration on yield was studied at the reaction time of 10 hrs. Amount of catalyst in relation to the amount of gur was varied from 0.01 percent to 0.20 percent. The experimental results as shown in Figs.5 and 6 revealed that the maximum yield could be obtained at a catalyst concentration of 0.08 percent.

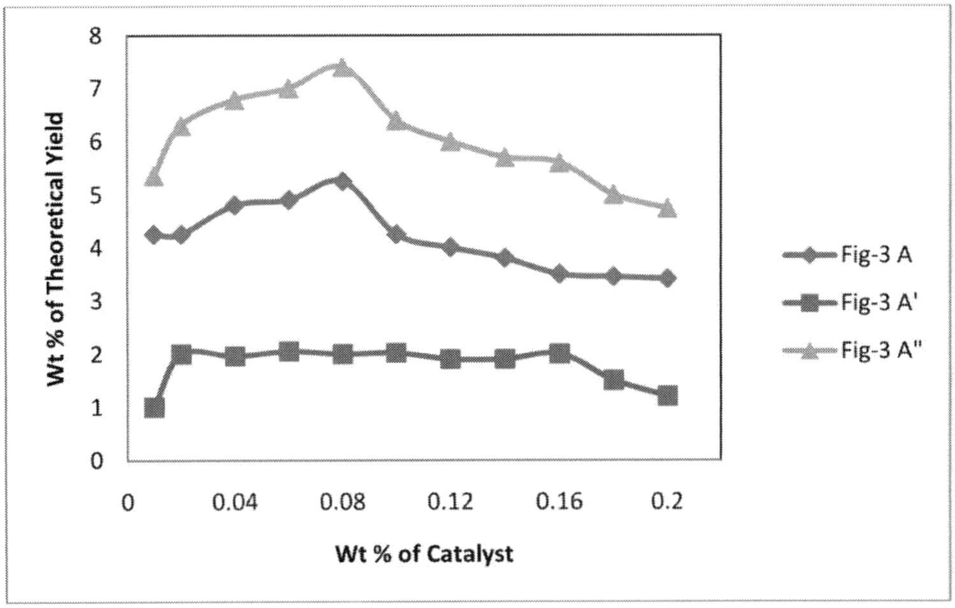

Fig. 5 Percent of theoretical yield of oxalic acid as a function of percentage of catalyst

Effect of Catalyst Promoters:

At the optimum amount of catalyst (0.08 wt. % V_2O_5), the effect of three promoters was studied using oxides of Tungsten (WO_3), Molybdenum (MoO_3) and Iron (Fe2O3). The amount was varied from 0.04 to 0.12 wt percent. The experimental results are shown in Fig. 6 reveals the following:

In case of Tungsten Oxide the yield gave a maximum at wt. 0.08% while in other two cases the yield decreased as the amount of promoters was increased. Furthermore, on comparing the results, it was observed that vanadium pentoxide alone used as catalyst gave better yield (74.4% of theoretical) than the maximum obtained in the case of Tungsten Oxide (66.0 wt% of the theoretical). It was also observed that the use of promoters added colour to the oxalic acid crystals such as reddish brown in the case of ferric oxide, and lost the aesthetic value of the product.

For sugarcane molasses the following additional experiments were performed.

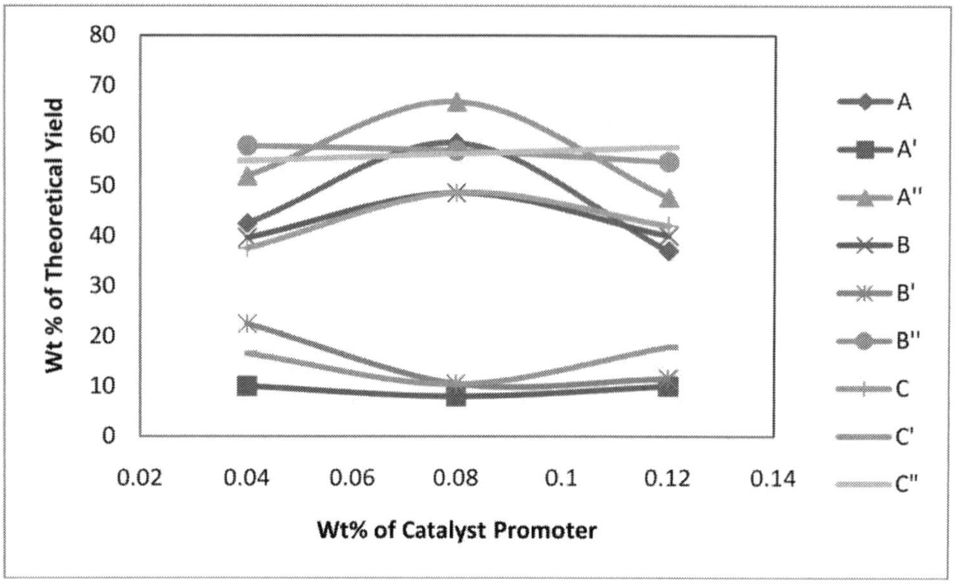

Fig.6 Percent of theoretical yield as a function of percentage of catalyst promoters.

Catalyst Recirculation: Vanadium pentoxide was impregrated in glass beads and alumina and used in place of the powder. It was observed that although at high acid concentrations there was no effect but at lower acid concentrations after the batch was over the oxide was soluble and hence it could not be recovered or recycled.

Gas Absorption: Nitrogen oxides evolved along with air and water vapor could be arrested only partially. As it is clear from the results, it is felt that a better absorber should be designed for complete removal of nitrogen oxides may cause air pollution and also offset the cost of oxalic acid production. The best way is to absorb it and recycle so as to reduce the nitric acid requirements.

Chemicals from Mother Liquor: Effluent liquid in the form of mother liquor contains unreacted acids, catalyst and other impurities originally present in the raw material (in the present case sugarcane molasses). Production of oxalic acid on industrial scale may pose the liquid effluent disposal problem and hence this treatment was tried. Effluent stream should be treated in such a manner that the pollution hazards are controlled as well as the useful chemicals are recovered to reduce the cost of production effluent treatment. Commercial grade soda ash was employed and the results are given in Table 3. It was noticed that no nitric or nitrates were present and about 73% conversion to glauber salt was obtained. In this way the left over sulfuric acid was recovered as well as the liquid effluent was treated.

Mother Liquor Recycling: Mother liquor analysis (after oxalic acid extraction) showed that a large portion contained unreacted dehydrant. Five batches were run using mother liquor. Make up nitric acid and fresh sample of molasses were added to start the batch. The results as discussed clearly show that an overall reduction in oxalic acid yield to an extant of 50% was obtained. Hence it is not instead it should be treated as discussed earlier.

CONCLUSIONS:

There are possibilities of establishing various molasses based acids in India such as lactic acid, oxalic acid, tartaric acid, ctric acid, aconitic acid, monosodium glucamate, dextran,L-Lysine, Xanthenes gum, glycerine, yeast, 2-keto gluconic, 5-keto gluconic , a-ketoglutaric, fumaric acids , glycolic acid,kojic acid and glucaric acid etc.These are mostly imported. Production are very scarce.

Projects regarding production of lactic acid, citric acid, tartaric acid and oxalic acid can be considered as competitive though there could be scope for further optimization of quality & cost standards. Also, more R&D initiatives are required to produce products of various grades, catering to the requirement of specific user. R& D work on the above acids are going on in various industries as well as research laboratories in India. Little R&D push, a unit would be able to produce oxalic acid from molasses. Producing the acid from cane molasses would be very cost-effective and India would be able to take a substantial share of the world market.

The following conclusions can be drawn from the results of investigation for producing oxalic acid dehydrates crystal using sugarcane jiggery and molasses as the raw materials.

1. The maximum yield, 74 wt percent of the theoretical could be obtained in 10 hrs. and increasing or decreasing the reaction time would decrease the yield.
2. Vanadium pentoxide at a concentration of 0.08wt% gave the maximum yield.
3. Addition of promoters like WO_3, MoO_3 and Fe_2O_3 decreased the yield and added colour to the final product, hence it is not recommended.
4. Maximum yield, 48.5 weight percent of theoretical could be obtained while in the case of gur (sugarcane jaggery) under similar condition it was 74.0 wt. percent.
5. Vanadium Pentoxide used as a catalyst has some solubility in dilute acids and hence cannot be re-cycled at this stage.
6. Nitrogen oxides can be absorbed in alkaline solution to give useful products (fertilizers) and control the air pollution.
7. Mother liquor re-cycling reduces the yield by 50% and is hence not recommended at this stage.
8. Mother liquor can be treated with alkalis to get useful products like glauber salt, ammonium sulphate etc.

REFERENCES:

1. http://search.conduit.com/?SearchSource=10&ctid=CT2776682
2. http://www.google.co.in/
3. *http://www.nandinichemical.com/publications/molassesbasedprojects.html*
4. Lourdes R. Fragante, Lactic acid from molasses, RDE abstract of completed research,Sugar and co-products utilization/molasses,ibid
5. TIFAC Report on "Technology for Production of Lactic Acid" 2001
6. Chemicals Weekly, July 5,2011 pp.239-273
7. Ali. S. et al (2002), Production of citric acid by *Aspergillus niger* using cane molasses in a stirred fermentor, Electonic Journal of Biotechnology, 5, Issue December 15th 2002

8. Ikram-ul-Haq et al (2004), Citric Acid production by selected mutants of *A. niger* from cane molasses, Bioresource Technology 93:125-130
9. Bhatt S.M. and S.K.Srivastava,Lactic Acid Production from Cane Molasses by *Lactobacillus delbrueckii* NCIM 2025 in Submerged Condition: Optimization of Medium Component by Taguchi DOE Methodology, Food Biotechnology, 22:115–139, 2008
10. Aksu Z. and T.Kutsal,Lactic acid production from molasses utilizing *Lactobacillus delbrueckii* and invertage together, Biotechnology Letters , Vol.8, No.3, pp.157-160,1986
11. James M. Church, Ruth Blumberg, Synthesis of Tartaric Acid, *Ind. Eng. Chem.*, 1951, *43* (8), pp 1780–1786
12. W. E. Barch, The Oxidation of 5-Keto-gluconic Acid with Nitric Acid in the Presence of Vanadium, J. Am. Chem. Soc., 1933, 55 (9), pp 3653–3658
13. R. Klasen, S. Bringer-Meyer and H. Sahm, Biotechnology and Bioengineering 40, 183, 1992.
14. Agrawal, Prasad, H. and Rao, M.B., 1979. "Influence of different processing conditions on the production of oxalic acid from cellulosic materials by nitric acid oxidation. Ind. J. Technol. 17, pp. 11–15.
15. Bretschneider, R.B.,Kopriva, J.. Markova and Z.Ondrackova, " Oxidation of Sucrose by nitric acid",Listy Cuker 1966-82,215-219,1967,36-39,and 227-232.
16. Bernsten,R.A.,and E.F. Degering," Oxidation of glucose with oxides of Nitrogen "Sugar Industry Abstracts,1949,11,87,258
17. Bose , S., S.Mukherjee,and A.N. Srivastava," Production of oxalic acid from sugarcane molasses", Proc.4th Joint Conv. S.T.A.I., D.S.T.A. and S.I.S.S.T.A.,G-17-21,1971
18. Chia, K.S. and C.C.Lin," Oxidation of Carbohydrates-synthesis of Glucosachharic Acid from sucrose",Chemistry, Taiwan,1962,98-104
19. Deshpande, S.D. and Vyas, S.N., 1979. Oxidation of sugar to oxalic acid and absorption of oxides of nitrogen to sodium nitrite. Ind. Eng. Chem. Prod. Res. Dev. 18 1, pp. 69–71.
20. Dube, S.K., Vasudevan, P. and Khandelwal, B., 1982. Oxalic Acid Manufacture. J. Chem. Tech. Biotechnol. 32, pp. 909–919.
21. Guo, W., Jiang, H. and Huaxue, Yu.N., 1997. Manufacture of oxalic acid from beet waste slag by hydrocarbonylation at atmospheric pressure. J. CA 45 4, pp. 240–241.
22. Gupta, K.C., Suri, S.K. and Bose, S., 1980. Oxalic acid from deionised sugarcane juice. Proc. STAI 44, pp. 119–125.
23. Gürü, M., et al., 1991. eker Pancarı Melasından Nitrik Asitle Okzalik Asite Yükseltgenmesinde Optimum artların Belirlenmesi. VII. Kimya ve Kimya Mühendisli i Sempozyumu, 2, 464, Gazima usa
24. Gürü, Metin, Ali Y. Bilgesüb and Vecihi Pamuka, Production of oxalic acid from sugar beet molasses by formed nitrogen oxides ,Bioresource Technology,Volume 77, Issue 1, March 2001, Pages 81-86
25. Hamel, R., Levasseur, R. and Appanna, V.D., 1999. Oxalic acid production and aluminum tolerance in *Pseudomonas fluorescens*. J. Inorg. Biochem. 76 2, pp. 99–104.
26. Mane, J.D., Jadhav, S.J. and Ramaiah, N.A., 1986. Production of oxalic acid from dry powder of parthenium hysterophorus L. J. Agricultural Food Chem. 34 6, pp. 989–990.
27. Mane, J.D., Modak, H.M., Ramaiah, N.A. and Jadhav, S.J., 1988. Utilisation of sugarcane trash and other cellulosic wastes for production of oxalic acid. Biological Wastes 25, pp. 171–176.

28. Nowak, P., 1986. Technology for obtaining oxalic acid from sugar by a pressure method. Przemysl Chemiczny 65 9, pp. 488–490.
29. Preen, F.H.E., "Oxalic acid from wood waste", Chemical age,1942-32,72,
30. Raha A.C., R.B.Nigam,and P.Sanyal : Production of oxalic acid from gur", Proc.S.T.A.I.,44(part 2),G-1-6,1976
31. Radulescu, G., Moise, M.-I., Cornos, L., Vapler, G., 1993. Preparation of oxalic acid by the thermal decomposition of sodium formate followed by cation exchange. Patent CA Section:45, Rom. RO 106878 B1 30: 4
32. Rujiter, G.J.G. et al., 1999. Oxalic acid production by *Aspergillus niger*: an oxalate-non-producing mutant produces citric acid at pH 5 and in the presence of manganese. Microbiology 145 9, pp. 2569–2576.
33. Sanyal, P., Ray, A.K. and Shukla, N.P., 1981. Oxalic acid from gur-effect of catalyst. Proc. STAI 45, pp. 21–26.
34. Sattar, A., Muhammad, D., Ashraf, M., Khan, S.A. and Bhatty, M.K., 1988. Oxidation of molasses to produce oxalic acid. Pakistan J. Sci. Ind. Res. 31 3, pp. 163–166.
35. Shukla, N.P., Ray, A.K., Sanyal, P., Oxalic acid from sugarcane molasses: process development and waste utilisation. Proc. 46th Convention STAI, pp. 93–101,1982
36. Sullivan, J.M., Williard, J.W., White, D.L. and Kim, Y.K., 1983. Production of oxalic acid via the nitric acid oxidation of hardwood (Red Oak) sawdust. Ind. Eng. Chem. Prod. Res. Dev. 22, pp. 699–709.
37. Scott., W.W.,Standard Methods of Chemical analysis, Fifth Ed., Vol.1, The Technical Press, London,644
38. Snell, F.D. and F.M.Biffen,Commercial Methods of analysis, First Ed., McGraw Hill Book Co., New York,1944,140
39. Tanz and Kao, Chemical Abstracts,1944,355466
40. Vishwalingam,K.,T.V.Suramaniam and B.Jagannadhaswamy," Process oxidation of ethylene glycol to oxalic acid in liquid in the presence of nitric acid" Ind. Chem.Engr.1980-22(4),11-13

(338f) Modeling the Hydrolytic Degradation and Distribution of Polymer Species In Biopolymers

Justin T. Kaffenberger, Huajiang Huang, Ulrike Tschirner, Ben Schroeder, Waleed Al-dajani and Shri Ramaswamy, Bioproducts and Biosystems Engineering, University of Minnesota, St Paul, MN

Modeling thehydrolytic degradation and distribution of polymer species in biopolymers

Justin Kaffenberger, Huajiang Huang, Ulrike Tschirner, BenSchroeder, Waleed Al-dajani, and Shri Ramaswamy

Department of Bioproducts and Biosystems Engineering,University of Minnesota, St. Paul, MN 55108

Abstract:

The US Department of Energy in its Plant/Crop-BasedRenewable Resources Report 2020 has put forth an initiative to have 10 percent ofchemical feedstock coming from plant-based renewable sources by 2020 and afurther increase to 50 percent by 2050.[1] Meeting these targets will require productionof current chemical feedstocks from renewable sources, such as the dehydrationof plant-based ethanol to produce ethylene.It will also require the development of alternative feedstocks that cansupplant the petroleum-based ones that currently predominate. One example of a chemical feedstock ofincreasing importance is lactic acid.

Lactic acid is a metabolite formed during glucosefermentation by many organisms. Due toits versatility, lactic acid has the potential to be a major renewable chemicalfeedstock. It can be esterified to formlactate esters, reduced to propylene glycol, or dehydrated to acrylic acid.[2] It can also be homopolymerized to formpolylactic acid (PLA) via condensation reaction. The resulting polymer exhibits mechanicalproperties similar to many petroleum-based polymers, including polyethyleneteraphthalate, polyethylene, and polystyrene.

The degradability of PLA is critical to its currentcommercial success. In disposable foodpackaging and service ware, the degradability and bio-absorptivity of thedegradation product of PLA makes it marketable as a green and environmentally-friendlyproduct. In biomedical applications, thedegradability and biocompatibility of PLA are essential. As PLA's ability to degrade is a major factorin its application functionality, understanding the fundamentals of thisprocess is critical to all stages of PLA manufacture and end-use application.

To this end, a mathematical model has been developed thatpredicts the time dependent change in the average molecular weight ofpolylactic acid (PLA) of varying crystallinity as it hydrolytically degrades ata given temperature. This model includes three possible hydrolysis reactionroutes: random chain scission, end scission (monomer removal from the acid endgroup, and monomer removal from the alcohol end group). Apparent rate constants for these reactionpathways were determined from experimental data. The resulting values suggest that,independent of the crystallinity of the starting PLA, acid end groups arehydrolyzed more rapidly than either alcohol end groups or random chainscission. This may help develop suitableapproaches to control the polymer properties and degradation and help developwider application for PLA products.

See more of this Session: Sustainability and Biobased Materials
See more of this Group/Topical: Forest and Plant Bioproducts Division

(338g) Experimental Study On Bast Fiber Reinforced Soy-Based Polyurethane

Shanshan Huo and Chad A. Ulven, Mechanical Engineering, North Dakota State University, Fargo, ND

Growing importance has been placed on the research of natural fibers and resins derived from natural resources as the need for reducing dependency on petroleum-derived products has increased. Natural fibers and biobased resins are potentially inexpensive, renewable resources capable of increasing the specific strength, specific stiffness, and damage tolerance of a variety of composites.

Polyurethane is one of the most versatile and intensively used industrial materials. Bio-based polyols developed from vegetable oils such as soybean oil, canola oil, palm oil and castor oil, are now being used in industry. Soy-based polyurethanes normally exhibit equivalent or improved physical and chemical properties due to the hydrophobic nature of triglycerides. Cellulosic fibers as an alternative for fiberglass have been studied by many researchers. Chemical modifications can be applied to reduce the amount of hydroxyl groups in cellulosic fiber to increase the hydrophobicity and improve the interfacial properties between fibers and matrices. Distillers dried grains with solubles (DDGS) is a by-product of corn-ethanol. Their potential use as reinforcing filler in biocomposites has been studied in thermoplastic composites, but not to a great extent in thermoset composites.

In this study, bast fiber reinforced soy-based polyurethane composites with DDGS filler are produced. The mechanical properties of soy-based PU matrix, soy-based PU matrix with DDGS, and cellulosic fiber reinforced soy-based PU composites with DDGS are investigated. The thermal properties of the matrix system and their composites are evaluated.

See more of this Session: Sustainability and Biobased Materials
See more of this Group/Topical: Forest and Plant Bioproducts Division

Improving the sustainability of biofuels through modification of the biomass production subsystem: Application to corn and lignocellulosic ethanol

Concerns about dwindling fossil fuel reserves and climate change, energy independence, and depletion of ecosystem goods and services have driven the development of alternative technologies such as biomass-based fuels (biofuels). Although biofuels may be a step towards alleviating the aforementioned concerns due to the renewable nature of the biomass feedstock, there are several drawbacks associated with them. Food crop-based fuels such as corn ethanol, which comprise the majority of the current biofuel mix in the US, require large amounts of fossil-based inputs purchased from the economy in order to grow the biomass. These inputs, such as fertilizers, pesticides, and fossil fuels greatly diminish the renewable nature of the plant itself and force the production of biofuels to be reliant on non-renewable fossil-based inputs. In addition to reliance on non-renewable inputs, biofuels are heavily reliant on ecosystem goods and services. Current large-scale farming practices result in the runoff of nutrients from fertilizers into rivers and streams, which can overwhelm the carrying capacity of the nitrogen (N) and phosphorous (P) cycles of these aquatic ecosystems with devastating results. Freshwater for irrigation is needed if natural precipitation is not sufficient. Topsoil, a vital local non-renewable resource, is also depleted through erosion in traditional monocultural row cropping. Ecosystem goods and services are also indirectly consumed through the life cycle of the inputs used in farming.

This research aims to improve the sustainability of biofuels, specifically corn ethanol, by modifying the corn farming system to use less purchased inputs and consume less ecosystem goods and services both directly and indirectly through the reduction of fossil-based inputs needed from the economy. The following design changes will be considered:

1) No-till/conservation tillage. This will reduce soil erosion and require less labor and fuel for preparing soil for seeding. Application of more herbicides may be necessary as tilling helps remove weeds.
2) Planting winter cover crops in the growing offseason, also known as "double cropping". This reduces bare soil exposure in the offseason, reducing erosion. Also, offseason weed growth can be reduced requiring less herbicide when the corn is planted. Depending on the species planted, soil nutrients can also be rebuilt in the offseason. However, additional labor, energy, and other inputs will be needed for planting and harvesting the cover crop.
3) Intercropping other species amounts corn during the growing season. This is often referred to as a polyculture. This can reduce soil exposure and hence reduce erosion and reduce fertilizer consumption due to fixation of nutrients by the intercropped species. However, separation of corn from the intercropped species may require additional inputs.
4) Constructed wetlands and/or riparian buffers for nutrient runoff capture. This can prevent nutrients from entering rivers and streams and reduce eutrophication.

5) Using treated sludge from wastewater treatment facilities (biosolids) for fertilizer, displacing the need for fossil-based artificial fertilizers. However, there is a limited supply of biosolids and may not be enough for large-scale application. Also, biosolids may be more labor and energy intensive to apply.
6) Fossil fuel substitution, i.e., replacing partially or fully the fossil fuels used in the farming system with biofuels to reduce the reliance on non-renewable inputs. Any disadvantage of biofuels will be compounded when fed back into the production of more biofuels.

The impact of these design changes will be quantified both directly and indirectly in the life cycle through the use of Ecologically Based Life Cycle Assessment (Eco-LCA) to determine the overall effect of the changes on the sustainability of corn ethanol. This LCA method will quantify traditional LCA impacts, such as emissions and energy consumption, but will also quantify the multitude of ecological goods and services consumed throughout the life cycle. This ecological accounting is vital to ensure the sustainability of any system since everything relies directly or indirectly on ecosystem services. Additionally, the resulting corn farming design is an example of the design of a technological-ecological network, where both technological and ecological variables are considered in the design of the overall system. Such systems are more sustainable than systems that use technological variables only, since technology alone cannot lead to sustainability due to second law constraints.

Although these design changes may improve the sustainability of corn ethanol, the scaling up of biofuel production can compete with food and lead to indirect land use change where native ecosystems such as forests are converted to farmland for biomass production, which may create a net greenhouse gas increase. Therefore, it is worth considering other biomass sources that can be produced on marginal lands. The case of producing ethanol from a low input, high diversity (LIHD) lignocellulosic grassland on degraded farmland will also be presented. LIHD biomass requires much fewer inputs than most monocultural biomasses and may be a sustainable source of liquid fuel. LIHD ethanol will also be compared with food crop-based ethanol as well as other cellulosic ethanol sources.

(339b) Agent-Based Modeling Analysis of Biomass Feedstock Production System Dynamics and Resilience

Yogendra Shastri[1], Luis Rodriguez[2], Alan Hansen[2] and K.C. Ting[2], (1)Department of Agricultural and Biological Engineering, University of Illinois at Urbana-Champaign, Energy Biosciences Institute, Urbana, IL, (2)Department of Agricultural and Biological Engineering, University of Illinois at Urbana-Champaign, Urbana, IL

An efficient and sustainable energy crop production system is essential for the success of the emerging biorefinery industry. However, there is currently a significant lack of understanding regarding the development and dynamics of the biomass feedstock market upstream of the biorefinery, which is a critical component of the value chain. This includes assessing the time-scale of its development as well as the impact of various engineering, technological, regulatory and social factors on the dynamics. Various uncertainties and disturbances can impact the functioning of this system, and understanding the resilience of the system in the presence of those is also important. The analysis is especially challenging due to the involvement of multiple stakeholders such as farmers, refinery, transportation companies and storage elevators that must participate and yet compete for a fraction of the profit.

The objective of this work is to conduct such as analysis using the complex adaptive systems theory approach. The work develops an agent-based model that incorporates various stakeholders (agents) within the feedstock production system. The model, developed in MATLAB®, uses an object-oriented programming approach by considering each agent-type as an independent class and a particular realization of that agent-type as an object (instance) of that class. The individual characteristics of an object are defined by a set of attributes. The behavior of each agent as well as its interaction with other agents is based on a set of rules (methods). Since the behavioral and social factors play an important role in the decisions made by individual agents, a mental model incorporating some of these factors is also developed for each agent. The set of decision rules are consequently based not only on the economic and technological rationale, but also on the unique mental models and cognitive processes. The learning/adaptive behavior exhibited by the agents is also modeled which enables the simulation of emergent behavior of the system.

The model has previously been used to study the development of the biomass market under different scenarios. The focus of this work is on understanding the impact of various disturbances and uncertainties on the dynamics of the system, which enables us to study the resilience of the system. Different possible disturbances such as loss of productivity, price fluctuations and infrastructure breakdown will be modeled. Desirable but sudden changes such as the availability of an efficient technology will also be considered. Simulation studies incorporating these disturbances will be conducted to study their impact on the dynamics of the feedstock production system. Preliminary studies using this model have illustrated that the development of this sector is gradual, nonlinear and may take up to 15 years. The impact of the time of disturbance along this development trajectory will therefore be critical and will also be studied. Simulations for different types of energy crops such as switchgrass and Miscanthus will also identify whether the dynamics are a function of the energy crop. The work will also enable us to evaluate different system resilience measures appropriate for analyzing the feedstock production system.

See more of this Session: Sustainable Biomass Feedstock Production and Supply for the Emerging Biorefinery Industry
See more of this Group/Topical: Sustainable Engineering Forum

(339c) Effect of Storage Method and Duration On the Bioprocessing of Lignocellulosic Biomass

Arun Athmanathan, ABE/Lorre, Purdue University, West Lafayette, IN and Nathan S. Mosier, Agricultural and Biological Engineering & Laboratory of Renewable Resources Engineering (LORRE), Purdue University, West Lafayette, IN

Lignocellulosic biorefineries require an efficient system of biomass storage between harvesting and processing. This study reports the effect of storage method and duration on the composition and processing performance of sweet sorghum bagasse. Sorghum bagasse was harvested at 85% moisture content, crushed to extract the juice, and then stored. Four storage conditions were used: frozen after crushing (control), ensiled after crushing (67% moisture content), or dried in the field and baled at high (27% by weight) or low (11% by weight) moisture contents. After 24 weeks of storage indoors at 20°C, 5.5% dry matter was lost from low moisture content bales, 29.4% from high moisture bales, and 7% from silage. Compositional analysis and mass balances determined that mass loss was primarily carbohydrates. Pretreatment and SSF at high solids loading (15% w/w) were then carried out for the materials. Distinct differences in performance were noted, with lower enzymatic digestibility of the glucan observed for dry storage compared to wet storage.

To examine non-compositional impacts of storage method, bagasse samples were enzymatically hydrolyzed at 1% w/w solids loading without drying, air-dried (20°C), and oven-dried (45 °C). Release of glucose stopped after <5 hours for dried bagasse with final yields being 70% (air-dried) and 57% (oven-dried) of total glucan. By comparison, non-dried samples continued to hydrolyze beyond 5 hours, resulting 87% glucan yield after 24 hours. These results indicate that hornification during drying may also negatively affect enzyme accessibility to the cellulose and lower bioprocessing performance.

See more of this Session: Sustainable Biomass Feedstock Production and Supply for the Emerging Biorefinery Industry
See more of this Group/Topical: Sustainable Engineering Forum

Determination of kinetic parameters of self-ignition wood pellets during storage

Wendi Guo[1], Xiaotao Bi[1], C. Jim Lim[1], Shahab Sokhansanj[1,2]

1. Chemical & Biological Engineering, University of British Columbia, Vancouver, BC V6T 1Z3
2. Environmental Sciences Division, Oak Ridge National laboratory, Oak Ridge TN 37831

1. Introduction

Accidental fires in wood pellets silos happen frequently (Channel 3000 2007; Aines 2008; Bartel 2008). It appears that the oxidative biological reactions are initially responsible for simultaneous gas release and temperature increase. (Walker and Harrison 1977; Lehtikangas 2000; Wihersaari 2005b, 2005a). The off-gassing and self-heating phenomena of wood pellet during storage have been investigated by several research institutions as well as in our Biomass & Bioenergy laboratory at UBC (Svedberg *et al.* 2004; Wadso 2007; Kuang *et al.* 2008; Svedberg *et al.* 2008; Kuang *et al.* 2009a; Kuang *et al.* 2009b; Guo *et al.* 2011). The results to date show that the rate of self-heating decreases with the age of pellets. The reaction rate is also highly dependent on storage temperature. A maximum heat release rate of 0.3 W/(kg pellet) has been recorded when a mass of pellets was stored in a sealed container maintained at 50 °C.

It is postulated that chemical oxidation dominates the self heating process when the storage temperature increases to 80 °C. This process leads to a higher temperature accelerating self heating when a self-ignition may occur (Meijer and Gast 2004). The self ignition of wood and other biomass was widely investigated experimentally under heat flux. Models were developed to predict the thermal ignition for those materials (Bilbao *et al.* 2001; Bilbao 2002; Pauner and Bygbjerg 2007)

The kinetic of wood pellets self heating at higher temperatures and the self-ignition phenomena haven't been fully understood. Experiments were carried out in this work at temperature range of 100 °C to 200 °C to study wood pellets chemical oxidation process. A self-heating model was developed to simulate the self-heating process using the results obtained from the experiment. A numerical method was developed to solve the energy balance and obtain the kinetic parameters for self-heating process.

2. Methodology

Experiments were carried out using a lab size oven (Thermal scientific, 1.7 cu. ft). Cylindrical containers were used as the self heating reactor and were put in the oven (constant temperature) with wood pellets loaded. Assuming the self heating process is a first order reaction and the self heating heat can be described as reaction heat (enthalpy of reaction, $\Delta_r h$) multiplying the reaction rate constant $[k=A\ \exp(-E/RT)]$.

The energy balance of the self-heating process inside a cylindrical container is expressed by

$$\rho C_P \frac{\partial T}{\partial t} = \lambda(\frac{1}{r}\frac{\partial}{\partial r}(r\frac{\partial T}{\partial r}) + \frac{\partial^2 T}{\partial z^2}) + \rho \cdot \Delta_r h \cdot A\exp(-\frac{E}{RT}), \quad (1)$$

where, C_p is the specific heat capacity of the pellets, kJ/kg; λ is the thermal conductivity of the pellets , W/(m K); ρ is the bulk density of the pellets, kg/m^3; T is the temperature of the pellets, K; r is the radius of the container, m; $\Delta_r h$ is the reaction enthalpy of the self-heating reaction, kJ/mol; A is a exponential factor; E is the activation energy of the self-heating reaction, J/mol; R is gas constant, 8.314 J/(molK); t is time of the self heating reaction, s.

A curve-fitting method was developed to extract the kinetic parameters (including $\Delta_r h$, A and E) using the experimental data (temperature vs. time). Eq.(1) was solved by MatLab using the thermal properties obtained from the previous studies along with the boundary conditions shown in Figure 1. The best fitted kinetic parameters were determined by minimizing the residuals between the numerical solution and the experimental temperature profiles.

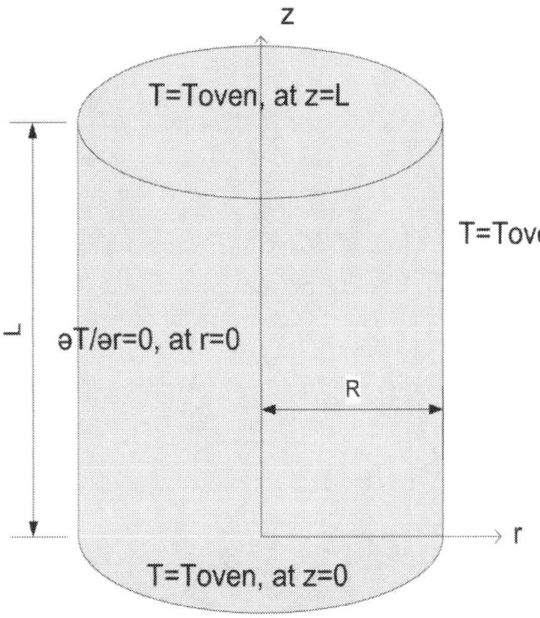

Figure 1 Boundary conditions and assumptions

Assumptions:

Bulk pellets can be treated as homogeneous materials with effective thermal properties (Cp, λ, ρ).

Initial condition t=0, $T=T_0$ (initial temperature)

No air movement inside the container

Fast heat conduction at container wall ($T_{wall}=T_{oven}$)

3. Experiment setup and materials

3.1 Sample

Softwood pellets produced in British Columbia (BC) were used in this research. Pellets with designation A, B and C in this research were manufactured from 3 different wood pellet plants in BC. They were made of whitewood mostly pine; both pellet A and pellet C were manufactured by Princeton Co-Generation Corp. (Princeton, BC, CanadaV0X 1W0). Pellet B was manufactured by Premium Pellet Ltd. (Vanderhoof, British Columbia Canada, V0J 3A0). Pellets were on average 6.2 mm in diameter and 10 to 25 mm in length. All pellets were made using whitewood with no bark. The pellets in this region are made from sawdust and shavings, a by-products of milling spruce, pine and fir for lumber.

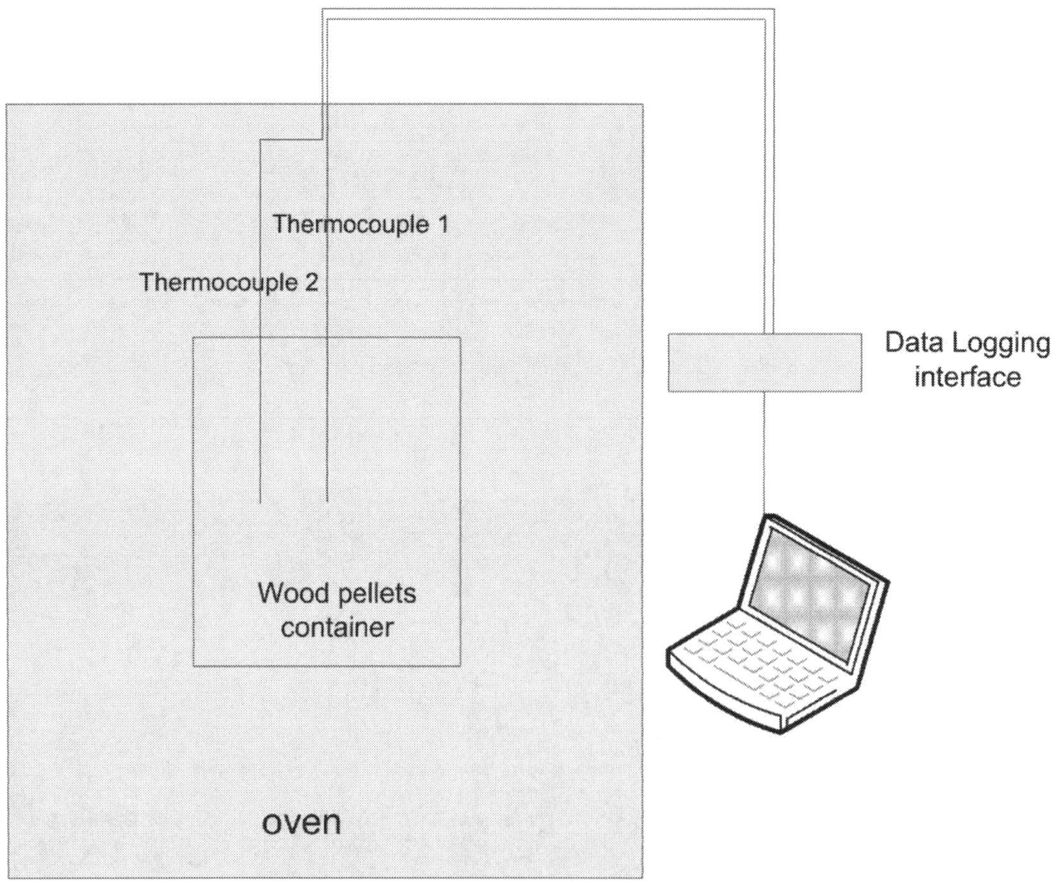

Figure 2 Experimental setup

3.2 Experimental setup

Figure 2 shows the experimental setup. Three sizes of containers made from steel were used in this experiment, small (H 106mm, D 120mm); Medium (H 166mm, D 190mm); Large (H 226mm, D 240mm). Top open short cylinders were used as reactor and two thermocouples were placed inside the container, one right in the middle (T1) and another one half way to the center (T2) (radial). The ends of two thermocouples were kept at the same axial level (same distances to the top and bottom). After placing the thermocouples, the containers were filled up with wood pellets (weight was recorded) and put inside of an oven (maintained at a constant temperature (between 120°C to 200°C); temperatures were recorded using a data acquisition system from the start till after the crossing point.

4 Results and Discussion

4.1 Oven temperature and self-heating

Figure 3 plots the experimental results of two different size containers at different oven temperature. The experimental results show that there is no self heating detected below an oven temperature of 100 °C at all lab size self heating containers which used in this study. Figure 3.a shows the temperature profiles of the large container (L)at 103 °C for 4 days and no self heating was detected. Self-heating was detected when the oven temperature was set at 140 °C for large containers (L), as shown in Figure3b where temperatures increased with time. When the center temperature (T1) crossed over the off-center temperature (T2) at 135°C (the crossing point), both temperature increased rapidly and reached a higher steady state temperature. Here, we consider the CP point (crossing point) as a signal of the detectable self-heating. When the oven temperature increased to 150 °C, the temperature T1 and T2 of the large container (L) continued increasing and reached 158 °C (the peak of T1) a steady state (decrease slightly), as shown in Figure 3.c. On the other hand, for the medium container, the crossing point temperature didn't appear until the temperature reached 147 °C (Figure 3.c.). For the large container, there is finally an uncontrollable thermal run-away which leads to a self-ignition when the oven temperature increased to 170 °C (shown in Figure3.d), while the temperature of medium container remains at a steady state (around 175 °C) . For the large container, temperatures kept running away after the oven heat was turned off and led to a self-ignition at 200 °C. Figure 3.f shows the burning wood pellets. The ignition happens at a point near the center.

It can be concluded that the heat release of the oxidization at temperature 100-200 °C is significant enough to be detected using a lab technique and to cause a self-ignition. The crossing point (detectable point of self-heating) and the thermal run-away temperatures are highly dependent on the size of the containers and the temperature of the oven.

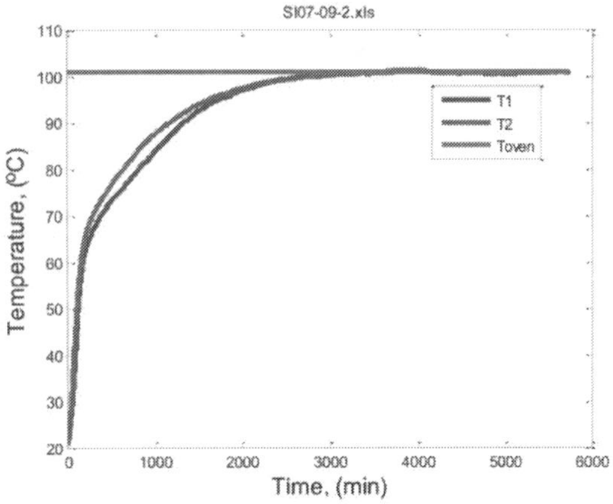

a.Oven temperature 103 °C, no crossing point, no thermal run away.

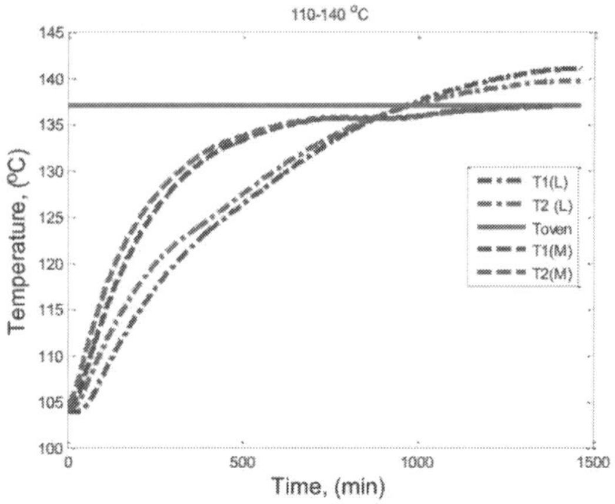

b. Oven temperature 140 °C crossing point after 10 minutes (Large and Medium container)

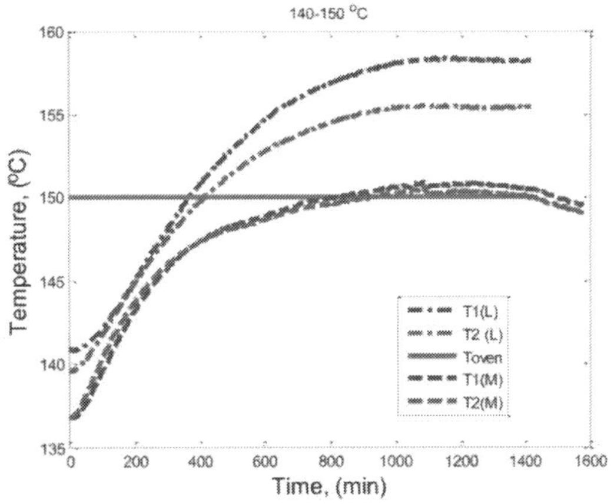

c. Oven temperature 150 °C (L&M)

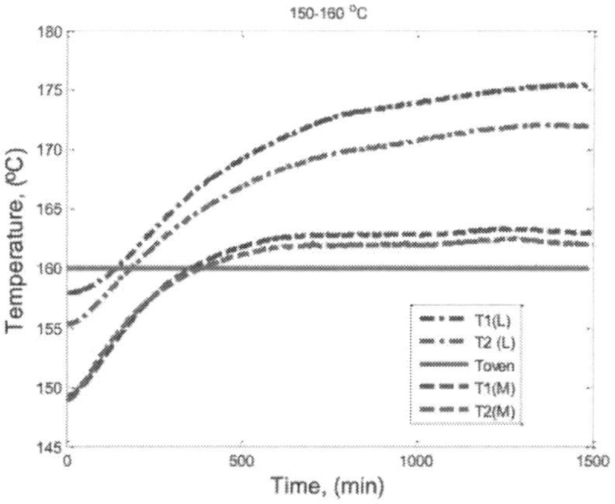

d. Oven temperature 160 °C (L&M)

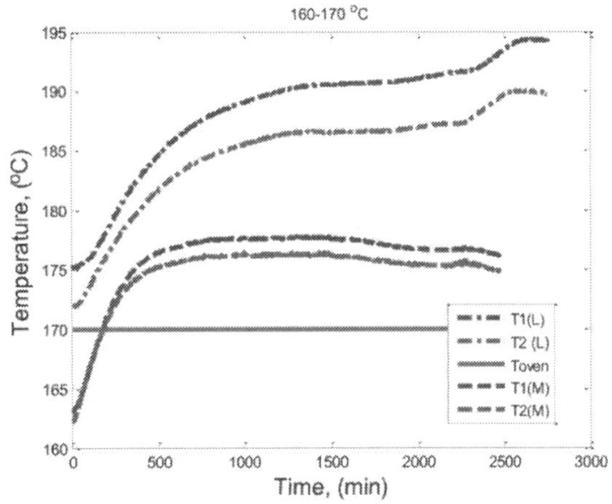

e. Oven temperature 170 °C and self-ignition (L&M).

f. Burning wood pellets.

Figure 3 Self-heating and self-ignition -Temperature profiles at different oven temperatures.

4.2 Kinetic parameters from curve fitting

A curve-fitting method was developed by solving the energy balance equation Eq (1) based on the boundary conditions shown in Fig. 1. The kinetic parameters were obtained by searching the best fit using the curve-fitting method and both data from large containers and medium containers. Figure 4 shows the typical best fitted curves in comparison with the experimental data.

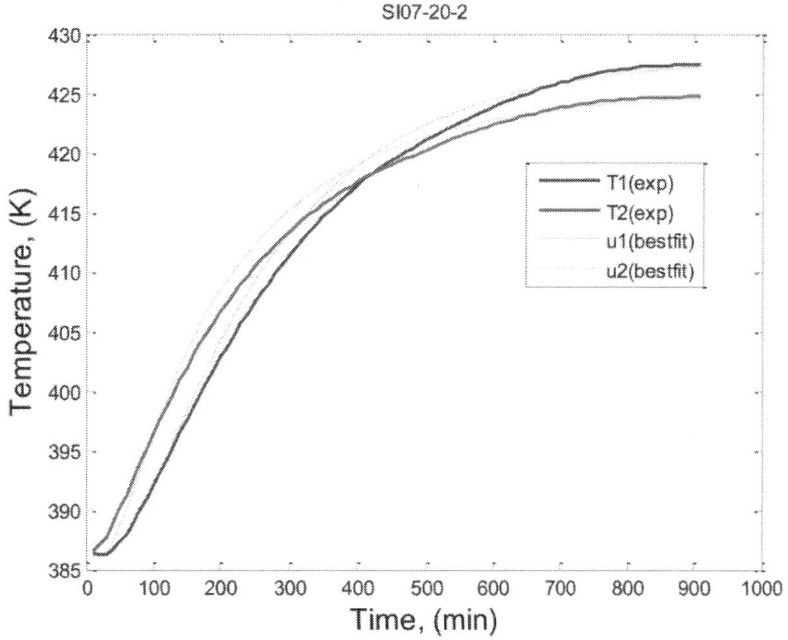

Figure 4 Comparison of the experimental data and best fitted curves of the kinetic parameters. (Experimental results of 'SI07-20-2' comparing to the numerical results of parameter E=62.5 kJ/mol, $\rho\Delta_r hA=5.7*10^7$ J.K/(m^3.s)

Further analysis is in progress.

5. Summary

Experiments were carried out to investigate the self-heating and self-ignition process of wood pellets at temperatures ranging from 100°C to 200°C. Three sizes of containers were used. The results show that the self-heating speed and thermal run-away temperature are highly dependent on the oven temperature and the size of the reactors. A curve-fitting method is developed to extract the kinetic parameters.

Aines, D. 2008. Fire reported for second time in six weeks at wood stove pellet plant. Retrieved 08 July 2010, from http://www.herald-mail.com/?cmd=displaystory&story_id=201753&format=html.

Bartel, R. 2008. Fire at westwood fibre products Retrieved 08 July 2010, from http://www.silk.fm/node/763122.

Bilbao, 2002. A model for the prediction of the thermal degradation and ignition of wood under constant and variable heat flux. Journal of Analytical and Applied Pyrolysis **62**(1): 63.

Bilbao, R., Mastral, J.F., Aldea, M.E., Ceamanos, J., Betr, M., Lana, J.A., 2001. Experimental and theoretical study of the ignition and smoldering of wood including convective effects. Combustion and Flame **126**(1-2): 1363-1372.

Channel 3000. 2007. 2 injured in explosion, fire at wood pellet warehouse. Retrieved 08 July 2010, from http://www.channel3000.com/news/10947450/detail.html.

Guo, W., Trischukb, K., Bi, X., Lim, C.J., Sokhansanj, S., 2011. Measurements of wood pellet self-heating kinetic parameters using isothermal calorimetry. Submitted to Biomass & Bioenergy. **Under Review**.

Kuang, X., Shankar, T.J., Bi, X.T., Lim, C.J., Sokhansanj, S., Melin, S., 2009a. Rate and peak concentrations of off-gas emissions in stored wood pellets—sensitivities to temperature, relative humidity, and headspace volume. Annals of Occupational Hygiene **53**(8): 789-796.

Kuang, X., Shankar, T.J., Bi, X.T., Sokhansanj, S., Jim Lim, C., Melin, S., 2008. Characterization and kinetics study of off-gas emissions from stored wood pellets. Annals of Occupational Hygiene **52**(8): 675-683.

Kuang, X., Shankar, T.J., Sokhansanj, S., Lim, C.J., Bi, X.T., Melin, S., 2009b. Effects of headspace and oxygen level on off-gas emissions from wood pellets in storage. Annals of Occupational Hygiene **53**(8): 807-813.

Lehtikangas, P., 2000. Storage effects on pelletised sawdust, logging residues and bark. Biomass Bioenerg. **19**(5): 287-293.

Meijer, R., Gast, C.H., 2004. Spontaneous combustion of biomass: Experimental study into guidelines to avoid and control this phenomena. KEMA Power Generation & Sustainables.

Pauner, M., Bygbjerg, H., 2007. Spontaneous ignition in storage and production lines: Investigation on wood pellets and protein powders. Fire and Materials **31**(8): 477-494.

Svedberg, U.R.A., Hogberg, H.E., Hogberg, J., Galle, B., 2004. Emission of hexanal and carbon monoxide from storage of wood pellets, a potential occupational and domestic health hazard. Annals of Occupational Hygiene **48**(4): 339-349.

Svedberg, U.R.A., Samuelsson, J., Melin, S., 2008. Hazardous off-gassing of carbon monoxide and oxygen depletion during ocean transportation of wood pellets. Annals of Occupational Hygiene **52**(4): 259-266.

Wadso, L., 2007. Measuring chemical heat production rates of biofuels by isothermal calorimetry for hazardous evaluation modelling. Fire and Materials **31**(4): 241-255.

Walker, I., Harrison, W., 1977. Self-heating of wet wood. 1. Exothermic oxidation of wet sawdust. New Zealand Journal of Science **20**: 191-200.

Wihersaari, M., 2005a. Evaluation of greenhouse gas emission risks from storage of wood residue. Biomass Bioenerg. **28**(5): 444-453.

Wihersaari, M., 2005b. Greenhouse gas emissions from final harvest fuel chip production in finland. Biomass Bioenerg. **28**(5): 435-443.

(339e) Coupled Optimization and Simulation for Multi-Biomass Source-to-Biorefinery Supply Chain Modeling and Analysis

William Faulkner[1], Joseph Amundson[1], Dr. Fazleena Badurdeen[2] and Dr. Jeffrey Seay[3], (1)University of Kentucky, Lexington, KY, (2)Institute for Sustainable Manufacturing, University of Kentucky, Lexington, KY, (3)Chemical and Materials Engineering, University of Kentucky, Lexington, KY

As cost associated with fossil fuels rise due to various environmental, social, and economic factors, it has become prudent to examine the possibility of exploiting alternative sources of raw materials for the production of energy and chemicals. Biomass, of various forms, has been proposed as an option to fill this need for alternative raw materials. Several problems with biomass as an alternative to fossil fuel derived materials instantly arise: low bulk energy density, the resulting high transport costs, seasonal variability of biomass supply, and the risks inherent to relying on a largely agriculturally based system are several examples. To begin to realize energy and chemical production with biomass as the feedstock on an economic scale, supply chain optimization and a robust means of simulating variation over long periods of time is critical.

Due to a relatively low energy density of the given biomass feedstocks, the majority of the biorefinery supply chain costs occur in the process of raw material transportation. By optimizing this aspect of the biorefinery supply chain, given the plant and potential feedstock source locations, the cost associated with raw material transportation can be minimized. In order to examine the potential longevity and general robustness of the output from the optimization model, simulation is a valuable tool. In this way, the effects of perturbations on the various feed stock supplies due to weather variability, crop yield variability, or other factors can be examined and analyzed.

This presentation focuses on the development of a coupled system comprised of an optimization model for the upstream portion of biorefinery supply chains in Western Kentucky and a discrete-event simulation program. The optimization model minimizes the transportation costs associated with moving biomass feedstock from centralized storage facilities to the biorefinery location. Multiple feedstocks available in the region under consideration including chicken litter, forest residue, and corn stover were taken into consideration. Various alternative shipping methods as well as a centralized versus a distributed production facility layout are considered as well. The discrete-event simulation model allows for the examination of the robustness of the determined optimal supply chain design. This combination of optimization and simulation for the examination of multi-biomass source-to-biorefinery supply chains could potentially serve as a template for future work in biorefinery supply chain model development and decision-making.

See more of this Session: Sustainable Biomass Feedstock Production and Supply for the Emerging

(339f) Improving the Quality of Spent Mushroom Substrate (SMS) As Feedstock for Biorefineries

Maria Nydia Ruiz-Felix, Chemical Engineering, Villanova University, Villanova, PA, Justin M. Yeash, Department of Chemical Engineering, Villanova University, Villanova, PA and Justinus A. Satrio, Chemical Engineering Department, Villanova University, Villanova, PA

In the current ecological climate there is a strong push to find a renewable energy resource to reduce the usage of fossil fuels which are depleting and are causing greenhouse gas emissions. Biorenewable materials, particularly lignocellulosic biomass, have received increased attention recently as the only carbon-containing resources to produce renewable energy, particularly liquid transportation fuels. Many biomass resources that have been explored in the past include wood chips, tree bark, corn stover, switch grass, bagasse, and other organic wastes. Some of these sources are byproducts of commercial agriculture and some provide opportunities for biomass to energy agribusiness, where the crop is grown just for energy harvesting. At Villanova University, we are interested in research on biomass to energy conversions because the school is situated in the proximity of some major agricultural areas in the Mid-Atlantic region, like Lancaster County in Pennsylvania, which produces numerous biorenewable materials that can be used for energy production. Another important region near Villanova University is Chester County, which is known to be "the mushroom growing capital of the world". Currently, Chester County accounts for around 37% of mushroom production in the U.S.

Mushroom growing produces a large amount of waste product called spent mushroom substrate (SMS). After the mushroom fungi grow and fruit the nutrients in the substrate are no longer favorable for mushroom growth and the substrate is considered waste. SMS poses a unique opportunity as a renewable energy resource since it is available in very large amounts, especially in Chester County and the surrounding areas in Pennsylvania. SMS is typically land applied in nearby agricultural fields or sold as compost. These avenues provide little benefit to the mushroom growing companies and are unsustainable due to the methane released through the decomposition of the substrate.

As is, SMS is not the most ideal feedstock for biorefining due to its high moisture and ash/inorganic materials content. Our research is to explore how to improve the quality of SMS to make it an ideal feedstock for biorefining by using torrefaction as a pretreatment method. Torrefaction is the process by which a biomass source is heated in the absence of oxygen to between 200 and 300°C for residence times ranging from 15 to 60 minutes. In this process the biomass will lose 30% of its mass, but only lose 10% of its initial energy which results in significant energy densification. This energy densification is important for biomass feedstocks which may require transportation from the source to further processing facilities. Additionally, torrefaction helps to make biomass properties uniform which enables processing facilities to expend less energy adjusting to various feedstock properties. One of the crucial steps in the conversion of biomass to energy is making sure that the biomass characteristics are within optimal specifications for pyrolysis and further thermal degradation.

To be reported are the results of the torrefaction of SMS at various process temperatures and lengths of processing time. Properties being evaluated are heating values, ash content, and elemental chemical composition. These properties will be analyzed to determine the effects that torrefaction and cleaning by mechanical separation have on the quality of the SMS. Results from fast pyrolysis of torrefied/cleaned SMS compared to untreated SMS by using pyro-GC/MS will be reported as well. Finally, assessment on the sustainability of the torrefaction and cleaning by mechanical separation process for improving the quality of SMS will be presented.

See more of this Session: Sustainable Biomass Feedstock Production and Supply for the Emerging Biorefinery Industry
See more of this Group/Topical: Sustainable Engineering Forum

(354a) Sustainability Evaluation of LNG Processing Technologies

Preeti Gangadharan, Anand Zanwar and Helen Lou, Dan F. Smith Department of Chemical Engineering, Lamar University, Beaumont, TX

Growing demand for clean fuels has led to a renewed urgency in the search for alternative fuels. With the recent shale gas development initiatives, natural gas is poised to capture a much larger share of the energy market. Being clean-burning, natural gas provides many environmental advantages over oil and is also a more economical source of energy.

LNG processing technologies are required for purification and transportation of natural gas before it can be used as a fuel. The LNG processing technologies include liquefaction followed by transportation and gasification of natural gas for use by a consumer. In this paper, all processes involved in LNG processing are simulated using the Aspen Plus software. The simulation is followed by a comprehensive sustainability assessment which covers safety and efficiency assessment, in addition to economical and environmental considerations.

The Enhanced Inherent Safety method used for the safety assessment can be a good tool in the early design stage. Energy and exergy efficiency of the processes involved in LNG processing technologies are evaluated. The assessment identifies the major factors contributing to the efficiency and safety concerns in the technology that is currently used, and provides guidance to designers for future improvements.

See more of this Session: Carbon Efficient Chemical Engineering Systems
See more of this Group/Topical: Sustainable Engineering Forum

(354b) Modeling and Optimization of Natural Gas Liquefaction Process for Reducing Energy Consumption

Jeongwoo Jeon, Kiwook Song, Chul-Jin Lee, Youngsub Lim and Chonghun Han, School of Chemical and Biological Engineering, Seoul National University, Seoul, South Korea

Because of limited fuel reserves, increasing prices of oil and growing concerns about the environment, natural gases are receiving attention as low-carbon, eco-friendly fuels. Natural gases have very large gas-to-liquid volume ratio, so the liquefaction process is essential to reduce volume of natural gases. Through the liquefaction process, natural gas becomes liquefied natural gas (LNG). LNG has a comparative advantage in terms of transport and safety. There are several kinds of natural gas liquefaction process. Among the liquefaction processes, propane pre-cooled mixed refrigerant (C3MR) process is the most widely used natural gas liquefaction process to date.

In this study, a steady-state simulation of the C3MR process was performed using ASPEN HYSYS™ and it was optimized to reduce energy consumption. Optimization formulation is developed to minimize operating cost of the process. Optimization variables are temperature, pressure, and flow-rate of refrigerant and composition of the mixed refrigerant.

This research was supported by a grant from the LNG Plant R&D Center funded by the Ministry of Land, Transportation and Maritime Affairs (MLTM) of the Korean government.

See more of this Session: Carbon Efficient Chemical Engineering Systems
See more of this Group/Topical: Sustainable Engineering Forum

(354c) Ethylene Oxide Production without CO2 As Byproduct: Is the Homogeneous H2O2-Based Catalytic Process Practically Viable?

Madhav Ghanta[1], Hyun Jin Lee[2], Daryle Busch[2], Darryl Fahey[3] and Bala Subramaniam[1], (1)Chemical and Petroleum Engineering, Center for Environmentally Beneficial Catalysis, University of Kansas, Lawrence, KS, (2)Department of Chemistry, University of Kansas, Lawrence, KS, (3)Center for Environmentally Beneficial Catalysis, University of Kansas, Lawrence, KS

Abstract

The novel liquid phase ethylene oxide (EO) technology developed at CEBC entails the transport of ethylene from the gas phase into the aqueous phase containing the dissolved H_2O_2 oxidant, methyl trioxorhenium catalyst, and pyridine N-oxide promoter in methanol co-solvent. The ethylene undergoes epoxidation to form corresponding epoxide with no detectable CO_2 formed as byproduct.[1] For, a batch residence time of 1 h, the EO yield and selectivity are 48% and 99+%, respectively. Further, the EO productivity 0.7-4.4 of [g EO/(g Re)/hr] in the CEBC process compares with the EO productivity of 2.2-4.1 of [g EO/(g Ag)/hr]. In this talk, we present a comparative techno-economic analysis benchmarked against the conventional EO technology. Two case studies are performed to study the effect of various parameters (such as source of the oxidant, mode of removing the heat of reaction and recovery/ decomposition of the H_2O_2 oxidant) on the economic viability of the CEBC process thus identifying the major economic impact drivers. Further, the environmental performance of the proposed process is compared against the conventional EO technology in a cradle to gate analysis performed using the GaBi software.[2] The reduction in GHG emissions and other adverse environmental effects due to the conservation of ethylene feedstock, elimination of CO_2 byproduct and easy separation of the products and byproducts from unreacted reactants will be quantified and compared with the environmental footprint of H_2O_2 production. These results along with process development issues such as immobilization of MTO catalyst onto a soluble polymer support, and its stability and durability will be presented and discussed.

References:

1. Lee, H-J., et. al., *Towards a CO_2 -free ethylene oxide process: Homogeneous ethylene oxide in gas-expanded liquids*. **2010**. 65, 128-134.
2. GaBi software, 4.0; PE International: Germany, 2011.

(354d) Use of Lifecycle Tools to Optimize Sustainable Energy Campuses

Earl R. Beaver, Practical Sustainability, LLC, Chesterfield, MO

This paper describes the use of GaBi (TM) software to optimize the design of a Sustainable Energy Campus based upon feed of scrap automobile and truck tires. Further, once the Sustainable Energy campus is defined, the software and related tools are used to compute the benefits of reduced lifecycle consumption of fossil fuels, emission of greenhouse gases and other wastes and the overall energy savings.

The Sustainable Energy Campus combines multiple technologies from different industries to create a truly sustainable "waste-to-profit" campus where the waste from one technology being used becomes the input feedstock to another technology that utilizes that waste to create a full spectrum of highly profitable fuel products. All the while, greenhouse gases are reduced, materials are recovered, energy is optimized and virtually no emissions are released to the environment. Further, the reality is highly profitable immediately and for the long term. Thus, genuine waste-to-profit.

Technologies from the rubber, chemical and petroleum industries are at play in the Sustainable Energy Campus. A Carbon Black recovery production line takes scrap automobile and truck tires and uses a vacuum pyrolysis process to convert those tires into a product called Carbon Black Alternative and a stream of pyrolysis oil. Carbon Black is present in 75-85% of all new tires produced worldwide. A new cracking and reaction technology combined with a sophisticated separation process converts the pyrolysis oil into raw materials for gasoline, kerosene, diesel, naphtha, jet A fuel base and liquified low molecular weight hydrocarbons (propane, butane and pentane). While each step can be profitable independently, combining them in a single site with cogeneration allows for energy and materials optimization as well as maximizing revenue. Further, minor components are captured and isolated which have extremely high value. The key materials are cyclic terpenes and high been recovered at high yields and high purities.

Since a single scrap tire weighs approximately 21 pounds and one tire is produced per year per person in industrialized countries, the Sustainable Energy Campus provides a means of producing massive amounts of valuable products which otherwise would be produced from incremental fossil fuels and raw materials at significant cost to society and the environment. A full scale tire pyrolysis unit utilizing this teechnology has been in operation for more than one year and has produced more than ten million pounds of carbon black along with the co-products cited above. The data related to mass of products, energy savings and net environmental impact will be presented and discussed.

See more of this Session: Carbon Efficient Chemical Engineering Systems

(354e) Cost Minimization Model for Heat Recovery for Hydrogen Production From Biomass Steam Gasification

Abrar Inayat, Murni M. Ahmad, Mohamed I. A. Mutalib and Suzana Yusup, Chemical Engineering, Universisti Teknologi PETRONAS, Tronoh, Malaysia

With the current energy and environmental crisis, hydrogen economy has now gained a positive outlook. This alternative form of energy becomes more attractive when produced from a sustainable source such as biomass, especially for an agricultural country such as Malaysia. However, the thermochemical conversion of biomass into hydrogen can be energy intensive due to its endothermic nature. In this work, a simple linear mathematical problem (LP) has been developed to design a heat integrated flowsheet for an energy efficient hydrogen production from oil palm empty fruit bunch via steam gasification with in situ CO_2 capture. The LP transshipment model targets to minimize the utility cost by considering the heat content of process streams at a fixed number of temperature intervals. Three constrained matches are imposed for the synthesized flowsheet for the hydrogen production with hot and cold utility requirements of 1420 hJ/h and 832 kJ/h respectively. The proposed model is able to predict a minimum utility cost of US$ 8.76/ kJ h. The minimum of heat load for hot and cold utilities solved are 631.20 kJ/h and 0.743 kJ/h respectively. This gives an energy savings of approximately 55% in hot utilities and 98% in cold utilities. Furthermore, the production cost of hydrogen is predicted to decrease from US$ 1.91 to 1.80 per kg of H_2 via the optimization approach. A result comparison is made with a previous heat integration work on the flowsheet that was based on pinch analysis.

See more of this Session: Carbon Efficient Chemical Engineering Systems
See more of this Group/Topical: Sustainable Engineering Forum

(354f) Completing the Cycle: Capturing Gaseous Waste Carbon As Fuel/Chemicals

Michael A. Schultz and Derek W. Griffin, LanzaTech Limited, Roselle, IL

The world currently faces an enormous energy challenge. LanzaTech responds to this challenge by capturing and beneficially reusing waste carbon to make low carbon fuels and chemicals. The International Energy Outlook, in its 2010 reference case, projects that world primary energy consumption will increase by 49 percent (1.4 percent per year) from 495 quadrillion Btu in 2007 to 739 quadrillion Btu in 2035.[1] In order to meet this challenge, LanzaTech provides a technology solution to improve process energy efficiency while simultaneously reducing greenhouse gas emissions.

Through gas fermentation, LanzaTech transforms *waste* into *resources*. Industrial waste gases or low value byproducts become the inputs for making valuable low-carbon fuels and chemicals. Industrial and chemical processes produce off-gases containing carbon monoxide and/or carbon dioxide that are either vented to the atmosphere, flared, or burned as fuel; a sub-optimal use of the carbon and energy contained in these off-gases. The LanzaTech process captures these waste gases with renewable resources and transforms them into energy rich fuels or chemicals such as ethanol, acetic acid, or 2,3-butanediol, among others. The LanzaTech process offers superior carbon conversion, energy efficiency, and greenhouse gas emission performance compared to conventional and emerging routes to the same products.

The LanzaTech gas fermentation process is described including potential applications in chemical, petrochemical, refinery, gas-to-liquids, and industrial plants (such as coal and steel). Key technology characteristics including gas conditioning, bioreactor design, and fermentation expertise are highlighted to demonstrate how the LanzaTech process can integrate with a wide range of industrial processes. An additional benefit of the LanzaTech process is the carbon capture efficiency illustrated in a life cycle analysis comparing LanzaTech's route to ethanol with conventional production of gasoline and corn ethanol.

[1]"IEA - 2010 International Energy Outlook - World Energy Demand and Economic Outlook." http://www.eia.doe.gov/oiaf/ieo/world.html (accessed February 24, 2011).

See more of this Session: Carbon Efficient Chemical Engineering Systems
See more of this Group/Topical: Sustainable Engineering Forum

(364a) Impact of Lignin Extraction and Cellulose III Conversion On the Enzymatic Digestibility of Corn Stover

Leonardo da costa Sousa, Shishir Chundawat, Nirmal Uppugundla, Bruce Dale and **Venkatesh Balan**, Chemical Engineering and Material Science, Great Lakes Bioenergy Center, Michigan State University, Lansing, MI

The benefits of delignification on plant cell wall digestibility by hydrolytic enzymes are well described in the literature. The presence of lignin increases non-productive binding of cellulases, resulting in enzyme inhibition and decrease in the sugar yields. Moreover, lignin acts as an anti-microbial agent, reducing cell viability and ethanol yields.

Recently, additional benefits by modifying the crystal structure of native cellulose (cellulose I) to cellulose III were also reported. After this modification, it was possible to double the rate of enzymatic hydrolysis of cellulose. However, this increase was observed for purified cellulose, native from algae (Cellulose Iα), which has a different crystal configuration than the cellulose native from plants (Cellulose Iβ). Therefore, it is not evident that this modification is possible to be obtained in plant cell walls and that it can represent the same kind of benefits observed for purified cellulose. From the literature, cellulose III is obtained by contacting cellulose I to supercritical ammonia or different types of amines in anhydrous conditions. Additionally, water is known to prevent the conversion of cellulose I to cellulose III, however the water concentration limit that allows cellulose III to be formed is not yet clearly defined. For biorefinery applications, plant biomass will contain a certain degree of moisture that could prevent stable cellulose III formation and therefore, it is important to verify in what extent is possible to convert native cellulose to cellulose III in plant biomass.

In this perspective, our goal was to test the benefits of cellulose III conversion under non-anhydrous conditions, using corn stover, and couple this modification with lignin extraction in a single stage process – Extractive AFEX (E-AFEX). These benefits were compared with the traditional non-extractive ammonia based pretreatment method - AFEX.

See more of this Session: Developments In the Pretreatment of Lignocellulosics for Bioconversion I
See more of this Group/Topical: Sustainable Engineering Forum

(364b) The Effects of Cationic Polyelectrolyte On the Pretreatment and Enzymatic Hydrolysis of Corn Stover

Shaowen Ji and Ilsoon Lee, Chemical Engineering and Materials Science, Michigan State University, East Lansing, MI

Lignocellulose is the most abundant renewable biomass and the potential substrate for ethanol production. The costly pretreatments and large consumption of enzymes are needed to reach high cellulose conversion. Lignin, as one of major cell wall components, is considered as the barrier to enzymatic hydrolysis by blocking cellulase from productively being adsorbed. The removal or effective modification of lignin is required for efficient pretreatment process. In this study, the effects of cationic polyelectrolyte, poly-(dimethyldiallylammonium chloride) (PDAC), on alkaline pretreated corn stover were investigated. Structural study by Scanning Electron Microscopy (SEM) imaging revealed particles morphology on cell walls of pretreated corn stover. These particles were confirmed to contain lignin by Confocal Laser Scanning Microscopy (CLSM) and Transmission Electron Microscopy (TEM) imaging. Chemical compositions of pretreated corn stover at different alkaline and polyelectrolyte concentrations were analyzed. Higher polyelectrolyte concentration in the pretreatment resulted in higher lignin retention. Evaluation of enzymatic hydrolysis showed the enhancement of cellulose and hemicellulose conversion after polyelectrolyte pretreatment. PDAC was proposed to facilitate cellulose disruption and lignin redistribution via its interaction with lignin, which significantly modified biomass morphological properties, such as increased porosity and improved accessibility of cellulose surface.

See more of this Session: Developments In the Pretreatment of Lignocellulosics for Bioconversion I
See more of this Group/Topical: Sustainable Engineering Forum

(364c) Novel Lignocellulosic Biomass Pretreatment Using a Modified Taylor-Couette Mixer for Biofuels and Biomaterials

Wei Wang and Ilsoon Lee, Department of Chemical Engineering and Materials Science, Michigan State University, East Lansing, MI

Bioethanol derived from lignocellulosic biomass can be a good candidate for energy alternative, not only because lignocellulosic biomass is abundant and sustainable, but also bioethanol production forms carbon cycle in a more timely fashion compared to fossil fuel, which leads to reduction of green gas emission. Lignocellulosic biomass pretreatment, which is one of the essential steps in bioethanol production, accounts for the majority portion of the overall bioethanol production cost. Hence an effective and efficient pretreatment method is highly demanded. Although this research area has been deeply investigated several decades ago, it has not attracted tremendous attention until recently. In our research, cost-effective and breakthrough biomass pretreatment was conducted with a novel modified Taylor-Couette mixing system, aiming to overcome the natural recalcitrance of lignocellulosic biomass in a short processing time and facilitate a better accessibility of enzymes to polysaccharides. Compositional analysis proved significant removals in both lignin and hemicellulose and rich glucan content left in the residue only after 2 minutes or less. Parameters like mixing speed, retention time, biomass loading and corresponding enzyme digestibility will be presented.

See more of this Session: Developments In the Pretreatment of Lignocellulosics for Bioconversion I
See more of this Group/Topical: Sustainable Engineering Forum

(364d) Modeling and Scale up of CO2-Water Pretreatment of Guayule Biomass

Ehsan Moharreri and J Richard Elliott, Chemical and Biomolecular Engineering, The University of Akron, Akron, OH

A carbon dioxide – water pretreatment process was developed for guayule biomass. The guayule biomass was mixed with water and then loaded to a reactor where it was exposed to a biphasic environment consisting of CO_2 rich vapor phase and water rich liquid phase. Then, the biomass was enzymatically hydrolyzed and the sugar content was measured at the end. The effects of temperature, pressure, moisture content, accessible volume and holding time in the pretreatment step were investigated. One hundred and twenty (120) experimental runs were carried out in the temperature range of 145 C to 210 C, pressure was between 3.4 MPa to 34 MPa, the moisture content ratio was between 0.6 to 6, the accessible volume was between 0.2 to 0.1 L and the holding time was between 20 to 50 minutes. A model was fit to the experimental data based on dissolved CO_2 in the liquid phase using CO_2-H_2O phase envelop and a temperature dependant constant. The model fits to one hundred and twenty (120) data points with an average absolute percent deviation(AAD%) of 13. Then, the same process was implemented to a pilot scale vessel with a 1 liter volume. The process was carried out in batch and semi-continuous methods and twenty (20) data points were produced to confirm the model. The model predicted the pilot plant runs with a 15% average absolute percent deviation (AAD%). In the lab scale of 190 C, 26 MPa, moisture ratio of 1.5, accessible volume of 0.22L, biomass load of 15 grams and 35 minutes in holding time resulted in 78% glucose yield and 66% hemicelluloses yield. In the pilot scale of 200 C, 10MPa, moisture ratio of 2, accessible volume of 0.88, biomass load of 60 grams and 35 minutes in holding time resulted in 78% total sugar yield.

See more of this Session: Developments In the Pretreatment of Lignocellulosics for Bioconversion I
See more of this Group/Topical: Sustainable Engineering Forum

(499g) Effect of Transition Metal Catalyst On Lignin Oxidation During Alkaline Hydrogen Peroxide Pretreatment

Zhenglun Li, Department of Chemical Engineering and Material Science, Michigan State University, East Lansing, MI and David Hodge, Chemical Engineering and Materials Science, Michigan State University, East Lansing, MI

Biomass lignocellulose can be used as feedstock for bioethanol production, and a pretreatment is necessary for achieving high ethanol yield and economical feasibility. As a method to increase enzyme digestibility of biomass by selective delignification, alkaline hydrogen peroxide (AHP) pretreatment has been proposed and studied. The lignin recalcitrance in lignocellulose biomass can be removed by H_2O_2 oxidation and the enzymatic hydrolysis efficiency of the polysaccharides can be greatly improved. However, a big issue with AHP pretreatment is the high cost of hydrogen peroxide during the pretreatment, and the selectivity of the oxidant need to be improved, i.e. to achieve highest pretreatment efficiency with low level of H_2O_2 loading.

In 1890s, Fenton discovered that iron ion can catalyze the oxidation of organic substrate by hydrogen peroxide. Since then, research has been done on the application of transition metal catalysts in catalyzing the phenol contaminant in wastewater and pulp bleaching. It is desired that catalyst can enhance selective electron transfer from lignin substrate to the oxidant and improve the selectivity of delignification reactions to unproductive side reactions.

See more of this Session: Developments In the Pretreatment of Lignocellulosics for Bioconversion I
See more of this Group/Topical: Sustainable Engineering Forum

(364f) Progress Toward Biological Pretreatment - Understanding Plant Cell Wall Degradation by Termite and White-Rot Fungi

Shulin Chen[1], Jing Ke[2], Dhrubojyoti D. Laskar[2] and Jijiao Zeng[2], (1)Department of Biological Systems Engineering, Washington State University, Pullman, WA, (2)Washington State University, Pullman, WA

In an effort developing more efficient and environmentally benign lignocellulosic biomass processing technologies, the authors' laboratory has been studying plant cell wall degradation process in termite and white-rot fungi. This presentation summarizes the results so far obtained from various experimental investigations of the lab. First, termite guts provide both aerobic and anaerobic environments, the former is required for lignin degradation. The termite guts also have close to neutral pH range, which suggest existence of lignin degrading enzymes that are different from that of fungi. The organization of the gut system allows the conversion of biomass in termite to occur as a sequential process in which the pretreatment of the biomass occurs in the foregut and midgut, whereas the hydrolysis of the cellulose occurs mainly in the hindgut. There are synergies between termite and its symbiont in deconstructing plant cell walls. The chewing action of the termite increases the surface area for enzymatic actions, the enzymes produced by the termite tissues modifies the bonds that forms the barriers for cellulase access and symbionts produce the enzymes that are mainly responsible for cutting the cellulose polymer to sugar units. The results shows the actual changes of lignin structure as the biomass material pass through the termite guts. The lignin structure changes are more modification than degradation in nature. Equally import is the deconstruction of the association of lignin with other components of plant cell walls such as hemicelluloses that occurs in the foregut and midgut. It becomes apparent by comparing with other biological systems that the enzyme systems and their actions in biological plant cell wall degradation systems are more complex than what was previously known. A generality has also been observed in that most of the biological systems studied tended to reduce G unit from the original lignin structure. Additionally, termite is a more effective plant cell wall degradation system than white rot-fungus because of the chewing action of the termite and the better-controlled environment in the termite guts. Termite is also more effective than the cow stomach as the latter lacks capability of lignin modification. Developing an artificial system that can surpass the natural biological system in major performance criteria may offer the ultimate solutions to lignocellulosic biomass processing and utilizations.

See more of this Session: Developments In the Pretreatment of Lignocellulosics for Bioconversion I
See more of this Group/Topical: Sustainable Engineering Forum

The Investigation of Aluminium Wastes Encapsulated in the Individual Phases of Ordinary Portland Cement

T.P. Neville[1], C. Kuenzel[2], P. Lettieri[1], J. Bensted[1], G.M. Photiadis[1], M. Tyrer[1] and S.J.R. Simons[1]
[1]Department of Chemical Engineering, University College London
[2]Department of Civil and Environmental Engineering, Imperial College London

ABSTRACT

The storage of expended radioactive wastes is one of the most costly aspects of nuclear energy due to the social and political concerns over environmental and safety issues. Immobilisation of radioactive wastes is carried out to allow them to be handled and stored safely. Currently, Ordinary Portland Cement (OPC), blastfurnace slag (BFS) and pulverised fuel ash (PFA) are used as an encapsulation matrix. The high pH of the cement system limits the solubility of many of the radioactive nuclides present. However, corrosion can occur under these conditions, particularly for aluminium. The products of corrosion, hydrogen gas evolution and expansive hydroxides can lead to poor wasteform quality which raises questions about the environmental and safety issues of cemented wasteforms. This work aims to gain further understanding of the corrosion of aluminium simulant wastes within cement matrices by separating the complex OPC system into easy-to-define pure systems. Therefore, the pure phases, tricalcium silicate, dicalcium silicate, tricalcium aluminate and tetracalcium aluminoferrite are synthesised and the corrosion of Al studied. As a result, a comparison between corrosion in OPC and pure phases can be drawn to provide fundamental information on the governing mechanisms of the corrosion reactions.

1. INTRODUCTION

The UK nuclear industry started in the 1950s, first with weapons production and then the opening of the first commercial power plant in 1956. In the UK there are currently 19 nuclear reactors. As of 2009 these provide 17.9% of the country's electricity supply [1]. With growing pressure on the UK Government to deal with climate change and ensure energy security new options of electricity generation, including nuclear, have to be examined. The Government recently reported that new nuclear builds can help meet environmental targets to cut CO_2 emissions [2]. However, the current fleet of nuclear power stations is entering the last years of design life, which will leave a gap in energy supply and may compromise both energy security and climate change goals.

One of the major questions about further use of nuclear power is what to do with the radioactive by-products of the processes. Currently these are classified as wastes and their storage and safe disposal is a major issue for the nuclear industry. In addition to the legacy waste from the first reactors, new wastes have to be treated and stored to stop any loss of material to the environment.

Immobilisation of radioactive wastes is carried out to allow them to be handled and stored safely. This technique is currently carried out using composite cements [3], which are blends of Ordinary Portland Cement, blastfurnace slag and pulverised fuel ash. The immobilised wastes are collectively known as wasteforms. The composite cements have low permeability and will incorporate various waste compounds into the hydrated structure of the cements, which minimise transport of the radionuclides into the environment. The hardened cement allows for the waste material to be stored as a solid which provides strength so that it is stable for long term storage.

This project aims to provide the nuclear industry with more information about the reactions and products formed in solidified waste systems. This will improve the understanding of wasteforms and act as underpinning support for waste management strategies. This project is a continuation of work carried out by the National Nuclear Laboratory, concerning the corrosion of Magnox wastes in OPC/BFS composite cements.

Issues surrounding the use of OPC systems include heat release during hydration and the high pH environment. The heat produced from the cement hydration can have several effects on the wasteform, including weakening of the structure. The high pH is important for limiting the solubility of some of the radionuclides [4]. However, metals like aluminium suffer corrosion under such conditions. Figure 1 is a Pourbaix diagram which indicates the stability of the Al at different pH values [5].Corrosion of the metals creates hydrogen gas and expansive metal oxides [6]. These have the potential to form cracks and pores within the cement structure, and thus, can lead to poor quality waste forms with potential routes for material loss to the environment [7].
The work reported here is focused on elucidating the fundamental interactions between pure cement phases and aluminium to determine whether the previously mentioned reactions can be controlled by the wasteform chemistry.

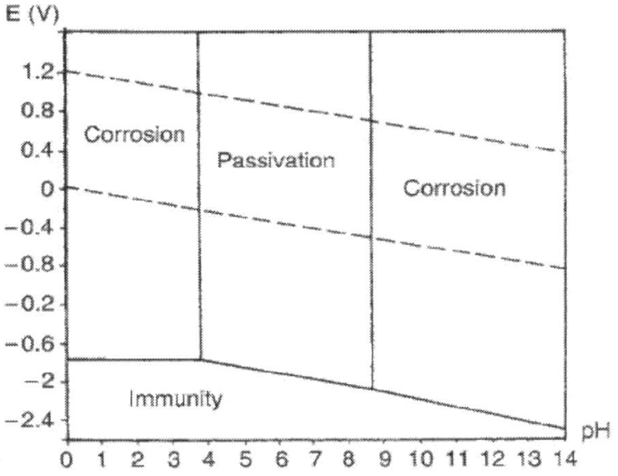

Figure 1. Pourbaix Diagram of aluminium [5].

2. METHODOLOGY

In order to address the issues of corrosion within cement matrices and following on from previous work into corrosion within OPC/BFS systems [8], this work looks at furthering the knowledge of the cement system by characterising and comparing simplified cements. The complex nature of the cement system due to the numerous compounds and contaminants make it difficult to establish specific causes contributing to corrosion processes.

The route taken by this project is to separate the composite cement system into its constituent parts, or phases, and carry out a systematic examination of these pure phases. The individual phases will be examined and the behaviour compared with the more complex systems, with the aim of deducing the causes and effects of corrosion. The four major phases are tricalcium silicate, dicalcium silicate, tricalcium aluminate and tetracalcium aluminoferrite [9], the percentage amounts are listed in Table 1 below [10]. There is a range of compositions of an OPC clinker for reasons such as the raw materials used, fuel types and production facilities [11]. When combining the potential differences of an OPC clinker with the unknowns of the waste material results in a very difficult system to define.

Table 1. Ordinary Portland Cement phase names, formulae and amounts.

Name	Composition	Cement Notation	Weight %
Alite	$3CaO \cdot SiO_2$	C_3S	55– 65
Belite	$2CaO \cdot SiO_2$	C_2S	15– 25
Aluminate	$3CaO \cdot Al_2O_3$	C_3A	8 – 14
Ferrite	$4CaO \cdot Al_2O_3 \cdot Fe_2O_3$	C_4AF	8 – 12

It is stated in the literature that the different phases have different effects on the hydration of the cement itself, Figure 2 shows the heat evolution of OPC hydration [12]. The different peaks show the different rates of reaction with water of the phases. The aluminate phase can be noted to react quickly and has significant heat evolution and give rise to flash setting if present in high quantity. Whilst tricalcium silicate reacts slower, and with lower heat evolution, it is known to contribute the early strength in cement pastes. Dicalcium silicate has a much slower reaction with water and develops compressive strength over a much longer period of time, with very little heat evolution from the reaction.

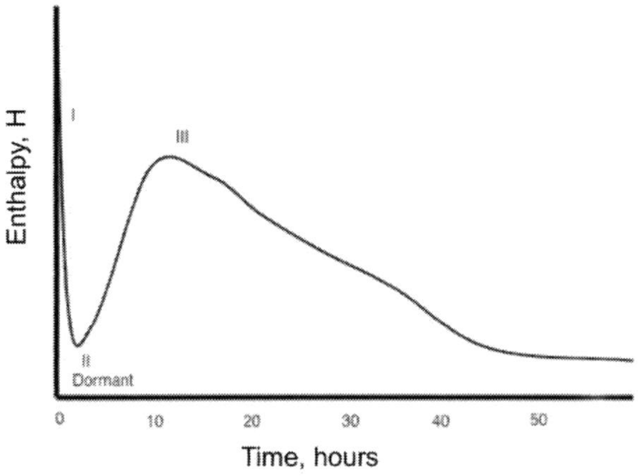

Figure 2. Heat evolution during OPC hydration. I aluminate peak, III silicate hydration [12].

It has been shown that the rate of corrosion is dependent upon pH of the medium. The hydration of the cement phases produces alkaline compounds. Hydrated OPC produces a pore solution with an approximate pH ~13.5. The pure phases will have minimal foreign ions and thus will undergo known hydration reactions which will allow for clearer assessment of the product matrix interactions with the metal.

3. EXPERIMENTAL

The first part of this work has been to break up OPC clinker into its 4 major phases. These phases are to be synthesised for this work using pure reactant materials, >99% $CaCO_3$ and SiO_2 from Sigma Aldrich, Doping agents were added to the mixture to ensure the correct crystal form, in small quantities ~1 %wt [13]. with expected phase purity around 98%. The powders were mixed in a tumble mill for 24 hrs and then pressed into pellets. The pellets were then sintered at 1450 °C for the required time followed by quenching to minimise phase transitions. The purification of the individual phases is achieved by a cyclic process of mixing and re-sintering. The current work looks at the calcium silicate phases as these represent the major part of the overall cement. Powder X-ray diffraction using a Philips PANalytical X'Pert was used to assess the product powders and ensure the desired crystals achieved. To ensure further product homogeneity, EDX maps were used to check Ca/Si ratios.

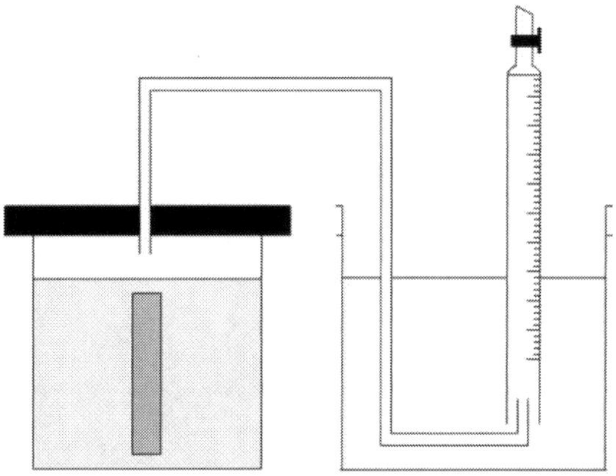

Figure 3. Hydrogen gas collection by water displacement flow from encapsulated aluminium rod on left.

The initial experiments undertaken to examine the corrosion of the system were hydrogen gas displacement measurements. High purity 99.99% aluminium rods, supplied from Goodfellows UK, 6mm in diameter and a known length, were encapsulated in the pure phases and sealed; for comparison rods were also placed in pH buffer solutions ranging from pH 11-13.5. The hydrogen was collected via a tube in to a measuring cylinder under a head of water. Figure 3 illustrates the experimental setup. All measurements were carried out at room temperature and readings corrected against a reference standard for volume fluctuations.

Hydrogen gas evolution provides an indication of corrosion rate. However, the method does not yield information about the internal structure and state of the cement matrix. Investigations of the internal structure were undertaken by measurements post hydration. SEM and EDX mapping of aged samples were undertaken to look at the porosity and the growth and movement of aluminium around the cement/metal boundary. A JEOL JSM6480LV SEM with EDX capacity was used to probe the cement/metal interface. The samples were run at low vacuum to attempt to retain moisture of the interface and avoid significant drying and cracking. Two methods were employed, encapsulation, then cutting open and setting rod against removable flat surface. The first method allows viewing of the effects of encapsulation upon corrosion of the metal, whilst the later enables a clearer picture of lateral corrosion growth into the cement.

4. RESULTS AND DISCUSSION

Figure 3 shows the XRD trace of tricalcium silicate synthesised for this work. The trace was compared to the reference to ensure the correct phase was achieved. The trace indicates that there is trace CaO present; this peak can be reduced with re-sintering, allowing combination with unreacted SiO_2. The purity of the samples is high enough for this work to demonstrate the effects of the individual phase.

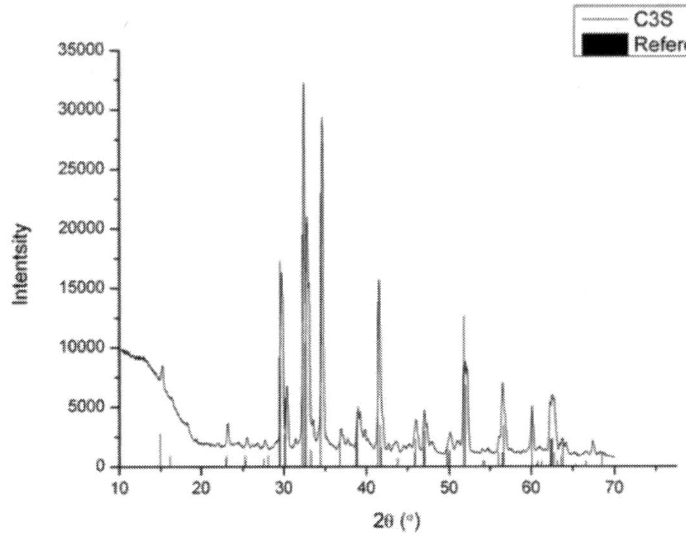

Figure 4. XRD Trace of tricaclium silicate sample after first sintering compared against reference peaks.

The expected driver of corrosion within a cement matrix is the pH of the pore solution, as demonstrated from the Pourbaix Diagram. At high pH the aluminium is more susceptible to corrosion as the oxide films that form on the metal surface are increasingly soluble and expose fresh metal to the electrolyte. Figure 5 shows the corrosion rate of aluminium rods in pH buffer solutions. The corrosion rate rises sharply with increasing pH.

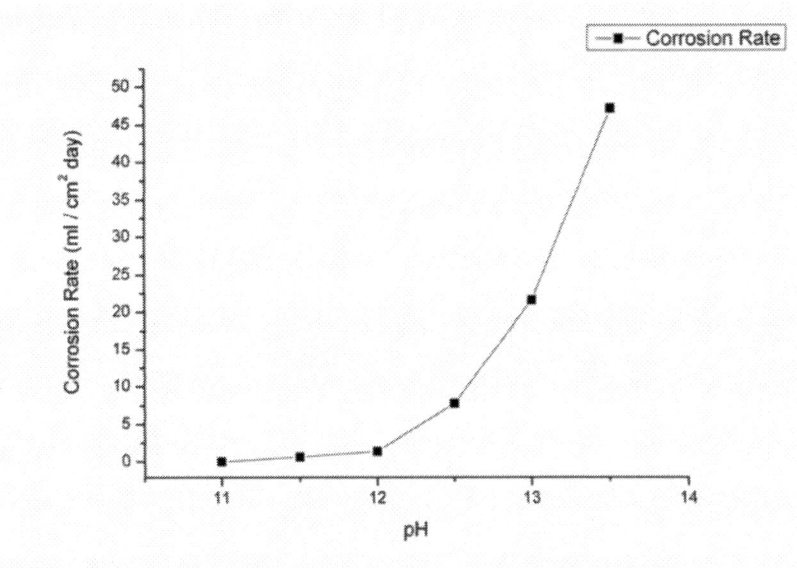

Figure 5. Aluminium corrosion rate in pH buffers.

The corrosion within a solid medium, however, demonstrates a different behaviour. Figure 6 shows the cumulative hydrogen volume evolved over 10 days. The rates of the evolution from the pure phases agree with the results of the buffer solutions; the initial rates of the evolution are seen to be higher for the tricalcium silicate. These results, however, tend to plateau and the corrosion rate falls significantly. It is noted that the lower pH dicalcium silicate evolves a total hydrogen amount of around twice that of the tricalcium silicate per unit area.

The chart also shows the pH change of 1 g of each phase in 100 ml of deionised water. The result shows the difference in hydration rate of the phases. The tricalcium silicate solution rapidly increases in to a maximum of around pH 12.5. The dicalcium silicate reaction is much slower and steadies out at around pH 12.3. This concurs with a mechanism of a pH driven corrosion rate. However, with regard to total hydrogen evolution and the plateauing of evolution the hydration of the cement phases must play a role.

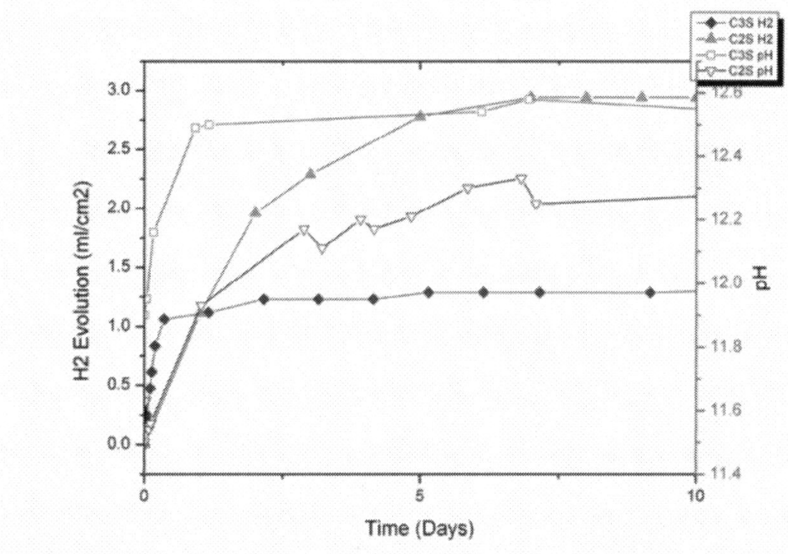

Figure 6. Calcium silicate corrosion extent and pH evolution, showing both silicates.

Figure 7 is a group of SEM images of aluminium rods within tricalcium silicate matrices. The left image shows the lateral corrosion of an aluminium rod; the rod is to the right of the image and the matrix to the left. It can be seen between the surfaces that a layer of corrosion product has grown. EDX mapping and line scans were taken of the section and it was demonstrated that the layer contains Al and Ca compounds. It can be seen in the cement matrix that there is formation of holes; these can be attributed to hydrogen gas evolution. The right hand image demonstrates clearly the corrosion within a matrix. The sample was cut in half through the metal. The metal was consequently removed exposing the cement surface. The revealed surface shows a significant bubbling structure formed by gas evolution. The formation of the bubbles significantly reduces the surface contact of the metal and the cement and the metal readily becomes detached from the matrix.

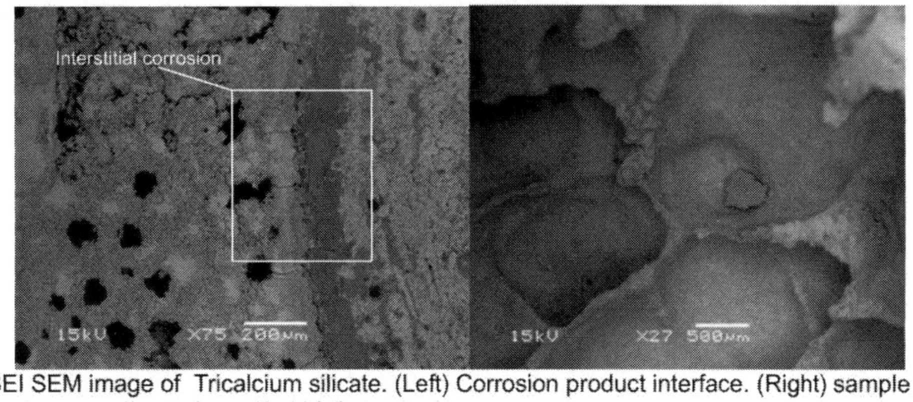

Figure 7. BEI SEM image of Tricalcium silicate. (Left) Corrosion product interface. (Right) sample with the aluminium rod removed, note large "bubble" structure.

Hydration of the cements consumes the liquid water present in the matrix, it is well documented that tri- and dicalcium silicates react at different rates and this has been shown in the work also. The consumption of water plays a significant role in corrosion as the loss of water removes the electrolyte necessary for corrosion. Secondly the interstitial phase generated between the cement and metal has been shown by EDX measurements to contain both Ca and Al atoms, forming a passive layer to minimise corrosion.

5. CONCLUSION

The work to date agrees with work for similar materials and clearly demonstrates questions and issues with aluminium encapsulation.

The resolution of the interstitial layer, the compounds that produce it and the effect it has on passivation are to be investigated.

Further work demonstrating water consumption and hydration effects on corrosion rates will be undertaken using electrochemical techniques such as impedance spectroscopy. These will produce an in-situ picture of within the matrix, to allow direct assessment of causes. The pure phases will also allow for modelling of reactions and predictive modelling of the corrosion.

6. ACKNOWLEDGEMENTS

The author would like to thank Prof. Chris Cheeseman and Tingting Zhang from the Department of Civil Engineering at Imperial College London for their advice, ideas and access to facilities throughout this work. Ed Butcher at NNL for his guidance with the direction of the project and supply of materials. Dr. Dan Brett and the members of the Centre for CO_2 Technology from the department of chemical engineering UCL for equipment and spare hands. The author would like to thank the DIAMOND consortium and the EPSRC for the funding of this project.

References

1. World Nuclear, A. Nuclear share figures, 1999-2009. http://www.world-nuclear.org/info/nshare.html.
2. Government, H. M. *The UK Low Carbon Transition Plan National strategy for climate and energy*; HM Government: 2009.
3. Ojovan, M. I.; Lee, W. E., Introduction to Immobilisation. In *An Introduction to Nuclear Waste Immobilisation*, Elsevier: Oxford, 2005; pp 1-8.
4. Glasser, F. P., Fundamental aspects of cement solidification and stabilisation. *Journal of Hazardous Materials* **1997,** *52* (2-3).
5. Vargel, C., *Corrosion of Aluminium*. Elsevier: 2004.
6. Setiadi, A.; Milestone, N. B.; Hill, J.; Hayes, M., Corrosion of aluminium and magnesium in BFS composite cements. *Advances in Applied Ceramics* **2006,** *105*, 191-196.
7. Atkins, M.; Glasser, F. P., Application of portland cement-based materials to radioactive waste immobilization. *Waste Management* **1992,** *12* (2-3), 105-131.
8. Setiadi, A. PhD Thesis University of Sheffield. 2006.
9. Lea, F. M., *The Chemistry of Cement and Concrete*. Edward Arnold Ltd: 1970; Vol. 3rd.
10. Taylor, H. F. W., *The Chemistry of Cements*. Academic Press: 1964.
11. (a) Bensted, J.; Barnes, P., *Structure and Performance of Cements*. Spon Press: 2002; Vol. 2nd; (b) Bye, G. C., *Portland Cement Composition,Production and Properties*. Pergamon Press: 1983.
12. WHD Cement hydration. http://www.understanding-cement.com/hydration.html.
13. Wesselsky, A.; Jensen, O. M., Synthesis of pure Portland cement phases. *Cement and Concrete Research* **2009,** *39* (11), 973-980.

(368b) Leaching Behavior of Hazardous Heavy Metals From Lime Fly Ash Cements

Ojas Chaudhari and Joseph Biernacki, Chemical Engineering, Tennessee Technological University, Cookeville, TN

Disposal of fly ash in landfills in large amounts results in sacrificing precious land space and potentially endangering neighboring communities and water resources. Productive utilization of fly ash is one approach to reduce the impact of this disposal problem. A major use for fly ash is in portland cement-based materials such as concrete. Unfortunately, not all fly ash is suitable for use as a partial cement substitute. One concern among concrete producers is the high residual content of carbon of some ashes, which interferes with other additives used to make modern concrete. Therefore, other large volume application for off-specification ash are sought. One possibility is to use high carbon bearing ash in combination with lime to provide cement for low strength non-structural applications such as in road base mix formulations. Incorporating such materials in road base formulations however, presents several environmental challenges. The main concern with this application is the possibility that the fly ash will come into contact with water and hence contaminate drinking resources or wildlife waters. Fly ashes, by virtue of their origin, can contain trace toxins and hazardous heavy metals. The goal of this research is to investigate the efficacy of hydrated ash to sequester heavy metal ions and discern if ash chemistry is a factor in the ability of the associated hydrates to bind the metals. Two ashes were investigated, one that is suitable for use as a cement substitute and the other having a high carbon content.

See more of this Session: Emissions and Waste Control From Emerging Industries
See more of this Group/Topical: Environmental Division

COMPARATIVE STUDY OF PERFORMANCE OF DIFFERENT CHELATING AGENTS FOR EXTRACTION OF NI FROM SPENT CATALYST

*Garima Chauhan, K.K.Pant, K.D.P.Nigam**
Indian Institute of Technology Delhi, New Delhi, India-110016

ABSTRACT

The mobilization of heavy metals from spent catalyst is of the essence due to increasing demand of complex metallic composites in industries and to reduce the metallic waste. Present study involves the development of environmental friendly methodology for extraction of nickel from spent catalyst of fertilizer industry by adopting the chelating agents. The concept of chelation explored here to bind metal from spent catalyst and subsequently to be de-chelated by acid addition to extract metal. All chelation experiments were conducted in atmospheric reflux conditions and process parameters (reaction time, molar concentration of solvent, solid-liquid ratio of reacting mixture, pH, reaction temperature, particle size) were optimized in order to maximize the metal extraction. Three different chelating agents (EDTA, DTPA, and NTA) were employed for extraction of Ni. EDTA has been used earlier in the literature for Ni extraction with solid to liquid ratio 1:50. To reduce the S:L ratio, experiments were conducted with S/L 1:20 and very less increment were seen with increasing S/L. It was also observed that at optimized reaction conditions, DTPA showed the highest (84%) Ni extraction among all three chelating agents. Recovery of DTPA was not more than 40% at atmospheric condition, while at refrigeration condition, DTPA recovery was improved and reached upto 80-86%. Recovery of EDTA solid (93%-97%) was highest among these chelating agents. Comparative evaluation of various chelating agents on the basis of extraction efficiencies concluded that complexing ability of DTPA is slightly higher than EDTA but poor biodegradability and less recovery of DTPA makes this chelating agent less feasible to use. NTA showed its maximum 65% Ni extraction. Less extraction of Ni using NTA is due to its quadridentate structure which causes lesser stability of Ni-NTA complex. Characterization of spent catalyst and chelating agents was performed to understand the structural and phyisco-chemical properties.

KEYWORDS

Chelation, spent catalyst, chelating agent, De-chelation

INTRODUCTION

The industrial era has seen a sharp increase in both the amounts and variety of catalysts that find applications in industry. Many catalysts, used in chemical, petrochemical, and pharmaceutical process, contain valuable precious metals. Most of these catalysts are supported on alumina/silica with varying percent of metal; metal concentration could vary from 2.5 to 30 % *(Siemens et al, 1986)*. Metals like Ni, Mo, Co, Rh, Pt, Pd, etc., are widely used in catalysts. These catalysts lose their activity gradually at the time of process reactions and after certain cycles, it has to be replaced by fresh catalyst. The deactivated catalyst is considered as 'spent catalyst'. The Environmental Protection Agency (EPA) added spent hydrotreating catalyst (K171) and spent hydrorefining catalyst (K172) to its hazardous waste list in August 1998 because of their self-heating behavior and toxic chemical content *(USEPA, 1998)*. Metals such as Cd, Cr, Pb, Cu, Ni and V are included in the list of potentially hazardous wastes published by the Environment Canada *(Rapaport, 2000)*. The mobilization of heavy metals from waste catalyst is, therefore, important to prevent environmental pollution and to conserve the resources. A variety of processing approaches have been proposed heretofore to seek solution for the aforementioned problem.

The metal recovery from spent catalyst can be achieved by chelation technology. Chelation, a hydrometallurgical method, is the formation or presence of two or more separate coordinate bonds between a polydentate ligand and a single central atom *(McNaught &Willkinson, 1997)*. The three most commonly used chelating groups are polyphosphate, hydroxycarboxylic acid, and amino polycarboxylic acid *(Peters, 1999)*. Amino polycarboxylic acid chelating agents (Ethylenediamine tetraacetic acid (EDTA), nitrilotriacetic acid (NTA), diethylene triamine pentaacetic acid (DTPA)) are used most frequently because of their strong binding capacity with metal ions than polyphosphates. These chelating agents maintain their sequestering ability over a wider pH range than hydrolytic acid types *(Pastor et al, 2007)*. Many researchers have studied the effect of chelating agent on metal extraction from soils and catalyst. *Lim et al, 2004* demonstrated better extraction of heavy metals (Pb and Cd) from the acidic soil using chelation technology than acid leaching. *Elliott & Brown, 1989* compared the ability of EDTA and NTA to solubilize Pb from a highly contaminated soil collected from a battery recycling facility. *Goel et al, 2009* studied the exchange behavior of nickel towards EDTA from primary reforming waste catalyst used in fertilizer industry and reported 95% extraction of Nickel with 0.8M concentration of EDTA, solid to liquid ratio 1:50 (g/ml), particle size 100µm, pH 11, 10 h of chelation time, 700rpm and 100°C. *Vuyyuru et al, 2010* developed an innovative combination of chelation and dechelation as a method for Ni recovery and achieved 95% Ni recovery in the nickel extraction carried out under hydrothermal condition in an autoclave at temperatures of 150°C over a 4h period. The major advantages of using chelation technique are reusability of chelating agent, no hazardous waste generation, no corrosive environment and eco-friendly process.

The present investigation deals with the utilization of aminoploycarboxylate chelating agents to extract nickel from primary reforming spent catalyst of fertilizer industry. Reusability of chelating agents and characterization of recovered chelating agent were also investigated.

EXPERIMENTAL SECTION

Spent Catalyst

The spent catalyst used in this study was provided by Shriram Fertilizers and Chemicals (SFCL), Kota, India. The spent catalyst contained almost 25-28% Ni. Catalyst was calcined in atmospheric oxygen flow in a temperature controlled muffle furnace for 5 h at 550°C. Characterization techniques were adopted for the physico-chemical evaluation of spent catalyst.

Experimental Procedure

Experiments were carried out in atmospheric reflux condition by stirring slurry of spent catalyst particles in the extractant solution (a solution of chelating agent with sodium hydroxide to make it alkaline) with gradual heating. After a certain reaction time, solution was filtered and the residue was sent for the analysis. Sulfuric acid (98%) was added dropwise in the filtrate to reach pH-2 for dechelation mechanism. Precipitated chelating agent was again filtered and solution of $NiSO_4$ was analysed. Effect of process parameters (reaction time, solid to liquid ratio, molar concentration of chelating agent, pH, reaction temperature and particle size) on metal extraction was studied.

$Na_2EDTA + Ni \longrightarrow [EDTA\text{-}Ni] + 2Na^+$ (Chelation)............ (1)

$[EDTA\text{-}Ni] + H_2SO_4 \longrightarrow NiSO_4 + H_2EDTA$ (Dechelation).......... (2)

RESULTS AND DISCUSSION

Characterization of spent catalyst

Scanning electron microscopy (SEM) was performed to ascertain the morphology of nickel on the surface of the alumina. Clustered and non uniform distribution of nickel was seen for spent catalyst due to coke formation at the surface of the catalyst whereas residue of the catalyst contained small number of clusters which explains the extraction of nickel. The same distribution of Ni was also brought out in the literature *(Goel et al, 2010).* Energy Dispersive X-Ray Analysis (EDX) confirms the presence of nickel in spent catalyst and the quantitative analysis showed around 26% Ni content in the sample. Predominance of alumina peaks and very small peak for Ni in EDX spectrum of residue confirms the metal recovery from spent catalyst. X-Ray Diffraction analysis showed sharp crystalline peaks of nickel oxide and nickel aluminate due to long term use of catalyst at high temperature in industries (*Stefanescu et al, 2007).* The surface area measurement was performed using BET surface area analyzer. Surface area for the spent catalyst was 2.68 m^2/g while, for residue, it was 3.80m^2/g. The large surface area of the residue explains the extraction of nickel and availability of more free sites.

Effect of Reaction time on Nickel extraction:

The effect of heating time on nickel recovery was observed by doing a series of digestion experiments with 0.8 M chelating agent and particle size 100 μ for a period of 2 to

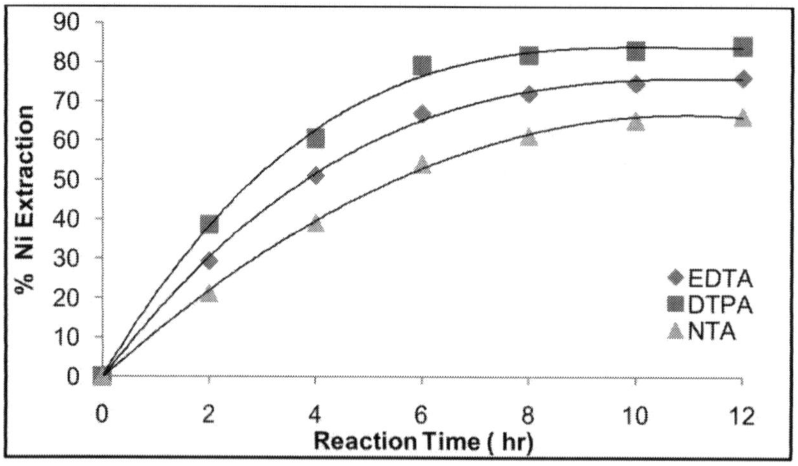

Figure: 1 Effect of Reaction time on metal extraction

12 hours. It was noticed that equilibrium time was 8 hr for EDTA while for DTPA, the equilibrium was obtained at nearly 6 hr reaction time. NTA achieved equilibrium condition at

9-10 hr reaction time. Prolonging the contact time is accompanied by a noticeable increase in the metal removal which means that nickel recovery is a rate process depending on the time of the reaction.

Effect of Molar concentration of Chelating Agent

The effect of concentration of different chelating agent on the recovery of nickel was also studied using different concentrations of chelating agents from 0.2M to 1M. It is indicated that excess reactant pushes the reaction forward due to reversibility of the reaction. From Fig. 2, it is observed that with the increase of chelating agent concentration

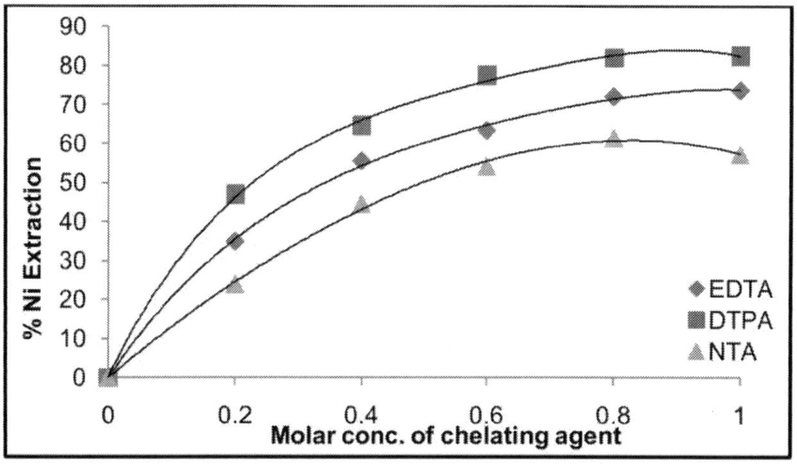

Figure: 2 Effect of molar concentration of chelating agent on metal extraction

from 0.2M to 1M, there is gradual change in the percentage extraction of nickel for each chelating agent. Maximum extraction of nickel was obtained using DTPA which could be explained by the number of binding sites of chelating agent. EDTA has six binding sites (4 acid sites and two amines sites) that make coordination bond with metal, while DTPA has eight sites (five acid and three amine sites), which shows stronger binding capacity and more extraction capacity than EDTA, but cannot be recovered easily due to a complicated process to break the coordination bond. Effect of less number of binding sites (3 acid sites, 1 amine site) and subsequently less number of rings in Ni-NTA complex could be seen in the form of lesser metal extraction.Extraction of Ni was not significant at lower concentration (0.2 M) of chelating agent. This can be illustrated by the low concentration of the ligand compared to the mobilized nickel *(Goel et al., 2010)*.

Effect of Catalyst to extractant solution ratio

The influence of solid-liquid ratio on Ni desorption was investigated for 1:10 to 1:50. It was noticed from the results that the removal efficiency of chelating agent was increased with an increase in solid:liquid ratio. With an increase of extractant volume, chelating agent concentration in the solution increases which leads to more Ni extraction and is in accordance with Le-Chatelier Principle. Decrease in Ni extraction with a stoichiometric excess of NTA could be explained by the quadridentate nature of NTA. Ni complexation leaves two metal coordination sites available for interaction with the catalyst surface thus readsorption of the NiNTA- could occur through the coordination sphere of the metal *(Elliot & Brown, 1989)*.

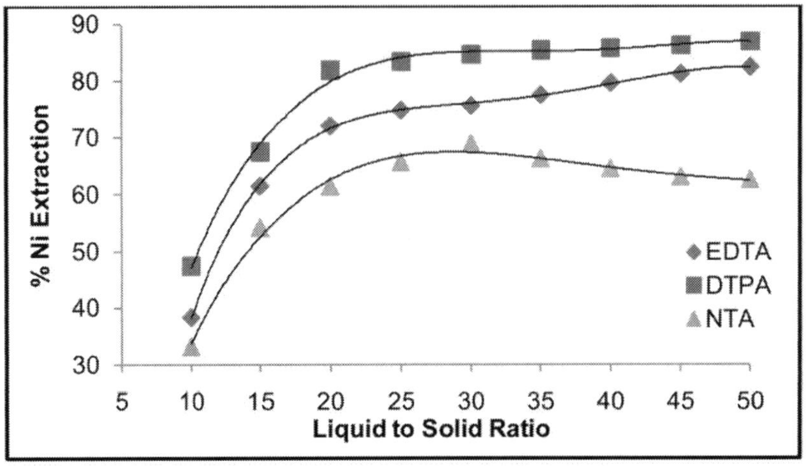

Figure: 3 Effect of catalyst to extractant volume on metal extraction

Effect of Reaction Temperature

In the present study, reaction was investigated first at room temperature and then the temperature of reactor varied from 50°C to 100°C. From Fig. 4 it is observed that, at room temperature, no any significant metal recovery could be seen. A steady rate of increase in extraction of nickel with increase in temperature suggests the kinetically controlled reaction. Results were in good agreement with *Vuyyuru et al, 2010* where the reaction was performed at autogenous condition using EDTA as chelating agent.

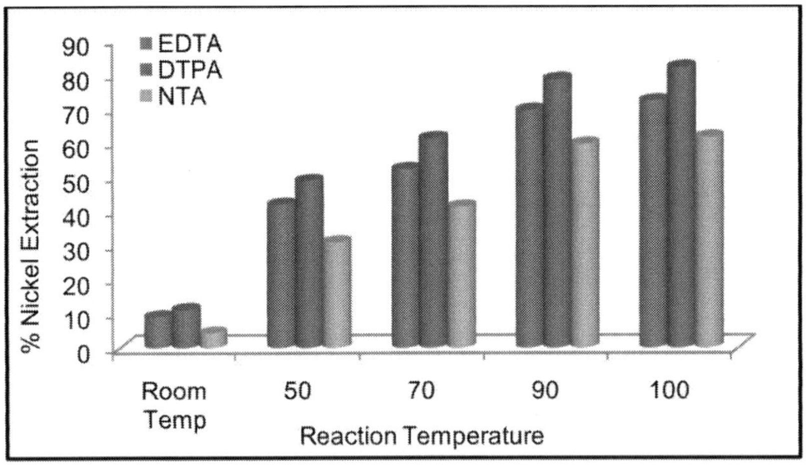

Figure: 4 Effect of Reaction Temperature on Metal Extraction

Effect of pH on Nickel Extraction

Experiments were conducted to investigate the effect of pH on metal extraction. For all the three chelating agent, effect of pH could be clearly seen in the Fig. 5. At pH 11, higher metal extraction was achieved for EDTA and DTPA than at lower pH. It could be explained by the protonated particle surface due to increase in H^+ ion concentration and thus sorption of metal-chelant complexes promoted (*Hauser et al, 2005)*. In case of NTA, decrease in metal extraction was observed at higher alkaline condition (pH 11) which illustrated the weaker complexing ability of NTA than other specified chelating agent to make it less able to prevent hydrolysis *(Elliot & Brown, 1989)*.

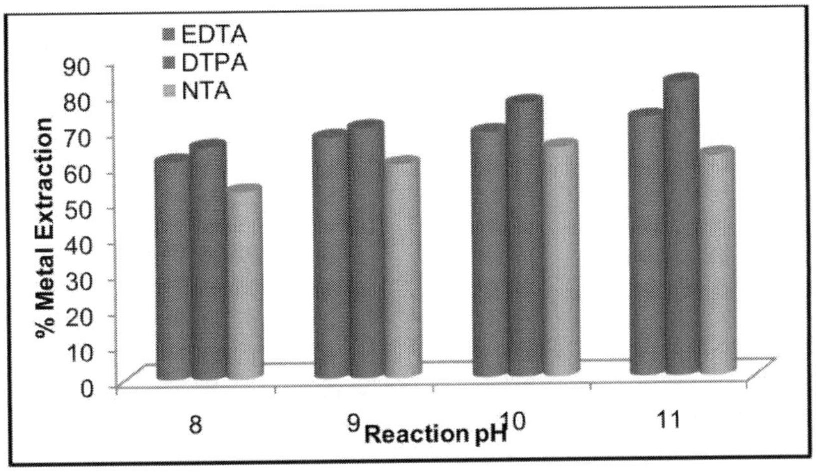

Figure: 5 Effect of pH on Metal Extraction

Effect of particle size on Nickel Extraction

The influence of particle size was studied using a particle range of 100–800µm. It was illustrated that higher % of Ni was extracted using smaller particle size for all three chelating agent. It was depicted from the Fig. 6 that at particle size lower than 300µm with stirring speed 700 rpm, external mass-transfer resistance can be ignored. Internal mass-transport resistance (diffusion in pores of the particles) can also be neglected, at lower particle size.

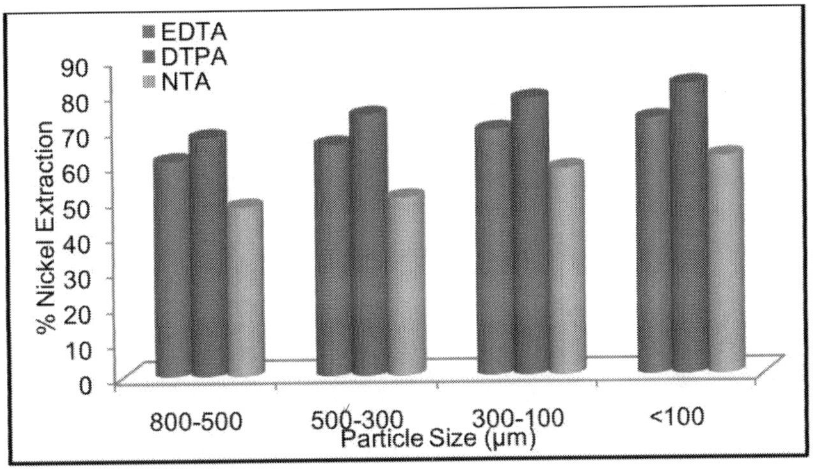

Figure: 6 Effect of particle size on Metal Extraction

Reusability of Chelating Agent

Chelating agents were recovered by addition of sulphuric acid to the metal-chelate complex. Recovery of EDTA and NTA were 93-98% in all cases in just 6 hours while recovery of DTPA was not more than 40% due to its higher complexing abilities. For better recovery of DTPA, dechelation was performed at 4°C for 3 days and recovery was improved upto 80-86%. These recovered chelating agents were recycled back to the process. Reusability of chelating agent was investigated and a consistent loss in performance was observed from about 72% to 57% for the fourth EDTA recovery cycle. The loss in efficiency of EDTA can be attributed to the change from disodium salt of EDTA (fresh) to protonated EDTA *(Vuyyuru et al, 2010)*. DTPA and NTA also showed loss in

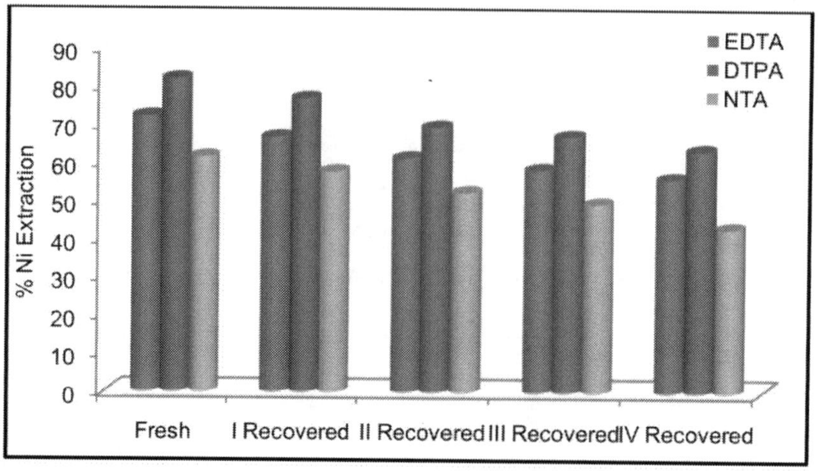

Figure: 7 Reusability of Chelating Agent

performance from 82% to 63% and 62% to 43% respectively due to difference in moisture content between fresh and recovered chelating agent.

Characterization of Chelating Agent

To investigate any possible change in the structure of the EDTA and recognize any impurities after regeneration different analysis methods were adopted, namely, (1) TGA (thermo gravimetric analysis), (2) EDX (energy-dispersive X-ray) analysis, (3) FT-IR (nuclear magnetic resonance) spectroscopy.

Thermogravimetric Analysis (TGA)

Thermo gravimetric analysis was performed in the range of 50-450 °C. Figure 8 shows the TGA curve for fresh and recovered EDTA. Fresh EDTA shows the initial weight loss with temperature until about 200 °C, which was due to water content (about 10% weight loss) whereas for the recovered EDTA, it is about 8%. Further loss in weight was due to oxidation of EDTA to CO_2, water, and nitrogenous compounds. The second

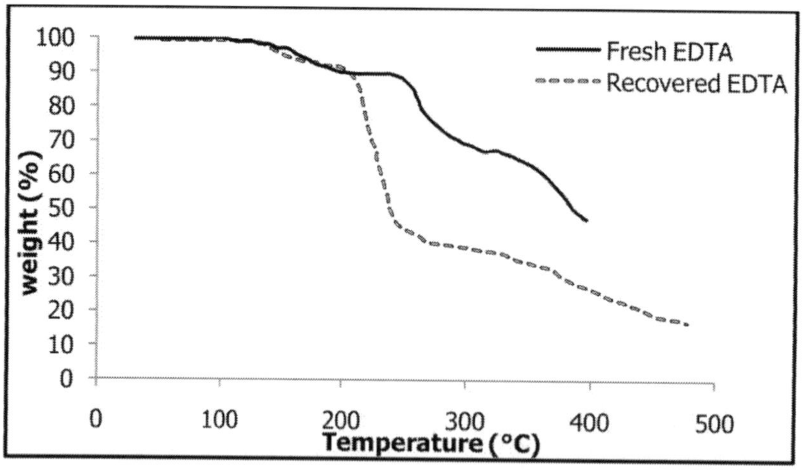

Figure: 8 Thermogravimetric Analysis for EDTA

transition commences at about 240 °C and ends at 320 °C, with a weight loss of an additional 20%, however, in case of recovered EDTA, the second transition commences at about 200°C and ends at 280°C, with a weight loss of an additional 52%. Similar thermogravimetric curve was also reported in literature *(Goel et al., 2009; Vuyyuru et al, 2010)*.The recovered EDTA is protonated and thus is expected to lose more weight at lower temperatures. The larger sulfur content in the recovered EDTA is also a major contributor to the extra weight loss, as the sulfur will be oxidized to SO_2. Higher content of sulphur in the case of recovered EDTA reduces the efficiency of chelating agent in the subsequent recovery steps. Similar kinds of observations were seen for NTA and DTPA also. (Figures for NTA and DTPA are not shown here due to space constraint.)

Energy Dispersive X-Ray Analysis (EDX)

EDX studies were performed for fresh and recovered chelating agent. The scanned graphs show the different elements present in EDTA and NTA. The graphs of the recovered chelating agent indicate an increasing amount of sulfur, as sulfuric acid was used for dechelation, leading to sulfur-containing groups on the surface. The differences in Ni extraction efficiency after subsequent chelating agent recovery could be correlated with the continued increase in sulfur content after each recovery step.

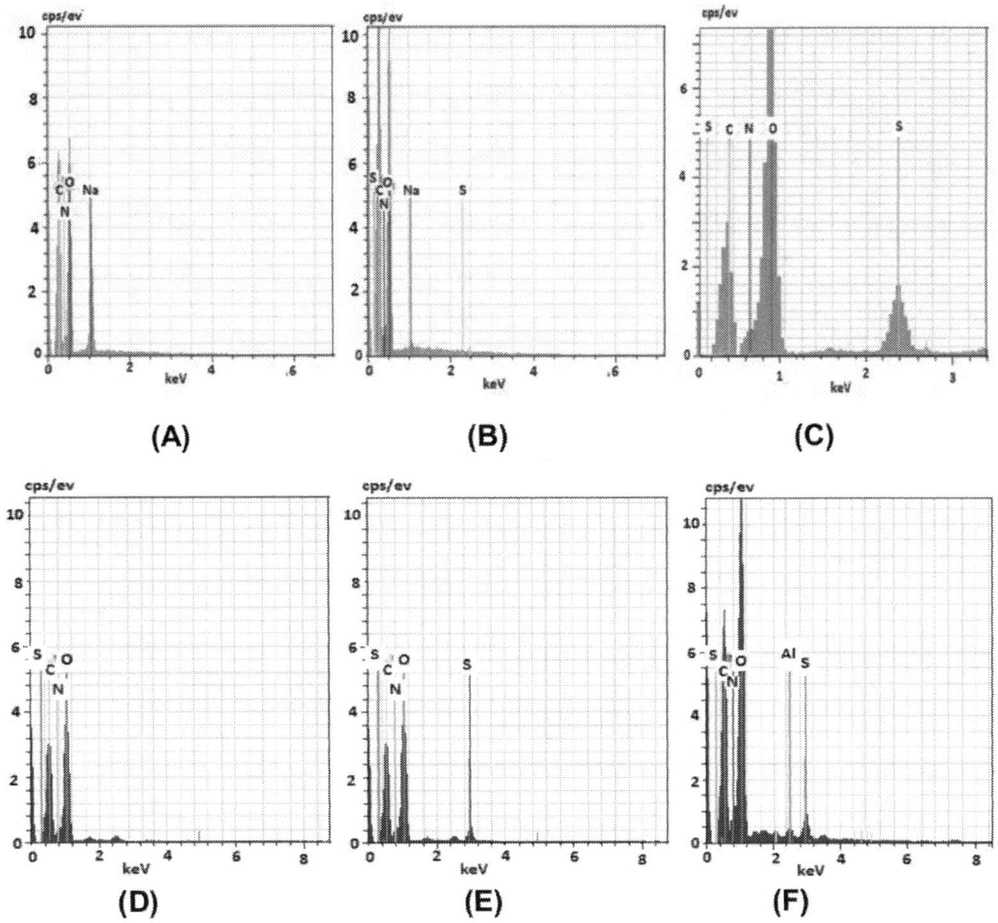

Figure: 9 EDX Analysis for (A) Fresh EDTA (B) I Recovered EDTA (C) II Recovered EDTA (D) Fresh NTA (E) I Recovered NTA (F) II Recovered NTA

Another factor that is clear from EDX analysis is that, after even one use, the Na content is trifling and in the second cycle, all of the Na available in the EDTA has disappeared. The protonated EDTA has a different molecular weight in comparison to the disodium salt of EDTA, thereby leading to differences in the measured molarity of the once-regenerated EDTA.

Fourier Transform Infrared Spectroscopy

Characterization of chelating agents and its recovered form was carried out by FT-IR spectroscopy also to identify structural characteristics of chelating agent. Aminopoly-carboxylate chelating agent was mixed with KBr and pellets were made for the analysis. The region between 3600 and 3200 cm^{-1} is generally attributed to O-H and N-H stretching vibrations. For hydrogen bonded dimeric carboxylic acids, a downward shift of O-H stretching bands was observed near 3000 cm^{-1} for fresh and recovered chelating agents. Bands between 3600 and 3000 cm^{-1} indicating the presence of hydroxyl groups as a result of hydrogen bonding and complexed water molecules as well as N-H stretching. These are shown by all the three spectrum. Broad vibrational bands were found between 2700 and 2500 cm^{-1} which were assigned to overtune and combination bands of C-O stretching and O-H bending enhanced. Hydrogen bonding tends to decrease the frequency of the carbonyl vibration of free carboxylic acid groups and increase the frequency of the anti-symmetric carboxylate stretching vibration. The band at 1711 cm^{-1} confirms the presence of saturated aliphatic carboxylic acid in fresh and recovered EDTA. For carboxylic acids and carboxylates, this region includes bands for O-H bending, C-O stretching, and C-H bending vibrations. (Figures are not shown here due to space constraint.)

CONCLUSION

The data reported here is significant demonstration of the use of chelation chemistry as an alternative to traditional acid leaching method, as applied to metal recovery from a spent catalyst. It was concluded that DTPA could be a considerable chelating agent for metal extraction from spent catalyst but poor biodegradability and less recovery due to high complexing abilities make DTPA less favorable chelating agent. Optimum condition established through one stage chelation approach could extract 72 %, 82% and 65% of Ni with EDTA, DTPA and NTA respectively. Characterization of catalyst and chelating agent confirms the extraction of nickel from catalyst successfully.

REFERENCES

1. Elliott, H.A., Brown, G.A., 'Comparative evaluation of NTA and EDTA for extractive decontamination of Pb-polluted soils', Water Air Soil Poll. 45 (1989) 361-369.
2. Goel, S., Pant, K.K., Nigam, K.D.P., 'Extraction of nickel from spent catalyst using fresh and recovered EDTA', Journal of Hazardous Materials 171 (2009) 253-261.
3. Goel, S., Gautam, A., 'Effect of chelating agents on mobilization of metal from waste catalyst', Hydrometallurgy 101 (2010) 120–125.

4. Hauser, L., Tandy, S., Schulin, R., Nowack, B., 'Column extraction of heavy metals from soils using the biodegradable chelating agent EDDS', Environ. Sci. Technol. 39 (2005) 6819–6824.

5. Kobayashi, J., Morii, F., Muramato, S., 'Removal of Cadmium from Polluted Soil with the Chelating Agent EDTA', Conference on trace substances in environmental health, Columbia, 8 (1974) 179.

6. Lim, T.T., Tay, J.H., Wang, J.Y., 'Chelating-Agent-Enhanced Heavy Metal Extraction from a Contaminated Acidic Soil', J. Envir. Engrg. 130 (2004) 59-66.

7. McNaught A. D., Wilkinson, A., IUPAC Compendium of Chemical Terminology, 2nd ed. the 'Gold Book', Blackwell Scientific Publications, Oxford (1997).

8. Means, J. L., Kucak, T., Crerar, D. A., 'Relative degradation rates of NTA, EDTA and DTPA and environmental implications', EnViron. Pollut. B *(*1980) 45–60.

9. Pastor, J., Aparicio, A. M., Gutierrez-Maroto, A., Hernandez, A. J., 'Effects of two chelating agents (EDTA and DTPA) on the autochthonous vegetation of a soil polluted with Cu, Zn and Cd', Sci. Total EnViron. *378* (2007) 114–118.

10. Peters, R. W., 'Chelant extraction of heavy metals from contaminated soils', *J.* Hazard. Mater. *66* (1999) 151–210.

11. Rapaport, D., 'Spent hydroprocessing catalysts listed as hazardous wastes', Hydrocarb Process 79 (2000) 11–22.

12. Siemens, R.E., Jong, B.W., Russell, J.H., 'Potential of spent catalysts as a source of critical metals', Conserv Recycling 9 (1986) 189–196.

13. Stefanescu, A., Veen, A.C.V., Brunel, E.D., Mirodatos, C., 'Investigation of a Ni-based steam reforming catalyst developed for the coating of microstructures' Chem. Eng. Sci. 62 (2007) 5092–5096.

14. United States Environmental Protection Agency (USEPA), 'Hazardous waste management system: Identification and history of hazardous waste; petroleum refining process waste', Federal Register 63 (1998) 151.

15. Vuyyuru, K.R., Pant, K.K., Krishnan, V.V., Nigam, K.D.P., 'Recovery of Nickel from Spent Industrial Catalysts Using Chelating Agents', Ind. Eng. Chem. Res. 49 *(*2010) 2014–2024.

(369a) Modeling Aggregation and Size Distribution of Nanoparticles In Aqueous Suspensions

Haoyang H. Liu, Chemical and Biomolecular Engineering, University of California, Los Angeles, Los Angeles, CA, Sirikarn Surawanvijit, Chemical&Biomolecular Engineering, University of California, Los Angeles, Los Angeles, CA, G. Orkoulas, Department of Chemical and Biomolecular Engineering, University of California, Los Angeles, Los Angeles, CA and Yoram Cohen, Chemical & Biomolecular Engineering, University of California, Los Angeles, Los Angeles, CA

The size distribution of the engineered nanoparticles (eNMs) and their aggregates in aqueous suspensions is crucial for predictions of the fate and transport of nanoparticles in the environment. Diffusion, sedimentation and interaction of nanoparticles with solid surfaces (including sedimentation), as well as living organisms are all affected by the size distribution of eNMs. In order to obtain the nanoparticle size distribution over the expected ranges of water chemistry and nanoparticle properties, predictive models are needed to assess the aggregation tendency of eNMs. In the present work, a constant-number kinetic Monte Carlo based model was developed to predict the aggregation rate and size distribution of nanoparticles. The modeling approach followed the "particles in a box" simulation method while allowing for particle sedimentation. The likelihood of nanoparticle aggregation was determined, based on the classical DLVO theory, via calculation of the aggregation probability of nanoparticle pairs. In developing the model, the reasonable minimum number of simulation particles and simulations instances were first evaluated to ensure convergence of the stationary statistics, demonstrating the ability to use parallel computing to dramatically decreasing simulation time. Model performance with respect to the predicted size distribution of eNMs was evaluated via comparison with experimental dynamic light scattering (DLS) measurements and with available literature data under varying conditions (e.g., pH, ionic strength, size distribution of the primary nanoparticles and nanoparticle surface charge). Model performance compared favorably with experimental DLS data demonstrating the potential of the present modeling approach for environmental analysis of the transport and fate of nanoparticles

See more of this Session: Environmental Implications of Nanomaterials: Fate and Transport
See more of this Group/Topical: Topical C: Environmental Aspects, Applications, and Implications of Nanomaterials and Nanotechnology

(369b) Stability of Engineered Nanoparticles In Aqueous Systems: Elucidating the Roles of Capping Agents and Natural Organic Matter

Jeffrey A. Nason, School of Chemical, Biological and Environmental Engineering, Oregon State University, Corvallis, OR

The rapid development of novel nanomaterials and their incorporation into consumer products has not been paralleled with an equal effort investigating the possible environmental implications. Increasingly, the properties of engineered nanoparticles (ENPs) are tailored for specific applications through the use of organic capping agents. Furthermore, when released into the environment it is probably that ENPs will interact with natural organic matter (NOM). The objective of this work is to develop an improved understanding of the roles of synthetic capping agents and NOM in controlling the aggregation of engineered nanoparticles (ENPs) in aquatic environmental systems. Using a suite of ligand-stabilized gold nanoparticles (AuNPs) and various NOM isolates, the specific aims of the project are to: (1) Correlate the physicochemical properties of capping agents with interactions between capped ENPs in aquatic systems; and (2) Correlate the physicochemical properties of capping agents and NOM with interactions between capped ENPs and NOM in aquatic systems. ENP stability with respect to aggregation is being assessed using the zeta potential and aggregation rates measured using time-resolved dynamic light scattering. Results indicate that NOM is adsorbed onto the AuNP surfaces, alterning surface charge and providing stability with respect to aggregation. Different classes of NOM stabilize AuNPs to different extents. The specific mechanisms of NOM-AuNP interactions are being probed using advanced surface and solution phases spectroscopic techniques (e.g., fluorescence, Raman, and EDX). By using a group of well-characterized ENPs that vary only with respect to their surface functionality, we aim to correlate environmental behavior with ENP properties in structure-activity type relationships. For example, results indicate that the stabilizing effects of NOM are related to the molecular weight of the specific NOM fractions. This work will provide an improved ability to predict environmental fate based on NP characteristics and yield the feedback necessary for the design of safer nanomaterials.

See more of this Session: Environmental Implications of Nanomaterials: Fate and Transport
See more of this Group/Topical: Topical C: Environmental Aspects, Applications, and Implications of Nanomaterials and Nanotechnology

(369c) Removal of Nanoparticles In Semiconductor Manufacturing Effluents Using Porous Media Filtration

Jeff Rottman, Reyes Sierra and Farhang Shadman, Chemical and Environmental Engineering, University of Arizona, Tucson, AZ

The use of engineered nanoparticles in industrial manufacturing has rapidly increased due to their unique electronic, optical, thermal and photoactive properties. Nanoparticles are used in various fields including textiles, biomedical applications, food and agriculture and semiconductor manufacturing. Frequently used nanoparticles include titania (TiO_2), which is commonly used in pigments and as a photocatalyst, and silica (SiO_2), which is commonly used as an additive in chemical mechanical planarization (CMP) slurries in semiconductor manufacturing. Along with the benefits these nanoparticles provide, numerous concerns into the environmental, safety, and health effects of the released particles arise. Understanding the fundamental interactions of these nanoparticles with typical treatment technologies will allow for improved design of abatement technologies. Porous media filtration is a common technique used to remove particulates from water. The simplicity and ubiquity of this process makes it ideal for the removal of nanoparticles in waste streams. The principles of nanoparticle transport in porous media also relate to the environmental fate of these nanoparticles in soil and ground water.

Custom nanoparticles of silica have been developed which have a fluorescent core and controllable final size. These particles, due to their fluorescence, can be measured at significantly lower concentrations than in current studies. Additionally, the ability to control the final size of the particles provides a true comparison of size effects on nanoparticle retention. A test bed, equipped with real-time and on-line spectrometry, was developed and used to study the dynamics of capture and removal of both TiO_2 and these SiO_2 nanoparticles from wastewater. Studies were conducted comparing the effect of key solution, nanoparticle, and bed characteristics, including pH, ionic strength, the presence of surfactants, heavy metals, and dispersants, as well as particle and collector sizes. A process model is under development for analysis of data from this novel experimental method and for elucidating the fundamental parameters affecting nanoparticle retention and removal in such systems.

See more of this Session: Environmental Implications of Nanomaterials: Fate and Transport
See more of this Group/Topical: Topical C: Environmental Aspects, Applications, and Implications of Nanomaterials and Nanotechnology

(369d) Adsorption of Gold Nanoparticles and Humic Acid On Activated Carbon Used In Drinking Water Treatment

Holly A. Stretz, Department of Chemical Engineering, Tennessee Technological University, Cookeville, TN, Vasanta Pallem, Department of Chemical Engineering and Center for the Management, Utilization and Protection of Water Resources, Tennessee Technological University, Cookeville, TN and Martha J. M. Wells, Department of Pharmaceutical Sciences, Midway College School of Pharmacy, Paintsville, KY

Granular activated carbon (GAC) is primarily used in drinking water treatment processes for removing dissolved organic matter comprising of humic and fulvic acids (DOM), and other pollutants. The increasing applications of gold nanoparticles (GNPs) in diagnostic imaging, chemical and biological sensing, also present potential concerns for adsorption and removal of GNPs in water treatment units, following post-clinical discharge. The current study employs spectroscopic separation to evaluate the extent of adsorption of citrate coated GNPs with core size 50 nm (GNPs-50), and HA separately and in mixtures. Modified Freundlich isotherms were used to describe the adsorption behavior of humic acid and gold in mixtures. The novelty of the approach was to examine the adsorption behavior for synergistic or antagonistic effects when all three materials (gold nanoparticles, HA and activated carbon) were present, and to carefully account for the fluorescence of the controls in determining fractional fluorescence prior to establishing the isotherms. The GNPs-50 were completely adsorbed on activated carbon when present individually, however in the presence of HA the adsorption of GNPs-50 decreased significantly. The adsorption of humic acid also decreased in the presence of GNPs-50 showing an antagonistic effect. Therefore, the mixtures of HA/GNPs-50 present significant implications for the fate and removal of GNPs and HA in water processing units.

See more of this Session: Environmental Implications of Nanomaterials: Fate and Transport
See more of this Group/Topical: Topical C: Environmental Aspects, Applications, and Implications of Nanomaterials and Nanotechnology

(369e) High Throughput Collection and Detection of Environmental Nanoparticles

Fanxu Meng[1], Serdar Ozturk[1], Maria D. King[2], Yassin A. Hassan[3] and Victor M. Ugaz[1], (1)Artie McFerrin Department of Chemical Engineering, Texas A&M University, College Station, TX, (2) Department of Mechanical Engineering, Texas A&M University, College Station, TX, (3)Departments of Nuclear and Mechanical Engineering, Texas A&M University, College Station, TX

The increasing prevalence of manufactured nanoparticles in the environment poses new risks to human health, pointing to the need for advanced nanoparticle sampling and analysis approaches capable of continuously monitoring room-sized volumes so that dynamic changes in exposure levels can be rapidly detected. Here we describe our efforts to address this need through development of a new technique that enables nanoparticles to be collected and concentrated at very high flow rates, so that they can be directly detected using a continuous-flow microfluidic approach coupled with particle sizing analysis. Nanoparticle detection is accomplished by harnessing interfacial fluorescence phenomena that uniquely emerge in a microfluidic environment. Fundamental parameters including sensitivity and specificity are characterized to establish detection limits for comparison with current-generation instrumentation. We then explore feasibility of coupling this component with an autonomous Wetted Wall Cyclone (WWC) collector developed in our laboratory and fabricated at Northrop Grumman Inc. This system is capable of performing continuous air sampling at 400 L/min with a cutpoint of 500 nm. Performance was characterized by sampling colloidal (anatase, 5-30 nm) and amorphous powder (5-30 nm) TiO_2 nanoparticles aerosolized and collected in a laboratory flowcell, and comparing the results against a standard glass fiber filter sampler. Environmental monitoring capability was assessed by sampling air in a freshly painted classroom for 24 hours at 100 L/min. Each hydrosol sample was weighed and quantitated alongside suspensions extracted from the reference filter using inductively coupled plasma mass spectrometry. This analysis indicated that the WWC system achieved collection efficiencies of 83.5% and 56.5% for aerosolized colloidal and amorphous power TiO_2 nanoparticles, respectively. The level of TiO_2 in the painted room air was around 0.2 ppt, however, SiO_2 was also detected at 1 ppb. Spectral analysis of the interfacial fluorescence offers potential to enable chemical fingerprinting to be rapidly performed in order to obtain a detailed picture of nanoparticle exposure levels while simultaneously providing the ability to respond to sudden environmental changes

See more of this Session: Environmental Implications of Nanomaterials: Fate and Transport
See more of this Group/Topical: Topical C: Environmental Aspects, Applications, and Implications of Nanomaterials and Nanotechnology

(394a) Integrating Nanocellulose Production with Biofuel

Junyong Zhu, USDA Forest Products Laboratory, Madison, WI

This study demonstrated the feasibility of integrating the production of nano-fibrillated cellulose (NFC), a potentially highly valuable biomaterial, with sugar/biofuel (ethanol) from wood fibers. Commercial cellulase enzymes were used to fractionate the less recalcitrant amorphous cellulose from a bleached kraft eucalyptus pulp resulting in a highly crystalline and recalcitrant cellulose (RC). The RC is difficult to hydrolyze to sugars but very suitable for producing biobased nanomaterials through mechanical homogenization. A range of fractionation yields of RC from 10–70% can be achieved by varying fractionation duration and enzyme dosage. The crystallinity of the RC was found to be as much as 24% higher than that of the original bleached pulp. The cellulase fractionation process facilitated mechanical homogenization by significantly reducing the degree of polymerization (DP) to about 400 and the length of the fibers to about 200 mm. The hydrolyzed sugars were found to be easily converted to ethanol through yeast fermentation with excellent efficiency of 91%. Films made from nano-fibrillated cellulose were found to be optically transparent with opacity as low as 12%. The NFC films were mechanically strong and stiff with tensile strengths and moduli approximately 10 and 6 times higher than film made from fibers that had not been nano-fibrillated.

See more of this Session: Nanocellulose-Based Materials and Composites
See more of this Group/Topical: Forest and Plant Bioproducts Division

SURFACE AND AGING CHARACTERIATICS OF PAPER COATED WITH NANO SIZE ZINC OXIDE PIGMENT

Sanjay Tyagi. Renu Tyagi* , B P Thapliyal* R M Mathur* and A K Ray***

**Central Pulp and Paper Research Institute, Saharanpur – 247001*
*** Indian Institute of Technology, (Saharanpur Campus), INDIA*

ABSTRACT

Quality of paper is improved on pigment coating in terms of surface, optical properties and printability characteristics. One of the major parameter for aqueous pigment coating is the type of pigment used and also its particle size and distribution. Zinc oxide is one of the pigment s employed for special coating application. It is offlet nano size particles are playing prominent roles in several areas ranging from medicine to electronics. In paper making there is no exception. Few specialty grades of paper are already using nano size calcium carbonates both as filler and also as coating pigments. Application of water based barrier coating in paper and board in food packaging industry have proven successful. It is also important due to substantial smaller size particles (nano range) compared to macro size solid particles (in micron range) can provide better functional properties of paper in some specialized applications. However, ZnO is present in various forms nano size ZnO particles can enhance the properties significantly to improve optical and surface properties and specially the effect of aging. In this present therefore an attempt has been made to study the potential technological applications of nano zinc oxide particles on paper surface made from hardwood furnish and bamboo using 100% nano ZnO and with admixture of coating grade China clay using laboratory rod coater. All the required properties were evaluated as per ISO standard and compared with the same base paper coated with micro size ZnO pigment particles. From the detailed study it is observed that paper coated with nano ZnO particles imparts better properties as compared to bulk ZnO in terms of brightness, whiteness, gloss, smoothness and printability. Aging properties were also thoroughly studied under artificial condition of varying temperature (50% – 120%)and humidity (10% - 90%). And also UV light, paper coated with nano ZnO showed better resistance to brightness reversion and UV protecting potential.

INTRODUCTION

Paper is pigment coated to improve the quality of paper in terms of surface, optical and printability characteristics. Paper coating ingredients which generally contains pigments, binders and additives is applied to the paper surface to improve the above properties of the paper. One of the major parameter for aqueous pigment coating is the type of pigment used and also its particle size distribution [1,2,3]. A commonly used pigment includes china clay, calcium carbonate, silica, zeolite, titanium dioxide, alumina trihydrate and polystyrene [2]. Zinc oxide is one of the pigments employed for special coating application [4]. Presently all above mentioned conventional pigments are available in the market have their particle size in the range of 1 – 10 micron [5,6]. The size of these particles is too big for the preparation of coating pigments to be used efficiently in coating machines. Evenness and uniformity in surface smoothness and thickness of coating layer which is one of the two prime requirements for producing good multi color print, generally lacks with conventional micron sized coating pigments [7,8,9]. Pore structure and contact angle wettability effect print quality by preventing ink spreading. In order to prevent wicking and feathering, it is important that the thickness of coating layer be homogeneous to a scale of few microns in depth which also helps in the absorption of successive droplets of ink at high delivery rates and any water present[9,10].

Application of nano size Zinc oxide in papermaking

While papermaking is normally a statistical process, nano technology makes use of the classic process where the position of every atom is specified and controlled. Yet despite this apparent discrepancy, between two technologies, the paper industry has been practicing nano particles over the conventional pigments long before it was marketed as latest technology trend [5,6].

Nano sized calcium carbonate has been used by paper industry since long as filler due to the advantage of high opacity, high brightness and whiteness [11]. At present, it is also used in special paper products such as baby diapers, sanitary napkins, etc.

Zinc oxide is also one of the pigments finding its use for special coating applications[4]. It is offlet nano size particles are playing prominent role in several areas ranging from medicine to electronics [12]. ZnO is essential for rubber and tier industry and also used as natural rubber, synthetic rubber and latex vulcanization active agents, reinforcing agent and coloring agents. In paint industry, ZnO in addition to color strength and hiding power outside but also a light coating of preservative and agents [13]. In medical and health and food industries, ZnO has a Sida, bleeding, hygienic convergence of functions. In electronics industry, ZnO varistor is the main raw material is magnetic, optical material such as the main additive. Prepared using ZnO varistor, not only has low sintering temperature, but also improve the performance of porous sensitive resistor such as the flux, the non linear coefficient [14,15].

In the present study, an attempt has been made to study the potential technological applications of nano zinc oxide particles on the paper surface made from the pulp of hardwood furnish and bamboo using 100% nano ZnO and with the admixture of coating grade china clay by taking trials on laboratory rod coater. The Zinc oxide soluble starch nano composite was prepared by wet chemical method using Zinc nitrate and sodium hydroxide as precursors and soluble starch as stabilizing and compared with the same base paper coated with micro size ZnO pigment particles.

EXPERIMENTAL

Coating formulation

The coating trials were carried out using Laboratory rod coater. The coated sheets were dried immediately in laboratory oven at the temperature of 105 C for maximum 2 minutes. The following two coating formulations were used for coating trials.

Ingredient	Composition I	Composition II
Pigment		
• Zinc Oxide (Nano or Ordinary)	500 parts	1000 parts
• China clay	500 parts	
Binder		

• Oxidized Starch	90 parts	90 parts
• Styrene Butadiene Latex (SBR Latex)	60 parts	60 parts
Dispersant		
• Anionic polymer	5 parts	5 parts
	Total solids maintained- 65±2%	

Preparation of coating colour

The preparation method for the coating slip is as under-

(a) Starch cooking

The oxidized starch solution was prepared separately by cooking under continuous agitation in hot water at temperature 90-95 ^{0}C, for about 15 minutes, i.e. upto the point when the whole solution became fully gelatinized.

(b) Pigment preparation

Pigment slurry was made in laboratory high-shear coating mixer. The dispersant was firstly added to desired quantity of water. When homogeneous solution was formed, weighed quantities of pigments were added, under constant stirring. A thick paste thus resulted, which became uniform at high shear. Time of agitation was, however kept at about 15 minutes.

(c) Addition of binders to pigments

The freshly prepared starch solution and synthetic binder SBR latex (in required amount) were added to the pigment paste in laboratory high-shear coating mixer, under constant agitation. The mixture was kept under constant agitation at high shear for about 30 minutes to ensure uniform mixing and dispersion of all the ingredients. Finally the slurry was taken out of the mixer.

(d) Screening

The coating slurry was next screened on a 300-mesh screen to eliminate foreign matters, coagulated coating and particles of dried coating slurry etc. Thus a uniform coating slurry; free from dirt, was ready for coating.

Calendering

Coated sheets after dried in laboratory oven were calendered using the laboratory calender with softnip mode, at 40 ^{0}C temperature & 30 bar pressure. The sheets were passed twice through the softnip.

Evaluation of optical characteristics of the pigments

The two pigments- Zinc Oxide (ordinary) and Zinc Oxide (nano particles), which were used in coating trials, were evaluated also for following optical characteristics; as an extension of the detailed study.

- Brightness (%)
- Whiteness (%)
- Yellowness (%)

Evaluation of base paper and coated paper

Base paper as well the coated & calendared papers were conditioned at 27±1 ^{0}C temperature and 65±2 % relative humidity prior to testing of following properties as per standard methods mentioned below.

Property	Standard methods
Grammage (g/m^2)	ISO 534
Brightness (% ISO)	ISO 2470

Gloss (%)	Tappi 480 om-99
Whiteness (%)	ISO 2470
PPS Roughness (micron)	ISO 8791/4
Picking velocity (cm/s)	ISO 3783
Print density (g/cm^3)	Using IGT printability tester Model AIC2-5
Print uniformity index	Using IGT printability tester Model AIC2-5
Stain length (mm), oil absorbency	Using IGT printability tester Model AIC2-5

RESULTS & DISCUSSION

Optical characteristics of base paper

The base paper used in the coating trials, was procured from commercial paper mill. The base paper was made from the pulp containing furnish of hardwood pulp and bamboo. Base paper was evaluated for different optical & printing properties for comparison purpose with coated & calendered paper. The result is tabulated in Table-I.

Optical characteristics of pigments

The optical characteristics of the pigments used in coating are given in Table-II. By comparing the values of the properties obtained for nano Zinc Oxide and ordinary Zinc Oxide, it was found that the starch-encapsulated nano Zinc Oxide particles had slightly lower brightness, whiteness and was relatively more yellowish in colour as compared to ordinary Zinc Oxide particles. This is probably due to the preparation process.

Characteristics of Coated & Calendered Paper samples

The coated paper samples, after softnip calendering, were evaluated for

(a) Optical characteristics
(b) Surface characteristics
(c) Printing characteristics
(d) Aging characteristics

(a) Optical characteristics

The effect of nano size zinc oxide particles on brightness of coated and calendered paper is given in Figure 1. Results show that nano zinc oxide gives better brightness as compared to conventional zinc oxide in both cases. i.e. when used alone and when used in combination with coating grade china clay. The same trend is true with whiteness value of coated paper (Fig. 3).

The value gloss of paper coated with nano ZnO also improved as compared to paper coated with conventional ZnO (Fig. 2). However, the improvement is low as compared to brightness and whiteness values of base paper, there is a fall in their values after coating with ZnO and China clay. The reduction in values is more in case when china clay is used in coating formulation. This is probably due to the lower brightness of china clay and binder additives which results in overall reduction of optical properties of coated paper.

(b) Surface characteristics

PPS Roughness

The surface smoothness of coated paper which is an important property in deciding the print performance of paper, improved with the application of nano ZnO. The PPS roughness in case of nano ZnO is relatively lower than conventional ZnO. The same trend was obtained when nano ZnO is used with china clay. A smooth surface was obtained when china clay is used in coating furnish. This is probably due to particle shape. China clay being platelet shape flattened more during the calendaring operation in compared to ZnO which is spherical in shape (Fig. 4).

(c) Printing characteristics

Picking velocity

Picking velocity of paper coated with nano ZnO was found to be higher as compared to conventional ZnO (Fig. 5) in both cases when used alone and in combination with china clay. This indicates that paper coated with nano ZnO can run easily on high speed printing presses or at the same printing speed, less binder is required in coating formulations with nano ZnO. The excess amount of binder always has negative effect on the printing characteristics.

Print Mottle (Print Non-Uniformity)

Print non uniformity leads to print mottle which is uneven appearance of colour density in a print. This is due to the surface factors to absorb ink evenly due to variation in base paper formation or coat weight. The non uniformity printing index is lower for paper coated with nano ZnO as compared to paper coated with conventional ZnO (Fig. 8).

The variation in sheet matrix formation formed by coating layer of conventional ZnO (due to non uniform size of ZnO particles) probably the reason for this. Paper coated with ZnO alone showed lower number as compared to paper coated with a mixture of china clay and ZnO since particle size of china clay is largely varied. The sheet matrix variation can lead to print mottle non uniform compression during calendering, resulting in variation in coating layer densification. Sheets that contain binder migration, coat weight variation, inferior coating profile and heterogeneous pigment blends can all cause mottle formation on the printed sheet.

Print density

Print density is affected by ink setting/ink holdout. For good density the vehicle of printing ink should stay on the top of paper together with ink pigment. The more the dried ink is retained on the outer surface of the paper, the higher will be the ink density, accordingly, the lower will be the ink demand. Paper with large pores in coating will have more penetration of printing ink leading the lower density. This is observed in case of papers coated with conventional ZnO coating (Fig. 7). The same is true with china clay in the coating formulation. This indicates that more ink will be required to print coated papers with conventional ZnO coatings.

Oil absorbency

Oil absorbency which is an indication of rate of penetration of oil was found to be more in the case of paper coated with nano size ZnO alone and in combination with china clay. (Fig 6)

(d) Effect of Aging

The effect of aging on the optical properties of coated and calendered paper is shown in the graph (Fig. 9-11) and tabulated in table (4-6). The results show that nano zinc oxide gives excellent UV protecting strength in both cases when used alone and when used in combination with coating grade china clay. The reduction in brightness and whiteness is minimum in the case of nano zinc oxide. The lower value of indicated the higher yellowness, which incase after aging maximum when china clay is used.

CONCLUSIONS

The use of nano zinc oxide as a coating pigment improves the paper properties in terms of optical, surface and printability as compared to conventional zinc oxide. This indicates that use of nano ZnO can be a good substitute for the replacement of conventional ZnO in the case of speciality grade papers. This opens a new scope for the use of nano Zinc oxide as a coating pigment for speciality grade of papers. The application of nano ZnO also exhibits excellent UV protection characteristics when tested at different condition of humidity and temperature. Paper represents good aging resistance properties which increases the longevity of paper. This opens a new scope of manufacturing paper with much longer life which lasts for longer years paper manufactured under connectional methods

The use of nano Zinc oxide as a coating pigment results in the paper with better properties as compared to conventional ZnO in terms of brightness, whiteness, gloss, smoothness, UV protection and printability characteristics i.e. picking velocity, print uniformity, print density, and oil absorbency .

Refrences:

1. 'Y.V.Sood, S.Tyagi. R.Tyagi, P.C.Pande, R.Tandon & T.K.Roy', 'Quality of indigenous coated paper and ways to improve it', *'J.IPPTA, 2008, 20 (4), pp'*
2. 'Coating pigments: from size and shape to nanotech', *'www.thefreelibrary.com/coating+pigments:+from+size+and+shape+to+nanotech.....'*
3. 'Surface characteristics of coating layers formed by coating pigments with different particle sizes and size distribution' *'J. Coat. Tech. and Res'*
4. 'N. Vatansever & S. Polat', 'Effect of zinc oxide type on aging properties of styrene butadiene rubber compounds', *'www.sciencedirect.com, 2009'*
5. Roxare, 'Nanotechnology in paper making', *'http://pulpandpapercircle.com/2008/07/nanotechnology –in-paper-making'*
6. 'J.Shen, Z.Song, X.Ojan, F.Yang & F.Kong', 'Nano fillers for papermaking wet end applications', *'www.bioresources.com, 2010, 5(3)'*
7. G. Chinga, 'Detailed characterization of paper surface structure for gloss assessment', *'JPPS, 2004, 30 (8), pp 222-227'*
8. G. Ström, S. E. Research, A. Englund & M. Karathanasis, 'Effect of coating structure on print gloss after sheet fed offset printing', *'J. Nord. Pulp & Paper, 2003, 18 (1), pp108-115'*
9. 'G. Chinga-Carrasco, M. Axelsson, Ø. Eriksen & S. Svensson', 'Structural characteristics of pore networks affecting print through', *'JPPS, 2008, 34 (1), pp 13-22'*
10. 'D.W. Donigian, J. N. Ishley & K.J. Wise, 'Coating pore structure and offset printed gloss', *'J.TAPPI, 1997, 80 (5), pp 163-172'*
11. T. Nypelö, M. Österberg & J. Laine, 'Ultra thin coating of paper by tailor made nanoparticles', *'TAPPI International Conf. on Nanotechnology for the Forest Product Industry, 2010'*
12. 'V. Bianchard & P. Blanchet', 'Color stability for wood products during use: Effects of inorganic nanoparticles', *'www.bioresources.com, 2011, 6(2)'*
13. 'B. Wang, N. Han, D. Meng, R. Yue, J. Yan & Y. Chen', 'Filter paper- templated preparation of ZnO thin films and examination of their gas sensing properties', *'www.sciencedirect.com, 2011'*
14. 'V. Prasad, A. J. Shaikh, A. A. Kathe, D. K. Bisoyi, A.K. Verma, N. Vigneshwaran', 'Functional behavior of paper coated with zinc oxide-soluble starch nanocomposites', *'J. Mat. Proc. Tech., 2010, 210 (14), pp 1962-1967'*
15. 'Surface characteristics of coating layers formed by coating pigments with different particle sizes and size distribution' *'J. Coat. Tech. and Res.'*

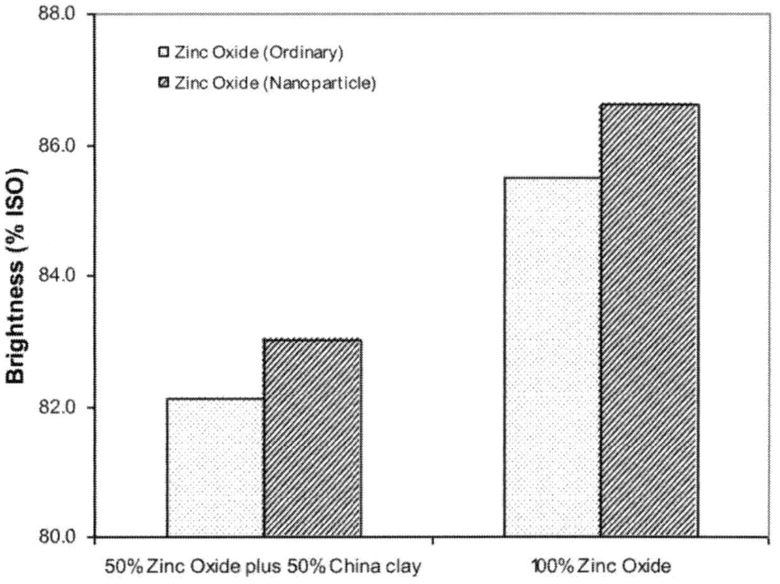

Fig. 1: Effect of Zinc Oxide nanoparticles on brightness of coated & calendered paper

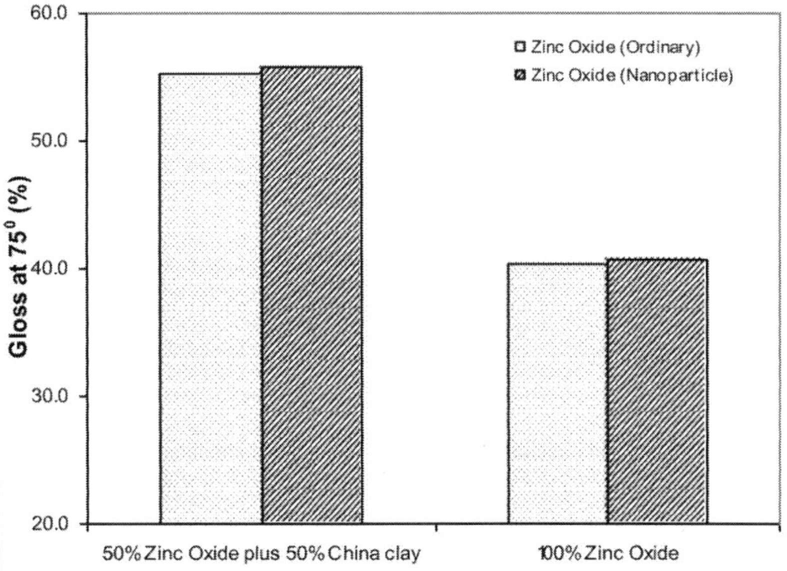

Fig. 2: Effect of Zinc Oxide nanoparticles on gloss of coated & calendered paper

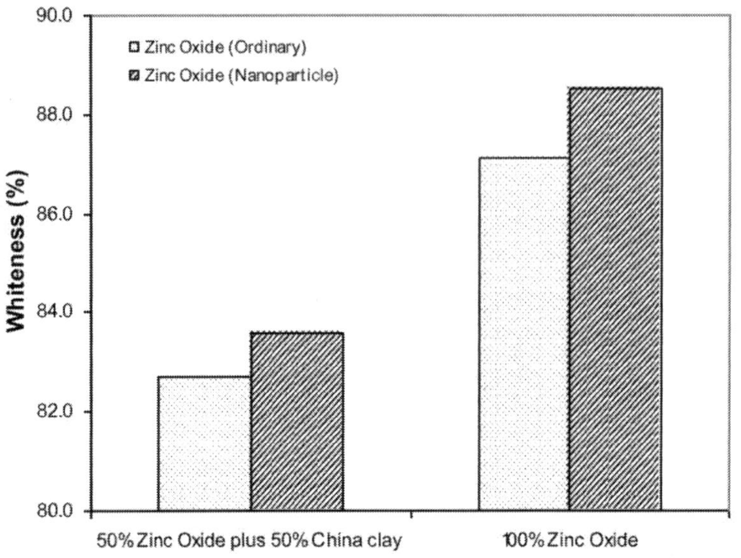

Fig. 3: Effect of Zinc Oxide nanoparticles on whiteness of coated & calendered paper

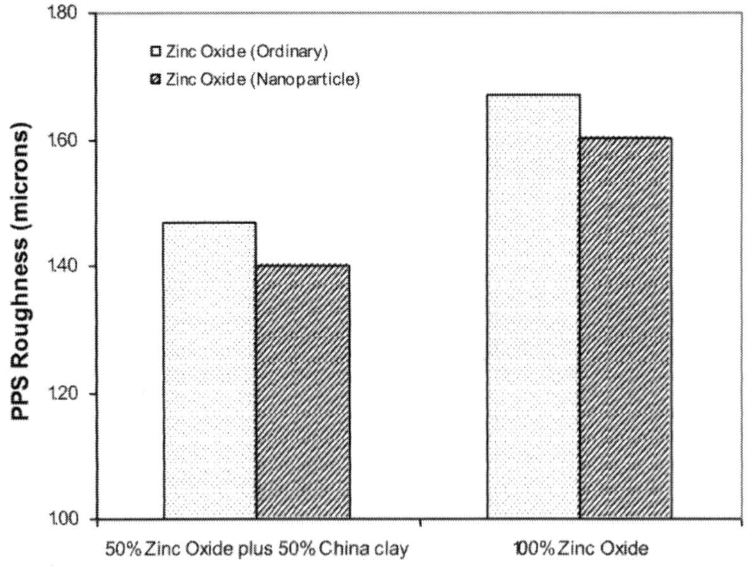

Fig. 4: Effect of Zinc Oxide nanoparticles on the roughness of coated & calendered paper

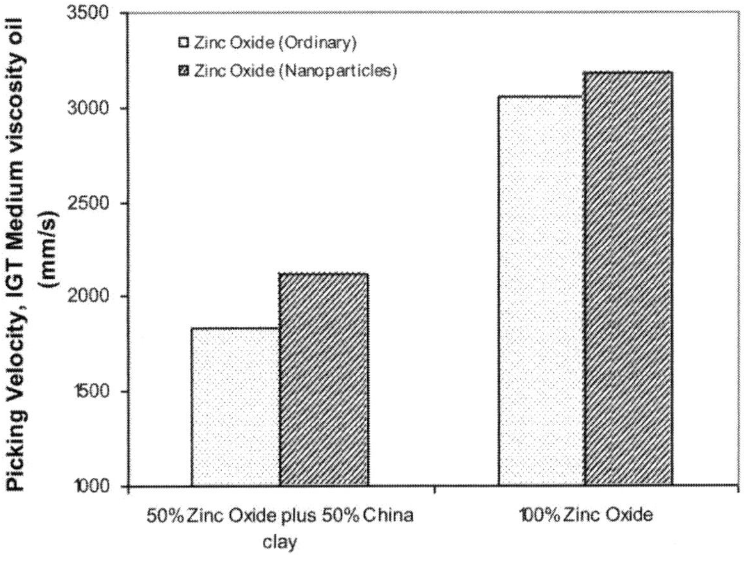

Fig. 5: Effect of Zinc Oxide nanoparticles on picking velocity of coated & calendered paper

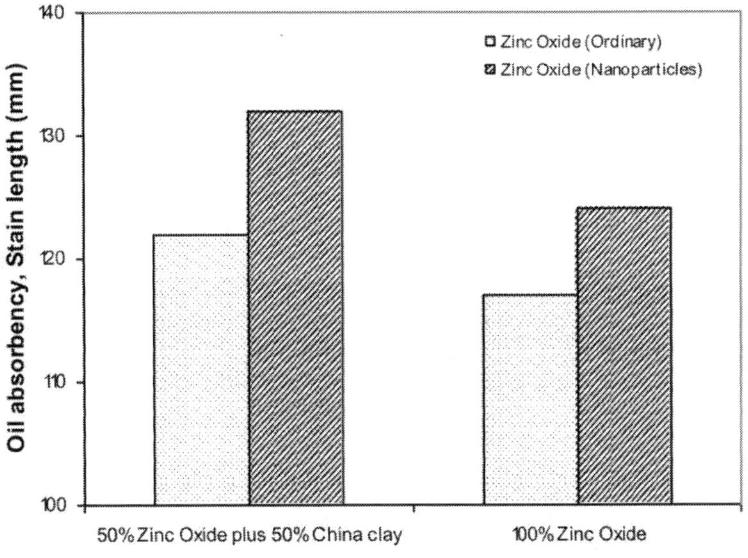

Fig. 6: Effect of Zinc Oxide nanoparticles on oil absorbency of coated & calendered paper

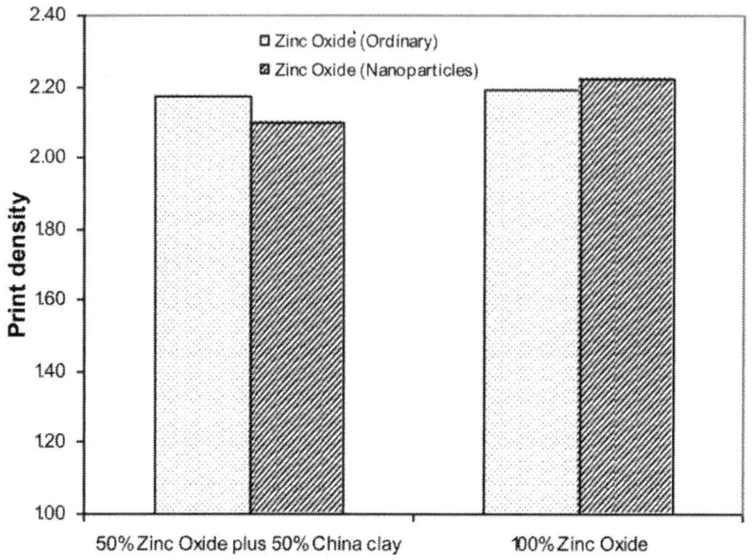

Fig. 7: Effect of Zinc Oxide nanoparticles on print density of coated & calendered paper

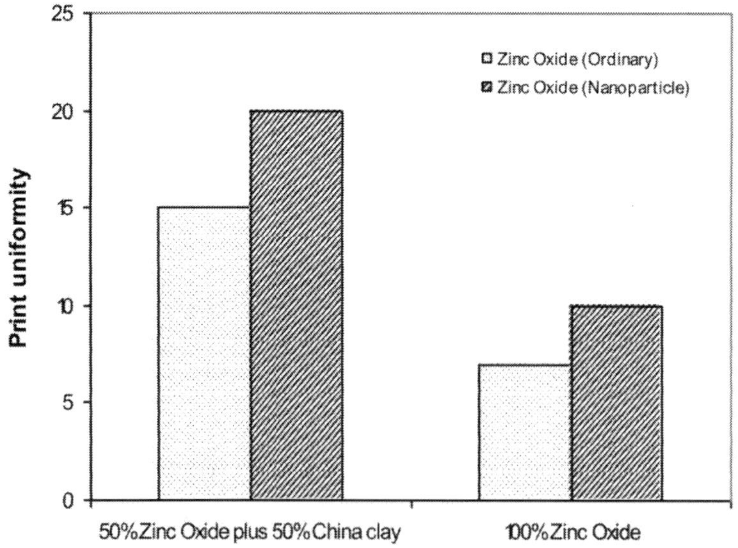

Fig. 8: Effect Zinc Oxide nanoparticles on print uniformity of coated & calendered paper

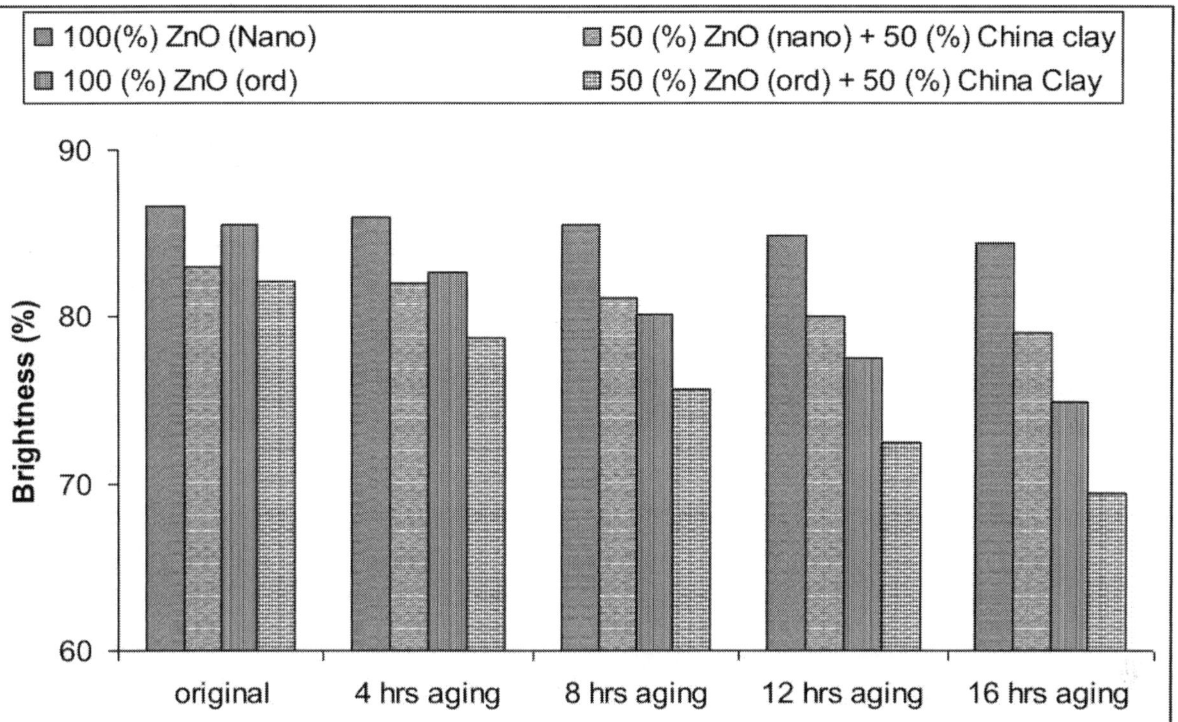

Fig. 9: Effect of zinc oxide nano particle on brightness reversion of coated and calendared paper

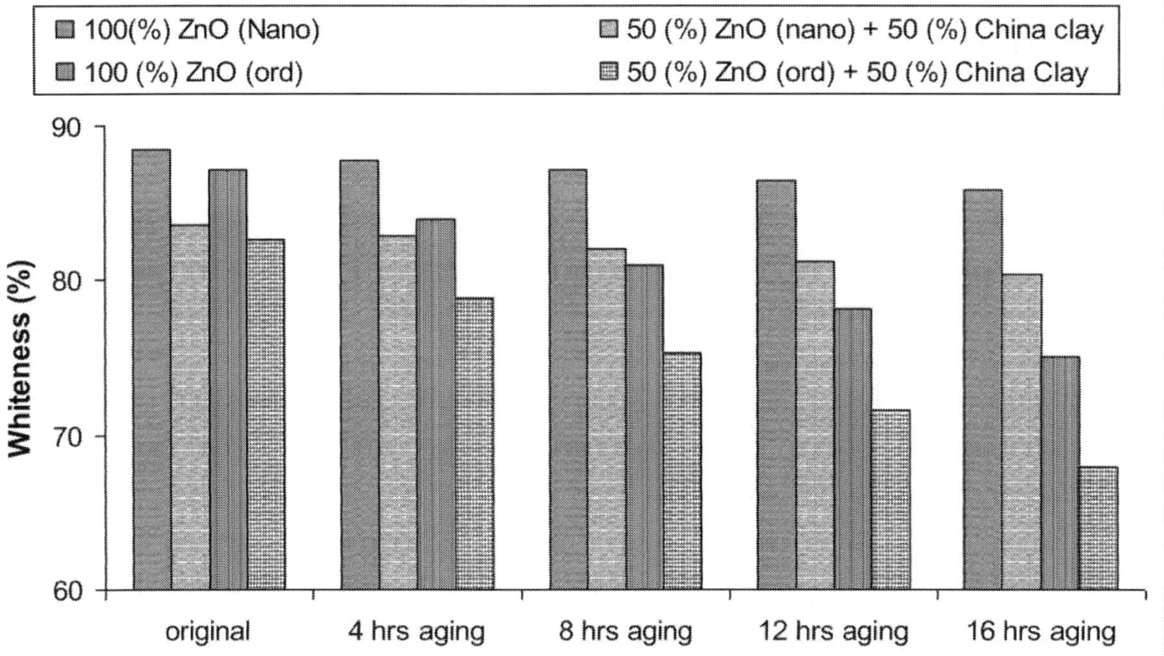

Fig. 10: Effect of zinc oxide nano particles on whiteness reversion of coated and calendared paper

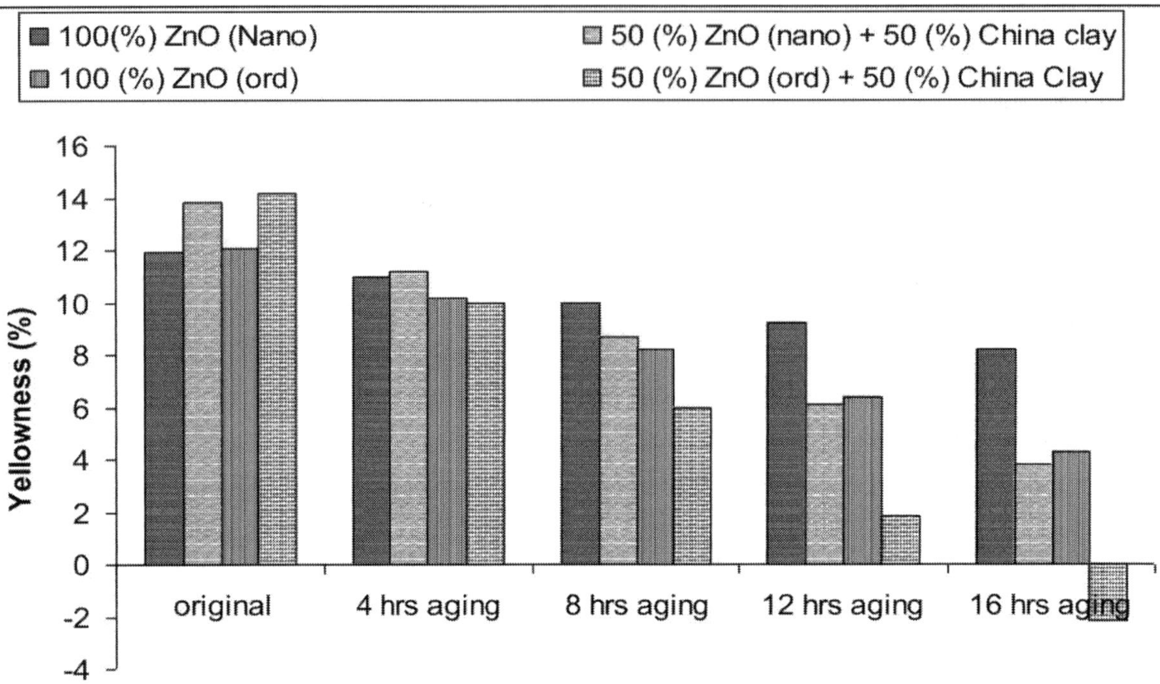

Fig.11: Effect of zinc oxide nano particle on yellowness reversion of coated and calendared paper

Table-I: Optical & printing characteristics of base paper used for coating

Property	Values obtained for the sample
Brightness (%)	88.5
Opacity (%)	80.0
Yellowness (%)	-5.13
Whiteness (%)	124.2
Gloss at 75^0 (%)	6.9
Porosity, Bendtsen (ml/min)	593

Table-II: Optical & printing characteristics of Zinc Oxide pigments used in coating

Property	Zinc Oxide (Ordinary)	Zinc Oxide (Nanoparticle)
Brightness (%)	97.8	96.2
Whiteness (%)	93.9	91.2
Yellowness (%)	1.64	1.70

Table-III: Different optical & printing characteristics of coated & calendared paper

Property	Pigment composition for coating			
	50% Zinc Oxide (Ord.) plus 50% China clay	50% Zinc Oxide (Nano) plus 50% China clay	100% Zinc Oxide (Ord.)	100% Zinc Oxide (Nano)
Coat weight (g/m^2)	12.7	12.6	12.8	12.4
Brightness (%)	82.1	83.0	85.5	86.6
Whiteness (%)	82.7	83.6	87.1	88.5
Gloss at 75°(%)	55.2	55.8	40.4	40.6
PPS Roughness (microns)	1.47	1.40	1.67	1.60
Oil absorbency, IGT Stain length (mm)	122	132	117	124

Picking velocity, IGT Medium viscosity oil (mm/s)	1830	2120	3060	3180
Picking Viscosity Velocity Product (VVP), IGT(Poise.m/s)	805	933	1346	1399
Print density	2.17	2.10	2.19	2.22
Print uniformity	15	20	7	10

Table-IV: Brightness reversion of coated and calendared paper after aging

Pigment Composition	Brightness (%)				
	Original	4 hrs aging	8 hrs aging	12 hrs aging	16 hrs aging
100(%) ZnO (nano)	86.6	86	85.5	84.9	84.4
50(%) ZnO (nano) + 50 (%) China Clay	83	82	81.1	80	79
100 (%) ZnO (ord)	85.5	82.7	80.2	77.5	74.9
50 (%) ZnO (ord) + 50 (%) China Clay	82.1	78.7	75.7	72.5	69.4

Table-V: Whiteness reversion of coated and calendared paper after aging

Pigment Composition	Whiteness (%)				
	Original	4 hrs aging	8 hrs aging	12 hrs aging	16 hrs aging
100 (%) ZnO (nano)	88.5	87.7	87.2	86.5	85.9
50 (%) ZnO (nano) + 50 (%) China Clay	83.6	82.9	82.1	81.2	80.4
100 (%) ZnO (ord)	87.1	84	81	78.1	75.1
50 (%) ZnO (ord) + 50 (%) China Clay	82.7	78.8	75.3	71.6	68

Table-VI: Yellowness reversion of coated and calendared paper after aging

Pigment Composition	Whiteness (%)				
	Original	4 hrs aging	8 hrs aging	12 hrs aging	16 hrs aging
100 (%) ZnO (nano)	11.9	11	10	9.2	8.2
50 (%) ZnO (nano) + 50 (%) China Clay	13.86	11.2	8.7	6.1	3.8
100 (%) ZnO (ord)	12.1	10.2	8.2	6.4	4.3
50 (%) ZnO (ord) + 50 (%) China Clay	14.2	10	6	1.8	-2.2

(394c) Investigation of Mass Transport Properties of Microfibrillated Cellulose (MFC) Films

Matteo Minelli[1], **Marco Giacinti Baschetti**[2], Ferruccio Doghieri[2], Mikael Ankerfors[3], Tom Lindstrm[3], David Plackett[4] and Istvn Sir[4], (1)Department of Chemical Engineering, Mining and Environment Technology, University of Bologna, Bologna, Italy, (2)Department of Chemical Engineering, Mining and Environmental technology, University of Bologna, Bologna, Italy, (3)Innventia AB, Stockholm, Sweden, (4)Solar Energy Programme, Ris National Laboratory for Sustainable Energy, Technical University of Denmark, Roskilde, Denmark

Investigation of masstransport properties of microfibrillated cellulose (MFC) films

M.Minelli, M. Giacinti Baschetti, F. Doghieri, M. Ankerfors, T. Lindstrm, I.Sir, D. Plackett

Cellulose-based nanocomposites are now emerging as promising newmaterials suitable for a variety of applications. One source for such newnanocomposites is microfibrillated cellulose (MFC), produced by delamination ofcellulosic fibers in high-pressure homogenizers. MFC is an organic andbiodegradable reinforcement in polymer nanocomposites because of its highaspect ratio, good mechanical properties and ability to form networks: promisingproperties that may make this material suitable for use in a number ofindustrial fields and products.

The structure and transport properties of a four different filmsbased on two different generations of microfibrillated cellulose (MFC), aloneor in combination with glycerol as plasticizer, were investigated throughFE-SEM analysis and sorption or permeation experiments. FE-SEM revealed the existenceof complex structures in the different samples. A porous, closely packed fibernetwork, more homogeneous in the samples containing glycerol, wascharacteristic of the surface of MFC films; while film cross-sections presenteda dense layered structure with no evidence of porosity. Water vapor sorptionexperiments confirmed the hydrophilic character of these cellulosic materialsand showed a dual effect of glycerol, which reduced the water uptake at lowwater activity while enhancing it at high relative humidity, as showed inFigure 1. The observed trends of water sorption can be described considering thatboth physical adsorption on fiber surfaces and absorption in the celluloseamorphous phase takes place and therefore a dual-model sorption model wasemployed.

The water diffusivity (in the panel in Figure 1) in dry samples wasremarkably low for a porous material (10^{-11}-10^{-12} cm^2/s), confirming the existence of complex structures below thefilm surface; however, when water is present in the system, *D* rapidlyincreases with an exponential trend. Diffusivity is also definitely affected byplasticization, being higher for glycerol-containing samples.

Similar behavior was observed in permeation experiments. Dry MFCfilms showed excellent oxygen barrier properties, comparable with those ofpolymers usually considered suitable for ultra-high barrier applications.Indeed, when dry, the cellulose network presents a particularly compact andstiff structure, causing the observed barrier effect. Furthermore, although MFCfilms are characterized by a certain porosity, FE-SEM images of sample surfacessuggests that the pores are not substantially interconnected

When the water content in the membrane is raised, however, adramatic decrease in these properties was observed (Figure 2).

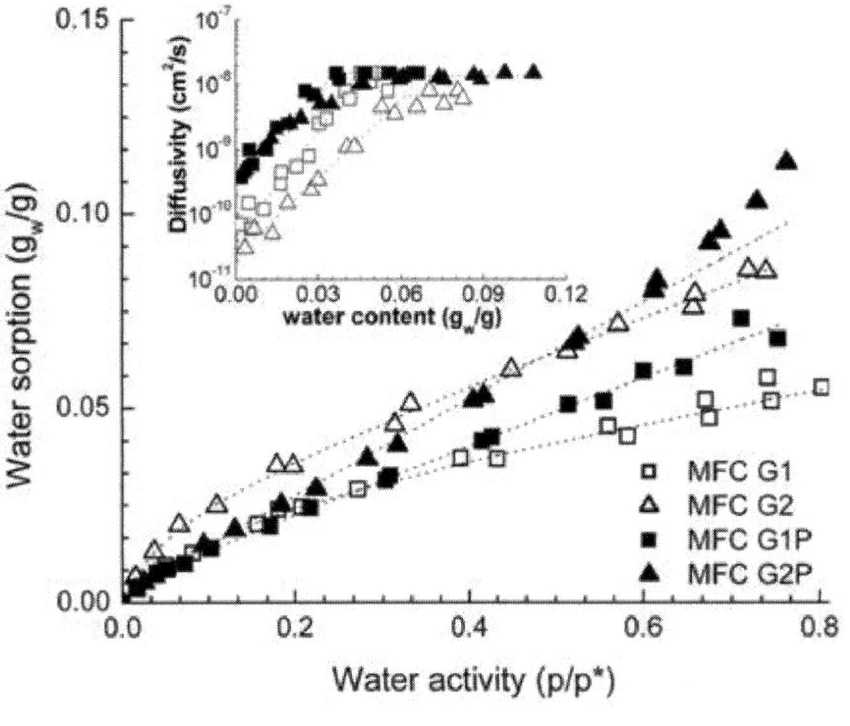

Figure 1. Water vapor sorption isothermsin MFC samples at 35°C, dotted lines are given by the best fitting of the data withthe dual-mode model equation. In the panel: water diffusivity at 35°C in the four different MFC samples as function of the average water concentration.

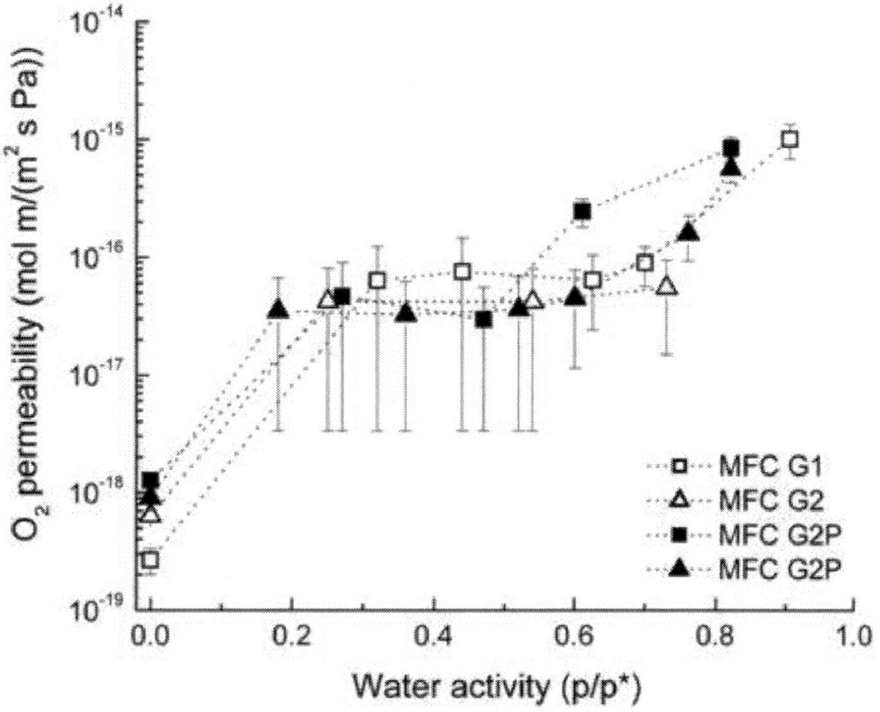

Figure 2. Oxygen permeability in thefour MFC samples at 35°C as a function of water activity.

See more of this Session: Nanocellulose-Based Materials and Composites
See more of this Group/Topical: Forest and Plant Bioproducts Division

(394d) Co-Products of Bioenergy System: Characterization of Saccharification Residuals

Han-Seung Yang, Shona Duncan, William T. Y. Tze and Jonathan Schilling, Department of Bioproducts and Biosystems Engineering, University of Minnesota, Saint Paul, MN

AIChE2011 Annual Meeting - Forest and PlantBioproducts Division

Co-Products of Bioenergy System: Characterizationof Saccharification Residuals

Han-Seung Yang[1], WilliamT. Y. Tze[1], Jonathan Schilling[1] and Shona Duncan[1]

[1]Departmentof Bioproducts and Biosystems Engineering

University of Minnesota

Saint Paul, MN 55108-6130 USA

Tel: 612-625-5200 Fax: 612-625-6286

e-mail: hsyang@umn.edu

The objective of this study is to investigate the effects ofdilute acid pretreatment and enzymatic saccharification on the reactivity,porosity, and crystallinity of solid residuals of biomass. Aspen wood flour was pre-treated with adilute acid treatment followed by enzymatic saccharification. The solid phase of the saccharificationprocess was analyzed using standard NREL protocols for amounts of lignin, pentosans,and hexosans. The functional groupsof solid residuals were analyzed using mid-infrared spectroscopy and hydroxylnumbers. The porosity andcrystallinity of the solids were also evaluated usingrespectively, nitrogen adsorption and X-ray diffraction. Cellulose nanocrystals were extractedfrom the solid residuals and verified using atomic force microscopy. The results will be discussed in thelight of reactivity, porosity, and crystallinity of saccharification residues topave the way for down-stream conversion into co-products of bioenergy(bio-alcohol) systems.

See more of this Session: Nanocellulose-Based Materials and Composites
See more of this Group/Topical: Forest and Plant Bioproducts Division

(394e) Bio-Nano Reinforcement of Polylactic Acid with Surface Modified Cellulose Nanocrystals

Jos Luis Orellana, Esteban E. Urea-Benavides and Christopher L. Kitchens, Department of Chemical and Biomolecular Engineering, Clemson University, Clemson, SC

Cellulose nanocrystals (CNC) have the potential to serve as a nanoscale reinforcement for polylactic acid (PLA), which has become a popular polymer but some of the deficiencies still need to be overcome. The main goal of this work is to improve the mechanical properties of the PLA films – especially toughness – by the addition of surface-modified CNC in the polymer matrix without sacrificing the thermal stability and processability.

CNC were isolated via acid hydrolysis from cotton and surface modified with an esterification process. Transmission Electron Microscopy (TEM) images demonstrated the isolation of the nanocrystals although aggregation can be evidently observed. CNC was further functionalized with different surfactants including decylamine (DA) and cetyltrimethylammonium bromide (CTAB). The dispersibility of the CNC was studied in different solvents such as THF, chloroform and ethyl acetate. Surface modified CNC were added in different concentrations to PLA. Tensile properties of the films were studied using an Instron universal testing machine.

It was found that the stability of the surface-modified CNC improved in organic solvents. DA greatly improved the stability of CNC in both THF and ethyl acetate; while CTAB did in chloroform. No settling was observed in these suspensions for at least three weeks. The toughness and elongation at break of PLA-CNC nanocomposites were improved by the addition of CNC by at least two-fold increase compared to plain PLA. Tensile strength and tensile modulus were not significantly affected by the addition of the crystals. The mechanical properties of the composites were compared and related to the CNC dispersion quality in the different solvents.

See more of this Session: Nanocellulose-Based Materials and Composites
See more of this Group/Topical: Forest and Plant Bioproducts Division

TREATMENT OF AMMUNITION WASTEWATER USING ULTRASONIC FENTON (US-FENTON) PROCESS

YangangLi[a, b], Wen-Pin Hsieh[a], RovshanMahmudov[a], and C.P. Huang[a]*

[a] Department of Civil and Environmental Engineering, University of Delaware, Newark, DE 19716, USA

[b] College of Water Resources and Architecture Engineering, Northwest A&F University, Yangling, Shaanxi, 712100,China

[*] Corresponding author. C. P. Huang (Fax: 1-302-831-8428; Email: huang@udel.edu)

Abstract

Ammunition wastewater contains several persistent compounds, such as 2,4,6 trinitrotoluene (TNT), 2,4 dinitroanisole (DNAN), and 3-nitro-1,2,4-triazol-5-one (NTO), that are recalcitrant during conventional treatment. In this study, the wastewater collected from a regional ammunition facility was screen treated with various methods, including UV-photolysis, Fenton, photo-Fenton and ultrasound-Fenton (US-Fenton). Results showed that US-Fenton was the best method among all screened and was selected for further study. The effect of pH, temperature, reaction time, US energy intensity, initial TOC concentration, and the molar ratio of iron to hydrogen peroxide were investigated. TOC, COD, and color removal increased with decreasing pH. TOC, COD, and color removal were increased with increase in temperature and US intensity. Color removal was very fast, reaching 80 % in 10 minutes. Compared to color removal, TOC and COD removal were slower, only about 20 % in 10 minutes, then approached 65 % and 93 % in 120 minutes, respectively. The optimal molar ratio of iron to hydrogen peroxide was 500. The carbon oxidation numbers were calculated for different treatment options. The results showed that the carbon oxidation number increased with increase in time. The toxicity of treated wastewater was evaluated by monitoring the respiration rate of *E. coli* exposed to the ammunition wastewater before and after treatment. Results indicated that sample treated for 300 minutes improved the respiration rate of *E. coli* compared to untreated samples.

Keyword: Ammunition wastewater, Ultrasound, Fenton, Total organic carbon (TOC), Chemical oxygen demand (COD), Toxicity.

1. Introduction

Ammunition wastewater from the wash-down of energetic materials handling equipment at military facilities has a complex chemical and physical characteristics and contains a number of harmful contaminants, such as metals, nitroaromatic compounds, and various oxyanions that require proper treatment before discharge [1, 2]. Nitroaromatic compounds have been widely used in the manufacture of explosives since World War II [3, 4]. Some metals and nitroaromatic compounds are toxic to humans and animals, and are generally very persistent in the environment if discharged without proper treatment. In addition, environmental transformation products of nitroaromatic compounds (i.e. azoxy- and azo-compounds) can be even more toxic than the parent nitroaromatics [5-7] .

Much research has been conducted in order to develop an effective ammunition wastewater treatment technology that would be applicable for wide changes in composition of these wastewaters. Several biological and physical-chemical treatment processes have been used. Yin et al. studied reductive transformation of TNT using aerobically grown *Escherichia coli* cultures [8]. Kang et al. investigated abiotic transformation of TNT via oxidative-coupling reaction using Mn oxide [9]. Oh et al. studied the removal of TNT and RDX using Fenton process as a subsequent treatment following iron pretreatment.[10]. Yardin et al. investigated the remediation of TNT contaminated soil using cyclodextrins coupled with Photo–Fenton treatment [11].

Advanced oxidation process (AOP) has been shown as promising to treat wastewaters containing recalcitrant contaminants. Particularly, AOP based on ultrasound irradiation has been found highly effective. Ultrasound manifests its physical and chemical effects through the phenomenon of cavitation, which can be defined as the cycle of formation, growth, and collapse of micro bubbles within the media. Transient collapse of cavitation bubbles results in concentration of energy and generation of extremely high local temperature and pressure. Temperature as high as 5000 °K and pressures up to 500 bar have been suggested during the transient collapse of the micro bubbles [12]. Under these conditions, entrapped molecules of dissolved gases (e.g. vaporized water and solutes) can be brought to an excited state and undergo hemolytic bond breakage to generate radicals; when water is the solvent, $OH^\bullet$, $H^\bullet$, and $HO_2^\bullet$ radicals are generated [13]. Hydroxyl radical is among the strongest oxidants with an E_H^o value of 2.33V and can react non-selectively with a host of organic and inorganic compounds [14]. Free radicals generated during water and oxygen dissociation combined with local high temperatures and pressures from formation and collapse of micro bubbles induce/accelerate oxidation reactions in the liquid medium [15].

$$H_2O \text{ ---))))}\rightarrow H^\bullet + OH^\bullet \quad (1)$$
$$O_2 \text{ ---))))}\rightarrow 2O \quad (2)$$
$$2O + H_2 \rightarrow 2OH^\bullet \quad (3)$$
$$H^\bullet + O_2 \rightarrow OH^\bullet + O \quad (4)$$

However, total mineralization of organic pollutants by means of ultrasound irradiation alone is very energy consuming, and in many cases has low efficiency, therefore its application for a full scale industrial plant is still impractical. To overcome these limitations, various combined ultrasound treatment systems have been proposed [16]. These treatment methods combine high energy confined within micro bubbles generated by ultrasound with chemical sources of hydroxyl radical formation. Over the past several years, there has been an increasing interest in the sonochemical process for the treatment of recalcitrant contaminants in aqueous solutions that are hardly amenable to conventional treatment [17-23]. Sonochemical treatment includes ultrasound treatment coupled with oxidants such as ozone [13, 14, 24], hydrogen peroxide (H_2O_2)[25, 26] , electrochemical methods [27, 28], Fenton's reagent[23, 29] , photocatalysts[29], and other catalytic oxidation processes [30]. Particularly, the combination of ultrasound with Fenton's reagent has evolved as a promising AOP for the treatment of ammunition wastewater. In Fenton oxidation process, the reaction of Fe^{2+} with H_2O_2 generates $OH^\bullet$ radicals as shown in Eq. 5.

$$H_2O_2 + Fe^{2+} \rightarrow Fe^{3+} + OH^- + OH^\bullet \quad (5)$$
$$RH + OH^\bullet \rightarrow H_2O + R^\bullet \quad (6)$$

Hydroxyl radicals react with organic compounds to generate other radicals as shown in Eq. 6; organic radicals are unstable and degrade rapidly. The results of several studies have shown that combined ultrasound and Fenton process were better than ultrasound or Fenton process alone.

In this study, ultrasound (US) assisted Fenton reaction was applied for the treatment of ammunition wastewater. The objectives of this study were (1) to determine the optimal conditions for the treatment of ammunition wastewater, and (2) to determine the effect of treatment on the short-term toxicity of ammunition wastewater.

2. Experiments

2.1 Materials

All chemicals used in this study were ACS (American Chemical Society) certified grade and obtained from either Fisher Scientific or Sigma Aldrich and used without further purification.

The ammunition wastewater samples generated from wash-down of the equipment handling energetic materials and occasional sump out operations was collected for use in this study. Table 1 shows the physical-chemical characteristics of the wastewater.

2.2 Analytical methods

Chemical oxygen demand (COD) was measured by the Hach colorimetric method at 600 nm using a Hach spectrophotometer (Hach DR/2000) according to The Standard Methods for the Examination of Water and Wastewater (1990). Residual hydrogen peroxide concentration in treated ammunition wastewater was analyzed using the titanium sulfate method [31]. Hydrogen peroxide interferes with COD measurements, and therefore the final COD values of treated ammunition wastewater were corrected for the residual hydrogen peroxide.

Total organic carbon (TOC) was measured with a TOC analyzer (Apollo 9000HS, Teledyne-Tekmar). Soluble ferrous iron concentration [Fe^{2+}] was determined using the 1,10-phenanthroline colorimetric method. To avoid the interference of ferric iron, ammonium fluoride was added as the masking agent. Total soluble iron concentration was determined by reducing ferric to ferrous ion by hydroxylamine under acidic conditions. Total iron concentration then was measured by the same method as Fe^{2+}concentration. Ferric ion concentration was estimated by subtracting ferrous ion from total iron.

The concentration of anions was analyzed using an ion chromatograph Dionex 500 equipped with a conductivity detector, AG-11 guard column, and AS-11 metal-free anion column (25 mm×4 mm i.d.). The eluent was a mixture of 87% deionized water, 10% 0.2 M NaOH and 3% acetonitrile. The effluent flow rate was 1 mL/min and the injection loop volume was 50 μL. Cations were determined using Inductively Coupled Plasma Spectrometer (ICP-OES) (Optima 7300 DV, Perkin-Elmer Corp). The mixed liquor and all other samples were digested with nitric acid before ICP-OES analysis for the determination of total metal concentration.

The toxicity of treated and untreated wastewater was determined with short-term toxicity test using *Escherichia coli*. The *E. coli* culture medium was prepared by dissolving 1.6 g KH_2PO_4, 1.6 g K_2HPO_4, 1.0 g NaCl, 2.5 g glucose, 4.0 g $(NH_4)_2SO_4$, 0.7 $MgSO_4$ and 0.5 g sodium citrate in 1 L of distilled water. The finial pH was adjusted to 7.0 using 4 M solution of NaOH. Culture medium was sterilized in an autoclave at 125°C for 15 min. Then 50 mLof sterilized culture medium was carefully transferred to 125 mL Erlenmeyer flask and 0.5 mL of *E. coil* seed solution was added. The flasks were placed on a reciprocal shaker and *E. coli* culture was allowed to grow at a constant temperature of 37°C for 12 h. To test the effect of the chemical treatment on the toxicity of ammunition wastewater, one mL of the following solutions were added to flasks containing *E. coli* cultures: original wastewater, original wastewater diluted 10 times, and wastewater treated for 60 and 300 minutes, respectively. After predetermined time intervals, one-mL aliquots of liquid culture medium were withdrawn and analyzed for carbon dioxide, according to the method proposed by Jardim et al. [32]. The flasks were shaken on a reciprocal shaker throughout the entire process.

2.3 Sonochemical reactor

Figure 1 shows the schematic diagram of the experimental apparatus. Sonochemical experiments were carried out using the ultrasoound generator (Cole-Parmer 600-Watt, 20 kHz,) equipped with a titanium probe transducer (Cole Parmer, Model CV 17). The experiments were conducted in a 40-mL borosilicate glass vial. The reactor was filled with 30 mL of ammunition wastewater and 5 mL of Fenton reagent. To maintain constant ionic strength, 0.5 mL of 0.7 M of stock NaCl solution was added to yield constant ionic strength of 10^{-2} M. The reactor was sealed with a Teflon-lined rubber stopper and immersed into a jacketed glass reactor containing DI water, in which the temperature was maintained at 25°C by a water circulation and temperature control system (Frigomix 1495, Fisher Scientific).

3. Results and Discussion

3.1 Screening of various treatment methods

Several methods were screened for treating the ammunition wastewater, including, UV-photolysis (UV $_{254\ nm}$), Fenton, photo-Fenton, and ultrasound-Fenton using TOC as the main performance indicator. The initial TOC concentration of the ammunition wastewater was 710±50 mg/L and pH of 2.8±0.2. The TOC concentration of the ammunition wastewater was monitored for 1 h and the results are shown in Figure 2. Results showed that the degradation of TOC was relatively low, less than 10%, when the ammunition wastewater was treated with only UV photolysis, $FeSO_4$, or H_2O_2. Low TOC degradation was also observed for Fenton and photo-Fenton processes. The highest TOC removal at 58% was observed when the wastewater was treated by a combination of ultrasound and Fenton regent with the concentration of [Fe^{2+}] and [H_2O_2] of 5×10^{-4} M and 0.25 M, respectively, and US intensity of 300 watts. The oxidation efficiency followed the order: ultrasound + Fenton > Fenton > photo+Fenton > H_2O_2 > $FeSO_4$ > UV. The ammunition wastewater had strong color that prevented the UV light from passing through the sample, which could affect the treatment efficiency of the UV and the photo-Fenton processes compared to US-Fenton process. The results clearly indicated the superiority of the ultrasound-Fenton process over other treatment processes studied. Therefore further studies ware focused on the ultrasound-Fenton method.

3.2 Effect of ultrasound intensity

The sonicator used in this research had a constant output frequency of 20 kHz. Higher power intensity results in the formation of cavities of higher temperature [33]. Figure 3 shows the degradation of color (measured as absorbance at 348 nm), TOC and COD as a function of power intensity. The color removal increased slightly from 83.7 to 87.7 % when the power intensity was increased from 0 to 300 watts. The removal of TOC and COD, however, improved significantly when the power density was increased. TOC and COD removal increased from 30.5 and 13.6 % to 73.6 and 55.7%, respectively, when the power intensity increased from 0 to 300 watts. Higher degradation percentage of ammunition wastewater was achieved at higher power intensity. Increase in sonic energy increases the number of generated cavities and the cumulative pressure [33]. This results in a greater $OH^{•}$ radical generation rate which will enhance the decomposition of organic compounds. On the other hand, collapse of cavities enhances the reaction of OH• radicals with organic compounds by greatly accelerating the mass transfer rates.

3.3 Effect of Hydrogen Peroxide to Ferrous Iron molar Ratio

Figure 4 shows the removals of COD, TOC, and color at fixed ferrous iron dose of 5×10^{-4} mole/L and varying the concentration of hydrogen peroxide to yield various molar ratios of hydrogen peroxide to ferrous iron from 10 to 1500. As the molar ratio was less than 500, the removal of TOC and COD increased with the molar ratio of H_2O_2/Fe^{2+}. The removal of TOC deteriorated when the molar ratio was increased from 500 to 1000. COD removal increased with the molar ratio of H_2O_2/Fe^{2+}. The maximum removal of TOC and COD at 69 and 54%, respectively, was achieved at H_2O_2/Fe^{2+} molar ratio of 500 and 1500. In Fenton oxidation process, the reaction of Fe^{2+} with H_2O_2 generates OH• radicals which are strong oxidants (Eq. 5). Reaction between hydroxyl radical and organic compounds produces organic radicals as shown in En. 6. The organic radicals are unstable and decompose rapidly. At low H_2O_2/Fe^{2+} molar ratios there is insufficient supply of hydrogen peroxide for the generation of hydroxyl radical (Eq. 1 and Eq. 2). Therefore increase in H_2O_2/Fe^{2+} molar ration will increase $OH^{•}$ generation and lead to better TOC removal. On the other hand, if the H_2O_2/Fe^{2+} molar ratio is too high, one or more side reactions would occur as shown in Eqs. 7 -10, i.e.,

$$H_2O_2+Fe^{3+}\rightarrow Fe^{2+}+HO_2^{•}+H^{+} \quad (7)$$

$$H_2O_2+OH^{•}\rightarrow HO_2^{•}+H_2O \quad (8)$$

$$Fe^{3+}+HO_2^{•}\rightarrow Fe^{2+}+O_2+H^{+} \quad (9)$$

$$Fe^{2+}+HO_2^{•}\rightarrow Fe^{3+}+HO_2^{-} \quad (10)$$

The above reactions result in the consumption of hydrogen peroxide as well as hydroxyl radical which impedes the degradation of organic compounds in question. It is interesting to note that color removal was independent on the molar ratio of hydrogen peroxide to ferrous ion.

3.4 Effect of pH

Fig. 5 shows the effect of the initial pH on the treatment of ammunition wastewater by US-Fenton process. TOC, COD, and color removal were significantly affected by pH. The removal was greater at low initial pH; at pH 2.0, TOC, COD and color removal were 64, 66, and 74%, respectively. In contrast, at pH 10.0, the degradation of TOC, CO and color were 48, 12, and 12%, respectively. Other studies have showed similar pH effect. Zhang et al. (2009) examined the effect of initial pH on the discoloration of C.I. Acid Orange 7 in the pH range of 2 to 5 and reported that the discoloration increased with decrease in pH and reached the maximum value at pH of 2. Our results agreed with those reported by Hermosilla et al. (2009) and Zhang et al. (2009) who studied the effect of pH on the treatment of landfill leachate in the pH range between 2.0 and 4.0 [34]. The optimal pH was found to be at 2.5. It is known that both Fe^{2+} and Fe^{3+} will precipitate into hydroxides at alkaline pH, which in turn will lead to decrease in the removal efficiency at high pH values. Iron precipitation at higher pH values could be visually observed by the deposition of brown precipitates at the bottom of the reactor. Moreover, the oxidation potential of hydroxyl radicals declined at higher pH values

3.5 Effect of Temperature

The effect of temperature on the treatment of ammunition wastewater was investigated and the results are shown in Figure 6(a). When the temperature was raised from 25 to 40°C, the removal of TOC, COD, and color increased and reached 76%, 87% and 93%, respectively. The highest increase was observed for COD removal and increase in carbon oxidation state.

In order to describe the relationship between of the organic matter decomposition rate and temperature, COD removal rate was used [35]. It was found that the COD removal rate followed first order rate equation as given by Eqn. (7):

$$-\frac{d[C]}{dt} = \kappa[C] \text{ or } C = C_0 \exp(-\kappa t) \qquad (7)$$

Where: C = the concentration of COD at time t, C_0 = initial concentration, κ = removal rate constant and t = reaction time. Figure 6(b) shows the effect of temperature on the first order rate constant for the COD removal. The rate of COD removal increased sharply when the temperature increased from 20 to 35°C followed by a gradual increase as the temperature increased from 35 to 60°C. Such behavior can be attributed to the fact that the rate of redox reactions is generally accelerated by increasing the temperature. At higher temperatures, the reaction of hydrogen peroxide and iron (Fe^{2+}/Fe^{3+}) proceeds faster, this can increase the rate of hydroxyl radical generation and consequently the degradation of organic compound[33]. From the degradation mechanism of US, it is known that when the cavitation bubbles explosively collapse, the temperature and pressure in the bubbles can reach up to several thousand Kelvin and several hundred atmospheres, respectively. Under these conditions, organic compounds decompose directly by pyrolysis inside the bubbles [36]. However, increase of temperature over a certain level may result in an adverse effect. In the US process, higher temperatures are likely to facilitate bubble formation due to an increase in the equilibrium vapor pressure; however, at elevated temperatures bubbles contain more water vapor that may cushion bubble implosion and consequently reduce the maximum temperature obtained during bubble collapse [37].

3.6 Effect of Reaction Time

Figure 7(a) shows TOC, COD, and color removal as a function of reaction time. Color removal was very fast and within the first 10 minutes the degradation reached 85%. In contrast within the first 10 minutes TOC and COD removals were significantly slower than color removal, at only 19 and 21%, respectively. However, after 10 minutes color removal exhibited no obvious change up to 300 minutes of reaction time reaching the maximum removal of 87%. TOC and COD removals increased gradually reaching 55% and 92%, respectively, after 120 minutes of the reaction time, after which the TOC and COD removal rates slowed down significantly reaching 60% and 95%, respectively, in 300 minutes. The removal efficiency of TOC was lower than that of COD. In order to explain better COD removal, the average carbon oxidation state was determined. The average oxidation state is defined as follows:

$$\text{Average oxidation state} = 4[1- (\text{COD/TOC})] \qquad (11)$$

Where: COD is expressed in moles of O_2 / L and TOC is in moles of C / L. Although the hydroxyl radicals and cavitations created by Fenton reagent and US have strong oxidability, they cannot oxidize all organic carbon into inorganic carbon, such as carbonate, bicarbonate or carbon dioxide [34]. The main reasons are the complicated composition of the ammunition wastewater and a complex Fenton oxidation mechanism. Therefore, only a portion of the organic carbon is mineralized into inorganic carbon. In addition, the decomposition of higher molecular weight organic matter generates lower molecular weight organic intermediates, which can lead to increase of the average oxidation state in the residual sample solution, but not the COD value. Increase in the average oxidation state means that COD to TOC ratio should decrease, just like Figure 7(a) shows, with time increasing from 0 to 300 minutes, the average oxidation state went up from -2.1 to 3.3.

Residual concentrations of ferrous iron, ferric iron, and hydrogen peroxide were measured during the reaction as well (Figure 7(b)). During the first 30 minutes, most of ferrous iron was consumed and its concentration decrease from 28 mg/L to 4.1 mg/L. At the same time, ferric iron was produced rapidly. The disappearance of Fe^{2+} and generation of Fe^{3+} occurs according to Eqn.(1) and (6). In the beginning of reaction, ferric iron rapidly reacted with H_2O_2 to generate $HO_2\cdot$and ferric iron. It is also possible that some amount of ferrous iron was consumed by a different side reaction because ferrous iron consumption rate was greater than the ferric iron generation rate. As the reaction continued, the concentration of ferrous iron in the solution increased slightly to 4.8 mg/L at the 60 minutes. This indicates that after 30 minutes of reaction time ferrous iron generation rate slightly exceeded the consumption rate. After 180 minutes, the concentration of ferrous iron became relatively steady at about 2 mg/L. Hydrogen peroxide showed different behavior than ferrous iron during the reaction. The concentration of hydrogen peroxide decreased rapidly in the first 60 minutes, and then went down slightly reaching the residual concentration of 21 mg/L in 300 minute.

3.7 Effect of Initial TOC Concentration

Because every compound in the wastewater has different removal efficiency, the overall removal efficiency can be affected by different initial concentration of the compounds. The effect of the initial concentration on the wastewater treatment efficiency was studied by diluting the original wastewater to different TOC values. Figure 7 shows the treatment results at different initial TOC values. It was very interesting to note that the observed results were opposite to what one would expect, e.g. better TOC removal at higher concentrations. TOC removal efficiency increased and reached the maximum value of 55% when the dilution factor increased to 10. Further increase in the dilution factor above 10 resulted in decrease in the TOC removal efficiency. When the initial wastewater was diluted 50 times the removal efficiency decreased to 23% and COD removal followed similar pattern. In order to explain such behavior, the average carbon oxidation number of the residual solutions was determined. When the samples were diluted 10 times, the treatment resulted in decrease of the oxidation state from -1.1 to -3.1, but at dilution factor of 50, carbon oxidation number slightly increased from -3.1 to -2.0 compared to the sample with the dilution factor of 10. At the same time, gradual accumulation of precipitate was observed at the bottom of the reactor at dilution factors greater than 10. Moreover, the quantity of precipitate increased proportionally to the dilution factor. The highest quantity of precipitate of 67 mg/L was reached after treatment by US- Fenton at the dilution factor of 50. The precipitate was collected, digested, and analyzed for TOC and iron concentration. The results showed that the precipitate contained TOC and iron in the amounts of 19.1 mg-C/L and 27.9 mg/L, respectively. The changes in the average carbon oxidation states indicated that possibly some new organic intermediates formed during the treatment process that caused TOC increase in the solution. With the initial TOC of sample decreasing, some intermediates formed slightly soluble ferrous iron complexes that easily precipitated and lead to the decrease of total iron concentration of solution and the efficiency of Fenton process, therefore TOC and COD removal became worse.

3.8 Toxicity

The toxicity of the wastewater and treated wastewater were checked by *E. coli.* The latter is easy to culture in laboratory conditions and shows a high production rate of CO_2. The final CO_2 concentration in the culture medium was measured by flow injection analysis with a conductmetric detector. Fig. 9 shows the *E. coli* responses to different sample. The results showed treatment by US- Fenton can reduce the toxicity as well as dilution. Due to hydrogen peroxide can also inhibit the respiration of *E. coli* and the concentration of hydrogen peroxide which was 1374 mg/L was still high in the treated wastewater after reaction for 60 minutes, the toxicity in the sample treated for 60 minutes was stronger than that diluted 10 times. In sample treated for 300 minutes, the inhibition in *E. coli* showed weakest. It is likely that some intermediates generated during the reaction were beneficial to *E. coli* growth.

4. Conclusion

Several methods, such as photolysis (UV), Fenton, photo-Fenton, and ultrasound-Fenton (US-Fenton), were studied for the treatment of ammunition wastewater. The results showed that US-Fenton process can be a promising method for ammunition wastewater treatment. The effect of operational parameters (ultrasonic intensity, pH, temperature, hydrogen peroxide to ferrous iron molar ratio, initial TOC concentration, and reaction time) on the TOC, COD, and color removal efficiencies were studied. Generally, the degradation of TOC, COD, and color increased with decrease in pH and increase in temperature and US intensity. During the treatment process, the optimal molar ratio of hydrogen peroxide to ferrous iron was 500. The effect of initial TOC concentration was complicated compared to other factors. The results showed that too low and high initial TOC concentration caused decline in the removal efficiency and the optimum initial TOC concentration of 70 mg-C/L.

Toxicity of the original wastewater, diluted wastewater, and treated wastewater has been investigated. The results showed that both dilution and treatment of ammunition wastewater by US-Fenton reduced the toxicity of the wastewater. The toxicity was affected by wastewater dilution factor and US-Fenton treatment duration. For shorter treatment times (i,e, 60 min), the toxicity of the wastewater treated by US-Fenton process increased compared to original wastewater diluted 10 times. This was due to the presence of residual unreacted hydrogen peroxide inhibiting *E. coli* respiration. However, after 300 minutes treatment period, the toxicity of wastewater almost disappeared.

References

[1] K. Ayoub, E.D. van Hullebusch, M. Cassir, A. Bermond, Application of advanced oxidation processes for TNT removal: A review, J. Hazard. Mater., 178 (2010) 10-28.

[2] M.J. Liou, M.C. Lu, J.N. Chen, Oxidation of explosives by Fenton and photo-Fenton processes, Water Research, 37 (2003) 3172-3179.

[3] J. Hawari, A. Halasz, T. Sheremata, S. Beaudet, C. Groom, L. Paquet, C. Rhofir, G. Ampleman, S. Thiboutot, Characterization of metabolites during biodegradation of hexahydro-1,3,5-trinitro-1,3,5-triazine (RDX) with municipal anaerobic sludge, Applied and Environmental Microbiology, 66 (2000) 2652-2657.

[4] P.M. Bradley, F.H. Chapelle, J.E. Landmeyer, J.G. Schumacher, MICROBIAL TRANSFORMATION OF NITROAROMATICS IN SURFACE SOILS AND AQUIFER MATERIALS, Applied and Environmental Microbiology, 60 (1994) 2170-2175.

[5] P.R. Gogate, S. Mujumdar, A.B. Pandit, Sonochemical reactors for waste water treatment: comparison using formic acid degradation as a model reaction, Advances in Environmental

Research, 7 (2003) 283-299.
[6] W.S. Chen, G.C. Huang, Sonochemical decomposition of dinitrotoluenes and trinitrotoluene in wastewater, J. Hazard. Mater., 169 (2009) 868-874.
[7] M.R. Hoffmann, I. Hua, R. Hochemer, Application of ultrasonic irradiation for the degradation of chemical contaminants in water, Ultrasonics Sonochemistry, 3 (1996) S163-S172.
[8] H. Yin, T.K. Wood, B.F. Smets, Reductive transformation of TNT by Escherichia coli: pathway description, Applied Microbiology and Biotechnology, 67 (2005) 397-404.
[9] K.H. Kang, D.M. Lim, H. Shin, Oxidative-coupling reaction of TNT reduction products by manganese oxide, Water Research, 40 (2006) 903-910.
[10] S.Y. Oh, P.C. Chiu, B.J. Kim, D.K. Cha, Enhancing Fenton oxidation of TNT and RDX through pretreatment with zero-valent iron, Water Research, 37 (2003) 4275-4283.
[11] G. Yardin, S. Chiron, Photo-Fenton treatment of TNT contaminated soil extract solutions obtained by soil flushing with cyclodextrin, Chemosphere, 62 (2006) 1395-1402.
[12] V.S. Moholkar, Mechanistic optimization of a dual frequency sonochemical reactor, Chemical Engineering Science, 64 (2009) 5255-5267.
[13] P. Riesz, T. Kondo, C.M. Krishna, SONOCHEMISTRY OF VOLATILE AND NONVOLATILE SOLUTES IN AQUEOUS-SOLUTIONS - EPR AND SPIN TRAPPING STUDIES, Ultrasonics, 28 (1990) 295-303.
[14] G.A.D. Briggs, ULTRASOUND - ITS CHEMICAL, PHYSICAL AND BIOLOGICAL EFFECTS - SUSLIK,KS, Interdisciplinary Science Reviews, 15 (1990) 190-191.
[15] R.A. Torres, F. Abdelmalek, E. Combet, C. Petrier, C. Pulgarin, A comparative study of ultrasonic cavitation and Fenton's reagent for bisphenol A degradation in deionised and natural waters, J. Hazard. Mater., 146 (2007) 546-551.
[16] Y.G. Adewuyi, Sonochemistry: Environmental science and engineering applications, Industrial & Engineering Chemistry Research, 40 (2001) 4681-4715.
[17] R. Kidak, N.H. Ince, Ultrasonic destruction of phenol and substituted phenols: A review of current research, Ultrasonics Sonochemistry, 13 (2006) 195-199.
[18] P.S. Bapat, P.R. Gogate, A.B. Pandit, Theoretical analysis of sonochemical degradation of phenol and its chloro-derivatives, Ultrasonics Sonochemistry, 15 (2008) 564-570.
[19] I. Gultekin, N.H. Ince, Ultrasonic destruction of bisphenol-A: The operating parameters, Ultrasonics Sonochemistry, 15 (2008) 524-529.
[20] O. Hamdaoui, E. Naffrechoux, Sonochemical and photosonochemical degradation of 4-chlorophenol in aqueous media, Ultrasonics Sonochemistry, 15 (2008) 981-987.
[21] Z.B. Guo, R. Feng, J.H. Li, Z. Zheng, Y.F. Zheng, Degradation of 2,4-dinitrophenol by combining sonolysis and different additives, J. Hazard. Mater., 158 (2008) 164-169.
[22] Z.L. Wu, B. Ondruschka, G. Cravotto, Degradation of Phenol under Combined Irradiation of Microwaves and Ultrasound, Environmental Science & Technology, 42 (2008) 8083-8087.
[23] Y. Segura, R. Molina, F. Martinez, J.A. Melero, Integrated heterogeneous sono-photo Fenton processes for the degradation of phenolic aqueous solutions, Ultrasonics Sonochemistry, 16 (2009) 417-424.
[24] I.K. Kim, C.P. Huang, P.C. Chiu, Sonochemical decomposition of dibenzothiophene in aqueous solution, Water Research, 35 (2001) 4370-4378.
[25] F. Chemat, P.G.M. Teunissen, S. Chemat, P.V. Bartels, Sono-oxidation treatment of humic substances in drinking water, Ultrasonics Sonochemistry, 8 (2001) 247-250.
[26] J.G. Lin, C.N. Chang, J.R. Wu, Y.S. Ma, Enhancement of decomposition of 2-chlorophenol with ultrasound/H2O2 process, Water Science and Technology, 34 (1996) 41-

48.
[27] R.B. Doppalapudi, G.A. Sorial, S.W. Maloney, Electrochemical reduction of simulated munitions wastewater in a bench-scale batch reactor, Environmental Engineering Science, 19 (2002) 115-130.
[28] D.K. Palaniswamy, G.A. Sorial, S.W. Maloney, Electrochemical reduction of 2,4,6-trinitrotoluene, Environmental Engineering Science, 21 (2004) 203-218.
[29] I. Grcic, M. Obradovic, D. Vujevic, N. Koprivanac, Sono-Fenton oxidation of formic acid/formate ions in an aqueous solution: From an experimental design to the mechanistic modeling, Chemical Engineering Journal, 164 (2010) 196-207.
[30] E.V. Rokhina, M. Lahtinen, M.C.M. Nolte, J. Virkutyte, The influence of ultrasound on the RuI3-catalyzed oxidation of phenol: Catalyst study and experimental design, Applied Catalysis B-Environmental, 87 (2009) 162-170.
[31] Y.W. Kang, M.J. Cho, K.Y. Hwang, Correction of hydrogen peroxide interference on standard chemical oxygen demand test, Water Research, 33 (1999) 1247-1251.
[32] W.F. Jardim, C. Pasquini, J.R. Guimaraes, L.C. Defaria, SHORT-TERM TOXICITY TEST USING ESCHERICHIA-COLI - MONITORING CO2 PRODUCTION BY FLOW-INJECTION ANALYSIS, Water Research, 24 (1990) 351-354.
[33] H.H. Sun, S.P. Sun, J.Y. Sun, R.X. Sun, L.P. Qiao, H.Q. Guo, M.H. Fan, Degradation of azo dye Acid black 1 using low concentration iron of Fenton process facilitated by ultrasonic irradiation, Ultrasonics Sonochemistry, 14 (2007) 761-766.
[34] H. Zhang, H.J. Choi, C.P. Huang, Effects of important reaction conditions on landfill leachate treatment by Fenton process, Fresenius Environmental Bulletin, 15 (2006) 43-47.
[35] I.K. Kim, C.P. Huang, Sonochemical degradation of polycyclic aromatic sulfur hydrocarbons (PASHs) in aqueous solutions exemplified by benzothiophene, Journal of the Chinese Institute of Engineers, 28 (2005) 1107-1118.
[36] J. Liang, S. Komarov, N. Hayashi, E. Kasai, Recent trends in the decomposition of chlorinated aromatic hydrocarbons by ultrasound irradiation and Fenton's reagent, Journal of Material Cycles and Waste Management, 9 (2007) 47-55.
[37] E. Psillakis, G. Goula, N. Kalogerakis, D. Mantzavinos, Degradation of polycyclic aromatic hydrocarbons in aqueous solutions by ultrasonic irradiation, J. Hazard. Mater., 108 (2004) 95-102.

Table 1. Physical chemical characteristics of ammunition wastewater

pH	Conductivity (μs/cm)	TOC	COD	TS	SS	TDS	Absorbance (348 nm)
		mg/L					
2.8±0.2	218.82	710±50	1450±150	2.24	0.01	2.16	3.873±0.01
Ionic composition, mg/L							
Cl^-	NO_3^-	SO_4^{2-}	Na^+	K^+	Ca^{2+}	Mg^2	NH_4^+
35.8	2503	46.7	157.5	13.6	29.9	0.02	219.9

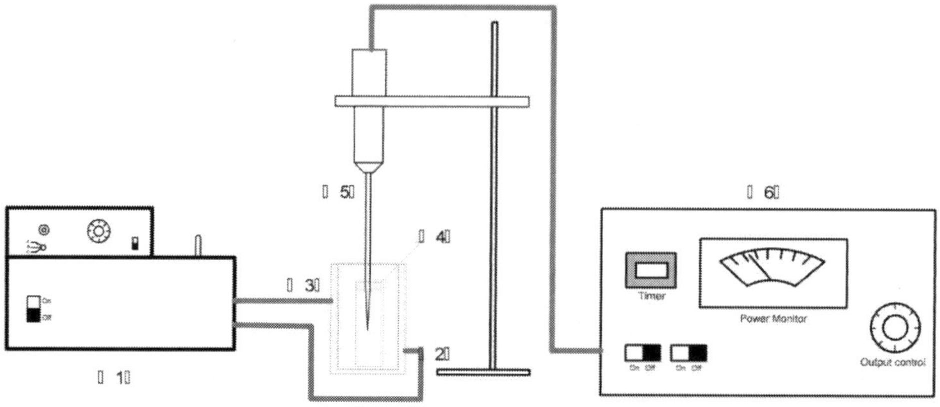

Figure 1. Schematic diagram of experimental setup for the sonochemical degradation of ammunition wastewater: (1) water circulation bath and temperature control system, (2) cooling water inlet, (3) cooling water outlet, (4) reactor, (5) converter and (6) ultrasonic generator

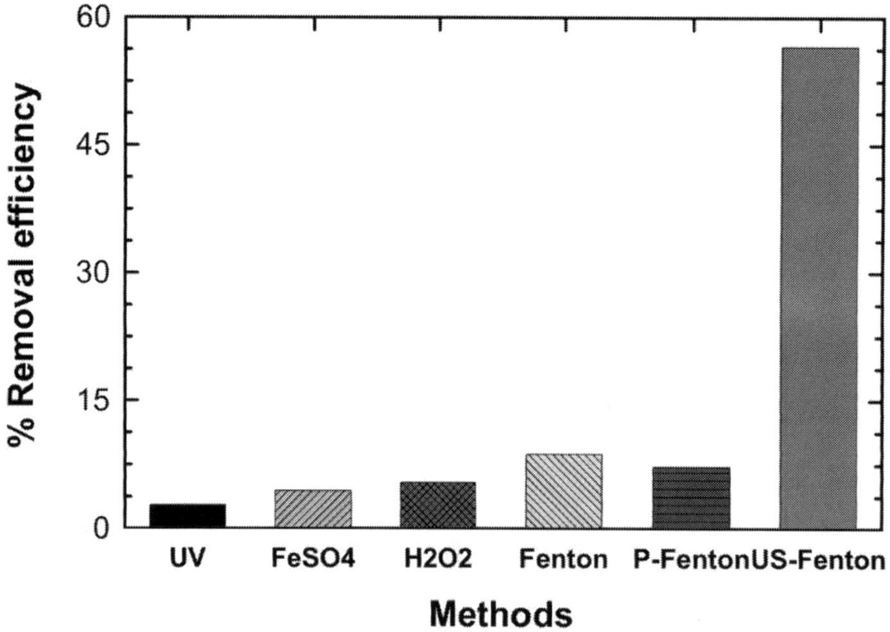

Figure 2. Comparing the treatment efficiency of various methods for the treatment of ammunition wastewater. Experimental conditions: $[TOC]_0$ = 710±50 mg/L, $[COD]_0$ = 1450±150 mg/L, Initial pH = 2.7±0.1, $[Fe^{2+}]$ = 5×10^{-4} M, $[H_2O_2]$ = 0.25 M, [NaCl] = 10^{-2} M , $UV_{254\ nm}$ irradiation = 14 watts, US intensity = 300 watts , Temperature = 25°C, Reaction time = 60 minutes.

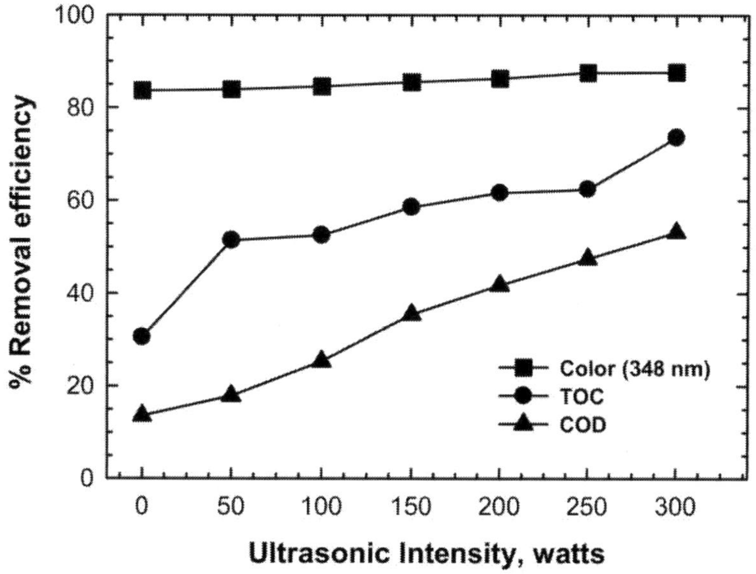

Figure 3. Effect of ultrasound intensity on the removal of TOC, COD, and color from the ammunition wastewater. Experimental conditions: $[TOC]_0$ = 710±50 mg/L, $[COD]_0$ = 1450±150 mg/L, Initial pH = 2.7±0.1, $[Fe^{2+}] = 5\times10^{-4}$ M, $[H_2O_2]$ = 0.25 M, [NaCl] = 10^{-2} M , Temperature = 25°C, Reaction time =60 minutes.

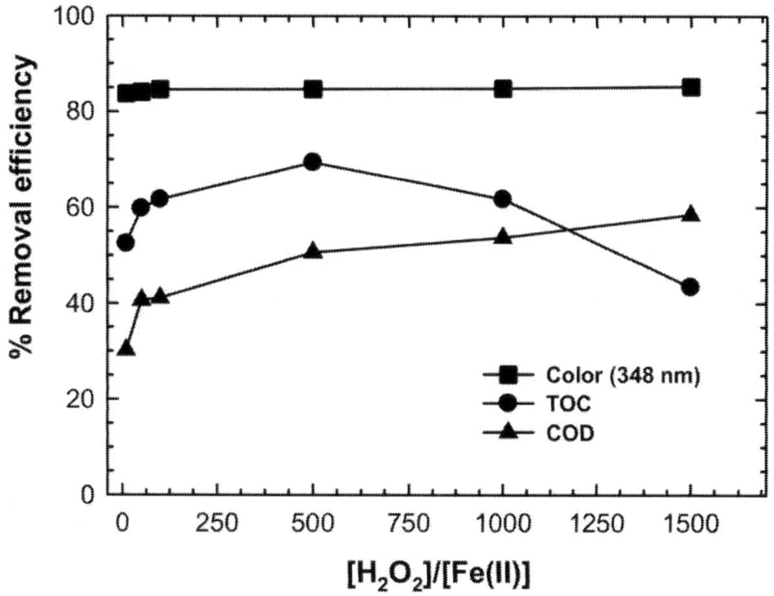

Figure 4. Effect of $[H_2O_2]/[Fe^{2+}]$ on the removal of TOC, COD, and color from the ammunition wastewater. Experimental conditions: $[TOC]_0$ = 710±50 mg/L, $[COD]_0$ = 1450±150 mg/L, Initial pH = 2.7±0.1, [NaCl] = 10^{-2} M, Temperature = 25°C, US intensity = 300 watts, Reaction time = 60 minutes.

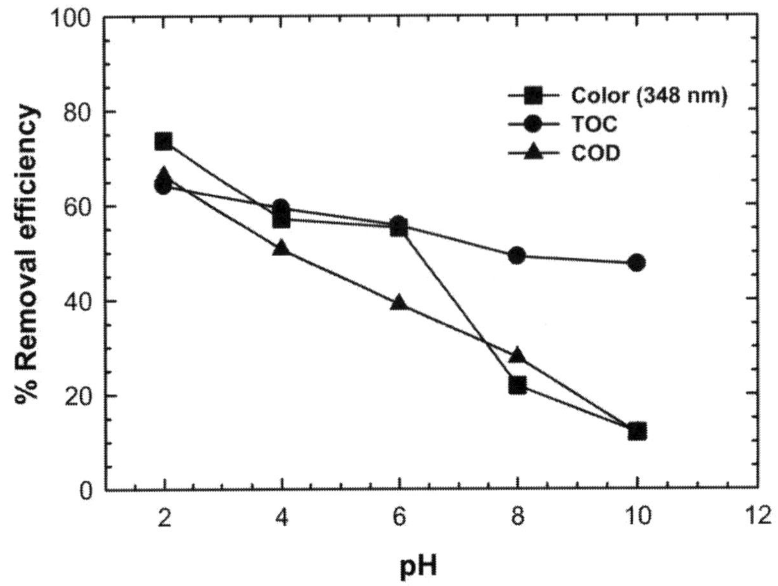

Figure 5. Effect of pH on the removal of TOC, COD, and color from the ammunition wastewater. Experimental conditions: $[TOC]_0$ = 710±50 mg/L, $[COD]_0$ = 1450±150 mg/L, $[Fe^{2+}] = 5\times10^{-4}$M, $[H_2O_2]$ = 0.25 M, [NaCl] = 10^{-2} M, Temperature = 25°C, US intensity = 300 watts, Reaction time = 60 minutes.

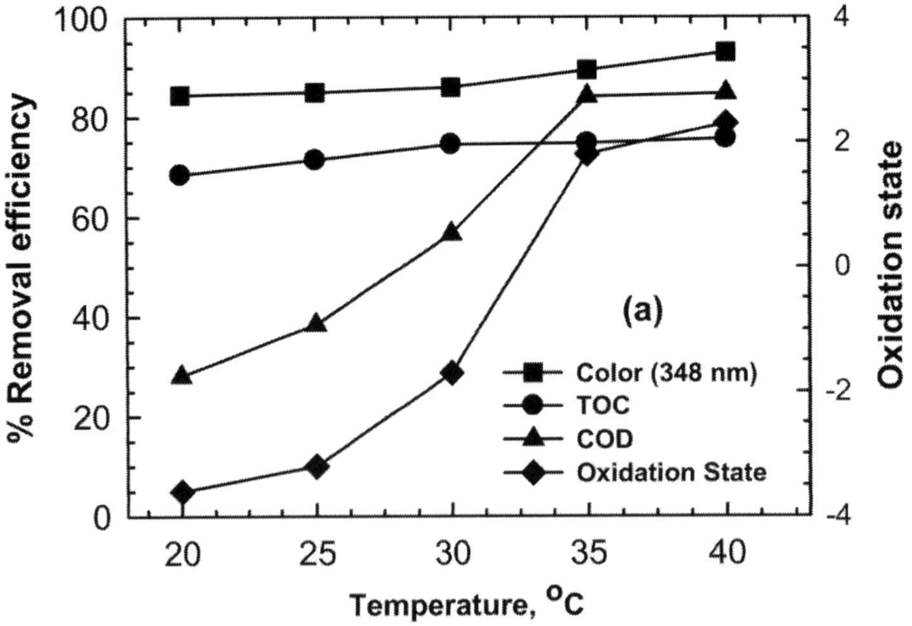

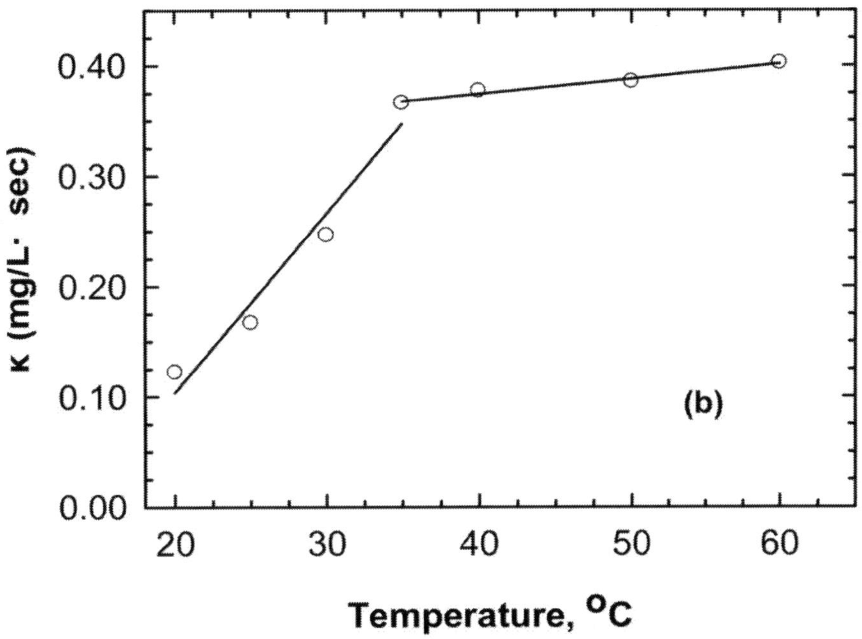

Figure 6. Effect of temperature on changes of carbon oxidation state and TOC, COD, and color removal efficiencies (a) and COD degradation rate constant for the ammunition wastewater (b). Experiment conditions: $[TOC]_0$ = 710±50 mg/L, $[COD]_0$ = 1450±150 mg/L, $[Fe^{2+}]$ = 5×10-4 M, $[H_2O_2]$ = 0.25M, [NaCl] = 10^{-2} M, Initial pH = 2.7±0.1, US intensity = 300 watts, Reaction time = 60 minutes.

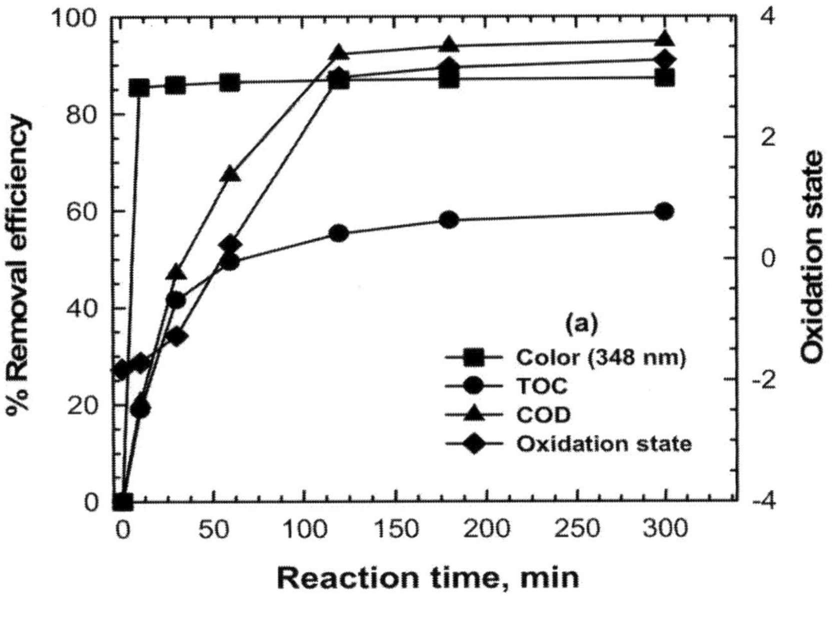

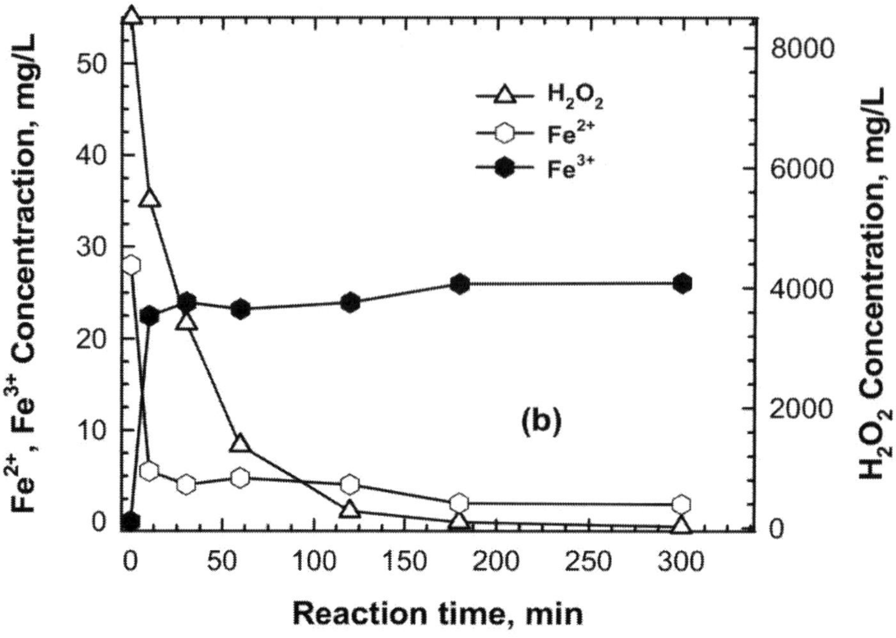

Figure 7. The removal of TOC, COD, and color from the ammunition wastewater (a) and change in the concentration of Fe^{2+}, Fe^{3+}, and H_2O_2 as a function of time (b). Experimental conditions: $[TOC]_0$ = 710±50 mg/L, $[COD]_0$ = 1450±150 mg/L, Initial pH = 2.7±0.1, $[Fe^{2+}]$ = 5×10^{-4} M, $[H_2O_2]$ = 0.25M, [NaCl] = 10^{-2} M, US intensity = 300 watts.

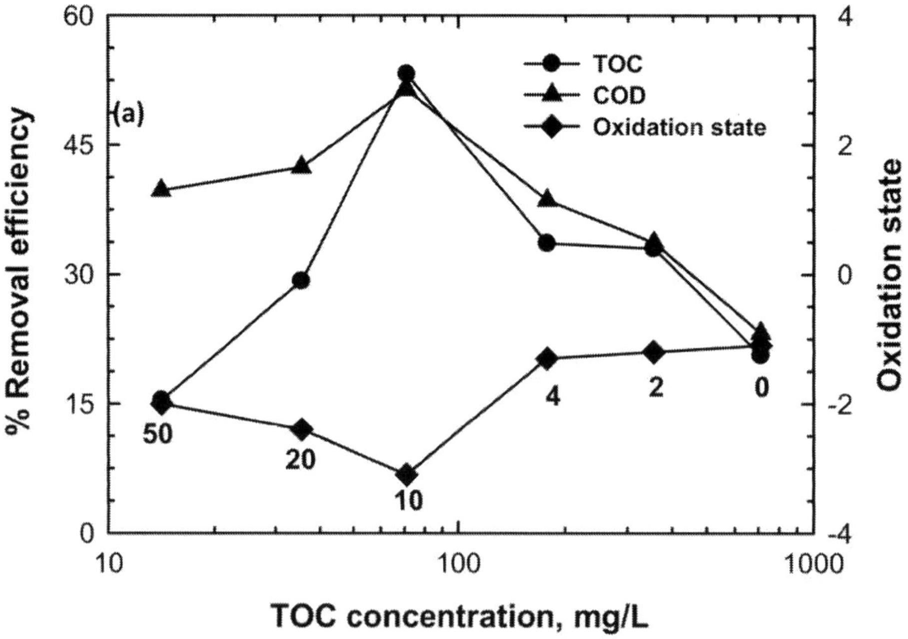

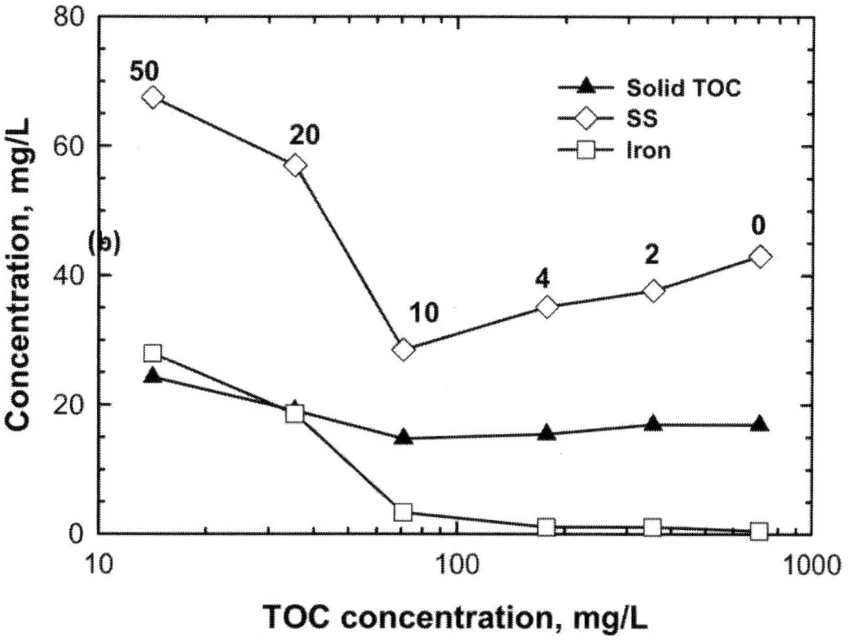

Figure 8. The effect of the initial concentration on the TOC and COD removal efficiency and carbon oxidation state during the treatment of the ammunition wastewater (a) and on the suspended solids (SS) concentration, iron concentration, and solid TOC (b). Numbers represent the dilution factor of the ammunition wastewater. Experimental conditions: $[TOC]_0$ = 710±50 mg/L, $[COD]_0$ = 1450±150 mg/L, $[Fe^{2+}]$ = 5×10^{-4} M, $[H_2O_2]$ = 0.25M, [NaCl]= 10^{-2} M, Initial pH = 2.7±0.1, Temperature = 25°C, US intensity = 300 watts, Reaction time = 60 min.

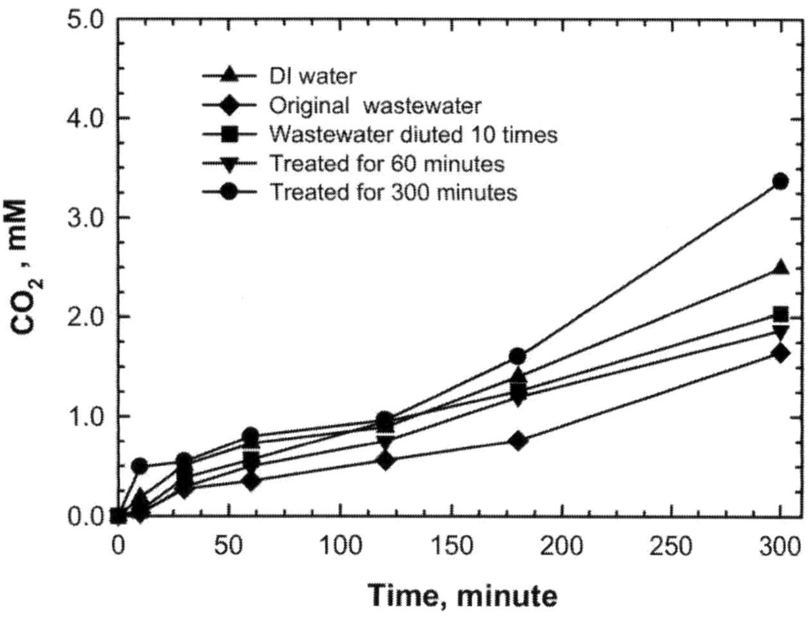

Figure 9. The toxicity test for the treated and untreated ammunition wastewater. Experimental conditions: pH = 7.0, Temperature = 37°C, Culture medium volume =50 mL.

(399b) COD Reduction From Leachate with An Advanced Oxidation Process (AOP)

Kashinath Banerjee, Veolia Water Solution & Tech, Moon Township, PA and Tapas Das, School of Engineering, Saint Martin's University, Lacey, WA

A landfill in the eastern part of United States was in operation between early to mid 1990s. The existing leachate treatment facility consists of pH adjustment, chemical coagulation, solid/liquid separation, sand filtration, carbon adsorption, and sludge dewatering. After the treatment, the water is discharged to the river, and is regulated by a National Pollution Discharge Elimination System (NPDES) permit. Occasionally, the discharge from the treatment facility exceeded the permit limitations for Chemical Oxygen Demand (COD), Biochemical Oxygen Demand (BOD), and Total Organic Carbon (TOC). The objectives of this study were to determine treatment methods which would enable compliance with the applicable discharge limitations; to establish the desired operating conditions of the process; and to investigate the effect of various parameters such as pH, catalyst dosages, reaction time, etc. on the COD destruction efficiency.

A series of influent and effluent samples were collected from the existing leachate treatment facility to characterize the wastewater, to determine the efficiency of the existing treatment system, and to perform treatability studies. All samples were preserved and handled in accordance with the U.S. EPA approved procedures. All analyses were performed in accordance with the procedures described in Standard Methods for the Examination of Water and Wastewater. Treatability studies were conducted using the following oxidation technologies: chlorine, potassium permanganate at different pH conditions, Fenton Reagent (hydrogen peroxide in presence of Ferrous ion), and Hydrogen peroxide/UV light.

The initial wastewater characterization results reveal that the COD, BOD, and TOC concentrations of the sample were approximately 150, 33, and 50 mg/L respectively. The sample had a pH of about 10.8, contained phenol ranging between 3 and 5 mg/L.

Chlorine, and potassium permanganate under acidic as well as under alkaline conditions were not effective in reducing COD to less than the discharge limitation of 35 mg/L. Treatability study results revealed Fenton Reagent or a combination of H2O2/UV which generated hydroxyl radicals (OH*) were capable to reduce the COD to below 35 mg/L within 30-minutes of reaction. The Fenton reaction was conducted at a pH of about 3.5 in presence of a hydrogen peroxide dosage of 1700 mg/L and about 200 mg/L of ferrous iron. After the reaction was completed, the residual hydrogen peroxide was removed from the reactor using a reducing agent, and the pH of the resultant mixture was raised to about 8.0 by the addition of sodium hydroxide. COD reduction with H2O2/UV is similar to that described above, except that in this process, the organic molecules will be oxidized using a combination of UV/H2O2/catalyst.

(399c) Photocatalytic Decomposition of Geosmin and 2-Methylisoborneol by M-InVO4/TiO2 with Visible Light In Recirculating Aquaculture Systems

Sandra L. Pettit[1], Louis J. Rufo[2] and John T. Wolan[2], (1)Dept of Chemical & Biomedical Engineering, University of South Florida, Tampa, FL, (2)University of South Florida, Tampa, FL

Geosmin and 2-methylisoborneol (MIB) are odor and taste contaminants often found in drinking water and aquaculture systems. They are tertiary alcohols produced by several classes of cyanobacteria and actinobacteria. They have a strong odor which can be detected by humans at concentrations as low as 4-10 ng/L. In addition, these compounds impart a gamey, off-flavor taste to aquaculture species. Although they are not a direct threat to public health, consumers frequently use smell as a primary indicator of safety.

Conventional water treatment (coagulation/ flocculation/ sedimentation/ filtration) and chlorination are not effective in the removal of these compounds. Powdered or granulated activated carbon is generally effective. However, this usually requires high dosing rates and lose efficacy with increased water turbidity. Ozonation, although effective in drinking water systems, has several limitations in recirculating aquaculture systems (RAS). Dissolved organic material present in RAS rapidly depletes ozone, reducing the residual ozone concentration. Furthermore, ozone residuals may cause gill adhesions and increased mortality rate in fish. Because of these drawbacks, many aquaculture systems use fresh water purge to reduce geosmin and MIB concentrations. Purge systems consume large quantities of fresh water and produce significant waste streams.

Ideally, an effective removal system would balance water consumption, fish mortality, and operating cost. TiO_2 is an effective photocatalyst for the decomposition of organic compounds. It is inexpensive, photostable, and chemically and biologically inert. However, only the ultraviolet portion of the solar irradiation (amounting to ~4% of the incoming solar energy on the earth's surface) is absorbed by TiO_2 due to its high intrinsic band gap (3.2 eV for anatase and 3.0 eV for rutile). When combined with M-$InVO_4$, where M is a transition metal, photoinitiation is moved to the visible spectrum range via electron transfer. In the present study, efficacy of M-$InVO_4/TiO_2$ is explored for visible-range photocatalytic decomposition of geosmin and MIB in recirculating aquaculture systems.

See more of this Session: Photo-Catalytic and Advanced Oxidation/Reduction Processes for Water Treatment

(399d) Advanced Oxidation for Water Processing In Microstructured Reactors

Kevin Caple, Nick Kusanto, Kenneth Williamson, **Jeffrey A. Nason**, Goran N. Jovanovic and Alexandre F.T. Yokochi, School of Chemical, Biological and Environmental Engineering, Oregon State University, Corvallis, OR

An important issue in potable water processing is the need for removal of trace organic chemicals. Conventional water treatment (coagulation/ flocculation/ sedimentation/ filtration) and chlorination is not always effective in the removal of these compounds. An approach that has been proposed for the purpose is advanced oxidation employing photochemical reactions at the surface of a photocatalyst such as TiO_2 [[i]].

Since the photochemical oxidation reaction is a surface reaction, to achieve sufficient flow rates it is imperative to increase the surface to volume ratio of the reactor, and we have investigated employing microstructured reactors implementing these advanced oxidation processes. An approach that has been employed to activate reactions in macroscale reactors is the use of electric discharge through the reaction medium which leads to the creation of highly reactive species as a consequence of the electrical discharge [[ii]] as an alternative to photocatalyzed approaches. We have therefore developed a microreactor employing a low voltage corona discharge for the disinfection and organic compound oxidation in water, a technique that moves the oxidation reaction from the surface of the reactor to the bulk of the fluid, thus increasing process efficiency.

In our presentation we will discuss recent results of the use of microstructured reactors for the advanced oxidation of organic contaminants in water.

[i] Ollis, D.F.; Pelizzetti, E.; Serpone , N. “Destruction of water contaminants” *Environ. Sci. Technol.* **1991**, *25*, 1522-1529.

[ii] Kirkpatrick, M. J.; Locke, B. R. “Hydrogen, Oxygen, and Hydrogen Peroxide Formation in Aqueous Phase Pulsed Corona Electrical Discharge” *Ind. Eng. Chem. Res.* **2005**, 44, 4243-4248.

See more of this Session: Photo-Catalytic and Advanced Oxidation/Reduction Processes for Water Treatment
See more of this Group/Topical: Topical 1: Water Technology for Developed and Developing Countries (see also Separations Division)

KINETICS OF OXIDATION OF DYES IN EFFLUENTS USING A POROUS ELECTRODE OZONATOR

Alexander P. Mathews[1] and Kishora K. Panda[2]
[1]Department of Civil Engineering, Kansas State University, Manhattan, KS 66506, USA
[2]HDR Engineering, Inc., 3230 El Camino Real, Suite 200, Irvine, CA 92604

ABSTRACT

Ozonation of dye Reactive Blue 19 wastewater was conducted using an *in situ* porous electrode ozone generator and reactor system. In this system, corona discharge is made at the gas liquid interface inside a porous barrier that separates the generator from the tubular reactor. The generated ozone from the outer generator diffuses through this porous barrier and reacts with the contaminants in the fluid that is being treated inside the reactor. Oxidation studies indicate that feed gas flow rate, and ozone concentration, pH, and initial dye concentration affects the kinetics of dye decolorization. Fast kinetic regime observed for dye oxidation followed by intermediate regime for the reaction of ozone with the byproducts. The specific power consumption for ozonation of wastewater in *in situ* reactor is significantly lower than that for conventional ozonators. High mass transfer and low power consumption for this reactor system makes it an efficient and cost-effective reactor for ozonation process.

INTRODUCTION

The treatment and disposal of dye wastewater is an environmental concern due to the inadequacy of the treatment methods in treating such complex wastewater, the aesthetic problem in the receiving water, and the effect on public health. Treatment alternatives available for dye wastewater include coagulation, chemical oxidation, activated carbon adsorption, and biological treatment. Coagulation and carbon adsorption may remove certain categories of dye up to 90% but possesses the drawback of generating large amount of sludge. Conventional biological processes are not successful since most dyes are resistant to biological degradation and toxic to the organisms used (Gardner et al., 1990; Zhou and Smith, 1997). Studies have reported that partial oxidation followed by biodegradation is a cost–effective treatment alternative for this kind of wastewater (Alvares et al., 2001; Zhou and Smith, 1997).

Ozone, an oxidant of choice for many reasons in comparison to others, has limited use at present for the reason of treatment cost. Ozonation efficiency is affected due to mass transfer limitations, and ozone decomposition and reaction kinetics in the reactors. The high costs associated with ozone application can be reduced by increasing the mass transfer rate in the dissolution of ozone gas, and by reacting ozone as it is produced with the contaminants in the solution by developing an *in situ* ozone generator (Mathews et al., 2004).

Ozone generation through the electrical discharge process produces, besides ozone, highly reactive short-lived species near the corona discharge region (Sano et al., 2002). The gaseous species that are generated include free electrons, negative ions, positive ions, and uncharged radicals (Tamon et al., 1995). The current ozonation systems are unable to utilize these short-lived species for ozonation process since the reactor is separated from the generator. Sharma et al., 2000 and Sano et al., 2002 studied the impact of these species on the degradation of organic matters. They reported that the rate of decomposition of organics was higher in comparison to ozone oxidation alone, and attributed this to the reaction of organics with the short-lived species formed near the corona discharge region.

In this research, a porous electrode ozonator has been designed in which the ozone generator and the reactor form a single unit (Mathews et al., 2004). The design consists of an outer peripheral ozone generation region and an inner tubular reactor separated by a porous tube. The corona discharge occurs when a pulsed-modulated AC HT (5kV – 10kV) is applied between the outer electrode and the inner porous electrode. Generated ozone from the corona region diffuses through the porous tube into the reactor through which treatment fluid passes. This design was envisaged to provide several benefits such as increasing ozone mass transfer rate by using a fine porous diffuser, employing the reactive species generated near corona region for degradation of organic matter, consuming less power for operation, and minimizing the loss of ozone by diffusing immediately to water upon generation. The specific goal of the present work was to study the decolorization kinetics of Reactive Blue 19 dye in the tubular *in situ* ozonator, and to determine the effects of process parameters on ozonation efficiency and power consumption in comparison with conventional ozonators.

METHODOLOGY

In situ ozone generator and reactor design

The porous electrode *in situ* ozone generator and reactor system was designed essentially in one unit wherein the inner tubular reactor is separated from the outer peripheral concentric ozone generator by a porous barrier. An outer glass tube of length 29.6 cm and internal diameter 2.62 cm was placed concentrically with an equal length inner porous ceramic tube of outer diameter 2.52 cm, and internal diameter 1.95 cm, and the peripheral gap was made airtight by sealing the ends. The volume of the inner tubular reactor was 88.4 cm_3. Details of the reactor system is described elsewhere (Mathews et al., 2004). Ozone is generated by electrical corona discharge process. The electrical circuitry was designed to supply less hazardous high frequency (15 to 20 kHz) pulsed modulated HT AC (5KV to 10 KV) from a regular 110 V wall terminal (Padmanabhan et al., 2000). The outer glass tube covered with aluminum foil was used as the hot electrode, and the porous tube with an embedded steel wire acted as the ground electrode. Corona discharge occurred in the peripheral gap when high voltage is applied between the two electrodes.

Experimental

Figure 1 shows the experimental set up for studying the oxidation of dye wastewater in the *in situ* ozone generator and reactor system. The dye oxidation process was studied under a closed loop water recirculation system wherein the water was continuously pumped from a recirculation unit of 1.1L capacity and circulated through the reactor. Oxygen (ultra high purity and extra dry grade, Air Products, KS, U.S.A) was supplied through a flow meter with needle valve to the generator for ozone generation. The ozone and oxygen mixture from the generator diffused through the porous ceramic tube and absorbed into the flowing wastewater inside the reactor.

The unreacted dye and the unabsorbed gas were carried to the recirculation unit where the gas was separated and escaped to the vent as off-gas. The unit was designed to separate gas in a disengagement section at the top of the unit while holding the liquid at the bottom. The contents of the recirculation unit were mixed with a magnetic stirrer. The recirculation unit had openings for taking water samples and adding make-up water, off-gas line, and temperature probe. Samples were taken from the recirculation unit and analyzed for dissolved ozone and dye concentration. The off-gas from the reactor was separated from the liquid in the recirculation unit and then sent to an online ozone gas analyzer (BMT 964 analyzer, BMT Messetechnik, Berlin, Germany) through a demister. The off-gas was absorbed in the 2% KI trap and the gas was vented out.

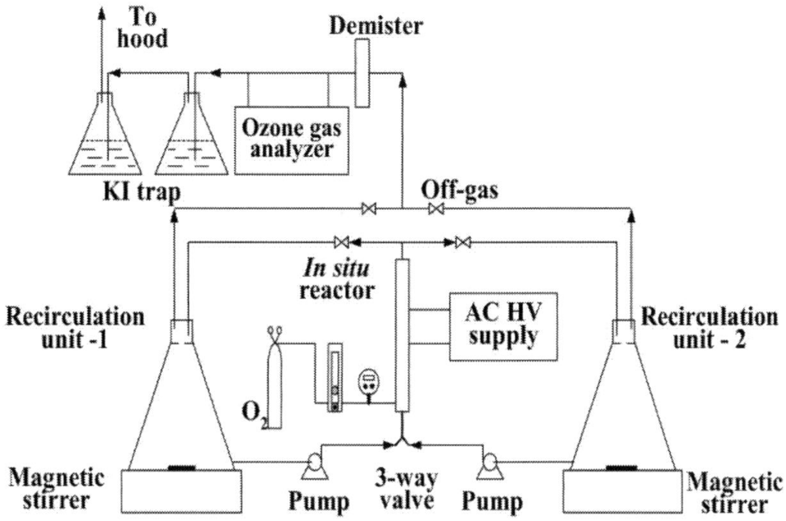

Figure 1. Experimental set-up for dye oxidation study using *in situ* reactor

Analytical Methods

Ozone concentration in the inlet gas was determined indirectly as the generated ozone diffuses and dissolves into the liquid immediately. Ozone generation rates were determined from the analyzed steady state dissolved and off-gas ozone concentrations. Eq.1 can be used to calculate the ozone generation rates.

$$G_{in} = G_{OSS} + \frac{kC_{USS}^{m}V}{Q_g} \tag{1}$$

Here G_{OSS}, C_{USS} are the steady state concentration of off-gas and dissolved ozone respectively, k the ozone self-decomposition rate constant, V the total volume of water in the recirculation unit and the reactor, Q_g the gas flow rate, and m the reaction order for the self-decomposition of ozone in water.

Prior to the dye oxidation experiment, 0.5L of distilled, deionized water at pH 2 was absorbed with ozone in the reactor by circulating the water from another recirculation unit placed in parallel (Fig.1). The experiment was carried out until dissolved ozone concentration in the recirculation unit reached a constant value. A voltage regulator (Variac Autotransformer, General Radio Company) was used to adjust the power supply to the circuitry to obtain a desired generation rate. The dissolved ozone concentration was determined by using the indigo colorimetric method (Clesceri et al., 1998).

Dye concentration in wastewater was determined from the standard curve developed for Reactive Blue 19 dye. Standard dye solutions at concentrations of 0.002 to 0.16 g/L were prepared, and their absorbance was measured at the maximum absorbed wavelength of 590nm on the spectrophotometer (Cary 50 Bio UV-Vis system, Varian, Walnut Creek, California). During dye ozonation experiments, water samples were removed from the recirculation unit and absorbance was read immediately on the spectrophotometer without any dilution. The sample water was returned into the recirulcuation unit to maintain a constant liquid volume inside the unit.

Distilled deionized (DDI) water (conductivity <0.06 μS/cm) was used to prepare dye wastewater. Oxidation of dye Reactive Blue 19 (Sigma-Aldrich Co., St.Louis, MO) was studied at pH 2 and 7. The pH was adjusted to 2 by addition of sulfuric acid, and to 7 by addition of calculated amounts of NaH_2PO_4 and Na_2HPO_4. All experiments were conducted at a temperature of 21 ± 0.5^0C, and ionic strength of 0.05 M (pH 2) and 0.1M (pH 7). Other chemicals used were reagent grade and were obtained from either Fisher Scientific or Sigma-Aldrich Co.

RESULTS AND DISCUSSIONS

Ozone generation rate

The ozone generation rate was calculated from dissolved and off-gas ozone measurements. It was assumed that the ozone generated in the corona discharge region was either absorbed into the liquid phase or escaped into the gas phase, and there is no ozone loss by reaction while diffusing through the porous ceramic electrode. Ozone generation rates at different gas flow rates and corresponding steady state off-gas ozone concentrations are shown in Table 1. As indicated in the Table and Eq. (1), the difference between G_{in} and G_{OSS} is due to the self-decomposition of ozone in water, and varies from 0.529% to 4.332%, increasing with a decrease in flow rate. The differences are higher at a higher pH of 7 due to the higher ozone decomposition rate by reaction with the hydroxyl ions. The generation rates were also verified by using the KI wet chemistry method (Rakness et al., 1996) and results were found to be within ±5% of the values (Table 1) calculated by the mass balance method. This deviation is acceptable considering the fluctuations in ozone generation using high frequency and high voltage corona discharge.

Ozonation of dye in in situ reactor

The concentration of dye, dissolved ozone, and off-gas ozone in recirculation unit was monitored during the ozonation of Reactive Blue 19 dye. Figure 2 shows the profiles for the three measured parameters for an ozonation experiment in *in situ* reactor. From the figure, it can be concluded that the Reactive Blue 19 ozonation can be divided into two phases. During the first 35 minutes, the process is governed by the oxidation of dye. The reaction regime is observed to be fast as the

Table 1 - Estimated ozone generation rate based on mass balance (pH = 2, and liquid velocity u_l= 20.1 m/h) and KI test (Panda and Mathews, 2004).

Feed gas flow Q_g (L/min)	Feed gas pressure (kPa)	Steady-state off-gas O_3 G_{OSS} (mg/L)	Ozone generated* G_{in}† (mg/L)	Steady-state off-gas O_3 flow rate (g/h)	Generation rate (g/h)†		Variation (%)
					Mass Balance	KI Test	
0.02	115.92	35.00	36.58	0.0493	0.0515	0.0542	0.84
0.03	119.37	31.25	31.88	0.0680	0.0694	0.0671	1.23
0.04	122.82	25.00	25.54	0.0747	0.0763	0.0737	1.70
0.05	133.16	33.80	33.98	0.1368	0.1375	0.1399	2.24

† not at constant voltage
* Calculated from mass balance

dissolved ozone concentration in water is zero. All the absorbed ozone in the water reacted with the dye either in the film or in the bulk water. The chemical reaction in the liquid increased the mass transfer rate which was observed from the difference between the feed gas and off-gas ozone concentration. During the second period i.e after 35 minutes, the oxidation of by-products governs the oxidation process. The reaction kinetics of by-products with ozone in this phase is low and not limited by the mass transfer rate. In addition, appearance of dissolved ozone indicates that there is no significant reaction of ozone with the by-products inside the film.

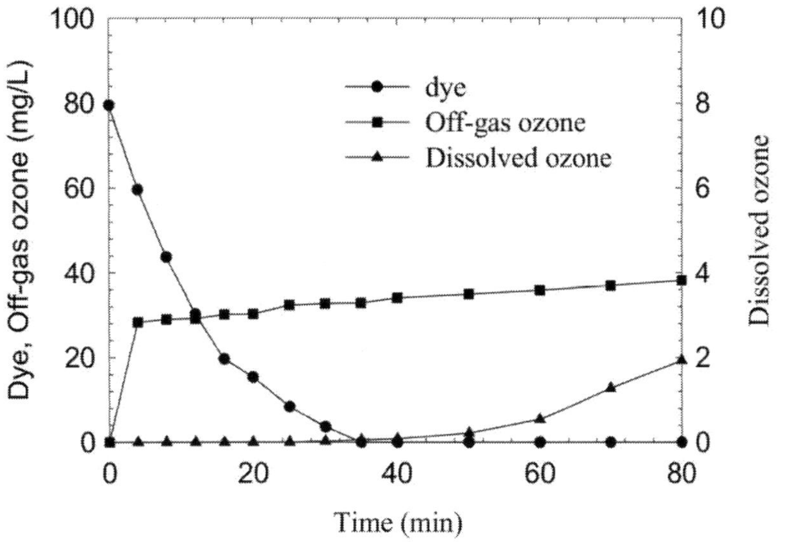

Figure 2. Concentration profiles for dye, dissolved, and off-gas ozone in *in situ* reactor (pH=2, G_{in} =50 g/Nm^3, C_o = 80 mg/L, Q_g=50 mL/min)

Effect of gas flow rate

The rate of dye oxidation was investigated by varying the gas flow rate. Figure 3 presents the dye decolorization at two different gas flow rates of 56 mL/min and 118 mL/min for an initial dye concentration of 80 mg/L. The rate of decolorization was higher at gas flow of 118mL/min than that at 56 mL/min. Since dye oxidation reaction is in fast kinetic regime until dye is present in water, the decolorization rate is mass transfer limited. So at a constant ozone generation rate, higher reaction rate is due to the increased mass transfer rate at gas flow rate of 118 mL/min in comparison to that at 56 mL/min. Gas flow rate has been reported to increase the specific interfacial area due to decrease in bubble size in stirred tank reactor (Sheffer and Esterson, 1982; Wu and Masten, 2001) and bubble columns (Roustan et al., 1996). In the *in situ* reactor, the mixing regime and the bubble size are possibly changed with increase in gas flow rate, and that affected the overall mass transfer rate by decreasing the bubble size and increasing the mixing intensity inside the reactor. The bubble diameters were difficult to measure in this small and closed reactor, but mixing was observed to increase with the increase in gas flow.

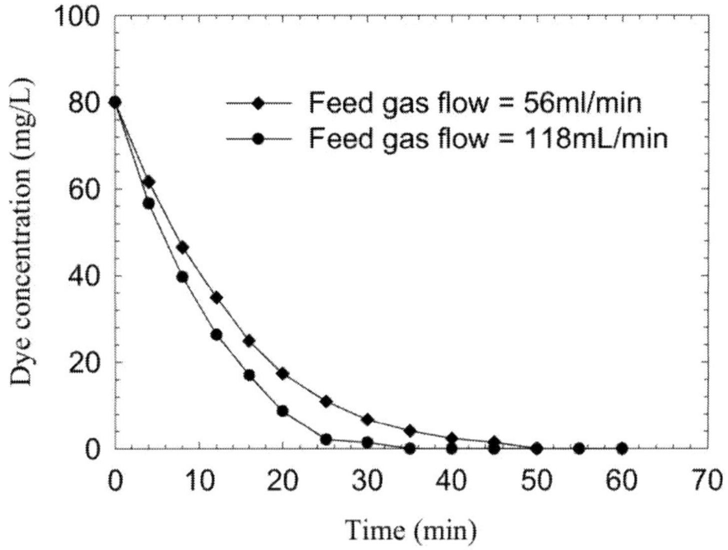

Figure 3. Effect of feed gas flow rate on dye decolorization
(C_o = 80 mg/L, G_{in} = 50 g/Nm^3, and pH = 2)

From the mass transfer studies in *in situ* reactor, a power law relationship, given in Equation 2 and 3, has been observed between superficial gas velocity and overall mass transfer coefficient (Panda and Mathews, 2004).

$$\text{pH} = 2\text{: } K_La = 0.4149\, u_g^{\,0.3878} \tag{2}$$

$$\text{pH} = 7\text{: } K_La = 0.3775\, u_g^{\,0.4495} \tag{3}$$

where K_La is the overall mass transfer coefficient (min^{-1}), and u_g the superficial gas velocities (m/h).

Effect of pH of water

The ozonation of dye at initial pH of 2 and 7 was examined, and the results for an experimental condition with initial dye concentration of 40 mg/L and ozone generation rate of 60.6 g/Nm^3 are shown in Figure 4. The rate of dye decolorization was observed to be higher at pH 2 than that at pH 7. Ozonation reaction pathways include direct oxidation by ozone or radical oxidation by hydroxyl radical. The hydroxyl radical reaction predominates under basic conditions and oxidation potential of the reaction is much higher than that for molecular ozone. So reaction rate at pH 7 is expected to be higher than at pH 2. For this particular anthraquinone dye, similar findings were reported in literature in relation to the effect of pH on dye oxidation (Chu and Ma, 2000). The characteristics of reactive blue 19 such as insensible structure for oxidation, leuco-form, and higher solubility in acidic water contribute to a higher decolorization rate in acidic conditions.

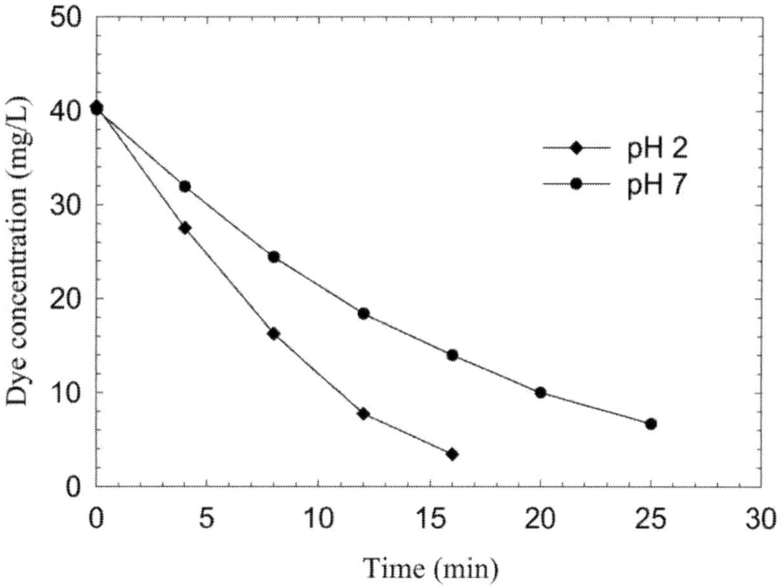

Figure 4. Effect of pH on decolorization rate of dye wastewater (C_o =40 mg/L, G_{in} = 60.6 g/Nm^3, and Q_g =33.6 mL/min)

Effect of feed gas ozone concentration

Figure 5 displays the effect of feed gas ozone concentration on the rate of decolorization at an initial dye concentration of 80 mg/L and pH 2. Three ozone concentrations investigated were

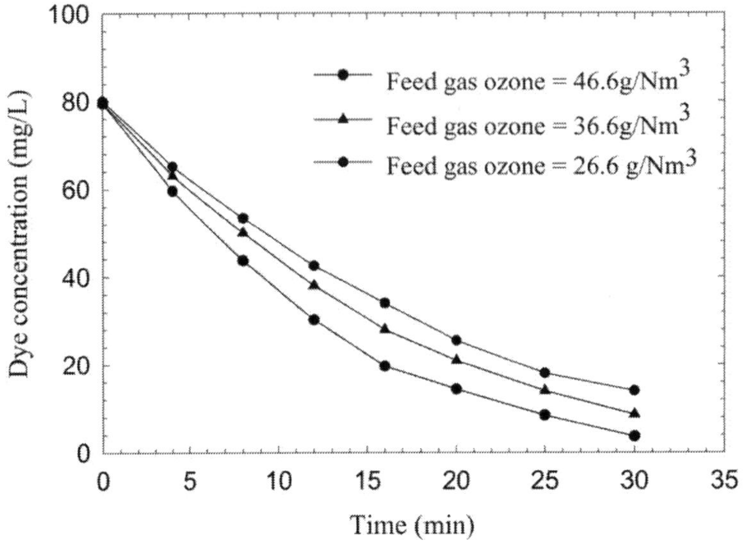

26.6 g/Nm^3, 36.6 g/Nm^3, and 46.6 g/Nm^3 at a constant feed gas flow of 50 mL/min. As shown in the figure, the decolorization rate increased with the increase in the feed gas concentration. Increasing feed gas ozone concentration increases the mass transfer rate, thus increasing the decolorization of dye for which reaction rate is mass transfer limited until complete dye decolorization. Shu and Huang, 1995 reported the increase in oxidation rate of dye Acid Orange 10 with an increase in mass transfer rate in the reactor. Decolorization of dye reactive blue 19 was studied by Lall et al., 2003, and observed an increase in reaction rate with the increase in concentration of ozone in inlet gas.

Figure 5. Effect of feed gas ozone concentration on dye decolorization (C_o = 80 mg/L, pH = 2, and Q_g = 50 ml/min)

Effect of initial dye concentration

The dye decolorization in *in situ* reactor was studied by varying the initial dye concentration in water. Results of dye decolorization at two initial concentrations of 40 mg/L and 80 mg/L are presented in Figure 6. The reaction rate decreased with the increase in the initial dye

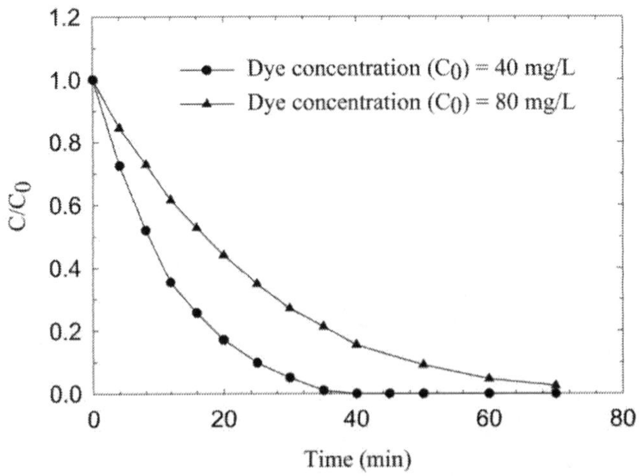

Figure 6. Effect of initial dye concentration on dye Decolorization (G_{in} = 50 g/Nm^3, pH 7 and Q_g = 50 mL/min)

concentration and this result is in agreement with the literature (Chu and Ma, 2000; Lall et al., 2003; Wu and Wang, 2001). These researchers used regular ozonators to study the dye decolorization. Hence the effect of initial dye concentration on kinetics of dye decolorization as observed in *in situ* reactor is similar to that in regular ozonators.

Power consumption

The *in situ* ozonator developed in this study is relatively small in size in comparison to regular ozonators. In comparision, *in situ* reactor needs significantly less power, required to circulate water through a small reactor. The specific power requirement is calculated for the *in situ* reactor to range from 0.002 to 0.004 kWm^{-3} (Panda and Mathews, 2004). The range of values for power requirement in bubble columns and stirred tank reactors reported in literature are from 0.01 to 0.1 kW m^{-3} and 0.5 to 10 kW m^{-3} respectively (Panda and Mathews, 2004).

CONCLUSIONS

Reaction efficiencies for dye decolorization are higher in this reactor system due to *in situ* ozone dissolution and reaction at the porous electrode gas-liquid interface, and involvement of reactive species formed near corona region in oxidation reaction. The reaction kinetics was observed to be in fast regime until dye was present in the water. Feed gas flow rate and ozone concentration, initial dye concentration, and pH of water affected the kinetics of oxidation. Kinetics of decolorization is directly dependent on the dye concentration. Higher reaction rate was observed in acidic medium than in neutral water due to the typical characteristics of this particular dye such as leuco-form, high solubility in acidic water, and insensible structure for oxidation. Small size of the reactor and absence of any stirring requirement makes this design less power intensive in comparison to the bubble column and the stirred tank reactor used in ozonation process.

ACKNOWLEDGEMENTS

This research was supported by a research grant and REU supplement from the National Science Foundation (BES – 0209343). We acknowledge and appreciate the assistance of Jamie Klein in the conduct of studies and laboratory analysis.

REFERENCES

Zhou, H.; Smith, D. W. (1997) Process Parameter Development for Ozonation of Kraft Pulp Mill Effluents. *Water Sci. Tech.*, **35** (2-3), 251.

Gardner, D. A.; Holdsworth, T. J.; Shaul, G. M.; Dostal, K. A.; Betowski, D. L. (1990) *Aerobic and Anaerobic Treatment of C.I. Disperse Blue* 791 EPA-600/S2-89/051; Office of Research and Development, U.S. Environmental Protection Agency, Cincinnati, Ohio.

Alvares, A. B. C.; Diaper, C., and Parsons, S. A. (2001) Partial Oxidation of Hydrolyzed and Unhydrolyzed Textile Azo Dyes by Ozone and the Effect on Biodegradability. *Trans IchemE,* **79** (B), 103.

Mathews, A. P.; Panda, K. K.; Ananthi, S.; Padmanabhan, K. (2004) Mass Transfer and Oxidation Kinetics in an *In Situ* Ozone Generator. *Water Sc. & Tech.*, **49** (4), 19.

Sano, N.; Kawashima, T.; Fujikawa, J.; Fujimoto, T.; Kitai, T.; Kanki, T. (2002) Decomposition of Organic Compounds in Water by Direct Contact of Gas Corona Discharge: Influence of Discharge Conditions. *Ind. Eng. Chem. Res.*, **41**, 5906.

Tamon, H.; Mizoata, H.; Sano, N.; Okazaki, M. (1995) New Concept of Gas Purification by Electron Attachment, *AIChE J.*, **41**, 1701.

Sharma, A. K.; Camaioni, D. M.; Josephson, G. B.; Goheen, S. C. (2000) Destruction of Pentachlorophenol Using Glow Discharge Plasma Process. *Env. Sci. Technol.*, **34**, 2267.

Padmanabhan, K.; Ananthi, S.; Patel, K. (2000) Portable Ozone Generator. *Electronics for You,* **32**, 46.

Clesceri, L. S.; Greenburg, A. E.; Eaton, A. D. (1998) *Standard Methods for the Examination of Water and Wastewater*. 20th Edn., Washington, DC.

Panda, K. K.; Mathews, A. P. (2004) Ozone Mass Transfer in an *In Situ* Ozone Generator and Reactor II. Experimental Studies. *Water Research,* (under review)

Sheffer, S.; Esterson, G. L. (1982) Mass Transfer and Reaction Kinetics in the Ozone/Tap Water System. *Water Research,* **16**, 383.

Wu, J. J.; Masten, S. J. (2001) Mass Transfer of Ozone in Semi Batch Stirred Reactor. *J. Env. Eng.*, **127** (12), 1089.

Roustan M.; Wang R.Y.; Wolbert, D. (1996) Modeling Hydrodynamic and Mass Transfer Parameters in a Continuous Ozone Bubble Column. *Ozone Sci. Eng.*, **18**, 99.

Chu, W.; Ma, C. (2000) Quantitative Prediction of Direct and Indirect Dye Ozonation Kinetics. *Water Res.,* **34** (12), 3153.

Shu, H. Y.; Huang, C. R. (1995) Degradation of Commercial Azo Dyes in Water Using Ozonation and UV Enhanced Ozonation Process. *Chemosphere*, **31**, 3813.

Lall, R.; Mutharaan, R.; Shah, Y. T.; Dhurajati, P. (2003) Decolorization of the Dye, reactive Blue 19, Using Ozonation, Ultrasound, and Ultrasound-Enhanced Ozonation. *Water Env. Res.*, **75**(2), 171.

Wu, J.; Wang, T. (2001) Ozonation of Aqueous Azo Dye in a Semi-Batch reactor. *Wat. Res.*, **35** (4), 1093.

(399f) Adsorptive-Reactive Nanoparticulate Systems for the In Situ Remediation of Chlorinated Hydrocarbons

Bhanukiran Sunkara[1], Jingjing Zhan[1], Yingqing Wang[1], Vijay T. John[2], Jibao He[3] and Gary McPherson[4], (1)Department of Chemical and Biomolecular Engineering, Tulane University, New Orleans, LA, (2)Chemical & biomolecular engineering, Tulane university, New orleans, LA, (3)Central Instrumental Facilities, Tulane University, New Orleans, LA, (4)Department of Chemistry, Tulane University, New Orleans, LA

Chlorinated hydrocarbons such as trichloroethylene (TCE) form a class of dense non-aqueous-phase liquid (DNAPL) toxic contaminants in soil and groundwater. The *in situ* injection of nanoscale zerovalent iron (NZVI) to reduce DNAPLs is a potentially simple, cost-effective, and environmentally benign technology that become a preferred method in the remediation of these compounds. However, unsupported NZVI particles exhibit ferromagnetism leading to particle aggregation and a loss in mobility through the subsurface. This work focuses on engineering submicron spherical carbon particles as effective carriers/supports for NZVI particles to address the *in situ* remediation of chlorinated hydrocarbons. We describe here a simple one-step process, demonstrating that it is possible to vary the placement of iron nanoparticles on carbon microspheres either on the surface of the microsphere or in the interior. These composites have high reactivities coupled with strong adsorption capabilities. The optimal size of these multifunctional particulate systems enables effective transport through sediments. The particulate systems are obtained from inexpensive precursors and through a semi-continuous method which allows for large scale synthesis of the composites necessary for eventual *in-situ* application. The detailed characterization of these multifunctional colloids through high resolution electron microscopy and their functional properties will be described. The mobility of these composites through model sediments using capillary and column transport experiments will be discussed.

See more of this Session: Photo-Catalytic and Advanced Oxidation/Reduction Processes for Water Treatment
See more of this Group/Topical: Topical 1: Water Technology for Developed and Developing Countries (see also Separations Division)

Considering Probability in Process Optimization Using Bayesian Network Approach: MTBE Production Case Study

Anahita Sadeghian[1], Farhang Jalali Farahani[2]
[1] M.Sc. student, [2] Professor, Chemical Eng. Dept., University of Tehran, Iran

Abstract

One of the major areas of engineering application of optimization is found in the tuning of existing operating plants. The chemical industry has undergone significant changes during the past decades due to the increased cost of energy, increasingly stringent environmental regulations, and global competition in product pricing and quality and one of the most important engineering tools for addressing these issues is optimization; but formal application of optimization is really not warranted because of the uncertainty that exists in the mathematical representation of the process or the data used in the model of the process, so optimizing the processes considering uncertainties has some merit. Uncertainty arises in many situations, for example experts may be uncertain about their own knowledge; there may be uncertainty inherent in the situation being modeled, or uncertainty about the accuracy and availability of information [*1*].

Since the field of optimization is still a very active research area and in recent years various new approaches to optimization have been proposed, this paper aims to use a new probabilistic approach allowing the uncertain parameter to be a random variable and lets us make probability statements about it, posterior to the data sets.

In this study, a Bayesian network, as a graphical model for reasoning under uncertainty is used as a model of the process operation. By building a probabilistic model, i.e. Bayesian network, the quantity of the plant product is maximized regarding to uncertainties like flow rates and temperatures.

Key words*: Optimization, Uncertainty, Bayesian Network, MTBE, Probability*

Introduction

The Bayesian approach will typically yield improved analyses, by better taking into account what is actually known and what is not known about the system to be optimized. We briefly review Bayesian methods for readers who are not familiar with this type of analysis in the coming sections and suggest ways in which these methods can be applied to optimization.

Bayesian methods are particularly appropriate for use in practice when optimizations are costly, or when input uncertainties are large. Our work provides guidance on the use of Bayesian methods for optimization analysis. We hope that it will stimulate readers to learn more about this important subject, and also encourage further research in this area.

Since we cannot completely eliminate uncertainty, we need to model it. In real life when we are faced with uncertainty, we use plausible reasoning. We adjust our belief about something, based on the occurrence or non-occurrence of something else. Statistics is the science that relates data to specific questions of interest. This includes devising methods to gather data relevant to the question, methods to summarize and display the data to shed light on the question, and methods that enable us to draw answers to the question that are supported by the data.

Data almost always contain uncertainty. This uncertainty may arise from selection of the items to be measured, or it may arise from variability of the measurement process.

Analysis of the results of probabilistic optimization rests fundamentally on the application of statistics. Historically, there have been two major schools of thought in statistics-Bayesian and classical. However, with few exceptions (e.g., [*2*, *3*]), only the classical approach has been significantly represented in the literature. The main advantage of Bayesian statistics is its ability to incorporate prior information into an analysis in a rigorous and coherent fashion-see, e.g., [*4*, *5*, *6*, *7*, *8*] for introductions to Bayesian statistics.

Bayesian belief networks offer consistent semantics for representing uncertainty and an intuitive graphical representation of the interactions between various causes and effects. They often produce very convincing results when the historical information in the conditional probability tables or the evidence known is inexact. A Bayesian belief network is a model that is especially useful when the information about the past and/or the current situation is vague, incomplete, conflicting, and uncertain. With the historical information stored in the conditional probability tables, they can be used to help make decisions, or as a way of automating a decision-making process. A quantified Bayesian Network represents a complete probabilistic model of the domain, here the process. One can easily use such a network to derive the joint probability distribution for all variables involved; and it is clear that the network contains sufficient information for computing answers to all queries regarding the variables [*9*].

Bayesian networks have been used in a wide variety of domains, including biology [*10*], computer games [*11*], data mining [*12*, *13*], diagnosis of failures [*14*, *15*], fault diagnosis in chemical processes [*16*], medicine [*17*, *18*, *19*, *20*, *21*], prediction [*22*, *23*], reliability analysis [*24*], students advising [*25*, *26*], forensic science [*27*] and weather forecasting [*28*].

Theoretical framework

Bayesian approach

There are two main philosophical approaches to statistics. The first is often referred to as the frequentist approach. Sometimes it is called the classical approach. The alternative approach that we apply is the Bayesian approach. It applies the laws of probability directly to the problem. This offers many fundamental advantages over the more commonly used frequentist approach. The Reverend Thomas Bayes first discovered a theorem that now bears his name.

Bayes showed how inverse probability could be used to calculate probability of antecedent events from the occurrence of the consequent event. The ideas that form the basis of this approach are:

- Since we are uncertain about the true value of the parameters, we will consider them to be random variables.
- The rules of probability are used directly to make inferences about the parameters.
- Probability statements about parameters must be interpreted as "degree of belief" [*29*].

Bayes' Theory

For any propositions A and B, the degree of belief that A is true, given that one assumes that B is true, is equal to the degree of belief that A and B are both true, given background information I, divided by the degree of belief that B is true, given background information I, provided that Pr(B|I) > 0:

$$\Pr(A|B,I) = \frac{\Pr(A,B|I)}{\Pr(B|I)} \qquad (\boldsymbol{eq}.1)$$

(eq. 1) is called Bayes' theorem [*27*].

Bayesian Networks

A Bayesian network is a graphical representation of an n-dimensional probability distribution. It is a directed acyclic graph (DAG) in which each node represents a variable of interest, and the arcs represent dependencies between the variables. The strengths of the dependencies are quantified by conditional probabilities. Thus, the architecture of a Bayesian network involves two components: (1) the structure of the network, which describes the direct association pairs of variables, and (2) the parameters of the network that represent the probability distribution [*31*].

Bayesian Networks, comparing to the other networks that consider uncertainty such as Markov Networks, use a richer language of directed graphs where the directions of the arrows permit us to distinguish genuine dependencies from spurious dependencies induced by hypothetical observations [*29*]. To represent causal relations, we identify the parents of each node (nodes incident to the tails of all arcs into the node) that are the direct causes of the node.

To benefit from Bayesian Networks in our work, an optimum structure should be determined which is relied to have taken into account all the data fed to it and also has the optimum number of arcs and logical order of the nodes.

Conditional probabilities

Once the topology of the BN is specified, the next step is to quantify the relationships between connected nodes – this is done by specifying a conditional probability distribution for each node. As we are only considering discrete variables in this work, this takes the form of a conditional probability table (CPT).

Root nodes also have an associated CPT, although it is degenerate, containing only one row representing its prior probabilities. The size of the CPT is, in fact, exponential in the number of parents [*29*].

Process Description

In this case study, a working plant producing MTBE in a petrochemical complex in Iran is considered. In this MTBE process methanol and isobutene react to produce the high-octane oxygenate, methyl tertiary butyl ether, MTBE, which is primarily used in gasoline blending as an octane enhancer to improve hydrocarbon combustion efficiency. The technology is shown in figure1.

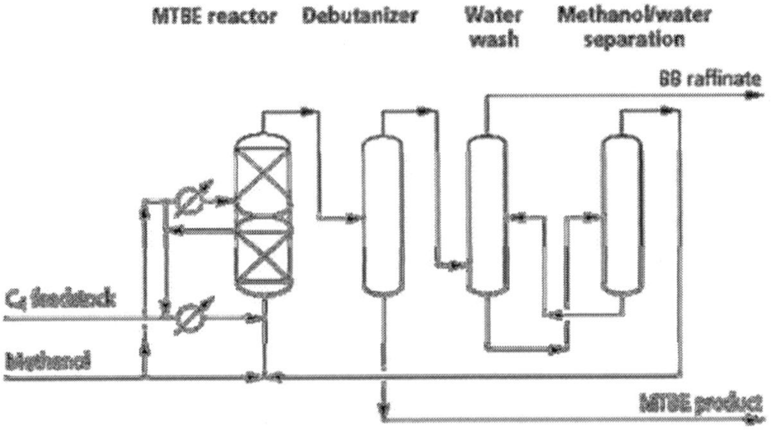

Figure1. Process Flow Diagram

Network Formation Using Industrial Data

Data Preparation

First off, a plant was targeted for acquiring appropriate amount of operational data. Before treating/pre-processing the data, each parameter which is considered to be an uncertainty in operation was plotted to reveal its fluctuation. Due to an approximate steadiness of some parameters around a level, the most fluctuating ones were selected to narrow down and simplify the overall procedure. Histograms are the most common way to show the distribution of data. There are some steps counted to construct a histogram in literature [*29*].

To have a uniform data in order to reduce errors arising from different orders of magnitude and also omitting the unit diversity, at first the data sets were normalized, by use of (eq. 2).

$$Data_{i,normalized} = \frac{Data_i - Data_{i,min}}{Data_{i,max} - Data_{i,min}} \qquad (\boldsymbol{eq}.\ 2)$$

$$i = 1, 2, \ldots, Number\ of\ data\ for\ parameter\ i$$

Finally according to the data distribution plots, all operating data were discretized to some bounds. By sure, the number of discretized bounds affects the calculation accuracy; it means the more the number of bounds we choose, the more precise answer we would get. The ultimate state can be of assuming a continuous distribution.

We have chosen an average number of bounds in this paper for a guaranteed result in a reasonable run time and a simple and easy to understand use of Bayesian Network approach in chemical engineering optimization. Table1 illustrates the discrete values.

Table1. Discrete values corresponding to each parameter

Parameter	**Function**	**Operating Range**	**Unit**	**Discrete Values**
A	iC_4 Flow Rate	[88.86-170.6]	m^3/hr	0.9, 0.7, 0.5
B	MeOH Flow Rate	[6.4-40]	m^3/hr	0.2, 0.5, 0.9
C	R1 Recycle Flow Rate to Plant Feed	[46-325]	m^3/hr	0.9, 0.7, 0.4, 0.1
D	MTBE Flow Rate	[13.8-206]	m^3/hr	0.1, 0.3, 0.8
E	Tower's Tray #10 Temperature	[78.2-166]	^oC	0.3, 0.75
F	Tower's Reflux Flow Rate	[150-152.8]	m^3/hr	0.9, 0.7, 0.5, 0.3, 0.1
G	Tower's Distillate Flow Rate	[54.68-150]	m^3/hr	0.25, 0.75

Network Structure

Two approaches are mainly available for determining the structure of a Bayesian network. The first can be a human expert who is an expert in the domain and dictates the arcs that connect the nodes in the network. Construction of Bayesian networks using this method is time consuming, and is limited by the subjective judgment of the expert consulted.

For those reasons, there has been substantial recent interest in identifying a computerized, objective methodology for inducing the graphical structure of the Bayesian network from a multivariate data set [*30*].

Here in this note, a free simplified wizard named BN PowerConstruction from a package, named Belief Network Powersoft, is used to determine the best network that meets the relation between data sets. This system is based on a three-phase belief dependency analysis algorithm.

Three-Phase Dependency Analysis

The three-phase dependency analysis algorithm, the procedure used to design Bayesian network, has three phases: drafting, thickening and thinning [*31*]. TPDA designers introduced their work as follows: "In the first phase, this algorithm evaluates mutual information of each pair of nodes as a measure of closeness and creates a draft of structure based on this information. In the second phase, the algorithm adds arcs when the pairs of nodes cannot be d-separated (directed separation of two nodes in a graph). The result of phase two is an independence map (I-map) of the underlying dependency model. In the third phase, each arc of the I-map is examined using conditional independence tests (CI tests) and will be removed if the two nodes of the arc can be de-separated. The result of phase three is the minimal I-map." Interested readers are referred to the mentioned reference for having a clue about the tests and their procedures.

For our case, as mentioned above, we took advantage of a piece of software that is available on the internet for free, Belief Network PowerSoft; this software is developed regarding TPDA algorithm [*32*]. By feeding the data sets and assigning some restrictions for some nodes, indicating which nodes are bound to be roots or leafs, the optimum network structure was determined. Figure2 shows the chosen Bayesian Network which has seven arcs of its possible twenty-one or $\binom{7}{2}$ arcs. The nodes are selected to function as isobutene feed flow rate, methanol feed flow rate, first reactor recycle flow rate, MTBE product flow rate, temperature of separation tower top section, reflux flow rate and distillate flow rate.

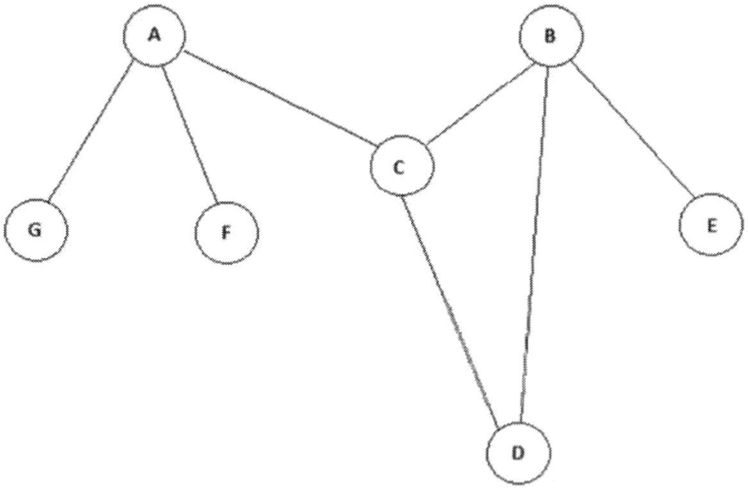

Figure2. Optimum network structure according to TPDA algorithm

The conditional probability tables (CPT) of all nodes were developed by a code written in MATLAB environment, and the network was also developed by FullBNT-1.0.4 software linked to MATLAB [*33*].

History Matching

After the network is developed as the model of a system, it should be checked for evaluation to make the user sure of its performance. Here, for Bayesian Network a history matching procedure should be done. In other words, the degree of reliability of the network is gained by checking its accordance with the data used to develop it; it means to what extent does the history of system, the recorded data, match the performance?

In order to do so, data sets of a certain prediction problem and ones of a certain decision-making problem were compared with each other and ninety and seventy percent of reliability was calculated respectively. This indicates that inferences made by the network in each of the modes are fairly acceptable, especially in prediction mode. Of course having more data, both quantitatively and qualitatively, will ensure a better network and higher degrees of reliability.

Results and discussion

Here in this case study, our objective is to maximize the production rate of MTBE by adjusting the optimum operating conditions. Production rate of MTBE is a function of operating conditions, so the problem would be:

$$D = f(A, B, C, E, F, G)$$

$$\begin{cases} \text{Maximize } f \\ \text{subject to } A \in [88.86, 170.6] \\ B \in [6.4, 40] \\ C \in [46, 325] \\ E \in [78.2, 166] \\ F \in [150, 152.8] \\ G \in [54.68, 150] \end{cases}$$

According to this fact that two of the considered variables are not adjustable and are imposed upon the plant from the previous units or any other imaginable sources, some scenarios are likely to occur. As each of the two imposed variables has three discrete values, the number of possible scenarios will be equal to nine. Conditional properties for all of the other nodes in the network are then extracted from the network by updating each case's conditions.

The table below only shows the computed probabilities of all discrete values of the objective variable, in response to the states which were determined for the two input variables. This is sort of prediction mode problem.

Table2. Conditional properties for node D in all possible scenarios

No.	A state	B state	D(1)	D(2)	D(3)
1	1	1	0.88234	0.0764958	0.0411641
2	1	2	0.325567	0.666633	0.00779922
3	1	3	0.5	0.25	0.25
4	2	1	0.935262	0.0529783	0.01176
5	2	2	0.342173	0.65077	0.00705673
6	2	3	0.5	0.25	0.25
7	3	1	0.893731	0.0625188	0.04375
8	3	2	0.325567	0.666633	0.00779922
9	3	3	1	0	0

The table indicates that the best, moderate and worst states of the objective, which is defined to be the amount of MTBE production, might happen in which case and with how much probability. Since more production flow rate is more desirable, third state of variable D is the best possibility to arise.

According to the results, in scenarios number three and six, third state of variable D, say D(3), has the highest probability to happen among all states. Computing a series of amounts, namely expected values, can end up in better understanding of the fact that what is going on. The expected value corresponding to each scenario is equal to the sum of its states' product of values and their associated probabilities.

Table below shows the expected value for all of the scenarios and the amount of its deviation percentage from average value. This percentage can be positive or negative regarding any increase or decrease in production flow rate.

Table3. Expected values and their deviation from average

Scenario Number	Expected Value for D=$\sum x.P(x)$	Deviation Percentage From Average Value
1	41.4987	-19.8093
2	59.6947	15.3521
3	76.265	47.372
4	36.6387	-29.2006
5	58.985	13.9807
6	76.265	47.372
7	41.3093	-20.1753
8	59.6947	15.3521
9	33.02	-36.1932

By comparing all scenarios' expected values with one another, it is found that the third and sixth ones predict the maximum production rate and in these two, occurrence of the best discrete value for this study's objective is more probable; also it can be stated that the amount of the positive deviation from average recorded value is higher than the other scenarios'. Other operating conditions in these two cases are reported in table4.

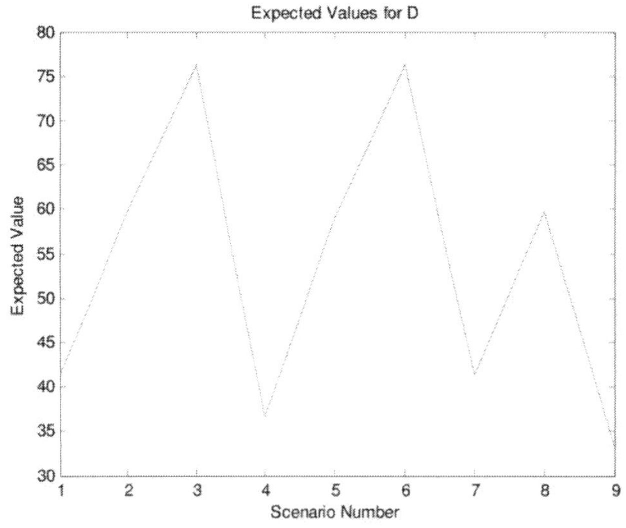

Figure3. Expected values for D

Table4. Optimum operating conditions' probabilities

<table>
<tr><th></th><th colspan="5">Scenario No. 3</th><th colspan="5">Scenario No. 6</th></tr>
<tr><td>C</td><td>0.25</td><td>0.25</td><td>0.25</td><td colspan="2">0.25</td><td>0.25</td><td>0.25</td><td>0.25</td><td colspan="2">0.25</td></tr>
<tr><td>E</td><td colspan="2">1</td><td colspan="3">0</td><td colspan="2">1</td><td colspan="3">0</td></tr>
<tr><td>F</td><td>0.2897</td><td>0.0561</td><td>0.2991</td><td>0.2991</td><td>0.0561</td><td>0.1183</td><td>0.0215</td><td>0.6129</td><td>0.2258</td><td>0.0215</td></tr>
<tr><td>G</td><td colspan="2">0</td><td colspan="3">1</td><td colspan="2">0</td><td colspan="3">1</td></tr>
</table>

In the above-mentioned case, production flow rate rises about fifty percent by comparison with the average production rate. Overall, it can be understood that in all scenarios, excluding the first, fourth, seventh and ninth, one can expect the plant production rate to be more than its typical values. These cases are more satisfactory and fitting to plant's optimum operation. As a suggestion a study can be performed on the previous plants' behavior or the feed stock storages of the under-study plant, to achieve a series of appropriate conditions for this plant, and to provide it with its desired requirements.

References

1. **T. F. Edgar, D. M. Himmelblau**, "The Nature and Organization of Optimization Problems," Chapter 1, in "Optimization of Chemical Processes," 2nd ed., McGraw-Hill, New York (1988).
2. **P.W. Glynn**, "Problems in Bayesian analysis of stochastic simulation," Proceedings of the 1986 Winter Simulation Conference, pp. 376-383 (1986).
3. **B.L. Nelson, *et al.***, "Approximation-assisted point estimation," Oper. Res. Lett., 20, pp. 109-118 (1997).
4. **J.O. Berger**, "Statistical Decision Theory and Bayesian Analysis," 2nd ed., Springer, Berlin (1985).
5. **D.R. Cox, and D.V. Hinkley**, "Theoretical Statistics," Chapman and Hall, London (1974).
6. **A. Gelman, and J.B. Carlin, and H.S. Stern, and D.B. Rubin**, "Bayesian Data Analysis," Chapman and Hall, London (1995).
7. **A.E. O'Hagan**, "Kendall's Advanced Theory of Statistics," Arnold, Paris (1994).
8. S.J. Press, "Bayesian Statistics: Principles, Models, and Applications," Wiley, New York (1989).
9. **Judea Pearl**, "Probabilistic Reasoning in Intelligent Systems - Networks of Plausible Inference" 1st ed., Morgan Kaufmann (September 15, 1988).
10. **N. Friedman, and M. Linial, and I. Nachman, and D. Pe'er**, "Using Bayesian networks to analyze expression data," Journal of Computational Biology, 7, pp. 601–620 (2000).
11. **J. Valadares, and G. Ramalho, and M. Ladeira**, "Modeling complex management games with Bayesian networks: The FutSim case of study," Proceedings of the Agents in Computer Games, Edmonton, Canada (2002).
12. **D. Margaritis, and C. Faloutsos, and S. Thrun**, "NetCube: A scalable tool for fast data mining and compression," Proceedings of the 27th Conference on Very Large Databases, VLDB, Rome, Italy (2001).
13. **J.R. Castelo**, "The discrete acyclic digraph Markov model in data mining," Thesis, presented to University of Utrecht, Utrecht, Netherlands, in partial fulfillment of the requirements for the degree of Doctor of Philosophy (2002).
14. **C.L. Chen, and S. Talukdar**, "Causal nets for diagnosis," Proceedings of the Expert Systems Application to Power Systems IV, CRL Publishing, pp. 379–386 (1993).
15. **S. Sarkar, and S. Sriram**, "Bayesian models for early warning of bank failures," Management Science, 47 (11), pp. 1457–1475 (2001).
16. **Haoran Liu, and Jean-Christophe Buvat, and Lionel Estel, and Isabelle Polaert**, "Bayesian Network Method for Fault Diagnosis in a Continuous Tubular Reactor," Chemical Product and Process Modeling, Published by The Berkeley Electronic Press, 5 (2010).
17. **S.M. Schwartz, and J. Baron, and J.R. Clarke**, "A causal Bayesian model for the diagnosis of appendicitis," Uncertainty in Artificial Intelligence, 2, pp. 423–434 (1988).

18. **I.A. Beinlich, and H.J. Suermondt, and R.M. Chavez, and G.F. Cooper**, "The ALARM monitoring system: A case study with two probabilistic inference techniques for belief networks," Proceedings of the Second European Conference on Artificial Intelligence in Medicine, pp. 247–256 (1989).
19. **S. Andrassen, and F. Jensen, and K. Olensen**, "Medical expert systems based on probabilistic causal networks," International Journal of Medical Computing, 28, pp. 1–30 (1991).
20. **S. Mani, and S. McDermott, and M. Valtorta**, "Mentor: A Bayesian model for prediction of mental retardation in newborns," Research in Developmental Disabilities, 18, pp. 303–318 (1997).
21. **W.-K. Wong, and A.W. Moore, and G. Cooper, and M. Wagner**, "Bayesian network anomaly pattern detection for disease outbreaks," Proceedings of the Twentieth International Conference on Machine Learning-ICML, Washington, DC (2003).
22. **B. Abramson**, "The design of belief network-based systems for price forecasting," Computers and Electrical Engineering, 20, pp. 163–180 (1994).
23. **G. Yigun, and D. Peiris, and J. Crawford, and J. NcNicol, and B. Marshall, and R. Jefferies**, "An Application of Belief Networks to Future Crop Production," IEEE Comput. Soc. Press., pp. 305 - 309 (1994).
24. **J.G. Torres-Toledano, and L. Sucar**, "Bayesian networks for reliability analysis of complex systems," Progress in Artificial Intelligence, IBERAMIA 98, Berlin, Springer, pp. 195-206 (1998).
25. **C. Conati, and K. VanLehn**, "Probabilistic plan recognition for cognitive apprenticeship," Proceedings of the 18th Annual Conference of the Cognitive Science Society, San Diego, CA, pp. 403-408 (1996).
26. **J. Royalty, and R. Holland, and J. Goldsmith, and A. Dekhtyar**, "POET: The online preference elicitation tool," Technical Report WS-02-13, AAAI (2002).
27. **F. Taroni, and C. Aitken, and P. Garbolino, and A. Biedermann**, "Bayesian Networks and Probabilistic Inference in Forensic Science," John Wiley & Sons, Chichester (2006).
28. **R. Kennett, and K. Korb, and A. Nicholson**, "Seabreeze prediction using Bayesian networks: A case of study," Proceedings of the Proceedings of the 5th Pacific-Asia Conference on Advances in Knowledge Discovery and Data Mining, Springer-Verlang, New York (2001).
29. **William M. Bolstad**, "Introduction to Bayesian Statistics," 2nd ed., John Wiley & Sons, (2007).
30. **Ana M. Martinez-Rodrigueza, and Jerrold H. Mayb, and Luis G. Vargas**, "An optimization-based approach for the design of Bayesian networks," Mathematical and Computer Modeling, 48, pp. 1265–1278 (2008).
31. **Cheng, J., and Greiner, and R. Kelly, and J. Bell, and D. Liu**, "Learning Bayesian networks from data: An information-theory based approach," Artificial Intelligence, 137, pp. 43-90 (2002).
32. "J Cheng's Bayesian Belief Network Software," available via http://webdocs.cs.ualberta.ca/~jcheng/bnsoft.htm. [Accessed 10/09/10.]
33. "Bayes Net Toolbox for Matlab," available via http://www.cs.ubc.ca/~murphyk/Software/BNT/bnt.html. [Accessed 10/09/10.]

A novel recuperative configuration for enhancement of ethylene oxide production via integration of ethylene oxidation process and cyclohexane dehydrogenation

School of Chemical and Petroleum Engineering, Department of Chemical Engineering, Shiraz University,

P.O. Box 71345, Shiraz, Iran

M.R. Rahimpour[1*], M.R.Gholipour, M. ShokrollahiYancheshmeh, M.R.Dehnavi, S. Amirabadi
D. Iranshahi

Abstract

In this article, the process intensification (PI) concept was studied for simultaneous hydrogen and ethylene oxide production. The most important advantages of Process intensification are reduction of equipment size and environmental impact, increase of energy utilization, and so it has attracted great attention. One example of process intensification is a novel reactor concept that facilitates direct coupling of exothermic and endothermic reactions. In this work, endothermic dehydrogenation of cyclohexane is supposed to occur in the shell side of conventional ethylene oxide synthesis reactor. This coupling has more importance in the others, for the reason that, hydrogen has been produced in endothermic side and also ethylene oxide in exothermic side. A one-dimensional heterogeneous model was used for steady state simulation of thermally coupled reactor (TCR). The simulation results are compared with the outputs of conventional ethylene oxide reactor (CR). The results demonstrate the performance of thermally coupled reactor is beneficial. However, the performance of the new configuration needs to be tested experimentally over a range of parameters under practical operating conditions.

Keywords: Ethylene oxide synthesis;cyclohexane dehydrogenation;Hydrogen production; thermally coupled reactor

1. INTRODUCTION

One of the fields of PI is coupling of reactions. Coupling exothermic and endothermic reactions can save energy and reduce capital and operational costs. This process can be done in the following reactor configurations: direct coupling, regenerative coupling, and recuperative coupling. In his study, we considered coupling of ethylene oxidation process with dehydrogenation of cyclohexane to benzene in a recuperative reactor as a thermally coupled reactor [1-6].

---

[*]Correspondingauthor. Tel.:+98 711 2303071 ; fax : +98 711 6287294.
E-mail address : rahimpor@shirazu.ac.ir(Prof. M. R. Rahimpour).

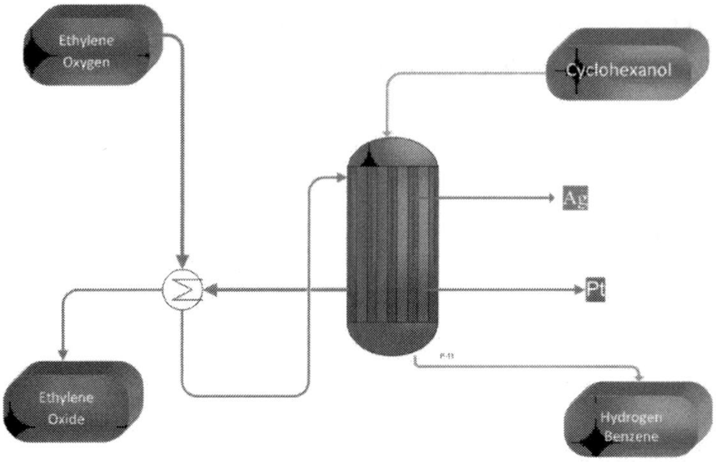

Figure 1. Process Flow Diagram of a thermally coupled reactor.

2. Reaction Scheme and Kinetics

The following main reactions take place in the ethylene oxidation process [7]:

$$C_2H_4 + 0.5O_2 \xrightarrow{Ag} C_2H_4O \qquad r_1 = \frac{k_1 P_E P_O^{n_1}}{(1+K_E^1 P_E)^2} \qquad (1)$$

ΔH_{298} = -105.39 Kj/mol

$$C_2H_4 + 3O_2 \xrightarrow{Ag} 2CO_2 + 2H_2O \qquad r_2 = \frac{k_2 P_E P_O^{n_2}}{(1+K_E^2 P_E)^2} \qquad (2)$$

ΔH_{298} = -1321.73 Kj/mol

The catalyst that is used affects on the kinetic parameters; consequently the kinetic parameters which reduce as much as possible the difference between calculated and experimental values of outlet molar flow rate of components should be found. The differential evolution (DE) strategy was used to optimize the following objective function:

$$OF = \sum_{i=1}^{m} (F_{i_{cal}} - F_{i_{exp}})^2 \quad (3)$$

The reaction scheme for the dehydrogenation of cyclohexane to benzene is as follows [8]:

$$C_6H_{12} \leftrightarrow C_6H_6 + 3H_2 \qquad r = \frac{-k(K_P P_C / P_{H_2}^3 - P_B)}{1+(K_B K_P P_C / P_{H_2}^3)} \qquad (4)$$

ΔH_{298} = +206.2 Kj/mol

3. Mathematical model

In this section, the mathematical model should be developed based on the assumptions below:

- Co-current flow regime
- One dimensional heterogeneous model
- Steady state condition
- Axial diffusion of heat and mass are dispensable
- The radial diffusion in the catalyst particles is dispensable
- The same bed porosity in both axial and radial direction
- The ideal gas mixture
- Heat loss to the surrounding is dispensable

The steady state mass and energy balance equations for solid phase are given below:

$$a_v c_j k_{gi,j}(y^g_{i,j} - y^s_{i,j}) + r_{i,j}\rho_b = 0 \tag{8}$$

$$a_v h_f(T^g_j - T^s_j) + \rho_b \sum_{i=1}^{N} r_{i,j}(-\Delta H_{f,i}) = 0 \tag{9}$$

In above equations, $y^s_{i,j}$ and T^s_j are the solid phase i[th] component mole fraction and temperature in j[th] side of reactor, respectively.

The following steady state mass and energy balance equations have been developed for the fluid phase:

$$-\frac{1}{AC}\frac{dF_{i,j}}{dz} + a_{vj} c_j k_{gi,j}(y^s_{i,j} - y^g_{i,j}) = 0 \tag{10}$$

$$-\frac{C^g_{pj}}{AC}\frac{d(F_j T^g_j)}{dz} + a_{vj} h_f (T^s_j - T^g_j) \pm \frac{\pi D_j}{AC} U(T^g_2 - T^g_1) = 0 \tag{11}$$

In above equations, $F_{i,j}$ and T_i^g are the ith component molar flow rate and temperature in j[th] side of the reactor for fluid phase, respectively.

The following boundary conditions are applied:

z=0; $F_{i,j} = F_{i0,j}$ $T_j^g = T_{0j}^g$ $P_j^g = P_{0j}^g$ (13)

Where Fi0,j , T0jg and P0jg are the molar flow rate of ith component, temperature and pressure at the entrance of jth side of the reactor, respectively.

4. Model Validation

The steady state model validation is performed between the simulation results and the industrial plant data for CR. The TCR modeling results and the corresponding observed data of the industrial plant are presented in Table 1. As it can be seen, the simulation results have an acceptable agreement with the plant data. Therefore, this mathematical model performs well under the industrial conditions.

Table 1. A comparison between simulation results and industrial plant data.

Table 4. A comparison between simulation results and industrial plant data

Flow rate (mol/s per tube)	Plant data	Model predicted	Relative error (%)
Ethylene	0.1870	0.197097	-5.41367
O2	0.0369	0.037739	-2.22322
Ethylene Oxide	0.0163	0.017188	-5.36781
CO2	0.0788	0.078908	-0.13768
H2O	0.0195	0.018833	3.339438
Temperature	504.35	513.2797	-1.77054

5. Results and Discussion

For better understanding of reactor performance, we investigated ethylene oxide yield and cyclohexane conversion inFig. 2(a)-(b). These two parameters are defined as follows:

$$Ethyleneoxideyield = \frac{N_{C_2H_4O}}{N_{C_2H_4}} \quad (14)$$

$$Cyclohexaneconversion = \frac{F_{C_6H_{12},in} - F_{C_6H_{12},out}}{F_{C_6H_{12},in}} \times 100 \quad (15)$$

As it can be seen in Fig.2 (a)-(b), maximum cyclohexane conversion is 34.6853%, and ethylene oxide reaches to 0.0866 and 0.0783 yields in the exothermic side of TCR and CR, respectively. The comparison of ethylene oxide yield in TCR with CR shows that the ethylene oxide yield in output of TCR is increased by 0.83%. The preference of thermally coupled reactor is obviously detected from this figure because the ethylene oxide yield reaches to the highest value in this configuration. The ethylene oxide yield enhances in TCR as a result of heat transfer progress between exothermic and endothermic sides.

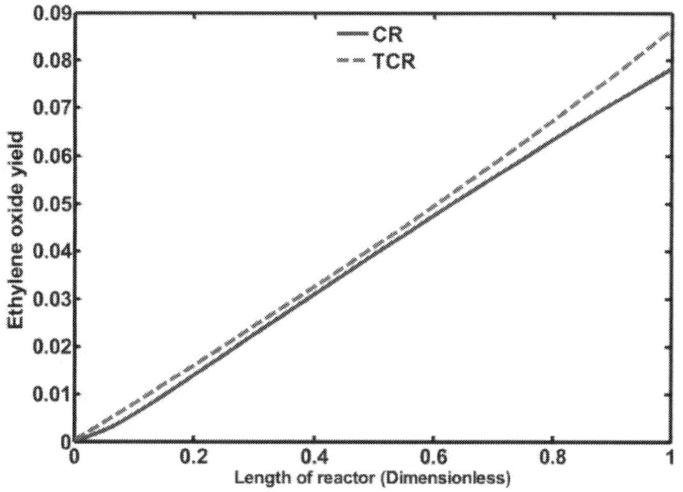

Figure 7(a). Cyclohexane conversion in endothermic side of TCR

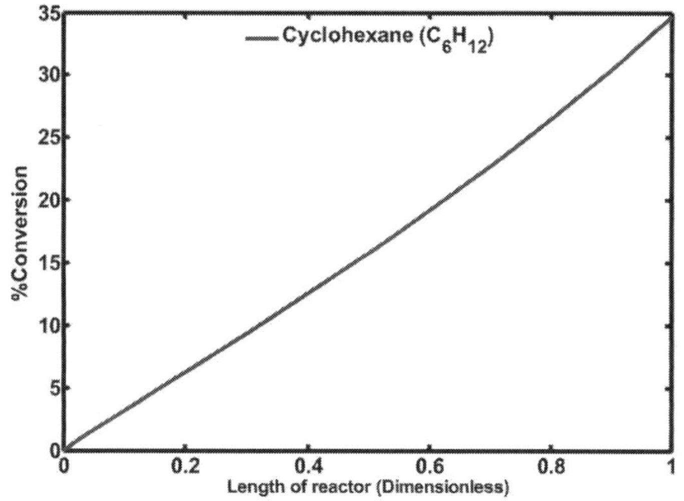

Figure 7(b). Ethylene oxide yield in exothermic side of TCR and CR along the reactor length.

The temperature profiles in the exothermic and endothermic sides of TCR and CR are illustrated in Fig.3. The endothermic side temperature decreases at the beginning of TCR and slightly increases at the rest of the reactor. The exothermic side temperature increases along TCR and reaches 528.29K at the end of the reactor.In addition, the highest temperature is observed at the exothermic side, since this is where heat is generated. Part of this heat is used to drive the endothermic reaction and the rest to heat the reaction mixtures in both sides. The temperature of the dehydrogenation side is always lower than that of the exothermic side in order to make a driving force for heat transfer from the solid wall. The abrupt increase in the temperature profile of CR takes place at the entrance of the reactor. The heat transfer does not perform well and the temperature is constant along CR which implies that the reactor temperature reaches the cooling water temperature and becomes constant.

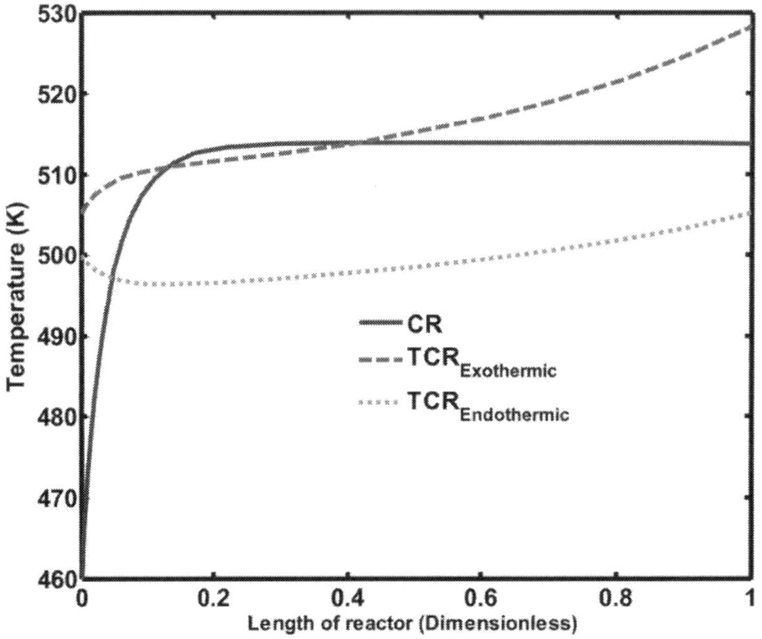

Figure 3. Temperature profile of both exothermic and endothermic sides of the TCR.

Fig.4 demonstrates the variation of the generated and consumed heat flux in the exothermic and endothermic side, respectively, and transferred heat from the solid wall along the reactor for TCR. At the inlet of the reactor, the generated heat in the exothermic side is lower than consumed heat in the endothermic side. Therefore the endothermic side temperature decreases at the beginning of TCR. At the rest of reactor, there is a reversed trend and consequently the temperature in both sides of reactor increase as illustrated by the temperature profile in Fig.3.

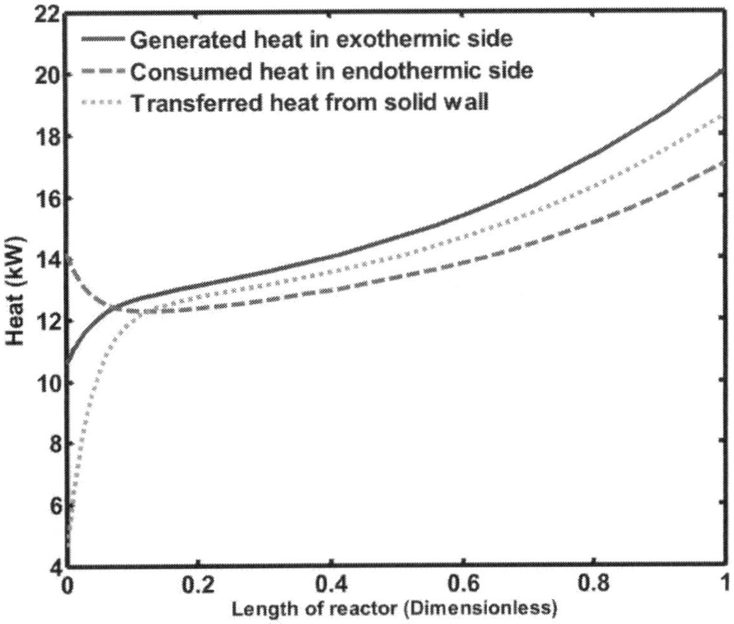

Figure 4. Fluctuation of the generated and consumed heat in the exothermic and endothermic sides of the TCR.

6. Conclusion

In this study, simultaneous production of ethylene oxide and hydrogen is verified in the thermally coupled reactor. Ethylene oxide as a main product is supplied in the exothermic side of reactor and hydrogen is procreated in the endothermic side of TCR. The steady state simulation of thermally coupled reactor was achieved by considering a one dimensional heterogeneous model. The simulation results of thermally coupled reactor were compared with the results of simulation for CR. The profiles of different parameters such as molar flow rates of components in the exothermic and endothermic sides, heat flux and temperature through the length of reactor were observed.

This comparison proved that this novel configuration can provided the needed heat to heat up the mixtures and to move the endothermic reaction better. The efficiency of heat transfer has been enhancedbecause of short distance between the consumed and transferred heat.Furthermore, the ethylene oxide yield as a main product has been increased in the TCR and reaches 0.0866 in the outlet of novel reactor.Finally, the results illustrate thatcoupling of ethylene oxidation process and cyclohexane dehydrogenation in a thermally coupled reactor is sane and beneficial.

References

[1] CANMET. Process integration.Varennes, Canada: CANMET energy technology centre (CETC). See also: http://cetc-varennes.nrcan.gc.ca/en/eb_o.html; 2009.
[2] P. Ras˘kovic´, A. Anastasovski, Lj. Markovska, V. Mes˘ko., Process integration in bioprocess indystry: waste heat recovery in yeast and ethyl alcohol plant, Energy 35 (2010) 704–717
[3] F. Friedler., Process integration, modelling and optimisation for energy saving and pollution reduction., Chemical Engineering Transaction 18 (2009) 1–26.
[4] Process integration for energy and water saving, increasing efficiency and reducing environmental impact, Applied Thermal Engineering 30 (2010) 2265–2269.
[5]Dautzenberg FM, Mukherjee M. Process intensification using multifunctional reactors. ChemEngSci 2001;56(10):251–67.
[6]Ramaswamy RC, Ramachandran PA, Dudukovic´ MP., Recuperative coupling of exothermic and endothermic reactions. ChemEngSci 2006;61:459–72.
[7]Lafarga D., Al-Juaied M.A., Bondy C.M., Varma A. Ethylene Epoxidation on Ag-Cs/r Al2O3 Catalyst: Experimental Results and Strategy for Kinetic Parameter Determination. Ind. Eng. Chem. Res. 2000,39, 2148-2156.
[8]Itoh N. A membrane reactor using palladium.AIChE J. 33 (1987) 1576-1578.

AUTHOR INDEX

AUTHOR INDEX

AUTHOR INDEX

AUTHOR INDEX

AUTHOR INDEX

AUTHOR INDEX

AUTHOR INDEX

AUTHOR INDEX

AUTHOR INDEX

AUTHOR INDEX

AUTHOR INDEX

AUTHOR INDEX

Environmental Division

Core Programming Topic at the 2011 AIChE Annual Meeting

Minneapolis, Minnesota, USA
16-21 October 2011

Volume 2 of 2

ISBN: 978-1-61839-730-0

Printed from e-media with permission by:

Curran Associates, Inc.
57 Morehouse Lane
Red Hook, NY 12571

proceedings.com

Some format issues inherent in the e-media version may also appear in this print version.

Printed by Curran Associates, Inc. (2012)

AIChE
3 Park Avenue
New York, NY 10016-5991

Phone: (203) 702-7660
Fax: (203) 775-5177

www.aiche.org

Additional copies of this publication are available from:

Curran Associates, Inc.
57 Morehouse Lane
Red Hook, NY 12571 USA
Phone: 845-758-0400
Fax: 845-758-2634
Email: curran@proceedings.com
Web: www.proceedings.com

TABLE OF CONTENTS

Volume 1

Volume 2

SIMULATION AND MULTIOBJECTIVE OPTIMIZATION OF A CONTINUOUS BIOCHEMICAL REACTOR USING MULTILAYER MODELING TECHNIQUE

Ghanim M. Alwan, Chemical Engineering, University of Technology, Baghdad, Iraq and Muthanna.H.Aldahhan, Chemical and Biological Engineering, Missouri University of Science and Technology, Rolla, MO

Abstract:

Biochemical reactors are essential unit operations in a wide variety of biotechnological process. As compared to conventional chemical reactors, bioreactors present unique modeling and control challenges due to complexity of the underlying biochemical reactions.

The biological processes are inherently very nonlinear and have frequently changing optimum operating conditions. Additionally, many available mathematical models for biological reactions are either in exact approximations or not suitable for a control design since no accurate biological law that can be exploited has been proposed. Optimization of biochemical reactor process requires a mathematical model that describes and predicts the process behavior.

The present biochemical reactor which used substrate (glucose) was simulated using multilayer modeling technique. The first layer is hybrid model was consist the framework of the dynamic mass balance equations supplemented with the sub models of kinetic reaction and the optimum problem represent the second modeling layer.

To overcome the limitation of experimental data, the accurate simulated model was used as a powerful tool to generate data which construct the objective functions.

The objective of the optimization technique was to maximize the conversion and minimize the cycle time of reaction. The decision (effective) variables were the dilution rate and the inlet concentration of the substrate. Optimization technique was the powerful tool to generate several new designs and sets of operating conditions. This can reduce the number of experimental runs and the cost consumed for the design and operation. Also it limits the best range of the operating conditions which improve the efficiency of the reactor. The yield of the biomass increased by 43% at the optimum conditions.

The stochastic search by Genetic algorithm with adaptive operators was the best to perform the multiobjective optimization of the present nonlinear interacted biochemical reactor.

Keywords: Biological Engineering, Genetic Algorithm, Multilayer Model, Optimization.

Nomenclatures:

D: Dilution rate, [hr^-1]

F: Flow rate, [liter/hr]

Km: Saturation constant,[gm/liter]

r1: Rate of cell generation,[gm/liter.hr]

r2: Rate of substrate consumption,[gm/liter.hr]

S: Substrate concentration into the reactor,[gm/liter]

Si: Inlet substrate concentration,[gm/liter]

V: Volume of the reactor,[liter]

X: Biomass concentration into the reactor,[gm/liter]

Y: Yield, [-]

μmax: Maximum specific growth rate coefficient,[hr^-1].

μ(s): Local specific growth rate coefficient, [hr^-1].

Introduction:

A biochemical reactor (Figure 1) is a vessel in which is carried out a chemical process involves organisms or biochemically active substances derived from such organisms. The CSTR biochemical reactor is widely used in the treatment of liquid wastes.The process kinetics can be characterized by the following reaction scheme:

$$\text{Substrate} \rightarrow \text{Biomass} + \text{Gas}$$

The goal of biochemical engineering research in industrial production of biologically based products such as food, pharmaceuticals, various commodity and specialty Chemicals (Lee, 1992).

Biochemical reactors are essential unit operations in a wide variety of biotechnological process. As compared to conventional chemical reactors, bioreactors present unique modeling and control challenges due to complexity of the underlying biochemical reactions (Henson, 2003).

The biological processes are inherently very nonlinear and have frequently changing optimum operating conditions.Additionally, many available mathematical models for biological reactions are either in exact approximations or not suitable for a control design since no accurate biological law that can be exploited has been proposed (Alhumaizi andAjbar, 2004 and Shimizu, 2006).

A novel roubst controller is proposed for a CSTR bioreactor that controls the culture dilution rate into the rector in order to maximize a cost function represnting the biomass yield (Kapadia etal.2008).

Optimization of chemical and related processes requires a mathematical model that describes and predicts the process behaviour.Optimization refers to finding the values of decision variables which correspond to and provide the maximum or minimum of one or more desired objectives. Without optimization of design and operations, manufacturing and engineering activities will not be as efficient as they are now (Rangaiah,2009).

Multilayer Model:

The present biochemical reactor was simulated using the multilayer modeling technique. The first layer is hybrid model was consist the framework of the dynamic mass balance equations supplemented with the sub models of kinetic reaction and the optimization problem represent the second modeling layer.

Hybrid model:

Dynamic modeling for optimization and control requires models that desribe the essential dynamic characteristics of the process under study.For complex nonlinear processes, simple and effective model may be difficult to derive.In such cases hybrid models may be attractive. These hybrid models consist of a framework of dynamic mass and energy balances, supplemented with sub-models describing additional equations, such as mass transformation and transfer rates (Roffel and Betlem, 2006).

In the present work, the following assumptions were adopted for the model:

1-Homogenous liquid-phase system.

2-First order irreversible reaction.

3-Constant hold-up.

4-Isothermal conditions.

5-Follow the Monod law.

6-No biomass in the feed stream.

7-Constant yield

The component material balances for biomass(X) and substrate(S) are:

$dX/dt = r1-(F/V)\,X$ (1)

$dS/dt = (F/V)\,Si - (F/V)\,S - r2$ (2)

And the reaction rate equations are:

$r1=\mu(s)\,X1$ (3)

The yield $Y = r1/r2$ (4)

Then, $r2 = r1/Y$ (5)

For Monod law;

$\mu(s) = (\mu max * S)/\,(Km+S)$ (6)

Equations (1&2) can be simplified to the following form:

$dX/dt = (\mu(s) - D)$ (7)

$dS/dt = D\,(Si - S) - (X/Y)\mu(s)$ (8)

Where $D=F/V$

Equations (7&8) and the supplementary kinetic equations (3 to 6) represent the hybrid model of the present biochemical reactor.The reactor can be viewed as nonlinear dynamic system and the dynamic simulation is very useful tool for model validation.

Description of the system:

The system was constructed in the laboratory of Energy, Environmental and Chemical Engineering Department-Washington University in Saint Louis as shown in Figure (1).

The main equipments are:

1-Cylinderical biochemical reactor manufactured from Plexiglas with capacity of 20 liters.

2-Polyethylene storage tank with capacity of 50 liters.

3-Rotary pump with discharge flow rate of (2-10) liters/hr.

4-Stirrer with steel blades of (0-100) rpm.

5-Rotameters with range of (0-10) liters/hr for water and air.

6-Digital pH, conductivity and thermometer with T-type thermocouple.

7-Manual steel needle valves.

The wastewater contains substrate (glucose) with the kinetic parameters (Cutlip and Shacham, 2007) of; maximum specific growth rate coefficient ($\mu max=0.3hr^{-1}$), saturation constant ($Km=0.5gm/liter$) and yield($Y=0.4$).

Optimization Technique:

Deterministic algorithms for function optimization are generally limited to convex regular functions.However, many functions are either not differentiable or need a lot of difficult mathematical treatment(discretization, sensitivitycompution….etc) for differentiating.Therefor, stochastic sampling methods could be used instead to optimize such functions. Traditional deterministic searchtechniques, use characteristics of the problem (e.g, Jacobin, Hessian) to determine the next sampling point while stochastic search techniques do not need such information.Instead,the next step in stochastic methods would be determined based on statistical sampling rules rather than a set of rigorous deterministic. It is mostly based on NSGA-II and Omni-optimizer (Palonen etal.2009).

In the present work, the Genetic algorithm (GA) has been applied to investigate the optimum conditions of the biochemical reactor process.GA is search algorithm based on mechanics of natural selection and natural genetics. Philosophically GA is based on Darwins′theory of 'survival of the fittest'.There are several genetic operators, such as; population, selection, crossover, mutation….etc.Eachchromosome represents a possible solution to the problem being optimized and each bit(or group of bits) represent a value of variable of the problem (gene).A population of chromosomes represents a set of possible solution. These solutions are classified by an evaluation function, giving better values, or fitness, to better solutions (Gupta and Srivastava, 2006).

Results and Discussion:

This section illustrate the experimental, simulated and optimum results of the process also the discussion of the biochemical process behaviors.

-Starting Operating Conditions:

It is better to limit the initial operating conditions of the system by combined the simulation model with the optimization technique.

The model equations (6, 7&8) have to be solved by the optimization search technique (Levenberg-Marquardt method) with the aid of the MATLAB computer program (fsolve code).The initial conditions for these variables are obtained by considering steady state operation of the reactor as shown in Table (1).

-Unsteady State Conditions:

The biochemical reactor model equations (6,7and 9) were solved numerically under unsteady state conditions using 5th order Runge-kutta method with the aid of the MATLAB program (ode45 code).Starting from steady state operating conditions (Table1). It was better to simulate the mathematical model to describe and predict the process behavior.

Figures (2-7) explain the behaviors of the process for different values of the process variables; dilution rate (D) and inlet substrate concentration (Si).The growth rate of the biomass was fast, so that the concentration of the biomass into reactor was increased with increasing Si or decreasing D as shown in Figure (2).This is due to that the Si was directly affect on the specific growth rate constant (μ) regarding to Monod law (equation 6).While the increasing of D tend to moderate growth rate of biomass then reduced the concentration of biomass and increased the concentration of the substrate(S) into the biochemical reactor (Henson etal.2001).As the result to this behavior of the process, the conversion would be decreased (Figure 3).Then the cycle time of the reaction was increased with decreased Si and increased D as shown in Figures (4-7).

The biomass concentration curves have S-shape and seem more sluggish when compared with the substrate curve which is faster and have exponential shape.This is due to that the growths of the biomass was starting with very low initial concentration, so it was needed to interval time to motivation and grow the cells of biomass. The delay time in the response of the biomass growth get it sluggish situation when compared to the fast substrate decays which starting with high concentrations (6 to 80 gm/liter).

The intersection point between two curves (Figures 4-7) indicates to the equilibrium point of the system. At this point the rate of cell generated equal to the rate of substrate consumption (Cutlip and Shacham, 2007).The time of equilibrium was depended upon the reaction rate (r1 and r2), D and Si.

Validity of the Process Model:

Agree results obtained when compared the experimental data with the simulated results (Figures 8 and 10).The deviation from the reality for two Figuers is less than

5%.This indicates that the hybrid model is accurate for the nonlinear unsteady state biochemical process.

Therefore, the simulated dynamic model based the optimization can be implemented to find the optimum process variables.

Optimization Problem:

The optimization objective functions represent the second layer model. In the present work, it is desired to maximize the conversion of the substrate and minimize the cycle time of the reaction.The major input process variables which affected on the objectives are Si and D.So that, it was better to correlate these objectives directly with the decision variables (Si and D).

Regarding to the high accuracy of the simulated model, the data for construct the optimization model could be then generated. By using the nonlinear least squares model estimation (Levenberg marquardt method)with the aid of the computer program (Statistica version 6), the nonlinear empirical equations were obtained:

Minimize Y1=7.987*D^0.1*Si^-0.827 (9)

Maximize Y2=0.813*D^ -0.036*Si^0.037 (10)

Subject to inequality constraints

$$0.01 \leq D \leq 0.8$$
$$6.0 \leq Si \leq 80.0 \quad (11)$$

The correlation coefficient(R) of equations (9&10) is 0.96.

Where:

Y1: Cycle time of reaction.

Y2: Conversion.

The objective functions (equations 9 and 10) are highly non linear and conflicted.So that the multiobjective optimization technique is suitable for solving this problem.

Decision Variables:

Equation (9) describes that the inlet concentration of the substrate (Si) is more effective which have power of (0.827) than the dilution rate (D) with power of (0.1).This means that the cycle time of reaction is controlled by Si .Figures (3-8) confirm this situation. While equation (10) indicates to that the power of D and Si are slightly difference, so that it have similar effect (Figure 2), then the conversion can be controlled either by D or Si.In the industrial applications the control system for conversion or concentration of the substrate into the biochemical reactors is easly manipulated by the dilution rate D (Kapadia etal.2008).

Multiobjective Optimization Technique:

In the present work, the objective was to minimize the cycle time of reaction (equation 9) and maximize the conversion (equation 10).These two equations are conflicted and highly nonlinear. The deterministic search was unuseful.So that, the advanced stochastic technique as Genetic Algorithm (GA) was suitable.The MATLAB Tool Box of the multiobjective GA was used.There would be many optimal solutions for the multi-objective problem. Figure (9) illustrate the output of the algorithm operators.The Pareto front indicate to nine chromosomes were obtained.Each chromosome represens a possible solution to the problem.This means that 9 sets of design and operating conditions for the biochemical reactor. The number of experimental runs and the cost consumption for design and operation can be reduced by optimization search. All of them get the optimum objectives within the operating conditions limit (equation 11). The researchers and decision makers can select the best solution.

Several experiments were applied and the operators of the GA were tunned at each experiment to get the best solution.The best operators of GA explained in Table (2). While Table (3) illustrate the optimum solution of the present optimization problem. Results of the optimization search show that the optimum results stay in acceptable limits (equation11).Figure (8) illustrate the improvement of the optimization technique on the biochemical reactor. The Score Histogram indicate the the cycle time of reaction can be minimized to(0.146 hr) and maximized the conversion to about (100%) within the limit of the operating conditions (equation 12).

Figure (9) confirm the improvement in the process when operated at the optimum conditions (Table 3).The steady state value of the biomass concentration into the reactor is equal to 5.0 gm/liter (Figuer 10).When compared this with Figure (2-a),the biomass concentration at (D = 0.727 hr^-1) is equal to 1.73 gm/liter. This indicates that the yield of the biomass into the biochemical reactor can be increased by 43% at optimal conditions.

The accuracy of the simulated model is clear when compared the experimental and simulated results at the optimum operating conditions (Figure 10).However the GA is suitable for the present nonlinear biochemical reactor process.Thesuccess of the optimization is strongly affected by the formulation of the objective functions and the selection of an appropriate optimization algorithm (Palonen etal.2009).

Conclusions:

1-The biochemical reactor was highly nonlinear system and the reaction parameters needed to adaption regarding to the operating conditions.

2-The effective decision variables were the dilution rate and the inlet concentration of the substrate.

3-Hybrid model was more accurate for the complex underlying biochemical reactions.

4-Agree results obtained by the simulated model when compared with the experimental results.

5-The accurate simulated model could be used to generate data and construct the optimization problem equations.

6- The simulated model based optimization implemented by multilayer modeling technique to find the optimum parameters of the reactor.

7-Multiobjective Genetic algorithm was the best stochastic search technique to solve the nonlinear conflicted objective functions. The optimization search can generate several new designs and sets of operating conditions. This reduced the number of experimental runs and the cost consumed for design and operation.

8- The efficiency of the reactor will be improved by the optimization technique.

9-The accuracy of the Genetic algorithm could be increased by good formulation of the objective functions and by using the appropriate Genetic operators.

References:

1-Lee, J.M.'Biochemical Engineering', Englewood Cliffs.NJ, Prentice-Hall, (1992).

2-Henson, M.A.,'Dynamic Modeling and Control of Yeast Cell Populations in

Continuous Biochemical Reactors, Computers and Chemical Engineering, 27, 1125-1199, (2003).

3-Alhumaizi, k.and Ajbar,AH,'Optimization of an Unstructed First Order Kinetic Model of Cyclically Operated Bioreactors', Department of Chemical Engineering, King Saud University, (2004).

4-Shimizu, k.,'an Overview on the Control System Design Of Bioreactors, Measurement and Control Journal, vol.50,no.1,pp.65-84,(2006).

5-Kapada, A.P, Nitendra, N. and Timothy, C.,'Lyapunov-Based Continuous-Stirred Tank Bioreactor Control to Maximize Biomass Production Using the Haldane and Monod Specific Growth Models', Electrical and Computer Engineering, Clemson University, Clemson, Sc, (2008).

6-Gade Pandu Rangaiah,'Multiobjective Optimization Technique and Applications in Chemical Engineering, vol.1, Ist edition, world scientific publishing co.Ltd, (2009).

7-Roffel, B. and Betlem, B.,'Process Dynamics and Control: Modeling For Control and Prediction', Jhon-wiley&sons Ltd, 3rd edition, (2006).

8- Cutlip.M.B. and Shacham, M.,'Modular and Multilayer Modeling Application to Biological Processes', 17th European Symposium on Computer Aided Processing Engineering-ESCAPE 17, pp.1-10,(2007).

9-Palonen, M, Hasan, A.and Siren, K.,'A Genetic Algorithm For Optimization of Building Envelope and HVAC System Parameters',11th International IBPSA Conference, Glasgow, Scotland, July 27-30, (2009).

10-Gupta, A.K and Srivastava, R.K.,'Integral Water Treatment Plant Design Optimization: Genetic Algorithm Based Approach", IE Journal, vol.8, septemper,(2006).

11-Henso, M.A and Zang,Y.,'Biofurcation Analysis Of Continuous Biochemical Reactor Models', Biotecol prog.17, pp.647-660, (2001).

Table (1)-Optimum initial operating conditions.

X(gm/l)	Si(gm/l)	D(hr^-1)	Y(-)	μ(hr^-1)
0.0	6.0	0.3	0.4	0.4

Table (2)-Best operators of GA.

Operator	Type and values
Population type	Double vector
Population size	80
Selection function	Tournament
Crossover function	Scattered
Crossover fraction	0.8
Mutation function	Adaptive feasible
Migration direction	Forward
Migration fraction	0.2
Hybrid function	Fgoalattain
Number of chromosome	9
Number of generation	183
Function tolerance	1.0E-6

Table (3)-The multiobjective optimum results.

Cycle time of reaction(hr)	Conversion(%)	Si(gm/l)	D(hr^-1)
0.146-1.77	0.87-1.0	6.0-80.0	0.01-0.8

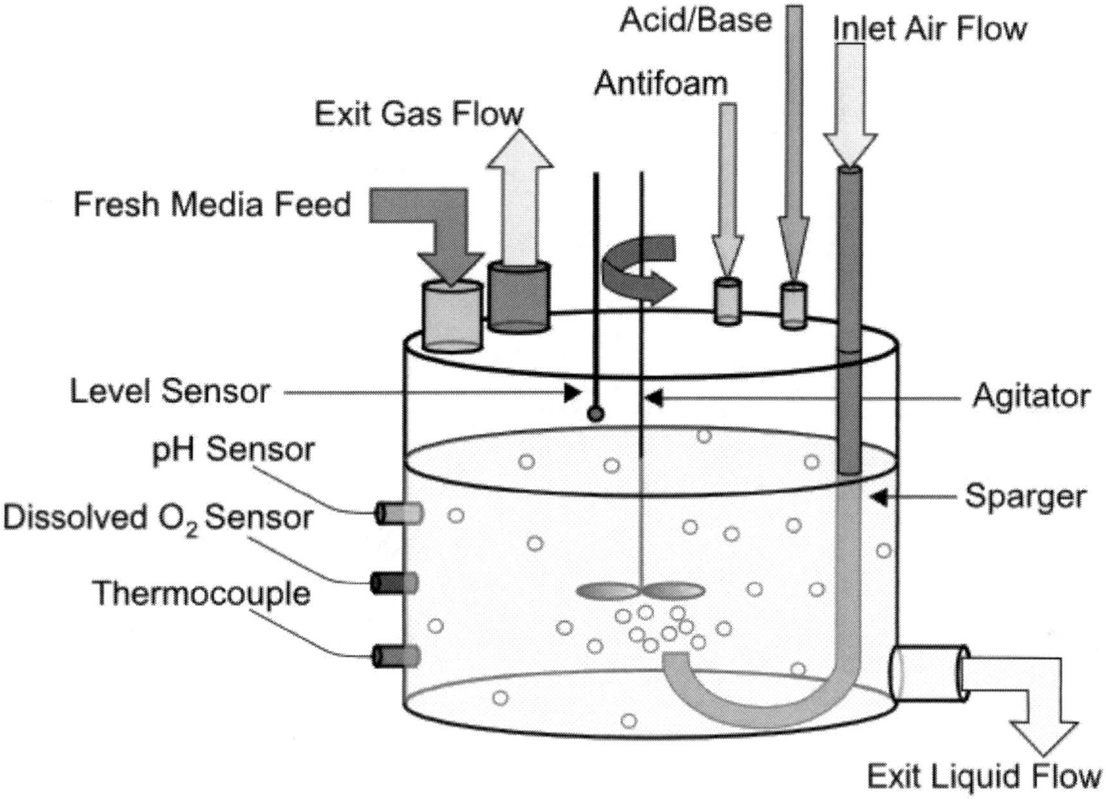

Figure (1): Schematic diagram of biochemical reactor.

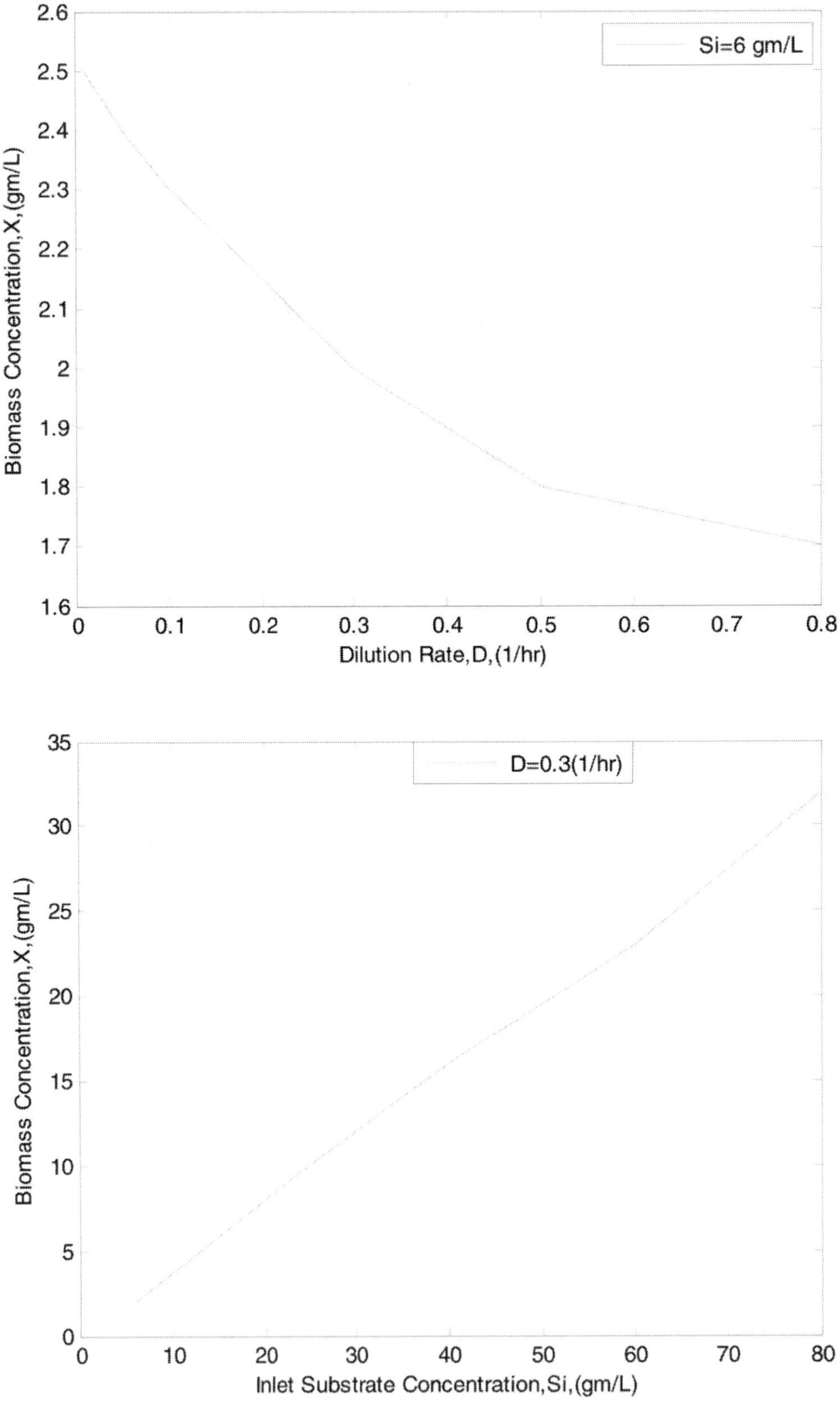

Figure (2): Biomass concentration for different value of ;(a)Dilution rate,(b)inlet substrate concentration.

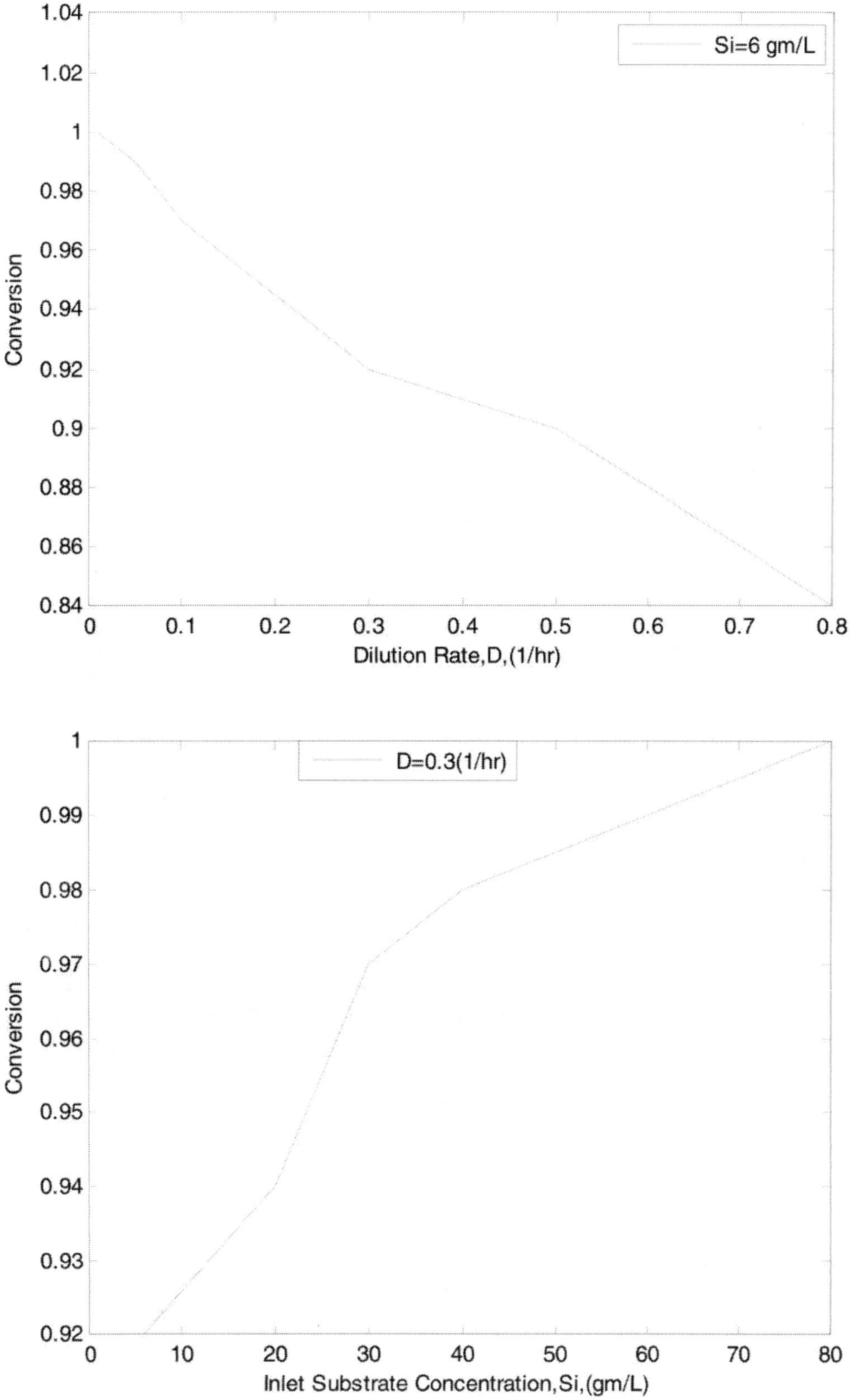

Figure (3): Conversion of substrate as a function of ;(a)dilution rate,(b)inlet substrate concentration.

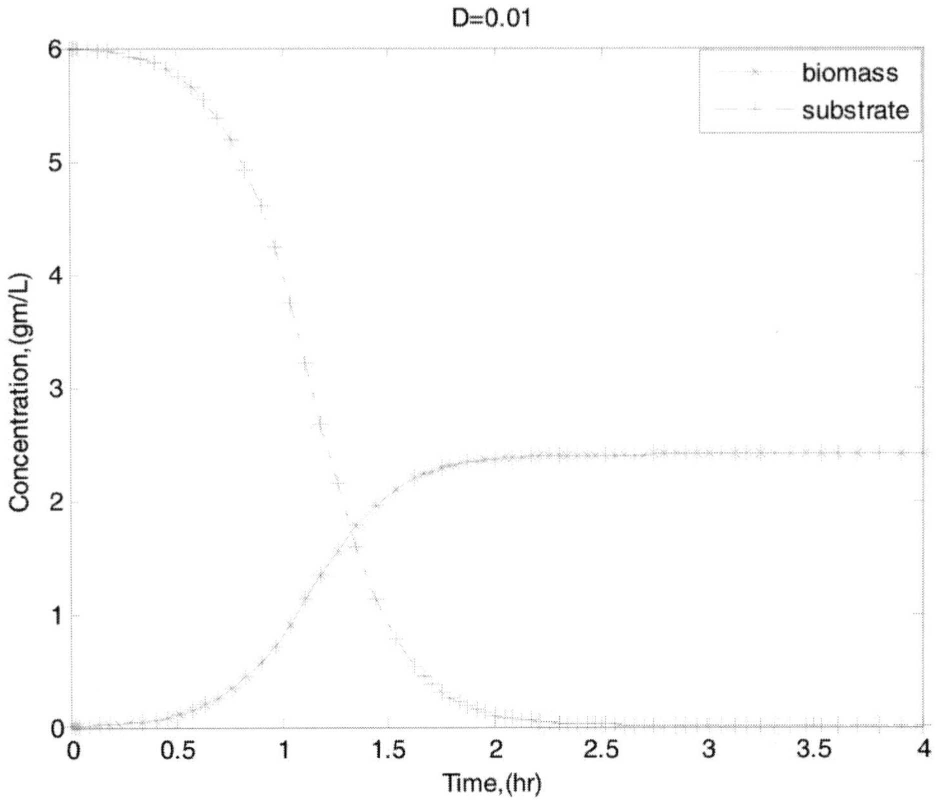

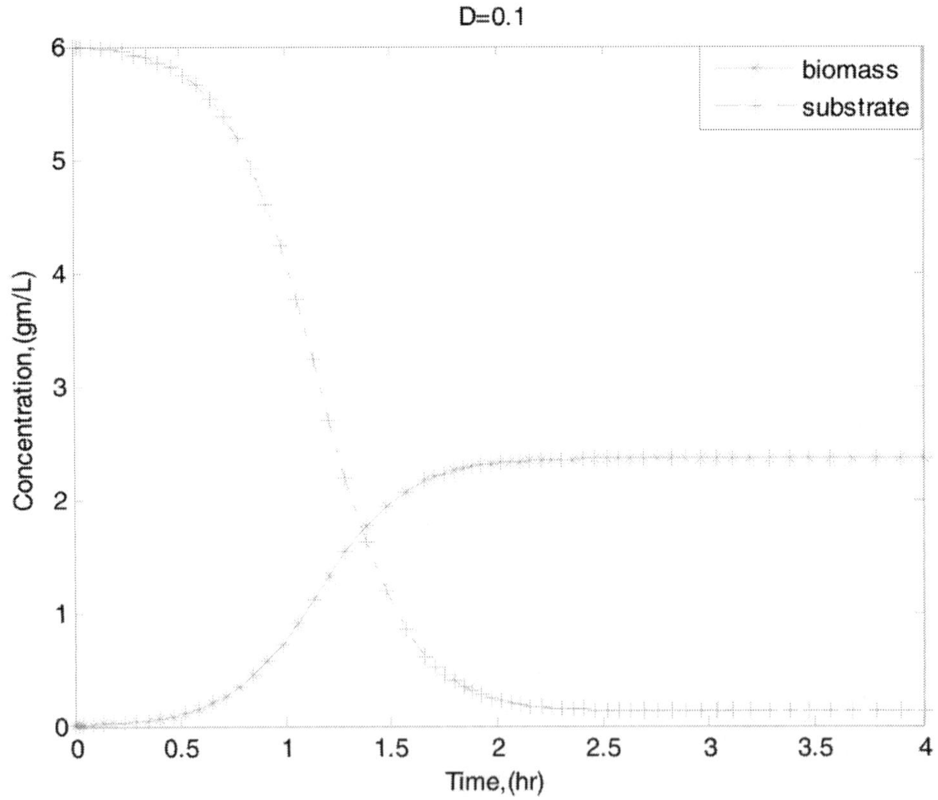

Figure (4): Concentration of biomass and substrate into reactor as a function of time for;(a)D=0.001,(b)D=0.1.

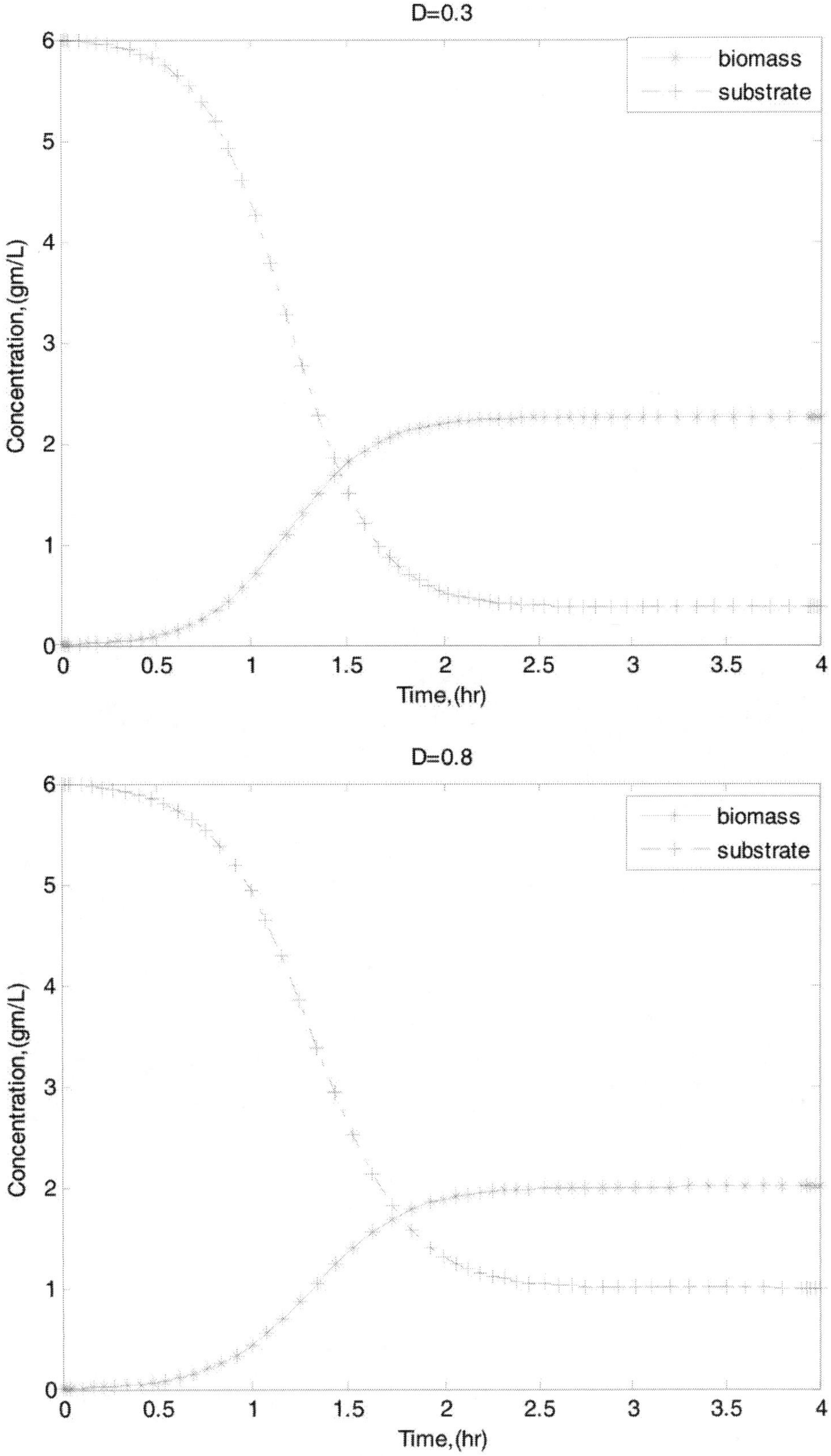

Figure (5): Concentration of biomass and substrate into reactor as a function of time for;(a)D=0.3,(b)D=0.8.

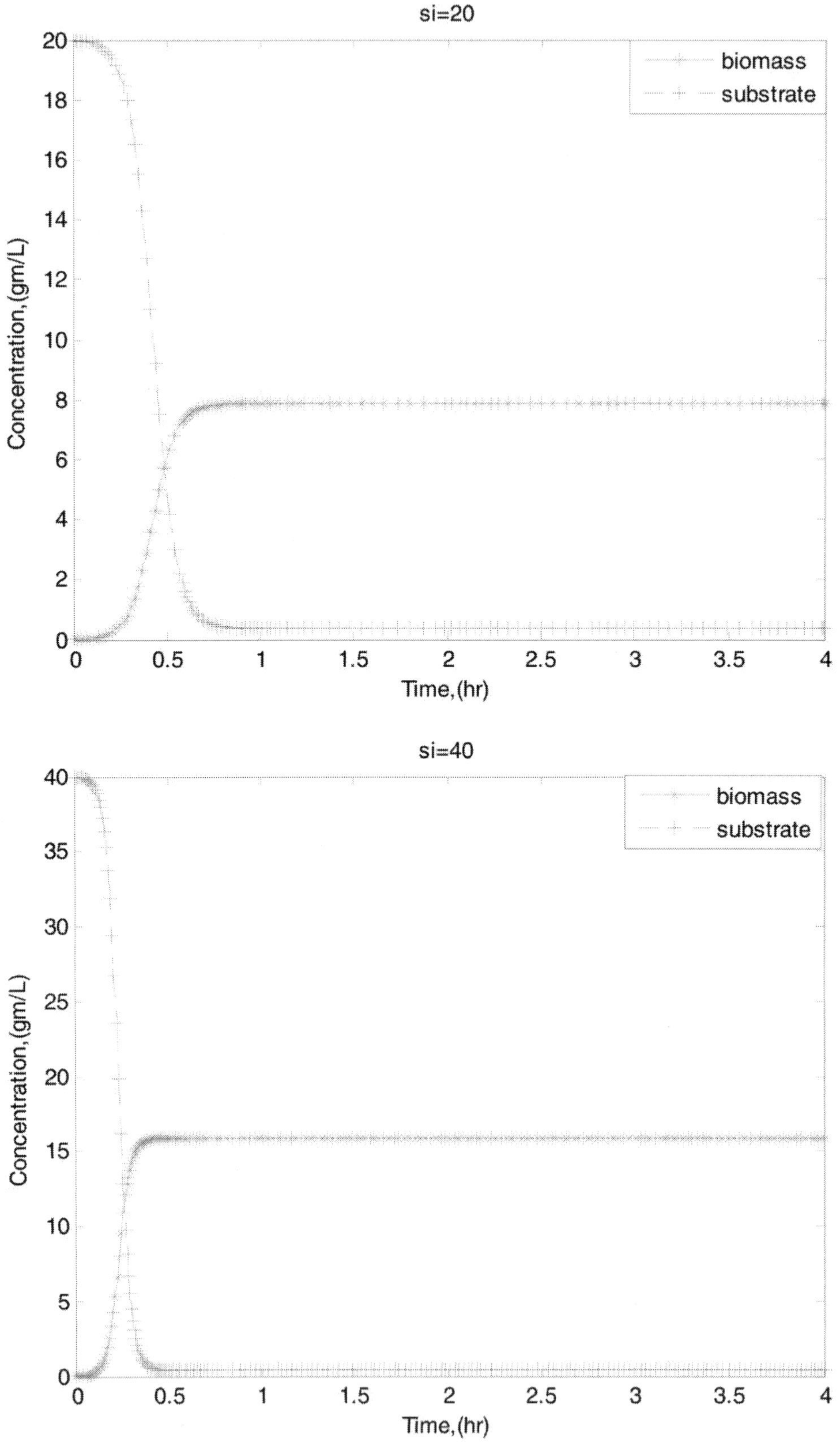

Figure (6): Concentration of biomass and substrate into reactor as a function of time for;(a)Si=20,(b)Si=40.

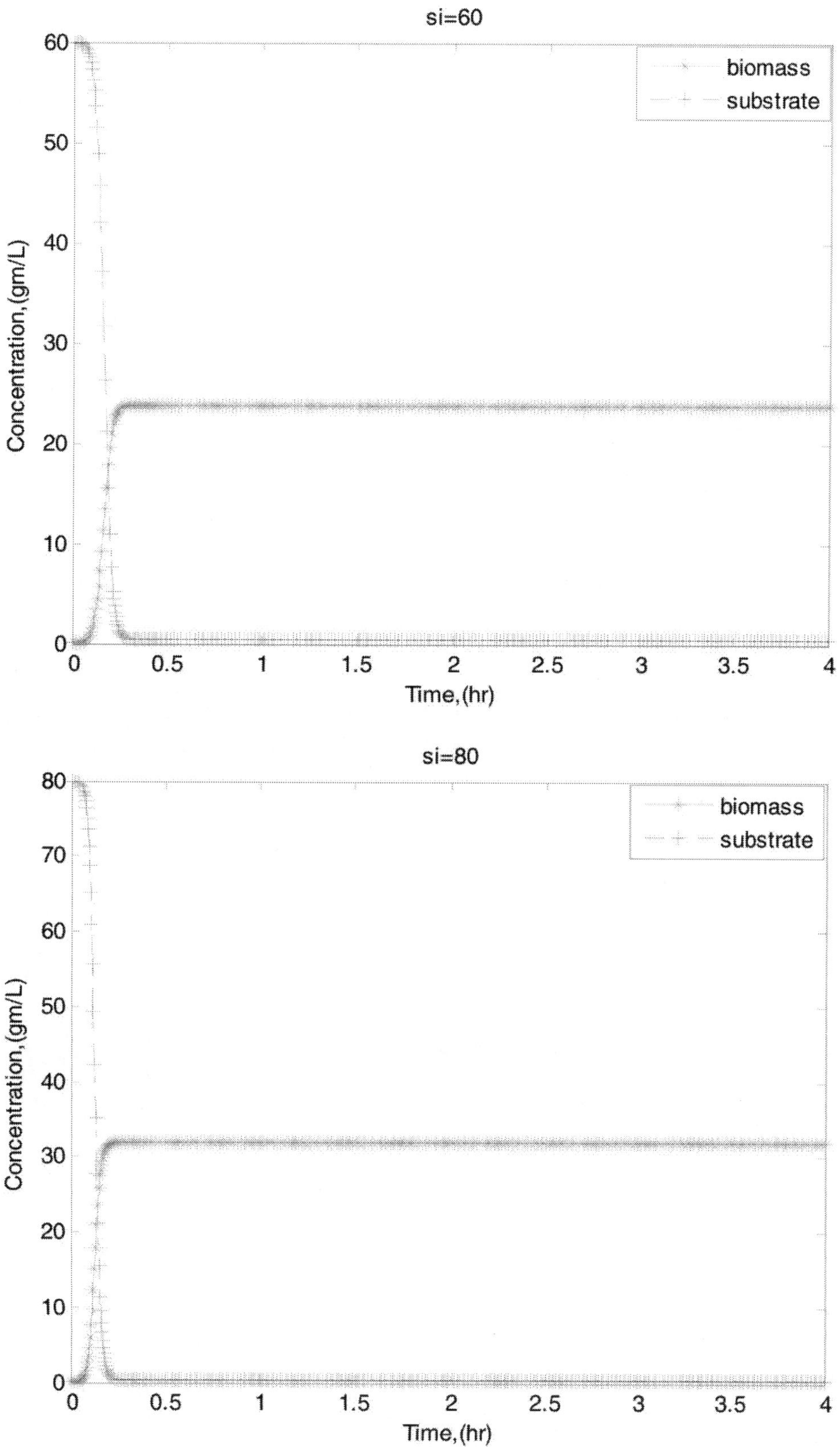

Figure (7): Concentration of biomass and substrate into reactor as a function of time for;(a)Si=60,(b)Si=80.

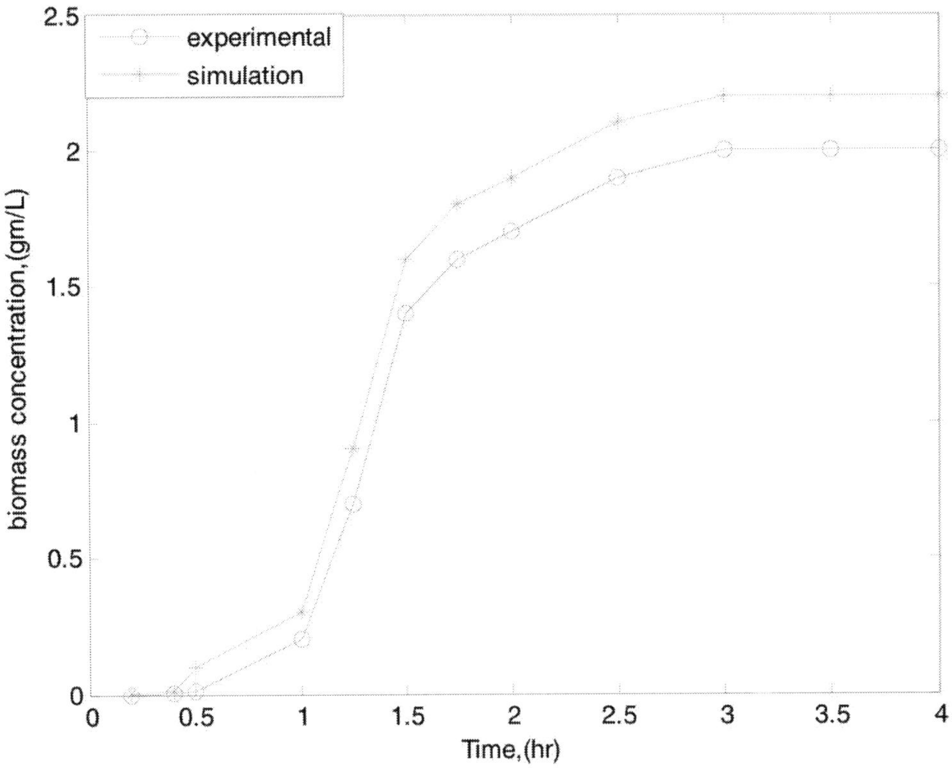

Figure (8):The comparison between the simulated and experimental results at(D=0.3,Si=6).

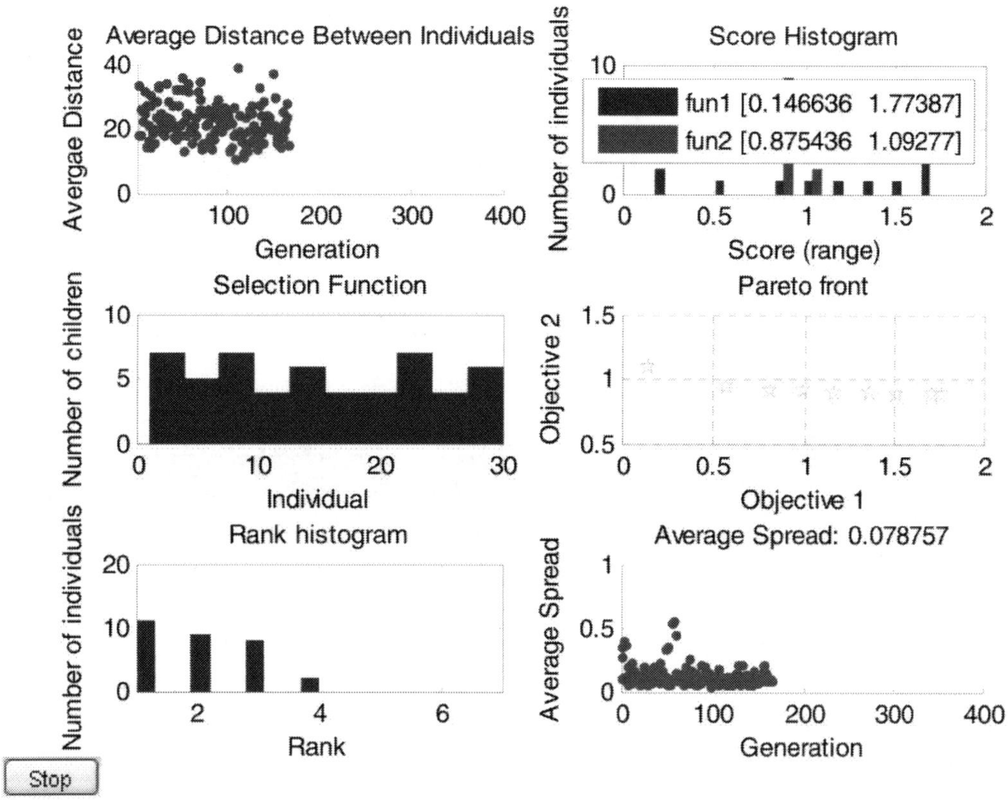

Figure (9):Optimum results by the multiobjective GA.

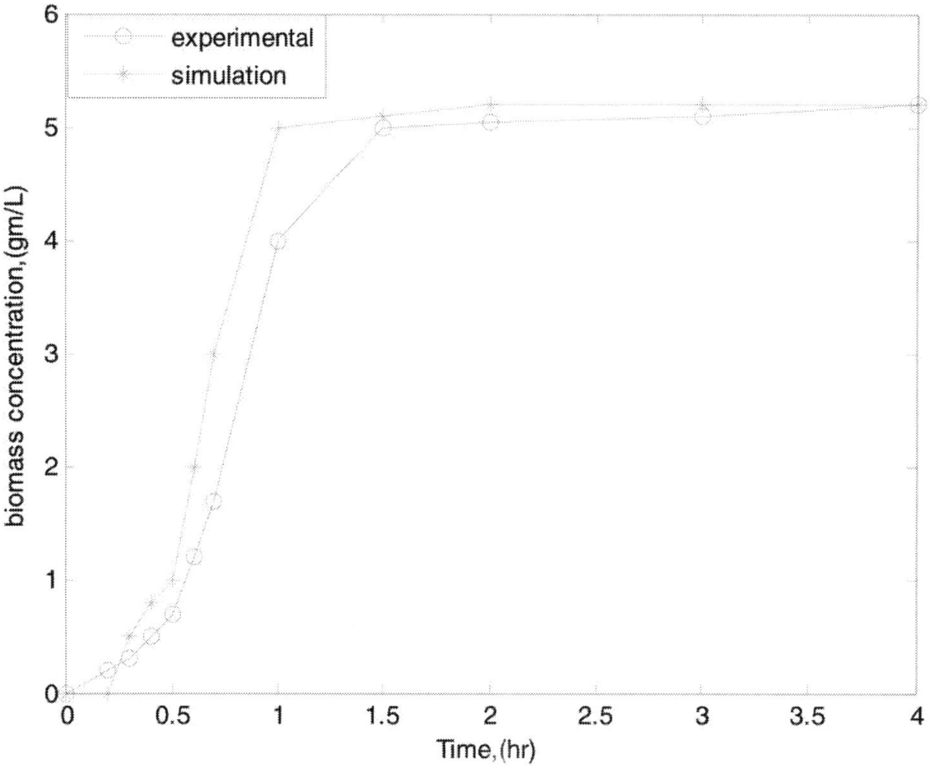

Figure (10): Experimental and simulated results at optimum operating conditions (D=0.727, Si=12)

(412d) HYDRODYNAMICS and OPTIMIZATION STUDIES IN AN Inverse Fluidized BED Bioreactor for Tannery Wastewater TREATMENT

Manivasagam Rajasimman and Chinnappan Karthikeyan, Chemical Engineering, Annamalai University, Chidambaram, India

In this work, the hydrodynamic characteristics and optimum conditions for the treatment of tannery wastewater in an inverse fluidized bed bioreactor were studied. Low-density particles made of polypropylene of density 870 kg/m^3 were used in this study. The behavior of air hold up and minimum fluidization velocity were studied for various bed heights and biomass grown on the particles. Experiments were carried out to find the optimum bed height and air flow rate for the maximum degradation of tannery wastewater. From the result it was observed that the air hold up increases with increase in air velocity and decreases after it reaches the critical velocity. Air hold up also increases with increase in bed height and decreases. The minimum fluidization air velocity increases with increase in bed height and decreases with increase in biomass weight. The maximum COD removal of 88.3% was achieved in this reactor at a bed height of 80 cm and for the air flow rate of 64.4 cc/s

See more of this Session: Sustainable Chemicals II
See more of this Group/Topical: Sustainable Engineering Forum

SUPERHYDROPHOBIC SURFACES PRODUCED BY SUPERCRITICAL FLUID TECHNOLOGY

Charlotta Turner[1], *Oskar Werner*[1,2] *and Lars Wågberg*[2]
(1) Department of Chemistry, Centre for Analysis and Synthesis, Lund University, Lund, Sweden
(2) Fibre and Polymer Technology, Royal Institute of Technology, Stockholm, Sweden

Introduction

Superhydrophobic surfaces are commonly described as surfaces that mimic the Lotus flower and its self-cleaning properties (*1;2*). A search in Web-of-Science Aug-11 on the topic "superhydrophobic surface" gives 2.340 hits of which 2.018 were published 2007 and later. An explanation to this trend could be that novel techniques combined with new theoretical understanding have made it possible to synthetically produce superhydrophobic surfaces.

More strictly speaking, the term superhydrophobicity (*3*) is used to describe the effect when a water droplet is rolling off a surface instead of sliding. This process is dependent on both the geometric structure and the intrinsic contact angle of the used material. In general, the geometric structure of the surface should be rough, containing micron-sized pillars with air pockets in-between, upon which water droplets will slide with extremely low friction. Surfaces can be considered superhydrophobic when they exhibit this property. As a rule of thumb contact angles towards pure water above 150 degrees indicate, but do not guarantee, superhydrophobic (strongly water-repellent), self-cleaning behavior.

There are many obvious applications of superhydrophobic and self-cleaning surfaces such as windows and environmentally exposed glass or lacquered surfaces (*4*) but if the technique could be applied to for example paper surfaces it is easy to envisage how new products can be prepared from this commodity product, for example, rain repellency coatings for boxes or breathable cover materials that will allow moisture penetration but prevent condensed water transport. Today most printing papers must be sized to an appropriate level so that the interactions between the paper and printing ink can be controlled (*5*), but this treatment only produces a lower degree of hydrophobicity with contact angles around 110 degrees for fully sized papers where the fibers have been saturated with sizing agent (*6*). These papers are non-absorbent but far from superhydrophobic and they will be wet through with water if the water drops are placed on the surface for a prolonged period of time.

There are several examples in the literature where superhydrophobic surfaces have been produced with wood fibers as a base substrate. In one of the existing examples (*7*), a highly fluorinated, tailor-made polymer has been grafted onto filter paper using organic solvents. However, fluorinated compounds are harmful to the environment, and hence more sustainable alternatives would be an important development. Furthermore, grafting to fibers is a relatively tedious procedure involving several chemical treatment steps often involving organic solvents. Another recently investigated technique is to deposit titanium dioxide nanoparticles onto the paper (*8;9*).

The methodology investigated in the presented research utilizes a common sizing agent used in the paper industry, alkyl ketene dimer (AKD). Since the described process does not require organic solvents but only supercritical carbon dioxide (scCO2) as solvent, it is also an environmentally more sustainable method. The particle formation and deposition technology used is the Rapid Expansion of Supercritical Solution (RESS) process. This technique could potentially enable a much more convenient way of making paper and basically any other solid substrate, superhydrophobic.

In our previous research, we have produced superhydrophobic self-cleaning surfaces by dissolving hydrophobic compounds in scCO2, followed by spraying the solution through a pressure nozzle employing the RESS technique (*10;11*). To the best of our knowledge, this combination of RESS and waxes has not been tested before and it was therefore considered interesting and important to investigate if this way of crystallizing AKD could be used to induce superhydrophobicity to solid substrates. The only processes developed so far based on scCO2 technology for making superhydrophobic coatings are relatively slow and they utilize fluorinated polymers (*12-14*). What we have developed is a rapid process (only a few seconds) using environmentally friendly coating chemicals. In the presented research, we have studied the effect of different pre-expansion pressure and temperature as well as distance between the nozzle and the treated surface, with regard to particle morphology and size distribution. Laser diffraction measurements have been conducted directly perpendicularly to the spraying cone from the RESS system, providing information about the particle size distribution as a function of spraying distance. Thermography was also used to obtain information about the phase transfer of CO2 from liquid to gas in the spray.

Materials and Methods

AKD granules (DR SF 300) with a mean particle diameter of approximately 4.5 mm containing a mixture of C_{18} AKD and C_{16} AKD were supplied by EKA Chemicals (Bohus, Sweden). Liquid carbon dioxide purchased from Air Liquide Gas AB (Eskilstuna, Sweden) was used in the RESS process. The surfaces used were 15mm×15mm pieces cut outs from polished silicon wafers (150 mm, CZ/1-0-0/BORON/P, MEMC Electronic Materials, Novara, Italy), which were cleaned with water/ethanol/water before use.

The RESS system used in this study was a simple in-house made 50 ml horizontal cylindrical pressure vessel connected to a sapphire nozzle of diameter 0.004" (Thar Technologies, USA) via a valve. The nozzle and the tubing were at all time heated to a temperature slightly above that of the vessel. The laser diffraction system used was a Malvern 1000 E with a 100 mm Fourier lens (Malvern Instruments, Ltd.). The RESS spray was directed through the laser beam. With the Fourier lens of 100 mm, optical diameter sizes between 0.2 and 120 µm can be detected. The evaluation software was Mastersizer X v2.19. See **Figure 1** for a schematic of the system.

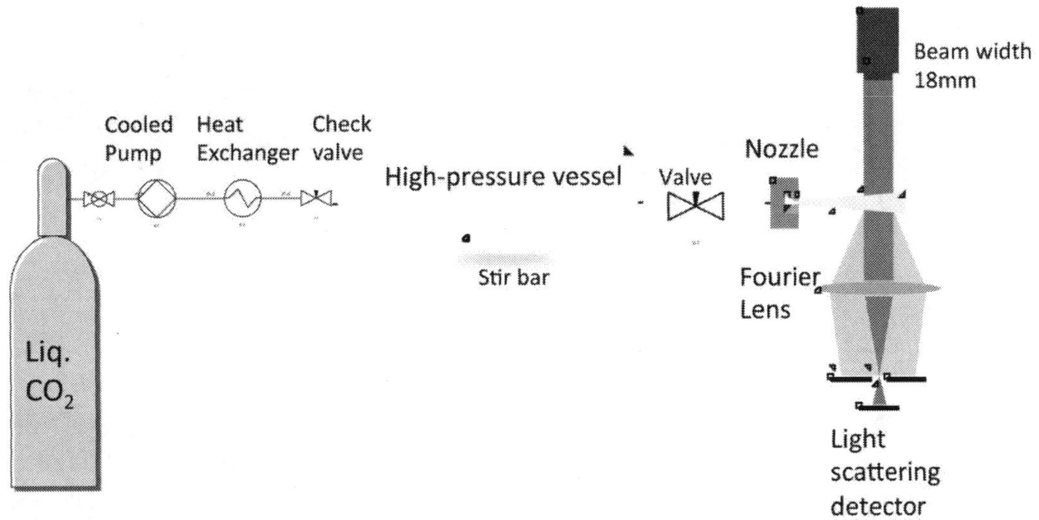

Figure 1*. Schematic of the RESS-laser diffraction system. SEM images were taken from surfaces produced at corresponding spraying distances, without the laser beam connected.*

Temperature has been measured in the spray cone. This has been done by spraying along a piece of thin sheet metal from which snapshots were taken with a Thermal Imager Ti30 by Fluke

(Everett, Washington, USA) with the possibility to read temperatures from -15°C to 200°C. The metal sheet has been coated with ordinary scotch tape, which according to the manual of the instrument has an emissivity of 0.93 rendering it fairly close to a black body.

The SEM system used in this system was a LEO 1530 and the surfaces were sputtered with a 5 nm thick layer of gold/palladium before measurement. Contact angles were measured to study the hydrophobic properties of the AKD surface with the aid of a semi-automatic CAM200 contact angle system (KSV Instruments Ltd, Helsinki, Finland).

Results and Discussions

A series sprayed at PET of 57°C and PET of 180 bar were measured at different distances from the nozzle using light scattering, as those conditions showed good repeatability and gave interesting results in initial tries. In order to separate the effect of the expanding CO_2 from formation of the particles, measurements were also made on CO_2 without dissolved AKD. **Figure 2** shows some of the obtained laser diffraction data as well as SEM micrographs.

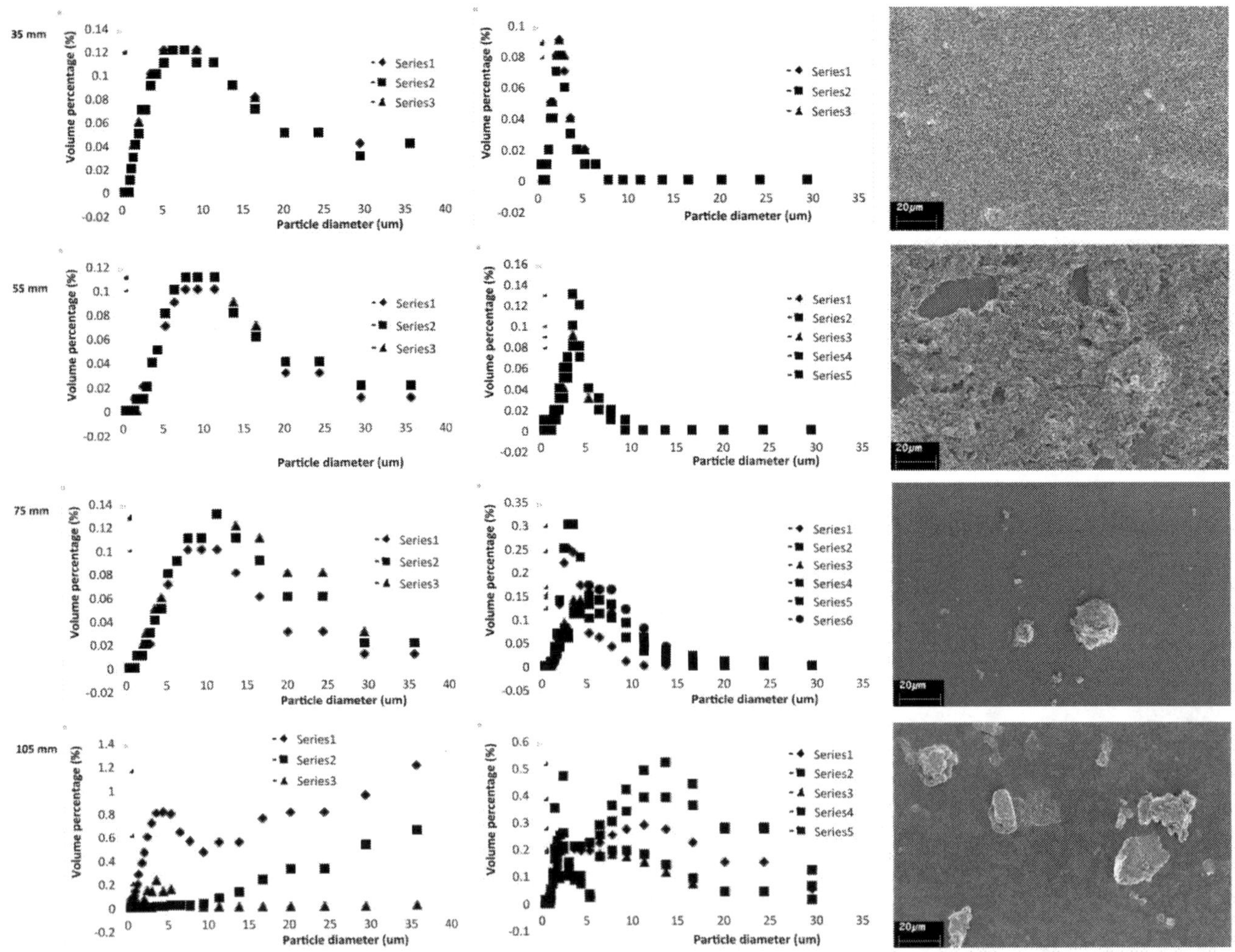

Figure 2*. Effect of spray distance on particle size distribution, as determined by laser diffraction. From top to bottom, the following spray distances were used: 35 mm; 55 mm; 75 mm; and 105 mm. To the right are SEM images showing representative parts of the respective sprayed surfaces.*

For the blank (CO_2 only) there is mainly a bell shaped peak at 10 µm. For the AKD samples a very sharp peak can be seen at around 2.5 µm for all the spray distances, but at longer distances a broader peak at around 6 µm is more pronounced. Thermography images revealed that the low temperature zone, for the selected pre-expansion state (57°C, 180bar) is approximately 60 mm long (data not shown). Since heavy cool down is depending on the phase transition of CO2 from liquid to gas, it is likely that 60 mm from the nozzle the phase transition is already complete. Hence, particles hitting the surface after traveling more than 60 mm will do so in a dry state while below 60 mm quite likely particles are still in liquid droplets of CO2.

Comparing the SEM micrographs and diffraction data, the continuous surface seen for 35 mm and 55 mm correlates with the dominant peak at 2.5 µm. At longer spray distances, on the other hand, there was no continuous surface being produced, but rather unevenly distributed particles and particle agglomerates. This could be due to that dry particles are more difficult to capture on a surface, and that the small particles will not even reach this far, and if they do, they are spread out. The later assumption is supported by the diffraction data, showing that the 2.5 µm peak is not as distinct or repeatable at those longer distances.

In **Figure 3**, peak position (particle size) has been plotted against different pre-expansion temperature and density (pressure was varied between 90 and 200 bar). It is obvious that even though none of these parameters, within the experiments settings, can alone be set to hinder the formation of small particles, low temperature and high density is a prerequisite for the formation of mainly large ones.

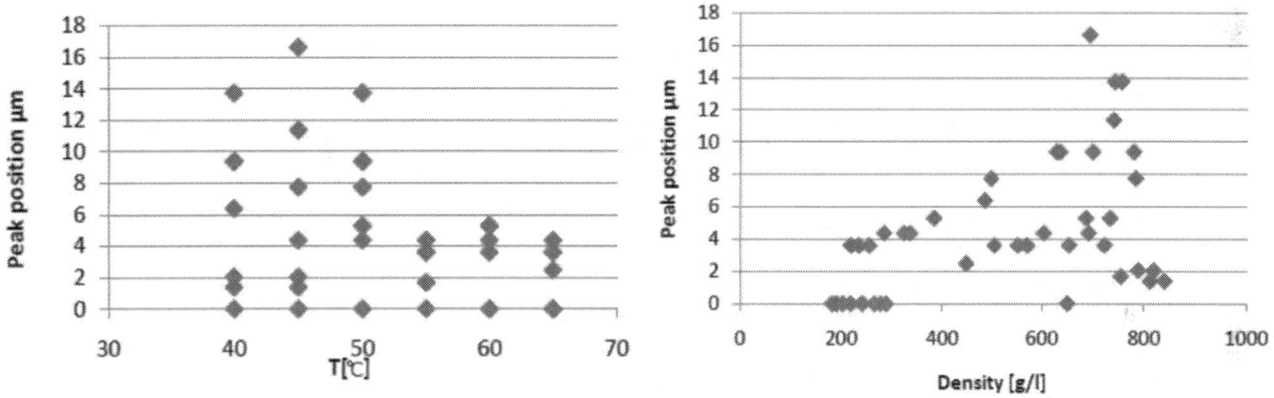

Figure 3. *Peak size by laser diffraction as a function of pre-expansion temperature (left) and density (right).*

Surfaces produced with high coverage of AKD crystals as discussed above are superhydrophobic, with static water contact angles of around 150-170°, see for example **Figure 4** below as well as our previous results in (*10*).

Figure 4. *AKD-covered cellulose surface produced by RESS at 300 bar and 40°C, 100 mm spray distance. Obtained static water contact angle is 153°.*

Conclusions

Light scattering data turned out to correlate well with SEM images when comparing the data obtained at different spray distances from the nozzle. A correlation could be seen for pre-expansion conditions producing smaller particles (as measured by diffraction) and a resulting evenly covered surface of small flakes (1-5 μm in diam.). By coupling data from SEM images, light scattering and thermography of the spray, conclusions could be drawn that at shorter distances CO2 is still in liquid phase and enables efficient transportation and attachment of AKD crystals to the surface, while at longer spray distances, CO2 is in gas phase and the formed particles are carried away aerodynamically without adsorbing to the surface.

Acknowledgements

This work was supported by the Swedish Foundation for Strategic Research (SSF, 2005:0073/13 and RMA08-0044), the Carl Trygger foundation (CTS 09:389), Bo Rydin and SweTree Technologies.

References

1. Barthlott, W.; Neinhuis, C. Purity of the sacred lotus, or escape from contamination in biological surfaces. *Planta* **1997**, *202*, 1-8.
2. Li, X.-M.; Reinhoudt, D.; Crego-Calama, M. What do we need for a superhydrophobic surface? A review on the recent progress in the preparation of superhydrophobic surfaces. . *Chem. Soc. Rev.* **2007**, 1350-1368.
3. Lafuma, A.; Quere, D. Superhydrophobic states. *Nat. Mater.* **2003**, *2*, 457-460.
4. Yoshimitsu, Z.; Nakajima, A.; Watanabe, T. *et al.* Effects of surface structure on the hydrophobicity and sliding behavior of water droplets. *Langmuir* **2002**, *18*, 5818-5822.
5. Shen, W.; Zhang, H. L.; Ettl, R. Chemical composition of 'akd vapor' and its implication to akd vapor sizing. *Cellulose* **2005**, *12*, 641-652.
6. Wågberg, L. Polyelectrolyte adsorption onto cellulose fibres – a review. *Nord. Pulp Paper Res. J.* **2000**, *15*, 586-597.
7. Nyström, D.; Lindqvist, J.; Östmark, E. *et al.* Superhydrophobic bio-fibre surfaces via tailored grafting architecture. *Chem. Commun.* **2006**, *34*, 3594-3596.
8. Huang, L. H.; Chen, K. F.; Lin, C. X. *et al.* Fabrication and characterization of superhydrophobic high opacity paper with titanium dioxide nanoparticles. *J. Mater. Sci.* **2011**, *46*, 2600-2605.
9. Teisala, H.; Tuominen, M.; Aromaa, M. *et al.* Development of superhydrophobic coating on paperboard surface using the liquid flame spray. *Surf. Coat. Technol.* **2010**, *205*, 436-445.
10. Quan, C.; Werner, O.; Wagberg, L. *et al.* Generation of superhydrophobic paper surfaces by a rapidly expanding supercritical carbon dioxide-alkyl ketene dimer solution. *J. Supercrit. Fluids* **2009**, *49*, 117-124.
11. Werner, O.; Quan, C.; Turner, C. *et al.* Properties of superhydrophobic paper treated with rapid expansion of supercritical co2 containing a crystallizing wax. *Cellulose* **2010**, *17*, 187-198.
12. Desbief, S.; Grignard, B.; Detrembleur, C. *et al.* Superhydrophobic aluminum surfaces by deposition of micelles of fluorinated block copolymers. *Langmuir* **2010**, *26*, 2057-2067.
13. Nikitin, L. N.; Gallyamov, M. O.; Said-Galiev, E. E. *et al.* Supercritical carbon dioxide: A reactive medium for chemical processes involving fluoropolymers. *Russ. J. Gen. Chem.* **2009**, *79*, 578-588.
14. Gallyamov, M. O.; Nikitin, L. N.; Nikolaev, A. Y. *et al.* Formation of superhydrophobic surfaces by the deposition of coatings from supercritical carbon dioxide. *Colloid J.* **2007**, *69*, 411-424.

Jet Fuel from Air, Water, and Uranium

Charles Forsberg and John Michael Galle-Bishop
Massachusetts Institute of Technology; Email: cforsber@mit.edu
Proceedings of the American Institute of Chemical Engineers; October 16-21, 2011; Paper 235073

Abstract

Liquid fuels can be produced from air and water. We developed flowsheets for production of 6400 barrels per day of JP5, a jet fuel that is useable as diesel fuel. The feed stocks are carbon dioxide extracted from air and water obtained from the ocean. Heat and electricity to convert the feed stocks into liquid fuel are provided by either a pressurized-water reactor (PWR) or a high-temperature reactor (HTR). PWRs are the most common existing commercial nuclear power technology with peak temperatures of ~285°C. The PWR thermal heat-to-liquid-fuel efficiency is ~22% versus the typical heat-to-electricity efficiency of 33%. HTRs are advanced reactors with peak temperatures of ~700°C. The heat-to-liquid-fuel efficiency is ~31% versus a heat-to-electricity efficiency of 45%. Because the supply of air and ocean water is effectively unlimited, the flowsheet provides an upper limit for the cost of liquid fuel relative to the cost of electricity—between 2 and 3 times the cost of electricity on a per unit heat basis. On a per unit energy basis, today liquid fuel prices in the U.S. are about equal to the price of electricity The study was done in the context of a navy nuclear tanker to manufacture JP-5 at sea at a rate sufficient to supply a carrier strike group under wartime conditions. The flow sheets are essentially identical for navy-at-sea or commercial-on-land applications.

I. Introduction

The design goal [1, 2] was a U.S. navy nuclear-powered tanker that could manufacture on-board at-sea all the liquid fuel as needed for a carrier strike group. This included fuel for both the carrier aircraft and all escort vessels. The aircraft carrier has nuclear reactors for propulsion but the escort vessels use marine diesel or jet fuel for power. Such a tanker would free the U.S. Navy from most of its logistical burdens. JP5 is a refined jet fuel for navy applications that is compatible for use in marine gas turbine and diesel engines and thus could meet all U.S. Navy at-sea liquid-fuel requirements. It differs from conventional jet fuel primarily in its lower flammability—an added requirement because of the risk of fire aboard ship.

The flow sheets for liquid fuels production on a navy tanker are essentially identical to those required to manufacture unlimited liquid fuels from air and water to meet commercial liquid fuels requirements. Uranium resources are sufficient for millennia [3]. The manufacture of fuels from air and water results in no net carbon dioxide releases to the atmosphere and thus no risk of climate change. Air and ocean water are available in unlimited quantities in most countries; thus, this is an option for long-term independence from foreign liquid fuel sources.

Today oil is the preferred low-cost feedstock for liquid fuels production. Alternative feed stocks (natural gas, coal, shale oil, oil sands, biomass) are beginning to be used for liquid fuels production. Liquid fuels from air and water are the only unlimited long-term liquid-fuels production option; thus, it provides an upper long-term estimate of the costs of liquid fuels. The plant size used in this case study is ~1% of a global-scale oil refinery.

The concept of converting carbon dioxide from air and water to liquid fuels is not new. Initial studies [4] were conducted in the 1970s and recent reviews [5] report on much of the progress since then. Advancing technologies however have improved the economics and provided a much sounder basis for estimating costs. These areas are described below

II. Flowsheets

The flowsheet is shown in Fig. 1. The major process steps include:

- *Carbon dioxide feedstock.* Carbon dioxide is extracted from air by absorption with regeneration using humid air. This energy-efficient technology is being tested in pilot plants [6]. This is the most technically immature process step. There are many other options to recover carbon dioxide from air [7].

  The concentration of carbon dioxide in seawater is about 40 times larger; however, in the ocean the carbon dioxide exists in several forms including bicarbonates. The equilibrium kinetics between these various forms and dissolved carbon dioxide are slow [8]. As a consequence, the size of the equipment to remove carbon dioxide from the ocean is too large for a ship-based system. However, extraction of gaseous carbon dioxide from the ocean may be an option for a plant sited on the coast.

- *Water feedstock.* Water is purified and converted to steam—a commercial technology.

- *Syngas production.* Water in the form of steam and CO_2 are fed to high-temperature coelectrolysis cells [9] to produce syngas (H_2, CO) and oxygen. Added hydrogen [10] is made by high-temperature steam electrolysis (HTE). The equipment operates at temperatures near 800°C, is at the pilot plant stage of development, and is based on solid-oxide electrolytic fuel cells operated in reverse. HTE uses about 75% of the energy of traditional alkaline-cell electrolysis.

  HTE is potentially a key technology for integration of nuclear-wind or nuclear-solar electricity hydrogen systems [11]. In these applications, the HTE system uses steam (heat) from nuclear plants and electricity from the grid to produce hydrogen that primarily supplies industrial markets but can also be used for peak electricity production. Such advanced systems may provide an economic method to use excess renewable electricity when electricity production exceeds demand. Consequently, there are expected to be significant incentives for development and ultimate commercialization of this technology.

- *Liquid fuels production.* The syngas is fed to a Fischer-Tropsch (FT) system to convert syngas to liquid fuels. Micro-channel chemical reactors are assumed [12, 13]. The FT process using traditional chemical reactors is a commercial process whereas micro-channel chemical reactors are in the pilot plant stage of testing.

  There are large efforts underway today to improve the FT process for conversion of natural gas to syngas and then to liquid fuels. Shell Oil recently completed its Pearl Project in Qatar, the first 19-billion-dollar global-scale natural-gas-to-diesel-fuel-plant that consumes the equivalent of ~2.5% of U.S. natural gas production. Given the large price differences between natural gas and liquid fuels per unit energy, there are large financial incentives to build more such plants, improve their efficiency, and reduce their capital costs. The rapid advances in natural-gas syngas to liquid-fuels plants is directly applicable to liquid fuels from air facilities.

- *Refining.* Standard refining technology is used to produce the final JP5 liquid fuel with unusable hydrocarbons recycled through an auto-thermal reformer.

  Because of the increased demand for jet and diesel fuel, there is significant work underway [14] to develop better flow sheets to convert FT liquids from coal liquefaction plants into primarily diesel and jet fuel. The normal FT process produces a spectrum of light, medium, and heavy hydrocarbons. Narrowing the product stream to produce more high-value gasoline, diesel, and jet fuel with smaller quantities of low-value light and heavy fractions has a large impact on profitability for existing and future FT plants.

- *Auto-thermal reforming.* The FT process produces a relatively wide distribution of hydrocarbons. Refining operations can convert many of these products to JP5 but there are significant light and heavy products that are not convertible. All of these are recycled back to an auto-thermal converter and burnt with oxygen to produce a syngas for recycle.

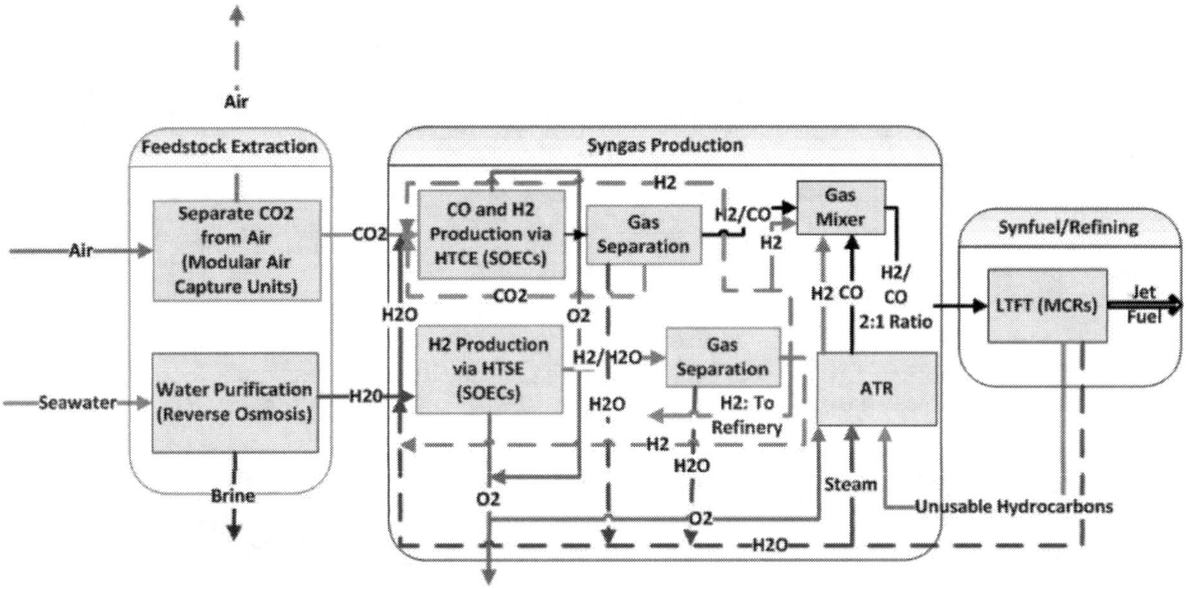

Figure 1: Top-Level Flow Sheet for Air and Water to Jet Fuel/Diesel

The technologies were chosen based on efficiency and space requirements. Space is a major constraint aboard a ship but would not be a constraint for a land-based facility. Each of the non-commercial technologies is at a point where a reasonable case for future commercial deployment can be made. For each non-commercial technology there is a backup commercial technology. For hydrogen production, it is conventional electrolysis, a technology that has been commercial for over a century. The least developed technology is the removal of carbon dioxide from air. Because most of the energy is consumed by high-temperature electrolysis and co-electrolysis systems, this is the critical technology in terms of overall plant efficiency and costs.

The process produces a single product (JP5). However, the flow sheets herein are identical to those required to manufacture unlimited gasoline, diesel, and jet fuel from air and water—except for minor changes in the last refining steps. The differences in energy consumption would be minor. The navy tanker output is ~1% of a global-scale oil refinery. Consequently, the work provides an estimate of what is required to produce liquid fuels with no net release of carbon dioxide to the atmosphere. A commercial plant would be slightly easier to build because (1) no need to restrict plant size to fit aboard ship and (2) somewhat smaller refinery subsection because there is no need to convert gasoline fractions to JP-5

III. Nuclear Reactor Choices

The process requires heat and electricity. All energy was assumed to be supplied by nuclear reactors. Two cases were evaluated. The base case assumed energy provided by pressurized water reactors (PWR)—the main commercial nuclear reactor technology today. The PWR peak temperature is 285°C. The alternative case assumed energy was provided by a salt-cooled high-temperature reactor with a peak temperature of 700°C. This is an advanced high-temperature reactor that may be available in 20 years. Resistance heating is used where required to obtain high temperatures.

IV. Energy Balances

The heat balances for a system to produce 6400 barrels per day of JP5 are shown in Table 1. The percentage of the total nuclear reactor power used in particular unit operations is shown in parenthesis. As would be expected, most of the energy input is for coelectrolysis and electrolysis that together require 590 MWe. That energy is to produce carbon monoxide from carbon dioxide and hydrogen from water for fuels production. The capture of carbon dioxide from the atmosphere is not a major energy consumer. The energy requirements for other operations are small. The smaller power requirements for ship propulsion are not shown.

If PWRs are the source of heat and electricity, the thermal heat-to-liquid-fuel efficiency is ~22% versus the typical heat-to-electricity efficiency of 33%. If HTRs are the source of heat and electricity, the heat-to-liquid-fuel efficiency is ~31% versus a heat-to-electricity efficiency of 45%.

V. Implications of Liquid Fuels from Air and Water

Converting carbon dioxide from air and hydrogen from water into liquid fuels places an upper limit on the cost of liquid fuels. By doing so, it defines the minimum technical and economic requirements for alternative technologies such as electric cars that might replace liquid fuels for specific applications. The technology could also be deployed under several different conditions.

- *Military Applications—Navy Tanker*. The original goal of this work [1] was a nuclear tanker for the U.S. Navy that could manufacture on-board at-sea all the liquid fuel as needed for a carrier strike group. This included fuel for both the carrier aircraft and all escort vessels. The carrier has nuclear reactors for propulsion. Such a tanker would free the U.S. Navy from most of its logistical burdens. The economics of navy fuel supply are significantly different than for commercial users because of the need to assure fuel is provided under war time conditions—a requirement that imposes the need for military protection of tankers.

Table 1. Heat and Energy Balances for Producing 6400 Barrels of JP5 Jet Fuel Per Day Using a Pressurized-Water Reactor or High-Temperature Reactor

System	PWR Parameter; (% Heat[1])	HTR Parameter: (% Heat[1])
Reactor Power		
Total Thermal Power (MWth)	2082 (100%)	1456 (100%)
Thermal Power to Power Cycle (MWth)	1973 (95%)	1285 (88%)
Electrical power from Power Cycle (MWe)	658 (95%)	650 (88%)
Thermal Power for Process Heat (MWth)	109 (5%)	171 (12%)
CO_2 Capture and Desalination (MWe)	21.7 (3%)	21.7 (3%)
High Temperature Coelectrolysis		
Temperature of heat input (°C)	27-285	27-655
CO_2 Feed Heating (MWth)	7.80	21.1
H_2O Feed Heating (MWth)	69.3 (3%)	102 (7%)
HTCE Heat Total (MWth)	77.1	123
HTCE Electricity (MWe)	534 (77%)	534 (73%)
High-Temperature Electrolysis		
H_2O Feed Heating (MWth)	12.0	17.6
HTE Total (MWth)	12.0	17.6
HTE Electricity (MWe)	55.9 (8%)	55.9 (8%)
HTE Electric Resistance Heating (MWe)	0.771	0.0
Auto thermal Reforming		
H_2O heating (MWth)	10.3	13.4
Hydrocarbon Heating (MWth)	2.96	7.86
ATR Heat Total (MWth)	13.2	21.2
ATR Electric Resistance Heat (MWe)	5.380	0.481
Product Upgrade		
Temperature of heat input (°C)	27-285	27-350
Product Upgrade (MWth)	6.24	8.59
Electric Resistance Heating (MWe)	2.49	0.0
Compression (MWe)	4.85	4.85
Compression Non-Refining (MWe)	34.3	34.3
Overall Cycle Performance		
Power cycle efficiency	33.1	44.6
Heat JP5 Combustion/Reactor Heat Input	21.8	31.1

[1]Percentage of total heat from the reactor

JP5 is a refined jet fuel for navy applications that is compatible for use in marine diesel and gas turbine engines and thus could meet all U.S. Navy at-sea liquid-fuel requirements. It differs from commercial jet fuel primarily in its lower flammability—an added requirement because of the risk of fire aboard ship. A nuclear tanker appears feasible given current directions in research and development. The largest gap is associated with the systems to remove carbon dioxide from the atmosphere where there is a need to reduce the system volume by about a factor of four. The system in the flow sheets is a system optimized for siting on land. No effort was made to develop a conceptual system with a smaller footprint.

- *Special commercial markets*. Most of the cost for liquid fuels production is associated with electricity to produce hydrogen via low-temperature or high-temperature water electrolysis. For parts of the world with very low hydro electricity costs and high liquid fuel costs (such as Iceland), this process could become commercially viable (particularly if there were restrictions on greenhouse gas releases). This option is available for any other stranded non-fossil energy source.

- *Normal commercial markets* If there are severe restrictions on greenhouse gas releases to the atmosphere, liquid fuels from air are one of the few options available and could become the economically preferred option. If atmospheric releases of carbon dioxide are limited[1], there are two feedstocks to produce liquid hydrocarbon fuels: biomass and atmospheric carbon dioxide. In both cases carbon dioxide from burning fuel is recycled through the air back to liquid fuels production.

  Biomass would be the preferred feedstock because biomass is a more reduced form of carbon ($\sim COH_2$) compared to carbon dioxide (CO_2). That implies less energy to convert it into a hydrocarbon fuel. However, total biomass resources are limited and may not be able to meet world liquid fuel demands [15].

  As discussed earlier, HTE is potentially a key technology for integration of nuclear-wind or nuclear-solar electricity hydrogen systems [11]. In these applications, the HTE system uses steam (heat) from nuclear plants and electricity from the grid to produce storable hydrogen for industrial markets at times of excess electricity production. Large-scale hydrogen storage is cheap and thus the system has the capability to match electricity and hydrogen production with demand when the production is determined by the energy source (wind or solar), not the customer demand. Production of liquid fuels from air could be done using that hydrogen with small modifications of the flowsheets.

VI. Conclusions

Because the supply of air and ocean water is effectively unlimited, this process when coupled to nuclear power plants provide an upper limit for the cost of liquid fuel relative to the cost of electricity—between 2 and 3 times the cost of electricity on a per unit heat basis because costs are dominated by the energy costs. On a per unit energy basis, liquid fuel prices today in the U.S. are about equal to the price of electricity ($4/gallon ~ $0.10/kWh) . The critical technologies are either commercial technologies or in the pilot plant stage of development. For each of the technologies in the pilot plant stage of development, more conventional technologies with somewhat lower efficiencies and higher capital costs are available.

[1] There is the option of producing liquid fuels from fossil fuels with removal and sequestration of carbon dioxide from the air. This option depends upon the economic and social viability of carbon dioxide sequestration. This option does not address the national security challenge of dependence on foreign fuels.

VII. References

1. J. M. Galle-Bishop, C. W. Forsberg, and M. J. Driscoll, *Nuclear Tanker Producing Liquid Fuels from Air and Water: Applicable Technology for Land-Based Future Production of Commercial Liquid Fuels*, MIT-NES-TR-13, Center for Advanced Nuclear Energy Systems, Massachusetts Institute of Technology (June 2011).
2. J. M. Galle-Bishop, *Nuclear Tanker Producing Liquid Fuels from Air and Water*, Thesis, Department of Nuclear Science and Engineering, Massachusetts Institute of Technology, Cambridge, Massachusetts (2011)
3. M. Kazimi, E. J. Moniz, C. W. Forsberg, et. al., *The Future of the Nuclear Fuel Cycle—An Interdisciplinary MIT Study*, Massachusetts Institute of Technology, Cambridge, Massachusetts (2011).
4. M. Steinberg, "Nuclear Power for the Production of Synthetic Fuels and Feedstocks," *Energy Policy*, **5**(1) pp. 12-24 (1977).
5. C. Graves, S. D. Ebbesen, M. Mogensen, and K. S. Lackner, "Sustainable Hydrocarbon Fuels by Recycling CO_2 and H_2O with Renewable or Nuclear Energy', *Renewable and Sustainable Energy Reviews*, **15**, 1-23 (2011)
6. K. S. Lackner, "Capture of Carbon Dioxide from Ambient Air," *The European Physical Journal (Special Topics),* **176,** pp. 93-106 (2009).
7. M. Ranjan, "M.S. Thesis: Feasibility of Air Capture," Massachusetts Institute of Technology, http://hdl.handle.net/1721.1/59782, Cambridge, MA. (2010)
8. S. Werner and J. Morgan, *Aquatic Chemistry: An Introduction Emphasizing Chemical Equilibrium in Natural Waters*, Wiley-Interscience, New York, pp. 780 (1970).
9. J. O'Brien, M. McKellar, and E. Harvego, *High-Temperature Electrolysis for Large- Scale Hydrogen and Syngas Production from Nuclear Energy – System Simulation and Economics*, Idaho National Laboratory, INL/CON-09-15563, Idaho Falls, ID.(2009)
10. J. O'Brien, C. Stoots, and J. Herring, J., *High Temperature Electrolysis for Hydrogen Production from Nuclear Energy – Technology Summary*, Idaho National Laboratory, INL/EXT-09-16140, Idaho Falls, ID. (2010)
11. C. W. Forsberg and G. Haratyk, "Nuclear Wind Hydrogen Systems for Variable Electricity and Hydrogen Production, " *Proc. American Institute of Chemical Engineers Meeting,* Minneapolis, Minnesota, October 16-21, 2011.
12. J. McDaniels, "Unlocking Biofuels with Microchannel Technology," Velocys, Plain City, OH., http://velocys.com/ (2010).
13. A. Tonkovich, D. Kuhlmann, and A. Rogers, "Microchannel Technology Scale Up to Commercial Capacity," *Chemical Engineering Research and Design*, **83**(A6) pp. 634-639 (2005).
14. A. De Klerk, A., *Fischer-Tropsch Jet Fuel Process*, US 2010/018568 A1pp. 1-14 (2010).
15. C. W. Forsberg, "Nuclear Energy for a Low-CO2-Emission Transportation System with Liquid Fuels," *Nuclear Technology*, **164**(3) pp. 348-367 (2008).

Energy Production From Municipal Waste Plastics using by Thermal Degradation
Moinuddin Sarker, PhD, MCIC
Vice President, Research & Development
Natural State Research, Inc.
37 Brown House Road (2nd Floor)
Stamford, CT-06902
E-mail: msarker@naturaltateresearch.com
www.naturalstateresearch.com
Natural State Research
AIChE Annual Meeting, 18th October, 2011

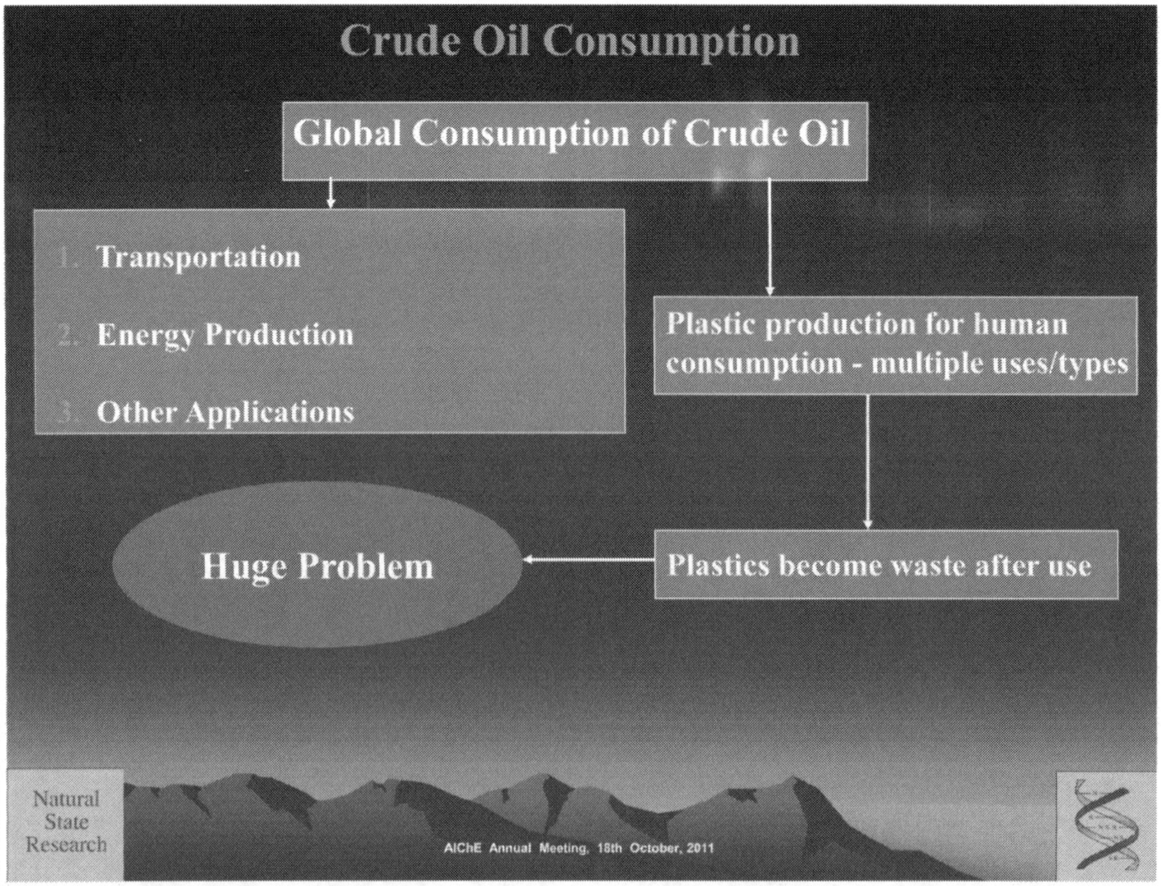

Crude Oil Consumption
Global Consumption of Crude Oil
1. Transportation
2. Energy Production
3. Other Applications
Plastic production for human consumption - multiple uses/types
Plastics become waste after use
Huge Problem
Natural State Research
AIChE Annual Meeting, 18th October, 2011

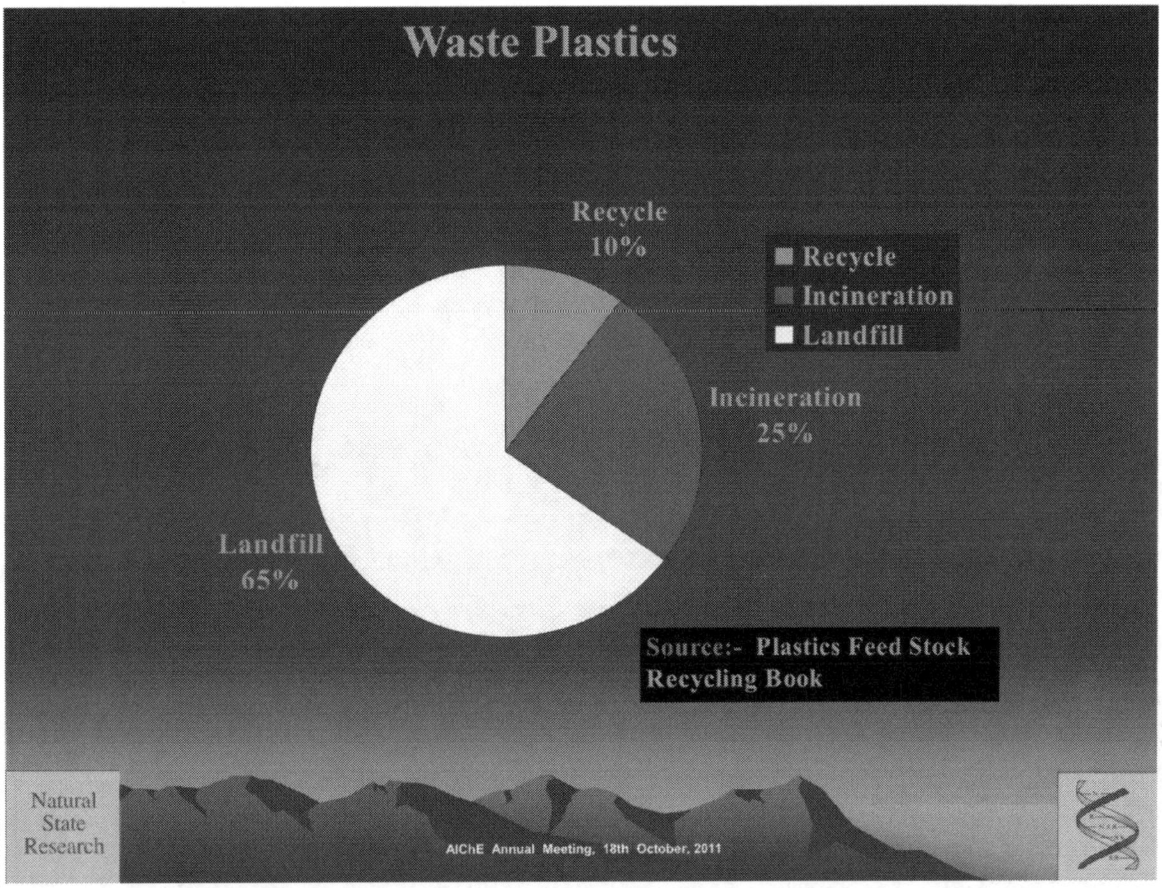

Waste Plastics
Recycle
10%
Incineration
25%
Landfill
65%
Recycle
Incineration
Landfill
Source:- Plastics Feed Stock
Recycling Book
Natural
State
Research
AIChE Annual Meeting, 18th October, 2011

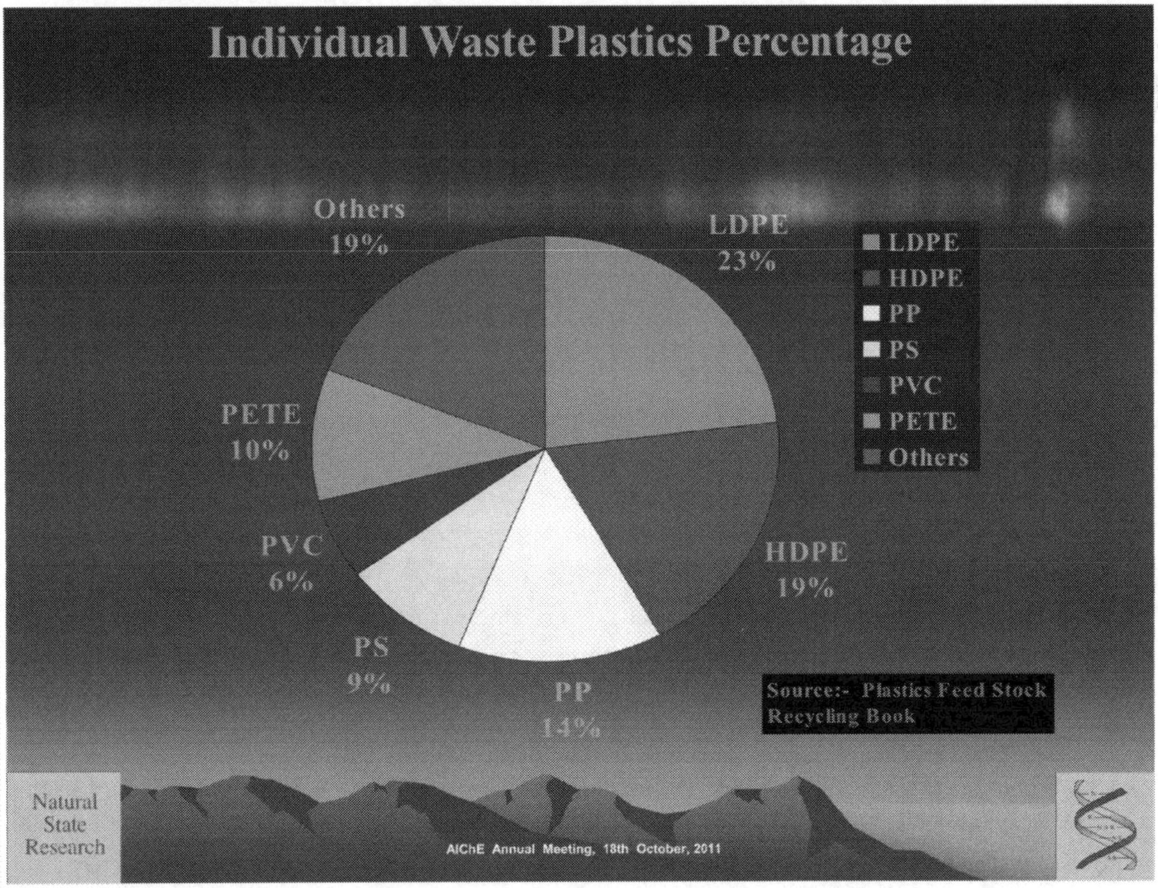

Individual Waste Plastics Percentage
Others
19%
LDPE
23%
PETE
10%
PVC
6%
HDPE
19%
PS
9%
PP
14%
LDPE
HDPE
PP
PS
PVC
PETE
Others
Source:- Plastics Feed Stock
Recycling Book
Natural
State
Research
AIChE Annual Meeting, 18th October, 2011

Plastics Name & Code

Plastic Identification Code	Type of Plastic Polymer	Common Packaging Applications
01 PET	Polyethylene Terephthalate (PET, PETE)	Soft drink, water and salad dressing bottles; peanut butter and jam jars
02 PE-HD	High Density Polyethylene (HDPE)	Milk, juice and water bottles; trash and retail bags.
03 PVC	Polyvinyl Chloride (PVC)	Juice bottles; cling films; PVC piping
04 PE-LD	Low Density Polyethylene (LDPE)	Frozen food bags; squeezable bottles, e.g. honey, mustard; cling films; flexible container lids.
05 PP	Polypropylene (PP)	Reusable microwaveable ware; kitchenware; yogurt containers; margarine tubs; microwaveable disposable take-away containers; disposable cups and plates.
06 PS	Polystyrene (PS)	Egg cartons; packing peanuts; disposable cups, plates, trays and cutlery; disposable take-away containers;
07 O	Other (often Polycarbonate or ABS)	Beverage bottles; baby milk bottles; electronic casing.

Natural State Research

AIChE Annual Meeting, 18th October, 2011

Plastics in Use

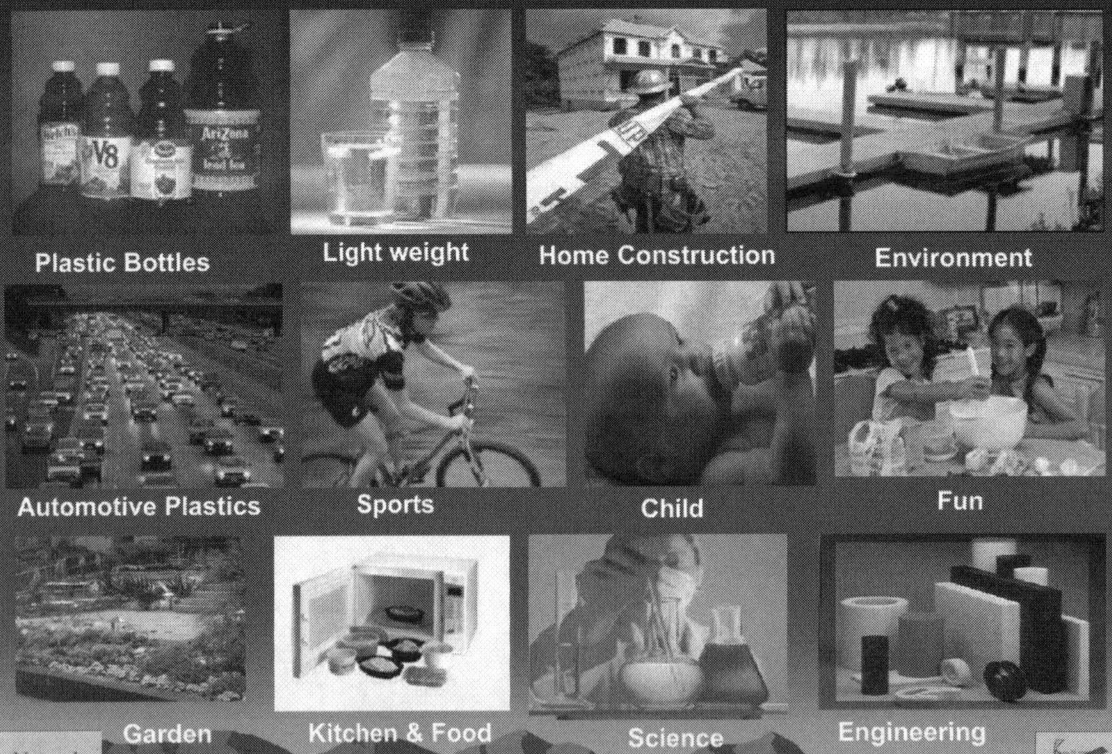

Natural State Research

AIChE Annual Meeting, 18th October, 2011

Waste Plastics
Baled Polythene Film
Mixed Plastic Films
Plastic Sheet
Mixed Plastic Bottles
Garbage Bags
Plastic Rope /String
Coat Hangers
Film on Rolls
CD's
Shredded Plastics
Granulated Plastics
EPS Briquettes
Car Bumpers
Natural State Research
AIChE Annual Meeting, 18th October, 2011

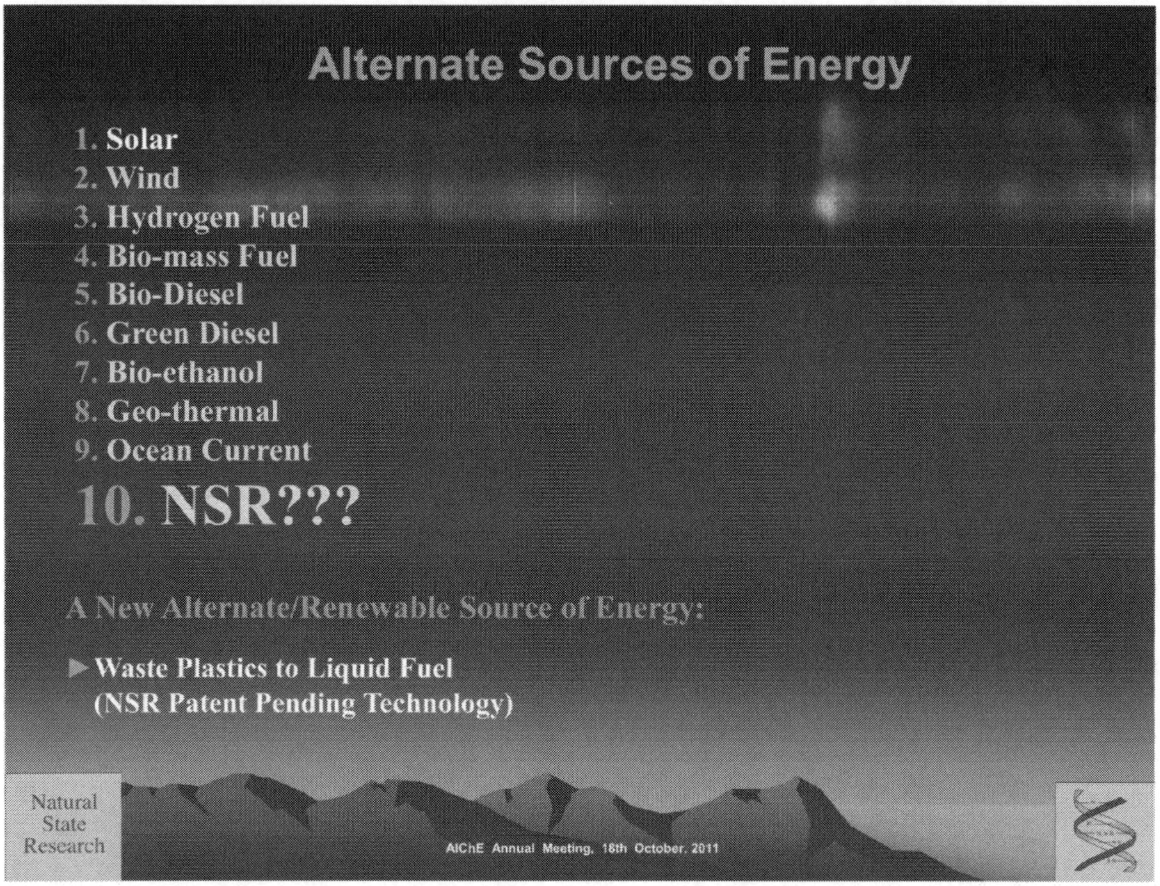
Alternate Sources of Energy
1. Solar
2. Wind
3. Hydrogen Fuel
4. Bio-mass Fuel
5. Bio-Diesel
6. Green Diesel
7. Bio-ethanol
8. Geo-thermal
9. Ocean Current
10. NSR???
A New Alternate/Renewable Source of Energy:
► Waste Plastics to Liquid Fuel
(NSR Patent Pending Technology)
Natural State Research
AIChE Annual Meeting, 18th October, 2011

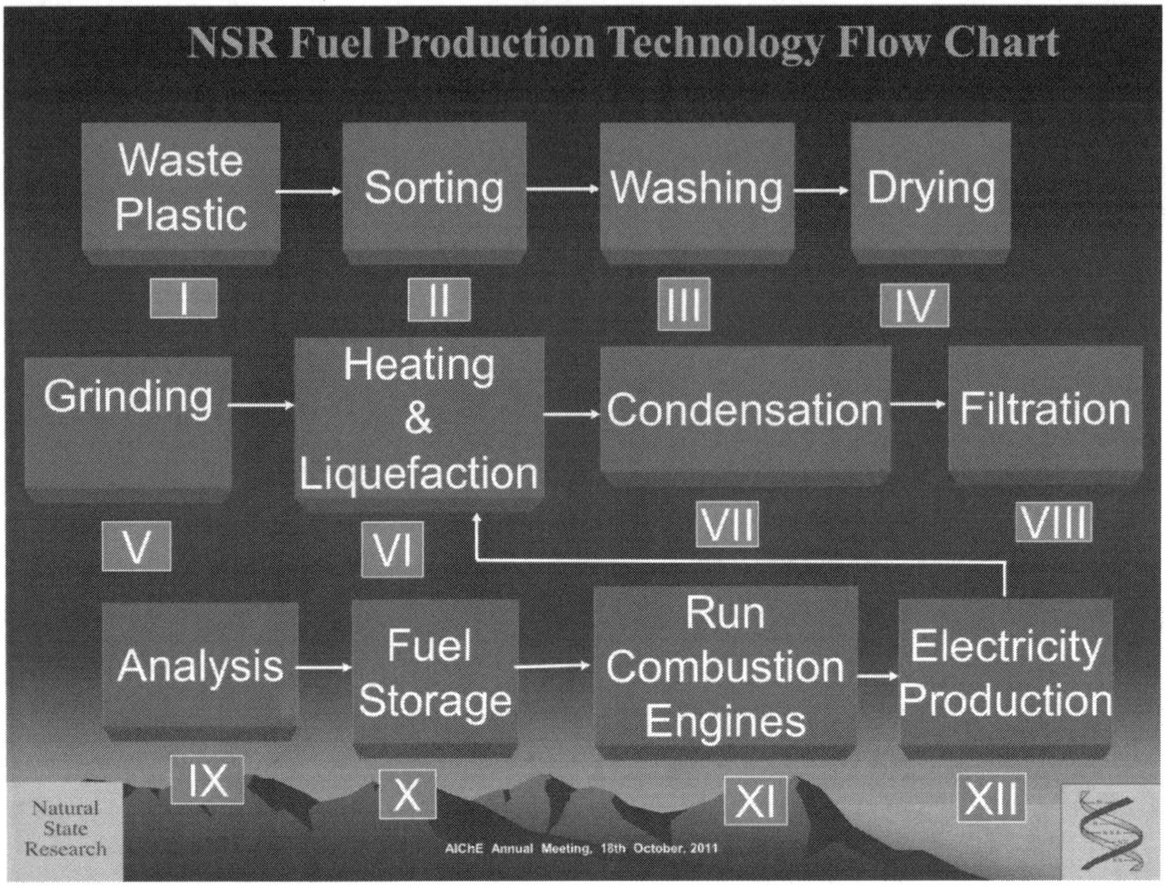
NSR Fuel Production Technology Flow Chart
Waste Plastic
I
Sorting
II
Washing
III
Drying
IV
Grinding
V
Heating & Liquefaction
VI
Condensation
VII
Filtration
VIII
Analysis
IX
Fuel Storage
X
Run Combustion Engines
XI
Electricity Production
XII
Natural State Research
AIChE Annual Meeting, 18th October, 2011

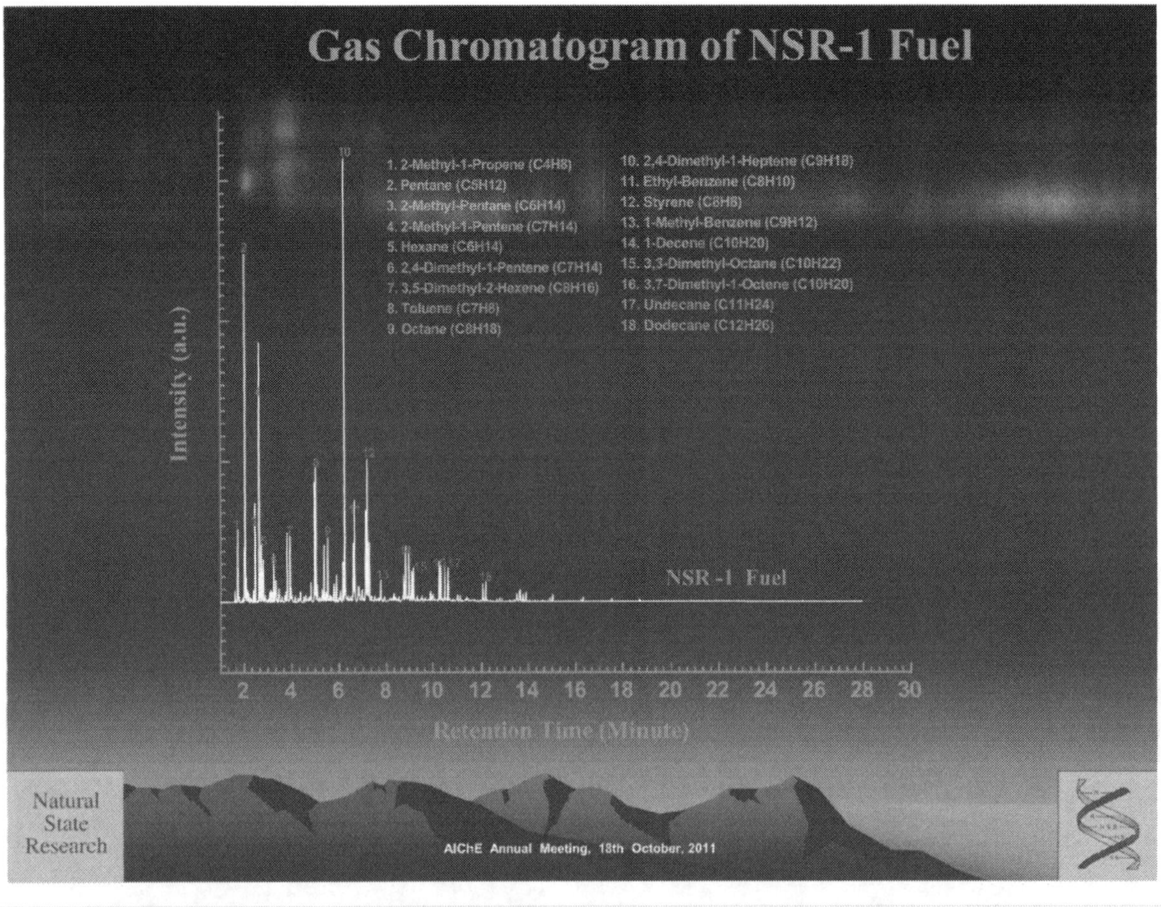
Gas Chromatogram of NSR-1 Fuel
1. 2-Methyl-1-Propene (C4H8)
2. Pentane (C5H12)
3. 2-Methyl-Pentane (C6H14)
4. 2-Methyl-1-Pentene (C7H14)
5. Hexane (C6H14)
6. 2,4-Dimethyl-1-Pentene (C7H14)
7. 3,5-Dimethyl-2-Hexene (C8H16)
8. Toluene (C7H8)
9. Octane (C8H18)
10. 2,4-Dimethyl-1-Heptene (C9H18)
11. Ethyl-Benzene (C8H10)
12. Styrene (C8H8)
13. 1-Methyl-Benzene (C9H12)
14. 1-Decene (C10H20)
15. 3,3-Dimethyl-Octane (C10H22)
16. 3,7-Dimethyl-1-Octene (C10H20)
17. Undecane (C11H24)
18. Dodecane (C12H26)
Intensity (a.u.)
NSR-1 Fuel
2 4 6 8 10 12 14 16 18 20 22 24 26 28 30
Retention Time (Minute)
Natural State Research
AIChE Annual Meeting, 18th October, 2011

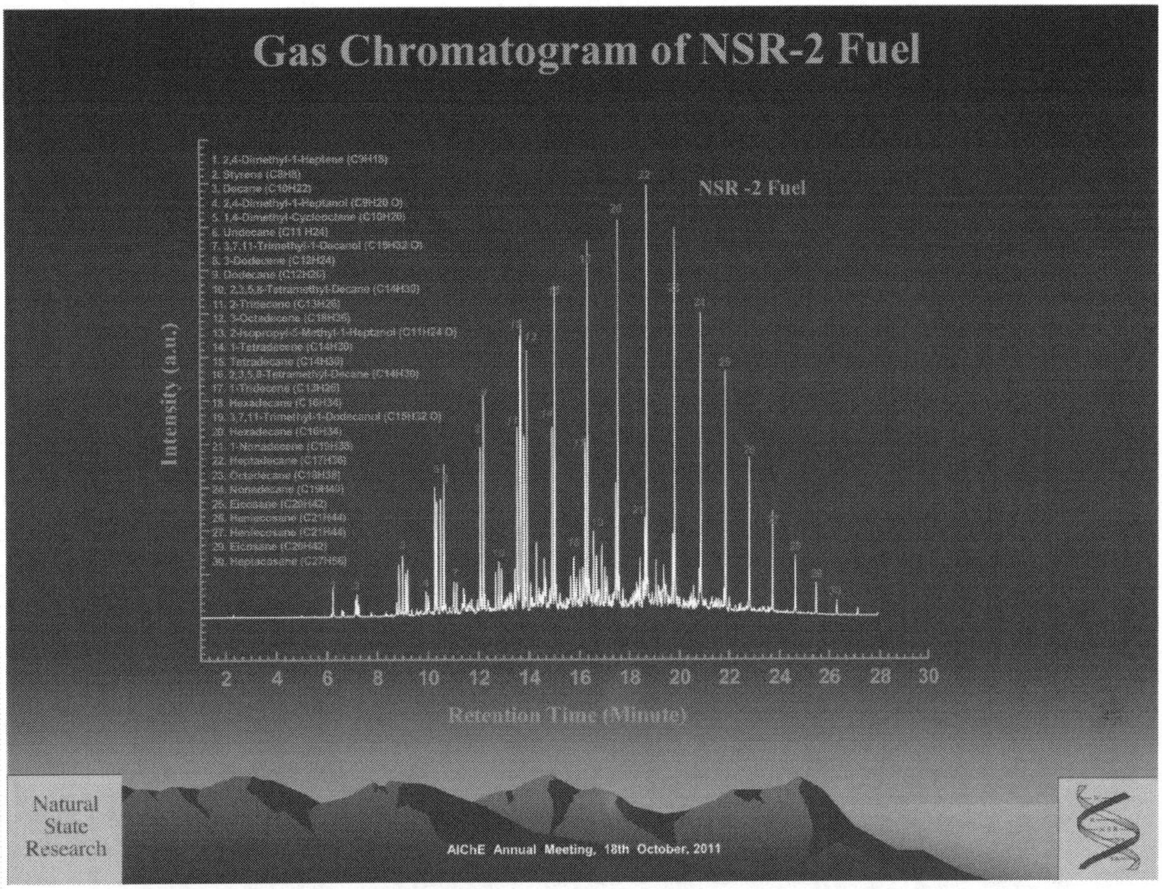
Gas Chromatogram of NSR-2 Fuel
NSR -2 Fuel
1. 2,4-Dimethyl-1-Heptene (C9H18)
2. Styrene (C8H8)
3. Decane (C10H22)
4. 2,4-Dimethyl-1-Heptanol (C9H20 O)
5. 1,4-Dimethyl-Cyclooctane (C10H20)
6. Undecane (C11 H24)
7. 3,7,11-Trimethyl-1-Decanol (C15H32 O)
8. 3-Dodecene (C12H24)
9. Dodecane (C12H26)
10. 2,3,5,8-Tetramethyl-Decane (C14H30)
11. 2-Tridecene (C13H26)
12. 3-Octadecene (C18H36)
13. 2-Isopropyl-5-Methyl-1-Heptanol (C11H24 O)
14. 1-Tetradecene (C14H28)
15. Tetradecane (C14H30)
16. 2,3,5,8-Tetramethyl-Decane (C14H30)
17. 1-Tridecene (C13H26)
18. Hexadecane (C16H34)
19. 3,7,11-Trimethyl-1-Dodecanol (C15H32 O)
20. Hexadecane (C16H34)
21. 1-Nonadecene (C19H38)
22. Heptadecane (C17H36)
23. Octadecane (C18H38)
24. Nonadecane (C19H40)
25. Eicosane (C20H42)
26. Heneicosane (C21H44)
27. Heneicosane (C21H44)
29. Eicosane (C20H42)
30. Heptacosane (C27H56)
Intensity (a.u.)
2 4 6 8 10 12 14 16 18 20 22 24 26 28 30
Retention Time (Minute)
Natural State Research
AIChE Annual Meeting, 18th October, 2011

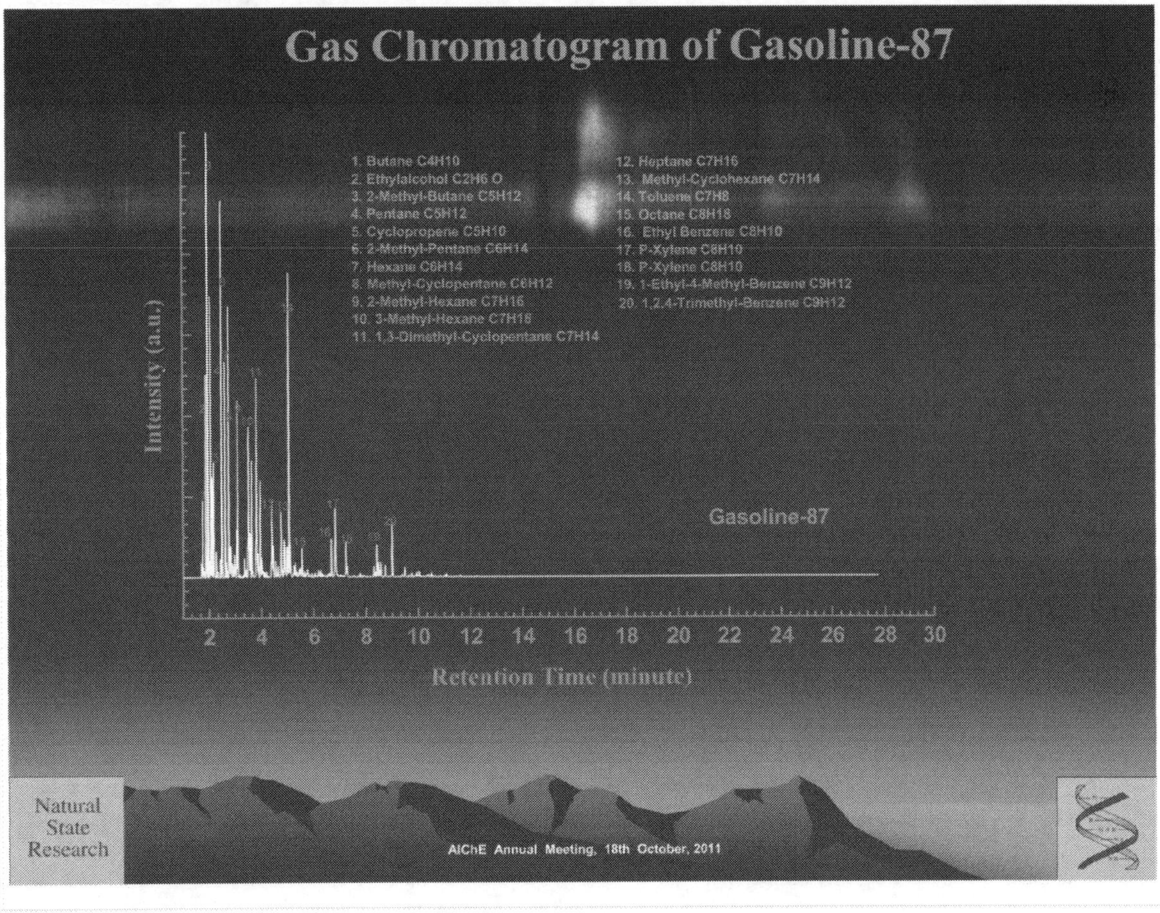
Gas Chromatogram of Gasoline-87
1. Butane C4H10
2. Ethylalcohol C2H6 O
3. 2-Methyl-Butane C5H12
4. Pentane C5H12
5. Cyclopropene C5H10
6. 2-Methyl-Pentane C6H14
7. Hexane C6H14
8. Methyl-Cyclopentane C6H12
9. 2-Methyl-Hexane C7H16
10. 3-Methyl-Hexane C7H16
11. 1,3-Dimethyl-Cyclopentane C7H14
12. Heptane C7H16
13. Methyl-Cyclohexane C7H14
14. Toluene C7H8
15. Octane C8H18
16. Ethyl Benzene C8H10
17. P-Xylene C8H10
18. P-Xylene C8H10
19. 1-Ethyl-4-Methyl-Benzene C9H12
20. 1,2,4-Trimethyl-Benzene C9H12
Gasoline-87
Intensity (a.u.)
2 4 6 8 10 12 14 16 18 20 22 24 26 28 30
Retention Time (minute)
Natural State Research
AIChE Annual Meeting, 18th October, 2011

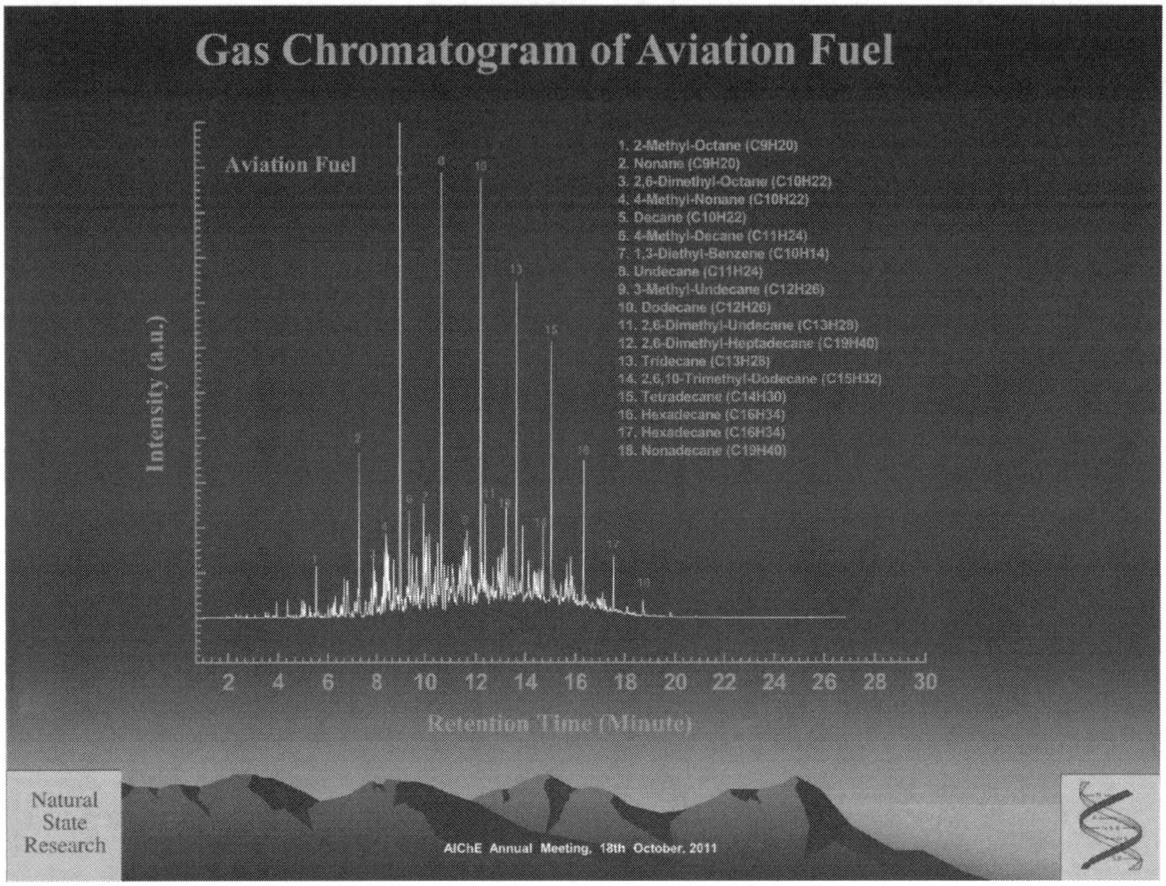

Gas Chromatogram of Aviation Fuel
Aviation Fuel
1. 2-Methyl-Octane (C9H20)
2. Nonane (C9H20)
3. 2,6-Dimethyl-Octane (C10H22)
4. 4-Methyl-Nonane (C10H22)
5. Decane (C10H22)
6. 4-Methyl-Decane (C11H24)
7. 1,3-Diethyl-Benzene (C10H14)
8. Undecane (C11H24)
9. 3-Methyl-Undecane (C12H26)
10. Dodecane (C12H26)
11. 2,6-Dimethyl-Undecane (C13H28)
12. 2,6-Dimethyl-Heptadecane (C19H40)
13. Tridecane (C13H28)
14. 2,6,10-Trimethyl-Dodecane (C15H32)
15. Tetradecane (C14H30)
16. Hexadecane (C16H34)
17. Hexadecane (C16H34)
18. Nonadecane (C19H40)
Intensity (a.u.)
2 4 6 8 10 12 14 16 18 20 22 24 26 28 30
Retention Time (Minute)
Natural State Research
AIChE Annual Meeting, 18th October, 2011

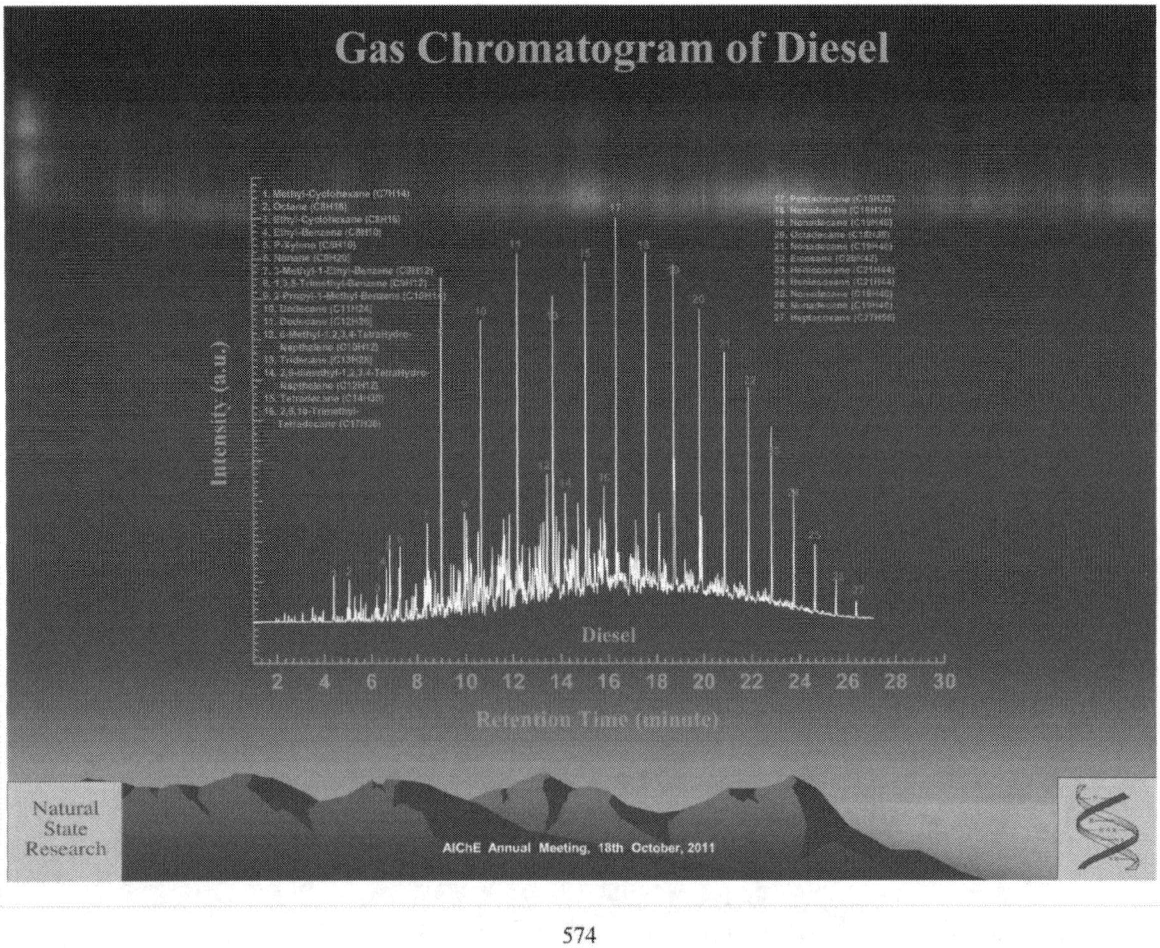

Gas Chromatogram of Diesel
Intensity (a.u.)
Diesel
2 4 6 8 10 12 14 16 18 20 22 24 26 28 30
Retention Time (minute)
Natural State Research
AIChE Annual Meeting, 18th October, 2011

Carbon Content of Various Fuels

Fuel Name	Carbon Content
NSR -1 Fuel	C_4-C_{12}
NSR -2 Fuel	C_9 - C_{27}
Gasoline (87)	C_4 - C_{11}
Aviation Fuel (Jet Fuel)	C_7- C_{19}
Diesel	C_6 - C_{27}

Natural State Research

AIChE Annual Meeting, 18th October, 2011

Differential Scanning Calorimetry (DSC) curves of several fuels illustrating different onsets of the boiling points

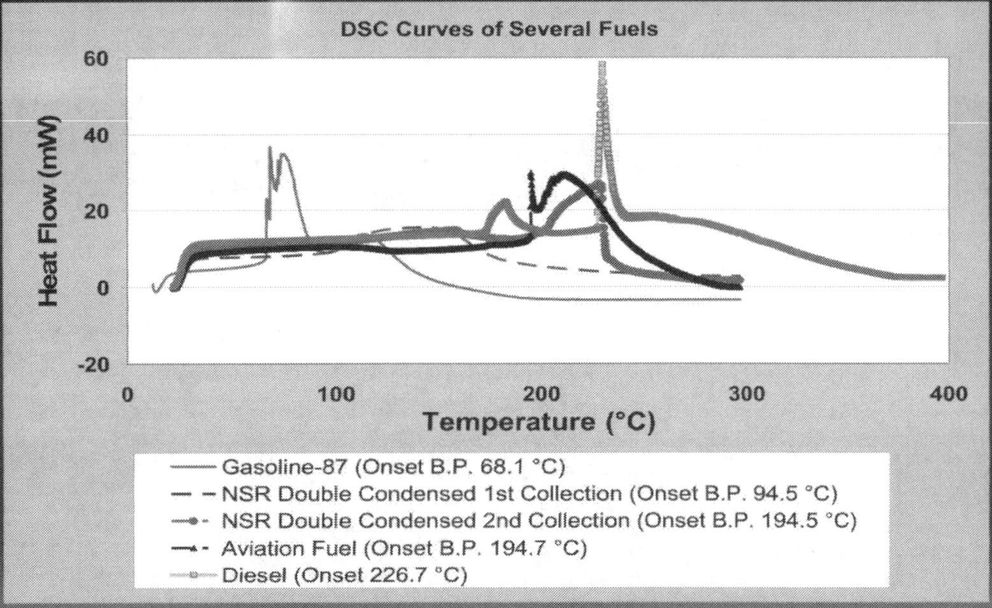

Natural State Research

AIChE Annual Meeting, 18th October, 2011

Energy Content of Various Fuels

Type of Fuel	Total BTU /Gallon	Mega Joules /Gallon
NSR -1 Fuel	113,934	120
NSR-2 Fuel	133,366	141
Gasoline - 87	125,000	132
Diesel	138,700	146
Aviation Fuel	120,200	127
Jet Fuel	135,000	142
Compressed Natural Gas	138,700	146
Distillate Fuel	138,700	146
Residual Fuel	149,700	158

1 kWh = 3,412 Btu (British Thermal Unit)
1 kWh = 3.6 MJ (Mega Joules)

Natural State Research

AIChE Annual Meeting, 18th October, 2011

Properties of Various Fuels

Sample	Boiling Point (^{0}C) (Initial)	Freezing Point (^{0}C)	Cloud Point (^{0}C)	Cetane Index	Octane Number	Density (g/ml)	Color / Appearance
NSR- 1	41.5	- 50.0	- 24.0	42.3	76	0.74	Pass – clear and bright
NSR -2	171.1	N/A	- 12.0	58.3	N/A	0.80	Pass – clear and bright
Gasoline-87	68.7-204.0	- 72.7	N/A	N/A	87,89,93	0.71-0.75	Pass – clear and bright
Jet Fuel	N/A	- 47.0	N/A	N/A	N/A	0.79-0.84	Clear straw and bright
Diesel	215.5	-10.0 (Gel Point)	- 5.0	42.5	N/A	0.85	Clear greenish and bright

All Preliminary test of NSR Fuels have done by www.intertek.com

Natural State Research

AIChE Annual Meeting, 18th October, 2011

Preliminary Emission Test of NSR Fuels and Comparison with Outside Fuels

DIPE = Di-isopropyl Ether
TAME = Tetra Amyl-Methyl Ether
MTBE = Methyl Tetra-Butyl Ether
DVP = Dry Vapor Pressure (Psi)
E150 = % volume evaporated at 1500C)

Reference Websites:

http://www.med.govt.nz/templates/page 10338.aspx

Sample	DVP (Psi)	E150	Sulfur (mg/kg)	Benzene (Vol. %)	Methanol (Vol. %)	Ethanol (Vol. %)	Oxygen (Wt %)	Olefins	DIPE	TAME	MTBE
NSR 1	4.54	5.6	6.2	0.15	0.00	0.00	0.00	>55	0.00	0.00	0.00
NSR 2	N/A	N/A	4.7	N/A	N/A	N/A	N/A	N/A	N/A	N/A	N/A
Gasoline	6.52	75	150	1	N/A	10	1	18	N/A	N/A	N/A
Diesel	N/A	N/A	50	N/A	N/A	N/A	N/A	N/A	N/A	N/A	N/A

All Preliminary test of NSR Fuels were done by www.intertek.com

Natural State Research

AIChE Annual Meeting, 18th October, 2011

NSR Fuel

NSR Fuel USA Patent Pending

International Patent / PCT (Patent Pending)

NSR Fuel is a combination of Gasoline, Diesel, Aviation and Jet Fuel

Preliminary research results indicate that NSR Fuel is highly flammable and substantially similar to Aviation and Jet Fuel

Natural State Research

AIChE Annual Meeting, 18th October, 2011

Driving an Automobile with NSR Fuel

NSR Car Test Experiment

Specification for test vehicle:

1984 Oldsmobile (V-8, 5.0 liters powered engine) and 4 doors Sedan

Experimental Conditions

Driver: Eugene Sheveron (180 lbs)
Along with 3 other researchers (Weight = 550 lbs)
5 Gallon Fuel Container (35 lbs)
Spare Tire (50 lbs)
Total Load : 815 lbs

Mileage obtained:

NSR Fuel (76) = 18 Miles/ Gallon

Gasoline (87) = 15 Miles/ Gallon

Cost Analysis

- **Thermal Process at ambient pressure and 370-420 (Degrees Celsius)**
- **NSR Fuel Energy input = 15-20 kWh**
  US$ 1.65 - 2.20
  (excluding overhead)
- **Energy output = 55-60 kWh**
  US$ 6.05 – 6.60
- **2.90-3.00 Kg of Plastics = 3.8 litter = 1 gallon in Lab Scale (Commercial scale is would be less than US$ 1.00 per gallon)**

Present Price:

- **Gasoline = US$ 2.50-3.00 per gallon, Energy output: 35-40 kWh**
- **Diesel: US$ 3.00-3.50 per gallon**
- **Aviation Fuel: US$ 8.00-10.00 per gallon**

Plans and Projection

2011-2012 → Pilot plant

2012 → Patent pending; Apply for franchise registration

2013 → 1st commercial plant

2014 → Consumers use NSR Fuel

2013-2015 → Franchise and license NSR technology throughout USA

2015-2020 → Franchise and license NSR technology worldwide

Natural State Research

AIChE Annual Meeting, 18th October, 2011

Advantage of NSR Technology and Products

1. Efficient
2. Economic (cost effective)
3. Environment friendly
4. 95-98% yield with minimum solid waste/ residue
5. No need to sort out waste plastics (except PETE and PVC)
6. Easy to setup (modular and mobile unit)
7. NSR fuels work with existing internal combustion engine
8. Creates jobs
9. Free/ little cost for raw materials
10. Easy to work with existing recycling and city waste facilities
11. Low investment with high Return

Natural State Research

AIChE Annual Meeting, 18th October, 2011

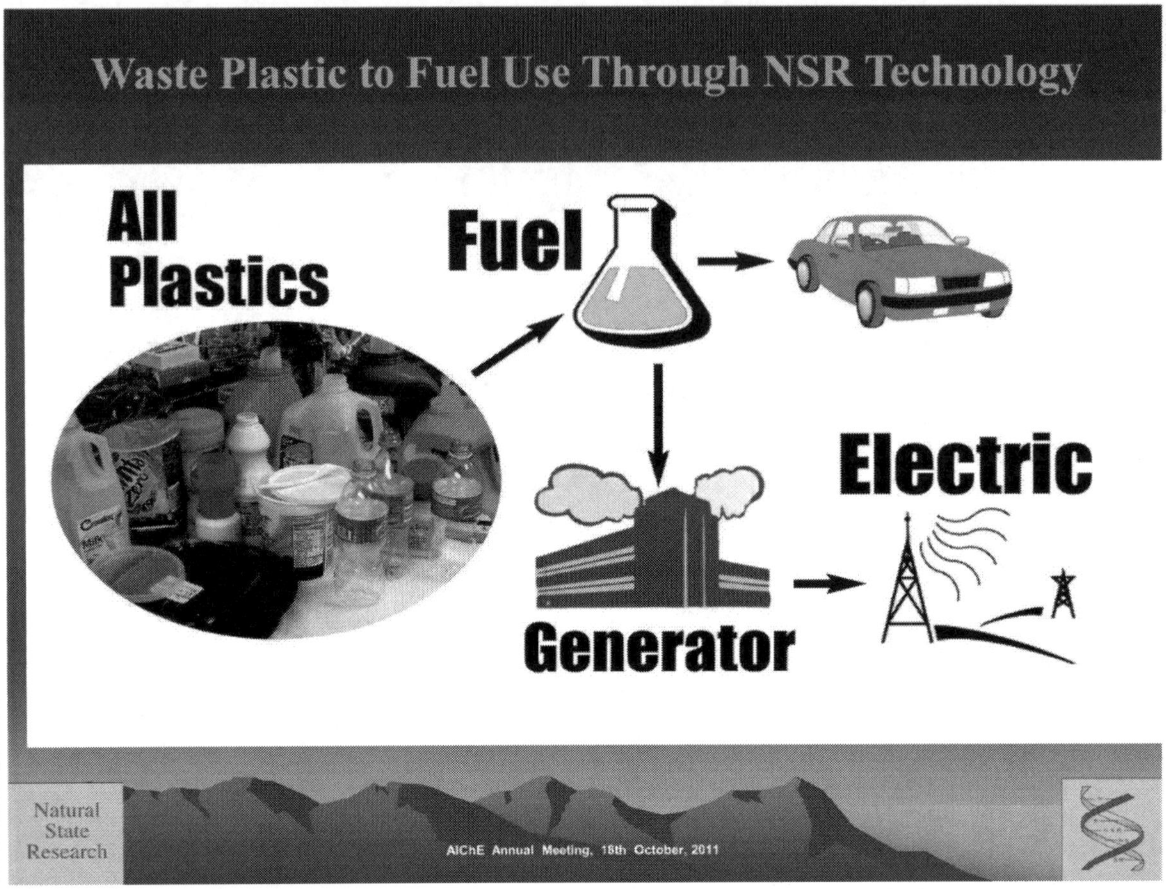

Waste Plastic to Fuel Use Through NSR Technology
All Plastics
Fuel
Electric
Generator
Natural State Research
AIChE Annual Meeting, 18th October, 2011

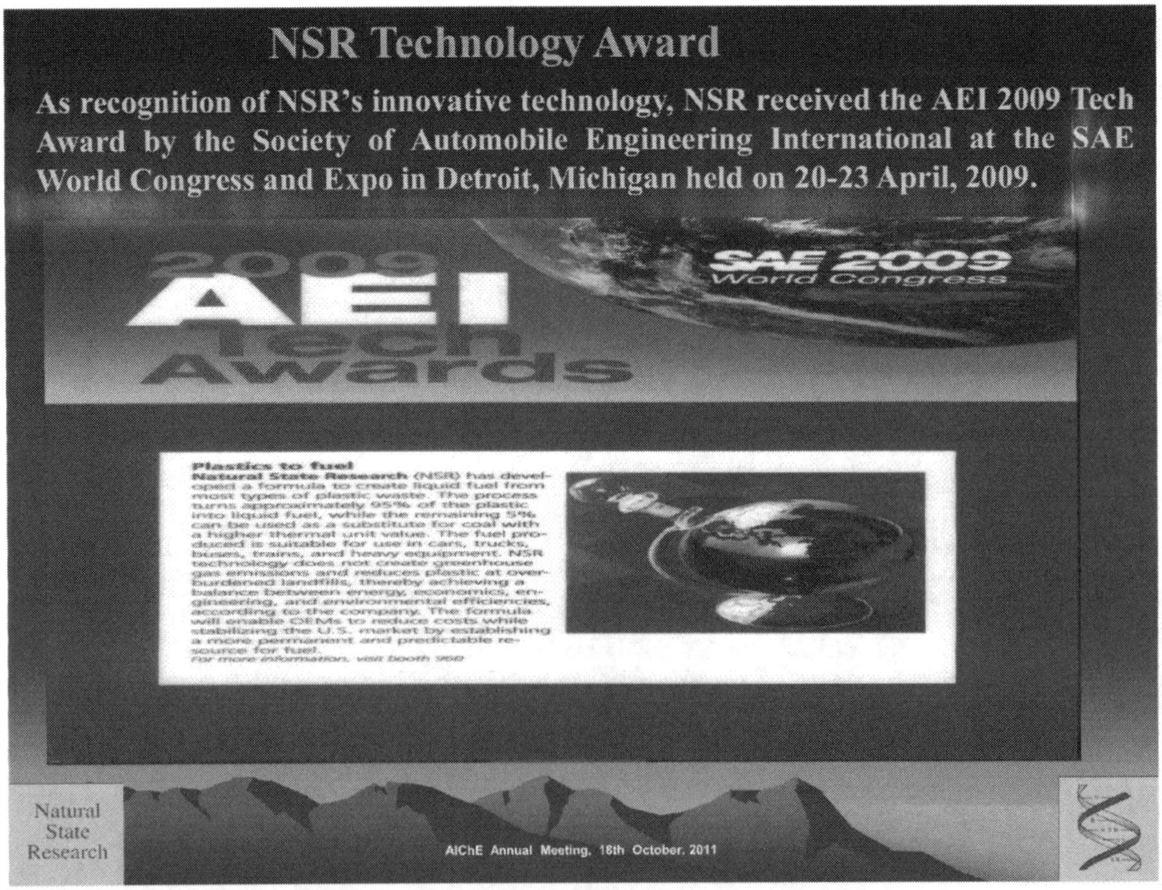

NSR Technology Award
As recognition of NSR's innovative technology, NSR received the AEI 2009 Tech Award by the Society of Automobile Engineering International at the SAE World Congress and Expo in Detroit, Michigan held on 20-23 April, 2009.
2009
AEI
Tech
Awards
SAE 2009
World Congress
Plastics to fuel
Natural State Research (NSR) has developed a formula to create liquid fuel from most types of plastic waste. The process turns approximately 95% of the plastic into liquid fuel, while the remaining 5% can be used as a substitute for coal with a higher thermal unit value. The fuel produced is suitable for use in cars, trucks, buses, trains, and heavy equipment. NSR technology does not create greenhouse gas emissions and reduces plastic at overburdened landfills, thereby achieving a balance between energy, economics, engineering, and environmental efficiencies, according to the company. The formula will enable OEMs to reduce costs while stabilizing the U.S. market by establishing a more permanent and predictable resource for fuel.
For more information, visit booth 960
Natural State Research
AIChE Annual Meeting, 18th October, 2011

Innovation Pipeline Award

For recognition for its outstanding achievement, NSR was awarded the Innovation Pipeline Award 2009 "companies to watch" at the Connecticut Technology Council at OMNI Hotel New Haven, CT, held on September, 17 2009 and 2010.

Natural State Research

AIChE Annual Meeting, 18th October, 2011

Conclusion

- **Convert waste plastics to hydrocarbon fuel**
- **Produce electricity using generator / turbine**
- **Reduce landfill by converting waste plastic to useful hydrocarbon fuel**
- **Save marine aquatic life by reducing water pollution**
- **Create local and sustainable jobs**

Natural State Research

AIChE Annual Meeting, 18th October, 2011

NSR Research Team

1. Karin Kaufman, PhD
   CEO and President
2. Moinuddin Sarker, PhD, MCIC.
   Vice President, Research and Development (Inventor)
3. Ellery Cressia
   Operation Manager
4. ASM Din Mohammad, M.Sc.
   Lab. Manager
5. Mohammad Mamunor Rashid, M.Sc.
   Plant Manager and Lab Coordinator
6. Nicholas Schiralli
   Come Back Manager
7. Tanya Stewart
   Office Coordinator
8. June Metz
   Business Manager

Natural State Research

AIChE Annual Meeting, 18th October, 2011

9. Mairim Vant
   Network Engineer
10. Aminul Kabir, M.Sc.
    Lab Chemist
11. Ashiquz Zaman
    Research Assistant
12. Mohammad Molla
    Intern
13. Kamrul Islam, M.Sc. MBA
    Lab Technician
14. Sadikur Rahman, M.Sc
    Lab Chemist (Junior)

Consultants:

1. Prof. Richard Parnas, University of Connecticut, USA
2. Sabir Majumder, PhD
3. Dennis Keehn
   Marketing Consultant
4. Fuad Rahman, PhD

Natural State Research

AIChE Annual Meeting, 18th October, 2011

For more information regarding NSR Products and Technology;

Please contact Dr. Moinuddin Sarker at NSR:

+1 (203) 406-0675

OR

E-mails: msarker@naturalstateresearch.com

OR

www.naturalstateresearch.com

Natural State Research

AIChE Annual Meeting, 18th October, 2011

Questions??

Natural State Research

AIChE Annual Meeting, 18th October, 2011

Thank You Very Much
Natural
State
Research
AIChE Annual Meeting, 18th October, 2011

Volatile Sulfur Compounds Removal in Biogas Using Thermo-Oxidative Pretreatment

Bipro Ranjan Dhar[1], Elsayed Elbeshbishy[2], Hisham Hafez[2], George Nakhla[1,2], Madhumita B. Ray[1,*]

[1]Department of Chemical and Biochemical Engineering, [2]Department of Civil and Environmental Engineering, University of Western Ontario, London, Ontario, Canada, N6A 5B9

Abstract

Thermo-oxidative pretreatment of municipal waste activated sludge was conducted using thermal pretreatment at 60°C in presence of oxidants (0.6 mg H_2O_2+1.5 mg $FeCl_2$/mg S^{2-}) prior to a continuous anaerobic digester operated at 10 days solid residence time (SRT). For the pretreated feed digester, the hydrogen sulphide (H_2S), dimethyl sulphide (DMS) and methanethiol (MT) concentrations in biogas decreased by 75%, 40%, and 10% compared to the control digester, respectively. Compared to the control digester, overall total suspended solids (TSS) and volatile suspended solids (VSS) removal efficiency were 10% and 11% higher, respectively, while methane production rate (L CH_4/day) increased by 20%.

Keywords: Anaerobic digestion, Biogas, Volatile sulfur compounds

* To whom all correspondence should be addressed: Department of Chemical and Biochemical Engineering, University of Western Ontario, London, Ontario, Canada N6A 5B9
Tel.: +1 519 661 2111x81273; Fax: +1 519-661-3498; E-mail: mray@eng.uwo.ca

1. Introduction

Biological treatment of wastewater produces a significant amount of Sludge. Although anaerobic (AD) digestion is an established and widely accepted sludge stabilization process, conventional anaerobic digestion has some limitations and operating challenges. Anaerobic digestibility of municipal waste activated sludge (WAS) is very poor due to the rate-limiting hydrolysis step. In addition, presence of volatile sulfur compounds (VSCs) such as hydrogen sulfide (H_2S), mercaptans in biogas diminishes the utility value of the biogas as they cause corrosion in combustion engines. Pretreatment of sludge prior to AD is a common approach to increase the solid reduction and biogas production in anaerobic digestion.

Various pretreatment techniques have been reported in literature including thermal, chemical, and mechanical pretreatments. Sometimes, different pretreatment technologies are combined together to improve the pretreatment performance. Although the effects of thermal-oxidative pretreatment of sludge have been reported in literature [1, 2], very limited information is available on the reduction of volatile sulfur compounds (VSCs) in biogas. This study investigated the impact of thermo-oxidative pretreatment on VSCs removal during anaerobic digestion. Thermo-oxidative pretreatment of municipal waste activated sludge was conducted using thermal pretreatment at 60°C in presence of 0.6 mg H_2O_2+1.5 mg $FeCl_2$/mg S^{2-} as oxidants with an objective to disintegrate sludge for enhancing anaerobic digestion as well as to remove volatile sulfur compounds generation potential in biogas.

2. Methodology

Thermo-oxidative pretreatment of sludge was conducted using conventional heating at 60°C in presence of oxidants (0.6 mg H_2O_2+1.5 mg $FeCl_2$/mg S^{2-}). The dosages used were based on the theoretical requirement of the chemical dosages for 100% dissolved sulfide (S^{2-}) removal in raw untreated sludge [3]. Thermal pretreatment of 750 ml sludge was done in a flask closed with a rubber cap fitted with a temperature probe and placed on a hot stirred plate (Corning Stirrer/Hot plate, Model PC-420, Corning Incorporated, USA), and heated at the set temperature for 30 minutes. Continuous anaerobic digestion of sludge (750 ml/d) was carried out using two identical completely mixed anaerobic bioreactors (10 L), with a working volume of 7.5 L and a solid retention time (SRT) of 10 days, and maintained at constant mesophilic temperature of 37±1 °C. During start-up the digesters were initially seeded with 7.5 L of anaerobically digested sludge collected from the Guelph wastewater treatment plant (Guelph, Ontario, Canada). Then, the reactors were separately fed with raw and pretreated WAS. Detailed analysis of both liquid and gas samples commenced after steady-state conditions were reached, approximately after three turnovers of the mean SRT. The influent and effluent sludge samples were analyzed for total suspended solids (TSS), volatile suspended solids (VSS), total chemical oxygen demand (TCOD), soluble oxygen demand (SCOD), ammonia nitrogen (NH_4-N), volatile fatty acids (VFAs), different protein and carbohydrate fractions, pH, and alkalinity. The biogas samples were analyzed for methane (CH_4), hydrogen sulfide (H_2S), dimethyl sulfide (DMS), and methanethiol (MT) concentrations.

3. Results & Discussions

3.1 Impact of Pretreatment

The impact of thermo-oxidative pretreatment on various WAS characteristics are shown in Table 1. After pretreatment TCOD remained almost constant, while SCOD significantly increased from 1448±109 mg/L to 3615±164 mg/L due to the release of various extra-cellular polymeric substances (EPSs) such as carbohydrates, proteins etc. Compared to raw WAS, the

TSS and VSS in pretreated sludge decreased by an average of 10% and 15%, respectively. The total volatile fatty acid (TVFA) in pretreated sludge was 17% higher than the raw WAS.

Table 1: Characteristics of raw and pretreated WAS.

Parameter	Raw WAS	Pretreated WAS
TCOD (mg/L)	21328±897	20881±722
SCOD (mg/L)	1448±109	3615±164
TSS (mg/L)	18253±714	16418±752
VSS (mg/L)	13237±609	11132±381
TVFA (mg COD/L)	360±20	420±25
Ammonia (mg/L)	143±10	134±17
TKN (mg/L)	1130±43	1209±125
STKN (mg/L)	700±15	756±19
Particulate protein (mg/L)	2644±275	1965±216
Bound protein (mg/L)	828±60	516±128
Soluble Protein (mg/L)	453±103	1229±166
Particulate carbohydrate (mg/L)	1733±249	1426±110
Soluble Carbohydrate (mg/L)	108±24	364±58
pH	7.1±0.1	6.9±0.1
Alkalinity (mg as $CaCO_3$/L)	3000±200	4500±250

Protein content in sludge is usually divided into three different fractions: cell, bound and soluble [4]. The cell protein represents the protein fraction inside the microbial cell, and the bound protein is the loosely attached protein to the microbial cell wall, while the soluble protein represents the protein fraction available in the aqueous phase. Particulate protein represents the combination of cell protein and bound protein. Bound protein has been identified as a major precursor for volatile organic sulphur compounds generation during digestion [5]. As shown in Table 1, due to thermo-oxidative pretreatment the particulate protein and bound protein decreased by an average of 26% and 38%, respectively, and the average soluble protein concentration in pretreated sludge was 2.7 times higher than the raw sludge. After pretreatment, particulate carbohydrate in pretreated sludge decreased by an average of 18%; while soluble carbohydrate increased by almost 3 times compared to the raw untreated sludge. These results suggest that the thermo-oxidative pretreatment has significantly released the EPSs through the solubilization of sludge. The reduction in total protein and particulate carbohydrate agrees well with the increase in soluble protein and soluble carbohydrate, respectively. No ammonia (NH_4-N) solubilization was observed after pretreatment.

3.2 Anaerobic Digestion

Table 2 summarizes the steady-state digested sludge quality and methane production for the control and pretreated feed digesters. Compared to the control digester, a significant increase of 20% in methane production rate to 3.1 L CH_4/day was observed for the digester fed with pretreated sludge (Table 2). The average TCOD concentrations in influent was almost the same for both digesters, while the SCOD was ~2.5 times higher in pretreated sludge compared to the raw untreated WAS (Table 1). Compared to the control digester, a slight enhancement in TCOD removal was observed for the digester processing the pretreated sludge, while a significant increase in SCOD removal was observed for the pretreated feed digester. The SCOD removal efficiency for control and pretreated feed digesters were 56% and 74%, respectively.

Although prior to the anaerobic digestion the thermo-oxidative pretreatment has significantly reduced the TSS and VSS concentrations compared to the raw WAS, steady-state TSS and VSS removal efficiency during anaerobic digestion for both reactors were almost same. At 10-days SRT, the VSS removal efficiencies for the control and test digester were 40% and 42%, respectively. However, compared to the digestion of raw WAS, the overall (digestion +

pretreatment) TSS and VSS removal in the test digester (including pretreatment) increased by 10% and 11%, respectively. The lack of improvement in suspended solids reduction during digestion due to the pretreatment suggests that the thermo-oxidative pretreatment did not improve the biodegradability of particulate matters.

Table 2: Digested sludge characteristics and methane production.

Parameter	Control Digester	Pretreated Feed Digester
TCOD (mg/L)	13134±320	10800±200
SCOD (mg/L)	630±27	926±84
TSS (mg/L)	12800±350	11300±630
VSS (mg/L)	8000±620	6500±350
TVFA (mg COD/L)	175±23	170±11
Ammonia (mg/L)	635±33	670±47
Particulate protein (mg/L)	1965±216	651±115
Bound protein (mg/L)	516±128	279±103
Soluble Protein (mg/L)	181±58	396±76
Particulate carbohydrate (mg/L)	1053±129	832±80
Soluble Carbohydrate (mg/L)	55±9	80±26
pH	7.5±0.15	7.6±0.15
Methane Production Rate (L CH_4/day)	2.6±0.2	3.1±0.15
CH_4 in biogas (vol. %)	63	66

For the control digester, the particulate protein, bound protein, and particulate carbohydrate removal efficiencies due to digestion were 68%, 62%, and 39%, respectively; while overall removal efficiencies achieved with pretreatment and digestion were 75%, 66%, and 52%, respectively. The results suggest that there was no enhancement in final degradation of bound protein after digestion of both raw and pretreated sludge. For both digesters, the effluent ammonia concentrations increased with solids reduction. As shown in Table 2, the TVFA removal efficiencies for the control and the pretreated feed digester were 51% and 55%, respectively.

3.3 Volatile Sulphur Compounds Removal

The dissolved sulfate and sulfide were analyzed in the raw and pretreated sludge to assess the impact of pretreatment on dissolved sulphide (Table 3). Dissolved sulfide can partition from sludge into the biogas as H_2S. Therefore, a reduction in dissolved sulfide prior to the anaerobic digestion may reduce the H_2S generation potential of sludge during anaerobic digestion. H_2O_2 and $FeCl_2$ can react with dissolved sulfide via a number of different pathways. The following reactions are possible [3, 6]:

$$H_2S + FeCl_2 \rightarrow FeS\downarrow + 2HCl \quad (1)$$

$$H_2S+H_2O_2 \rightarrow S+2H_2O \quad (2)$$

$$H_2S+4H_2O_2 \rightarrow 4\ H_2SO_4+4\ H_2O \quad (3)$$

$$2FeS + 3H_2O_2 \rightarrow 2S+2\ Fe\ (OH)_3 \quad (4)$$

However, the reaction pathways at elevated temperature due to thermal pretreatment are still unknown. After thermo-oxidative pretreatment, dissolved sulfide concentration decreased from 80 mg/L to 65 mg/L, while dissolved sulfate concentration remained almost constant. Dissolved sulfide might be converted to ferrous sulfide (FeS) or colloidal sulfur according to reactions (1) and (4). Compared to the raw WAS, 19% removal in dissolved sulfide was observed due to thermo-oxidative preteratment, while the oxidant might react with other compounds in sludge during pretreatment resulting in lower than expected degree of removal of sulfide. The dissolved sulfate and sulfide removal efficiencies were same for both digesters. The very low sulfate removal (<10% for both digesters) during anaerobic digestion indicates the low numbers of sulfate reducing bacteria (SRB) in the anaerobic digesters.

In this study, concentrations of three major VSCs (H_2S, MT and DMS) were regularly monitored in the biogas for both the digesters. The average H_2S, MT and DMS concentrations for control digester were 15, 4 and 0.16 ppm, respectively. H_2S, DMS and MT concentrations in pretreated digester biogas decreased by an average of 75%, 40% and 10% compared to the control digester, respectively (Figure 1). The statistical paired student's t-test indicates H_2S and DMS removal efficiencies in biogas were statistically significant at the 95% confidence level, while MT removal efficiency was insignificant at 95% confidence level. The observed decrease in VSCs concentration in pretreated feed digester might be due to the reduction of two major precursors of volatile sulphur compounds (dissolved sulphide and bound protein) removal through pretreatment.

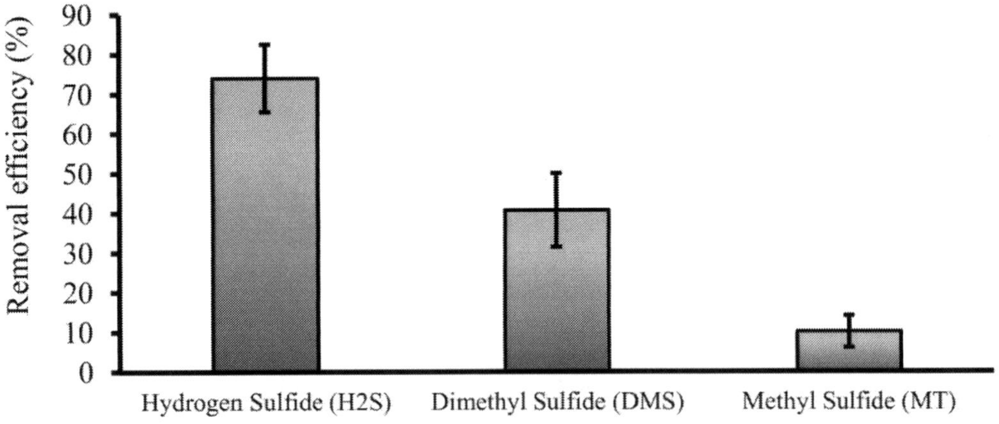

Figure 1: Average volatile sulfur compounds (VSCs) removal in biogas compared to control.

4. Conclusions

The thermo-oxidative pretreatment has shown significant improvement on sludge solubilization along with a significant release of EPSs (protein and carbohydrate). However, solids removal efficiencies during digestion for both digesters were almost the same, although overall (pretreatment and subsequent digestion) TSS and VSS destruction efficiencies with pretreatment were 10% and 11% higher than the control, respectively. Methane production rate for pretreated feed digester increased by 20% compared to the control digester. For the pretreated feed digester, the H_2S and DMS concentrations in biogas decreased by 75% and 40% compared to the control digester, respectively.

References

[1] J. A. Cacho, N. Madhavan, M.T. Suidan, P. Ginestet, P., J.M. Audic, Oxidative co-treatment using hydrogen peroxide with anaerobic digestion of excess municipal sludge, Water Environ. Res. 78 (2006) 691-700.

[2] C. Eskicioglu, A. Prorot, J. Marin, R.L. Droste, K.J. Kennedy, Synergetic pretreatment of sewage sludge by microwave irradiation in presence of H_2O_2 for enhanced anaerobic digestion, Water Res. 42 (2008) 4674-4682.

[3] J.R. Walton, M.S. Velasco, E. Ratledge, Peroxide Regenerated Iron-Sulfide Control (PRI-SC)™: Integrating Collection System Sulfide Control with Enhanced Primary Clarification by Adding Iron Salts and Hydrogen Peroxide, Water Environment Federation Technical Exhibition and Conference (WEFTEC), Los Angles, California, 2003.

[4] R. Dimock, E. Morgenroth, The influence of particle size on microbial hydrolysis of protein particles in activated sludge, Water Res. 40 (2006) 2064-2074.

[5] M. Higgins, D. Glindemann, J.T. Novak, S.N. Murthy, S. Gerwin, R. Forbes, Standardized biosolids incubation, headspace odor measurement and odor production consumption cycles, in: Proceedings Water Environment. Federation and AWWA Odors and Air Emissions Conference, Bellevue, Washington, 2004.

[6] ASCE, Odor control in wastewater treatment plants, American Society of Civil Engineers Publications, USA, 1995.

[7] B.R. Dhar, E. Youssef, G. Nakhla, M.B. Ray, Pretreatment of waste activated sludge for volatile sulfur compounds control in anaerobic digestion, Bioresour. Technol. 102 (2011) 3776-3782.

(413d) Multidisciplinary Approach In Developing Region Specific Models for Sustainable Biorefining

Sumesh Sukumara[1], Jeffrey Seay[1] and Fazleena Badurdeen[2], (1)Chemical and Materials Engineering, University of Kentucky, Lexington, KY, (2)Mechanical Engineering, University of Kentucky, Lexington, KY

Use of foreign fossil fuel for transportation and producing energy has led to energy dependence and growing environmental concerns. Among the various choices of renewable energy resources, biomass intrigues many researchers due to its widespread availability, cost effectiveness and its applicability as a sustainable energy source. Large scale biomass production and improved manufacturing techniques are some of the major areas where researchers are working on. However, because of high variations in the availability of biomass from one region to another there is a necessity to develop region specific economic models for biorefining. Developing economic models requires a multidisciplinary approach as the variation in biomass results in significant change in collection, storage techniques, supply and processing. This presentation aims to show how different disciplines of engineering and sciences can be combined to develop a region specific economic model for sustainable biorefinery based on Geographic Information Systems (GIS), supply chain optimization and process systems engineering. The Jackson Purchase in western Kentucky is the region where this model will be applied first. The Jackson Purchase region is not only rich in coal and biomass but also has a large number of poultry farms. Hence poultry litter can be a potential feedstock for integrated biorefinery in this region. Forest residue, corn stover and chicken litter are some of the feedstocks that are considered in this model for producing liquid transportation fuels. However, the distribution of lignocellulosic biomass in the country is very uneven and the first challenge is to locate the biomass. GIS is a technique by which biomass in a region can be located and by coupling it with supply chain optimization, location of potential biorefinery can be suggested. ILOG OPL is the tool used for optimizing the transportation cost of biomass from supply points to the potential biorefineries. Process systems engineering can be used to develop and optimize the processes. Aspen plus V-7.2 process engineering suite is the process simulation software used for this purpose. Gasification is a proven thermochemical process that is optimized in this model to produce synthesis gas and liquid hydrocarbons from forest residue and chicken litter. Also, a biochemical process is optimized in which corn stover is used to produce ethanol which can be marketed as transportation fuel. This model also looks into costs associated with storage and preprocessing of biomass and chicken litter. Detailed ash analysis will also be done in order to suggest most environmentally friendly conversion techniques among the various available choices. Finally, in future all the above optimized models will be combined to estimate the total production cost of various marketable products from different locally available feedstocks. The results, thus, obtained from these models will not only determine the location and number of biorefineries required to sustain a given region, but will also estimate the adequate monetary investment on sustainable biorefining required for that specific region.

(413e) Environmental Releases In the Fuel Ethanol Industry

Debalina Sengupta, Chemical Engineering, US Environmental Protection Agency, Cincinnati, OH and Raymond L. Smith, US EPA, Cincinnati, OH

Corn ethanol is the largest produced alternate biofuel in the United States. More than 13 billion gallons of ethanol were produced in 2010. The projected corn ethanol production is 15 billion gallons by 2015. With increased production of ethanol, the environmental releases from ethanol industry are going to increase.

The common studies with fuel ethanol plants typically report criteria air pollutants and greenhouse gas emissions. This prompted the need to evaluate the hazardous air pollutants released from ethanol plants and an overview of these studies will be presented in this paper.

This research shows a relative comparison of toxic releases in the United States, and particular to the corn-ethanol industry. Data is collected from the TRI reporting database and compiled for emissions for the toxic chemicals. Ethanol is produced industrially by two processes, wet milling, and dry grinding. In the wet milling process, there is a range of products pertaining to the food and feed industry other than ethanol. In the dry grind process, the products are ethanol and distillers grains with solubles. The processes produce ammonia, acetaldehyde, formaldehyde, acrolein, etc. as toxic releases. These releases are reported to the TRI database, and the releases for the ethanol industry are worth studying for the environmental effects and trends for future plants.

The present work attempts to capture these trends and present them for further evaluation.

See more of this Session: Sustainable Fuel From Renewable Resources II
See more of this Group/Topical: Environmental Division

(413f) Membrane Separation of Ethanol From Bio-Ethanol Containing Gasoline

Akihiro Yamasaki[1], Yohei Ueda[2], Naoki Takahashi[2], Yuka Sakai[2], Shigeo Satokawa[2] and Atsushi Iizuka[3], (1)Department of Materials and Life Science, Seikei University, Musashino, Tokyo, Japan, (2) Seikei University, Musashino, Tokyo, Japan, (3)Research Center for Sustainable Materials & Engineering, Institute of Multidisciplinary Research for Advanced Materials, Tohoku University, Sendai, Miyagi, Japan

Polymeric membranes for enriching ethanol from bio-ethanol containing gasoline (BEG) have been developed for the on-board separation process. Poly (vinyl alcohol) (PVA)-base membranes with controlled hydrophilicity were prepared and their separation performances were examined based on the pervaporation experimental studies. The hydrophilicity of the PVA-based membranes was controlled by changing the degree of saponification (the fraction of hydroxyl groups). The PVA-based membranes showed high selectivity for ethanol, of which the concentration in the permeate side was higher than 90% for a wide range of the feed composition studied. Increasing the hydrophilicity of the membranes increased the ethanol selectivity over a model hydrocarbon (n-heptane), and the flux decreased. The highest flux observed was as high as 100 $g/m^2/h$. The results suggest that the membranes can be applicable for the separation processes of ethanol from BEG.

See more of this Session: Sustainable Fuel From Renewable Resources II
See more of this Group/Topical: Environmental Division

(413g) Life Cycle Analysis of Geothermal Energy Extraction

Victor R. Vasquez and O. Hanbury, Chemical Engineering, University of Nevada, Reno, Reno, NV

As petroleum resources dwindle, many alternative transportation fuels have been proposed as replacements. One of the most versatile of these is electricity, which is source agnostic from the perspective of the vehicle, allowing any power generating process, either traditional or renewable to be used. It is then important to assess the externalities of these sources to meet modern emissions standards. In this work, we use a cradle-to-grave LCA approach to evaluate the life cycle issues of geothermal energy extraction (hydrothermal reservoirs) and compare it to the environmental impact of competing energy and fuel sources.
For comparisons purposes, we use the results of modeling frameworks such as GREET to estimate the baseline environmental impact of more traditional fuel pathways (i.e., petroleum or biomass-based). These are used to evaluate other fuel candidates or validate the potential renewable energy sources from an environmental impact perspective.
This work follows ISO 14040 standards for life cycle analysis applied to geothermal energy. This breaks down energy production into the discrete stages of drilling, production, and site reclamation and allows for further extensive modularization. The environmental impacts from geothermal energy depend heavily on the chemistry of the reservoir, the power facility equipment, and the manner in which the spent fluids are handled.
These factors are summed over the life of a facility using Monte Carlo simulation, to give an estimation of the total impact that a geothermal facility will have in its lifetime. The results show that while the impacts of even renewable energy sources are non-negligible, their use can reduce the influence transportation has on the environment.

See more of this Session: Sustainable Fuel From Renewable Resources II
See more of this Group/Topical: Environmental Division

(426a) Degradation by PULSE Corona Discharges: STUDY ON the POSSIBLE Chemical Degradation Pathway for Diclofenac

Cynthia M. Torres-Godoy, Chemistry, TENNESSEE TECHNOLOGICAL UNIVERSITY, Cookeville, TN, Dr ANDREW Callender, CHEMISTRY, TENNESSEE TECHNOLOGICAL UNIVERSITY, Cookeville, TN, Dr PEDRO E. Arce, Chemical Engineering, TENNESSEE TECHNOLOGICAL UNIVERSITY, Cookeville, TN and Dr DENNIS B. George, WATER Resources CENTER, TENNESSEE TECHNOLOGICAL UNIVERSITY, Cookeville, TN

DEGRADATION BY PULSE CORONA DISCHARGES: STUDY ON THE POSSIBLE CHEMICAL DEGRADATION PATHWAY FOR DICLOFENAC

Torres-Godoy, C., Tennessee Technological University

Dr. Callender, A., Tennessee Technological University

Dr Arce, P, E., Tennessee Technological University

Dr George, D., Tennessee Technological University

The removal of hazardous organic pollutants, such as, polychlorinated biphenyls (PCB's), polycyclic aromatic hydrocarbons (benzopyrene, dibenzo anthracene), pesticides (DDT, lindane), pharmaceutical products (Diclofenac, Ibuprofen, Carbamazepine) from waste water is a growing issue in environmental science and technology. Innovative methods for water remediation have been examined for removing contaminants from waste sites.

Pulsed corona discharge (PCD) is a non-thermal plasma technology characterized by low gas temperatures and high electron temperatures. This technology has the potential of effective water decontamination and the capability of efficient removal or degradation of organic contaminants. Two key aspects (chemical and physical) exist during a pulsed corona discharge process, and are thought to occur simultaneously in synergism for direct or indirect degradation of organic compounds. The chemical aspect of the process, which is the main focus of this research, mainly involves the formation of various active radicals, such as, ·OH. ·H, ·O, $\cdot O_2$ and $\cdot HO_2$ and active molecules, such as, H_2O_2 and O_3 and their chemical reactions and pathways. All these active species and reactions involve a high oxidation potential to break the molecular bonds of organic contaminants; and, therefore, to lead to effective decontamination of water fluent.

Using the fundamentals of organic chemistry with pulsed corona discharge principles, this research explores the degradation pathway of diclofenac in PCD using LC-MS to identify the intermediate degradation products. The rate of diclofenac degradation is confirmed through the use of ion-exchange chromatography to determine the amount of chloride released. The role of parameters such as electrode distance (gap) and solution conductivity are also assessed.

See more of this Session: Advances in Water Remediation and Treatment Systems
See more of this Group/Topical: Environmental Division

Impact of ultrasonication on degradation of estrone in pure water and waste activated sludge

Charu Chawla[1a], *George Nakhla*[1b] *Madhumita B. Ray*[1c*]
[1] *The University of Western Ontario, London, Ontario, Canada*
[a]*cchawla2@uwo.ca* , [b]*gnakhla@uwo.ca*, [c]*mray@eng.uwo.ca*

Introduction

Natural estrogens such as estrone (E1), 17-α, β estradiol (E2) and estriol (E3) and synthetic estrogen17-α-ethinylestradiol (EE2) have been categorized under “emerging contaminants” or endocrine disrupting chemicals (EDCs) owing to their potential for causing severe health related risks. These compounds have been detected in trace concentrations in surface waters possibly due to effluents from the conventional wastewater treatment plants, which have not been designed for their removal. One of the primary studies conducted on the presence of estrogens in wastewater treatment plant (WWTPs) effluents was carried out by Ternes et al.[1], who reported concentrations of 0.015μg/l and 0.027μg/l for 17β-estradiol and estrone, respectively. The presence of EDCs in WWTP effluents in trace amounts has been observed to affect the regional fauna and induce hormonal changes in aquatic organisms such as male rainbow trouts [2]. Besides wastewater effluents, natural estrogens have been detected in considerable amounts in anaerobically digested biosolids generated from WWTPs. Due to the affinity of estrogens towards the solid phase, they tend to accumulate on the sludge particulates in the activated sludge process and finally end up in the anaerobic digestion unit. Sludges generated in WWTPs are processed either by anaerobic digestion or other means of stabilization, prior to their disposal or further use as biosolids. Presence of recalcitrant organics such as natural estrogens in biosolids restricts their disposal options. The variations observed in the amount of estrogens detected in biosolids could be attributed to inherent wastewater characteristics or background concentrations in the raw influent entering WWTPs.

Studies conducted on the anaerobic digestibility of estrogens so far have yielded contrasting results. Carballa et al. [3] reported 85% removal of E1+E2 and EE2, at influent concentration comprising background and spiked concentrations, ranging from 4-400 μg/L, in a continuous sludge digestion process using a mixture (70:30 v/v) of primary and waste activated sludge as feed to a pilot scale anaerobic digester operated at sludge retention times of 20 and 10 days at mesophillic conditions. Contrary to the above findings, Czajka and Londry [5] observed no anaerobic degradation of EE2 added at 5 mg/L over an incubation period of 3 years in a lab scale study.

Many pretreatment methods such as chemical, thermal, mechanical, ultrasonic disintegration

---

* Corresponding author : Tel: +1 5196612111x81273
mailto:mray@eng.uwo.ca (M Ray)

or the combination of above pretreatment methods have been studied to improve anaerobic digestion by enhancing COD solubilization thereby reducing the residence time in the digester. Ultrasonication has been demonstrated to improve volatile solid (VS) reduction (38% to 50%), gas production, and dewaterability of sludges in a full scale study [4]. Besides particle disintegration, ultrasonication is known to degrade organic compounds in solution [5]. In the present study, effect of ultrasonication as a pretreatment method on the anaerobic digestibility of estrone (E1) in waste activated sludge (WAS) is studied.

Experimental

Materials

Estrone was obtained from Sigma Aldrich (Oakville, CA) with 98% purity. All organic solvents such as acetone (distilled-in-glass grade), methanol (distilled-in-glass grade), ethyl acetate (HPLC grade) and dichloromethane (HPLC grade) were obtained from Caledon Laboratories (Georgetown, Ontario, CA). Derivatization agents N, O-Bis (Trimethylsilyl) trifluoroacetamide (BSTFA) and pyridine were obtained from Supelco (Oakville, CA) and Caledon Laboratories (Georgetown, Ontario, CA), respectively. Ultra-pure water (conductivity of 18Ω) was obtained from Millipore water systems (Billerica, MA). Supelclean™ ENVI-18 SPE cartridge (from Sigma Aldrich, CA) tube with bed wt. of 500 mg with a volume of 3 mL was used for extraction of estrogen.

Waste activated sludge was collected from the aeration tank of the Adelaide Pollution Control plant located in London, Ontario, Canada and stored at a temperature of 4°C prior to use. Anaerobic inocula collected from primary anaerobic digester at the St Mary's (Ontario, CA) wastewater treatment plant, was used as microbial consortium for the anaerobic batch studies.

Analytical methods

Water Quality parameters

All water quality parameters were analyzed according to the standard APHA methods [6].Soluble parameters were analyzed after filtering the samples through 0.45 µm membrane filters. HACH vials were used to measure the total and the soluble chemical oxygen demand (TCOD and SCOD) in a HACH reactor. Biogas production was measured by releasing gas pressure in the vials using glass syringes (Perfektum; Popper & Sons Inc. NY,USA) to equilibrate with the ambient pressure. Biogas composition was determined by a gas chromatograph (Model310, SRI Instruments, Torrance, CA) equipped with a thermal conductivity detector (TCD) and a molecular sieve column (Molesieve 5A, mesh 80/100, 182.88X0.3175 cm). The temperatures of the column and the TCD detector were 90 and 105 °C, respectively. Argon was used as carrier gas at a flow rate of 30 mL/min, pH and ORP were measured using Oakton pH meter and ORP meter (Eutech instruments, IL, USA), respectively.

Estrogen analysis

100 mL of the sample volume was used for estrone analysis using the solid phase extraction

(SPE) method. The sludge samples were centrifuged at 5000 rpm for 10 minutes (Sorvall RC 5B Superseed Centrifuge, DU Pont Instruments, and USA). The samples were then filtered through glass microfiber binder free filter papers GF/D 2.7μm and GF/C 1.2 μm (Whatman, GE Health care, USA), sequentially. The filtrate was passed through Envi18 SPE cartridges at flow rates less than 4mL/min by applying vacuum in a side arm filtration flask (VWR, CA). The cartridges were activated by passing 3x 2mL acetone, 3x 2mL methanol and rinsed with 5x 2mL MiliQ water prior to loading of samples. After the samples were passed through SPE column, the tubes were rinsed with MiliQ water. The tubes were dried for 30 minutes using vacuum and thereafter the analyte was extracted with 15 mL of acetone. The extract was concentrated in concentration station (Turbo Vap II, Caliper Life Sciences, MA, and USA) under a gentle stream of nitrogen at 45 ^{0}C and 11-12 psi. The extracted sample was subjected to solvent exchange with dichloromethane (DCM) and concentrated further to 0.5 mL which was reconstituted to 1 mL. The concentrated sample contents were transferred in 2 mL GC vials (Agilent Technologies, USA); rinsed with 1 mL DCM, prior to analysis 100μL of BSTFA and 50 μL of pyridine were added. The sample contents were incubated at 55 0C for 30 minutes in an oven (ED 53-LL, Binder, Tuttingen, Germany).

Analysis was performed using a gas chromatograph (GC-2010, Shimadzu) coupled with a quadrupole mass spectrometer (GCMS-QP2010S, Shimadzu) equipped with an auto injector (AOC-5000, Shimadzu). Chromatographic separation was achieved using a BPX50 (50% phenyl polysilphenylene-siloxane) type capillary column of 30 m × 0.25 mm i.d. (film thickness of 0.25 μm) obtained from SGE (Austin, TX). Helium was used as the carrier gas at a column flow rate of 1.5 mL/min. The injector temperature was maintained at 320^{0}C. The GC oven temperature was programmed as follows: 2 min at 100 °C, first ramp 20°C/min to 150°C, held for 5 min, second ramp 10°C/min to 240°C held for 1 min and third ramp 20°C/min to 300°C and held for 14 min. The mass spectrometer was operated in selected ion monitoring (SIM) mode with positive ionization by electron impact (EI). The ion source and interface temperatures were maintained at 250 ^{0}C and 320 ^{0}C respectively with detector voltage relative to the auto tuning. The retention time for E1 was 22.9 minutes and ions monitored had mass charge ratios (m/z) of 257 and 342.

Results and Discussion

Adsorption study

Secondary sludge was thickened by gravity and treated thermally at 80^{0}C for 45 minutes in order to inhibiting biological activity. Prior to inactivation background concentration of E1 in WAS was measured. An average of 20-22 ng/L of E1 was found to be in the sludge. Thereafter, waste activated sludge was spiked with different concentrations (1- 50 μg/L) of E1 and was stirred at 200 rpm for 20 hours on magnetic stirrer bar plate (Henry Troemner LLC, USA). Samples were collected at regular intervals of 10 min, 30 min, 1 hr, 5 hrs and 20 hrs, to determine equilibration time and concentration. The data were plotted to establish the adsorption isotherm ($q= 0.103\ C_e^{0.83}$) of estrone on WAS.

Ultrasonication pretreatment

A lab scale horn type sonicator (Sonics, Vibra Cell USA) with an applied frequency of 20 KHz and rated power 0.75 kW was used as the pretreatment device. Sonication was carried out in a beaker containing 200 mL of ultrapure water spiked with E1 at a concentration of 200 µg/L.. Temperature was monitored during the sonication time and was not allowed to exceed the room temperature i.e. 25±2 ^{0}C. Four different sonication energies of 300, 500, 1000 and 1600 kJ/L were used for the experiments to determine the degradation of E1 under sonication only. Sonication showed >50% degradation of estrone in ultra pure water at 1600 KJ/L. The degradation was observed to follow first order kinetics.

Anaerobic batch study

Batch anaerobic studies were conducted in 250 mL serum bottles (Wheaton Science Products, USA). It was conducted to assess the anaerobic digestibility of sonicated estrone and to investigate the effect of estrone on methanogenic activity. Each bottle contained a 1:1 v/v mixture of anaerobically digested sludge as inocula and sonicated estrone. The bottles were purged with ultrapure nitrogen to remove any dissolved oxygen inside the bottles and capped tightly with rubber stoppers. The bottles were subjected to mechanical agitation at a speed of 180 rpm (MaxQ 4000, Thermo Scientific, Fremont, CA) and maintained at a mesophillic temperature of 37 ^{0}C. Anaerobic digestion studies conducted for both ultrasonic treated and untreated E1 showed no significant improvement in E1 removal from sludge indicating recalcitrant nature of both E1 and its intermediates by ultrasonication.

References

[1] T.A. Ternes, M. Stumpf, J. Mueller, K. Haberer, R.-. Wilken, M. Servos, Behavior and occurrence of estrogens in municipal sewage treatment plants — I. Investigations in Germany, Canada and Brazil, Sci. Total Environ. 225 (1999) 81-90.

[2] C. Desbrow, E.J. Routledge, G.C. Brighty, J.P. Sumpter, M. Waldock, Identification of estrogenic chemicals in STW effluent. 1. chemical fractionation and in vitro biological screening, Environ. Sci. Technol. 32 (1998) 1549-1558.

[3] M. Carballa, F. Omil, A.C. Alder, J.M. Lema, Comparison between the conventional anaerobic digestion of sewage sludge and its combination with a chemical or thermal pre-treatment concerning the removal of pharmaceuticals and personal care products, Water Sci. Technol. 53 (2006) 109-117.

[4] F. Hogan, S. Mormede, P. Clark, M. Crane, Ultrasonic sludge treatment for enhanced anaerobic digestion, Water Sci. Technol. 50 (2004) 25-32.

[5] M. Goel, H. Hongqiang, A.S. Mujumdar and M. B. Ray, Sonochemical Decomposition of Volatile and Non-volatile Organic Compounds-A Comparative Study, Water Research, 38 (19) (2004) 4247-4261.

[6] Standard methods for the examination of water and wastewater, 20 ed., APHA, Washington DC, USA, 1998.

(426c) Advanced Oxidation Processes with DC-Corona Discharge

Muhammad suleman Tahir, Institute of Chemical engineering and envirinmental technology, Graz University of Technology, Graz, Austria, Mahmood Saleem, Institute of Chemical Engineering and Technology, University of the Punjab, Lahore, Pakistan, Lahore, Pakistan and Matthus Siebenhofer, Intitute of chemical engineering and environmental technologies, Graz University of Technology, Graz, Austria

Abstract

Thiscontribution presents the results of degradation of organic constituents and non-biodegradablehazardous pollutants from wastewater and off-gas with DC-Corona discharge. A coronadischarge reactor equipped with brush type discharge electrode was investigated.The set up consisted of a continuously rinsed tube- type electrostaticprecipitator made of glass. The tube of 1 m net length and 76 mm inner diameterwas equipped with brush electrodes of different brush diameter and wirediameter. During operation the water film acted as earthed counter electrode.

Thecurrent/voltage data was modeled to develop an adapted correlation for brushdischarge at specific operation conditions. The effect of brush discharge coronacurrent on formation of oxidizing reagents was investigated and modeled and generationof the oxidizing reagent ozone, formed through brush discharge was optimized. DC-Coronadischarge with brush type discharge electrodes generates UV-radiation at elevatedoperation corona current. The radiation intensity of UV-C radiation was determined.

Degradationof selected constituents like acetone, EDTA, phenol etc. was recorded for specificcorona current. Degradation rate of selected constituents was higher in thedomain of ozone formation. Complete elimination of acetone and partialdegradation of EDTA and phenol was recorded at a corona current of 0.3-0.45mA.

See more of this Session: Advances in Water Remediation and Treatment Systems
See more of this Group/Topical: Environmental Division

(426d) Electrochemical Oxidation of Landfill Leachate; Performance of Boron-Doped Diamond Anodes

Dieter Woisetschlger[1], Bernd Humpl[2], Michael Koncar[2] and Matthus Siebenhofer[1], (1)Insitute of Chemical Engineering and Environmental Technology, Graz University of Technology, Graz, Austria, (2)VTU Technology, Grambach/Graz, Austria

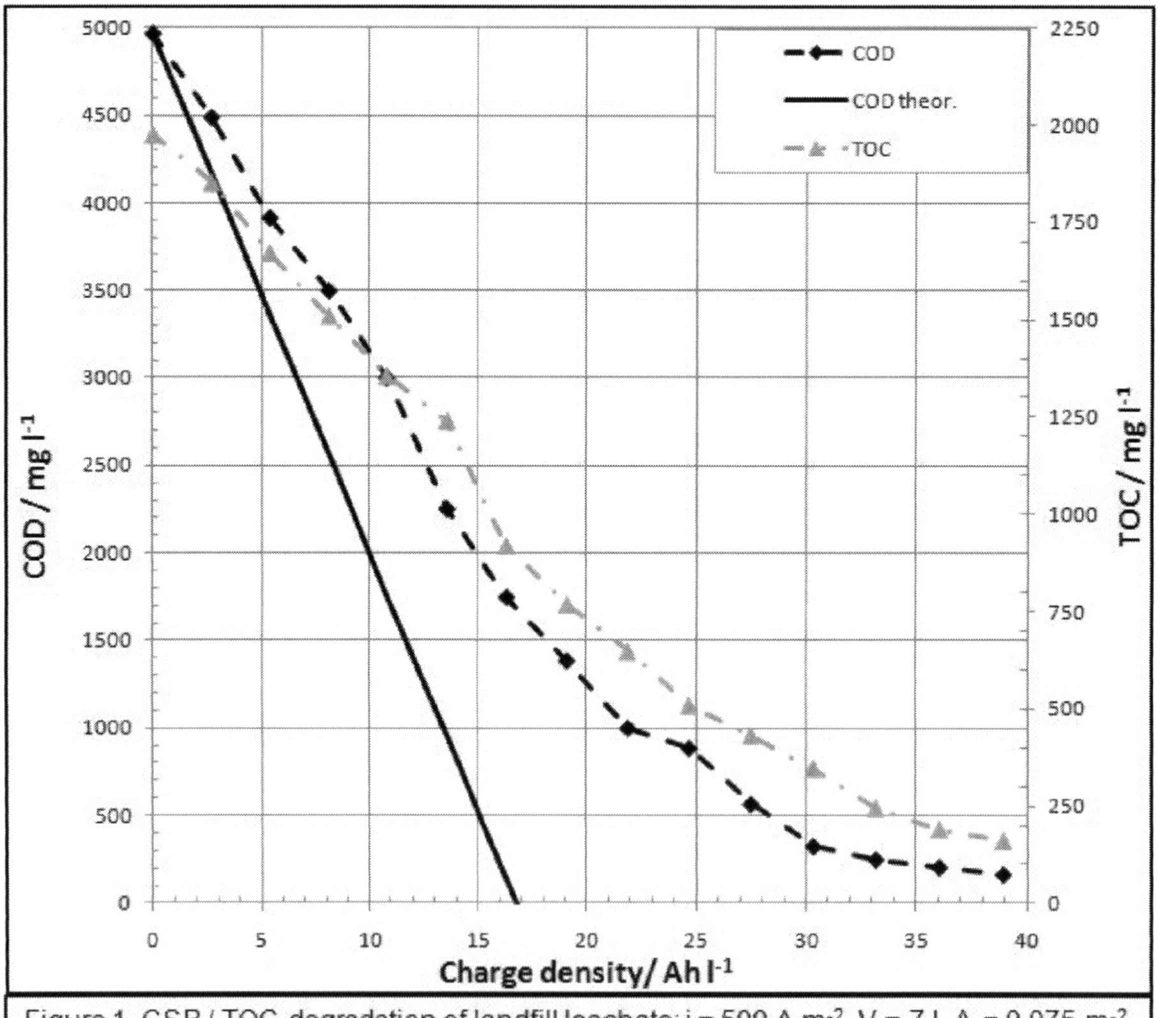

Figure 1. CSB / TOC degradation of landfill leachate: $j = 500\ A\ m^{-2}$, $V = 7\ l$, $A = 0{,}075\ m^{-2}$

See more of this Session: Advances in Water Remediation and Treatment Systems
See more of this Group/Topical: Environmental Division

(426f) Phosphate Removal In Wastewater by Functionalized Mesoporous SBA-15

Seung-yeon Lee[1], Jae-woo Choi[1], Seung-gun Chung[1], Ki-bong Lee[2] and **Sang-hyup Lee**[1], (1)Water Research Center, Korea Institute of Science and Technology, Seoul, South Korea, (2)Chemical and Biological Engineering, Korea University, Seoul, South Korea

Abundant phosphorous (P) in water causes eutrophication which threatens human health. As the regulation to control the level of phosphate in wastewater becomes much stricter, more efficient phosphate removal method should be developed. Existing phosphate removal methods such as biological, physical and chemical treatments are limited to remove phosphate efficiently. For this reason, mesoporous silica (SBA-15) was applied to remove phosphate in wastewater. SBA-15 has large surface, abundant reactive ionized surface, the simplicity to control their pore size, highly uniformed pores and application of functional groups. However, due to its low affinity for phosphate itself, SBA-15 was modified with several functional groups. These functional groups were chosen by their affinities for phosphate. Amine, aluminum, titanium, and thiol groups were adopted as the phosphate affinitive functional groups. Functionalized SBA-15 samples were synthesized by the post-synthesis method and analyzed with XRD, TEM, EDX and BET. To determine their adsorption capacities, equilibrium and kinetic tests were conducted and their adsorption capacities were determined by time and concentration. Each sample was analyzed by ion chromatography. According to each data, all samples were synthesized well. From the adsorption capacity tests, amino-functionalized SBA-15 showed the best adsorption capacity (66.57 mg/g) and possibility of its practical applications to real wastewater among these functionalized samples. Also Al-functionalized SBA-15 also had good adsorption capacity for phosphate (39.60 mg/g). Based on this study, it is proven that functionalized SBA-15 is a promising adsorbent for phosphate effectively than existing treatment methods.

See more of this Session: Advances in Water Remediation and Treatment Systems
See more of this Group/Topical: Environmental Division

(453a) Removal of Selenite with Microbial Fuel Cells Utilizing Shewanella Oneidensis MR-1

Prithiviraj Chellamuthu, Department of Biological Sciences, University of Southern California, Los Angeles, CA, Mandy Ward, Department of Earth sciences, University of Southern California, Los Angeles, CA, Kenneth Nealson, Department of Earth Sciences, University of Southern California, Los Angeles, CA and Miriam Kronen, Department of Microbiology, University of Heidelberg, Heidelberg, Germany

Selenium is a trace element with important physiological roles in immune function and oxidative stress; however, at elevated levels it is known to be toxic to all forms of living organisms. Apart from naturally occurring selenium in rocks, selenium compounds are widely used in electronics, oil production and agricultural applications, which has resulted in the contamination of ground water. Selenite (+4 oxidation state) and selenate (+6 oxidation state) are the oxidized forms of selenium. Both are soluble, mobile and toxic in soil and water systems. Even at low levels of contamination these elements have deleterious effects on fauna, where bio-magnification plays a significant role in inducing toxicity. The EPA has set a limit of .05 parts per million (ppm) in drinking water. Hence remediation is important and a variety of methods have been proposed, including activated alumina, coagulation/filtration, reverse osmosis and bioremediation. Recent investigations have shown that bioremediation, which utilizes indigenous or engineered bacteria for the removal of selenium compounds from the soil and/or water systems could be a viable option.

Microbial fuel cell (MFC) systems utilize the catalytic ability of electrogenic bacteria, unique organisms that can convert energy from organic compounds to electricity through the respiration of solid electrodes. There has been much research into the applications of MFCs, including recent investigations that suggest that the cathode of a MFC can be used for bioremediation and the extraction of valuable metals. Bioremediation results from the ability of the bacteria to convert toxic compounds into more manageable, nontoxic forms. In a MFC, transformation of toxic compounds occurs when used as electron acceptors or oxidants in the cathode compartment. Researchers have shown the applicability of MFC bioremediation technology for the removal of various metals including copper, chromium, selenium and arsenic. In order to fully exploit the potential of this technology it is important to understand the electron transfer mechanism involved in this reduction/ remediation process.

Shewanella oneidensis MR-1 is a facultative anaerobic bacterium known for its versatility in using a variety of soluble and insoluble electron acceptors. This versatility stems from an array of multi-heme cytochromes which makes them an interesting candidate for bioremediation of toxic heavy metals such as uranium, selenium and chromium. Understanding respiratory electron transfer pathways will help in the design of engineered organisms for better removal of toxic contaminants from water. Using a MFC, we have identified the key cytochromes involved in the reduction of selenium under fuel cell conditions. We have also optimized the maximum concentrations for optimal removal rates without affecting the performance of the MFC. Using various electrochemical techniques, the operating parameters, such as internal resistance and maximum power output, have been determined. Our results indicate this technology could satisfy the stringent EPA water quality standards for removal of selenium.

See more of this Session: Fundamentals of Biomass Utilization
See more of this Group/Topical: Environmental Division

(453b) Reaction Mechanism of Catalytic Oxidative Depolymerization of Cellulosic Biomass

Hongfei Lin, Chemical and Materials Engineering, University of Nevada, Reno, NV

While biomass recalcitrance is one of the biggest challenges in biofuel industry, Nature has the key to unlock this puzzle. When a plant grows, the cell wall undergoes re-structuring. Recent biological studies found that hydroxyl radical (•OH), a very short-lived and strongly oxidative radical, is harnessed by the growing plant cells to loosen their cell walls by nonenzymatically cleaving the wall polysaccharides, although the physiological significance of this loosening mechanism is unclear. This nonenzymatic process sheds light on developing chemical oxidation methods for biomass deconstruction under mild reaction conditions. We are developing cost competitive catalytic biomass deconstruction strategies using aqueous phase partial oxidation (APPO) for releasing platform intermediates from solid biomass with a focus on the production of organic acid streams that can be readily upgraded and converted to commodity chemicals or infrastructure compatible liquid hydrocarbon fuels. The reaction mechanism of the APPO process has been investigated. A free radical route is presented to interpret the pathway of cleaving cellulosic biomass via the APPO process and the functionalities of oxidation catalysts are correlated to the reaction mechanism. By understanding the reaction mechanism, we are able to optimizing the catalytic system to realize high yield of the organic acids and to manipulate the selectivity of the APPO process.

See more of this Session: Fundamentals of Biomass Utilization
See more of this Group/Topical: Environmental Division

(453c) Infrared Spectroscopy and Thermo Gravimetric Analysis for Cellulose and Residue by Hydrothermal Decomposition

Lidong Geng and Zhongjun Zhu, Chinese Society for Environmental Sciences, Beijing, China

Abstract: Under the background of energy and resource shortage and the environment pollution, the utilization and research of biomass wastes as a kind of resource attracting worldwide attention. Hydrothermal reaction provides an effective method for the reutilization of biomass wastes. The disposal and reutilization of biomass wastes were realized by hydrothermal reaction in batch hydrothermal reactor. By using Fourier transform spectrum analysis, the microcosmic structure of micro crystallite cellulose and its products were compared. The pyrolysis behavior of cellulose was studied by thermo gravimetric analysis (TGA). The experimental results showed that the molecular structure of micro crystallite cellulose changed obviously and the decomposition of micro crystallite cellulose mainly occurred at the temperature range of 250-400°C. The weight loss(TG) and the rate of weight loss (DTG) shift to higher temperature region with the increase of heating-up rate. The thermal dynamics for biomass wastes was coincidence to the first order reaction dynamics by Coat-Redfern analysis. The FTIR and TGA analysis indicated that the hydrothermal decomposition of micro crystallite cellulose was composed by low temperature hydrolysis and high temperature pyrolysis.

See more of this Session: Fundamentals of Biomass Utilization
See more of this Group/Topical: Environmental Division

(453d) Landfill Gas Utilization In Gas Turbines: Experimental Results for Combustion of Landfill Impurities In Premixed Flames

Aydin Jalali[1], Mirmohammadyousef Motamedhashemi[2], Fokion N. Egolfopoulos[1] and Theodore Tsotsis[2], (1)Department of Aerospace and Mechanical Engineering, University of Southern California, Los Angeles, CA, (2)Mork Family Department of Chemical Engineering and Material Science, University of Southern California, Los Angeles, CA

Gas turbines and other conventional internal combustion engines can combust landfill gas (LFG), which is an important renewable energy source, to generate electricity. However, the corrosion and damage that can be caused by its impurities may reduce the operating life of these combustion engines when operating on LFG. One class of major impurities present in LFG is Siloxanes. In our research, the decomposition of Hexamethyldisiloxane (also referred to as L_2) present in LFG is experimentally investigated in order to provide a better knowledge of the way it is degraded during the combustion of LFG. Several experimental techniques were utilized in order to study premixed flat flames, and to gain a better understanding about the durability of L_2 in the flame environment and the kinetics of its decomposition. In our experiments, we have utilized the counterflow technique, which allows one to accurately probe and analyze the mechanisms leading to the formation of silica microparticles resulting from the decomposition of L_2. We also applied laser extinction techniques for the *in situ* monitoring of particle formation, as well as GC-MS techniques for measuring the kinetics of siloxane decomposition.

It was found that microcrystalline silica particles were generated during LFG combustion, leading to a rapid coverage of the surface of Ni/Cr sheets placed in the flame environment, and forming a light white layer of solid particles, whose surface was analyzed by atomic force microscopy (AFM). The size of these particles was then estimated using SEM/EDAX. The effect of L_2 concentration on its conversion along the flame was found for a number of different feed concentrations. Volume fractions of particles within the flame were also measured *via* the laser extinction method indicating a linear relationship between the concentration in the fuel and the corresponding volume of particles in the flue-gas. The temperature profile has been measured experimentally in order to identify the kinetics of burning of the siloxane compounds. This fundamental insight is important in terms of being able to accurately determine the maximum allowable siloxane content for biogas that is safe to use without leading to deposits and microparticle formation.

See more of this Session: Fundamentals of Biomass Utilization
See more of this Group/Topical: Environmental Division

(453e) New Modeling Framework of Enzymatic Hydrolysis of Cellulose

Yang Zhang, Deparment of Chemical Engineering, Michigan Technological University, Houghton, MI and Wen Zhou, Chemical Engineering, Michigan Technological University, Houghton, MI

Cellulose has long been recognized as a renewable carbohydrate source for human energy use. It is available in large quantities in plant biomass. In recent years, for both economic and environmental reasons, intense research efforts have been directed at the utilization of cellulose to produce short-chain soluble sugar oligomers, such as glucose or cellobiose, which are subsequently or concurrently metabolized by microorganisms to produce biofuels such as bioethanol. The hydrolytic conversion of cellulose into soluble, fermentable sugars is achieved by means of cellulase systems which contain three major types of enzymatic activities: (1) endoglucanases which cut beta(1,4) glucosidic bonds randomly at all internal bond sites of insoluble cellulose chains in solid substrates; (2) exoglucanases which cut bonds only at the ends of insoluble cellulose chains; and (3) b-glucosidases which hydrolyze soluble oligomer sugars into glucose. The endoglucanases and exoglucanases thus depolymerize long solid cellulose chains into short soluble sugars and this is believed to be the rate-limiting step of the cellulose hydrolysis process.

In this work, we develop a general framework for a realistic rate equation modeling of cellulose hydrolysis using noncomplexed cellulase. Our proposed formalism, for the first time, takes into account explicitly the time evolution of the random substrate morphology resulting from the hydrolytic cellulose chain fragmentation and solubilization. This is achieved by integrating novel geometrical concepts to quantitatively capture the time-dependent random morphology, together with the enzymatic chain fragmentation, into a coupled morphology-plus-kinetics rate equation approach. In addition, an innovative site number representation, based on tracking available numbers of beta(1,4) glucosidic bonds, of different site types, exposed to attacks by different enzyme types, is presented. This site number representation results in an ordinary differential equation (ODE) system, with a substantially reduced ODE system size, compared to earlier chain fragmentation kinetics approaches. This formalism enables us to quantitatively simulate both the hydrolytically evolving random substrate morphology and the profound, and heretofore neglected, morphology effects on the hydrolysis kinetics. By incorporating the evolving morphology on an equal footing with the hydrolytic chain fragmentation, our formalism provides a framework for the realistic modeling of the entire solubilization process, beyond the short-time limit and through near-complete hydrolytic conversion.

See more of this Session: Fundamentals of Biomass Utilization
See more of this Group/Topical: Environmental Division

(453f) Theoretical Analysis of Biodiesel Production Under High Temperature and Pressure Conditions

Fernando Olmos, Chemical and Biomolecular Engineering, UCLA, Los Angeles, CA and Vasilios Manousiouthakis, Chemical Engineering Department, University of California at Los Angeles, Los Angeles, CA

Typically, biodiesel is synthesized by the transesterification of vegetable oil in heterogeneous fashion, i. e. in multiple liquid phases since the reactants, methanol and vegetable oil, are immiscible. In this work, we propose that the synthesis of fatty acid methyl esters (FAME), which after purification become biodiesel, take place in a solution where methanol, vegetable oil, and FAME are all homogeneously dissolved. Carrying out the reaction scheme in this manner eliminates any mass transfer effects between the oil and methanol phases. In addition, the overall reaction rate is further enhanced by the gravity based removal of the glycerol final product from the reaction mixture. The proposed process is modeled as a multiphase reactor with three phases (vapor, FAME, glycerol) in equilibrium with each other. Reactions are only considered to be carried out in the FAME phase. The mathematical model constitutes a Differential-Algebraic-Equation (DAE) problem that is solved through the implementation of a hybrid Runge-Kutta/Newton-Raphson solver. Several temperature and pressure conditions are considered to explore the operational characteristics of this process. Comparisons with the original heterogeneous biodiesel production process are carried out.

See more of this Session: Fundamentals of Biomass Utilization
See more of this Group/Topical: Environmental Division

New design for Advanced Zero Emission Power Plant concept

Wei Chen[1], Louis Winnubst[1*], Arian Nijmeijer[1]

[1] Inorganic Membranes, IMPACT Institute for Energy and Resources, University of Twente,

P.O. Box 217, 7500 AE Enschede, The Netherlands.

a.j.a.winnubst@utwente.nl

The most significant anthropogenic greenhouse gas is CO_2, which arises mainly from use of fossil fuels. A new technique called Advanced Zero Emission Power Plant (AZEP) has been developed these years targeting CO_2 capture and storage (CCS). In AZEP concept, CO_2-diluted O_2 is used to combust the fossil fuels, which produces a concentrated CO_2 gas stream and thus enables efficient CO_2 capture. The O_2/CO_2 mixture, necessary for combustion, could be supplied by using CO_2 to transport the oxygen as separated by oxygen permeable membrane from the air. After combustion, the concentrated CO_2 is partly recycled to the membrane as sweep gas, and the rest is captured.

Mixed Ionic and Electronic Conductor (MIEC) membrane is one of the most promising oxygen permeable membranes for AZEP. However, most of the MIEC materials, which usually have the oxygen deficient perovskite structure, are not stable in CO_2 containing atmosphere because of the reaction between CO_2 and the MIECs. In our work, we have found that the CO_2 tolerance of some MIEC materials can be greatly enhanced by increasing the oxygen partial pressure in CO_2 and the presence of oxygen can effectively protect the MIEC membranes. Inspired by this phenomenon, we have developed new concept of AZEP by adding certain amount of oxygen to the inlet of the separation unit to protect the membranes, following oxygen permeation measurements have shown that the protective effect of oxygen for the membranes is quite remarkable.

(461b) Tunable Reversible Ionic Liquids for CO_2 Capture

Elizabeth J. Biddinger[1], Amy Rohan[1], Jackson Switzer[1], Swetha Sivaswamy[1], Kyle Flack[2], Manish Talreja[1], Manjusha Verma[2], Pamela Pollet[2], Charles L. Liotta[2] and Charles A. Eckert[1], (1)School of Chemical & Biomolecular Engineering, Georgia Institute of Technology, Atlanta, GA, (2)School of Chemistry & Biochemistry, Georgia Institute of Technology, Atlanta, GA

Regulation of CO_2 emissions is likely inthe near future. The technology for CO_2 capture from largepoint-source CO_2 generators, such as coal-fired power plants, needsto be developed before CO_2 emission regulation is mandated. Effective CO_2 capture systems must produce pure CO_2streams without becoming an overly parasitic burden to the energy productionprocess. Monoethanol amine (MEA) and otheraqueous-based systems have a record of high CO_2 capture capacitiesbut also high regenerative energies resulting in an increased parasitic load tothe plant. Our research uses tunable reversible ionic liquids (RevILs) as the CO_2 capture system. Here,we maintain the benefit of high CO_2 capture capacities throughchemisorptions of CO_2 without the water excesses resulting in thepossibility of a less parasitic regeneration of the absorption system.

The RevILs that will be discussed here areone-component silylamines. Before CO_2 capture they are silylamine molecular liquids. Upon reaction with CO_2,the ionic form of the silylamine is producedreversibly. The silyl group was incorporated with areactive primary amine to form the RevILs withreduced viscosities. The structure of these silylaminescan be tuned to examine and ultimately enhance the desired properties of theabsorption system. The impact of these structure modifications on theproperties of the silylamines have been studied withtwo objectives in mind. The first objective is to establish anunderstanding of what the structural modifications do to the overall behaviorof the RevILs, including CO_2 capturecapacity, energy of regeneration and viscosity. Once thesestructure-property relationships are developed, the second objective is to usethese structure-property relationships to design a commercially viable RevIL system. Additional factors that are consideredin this second objective include ease of synthesis of the silylamineand stability of the silylamine during CO_2capture and release cycling. Promising RevILsidentified will be compared to conventional CO_2 capture systems.

See more of this Session: Ionic Liquids and Membranes for Carbon Capture
See more of this Group/Topical: Sustainable Engineering Forum

MEMBRANE CARBON CAPTURE AND THE CRITICAL ROLE OF PROCESS SYSTEM ENGINEERING

Rajab Khalilpour, Ali Abbas

School of Chemical and Biomolecular Engineering,
The University of Sydney, Sydney, Australia
Tel: +61 4100 91600; Fax: +61 2 9351 2854; E-mail: r.kh@usyd.edu.au

2011 AIChE Annual Meeting: Minneapolis, MN, October 16-21

Abstract

There is general consensus that when the concentration of target component (TC) in the membrane inlet feed is comparatively low, membranes tend to become an infeasible option for separation of TC. For this reason, there is a controversy in the literature on the feasibility of membranes for post-combustion carbon capture where the concentration of CO_2 in the power plant's flue gas is usually less than 15%. Our recent study shows that, the main reason for inconsistency in analyses is due to the application of simple membrane models. Such simplifications become critical when the TC concentration is at threshold values (as for CO_2 in the flue gas) [1].

In this presentation, we will present a systematic methodology for analysis of membrane systems considering multicomponent flue gas with CO_2 as TC. We avoided simplifying assumptions namely multicomponent flue gas represented by CO_2/N_2 binary mixture or considering the co/counter current flow pattern of hollow-fibre membrane system as mixed flow. We found, via parametric analyses, optimal regions of flue gas pressure and membrane area within which a techno-economical process system design could be carried out. High selectivity was found to not necessarily have notable impact on PCC membrane performance; rather, a medium selectivity combined with medium or high permeance could be more advantageous.

[1] Khalilpour, R., Abbas, A., Lai, Z., Pinnau, I. 2011. Modeling and Parametric Analysis of Hollow Fibre Membrane System for Carbon capture from Multicomponent Flue Gas. AIChE Journal.

(461d) A-Priori Estimation of Reactive Amino Acid Based Ionic Liquids for CO2 Absorption

Johannes Voelkl, Karsten Mller, Liudmila Mokrushina and Wolfgang Arlt, Chair of Separation Science and Technology, University of Erlangen-Nuremberg, Erlangen, Germany

The separation of CO_2 is an important topic in thermal separation science. The scrubbing of carbon dioxide from exhaust gases has gained more interest in the last years to reduce the emissions of green for counteracting global warming but is also an important process step in biogas upgrade to produce methane from regenerative sources. A commonly used process is absorption with following regeneration. Often amine solutions are used as solvents, which react with carbon dioxide and thus can bind CO_2 with a high capacity and selectivity. This comes along with a huge energy demand for regeneration and losses of solvent due to irreversible reaction products.

Novel solvents should be able to overcome these disadvantages but provide comparable separation performance to optimize the carbon dioxide scrubbing. Ionic liquids are a promising solvent family. Their modular setup of anion and cation makes the development of a task specific solvent possible.

In this work ionic liquids with an amino acid anion are investigated systematically in their ability as CO_2 absorbents. They can bind carbon dioxide physically as well as chemically due to the reaction of the amine functional group of the amino acid with CO_2. Both influences were screened with a-priori quantum chemical based methods. The change of free Gibbs enthalpy and reaction enthalpy of the reactions were predicted with DFT-calculations. Thereby the different equilibrium conversion and the tendency towards consecutive reactions were investigated. The anions were coupled with various cations in a screening of the physical solubility of CO_2 in the ionic liquids. The solubility is predicted with the COSMO-RS model. The combination of the chemical and phase equilibrium leads to the CO_2 loading as a function of the pressure. It can be seen that carbon dioxide is mainly bound chemically but the variation of the cation leads to a difference in the CO_2 loading of up to 10% to 15 % only due to different physical solubility. From these results the hypothetical solvents were benchmarked based on their predicted properties, whereas high carbon dioxide capacity and low reaction energy for an energetic favorable regeneration are desired. The most promising candidates were synthesized to measure the carbon dioxide solubility and validate the predictions. This approach shows a systematic solvent development based on quantum chemical methods which reduces experimental time and saves development costs.

See more of this Session: Ionic Liquids and Membranes for Carbon Capture
See more of this Group/Topical: Sustainable Engineering Forum

NANOMATERIALS: THE LATEST EMERGING ENVIRONMENTAL CONTAMINANT IDENTIFICATION AND MEASUREMENT

Presented by:
Katrina E. Varner
US Environmental Protection Agency/Office of Research & Development
Las Vegas, NV, USA

BIOGRAPHY:

My primary research focus as an environmental research scientist is method development and evaluation of parameters monitoring for contaminants and hazardous waste. Method and monitoring development involve the detection, speciation, and identification of nanomaterials in various media. This includes the development of screening techniques for groundwater and soil. My goal is to advance analytical techniques for the characterization and remediation of nanomaterials in various media including water, soil, and tissue to estimate the daily exposure of animals and humans to these materials.

Introduction

The basic research presented is a novel tool for the detection, characterization, quantification, and monitoring of nanomaterials in the environment. EPA is becoming more proactive concerning environmental issues. Emerging contaminants is an area EPA is investigating to determine if new products introduced by man are harmful to people or the environment. Nanomaterials are the latest emerging chemical group. These materials are being incorporated into commercial products at a faster rate than the development of knowledge and regulations to mitigate potential environmental impacts. Product technology is rapidly advancing, but are we, scientists and engineers, keeping pace with understanding and communicating information about potential health risks and effect to the public from these products? EPA's Office of Research and Development charged its laboratories to perform research on nanomaterials to gain understanding of the transformation, transportation, and longevity of these materials. The National Exposure Research Laboratory is specifically responsible for conducting nanotechnology studies.

EPA's National Exposure Research Laboratory-Environmental Sciences Division is researching the detection, characterization, quantification, and monitoring of nanomaterials in the environment. Before the possible risks of these materials in the environment can be determined, methods are needed for their collection, separation, detection, identification and quantification. The areas of research encompass the sources, fate and transport, exposure, and mitigation or prevention of the risks associated with nanomaterials. This research requires technology be applied to detect, identify, quantify and minimize exposure to engineered nanomaterials in the environment. Currently, nano-scale metallic and carbon based particles are not detectable in water or soil.

EPA is concerned with the transport of nanomaterials in the environment which requires understanding transfer mechanisms within environmental media, and measurement of various environmental parameters and nanomaterials parameters. Nanomaterial parameters include particle size and charge, chemical or elemental composition and surface modifications. Environmental media parameters include pH, ionic strength, flow rate, composition, and the presence of naturally occurring (such as dissolved organic carbon) as well as anthropogenic contaminant. Understanding interactions between nanomaterials and environmental media is also critically important.

The goals of this research are to develop and test novel monitoring and characterization methods for engineered nanomaterials (ENMs) with the initial focus on carbon nanoparticles (e.g. fullerenes) and metal nanoparticles (silver and gold) using electrochemical analyses. Increasing ENM content in consumer products may have a negative environmental impact. Many nano-enabled products including fullerene lotions and cosmetics contain reactive oxygen species that are readily introduced as non-point source pollution into the environment. Application onto the skin might pose more problems that may include immediate health concerns. Measurement of these materials is challenging because of their low concentrations and the presence of natural and anthropogenic nanomaterials containing the same elements. We are investigating the use of electrochemical probes with Quartz Crystal Microbalance (QCM) to quantitatively analyze for C_{60} versus C_{70}. The initial environmental matrix of water was utilized in the laboratory and further environmental matrices analyzed will include ground and surface water, soil and sediment.

Approach

A QCM gold electrode is modified with gold nanoparticles (AuNPs) prior to fictionalization using N-dicyclohexylcarbodiimide/N-hydroxysulfosuccinimide (DCC/NHS) chemistry (Figure 1). Decrease in frequency or mass increase is used to measure the adsorption/enumeration of C_{60} by the beta-cyclodextrin amine (ß–CD-NH_2) cavity attached on the QCM.

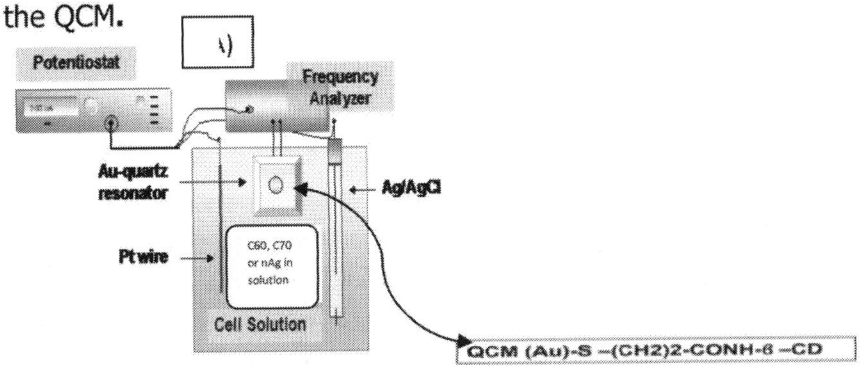

(B)

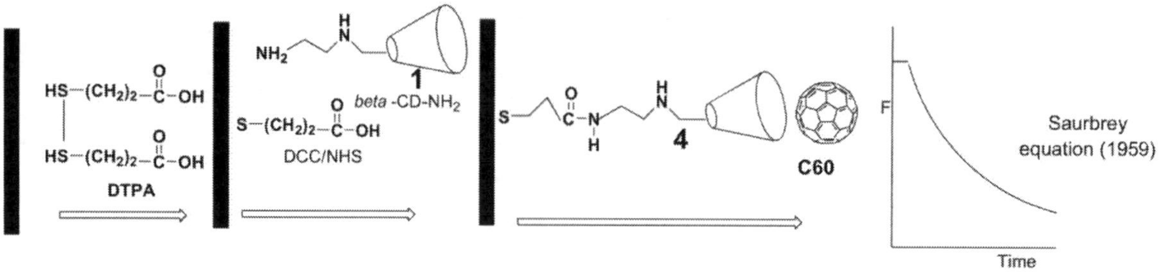

Δf is the change in frequency to coating (Hz), f0 is the frequency of quartz plate (MHz), Δm is the mass of engineered nanomaterial (in pico-grams), A is the area of quartz plate or electrode

Figure 1: *Schematic of modified QCM (A); Sensor design and surface modification using N-DCC/NHS chemistry (B).*

Discussion

The QCM is used to measure mass changes as well as the in-situ measurements of charge, current, potential and the resistance changes following the uptake of the particles. The unique sugar complex bucket-like structure is used to capture the spherical soccer ball-like structured C_{60}. Using amino modified ß–CD (a water-soluble polymer with synthetic receptor), the interactions of selected nanostructured materials (e.g. gold, silver, and fullerenes) are directly monitored by measuring the time-dependent changes in series resonant frequency by the QCM sensor. QCM provides a relationship between the resonance oscillation frequency of piezoelectric crystal and the interfacial mass change. The data are used to determine the quantity of the nanomaterial and comparisons are made with conventional methods such as Ultra Visible Spectroscopy (UV-Vis), Transmission Electron Microscopy (TEM) and Field Emission Scanning Electron Microscopy (FESEM). When the QCM is saturated with the nanomaterial, the attached nanomaterials are quantitatively released from the probe during a regeneration step washing with acid or alkaline hydrogen peroxide due to its selective association and dissociation behavior with C_{60}. And measure the concentration levels of C_{60} in water using simple QCM mass changes Application of a voltage between electrodes results in a shear deformation of the crystal

Results

Reproducible results are obtained using AuNPs in the immobilization chemistry. Figure 2A shows a plot of frequency versus time for the control, the QCM (AuNPs)-S–$(CH_2)_2$-CONH-ß–CD-NH_2 electrode in toluene. The figure demonstrates that no adsorption was noted on the modified QCM. In contrast, the frequency decreased (evidence of mass increase according to the Saurbrey equation) in the presence of 0.14 mg of C_{60} ml toluene (Figure 2B). These results suggest that the sensor can detect C_{60} in solution.

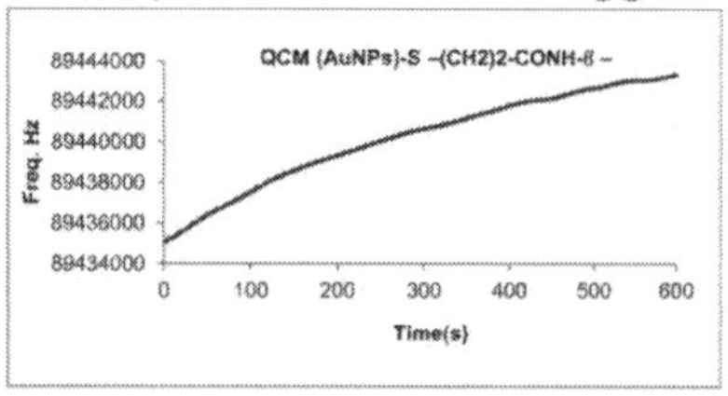

(A)

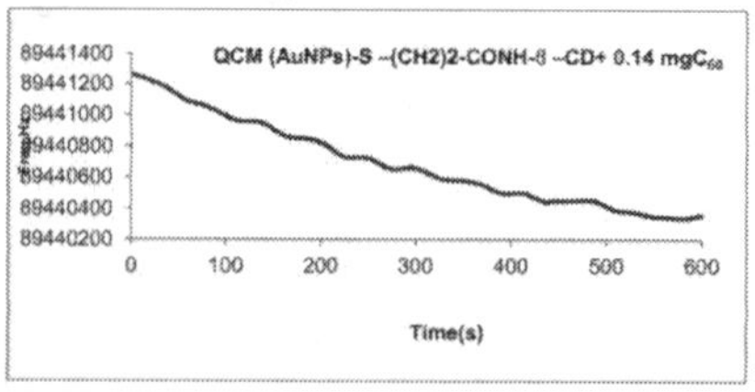

(B)

Figure 2: *Frequency vs. time for QCM (Au)-S-(CH$_2$)$_2$-CONH-ß-CD in presence of (A) toluene only (B) 0.14 mg C 60 + toluene (run time = 600 seconds)*

Figure 2B shows that the decrease in frequency continued up to 600 seconds. Additional frequency vs time change data showed that at 700s the maximum adsorption of C_{60} from 0.14 mg C_{60} /ml toluene was reached. This is observation indicated that all available active sites on the sensor were saturated. Sensor calibration data also showed that at much lower concentrations, the sensor can be calibrated up to 700 seconds. The QCM sensor has been tested for varying C_{60} concentrations.

Characterization

SEM images show the expected gold SEM micrograph. SEM micrographs of a clean modified QCM electrode (Figure 3A) and after exposure to C_{60} in toluene are shown. In the presence of C_{60} the SEM micrograph showed evidence of near spherical nanoparticles attributed to C_{60} clusters (Figure 3B), which were not observed on the pre-cleaned bare QCM.[1, 2] The presence of fullerene clusters of about 200 nm on the sensor surface suggests the aggregation of fullerenes since they exist as fractal systems which is consistent with previous studies.[3]

(A) (B)

Figure 3: *Pre-cleaned QCM, image @ 200nm (A) and QCM after evaporating drop of 2.8 mg C_{60}/toluene solution, image@ 200nm (B)*

Conclusions

The outcome of this research is to provide quantitative data to exposure risk assessors that make possible predicative estimates of compounds most likely to be of concern for the protection of human health and the environment. The ultimate goal, to discover and define previously unknown or poorly understood vulnerabilities of nanomaterials, is being accomplished. The successful demonstration using the synthesis of ß-CD-NH_2, its attachment and modification of QCM transducer via DCC/NHS chemistry is a promising new tool to address that goal. The QCM sensor showed specificity for C_{60} extraction when compared to the control solution by measuring the mass changes as well as the in-situ measurements of charge, current, potential and the resistance changes following the absorption and isolation of the particles to determine the presence of nanomaterials. A new practical measurement method for engineered, C_{60}, fullerenes is demonstrated as a new analytical tool; the first nanosensor to measure ENMs utilizing a size exclusive approach for the quantitative measurement of fullerenes.[4]

References:

1. Sadik, O.: Kikandi, S.; Wang, Q.; Varner, K. "Sensors as Tools for Quantitation, Nanotoxicity and Nanomonitoring Assessment of Engineered Nanaomaterials", *J. Environ. Monit.* **2009**, 11, 1782-1800.

2. Liu, Y.; Yang, W.; Chen, Y. Chemical Communications, **2005**, 4208-4210

3. Kikandi, S; Okello, V.; Wang, Q.; Sadik, O.; Varner, K.; Burns, S. "Size-Exclusive Nanosensor for Quantitative Analysis of Fullerene C_{60}", *Environ Science and Tech.* Accepted **2011**.

4. Bulavin, L.; Adamenko, I.; Yashchuk, V.; Ogul'chansky, T.; Prylutskyy, Y.; Durov, S.; Scharff, P. "Self-Organization C60 Nanoparticles in Toluene Solution", J. *of Molecular Liquids* **2001**, 93, (1-3), 187-191.

Notice: Although this work was reviewed by EPA and approved for publication, it may not necessarily reflect official Agency policy.

Work performed by:
Samuel Kikandi (Student Services Contract #RFQ-RT-08-00130), Qiong Wang, Veronic Okello under guidance and mentorship of Omowunmi Sadik, Director, Center for Advanced Sensors & Environmental System, State University of New York-Binghamton, Department of Chemistry, POB 6000, Binghamton, NY 13902-6000

(462b) Enabling In Situ Real-Time Characterization of Interfaces with Quartz Crystal Microbalance with Dissipation Monitoring

Mark A. Poggi, Q-Sense and Farfield, Biolin Scientific, Linthicum Heights, MD, Archana Jaiswal, Q-Sense, Biolin Scientific, Linthicum Heights, MD and Matthew Dixon, Biolin Scientific, Linthicum Heights, MD

There is a growing need for new technologies to quantitatively measure the surface properties of biological and material interfaces. One technique in particular, the Quartz Crystal Microbalance with Dissipation Monitoring (QCM-D), fulfills the need for monitoring real-time dynamic adsorption and desorption phenomena. Capable of operating in liquids and in real-time, QCM-D provides a powerful approach to analyze the *in situ* thickness, structural, and viscoelastic properties of molecules and materials at interfaces. This presentation will focus on the application of QCM-D to monitoring self assembly processes. Examples to be discussed will be recent research done in the areas of: protein deposition or fouling, polyelectrolyte multilayer build-up, nanoparticle/nanotube assembly, lipid membrane fuel cell interfaces. The QCM-D technique has allowed scientists to achieve a more fundamental understanding of how dynamic materials behave on the molecular scale and current research in several of these areas will be highlighted.

See more of this Session: Life Cycle Studies Associated with Nanomaterials and Nano-Enhanced Products
See more of this Group/Topical: Topical C: Environmental Aspects, Applications, and Implications of Nanomaterials and Nanotechnology

(462c) Mobility and Deposition of Silver Nanoparticles On Silica Surfaces Under Environmentally Relevant Conditions

B. Reginald Thio, Engineering Product and Design, Singapore University of Technology and Design, Singapore, Singapore, **Milka O. Montes**, Bren School of Environ. Sci & Mgmt, University of California, Santa Barbara, Santa Barbara, CA, Mahmoud A. Mahmoud, School of Chemistry and Biochemistry, Georgia Institute of Technology, Atlanta, GA and Arturo A. Keller, Bren School Environ. Sci & Mgmt, University of California, Santa Barbara, Santa Barbara, CA

Increasing use of nanomaterials in commercial products has led to widespread concerns about their potential environmental and health impacts. To better understand the transport, fate and behavior of nanoparticles in aquatic systems, it is essential to understand their interactions with different components of natural waters including natural organic matter (NOM) over a broad range of physicochemical conditions. Ag nanoparticles are widely used in many consumer applications such as antibacterial socks, and pigments. Deposition and mobility rates and the effects of NOM on particle deposition, as well as transport behavior of Ag nanoparticles onto a silica surface were evaluated using thin layer chromatography (TLC), and quartz crystal microbalance with dissipation (QCM-D) over a broad range of solution conditions.

See more of this Session: Life Cycle Studies Associated with Nanomaterials and Nano-Enhanced Products
See more of this Group/Topical: Topical C: Environmental Aspects, Applications, and Implications of Nanomaterials and Nanotechnology

(462d) Effect of the Aggregation of TiO2 Nanoparticles On Their Fate and Transport In Natural Waters

Mara Marta Fidalgo[1], Marina B. Romanello[1], Liliana M. Bertini[1], Linna Du[2] and Jianhong Ren[2], (1) Department of Chemical Engineering, Instituto Tecnolgico de Buenos Aires, Buenos Aires, Argentina, (2)Department of Environmental Engineering, Texas A&M University Kingsville, Kingsville, TX

Nanotechnology is a growing area with many promising and revolutionary advances in a variety of fields, from medicine to sporting good design and personal care products. One of the most widely applied nanomaterial is nanoTiO_2. Although it is a well-known material at the macroscale, the environmental implications of the special properties exhibited by TiO_2 in its nanoform are not fully understood. Of particular interest are the aggregation processes in natural waters, the interaction with natural organic matter, and the effect of the presence of divalent ion such as calcium and magnesium. The different conditions of the water matrix that TiO_2 nanoparticles may encounter in surface and groundwater will influence their physico-chemical properties, and thus modify their surface charge, aggregation, and attachment characteristics. These properties are key to determine the fate and transport of nanomaterials in the environment, and therefore a complete understanding of the transformations that TiO_2 undergo in a natural setting is necessary to assess the risk posed to the environment by products containing this nanomaterial.

The size and surface charge of TiO_2 nanoparticles was investigated as a function of pH for a variety of relevant conditions, including different levels of ionic strength, concentration of natural organic matter and non-indiferent cations by Dynamic Light Scattering (DLS) using a Nanosizer Z90 (Malvern, UK) and by means of settling column experiments for those cases when DLS was not applicable. By size the authors refer to the measured equivalent hydrodynamic diameter of the aggregates formed by the nanoparticles. DLVO theory was applied to model the experimental conditions explored, and generally good agreement was observed when aggregate size is taken as effective particle size for the calculations.

Aggregate size and morphology were investigated by Scanning Electron Microscopy using a FESEM (JEOL JSM-6701F). To preserve the aggregates conditions as close as possible to their original form in suspension, a substrate was placed vertically in each suspension and the sample to be imaged was formed by aggregates that attached during the descent of the liquid level as the suspension evaporates. Special care was taken so that the change in concentration of the constituents in the solutions did not vary significantly during this process, taking into account the ratio of the relative change of volume to the initial volume of sample suspension and interrupting the process accordingly. Prior to imaging, the samples were gold coated with a sputter coater (Denton Vacuum Desk IV) to provide a conductive surface.

Sand column transport experiments were conducted for some characteristic conditions to investigate the mobility of nanoTiO_2 in porous media. The results were modeled applying filtration theory.

It was observed that aggregation is dependent on pH, shear (mixing conditions), presence of natural organic matter and divalent ions through their effect in the zeta potential of the particles although this parameter alone was insufficient to explain the different size of the aggregates. Attachment to the sand grains during flow in the porous media followed a similar trend.

This work provides critical insights on the physico-chemical properties of TiO_2 nanoparticles suspended in environmentally relevant aqueous matrixes. The information presented is key to understand the interaction with the natural environment: their fate and transport on groundwater and surface water, their bioavailability and their subsequent effect on the biota.

See more of this Session: Life Cycle Studies Associated with Nanomaterials and Nano-Enhanced Products
See more of this Group/Topical: Topical C: Environmental Aspects, Applications, and Implications of Nanomaterials and Nanotechnology

(462e) Application of Screening-Level Life Cycle Assessment to Emerging Nanoproducts: Nanosilver Textiles and CNT Electronics

David E. Meyer, National Risk Management Research Laboratory, U. S. Environmental Protection Agency, Cincinnati, OH and Venkata K. K. Upadhyayula, Oak Ridge Institute of Science and Education (ORISE), Oak Ridge, TN

As global concerns for environmental sustainability grow, emerging technologies must be scrutinized early in deployment to insure they represent the most environmentally conscious option. Screening-level Life Cycle Assessment (LCA) can provide a quick tool to identify the life cycle hot-spots and focus research efforts to help minimize the burdens of a technology while maximizing its benefits. This talk examines the application of LCA to emerging nanotechnologies using case studies of nanosilver socks and a field emission display device (FED) containing carbon nanotubes (CNTs). The use of nanoscale silver in consumer products has exploded in popularity because of antimicrobial effects, making it the largest nanocomponent in use. The use of CNTs in nanoproducts is expected to offer numerous environmental benefits such as energy efficiency and the waste reduction. Although these products are considered beneficial to society, some attention must be given to the potential environmental impacts arising throughout the life cycle as production demands escalate.

See more of this Session: Life Cycle Studies Associated with Nanomaterials and Nano-Enhanced Products
See more of this Group/Topical: Topical C: Environmental Aspects, Applications, and Implications of Nanomaterials and Nanotechnology

(480a) ISEE: A Computational Tool for Industrial Sustainability Assessment and Decision Making

Zheng Liu, Department of Chemical Engineering and Materials Science, Wayne State University, Detroit, MI and Yinlun Huang, Dept. of Chemical Engineering & Material Science, Wayne State University, Detroit, MI

In the study on industrial sustainability, a major challenge is how to pursue effective sustainability assessment and decision making for the short- to long-term sustainable development of industrial systems. A sustainable system should demonstrate high efficiency of material and energy utilization, minimum waste generation, assured safety, high level social responsibility, etc.

Sustainability assessment and decision making of industrial systems is a multi-objective and interdisciplinary task, which has been greatly challenged due to the inherent complexity and uncertainty carried by the sustainability essential. To facilitate industrial practice on engineering sustainability, a computational tool, namely ISEE (Industrial Sustainability Evaluation and Enhancement), has been designed, where basic sustainability principles are embedded in a systems approach for sustainability assessment and decision making. The tool is featured by its capability of processing uncertain data and information, assessing sustainability status quo and predicting its future performance, and evaluating design alternatives using various sustainability metrics. Based on the assessment, the tool is capable of identifying the most desirable design for sustainability improvement. As a case study, biodiesel manufacturing in a large geographical region was illustrated.

See more of this Session: Sustainability Metrics, Indicators and Life Cycle Assessment
See more of this Group/Topical: Sustainable Engineering Forum

(480b) Engineering Involvement In Green School Development

Cory D. Jensen, Division of Engineering, Colorado School of Mines, Golden, CO

The emerging term of a "Green School" includes facets such as; technologies that would make institutional learning more energy efficient, are inherently environmentally friendly, and are envisioned to pass on concepts of stewardship and sustainability to learners. It is proposed that relatively simple and behavior based teaching concepts such as recycling, gardening, and touring of "green" social centers has the potential to impact long term knowledge of youth that would promote such practices. We have developed innovative teaching programs at the Cole Arts and Science Academy (CASA) via innovative infusion programs to promote such learning practices and hopefully instill concepts of stewardship. Such programs have been made possible by state (Colorado) education grants. To date, three classes have been planned and implemented: 1) "Garden Extravaganza" and 2) Green House Gardening, and 3) Holistic Garden Management. One of the goals of this unique urban agriculture based curriculum is to impart first principle applied concepts such as mass and energy balances that also carry tones of stewardship and community. As an extension to this public school outreach and education program, evaluation of the program, applied technology and study thereof, are directly related to current sustainability science practices that are gaining acceptance. This talk will highlight our efforts to develop Green School education programs and sustainability science & research. The extension of this work via research could lead to; 1) the refinement of engineering models in coordination with urban green school concepts (e.g. green house, energy flows, systems), 2) provide for engineering design projects that incorporate life cycle analysis (LCA) perspectives, 3) provide a basis for entrepreneurship, and 4) could be used as a basis for extensions to larger social implications (e.g. lifelong stewardship behavior) or novel extensions of life cycles.

See more of this Session: Sustainability Metrics, Indicators and Life Cycle Assessment
See more of this Group/Topical: Sustainable Engineering Forum

(480c) Towards Improved C and N Footprints and Understanding Their Nexus

Shweta Singh and Bhavik R. Bakshi, William G. Lowrie Department of Chemical and Biomolecular Engineering, The Ohio State University, Columbus, OH

Carbon and Nitrogen are the two most disrupted bio-geochemical cycles over their natural states. There have been several works that relate the impact of C on various components of N cycle such as N stocks in soil and vice versa. Similarly, the role of anthropogenic activities on different components of these natural cycles has also been quantified using the Ecologically Based Life Cycle Assessment (Eco-LCA). The realizations of extent of anthropogenic impact on the C and N flows have led to development and extensive use of C and N footprint measures. This is mainly due to the fact that footprint measures have become standard and easy way of quantifying environmental impacts.

Carbon footprint has become an essential metric in any Life Cycle study along with several studies that have indicated efforts to enhance the strength of this metric. On the other hand, Nitrogen footprint is a new entrant in the group of footprint measures but is rapidly gaining importance. However, the current C and N footprint metrics are not without limitations since these miss several components of flow of the natural C & N cycles. In this work, these limitations of current metrics are realized and improved C and N metrics are proposed to study the impacts of any process or product. The development of new metrics will be explained based on several flows that are not accounted in current metric calculations. Even though both the C footprint and N-footprint can be calculated according to the present definitions of these metrics there have been hardly any studies that have looked into the relation of these metrics. Hence, in this work along with the comparison of new and old C & N footprint measures, the relation between the C & N footprint will also be studied. A case study on bio-fuels will be presented to study the utility of using newly defined metrics for C & N over old C & N metrics and the relation among C & N metrics itself.

See more of this Session: Sustainability Metrics, Indicators and Life Cycle Assessment
See more of this Group/Topical: Sustainable Engineering Forum

(480d) Life Cycle Assessment of Waste-to-Energy Technologies

Zac Coventry, Civil Engineering, University of Colorado Denver, Denver, CO and **Arunprakash T. Karunanithi**, Department of Civil Engineering, University of Colorado Denver, Denver, CO

The United States generated 243 million tons of municipal solid waste (MSW) in 2009. With household waste generation continuing to grow at a rate of 3% per annum (Baidoo, Ferguson et al. 2009), the annual United States MSW generation is expected to reach 400 million tons by the year 2030. The current trend for the treatment of MSW is for 54% to be landfilled, 33% recycled and 12% combusted with energy recovery (United States Environmental Protection Agency 2010).

The disposal of MSW through landfilling causes direct and indirect environmental harm through increased greenhouse gas (GHG) emissions. For example, U.S. methane emissions from landfills were close to 130 $TgCO_2e$ in 2007 or 32% of all human generated methane emissions for that year (Jaramillo and Matthews 2005). In the larger context, such emissions contribute to climate change, smog formation and human health degradation. In light of these findings, a number of experts within the municipal solid waste management community are calling for a change in the waste management paradigm. For example, the City of Austin, Texas recently ratified a zero-waste master plan that aims to achieve a 20% reduction in solid waste disposal to landfills by 2012 and 90% reduction by 2040. Such dramatic shifts in the MSW waste flow will likely call for new, or different, MSW treatment options.

One of the primary options under consideration is waste-to-energy (WTE). In general terms, WTE technologies seek to convert post-recyclable MSW into a useful form of energy. These approaches range from "tried and true" technologies such as advanced thermal recycling (i.e. incineration with a highly effective emissions control system) to edge-of-the envelope technologies such as plasma-arc reduction. Many opinion leaders in the solid waste management community have lauded praise on WTE technologies claiming that they offer the dual benefits of decreasing net GHG emissions and helping ameliorate the United States' present energy crisis. However, a realistic understanding of the potential benefits of WTE technologies demands an accurate accounting of each technology's environmental impact over its entire life cycle. Moreover, those impacts must be reported in such a way as to promote meaningful comparison with the current landfill-based approach for solid waste disposal.

We present a Life Cycle Assessment study of a gasification WTE technology and compare its environmental impact with other waste to energy technologies and landfill emissions. Although individual life cycle assessment studies have been performed for most or all of the viable WTE technologies (Kalogo, Habibi et al. 2007; Velumani and Meenakshi 2007; Datiz 2008; Chester and Martin 2009; Ducharme 2010; Zaman 2010) each of these LCAs are based on different process boundaries (in terms of number of upstream and downstream processes included), geographical differences (e.g. European data vs. U.S. data), different emission factors and varying sets of assumptions. Such disparities disallow meaningful comparison between the various MSW management strategies. This study's LCA encompasses emissions from multiple upstream processes thereby "capturing" the majority of energy use and emissions. For example, it considers energy use and emissions associated with a facility's construction and demolition averaged over its productive life. The LCA begins with a detailed process flow diagram for each technology that leads to the determination of relevant unit process data modules. Next an inventory life cycle impact assessment is carried out. This assessment focuses on mid-point indicators that fall under four major categories: climate change impact, human health impact, ecosystem impact, and non-renewable energy used. Special emphasis is placed on the global warming potential indicator since it accurately reflects greenhouse gas (GHG) emissions represented as CO_2 equivalents. Finally, the paper explores the usefulness of LCA studies in informing policy decisions on the adoption of waste-to-energy technologies.

See more of this Session: Sustainability Metrics, Indicators and Life Cycle Assessment
See more of this Group/Topical: Sustainable Engineering Forum

(480e) Review of Life-Cycle and Techno-Economic Analyses for Algae Systems: Bottlenecks to Sustainability

Melissa Rickman, Chemical and Biological Engineering, University of Colorado, Boulder, CO, John Pellegrino, Department of Mechanical Engineering, University of Colorado, Boulder, CO and Stephanie Shaw, Air Quality, Environment Sector, EPRI, Palo Alto, CA

It is increasingly suggested that bioremediation of atmospheric carbon dioxide (CO_2) can be achieved by fixing CO_2 as algal biomass. This topic has been the subject of numerous life-cycle analyses (LCAs) in recent years, which have attempted to assess the environmental impacts of such processes, with mixed conclusions. Authors of techno-economic analyses (TEAs) have endeavored to determine the practical feasibility of algae production systems, but it is difficult to draw conclusions from their results due to uncertainty surrounding the wide variety of process designs proposed for this immature technology. To date, no comprehensive review of publicly-available LCAs and TEAs has been published.

Herein we have assembled a critical review of studies that have been published, and provide some illustrative benchmarking calculations based on fundamental relationships. It is clear that analysis of the detailed process designs and operations is necessary to quantify the potential sustainability of a system: in short, sustainable production of algal biomass is not an *a priori* result of all process scenarios. Good LCA studies identify major technical and economic lever points, but this step is often overlooked. It appears likely that co-production of high-value products (in addition to biofuels) is the most economically efficient way to lower carbon footprints of industrial CO_2 emitters.

See more of this Session: Sustainability Metrics, Indicators and Life Cycle Assessment
See more of this Group/Topical: Sustainable Engineering Forum

(480f) Assessment Processes for Sustainability and Sustainable Processes

Larry Erickson[1], Ryan Green[2], Blase Leven[2], Oral Saulters[2] and Terrie Boguski[2], (1)Chemical Engineering, Kansas State University, Manhattan, KS, (2)Center for Hazardous Substance Research, Kansas State University, Manhattan, KS

There is a growing emphasis to include sustainability in decisions that are made, especially in capital investments, energy intensive processes, and strategic planning. The interdisciplinary tools and approaches of sustainability science are providing an essential framework for strategic planning, process systems design, problem solving and decision making across systems, sectors, and scales. Sustainability assessments that encompass quantifying ecosystem services, product life cycles, and potential impacts are being integrated into leadership decision making. Governments, corporations, organizations, schools, and communities have all endeavored to establish indicators and metrics for measuring progress toward sustainability goals. Although sustainability initiatives may have moved from the voluntary realm into the regulatory universe in some capacities with demands from customers, shareholders, stakeholders, insurers, and government agencies, there are many unknowns in this fertile transdisciplinary field of study. The results of a literature review of sustainability indices and assessment processes will be presented.

See more of this Session: Sustainability Metrics, Indicators and Life Cycle Assessment
See more of this Group/Topical: Sustainable Engineering Forum

(482a) Heterogeneity Analysis of the Dry FGD Process At Medium Temperature In CFB Reactor Based On Eulerian Simulation

Fei Li, Institute of Process Engineering, Chinese Academy of Sciences, Beijing, China

The flow field structure and the desulfurization reaction of the dry FGD process in CFB reactor was studied by numerical method with Eulerian two-fluid model. By using the Syamlal & O'Brien drag force model with particle clustering effect correction and the improved sulfur capture model for the T-T sorbent, the relative errors of the computed bed inventory and desulfurization efficiency are lower than 5% and 6% respectively in comparison with experimental data. The simulated results show that from the bottom to the top of the bed, the flow field changes from dense region to transition region and then to dilute region; the PDDs of particle volume fraction, gas and particle-phase velocities changes from monotonous single-peak distribution to heterogeneous two-peak distribution and finally recover monotonous single-peak distribution corresponding to the axial distribution of heterogeneity of cross-sectional particle volume fraction distribution. Analysis of the heterogeneity of the particle volume fraction implies that the heterogeneity reaches maximum at the transition region while the sub-peak appears near the outlet under the hat chamber structure. The heterogeneity of SO2 concentration distribution increases with the increasing of the product of the particle volume fraction and the heterogeneity of its distribution. Finally, it was found that The heterogeneous axial distribution of sorbent particles shortens the effective sulfating reaction time thus the Calcium utility efficiency and with a certain particle volume fraction, the sulfating reaction rate as well as the desulfurization efficiency increase as the heterogeneity of the particle volume fraction distribution decreases. This work ascertains the optimization method for improving the Calcium utilization efficiency and desulfurization efficiency apparently that is to improve the uniformity of the distribution of sorbent particles.

See more of this Session: Technologies for Reductions of Emissions From Stationary Combustion Sources
See more of this Group/Topical: Environmental Division

DRY-TYPE OFF-GAS TREATMENT TECHNOLOGY OF EXHAUST GASES FROM THE SEMICONDUCTOR INDUSTRY FOR FLUORINE RECYCLING

Dr. Shinji Yasui, Mr. Syunpei Yamaji, Tsukuri College, Nagoya Institute of Technology, Nagoya, Japan, Mr. Tadashi Shojo, Mr. Goichi Inoue, Dr. Kunihiko Koike, Iwatani Corp., Shiga, Japan, Dr. Akihiro Takeuchi, Mr. Yoshio Iwasa, Chubu Electric Power Co., Inc., Nagoya, Japan

Introduction

Fluorocarbons and perfluoro compounds (PFCs) contribute to global warming and are used in large quantities in the semiconductor industry, which must reduce the emission of these gases to the atmosphere to achieve the requirements of the Kyoto Protocol. In the semiconductor industry, voluntary reduction goals for PFCs were set at the World Semiconductor Council held in April 1999 and ongoing reduction efforts have been made. However, since HFCs and PFCs are used in critical processes including chamber cleaning and etching during the manufacturing of semiconductors such as LCDs and solar panels, the consumption of these chemicals increases every year. Therefore, in spite of the reduction efforts, the emission of fluorocarbons and PFCs has increased in recent years.

The technology for the treatment of PFCs, which has been supplied in the semiconductor industries as an inexpensive process, is a combination of combustion to decompose PFCs to acid components such as HF and scrubbing to neutralize the acid components. However, because the combustion method has a low decomposition ratio and the treatment of the resultant wastewater has a high energy load, a new dry treatment technology is desirable. Consequently, we are developing a new dry-type treatment technology for PFCs that consists of an electric furnace for decomposition of the PFCs and a dry-fixation furnace for absorption the acid components, which enables fluorine recycling. Fluorine is a precious resource for Japan, which imports most fluorine from China or Mexico. The recovery of fluorine resources from exhaust gas containing fluorine is also an important issue for Japan. This paper describes the details of the treatment process and the developed technologies for the fluorine recycling system.

Dry-type treatment system

In the semiconductor industry, silane (SiH_4) is used for Si deposition and NF_3 is mainly used as the chamber cleaning gas. These gases are used alternately in the chamber; the exhaust gases from the process chamber include SiH_4, NF_3, and SiF_4 diluted with N_2 through the vacuum pump. The processing flow of the developed treatment system is shown in Fig. 1. Initially in the treatment process, NF_3 and SiH_4 are decomposed in the electric furnace to form HF, SiF_4, and SiOx. Then the SiOx powder is removed using a Teflon filter and the remaining HF and SiF_4 are absorbed by solid absorbents in the subsequent dry-fixation furnace. The dry-fixation furnace is a switching system with two columns filled with an upper layer of calcium carbonate and a lower layer of sodium bicarbonate, respectively; the calcium carbonate in the upper layer absorbs HF and converts it to CaF_2 and the sodium bicarbonate in the lower layer absorbs HF and SiF_4 to converts it to Na_2SiF_6.

There are two reasons for selecting an electric furnace for the decomposition of the PFCs. One is that the decomposition temperature is lower (800 °C) than that required for the combustion method (1500 °C). Since the Teflon filters and dry-fixation furnace are processed at temperatures of 150 °C, it is desirable for the temperature of the decomposition furnace for PFCs at the preceding stage to be lower. The second reason is that the electric furnace does not require excessive gas such as air for cooling and decomposition of the PFCs; therefore, the off-gas from the furnace contains a higher concentration of HF, which improves the gas-solid reactivity of HF and $CaCO_3$ in the fixation furnace and enables the recovery of high-purity calcium fluoride. The treatment flow rate in the developed system is 300 L/min (3% NF_3 and 0.5% SiH_4 in N_2) from emissions from the three chambers of the semiconductor-processing device.

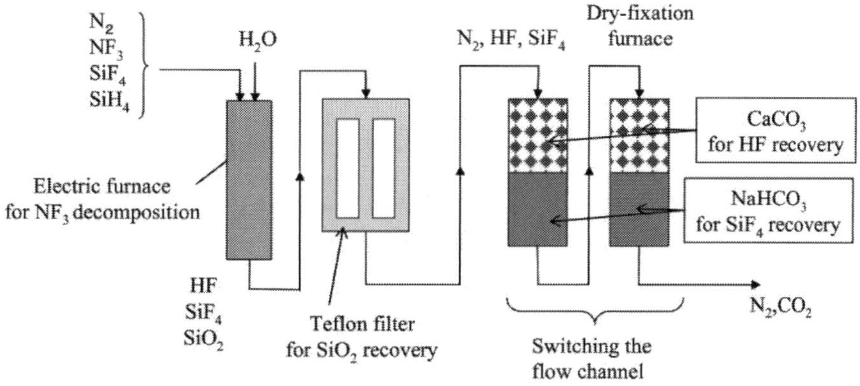

Fig. 1. Schematic of the dry-type gas treatment system.

Electric furnace for PFC decomposition

Important parameters for the design of the electric furnace for NF_3 decomposition are the reaction temperature and the volume of the reaction tube. Therefore, the decomposition properties of NF_3 were investigated. The decomposition reaction is expressed by the following equation:

$$NF_3 + (3/2)H_2O \rightarrow 3HF + (1/2)N_2 + (3/4)O_2 \qquad (1)$$

From the decomposition experiments, this decomposition reaction was determined to follow the first-order reaction formula and apparent rate constants for the temperature were thus obtained. The results are shown in Fig. 2. From these results, the frequency factor (k_0) and the activation energy (E) of the decomposition reaction were obtained as follows:

$$k = k_0 \exp(-E / RT) \qquad (2)$$

$k_0 = 2.69 \times 10^4$ (s^{-1})
$E = 6.55 \times 10^4$ (J/mol).

By using the obtained apparent rate constants, the residence time required for the decomposition of NF_3 was calculated by setting the decomposition ratio and reaction temperature. From the treatment system specifications, the reaction temperature and decomposition ratio were set at 800 °C and of 99.9997% (1 ppm of HF in the exhaust gases), respectively, and the required residence time was calculated to be 1.7 seconds. Using the calculated residence time, the volume of the reactor tube can be determined from the flow rate of the process gas.

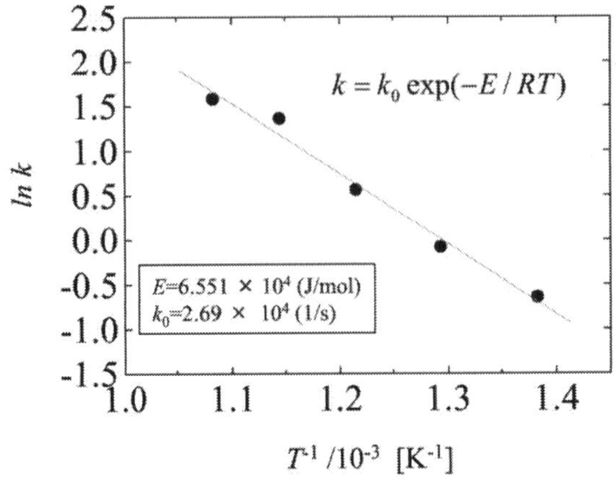

Fig. 2. The experimental results of the apparent reaction rate constants.

Similarly, the decomposition properties of SiH_4 were confirmed by decomposition experiments. The decomposition reaction of SiH_4 with H_2O occurs according to the following equation:

$$SiH_4 + 2H_2O \rightarrow SiO_2 + 4H_2 \qquad (3)$$

From the results of the decomposition experiments, a decomposition ratio of 99.8% was achieved with reaction conditions of 700 °C and a residence time of 0.7 s. This result shows that the conditions required in the electric furnace for the decomposition of NF_3 are also effective for the decomposition of SiH_4.

Dry-type fixation furnace

In a conventional dry-type gas treatment system, recycling the solid absorbent after the absorption reaction is not considered. In our gas treatment system with two types of solid absorbents and a fixation device connected to a series of two columns, over 97% of the pure calcium fluoride can be recovered and recycled to obtain fluorine resources. In addition, a continuous gas treatment is achieved by switching these columns. To obtain the high-purity CaF_2, sodium absorbents were used to absorb the HF and SiF_4 leaking from the $CaCO_3$ layer. This section shows the gas-solid reaction properties of this fixation furnace.

HF absorption of $CaCO_3$

1) Experimental conditions and method

To recover the high-purity CaF_2, it is necessary to clarify the gas-solid reaction properties of HF and $CaCO_3$. In other studies, the gas-solid reaction properties of $CaCO_3$ and HF at a high concentration in the reacted gases, which was generated by the decomposition of pure fluorocarbons, were investigated [1, 2]. However, the reaction properties at low concentrations of HF have not yet been clarified. Therefore, the reaction properties of HF and $CaCO_3$ were investigated in this experiment. The experimental conditions are shown in Table 1. The important parameters for the design of reactors for gas-solid reactions are the particle size of the solid absorbents, reaction temperature, and superficial velocity. The experimental conditions are consistent with the design specifications of these parameters for a practical device.

Table 1. Experimental conditions for the gas-solid reaction.

	Practical system	Experimental conditions
Processing flow rate	400 L/min	1.6 L/min
Inner diameter	45 cm	2.8 cm
Superficial velocity	6 cm/s	6 cm/s
L/D ratio	1.2	1.2
$CaCO_3$ weight	120 kg	31 g
Space velocity (SV)	400 h^{-1}	6700 h^{-1}
Absorbent size	1～2 mm, 2～5 mm	1～2 mm, 2～5 mm
Fixation temperature	150 °C	150 °C

The experiment flow is shown in Fig. 3. The reaction gas including HF was obtained by thermal decomposition of HFC134a ($C_2H_2F_4$) with water vapor and diluted at the entrance of the fixed-bed reaction furnace. The fixed-bed reactor filled with calcium carbonate is made of SUS, which is heated uniformly in a vertical electric furnace. The temperature of the vertical electric furnace was set at 150 °C. HF gas concentrations of 1, 5, and 10% and $CaCO_3$ grain sizes of 1~2 mm and 2~5 mm were investigated. The breakthrough properties for each experimental condition were investigated by measuring the concentration of F ions in the impinger. The experimental conditions are summarized in Table 2.

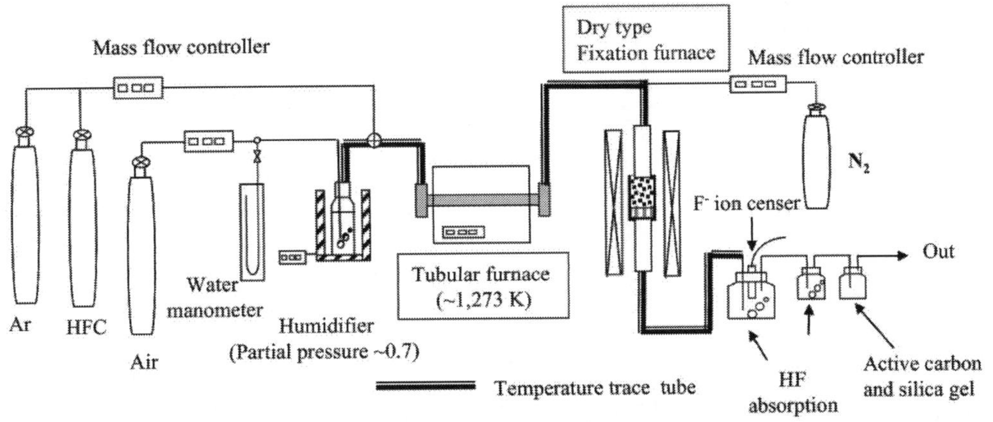

Fig. 3. Experiment flow of the gas-solid reaction of HF and $CaCO_3$.

Table 2. Experimental conditions of the gas-solid reaction of HF and $CaCO_3$.

	Frow rate of the treated gases						Pyrolysis gas conditions		Fixation conditions						
									Absorbent						
Test no.	$C_2H_2F_4$ [ml/min]	Ar [ml/min]	H_2O [ml/min]	Air [ml/min]	N_2 [ml/min]	Total [ml/min]	HF frow rate [ml/min]	HF concentration [mol/m^3]	Type	Weight [g]	grain size [mm]	Reaction temperature [°C]	Space velocity [cm/s]	SV [h^{-1}]	L/D
1	5	300	50	100	1100	1555	20	0.574	$CaCO_3$	31	1～2	150	6.1	4408.4	1.2
2	20	300	75	150	1000	1595	80	2.239	$CaCO_3$	31	1～2	150	6.2	4295.5	1.2
3	40	300	150	300	700	1590	160	4.492	$CaCO_3$	31	1～2	150	6.2	4394.3	1.2
4	5	300	50	100	1100	1555	20	0.574	$CaCO_3$	31	2～5	150	6.1	4408.4	1.2
5	20	300	75	150	1000	1595	80	2.239	$CaCO_3$	31	2～5	150	6.2	4295.5	1.2
6	40	300	150	300	700	1590	160	4.492	$CaCO_3$	31	2～5	150	6.2	4394.3	1.2

2) Experimental results

The results of the breakthrough properties are shown in Fig. 4. The HF concentration in the off-gas from the fixed-bed reactor increased gradually after the breakthrough time and finally reached the initial concentration at the entrance of the fixed-bed reactor. The results shown in the vertical axis in Fig. 4 were normalized by the initial HF concentration at the entrance.

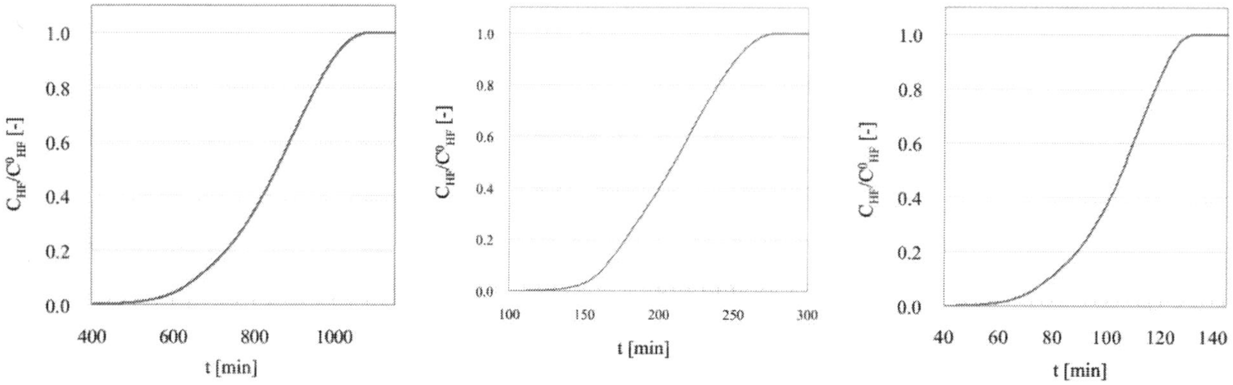

(a) HF conc.: 1% (b) HF conc.: 5% (c) HF conc.: 10%

(1) Grain size: 1~2 mm

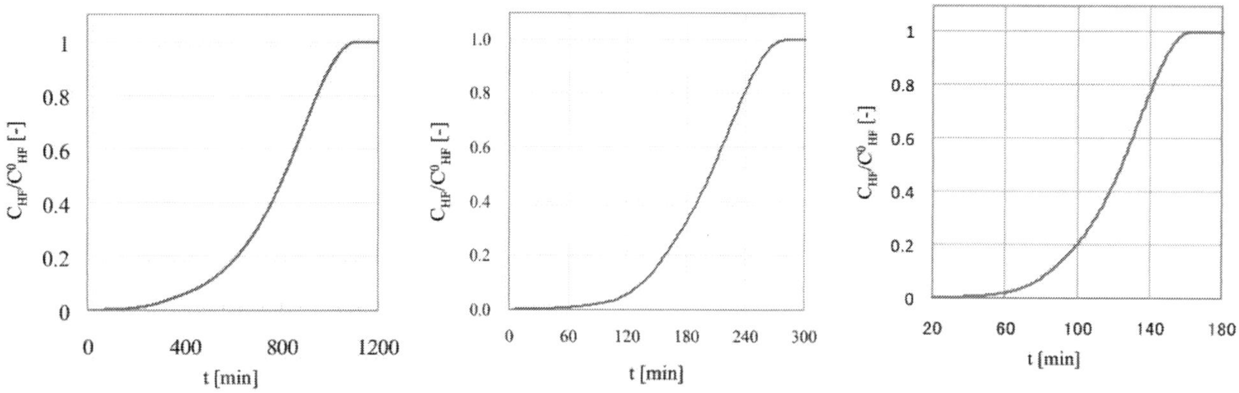

(a) HF conc.: 1% (b) HF conc.: 5% (c) HF conc.: 10%

(2) Grain size: 2~5 mm

Fig. 4. HF breakthrough properties of $CaCO_3$.

The elapsed time from the breakthrough point to the complete breakthrough time (C_{HF}/C^0_{HF} = 1) increased at under low HF concentration conditions. For a fixed-bed reactor, the slope of the breakthrough curve ($d(C_{HF}/C^0_{HF})/dt$) can be regarded as equal to the reaction rate (dX/dt) of the $CaCO_3$ filling at the outside edge of the fixed-bed reactor. Therefore, the slopes of the breakthrough curves of these results were evaluated and plotted against the HF concentration at the slope points, as shown in Fig. 5. From these results, the reaction rate of $CaCO_3$ can be determined using a first-order reaction formula of the HF concentration.

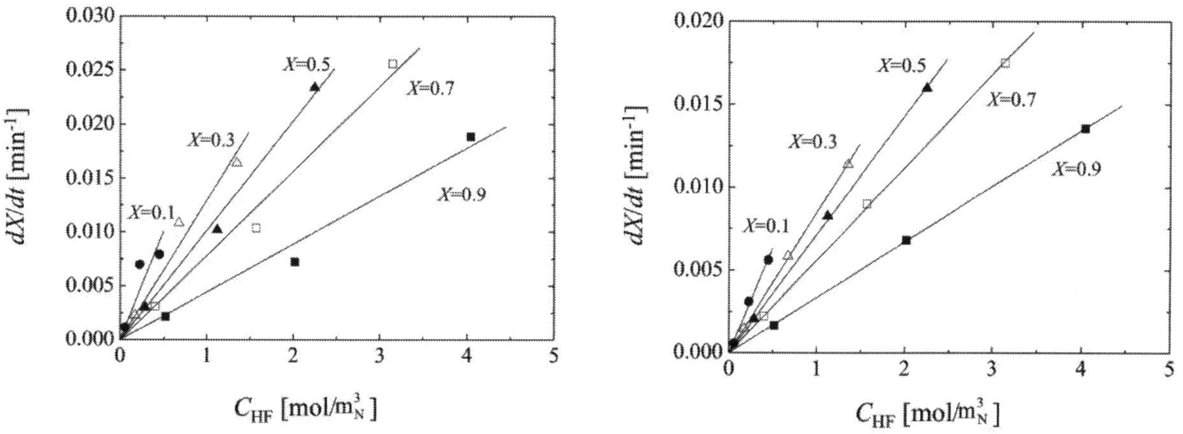

(a) Grain size: 1~2 mm (b) Grain suze: 2~5 mm

Fig. 5. Reaction rate of $CaCO_3$ plotted against the HF concentration.

The results of the reaction rates of $CaCO_3$ were investigated using the gas-solid reaction model of a shrinking unreacted core system. The reaction rate of a shrinking unreacted core system is expressed using the following equation[3]:

$$\frac{dX}{dt} = \frac{C_{HF}}{f_S^{-1}\cdot(1-X)^{-2/3} + f_p^{-1}\cdot\left[(1-X)^{-1/3}-1\right] + f_g^{-1}}. \quad (4)$$

Here, f_S is the rate constant for the particle surface chemical reaction, f_p is the rate constant for diffusion through the product layer of a spherical particle, and f_g is the rate constant for the external mass transport. By evaluating the results of Fig. 5 using equation (4), the overall reaction rate of $CaCO_3$ could be expressed in terms of the rate of diffusion through the product layer and the rate of external mass transport; each rate constant was obtained, as shown in Table 3.

Table 3. Rate constants of the gas-solid reaction of HF and $CaCO_3$.

Grain size ~1–2 mm	$f_p = 6.67\times10^{-3}$ [m³/(mol·min)]	$f_g = 1.88\times10^{-2}$ [m³/(mol·min)]
Grain size ~2–5 mm	$f_p = 5.37\times10^{-3}$ [m³/(mol·min)]	$f_g = 1.17\times10^{-2}$ [m³/(mol·min)]

By using each obtained rate constant, it was possible to calculate the elapsed time from the breakthrough point to the complete breakthrough by setting the initial concentration of HF (C^0_{HF}) and the reaction ratio of $CaCO_3$ (e.g., 0.95). From this calculated time, the amount of HF gas leaking from the $CaCO_3$ layer until the $CaCO_3$ converted into high-purity CaF_2 was calculated from the treatment flow rate. This amount of HF gas needs to be absorbed in the subsequent $NaHCO_3$ layer.

SiF_4 absorption for $NaHCO_3$

The role of the $NaHCO_3$ layer is to absorb the HF and SiF_4 emitting from the $CaCO_3$ layer. To determine the amount of $NaHCO_3$, it is necessary to calculate the absorption capacity of each gas. Here, the amount of $NaHCO_3$ for HF absorption was confirmed to supply the equimolar amount of HF leaked from the $CaCO_3$ layer, because the reaction of $NaHCO_3$ and HF at 150°C generate not only NaF and $NaHF_2$, HF can be absorbed completely by an equimolar amount of $NaHCO_3$. Therefore, this section describes the results of the experiments for examination of the absorption capacity of SiF_4.

1) Experimental conditions and method

The experimental flow is shown in Fig. 6. At first, HF was generated by thermal decomposition of HFC134a ($C_2H_2F_4$) with water vapor and air; SiF_4 was generated by the reaction of the HF and pieces of quartz. After that, these reaction gases were introduced into the fixed-bed furnace with N_2 dilution gas to set the SiF_4 concentration and then reacted with sodium-based absorbents in the furnace. We used two types of sodium-based absorbents: $NaHCO_3$ and NaF. NaF was used because it is generated by the reaction of HF and $NaHCO_3$ in the fixed-bed furnace. The concentrations of HF and SiF_4 in the exhaust gas from the fixed-bed furnace were analyzed by FT-IR and the amount of SiF_4 absorbed by each absorbent until the breakthrough time was determined. Table 4 shows the gas-solid reaction conditions of this experiment.

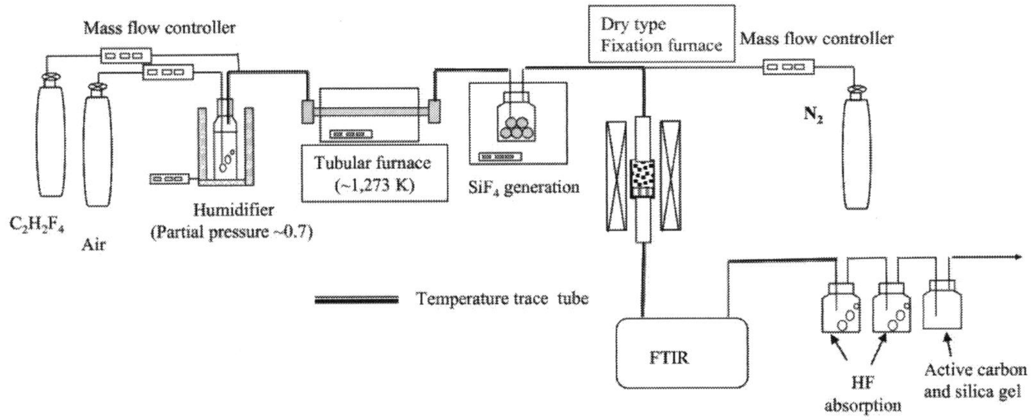

Fig. 6. Experimental flow of the gas-solid reaction of SiF_4 and sodium absorbents.

Table 4. Experimental conditions of the fixed-bed furnace.

	W	*T*	*V*	*L*	*L/D*	*v*	*SV*
	g	°C	cm^3	cm	-	cm/s	h^{-1}
$NaHCO_3$ or NaF	12	150	10.4	1.68	0.60	11.7	25066

2) Experimental results

Figure 7 shows the results of the breakthrough properties of each absorbent. The breakthrough times of SiF_4 in both absorbents were more than 200 min from after the start of the gas-solid reaction. As the concentration of SiF_4 was less than 3 ppm, which is the TLV value of SiF_4, until the breakthrough time, it was confirmed that both sodium-base absorbents can completely absorb SiF_4. The breakthrough times for the $NaHCO_3$ and NaF absorbents were 230 min and 250 min, respectively. The absorption capacity of each absorbent was then calculated using that obtained time to be 0.133 g/g $NaHCO_3$ and 0.145 g/g NaF. $NaHCO_3$ can absorb both HF and SiF_4 that leaks from $CaCO_3$ layer. The amount of

$NaHCO_3$ filling in the fix-bed furnace can be calculated using the obtained absorption capacities for each gas by the amount of $CaCO_3$ loading in the fixation furnace.

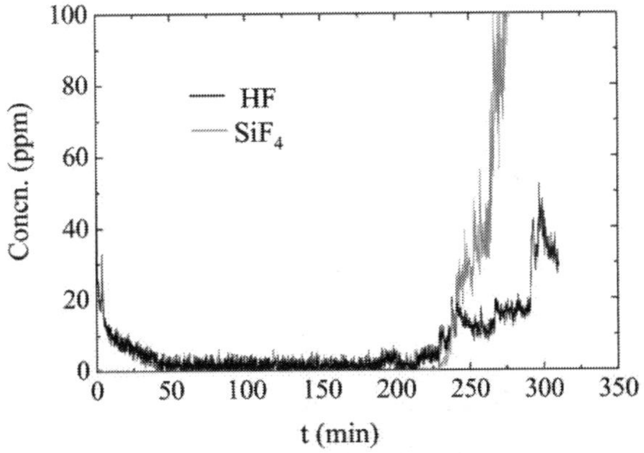

(a) $NaHCO_3$

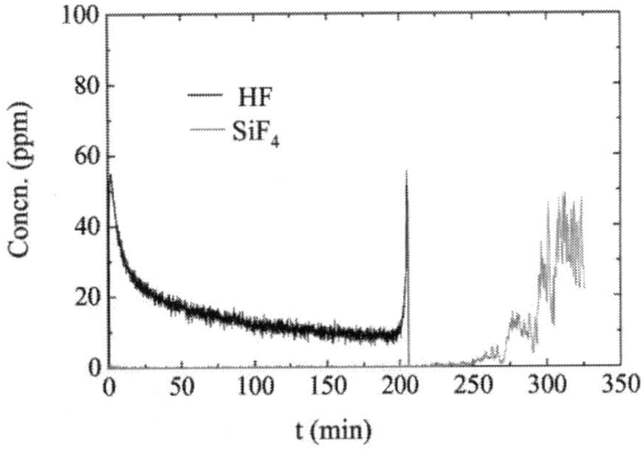

(b) NaF

Fig. 7. Breakthrough properties of HF and SiF_4.

Chemical properties of obtained fluorides

The major advantage of this dry treatment system is the ability to recycle the obtained fluorides. This section shows the chemical properties of the obtained fluorides. Figure 8 shows the EPMA photographs of the cross-section of the calcium absorbents before and after the reactions. It is confirmed that the reaction with fluorine proceeds to the core of the calcium absorbent. Mg and Si impurities are detected at about several hundred or several thousand ppm in the obtained calcium fluorides. These impurities are contained originally in the raw calcium carbonate materials as dolomite ($CaMg(CO_3)_2$) and silica. However, the purity of the obtained CaF_2 is over 97%, which is sufficient for it to be recycled as fluorine resources.

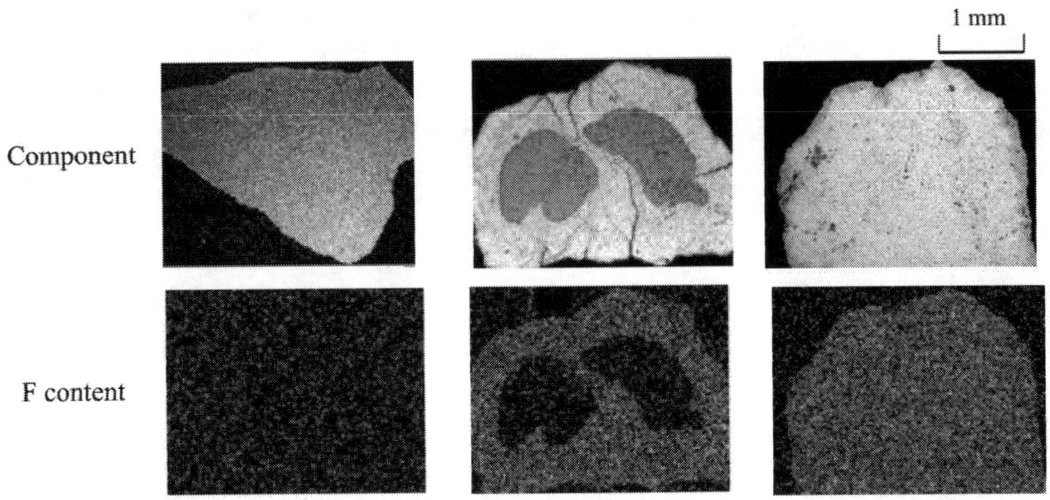

(a) Before reaction (b) During reaction (c) After reaction

Fig. 8. EPMA photographs of the $CaCO_3$ absorbents before and after experiments.

The EPMA photographs of the cross section of the $NaHCO_3$ absorbents before and after the experiments are shown in Fig. 9. Similar to the $CaCO_3$ absorbent, the fluorine and silicon reaction proceeded to the core of the absorbents. Silicon was detected as Na_2SiF_6 by X-ray diffraction analysis. This sodium-based material can be also reused as a special material by the fluorine industry.

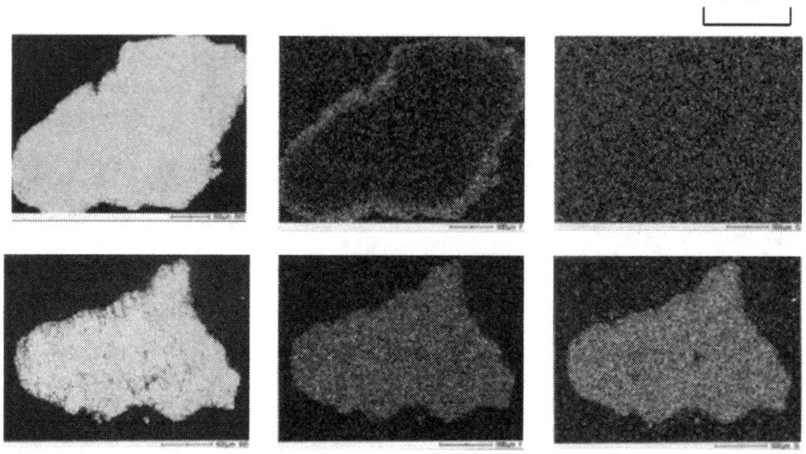

Fig. 9. EPMA photographs of the $NaHCO_3$ absorbents before and after experiments.

Conclusions

A dry-type gas treatment system was developed to enable fluorine recycling from the off-gas containing NF_3 and SiH_4 from semiconductor facilities. NF_3 and SiH_4 are decomposed in an electric furnace and two types of the solid absorbents, $CaCO_3$ and $NaHCO_3$, absorb the HF and SiF_4 in the decomposition gases. In this study, the decomposition properties of NF_3 and SiH_4 were clarified and the reaction tube in the electric decomposition furnace was designed. In addition, the HF and SiF_4 breakthrough properties of each absorbent were investigated. From the experimental results of the gas-solid reaction rate constants and the absorption capacities, the optimum dry-type fixation furnace can be designed. Furthermore, by examining the chemical compositions of the obtained fluorides, it was confirmed that the calcium fluoride could be recycled as a fluorine resource.

Reference

1. S. Yasui, K. Ikeda and H. Shirai (2007), "Research on dry-processing technology for reconverted resources fluorine from waste chlorofluorocarbons," AIChE Annual Meeting 2007, Salt Lake City, Paper no.173d.
2. S. Yasui, S. Nakai, A. Ueno, T. Utsumi, T. Imai and T. Murakami (2008), "Low temperature dry processing technology for exhaust gases containing fluorine using calcium absorbents," CHISA 2008, Praha.
3. J. Szekely, J. W. Evans and H. Y. Sohn (1976), Gas-solid reaction, Academic press, NY.

(482c) Characterization of Cupric Chloride-Impregnated Sorbents for Elemental Mercury Capture From Coal Combustion Flue Gas

Xin Li and Joo-Youp Lee, School of Energy, Environmental, Biological, and Medical Engineering, University of Cincinnati, Cincinnati, OH

Carbon sorbent injection is one of the most promising mercury control technologies from coal-fired power plants. $CuCl_2$-impregnated activated carbon sorbent has demonstrated excellent mercury oxidation and adsorption performance in our previous studies. In addition, when $CuCl_2$ is impregnated onto noncarbonaceous substrates, the material has shown excellent elemental mercury oxidation, but not adsorption capability. Our previous study using XAFS shows that mercuric chloride ($HgCl_2$) is a major resultant oxidized mercury compound generated over the $CuCl_2$-impregnated activated carbon sorbent. This study focused on investigating the structure and electronic state of $CuCl_2$-impregnated sorbent supported on activated carbon and alumina.

$CuCl_2$-impregnated activated carbon and alumina sorbents were extensively characterized using TGA-MS, XRD, XPS, XAFS and SEM/EDS techniques. XAFS results suggested that $CuCl_2$ supported on AC be reduced to Cu+ species during impregnation, probably by the reduction sites on the support surface. Copper was still coordinated with Cl^- ions on carbon surface at relative high copper loading sorbent, whereas it might be mainly connected to oxygen functional group on carbon surface at low copper loading sorbent (i.e. 1% $CuCl_2$-AC), but Cu-C bonding was not observed on the sorbent. High Cl concentration on the sorbent enhanced the elemental mercury oxidation and organic polar solvent (isopropyl alcohol) could impregnate higher Cl loading on the carbon surface than water. The valence change of copper between Cu^{2+} and Cu^+ was probably responsible for the mercury oxidation reaction. Overall this presentation will address copper and mercury speciations over the carbon and alumina surfaces.

See more of this Session: Technologies for Reductions of Emissions From Stationary Combustion Sources
See more of this Group/Topical: Environmental Division

(482d) Organic Sulfur Removal From Refinery Fuel Gas

Vasilis Papavassiliou, Raymond Drnevich, Ramchandra Watwe and John Scalise, Praxair Technology Center, Tonawanda, NY

Organic Sulfur Removal fromRefinery Fuel Gas

Raymond F.Drnevich, Praxair & Vasilis Papavassiliou, Praxair, Tonawanda, NY; JohnScalise Burr Ridge IL; Ramchandra Watwe, Praxair, Houston TX

Refinery fuel gas has beenrecognized as one of the two major sources of sulfur emissions from refineries (the other being the Fluid Catalytic Cracking Unit or FCCU). Refinery fuel gas is amine treated to removeH_2S but refinery gas streams, especially those derived from cokerscontain other sulfur compounds such as mercaptans, thiophenes and sulfideswhich are not effectively treated by an amine system. Since refinery fuel gas usually representsthe majority of fuel that is used in refinery heaters and boilers any remainingsulfur in refinery gas will convert to SO_x during the combustionprocess and end up as emissions from the refinery stacks. Praxair has developed a new technology basedon a short contact time catalyst that can reliably convert organic sulfur to H_2Swhich can then removed in a conventional amine treater. The Praxair technology can be applied to anytype of refinery gas. The key to thisconcept is the development of a cost effective short contact time reactor(Refinery Gas Processor or RGP). RGPcontains a precious metal catalyst that can operate over a wide temperaturerange. Non-H_2S sulfurconversion occurs optimally around 700 F butthe reactor can operate as high as 1100 Fwith no need for feed dilution. Noexternal energy source is required. Theenergy demand is satisfied with either olefin hydrogenation or addition ofoxygen if olefin levels in the refinery gas are low. RGP can operate with high olefin levels andin that case the heat required to convert organic sulfur components to $H_2S$is easily provided by the olefin hydrogenation reactions. However, if the olefin concentration in thefeed is low, some oxygen can be added to the RGP reactor to provide make-upheat. The oxygen can be in the form ofpure oxygen or air if dilution of nitrogen is acceptable. The oxygen will react selectively with hydrogento produce water and heat, which will enable the catalytic conversion oforganic sulfur to H_2S. Noother oxygen side reactions occur.Praxair estimates that this technology is 25% less expensive thancaustic treating. The details thereactor design, experimental results and the process integration in a refineryfuel gas system will be discussed in detail.

See more of this Session: Technologies for Reductions of Emissions From Stationary Combustion Sources
See more of this Group/Topical: Environmental Division

NEW METHOD FOR THE PRODUCTION OF BTX STANDARD GAS SAMPLING WITH SPME AS AN ALTERNATIVE TO AIR MONITORING, ADVANTAGES AND LIMITATIONS

Bruna C. Queiroz; M. Veronica C. Oropeza;; Edmilson P. da Silva; Maria A. Mendes; Osvaldo Chiavone-Filho; Paulo F. Moreira Junior; Reinaldo Guidici
Federal University of Rio Grande do Norte, Natal, Brazil

INTRODUCTION

This paper aims the development of an experimental device to determine quantitatively the content of benzene, toluene and xylenes (BTX) in the atmosphere. BTX are extremely volatile solvents, and therefore play an important role in atmospheric chemistry, being precursors in the tropospheric ozone formation. In this work a BTX new standard gas was produced in nitrogen for stagnant systems. The BTX (benzene, toluene and xylene) are involved in the composition of various products, among them are: plastics, gasoline, paints, cosmetics and even some medicines, hence they are found as water, soil and air pollutants. These compounds as air pollutants play an important role in the formation of tropospheric ozone because they are classified as volatile organic compounds, and pass easily from solid or liquid state for the gas phase [1]. Due to the toxicological properties of volatile organic compounds (VOCs), and particularly benzene among the BTEX compounds [2] (benzene, toluene and xylenes), various field or laboratory techniques have been developed for continuous or discrete monitoring of these pollutants in air: automatic gas chromatography, passive or active sampling on sorption tubes or in canisters followed by gas chromatography in the laboratory after desorption. However, continuous on-site monitoring is very expensive and difficult to calibrate [2]. Analytical methods for volatile organic air pollutants suffer three fundamental problems [2]: (a) the need to analyse very low concentrations; (b) the need for analytical standards for method validation, which should contain the analytes of interest at the same concentration levels that are present in real-world samples; (c) care must be taken not to add to environmental pollution during sample preparation. Solid-phase microextraction (SPME) has been widely accepted as an innovative technique for the determination of organic pollutants in environmental matrices. Analytes are absorbed into or adsorbed onto the SPME fiber in the aqueous solution or in the headspace over the solution, and then directly desorbed from the SPME fiber within the GC injector at high temperature . SPME requires no solvents and provides a linear response over a wide range of analyte concentrations. It consists of exposing the sample to a small fiber with a silica core coated with a stationary phase in which the target compounds are trapped. The amount of analyte extracted by the fiber depends on the type of fiber and is proportional to the initial analyte concentration in the matrix. Sampling and pre concentration are combined in one step. Consequently, SPME has the advantage of reducing interference from complex constituents in the samples. Moreover, high resolution and low cost can be obtained through SPME in comparison with conventional methods like headspace or purge and trap. To date, the poly(dimethylsiloxane) (PDMS) coating is one of the most widely used coatings for extracting volatile analytes from environmental samples via absorption [3]. The relative ease of the extraction process control and the fact that the absorption process is not affected by competition between the absorbed analytes prompted the development of air sampling methods based on the PDMS coating [2,4]. In addition, the theory behind the equilibrium and nonequilibrium extraction process for absorptive (liquid-phase) PDMS coatings is well understood and described in the literature [2,3,7]. The ability of the absorbent fiber retain certain amounts of analytes depends mainly on the thickness of the liquid layer and the size of the analyte. The polarity of the fiber can increase the attraction of a particular analyte, but is the thickness of liquid film on the fiber that retains the analyte. Fiber adsorbents extract the analytes by physical interactions. The collection is by retention of the analytes within the internal pores. The micro-and mesopores are ideal for intermediate and small analytes and analytes usually will be withheld until a thermal or solvent desorption.In these types of fibers there is a limited number of adsorption sites, so a competition between the analytes take place, unlike the absorbent fiber [5,6]. An attempt of establish a theoretical description for porous polymer coatings, Gorecki at al based on Langmuir isotherm model. However, this model cannot be applied to Carboxen/PDMS fiber, since the porous in Carboxen/PDMS are so small that capillary condensation as well as simple adsorption can occur [16,17].

EXPERIMENTAL

Apparatus

The device used to perform all the experimental trials is shown in Figure 1. This unit consists of: 1- nitrogen cylinder, 2- mass flow controller, 3- pressure meter, 4-Vaccum pumper, 5- water pump, 6- heater, 7- SPME device, 8- mariotte bottle, 9- coil pipe, 10- thermostatic bath

Material and methods

Reagents: All the reagents (Benzene, toluene, p-xylene and methanol) were P.A. grade and purchased from Aldrich and Merck.
Stander Solution: The standard liquid was produced by make a solution contain 54.17% methanol and 45.83% of benzene, toluene and p-xylene in a ration of 1:1:1 for mass, whereas the density of the three compounds is almost the same (0,876 g/cm^3, 0,867 g/cm^3 and 0,864 for Benzene, Toluene and Xylene respectively) We decided to use the value of the average density of the liquid mixture of three compounds. All calculations were performed using the density value equal to 0,87 g/cm^3, The standard solution was store at freezer conditions.
Gas Production: The 20 liters glass mariotte was immersed into the bath and closed, until the bath temperature reaches 22,1 °C, then, vacuum was applied, and then nitrogen was filled into the container until it reaches atmospheric pressure. After the temperature of the system stabilize at 22.1 °C, was carried out a background, to ensure there were no contaminants in the system. The liquid standard was inserted the closed marriote at 19 °C by using a standard micro syringe. Since tests showed that temperatures higher then 22 °C produce losses through evaporation and instability of the system also occurs. Considering that the three compounds are very volatile, the evaporation process starts quickly, following the diffusion of the BTX molecules in the diluents gas, forming a gas solution with known concentration. The samples were collected manually injecting the SPME through from silicone septum, the fiber stayed in contact with gas solution during 10 minutes. After 10 minutes the fiber was withdrawn from the marriot and immediately introduced into the gas chromatograph injector. The compounds were then thermally desorbed and analyzed by mass spectrometry. Please use this example paper as your primary reference in completing your paper. Your assistance in helping follow the styles

and guidelines outlined for the following parts of your paper will help your association publish a more "professional looking" publication in the most cost-effective and timely manner possible. A manual sampling holder for SPME equipped with a 75-µm Carboxen-PDMS fiber (Supelco, Bellefonte, PA, USA) was used in all the experiment.

Instrumentation

The gas chromatograph mass spectrometer, GCMS-QP2010 plus (SHIMADZU) system, with a split-splitless injector, equipped with a Agilent (DB5-MS) 30M x 0,25 mm (Narrowbore) x 0,25 µm column was used for chemical analyses. Carrier gas was He and column flow was maintained at 0.88ml/mim. Initially the oven temperature was held at 40°C for 4 min, and then increased to 160°C at 20°C/min, and held at 250°C for 2 min. The temperature of injector was 300°C the temperature of the interface 310°C. The mass spectrometer instrument was equipped with a electron ionization ion source and a quadrupole mass analyzer. The analyses were performed in full scam mode with mass range from m/z 40 to m/z 500. The compounds were ionized into the electron ionization ion source and separated by ion trap mass analyzer Vacuum pump with motor single-phase WEQ, to remove air from the system. A thermostatic bath of Tecnal TE 184 to control de system temperature. Gauge Pressure Transmitter-Taylor electronic 4-20ma output. A 20 liters glass mariotte acquire from Vibrolab.

RESULTS AND DISCUSSION

Gas Production

Two basic methods can be to calibrate SPME for air sampling: static gases and dynamic gases [2,3,4]. In this work we are proposing a static method for the standard gas generation, as a new alternative way of being independent of the suppliers of gas pattern; we can produce the gas based on two theories: the theory of the equation of state of ideal gases, where pressure, temperature and volume were kept constant for all experiments, approximately 1 atm pressure, temperature of 22.1 ° C and the volume of 20 liters (volume of container where the gas was produced). And also based on the theory of diffusion in a stagnant gas. The profile of evaporation observed follows a parabolic function, from minimum values in the first 5 minutes, reaching maximum values between 20 minutes and 100 minutes and then concentration of the compounds decreases after 120 min. These results are shown on the Figure 2. This could be explained by the fact that the first few minutes the liquid mixture had not completely evaporated, after complete evaporation, the maximum concentration of the compounds in the fiber is reached, and finally the system starts to lose gas to the atmosphere because the system is not fully closed, this happens after two hours. The maximum signal intensity observed in the GCMS using SPME carboxen/PDMS are proportional to the concentration of gas, since the minimum values of concentration are considered as regions of instability.

Exposure Time

To determine the response profiles for these compounds, a standard gas mixture of BTX were prepared according to section 3.1 of this paper, on the concentration of 0,3 □l/ml. It was sampled with the Carboxen/PDMS fiber for different time intervals (1, 5, 10, 15 and 25 min). Plots of the GC-MS TIC intensity versus exposure time (Figure 1) show that, the sorption process reached an equilibrium state at the 15 min of exposure. Even so, it was decided that sampling with Carboxen/PDMS would be performed under non-equilibrium conditions sampling at 10 min fiber exposition. Because in equilibrium there is the competition of the pores of the fiber by the compounds analyzed. This would make it difficult to quantify the BTX. Popp and Paschke proved that the Carboxen/PDMS had an even better extraction capability than the PDMS fiber, though repeatability was slightly worse equilibrium time was prolonged [2], see fig. 1

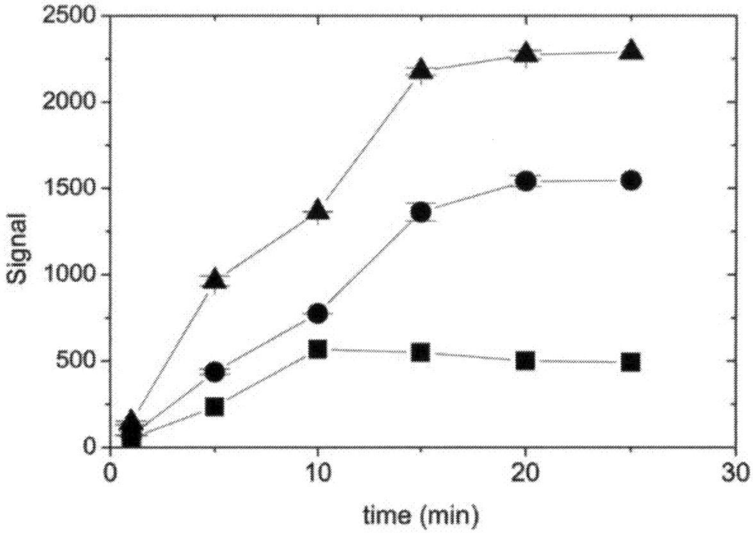

Figure 1- Profile of exposure time by plotting GC/MS total ion current intensity versus exposure times for each BTX, obtained using a Carboxen/PDMS fibre

Sampling Time

Sampling was performed in the time interval containing the points called for significant points. Significant points are those located after the time of initial diffusion, and before starting to experience loss of analyte. These points are located outside the region of instability and error. There is a time for the progressive diffusion of the injected liquid A series of experiments showed that the time interval to reach the maximum, stable and homogeneous diffusion within the Mariot. Significant points are located within the time interval between 20 minutes and 100 minutes after injection of the standard liquid. This interval corresponds to a ΔT of 80 minutes at which sampling can be performed. The concentration of the gas standard was calculated from the standard volume of standard liquid injected. All experiments were performed in triplicate. The method validation was performed by constructing a graph, plotting the concentration of the analyte versus signal intensities, see Fig. 2.

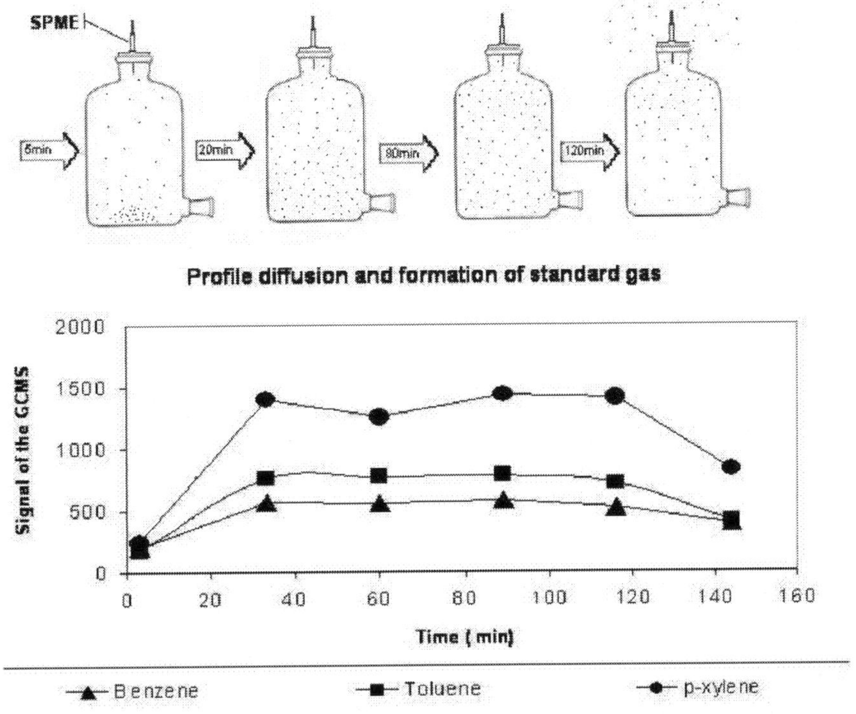

Figure 2- Profile diffusion triplicate: Benzene, Toluene and p-xylene. Concentration versus time. Analyses were performed at GCMS instrument

Calibration System

The calibration curve of BTX using of Carboxen/PDMS SPME fibers was obtained under non-equilibrium conditions and 10 min sampling. Consequently, severe control of the sampling time was necessary. The curve was obtained by plotting the values of concentration versus GC MS signal intensity, for each compound. Three replications were carried out for each of the five calibration levels (6778.01, 13538.26, 27120.9, 33904.832, 40685,8 μg/m³) and for the blank. These results are shown on the Figure 4. Least square linear regression analysis was used to estimate the slopes and intercepts of the calibration lines. The linear correlation coefficients were considered good for Xilene (R^2= 0,9854) and Toluene (R^2 = 0,9582). For benzene the value was a little lower, (R^2 = 0,8867). But searching the literature we found data showing that this value can be considered satisfactory for this compound.

Competition of the Compounds by the Porous of the Fiber

The fiber Carboxen/PDMS has a high selectivity, its porosity is 0.78 (mL g^{-1}) less than other fibers, such as divinylbenzene, which is 1.54 (mLg^{-1}), this means fewer sites available for adsorption, this fact explains its high selectivity [5,6,7]. Adsorption occurs on the surface of the pores of the fiber coating, and the driving force that determines the adsorption of compounds by the fiber are factors such as molecular weight, amount of analytical, vapor pressure, boiling point and polarity [7]. The experimental results led to two regions: a linear and a nonlinear range. In the linear range was observed that the increase of the signal was proportional to increasing concentrations, and the most influential factor in the responses of the signal was associated with small temperature changes (thermostatting system). However with the increasing concentration of the signal response of benzene decreased significantly, dropping to less than half. Adding to the fact that benzene has less affinity for the fiber, this leads us to propose that there is a certain concentration where fiber pores not have enough to absorb all compounds equally, then the compound with lower affinity, is replaced by the compound with higher affinity. This was observed in the concentration of gas than 122057.45 µg / m³ for each BTX. To corroborate this result, an experiment was performed where we made measurements of benzene adsorbed by the fiber: the first in a solution containing only benzene and, then in a solution containing the three compounds: benzene, toluene and p-xylene, benzene in both solutions was at the same concentration. Comparing the results it was possible to observe that there was a significant difference in the intensities of the signals, the signal of benzene when it was in solution with toluene and p-xylene was approximately 30% smaller than the intensity of the signal when there was only benzene in the solution. See Fig. 3. Thus conclude that there is a competition of the compounds by the pores of the fiber, and at high concentrations, when the fiber reaches a saturation state, the intensities of the measured signals do not represent the true concentrations of the compounds in solution.

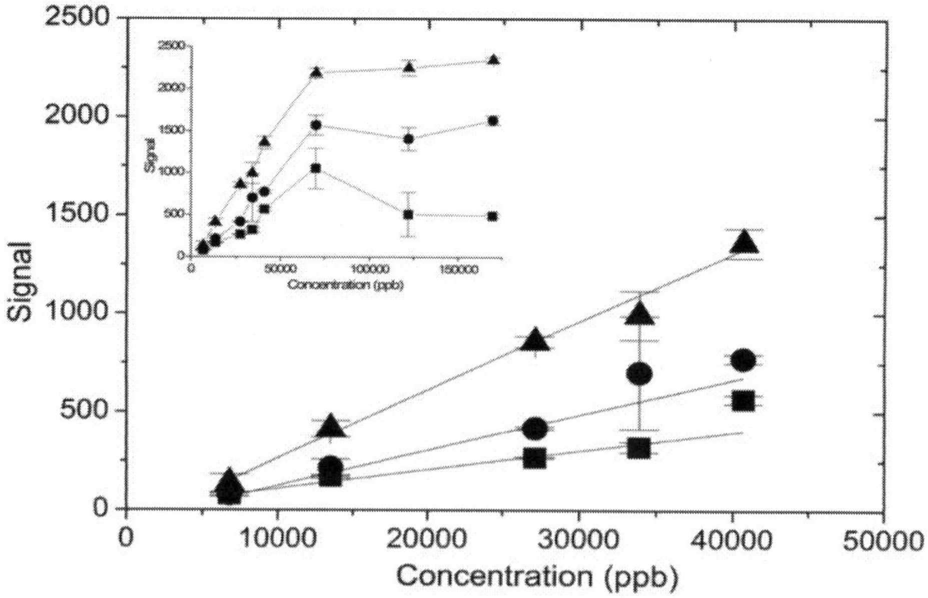

Figure 3- Linear and nonlinear profiles of a adsorption BTX in Carboxen/PDMS fiber

ACKNOWLEDGEMENTS

The financial support from FAPESP (Fundação de Amparo à Pesquisa do Estado de São Paulo) and CNPq (Conselho Nacional de Desenvolvimento Científico e Tecnológico) is gratefully appreciated.

REFERENCES

[1] Seinfeld, John H. et al. (2006), "Atmospheric chemistry and physics", Second edition, John Wiley & Sons, INC.

[2] Tumbiolo, S.; Gal, J-F.; Maria, P-C.; Zerbinati, O. Determination of benzene, toluene, ethylbenzene and xylenes in air by solid phase micro-extraction/gas chromatography/mass spectrometry, *Anal. Bioanal. Chem.*, 380: 824 (2004).

[3] Chen, Y .; Koziel, J.; Pawliszyn, J. Calibration for On-site analysis of hydrocarbons in aqueous and gaseous sample using solid-phase microextraction. *Anaytical.Chemistry.* 2003, 74,6485-6493.

[4] Arthur, C.L.; Pawliszyn, J. (1990), Solid phase microextraction with thermal desorption using fused silica optical fibers , *Anal Chem*, 62, pp. 2145.

[5] Chen, J.; Pawliszyn J. Solid Phase Microextraction Coupled to Highl Performance Liquid Chromatography. *Analytical chemistry.* **1995,** 76, 2530

[6] Gorecki, T.; (1999), *Applications of Solid Phase Microextraction*; First ed; RSC: Waterloo,

[7] Cho, Hyun-Jeong; Beak K.; Lee, Hyun-HO; Lee, Sang-Hyun; Yang, JI-Won Competitive extraction of multi-component contaminants in water by carboxen-polydimethylsiloxane fiber during solid-phase microextraction. *Journal of chromatography A.* 988, 177-184, 2003

(484a) Modeling of Long-Time Flux Decay In Cross-Flow Filtration

Philip P. Schonewill, Richard C. Daniel, Rick W. Shimskey, Justin M. Billing and Reid A. Peterson, Pacific Northwest National Laboratory, Richland, WA

Recent work at Pacific Northwest National Laboratory has been concerned with understanding how cross-flow filter flux decays with time. Understanding the cause(s) of flux decay is important for assessing long-term behavior of this separation technology, especially when used in challenging processing environments e.g. nuclear waste treatment. Fouling in the filter is a complex issue that is affected by many parameters: the chemistry of the solution being filtered, particle size and morphology, operating conditions, and cleaning methods. Based on filtration work performed at PNNL over the past several years, it is clear that a portion of the fouling can be "reversed" by back-pulsing the filter and another portion is irreversible (usually only rectified via cleaning protocols). However, the sheer number of parameters requiring consideration complicates effective modeling of both the long-time behavior of the cross-flow filter and its response to back-pulses.

To combat this, simple models were developed from first principles based on different suspected fouling mechanisms. The models were tested against long-term data (100-hour tests at a constant solids concentration) collected with PNNL's bench-scale filtration system. The flux was found to be well described by a cake mechanism model at intermediate times (starting at approximately 3 hours elapsed time); at longer times (approximately greater than 20 hours of filtration time) the flux behavior transitions to an intermediate mechanism behavior. The decay constant for these models can be predicted using a single dimensionless parameter, the ratio of the transmembrane pressure to the inertial pressure. Additional testing suggested that the filter could be conditioned by back-pulsing. These observations were used to build a comprehensive model to assist in determining a back-pulse strategy for cross-flow filters that will experience long-term decays in the filter flux.

See more of this Session: Theories and Applications of Cross Flow Filtration
See more of this Group/Topical: Separations Division

(484b) Control of Packed-Particle Flow Resistances During Cross-Flow Filtration In Microchannels

Nopphon Weeranoppanant, Levy Amar, **Michael Hill** and Edward F. Leonard, Chemical Engineering, Columbia University, New York, NY

Cross-flow filtration of fine suspensions through microsieves is a logical, sometimes necessary step in micro-processing. The observed limitation on steady-state filtrate flux accompanying this operation is generally attributed to particle accumulation upon the sieve. We consider two mechanisms: (1) blockage of filter pores by particles in contact with the filter surface, and (2) formation of a packed bed of particles that intrudes upon and is regulated by shear flow within the microchannel. This study examines the conditions under which a packed bed forms on a sieve, and provides a simple model for predicting its thickness when formed by hard non-interacting microspheres. Extensive experimental data are reported that corroborate the model. The good fit of these data lay a firm basis for the semi-quantitative analysis of the behavior of more complex suspensions.

See more of this Session: Theories and Applications of Cross Flow Filtration
See more of this Group/Topical: Separations Division

(484c) Cross-Flow Filtration of Plasma From Blood From a Microchannel

Xiangnan Ye[1], Levy Amar[1], Michael Hill[1], James Jones[2], Jacob Baggerman[3], Cees van Rijn[3] and **Edward F. Leonard**[1], (1)Chemical Engineering, Columbia University, New York, NY, (2)Department of Medicine -- St. Lukes'/Roosevelt Hospital, Columbia University, New York, NY, (3)Aquamarijn Micro Filtration, Zutphen, Netherlands

Steady cross-flow filtration of plasma through microsieves from blood in a microchannel is an important part of several biomedical applications under development, including a new artificial kidney. Erythrocytes, which comprise the largest volume fraction of all particulates in blood, are believed to create the controlling resistance at the sieve surface. Maximum steady-state filtration flux has been observed to be a function of wall shear rate, as predicted by any conventional cross-flow filtration theory, but to show weak dependence on erythrocyte concentration, contrary to theory based on convective diffusion. Maximum filtration flux has also been observed to be a function of sieve pore size and shape (in spite of their very low flow resistance) as well as sieve surface coating.

Recent experiments have correlated macroscopic measurements (filtration rates, transmembrane pressures) with direct observation of erythrocyte behavior at the filtering surface. At low filtration rates (low transmembrane pressures), erythrocytes roll across the filter surface, but at higher filtration rates (higher transmembrane pressures), erythrocytes are observed sticking to the sieve surface. Post-filtration SEM's, even those obtained at very low transmembrane pressures, reveal significant capture and deformation of erythrocytes in the filter pores. Careful balances among sieve design (which can be precisely defined by photolithography), surface treatments of sieves, and flow conditions, allow filtration fluxes at rates exceeding 0.1 cm^3/cm^2-min.

See more of this Session: Theories and Applications of Cross Flow Filtration
See more of this Group/Topical: Separations Division

An electrokinetic model for combined fouling during cross flow filtration of electrolyte solution and charged Colloids

Md Abdullaha Al Mamun, Reeshav Chatterjee, Subir Bhattacharjee,
Department of Mechanical Engineering, University of Alberta, Edmonton, AB, Canada T6G 2G8
Email: subir.b@ualberta.ca
Sirshendu De
Department of Chemical Engineering, Indian Institute of Technology, Kharagpur 721 302, India

Abstract

Colloidal fouling, formation of a cake due to deposition of colloids on to a membrane, is commonly encountered in pressure driven filtration processes, such as nanofiltration and reverse osmosis. The influence of these deposited charged colloids on the ion concentration polarization, ion rejection and the overall permeate flux is the focus of the present work. A common situation encountered during membrane filtration is the combined influence of charged colloids and the associated electrolyte on the flux decline and rejection. In this context, coupled model of concentration polarization (CP), cake layer formation by charged colloidal particles and ion transport through the interstitial spaces of the cake layer has been developed to predict the permeate flux decline and observed salt rejection during cross flow filtration of an electrolyte and colloidal mixture. The suspension structure of the cake layer is considered as a swarm of colloidal particles, represented using the Kuwabara cell model. For this system, the electroosmotic flow is calculated analytically using the Levine-Neale model. The analysis is then combined with the standard film theory of cross flow membrane filtration to predict the flux decline, observed rejection, cake layer thickness and resistance, and osmotic pressure difference. The model illustrates the combined influence of electrical and physical properties of colloids and electrolytes, mass transfer coefficient or crossflow condition and membrane resistance on the separation process. This approach provides a comprehensive insight into the mechanism of cake enhanced concentration polarization for charged colloids.

Keyword: Cake Enhanced Concentration Polarization, Kuwabara Cell Model, Levine-Neale Electrophoretic Model, Film Theory

1 Introduction

Membrane separation processes are increasingly being used in industries as a treatment process for separation of colloids, macromolecules, solutes, and ions. This is often achieved through processes like nanofiltration, whereby a single unit operation is used to remove multiple molecular weight and size species. The major operating costs associated with membranes include creating the pressure driving force for permeation, and mitigation of fouling. Both these operating costs are directly related to the overall energy consumption of membrane filtration. Several methods for fouling prevention and minimization has been suggested over the last few decades. But, most studies on fouling have typically ignored the concept of microscale manipulation of the fouling deposit itself to modify its permeability, or its ability to selectively adsorb ions or small organic contaminants. Therefore study of the fouling mechanism is important to develop a better understanding of the relation between the permeate flux, salt rejection, cake formation and to manipulate the fouled layer to enhance the preformece of the filtration process.

Hoek and Elimelech [1] conducted colloid fouling experiments during cross flow RO and NF, and reported cake enhanced osmotic pressure (CEOP) as the mechanism of flux decline. They proposed a model of CEOP, where hindered back-diffusion of ions within colloidal deposit layers was mentioned as the main mechanism of concentration polarization, resulting in an increase of osmotic pressure and decrease of pemeate flux.

Kim and co-workers [2] presented a detailed model of three component combined fouling to represent flux decline behavior of the system during dead end filtration. The model however introduces several parameters that need to be determined independently, such as, cake layer tortuosity, diffusive tortuosity, and hindered diffusivity of electrolyte through the cake layer. Additionally the model does not incorporate the effect of zeta potential of colloids and electrolyte solution conductivity in the flux decline behavior. However, the model presents a comprehensive mechanism for multi-component fouling from classical filtration concept.

Sim and co-workers [3] investigated the coupled effect of cake enhanced osmotic pressure and colloidal fouling in RO, using cross flow sampler-modified fouling index ultrafiltration (CFS-MFI$_{UF}$). They used CFS-MFI$_{UF}$ measurement to determine the deposition rate of colloidal mass and cake porosity, which then incorporates the CEOP effect in RO fouling. They also considered the hindered diffusivity of ions within the cake layer to model CEOP. Though, the technique provides good estimation of RO fouling profile, the model depends on the CFS-MFI$_{UF}$ or dead-end RO experimental data.

De and Bhattacharjee [4] proposed an electrokinetic model for flux decline during combined filtration of charged colloids and electrolyte. They assumed the cake layer as a bundle of charged cylindrical nano-pores similar to the membrane. Their model considered the electroviscous effect created by the charged nano-pores, which results in additional resistance of the cake layer. Though their model represents an improvement, the actual flow model should consider the flow of electrolytes around the charged colloids in a cake layer to be more realistic.

In this paper, a coupled model of concentration polarization, cake layer formation by charged colloidal particles and ion transport through the interstitial spaces of cake layer has been developed to predict the permeate flux decline and observed rejection due to fouling during cross flow nanofiltration of an electrolyte and colloidal suspension. The model cosiders the cake layer as a swarm of charged spherical particles instead of cylindrical nanopores, which is more realistic, and it can predict the flux decline and observed rejection without considering any correction parameter except the cake tortuosity. The model illustrates the combined influence of colloid mass transfer coefficient, electrical and physical properties of colloids and electrolytes, crossflow geometry and membrane resistance on the cake enhanced concentration polarization of the electrolyte.

2 Mathematical Model

During combined fouling of colloidal particles of radius a, surface potential ψ_c and electrolyte feed concentration of $C_{i,f}$ in a crossflow membrane filtration operation, the colloidal particles form a cake layer on the membrane and the electrolyte solution forms a concentration polarization layer on the cake surface as shown in Fig. 1. The dimensionless permeate flux through the cake layer and membrane can be represented by the following modified Darcy's equation [4]:

$$v_w^* = \frac{v_w}{v_w^0} = \frac{1-\Delta\pi_m^*}{1+R_g^*} \tag{1}$$

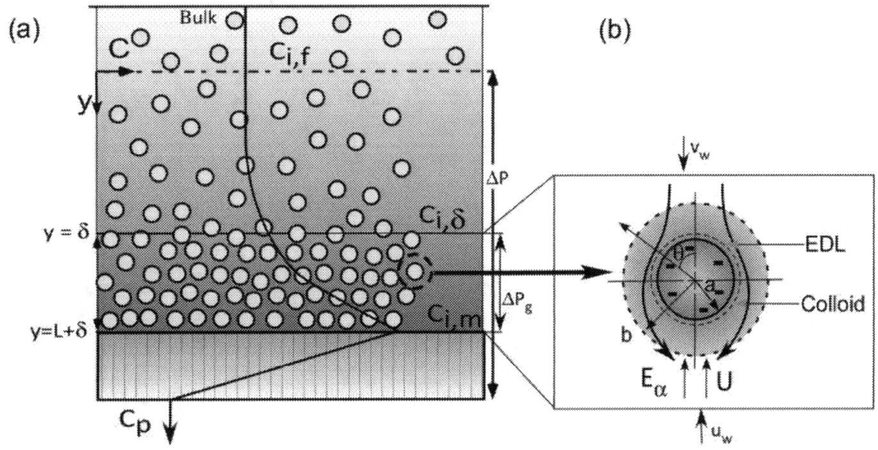

Figure 1: (a) Cake enhanced concentration polarization (b) Electrophoretic flow model around a charged colloid according to the Levine-Neale and Kuwabara cell model

Here, $v_w^o = \Delta P/\mu R_m$ is the pure water flux, $\Delta\pi_m^*$ is the dimensionless osmotic pressure across the membrane and R_g^* is the dimensionless resistance. The cake layer resistance R_g^* has been formulated based on the Kuwabara cell model and Levine-Neale electrophoretic flow model [5] of charged particles in a concentrated colloidal suspension as shown in Fig. 1b. The model considers the combined influence of hydrodynamic and electrokinetic resistance to the transport through the cake layer. According to the Levine-Neale model, electroosmotic flow (U) through the interstices of the cake layer due to induced streaming potential is given by [5]:

$$U = E_\infty \left(\frac{\varepsilon\psi_c}{\mu}\right) g_2(\kappa a, \varphi_p) \tag{2}$$

The expression of induced electric field (E_∞) in Eq. 2 at zero current condition, in accordance with Onsager principle of reciprocity for irreversible phenomena and for thin Debye layer, is [5]:

$$E_\infty = \frac{\Delta P_g}{L}\left(\frac{\varepsilon\psi_c}{\mu\sigma^\infty}\right) \tag{3}$$

Thus, the expression of electroosmotic flux (u_w) from the continuity assumption using Eqs. 2 and 3 is given as follows:

$$u_w = U(1-\varphi_p) = \frac{\Delta P_g}{L\sigma^\infty}\left(\frac{\varepsilon\psi_c}{\mu}\right)^2 (1-\varphi_p)g_2(\kappa a, \varphi_p) \tag{4}$$

Therefore, the total flux across the cake layer at no current condition, due to effective pressure across the cake ΔP_g and reverse electroosmotic flow is given as follows:

$$\begin{aligned}
v_w &= \frac{\Delta P_g}{\mu r_g L} - u_w \\
\text{or, } v_w &= \frac{\Delta P_g}{\mu r_g L}\left[1 - 9\varphi_p(1-\varphi_p)A_K \frac{\beta g_2(\kappa a, \varphi_p)}{2(\kappa a)^2}\right] \\
\text{or, } v_w^* &= \frac{\Delta P_g^*}{L^*/Mg} = \frac{\Delta P_g^*}{R_g^*}
\end{aligned} \tag{5a}$$

$$r_g = \frac{9\varphi_p A_K}{2a^2}; \quad M = \frac{2a^2 R_m}{9\varphi_p A_K d_e} \quad \text{and} \quad g = \left[1 - 9\varphi_p(1-\varphi_p)A_K \frac{\beta g_2(\kappa a, \varphi_p)}{2(\kappa a)^2}\right] \tag{5b}$$

Here, r_g is the hydrodynamic resistance of the cake layer, $A_K = 1/(1-9\varphi_p^{1/3}/5+\varphi_p-\varphi_p^2/5)$ accounts for the presence of neighbouring particles, $\beta = (\varepsilon\psi_c\kappa)^2/(\mu\sigma^\infty)$ and $L^* = L/d_e$. The function g accounts for the electroosmotic effect in a swarm of colloidal particles, depends on dimensionless Debye length κa, surface potential of colloidal particles ψ_c, solution conductivity σ^∞ and volume fraction of the cake layer, φ_p.

The dimensionless osmotic pressure $\Delta\pi_m^*$, for simple electrolyte (1:1) can be written from the vant Hoff equation:

$$\Delta\pi_m^* = \frac{2R_{ug}T}{\Delta P}(C_{i,m} - C_{i,p}) = \alpha\bar{C}_{i,m}R_r \tag{6}$$

where, $\alpha = 2R_{ug}TC_{i,f}/\Delta P$ and $\bar{C}_{i,m}$ is dimensionless salt concentration at the membrane surface. The expression of $\bar{C}_{i,m}$, derived by integrating the mass balance of electrolytes within the concentration polarization and cake layer with appropriate boundary conditions is:

$$\bar{C}_{i,m} = \frac{C_{i,m}}{C_{i,f}} = \frac{\exp\left[v_w^*(R_i + L^*R)\right]}{R_r + (1-R_r)\exp\left[v_w^*(R_i + RL^*)\right]} \tag{7}$$

Here, $R_i = v_w^o/k_i$, $R = v_w^o d_e/D_i^*$. The observed rejection of the membrane can be written as:

$$R_o = 1 - \frac{C_{i,p}}{C_{i,f}} = 1 - \bar{C}_{i,m}(1-R_r) \tag{8}$$

The dimensionless governing equation for colloidal cake layer growth, obtained by integrating the mass balance equation of colloids in the cake layer with appropriate boundary conditions, is given by [6]:

$$\frac{dL^*}{d\tau} = v_w^* \frac{C_{c,m}^* - \exp(v_w^* R_c)}{1 - \exp(v_w^* R_c)} \tag{9}$$

Here, $\tau = v_w^o C_{c,f} t/(d_e \rho_c)$, $R_c = v_w^o/k_c$, $C_{c,m}^* = C_{c,m}/C_{c,f}$. The dimensioless pressure drop with respect to operating pressure can be written as:

$$\Delta P_m{}^* + \Delta\pi_m^* + v_w^* R_g^* = 1 \quad (10)$$

Here, $\Delta P_m{}^*$ is dimensionless pressure drop across the membrane.

3 Numerical Methodology

Equations 1 and 9 are solved simultaneously in conjunction with the Eqs. 6 and 7, using a differential algebric equation solver DDASSL. For a 50 nm particle with surface potential -45 mV and cake volume fraction 0.64 for random-closed-packing, in 0.01 M NaCl solution, the value of g is 0.67. The value of M depends on particle size, cake volume fraction and hydrodynamic radius of crossflow channel and is less than unity for RO or NF. The volume fraction can be as high as 0.72 for hexagonal packing. The viscosity and relative permitvity of water is 0.89 mPa.s and 80.0, respectively. The electrolyte flow turbulence parameter R_i, is considered to be in the order of 1.0 becuase pure water flux and mass transfer coefficient of electrolyte both are of the order 10^{-5}. The corresponding value of R_c is considered to be of the order 10.0, because the diffusivity of colloidal particle is at least one order magnitude less than the electrolytes. For the silica colloidal particles, the bulk concentration is 0.2 kg/m^3, whereas the cake concentration is about 173 kg/m^3 [7]. Typical equivalent diameter of the cross flow channel is 10^{-3} m. For colloid feed concentration 0.2 kg/m^3, with particle density 2317 kg/m^3, τ= 0.01 represents 6.5 hr of operation in real time. The range of dimensionless parameters are listed in table 1.

Table 1: Range of dimensionless parameters used in simulations

Parameter	Range
κa	1 – 100
g_2	0 – 0.9
β	1.0 – 5.0
g	0.05 – 0.95
M	0.005 – 0.5
R_i	0.05 – 1.5
R_c	10.0 – 60.0
R	100 – 1000
α	0.09 – 0.9
$C_{c,m}^*$	800 – 900
R_r	0.8 – 0.9

4 Results and Discussion

Each of the dimensionless parameters listed in Table 1 affect the flux decline and observed rejection during cross flow filtration process. For the present parametric study, the effect of four parameters M,

g, R and R_c are considered and explained. The physical properties of the colloidal particle, cake layer and membrane are incorporated in the membrane and colloidal system interaction parameter M. The value of M is proportional to the square of particle size, i.e., for small particles, the value of M is small resulting in higher resistance since specific resistance is inversely proportional to M. As a result, the permeate flux decline is higher as shown in Fig. 4a. The observed rejection also increases when M is small, this is due to lower convection of salt compared to back diffusion across the cake layer. Therefore, concentration at the membrane surface is less significant for M=0.05 resulting in lower osmotic pressure as shown in Fig. 4a. Therefore, permeate flux and observed rejection can be changed by changing the size of the cake forming particles.

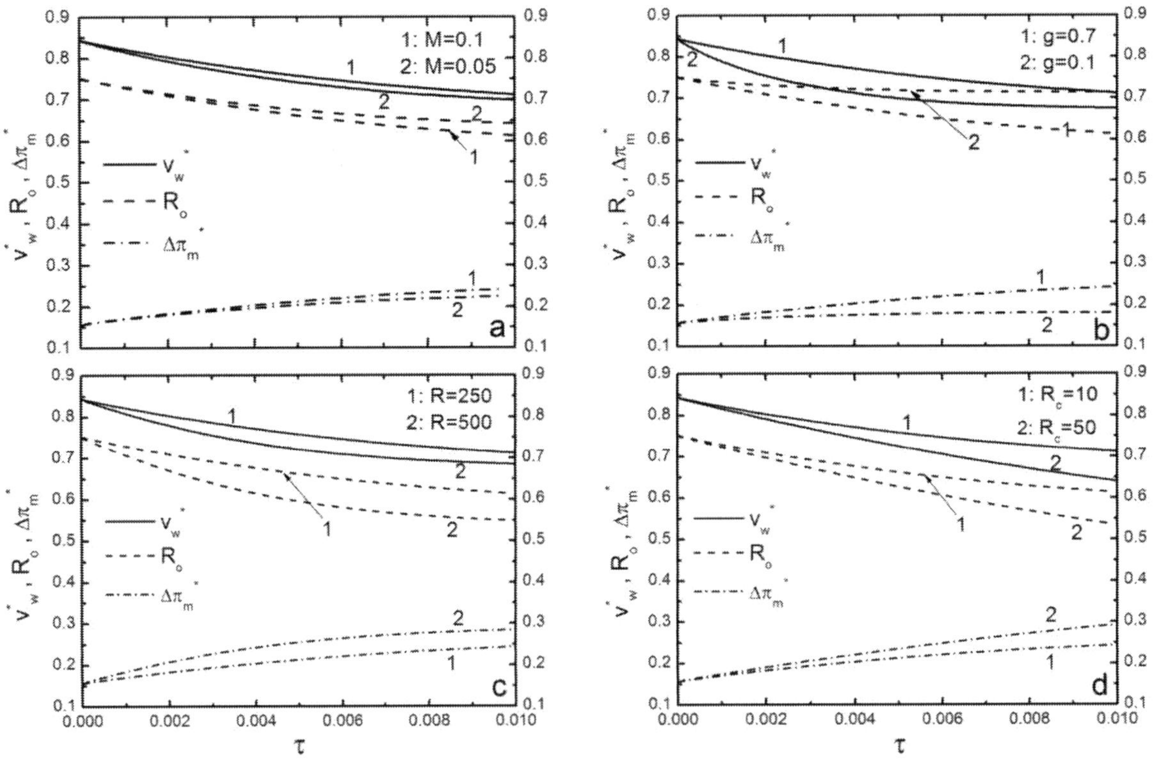

Figure 2: Variation of normalized permeate flux (v_w^*), observed rejection (R_o) and osmotic pressure $\Delta\pi_m^*$. (a) g=0.7, R=250.0, R_i=1.0, R_c=10.0, R_r=0.875, α=0.09 and $C_{c,m}^*$=850.0; (b) M=0.1, R=250.0, R_i=1.0, R_c=10.0, R_r=0.875, α=0.09 and $C_{c,m}^*$=850.0; (c) M=0.1, g=0.7, R_i=1.0, R_c=10.0, R_r=0.875, α=0.09 and $C_{c,m}^*$=850.0; (d) M=0.1, g=0.7, R=250.0, R_i=1.0, R_r=0.875, α=0.09 and $C_{c,m}^*$=850.0

The effect of electroosmotic parameter g, induced due to the presence of charged colloids and electrolyte solution, is shown in Fig. 4b. In absence of surface potential, the value of g goes to unity, implying absence of any induced electroosmotic flow. In presence of charged particles and salt solution the value of g is always less than 1. The parameter g is a strong function of κa and it decreases with increase of κa or with increase of electrolyte feed concentration. The parameter g is also a function of surface potential ψ_c and decreases with increase of ψ_c. Therefore, increase of surface potential of colloidal particles or increase of feed electrolyte concentration causes decrease of g. Since cake resistance is inversely proportional to g, increase of surface potential and feed concentration increases the electroviscous resistance of cake layer. As a result, permeate flux decline is more severe due to higher reverse electroosmotic flow induced from charged colloids. Due to the reverse electroosmotic flow of electrolyte, electrolyte concentration is less at the membrane surface, increasing the observed rejection and decreasing the transmembrane osmotic pressure as shown in Fig. 4b. Therefore, smaller value of g or higher value of κ and ψ_c, decreases the permeate flux and increases the salt rejection siginificantly. Thus, formation of cake layer with nano colloidal particles of high surface potential in concentrated salt solution can enhance the observed rejection of nanofiltration membranes or can enhance concentration polarization on cake surface significantly, which means the cake layer can be modeled as a secondary low salt rejection membrane of nano size pores with high surface potential.

The effect of parameter R on permeate flux and observed rejection is shown in Fig. 4c. The parameter R is defined as the ratio of pure water flux through the membrane to the diffusive flux through the cake layer. Higher value of R indicates that the convective flux is dominant compared to back diffusion through the cake layer which results in enhanced accumulation of electrolyte within the cake layer and increase in transmembrane osmotic pressure. As a result, the permeate flux decline is comparatively higher and observed rejection is lower for large value of R, according to Fig. 4c. Therefore, higher value of R enhances salt concentration at membrane surface and declines the permeate flux and observed rejection.

The parameter R_c is defined as the ratio of pure water flux to colloids mass transfer coefficient. The mass transfer coefficient solely depends on the geometry of cross flow channel and velocity. Since the model is for cross flow filtration lower value of R_c indicates higher mass transfer coefficient or, higher degree of turbulence and shear at the cake surface. Due to higher shear rate at cake surface, the cake layer can not grow much which leads to lower cake layer thickness and resistance, resulting in higher permeate flux as shown in Fig. 4d. Therefore, convective flux of electrolyte is higher for small value of R_c. Though the convective flux is higher, observed rejection is higher and transmembrane osmotic pressure is lower, due to smaller of ratio of convective flux to diffusive flux within the thinner cake layer for small R_c. Thus, higher mass transfer coefficient or small R_c enhances the permeate flux and observed rejection.

5 Concluding Remarks

The electrokinetic model presented in this paper is a simple coupled model for quantifying the combined effect of colloid-electrolyte system during crossflow nanofiltration. Due to reverse electroosmotic flow, the permeate flux and salt transport through the cake layer decrease. The salt concentration at the cake surface increases, which enhances the concentration polarization on cake surface. The cake enhanced concentration polarization will be severe for nano-colloids with higher surface potential. A detailed electrokinetic model of ion transport through the cake layer of nano-colloids and membrane will lead to a better understanding of the combined fouling process for high salt rejection membranes. Consideration of the electrokinetic interaction between the electrolyte and the colloid is important to develop a detailed model of the inter-relations between the permeate flux, salt rejection and cake formation. The study also includes that manipulation of the meso-scale structure of the cake will allow us to alter the porosity of cake layer, extent of adsorption of different species, and alter the salt rejection of the membranes.

Nomenclature

ΔP	Applied operating pressure, Pa
R_m	Membrane resistance, 1/m
R_r	Real rejection of membrane
d_e	Hydraulic diameter of cross flow channel, m
μ	Solvent viscosity, Pa.s
ε	Solvent electrical permitivity
R_{ug}	Universal gas constant, J/mole K
T	Absolute temperature, K
C_f	Solute concentration at feed (i.e., i= ions, c=colloids)
C_m	Solute concentration at membrane surface (i.e., i= ions, c=colloids)
ρ_c	Density of colloids, kg/m^3
L	Cake thickness, m
$D = \frac{k_B T}{6\pi\mu r}$	Brownian diffusion coefficient of solutes (i.e., i= ions, c=colloids), m^2/s
$D_i^* = \frac{(1-\varphi_p)*D_i}{[1-\ln(1-\varphi_p)^2]}$	Hindered diffusivity of ions within the cake layer, m^2/s
$k = 1.86\,\frac{D}{d_e}\left(\frac{u_c d_e^2}{D L_c}\right)^{\frac{1}{3}}$	Mass transfer coefficient co-relation of solute (i.e., i= ions, c=colloids) for laminar flow in thin channel [8], m/s

References

[1] E.M.V. Hoek and M. Elimelech. Cake-enhanced concentration polarization: A new fouling mechanism for salt-rejecting membranes. *Environmental Science and Technology*, 37:5581–5588, 2003.

[2] A.S. Kim, A.E. Contreras, Q. Li, and R. Yuan. Fundamental mechanisms of three-component combined fouling with experimental verification. *Langmuir*, 25:7815–7827, 2009.

[3] L.N. Sim, Y. Ye, V. Chen, and A.G. Fane. Investigations of the coupled effect of cake-enhanced osmotic pressure and colloidal fouling in ro using crossflow sampler-modified fouling index ultrafiltration. *Desalination*, 273:184–196, 2011.

[4] S. De and S. Bhattacharjee. Flux decline during cross flow membrane filtration of electrolytic solution in presence of charged nano-colloids: A simple electrokinetic model. *Journal of Colloid and Interface Science*, 353:530–536, 2011.

[5] J. Masliyah and S. Bhattacharjee. *Electrokinetic and Colloid Transport Phenomena*. Wiley Interscience, 2006.

[6] S. De and P. K. Bhattacharya. Modeling of ultrafiltration process for a two-component aqueous solution of low and high (gel-forming) molecular weight solutes. *Journal of Membrane Science*, 136:57 – 69, 1997.

[7] S.K. Zaidi and A. Kumar. Experimental analysis of a gel layer in dead-end ultrafiltration of a silica suspension. *Desalination*, 172:107–117, 2005.

[8] S. Kim and E.M.V. Hoek. Modeling concentration polarization in reverse osmosis processes. *Desalination*, 186:111–128, 2005.

(484e) Comparison of Membrane Performance: Analysis with Fouling Models

Mingyan Zhou[1], James E. Kilduff[1] and **Georges Belfort**[2], (1)Department of Civil and Environmental Engineering, Rensselaer Polytechnic Institute, Troy, NY, (2)Department of Chemical and Biological Engineering, Rensselaer Polytechnic Institute, Troy, NY

Membrane research and engineering applications often involve optimizing flux and rejection for a particular application. This process frequently presents the need to compare the performance of membranes having different inherent permeabilities; an important performance criterion is resistance to fouling. To identify the most appropriate (least biased) approaches to compare membrane performance, an analysis using fouling models, with an emphasis on the combined pore blockage and cake filtration model, was performed. We illustrate how the models can be used to provide a general framework for assessing fouling and developing ways to compare membrane performance. Comparisons at constant transmembrane pressure using the same initial flux and the same initial pressure were emphasized. This analysis demonstrates that membrane resistance has a significant effect on flux decline - flux decline is smaller for membranes having greater resistance, especially when the feed solution and membrane-foulant interactions are similar. This effect cannot be normalized regardless of whether flux is plotted as a function of filtration time and volume throughput. A better approach is to compare the effects of fouling by evaluating fouling resistance as a function of volume throughput. Alternatively, comparing membranes based on total resistance provides some advantages because it can be related to permeate production and energy cost. When comparing membranes having different resistance and rejection, the effects of fouling should be compared by evaluating fouling resistance as a function of mass accumulation on the membrane surface. The effects of cake compressibility are also discussed.

See more of this Session: Theories and Applications of Cross Flow Filtration
See more of this Group/Topical: Separations Division

(491a) Glycome Profiling Using Glycan-Directed Monoclonal Antibodies: Applications to Plant Cell Wall/Biomass Characterization

Sivakumar Pattathil[1], Michael G. Hahn[2], Heather McCormick[1], Jeffrey Miller[2], Virginia Brown[2], Ajaya Biswal[2], Trina Saffold[2], Malcolm O'Neill[2], Debra Mohnen[2], William S. York[2], Jaclyn D. DeMartini[3], Charles Wyman[4], Michael W.W. Adams[5], Irina Kataeva[6], Richard A. Dixon[7], Huanzhong Wang[7] and Fang Chen[7], (1)Complex Carbohydrate Research Center, University of Georgia, Athens, GA, (2)University of Georgia, Athens, GA, (3)Chemical and Environmental Engineering, University of California Riverside, Riverside, CA, (4)University of California Riverside, Riverside, CA, (5) Biochemistry and Molecular Biology, University of Georgia, Athens, GA, (6)Department of Biochemistry, University of Georgia, Athens, GA, (7)Samuel Roberts Noble Foundation, Ardmore, OK

Plant biomass is the major source for lingo-cellulosic bioenergy. Plant cell walls form the major part of the plant biomass. Research on plant cell wall/biomass is challenging owing to the complexity of the cell walls. Thus, one of the focus areas of bio-energy research has been the development of advanced tools that enable easy, rapid, high-throughput and reliable characterization of plant cell walls/biomass. Cell wall-directed monoclonal antibodies have now emerged as powerful probes to assist in the detailed structural analysis of plant cell walls. The worldwide collection of plant cell wall glycan-directed monoclonal antibodies is now sufficiently large and comprehensive to be used to monitor the majority of cell wall glycans. Here we report an ELISA-based method, Glycome Profiling, that employs a toolkit of 155 monoclonal antibodies (recognizing most major classes of plant polysaccharides) for characterizing diverse plant cell walls. The method involves sequential extraction of cell wall materials with increasingly harsh reagents to obtain fractions enriched in diverse wall polysaccharides, such as pectins, arabinogalactans, xylans, xyloglucans, and mannans. Each of these fractions is subjected to ELISA against the toolkit of antibodies and the binding responses are depicted as heat maps. The studies presented here outline multiple applications of Glycome Profiling to plant cell wall/biomass characterization and utilization. [This work is supported by a grant from US Department of Energy BioEnergy Science Center and grant DE-PS02-06ER64304. The BioEnergy Science Center is a U.S. Department of Energy Bioenergy Research Center supported by the Office of Biological and Environmental Research in the DOE Office of Science. Generation of the monoclonal antibodies used in this work was supported by the NSF Plant Genome Program (DBI-0421683)]

See more of this Session: Advances In Biofuels: DOE Bioenergy Research Centers I
See more of this Group/Topical: Sustainable Engineering Forum

(491b) Progress In Biomass Trait Screening Using a High Throughput Digestibility Platform

Nick Santoro[1], Shane Cantu[2] and Jonathan Walton[2], (1)Great Lakes Bioenergy Research Center, Michigan State University, East Lansing, MI, (2)Michigan State University, East Lansing, MI

A key step in the development of a viable lignocellulosic ethanol industry is the improvement of the plants that are to be used as biomass feedstocks. We have previously described a high throughput cell wall digestibility platform (HTDP) to facilitate the identification and evaluation of feedstock plants. Cell wall digestibility is assayed by the amount of free Glc and Xyl released after chemical pretreatment and incubation with a commercial cellulase mixture (Accellerase). One key component of the HTDP is a custom-designed workstation that grinds and dispenses individual plant biomass to within 0.1 mg accuracy. Combining this workstation with a second key component, a laboratory information management system (LIMS), provides the basis for generating a quantitative measure of cell wall digestibility on a vast number of samples.

The results presented in this talk summarize cell wall digestibility screening projects carried out in the two years since the HTDP opened. A collection of approximately 7,000 *Brachypodium* EMS mutant M2 lines have been screened. Classes of outliers are identified and studies are underway to further characterize the mutant lines. A collection of approximately 1,200 *Arabidopsis* T-DNA lines with insertions in cell wall biosynthesis-related genes have been screened for differences in cell wall digestibility. The chemical and genetic characterization of these lines is presented. Corn lines comprising a University of Wisconsin Diversity Set have been screened recently using corn stalk core samples. An initial survey of outliers is presented. Results for transgenic poplars containing knock-down insertions for decreasing the expression of the CCR gene are also presented. The poplar screening studies compare the cell wall digestibility of the transgenic and wild type poplar in response to a variety of pretreatment conditions as well as to geographical field location. Substantial differences in cell wall digestibility of the transgenic and wild type poplar are shown as the above parameters are varied.

In Summary, the HTDP is an extremely useful tool, demonstrating utility as a screening tool to identify improved plants from large sample populations (Brachypodium). In addition, the HTDP is flexible to accommodate novel sampling methods (corn stalk cores). Furthermore, once interesting plants are identified, they can be further tested in the HTDP using a number of varied pretreatment conditions (poplar).

See more of this Session: Advances In Biofuels: DOE Bioenergy Research Centers I
See more of this Group/Topical: Sustainable Engineering Forum

(491c) Rheological Properties of Biomass Deconstruction Process Using Ionic Liquid

Alejandro G. Cruz[1], Jeff Mentel[2], Seema Singh[3] and Blake Simmons[3], (1)Deconstruction, Joint BioEnergy Institute, Emeryville, CA, (2)Malvern Instruments LTD, MA, (3)Joint BioEnergy Institute, Emeryville, CA

Lignocellulosic biomass must be pretreated prior to enzymatic hydrolysis to improve sugar yields and saccharification efficiency. Ionic liquid (IL) pretreatment has shown great potential as a novel pretreatment technology with sugar yields as high as ~100%. However, to improve process economics, higher biomass loading is necessary. Biomass slurry in IL are highly viscous. To decrease production cost, the biomass slurry should be pumpable and exhibit least resistance to flow. Here we investigate rheological properties of neat IL, biomass slurry in ionic liquid through dissolution and regeneration processes. Complex viscosities and modulus (both elastic and viscous) measurements provide a detail insight of biomass solubilization and cellulose regeneration process and pinpoint the onset of cellulose regeneration event. By generating small-angle oscillatory frequency sweeps of pretreated switchgrass at various increased loading percentages we can characterize the nature of the slurry where the material has strong intermolecular association. For a viscous material such as in the 3% pretreated switchgrass sample, the viscous modulus is dominant over the elastic modulus. Just a slight increase in percent loading makes the slurry behave like a gel like structure and the elastic modulus is dominant over the viscous modulus. All percent loading pretreated switchgrass show frequency dependence and shear thinning behavior.

See more of this Session: Advances In Biofuels: DOE Bioenergy Research Centers I
See more of this Group/Topical: Sustainable Engineering Forum

(491d) Molecular Simulation of Lignocellulosic Biomass Components

Loukas Petridis, Center for Molecular Biophysics, Oak Ridge National Laboratory, Oak Ridge, TN

Lignocellulosic biomass comprises the vast majority of biomass on Earth and has the potential to play a major role in generation of renewable biofuels if cost-effective conversion can be achieved. Largely composed of plant cell walls, lignocellulosic biomass is a complex biological composite material that shows significant recalcitrance towards the structural deconstruction and enzymatic hydrolysis into sugars that is necessary for fermentation to bioethanol. Here, we report on studies combining molecular dynamics simulation and small angle neutron scattering to better understand structural properties of biomass components cellulose and lignin. Lignin, a major polymeric component of plant cell walls, forms aggregates in vivo after pretreatment of lignocellulosic biomass for ethanol production. The aggregates are thought to reduce ethanol yields by inhibiting enzymatic hydrolysis of cellulose. Extensive atomistic molecular dynamics simulations were used to demonstrate that the surfaces of the aggregates are highly irregular and characterized by a surface fractal dimension that is invariant across length scales from ~1–1000Å. The simulations reveal extensive water penetration of the aggregates and heterogeneous chain dynamics corresponding to a rigid core with a fluid surface. The detailed multiscale structure revealed here should aid in understanding biomass recalcitrance to hydrolysis and in feedstock engineering efforts to improve biofuel yield.

See more of this Session: Advances In Biofuels: DOE Bioenergy Research Centers I
See more of this Group/Topical: Sustainable Engineering Forum

(491e) Microfluidic Assays for Biofuels Research

Rajiv Bharadwaj[1], Aarthi Chandrasekaran[2], Chieh Chang[3], April Wong[4], Blake Simmons[2], Paul D. Adams[2] and Anup Singh[2], (1)Sandia National Laboratories, Joint BioEnergy Institute, Livermore, CA, (2)Joint BioEnergy Institute, Emeryville, CA, (3)Sandia National Laboratories, Joint BioEnergy Institute, Livermore, (4)Sandia National Laboratories, Livermore, CA

We are developing next-generation assays for biofuels research using microfluidic technologies to provide significant improvement over conventional platforms in throughput, sensitivity, multiplexing and speed of analysis. To develop these microscale assays we utilize technologies covering a wide spectrum of complexity based on the requirements of the assays. For example, using off-the-shelf components we have developed a microtiter plate-based assay for screening cellulases against heterogeneous substrates. This approach relies on in-situ biomass

regeneration in micro-volumes to volumetrically meter biomass (sub-mg loading) and also precisely control the amount of residual IL for screening activities of novel IL-tolerant cellulases. We have also developed more sophisticated platforms for screening of lignocellulolytic enzymes using microfluidic chip for electrophoretic analysis of glycans. These chips are being used to screen genetically-engineered enzymes, assess performance of pretreatment processes, and for discovering cellulase activities in microbial communities. We are also developing high throughput microchips to perform combinatorial DNA assembly for optimization of biofuel synthesis pathways.

See more of this Session: Advances In Biofuels: DOE Bioenergy Research Centers I
See more of this Group/Topical: Sustainable Engineering Forum

(491f) The Binding Properties of Cellulases On Allomorphs Cellulose and Pretreated Biomass During Enzymatic Hydrolysis

Dahai Gao, **Shishir Chundawat**, Venkatesh Balan and Bruce Dale, Chemical Engineering and Material Science, Michigan State University, Lansing, MI

The depolymerization of cellulose to monomeric glucose is the key step for cellulosic ethanol biorefinery process. Cellulases binding on biomass is the essential first step for effective breakdown of biomass to sugars. A new fast flow liquid chromatography (FPLC) based method has been developed to quantify endoglucanase I (EG I, GH family 7B), cellobiohydrolase I (CBH I, GH family 7A) and cellobiohydrolase II (CBH II, GH family 6A) individually in the heterogeneous reaction environment.

The binding level of individual enzymes during 48 hours hydrolysis was studied on different cellulose allomorphs: micro crystal cellulose Avicel (cellulose Iβ), liquid ammonia treated cellulose (cellulose III), sodium hydroxide treated cellulose (cellulose II) and phosphoric acid swollen amorphous cellulose (AC). The digestibility rank is AC>cellulose III>cellulose II>cellulose I. However, the AC has the highest initial binding capacity while cellulose III has the lowest. After the substrate solubilized, most of the cellulase returned to supernatant. But CBH II is less stable. Only around 60% CBH II can be recovered.

The time course binding studies were also performed under pretreated biomass. Ammonia Fiber Expansion (AFEX) treated corn stover (CS), dilute acid treated CS and ionic liquid pretreated CS are compared. The results show the existence of lignin is responsible for a significant amount of unproductive binding. For AFEX-CS, the initial binding level for those cellulases are >80%. However, after 48 hours digestion, only around 50% of CBH I, 28% of CBH II and 19% of EG I are detected in the supernatant. The remaining cellulases are believed to be bound unproductively onto lignin. Similar results were found on ACID-CS. The results are critical for improving our understanding of enzyme synergism, productive/unproductive enzyme binding and the role of pretreatment on enzyme accessibility to lignocellulosic plant cell walls, as well as help to engineer novel low unproductive binding enzymes and develop an economic enzymes recycle system.

See more of this Session: Advances In Biofuels: DOE Bioenergy Research Centers I
See more of this Group/Topical: Sustainable Engineering Forum

(493a) Effectiveness of Activated Carbon Adsorption for the Removal of Endocrine Disrupting Chemicals In Drinking Water

Varadarajan Ravindran, University of Southern California, Los Angeles, CA and **Massoud Pirbazari**, Sonny Astani Department of Civil and Environmental Engineering, University of Southern California, Los Angeles, CA

The US Environmental Protection Agency (USEPA) defines an EDC as an exogenous agent that interferes with the synthesis, secretion, transport, binding, action, or elimination of natural hormones in the body that are responsible for the maintenance of homeostasis, reproduction, development, and/or behavior. Such chemicals as those above have been found existing in wastewater, surface waters, sediments, groundwater, and even drinking water. As they exist in extremely low concentration (μ g/L or ng/L levels), and so highly sensitive measurement techniques are employed to detect them. The EDCs have been attributed as a cause of reproductive disturbance in humans and wildlife even at low concentrations. Human exposure to these chemicals in the environment is a critical concern with unknown long-term impacts. Natural and synthetic EDCs are released into the environment by humans, animals and industry, mainly through sewage treatment systems before reaching the receiving waters bodies including soil, surface water, sediment and ground water.

The EDC removal methods fall into three categories, namely, physical removal, biodegradation and chemical advanced oxidation. There is much research data on membrane processes for the effective removal of organic micro-pollutants; from pesticides to pharmaceuticals to personal care chemicals. In recent years, research on EDC removals by the membrane processes have greatly increased. Studies have discovered that the rejection efficiency of EDCs by membranes strongly depended on their physico-chemical properties, such as molecule weight, hydrophobicity (octanol-water partition coefficient or log K_{ow}), aqueous solubility, electrostatic properties, and other characteristics. Nevertheless, membrane technologies are very expensive as compared to other technologies such as advanced oxidation and activated carbon adsorption processes. Advanced oxidation technologies such as various combinations of ozone, hydrogen peroxide and UV radiation have been found effective in the destruction of most EDCs to meet the treatment goals. However, residuals have of these chemicals have been found after extensive treatment, and in some cases complete oxidation of EDCs had not been achieved besides yielding disinfection byproducts including aldehydes, ketones and haloacetic acids.

Under the circumstances activated carbon adsorption is a well-known, efficient and cost-effective technology for removing various organic contaminants including the EDCs. Ggranular activated carbon (GAC) is commonly employed in packed bed adsorbers for the removal of trace organic pollutants from water, although powder activated carbon (PAC) has also been used in some cases.. During the last few years, several researchers have demonstrated that activated carbon has a strong capability of removing a broad range of representative EDCs for artificial and real waste water in the laboratory and pilot-scale and full-scale operations. Studies on EDCs removal by GAC mainly focus on the removal efficiencies for EDCs in different water systems and the influence of factors such as the physicochemical properties of EDCs and the types of GAC produced from different materials. The effects of some fundamental environmental parameters including adsorbent concentration, pH, salinity and the presence of humic acid and surfactant on adsorption. A major concern in the application of fixed-bed GAC adsorbers is that that adsorption capacity of the activated carbon was greatly impaired by the presence of natural organic matter (NOM) such as humic and fulvic substances present in most nature water sources, and the problem is exacerbated in the case of EDCs regarding their removal at low concentrations (micro-grams or nano-grams per liter level). Furthermore, water quality parameters including alkalinity, hardness and presence of NOM influence the effectiveness of activated carbon adsorption. The presence of divalent cations in natural waters such as calcium influence the properties of NOM such as those of the macromolecules of humic substances. The mechanisms of complexation of humic substances with divalenet calcium ions are important within the context of EDC adsorption at trace levels. Humic substances have a strong tendency to form complexes with metal ions and EDCs, thereby altering their transport and adsorption characteristics. The paper will address the adsorption of model EDCs including alachlor, aldicarb, dieldrin, heptachlor, heptachlor epoxide, toxaphene, and polychlorinated biphenyls (congeners Arochlor 1016 and 1254) at low concentrations from water, in the presence and absence of humic substances. Particular emphasis will placed on the mechanisms involved in the transport and adsorption phenomena including those of competitive adsorption among humic acid, humic acid-EDC complexes, and the free EDC molecules. The mechanisms of compaction of humic molecules in the presence of calcium ions and the influence of this phenomenon with respect to competitive adsorption will be addressed. Additionally, the effectiveness of activated carbon technology will be discussed with special emphasis on meeting the treatment objectives for waters containing NOM such as humic substances. The presentation will address the relevant complexation, transport and competitive sorption mechanisms, and the means for improving the efficiency of the process for removal of the EDCs of interest.

See more of this Session: Advances in Drinking Water Treatment and Purification Technologies
See more of this Group/Topical: Environmental Division

(493b) Fluorescence Spectroscopy As a Tool for the Characterization of Natural Organic Matter and the Control and Optimization of Drinking Water Treatment Processes

Nancy P. Sanchez and Christopher M. Miller, Civil Engineering, The University of Akron, Akron, OH

Natural Organic Matter (NOM) present in source waters is a complex and not fully characterized mixture of humic, fulvic and protein like compounds which determine its characteristics and chemical properties. The unique character of the NOM in each water source is determined by biological, geographical and seasonal factors which cause dynamic behavior. Water treatment processes are highly influenced by the nature and concentration of NOM in the water under treatment. Coagulation efficiency and the extent of the Disinfection by-Products (DBP) formation are greatly determined by the specific character of the Natural Organic Matter in the source water. Fluorescence spectroscopy represents a tool for characterizing the nature of the NOM and assessing its dynamics, transformations and interactions under treatment. The analysis of the Excitation-Emission Matrices (EEM) requires the application of mathematical techniques that allow extracting the important amount of information present in these matrices. Parallel Factor Analysis (PARAFAC) has emerged as an integral approach for the analysis of the fluorescence matrices and its use has been widely reported in diverse environmental analyses. In this work, water samples from different water sources serving four different water treatment plants located in Ohio were collected during periods of over one year. Samples of raw water and water after coagulation were obtained three times a week. The continuous monitoring of the NOM levels represented as Total Organic Carbon (TOC) concentrations and the measurement of the fluorescence levels in the waters were carried out. The application of the PARAFAC technique allowed obtaining the distinctive fluorescence components in each water before and after treatment, offering insight not only about the particularities of each water source but also about the efficiency of the specific coagulation process applied in each facility. Results indicate that fluorescence analysis coupled with PARAFAC application may represent a practical tool to be used in the control and optimization of the water treatment operations increasing the efficiency of the processes and assuring the desired quality characteristics in the drinking water being supplied.

See more of this Session: Advances in Drinking Water Treatment and Purification Technologies
See more of this Group/Topical: Environmental Division

(493c) Decontamination of Arsenite Polluted Water by a Zirconium Oxide Based Sorbent

Yuming Zheng, Dan Wu, Ling Yu and **J. Paul Chen**, Department of Civil & Environmental Engineering, National University of Singapore, Singapore, Singapore

Pollution of Arsenite As(III) in the water is an issue of serious concern because of its seriously negative impacts on human health. In this study, the adsorption behaviors and mechanism of arsenite onto a zirconium oxide based sorbent are investigated. To explore the adsorption behaviors of arsenite onto the sorbent, the effect of pH, natural organic matters and coexisting ions on the adsorption of arsenite, the adsorption kinetics, as well as the adsorption isotherm were investigated. The pH effect studies showed that the sorbent exhibited high As(III) adsorption capacity at a wide pH range, and the optimal pH value for As(III) removal is between 9 to 10. The kinetics study illustrated that most of the uptake of As(III) onto the sorbent took place in the first 20 hours, and the adsorption equilibrium could be obtained within 48 hours. The adsorption isotherm study demonstrated that the sorbent was efficient for As(III) removal from aqueous solution, and the maximum adsorption capacity for As(III) was around 1.8 mmol/g at optimal pH. The adsorption isotherm was found to be well described by Langmuir model. The existence of humic acid, silicate, sulfate, phosphate or fluoride does not obviously effect the adsorption of As(III) onto the sorbent. Fourier transform infrared spectroscopy and X-ray photoelectron spectroscopy analyses together with the adsorption studies indicated that ion exchange played an important role in the adsorption process, and the hydroxyl functional group was involved.

See more of this Session: Advances in Drinking Water Treatment and Purification Technologies
See more of this Group/Topical: Environmental Division

BIO-REFINERY BASED ON INDIAN PAPER INDUSTRY WASTES

Sanjay Tyagi*, Narayan Chandra Mishra, and A.K.Ray
Indian Institute of Technology, Saharanpur Campus, Saharanpur, India
*Central Pulp and paper Research Institute, Saharanpur, India

ABSTRACT

Paper Industry is one of the major industries which discharge some wastes that can produce many valuable by-products. These are black liquor, steam and electrical energy, wood/ bark wastes, sawdust/ bamboo dust, solid wastes like lime mud, grits, dregs, ETP sludges and silica. These wastes can be converted to many value added chemicals, such as activated carbon from black liquor, sawdust, bamboo dust, broken chips and partly pulverized bark.

Black liquor /sulphite liquor produces ligno sulphones from rice straw, wheat straw, bagasse, bamboo, sarkanda and hardwood. It has diversified uses e.g. as a base for industrial detergents(washing powder and liquid soaps), oil-well drilling mud, cement and concrete additives, dispersants for a variety of chemicals, emulsifiers and stabilizers, brightening agents for electrolytic refining during electroplating, binders and adhesives for pellets and briquettes chemicals such as clay, calcium carbonate and titanium di-oxide resin ingredients, lignin-phenol-formaldehyde mixtures as adhesives for ply wood and other wood based panels, rubber additives, adsorbent/desorbent, micronutrient for agricultural applications, asphalt emulsions. Besides black liquor can also be converted to lignin-poly isocyanate foam, lignin reinforced polymers, lignin adducts, chlorobrominated lignins as effective fire retarding agents, composites of alkali lignin and biologically active agents such as fungicides, insecticides, nematocides, herbicides etc. ,activated carbon by coking or gasification of black liquor, hydrolytic degradation of ligno-cellulosic waste to sugars such as xylose and xylitol used as food stuff as sweeteners, synthetic fuel using solar furnace, silica from straw black liquor, hydrogen gas by electrolysis, vanillin, DMSO, methanol, acetate, formate, actate, saccharinic acids, lactones, turpentine and tall oil, alkali lignin, guaiacol, pyrocatechol, acetoguaiacone, biogas(methane) etc.

Cellulosic wastes can also produce cellulosic bio-ethanol using thermophilic bacterial species like Clostridium thermocellum. In the similar manner, the sludge can produce glucose by treatment with cellulase enzymes. Syngas from gasification of pith of some nonwood based plants can also be produced.

From straw soda black liquor, silica can be precipitated for industrial applications such as amorphous silica (for high quality synthetic pigments as paper fillers which can replace soap stone and TiO_2), filler, retention aid and as flocculating agent, for production of ceramics. The later can be employed for household china, fine earthen ware, concrete block, electric porcelain, silica light weight brick, clay brick, calcium silicate thermal insulating materials, reinforcement in rubber, in paint manufacture, adsorbent, grease thickener, insecticide and cosmetic industry.

An attempt has been made in this paper to review the existing status of Indian industry to examine up to what extent they are able to exploit the potential of aforesaid cellulose and lignin by-products. Further exploratory investigations were made to look into the feasibility of their manufacture in plant practice in India for its sustainability.

INTRODUCTION:

Pulp and paper Industry is the World's largest non-food biomass collection system. India is the 15th largest paper producer in the World. The installed capacity of Indian paper industry is 12 million tonnes and 11 million tonnes is in operation, however annual production is around 10 million tonnes with per capita consumption of 10.5 million tonnes. Now the Indian paper industry is growing at 9.5%CAGR per annum along with booming economy with GDP growth rate 8.6 % .It is projected a demand of paper and paper board in the country will be 15 million tonnes by the year 2015 and 22 million tonnes by the year 2020[1]. On the utilization of recovered paper side, India imports around 4.5-5 million tonnes of waste paper as the recovery rate at India rests at meager value of 20-22% [1]. In the coming decade Asia will be major hub for paper production and trade, particularly in a time when drastic paper production cut is taking place in Europe (minus 0.3%),North America (minus 0.5%) and other parts of the World**[2].**As a major economy of Asia, compounded with the booming economy of India and increasing demand of paper and paper board, India will provide consistent growth for decades to come. Together India and China are estimated to make up 59% of global incremental paper and paper board demand during 2009-2020. At the same time , North America and Western Europe is forecast to decline from 46% to 38% of global production by 2020.However, Indian manufacturers have to be more conscious on the emerging opportunities and growing market size as stringent competitions are ahead for them which have to be met by adopting clean technologies and quality products [2].

Pulp and paper industry is capital-intensive industry, intensive in terms of energy, chemical, water, labor and raw materials concomitant with high pollution load generation. Besides it demands various large expensive pieces of equipments for its manufacture to produce quality products. Because of its capital intensiveness, investment in the paper industry is always a long-term proposition. The technology of operations is also very much intricate. Today's plant personnel need the concept of production integrated with energy-environmental-productivity-quality issues. The discharges from paper industry include solid waste, liquid effluent and gaseous emissions. There are regulatory limits enforced on the discharge of these pollutants by the regulatory authorities and the industry is expected to treat its effluents to meet the standards. In some of the states zero effluent concepts are being implemented in waste paper based mills. The new process must induct lower energy, water consumption along with little or no impact on environment. In order to follow this strategy the technology has been continuously passing through many revolutionary changes over the last decades to achieve low manufacturing costs, to earn profits derived from the ability of the producer and to face international competition for quality standards in the global market.

PAPER MILL BASED BIOREFINERY:

The pulp and paper industry is at the forefront of the concept of bio-refinery since it is already a biorefinery -1st generation of Biorefinery –enabling sustainable production of pulp/ paper enhanced by the coproduction of bioenergy and biofuels and bioproducts, through matured technological processes that maximizes the value chain of pulpwood **[3].** However, the decreasing profitability of standard pulp and paper mills and the world's demand for a non-fossil Carbon Economy resulted a new biorefinery platform based upon complex technology concepts with higher revenue potential [3]. Biorefinery is an overall concept of an integrated and diversified processing plant where biomass feedstocks are converted into a wide range of valuable products, much likewise to petroleum refineries. Integrated birefinery is a processing facility that extracts carbohydrates, oils, lignin, and other materials from biomass, converts them into fuels, high value chemicals and other materials, with a near zero waste draw [4]. Paper Industry is one of the major industries which discharges some wastes as given below that can produce many valuable by-products [5-26].These are black liquor, steam and electrical energy, tall oil, wood/ bark wastes, sawdust/ bamboo dust, solid wastes like lime mud, grits, dregs, ETP sludge and silica. These wastes can be converted to many value added chemicals, such as activated carbon from black liquor, sawdust, bamboo dust, broken chips and partly pulverized bark, turpentine and tall oil. Important chemicals like methanol, acetate, formate, lactate, saccharinic acids, lactones, alkali lignin, guaiacol, pyrocatechol, acetoguaiacone can also be prepared. These are discussed as under:

Solid waste from big paper mills[12]

(Avg. data based on data collected from 8 paper mills)

Organic Waste: (T/T of Paper)

1.	Raw Material Preparation (Wood /bamboo dust)	0.1
2.	Pulp mill rejects	0.01
3.	Paper machine cleaner rejects	0.1
4.	ETP Sludge	0.2
	Sub-Total	0.41

Inorganic Waste: (T/T of Paper)

1.	Ash (Cinder + Fly ash) from boilers	0.80
2.	Lime sludge and grits (Recovery + Bleach liquor preparation)	0.70
	Sub-Total	1.50
	Total	1.91

Utilization of wastes from a paper industry can have enough potential for a perfect biorefinery in India. Production of many value added products can be explored. There are many Indian industries engaged in establishing biorefineries for producing multiple products. Some are mentioned below:

UTILIZATION OF WASTES FOR BIOREFINERY IN INDIA

1. Utilization of wood/nonwood wastes:

Large amounts of ligno cellulosic waste are generated through forestry and agricultural practices , pulp and paper industries, timber and agro-industries.

1.1 Production of bioethanol:

Cellulosic wastes, solid wastes like wood dust, bagasse pith, waste paper and even sludge can produce cellulosic bioethanol [8]. Alcohol can be produced from cellulosic wastes using thermophilic anaerobic bacteria species like Clostridium thermocellum at 60°C and pH of 7.5 under control anaerobic conditions [14]. Bioethanol from bagasse pith has been produced in the laboratory using thermophilic ethanologen, yeast strain at 50°C through sachharification both chemically and enzymetically (using hydrolytic enzymes like celluloses, laccases and xylanase) followed by fermentation of the hydrolysate. It is reported that almost 80% glucose (30 g/l)of enzyme hydrolysate was fermented to ethanol(9.5gm/l) within 40 h and thereafter some of the oligosachharides also converted to ethanol bringing the final ethanol concentration 11.5 gm/l in the broth [6,7]. Naithani et al. [9] describes utilization of Lantana Camara and pine needle as lignocellulosic wastes for ethanol production with a yeast culture of Saccharomyces Cerevisiae NCIM-3570 with final pH to 4.5 at a temperature of 30°C. The acid hydrolysate of hemicelluloses extracted bagasse with the yeast S.Cerevisiae for fermentation can produce ethanol after concentration and supplementation can produce 8.3% alcohol with fermentation efficiency of 83.3% compared to 7.8% for molasses medium of 28°Bx with FE of the order of 83.3%[25].

2. Utilization of fiber rich Sludge:

In the similar manner, the sludge can produce glucose by treatment with cellulase enzymes [10]. Treatment of cellulosic sludge with cellulase enzymes rapidly shortens the fiber length and then dissolves the fibers in to glucose. The glucose can then be sent to an ethanol plant for fermentation to ethanol. The economics are based on sludge containing 50 tonnes per day of cellulose. The enzyme cost is projected at $6,000 per day. Water treatment cost is $1,000 per day (including capital and O &M) assuming reverse osmosis will be needed. Thus the total costs are $7,100. The value of the glucose produced (assuming 70% conversion from fiber) is $13,000 per day. The value of sludge disposal avoidance is taken at $3,000 per day for a total benefit of $16,000 per day. Addition of c-PAM accelerates the rate of fiber hydrolysis. The other costs considered for investment are due to some equipment to handle the residual solids,acid and power. The cost-benefits have been found favorable [10]

3. Biomass Gassification:

Syngas from gasification of pith of some nonwood based plants such as bagasse can be produced [11]. Biomass thermo-conversion for energy production from bagasse has been attempted by Umrao et.al. [11] for pyrolysis in a free fall reactor followed by gasification in a quartz reactor to obtain a more efficient and complete conversion of the bagasse as well as production of a gas with high heating value or with a high H_2 + CO (Syngas) content to be used as a fuel for chemical industry. Pyrolysis of bagasse is done on a fluidized bed at high temperature due to which charcoal is formed and the volatile components are released. The charcoal is sent to the two stage gasifier. As a result solid ashes, and a gas mixture which is composed of CO_2,CO,H_2, water vapour, pyrolysis product , tar and hydrocarbons. This process saves one liter of liquid fuel by using 4-5 kg of bagasse.

4. **Bio-gas production [21, 24]**

Primary and secondary (biological) sludges generated in pulp and paper industry can be digested anaerobically to produce bio-gas through acid fermentation followed by conversion to methane and carbon di oxide by methane forming bacteria. Benefits of anaerobic digestion are:

- Production of Bio-gas which can be utilized as a fuel
- Reduction of organic solids (and stabilization)
- Sludge can be used as a soil conditioner or fertilizer.
- Sludge can be treated without pre-dewatering.

Typical values of bio-gas production and organic solid waste reduction using anaerobic digestion of pulp and paper mill sludges are:

Reduction of organic solids : 45-75%)

Bio-gas produced : 0.5-0.7 m^3/kg organic solids

The process details for a typical 30TPD bagasse based soda pulp mill in India producing 0.45 - 0.50 m^3/kg COD gas (75% CH_4 + 25% CO_2) with calorific value of 5400 kcals/m^3 is available. The process removes 90% BOD, and 70% COD at temperature range 35-37°C with saving of 4000kWh /day power and 5 tons per month nutrients [24].

5. **Particle board**: Pith + Lignin can make particle board.
6. **Other important value added products:**

Many valuable chemicals can be manufactured from paper industry wastes. Majority of the processes for chemical manufacture from paper industry wastes are patented [22].These are excellently reviewed by a number of investigators [19-23]. Some of the patented processes are highlighted below:

6.1 Activated carbon:

Ligno-cellulosic material such as saw dust, bamboo dust, broken chips, and partly pulverized bark is reported to have been converted into activated carbon.

Method I: The material was first treated with inorganic acid to remove sugars at low temperature range. Acid extracted material was then heated at **450°C for nine hours and** for one hour at 800°C. While steam was being injected, the tar was removed as condensate in this high temperature range. About 22 % yield of activated carbon was reported.

Method II:

Continuous production of activated carbon can be possible by carbonization of cellulosic material impregnated with activating agent such as zinc chloride. The process involved continuous supply of activating agent-impregnated material into carbonizing zone containing previously carbonized cellulosic material at such constant flow rate as to allow the impregnated material to reside for fifteen minutes in the carbonizing zone. The impregnated material was carbonized with the carbonized cellulosic material heated to 350°C by counter currently supplied hot gases and which formed an agitated fluidized bed by being fluidized with hot gas stream and being stirred by a mechanical means. The total weight of the product was two third of the original weight of the wood material supplied.

6.2 Chemicals from Bark:

Bark along with other refuse creates a disposal problem for paper mills using hard wood as raw material. Utilization of bark can be developed on following lines.

I. As a source of fuel for steam generation
II. As a source of chemicals, particularly tannins and tannin like compounds, which would be recovered by water/acid/alkali extraction of bark, prior to burning.
III. Marginal utilization as soil conditioner, extender for thermosetting resins' oil well mud additive, reinforcing and stabilizing agent for polypropylene moulding compounds.

6.3 Hydrolytic Degradation of Lignocellulosic waste to sugars, Xylose and Xylitol:

During acid hydrolysis of lignocellulosic biomass a number of chemicals are produced. These are sugars (Xylose), aliphatic acids (acetic, formic and levulinic), furan derivatives, furfural and 5-hexamethyl furfural(HMF) and phenolic compounds. Hydrolytic degradation of lignocellulosic waste to sugars such as xylose and xylitol used as food stuff as sweeteners can be potential chemicals. Hardwood and agricultural residues are rich in pentosans which mainly contain xylan. Waste and refuse generated from hard woods/straw/bagasse etc. while preparing these materials for pulping, could be degraded hydrolytically to D (+) Xylose and Xylitol, which are used in food stuff as sweeteners.

Hydrolysis of waste was carried out in first stage with sodium hydroxide of 4% concentration which removes acetic acid chemically bonded in the wood as sodium acetate. The liquid extract of this stage can be re-used after it has been brought to the alkali concentration to be used.

In the second stage of processing the residue from the first stage containing pentosan and cellolignin was further hydrolyzed with mineral acid e.g. sulfuric acid. When working without pressure, 1.5 to 6 % strength of H_2SO_4, and a ratio of liquid to solid of 3 to 6 parts by volume was preferred. Temperature was preferably 50 to 125°C and time was 2-4 hours. After completion of the second stage, the batch was filtered; the filtrate could be worked to Xylose or directly to Xylitol.The insoluble residue from the second stage could next be used to obtain organosolve lignin which was thermoplastic and very reactive; the material, this lignin preparation could be used as a base material for dyestuffs and pesticides. The residue remained after removal of lignin, on treatment with sodium chlorite followed by sodium hydroxide could be converted to practically white cellulose which could be used for industrial applications. Thus all wood waste could be converted to useful chemicals.

6.4 Synthetic Fuel using Solar Furnace:

A process of producing gases such as CO_2, CO, CH_4 and H_2 from pyrolysis and gasification of solid organic waste contained within a reactor has been patented [22]. The process comprises solar heating of working fluid (steam, CO_2 or a mixture of gases) to 600°C to 700°C by means of solar top furnace. The working fluid was contained in the lower section of the reactor situated at the focus of the solar top furnace. The heated working fluid was injected under pressure into porous reactant bed on which the catalyst ($NaHCO_3$) - mixed organic waste was situated. The catalyst-mixed organic waste was pyrolyzed causing the formation of CO_2, CO, CH_4 and H_2 gases and char residue.The hydrogen produced could be utilized to replace natural gas in a hydrogen economy, to make fertilizer or as a chemical feed stock. It is claimed that process did not require a source of pure oxygen, and provided an attractive method for using solar energy.

7.0 Basic chemicals from wastes

Wood waste has a complex carbohydrate lignin structure requiring severe conditions to reduce it to such basic chemicals as Methanol, Ethanol, Furfural and Phenols. Chemical production costs are directly related to scale and complexity of operation **[22].** Utilization of pine needles and bagasse as bed material in solid state fermentation for production of lactic acid by Lactobacillus strains has been an active research in IITRoorkee[26].

8.0 Cellulose nano crystals and nanocomposites:

Cellulose nanocrystals have been largely applied as reinforcing fillers in the preparation of nano composites materials with improved mechanical and barrier properties.

9.0 Utilization of black liquor and sulphite waste liquor

Pulp and Paper industry has a very huge lignin potential. According to rough estimate for every ton of bleached pulp produced, about 0.5 ton of lignin is either wasted to sewer or used as a low value fuel. If the means could be found to isolate it from spent pulping liquor and to convert it to usable chemicals, the financial returns might be sufficient to support separation and processing lignin preparations from spent liquor, thus helping abatement of stream-pollution. In many ways Kraft lignin is a versatile raw material. It can be easily isolated by acidification, preferably carried out with carbon dioxide (stack gas), whereby phenolic group are liberated and lignin precipitates. The precipitated lignin can be removed by filtration at elevated temperature and purified further by re-dissolution in water and re-precipitation by sulphuric acid.

9.1 Lignin Isolation and separation:

Lignin compounds from sulphate black liquor could be precipitated by bubbling CO_2 into it at about 60°C. The precipitated lignin was separated by filtration and purified by dissolving in dry hexane and re-precipitating in ether. Alkali lignin could be isolated, from soda process black liquor by neutralizing it with CO_2 from stack gases to pH 8.3 to 8.5; about 40% of the lignin was reported[9] to have been precipitated. The mix was heated to 85°C – 90°C to coagulate the lignin, was filtered. Residual lignin could be further precipitated by addition of sulfuric/hydrochloric acid. Isolated lignin was washed first with cold water and then with dilute sulfuric acid and dried on double drum drier.

The slimy, gelatinous and hard to filter form of the precipitated lignin from black liquor could be changed to a crystalline, easily filterable and quick settling precipitate by mixing small amount of water-immiscible solvent such as chloroform, methelyene bromide, l-nitropropane or the like, with lignin containing liquid such as Kraft/soda black liquor. The chloroform, or other agent, may then be recovered from the filtered liquid by vacuum- or steam distillation.

Chemical modification of Kraft lignin drastically changes its properties and product obtained are of commercial value. Common modifications can be summarized as follows:

1. Formation of ethers by reaction with alkyl/aryl halides
2. Formation of esters by, reaction with carboxylic acid
3. Formation of polymeric condensation product with phenols, aldehydes and amines
4. Conversion to sparingly soluble phenol derivatives in reaction with various salts of heavy metals
5. Sulphonation to Kraft lignosulphonates
6. Nitration carried out with nitric acid
7. Chlorination with aqueous chlorine
8. Oxidation, with air or oxygen produces derivatives with changed solubility and melting point
9. Demethylation of lignin can be accomplished by acidic and alkaline treatment and results in the formation of catechol group increasing the reactivity in condensate e.g. with formaldehyde.

9.2 Lignin derivatives:

Lignin thus isolated could be converted to large surface area. Lignin is useful in drug formulations for adsorbing acids in the digestive system, due to their extremely high surface area and their acid-alkali resistance. Other suggested uses include carriers for adhesives, catalyst substrate, reinforcement for plastics and carriers for insecticides and herbicides.

A method to produce biodegradable uncolored lignin solutions was described wherein aqueous lignin solutions such as pulp mill effluent were treated with polyvalent metal cations. The amount of cations should be sufficient to react with lignin, and the aromatics to produce a water insoluble precipitate which was separated from remaining liquor. The precipitate was subsequently acidified with acid solution and was allowed to stand for sufficient time to permit digestion. The precipitate thus formed was separated by centrifugation; and was treated with alkali to produce aqueous basic solution. This solution was found to be effective in controlling Scale formation in boiler water systems, when added to these systems in amount of 5-50 ppm. Thus method not only recovers valuable product from effluent but also reduces the amount of non-biodegradable and toxic materials prior to discharge. In addition color of the final discharge is also reduced.

Sulpho methylated alkali lignins mixed with aqueous ferrous salt and dichromate and spray dried to a powdered product was reported to be useful as drilling mud thinner.

9.3 Lignin Reinforced Polymers:

A method for forming lignin reinforced polymers was developed by co-precipitating lignin/latex slurry and gas drying the slurry to leave a lignin reinforced polymer free of inorganic salts. In a process, alkali lignin was dissolved in ammonium hydroxides; and the ammonium lignate thus formed was then modified by adding hexamethylenetetramine to the lignate and heating to 90°-110°C for two: hours. Thus treated lignin was mixed with latex to give emulsion, which was precipitated by CO_2. When the precipitate was dried at 300°F or below, inorganic salts formed in precipitation, sublimed leaving lignin/latex polymer.

9.4 Lignin- polyisocyanate/polyurethane foam:

These foams are useful as plastics and adhesives. Lignin-polyisocyanate foam can be prepared from polyethylene glycol solution of lignin when the later is treated with toluene di-isocyanate at 115°C which afforded foam having good compressive strength and 74% recovery. Polyurethane type foam from lignin preparations and organic isocyanates are also reported.

9.5 Lignin as dispersants:

Method: preparation of ozone treated lignin from Kraft black liquor by treating lignin with oxygen containing 3% ozone at 50°C for two hours; and spray drying the resulting solution has been reported. Oxidized lignin was found to be useful as dispersant for a variety of inorganic materials such as Clay, Cement, Calcium Carbonate, and Titanium Dioxide.

9.6 Lignin Adducts as surfactants:

Lignin adducts with 3-chloro-2-hydroxypropane-l-sulphonate, chloromethane sulphonate, acrolein and the like compounds are also reported to be useful as surfactants in dyestuff compositions.

9.7 Adhesive from lignin: Lignin-phenol-formaldehyde mixtures as adhesives are used for ply wood and other wood based panels:

Method: An adhesive from Kraft or Soda black liquor was prepared by reacting the black liquor (dry matter more than 25%) with formaldehyde and combining the product with separately prepared phenol-formaldehyde adhesive.

9.8 Lignin as coagulants and flocculants:

Lignin preparations from pulping liquor chemically modified by reaction with compounds such as formaldehyde, triazines, alkylamines were found [to] be useful as coagulants in treatment of sewage and other wastes. Cationic water soluble lignin amines prepared by reacting Kraft/Soda lignins with the reaction product of epichlorohydrin and a tertiary amine were suitable for a variety of uses such as flocculating agents, coagulants, retention aids, dispersing agents, asphalt emulsifiers and emulsion stabilizers.

9.9 Fire retarding agents:

Chlorobrominated lignins were highly effective as fire retarding agents. In one method, the inflammable material was coated or soaked with an aqueous slurry or alkaline solution of chlorobrominated lignin followed by drying. In addition chlorobrominated lignins could be combined with bonding compositions used for the production of bonded articles from particulate material, such as fibers, chips, granules, saw dust etc. eg. in the formation of chip board, particle board and the like. Chlorobromination was affected by adding chlorine and bromine concurrently or in two step operation. After completion the reaction, the chlorobrominated lignin was filtered, washed and dried.

9.10 Composites

Composites of alkali lignin and biologically active agents such as fungicides, insecticides, nematocides, herbicides and the like are useful as controlled release pesticides. These composites were prepared by intermixing ammonium salt of lignin with active agent under conditions of high shear mixing to form stable emulsion; acidifying the emulsion with mineral acid at pH 4-5 to precipitate the active agent/lignin emulsion with the active agent being physically absorbed within the lignin and drying the precipitate if desired.

9.11 Coke and activated carbon from black liquor

Process for activated carbon by coking or gasification of black liquor: In a process black liquor was heated in absence of free oxygen to an elevated temperature (450°-625°F) and pressure to prevent vaporization of the water, for a period sufficient to produce granular coke. The coke was separated from the aqueous phase to yield an aqueous effluent of low COD. The coke produced might be burned to supply heat, or used as a soil conditioner or carbonized to produce high grade coke for metallurgical purposes or used to produce activated carbon.

Organic values of black liquor have been converted into heat energy and activated carbon by a method which eliminates the recovery boiler from pulping liquor regeneration systems. Black liquor containing 16% solids or even lower or at any solids content up to 65% was hydropyrolyzed under very high pressure at 550°-650°F. The organic matter separates as a water-insoluble precipitate along with inorganic constituents such as Na_2CO_3, and minor amounts of Na_2SO_4, Na_2S. The organic matter could be utilized to produce heat energy or activated carbon which itself could be used for purifying effluents from various section of paper mill.

10.0 By-products from lignin

Generally by-products from lignin can be categorized as follows:

1. Low molecular weight compounds
2. Polymeric products

Table 1 gives a detailed picture of by-products (both commercial and non-commercial) obtained from Kraft cooking of wood using different methods of isolation. It can be seen from the table that the main by-products of Kraft cook are the degradation products of carbohydrates and lignin. The most consistent efforts on dry distillation of the alkaline black liquor powdered together with excess lime or alkali were made to obtain methanol, acetone, methylethylketone, higher ketones and other components of light and heavy oils in appreciable quantity (60-70 kg/ton of pulp each of the low molecular and high molecular compounds).

Low molecular weight compounds

The most practical approach to synthesis low molecular weight compounds from lignin is to heat the black liquor for 10-20 minutes at 250-290°C together with small amount of alkali and sodium sulphide. Table 2 covers the chemicals obtained during this process with their corresponding yield in kg/PTP. Under optimum conditions (291°C, 20.4 % of organic substance, 3.2 % Na_2S and 1.6 % of the NaOH based on total liquor) a maximum yield of 33 % of the organic products can be obtained as ether soluble phenols.

Of all chemicals listed in Table 2 the following chemicals are most important.

Use of lignin and lignin derivatives: Lignin Sulphonates and Kraft Lignin, Vanillin, Dimethyl sulphonates, Dimethyl Sulphoxide, Furfural, Methyl Mercaptan, Phenolics,thermosetting resins and graft polymers including polyurethanes, polyesters, polyamines and epoxies[13].

1. Vanillin
2. Furfural
3. Organosulphur compounds
    a) Dimethyl sulphide (DMS)
    b) Methyl mercaptans (MM)
    c) Dimethyl sulphoxide (DMSO)
    d) Dimethyl sulphone ($DMSO_2$)
4. Low molecular weight Phenolic compounds

10.1 Vanillin

Vanillin has a wide range of industrial application. Vanillin is the most widely flavoring agent,, commonly known as vanilla, due to its pleasant aroma and its property of enhancing other flavors.. It is universally used as flavoring agent with a property to mask other odors. Vanillin also finds use as a reductant and as material absorbing ultraviolet light. It can be incorporated into various plastics such as regenerated cellulose, polyvinyl ethers and polyamides to protect them against ultraviolet radiation. Ointments for protection against sunburn on the skin can be prepared with vanillin as active agent. Polymeric materials prepared from vanillin and vanillic acid have potential as fibres and films.

Generally vanillin is synthesized from Guaiacol or Eugenol.

It can be produced from lignin through oxidation of spent liquor with suitable oxidising agent or by extraction by a suitable solvent. There are a few methods for manufacturing Vanillin.

Method I: Catalytic oxidation under pressure in pressure vessel by prior raising pH of black liquor from 9.5 to 13 by sodium hydroxide and then heating for about 2 hours followed by extraction with chloroform. The aqueous layer is acidified with sulphuric acid to bring down the pH to 3.0 and the filtrate is again extracted with chloroform.

Commercial process based on alkaline oxidation of lignin with oxidising agent as cupric oxide (2% wt/wt.), various nitrobenzene compounds, mercuric oxide etc. have reported a better yield of vanillin from lignosulphonates. For example cupric hydroxide oxidation of lignin gives 22 % of vanillin at 160°C and 12.5 % at 102°C. In comparison a maximum yield of 23 % is obtained from nitrobenzene oxidation under optimum conditions of 1 hr at 180°C. Table No.2 gives the product obtained by alkaline oxidation of lignin. While vanillin is the most commonly reported simple compound obtained by alkaline oxidation of lignin, there are a number of other products that are formed which may eventually prove to be of commercial interest.

Method II: The black liquor is dissolved in chloroform and then extracted with saturated sodium bisulphate solution.

Purification of vanillin is an essential step when synthesized from lignin. Most practical way to purify it is to distillate under reduced pressure or co-distillation with saturated hydrocarbon.

10.2 Organo sulphur compounds from lignin

The most important products are DMS and DMSO.

DMSO is used as a catalyst and organic solvent for dyes and other chemical reactions. Since DMSO is an excellent solvent for many carbohydrate materials, it has found application in preparation of sugar derivatives. DMSO also has found utility as a material to protect cellulose tissue from freeze damage. In this case, it has been shown to penetrate-the cell walls better than glycerin and thus prevents excessive concentration of electrolytes in the cell during freezing. It has also been used successfully to protect red blood carpusals against freezing. Widespread interest in DMCO as a solvent for carrying drugs across cell membrane and as a drug itself have been developed.

Depending upon the source of lignin whether softwood or hardwood, the methoxyl contents of Kraft lignin varies from 14-20 %. Since methyl group represents about half of the weight of the methoxy group, it is theoretically possible to recover 7-10 % of lignin calculated as methyl radicals.

Synthesis of DMS

Method of production of DMS and MM is by heating Kraft spent liquor with sodium sulphide at 200°C and to purify the product obtained. The yields of DMS are about 3 % and MM 0.2-3 per cent on lignin solids.

Commercial process consist of heating concentrated spent liquor with elemental sulphur to a temperature of 200-250°C. The DMS formed is flashed from the reactor along with some water, MM and H_2S and collected by means of a condenser. The crude DMS is scrubbed by passing it through Sodium hydroxide and distilled.

DMS itself does not find any application but is a precursor for its halogenated and oxidized derivatives.

Synthesis of DMSO

This is a compound obtained from lignin with a variety of applications in chemical and pharmaceutical industries. This can be commercially obtained by oxidizing DMS by nitrogen-dioxide under optimum conditions. The reaction can be summarized as

$$3Me_2S + 2HNO_3 \rightarrow 3Me_2SO + 2NO + H_2O$$

The application of DMSO has extended in many fields. It has been used as a outstanding medium for promoting organic reactions and as a reactant solvent itself. The following points outline some of the most potential application of DMSO.

10.3 Polymeric compounds from lignin:

Lignosulphonates:

Bulk of the materials are imported from Russia and China **[19]**

This class of lignin polymers has attracted a great deal of interest because of their diversified uses. They can also be utilized as the base for-industrial detergents such as washing powders and liquid soaps.

10.3.1 Synthesis of lignosulphonates

Lignosulphonates can easily be prepared from purified Kraft lignin by sulphonation with solutions of alkali sulphites at elevated temperature and pressure. Sulphite spent liquor contains lignosulphonates which are formed during pulping reactions. Practical approach to isolate and purify this compound is to remove most sulphite and sulphate by lime addition upto pH of 10.5, more lime is added to the filtrate to give basic calcium lignosulphonate which precipitates in the pH interval of 10.5 - 12.2. More than half the lignosulphonate is thereby removed and separated by filtration. By the addition of sulphuric acid and magnesium and sodium lignosulphonates can be obtained.

10.3.2 Application of Lignosulphonates:

Black liquor /sulphite liquor produces lignosulphones from rice straw, wheat straw, bagasse, bamboo, sarkanda etc. and also from hardwood. Fundamental studies on the colloidal properties of purified lignosulphonates demonstrated their specific value for emulsification, defloccuation, chelation, adsorption and adhesion uses. It has diversified industrial applications; some major ones are as under:

i. Oil well drilling muds: In the drilling of oil wells, water and dispersants are continuously fed into the hole, to form, together with suspended fine clay, called as 'drilling mud'. Oil in water emulsions in a drilling fluid has been in use since the 1940's. Lignin sulphonates are valuable for stabilizing the oil in water emulsions which give the mud improved properties in control of water loss, reduction of torque on the drill stem, increase of bit life and general improvement of bore hole condition.

ii. Cement and concrete additives: Lignosulphonates are used as grinding aids in Portland cement and concrete. Their function is to reduce the agglomeration of the-ground particles, to reduce surface-tension at the solid-air interference of the particles-and to keep the surface of the grinding media clean & free. It reduces settling time; water reduction and these can also be used as air entraining agents in concretes. The entrained air gives increased resistance to spalling and cracking of the finished concrete. The dispersive action of ligno-sulphonates is useful in mixing, and placing concrete. Incorporation of 0.07 - 0.15 kg of lignosulphonates-concrete admixture per sack of cement will allow water reduction up to 20%. This results in better compressive strength and durable concrete. In oil well cementing it thus controls settling time and strength.

iii. Dispersants for variety of Chemicals: Lignin sulphonate dispersants usually have a molecular weight in the range of 10,000 - 40,000. The colloidal lignin coats the surface imparting a negative charge to the particle. The similarly charged particles now repel one another and a good dispersion is obtained. This is used as dispersants in refractories, tiles, absorbants, cosmetics, crayons, medicines, paint, paper coatings, tooth powder and dyes. It can also be base for industrial detergents (washing powder and liquid soaps

iv. Ore floatation: Grinding aids, refining of ores as floatation agent, improves filtering properties in Pb-Zn system.

v. Emulsifiers and stabilizers: Lignin emulsifiers are quite competitive with other commercial emulsifiers stabilizing oil-in-water emulsions. Emulsions formed are very stable even under frozen or boiling stage. The stabilization of fire fighting foams that results from the interaction of sodium bicarbonate solution and alum is achieved by an addition of sodium salt of Kraft lignin. Better results can be obtained by using lignosulphonates.

vi. Electrolytic Refining: Lignosulphonates have been successfully used as 'Brightening' agent for the plated surface preventing pitting and deposition of rough surface in electroplating.The electrolytic refining of copper , zinc etc. are some examples.

vii. Binders and adhesives: Lignosulphonate enhances the properties of the specific products e.g., pellets, briquettes. Large tonnage of spent liquor solids are used as binders for pellets and briquettes. 2% calcium lignin sulphonate in iron ore gives a pellet ten times as hard as that made with bentonite binder. The most familiar adhesive product is linoleum cement containing principally lignin sulphonates and clay. These cements are also used as adhesive for ceiling tiles, road surfacing, hard durable surface and laying of dust, chemicals such as clay, calcium carbonate and titanium di-oxide resin ingredients,

viii. Resin ingredients: Combination of lignin sulphonates, with low molecular weight urea and phenol formaldehyde resins have received attention as adhesive or binding agent for fibrous materials such as plywood, particle and chip board and paper laminates. Many good quality glues can be produced by incorporating lignin sulphonates with polyvinyl acetate or alcohol.

ix. Rubber additives: As extender, modified reinforcing pigment. Many types of lattices such as neutral nitrile and neoprene are compatible with lignin either in the acid or sodium salt form. Mixture of rubber with Kraft lignin is also suitable for use as rubber to glass adhesives.

x. Protein precipitants: Forms insoluble complex with proteins.

xi. Tanning agents: Forms more soluble tanning solution which facilitates recovery procedures.

xii. Sequestering agents: Ca, Ni, Sn, Zn & Al can be sequestered by lignosulphonates above pH 7.

xiii. Miscellaneous uses of lignosulphonates:

Other uses of lignosulphonates are lime plastering medium, storage battery plates-as expander and component of separator., lime plaster-use of less water with improved settling properties, crystal growth inhibitor, Ingot mold wash, and many more.

11.0 Other uses of black liquor: **Hydrogen gas by electrolysis,** Gasification of black liquor, Biogas from black liquor, Preparation of rubber additives, adsorbent/desorbent, micronutrient for agricultural applications, asphalt emulsions,

12. Utilization of lime sludge:

As lime sludge contain high silica content, so its burning is not considered economical. However, the following approach can be considered.

a. Burning to get low grade lime
b. In the cement industry for partial substitution of lime stones
c. As a conditioner for soil especially acidic one.
d. Burning after desilication
e. Lime mud comprising calcium carbonate was found to be useful in removing colour from pulping liquid waste effluents. In a process[32] effluents were treated with lime mud followed by clarification, separation of effluents from classifiers to remove colour bodies.
f. Puzzulanna Cement

13.0 Silica recovery from from straw black liquor:

Amorphous silica (for high quality synthetic pigments as paper fillers which can replace soap stone and TiO_2), filler, retention aid and as flocculating agent, for production of ceramics. The later can be employed for household china, fine earthen ware, concrete block, electric porcelain, silica light weight brick, clay brick, calcium silicate thermal insulating materials, reinforcement in rubber, in paint manufacture, adsorbent, grease thickener, insecticide and cosmetic industry.

Silica from boiler Fly ash, and Sludge are also used for manufacturing light weight bricks.

14. Industry Status in India:

- **Mission New Energy** : Uses agri-residues using enzymatic hydrolysis process produces 0.07Ml/year of cellulosic ethanol[3]
- **Utilization of bamboo dust:** Gasfication, Syn Gas by Hindustan Paper Corporation(HPC), **Nogaon,** Assam
- **Pith + Lignin to particle board**: Kailash Paper Mill , Awagwanpur, Moradabad, India
- **Biomethanation Process**: Satia Paper Mills, Pudumjee Paper Mills
- **Biomass cogeneration:** ABC paper mill

- **Sludge and Boiler Fly Ash to light weight Bricks**: ITC, Bhadrachalam, Rainbow Papers, Ahmedabad.
- **Corrugated roof sheets and Diesel from waste plastics in recycled paper:** Rainbow Papers, Damanganga
- **Producer gas plant using wood chips :** Andhra Pradesh Paper Mill (APPM), now International Paper, USA
- **Biogas, cogeneration and biomass gasification:**TNPL
- **PAFC :** BHEL, India[15].
- **Installation of biomethanation plant in pulp and paper industry with UASB reactor:** United Envirotechier
- **Biomass gassifier :** W2E projects and Systems
- **Lignosulphonates:** ABC paper mill, SailaKhhurd, Punjab

15. Status in Research Institutions

ETP sludge, pith and paper to Bio ethanol: Indian Institute of petroleum (IIP, Dehradun), Central pulp and paper research institute (CPPRI), Cellulose and paper branch,Forest research institute(FRI), National Sugar Institute , Kanpur,Basant Dada Patil Sugar Institute, Pune.

Cellulosic waste to alcohol: College of technology,Osmania University, Hyderabad, National Sugar Institute, Govt. of India, Indian Institute of Petroleum, Dehradun, Vasant Dada Sugar Institute, Pune, IITRoorkee.

Utilization of lime sludge,Fly ash :Central building research Institute, National thermal power corporation(NTPC)

Lignin grafted copolymers:Regional Research Laboratory(RRL at Bhubaneswar,Jammu, Jorhat)

Biogas –fuel cell based co-generation: Tata Energy Research Institute (TERI)

Production of lactic acid from waste sugarcane bagasse derived cellulose,cellulase production with Tricoderma citriviridae on solid bed ,use of acid hydrolysates of pine needles,and bagasse for lactic acid production using various strains such as Lactobacillus delbrueckii or lactobacillus pentosus have been explored in Department of paper technology, IITRoorkee( ), National sugar Institute , Kanpur, H.B.T.I., Kanpur

Ministry of New and Renewable Energy(MNRF) is keen to provide technical and financial assistance for setting up of energy generating plant based on biogas, cogeneration and biomass gasification

The technical parameters of a typical HRAD-PAFC (high rate anaerobic digester with phosphoric acid fuel cell) cogeneration system for a 50 TPD paper mill have been estimated by

TERI and development of a prototype PAFC have been explored by Bharat Heavy Elecricals Ltd.[15].

Table 1: By-products (both commercial & non-commercial) of the kraft cook

Compound	Origin	Quantity kg per ton of pulp
Methanol	Methoxyl groups of lignin & possibly glucuronoxylan	5
Acetate	Acetyl groups of the xylan as well as degraded carbohydrates	100-200
Formiate	Degradation of carbohydrates	40
Lactate	Degradation of carbohydrates	100
Saccharinic acids & lactones	Degradation of carbohydrates	250
Aliphatic sulfur compounds (methyl mercaptan, sulfide, disulfide)	Methoxyl groups of lignin and possibly glucuronoxylan	1
Turpentine	Terpenes	8-10
Tall oil (fatty & resin acid salts, phytosterols, etc.)	Wood resin components	20-100
Alkali lignin	Lignin	400-600

Table 2: By-products of the Kraft cook and from subsequent pressure heating of the black liquor with alkali and sulfide at 250-285°C (48 atm)

Before Pressure Heating	kg ptp	After Pressure Heating	kg ptp
Tall Oil	50	Acetic acid	60
Turpentine	17	Formic acid	60
Acetic acid	40	Methyl sulfide	40
Formic acid	40	Methyl mercaptan	8

Methanol	5	Pyrocatechol	30
Vanillin	2	Methyl catechol	11
Acetoguaiacone	2	Ethyl catechol	9
Guaiacol	2	Other ether-soluble phenols	150
Pyrocatechol	1	Homoprotocatechuic, protocatechuic and other phenol carboxylic acids, perhaps	150
Other phenols	42	Ether soluble acids of carbohydrate origin	150
Phenol carbonic acids	45	Demethylated lignin of novalac properties	100
Non-volatiles and lactic acids	145	Butanol-soluble lactones	35
TOTAL	**391**	**TOTAL**	**803**

CONCLUSIONS

Paper industry wastes can be properly utilized to produce value added products in order to act as second generation biorefinery. Lignin in Black Liquors has excellent potential for reuse in Fine Chemical Manufacture. The Indian Paper Industry is not yet ready for such a drastic transformation, since it will jeopardize the Chemical Recovery Processes. The Paper Industry is totally dependent upon the lignin black liquors as a source of energy fuel in Recovery Boilers. Serious attempts have to be made for medium and small scale manufacturing process. Some initiatives have been taken by Indian pulp and paper industry in recent years. A few research institutions have started research activities for establishing biorefinery based on paper/ sugar mill wastes.

It is necessary to determine the quantum of the wood waste/ agri-residue facility for the maximum size consistent with availability of raw material. A single product facility appears not to be competitive. In order to be economic utilization of the wood waste, multiple product facility is essential.

REFERENCES

1. Mukundan, P.G., Editorial, Inpaper International, Vol.12, No.2, Jan.-March 2010, Vol.13, No.1, Oct.-No.2010, www.inpaper.com.
2. Inpaper India, Vol.13(3), 14(2), March-April,2011, pp.24, 30.

3. Gabriel D.A., Sousa, Biorefinery development pathways: A survey for the pulp and paper industry, Proceedings of XXI Encontro Nacional da TECNICELPA/VI CIADICYP 2010, 12-15 Outubro 2010-Lisboa , Portugal.
4. Florbela Carvalheiro, Luis C.Duarte and Francisco M Girio, Hemicellulose biorefineries: a review on biomass pretreatments, JSIR, Vol.67, November, 2008, pp.849-864.
5. 56th Chemical weekly, July 05, 2011.
6. Thakur Vasanta V, Diwakar Pandey, R.K.Jain, R.M.Mathur, Second generation biofuels: Bioethanol from lignocellulosic materials, Inpaper India, Vol. 14(2), March-April, 2011, pp.4-12.
7. Jain, R.K., Thakur Vasanta V., Pandey Diwakar, Adhikari, Dixit A.K. and Mathur R.M., Bio ethanol from Bagasse Pith-A lignocellulosic waste biomass from paper / sugar Industry, IPPTA J., Vol.23, No.1, Jan-March, 2011, 169-173.
8. Mark E. Reno, Monique STREFF, Andrew Hird and Anjan Ray, Conversion of biomass to fuel as an environmental impact reduction opportunity for the pulp and paper industry, Inpaper International, Vol.13, No.3, April-June, 2011, pp.18-20.
9. Naithani Sanjay, Ganesh Joshi and Alok Kumar Dubey, Detoxification of Acid Catalyzed Lignocellulosic Hydrolyzate to enhance Fermentation Efficiency for Ethanol Production, INPAPER INDIA, Vol.14(2), March –April 2011, pp.14-22.
10. Banerjee Sujit, Energy from Pulp Mill Sludge,INPAPER INTERNATIONAL, Vol.13, No.1, Oct.-Nov. 2010, pp.10-12.
11. Umrao S., Pankaj Kumar and Sadhana Sachan, Method of producing Syngas from gasification of bagasse, Proceedings on "Advances in Chemical Engineering"-MacMillan, Advanced Research series February 27-28, 2009, Patiala, pp. 445-450.
12. Dhingra H.K. and Mohendru V.K., Status of Solid Waste Pollution From Indian Paper Industry and Future Prospects of its Management, IPPTA Vol.8,No.1, March 1996, pp.49-55.
13. Sahoo N, Rohella R.S.,Choudhury S.,Paul S.C.,and Chakravortty, V., Preparation of Value-Added Products From Black Liquor,IPPTA, Vol.8(3), September, 1996,pp.45-50.
14. Padmaja A, Venkateshwar S,Seenayya G and Bhagavanth Rao M, Evaluation of cellulosic substrates in the conversion of biomass to ethanol,Indian Chemical Engr., Section A . Vol.37 No.3, Jul-Sep(1995),pp.140-144.
15. Kurup ,G. Rajendra Chawla S.K.,Mathur A., Biogas –fuel cell based cogeneration in the pulp and paper Industry, IPPTA convention issue ,1992,pp.25-30.
16. Singh S.P., Recent developments in characterization and utilization of black liquor lignins from small pulp and paper mills, IPPTA, Vol.2 (3), 1990, pp.11-15.
17. Adsul, M.G., Verma A.J. and Ghokle, D.V., Lactic acid production from waste sugarcane bagasse derived cellulose, Green chemistry, 9, 58-62, 2007.
18. Guerra, G.,Casado, M.R.L.G., Arguelles,M.I., Sanchez M.I.,Manzano, A.M., and Guzman,T., Cellulase production with sugarcane straw by Trichoderma citrinoviridae on solid bed, sugar Tech.,8(1),30-35.

19. Managaonkar , N.D. and Kulkarni, Y.S.,Preparation and utilization of lignin by-products and lignosulphonates obtained from wood black liquor,IPPTA convention Issue,1987 pp.40-52
20. Ramaswamy, V. B iotechnology –applications in waste utilization and pollution abatement, ibid, pp.1-13.
21. Jain Sailendra, Environmental protection in Paper Industry, ibid, pp.53-61.
22. Dhake, J.D., Utilization of pulp and paper industry waste for chemicals,ibid, pp.19-29.
23. Trivedi, M.K., Lignin recovery from alkaline pulping of bagasse, ibid, pp.30-35.
24. NIEM (UNEP) Project Under Phase II- Activity III, October, 1990, Institute of Paper Technology (University of Roorkee), Saharanpur-247001 (India), pp.64-65, 66-67.
25. Lehri Madhu, Vartika Saxena and P.K. Agarwal, Preliminary studies on ethanolic fermentation of acid hydrolysates of hemicelluloses-extracted bagasse,pp.G587-G593, , Proceedings of the 57th Annual Convention of STAI, New Delhi, 13th -15th October,1995
26. Manoj Kumar Ghosh, Uttam K. Ghosh, Utilization of pine needles as bed material in solid state fermentation for production of lactic acid by Lactobacillus strains, Bio Resources, Vol.6, No.2,2011, pp.1556-1575.

Lignin – A Renewable Precursor for Carbon Nanofibers

John F. Kadla,[1] Ian Dallmeyer,[1] and Frank Ko[2]
[1]Biomaterials Chemistry Lab, Dept. of Wood Science, Faculty of Forestry
[2]Advanced Fibrous Materials Lab, AMPEL, Dept. of Materials Engineering
[1,2]University of British Columbia, Vancouver, BC

Abstract

Here we report on our research efforts aimed at developing carbon nanofibers through electrospinning of Kraft lignin. The research described herein focuses on understanding the effect of polymeric additives in influencing viscoelastic fluid properties which control the fiber morphology and diameter of fibers electrospun from blend solutions of softwood Kraft lignin (SKL) and poly(ethylene oxide) (PEO). To characterize the relevant fluid properties we report on the solution shear properties as well as the extensional properties of spinning solutions. We demonstrate that viscoelasticity can be varied in order to control the spinning process in our system by changing the concentrations of SKL and PEO as well as the PEO molecular weight. We also report on the mechanical properties of CNF precursor fibers as a function of thermal processing parameters. The results of these studies indicate that electrospinning is a promising platform for the development of micro- and nanofibrous carbon materials from technical lignins.

Introduction

Lignin is a naturally abundant biopolymer which reinforces the carbohydrate matrix, facilitates water and nutrient transport, and protects against invasion by microorganisms in the stems of trees and other terrestrial plants.[1,2] In terms of its chemical structure, lignin is an aromatic heteropolymer consisting of an irregular sequence of phenylpropane units connected by a variety of interunit linkages.[1,2] The renewable nature, abundance, and availability of lignin from the pulp and paper industry[3] are driving forces for the development of value-added products from lignin. Carbon fibres (CFs) are one example of value-added products that can be derived from lignin.[4] CFs have numerous applications including reinforcement in advanced light weight composite materials,[5] catalyst substrates,[6] adsorbents,[7] electrodes,[8] and tissue engineering scaffolds.[9] The most common precursors for commercial CFs are petroleum pitch, synthetic copolymers based on polyacrylonitrile (PAN), and rayon.[5] While PAN-based CFs have superior tensile strength, the high cost of PAN and non-renewable nature of petroleum-derived precursors are disadvantages.

Lignin is a potential renewable precursor for CFs, but the complex, heterogeneous chemical structure, the variability in properties among different types of isolated lignins, and the comparably poorer mechanical properties of lignin-based CF are barriers to their application.[4,10,11] Nevertheless, significant research efforts are being devoted to the production of CFs based on lignin, for example, at Oak Ridge National Laboratory[12]. Researchers in our lab have reported on thermal extrusion of lignin and lignin-polymer blends and conversion of these precursors to CFs by thermal treatment during the last decade.[4,13,14] Polymer blending was employed as a strategy to improve the spinnability of lignin[4], increase the tensile strength[13], or modify the surface porosity[14], depending on the blending polymers and their extent of interactions with lignin.[15-18] While studies conducted to date have reported improvements in fibre properties, a deeper understanding of the relationship between the complex macromolecular structure of lignin, interactions with additives, processibility, and material properties is needed to expand into a wider range of applications for lignin-based CFs. In addition, novel processing strategies must be explored to improve properties for specific applications and overcome some of the difficulties associated with lignin mentioned above.

One strategy to improve the properties of lignin-based CFs is to reduce the fiber diameter, since the tensile strength of CFs is known to increase with decreasing fibre diameter,[4,19,20] and applications such as filtration and catalysis depend on high specific surface area, which increases with decreasing diameter. Electrospinning is a robust strategy to produce fibers with diameters in the range of a few hundred nanometers or less.[21-23] Electrospinning of CF precursors such as PAN,[24-26] phenolic resin,[27] and pitch[28] has been reported. Electrospinning of organosolv lignin[29,30] and blends of sulfonated alkali lignin with PAN[31] have also been reported in the literature. However, lignin in general has been observed to display a high tendency to electrospray into droplets instead of forming fibers. To overcome this limitation, we have reported on the electrospinning of blends of several technical lignins with small amounts of poly(ethylene oxide) (PEO).[32] The advantages of our system compared to other reported systems are applicability to different technical lignins, the ability to spin fibers with greater than 99% lignin content with a single spinneret, and the ability to control the fiber diameter by varying the viscoelasticity of the spinning solution. Our current research in the area of lignin electrospinning is aimed at two main objectives: 1) To understand the role of lignin structure and interactions with additives in determining the viscoelastic fluid properties which govern the spinning process, and 2) To thermally process the electrospun fabrics and characterize the material properties as a function of processing parameters. In the following sections we provide an overview of our research efforts in the areas of rheology and its effect on the electrospinning process and processing and characterization of nonwoven fabrics produced from fractionated softwood Kraft lignin.

Materials and Methods

Lignin and solution preparation:

Softwood Kraft Lignin (SKL, Indulin-AT, Meadwestvaco, Glen Allen, VA, USA) was washed with dilute HCl (pH 2) to remove Na^+ ions and fractionated by sequential extraction with organic solvents according to a procedure based on that of Morck et al.[33] The F_4 fraction (according to the nomenclature of Morck et al[33]) was selected for rheological and electrospinning experiments. Polyethylene oxide (PEO) with M_v of 1000 and 5000 kDa (denoted PEO1M and PEO5M) were obtained from Sigma-Aldrich. N,N'-dimethylformamide (DMF, ACS Reagent grade) was obtained from Fisher Scientific (Ottawa, ON) and used as received.

Solution preparation:

SKL and SKL:PEO solutions in DMF were prepared as follows: SKL solutions were first prepared by adding dry lignin powder to the solvent. In the case of SKL:PEO solutions, PEO was first dissolved by briefly heating the solution at 80°C for 10-15 minutes, then SKL was added to the PEO solution. The mixture was heated at 80°C for 2 hours with intermittent vortexing every 30 minutes for 1-2 minute intervals. Solutions without PEO were prepared at concentrations of 25, 30, 35, 40, 45, and 50% by mass. Solutions with PEO were prepared with 25-45% SKL and 0.1-0.4% PEO (M_v = 1 or 5 x 10^6 g/mol).

Rheology:

The shear viscosity η was obtained from small amplitude oscillatory shear as the limit of $G''(\omega)/\omega$ at low ω or from steady shear measurements increasing shear rate from 1-1000 s^{-1} using an AR2000 torsional rheometer (TA Instruments, Grimsby, ON) with 40 mm diameter cone geometry and cone angle of 2°. A capillary breakup extensional rheometer (Haake CaBER 1, Thermo Fisher Scientific, Waltham, MA) was used to characterize the extensional fluid properties. In CaBER experiments, plate diameters of 4 mm, initial plate separation of 2 mm, final plate separation of 12 mm, and plate separation time (strike time) of 50 ms were used. Where applicable (for elastic solutions), the characteristic time scale of viscoelastic stress growth in uniaxial extensional flow λ (herein referred to as the relaxation time) was obtained by fitting the portion of the thinning profile where exponential thinning was observed as described by Rodd et al[34] using the equation:

(1) $D_{mid}(t) = D_1((G^*D_1)/4\sigma)^{1/3}\exp(-t/3\lambda)$

Where $D_{mid}(t)$ is the midpoint filament diameter, D_1 is the filament midpoint diameter just after the upper plate stopped moving, G is the elastic modulus, and σ is surface tension.

Electrospinning:

Electrospinning experiments were conducted using two different apparatuses. One was a self-built apparatus previously described[32] with a flat-tip needle as spinneret (gauge 18, 21, 22, or 25), operating voltage of 15 kV, flow rate of 0.03 mL/min, and spinneret-to-collector distance of 15-20 cm. The second was a Kato Tech Nanofiber Electrospinning Unit (NEU) (Kato Tech Co., LTD, Kyoto, Japan) using a horizontal configuration, rotating drum collector, and the same operating parameters listed above.

Thermal processing of electrospun fabrics:

Electrospun fabrics were heated under different conditions in a Hewlett Packard 5890 Series II gas chromatography oven under air. The samples were mounted on a Teflon-wrapped glass sample holder and heated from 25°C at a specified heating rate up to a specified final temperature and holding at the maximum temperature for 1 hour. Heating rates of 0.5, 5, and ~40 °C/min and final temperatures of 150, 200, and 250°C were used. Each condition was repeated three times and tensile testing was carried out for each of the three runs using 20 samples per run. The results were reported as the average of the three runs +/- the standard deviation. Following thermostabilization, carbonization was carried out in a GSL 1100X2 tube furnace (MTI Corporation, Richmond, CA) under N_2 heating at a rate of 10°C/min to 1000°C and holding at 1000°C for 1 hour.

Characterization of fibers and fabrics:

Optical and scanning electron microscopy was used to characterize fiber appearance using and Olympus BX41 Microscope (Richmond Hill, ON) and Hitachi S3000N SEM using gold-coated samples and accelerating voltage of 5 kV. Fiber diameters were measured using ImageJ analysis software counting a minimum of N = 100 fiber diameters per sample. The tensile properties of electrospun nonwovens fabrics were measured on a KES-G1 Kawabata microtensile tester as described elsewhere.[35] A density of 1.3 g/mL was obtained from an MSDS for alkali lignin and used to convert tensile data from units of N/tex to units of MPa.

Results and Discussion

Shear rheometry:

Shear rheometry has been used to characterize the viscosity of electrospinning solutions in the literature and correlations between shear viscosity, electrospinnability, and fiber diameter have been reported for a variety of different systems.[36-38] We first reported on the shear properties of lignin solutions with and without PEO in DMF in a published study.[32] In general, the shear properties of lignin solutions was Newtonian, with a viscosity independent of shear rate. However, it was clear in previous work that different lignins have different viscosities.[32] The observed differences underscore the inherent variability in lignins obtained by different isolation procedures or from different botanical origins. To focus our study, we selected to characterize SKL in greater detail.[32] The viscosity of SKL solutions varied over roughly 3 orders of magnitude when increasing from 25-50%. The rather strong dependence of viscosity on SKL concentration $\eta \sim c^7$ indicated that intermolecular interactions between SKL molecules increased the viscosity. Interestingly, solutions with PEO displayed a similar dependency of viscosity with concentration, with slightly higher values of viscosity but overall showing the same trend. However, SKL:PEO solutions were electrospinnable at shear viscosities far lower than those of SKL solutions. This result indicated that shear viscosity was not the key fluid property influencing the electrospinning behavior, since typically higher viscosity promotes fiber formation.

CaBER:

While viscosity affects the electrospinning process, it is also known that the elastic properties of polymer solutions strongly affect the ability to form fibers and the fiber diameter.[39-43] Based on comparison of our results with reports from the literature, we hypothesized that elasticity resulting from PEO addition may be responsible for the improved electrospinnability of lignin:PEO solutions. To test our hypothesis we varied the elasticity by varying the PEO molecular weight and the concentrations of SKL and PEO and investigated the ability to form fibres and the fibre diameter. Furthermore, we characterized the extensional properties of our solutions using a capillary breakup extensional rheometer (CaBER).

Figure 1 shows the thinning profile of the SKL solutions without PEO for concentrations of 40, 45, and 50%. Time is on the abscissa and D_{mid} as recorded by a laser micrometer is shown on the ordinate normalized by the initial filament diameter D_1. In general SKL solutions at or below 35% concentration could not be measured using the CaBER due breakup occurring before the end of the initial step. In Figure 1 it can be observed that the filament thinning profiles were linear in time, consistent with the behavior of a Newtonian fluid, in agreement with shear measurements.[44]

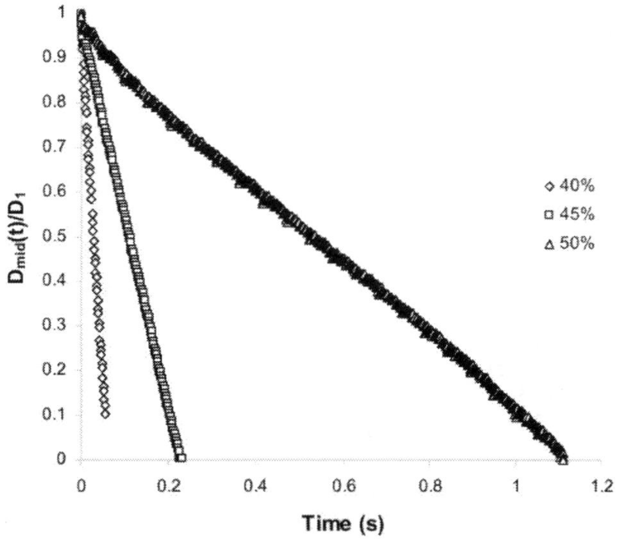

Figure 1: Thinning profiles of SKL solutions without PEO from CaBER experiments

In contrast, solutions with PEO displayed very different thinning behavior. Figure 2 shows the thinning profiles of several solutions with different SKL concentration, PEO concentration, and PEO molecular weight. Solutions with PEO display exponential thinning behavior, which indicates elasticity.[44] Elasticity implies that the extensional viscosity η_{ext} is not constant during thinning but increases with decreasing filament diameter (strain hardening). The relaxation time λ was determined from the CaBER data. Table 1 summarizes the compositions of the fluids and the values of λ obtained from the data. It can be seen in the table that λ increases with SKL concentration, PEO concentration, and PEO molecular weight, in agreement with published results.[45] Taken together these results show that the viscoelasticity of the SKL:PEO spinning solution can be varied by changing the relative amounts of SKL and PEO as well as the PEO molecular weight. The results also suggest that strain hardening occurring in extensional flow (which is absent in solutions without PEO) may provide the stability previously observed for lignin:PEO solutions in electrospinning.[32] We now discuss the effect of viscoelasticity on electrospinning.

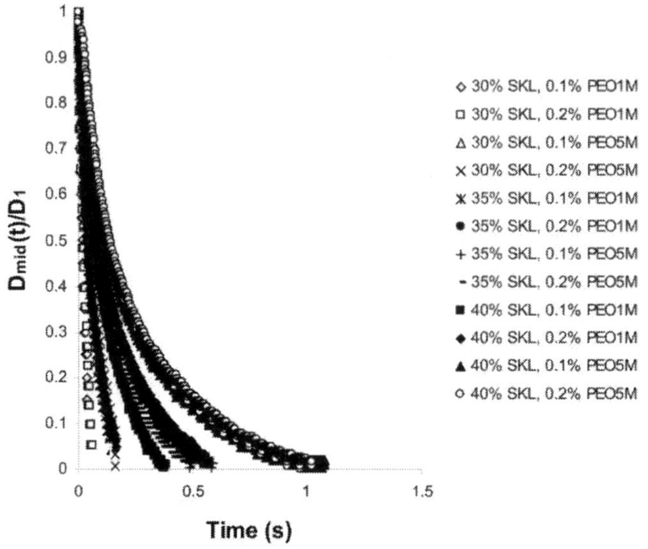

Figure 2: Thinning profiles of SKL:PEO solutions from CaBER experiments

Table 1: Fluid compositions and relaxation times

SKL concentration (%)	PEO concentration (%)	PEO mol. weight (g/mol)	λ (ms)
30	0.1	1×10^6	7.3
30	0.2	1×10^6	11.4
30	0.1	5×10^6	21.7
30	0.2	5×10^6	24.7
35	0.1	1×10^6	19.1
35	0.2	1×10^6	21.1
35	0.1	5×10^6	50.6
35	0.2	5×10^6	49.5
40	0.1	1×10^6	39.9
40	0.2	1×10^6	63.1
40	0.1	5×10^6	92.3
40	0.2	5×10^6	99.3
45	0.1	1×10^6	134.8
45	0.2	1×10^6	150.6
45	0.1	5×10^6	240.8
45	0.2	5×10^6	408.8

Electrospinning:

Example SEM images are shown in Figure 3 to illustrate the results of spinning experiments. Similar to our previous observations, only electrospray was formed without PEO below a concentration of 45-50%, with beaded fibers forming at 45%. In contrast, nearly all of the solutions with PEO were capable of forming fibers even though the amount of PEO in solution was less than 1% relative to SKL, producing fibers with SKL contents greater than 99%. The ability to achieve processibility with low PEO content is an advantage of the SKL:PEO system. Comparing rheology and electrospinning experiments, it can clearly be seen that fiber diameter increases with increasing λ. Figure 4 shows a plot of diameter vs. λ. The mean fiber diameter correlates well with λ, indicating that elasticity is correlated with the fiber diameter. Furthermore, the error bars show the standard deviations of the measured fiber diameters. It is also clear from the plot that the width of the diameter distribution tends to increase with increasing λ. This is partially due to the flattened, twisted morphology which tended to occur in solutions with higher viscoelasticity. The flattened morphology may be due to unevaporated solvent trapped in the core of the fiber followed by collapse of the fiber during the bending instability. The results presented here demonstrate that by controlling the viscoelasticity of the spinning solution, the mean diameter and width of the diameter distribution can be controlled.

Figure 3: SKL:PEO fibers produced from solutions with different viscoelasticities (Scale bar = 10 μm)

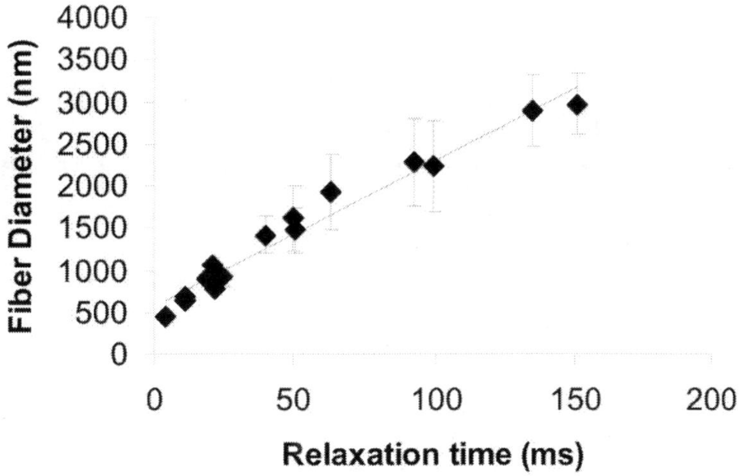

Figure 4: Fiber diameter vs. relaxation time for SKL:PEO solutions

Thermal Processing:

To prepare carbon fibers from lignin, an oxidative thermostabilization stage is necessary to prevent fibers from fusing during carbonization.[4] In previous work a slow heating rate was required to prevent fusion of melt-spun hardwood Kraft lignin (HKL) fibers.[4] The necessity of using a slow heating rate can be considered a disadvantage since it increases the time and energy required to generate carbon fibers. Due to its different chemical structure, SKL is expected to be more thermally stable.[4] However, previous attempts at melt spinning SKL were unsuccessful due to the tendency of SKL to char, so the thermostabilization of SKL fibers has not been reported in the literature. An SEM of a sample carbonized with and without thermostabilization is shown in Figure 5, showing fusion of fibers on the left and non-fused fibers on the right. To investigate thermostabilization of SKL fibers, we subjected our electrospun fabrics to different heating rates and final heating temperatures. It was observed that in contrast to melt-spun fibers or non-thermostabilized samples, none of the samples fused, indicating good stability. These samples were therefore suitable for conversion to carbon nanofibers. An SEM of carbonized 99% SKL fibers is also shown in Figure 5.

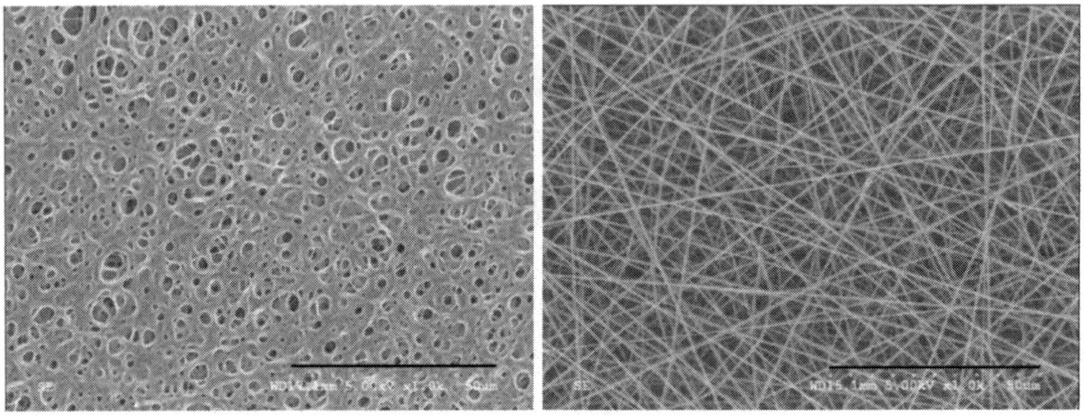

Figure 5: Carbonized 99% SKL fibers without thermostabilization (left), and after thermostabilization at 5°C/min to 250°C (right). Scale bar = 50 μm

Tensile testing:

To investigate the mechanical properties of SKL nonwoven mats, samples thermostabilized under different heating rates and temperatures but with similar diameters on the order of 1 μm were subjected to microtensile testing. The modulus and tensile strength values for different heating rates and thermostabilization temperatures are shown in Table 2. A steady increase in tensile strength was observed as the maximum treatment temperature increased from roughly 2-4 MPa for fibers heated to 150°C to values of more than 20 MPa for samples heated to 250°C. A similar trend was observed for elastic modulus, which increased from roughly 300 MPa to values of 1 GPa for fibers treated at 150 and 250°C, respectively. Interestingly, fibers heated at the lowest heating rate had a lower tensile strength, while comparable tensile strength was observed for samples heated at 5°C/min or faster. This could indicate that longer heating times lead to degradation of the mechanical properties of SKL fibers. The results presented here demonstrate the feasibility of generating precursors for carbon nanofibers from SKL by electrospinning.

Table 2: Mechanical properties of thermostabilized SKL:PEO nonwoven fabrics

Heating rate (°C/min)	T_{max} (°C)	Tensile strength (MPa)	Standard Deviation (MPa)	Elastic Modulus (MPa)	Standard Deviation (MPa)
As-spun	n/a	4.0	1.1	327.6	67.0
0.5	150	2.9	1.0	324.8	147.3
0.5	200	12.4	3.4	938.9	199.0
0.5	250	14.5	2.9	818.2	160.1
5.0	150	1.7	0.6	231.7	117.1
5.0	200	13.3	3.0	966.1	147.0
5.0	250	24.0	5.9	1055.9	269.5
47.0	150	3.7	1.0	322.6	121.6
41.0	200	10.7	1.9	849.4	101.6
39.0	250	22.1	3.7	1131.4	149.9

Summary:

The results from rheological experiments have demonstrated for the first time that while SKL solutions without PEO behave like Newtonian fluids while undergoing extensional flow, SKL:PEO solutions display non-Newtonian elastic behavior. We have demonstrated that SKL:PEO solution viscoelasticity strongly influences the fiber diameter and the width of the diameter distribution using shear rheometry and CaBER coupled with electrospinning experiments. By varying the SKL concentration, PEO concentration, and PEO molecular weight, we have demonstrated the ability to vary the viscoelastic properties over a wide range with less than 1% PEO added to the spinning solution (relative to SKL mass). With a sufficient understanding of SKL:PEO solution viscoelasticity, we then carried out thermostabilization experiments and measured the tensile properties of nonwoven SKL fabrics as a function of temperature and heating rate. In contrast with some previous reports on melt-spun fibers produced from HKL, electrospun SKL fibers did not fuse under any of the thermostabilization conditions tested, whereas carbonization performed without thermostabilization resulted in fiber fusion. In future work we hope to further optimize the process of lignin electrospinning to move toward toward carbon nanofiber precursors with properties suitable for real-world applications.

Acknowledgements

We would like to acknowledge North Carolina State University, the U.S. Department of Agriculture, and the Natural Sciences and Engineering Research Council of Canada (NSERC) for supporting this research.

References

1) Boerjan, W., Ralph, J., Baucher, M. (2003) Lignin Biosynthesis. *Annu. Rev. Plant Biol.* 54; 519-546

2) Ralph, J., Lundquist, K., Brunow, G., Lu, F., Kim, H., Schatz, P.F. Marita, J.M., Hatfield, R.D., Ralph, S.A., Christensen, J.H., Boerjan, W. (2004) Lignins: Natural polymers from oxidative coupling of 4-hydroxyphenylpropanoids. *Phytochemistry Reviews* 3; 29-60

3) Brodin, I., Sjoholm, E., Gellerstedt, G. (2009) Kraft lignin as feedstock for chemical products: The effects of membrane filtration. *Holzforschung*: 63, 290-297.
4) Kadla, J.F., Kubo, S., Venditti, R.A., Gilbert, R.D., Compere, A.L. (2002) Lignin-based carbon fibers for composite fiber applications. *Carbon* 40: 2913–2920
5) Morgan, P (2005). Carbon fibers and their composites Taylor and Francis, Boca Raton
6) Im, J.S., Park, S.J., Kim, T., Lee, Y-S. (2009) Hydrogen storage evaluation based on investigations of the catalytic properties of metal/metal oxides in electrospun carbon fibers International Journal of Hydrogen Energy; 34: 3382-3388
7) Osmond, N.M. (2000) Activated carbon fibre adsorbent materials. *Adsorption science and technology* 18, 6, 529-539
8) Sharma, C.S., Katepalli, H., Sharma, A., Madou, M. (2011) Fabrication and electrical conductivity of suspended carbon nanofiber arrays. *Carbon*; 49: 1727-1732
9) Aoki, K., Usui, Y., Narita, N., Ogiwara, N., Iashigaki, N., Nakamura, K., Kato, H., Sano, K., Ogiwara, N., Kametani, K., Kim, C., Taruta, S., Kim, Y.A., Endo, M., Saito, N. (2009) A thin carbon-fiber web as a scaffold for bone-tissue regeneration. *Small* 5: 13, 1540-1546.
10) Kubo S, Ishikawa N, Uraki Y, Sano Y. (1997) Preparation of lignin fibers from softwood acetic acid lignin—Relationship between fusibility and the chemical structure of lignin. *Mokuzai Gakkaishi* 43, 8, 655–662.
11) Dave,V., Prasad, A., Marand, H., Glasser, W. (1993) Molecular organization of lignin during carbonization. *Polymer* 34, 15, 3144-3154.
12) http://www.ornl.gov/ornlhome/news_items/news_100920.shtml
13) Kubo, S., Kadla, J.F. (2005a) Lignin-based Carbon Fibers: Effect of Synthetic Polymer Blending on Fiber Properties. *Journal of Polymers and the Environment*, Vol. 13, No. 2. 97-105.
14) Kubo, S., Yoshida, T., Kadla, J.F. (2007) Surface Porosity of Lignin/PP Blend Carbon Fibers. *J. Wood Chemistry and Technology* 27; 257-271.
15) Kubo, S., Kadla, J.F. (2005b) Kraft Lignin/Poly(ethylene oxide) Blends: Effect of Lignin Structure on Miscibility and Hydrogen Bonding. *Journal of Applied Polymer Science* 98; 1437–1444
16) Kadla, J.F., Kubo, S. (2003) Miscibility and Hydrogen Bonding in Blends of Poly(ethylene oxide) and Kraft Lignin. *Macromolecules 36:* 7803-7811
17) Kadla, J.F., Kubo, S. (2004) Lignin-based polymer blends: analysis of intermolecular interactions in lignin-synthetic polymer polymer blends. *Composites, Part A: Applied Science and Manufacturing* 35; 395-400
18) Kubo, S., Kadla, J.F. (2006) Effect of poly(ethylene oxide) molecular mass on miscibility and hydrogen bonding with lignin. *Holzforschung*. 60; 245–252.
19) Tagawa, T., Miyata, T. (1997). Size effect on tensile strength of carbon fibers. *Materials Science and Engineering* A238, 336–342
20) Tanaka, T., Nakayama, H., Sakaida, A., Horikawa, N., (1999) Estimation of tensile strength distribution for carbon fiber with diameter variation along fiber. *Materials Research Science International* 5, 2, 90-97.
21) Rutledge, G.C., Fridrikh, S.V. (2007) Formation of fibers by electrospinning. *Advanced Drug Delivery Reviews* 59, 1384-1391.
22) Reneker, D.H., Yarin, A.L. (2008) Electrospinning jets and polymer nanofibers. *Polymer* 49, 2387-2425

23) Greiner, A., Wendorff, J.H. (2007) Electrospinning: A fascinating method for the preparation of ultrathin fibers. *Angew. Chem. Int. Ed.* 46, 5670-5703.
24) Joshi, P., Zhang, L., Chen, Q., Galipeau, D., Fong, H., Qiao, Q. (2010) Electrospun Carbon Nanofibers as Low-Cost Counter Electrode for Dye-Sensitized Solar Cells. *ACS Applied Materials and Interfaces*; 2: 12, 3572-3577.
25) Arshad, S.N., Naraghi, M., Chasiostis, I. (2011) Strong carbon nanofibers from electrospun polyacrylonitrile. *Carbon*; 49: 1710-1719
26) Moon, S., Farris, R. (2009) Strong electrospun nanometer-diameter polyacrylonitrile carbon fiber yarns. *Carbon* 47: 2829-2839
27) Imaizumi, S., Matsumoto, H., Suzuki, K., Minagawa, M., Kimura, M., Tanioka, A. J. (2009) Phenolic Resin-Based Carbon Thin Fibers Prepared by Electrospinning: Additive Effects of Poly(vinyl butyral) and Electrolytes. *Polymer Journal* 41: 12, 1124-1128.
28) Park, S.H., Kim, C., Yang, K.S. (2004) Preparation of carbonized fiber web from electrospinning of isotropic pitch. *Synthetic Metals* 143: 175-179.
29) Lallave, M., Bedia, J., Ruiz-Rosas, R., Rodríguez-Mirasol, J., Cordero, T., Otero, J.C., Marquez, M., Barrero, A., and Loscertales, I.G. (2007) Filled and Hollow Carbon Nanofibers by Coaxial Electrospinning of Alcell Lignin without Binder Polymers. *Advanced Materials* 19; 4292-4296
30) Ruiz-Rosas, R., Bedia, J., Lallave, M., Loscertales, I.G., Barrero, A., Rodrı´guez-Mirasol, J., Cordero, T. The production of submicron diameter carbon fibers by the electrospinning of lignin. *Carbon* 48 (2010) 696–705
31) Seo, D.K., Jeun, J.P., Kim, H.B., Kang, P.H. (2011) Preparation and characterization of the carbon nanofiber mat produced from electrospun PAN/lignin precursors by electron beam irradiation. *Rev. Mater. Sci.*; 28: 31-34.
32) Dallmeyer, I., Ko, F., Kadla, J.F. (2010) Electrospinning of technical lignins for the production of fibrous networks. *Journal of Wood Chemistry and Technology* 30: 315-329.
33) Morck, R., Yoshida, H., Kringstad, K.P., Hatakeyama, H. (1986) Fractionation of Kraft lignin by successive extraction with organic solvents. *Holzforschung* 40(suppl.): 51-60.
34) Rodd, L.E., Scott, T.P. Cooper-White, J.J. McKinley, G.H. (2005) Capillary breakup rheometry of low viscosity elastic liquids. *Applied Rheology* 15: 12-27.
35) Gandhi,M., Yang,H., Shor, L., Ko, F. (2009) Post-spinning modification of electrospun nanofiber nanocomposite from Bombyx mori silk and carbon nanotubes. *Polymer* 50: 1918-1924.
36) Gupta, P., Elkins, C., Long, T.E., Wilkes, G.L. Electrospinning of linear homopolymers of poly(methyl methacrylate): exploring relationships between fiber formation, viscosity, molecular weight and concentration in a good solvent. *Polymer* 46 (2005) 4799–4810
37) McKee, M.G., Wilkes, G.L., Colby, R.H. Long, T.E. (2004) Correlations of solution rheology with electrospun fiber formation of linear and branched polyesters. *Macromolecules* 37, 1760-1767
38) Shenoy, S.L., Bates, W.D., Frisch, H.L., Wnek, G.E. (2005). Role of chain entanglements on fiber formation during electrospinning of polymer solutions: good solvent, non-specific polymer–polymer interaction limit. *Polymer* 46, 3372–3384
39) Feng, J.J. (2002) The stretching of an electrified non-Newtonian jet: A model for electrospinning. *Physics of Fluids* 14, 11, 3912-3926.
40) Feng, J.J. (2003) Stretching of a straight electrically charged viscoelastic jet. *J. Non-Newtonian Fluid Mechanics* 116, 55-70.
41) Carroll, C.P. Joo, Y.L. (2006) Electrospinning of viscoelastic Boger fluids: modeling and experiments. *Physics of Fluids* 18, 053102-1-14.

42) Yu, J.H., Fridrikh, S.V., Rutledge, G.C. (2006). The role of elasticity in the formation of electrospun fibers. *Polymer* 47, 4789–4797

43) Helgeson, M.E., Grammatikos, K.N., Deitzel, J.M. Wagner, N.J. (2008) Theory and kinematic measurements of the mechanics of stable electrospun polymer jets. *Polymer* 49, 2924-2936

44) Entov, V.M., Hinch, E.J. (1997) Effect of a spectrum of relaxation times on the capillary thinning of a filament of elastic liquid. *J. Non-Newtonian Fluid Mechanics* 72, 31-53.

45) Clasen, C., Plog, J.P., Kulicke, W.-M., Owens, M., Macosko, C., Scriven, L.E., Verani, M., McKinley, G.H. (2006) How dilute are dilute solutions in extensional flows? *Rheology* 50: 6, 849-881.

(499d) Development of Nylon Biocomposites Through the Torrefaction of Waste Stream Agricultural by-Products

Jessica L. Lattimer and Chad A. Ulven, Mechanical Engineering, North Dakota State University, Fargo, ND

In the last several years the use of biobased materials for fillers in thermoplastics has seen a remarkable increase. The desire to produce lighter, stronger, and more ecologically friendly materials being key driving forces behind the movement. While the focus has thus far been on producing biobased composites out of commodity polyolefins or bio-derived resins, little work has been done in the realm of engineering thermoplastics. For engineering thermoplastics, the increased processing temperatures are unfavorable for biocomposite production, causing significant damage to the biobased fillers. As the automotive industry is one of the largest consumers of nylon and the ever increasing push to make cars green, there is a natural fit for research into producing biobased nylon composites. Finding a way to drive off the constituents within the biomass that degrade above the traditional processing temperatures of polyolefins would provide a new biomass that could withstand the processing of engineering thermoplastics.

From previous studies into the use of biobased materials for energy production it is known that torrefaction, conducted between 220-300 °C, will drive off the hemicellulose, fats, waxes, and other constituents yielding a biomass consisting of mostly cellulose, degrading between 300-375 °C, and lignin, degrading slowly over 250-500 °C. As cellulose is the constituent in biomass that acts as a reinforcing agent in the biocomposite, this is a potential method to prepare biobased fillers for introduction in to engineering thermoplastics. This study describes the development of nylon based biocomposites using torrefied sunflower hulls (TSFH), oat hulls (TOH), and flax shive (TFS).

To prepare the biocomposites, nylon was blended with each of the torrefied biomasses analyzed based on weight percentage. This study looks at three grades of the biocomposite; neat nylon-6, 16 wt% filler in nylon-6, and 20 wt% filler in nylon-6. The biocomposite samples prepared were tested for static and dynamic properties in addition to characterizing their resistance to moisture uptake as a function of torrefied biomass loading. Correlations between biomass input and loading are discussed.

See more of this Session: Biobased Materials II: Lignin-Based Materials
See more of this Group/Topical: Forest and Plant Bioproducts Division

(499f) Lignin-Based Carbonaceous Nano-Fibrous Felts

Chuilin Lai[1], Lifeng Zhang[1], Hao Fong[1] and **Lew P. Christopher**[2], (1)Chemistry, South Dakota School of Mines & Technology, Rapid City, SD, (2)Center for Bioprocessing Research and Development and Department of Chemical and Biological Engineering, South Dakota School of Mines & Technology, Rapid City, SD

Lignin is a naturally existing biopolymer widely found in lignocellulosic plant biomass at an average of 20-30%. It is also a high-volume byproduct of wood pulping which exceeds 15 million tons per year in the US. Nonetheless, lignin has been utilized mainly as a low-grade fuel for production of heat and power to date. There is however a growing demand to identify applications with high economic value for lignin. Although production of carbon fibers from lignin has been reported in literature, the low mechanical properties obtained so far have prevented their commercial use in reinforcement and/or composites applications. Nano-fibrous felts consisting of porous carbon nanofibers derived from lignin, however, have potential applications that do not rely on mechanical properties but on large specific surface area. The use of low-cost lignin as the precursor to develop carbonaceous nano-fibrous felts is anticipated to have a great impact on these applications. The nanomaterials processing technique of electrospinning provides a straightforward approach for convenient preparation of fibers with diameters ranging from tens to hundreds of nanometers. In this research, innovative carbonaceous nano-fibrous felts with high specific surface area were developed by electrospinning aqueous mixtures of alkali lignin and poly(vinyl alcohol) with varied mass ratios followed by heat treatments for stabilization and carbonization. The developed carbonaceous nano-fibrous felts had a large pore volume and specific surface area and are therefore expected to find important applications as catalyst support, supercapacitor electrode, and hydrogen storage materials.

See more of this Session: Biobased Materials II: Lignin-Based Materials
See more of this Group/Topical: Forest and Plant Bioproducts Division

(519a) Incorporating Sustainability Into the Design of Chemical Process: From Reaction Pathway Selection to Process Design

Kailiang Zheng and Helen Lou, Dan F. Smith Department of Chemical Engineering, Lamar University, Beaumont, TX

It is important to incorporate sustainability issues into the design of chemical process, as more and more social and environmental concerns arise, and natural resources become costly and less available. Sustainability is multi-dimensional and complex in nature, as economic, societal and environmental issues need to be considered together. Thus a structured approach is needed to assist in incorporating sustainability into the new process design, as well as the retrofit of an existing design. In this paper, a sequential approach from reaction pathway selection to process design is proposed, in light of sustainability. At the early design stage, the sustainability performance of each potential reaction pathway is evaluated in terms of possible profit, driving force of the pathway (Gibbs energy), inherent safety index, environmental analysis, mass efficiency, reaction selectivity and conversion, respectively. This can help the designers to screen out the inferior reaction alternatives systematically and efficiently, thus reduce the complexity and labor in the following design stage. Then in the detailed process design stage, the sustainability performance of each process is assessed sequentially considering economic, societal, environmental and efficiency issues.

Root cause analysis is a powerful tool to identify and illustrate the fundamental cause of a problem. Integrating root cause analysis into the presented sequential sustainable design approach can help identify the root cause and improvement opportunity for a specific design. The idea of this approach is illustrated via several case studies.

See more of this Session: Design and Operation for Sustainability
See more of this Group/Topical: Sustainable Engineering Forum

(519b) Optimization of CO_2 Capture Process with Aqueous Amines - a Comparison of Two Simulation-Optimization Approaches

Aroonsri Nuchitprasittichai, Chemical Engineering, The University of Tulsa, Tulsa, OK and Selen Cremaschi, Department of Chemical Engineering, The University of Tulsa, Tulsa, OK

Aqueous amine is a solvent considered for CO_2 recovery from the flue gas of a refinery gas turbine by chemical absorption/desorption process [1]. The performance and the economics of this process depends on the choice of amine absorbent and its concentration, the number of stages in absorber and stripper columns, and the operating conditions.

In this work, we used two different simulation-based optimization approaches to minimize the CO_2 capture cost for post-combustion CO_2 removal. The first simulation-based optimization technique uses local search space to estimate an appropriate direction to reduce the objective function, i.e., response surface methodologies. The optimal solution obtained using response surface methodologies will generally be a local optimum. In the second optimization approach, the simulation was used to build a surrogate model, i.e., an artificial neural network (ANN), of the objective function over the whole decision space, and the optimization was performed using the surrogate model. Depending on the accuracy of the surrogate models, the solutions obtained using this approach can be shown to be global within the bounds of the data used to generate the surrogate models.

The impact of different amines absorbents, design parameters and operating conditions on the CO_2 capture cost were analyzed using both optimization techniques. Monoethanolamine (MEA), diglycolamine (DGA), diethanolamine (DEA), methyl diethanolamine (MDEA), triethanolamine (TEA), and blended aqueous solutions of these amines have been considered in our analyses. The required specification for the CO_2 capture process was 96 mole% CO_2 purity at the product for all simulations. The results are the optimum number of stages for the absorber and stripper columns, and the optimum absorbent concentration and operating conditions, i.e., the ones that give the minimum cost for CO_2 capture process. In this presentation, the comparison of the results obtained from both simulation-optimization techniques will be discussed. This comparison will provide the accuracy and the validity of the response surface technique results for the CO_2 capture process optimization. It should be noted that response surface methodology is simpler to develop and requires less expert knowledge compared to the ANN approach. Therefore, the response surface methodology can be used for CO_2 capture cost optimization if the results are closer to the global optimum obtained by the ANN approach.

Reference:

[1] Mofarahi, M., Khojasteh, Y., Khaledi, H., & Farahnak, A. (2008). Design of CO_2 absorption plant for recovery of CO_2 from flue gases of gas turbine. Energy, 33, 1311-1319.

See more of this Session: Design and Operation for Sustainability
See more of this Group/Topical: Sustainable Engineering Forum

(519c) A New Process for TiO2 Production

David R. Corbin[1], Keith W. Hutchenson[1], Stephen E. Lyke[2], Eugene M. McCarron III[2], Mark B. Shiflett[1], Charlie C. Torardi[1] and Joseph J. Zaher[1], (1)Central Research and Development, DuPont Company, Wilmington, DE, (2)Titanium Technologies, DuPont Company, Wilmington, DE

Because of its high refractive index, stability and non-toxicity, TiO_2 has found application as the primary opacifying agent in paint, paper and plastics. Annual world consumption of TiO_2 is estimated to be about 10MMM lbs. With an annual growth rate of about 3%, significant new capacity is needed each year to meet the growing demand. This presentation will summarize our efforts at developing a novel next generation process for TiO_2 production based on ammonium binoxalate digestion of ilmenite ore.

See more of this Session: Design and Operation for Sustainability
See more of this Group/Topical: Sustainable Engineering Forum

(519d) Low Temperature Hydrothermal Crystallization for a New TiO2 Process

Keith W. Hutchenson[1], Sheng Li[2], David R. Corbin[1], Charles C. Torardi[1] and Eugene M. McCarron[1], (1)DuPont, Wilmington, DE, (2)Chemical and Biological Engineering, Princeton University, Princeton, NJ

This presentation summarizes the development and optimization of a novel low temperature hydrothermal route to recover base TiO_2 product from an intermediate titanyl hydroxide species formed in the TiO_2 process following digestion of the iron titanate ore feed material. This work was part of an effort to develop a new TiO_2 production process based on ammonium hydrogen oxalate digestion that can process low-grade ilmenite ores with lower energy and capital investment intensity and a smaller environmental footprint than conventional processes.

Statistically designed experiments have illustrated the feasibility of producing both pigment-grade and nano-size TiO_2 base materials at mild hydrothermal process conditions. Small scale screening experiments on synthetic intermediates were carried out to examine the effects of various reaction variables on the final TiO_2 properties. The screening study identified reaction time, hydrochloric acid to titanium ratio, and the concentration of a particular mineralizing salt as the three key factors most influential to the formation and growth of rutile TiO_2. Based on the results of the screening experiments, empirical process models, which express TiO_2 product properties as a function of the critical process variables, were developed using statistical response surface design techniques. The models were used to optimize the reaction conditions for the production of either pigmentary or nano rutile base products. The predicted reaction conditions were then validated at the same 10 mL scale.

The empirical model results identified a relatively narrow operating window for the production of pigmentary rutile TiO_2 at relatively low temperature (235-250 C) hydrothermal crystallization conditions. The process was successfully repeated on a 1L reactor scale and resulted in the production of pigmentary rutile base. This material was then successfully surface treated via the standard chloride wet treatment technology. Subsequent work validated the application of this process to an authentic ore-derived titanyl hydroxide intermediate and demonstrated the formation of pigmentary rutile TiO_2.

See more of this Session: Design and Operation for Sustainability
See more of this Group/Topical: Sustainable Engineering Forum

(519e) Integration of Electroplating Process Design and Operation for Simultaneous Productivity Maximization, Energy Saving, and Freshwater Minimization

Chaowei Liu, Chuanyu Zhao and Qiang Xu, Dan F. Smith Department of Chemical Engineering, Lamar University, Beaumont, TX

Electroplating annually produces numerous plated workpieces for U.S. pillar industries. However, it is also a major environment polluter in forms of wastewater, sludge, and spent solution. Thus, waste source reduction for the electroplating industry is a must. In practice, when productivity and energy use efficiency are also taken into account, the design and operation of an electroplating process become very complicated and need an in-depth study. In this paper, productivity maximization, energy saving, and freshwater/wastewater minimization are simultaneously addressed for the optimal design and operation of electroplating processes, which generates a triple-objective mixed-integer dynamic optimization (MIDO) model. The MIDO model is iteratively solved by a developed methodology to obtain the 3D Pareto frontier of the optimization problem, which provides important technical supports for the design and operation of electroplating processes. The efficacy has been demonstrated with a case study on an electroplating process.

Keywords: Hoist scheduling, Water network synthesis, Energy saving, Optimization

See more of this Session: Design and Operation for Sustainability
See more of this Group/Topical: Sustainable Engineering Forum

(519f) Design of Biofuel Additives Using Chemometric Modeling and Molecular Design Techniques

Subin Hada, Charles C. Solvason and Mario Richard Eden, Department of Chemical Engineering, Auburn University, Auburn, AL

Any type of feedstock that contains free fatty acids and/or triglycerides such as vegetable oil, waste oil, animal fat, and waste grease can be converted into biodiesel. Because of its biological origin, biodiesel has many advantages compared to its petroleum derived counterparts. However, it does exhibit several limitations like oxidative stability and low temperature operability that have impaired its wide use and commercialization. Simultaneous solution of these problems has proven difficult as improvements in one area tend to impair another. As a result, design of an optimal fuel for all environments can be a quite difficult task. In addition, feedstock origin will impact the final quality of the biodiesel product. Therefore, to produce fuel grade biodiesel, the characteristics of the feedstock are very important during the initial research and production stage.

Several approaches are possible for improving the fuel properties of biodiesel. Modification of the fatty acid composition through physical processes or uses of additives are the most prevalent. Fuel additives, such as antioxidants, cetane enhancers, or cold-flow improvers have become indispensable tools not only to alleviate the drawbacks described above, but also to assure that any fuel blend will meet international and regional standards regardless of origin. However, an additive solution to one problem often aggravates another problem. Furthermore, the questions of additive compatibility, required additive levels, the effect on other properties, and whether these additives function as designed for biodiesel fuels with differing fatty acid profiles still remain challenges and that need further investigation.

Therefore, it is desired to molecularly design biodiesel additives that simultaneously account for the unintended effect on other fuel properties in the neat and the blend fuel in order to achieve the performance properties of the petroleum based fuel. In this way, biofuels can be formulated that are adaptable to a range or blend of feedstocks and the desirable fuel characteristics like oxidative stability and wide operating temperature range.

In this work, we aim to identify all possible compounds which, when added to off spec biodiesel, results in a fuel that satisfies performance standards such as ASTM D6751 in the US and EN14214 in Europe. To meet this end, multivariate characterization data obtained from IR spectroscopy of common additives were combined with decomposition and property clustering techniques in a reverse problem formulation. In this approach, the fuel additive property targets are identified in the first reverse problem followed by molecular design to match the targets. The characterization data consists of multitude of properties of interest (such as cetane number, melting point, and kinematic viscosity) to ensure adequate performance. To facilitate an efficient design we consolidated these various properties into a latent property domain using principle component analysis (PCA) techniques. Finally, characterization based molecular design using group contribution parameters are then used to build novel additives that match the fuel specifications in the latent property space.

See more of this Session: Design and Operation for Sustainability
See more of this Group/Topical: Sustainable Engineering Forum

Rule of Thumb for Simulating Biomass Pyrolysis in Packed Bed Reactor

Widya Wijayanti, Ken-ichiro Tanoue Takahiro Suetomi, Tatsuo Nishimura,
Yamaguchi University, Ube, Japan

Miki Taniguchi and Ken-ichi Sasauchi,
Chugai Ro Co.ltd, Sakai, Japan

Introduction

Compared to other countries, Japan is particularly rich in forests; the percentage of Japan's landmass that is forested is 67%, which is the third highest in the world. Only 20% of all wood is effectively utilized as timber; the rest, including waste wood (4 million tons/year) and natural forests (10 million tons/year), is unused. In recent years, there has been increasing demand for quickly produced, environment-friendly energy resources; in particular, there is wide interest in the production of energy from woody biomass, using wood resources which are currently underutilized. The use of biomass to provide energy has many merits: it is renewable, abundant, easy to store, and carbon-neutral. As Japan has many mountains with steep slopes, it is difficult to transport the felled trees. So that, the development of a compact gasifier of high quality, which can cope with variations both in the amount of biomass collected and in energy demand, is necessary. Furthermore, in order to ensure that the system is used in the most efficient manner, it is necessary to increase our understanding of the reaction mechanisms involved in both decomposition of biomass and heat transfer in the biomass layer.

It has been well known that the woody biomass decomposed to char, gas with having the middle heating value and tar undergoing the pyrolysis. There are many chemical reaction models of pyrolysis[11)].

The heat transfer with including the chemical reactions undergoing the pyrolysis has been also investigated by Koufopanos et al [7)], Won et al[18)] and etc. The roughly mechanism of the heat transfer could be predicted by the numerical simulation [4)]. However, most of reports assumed that the volume didn't change undergoing the pyrolysis and there is no report how the thermal conduction in the bed changes with the effect of the volume reduction.

In this report, the numerical simulation of the unsteady thermal conduction without the chemical reaction has been conducted and compared with the experimental one. The yield of solid component undergoing the fast and slow pyrolysis in the packed bed of biomass particles has been also compared with the analyzed one by the different three kinetic models (model-1: Thurner et.al[16)], model-2: Mohan et.al[8)], model-3: Miller et.al[9)]). Furthermore, the relationship between the volume reduction process in the packed bed and the chemical reaction undergoing pyrolysis was elucidated by the formularization.

Numerical simulation for thermal conduction without considering chemical reaction

Two dimensional numerical simulation in the packed bed of the biomass particles for heat transfer has been conducted. The size of the calculation domain is r = 53mm and z = 90mm. The regular mesh is utilized. The number of the mesh along the radial component and along the axial component are 140 and 234, respectively. Specific heat in the packed bed can be expressed by using the specific heat of the biomass and air and volume fraction. Effective thermal conductivity in the packed bed is adapted by the equation, which was proposed by Yagi et al[19)]. Density in the packed bed is treated as experimental value.

The equation and the boundary conditions are descretized over a control volume using the finite difference method. The calculation program is made originally by using fortran language. The calculation has been conducted by using not only the dependence of the physical properties on the temperature bat also the change of the position of the top in the packed bed. The temperature profile in the packed bed and the temperature at the top of the packed bed are solved by the SOR method and Newton-Raphson method, respectively.

Experimental set up and procedure

The rig consists of the supplying of nitrogen gas, a tubular reactor, the cold traps for the tar and water, gas-sampling bag and gas flow meter[12-15]. The furnace has the maximum output of 1.5 kW and can set the wall temperature of the reactor up to about 1073 K. The cooled traps in the iced bath consist of two erlenmeyer flasks of 500cc that are filled with glass wool and solid of $CaCl_2$. Both the tar and water are adsorbed by the traps. Generated gas through the traps fills with the sampling bag (GL Science Co., Ltd) for measurement of gas component. Gas flow rate is measured by the wet type gas meter (Shinagawa Co., Ltd, W- NK-2B). O_2 monitor is set at the exit of the reactor. Air in the reactor is exhausted by the change of the nitrogen gas. The nitrogen gas introduces into the reactor until the concentration of O_2 at the exit of the reactor is less than 1 %. After that, the supplying of the nitrogen gas stops. The slow pyrolysis whose heating rate is 400 K/hr has been applied conducted to investigate the heat transfer.

Pyrolysis in the packed bed of biomass particles[13-15]

For the slow pyrolysis, woody biomass particles of W_0 = 100 g and D_{p50} = 1.1 mm are introduced into the tube reactor. The wall of the reactor is heated by the electric furnace with the heating rate at 400 K/hr up to 1073 K. The temperature of the wall is maintained at 1073 K for about 60 min. The temperatures of the packed bed of biomass particles and the wall of the reactor are measured by the K-type thermocouples (Yamari Co., Ltd., ϕ= 1mm) every 1 minute and recorded in a memory of the personal computer through the data logger (Keyence Co., Ltd, NR-1000). Heat flow rate can be calculated by the change of the average temperature in the biomass layer for 1 minute.

Shrinking process in the packed bed of biomass particles[14, 15]

Dry biomass is set in the quartz beaker (63mm in height and 50mm in internal diameter). The weight of the layer and height of the layer show about 9 g and about 30mm, respectively. The beaker with having three thermocouples is set in the test rig. After the switch on of the heater, visualization of the biomass layer and temperature measurement are started. In order to avoid the tar-adsorption on the visualized window, the nitrogen gas gets continuously into the reactor. *Pseudotsuga menziesii* and wood

Table 1. Properties of biomass

True density of wood[10] (g/cm^3)	ρ_T	1.5							
Kind of biomass	–	*Pseudotsuga menziesii*				Wood bark			
Size of biomass particles (mm)	Dp	1				0.5			
Component	–	Hemicellulose	Cellulose	Lignin	Ash	Hemicellulose	Cellulose	Lignin	Ash
Mass fraction[1] (wt%)	w_i	24.4	45.8	28.0	1.7	29.8	24.8	43.8	1.6

bark are used as the sample of biomass. The components of the sample are listed in table 1.

Numerical simulation for chemical reaction and shrinking process in the packed bed of biomass particles

Chemical reaction models undergoing pyrolysis of biomass

There are many models for pyrolysis of biomass. In this study, three models were chosen in these models. The model-1 shows the representative model undergoing the pyrolysis[16]. In the model-1, gas, tar and char are generated by three- parallel reactions. The parameters of the rate constant, which were reported by Di Blasi et al[5]., were used. The model-1 doesn't consider the effect of main components of biomass.

On the other hand, The model-2[10] and model-3[11] shows that Hemicellulose, Cellulose and Lignin decompose parallel to gas, tar and char. For the model-2, the parameters of the rate constant, which were reported by Varhegyi et al[17], Capart et al. [3] and Chan et al. [2], were used. For the model-3, the parameters of the rate constant, which were reported by Miller et al[9], were used.

The effect of the char formation due to the secondary decomposition by tar was ignored. The yield of solid component every model and every setting temperature were calculated by the Runge Kutta method. The mass fraction for Hemicellulose, Cellulose and Lignin in the table 1 was utilized as the initial weight of for the model-2 and –3.

Shrinking process in the packed bed of biomass particles[15]

The volume of the packed bed is given by

$$V_{PB} = V_G + V_B \quad (1)$$

Where a V_G and a V_B show the volume in the gas phase and the volume in the solid phase, respectively. If the equation 1 is divided by $V_{PB,0}$,

$$\frac{V_{PB}}{V_{PB,0}} = \frac{V_G}{V_{PB,0}} + \frac{V_B}{V_{PB,0}} \quad (2)$$

Firstly the V_G was decided experimentally using the volume fraction ε_{PB} and the equilibrium volume of the packed bed. The dependence of the V_G on the setting temperature were shown by the approximated lines[15]. Secondary the V_B was solved by the numerical simulation for the chemical reaction. A weight of the solid component of biomass particles $[S_B]$ is given by

$$[S_B] = \rho_t V_B \quad (3)$$

Where ρ_t shows the true density of woods and its value[10] is about 1.5 g/cm^3. If the true density doesn't change undergoing the pyrolysis, the relationship between the volume and the weight of the solid component of biomass is given by,

$$\frac{V_B}{V_{B,0}} = \frac{[S_B]}{[W_{B,0}]} \quad (4)$$

Where a V_{B0} and a $[W_{B,0}]$ shows the initial volume and the initial weight of the biomass, respectively. From these procedures, the volume of the packed bed can be expressed by

For Pseudotsuga menziesii

$$\frac{V_{\mathrm{PB}}}{V_{\mathrm{PB,0}}}=\begin{cases}\dfrac{V_{\mathrm{B,0}}}{V_{\mathrm{PB,0}}}\dfrac{[S_B]}{[W_{\mathrm{B,0}}]}+0.899 & (T_{\mathrm{S}}\leq 532\,K)\\[2ex] \dfrac{V_{\mathrm{B,0}}}{V_{\mathrm{PB,0}}}\dfrac{[S_B]}{[W_{\mathrm{B,0}}]} & \\ +0.344+12.6*\exp\left(-\dfrac{T_{\mathrm{S}}}{171}\right) & (T_{\mathrm{S}}\leq 532\,K)\end{cases}$$

(5)

For bark

$$\frac{V_{\mathrm{PB}}}{V_{\mathrm{PB,0}}}=\begin{cases}\dfrac{V_{\mathrm{B,0}}}{V_{\mathrm{PB,0}}}\dfrac{[S_B]}{[W_{\mathrm{B,0}}]}+0.833 & (T_{\mathrm{S}}\leq 665\,K)\\[2ex] \dfrac{V_{\mathrm{B,0}}}{V_{\mathrm{PB,0}}}\dfrac{[S_B]}{[W_{\mathrm{B,0}}]} & \\ +0.672+108*\exp\left(-\dfrac{T_{\mathrm{S}}}{102}\right) & (T_{\mathrm{S}}\leq 665\,K)\end{cases}$$

The solid component $[S_{\mathrm{B}}]$ was calculated by solving the equation of the pyrolysis using the Runge Kutta method. By substituting the measured temperature in the packed bed at $t = t$ T_{PB} (t) to T_{S}, the volume of the packed bed was solved by equation 5.

Results and discussion

Experimental and calculated results for the heat transfer undergoing the pyrolysis [13, 14]

Fig. 1 shows the time course of the average temperature (T_{av}) in the packed bed of the biomass particles. Plots shows the experimental results while the solid line shows the calculated average temperature without the effect of chemical reaction. For $T_{\mathrm{av}} < 573$K, the experimental results can be reproduced by the calculation. However, the experimental results for 573K $< T_{\mathrm{av}}$ didn't agree with the calculated results because the packed bed of biomass is shrinking dramatically due to the formation of tar from cellulose. The temperature beyond the biomass layer rather than the temperature in the biomass layer was measured by the thermocouples at $z = 50$mm. Therefore, in order to reproduce the heat transfer in the biomass layer, it is necessary to investigate the shrinking process in the packed bed of biomass particles.

Relationship between the Formation Rate of Solid Component and Volume Reduction of Packed Bed of Woody Biomass Particles during Pyrolysis [14,15]

Fig 2 shows the relationship between the yield of the solid component and the setting temperature. Fig. 2a shows the results for *Pseudotsuga menziesii*. ○ shows the experimental results while the lines shows the calculated results for three models. The dependence of the experimental yield on the setting temperature is divided by almost two zones. For the zone 1 at $T_{\mathrm{S}} < 623$ K, the yield decreased

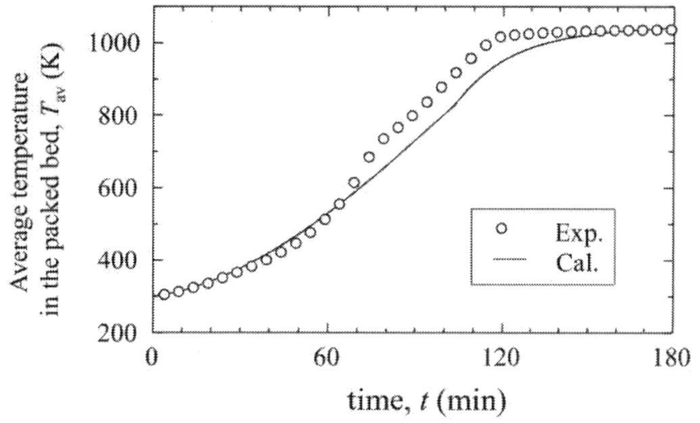

Fig.1 Comparison of the measured average temperature in the packed bed of biomass particles with the calculated one.

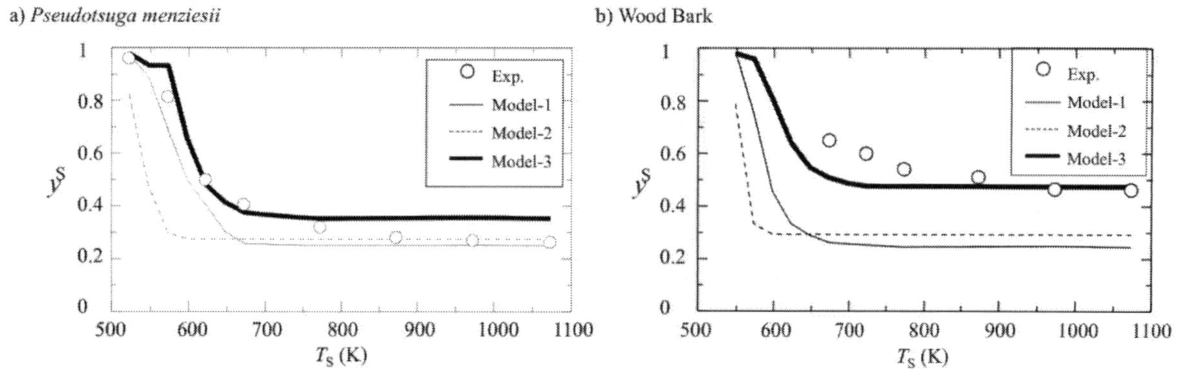

Fig. 2 Comparison of the experimental solid yield with the predicted one by three different kinetic models.

dramatically. In this zone, all components in the wood decomposed to tar and gas [6]. For zone 2 at $T_S >$ 623 K, the yield decreased slightly. In this zone, Lignin in the wood mainly decomposed to tar and gas[6]. The experimental yield for *Pseudotsuga menziesii* could be reproduced by the model-1 and model–3 except for the model-2. On the other hand, in fig 2b there was only 1 zone which the yield decreased slightly. As the experimental yield for Wood bark could be reproduced by the model–3, it was important to consider the lignin contents.

Fig.3 shows the time course of the volume in the packed bed for various setting temperatures. Fig. 3a shows the results for *Pseudotsuga menziesii.*○ shows the experimental results while the lines shows the calculated results. At $t <$ 50min, the average temperature in the packed bed of biomass was less than 523K and the volume of the layer changes hardly. At $t >$ 55min, the packed bed of biomass becomes carbonizing and then shrinking dramatically at $t >$ 65min. The experimental volume for the sawdust of *Pseudotsuga menziesii* could be reproduced by the model-1 and model–3. On the other hand, in fig 3b the volume decreased slightly with time and setting temperature. The experimental volume for Wood bark could be reproduced by the model–3.

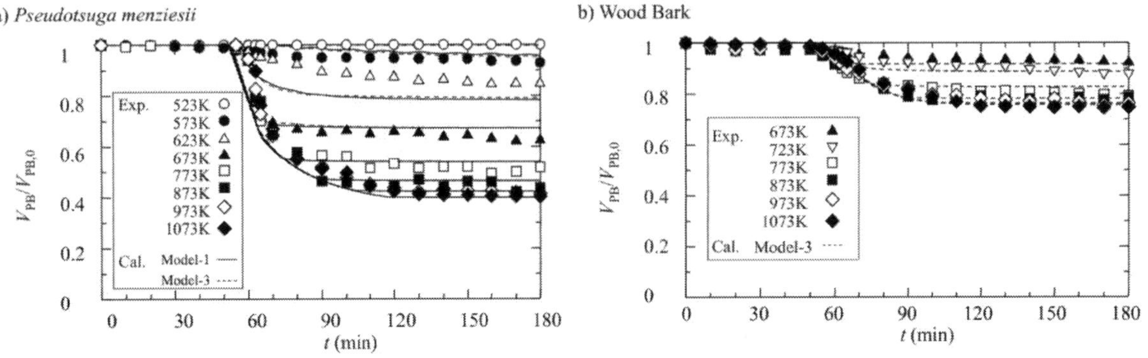

Fig. 3 Time course of the volume reduction process undergoing the pyrolysis.

Therefore the volume in the packed bed of the biomass undergoing the pyrolysis can be estimated by using the model-3 and the dependence of gas volume in the bed on the temperature.

The effect of the shrinking process on the heat transfer will be reported on the day of the presentation.

Conclusions

Transport phenomena in the pyrolysis of woody biomass has been investigated experimentally and numerically. Main conclusions are summarized as follows;

1) For $T_{av} < 573$K, the time course of the average temperature in the packed bed can be reproduced by the calculation. However, the experimental results for 573K $< T_{av}$ didn't agree with the calculated results. The packed bed of biomass is shrinking dramatically due to the formation of tar from cellulose. The temperature beyond the biomass layer rather than the temperature in the biomass layer was measured by the thermocouples at $z = 50$mm. Therefore, in order to reproduce the heat transfer in the biomass layer, it is necessary to investigate the shrinking process in the packed bed of biomass particles.
2) The yield of solid component undergoing the fast and slow pyrolysis in the packed bed of biomass particles has been compared with the analyzed one by the different three kinetic models (model-1: Thurner et.al, model-2: Mohan et.al, model-3: Miller et.al). From the results, the yield of solid component could be reproduced by the analysis using the model-3 even if the heating rate and the lignin content changed. Furthermore, the relationship between the volume reduction process in the packed bed and the chemical reaction undergoing pyrolysis was elucidated by the formularization. From the result, the volume in the packed bed of the biomass undergoing the pyrolysis can be estimated by using the model-3 and the dependence of gas volume in the bed on the temperature.

Refernces

1. Ayhan D. (1997)," Calculation of higher heating values of biomass fuels," Fuel, **76-5**, 431-434.
2. Chan, R. W. C., Krieger, B. B. (1981), "Kinetics of dielectric loss microwave degradation of polymers: lignin," J. Appl. Poly. Sci., 26-5, 1533-1553.
3. Capart, R., Khezami, L., Burnham, A. K., (2004)," Assessment of various kinetic models for the pyrolysis of a microgranular cellulose," Thermo. ACTA, 417-1, 79-89.
4. Di Blasi, C. (2008), "Modeling chemical and physical processes of wood and biomass pyrolysis," Prog. Energy Comb. Sci., **34**, 47-90.
5. Di Blasi, C. and Branka, C. (2001)," Kinetics of Primary Product Formation from Wood

Pyrolysis," Ind. Eng. Chem. Res., **40-23**, 5547-5556.
6. Hasegawa, I., Fujisawa, H., Sunagawa, K., Mae, K. (2005)" Quantitative Prediction of Yield and Elemental Composition during Pyrolysis of Wood Biomass", J. Jpn. Inst. Energy, **84** (1), 46-52 (in Japanese).
7. Koufopanos, C. Papayannakos, A., N., Maschio, G. and Lucchesi, A.(1991), "Modelling of the Pyrolysis of Biomass Particles. Studies on Kinetics, Thermal and Heat Transfer Effects," Can. J. Chem. Eng., 69, pp.907-915.
8. Mohan, D., Pittman Jr., C. U. and Steele, P. H. (2006)," Pyrolysis of Wood/Biomass for Bio-oil: A Critical Review," Energy Fuels, **20-3**, 848-889.
9. Miller, R. S., Bellan, J. (1997)" A Generalized Biomass Pyrolysis Model Based on Superimposed Cellulose, Hemicelluloseand Liqnin Kinetics" , Combust. Sci. Technol., 126, 97-137.
10. Okano, K. and Sofue, N., Mokuzai kagaku handbook, Asakura shoten, 117(2006).
11. Okumura, Y., (2011)," Modeling of Pyrolysis and Gasification Reactions of Many Types of Biomass" , J. Jpn. Inst. Energy, **90** (2), 122-131 (in Japanese).
12. Tanoue, K., Nishimura, T., Hinauchi, T., Uchida, K., Taniguchi, M. and Sasauchi, K.(2007), "Analysis of pyrolysis of woody biomass with heating rate," Energy and Resources, 28-3, pp.186-192 (in Japanese).
13. Tanoue, K., Hinauchi, T., THAUNG, O., Nishimura, T., Taniguchi, M. and Sasauchi, K.(2007), "Modeling of heterogeneous chemical reactions caused in pyrolysis of biomass particles," Adv. Powder Technol., 18-6, pp.825-840.
14. Tanoue, K., Widya, W., Yamasaki, K., Kawanaka, T., Yoshida, A., Nishimura, T., Taniguchi, M., Sasauchi, K., (2010) "Numerical Simulation of the thermal conduction of packed bed of woody biomass particles accompanying volume reduction induced by pyrolysis", J. Jpn. Inst. Energy, **89** (10), 948 (in Japanese).
15. Tanoue, K., Yamasaki, K., Nishimura, T., Taniguchi, M., Sasauchi, K., "A Relationship between the Formation Rate of Solid Component and Volume Reduction of Packed Bed of Woody Biomass Particles during Pyrolysis", J. Jpn. Inst. Energy, in press (in Japanese).
16. Thurner, F., Mann U. (1981)" Kinetic investigation of wood pyrolysis" , Ind. Eng. Chem. Proc. Des. Dev., **20**, 482-488.
17. Varhegyi, G., Antal Jr., M.J., Szekely, T. and Szabo, P. (1989)"Kinetics of the thermal decomposition of cellulose ,hemicellulose, and sugar cane bagasse", Energy Fuels, 3-3, 329-335
18. Won, C. P, Arvind, A. and Howard, R. B. (2010)"Experimental and theoretical investigation of heat and mass transfer processes during wood pyrolysis" , Combust. Flame, **157**, 481-494.
19. Yagi,S and Kunii, D., International Deveropment in Heat Transfer, Part IV, 742&750 (1961).

Acknowledgements
This work was supported in part by a Grant-in Aid for Scientific Research C (No. 19560174) from the Japan Society for the Promotion of Science. Our special thanks are extended to Dr. Morihisa Yokota of Ube Industries, Ltd., for helpful discussions on the work.

SYNTHESIS OF SUSTAINABLE PROPERTY- BASED WATER NETWORKS

Luis Fernando Lira-Barragán,[a] José María Ortega-Ponce,[a] Medardo Serna-González,[a] Mahmoud M. El-Halwagi[c]

[a] *Universidad Michoacana de San Nicolás de Hidalgo, Morelia, Mich., México*

[b] *Chemical Engineering Department, Texas A&M University, College Station, TX, USA*

Introduction

Water is one of the most important natural resources that support the life in the world. The chemical and process industry is one of the biggest consumers of the available clean water, and at the same time the industry is one of the highest producers of polluted wastewater. In this regard, recently there have been proposed a set of techniques to economically reduce the clean water consumption in the process industry considering the possibility to recycle, reuse and regenerate the wastewater produced in the industrial facilities. The proposed techniques have addressed the problem from different perspectives (i.e., minimizing the fresh water consumption, minimizing the wastewater discharge, minimizing the regeneration costs, etc) and using diverse techniques (i.e., graphical, algorithmic and mathematical programming based approaches). The heuristic approaches allow to easily identify targets ahead of the design and to obtain a graphical representation of the problem (see for example Wang and Smith, 1994; El-Halwagi and Spriggs, 1998). All previous methodologies only have considered the processes that happen inside the industrial facilities, without taking into account the processes that occur outside the plant for the wastewater streams discharged to the environment.

Therefore, the goal of the proposed work is to overcome the drawbacks of the previously reported methods for the synthesis of water integration networks, considering the sustainability of the surrounding watershed that is based on properties for the cases that multiple pollutants are presented and that make difficult to characterize the streams in terms of the composition. The optimal quantity and quality for the wastewater discharged from the new plant depends on the location selected to install the plant because the impact in the watershed depends strongly on the location selected (considering the integration with other streams and the chemical and biochemical reactions that occur in the watershed system); and therefore, the recycle and reuse network is highly influenced by the location of the new plant.

Model Formulation

The main stream of the river is fed by several streams or tributaries ($FT_{r,t}$). As it can be seen in **Figure 1**, the watershed system exchanges water during its trajectory (i.e., water for agricultural use, wastewater discharged to the river with and without treatment, industrial and residential effluents, etc.) and finally the watershed discharges to the final disposal. Besides, natural phenomena like precipitation, filtration and vaporization are taken into account in the model. These processes modify significantly the characteristics of the river; therefore, to track adequately the mean properties, the river is sectioned in parts where the properties can be considered constant (these sections are represented in **Figure 1** with ovals). These parts o sections are called reaches and these identify zones where no big intermediate effluents are discharged and/or extracted. In this sense, the model also indentifies the main effluents discharged in each reach; these effluents are called tributaries. The tributaries can be channels or branches of the river, which may contain discharges with/without treatment, industrial discharges, etc., and their flowrates ($FT_{r,t}$) and properties ($P_{p,r,t}^{Trib}$) affect the reaches where they are discharged. The discharges to the watershed contain compounds that interact chemically and biochemically with the system thought the flora and fauna established in the watershed.

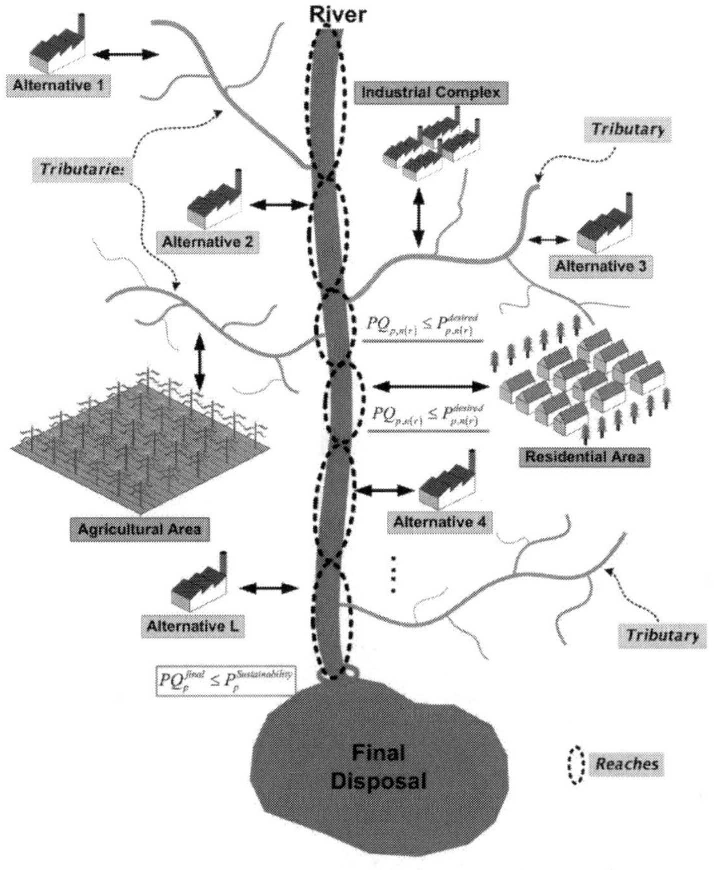

Figure. 1 Alternatives to locate the new industrial plant in a watershed system.

To ensure the sustainability of the surrounding watershed system, it is required to carry out an analysis prior to the optimization respect to the capacity of the final disposal to decompose chemically and biochemically the pollutants, and as consequence it is necessary to quantify these processes in terms of properties. The analysis yields the sustainability constraints which ensure that the pollutants will not be accumulated in the system and to maintain the properties of the watershed under reasonable limits.

In this sense, to consider the performance function of the interceptors, it is used a correction factor ($\gamma_p^{int^1,\ldots,int^N}$) that can be calculated prior to the optimization process, obtaining a linear function to model the interceptors. The correction factor must consider several aspects like the value of the property at the inlet conditions of the interceptor, the flowrate and design and operating parameters.

Thus, the correction factor is a very useful tool in this work, where the available technologies considered in the property interception network can be simulated (i.e., using ASPEN) to get the efficiency factors. It is worth of notice that the model includes a set of fictitious interceptors for each property to be intercepted with conversion factors and costs equal to zero. Besides, the property operators values for the inlet condition of the process sources are also known prior to the optimization process, $\psi_p\left(P_{i,p}^{InSource}\right)$, then the property values at the exit conditions of the property interception network can be calculated prior to the optimization process as follows:

$$\psi_p\left(P_{i,p}^{OutPIN,\mathrm{int}^1,\ldots,\mathrm{int}^N}\right)=\gamma_p^{\mathrm{int}^1,\ldots,\mathrm{int}^N}\psi_p\left(P_{i,p}^{InSource}\right),\quad i\in I, p\in P, \mathrm{int}^1\in INT^1,\ldots,\mathrm{int}^N\in INT^N$$

MFA Model based on properties

$$Q_r=Q_{r-1}+P_r+D_r+H_r+\sum_{t=1}^{N_{t(r)}}FT_{r,t}+QPNEW_{r(l)}-L_r-U_r,\quad \forall r\in R \tag{1}$$

$$\begin{aligned}\psi_p\left(P_{p,r}^{\mathrm{Reach}}\right)Q_r&=\psi_p\left(P_{p,r-1}^{In\mathrm{Reach}}\right)Q_{r-1}+\psi_p\left(P_{p,r}^{\mathrm{Prec}}\right)P_r+\psi_p\left(P_{p,r}^{Ind}\right)D_r+\psi_p\left(P_{p,r}^{\mathrm{Resid}}\right)H_r+\sum_{t=1}^{N_{t(r)}}\psi_p\left(P_{p,r,t}^{Trib}\right)FT_{r,t}\\&+\psi_p\left(P_{p,r(l)}^{PNEW}\right)QPNEW_{r(l)}-\psi_p\left(P_{p,r}^{Loses}\right)L_r-\psi_p\left(P_{p,r}^{Uses}\right)U_r-\int_{V=0}^{V_r}r_{p,r}dV_r,\ \forall p\in P, r\in R\end{aligned} \tag{2}$$

$$FT_{r,t} = S_{r,t}^{Untreat} + S_{r,t}^{Treat} + I_{r,t} + P_{r,t} + D_{r,t} + QPNEW_{r,t(l)} - L_{r,t} - U_{r,t}, \qquad \forall r \in R, t \in T \tag{3}$$

$$\psi_p\left(P_{p,r,t}^{Trib}\right)FT_{r,t} = \psi_p\left(P_{p,r,t}^{Untreat}\right)S_{r,t}^{Untreat} + \psi_p\left(P_{p,r,t}^{Treat}\right)S_{r,t}^{Treat} + \psi_p\left(P_{p,r,t}^{Ind}\right)I_{r,t} + \psi_p\left(P_{p,r,t}^{Prec}\right)P_{r,t} + \psi_p\left(P_{p,r,t}^{Direct}\right)D_{r,t}$$
$$+\psi_p\left(P_{p,r,t(l)}^{PNEW}\right)QPNEW_{r,t(l)} - \psi_p\left(P_{p,r,t}^{Loses}\right)L_{r,t} - \psi_p\left(P_{p,r,t}^{Uses}\right)U_{r,t} - \int_{V=0}^{V_{r,t}} r_{p,r,t}\,dV_{r,t}, \qquad \forall p \in P, r \in R, t \in T \tag{4}$$

$$\int_{V=0}^{V_r} r_{p,r}\,dV_r = k_p\left[\psi_p\left(P_{p,r}^{Reach}\right)\right]^{\sigma_p} V_r, \qquad \forall p \in P, r \in R \tag{5}$$

$$\int_{V=0}^{V_{r,t}} r_{p,r,t}\,dV_{r,t} = k_p\left[\psi_p\left(P_{p,r,t}^{Trib}\right)\right]^{\sigma_p} V_{r,t}, \qquad \forall p \in P, r \in R, t \in T \tag{6}$$

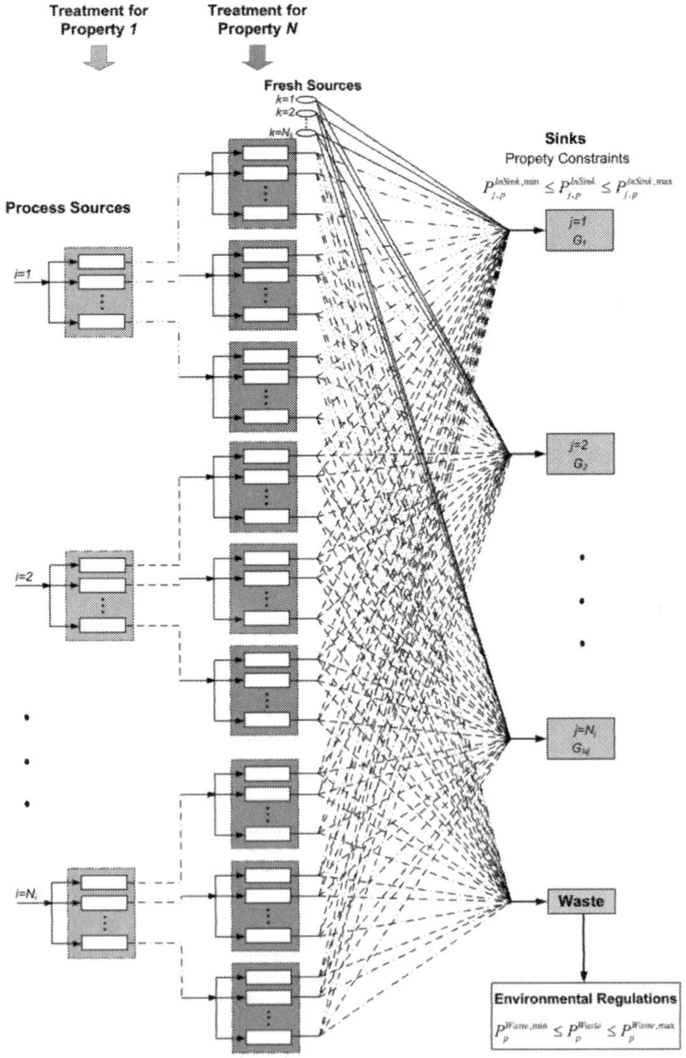

Figure 2. Superstructure for the recycle and reuse network

Location of the new plant

This work considers a set of alternatives to locate the new industrial facility (L), these options are generated prior to the optimization process. Then, for each alternative, it is necessary to consider the possible discharge of the new wastewater flowrate ($QPNEW_l$) and their property operators ($\psi_p\left(P_{p,l}^{PNEW}\right)$), thus, if the new industrial plant is located in the alternative 1, the wastewater flowrate and their property operators associated to this site must be bigger than zero; otherwise, if the new plant is not located in the position 1, the wastewater flowrate and their property operators discharged would be zero. This is included in the following disjunction:

$$\begin{bmatrix} Y_1 \\ \psi_p\left(P_{p,1}^{PNEW}\right)\geq 0, QPNEW_1 \geq 0 \\ \psi_p\left(P_{p,2}^{PNEW}\right)=0, QPNEW_2 = 0 \\ \vdots \\ \psi_p\left(P_{p,l}^{PNEW}\right)=0, QPNEW_l = 0 \end{bmatrix} \vee \begin{bmatrix} Y_2 \\ \psi_p\left(P_{p,1}^{PNEW}\right)=0, QPNEW_1 = 0 \\ \psi_p\left(P_{p,2}^{PNEW}\right)\geq 0, QPNEW_2 \geq 0 \\ \vdots \\ \psi_p\left(P_{p,l}^{PNEW}\right)=0, QPNEW_l = 0 \end{bmatrix} \vee \cdots \vee \begin{bmatrix} Y_l \\ \psi_p\left(P_{p,1}^{PNEW}\right)=0, QPNEW_1 = 0 \\ \psi_p\left(P_{p,2}^{PNEW}\right)=0, QPNEW_2 = 0 \\ \vdots \\ \psi_p\left(P_{p,l}^{PNEW}\right)\geq 0, QPNEW_l \geq 0 \end{bmatrix}, \quad \forall p \in P$$

The reformulation includes the transformation of the Boolean variables (Y_l) into a set of binary variables (y_l), when the Boolean variables are true, the associated binary variables must be set as one; otherwise, when the Boolean variables are false, the associated binary variables are zero. Therefore, the following relationship is used to select only one location to install the plant:

$$\sum_{l\in L} y_l = 1 \tag{7}$$

$$QPNEW_l \leq QPNEW^{Up}\, y_l, \qquad \forall l \in L \tag{8}$$

$$\psi_p\left(P_{p,l}^{PNEW}\right) \leq \psi_p\left(P^{PNEW}\right)^{Up} y_l, \qquad \forall p \in P, l \in L \tag{9}$$

where $QPNEW^{Up}$ and $\psi_p\left(P_{p,l}^{PNEW}\right)^{Up}$ are the upper limits for $QPNEW_l$ and $\psi_p\left(P_{p,l}^{PNEW}\right)$, respectively. Notice that $\psi_p\left(P_{p,l}^{PNEW}\right)^{Up}$ is an upper bound for all properties p. These constraints are required to ensure that $QPNEW_l$ and $\psi_p\left(P_{p,l}^{PNEW}\right)$ take values bigger than zero only when the new plant is located in a position l in particular.

$$\sum_{l\in L} QPNEW_l = Waste \tag{10}$$

$$\sum_{l\in L} \psi_p\left(P_{p,l}^{PNEW}\right) = \psi_p\left(P_p^{disc}\right), \qquad \forall l \in L \tag{11}$$

Superstructure for the water network

$$F_k = \sum_{j\in J} f_{k,j}, \quad \forall k \in K \tag{12}$$

$$W_i = \sum_{int^1 \in INT^1} w_i^{int^1}, \quad \forall i \in I \tag{13}$$

$$w_i^{int^1} = \sum_{int^2 \in INT^2} w_i^{int^1,int^2}, \quad \forall i \in I, int^1 \in INT^1 \tag{14}$$

$$w_i^{int^1,int^2,\ldots,int^{N-1}} = \sum_{int^N \in INT^N} w_i^{int^1,int^2,\ldots,int^N}, \quad \forall i \in I, int^1 \in INT^1,\ldots,int^{N-1} \in INT^{N-1} \tag{15}$$

$$w_i^{int^1,\ldots,int^N} = \sum_{j\in J} g_{i,j}^{int^1,\ldots,int^N} + g_{i,Waste}^{int^1,\ldots,int^N}, \quad \forall i \in I, int^1 \in INT^1,\ldots,int^N \in INT^N \tag{16}$$

$$G_j = \sum_{i \in I} \sum_{int^1 \in INT^1} \cdots \sum_{int^N \in INT^N} g_{i,j}^{int^1,\ldots,int^N} + \sum_{k \in K} f_{k,j}, \quad \forall j \in J \tag{17}$$

$$\psi_p\left(P_{j,p}^{InSink}\right) G_j = \sum_{i \in I} \sum_{int^1 \in INT^1} \cdots \sum_{int^N \in INT^N} \left[\psi_p\left(P_{i,p}^{OutPIN,int^1,\ldots,int^N}\right) g_{i,j}^{int^1,\ldots,int^N}\right] + \sum_{k \in K}\left[\psi_p\left(P_{k,p}^{Fresh}\right) f_{k,j}\right], \quad \forall j \in J, p \in P \tag{18}$$

$$Waste = \sum_{i \in I} \sum_{int^1 \in INT^1} \cdots \sum_{int^N \in INT^N} g_{i,Waste}^{int^1,\ldots,int^N} \tag{19}$$

$$\psi_p\left(P_p^{Waste}\right) Waste = \sum_{i \in I} \sum_{int^1 \in INT^1} \cdots \sum_{int^N \in INT^N} \left(\psi_p\left(P_{i,p}^{OutPIN,int^1,\ldots,int^N}\right) g_{i,Waste}^{int^1,\ldots,int^N}\right), \quad \forall p \in P \tag{20}$$

$$\psi_p\left(P_{j,p}^{InSink,\min}\right) \le \psi_p\left(P_{j,p}^{InSink}\right) \le \psi_p\left(P_{j,p}^{InSink,\max}\right), \quad j \in J, P \in P \tag{21}$$

Environmental constraints

$$\psi_p\left(P_p^{Waste,\min}\right) \le \psi_p\left(P_p^{Waste}\right) \le \psi_p\left(P_p^{Waste,\max}\right), \quad \forall p \in P \tag{22}$$

Constraints for the quality of specific reaches

$$\psi_p\left(P_{p,m(r)}^{DesiredMIN}\right) \le \psi_p\left(P_{p,m(r)}^{Reach}\right) \le \psi_p\left(P_{p,m(r)}^{DesiredMAX}\right), \quad \forall p \in P, m(r) \in M(R) \tag{23}$$

Sustainability constraints

$$\psi_p\left(P_p^{SustainableMIN}\right) \le \psi_p\left(P_{p,final}^{Reach}\right) \le \psi_p\left(P_p^{SustainableMAX}\right), \quad \forall p \in P \tag{24}$$

Objective Function

The objective function consists in the minimization of the total annual cost (*TAC*), which is constituted by the installation costs for the new plant, the fresh sources costs, the treatment cost (including operational cost, variable and fixed capital cost) and the piping costs. The installation cost ($Cland_l$) takes into account the land cost, the transportation costs for raw materials, products and services. Fresh sources cost ($FreC_k$) is caused by the required flowrate of fresh sources to satisfy the inlet flowrate in the process sinks as well as the properties constraints imposed on them. Also, process sources need to be treated prior to be sent to the process sinks, as well as prior to be discharged to the environment, in this sense; a set of technologies must be installed to treat the process sources yielding capital and operational costs for each one of the interceptors required. The operational cost ($VarC_{op}^{int^N}$) for the interceptors depends on the treated flowrate, while the capital cost contains a variable cost ($VarC_{cap}^{int^N}$), which also depends on the treated flowrate and a fixed component ($FixC_{cap}^{int^N}$). Finally, the piping costs for the four sections showed in the superstructure are considered. The first section are the pipes requires when the process sources are sent to treatment ($TrPipC$), once the process sources are treated they can be sent to wastewater stream ($WaPipC$) or process sinks ($EqPipC$), the next part covers the pipes to connect the fresh sources with process sinks ($FrPipC$).

$$
\begin{aligned}
Min\,TAC = \sum_{l\in L} Cland_l y_l + H_Y \sum_{k\in K} FreC_k F_k \\
+H_Y \sum_{i\in I}\left[\sum_{int^1\in INT^1} VarC_{op}^{int^1} w_i^{int^1} + \cdots \sum_{int^N\in INT^N}\left(VarC_{op}^{int^N} w_i^{int^1,\ldots,int^N}\right)\right] \\
+H_Y \sum_{i\in I}\left[\sum_{int^1\in INT^1} VarC_{cap}^{int^1} w_i^{int^1} + \cdots \sum_{int^N\in INT^N}\left(VarC_{cap}^{int^N} w_i^{int^1,\ldots,int^N}\right)\right] \\
+k_f \sum_{i\in I}\left[\sum_{int^1\in INT^1} FixC_{cap}^{int^1} z_i^{int^1} + \cdots \sum_{int^N\in INT^N}\left(FixC_{cap}^{int^N} z_i^{int^1,\ldots,int^N}\right)\right] \\
+H_Y \left[\begin{array}{l} TrPipC\sum_{i\in I}\left(\sum_{int\in INT} w_i^{int^1} + \cdots \sum_{int^N\in INT^N} w_i^{int^1,\ldots,int^N}\right) + WaPipC\sum_{i\in I}\sum_{int^1\in INT^1}\cdots\sum_{int^N\in INT^N} g_{i,Waste}^{int^1,\ldots,int^N} \\ +EqPipC\sum_{i\in I}\sum_{int^1\in INT^1}\cdots\sum_{int^N\in INT^N}\sum_{j\in J} g_{i,j}^{int^1,\ldots,int^N} + FrPipC\sum_{k\in K}\sum_{j\in J} f_{k,j} \end{array}\right]
\end{aligned}
\quad (25)
$$

Results and discussion

Case of study

This case considers the Balsas watershed that is one of the biggest systems in Mexico. This system transports several types of discharges until the Pacific Ocean. CONAGUA 20010a, 2010b reported the information for the system sowed on **Figure 6**. In this case, the properties considered were the composition, toxicity, density, COD, odor and color; however, there is a set of properties restricted in the environment and other one restricted in the process sinks. Thus, the properties restricted for the stream discharged to the environment are composition, toxicity, COD, odor and color, while the process sinks limit the composition and density. The system offers twenty possible locations to install the new industrial facility and their installation costs are presented in **Table 6**

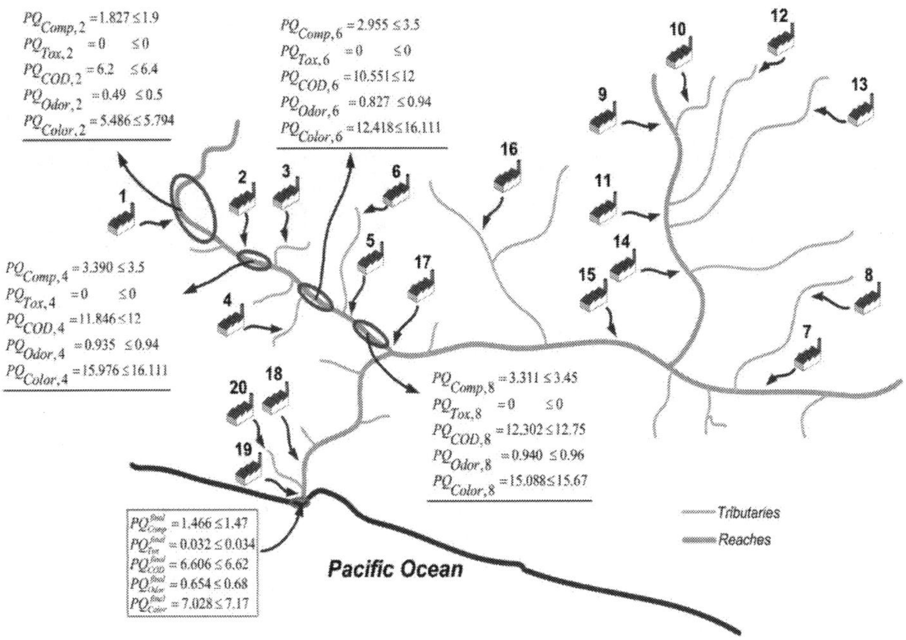

Figure 6. Balsas watershed system (possible locations and the constraints on the river)

Table 6. Installation costs for Example 2

Site	Annualized installation cost, $/year
1	8,000,000
2	8,500,000
3	9,000,000
4	10,500,000
5	9,500,000
6	9,000,000
7	15,500,000
8	14,500,000
9	17,700,000
10	22,000,000
11	18,000,000
12	17,500,000
13	24,000,000
14	28,000,000
15	23,500,000
16	17,000,000
17	20,000,000
18	19,000,000
19	16,000,000
20	17,000,000

Table 7. Fresh and process sources for Example 2

Sources	*Flowrate, m^3/s*	*Composition, ppm*	*Toxicity, %*	*COD, mgO_2/lt*	*Odor*	*Color*	*Density, g/ml*
			Process				
1	0.363	64.6	0.11	149	6.3	250	0.94
2	0.241	58	0.04	88	7.7	234	2.41
3	0.046	45.3	0	108	5.1	298	1.14
4	0.447	60	0.13	185	4.8	202	1.20
5	0.196	52	0.21	120	7.6	194	2.70
6	0.204	62	0.25	76	8.2	164	1.64
7	0.281	54.5	0	243	3.7	179	1.85
8	0.329	47.8	0.13	164	7.4	346	1.75
9	0.282	46	0.09	297	4.8	264	2.45
10	0.266	82.1	0.7	326	8.5	394	0.89
11	0.215	26.4	0.3	146	5.7	195	2.15
			Fresh				
1	-	0	---	---	---	---	2.3
2	-	1	---	---	---	---	2.45
3	-	2	---	---	---	---	1.8

Table 7 shows the characteristics for the fresh and process sources. **Table 8** shows the requirements for the flowrate, maximum composition as well as the minimum and maximum density for each process sink. The cost for fresh source 1 is $0.128/m^3, for fresh source 2 is $0.113/m^3, for fresh source 3 is $0.097/m^3. **Table 9** shows the characteristics for the property interceptors. For this case, the variables costs are included in just one term.

Table 8. Requirements for process sinks for Example 2.

Process Sink	*Flowrate, m^3/s*	*$Comp^{max}$*	*$Density^{min}$*	*$Density^{max}$*
1	0.348	16	2.2	2.4
2	0.428	19.4	1.95	2.1
3	0.406	23.3	2.35	2.55
4	0.312	28.5	1.7	1.9
5	0.336	18	2.2	2.3
6	0.294	14	1.8	2
7	0.196	22.5	2.45	2.55

Table 9. Available interceptors for Example 2.

Interceptor	Efficiency Factor	Variable Cost, $/m^3
	Composition	
Rec^1	0.89	0.026
Rec^2	0.77	0.021
	Toxicity	
Tox^1	1	0.054
Tox^2	0.92	0.049
	COD	
Aer^1	0.82	0.022
Aer^2	0.57	0.018
	Odor	
Ads^1	0.86	0.013
Ads^2	0.61	0.011
	Color	
Dec^1	0.90	0.019
Dec^2	0.80	0.015

On the other hand, the environmental legislation imposes the next constraints for the wastewater stream: $P_{Comp}^{Sustainable} = 50$, $P_{Tox}^{Sustainable} = 0.1$, $P_{COD}^{Sustainable} = 75$, $P_{Odor}^{Sustainable} = 2.7$, and $P_{Comp}^{Sustainable} = 100$. While the piping cost for each section in the superstructure are: $TrPipC = 0.001\,\$/m^3$, $WaPipC = 0.002\,\$/m^3$, $EqPipC = 0.005\,\$/m^3$, and $FrPipC = 0.003\,\$/m^3$.

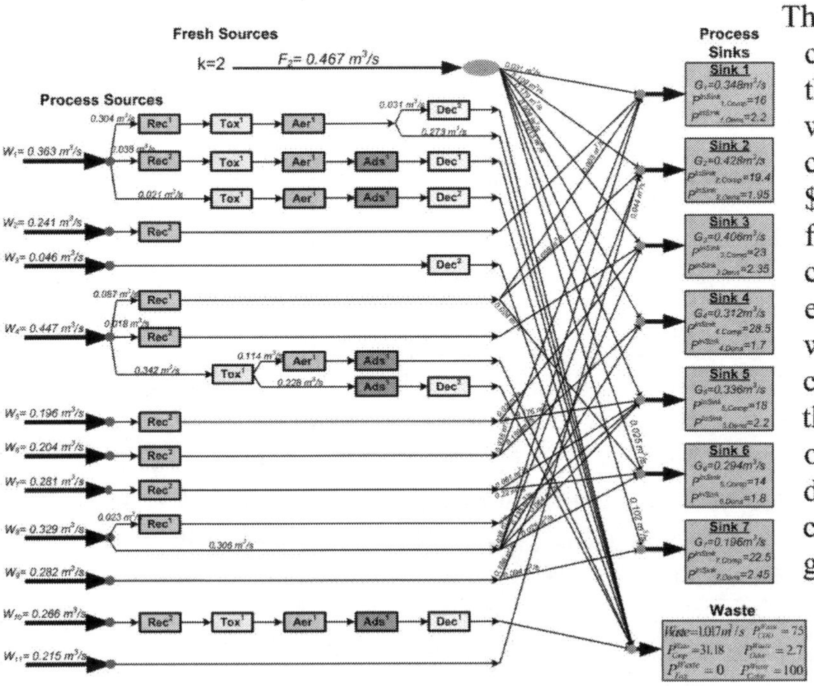

Figure 7. Mass integration for Example 2

Then, the proposed model was applied for this case of study, giving an optimal solution where the new facility must be placed in alternative 6 with a total annual cost of \$14.9x10^6/yr, composed by \$9x10^6/yr of installation cost, \$3.6x10^6/yr of treatment cost, \$1.5203x10^6/yr of fresh source cost and \$779,040/yr of piping cost. **Figure 7** shows the water network this example, notice that the value of toxicity in the wastewater stream is zero, satisfying the constraint in the reach 8, where is not allowed the existence of this property. Notice that most of the properties in the wastewater streams are discharged in the limit of the environmental constraints; however, the composition requires a greater reduction to satisfy the constraints imposed over the reach 8 ensuring the quality of the water extracted for human activities.

Conclusions

This paper proposes a new water integration approach to consider simultaneously the integration inside the plant and the sustainability of the surrounded watershed. The model is based on the properties of the streams, which is suited for the wastewater streams that are constituted by several pollutants. The proposed approach considers the integration of the wastewater discharged by the industrial facility to the other discharges and uses through the optimal selection of the location of the industrial facility. In addition, the model considers the natural degradation of the pollutants through the watershed and the effect to the specific properties that are constrained in the system. The proposed model is applied to one case from Mexico. The results show that the traditional water integration schemes that do not consider the surrounding watershed yield inappropriate solutions, because the environmental regulations are not enough to satisfy the sustainability of the system.

References

CONAGUA (2009a). Mexican National Water Commission. Water Statistics 2008. http://www.conagua.gob.mx/OCB07/Temas/EstadisticasBALSAS-FINALago09.pdf

CONAGUA (2009b). Mexican National Water Commission. Water Atlas in Mexico 2009.

El-Halwagi, M. M., and Spriggs, H. D. (1998). Solve design puzzles with mass integration. *Chemical Engineering Progress*, *94* (8), 25-44.

Wang, Y. P., and Smith, R. (1994). Wastewater minimization. *Chemical Engineering Science*, *49* (7), 981-1006.

(521b) Comparing Continuous and Batch Process Design Under Uncertainty for Biodiesel Production

Pahola T. Benavides[1], Urmila Diwekar[2] and Juan M. Salazar[2], (1)VRI- Custom, Chicago, IL, (2) Vishwamitra Research Institute, Center for Uncertain Systems: Tools for Optimization and Management, Clarendon Hills, IL

Abstract:

The continuing depletion of fossil fuel reserves and the increasing environmental concerns has encouraged engineers and scientists to look for an alternative, clean and renewable fuel that can reduce the negative environmental impact. Biodiesel has been considered as one of the best candidate of one of these renewable fuels. One of the pathways to biodiesel production is the transesterification reaction of triglycerides from vegetable oils and short-chain alcohols. A batch reactor is employed for the production of biodiesel. However, most of the other unit operations like decanters and distillation columns are operated continuously. The success of biofuel introduction in economy depends on the economics of these fuels. One of the factors which are problematic in this is the seasonal variation of feedstock, and transportation and storage costs of material for continuous processing. In this work, we compare batch and continuous processing options for biodiesel production under feedstock variation. These models will allow us to choose an optimal process configuration, depending on the size of the market, distance to market, and feedstock availability,

Key words: *batch process, continuous process, biodiesel, stochastic simulation, economical analysis.*

See more of this Session: Design of Sustainable Processes
See more of this Group/Topical: Computing and Systems Technology Division

(521c) Simultaneous Optimization of Flowsheet, Heat Recovery, and Water Network

Linlin Yang and Ignacio Grossmann, Chemical Engineering, Carnegie Mellon University, Pittsburgh, PA

Significant research work has been done in the area of heat integration in the past thirty years due to the incentive of improving energy recovery. Also, stricter environmental regulations have caused an increasing demand for minimization of water consumption and effluent generation for industrial processes. Various techniques have been developed in response to this need for process integration and they are presented in these reviews[1][3][4]. However, most of this research has been focused on minimization of energy and water separately. Furthermore, traditionally, process synthesis design is performed sequentially - the flowsheet is optimized first, then the resulting operating parameters are used in heat exchanger network synthesis (HENS) and water network (WN).

This presentation describes a novel and effective approach for the simultaneous optimization of process flowsheet subject to heat and water integration. Through simultaneous optimization approach, the idea is to represent the HENS and the WN through design targets such as energy consumption and water consumption to avoid solving a very large MINLP problem. The advantage of performing simultaneous optimization is that proper trade-offs of operation cost and investment cost can be taken into account.

While effective NLP targeting formulation for simultaneous process optimization with heat integration has already been presented by Duran & Grossmann[2], this is not the case for targeting minimum water use. Typical WN problems are formulated using NLP or MINLP models using a detailed superstructure as reported in many references[5]. Freshwater is first used for process units, and then wastewater generated from these processes is treated in various treatment units. Process units are defined by their allowable contaminant concentration level, whereas wastewater treatment units are defined based on their contaminant removal ratio. The superstructures consider systematic alternatives for water recycle and water reuse with the objective of minimizing the freshwater consumption, or more generally the total cost. In general, single-contaminant systems can be modeled using LP formulations, while multi-contaminant systems take the form of NLP models.

In order to incorporate water integration in the simultaneous optimization framework, a new LP formulation is presented to determine the target for a multi-contaminant water network with process units only. In particular, the bilinear equations of the mixer are replaced by a set of inequalities that are evaluated at the limits of inlets and outlet concentrations of the units. This relaxation, which follows the optimality conditions proved by Bagajewicz & Savelski[6], yields an LP that can be shown to predict the exact minimum amount of freshwater that is required by the process units. The LP targeting model is incorporated in the simultaneous optimization of a methanol synthesis process flowsheet with heat and water integration and resulted in an improvement of the economical objective function. The important observation here is that in the simultaneous approach, heating and cooling are integrated as part of the optimization so that improved operating conditions can be achieved. It is shown for a specific instance that the resulting flowsheet from the simultaneous method improves the profit by 17%.

For the case of multi-contaminant water network problems including wastewater treatment units with process units, the proposed LP targeting formulation is shown to predict a rigorous minimum freshwater requirement for special cases of the WN problem. For the general case, the lack of exact upper bound of contaminant concentration in the treatment units complicates the interactions in the mixer units. To circumvent this problem, a number of linear relaxations are used to obtain approximate but tight water targets for general WN problems. The proposed targeting model yields an MILP problem. The application of this model is also applied in the simultaneous optimization, heat and water integration of the methanol process yielding further improvements in the profit.

References:

[1] Bagajewicz, M., (2000). A review of recent design procedures for water networks in re0neries and process plants. *Computers & Chemical Engineering,* 24 (9), 2093-2115.

[2] Duran, M.A. & Grossmann, I.E. (1986). Simultaneous optimization and heat integration of chemical processes. *AIChE Journal*, 32, 123-138.

[3] Foo, D.C.Y. (2009). State-of-the-art review of pinch analysis techniques for water network synthesis. *Ind. Eng. Chem. Res.* 48, 5125–5159.

[4] Furman, K.C. & Sahinidis, N.V. (2002). A critical review and annotated bibliography for heat exchanger network synthesis in the 20th Century. *Ind. Eng. Chem. Res.*, 41, 2335-2370.

[5] Karuppiah, R. & Grossmann, I. E. (2008). Global optimization of multi-scenario mixed integer nonlinear programming models arising in the synthesis of integrated water networks under uncertainty. *Computers & Chemical Engineering*, 32(1-2), 145-160.

[6] Savelski, M. & Bagajewicz, M. (2003). On the necessary conditions of optimality of water utilization systems in process plants with multiple contaminants. *Chemical Engineering Science*, 58(23-24), 5349-5362.

See more of this Session: Design of Sustainable Processes
See more of this Group/Topical: Computing and Systems Technology Division

(521d) Incorporating Safety and Efficiency Considerations In Sustainability Assessment of Poly-Generation Systems

Preeti Gangadharan[1], Kailiang Zheng[2], Abhishek V. Jayswal[3], Anand Zanwar[1] and Helen Lou[1], (1) Dan F. Smith Department of Chemical Engineering, Lamar University, Beaumont, TX, (2)Department of Chemical Engineering, Lamar University, Beaumont, TX, (3)Dan F. smith Department of Chemical Engineering, Lamar University, Beaumont, TX

Sustainability assessment determines the most sustainable pathway, taking into consideration the economic, environmental, efficiency and societal aspects. Embedding exergy analysis and novel safety evaluation approach to complement the widely-accepted economic assessment and environmental impacts assessment in a decision tree evaluation framework provides a more comprehensive, yet faster methodology to compare related processes in terms of sustainability. Considering their promise to provide high material and energy conversion from natural resources, poly-generation systems are recognized as promising technologies for future chemical and power industries.

The commonly used inherent safety approach considers chemical and process safety, but does not give due emphasis to complexity of the process flow sheet. This becomes important in the case of poly-generation systems that accept multiple feeds and produce multiple products through many pathways for chemical and energy conversion. Since the simultaneous production of chemicals and electricity offers both economical viability and thermodynamic efficiency, exergy efficiency analysis is employed for thermodynamic comparison of proposed processes.

In this paper three different poly-generation systems, which use coal and natural gas as feed to produce DME and power are compared using a comprehensive sustainability assessment methodology, which takes into account the economic, environmental, societal and efficiency factors.

See more of this Session: Design of Sustainable Processes
See more of this Group/Topical: Computing and Systems Technology Division

(521e) Technology Evaluation and Decision Making for Sustainability Enhancement of Industrial Systems Under Uncertainty

Zheng Liu, Department of Chemical Engineering and Materials Science, Wayne State University, Detroit, MI and Yinlun Huang, Dept. of Chemical Engineering & Material Science, Wayne State University, Detroit, MI

Depletion of natural resources, environmental pressure, economic globalization, etc., demand seriously industrial organizations to ensure that their development is sustainable. Technology based sustainability enhancement is one of key approaches for industrial sustainability realization. However, identification of effective technologies for a given industrial system could be a combinatorial problem; this is especially true when multiple technologies are considered for adoption together. On the other hand, if the available data and information about the industrial system and the known technologies are incomplete, imprecise, and uncertain, the technology identification could be even more difficult.

In this paper, we introduce a simple approach for systematic sustainability assessment of industrial systems and technologies, and effective system sustainability enhancement under uncertainty. An industrial system can be an industrial zone, corporate, company, or plant of any complexity. We assume that the industrial organization has set a strategic plan that includes detailed development goals of economic, environmental, and social sustainability with a budget set aside for goal achievement. It is known that a number of technologies are seemingly promising for sustainability enhancement. A challenge is how to derive efficiently the most suitable solutions for identification of superior sustainability technologies.

The technology-assisted sustainability enhancement methodology to be introduced will consist of a sustainability assessment method for evaluating the performance of industrial systems as well as technologies, and a sustainability decision making and analysis method for identifying superior technology combinations for achieving the sustainability goals of industrial systems. The interval parameter based information processing and decision making permit the acceptance of uncertain information that could be available at the best. The logically designed solution identification procedure can make the combinatorial problem solved efficiently through reducing the solution space stage-wisely using different criteria set by the strategic plan. The identified solutions are also sufficiently detailed which can greatly facilitate the industrial organization in final decision making. The methodology is general that can be applied to the sustainability enhancement problems of any size and scope, and the efficacy of the approach is illustrated through a sustainability development problem about regional biodiesel manufacturing improvement.

See more of this Session: Design of Sustainable Processes

(521f) Integration of Production Processes, Waste Management and Energy Utility Generation In Chemical Batch Plants

Stavros Papadokonstantakis, Claude Rerat and Konrad Hungerbuhler, Chemistry and Applied Biosciences, Institute for Chemical- and Bioengineering, Swiss Federal Institute of Technology, Zurich (ETHZ), Zurich, Switzerland

The production of chemicals on industrial scale heavily relies on the availability of utilities such as energy in different forms e.g., steam, electricity, cooling water, brine and others. These utilities represent a significant share of the production cost of chemicals and define to a major degree the environmental impact resulting from chemical production. The efficiency of utility usage in chemical industry is primarily defined by two major factors, the efficiency of the utility generation (e.g. steam production) and the efficiency in using the generated utilities in achieving a productivity of chemicals as high as possible.This description evokes the idea of two separated problems and unfortunately nowadays the two fields are often treated like this and consequently substantial integration potentials are overlooked resulting in increased cost and environmental impact. However, the generation and the usage of utilities in chemical production are highly interlinked and integration potentials must be identified during early design stages of both utility generation systems and chemical production processes. Major aspects in this context are waste treatment in general and solvent management in particular. Increased process integration, e.g., recycling of material flows, reduces the amount of waste to be treated and therefore the additional utility input required by many waste treatment technologies (e.g., stripping, sewage treatment). However, waste can also be a source of utility generation, e.g., by generating steam through waste incineration. This is particularly an issue for the huge amount of waste solvents that are generated in pharmaceutical and fine chemical production typically operating in batch mode.

Chemical batch plants are characterized by frequent changes in production portfolio which implies dynamic patterns in energy demand and waste generation. Since waste treatment plays an important role in the energy system of a chemical plant, these dynamic patterns can increase energy demand and simultaneously offer opportunities for primary energy savings. In constrast, utility generation systems cannot be changed frequently because high investments are associated and high efficiency can only be achieved for a well-defined utility mix (e.g. the proportion of electricity to steam in Combined Heat and Power devices).

In order to study the potential of an integrated design in this three-component system (i.e., production processes, waste management and energy utility generation) suitable models are needed. In the present study the energy demand of batch chemical production processes was based on bottom-up unit operation models adjusted to real process measurements using calibrated control valve openings. The difference between theoretical and real utility consumption was used to develop empirical parametric equations describing thermal losses due to inefficient heating/cooling system operations. Validation of these bottom-up models was performed at different aggregation levels, including equipment specific unit operations, production lines and overall batch plant energy consumption. The results indicated that for all types of energy utility, equipment and unit operation the consumption could be estimated with a relative error between 5% and 35% depending on the aggregation level.

Typical batch plant waste treatment facilities were considered like incineration, wastewater treatment and wet air oxidation, while solvent recovery was treated with continuous and batch distillation units. This waste management system can make full use of the dual properties of waste solvents which can be an energy source as combustible for incineration or energy sink as material for recycling in distillation units. In order to include life cycle analysis aspects in the investigation, multi-input allocation models were used for the waste treatment units to include energy requirements and pollutant emissions of waste handling. These models assess emissions and utility consumption with few waste stream data using linear transfer coefficients derived by extensive industrial studies of the respective operations.

The energy utility generation system includes steam boilers and the associated steam network, gas turbines, a refrigeration cycle producing brine and of course the waste incineration plant. The heat load provided by utilities corresponds to the heat demand of the process taking into account an average efficiency for heat exchangers. Concerning the process streams temperatures, it is possible to assume that they are ΔTmin lower (or higher, if the utility is a cold stream) than the utility lower (higher) temperature. The size of the corresponding heat exchanger network and the amount of necessary primary energy sources can be optimized using Pinch analysis.

For the optimization of the integrated system in a modular form, three nested optimization problems were solved: the production planning of the batch processes (MILP), the mixing problem of the generated waste streams (MINLP), and the heat integration problem for energy utilities geneartion (LP). The operational costs and diverse life-cycle impact oriented environmental indicators were used as objective functions. A synthetic case study plant consisting of a mono-product and a multi-product building was used for investigating the efficiency of the integrated design. The case study plant reflects the complexity of a real batch plant that provided the data for the development of the bottom-up energy consumption models, the desired production capacities, the waste management constraints and the utility generation system. In this way, it was possible to define a base case corresponding to the current plant operation, to which the optimization results were compared. It is shown that an optimal waste mixing together with a suitable production planning (i.e., more homogeneously distributed production over the time horizon) can result in a significant reduction of ancillaries in waste treatment operations and primary sources for energy utility generation (approx. up to 50% less energy consumption related operational costs and 20% less environmental impacts expressed in Eco-Indicator-99 points).

See more of this Session: Design of Sustainable Processes
See more of this Group/Topical: Computing and Systems Technology Division

(521g) Economic Analysis of Municipal Wastewater Utilization for Thermoelectric Power Production

Iman Safari[1], Michael E. Walker[1], Javad Abbasian[1], Hamid Arastoopour[2], Ming-Kai Hsieh[3], Ranjani B. Theregowda[3], David A. Dzombak[3] and David C. Miller[4], (1)Chemical and Biological Engineering, Illinois Institute of Technology, Chicago, IL, (2)Department of Chemical and Biological Engineering, Wanger Institute for Sustainable Energy Research (WISER), Illinois Institute of Technology, Chicago, IL, (3)Department of Civil and Environmental Engineering, Carnegie Mellon University, Pittsburgh, PA, (4)U.S. Department Of Energy, National Energy Technology Laboratory, Morgantown, WV

The thermoelectric power industry in the U.S. uses a large amount of freshwater. The large water demand is increasingly a problem, especially for new power plant development, as availability of freshwater for new uses diminishes in the United States. Reusing non-traditional water sources, such as treated municipal wastewater, provides one option to mitigate freshwater usage in the thermoelectric power industry. The amount of freshwater withdrawal that can be displaced with non-traditional water sources at a particular location requires evaluation of the water management and treatment requirements, considering the quality and abundance of the non-traditional water sources.

This paper presents the development of an integrated costing model to assess the impact of degraded water treatment, as well as the implications of increased tube scaling in the main condenser. The model developed herein is used to perform case studies of various treatment, condenser cleaning and condenser configurations to provide insight into the ramifications of degraded water use in the cooling loops of thermoelectric power plants. Further, this paper lays the groundwork for the integration of relationships between degraded water quality, scaling characteristics and volatile emission within a recirculating cooling loop model.

See more of this Session: Design of Sustainable Processes
See more of this Group/Topical: Computing and Systems Technology Division

Considerations for Research, Development and Commercialization/Scale-up of Biomass Pretreatment Processes

Dale Monceaux[1], *Richard C. Agar*[1] *and Charles E. Wyman*[2], [(1)]*AdvanceBio, LLC, Milford, OH,* [(2)]*College of Engineering-Center for Environmental Research and Technology, University of California, Riverside, Riverside, CA*

Global efforts to cost effectively convert biomass-based polysaccharides to sugars via enzymatic hydrolysis and subsequently to fuels and chemicals have been ongoing since the early discoveries by the US Army Natick Research & Development Command. These, and other early developments related to enzymatic hydrolysis of cellulose were followed by advancements in microbial genetic engineering. Genetically modified microorganisms capable of utilizing xylose-rich hemicellulose fractions of biomass feedstocks in addition to the cellulose derived glucose, promised improved ethanol yields and process economics. The past fifteen years of research has led to other major advancements in cellulase protein chemistry as well as an improved understanding of the microstructure of lignocellulose, moving the world ever closer to commercially viable biomass-based fuels and chemicals.

Early research ultimately revealed the sensitivity of the feedstock, pretreatment, enzyme and organism regime. This, in turn has exposed the need for consistency in methodology as research proceeds from concept to commercial-scale operations.

Much current activities focusing on the commercialization of biomass based fuels and chemicals hinge on the development of cost-effective biomass pretreatment processes. This "first step" in the process of converting biomass to end-products is critical to successful commercialization efforts for a number of reasons. First, it provides reactive cellulose that can be efficiently enzymatically hydrolyzed to oligomeric or monomeric glucose. Second, it efficiently solubilizes hemicellulose fractions of the biomass, producing oligomeric and or monomeric xylose, mannose, galactose, arabinose, etc. Third, it produces minimal reaction by-products that inhibit enzymatic hydrolysis and fermentation processes. Last, it is mechanically and economically scalable.

Many pretreatment technologies employ thermo-chemical reactions that reduce the recalcitrance of ligno-cellulosic matrices, solubilize fractions of the biomass and expose reactive sites to subsequent reactions. Various pretreatment technologies are under development, each with a range of design basis conditions. **Table 1** illustrates general conditions for six common pretreatment chemistries.

Pretreatment	Temp (°C)	Residence Time (min)	Water Loading*	Catalyst	Catalyst Loading**	Catalyst Recyclable
AFEX	150	30	200	Anhydrous NH_3	152	Yes (95-95% of loaded)
Dilute Acid	140	40	891	1% H_2SO_4	9	No
Lime	120	240	1500	$CaOH/O_2$	100/100 psi	Maybe/Ne
Liquid Hot Water	200	10	663	Water	N/A	N/A
SAA	160	60	765	15% Aqueous NH_3	135	Yes, ecess
SO_2	180	10	900	5% SO_2/Steam	5/?	Yes, excess

* Water Loading = L/100 kg of dry biomass input to pretreatment

** Catalyist Loading = kg/100 kg of dry biomass input to pretreatment

Table 1

Typical Thermochemical Pretreatment Chemistries

Reaction kinetics varies significantly with feedstock type, mechanical pretreatment, catalyst type, catalyst loading, water loading, residence time and temperature. Tight control of key parameters is required to maximize conversion efficiency. Elevated catalyst concentration and residence times increase reaction rate but also xylose destruction as illustrated in **Table 2**. Consequently, optimum reaction conditions require mechanical designs that deliver controlled residence times under near-plug flow conditions.

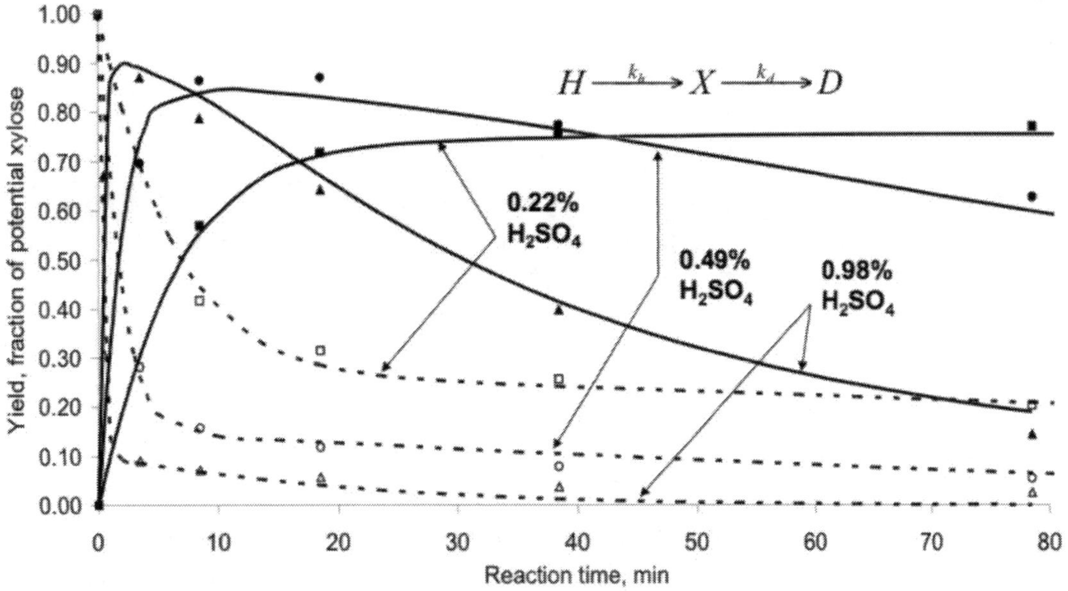

Table 2

Xylose Yield in Pretreatment vs Time at 160°C

Resource limitations, including feedstock supply, workspace, staffing and funding constrain the research and development efforts at each stage. Real-world feedstock properties such as foreign matter content, chemical composition variability, recalcitrance, bulk density, etc. introduce constraints that further complicate research and commercialization efforts. Feedstock selection dramatically impacts capital investment and production cost. Feedstock recalcitrance in conjunction with chemical composition effectively determines yield, and therefore the quantity of feedstock required per unit of fuel or chemical produced. Feedstock chemical composition and bulk density, in conjunction with reactor residence time establishes the size of the major components of the capital and energy intensive pretreatment reactor unit operation, as illustrated in **Table 3**.

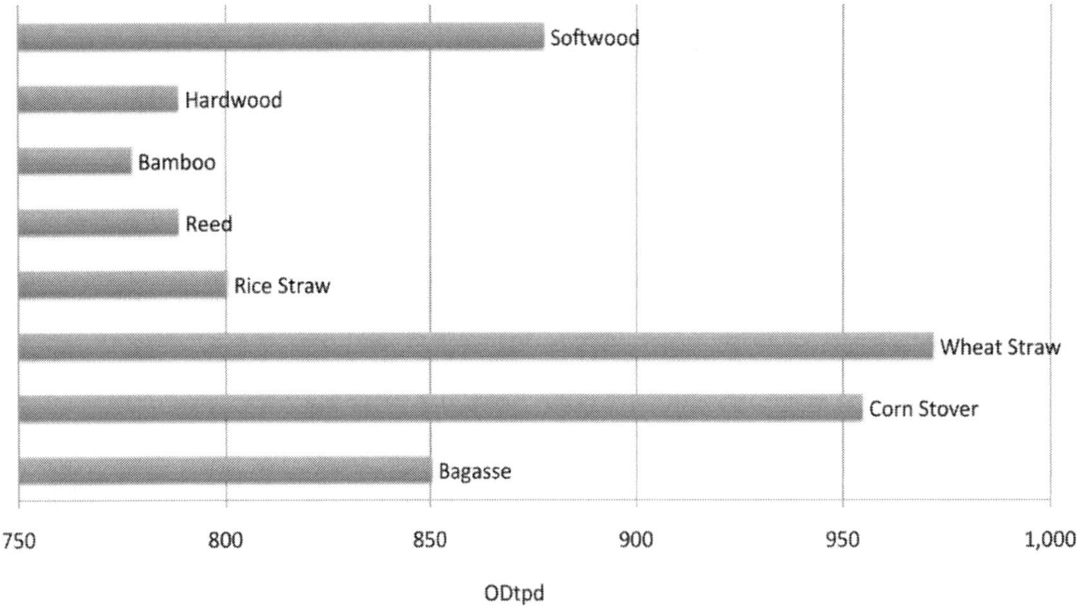

Table 3

Estimated Feedstock Tonnage per 100 Megaliters of Ethanol

High yields are critical to distribute both operating and capital costs. Low operating costs, including chemicals, energy and labor, are essential to providing adequate margins for return on capital. General operating cost categories must be lower for a new process than for established technologies making conventional products since low capital costs are essential to minimize financial exposure. The driving forces to contain capital investment include general technologies and processes that minimize equipment sizes, operate at lower pressures, operate at neutral or higher pHs as well as thermal processes that operate at lower temperatures and biochemical processes that operate at elevated temperatures.

Technical compromises frequently lead to approaches that ultimately expand the research and development program effort, schedule and budget. The impact of decisions made early in the technology development phase of a project may not fully materialize until late in the project after significant design and engineering expenditures have been made. The time period between research decision-making and engineering-based capital cost estimating delays the commercialization or leads to technology dead-ends. Visualization of the expected commercial scale process should occur very early in the technology research and development effort. A process "scale-down" approach to optimize research and technology development, scale-up and commercialization efforts has proven to be and effective strategy to accelerate project development and finance.

(522b) Improving Corn Stover Reactivity In Low Severity Pretreatment Through Deacetylation and Disc-Refining

Xiaowen Chen[1], Joseph Shekiro[1], mary Ann Franden[1], Min Zhang[1], Bonwook Koo[2], Sunkyu Park[2] and Melvin Tucker[1], (1)National Bioenergy Center, National Renewable Energy Laboratory, Golden, CO, (2) Department of Forest Biomaterials, North Carolina state university, Raleigh, NC

The current "state-of-art" dilute acid pretreatment to produce bio-ethanol from corn stover requires severe conditions to overcome biomass recalcitrance, which, in turn, offers 70~80% xylose monomer yield and 5~7% furfural yield. However, high acid loadings and high reaction temperatures harm the economic feasibility of the process. In addition, the amount of acetate and furfural produced during pretreatment is in a range that strongly inhibits cell growth and impedes ethanol fermentation. To decrease the impact of these issues, deacetylation, pretreatment with lower acid concentrations, lower temperatures, and integration of disc-refining was carried out in this work. The results show that deacetylation with dilute NaOH (0.1M) combined with low severitydilute-acid pretreatment increased corn stover reactivity by increasing the xylose monomer yield by 20% in pretreatment while degradation of xylan to furfural remained under 2%. Glucose yield increased by 10% and xylose yield by 20% during dilute solids enzymatic hydrolysis as a result of deacetylation of the feedstock prior to pretreatment. Particle size reduction by mechanical refining after pretreatment is shown to assist achieving> 90% cellulose yield in high solids (20 wt% total solids) enzymatic hydrolysis. Bioscreen C growth assays and mini fermentation at high solids (20wt%) showed less toxicity of deacetylated pretreated corn stover and enhanced xylose utilization by 20%.

See more of this Session: Developments In Biobased Alternative Fuels I
See more of this Group/Topical: Sustainable Engineering Forum

Reducing Process Complexity for Cellulosic Ethanol

Lonnie O. Ingram, Department of Microbiology and Cell Science
University of Florida, Box 110700, Gainesville, FL 32611 Ingram@UFL.edu

Lignocellulosic biomass represents a potential source of carbohydrates (cellulose and hemicellulose) for microbial fermentation into fuels and chemicals. Steam pretreatment with dilute mineral acids serves as an efficient approach to depolymerize hemicellulose and enhance cellulose digestion with enzymes. However, side products (furfural, 5-hydroxymethyl furfural, acetate, phenolics,and others) formed during pretreatment inhibit microbial growth and retard fermentation. Potential engineering solutions to this problem (separation of hemicellulose hydrolysate, counter-current-washing of fibers, and mitigation of toxins) add complexity and increase cost. The level of inhibitors produced during acid pretreatment is affected by the severity of conditions and the choice of acid. Weaker acids such as phosphoric acid produce lower levels of toxic side products and reduce the need for exotic metal alloys. Since mineral acids are not consumed by the pretreatment process, the higher cost of phosphoric acid as compared to sulfuric acid could be offset in part by recovery and reuse as a dilute fertilizer when crops are nearby. Recent progress has been made in the development of more robust *E. coli*-based biocatalyst with increased resistance to dilute acid hydrolysates. Using this biocatalyst and phosphoric acid pretreatment, we have been able to co-ferment sugarcane bagasse hemicellulose hydrolysate and enzyme-hydrolyzed cellulose in a single vessel using a variation of the simultaneous saccharification and co-fermentation process (SScF; Fig. 1). Hexose and pentose sugars from phosphoric acid pretreated sugarcane bagasse were fermented in a single vessel, eliminating process steps for solid–liquid separation and sugar cleanup. After pretreatment, a liquefaction step (L) with cellulase was included to improve mixing and saccharification (L + SScF), analogous to a corn ethanol process. Bagasse slurries containing 10% and 14% dry weight (fiber plus solubles) were tested using pretreatment temperatures of 160–190 C (1% phosphoric acid, 10 min). Enzymatic saccharification and inhibitor production both increased with pretreatment temperature. The highest titer (30 g/L ethanol) was obtained after incubation for 122 h using 14% dry weight slurries of pretreated bagasse (180 C). The highest yield (0.27 g ethanol/g dry bagasse) was 82 gal ethanol per dry ton of bagasse using this L+SScF process and 10% dry weight slurries.

Comparison of conversion processes

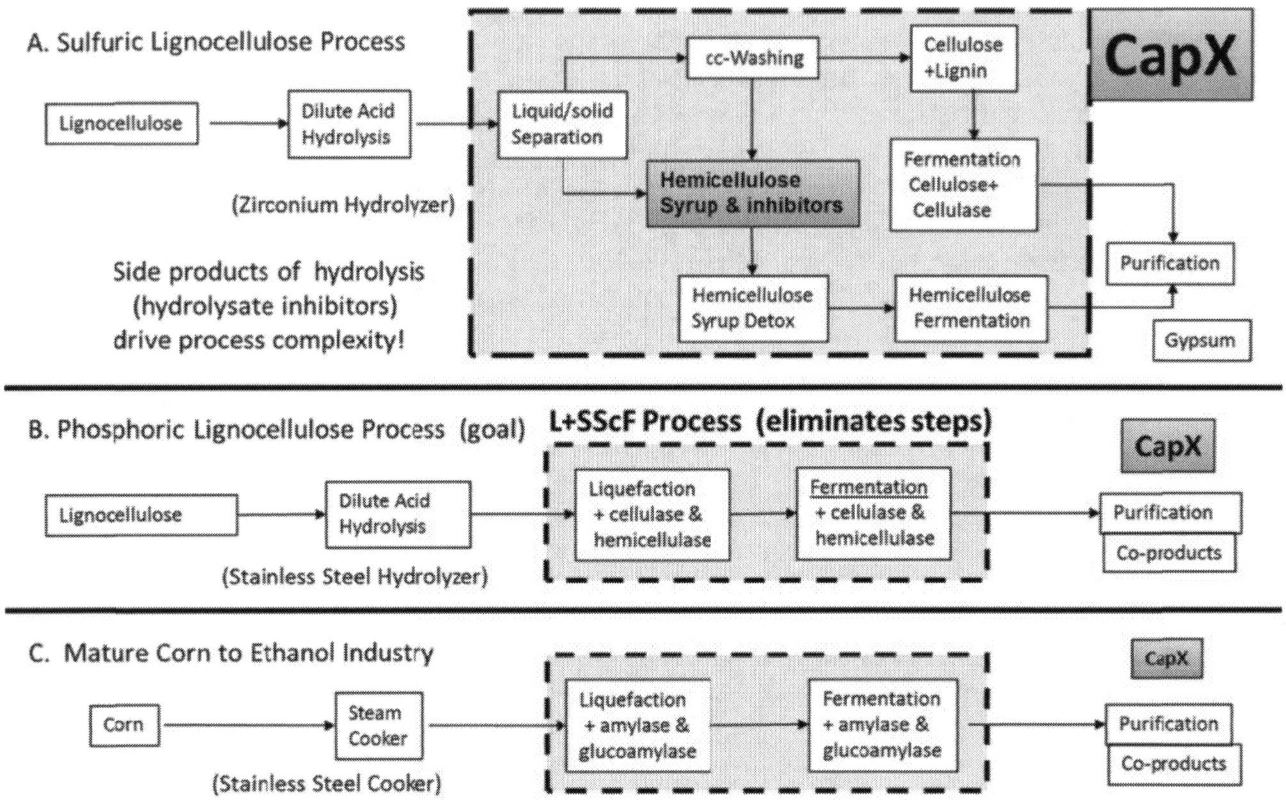

(522d) Co-Utilization of Methane In Steam-Biomass Gasification Using Concentrated Solar Energy

Aaron W. Palumbo, Erica L. Jorgensen and Alan W. Weimer, Chemical & Biological Engineering, University of Colorado at Boulder, Boulder, CO

Presented is a novel gasification application where biomass is steam-gasified and natural gas is steam-reformed simultaneously in a solar reactor. Solar-driven gasification is a proven technology, and can augment the biofuel industry as a domestic fuel source. Gasification performed in an indirectly heated gasifier can achieve temperatures greater than 1150 °C [1], mitigating or eliminating the production of CO_2, tar, char, and light hydrocarbon gases in the product gas. This reduces the need for gas clean-up technology and will not require a downstream water-gas shift reactor to adjust H_2:CO. Thermodynamics favor the formation of only hydrogen and carbon monoxide in the product synthesis gas. Therefore, a predictable H_2:CO can be controlled and tuned for downstream fuel conversion. Since the majority of biomass carbon is converted to CO, more than twice as much biomass is converted to a useable liquid fuel than in a traditional gasification process. Preliminary experiments show that reactor temperatures must exceed 1350 °C in order to reduce the effects of the water-gas shift reaction and reform practically all light hydrocarbon gases, including CH_4. This operating temperature is consistent with realizable concentrated solar energy systems.

1. Piatkowski, N., et al., *Solar-driven gasification of carbonaceous feedstock-a review.* Energy & Environmental Science, 2011. **4**(1): p. 73-82.

See more of this Session: Developments In Biobased Alternative Fuels I
See more of this Group/Topical: Sustainable Engineering Forum

(522e) Poly(oxymethylene) Dimethyl Ethers As Components of Tailored Diesel Fuel - Properties, Synthesis and Purification Concepts

Jakob Burger, Laboratory of Engineering Thermodynamics, Technische Universitt Kaiserslautern, Kaiserslautern, Germany

Poly(oxymethylene) Dimethyl Ethers as components of tailored diesel fuel - properties, synthesis and purification concepts

Jakob Burger [1*], Markus Siegert [2], Eckhard Ströfer [2], Michael Nilles [2], Hans Hasse [1]

[1] Laboratory of Engineering Thermodynamics, University of Kaiserslautern, Germany;

[2] BASF SE Company, Ludwigshafen, Germany
*Corresponding author: Jakob Burger, Laboratory of Engineering Thermodynamics, University of Kaiserslautern, Germany. Email: jakob.burger@mv.uni-kl.de

Keywords: Fuel additive, Soot reduction, Process development

Poly(oxymethylene) Dimethyl Ethers (POMDMEs), oligomers of the general structure CH_3-O-(CH_2-O)$_n$-CH_3, open a new route for tailoring diesel fuels. POMDMEs belong to the group of oxygenates which reduce soot formation in the combustion when added to diesel fuels. They can be produced on a large scale based on Methanol which can be produced from renewable resources. This contribution presents the process chain, possible synthesis routes and some fuel related properties of POMDMEs. Comparisons to other oxygenates, e.g. Dimethyl Ether, are given.

A particularly favourable route for the large scale production of POMDMEs in which they are formed from Methylal and Trioxane is proposed. The chemistry behind this route is given in detail. Measurements of the reaction kinetics and the reaction equilibrium were carried out in a batch reactor using acidic ion exchange resin as heterogeneous catalyst. The parameters of a pseudo-homogeneous model were fitted to the results of these experiments. The formation of side-products was studied experimentally and concepts to minimize them are discussed. Based on the reaction model a process concept was developed. It consists of a reactor and a downstream distillation sequence to yield pure POMDMEs of the most favourable chain length n=3-5.

The use of distillation technology allows an easy scale-up to huge single train plants which can potentially fulfil the need for millions of tonnes of the innovative oxygen carrier POMDME needed for the supply of clean diesel fuel in the future.

See more of this Session: Developments In Biobased Alternative Fuels I
See more of this Group/Topical: Sustainable Engineering Forum

(522f) Synthesis of Green Diesel Fuel From Fatty Acid Feedstocks Via Electrochemical Hofer-Moest Decarboxylation

Cody Wagner, Charles Coronella and Rainer Busch, Chemical Engineering, University of Nevada, Reno, Reno, NV

Production of biodiesel by esterification or transesterification of long-chain fatty acids is a well known process, but the fuel product has characteristics making it not fully compatible with petroleum diesel engines. A better option is to convert fatty acid molecules to hydrocarbons by removing their carboxyl group. Decarboxylation of fatty acid feedstocks was performed by electrolysis using graphite electrodes, which selectively causes a Hofer-Moest (or non-Kolbe)-type reaction. Methanol was used as a solvent for a robust selection of fatty acid feedstocks.

The liquid product created by the electrochemical reaction was composed of mainly alkenes with at least one double bond, ethers, and methyl esters. This fuel product has advantageous cold-flow properties, no sulfur content, heating values close to those of diesel fuels, and is capable of running in current diesel engines as a drop-in replacement for petroleum diesel, unlike biodiesel. The temperature, pH, concentrations of ions, and type of base used were all varied to optimized the current density, which limits the production rate. A separation method was tested for removal of the fuel product from the electrolyte. An efficiency was then calculated for several varieties of fatty acid feedstocks, comparing the electrical energy input to the product generated from the electrolysis. Overall first-law thermodynamics and economics were also evaluated, examining production rates and the large-scale viability of the process.

See more of this Session: Developments In Biobased Alternative Fuels I
See more of this Group/Topical: Sustainable Engineering Forum

Quantitative Comparison of Chemical and Biological Cardboard's Lignin Degradation

E. Isaza
Department of Chemical Engineering, Universidad de Los Andes, Bogotá, Colombia
R. Sierra R.
Advisor, Universidad de Los Andes, Bogotá, Colombia

ABSTRACT: An efficient way to increase biomass digestibility, thereby, make it usable in the production of second generation biofuels, is pretreatment. However, it demands important energy inputs which make it one of the most expensive stages in biofuels processing. Comparisons between chemical (oxidative with lime) and biological (with commercial *Laccase* extract) pretreatments were performed both by experimental measurements and by simulations. All experiments were performed following a full factorial experimental design, with lignin and sugar yields as response variables. The influence of the type of pretreatment (i.e., pretreatment yields) on the process that leads to biogasoline and jet fuel production (MixAlco®) was simulated in AspenPlus®. This simulation showed higher productivity and energy efficiency (15 and 60% respectively) for the chemical pretreatment over the biological pretreatment.

1 INTRODUCTION

First generation biofuels have emerged as an alternative to fossil fuels, whose price and environmental impacts have been subject of intense debate over the last years. From all biofuels, biodiesel and bioethanol are the most popular. They are currently produced using high energy crops such as soy, sugar cane, oil palm and corn starch as feedstocks (U.S. State Department, 2010). These resources, however, are characterized by such high commercial and alimentary value that they become restrictive for biofuels and generate issues of alimentary sustainability (Lee, Biological conversion of lignocellulisic biomasss to ethanol, 1997).

The utilization of more inexpensive feedstock in fermentative process such as lignocellulosic residues is a good baseline for the optimization of this technology (Lee, Biological conversion of lignocellulisic biomasss to ethanol, 1997). Good feedstock resources are agricultural, industrial, and municipal wastes among others. The *MixAlco®* process, created by Dr. Mark Holtzapple in Texas A&M University and patented by Terrabon L.L.C, uses this kind of waste to produce ethanol and higher energy alcohols.

In this process, four main stages may be differentiated: pretreatment, fermentation, chemical dewatering, gasification and cogeneration, thermal conversion involving hydrogen, and final conversion into biofuels. Among these, pretreatment is the most expensive (Gousamme, 2011).

In biological and some chemical pretreatments (such as lime pretreatment), biomass compositional changes are sought. These changes, include but are not limited to delignification. As a result, an increased biomass digestibility is observed. Other effects of pretreatment are: disruption of the crystalline structure of cellulose, increased pore size to enhance enzyme access to carbohydrates, and some degree of separation of the main lignocellulose constituent polymers.

Lignin is a highly heterogeneous tridimensional aromatic polymer, formed by three different units of phenylpropylene: coumaric alcohol, coniferyl alcohol and sinaphyl alcohol; bounded randomly through C-C and C-O-C type bounds (Sylvia, Fuhrman, Hartel, & Zuberer, 1998), (Lee, Biological conversion of lignocellulisic biomasss to ethanol, 1997).

Widely extended delignification processes such as Kraft pulping, (Lachental, Mortha, Sevillano, & Zaroubine, 2004) are controversial due to the use of environmental harmful chemicals. Delignification in alkali with possible oxygen addition is preferred because comparable delignification levels are obtained with less toxic waste products. Pretreatments such as oxidative and non-oxidative lime (i.e., calcium hydroxide ($Ca(OH)_2$), have been reported to promote lignin degradations close to 70% of the compound (Chang, Nagwani M, Kim, & Holtapple, 1999).

On the other hand wild occurring delignification processes are promoted by extracellular enzymes released by fungi, bacteria (Zhang, Kjonas, & Flurkey, 1999) or even plant cells; however, white rot fungi are the degraders by excellence (Lee, Biological conversion of lignocellulisic biomasss to ethanol, 1997). They generate extracellular enzymes such as *Laccase*, Manganese peroxidase and lignin peroxidase. Laccase (EC. 1.10, 32), benzendioloxigenoxidoreductase, is a copper containing enzyme capable of lignin degradation in the presence of oxygen. The only product of this transformation is water (Rodriguez Cuoto & Toca-Herrera, 2006). Because of this reason, this enzyme has been widely studied and used in industrial applications.

Since there are many advantages and drawbacks for both chemical and biological delignifications, it is desirable to formally compare these processes. However, a systematic comparison on the same basis (same feedstock, analytical procedures and general methods) was not found in the literature. This study aimed to make such comparison. To accomplish this goal, a full factorial design was implemented to both: oxidative lime delignification of cardboard (high in lignin content) and delignification of the same material using purified lacasse (provided by Sigma Aldrich)

The experimental data obtained from the best global conversion of carbohydrates was loaded to an existing preliminary simulation of the *MixAlco®* process (Gousamme, 2011) to determine which of the delignification processes is best on the basis of highest productivity and energy efficiency.

2 METHODOLOGY

2.1 *Chemicals and reagents*

2.1.1 Feedstock for pretreatments and substrate for enzymatic hydrolysis

Recycled cardboard is commonly used in Colombia to transport eggs. This biomass is characterized by high lignin content; a desirable feature for a delignification study. Untreated cardboard was the feedstock for pretreatment. The samples were obtained from *Kikes eggs*, Leader Company in the egg market at Colombia. Pretreated cardboard was used as substrate for enzymatic hydrolysis.

2.1.2 *Reagents for enzymatic delignification*

The selected enzyme was *Laccase Suberase* brand *Novartis*. It was activated in a phosphate buffer at pH 3, 4.5, 5, 5.5, 6, 6.5 and 7.

2.1.3 *Chemical pretreatment*

Hydrogen peroxide was obtained from *Panreac* at (%w/w) 0.5, 1, and 2. Calcium hydroxide $Ca(OH)_2$ also was obtained from Panreac.

2.1.4 *Enzymatic saccharification*

It was performed using cellulase 15 FPU obtained from *Sigma Aldrich.*

2.2 *Characterization of the selected biomass*

Quantification of biomass components (i.e., lignin, structural carbohydrates, and ash), was obtained following the laboratory analytical procedures (LAP) produced by the *National Renewable Energy Laboratory (NREL)* for lignocellulosic biomass (Sluiter A. , Hames, Ruiz, Scarlata, Sluiter, & Templeton, 2004). A summary of experimental protocols follows.

2.2.1 *Biomass conditioning*

In accordance to NREL Lap entitled "Preparation of Samples for Compositional Analysis" (Hames, Ruiz, Scarlata, Sluiter, Sluiter, & Templeton, 2008), all biomass was conditioned in a step previous to experimentation, by reducing: particle size (the fraction taken for experimentation was +40/-80 Thyler mesh) and the moisture content was reduced to less than 10%, by drying in a convection oven at 45°C for >24 h.

2.2.2 *Ash content determination*

The ash content of raw biomass was determined by exposing a known dry weight of biomass (7 replicates) to calcination during four hours in a muffle at 575°C. The weights of the resulting ashes were measured and related to the initial dry weight of each sample. (Sluiter A. , Hames, Ruiz, Scarlata, Sluiter, & Templeton, 2008).

2.2.3 *Structural sugars and lignin*

These components were measured in accordance to NREL laboratory analytical procedure entitled "Determination of Structural Carbohydrates and Lignin in Biomass" which involves two acid hydrolysis steps with H_2SO_4. The first stage is concentrated at 72% and the second diluted with 4%. This procedure results in solubilization of structural carbohydrates while lignin is precipitated (Scarlata, Slutier, Termpleton, Crocker,

Hames, & Ruiz, 2008). Sugar concentration is determined by the colorimetric method dinitrosalicilic acid or DNS, and lignin is separated and gravimetrically measured.

2.3 *Chemical pretreatment*

Cardboard (1 g), lime (0,5 g) and 45 mL of hydrogen peroxide were well mixed into a stainless steel reactor (Parr 500). Afterwards, the reactor was tightly closed, and submitted to a pretreatment at the temperature, hydrogen peroxide concentration and time of pretreatment that corresponds to the experiment according to the factorial design 2^3 summarized in Table 1.

Table 1. Factorial design of chemical pretreatment.

Factor	Leve l 1	Lev el 2
Hydrogen peroxide concentration%	0.5	2
Operating time (min)	30	120
Operating temperature (°C)	80	150

Pretreated samples were neutralized using 5 N HCl, washed with abundant distilled water, filtered through Whatman #1 filter paper and weighted. Afterwards, the harvested biomass was divided into three parts. The first part was kept frozen to be submitted to enzymatic hydrolysis later. Another part was put to dry in a 45°C convection oven, to be used in lignin and sugar quantification. The last part was used to determine moisture content.

A parameter was calculated to determine total solids yield after pretreatment (Y_T) as follows:

$$Y_T = \frac{\text{recovered biomass (g)}}{\text{biomass before pretreatment (g)}} \qquad (1)$$

With this parameter, it is possible to determine structural carbohydrates and lignin yields (Y_s) (Y_L) as defined in equations (2) and (3),

$$Y_s = \frac{\text{recovered carbohydrates (g)}}{\text{carbohydrates before pretreatment (g)}} = Y_T \frac{C_0}{C_t} \qquad (2)$$

$$Y_L = \frac{\text{recovered lignin (g)}}{\text{lignin before pretreatment(g)}} = Y_T \frac{L_0}{L_t} \qquad (3)$$

Where, (L)=Lignin; (C)= Sugars; (0)=initial; and (t)=at the end of the pretreatment.

For lime pretreatment, kinetics of lignin and sugar degradation were modeled by standard power law (Eq. 4), accounting for two reacting moieties (fast and slow) as proposed by (Sierra, et al., 2011). Model parameters are obtained by minimization of the sum of squared residuals.

$$-r_i = -\frac{dY_i}{dt} = k_{if} M_{O_2}{}^{\beta if} Y_{if} + k_{is} M_{O_2}{}^{\beta is} Y_{is} \qquad (4)$$

Where, (i)=Lignin or sugars; (j)=slow or fast reaction; k_{ij}= rate constant for substance; R=universal constant law for ideal gases (8.314×10^{-3} kJ/(mol K); M_{O_2}=oxygen loading (kg initial oxygen/kg dry biomass); and β_{ij}= reaction order for hydrogen peroxide.

2.4 *Enzymatic pretreatment*

2.4.1 *Activity determination*

Activity determination of the enzyme, *Laccase Suberase* brand *Novartis*, was made according to the *Sigma Aldrich* protocol (Aldrich, 1997). This protocol recommends the use of 0.216 mM Syringaldazine solution, water and *Laccase Suberase*, to determine the activity by a continuous spectrophotometric rate determination (Aldrich, 1997). Because experiments were run at several different pH (Table 2), one result of this experimentation was the recommended pH.

2.4.2 *Selectivity*

Enzymatic delignification is supposed to be highly (if not completely) selective. Nevertheless, a selectivity experiment was performed in order to ratify that the enzyme was free of cellulases. For this purpose, the enzyme (20 µL of *Laccase*) was put in contact with 0.1 g of cellulose at pH of 6, 6.5 y 7, and room temperature and 55°C for a day.

2.4.3 *Experimental Design*

A full factorial design 3^3 was made in order to determine enzyme loading, substrate concentration, residence time and temperature that would maximize enzyme performance at optimum pH of 6.5 and optimum temperature of 55°C. (Baldrian, Fungal Laccasses-ocurrance and properties, 2006). It is summarized in Table 2.

Table 2. Factorial design for enzymatic pretreatment

Factor	Level 1	Level 2	Level 3
Enzyme loading (mL/g substrate)	0.0 1	0.1	0.5
Substrate concentration (w/w)	1	5	10
Residence time (min)	60	360	1440

Immediately after pretreatment, enzyme pretreated samples were submitted to thermal shock to denature the enzyme and therefore stop the reactions. Subsequently, they were profusely washed with distilled water.

Based on the obtained results a decrease on the number levels of the factorial design is proposed. Reducing: residence time(0-1 h), enzymatic loading (20,150 µL) and substrate concentration (2 g). For a better compression of the degradation of lignin it is modeled by Michaelis Menten:

$$-r_S = \frac{V_{max} \times (S)}{K_m + (S)} \quad (5)$$

Where (S)=substrate concentration; K_m=affinity constant; and V_{max}=maximum degradation rate of substrate.

2.5 *Enzymatic hydrolysis*

To determine the effect of pretreatment in biomass digestibility, an enzymatic hydrolysis is made quantifying the cellulose fraction that is hydrolyzed by cellulase at optimum conditions.

The procedure is the *NREL Lap* for *Enzymatic Saccharification of Lignocellulosic Biomass* (Brown & Torget, 1996). It consists on exposing pretreated biomass with an excess of cellulase (15 FPU/g biomass) at 50ºC and pH 4.8 for 3 days. The protocol is modified adding 1g of biomass instead of 1 g cellulose equivalent in biomass.

The monomers of glucose and xylose are determined later by DNS, and digestibility is calculated with relation (5) and (6):

$$Y_{saccharification} = \frac{\text{digeted celulose (g)}}{\text{added celulose (g)}} \quad (5)$$

$$Y_{saccharification} = \frac{\text{sugars}_e \times V_T}{\text{biomass(g)} \times \text{SST} \times \text{sugars}_p} \quad (6)$$

Where, $sugars_e$=sugar content of saccharification extract (mg sugars/ml); V_T=total volume of saccharification (10.6 ml), biomass =pretrated biomass= 0.1 g; SST= total solid content in pretreated biomass; and $sugars_p$ =sugar content of pretreated biomass (g sugars/g pretreated biomass).

2.6 *Global conversion*

The combined effect of pretreatment and enzymatic hydrolysis a global recuperation factor is found:

(6)

Mentioned relation allows choosing the conditions of chemical and enzymatic pretreatment to load in the *MixAlco®* process simulation, where higher values of correspond to superior efficiency in pretreatment, therefore, preservation of sugars, breakage of crystalline structure of celluloses and hemicelluloses and higher lignin degradation.

2.7 *MixAlco® process simulation*

The simulation of the *MixAlco®* process, is modified from a existing one made by (Goussame,2011) where experimental data of both pretreatments are loaded. Accomplishing finally, the comparison of the two pretreatments, at a preliminary stage, because the simulation is an ideal modeling of the process.

The energy index was defined by (Gousamme, 2011) as,

(7)

3 RESULTS AND DISCUSSION

3.1 *Biomass characterization*

The obtained results of biomass characterization protocols are, that non pretreated biomass is composed in dry weight: 43% by lignin, 3% by ash, and a 44% of sugars, Illustration 1.

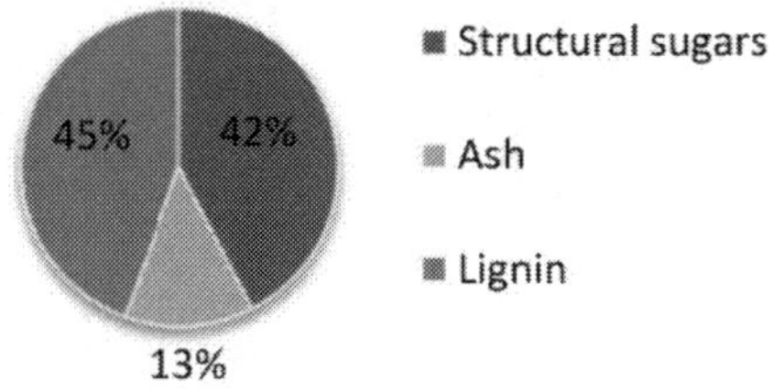

Ilustration1. Composition of cardboard in dry weight.

3.2 *Chemical pretreatment*

3.2.1 *Lignin degradation*

The average of the experimental results obtained for factorial design 2^3, for chemical pretreatment are presented in Illustration 2 with respective standard deviation.

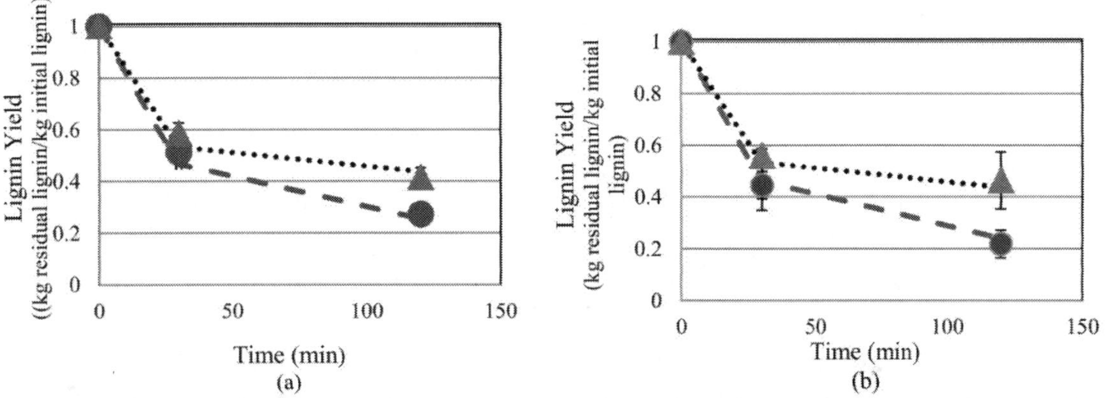

Illustration 2. Lignin yield for chemical pretreatment with lime Ca(OH)2 and respective kinetic model. (a) 80ºC(b)150ºC. (▲) 2% hydrogen peroxide, (●)0.5% hydrogen peroxide.

According to Illustration 2 at higher peroxide concentration, temperature and time, greater lignin degradation. This can be explained because higher hydrogen peroxide concentration generates more quantity of superoxide ions responsible of breaking C-C and C-O-C bounds.

As shown in Ilustration2, the time that superoxide ions take in breaking the lignin chain ends, where the degradation begins, is half hour. With higher residence time an increase of the lignin degradation is measurable, because the numerical relation between superoxide ions and attack sites become more efficient.

The highest delignification is at two hours, with lime ($Ca(OH)_2$) at 150°C and hydrogen peroxide at 2%.

Calculated lignin yields are presented as cutter lines in the operating rank of each pretreatment. Data is obtained by least squares minimization of the model presented by Sierra ET all, 2009, whose solution generates the parameters show in Table 3.

Table 3. Achieved values for parameters of lignin degradation with lime $Ca(OH)_2$.

Paramet er	Value	Units
Y_{if0}	0.545	g residual lignin/g initial lignin
a_{iif}	0.0597	min^{-1}
B_f	0.0257	
E_{iif}	0.0825	kJ/mol
a_{ifs}	4.96E-07	min^{-1}
B_s	2.10	
E_{iis}	0.878	kJ/mol

3.2.2 *Sugar Yields*

Average of experimental data obtained for sugar yield in chemical pretreatment are presented as marks in Illustration 3.

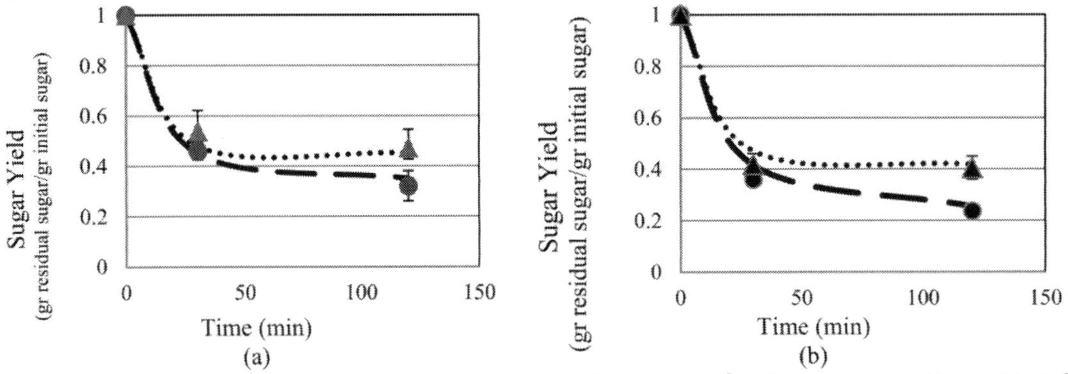

Illustration 3. Sugar yield for chemical pretreatment (a) 80°C, (b) 150°C. (··▲··) hydrogen peroxide at 2%, (−●−)hydrogen peroxide.0.5%

The behavior of sugar yield is similar to lignin yield, because at higher operation conditions lower sugar yield, this according to the presence of superoxide ions and harder thermal conditions.

Comparing part (a) and (b) of the Illustration 3, can be observed that sugars are sensitive to temperature degradation because at every pretreatment with equal time and peroxide concentration and higher temperature lower sugar yield is obtained.

The effect of time in sugar degradation is least at lower hydroxide concentration.

The kinetic modeling is made same as the lignin degradation, the obtained parameters are:

Table 4. Obtained values of kinetic parameters for sugar degradation with lime Ca(OH)2.

Parameter	**Value**	**Units**
Y_{if0}	0,489	g residual sugar/g initial sugar
a_{iif}	0,00156	min^{-1}
B_f	1,02	
a_{ifs}	911×102	min^{-1}
B_s	4,315	
E_{iif}	11,8	kJ/mol
E_{iis}	0,869	kJ/mol

3.3 *Enzymatic pretreatment with Laccase*

3.3.1 *Activity determination and enzyme selectivity*

The higher activity is 864 U ml^{-1} at pH 6.5, Illustration 4.

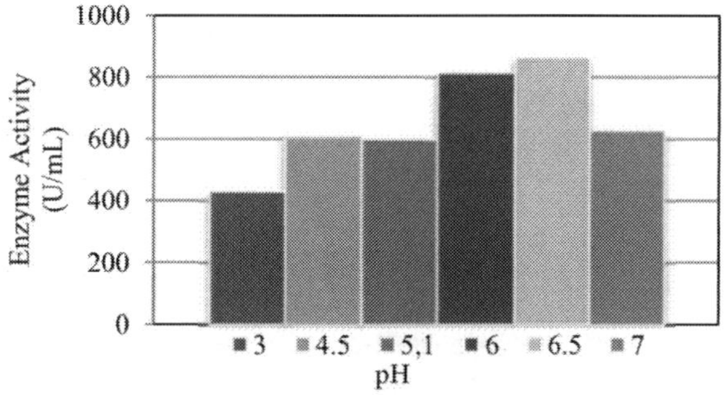

Illustration 4. Activity of the enzyme.

Obtained results for enzyme selectivity showed changes of 10^{-3} g, attributed to moisture content in the samples.

3.3.2 *Enzymatic lignin degradation*

Enzymatic experimental average results of the first substrate level (2 g) of the factorial design 3^3 are presented in Illustration 5. Only this level is presented because the behavior obtained for the other two levels of substrate, 1.5 and 1.0 g are very similar; (where at every enzymatic loading there is an increase of the lignin

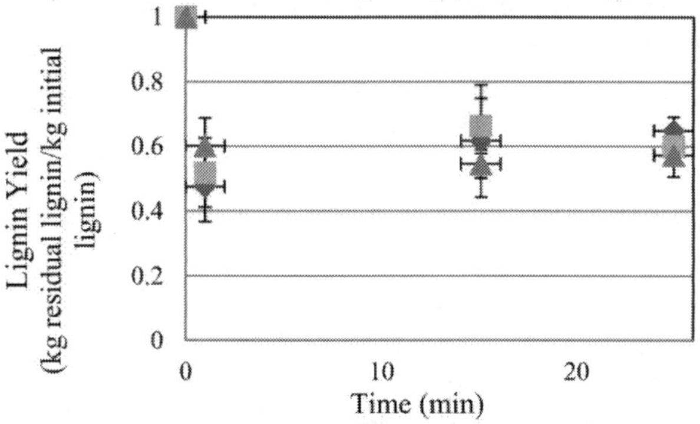

Illustration 5. Enzymatic delignification (◆) 20 μL, (■) 150 μL, (▲) 1000 μL.

For this reason the experimental conditions were cut to: a maximum of an hour, for residence time, and two levels of enzymatic loading, 20 and 150 μL, with a single substrate concentration, 2 g.

In Illustration 6, the results of the last Laccase experimentation are presented where the higher lignin degradation is 0.45(g residual lignin/g initial lignin), accomplish by 150 μL of enzyme at one hour. While the minimum yield is 0.75(g residual lignin/g initial lignin) obtain with 20 μL of Laccase at 30 min.

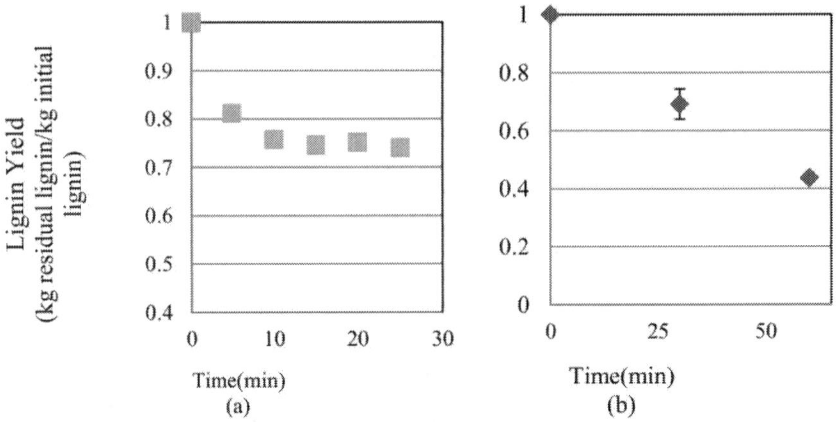

Illustration 6. Enzymatic delignification of 2 g of substrate (a) 20μL enzyme, (b) 150 μL enzyme.

The kinetic modeling for the enzymatic lignin degradation was made by the equation of Michaelis Menten as a first approximation.

The Lineweaver-Burk linearization was used to model the kinetic parameters, for 20 y 150 μl of enzyme data.

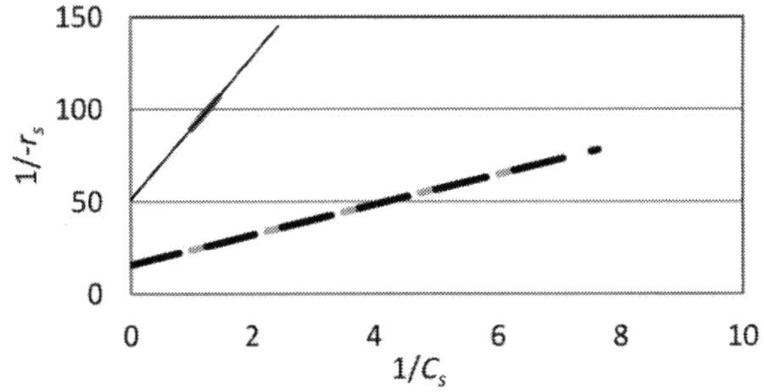

Illustration 7. Linewever-Burk for 2 g of biomass. (- -) 20 μL, ( ) 150 μL.

The values obtained for the kinetic parameters are presented at the Table 5.

Table 5.Obtained values for the enzymatic degradation of lignin.

Parameter	Level	Value	Units
V_{max}	20μl	0,0646	U mg^{-1}
K_m	20μl	0,529	g cm^{-3}
V_{max}	150μl	0,0195	U mg^{-1}
K_m	150μl	0,752	g cm^{-3}

3.3.3 *Sugar Yield*

As shown at Illustration 8, it can be identified that the higher sugar yield is obtained with enzymatic pretreatment at ten minutes and 20 μL of enzyme, while the lowest is also obtained with 20μL at 20 minutes.Thereby time is an important factor at this enzymatic loading level.

At 150 μL of enzyme, sugar yield keeps near 0.6 (g recover sugar/g initial sugar) in the selected time range.

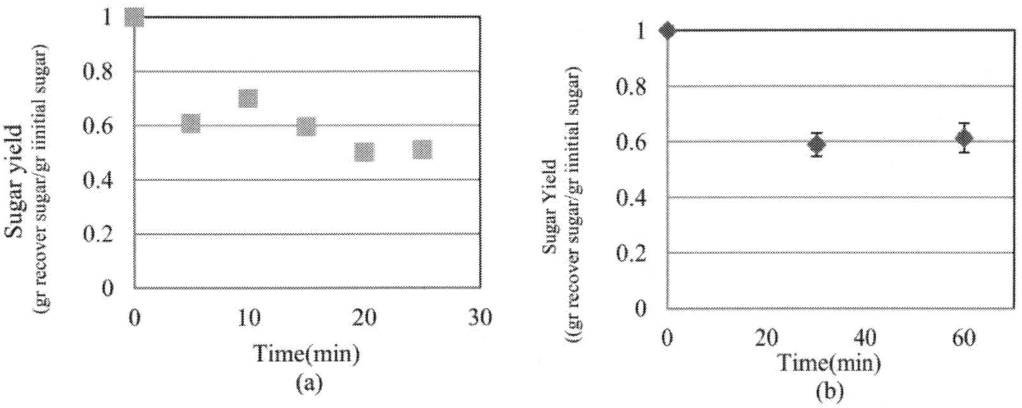

Ilustration8. Sugar yield for 2 g of substrate (a) 20 µL of enzyme (b) 150 µL of enzyme.

3.4 *Enzymatic saccarification*

Higher digestibility values were achieved for chemical pretreatment with the higher level of: time (2 h) and hydrogen peroxide concentration(2%), but since results are similar with different temperatures, Illustration 9, it is necessary to calculate global sugar conversion to select the best chemical pretreatment conditions.

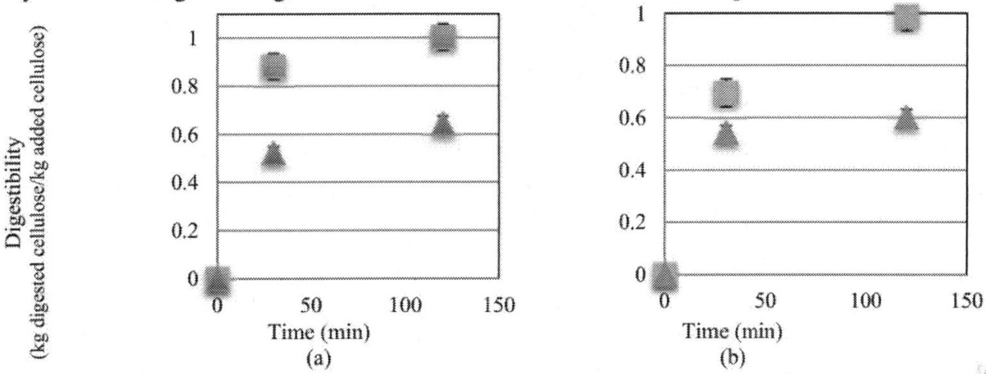

Illustration 9. Digestibility of chemical pretreated biomass.(a) 80ºC, (b) 150ºC. (▲) 0.5% peroxide,(■)2% peroxide.

Highest digestibility for enzymatic pretreatment is at one hour of pretreatment with *Laccase,* with 150 µL of enzyme loading, Illustration 10.

The biomass, pretreated with 20 µL of *Laccase* obtained the highest digestibility at 10 minutes of pretreatment, which can be justified with the fact that low enzyme concentration triggers quicker enzyme action because there is lower enzyme competition, but also is less efficient compared to 150 µl of enzyme where are more competition but higher lignin degradation.

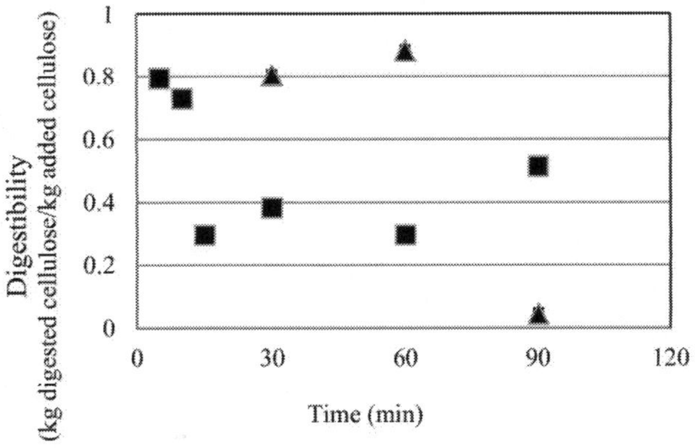

Illustration 10. Digestibility enzymatic pretreated biomass (■) 20µL and (▲)150 µL de enzyme.

The calculation of global conversion is necessary to determine which chemical and enzymatic pretreatment is the proper one to implement at Aspen® simulation of the *MixAlco®* process.

The global conversion results are presented in a contour graph, made using *MiniTab Statistical Software©*.

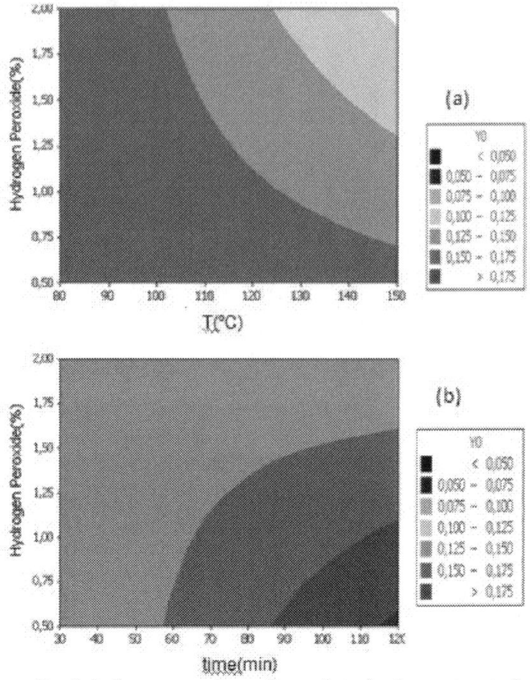

Illustration 11.Contour graph of global conversion for chemical pretreated biomass. (a) %Peroxide vs. temperature(°C),(b) %peroxide vs. time(min).

The global sugar conversion, Y_0, of chemical pretreatment samples are shown in Illustration 11, where at less peroxide concentration, lower temperature y higher time results in the best sugar conversion which can be explained by sensibility of sugar degradation to temperature and lower superoxide ions concentration.

The global sugar conversion for enzymatic pretreated biomass is presented in Illustration 12, where the zone with higher sugar conversion correspond to one hour, 150 μL of enzyme loading which can be explained due to enzyme selectivity to lignin degradation and the inverse relation between enzyme competivity and residence time.

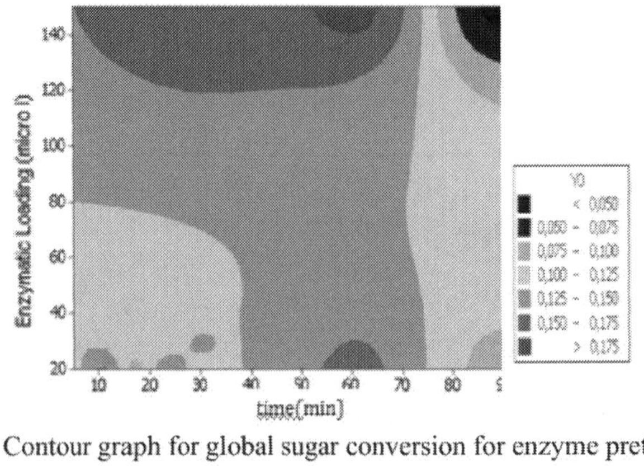

Illustration 12. Contour graph for global sugar conversion for enzyme pretreated biomass.

The best global conversion value was achieved with enzyme pretreated biomass at one hour of pretreatment and 150 μL of enzyme because the recuperation factor Y_t was close to one and the sugar degradation was low compare to chemical degradation recuperation factor close to ,5 and higher sugar degradation.

3.5 *Aspen Plus® Modeling*

The pretreatment conditions chosen to be loaded at the *MixAlco®* process simulation, design by (Gousamme, 2011) are the corresponding to chemical pretreatment 0.5% peroxide, 80ºC y 120 minutes and the corresponding to enzymatic pretreatment with 150 μL and one hour of residence time.

The higher biofuel production was achieved with chemical pretreatment with an energy index of 4,477 vs. 4,189 (kJ out/kJ in) of the enzymatic pretreatment.

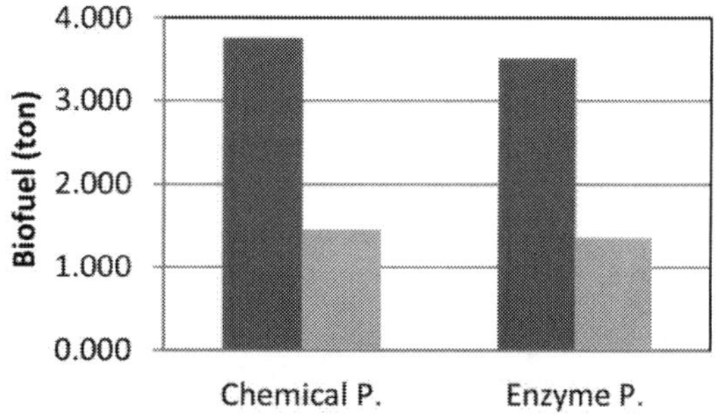

Illustration 13. Bar graph for biofuel production with chemical and enzyme pretreatment. ( ) Jet biofuel, ( ) gasoline.

4 CONCLUSIONS

Comparing chemical with enzymatic pretreatment, the enzymatic is more sensitive to time and pH, which complicates the implementation of this kind of pretreatment in the industry.

Enzymatic pretreatment also achieved the best values for digestibility and global sugar conversion.

The simulation, final step of the comparison between chemical and biological pretreatment, gives better results for the chemical pretreatment in biofuel generation and energy index.

5 BIBLIOGRAPHY

1. Aldrich, S. (1997 11-03). *Enzymatic Assay of LACCASE.* Retrieved 2011 12-03 from http://www.sigmaaldrich.com/etc/medialib/docs/Sigma/General_Information/2/laccase.Par.0001.File.dat/laccase.pdf
2. Baldrian, P. (2006). Fungal Laccasses-ocurrance and properties. FEMS Microbial Review 30 , 215-242.
3. Brown, L., & Torget, R. (1996). *Enzymatic Saccharification of lignocellulosic Biomass.* NREL.
4. Carvajal Montoya, S., & Giraldo Garcia, D. (2010). Simulación del proceso MixAlco en Aspen Plus y análisis de sensibilidad de los parámetros involucrados en las principales etapas del proceso. Bogota D.C: Proyecto de grado, Universidad de los Andes.
5. Chang, V., Nagwani M, M., Kim, C., & Holtapple, M. (1999). *Oxidative Lime Pretreatment of High-Lignin Biomass.* College Station: Texas A&M.
6. David, C., Salvador, S., Dirion, J., & Quintard, M. (2003). Determination of reaction scheme for cardboard thermal degradation using thermal gravimetricanalysis. *Journal of Analitycal an d Applied Pyroysis* , 307-232.
7. Fogler, S. H. (2008). Elementos de ingeniería de las reacciónes químicas. Naucalpan de Juarez, mexico: Pearson Prentice Hall.
8. Gandi, J., Holtzapple, M. T., Ferrer, A., Byers, M. F., Turner, N. D., Nagwani, M., et al. (1997). Lime treatment of agricultural residues to improve rumen digestibility. *Animal Feed and Science Technology* , 195.
9. Gousame, P. (2011). Simulation of the MixAlco Process. *Special Project* .
10. Granda, C. B., & Holtzapple, M. T. (2008). Part 1: Comparison of Three Biomass Conversion Platforms. In *Carboxylate Platform: The MixAlco Process.* Texas: Department of Chemical Engineering Texas A&M University.
11. Hames, B., Ruiz, R., Scarlata, C., Sluiter, A., Sluiter, J., & Templeton, D. (2008). *Preparation of Samples for Compositional Analysis.* Cole Boulevard, Golden, Colorado : National Renewable Energy Laboratory.
12. Hayatsu, R., Winans, R., McBeth, R., Scott, R., Moore, L., & Stuider, M. (1979). Lignin-like polymers in coals. *Nature* , 278.

13. Holtzapple, M., Davison, R. R., Ross, M. K., Aldrett-Lee, S., Nagwani, M., Lee, C.-M., et al. (1999). Biomass Conersion to Mixed Alcohol Fuels Using the MixAlco Process. *Applied Biochementry and Biotechnology* , 609-630.
14. Kaar, W. E., & Holtzapple, M. T. (2000). Using lime pretreatment to facilitate the enzymic hydrolysis. *Biomass and Bioenergy* , 189.
15. Lachental, D., Mortha, G., Sevillano, R.-M., & Zaroubine, M. (2004). Isolation of residual lignin from softwood kraft pulp. Advantages of the acetic acid acidolysis method. *Compites Rendus Biologies* , 911-916.
16. Lee, J. (1997). Biological conversion of lignocellulisic biomasss to ethanol. *Journal of Biotechnoly* , 3-4.
17. Lee, J. (1997). Biological conversion of lignocellulosic biomass to ethanol. *Journal of Biotechnology* , 4.
18. Mandels, M., Hontz, L., & Nystrom, J. (1974). Enzymatic Hydrolysis of Waste Cellulose. *Biotechnology and Bioengineering* , 1471-1493.
19. Morozova, O., Shumakovich, G., Gorbacheva, M., Shleev, S., & Yaropolov, A. (2007). "Blue" Laccases. *Biochemestry (Moscow)* , *73* (10), 1396-1412.
20. Mosier, N., Wyman, C., Dale, B., Elander, R., Lee, Y., Holtappple, M., et al. (2005). Features o promising technologies for pretreatment of lignocellulosic biomass. *Bioresource technology* , 673.
21. Pham, V., Holtazpple, M., & El-Halwagi, M. (2010). Techno-economic analysis of biomass to fuel ocnversion. *Journal of Microbiology and Biotechnology* .
22. Phillips, T. (2006 йил June). Structural Formula of Ethanoic Acid reacting with Propanol to form Propyl-Ethanoate.
23. Robert J. Wooley, V. P. (1996 йил Abril). Development of an ASPEN PLUS Physical Property Database for Biofuels Components. Golden, Colorado, Estados Unidos.
24. Rodriguez Cuoto, S., & Toca-Herrera, J. L. (2006). Lacasses in the textile industry. *Biotechnology and Molecular Biology review* , 115-120.
25. Scarlata, C., Slutier, J., Termpleton, a. D., Crocker, D. A., Hames, B., & Ruiz, R. (2008). *Determination of Structural Carbohydrates and Lignin in Biomass LAP.* Cole Boulevard, Golden, Colorado: National Renewable Energy Laboratory.
26. SE HOON, K. (2004). *LIME PRETREATMENT AND ENZYMATIC HYDROLYSIS.* Texas, USA: Texas A&M University Dissertation.
27. Sierra, R., & Holtzapple, M. T. (2009). Selectivity and Delignification Kinetics for Oxidative and Non-Oxidative Lime Pretreatment of Poplar Wood - Part III: Long-therm. *(Under Review)* .
28. Sierra, R., & Holtzapple, M. T. (2009). Selectivity and Delignification Kinetics for Oxidative Short-therm Lime Pretreatment of Poplar Wood - Part I: Constanr Pressure. *(Under Review)* .
29. Sierra, R., Smith, A., Granda, C., & Holtzapple, M. (2008). Producing fuels and chemicals from lignocellulosic biomass. *Chemical Engineering Progress* , 510-518.
30. Sluiter, A., Hames, B., Ruiz, R., Scarlata, C., Sluiter, J., & Templeton, D. (2008). *Determination of Ash in Biomass (LAP).* Cole Boulevard, Golden, Colorado: National Renewable Energy Laboratory.
31. Sluiter, A., Hames, B., Ruiz, R., Scarlata, C., Sluiter, J., & Templeton, D. (2004). *Determination of Total Solids in Biomass LAP.* Cole Boulevard, Golden, Colorado: National Renewable Energy Laboratory.
32. Smufit Kappa. (2011). *Preguntas frecuentes.* Retrieved 2011 6-Mayo from http://www.smurfitkappa.com.co/DropdownMenu/Contact/FAQ/Recycling/
33. Sylvia, D., Fuhrman, J., Hartel, P., & Zuberer, D. (1998). Decomposition of Naturraly Occuring Organic Materials. In *Principles and applications of soil microbiology.* (pp. 228-234). USA: Prentice Hall.
34. Tchobanoglous, G., Theissen, H., & Eliassen, R. (1982). Producción de desechos sólidos. In *DESECHOS SÓLIDOS PRINCIPIOS DE INGENIERÍA Y ADMINISTRACIÓN* (pp. 67-82). Ambiente y los Recursos Naturales Renovables.
35. Terrabon, L. (2008). *Terrabon Renewable Energy Solutions.* Retrieved 2009 30-Octubre from MixAlco Process: http://www.terrabon.com/mixalco_overview.html
36. Tisma, M., Znidarsic-Plazl, P., Plazl, I., Vasic-Racki, D., & Zelic, B. (2009). *Oxidation of Coniferyl Alcohol Catalized by catalized from Trametes versicolor.* Croacia, Eslovenia.
37. Tuomela, M., Vikman, M., Hatakka, A., & Itavaara, M. (2000). Biodegradation of lignin in a compst enviroment: a review. *Bioresource Technology* , 169-183.
38. U.S. State Department. (2010 йил 15-March). *Raw materials of biofuels.* Retrieved 2011 йил 11-March from http://www.america.gov/st/develop-spanish/2010/March/20100409143620CMretroP3.654116e-02.html
39. UNL. (2004). *Capítulo 2. Aspectos Generales.* From bibliotecavirtual.unl.edu.ar:8180/tesis/.../1/.../07-Capítulo%202.pdf
40. Van Soest, P. J. (1973). Lignin. In *Nutritional ecology of the ruminant* (p. 193). Cornell, USA: Cornell University Press.
41. Yamauchi, K., & Fukushima, K. (2004). The regulation from guaicyl to syringyl lingnin in the differenciating xylem of Robinia pseudoacacia. *Comptes Rendus Biologies* , 791-797.

42. Zhang, J., Kjonas, R., & Flurkey, W. (1999). Does N-hydroxyglycine inhibit plant and fungal laccases? *Phytochemestry* , 775-783.

(523b) Isoconversional Methods Applied to Kinetic Modeling Oxidative Alkaline Pretreatment of Sugarcane Bagasse

Rocio Sierra, Cuellar Gerardo, Simbaqueba Andres, Auza Nicolas and Copete Diego, Chemical Engineering, Universidad de los Andes, Bogota, Colombia

Delignification (decrease of lignin content) has shown to be a valid method to significantly increase lignocellulosic feedstocks digestibility. This operation (widely known as pretreatment) makes these abundantly available and inexpensive feedstocks usable in fermentation processes whose main goal is to produce biofuels and other chemicals. Because pretreatment is one of the most expensive processes in biofuels production plants, it has been extensively studied; however, a throughout literature search shows that development of kinetic models to better understand, optimize and control pretreatment, are very scarce if available at all.

This study uses sugarcane bagasse which is a residue from the sugar industry abundantly produced particularly near and in the Equator zone. This lignocellulosic resource is traditionally used in combustion processes to supply energy to the production plant; however life cycle analysis have shown that if lignocellulose is used as substrate in fermentation processes, the energy yield would be much closer to the theoretically available energy.

In this work, sugarcane bagasse is submitted to an alkaline pretreatment that uses an oxidative agent, systematically applied at widely varied ranges of temperature and time using a full factorial experimental design

Model fitting methods do not achieve a clean separation between the temperature dependent Arrhenius constant and the reaction model. On the contrary, the isoconversional methods evaluate the activation energy values at progressive degree of conversion without modelistic assumptions. This work applies this methodology to calculate activation energy for alkaline pretreatment of sugarcane bagasse and compares with previous work where the traditional curve fitting is employed.

See more of this Session: Developments In the Pretreatment of Lignocellulosics for Bioconversion II
See more of this Group/Topical: Sustainable Engineering Forum

(523c) Alkaline Deacylation of Corn Stover to Reduce Toxicity of Prehydrolyzates and Improve Enzymatic Hydrolysis

Urvi D. Kothari and Y. Y. Lee, Chemical Engineering, Auburn University, Auburn, AL

Hydrolysates from dilute acid pretreatment of biomass contain various biomass degradation products which are inhibitory to enzymes and toxic to microorganisms. One of the main inhibitory compounds is acetic acid which is released from the lignin-hemicellulose matrix in the biomass. Acetic acid at the level of 1 g/L inhibits cellulase activity by 10%. In most pretreatment liquors its concentration is in the range of 3-10 g/L and inhibition of cellulase activity by acetic acid at these concentrations is 20-60%. In this study, sodium carbonate was investigated as a deacylation reagent that can selectively remove part of the acetyl components in corn stover and reduce the acetyl concentration in the pretreatment hydrolysate. Sodium carbonate was chosen since it is a mild alkali that can preserve carbohydrates under the deacylation process conditions. It is inexpensive and can easily be recovered through established Kraft recovery process. Sodium carbonate of 5-10 wt% was used over 30-60°C for deacylation. A dilute-acid pretreatment was then applied under the conditions of 0.5-1% H_2SO_4 at 60-110°C in order to assess the effects of deacylation. The two-step process operated under optimized conditions led to reduction of the final acetyl concentration by 50-80%, and minimal loss of hemicellulose at 2-12%. The combined process further improved the glucan and xylan hydrolysis yields of the pretreated corn stover over those of the dilute-acid pretreatment alone.

See more of this Session: Developments In the Pretreatment of Lignocellulosics for Bioconversion II
See more of this Group/Topical: Sustainable Engineering Forum

(523d) Deconstruction of Wheat Straw Lignin by Streptomyces Viridosporus As Insight Into Biological Degradation Mechanism

Jijiao Zeng[1], Department of Biological Systems Engineering, Washington State University, Pullman, WA, Deepak Singh, Edeniq Inc, Visalia, CA, Dhrubojyoti D. Laskar, Washington State University, Pullman, WA and Shulin Chen, Biological Systems Engineering, Washington State University, Pullman, WA

Deconstruction of wheat straw lignin by *Streptomyces viridosporus* asinsight into biological degradation mechanism

JijiaoZeng[1], Deepak Singh[2],Dhrubojyoti Dey Laskar[1]and Shulin Chen[1*]

1. Department of Biological Systems Engineering, Washington State University,Pullman, Washington 99164-6120, USA

2. Edeniq Inc, 2505 N Shirk Road, Visalia, 93291 CA

[*]Corresponding author. Tel: +1 509 335 3743; fax: +1 509 335 2722.

E-mail address: chens@wsu.edu

Abstract:

Lignins are secondary natural organic compositions of plant cell wall besides cellulose and hemicelluse. Structurally lignins are linked with hemicelluloses and cellulose to form recalcitrant complex that resist the environmental stress. Biological systems created by nature over millions of years of evolution can be capitalized for targeted deconstruction of plant cell walls through selective cleavage and structural modification of lignin. Such modification causes changes in cell wall structure for facilitating enzymatic attack to cellulose for sugar release. The study of biological degradation mechanism on lignocellulosic materials could provide the new knowledge to convert plant biomass to biofuel and high value co-products. In this study, we have investigated the effect of wheat straw lignin structural changes by *Streptomyces viridosporus* on downstream enzymatic hydrolysis. ^{13}C CPMAS NMR revealed that the rising of carbonhydrate/lignin ratio accompany with enhanced intensity of methoxyl group in bio-treated wheat straw. The degraded lignin were isolated and analyzed by Fourier Transmission Infrared (FTIR), Pyrolysis Gas Chromatography/Mass Spectrometry (Py-GC/MS) and two dimenssional heteronuclear multiple quantum coherence nuclear magnatic resonance (2D HMQC NMR) to elucidate their structural change. The results indicate that the biological pretreated lignin were associated with critical change and/ or modification on their structure implying selective removal of guaiacyl (G) lignin unit which is more condense than syringyl (S) unit. Moreover, the treated lignin also dramatically reduced the inhibition of cellulose hydrolysis. This information provides new insights for understanding the biological degradation mechanism on lignin for cellulose utilization and exploring a possible route to convert lignin derived co-products.

See more of this Session: Developments In the Pretreatment of Lignocellulosics for Bioconversion II
See more of this Group/Topical: Sustainable Engineering Forum

Catchlight Energy LLC (CLE)

A 50/50 Chevron and Weyerhaeuser joint venture.

Our Mission: To commercialize the large-scale production of liquid transportation fuels from sustainable forest-based resources

Weyerhaeuser — Catchlight Energy — Chevron

Feedstocks at scale → Conversion Technology → High quality fuels to customers

A Unique End-to-End Value Chain Solution
Leveraging the strengths of two natural resource leaders

Catchlight Energy (CLE) brings strengths not found elsewhere - feedstocks development and procurement, fuels testing and refining, operations, conversion engineering.

CLE drivers aligned with national interests to develop renewable energy platform, increase energy independence & security and promote rural development.

October 15, 2011

1

Softwood – High Availability/High Recalcitrance

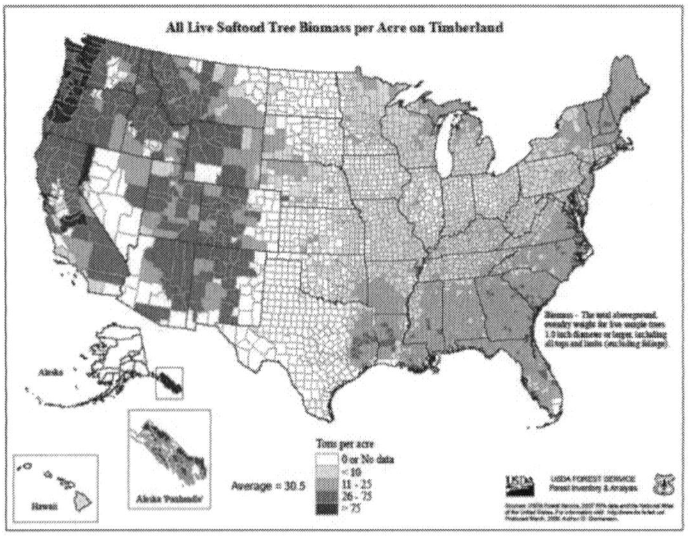

The use of softwood is of high critical value to the biofuels industry, but it remains a difficult feedstock for most pure bio-conversion processes.

October 15, 2011

2

Pretreatment Issues for Softwood and other Recalcitrant Biomass

Catchlight Energy

- **Dilute acid and base systems ineffective at reasonable levels.**
- **Known and proposed "Fractionation" pretreatments are excessively expensive, and not commercial scale.**
- **Pulping processes effective because of removal of lignin, but result in high and unsupportable costs.**
  - ➢ High capex (unless repurposed mill that is obtained at very low cost).
  - ➢ High opex (chemical costs or chemical recovery costs, wastewater treatment).
  - ➢ Fuels value can't compete with pulp value.

October 15, 2011

3

CLE Pretreatment

Catchlight Energy

- **Pulping, especially sulfite pulping, is an excellent but expensive pretreatment**
- **CLE Pretreatment has**
  - ➢ Higher throughput
  - ➢ Lower chemical consumption
  - ➢ Lower inhibitors
  - ➢ Excellent enzymatic hydrolysis results
  - ➢ Higher sugar yield

October 15, 2011

4

CLE Pretreatment Characteristics

Catchlight Energy

- Adaptable to existing infrastructure
- Scalable from pilot operations
- Feedstock flexible

October 15, 2011

CLE Pretreatment and Process to Sugar

Catchlight Energy

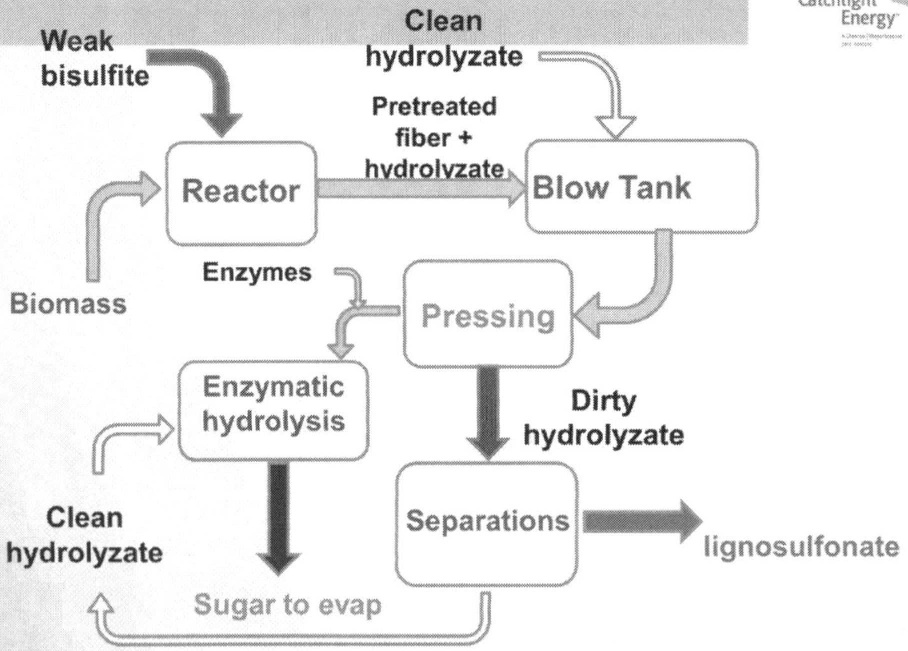

October 15, 2011

Softwood Sugar

Pretreated Softwood-Mill Residuals

Catchlight Energy

Softwood Residuals

Calcium Bisulfite Pretreatment

Pretreated Refined Pulp

Pretreated Residual Sugar Composition:

Arabinan	0.21%
Galactan	0.55%
Glucan	49.9%
Xylan	3.42%
Mannan	3.63%
Total	57.7%

Softwood Residual Sugar Composition:

Arabinan	1.01%
Galactan	2.09%
Glucan	39.5%
Xylan	6.05%
Mannan	10.5%
Total	59.2%

Pretreatment Liquor

Liquor Sugar Composition:

Arabinose	0.07%
Galactose	0.17%
Glucose	0.36%
Xylose	0.48%
Mannose	1.10%
Total	2.18%

October 15, 2011

7

Softwood Mill Residual Chips

Pilot Pretreatment Results

Catchlight Energy

Pilot Pretreatment Size: 3 kg OD chips/1 cu ft digester	
Overall Pretreatment Sugar Yield on Wood Sugar	88-95%
Furan Formation on Wood Sugar	<1.5%
Acetic Acid on Wood	~1.0%
Initial Cook Liquor pH	2.0-4.0
Pulp Yield on Wood	65%-80%

CLE pretreatment produces high yield of pulp and ultimately sugar.

October 15, 2011

8

Enzymatic Hydrolysis

Catchlight Energy

- Near 80% conversion of glucan up to 17% solids in the hydrolysis step using low dose of enzymes.
- Overall sugar yields of 87% attained on clean softwood with economic enzyme loading.
- Overall sugar yields of 80% attained on softwood forest residuals.
- Switchgrass and hardwood behave similarly although have higher C5 sugar in the product.

October 15, 2011

9

CLE Pretreated Hydrolysis

Glucose Yield vs. CTec2 Enzyme Dosage

Catchlight Energy

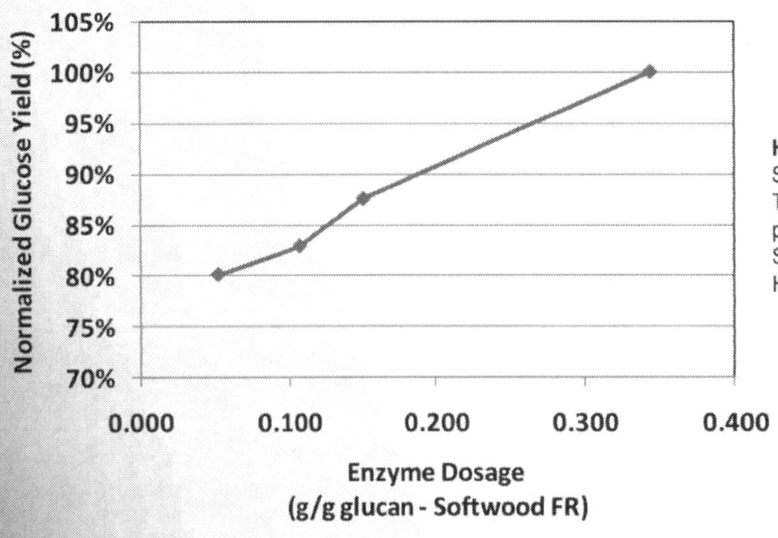

Hydrolysis Conditions:
Solid: 17%, unwashed
Temperature: 50°C
pH: 4.8-5.3
Shaking Speed: 200 rpm
Hydrolysis Time: 96 hrs

October 15, 2011

10

CLE Pretreatment Works on Multiple Feedstocks
Hydrolysis results at 4 days

Catchlight Energy

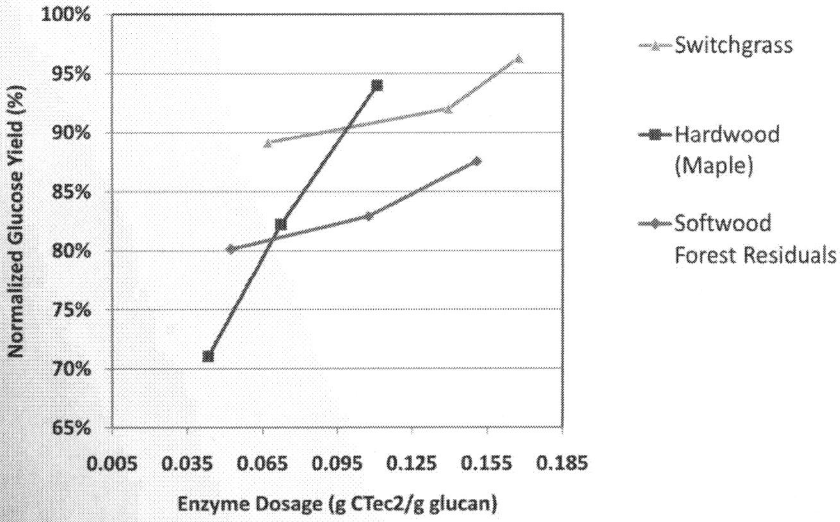

October 15, 2011

11

Commercial Data

CTec2 Dosage on Pretreated Corn Stover (Dilute Acid)

Catchlight Energy

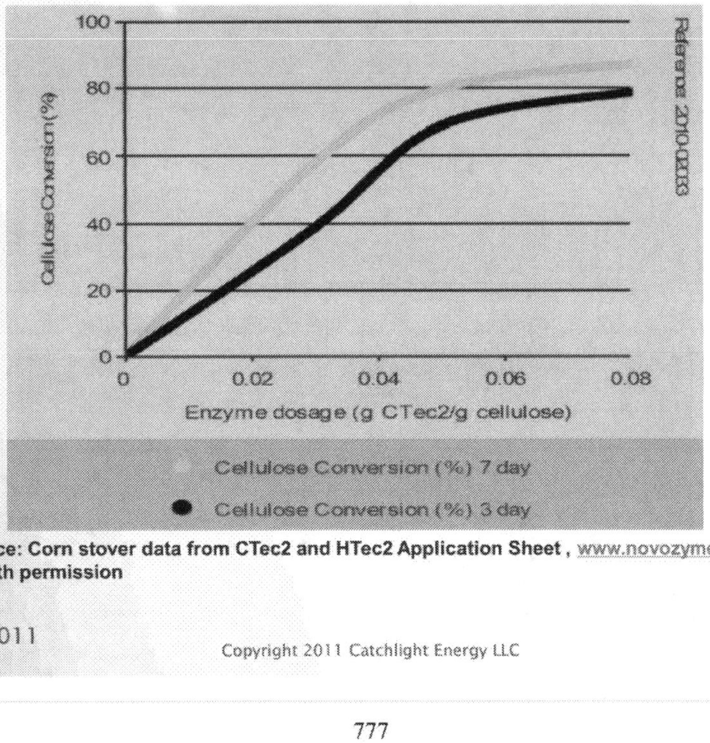

Reference: Corn stover data from CTec2 and HTec2 Application Sheet , www.novozymes.com
Used with permission

October 15, 2011

12

CLE Pretreatment Makes Softwood Hydrolyze Like Acid-Pretreated Corn Stover

Catchlight Energy

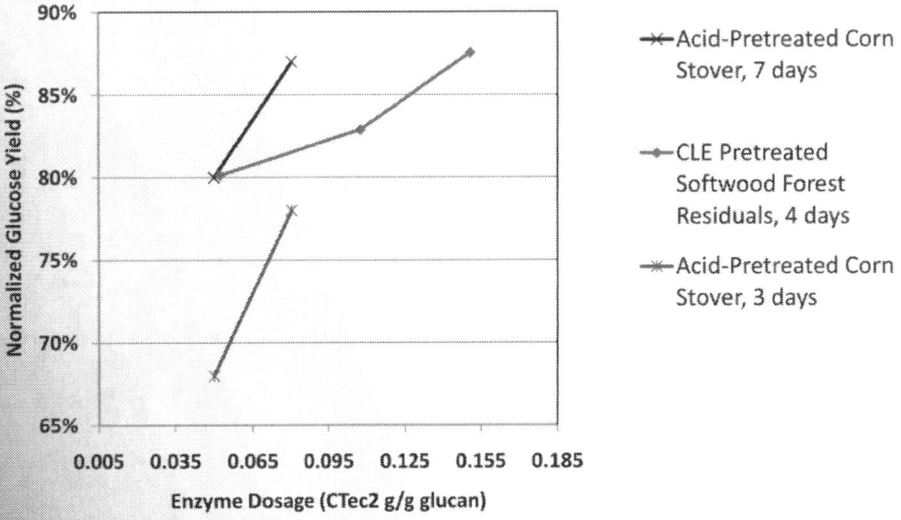

October 15, 2011

13

CLE Sugar: Relative Cost Components

Catchlight Energy

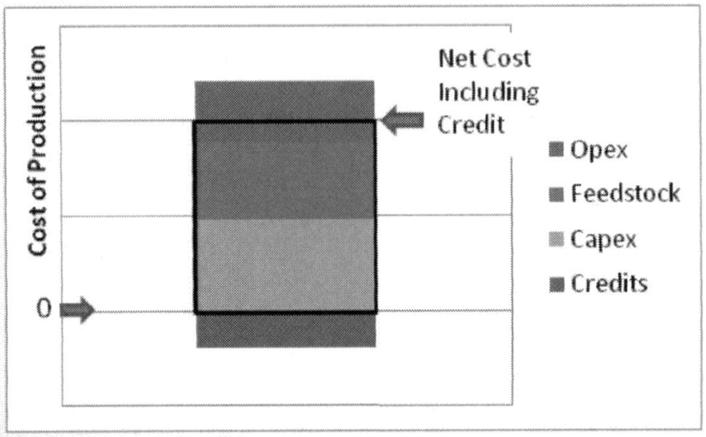

October 15, 2011

14

Conclusions

- **CLE's pretreatment technology is advantageous for difficult to hydrolyze feedstocks, such as softwood residuals which are available today at commercial scale.**
- **The technology is proven, scalable, and cost effective.**
  - ➢ An engineering design has been developed for a large scale greenfield biofuels with downstream processing and separations for sugar, lignosulfonate, and energy production.
- **CLE is actively working with select partners and welcomes interest in the production of sugar, ethanol, and drop-in biofuels.**

October 15, 2011

15

(523f) Economics of Pretreatment for Biological Processing

Ling Tao, Andy Aden and Richard T. Elander, National Bioenergy Center, National Renewable Energy Laboratory, Golden, CO

Economics of Pretreatment for Biological Processing

Ling Tao[1], Andy Aden[1] and Richard T. Elander[1]

[1] NationalRenewable Energy Laboratory, 1617 Cole Blvd., Golden, CO 80401, USA

Abstract:

A broad understanding of pretreatment economics within anoverall conversion context will be presented. The importance of pretreatmentin the overall economics of the biorefinery is addressed along with the keyparameters that influence the pretreatment economics. A large array ofpretreatment technology options exist and are currently being researched anddeveloped in this study. There appear to be many performances and costtradeoffs for these varying technologies. Only through continued R&D willthese tradeoffs be more fully understood. Process economics, especially thepromising pretreatment technologies studied in the Consortium for AppliedFundamentals and Innovation (CAFI) projects, are analyzed and compared indetails in this work. The six pretreatment technologies vary greatly in termsof their process design and projected total capital investment. Pretreatmentarea capital costs are based on past research, existing databases andengineering judgments, and include not only pretreatment reactor costs, butalso pretreatment chemical recovery and recycle equipment costs for theprocesses that use high pretreatment chemical loadings. More accurate costestimates for each pretreatment process will require specific vendor quotationsbased on pretreatment process conditions and any pretreatment chemical recoveryrequirements. Overall ethanol yield, which is largely based on achieved overallsugar yield achieved in pretreatment and subsequent enzymatic hydrolysis steps,is one of the single-most important factors in determining projected ethanolcosts for each process. Many key variables and parameters can have significantimpact on the pretreatment economics. While many are discussed within this talk, continuing R&D will generate data that further guide the development of thebiomass conversion industry.

See more of this Session: Developments In the Pretreatment of Lignocellulosics for Bioconversion II
See more of this Group/Topical: Sustainable Engineering Forum

(529a) Integrated Biogeochemical Modeling for Sustained Uranium Removal

Jiao Zhao and Radhakrishna Mahadevan, Department of Chemical Engineering and Applied Chemistry, University of Toronto, Toronto, ON, Canada

Improper management or accidental leakage ofhazardous wastes may lead to contaminated air, soil and water, which is notonly hazardous to public health but also to the ecosystem. For example, uranium contamination is aserious concern at several sites in theU.S. The soluble hexavalent uranium [U(VI)] can migrate to groundwater insurrounding areas, posing a problem for the drinking water safety. Whileevidences have shown that the reduction of soluble U(VI) to insoluble tetravalenturanium [U(IV)] is mainly a biotic process (microbial dissimilatory metalreduction), some bench-scale experiments indicate that U(VI) is readily reducedby some abiotic process (e.g. natural reductants such as ferrous iron and hydrogensulfide). Since experimental investigation of biotic and abiotic reduction wasconducted under different environmental conditions, it is necessary to developmodels that combine both biotic and abiotic factors for predicting establishmentof sustained uranium removal at actual environmental interfaces.

In this work, we developed a comprehensive 3-Dreactive-transport model for a recent column experiment (Moon et al., 2010)packed with the sediment from Old Rifle, CO. Groundwater from the Old Riflesite (which contains ca. 9 mM of sulfate but was amended with 20 μM U(VI)) waspumped upflow through the column at a rate of 0.035 ml/min. The biotic processsuch as genome-scale and kinetic models for microbial communities (*Geobacter*,*Rhodoferax* and *Desulfovibrio*) and electron capacitance for *Geobacter*(Zhao et al., 2010) has been coupled to the abiotic process in HYDROGEOCHEM(Yeh et al., 1998), where the finite element method was used to iterativelysolve governing equations by discretizing the column with 9 elements and 36nodes (Fig. 1A). The abiotic reduction of U(VI) is based on two publishedmechanisms -- uranium-hydroxyl species are the ones being reduced by sulfide(Hua et al., 2006) and the surface-adsorbed uranium species can be reduced bysurface-bound Fe(II) (Rosso et al., 2006 and Liger et al., 1999).

The model predicts that the biotic process plays amajor role in the U(VI) reduction under actual sediments (Fig.1B), which isconsistent with field-experiments (Anderson et al., 2002). Due to the mechanismof electron capacitance through which the reduced cytochromes can transferelectrons directly to oxidized metal ions with high rates (Wang et al., 2008),the uranium bioremediation process is still efficient and effective even whenthe sediments has been dominated by sulfate-reducing bacteria after 40 days ofacetate amendment. This simulation supports the recent findings (Williamset al., 2011) suggesting that following prolongedacetate addition, the small fraction of planktonic *Geobacter* species maystill be largely responsible for the vast bulk of U(VI) immobilization.Further, the model provides an explanation for the insignificant effect exertedby the abiotic process on the U(VI) reduction. Due to the dominant uranium-carbonatespecies present in the carbonate-containing system, the calculated total concentrationsof uranium-hydroxyl species are very low, leading to an extremely slow sulfide-drivenU(VI) reduction. Similarly, the low concentration of the surface-bound U(VI) (Moonet al., 2007) results in the low rate of surface catalysis of U(VI) reductionby Fe(II).

The model results suggest that themicrobially-generated hydrogen sulfide by sulfate-reducing bacterium may behelpful for the U(VI) reduction if the carbonate concentration can be reducedin the column amended with 20 μM U(VI). Fig. 2 indicates that by reducing theinfluent bicarbonate concentration from 8 mM to 3.5 mM, the total concentrationof uranium-hydroxyl species is significantly increased, generating a morethorough removal of the soluble U(VI) for long-term uranium immobilization. Itshould be noted that such abiotic process is dependent upon certaincombinations of experimental conditions, including the concentrations of U(VI),sulfide, bicarbonate and Ca(II). For example, if U(VI) concentration is low(e.g. typical values for Rifle groundwater), it might be possible that thebicarbonate amendment is beneficial to U(VI) reduction, since it will improvethe U(VI) desorption process, and hence increase the probability of the bioticreactions in groundwater.

Thecomprehensive model may provide auseful tool for investigating mechanisms and dynamics of abiotic and bioticinteractions under actual environmental conditions and subsequently for designingan optimal strategy for sustained uranium removal.

Reference

AndersonRT, Vrionis HA, Ortiz-Bernad I, Resch CT, Long PE, Dayvault R, Karp K, MarutzkyS, Metzler DR, Peacock A, White DC, Lowe M, Lovley DR. 2003. Stimulating the insitu activity of Geobacter species to remove uranium from the groundwater of auranium-contaminated aquifer. Appl Environ Microbiol 69(10):5884C5891.

Hua B, XuH, Terry J, Deng B. 2006. Kinetics of uranium(VI) reduction by hydrogen sulfidein anoxic aqueous systems. Environ. Sci. Technol. 40, 4666-4671

Liger E,Charlet L, Cappellen V P. 1999. Surface catalysis ofuranium (VI) reduction byiron(II). Geochim. Cosmochim. Acta 63,2939C2955.

Moon H,Komlos J, Jaffe PR. 2007. Uranium reoxidation in previously bioreduced sedimentby dissolved oxygen and nitrate. Environ Sci Technol 41:4587C4592.

Moon H,McGuinness L, Kukkadapu RK, Peacock AD, Komlos J, Kerkhof LJ, Long PE, JaffePR. 2010. Microbial reduction of uranium under iron- and sulfate-reducingconditions: Effect of amended goethite on microbial community composition anddynamics. Water Res 44:4015C4028.

Rosso KM. 2006. Mechanims and dynamics of abiotic and biotic interactions atenvironmental interfaces. ERSD Annual Report.

Wang Z,Liu C, Wang X, Marshall MJ, Zachara JM, Rosso KM, Dupuis M, Fredrickson JK,Heald S, Shi L. 2008. Kinetics of reduction of Fe(III) complexes by outermembrane cytochromes MtrC and OmcA of Shewanella oneidensis MR-1. Appl EnvironMicrobiol 74(21):6746C6755.

WilliamsKH, Long PE, Davis JA, Steefel CI, Wilkins MJ, N'Guessan AL, Yang L, NewcomerD, Spane FA, Kerkhof LJ, McGuinness L, Dayvault R, Lovely DR. 2011. Acetateavailability and its influence on sustainable bioremediation ofuranium-contaminated groundwater. Geomicro J. In press.

Yeh GT,Salvage KM, Gwo JP, Zachara JM, Szecsody JE. 1998. HYDROBIOGEOCHEM: A CoupledModel of HYDROlogic Transport and Mixed BIOGEOCHEMical Kinetic/EquilibriumReactions in Saturated-Unsaturated Media. Oak Ridge National Laboratory, Centerfor Computational Sciences, ORNL/TM-13668, Oak Ridge, TN37831.

Zhao J,Scheibe TD, Fang Y, Lovley DR, Mahadevan R. 2010. Modeling and sensitivityanalysis of electron capacitance for geobacter in sediment environments. JContam Hydrol 112(1C4):30-44.

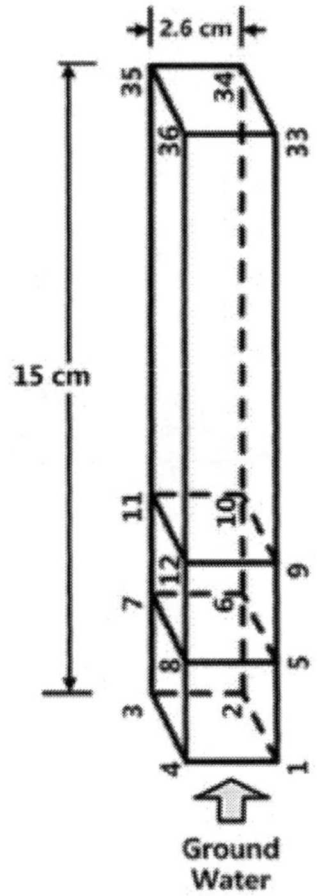

(A)

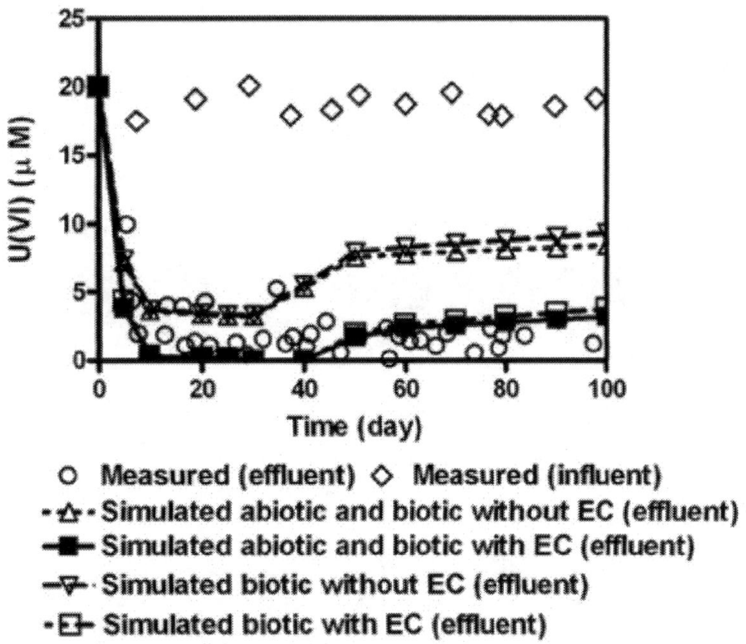

(B)

Fig. 1. Experimental3-D domain for the flow-through column (A) and the effect of abiotic and bioticprocess on U(VI) reduction (B). EC C Electron Capacitance for *Geobacter*.

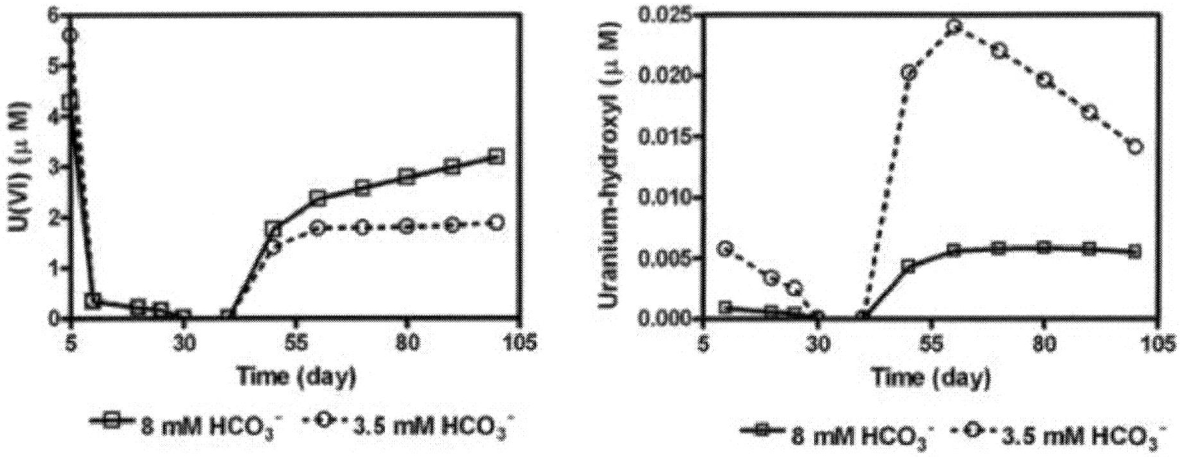

Fig. 2.Effect of carbonate concentration on total concentration of U(VI) species (A)and U(VI)-hydroxyl(B) in effluent.

See more of this Session: Fundamentals of Environmental Biotechnology
See more of this Group/Topical: Environmental Division

(529b) The Role of Geobacterial Cytochromes In Uranium Reduction

Bhushan P. Awate, Chemical engineering and Materials science, Michigan State University, East Lansing, MI, Gemma Reguera, Microbiology and Molecular Genetics, Michigan State University, East Lansing, MI and R. Mark Worden, Chemical Engineering and Materials science, Michigan State University, East Lansing, MI

Anthropogenic use of uranium has led to environmental contamination. Leaching of soluble uranium into the groundwater can lead to widespread human exposure and serious health hazards. One strategy to arrest groundwater contamination is to augment uranium precipitation via subsurface bacteria. *Geobacter* species have been shown to electrochemically reduce uranium from a soluble form U(VI) to an insoluble form U(IV). *Geobacter* cells use electron-carrying proteins to transfer electrons generated during respiration from the cytoplasm, through the cell membrane, to electron acceptors (mostly metal oxides) located outside the cell. Although *Geobacter's*'role in uranium reduction is widely recognized, the electron-transfer mechanism(s) used by these bacteria is not well understood. One distinctive feature of *Geobacter* species is the presence of more than 100 *c*-type cytochromes. These cytochromes are postulated to function as a heme network that can temporarily store electron produced during respiration and then shuttle the electrons to external electron acceptors, possibly with the involvement of conductive pili. Elucidation of *Geobacter*'s electron transfer mechanisms is particularly challenging due to the inherent complexity of *Geobacter's* heme network and the possibility of pleiotropic effects following gene knockouts in whole cells. In this study, we describe a biomimetic strategy, in which molecular architectures that mimic the structure and composition of *Geobacter*'s heme network are assembled on an electrode, and the resulting biomimetic interface is electrochemically characterized. This "bottom-up" approach enables hypotheses about structure-function relationships among key components of the electron transfer system to be tested. To illustrate the concept, a biomimetic interface was assembled that contained the periplasmic cytochrome PpcA, which is the only cytochrome that is conserved in all the members of the family *Geobacteraceae*. Recombinant PpcA from *Geobacter sulfurreducens* was overexpressed in *Escherichia coli*, purified, and immobilized on an alkanethiol self assembled monolayer bound to a gold electrode. Redox activity of the PpcA-containing interface was characterized using cyclic voltammetry, amperometry and chronocoulometry. The relative rates of PpcA-mediated reduction of soluble iron and uranium species were determined. The results are discussed in terms of the PpcA's possible role in groundwater uranium fixation.

See more of this Session: Fundamentals of Environmental Biotechnology
See more of this Group/Topical: Environmental Division

(529c) Evaluation of Non-Growth Biokinetic Models for Chromate Reduction with *Shewanellae*

Lewis Hsu[1], Kenneth H. Nealson[2] and Massoud Pirbazari[1], (1)Sonny Astani Department of Civil and Environmental Engineering, University of Southern California, Los Angeles, CA, (2)Department of Earth Sciences, University of Southern California, Los Angeles, CA

Hexavalent chromium is a carcinogenic pollutant oftenassociated with leather tanning and electroplating industries. Additionalapplications have included its use as a corrosion inhibitor in cooling waterand as a colorant in dye and pigment compounds. Its widespread use inindustrial application has often been met with poor waste management practices,leading to its release into the natural environment.

Bioremediation has been proposed as a viable remediationtool to reduce hexavalent chromium to its trivalent state. The rationale behindthis approach is that while the hexavalent form is very soluble and mobile inthe environment, the trivalent form is known to form insoluble precipitatesunder most environmental conditions. Such a transformation results in chromium immobility.Moreover, by utilizing microbes encountered in the subsurface environment, thepotential of designing an *in situ* treatment approach is much moreappealing when compared to more invasive *ex situ* approaches.

In particular, *Shewanella* species have beenproposed as model organisms capable of hexavalent chromium reduction. Thewidespread abundance of *Shewanella* in nature combined with the breadthof genomic and physiological knowledge make it an ideal organism to study inorder to understand mechanisms of chromium reduction in the environment. Tothis end, several *Shewanella* species were evaluated for their ability toreduce chromium under anaerobic conditions in batch reactors.

Data generated from these experiments was used toestimate biokinetic parameters for models proposed in literature for remediationof other types of pollutants. In this manner, we have evaluated the applicationof these models for hexavalent chromium reduction. A transformation capacity modelwas selected based on its agreement with experimental data. Furthermore, thesensitivity of this model to changes in biokinetic parameters was determined,as well as the effects of pH, temperature, and electron donor. The data generatedfrom these studies suggests that biomass, pH, and electron donor are among the keyparameters to consider when using designing bioremediation based treatmentsystems.

See more of this Session: Fundamentals of Environmental Biotechnology
See more of this Group/Topical: Environmental Division

Biotransformation of Coffee Pulp/Husk by Ligniculous Filamentous Fungi

María del Pilar Sierra-Gómez[1], and Patricia Ortiz-Bermúdez[1]
[1]Department of Chemical Engineering, University of Puerto Rico Mayagüez Campus
E-mails: maria.sierra@upr.edu, patricia.ortiz3@upr.edu

Abstract - Agro-industrial residues have been increasingly proposed as substrates in several bioprocesses. This is mainly because the recycling of these residues contributes to solve prevailing pollution waste management problems. Coffee pulp is the main byproduct in the wet processing of coffee; one of Puerto Rico's important agricultural commodities. Coffee pulp is close to 94% of the total weight of the coffee cherry, therefore innovative and environmentally friendly methods to reutilize this waste are necessary. The bioavailability of coffee pulp is restricted by the presence of recalcitrant and toxic compounds such as lignin, caffeine, and tannins. Thus, we propose the utilization of ligniculous filamentous fungi to detoxify coffee pulp, and convert it into an amenable source of carbohydrates and protein.

We have screened twelve strains of ligniculous filamentous fungi for caffeine tolerance. These strains were inoculated in caffeine concentrations of 0.2, 0.5, 1.0 and 4.0%. Caffeine degradation by candidate strains was assessed in liquid medium with defined amounts of pure caffeine. Experiments were conducted on minimal nutrient conditions at pH 4.5-5.5 and 60-70% moisture. Identification of caffeine reduction and its metabolites was done by HPLC analysis. Solid-state fermentations (SSFs) were carried out to assess the activity of ligniculous filamentous fungi on the coffee pulp. Chemical analysis of the degraded material was conducted to assess whether caffeine detoxification has occurred under SSF conditions. These analyses determined the extent of detoxification carried out by candidate strains on the actual coffee pulp.

Our preliminary results indicate that *Alternaria alternata* showed an unreported tolerance to caffeine of up to 1.0% on plates in rich medium. Experiments under defined liquid medium conditions with caffeine as a sole nitrogen source revealed that *A. alternata* detoxified up to 50% of the caffeine within 70 days. It is noteworthy that reduction of caffeine was also observed in experiments using caffeine as a sole carbon and nitrogen source although the percentage of reduction of caffeine was less. *Aspergillus tamarii*, a well-known caffeine detoxifier was grown in parallel with and its performance was similar to that of *A. alternata* under our culture conditions. SSF experiments demonstrated that coffee pulp supported the growth of *A. alternata,* preferentially at the larger particle size (i.e. #14 sieve supported growth better than #30 sieve). Preliminary extraction of caffeine from treated coffee pulp suggests that reduction of caffeine is possible under the SSF conditions we tested. Experiments are currently underway to assess caffeine detoxification of coffee pulp quantitatively, using *A. alternata* and *A. tamarii* as model strains.

(529e) Detecting Biological Networks In Clostridium Thermocellum Using Computational Approaches

Wen Zhou and Vinay Patil, Chemical Engineering, Michigan Technological University, Houghton, MI

Lignocellulosic plant biomass has long been recognized as a potential low-cost and sustainable source of mixed sugars for production of biofuels and other value-added chemicals. Lignocellulosic biomass comes from plant cell wall which is a composite structure with crystalline cellulose, hydrated hemicellulose, and lignin as major components. Corn stover, switchgrass, miscanthus, woodchips and the byproducts of lawn and tree maintenance are some of the more popular cellulosic materials for biofuel production. Production of biofuels from lignocellulose has the advantage of abundant and diverse raw material compared to sources like corn and cane sugars, but requires a greater amount of processing to make the sugar monomers available to the microorganisms that are typically used to produce ethanol by fermentation. Plants have evolved superb mechanisms for resisting assault on their cell wall structural sugars from the microbial and animal kingdoms, collectively known as biomass recalcitrance. Current technologies for the biological conversion of lignocellulosic biomass generally include pretreatment of biomass, hydrolysis of polysaccharides, and fermentation of C5 and C6 sugars.

Clostridium thermocellum is an anaerobic, thermophilic bacterium. C. thermocellum has garnered great research interest due to its cellulolytic and ethanologenic abilities, being capable of directly converting a cellulosic substrate into ethanol. This makes it useful in converting biomass into a usable energy source. The degradation of the cellulose is carried out in the bacterium by a large extracellular cellulase system called a cellulosome, which contains nearly 20 catalytic subunits. The cellulase system of the bacterium significantly differs from fungal cellulases due to its high activity on crystalline cellulose. In this work, I will present our recent work on understanding and improving the process of biologically converting biomass into biofuels through computational systems biology approaches. Specifically, I will focus my presentation on elucidating of a number of key regulatory pathways in cellulolytic organisms relevant to bioenergy production.

See more of this Session: Fundamentals of Environmental Biotechnology
See more of this Group/Topical: Environmental Division

(529f) Investigating the Mechanisms of Arsenic Removal by Microbial Layer In a Bio-Sand Filter

Hannah Gray[1], Charlotte Chan[1], **Varadarajan Ravindran**[2], Lewis Hsu[1] and Massoud Pirbazari[1], (1) Sonny Astani Department of Civil and Environmental Engineering, University of Southern California, Los Angeles, CA, (2)University of Southern California, Los Angeles, CA

The present study addresses the enormous global demand for safe drinking water in developing countries, and focuses on the removal of arsenic and bacterial pathogens from potable water supplies using a bio-sand filter. The bio-sand filter was constructed of a transparent polyethyl resin to facilitate visual observation of microbial activity and reactor flows. The filter medium essentially consisted of a supporting gravel layer, large sand layer, and biofilm layer containing a diverse microbial population, acclimated to a natural surface water source. Our work showed that the bio-sand filter r in-tandem with a pre-treatment column of iron-oxide-coated sand were effective in eliminating these contaminants. The present work focuses on determining the actual mechanisms of arsenic removal within the filter medium. In this study, the previous filter design was adapted to include a layer of iron-oxide-coated sand. Four sampling ports were strategically located to separately quantify arsenic removals in the biofilm and iron-oxide-coated-sand layers. Both sodium arsenate and sodium arsenite were spiked in the filter influent to determine whether arsenic removal was attributable to redox reactions.

The study is also directed at determining the actual mechanisms of arsenic removal within the filter, and the microbial strains primarily responsible for arsenic removal. Four sampling ports were strategically located in the bio-sand filter to separately quantify arsenic removals in the biofilm and iron-oxide-coated-sand layers. Both sodium arsenate and sodium arsenite were spiked in the filter influent to determine whether arsenic removal was attributable to redox reactions. The total arsenic removal was determined using the inductively coupled plasma –mass spectrometry (ICP-MS) technique. The bacterial removals were determined by the If arsenic removal occurred within the biofilm layer, the implications would be important, highlighting the ability of bacteria to store or process harmful forms of arsenic. Samples of schmutzdecke (biofilm layer) were obtained from the filter for microbial community analysis. Identification of dominant bacterial strains was accomplished by DNA isolation, PCR techniques, and DGGE fingerprinting. The DNA materials were isolated from these bio-layer samples, and the bacterial rRNA sequences were amplified by polymerase chain reaction (PCR) techniques. The PCR products were subsequently separated by denaturing gradient gel electrophoresis (DGEE), and sequenced for fingerprinting identification of dominant bacterial strains, before and after exposure to arsenic. Similar samples were also collected for scanning electron microscopy analysis to morphologically distinguish the microbial species associated with arsenic removal. Elemental analysis was undertaken to determine the locations of any immobilized arsenic species. The removals of bacterial pathogens (coliform used as surrogates for bacteria) in the bio-sand filter easily exceeded 5-logs (99.999%). The presentation would discuss the removals of arsenic and bacteria in the bio-sand filter at various depths of the filter. More importantly, it would address the mechanisms and interactions and mechanisms of arsenic removal in the bioactive filter.

See more of this Session: Fundamentals of Environmental Biotechnology
See more of this Group/Topical: Environmental Division

(532a) Ultra-Rapid Elimination of Biofilms On Metal and Ceramic Surfaces Via the Combustion of a Nanoenergetic Coating

Shramik Sengupta[1], Rajagopalan Thiruvengadathan[2], Byung-Doo Lee[1], Brandon M. Smith[1], Sachidevi Puttaswamy[1], Keshab Gangopadhyay[3] and Shubhra Gangopadhyay[3], (1)Biological Engineering, University of Missouri, Columbia, MO, (2)University of Missouri Columbia, Columbia, MO, (3) Electrical and Computer Engineering, University of Missouri, Columbia, MO

In Heat Exchangers and Cooling-Water Systems, which are an integral part of a wide variety of industrial processes, the presence of biofilms can reduce the effective heat transfer coefficient by as much as 50%. Biofilms consist of microbials colonies embedded on the surface of the equipment, and protected by a layer of extra-cellular polymeric substances secreted by the same microbes.

Removal of the biofilms requires that the equipment be taken "offline" and subjected to chemical and physical treatments that not only take a long time, but also require the use of toxic chemicals. Reducing the down-time for the equipment and the development of processes that do not require the use and disposal of toxic chemicals, is thus highly desirable from both economic and environmental points of view.

We present here a novel process that uses nano-energetic materials to eliminate biofilms in just a few seconds. The process involves spray-coating of an approximately 100 μm thick layer of nano-energetic material on top of the biofilm, allowing it to dry, and igniting the dried coating to incinerate the biofilm. The nanoenergetic material used is a two component mixture consisting of aluminum nanoparticles (fuel) dispersed in a THV220A (fluoropolymer oxidizer) matrix. Upon ignition, the Al nanoparticles react with THV 220A exothermically, producing high temperatures (> 2500K) for an extremely brief period (~100mS) that destroys the biofilm underneath, along with the bacteria harbored within. However, due to the small mass of nanoparticles present (~1 mg/cm^2 of coating), the total *amount* of heat produced is low (~0.1 KJ/cm^2) and the underlying metal or ceramic surface does not suffer observable damages. Since the byproducts of the process consist only of carbonaceous residue (that can be easily washed away) and gaseous combustion products, the process is not extremely fast, but also environmentally friendly.

See more of this Session: Green and Renewable Energy Aspects of Nanotechnology
See more of this Group/Topical: Topical C: Environmental Aspects, Applications, and Implications of Nanomaterials and Nanotechnology

(532b) Life Cycle Energy Analysis and Midpoint Assessment of Multimegawatt Wind Turbines with Polymer Nanocomposite Blade Material

Laura A. Merugula[1], Bhavik R. Bakshi[1] and Vikas Khanna[2], (1)Lowrie Department of Chemical and Biomolecular Engineering, The Ohio State University, Columbus, OH, (2)Department of Civil and Environmental Engineering, University of Pittsburgh, Pittsburgh, PA

Life cycle analysis ofvapor-grown carbon nanofiber (VGCNF or CNF) manufacturing shows production ishighly energy-intensive with a cumulative energy demand that is influenced byfeedstock selection and dominated by processing energy needed to maintainreactor temperature for the catalytic pyrolysis of the hydrocarbon feed.However, at relatively low mass percentages, CNFs can reinforce polymercomposites yielding significant improvements in properties such as mode-1delamination resistance, vibrational damping, tensile strength, fatigueresistance and Young's modulus. A new method termed "pre-binding" has beendeveloped to reinforce polymer composites with CNFs at the interface of longfibers and resin. This method improves the manufacturability of large pieceswith CNF-reinforced polymer nanocomposites (PNCs) by overcoming processingissues inherent to previous methods of PNC fabrication. Materials scientistsare considering use of this emerging material for wind turbine blade productionto meet the demand for larger, more resilient utility-scale wind power plants.

Wind energy converters (WECs)on-line today are a relatively new technology with installations of multimegawattratings having been introduced to the modern market only in the last 15-20years. The largest 3-blade horizontal axis wind turbine (HAWT) rotor size inoperation is 126 meters in diameter with generator ratings from 5 to 7 MW whilethe growing market is dominated by new installations from 1 to 3 MW. Apparentlimits of fiberglass technology have been reached as rotor blades beyond about40 meters generally require some inclusion of carbon fiber to address stiffnessand edgewise fatigue, which become critical at this length. The less densecarbon fiber also leads to a positive deviation in trends of weight toincreasing size. Recent press releases communicate progress in R&D formeeting industrial-governmental collaborative goals for 20 MW plants possiblyby 2020 using available technologies with rotors expected to have diametersnear 200 meters.

In an even shorter timespan, numerous nanotechnology-enabled products have been introduced to theconsumer market with over 1,300 manufacturer-identifiedproducts populating The Project on Emerging Nanotechnologies consumer productinventory today. These products include automobile body parts with carbon nanotubes(CNTs) for enabling online electrostatic painting and a personal watercraftusing nanomaterials for reinforcement of the polymer composite for alightweight, stronger material.

While life cycle energyconsumption of CNF-reinforced PNCs for use in automobile body panels has beenevaluated, similar trends were not expected for life cycle energy in windturbines due to differences in operational phase energy consumption. Also,previous assessments of PNCs have not considered the method of pre-binding,which uses acetone as a vessel for dispersion and distribution of the CNFs ontoglass or carbon fiber matting.

Preliminary studies thatassumed replacement of polymer composite blade materials on a 2 MW and 5 MWplant with CNF pre-bound PNCs indicated a range of parameterization valuesunder which the energy return on investment (EROI) would not decrease. Thesestudies have been updated in a process-based life cycle assessment (LCA) toinclude more relevant process information and have been extended to a midpointanalysis using the CML and Traci methods to determine areas of focus foroptimization. An energy analysis was extended to a 5 MW WEC using cumulativeenergy demand and power generation figures from published LCA literature.

Impacts are reported perkilowatt-hour generation of electricity by the wind power plant and compared tothe base case, which is a 2 MW WEC in Middelgrund that is inventoried in the proprietaryEcoinvent database. In every feasible scenario, the EROI of wind power plants withthe PNC blade material remained competitive to thermoelectric power generation,and life cycle water consumption remained favorable. Where CNF production wasassumed to be based on benzene feedstock, energy analysis results were muchmore favorable than methane. Midpoint analysis, however, was assessed with methaneas the preferred hydrocarbon source for pyrolysis in nanofiber manufacture. Similartrends emerged under extension to the 5 MW system energy analysis.

Assuming a masscontribution of CNFs that ranges from 1 to 5 % by weight of each blade, theoverall mass of the nanomaterials is approximately 0.01-0.06% of the overallmass of the WEC system, yet there is direct correlation between the massfraction of the nanomaterials to nontrivial impacts per kWh compared to thebase case. Where studies have shown manufacturing of virgin nanomaterials tobe solvent-intensive, this study finds that the use phase of the nanomaterial,where CNFs are integrated into a product system, is highly solvent-intensive aswell. Current laboratory scale ratios of solvent spent to nanofibers dispersedare on the order of $5x10^2$ g solvent per g CNF. Maturation of thismaterial development must include a reduction of solvent use by optimization ofthe dispersion process and evaluation of solvent recovery systems for reusethat considers effects of nanomaterials in the operation. Photochemicaloxidation is a concern in terms of acetone emissions, the degree to which willbe dependent on recovery, reuse, and emissions controls. Also, the degree ofsolvent fresh feed required has an impact on the overall system's globalwarming potential and eutrophication potential due to the manufacturing phaseof the acetone.

Evaluating impacts of emergingtechnologies is complicated by the fast-pace of development and deployment,significant gaps in data and uncertainties about the future. Compilation oflife cycle data is unwieldy at best due to complex supply chains and regionalvariations in categories as energy supply and transportation. This is furthercomplicated by incomplete understanding of ecosystem service impacts, thoughwork is underway to quantify and compile ecosystem services for inclusion in LCA.System performance improvements will not occur with PNC integration inisolation of other technological changes within WEC design and operation, yet analysesas this are meant to give insight to areas of concern by identifying impactsand order of magnitude effects associated with this emerging material for theassumed use.

See more of this Session: Green and Renewable Energy Aspects of Nanotechnology
See more of this Group/Topical: Topical C: Environmental Aspects, Applications, and Implications of Nanomaterials and Nanotechnology

(532c) Advanced Oxidation Processes with Carbon Nanotubes: Surface-Promoted Formation of Hydroxyl Radical During Ozonation

Rebekah L. Oulton[1], Michael J. Nalbandian[1], Howard A. Fairbrother[2], Kevin Wepasnick[2] and David M. Cwiertny[1], (1)Chemical & Environmental Engineering, University of California, Riverside, Riverside, CA, (2)Department of Chemistry, Johns Hopkins University, Baltimore, MD

Advanced Oxidation Processes (AOPs), which utilize hydroxyl radical (•OH) to chemically degrade contaminants, are becoming increasingly utilized in water and wastewater treatment for disinfection and removal of organic micropollutants via chemical oxidation reactions. Catalytic ozonation represents an alternative AOP in which a solid substrate such as activated carbon or common metal oxides are used to promote •OH formation from ozone. This project explores the efficacy of using multi-walled carbon nanotubes (MWCNTs) as new materials for promoting •OH production during ozonation. Using para-chlorobenzoic acid (*p*-CBA) as a hydroxyl radical probe, aqueous suspensions of functionalized (i.e., surface oxidized) and nonfunctionalized MWCNTs increased •OH concentrations in bulk solution by an order of magnitude or more over levels observed with ozone alone and resulted in •OH levels comparable to more traditional ozone-based AOPs (e.g., hydrogen peroxide with ozone). Production of •OH was greatest on oxidized MWCNTs, and evidence suggests this enhanced •OH production cannot be explained solely by the increased dispersivity of oxidized MWCNTs in aqueous suspensions. Chemical routes for MWCNT functionalization were used to systematically vary the surface oxygen content of MWCNTs, and the ability of these functionalized materials to generate •OH was subsequently determined in batch ozonation experiments, again using *p*-CBA as a radical probe. Complementary surface analysis with X-ray photoelectron spectroscopy (XPS) combined with chemical derivatization was used to correlate •OH production not only to total MWCNT surface oxygen content, but also to the concentration of specific functional groups (hydroxyl, carbonyl and carboxylic acid groups) resulting from oxidation. Results to date show that •OH production increases with MWCNT surface oxidation, and that carboxyl surface groups appear largely responsible for this increase in •OH production.

See more of this Session: Green and Renewable Energy Aspects of Nanotechnology
See more of this Group/Topical: Topical C: Environmental Aspects, Applications, and Implications of Nanomaterials and Nanotechnology

(532d) Evaluating the Dynamics of An Integrated Electrokinetic and Zero-Valent Iron Nanoparticle System for Treatment of Hexavalent Chromium In Groundwater

Ryan Thacher and Massoud Pirbazari, Sonny Astani Department of Civil and Environmental Engineering, University of Southern California, Los Angeles, CA

Electrokinetic technology is a promising groundwater treatment strategy capable of transporting ionic contaminants through groundwater by application of an electric potential across a contaminated area. While this is effective in the transport and concentration of contaminants at an electrode, complimentary remedial action may still be required to enhance treatment of the contaminants. Zero-valent iron nanoparticles (nZVI) may potentially offer an ideal solution for this genre of *in-situ* environmental remediation. Their small size coupled with their high surface area to volume ratio as well as reductive capacity, make nZVI a seemingly ideal medium for subsurface remediation where other classical remediation strategies such as excavation are not possible. This may be in deep aquifers, urban areas, or under existing infrastructure.

A comprehensive understanding of the fate and transport of nZVI in the subsurface is just as critical to *in-situ* remediation as the high reactivity. Upon introduction to an aquifer environment, nZVI must be readily dispersible and remain stable so that they reach the targeted contamination zone. Several researchers have investigated transport of nZVI through porous media, however most of the published work is conducted in idealized batch studies. These investigations have indeed provided a good knowledge base, but limited information exists concerning the fate, transport, and toxicity of nZVI in the natural environment. The challenging issues at hand include: nanoparticle agglomeration, nZVI storage and age, components of soil matrices, aquifer hydraulic properties, presence of natural organic matter, pH, and redox potential, among others.

This study investigates the treatment of simulated groundwater containing hexavalent chromium (Cr^{VI}), a known carcinogen, in a dynamic system using combined electrokinetic and zero-valent iron nanoparticle technology. Continuous flow columns are charged with simulated natural soils, followed by saturation with solutions containing Cr^{VI}, variable humic acid concentration, and at multiple pH values. Upon saturation, electric potential is applied, and dilute solutions of nZVI are added at regular intervals for the entirety of the experiment.

This presentation will discuss the influence of natural aquifer characteristics such as pH, natural buffer, and natural organic matter content on the transport of nZVI through subsurface environments, and how this impacts treatment of contaminants. Furthermore, this research will evaluate the dynamics of coupling electrokinetic and nZVI technologies in decontamination of the subsurface environment.

See more of this Session: Green and Renewable Energy Aspects of Nanotechnology
See more of this Group/Topical: Topical C: Environmental Aspects, Applications, and Implications of Nanomaterials and Nanotechnology

(532e) Biodegradation of Carbazole by Microbial Cells Coated with Magnetite Nanoparticles

Jianmin Xing, National Key Lab of Biochemical Engineering, Institute of Process Engineering, Chinese Academy of Sciences, Beijing, China

Microbes can remove nitrogen from carbazole through a series of enzyme-catalyzed reactions. Microbial cells of *Klebsiella* sp. LSSE-H2 were coated with magnetic Fe_3O_4 nanoparticles, which were synthesized by a coprecipitation method followed by modification with ammonium oleate. The average size of the magnetic particles was found to be in the range from 10 to 15 nm. Transmission electron microscopy (TEM) cross section analysis of the cells showed further that the Fe_3O_4 nanoparticles were strongly absorbed by the surfaces of the cells and coated the cells. The coated cells had superparamagnetic properties. The coated cells not only had the same activity as free cells but could also be reused. Biodenitrogenation rate was increased with the addition of Tween 80. The activity decreased less than 23% after three recycles.

See more of this Session: Green and Renewable Energy Aspects of Nanotechnology
See more of this Group/Topical: Topical C: Environmental Aspects, Applications, and Implications of Nanomaterials and Nanotechnology

Developing Experimental and Computational Nuclear Forensics Signals of Fast Neutron Irradiation Facility at TRIGA Reactor

Ian Schwerdt, Joseph Levinthal, Chris Dances, Luis Ortiz-R,
Dong-OK Choe and Tatjana Jevremovic
University of Utah Nuclear Engineering Program (UNEP)
The University of Utah, Salt lake City, UT

INTRODUCTION

Over 230 research reactors are operating on campuses and labs across the globe. These research reactors are diverse in their design, fuel type, power, and fuel enrichment. Most TRIGA reactors use Uranium Zirconium Hydride (UZrH) fuel that initially was designed to contain High-Enriched Uranium (HEU). A majority have been converted into low-enriched uranium (LEU) fuel (General Atomics)[1]. Fast neutrons can be used in several applications such as neutron hardness experiments for electronics and material tests. Accurate neutron flux and radiation dose measurements and computational estimates in reactors' irradiation facilities are of great importance for producing accurate nuclear signals of interest to nuclear forensics.

The University of Utah houses a 100 kW TRIGA reactor (UUTR), with a few irradiators, the Fast Neutron Irradiation Facility (FNIF), thermal irradiation facility (TI) central irradiation facility. The function of the FNIF is to provide a location for samples to be irradiated by fast neutrons coming from the reactor core. The FNIF is located directly next to the exterior of the reactor core. It consists of an aluminum shell filled with stacked lead bricks. The lead bricks are placed along the walls of the aluminum structure leaving a sizeable gap in the center for samples to be placed. The lead bricks block gamma radiation from the reactor core, while minimizing the amount of energy lost from scattered neutrons. The initial amount of energy of a neutron produced by fission follows a predictable curve called the neutron fission spectrum. The energy spectrum is a bell shaped curve that has a peak at about 1 MeV, and an average value of 2 MeV. As neutrons are produced in the reactor core, they lose energy by interacting with the matter in between the emission point and the FNIF. The vast majority of incident neutrons are produced from the two closest rows of fuel elements in the reactor core. The neutrons are emitted at a random trajectory from the fuel element, and form a cosine flux distribution in each axis direction. The profile within the FNIF is sum of the profiles from the first two rows. This project is largely based on the work presented at the Advancing Tools and Solutions for Nuclear Material Detection 2nd annual national conference by the authors. One of the goals of the previous project was to characterize this distribution of the neutrons with respect to energy and position by using a combination of experimental data and computer simulation. The objective of this project was to refine the simulation, and expand to new experimental techniques.

DESCRIPTION OF EXPERIMENTAL PROCEDURE

One way to map the neutron flux distribution is through the neutron activation of a material and measuring the induced radioactivity, which can be used to develop a profile of the fast neutron flux. For our experiment, we utilized the activation of Sulfur-32 to Phosphorous-32 through a charged particle emission (n, p) reaction. When a moderately fast neutron is captured by the nucleus of Sulfur-32, the nucleus will then emit a proton to become Phosphorous-32. The charged particle emission reaction (n, p) of Sulfur-32 is a threshold reaction requiring a fast neutron with energy in excess of 2 MeV. The Phosphorous-32 has one mechanism of decay, the emission of a detectable beta particle that is used to quantify the amount of neutron flux experienced while in the FNIF.

84 Sulfur-32 pellets were arranged in a 20 cm x 31cm grid pattern and irradiated in the FNIF for 15 minutes at 1kW (Fig. 1. a)). The samples were allowed to decay for seven (7) days so that the activity was low enough for safe handling. The activity of the samples was measured utilizing a Geiger-Mueller tube. The activities were standardized to a reference time corresponding to when the first sample was measured. A general fast neutron flux profile was obtained by plotting the intensity of the activities on a contour map (Fig. 1. b)). The FNIF was then modeled in GEANT4[2], in order to verify that physics modeling could replicate the general spatial variations of the fast neutron flux profile elucidated from the neutron activation of Sulfur-32.

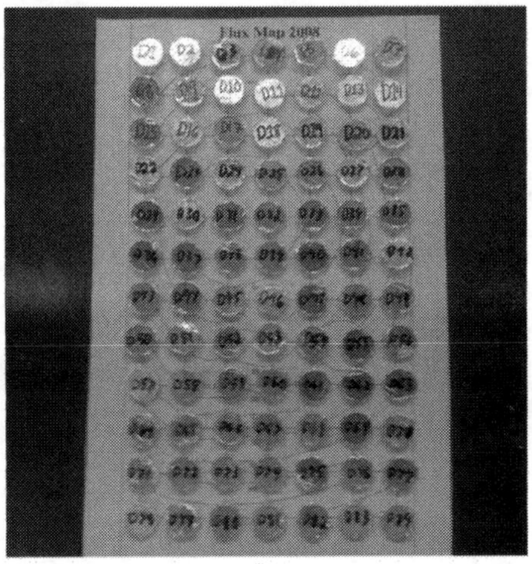

a)

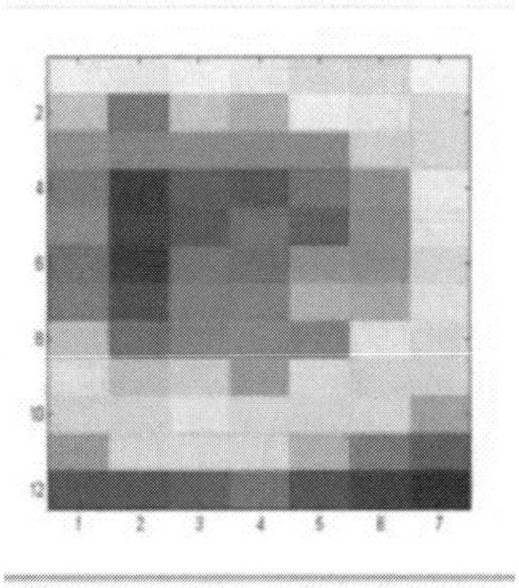

b)

Figure 1. a) Sulfur pellets laid out into a grid before irradiation; b) Experimental measurements

In addition to this initial neutron flux mapping of the FNIF, a separate experiment may be able to contribute to this project. A rhodium detector was used to measure the amount of neutron flux in the FNIF. The detector was lowered into the FNIF, and a flux value was determined based on the amperage induced in the detector. An initial flux of 4.6×10^{10} n/cm^{2}sec, was determined in this way (Fig. 2). This rough value was independent of the position of the detector. The value read by the detector represents a fraction of the total flux in the FNIF, at an arbitrary position. Future work with this experiment would be in integrating the detector into the GEANT4 model to aide in benchmarking the detector. The total neutron flux in the FNIF could be scaled from the neutron flux distribution knowing the value of the detector at its position. The FNIF facility offers an opportunity to benchmark the detector, and the computer simulation would aide in confirming the benchmark's validity. This potential advantage in understanding a detector ultimately contributes to the nuclear forensic field. These new detectors under development at UNEP will be optimized for fast detection at site of low and high energy neutrons of interest to nuclear forensics.

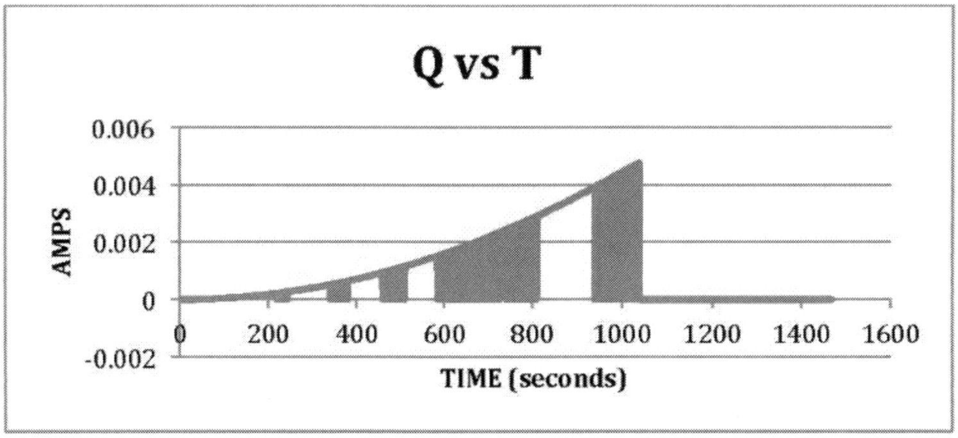

Figure 2. Charge induced in the rhodium detector vs. time to get amperage

DESCRIPTION OF GEANT4 SIMULATION

A computer code developed at CERN called GEometry And Tracking (GEANT4 Collaboration)[2] is used for simulation of FNIF neutron flux distribution. GEANT 4 is an open source code written in C++ designed to model the physics processes involved with particles passing through matter with a wide variety of applications. A simulation is constructed from a number of subclasses and then compiled as a single executable. The most basic simulation requires classes that define the geometry, the particle source, and the physics processes involved. A class called *G4DetectorConstruction* establishes the geometry. After establishing a given shape, it is placed into a *mother* or *world* volume and then is assigned different material properties. For this simulation, a trapezoid constitutes the aluminum casing of the first two rows of the reactor. The FNIF is represented by an aluminum shell encompassing a hollow lead block and is placed next to the aluminum trapezoid (Fig. 3).

A class called *G4particlesource* establishes the generation of any subatomic particles. Each neutron is generated in what is called *an event* with a number of attributes. For each *event* each neutron needs to be assigned a specific initial position, direction, and energy. The position is assigned randomly to a fuel element position in the modeled reactor core. The neutron is then assigned a random position along the surface of the fuel element. The *watt rejection* method is used to assign the neutron an energy level according to the probability of the fission energy spectrum. Finally, each neutron was given a random direction in all three dimensions. A standard physics library was used to simulate the neutron transport established in the physics library class. Additionally, a *scoring method* was used to count the total number of neutrons that pass through the inner chamber of the FNIF. A matrix of cells was established in all three dimensions, and at different energy levels. Each cell then counts the total number of neutrons that pass through indexed cells. From this data, a flux value is assigned to each cell at each energy level. This produces a three dimensional map of the flux at the FNIF and at three different energy levels.

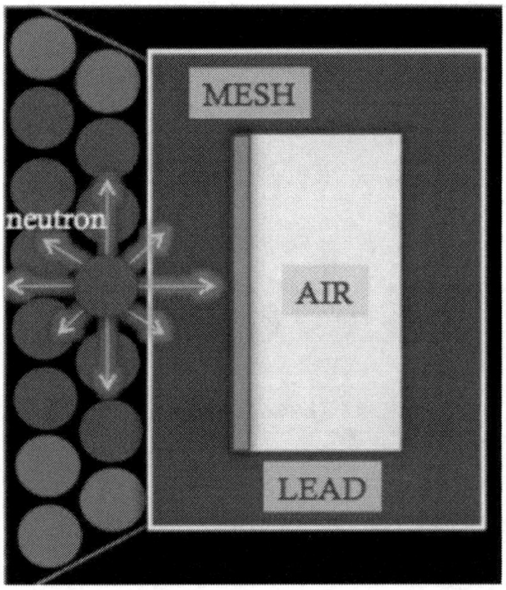

Figure 3. FNIF as modeled in GEANT4. The red block represents the stacked lead bricks in the FNIF. The yellow box in the middle represents the chamber where samples can be placed. The blue circles represent fuel elements, and the green empty slots. As in the diagram, neutrons are generated in the fuel elements and then enter the FNIF.

The results obtained from the simulation were then run through Matlab to graphically display the flux distribution. The three-dimensional flux distribution can be viewed from the top as in Fig. 3, or from the side or front of the FNIF. The side view would be the in the direction from the bottom of Fig. 3. The front view would be from the direction to the right of the image. Each perspective gives the neutron flux distribution along two different axis, leaving a third free. The intensity of the flux distribution on one axis, and the two other represented the cross section image at a particular distance along the third position axis (Fig 4.a). The same could be done with the different energy levels. Instead of altering the third position axis, one could vary the energy range of the distribution. However the profile of the cross sections varies only in the magnitude with respect to energy level. Instead, the energy spectrum inside the FNIF is shown in Fig 4.b). The graph in the top right of the image of Fig. 4. a) and the experimental graph of the data from Fig. 1.b) both depict a similar trend for the neutron flux distribution. Discrepancies arise from the mesh sizes used in plotting the experimental and simulation data. The differences in resolution between the two graphs adds to a bit distort image when comparing experimental with simulation distributions.

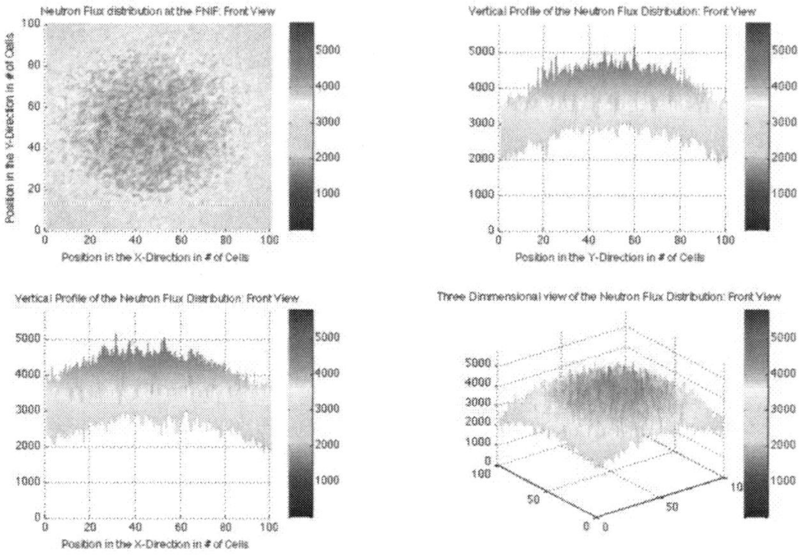

a)

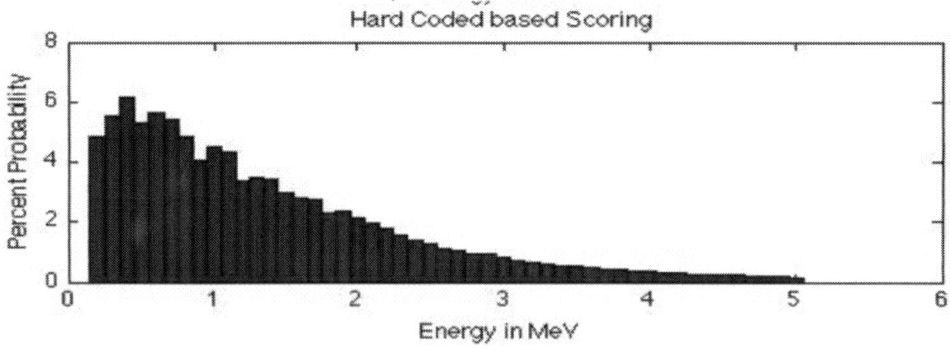

b)

Figure 4. a) The neutron flux distribution from a "forward perspective". That is from the point of view from the reactor towards the FNIF. The flux relative to both the x and y dimensions is shown. The other views are from other angles, of the same 3-D flux map. This same map is one cross section within the facility, and can be taken for any direction; b) The total energy spectrum in the FNIF. While the shape of the curve deeply resembles the fast fission spectrum, the peak has dropped from 2 MeV to just under .5 MeV. This shift also moves the average closer to approximately 1 MeV.

CONCLUSION

Recent development of simulation and experimental models regarding beam ports (irradiation facilities) at the UUTR includes a number of undergraduate and graduate students in a team work. With this effort we are now systematically analyzing each of the irradiation facilities in recording experimental measurements used for calibrating radiation detectors and benchmarking GEANT4 as well as MCNP5/X simulation models. In this paper we presented neutron flux distribution in one of UUTR irradiation ports that was mapped experimentally and modeled using GEANT4. Experiment was based on using the sulfur tablets that where placed on a grid, and then irradiated; the activity of the tablets was then measured. While the coarse grid was used, the experimental data provided accurate cosine flux distribution. This trend was shown to coincide with the computer simulation in GEANT4 reasonably well. GEANT4 simulation provided a higher resolution of the neutron flux distribution and also suggests an energy spectrum within the FNIF. The first measurements using rhodium detector has provided initial data for further comparisons. Future work would lie in integrating the rhodium detector measurements with GEANT4 simulations; this comparison will provide a better understanding of the detector accuracy.

References

1. General Atomics, "TRIGA Nuclear Reactors".

   Accessed: 9/10/2011 at http://www.ga-esi.com/triga/

2. GEANT4 Collaboration, "GEANT4 User's Application Guide for Application Developers"

   Accessed: 9/10/2011 at

   http://geant4.web.cern.ch/geant4/UserDocumentation/UsersGuides/ForApplicationDeveloper/html/index.html

Monte Carlo Simulation of Tritium Generation in The University of Utah TRIGA Reactor

Todd Sherman, Dong-OK Choe, Haori Yang, and Tatjana Jevremovic

Nuclear Engineering Program

The University of Utah, Salt Lake City, UT

Extended Abstract

INTRODUCTION

The University of Utah Nuclear Engineering Program houses a Mark I TRIGA Reactor, UUTR (University of Utah TRIGA Reactor). This reactor is a pool type reactor with water used as both the coolant and neutron moderator. Regular city water is filtered and then dispensed into the tank. UUTR core uses low enriched uranium fuel.

The UUTR is licensed by the Nuclear Regulatory Commission to operate at 100 kW. However, the usual operating power is 90 kW. At this power level a maximum neutron flux of about 3.6 $x10^{12}$ neutrons/cm^2-sec is produced at the center of the core. This flux decreases by about an order of magnitude toward the periphery of the core. **Figure 1** shows the layout of the UUTR, irradiation facilities, and the location of some of the heavy water elements.

The UUTR is equipped with various experimental irradiation facilities for neutron activation, fast neutron irradiation and gamma irradiation experiments, as indicated in **Figure 1**. The UUTR's experiments mainly include investigations of radiation hardening, neutron activation analysis, biomedical analysis and isotope production.

TRITIUM (T) GENERATION IN THE UUTR

Where T is generated in UUTR. In the outer ring of the UUTR core there are heavy water elements serving as both moderator and reflector. Neutrons that have reached the heavy water elements will likely be reflected back towards the core instead of escaping from the core. These heavy-water elements are filled with 68% D_2O and 32% H_2O. The TI (**Figure 1**) is also filled with heavy water of the same concentration. The heavy water elements and the TI are the primary sources of tritium production in the UUTR.

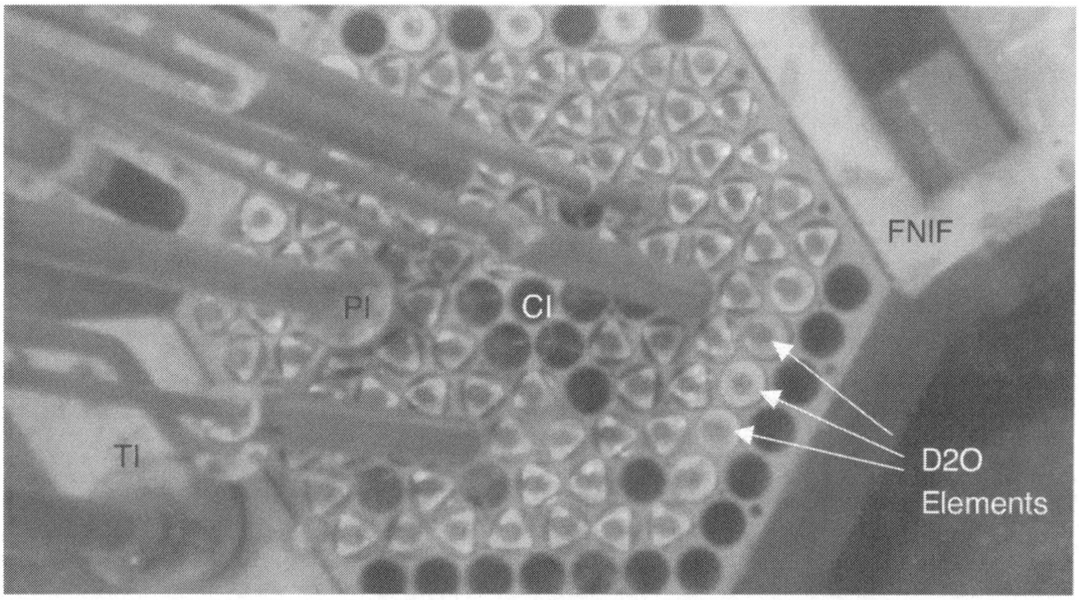

Figure 1 – The University of Utah TRIGA Reactor

[CI – Central Irradiation Facility; TI – Thermal Irradiation Facility; FNIF – Fast Neutron Irradiation Facility; PI – Pneumatic Irradiation Facility]

How is T generated? Hydrogen captures a neutron to become deuterium. Deuterium captures a neutron to become tritium, as illustrated by **Equation 1** and **Figure 2** (Dyadin, 1979). Tritium has a smaller scattering cross section than the other hydrogen isotopes. Tritium is unstable and undergoes a beta decay to become a stable helium-3. The half-life of tritium is 12.33 years.

Why it is important knowing the quantity of generated T?[1] Identifying and managing tritium quantity is important because of its potential adverse characteristics. Tritium, in large doses, can have serious biological consequences, both physically and radiologically. Tritium is also a key component in nuclear weapon proliferation, which makes it important to control. The EPA Safe Drinking Water Act limits the tritium concentration to 0.74 Bq/mL or 20,000 picocurie/ liter. This is the equivalent to 4.15 x10^{11} tritium atoms/ mL in drinking water.

---

[1] "Tritium is used by all nuclear weapons states to increase the explosive yield of atomic bombs and to miniaturize them. However, this radioactive material has not yet been put under appropriate international control comparable to the nuclear safeguards applied for plutonium and uranium. It is a neglected material in efforts to control the spread and further development of nuclear weapons." (Kalinowski, 2005)

$$ {}_{1}^{1}H + {}_{0}^{1}n \rightarrow {}_{1}^{2}H + {}_{0}^{1}n \rightarrow {}_{1}^{3}H \qquad (1) $$

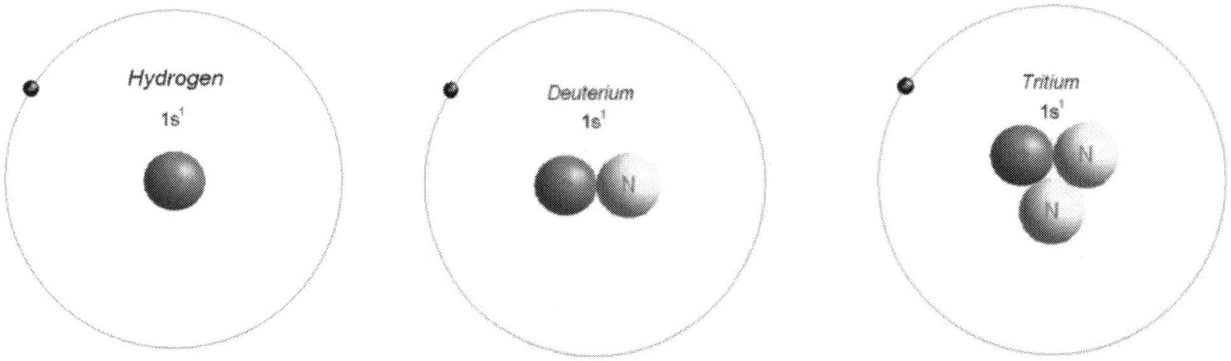

Figure 2 - Three different hydrogen isotopes

ANALYTICAL ESTIMATES OF TRITIUM GENERATION IN UUTR

The tritium generation in several heavy water reactors can be estimated analytically. By using simple neutron activation equations with a measured thermal neutron flux (Erykalov, 2005) we may derive tritium generation based on a very simplified reactor model. The following equation can be used to calculate the tritium generation rate in the UUTR heavy water elements:

$$ N_{Tritium} = N_{deuterium}\, \sigma\phi \qquad (2) $$

where:

N = number of atoms

ϕ = neutron flux (neutrons/cm^2-sec)

σ = Thermal neutron absorption cross section (cm^2)

An approximated thermal neutron flux of 3.6 x10^{11} neutrons/cm^2-sec was used to estimate tritium generation in the heavy water elements. The flux in each of the heavy water elements could be measured experimentally; such experiments have not yet been performed but are planed to be completed in 2012. The calculated tritium generation rate using Equation 2 in every heavy water element, at 90 kW, was estimated roughly to be in order of 4.758x10^9 atoms/ sec.

MCNP5 MODELING OF TRITIUM GENERATION

MCNP5 Model. The MCNP5 model of the UUTR core including all details is often used in predicating the outcomes of various experiments and operational settings that would otherwise be difficult to do experimentally. Key parameters (such as neutron flux inside the various irradiation facilities) were already experimentally measured and used to benchmark the MCNP5 model, so that any other modeling of different configurations or the experiments give the known confidence. The measured value of the neutron flux in for example the CI (**Figure 1**) of 3.6 $\times 10^{12}$ neutrons/cm^2-sec was used to scale the tallied MCNP5 flux values for this model. The CI is where the neutron flux is measured and well known at our facility. So the CI flux values were chosen to scale all the MCNP5 results to the measured values. This scaling factor was produced by dividing the measured flux by the MCNP5 flux tally. The purpose of the scaling factor is to normalize all of the MCNP5 results to same ratios.

The MCNP5 simulation was ran using 100,000 particles, which was enough to bring the relative errors under 0.1, which MCNP5 community considers as reliable. The tritium production was tracked and tallied using the track length estimation. In MCNP5, the average neutron flux in a certain cell can be calculated using track length estimation (F4 tally). In order for tritium to be produced in the UUTR it must come from a deuterium nucleus absorbing a neutron. So the neutron flux is imperative to track and record. The MCNP5 track length estimation tallies the relative average neutron flux. The average neutron flux in a cell can be written as:

$$\phi v = \frac{1}{V}\int dE \int dt \int dV \int d\Omega\, \Psi(\vec{r},\Omega,\mathrm{E},\mathrm{t})$$

$$= \frac{1}{V}\int dE \int dV \int d\Omega \int dt\, vn(\vec{r},\Omega,\mathrm{E},\mathrm{t})$$

$$= \frac{1}{V}\int dE \int dV \int dt\, vN(\vec{r},\mathrm{E},\mathrm{t})$$

The quality $dt \cdot v \cdot N(\vec{r},E,t)$ can be thought of as a track length density. Using F4 tally, MCNP5 estimates the average flux based on the accumulation of weighted track length for all particles tracks in the cell. Based on neutron flux calculated using F4 tally, the FMn card can be used to calculate any quantity of the form

$$C\int \phi(E) R_m(E) dE$$

$R_m(E)$ can be any response function from the MCNP cross-section library. The tritium generation rate in a heavy water element can be written as

$$\int \phi(E) \Sigma(E) dE$$

where $\Sigma(E)$ is the macroscopic cross section for deuterium capture reaction. It can be rearranged as

$$N\int \phi(E)\sigma(E)dE$$

where N is the number of density of deuterium. The reaction rate per unit volume is then

$$\frac{N}{V}\int \phi(E)\sigma(E)dE$$

Thus we can set the multiplicative constant on the FMn card as the deuterium density per cm^3

$$C = \frac{N}{V}$$

The MCNP5 F4 tally gives a unit neutron flux, ϕ, per unit time (neutrons/ cm^2). The FMn card simply multiplied ϕ with the microscopic neutron absorption cross section (barns) σ. The FMn output, therefore, was obtained as:

$$\phi\left(\frac{neutrons}{cm^2}\right)\sigma(barns) = \phi\sigma\left(\frac{neutrons}{cm^2}(barns)\right) \quad (3)$$

To convert this value into an actual tritium generation rate per cm^3 the results were multiplied by the scaling factor, the constant C, the conversion from $1x10^{-24}$ cm^2 = 1 barn and the volume of each heavy water element; these are shown in **Table 1**. The following relation explains this final calculation step:

$$SF\left(\frac{\frac{neutrons}{cm^2 \cdot sec}}{\frac{neutrons}{cm^2}}\right)\phi\sigma\left(\frac{neutrons}{cm^2}(barns)\right)C\left(\frac{N_{deuterium}}{cm^3}\right)\left(\frac{1capture}{neutron}\right)\left(\frac{1x\,10^{-24}cm^2}{1\,barn}\right)V(cm^3) = \frac{tritium\ generations}{sec} \quad (4)$$

where:

SF = scaling factor obtained by taking ratio of the measured CI flux and the MCNP5 simulation results
$\phi\,\sigma$ = FMn card results from MCNP5 simulation
C = multiplication constant used on the FMn card in the MCNP5 simulation
V = volume of a heavy water element

Tritium quantities in UUTR obtained by MCNP5. The MCNP5 gives the unit interactions. Thus all the MCNP5 values where multiplied by the scaling factor as well as the heavy water element volume (cm^3) and deuterium atom density (atoms/ cm^3). These results are summarized in **Table 1**. The unit tritium generation activities are also shown in **Table 1** as calculated from:

$$A = \lambda N \quad (5)$$

A = Activity (Bq)

λ = decay constant (sec^{-1})

N = Deuterium density per unit volume (atoms/ $cm^{3)}$

As shown in **Table 1**, the tritium generation rate in twelve heavy water elements at the UUTR is far below the EPA regulation even when the reactor is running at its full licensed thermal power. The tritium generation rate inside the TI is also well below the EPA regulation. Thus, the amount of tritium generated at the UUTR due to normal operation will not pose any radiological hazard to the staff members or the public if by any unexpected event is released to the environment.

Table 1 - Tritium Generation in the UUTR based on MCNP5 Model

Heavy Water Element	MCNP5 Tritium generation rate (Atom/sec)	Unit tritium activity generation rate (Bq/cm^3)
TI	$3.2899 \times 10^{10} \pm 0.0187$	4.77×10^4
1	$1.3294 \times 10^9 \pm 0.0369$	4.21×10^3
2	$1.5884 \times 10^9 \pm 0.0346$	5.03×10^3
3	$1.5543 \times 10^9 \pm 0.0342$	4.92×10^3
4	$1.3287 \times 10^9 \pm 0.0368$	4.21×10^3
5	$1.5168 \times 10^9 \pm 0.0330$	4.80×10^3
6	$1.4818 \times 10^9 \pm 0.0341$	4.69×10^3
7	$1.7073 \times 10^9 \pm 0.0336$	5.41×10^3
8	$1.6108 \times 10^9 \pm 0.0347$	5.10×10^3
9	$1.4987 \times 10^9 \pm 0.0348$	4.75×10^3
10	$1.3438 \times 10^9 \pm 0.036$	4.26×10^3
11	$1.5638 \times 10^9 \pm 0.0338$	4.95×10^3
12	$1.3477 \times 10^9 \pm 0.0363$	4.27×10^3

CONCLUSION

The tritium generation inside of the UUTR is investigated using a detailed model of the UUTR based on MCNP5. In order to obtain a tritium generation profile in the existing twelve heavy water elements and the TI, the MCNP5 values are normalized to a single source particle. To get the overall tritium generation rate, a scaling factor was calculated by taking ratio of the measured and simulated neutron flux at the CI (**Figure 1**). The MCNP5 simulation model as presented provided reliable values for tritium generation rate and activity. This model now serves as a base-line for our feasibility study of the UUTR power upgrade.

References

Dyadin, Y. V. (1979). Measurement of tritium concentration in heavy-water reactor. *Atomic Energy* , *47* (5), 916-918.

Erykalov, A. N. (2005). Computed trtium concentration in a heavy-water reactor. *Atomic Energy* , *99* (3), 210-214.

"MCNP-A General Monte Carlo N-Particle Transport Code, Vesion 5". LA-UR-03-0245, Los Alamos Nation Labortory (2003).

M. Kalinowski, "International Control of Tritium for Nuclear Nonproliferation and Disarmament", CRC Press, (2005).

GEANT4 SIMULATION OF IRRADIATION FACILITIES AND NEUTRON SOURCES AT UNIVERSITY OF UTAH TRIGA FOR NUCLEAR FORENSICS AND DETECTION

[1]Andrew Steven Voyles, [2]Haori Yang and [3]Tatjana Jevremovic
Utah Nuclear Engineering Program (UNEP), The University of Utah
[1]Undergraduate student in Nuclear Engineering Minor, [2]Assistant Professor, [3]Chair Professor and UNEP Director
[1]andrew.voyles@utah.edu, [2]haori.yang@utah.edu, [3]tatjana.jevremovic@utah.edu

Introduction

GEANT4 (**GE**ometry **AN**d **T**racking, release 4) high-energy physics simulation toolkit was used to simulate neutron source signatures, as well as various irradiation ports of the University of Utah TRIGA facility (UUTR): central irradiator, thermal irradiator, and some peripheral ports and components. Each of these ports has unique geometry and different neutron and gamma flux. Preliminary results for thermal irradiator and central irradiator, as well as for neutron sources are discussed in this paper, while the modeling of the remaining ports is underway.

Purpose

Being able to quickly and accurately determine the source of various nuclear materials through known nuclear signatures is an essential requirement for successful nuclear forensics. As a part of an ongoing project at UUTR, using GEANT4, we are developing comprehensive nuclear signature data sets and benchmarks against corresponding nuclear experiments. This paper discusses signatures of the various irradiation ports and samples irradiated in such ports at UUTR, signatures for the UUTR fuel element assemblies, and signatures of shielded and unshielded neutron and gamma sources.

Description of the actual work

GEANT4 program flow.

For a general GEANT4 simulation, a few files are adapted to fit the end user's run-time parameters, namely, the geometry of the detector and solids involved in the simulation, how particles are generated, and with what physical properties, and how the particles will propagate through space. At the time of compilation, these header files are read into the execution script, and complied to produce the executable of the simulation.

Once executed, GEANT4 provides two routes for program flow: interactive and batch modes. For batch mode, a macro file, written ahead of time, is read into the idling command prompt in the simulation, and executed lien-by-line. This is convenient for simulations that are compiled ahead of time, and simply running the macro file will carry out the entire simulation autonomously, without any further input. Another alternative is interactive mode. In this mode, the user can execute commands into the idling command prompt to achieve the exact same results as batch mode, for a macro containing the same commands. This mode is useful for debugging a simulation, or for just experimenting. However, both modes can be used to attain the same desired results. Once the simulation is completed, the user can then use the commands of either OpenGL or HepRep to output the visualization of their simulation. The simulation is designed to be general for all common simulation applications. End users will only need to supply data on the geometry and composition of their setup and samples in the *GeometryConstruction.cc* file, as well as the neutron flux values at runtime, and the simulation will be carried out in batch mode, and visualized.

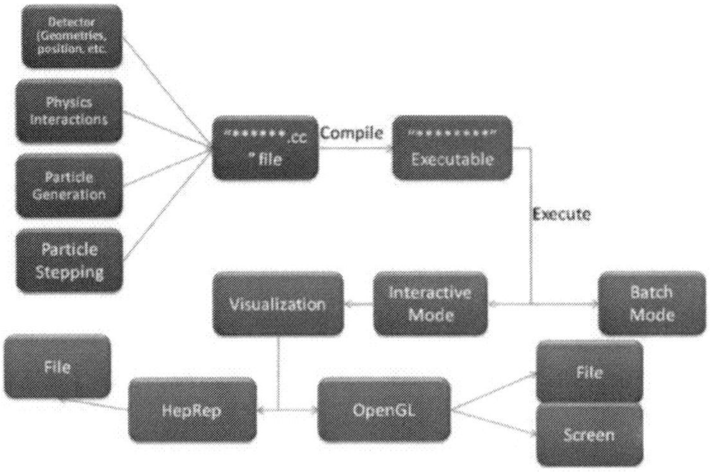

Figure 1. Program flow for generalized GEANT4 simulation.

GEANT4 detection and scoring.

Calculations of absorbed dose at each distance from the source were accomplished via use of multiple sensitive detectors in GEANT4. For example, 5cm thick germanium detectors were constructed at 25, 50, 75, 100, 125, 150, 175, and 200 cm away from the center of the sample source. These detectors were assigned logical scorers in GEANT4, made sensitive only to gamma photons. At the end of each simulation, the detectors calculate the accumulated gamma dose in each detector, and these results are printed to the command line, as well as to a file, designed for easy importing of data for later analysis and visualization. An example of the detector array, constructed around the UUTR central irradiator can be seen in Figure 2.

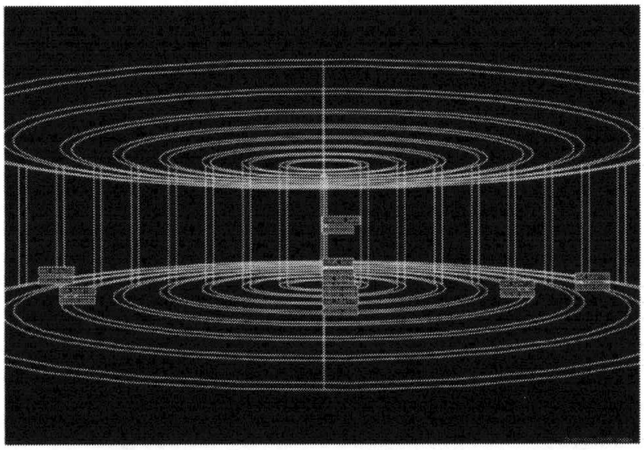

Figure 2. Example of detector array, constructed around central irradiator model. The same array of detectors was used for both the thermal and central irradiator models. Central model used here for illustration purposes.

GEANT4 model of neutron energy spectra.

The Watt model of fission neutron energies was used to model the thermal neutron flux in the various port models. This model is described in Equation 1, and a plot of the fraction of fission neutrons per MeV as a function of the neutron energy is seen in Figure 3 below.

$$\chi(E) = 0.453e^{-E}\sinh\,(2E)^{0.5}$$

Equation 1. Watt Model of Fission Neutron Energy distributions.

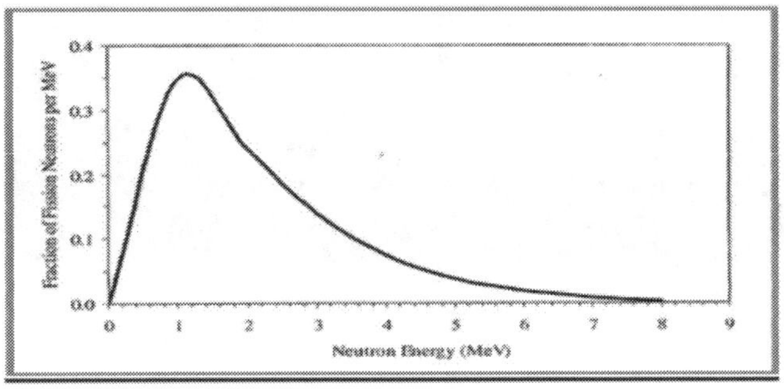

Figure 3. Watt Fission Neutron Energy distribution. Image taken from Jevremovic, "Nuclear Principles in Engineering" [4].

GEANT4 model of UUTR central irradiation port.

Figure 4 shows the model of a 1"-diameter, 5"-tall aluminum cylinder in the center of a 15.24 cm-diameter, 38.1 cm-tall cylindrical aluminum air shaft, subjected to a flux of 3.6 $*10^{12}$ $^{\text{neutrons}}/_{\text{cm}^2 * \text{s}}$. MCNP5 was used to simulate this same geometry, to obtain neutron flux-per-source-particle, to calculate the number of neutrons necessary to be generated in order to simulate a desired flux in GEANT4.

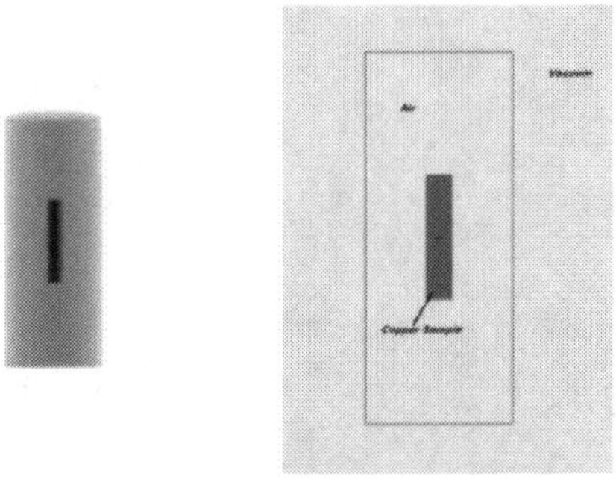

Figure 4. MCNP5 Model of a sample placed in central irradiator (CI); flux data used to run the corresponding GEANT4 model

GEANT4 model of UUTR thermal irradiation port.

Figure 5 shows the model of a 1″-diameter, 5″-tall aluminum cylinder 1″-diameter, 5″-tall aluminum cylinder in center of a 1″-diameter, 35.08-cm tall aluminum air shaft. Air shaft nested at the center of 11.4-cm diameter, 35.08*cm tall shaft of heavy water. Shafts embedded in aluminum trapezoidal prism with bases of 29.8cm and 47.4cm, 15.2cm wide, and 60cm tall, filled with heavy water. Aluminum volumes are 1/16″ thick. Sample was subjected to a flux of $3.6 * 10^{11}\ \mathrm{neutrons}/_{\mathrm{cm}^2 * \mathrm{s}}$.

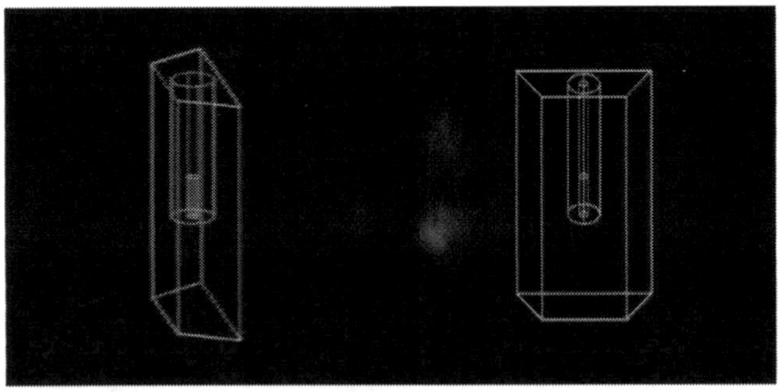

Figure 5. GEANT4 Model of an aluminum sample placed in thermal irradiator (TI)

GEANT4-based UUTR nuclear signatures.

Calculations of average neutron and gamma photon energy spectra were accomplished via tracking neutrons and gamma photons in desired volumes, specific to each of the above presented models. For the thermal irradiator and central irradiator models, the average neutron energy spectra were calculated from the neutrons in the air shafts leading up from the irradiation ports, and the average gamma photon energy spectra were calculated from the gamma photons in the samples in the irradiation ports. For the shielded and unshielded Pu-Be neutron source models, the average neutron energy spectra and the average gamma photon energy spectra were calculated from the neutrons and gamma photons, respectively, penetrating the inner surface of the inner, gamma photon detector.

Results

GEANT4 visualizations of UUTR central and thermal irradiation ports.

In the GEANT4 simulation model, the aluminum cylinder was exposed to 3.6 $*10^{12}$ $\text{neutrons}/_{\text{cm}^2 * \text{s}}$, homogeneous cylindrical flux, using 1.07745$*10^{-3}$ $\text{neutrons}/_{\text{cm}^2 * \text{s}}$ per neutron. Figure 6 illustrates the homogeneous neutron flux for an irradiated aluminum sample, simulated in GEANT4.

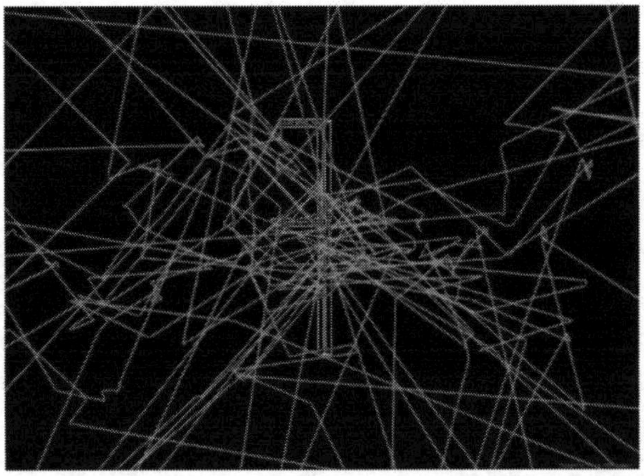

Figure 6. GEANT4 simulation of central irradiator irradiation of aluminum sample. Green rays indicate neutron and gamma tracks; red tracks indicate negatively-charged particles; blue tracks indicate positively-charged particles; white points mark stepping points for particles.

In the GEANT4 simulation model, the aluminum cylinder was irradiated with 3.6 $*10^{11}$ $\text{neutrons}/_{\text{cm}^2 * \text{s}}$, homogeneous planar flux. Figure 7 illustrates the homogeneous neutron flux for an irradiated aluminum sample, simulated in GEANT4.

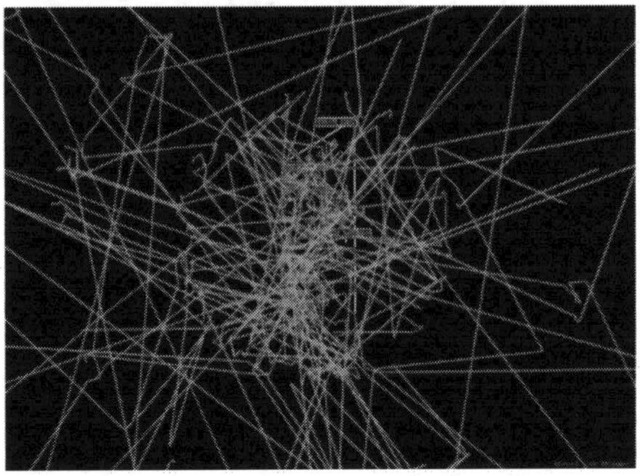

Figure 7. GEANT4 simulation of thermal irradiator irradiation of aluminum sample. Green rays indicate neutron and gamma tracks; red tracks indicate negatively-charged particles; blue tracks indicate positively-charged particles; white points mark stepping points for particles.

GEANT4 dose calculations for UUTR central and thermal irradiation ports.

In GEANT4, a particle flux is simulated by generating various large (>>10,000) numbers of source particles. The dependent variable, in this case, the dose rate at various distances from the irradiated sample sources, is then calculated for each run of particles. If a linear or some other relationship can be drawn between the number of source particles generated and the dependent variable, it is reasonable to be able to scale up the results of the dependent variable with respect to the number of source particles generated. In this case, the absorbed dose rate varied linearly with respect to the number of source particles for both the CI and TI. Thus, many runs of particles were simulated at various numbers of source particles between 100 and 1,000,000 source neutrons. This data was plotted to generate a best-fit regression line between the numbers of source particles and the absorbed dose rate. To compare results, the best-fit dose for 100,000 neutrons was used in the results for both the CI and TI models, discussed below. The dose vs. numbers of source particles plots for the TI model can be seen in Figure 8. From Figure 8, the mean dose for 100,000 neutrons generated was calculated for each distance from the source. This dose rate result is then scaled up by factor of $\frac{\text{desired actual flux}}{\text{flux for 100,000 neutrons}}$, to give results comparable to what should be seen experimentally. These absorbed dose rates were tabulated at each distance to yield a surface plot of dose rate as a function of distance from the source, seen in Figures 9 and 10.

As an example calculation, at 25cm from the irradiated sample, the absorbed dose for 0.154 μs irradiation time of aluminum sample from thermal irradiator is $5.995 \cdot 10^{-15}$ Gray. For a 2 hour irradiation time common for NAA, this equals a dose rate of $1.39 \cdot 10^{4}\ ^{mrem}/_{hr}$. For comparison, at University of Utah TRIGA, Area Radiation Monitors SCRAM reactor for detection of > 10 $^{mrem}/_{hr}$ dose rates. In this case, samples are suspended at half depth in reactor pool to cool until < 10 $^{mrem}/_{hr}$. For 1 $^{mrem}/_{hr}$ < dose rates < 10 $^{mrem}/_{hr}$, samples are cooled inside 4"-thick lead box until < 1 $^{mrem}/_{hr}$. Samples < 1 $^{mrem}/_{hr}$ are deemed safe. The central irradiator model displays the similar results, for a cylindrical neutron source.

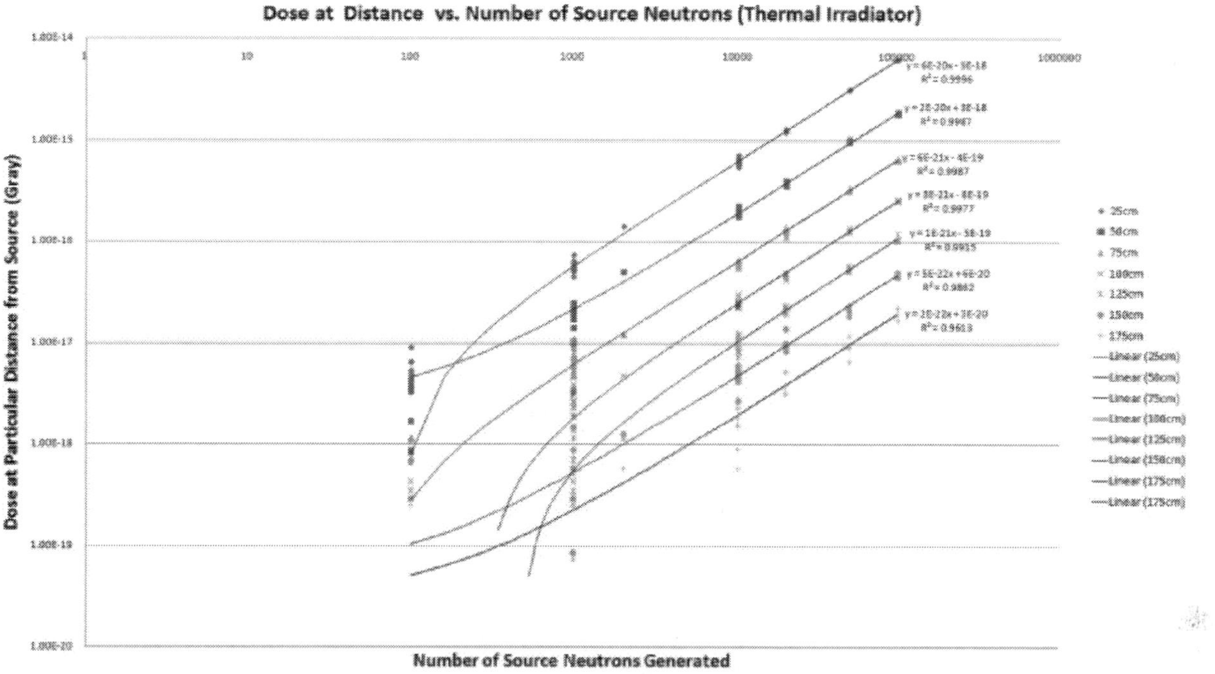

Figure 8. GEANT4 dose at various distances from sample as a function of the number of source particles generated for the Thermal Irradiator model. At each distance, for > 100 neutrons generated, there is a strong linear relationship between the independent and dependent variables.

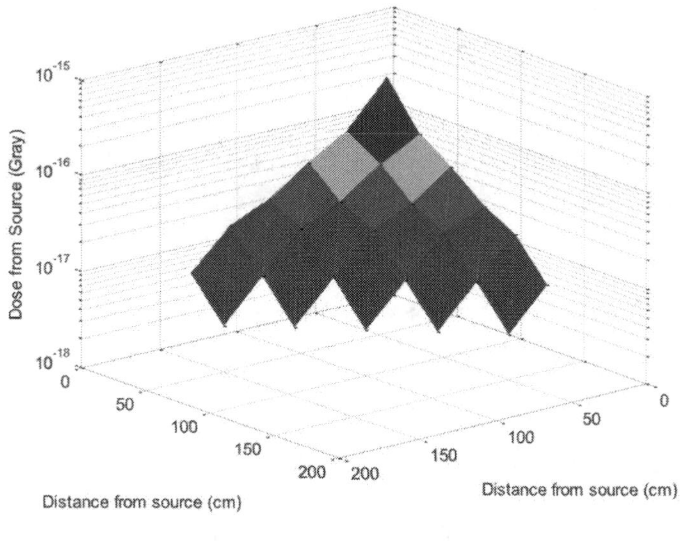

Figure 9. Dose plot for the thermal TRIGA irradiator model, as a function of distance from the source, corresponding to a 0.154 µs irradiation time. Image taken from Voyles et al. [7].

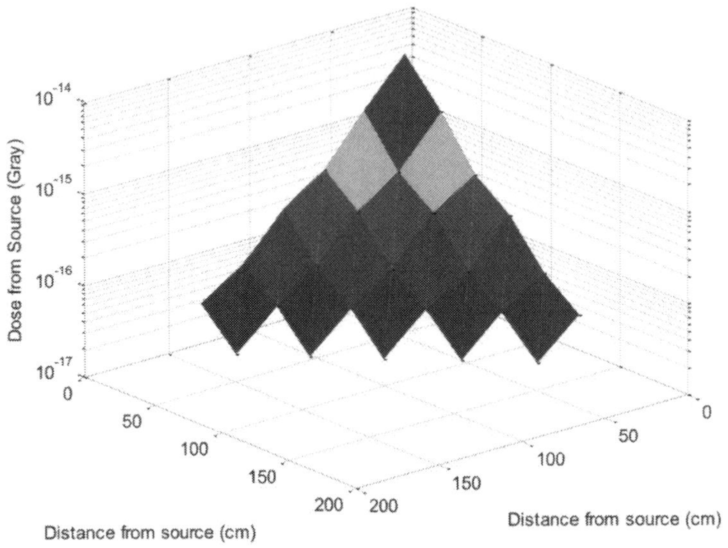

Figure 10. Dose plot for the central UUTR irradiator model, as a function of distance from the source, corresponding to a 0.0141 μs irradiation time. Image taken from Voyles et al. [7].

GEANT4 neutron source simulations.

GEANT4 simulations of an unshielded Pu-Be neutron source, as well as simulations of the same source, shielded with paraffin and lead, have been created. Figure 12 shows that very few of the green (neutrally-charged) tracks make it outside the paraffin shell. These tracks represent the neutrons and gamma photons, and so it is clear that the paraffin is a good moderating material for the neutrons (and slightly so for the gamma photons as well).

GEANT4 neutron and gamma signatures.

Average neutron and gamma photon energy spectra were created for the thermal irradiator, central irradiator, shielded Pu-Be neutron source, and unshielded Pu-Be neutron source models, using the methods described above, in "GEANT4-based UUTR nuclear signatures". After tracking gamma photons and neutrons being transported in the volumes specific to each model (described in the aforementioned earlier section), the kinetic energies of these particles were stored, and compiled into histograms for each of the thermal irradiator, central irradiator, shielded Pu-Be neutron source, and unshielded Pu-Be neutron source models. These histograms are shown in Figures 12, 13, 14, and 15.

For the central irradiator, as well as the shielded and unshielded Pu-Be neutron source models, gamma photon and neutron energy spectra for various simulation models, no experimental data were available for benchmarking. However, it is very obvious that in most of the models, excessive thermalization of neutrons, as well as somewhat for gamma photons, generated very indistinct gamma photon and neutron energy spectra. Preliminary benchmarking has been accomplished, using a 150-mg sample of limestone ($CaCO_3$), irradiated in the thermal UUTR irradiator for 2 hours, and then allowed to decay for 72 hours.

Figure 16 shoiws the experimental gamma photon energy spectrum, as well as the gamma photon energy spectrum for the simulation of this irradiation.

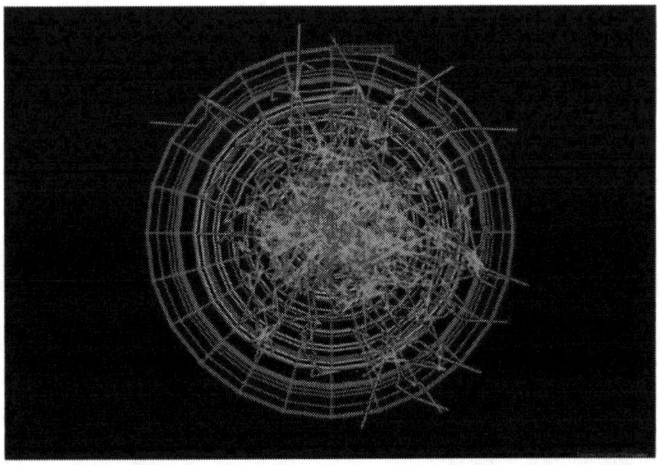

Figure 11: Top-down view of experimental setup for 6.700 · $10^6 \,\mathrm{neutrons}/\mathrm{cm}^2 * \mathrm{s}$, with particle tracks. Red lines indicate negatively-charged particle trajectory, green lines indicate neutrally-charged particle trajectory, blue lines indicate positively-charged particle trajectory, yellow dots indicate the actual step points used by Geant4.

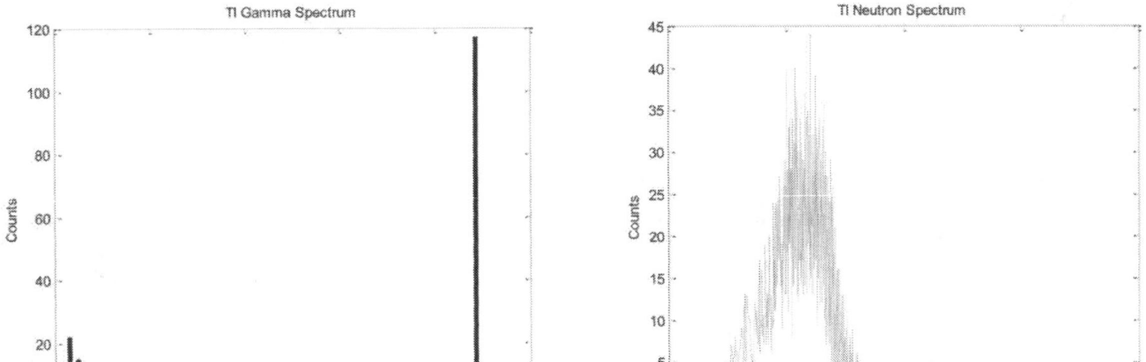

Figure 12. Average gamma photon and neutron energy spectra for the thermal TRIGA irradiator model of irradiation of aluminum sample, seen n Figure 7.

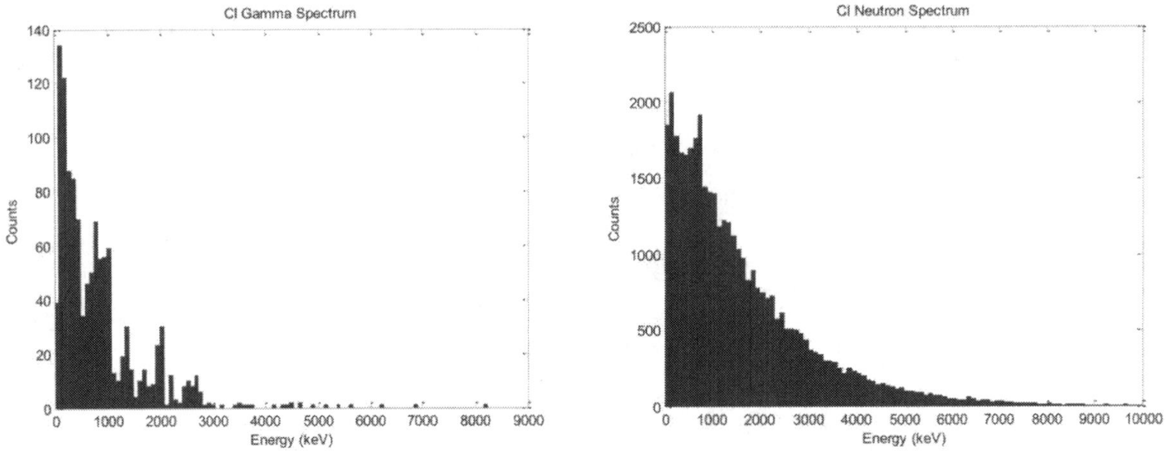

Figure 13. Average gamma photon and neutron energy spectra for the central TRIGA irradiator model of irradiation of aluminum sample, seen in Figure 6.

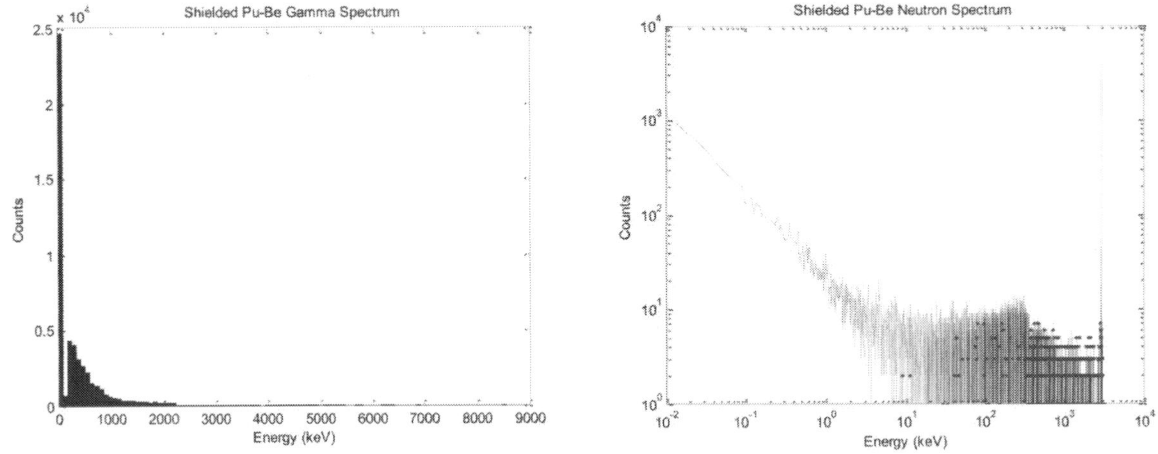

Figure 14. Average gamma photon and neutron energy spectra for the shielded Pu-Be neutron source model of irradiation of aluminum sample, seen in Figure 12.

As can be seen, there is some basic correlation between the experimental and simulation spectra, in Figure 17. The largest discrepancy comes in the large peak at ~2200 keV. This most likely comes from the presence of hydrogen nuclei in the simulation, and can be seen distinctly in the gamma photon spectrum for the thermal irradiator model, seen in Figure 13. This issue with such neutron activation experiments has been documented in literature, by Evans et al. [1]. Future work in this simulation will address and correct this issue.

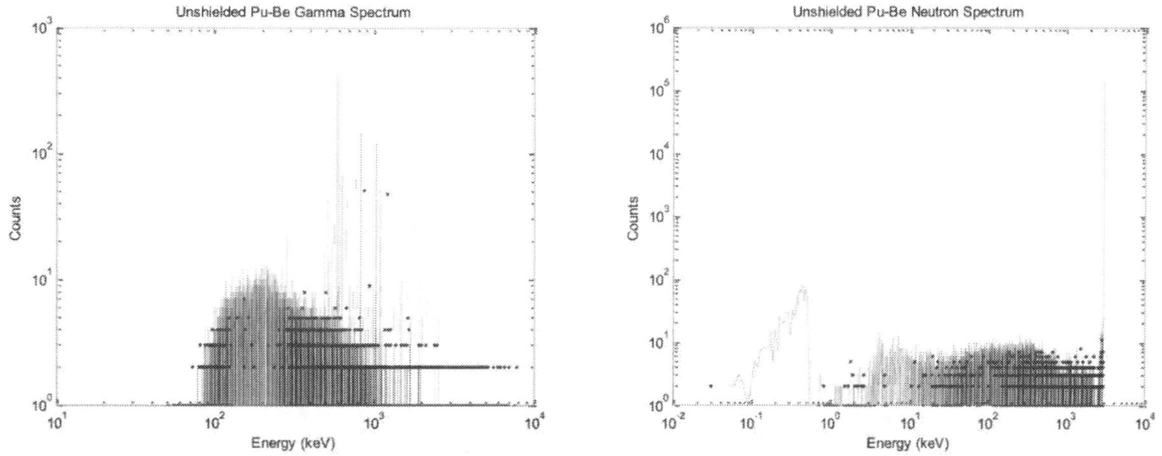

Figure 15. Average gamma photon and neutron energy spectra for the unshielded Pu-Be neutron source model of irradiation of aluminum sample.

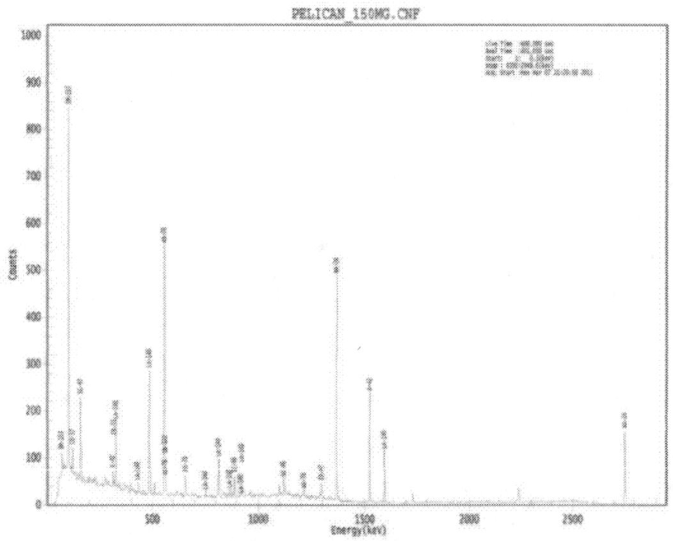

Figure 16: Experimental gamma photon energy spectrum for 150-mg limestone sample irradiated in UUTR TRIGA thermal irradiator for 2 hours.

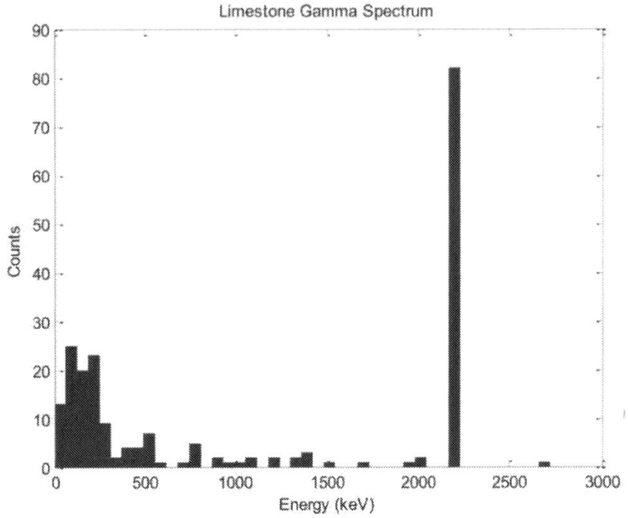

Figure 17: Experimental gamma photon energy spectrum for 150-mg limestone sample irradiated in UUTR TRIGA thermal irradiator for 2 hours.

Conclusion

A preliminary GEANT4 model (coupled with MCNP5 simulation for accurate representation of the neutron flux) was developed for a central irradiation port and thermal irradiation port at Utah TRIGA facility, as well as shielded Pu-Be neutron and gamma sources. Using MCNP5 to calculate flux-per-particle, our current GEANT4 model can simulate the irradiation of various samples. The next step is to expand the current model to include the fast neutron irradiation facility at Utah TRIGA facility. In addition, simulation results will be benchmarked against experimental results, to help make this application suitable for export, in the use of helping benchmark experiments in new facilities.

Acknowledgement

This research is partially funded by the DHS/DOE Nuclear Forensics Education Award Program.

References

1. Evans, C. J., Ryde, S. J. S., Hancock, D. A., and Al-Agel, F. "Monte Carlo Simulation of Prompt Gamma Neutron Activation Analysis Using MCNP Code". Applied Radiation and Isotopes, Vol. 49, No 5/6, pp. 541-543, 1998.

2. Geant4: a toolkit for the simulation of the passage of particles through matter. *CERN* - the European Organization for Nuclear Research, 2011. http://www.geant4.org/geant4/
3. HepRep: a Generic Interface Definition for HEP Event Display Representables. Joseph Perl, Stanford Linear Accelerator Center, 2009. http://www.slac.stanford.edu/~perl/heprep/index.html
4. Jevremovic, Tatjana. "Nuclear Principles in Engineering." New York: Springer Science+Business Media, Inc., 2008.
5. MCNP - A General Monte Carlo N-Particle Transport Code - Version 5. Monte Carlo Codes Group, Los Alamos National Laboratory, 2006. http://mcnp-green.lanl.gov/
6. OpenGL - The Industry's Foundation for High Performance Graphics. Khronos Group, 2010. http://www.opengl.org/
7. Voyles, Andrew S., Yang, Haori, and Jevremovic, Tatjana. "GEANT4 Simulation of Irradiation Facilities at University of Utah TRIGA." Unpublished paper presented at American Nuclear Society Student Conference, Atlanta, GA, 2011.

GEANT4 MODEL FOR NUCLEAR FORENSIC ANALYSIS OF NUCLEAR DETONATION IN URBAN ENVIRONMENTS

Jeffrey Taylor, Haori Yang, Xuesong Zhou, and Tatjana Jevremovic
University of Utah, Salt Lake City, UT, USA

Introduction

Nuclear forensics techniques are commonly applied in pre-detonation circumstances to help identify the material characteristics and source of intercepted material. Just as important, the field of post-detonation forensics applies similar techniques for device attribution. This paper summarizes research and work relevant to creating a simple model in GEANT4 for post-detonation nuclear forensics applications in urban environments. The simulation model incorporates an intense neutron and gamma ray pulse source located in an urban canyon landscape in downtown Salt Lake City with approximate building dimensions to simulate a low-yield nuclear device detonation. Following a quick review of nuclear forensics in urban environments, this paper characterizes model inputs, such as prevalent material types and the radiation source, describes the modeling approach in GEANT4, and discusses simulation results.

Nuclear Forensics in Urban Settings

A report published by the American Physical Society and the American Association for the Advancement of Science (AAAS) defines nuclear forensics as "... the technical means by which nuclear materials, whether intercepted intact or retrieved from post-explosion debris, are characterized (as to composition, physical condition, age, provenance, history) and interpreted (as to provenance, industrial history, and implications for nuclear device design)" (9). The primary objective of nuclear forensic analysis is to identify the material characteristics to help determine the material's origin, intended use, and the method of acquisition, with the ultimate goal being attribution. Since government agencies often focus their search for illicit nuclear material away from major urban areas (ports, borders, etc.), nuclear forensics in urban environments is often more focused toward post-detonation attribution.

In an urban detonation scenario, it may be assumed that a low-yield fissile device detonates in an urban canyon environment, with many large buildings in the immediate vicinity. Depending on the size and type of the device, there may be sufficient residual material for forensic analysis, but there are situations which may preclude this option (i.e. small material source experiencing significant dispersion after detonation). In these situations, attempting to identify the source material may rely upon measuring the prompt gamma rays and neutrons, which are released within the first seconds after the detonation, and comparing the observed spectral distributions with simulation results for a similar environment. This approach requires simulating the source and the surrounding environment, which obviously increases the uncertainty associated with the resulting source characterization estimates compared to physical forensic analysis.

Source Characterization Simulation

Simulation tools used for source characterization typically face a significant challenge modeling the source because they are described as a device. Modeling the source as a device requires consideration for the geometry, material construction, method of detonation, and other characteristics of the weapon, all of which are unknown and significantly impact the simulation results. Additionally, this approach requires modeling many potential combinations of device characteristics, increasing analysis time and further complicating the simulation process.

Since the source is the largest unknown element, the best option for limiting the remaining sources of additional uncertainty requires creating a more realistic simulated environment. Interactions with structures and other materials in the built environment affect the observed radiation measurements, so these effects should be considered in the simulation. As a result, creating this type of simulation requires knowledge of the built environment near the detonation site, including the dimensions and materials used in construction. This level of detail, which is greater than what is typically available for emergency response and analysis simulations, is particularly well-suited for implementation with a Geographic Information System (GIS). A GIS is a system for managing and analyzing spatially-referenced data which can associate attributes with objects with a defined shape and geographic location.

Modeling Approach

To address the issues associated with source characterization simulations described above, the proposed modeling approach in this paper simplifies the source model while utilizing a higher-detail environment model. The source material is simplified to a single point source with a user-defined spectral energy distribution and time-dependent intensity. This reduces modeling complexity by bypassing most device characteristics, reducing the size of the potential solution set (at the cost of increasing uncertainty). The surrounding environment is constructed from input data stored in a GIS, containing building geometries and object attributes (i.e. building with dimensions, shape, material, etc.). This provides the capability to model highly detailed urban areas. Lastly, the simulation is constructed within GEANT4 so that the simulation can fully incorporate all environment characteristics, such as particle interactions and shielding effects, which may affect measurement results. The remaining portion of this paper describes the process of building a simple model for simulation in GEANT4.

Model Inputs

As described above, the primary inputs for this simple GEANT4 model can be divided into the source characteristics and built environment characteristics. The source characteristics are defined by the time-based intensity of the source and the spectral energy distribution for prompt gamma rays and neutrons from a fission event. The built environment characteristics can be defined primarily by the structures in the urban environment, where each building has a center location, dimensions, and an assigned material composition.

Gamma Ray and Neutron Source Characterization

Research related to nuclear weapons testing in the 1950s was referenced to estimate the spectral energy distributions and time-based intensity of prompt gamma rays and neutrons from a small detonation. As seen in Figure 1, a peak in gamma rays occurs almost immediately after detonation (within a microsecond), followed by a decreasing trend in intensity. Fission products decaying after detonation, as well as neutron interactions with the atmosphere, contribute to more gamma rays being emitted after the initial prompt gamma rays when compared to a high altitude detonation (dashed line in Figure 1). Glasstone and Dolan describe a similar peak in neutrons emitted after detonation, where "... 99% of fission neutrons are produced almost immediately, within less than a millionth of a second of the initiation of the explosion" (340). From these observations, modeling the source as an intense pulse from a point source is a reasonable approximation for this model.

Characterizing the source spectral energy distribution for prompt gamma rays and neutrons, Glasstone and Dolan also provide very rough energy distribution histograms, as seen in Figure 2 and Figure 3. The gamma ray spectral distribution in Figure 2 shows nearly 70% below 750 keV, producing a large peak at the low side of the energy spectrum, with a relatively flat distribution at the remaining higher energies. The neutron spectral distribution in Figure 3 shows a relatively flat distribution at energies below 1 MeV, followed by a trend generally showing a nearly-linear decrease in the distribution at higher energy levels until a sharp decrease at 10 MeV. These spectral distributions were identified as examples in Glasstone and Dolan because each device is slightly different, so these distributions should not be considered characteristic of all detonations from fission events. In this case, they are only used to inform the creation of the initial energy spectrum for the model.

Urban Building Material Characterization

Materials in the urban environment vary from the steel and concrete buildings to the trees, cars, and people passing through the city. Since a static geometry model was used in this GEANT4 model, attention was drawn particularly to buildings, since they are expected to act like shielding materials within the environment and thus greatly influence the simulation results. Typical building materials in modern cities often include combinations of concrete, masonry, steel, wood, and glass. This model focused primarily on the major structural components of most buildings, specifically concrete, stone, brick, and steel. These four materials are composite materials, composed of several compounds with various types and compositions depending on the site of their creation, their purpose, and the personnel mixing the materials, requiring definition in the GEANT4 DetectorConstructor source file. Several composite materials are pre-defined in the GEANT4 materials database, but of the four materials defined in the model, only concrete was found in the database. The remaining three materials required definition.

Defining a composite material in GEANT4 requires the fractional mass (or percent composition) of the individual elements or compounds found in the composite material and the overall density of the material. Table 1 describes these general characteristics for each of the four materials. Since there are multiple types of stone, brick, and steel, more common building materials such as granite, red brick and A36 structural steel were chosen for their use in the model.

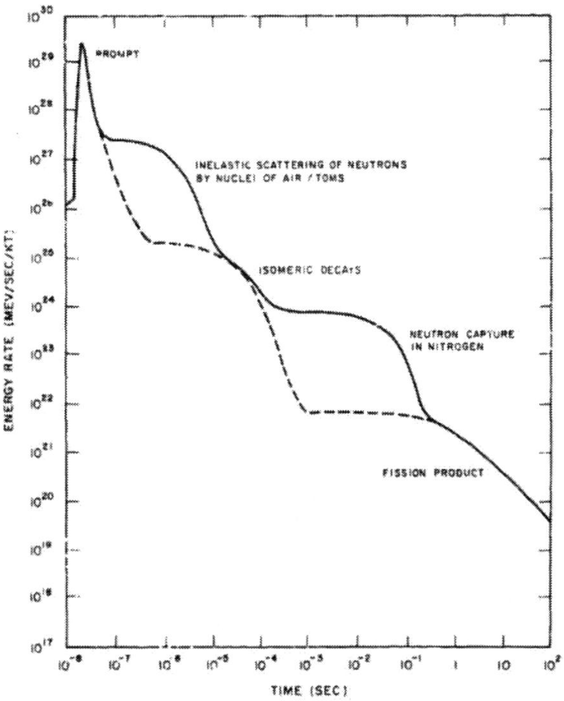

Figure 1. Temporal gamma ray intensity distribution after hypothetical detonation (Glasstone and Dolan 328)

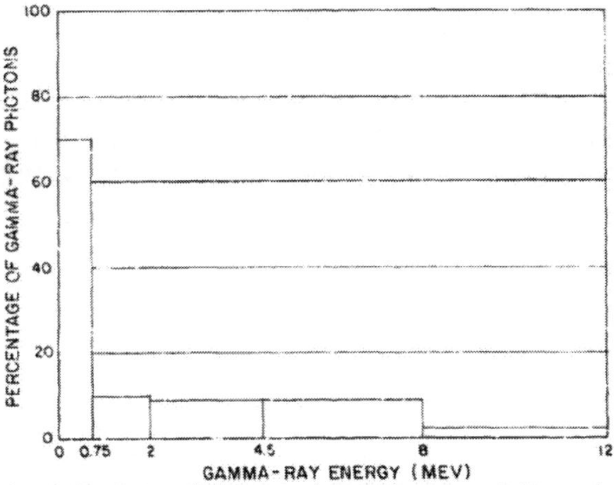

Figure 8.106. Spectrum of initial gamma radiation 2,000 yards from a 20-kiloton explosion.

Figure 2. Example of initial gamma ray spectral energy distribution (Glasstone and Dolan 361)

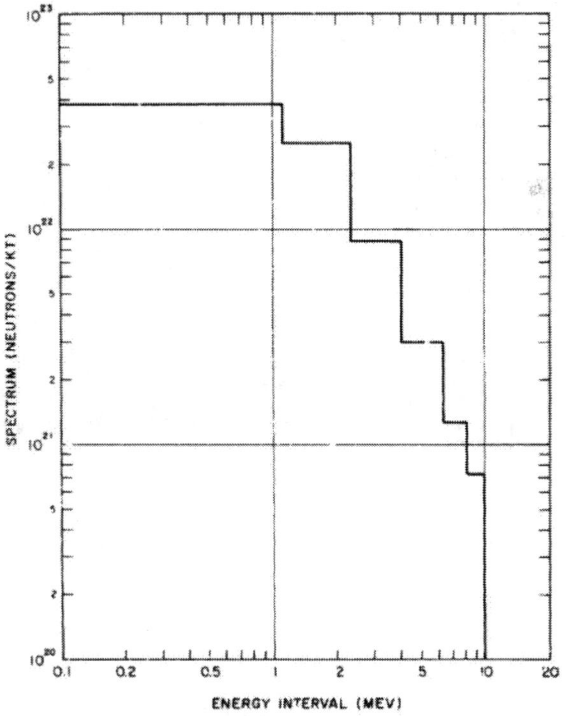

Figure 8.116a. Neutron spectrum for a fission weapon per kiloton total energy yield.

Figure 3. Example of initial neutron spectral energy distribution (Glasstone and Dolan 364)

Table 1. Fractional Mass Composition & Densities for Concrete, Stone, Brick, and Steel

Concrete[1]		**Stone (Granite)[2]**		**Red Brick[3]**		**Steel (A36 Stuct.)[4]**	
Element	% Comp.	Element	% Comp.	Element	% Comp.	Element	% Comp.
O	0.529	O	0.484	O	0.491	Fe	0.982
Si	0.337	Si	0.337	Si	0.280	Mn	0.010
Ca	0.044	Al	0.076	Al	0.159	C	0.003
Al	0.034	K	0.034	Fe	0.035	Si	0.003
Na	0.016	Na	0.027	Ca	0.032	Cu	0.002
Fe	0.014	Fe	0.022	Mg	0.003	Density = 7.85 g/cm^3	
K	0.013	Ca	0.013	Density = 2.0 g/cm^3			
H	0.010	Mg	0.007				
Mg	0.002	Density = 2.7 g/cm^3					
C	0.001						
Density = 2.3 g/cm^3							
Sources: [1] Geant4 User's Guide for Application Developers; [2] Blatt and Tracy; [3] Punmia et al.; [4] MatWeb							

Building Geometries and Locations

Most of the building locations and their dimensions were acquired by taking measurements in ESRI's ArcGIS using aerial photography available from the GIS Portal of the Utah Automated Geographic Reference Center. After determining the world environment's center location in the X-Y plane, building height information was acquired from the Emporis website, featuring lists of the tallest buildings in major cities with their heights. This manual process of connecting a building with attribute data in a GIS is visualized in Figure 4, where the green blocks represent the buildings with height data available, and the shape dimensions were defined by the spatially- and size-accurate aerial photography. The green blocks with light blue highlighting identify the buildings added to the GEANT4 DetectorConstruction file for this analysis. The building position can be estimated by measuring between the world and shape centers. The end result is a table of half-width sizes in the X, Y, and Z directions, as well as vector positions for the center of each building in reference to the world center, as seen in Table 2.

Predominant Physics in Urban Environments

The GEANT4 simulation is modified to consider particle interactions through attenuation, scattering, and absorption for the composite materials in the model. Using data available from the National Institute of Standards and Testing (NIST) and the Korean Atomic Energy Research Institute (KAERI), attenuation coefficients and cross sections were plotted for

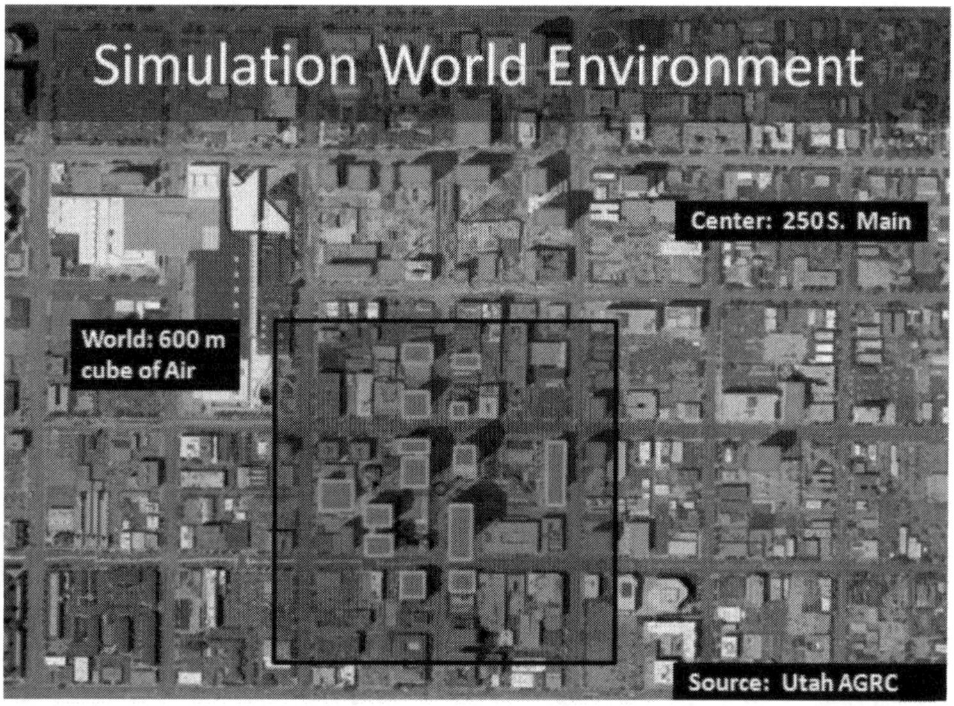

Figure 4. Small area in Salt Lake City as the basis for the simulation world environment

Table 2. GEANT4 Model Building Geometry Characteristics

Name	Half_Width (X)	Half_Length (Y)	Half_Height (Z)	X_Center	Y_Center	Z_Center
Wells Fargo	20	50	65	35	-75	65
One Utah Center	20	18	54	45	58	54
American Tower North	23	18	50	-113	-48	50
American Tower South	23	18	50	-108	-103	50
Walker Center	15	13	34	30	138	34
Hotel Monaco	25	10	33	-50	75	33
Tribune Bldg.	23	10	22	38	225	22
Judge Bldg.	20	18	16	35	-163	16
US Bank	23	23	23	-48	148	23
Kearns Bldg.	25	15	22	-45	240	22
222 Main	23	25	48	-48	30	48
Marriott Center	18	58	19	208	33	19
Hilton Center	30	25	28	-185	-10	28
JC Penney	23	23	23	-48	-168	23

gamma rays and neutrons, respectively, across the energy spectrum. These plots help to indicate which interactions are most likely to take place when gamma rays and neutrons with various energies come in contact with various materials in the urban environment.

Gamma Ray Interaction

Gamma attenuation coefficients from the NIST for concrete and steel are plotted in Figures 5 and 6, where the plots identify different coefficients for total attenuation, coherent and incoherent scattering, photoelectric absorption, and pair production between 1 keV and 20 MeV. Since the composition of brick, stone, and concrete are quite similar, and upon inspection were nearly identical, concrete was used here to visualize the attenuation coefficients representative of all three materials. As seen in Figure 5 below for concrete, the photoelectric effect is expected to dominate below 30 keV and incoherent scattering should be dominant between 30 keV and approximately 20 MeV, with pair production taking precedence at energies above 20 MeV. In Figure 6, the same interactions are expected to dominate in nearly the same energy levels, with the upper bound for photoelectric absorption slightly higher than for concrete, and pair production's lower bound is closer to 10 MeV.

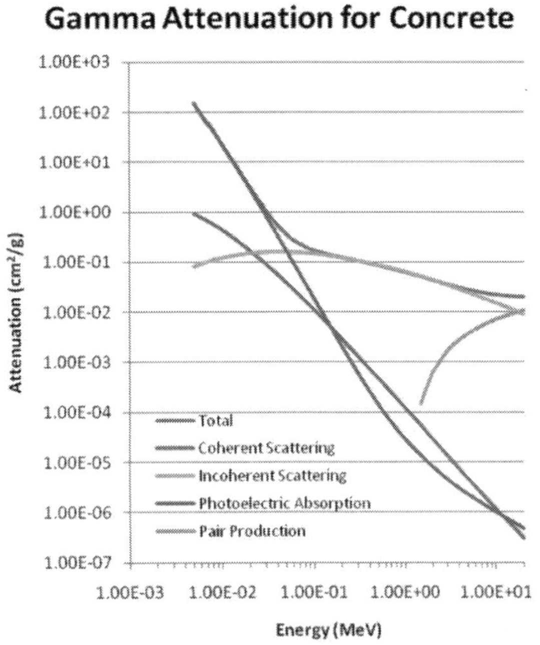

Figure 5. Gamma attenuation coefficients for concrete (Bergen)

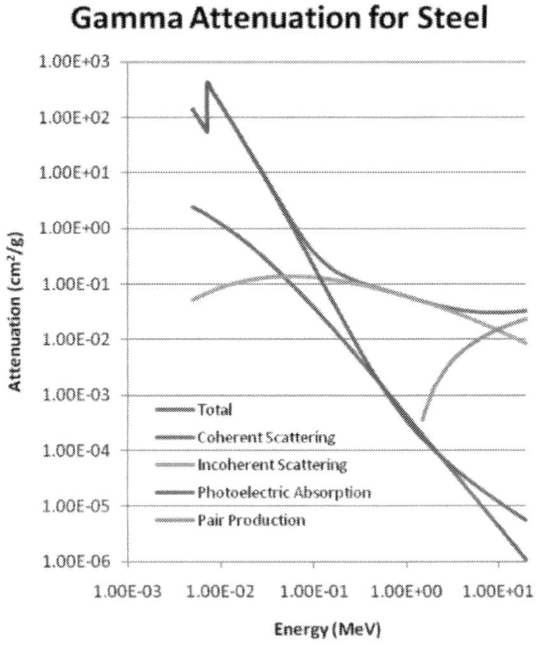

Figure 6. Gamma attenuation coefficients for steel (Bergen)

Neutron Interactions

Neutron cross section data from KAERI for Silicon, Oxygen, Aluminum, and Iron, the primary components of the four materials modeled, are plotted in Figures 7 through 10. These plots identify the total, elastic scattering, absorption, and radiative capture cross sections for neutron energies between approximately 10^{-11} MeV and 20 MeV. In each figure, it is readily apparent that elastic scattering should dominate all other interactions across nearly the entire energy spectrum. Radiative capture plays a much bigger role at lower energies for each

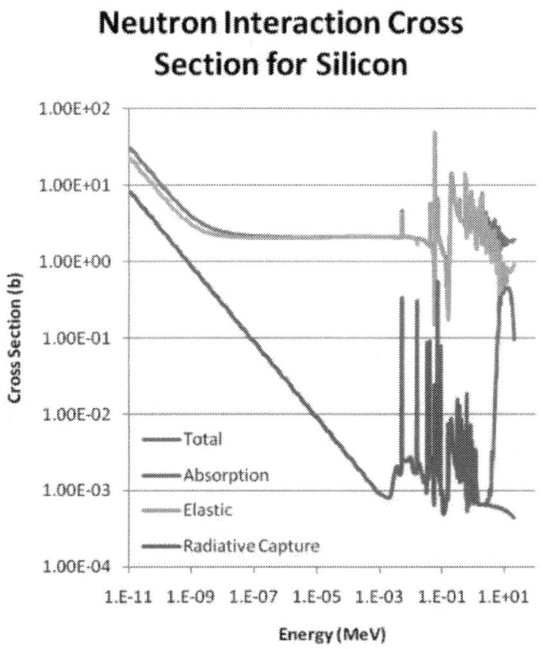

Figure 7. Neutron cross sections for silicon (KAERI)

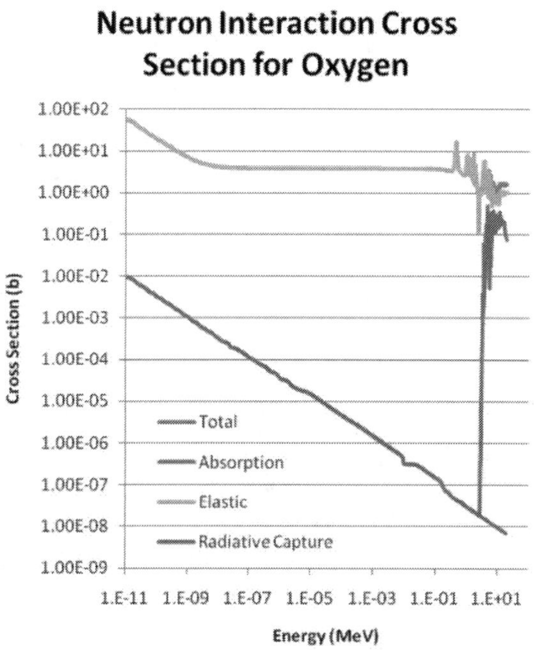

Figure 8. Neutron cross sections for oxygen (KAERI)

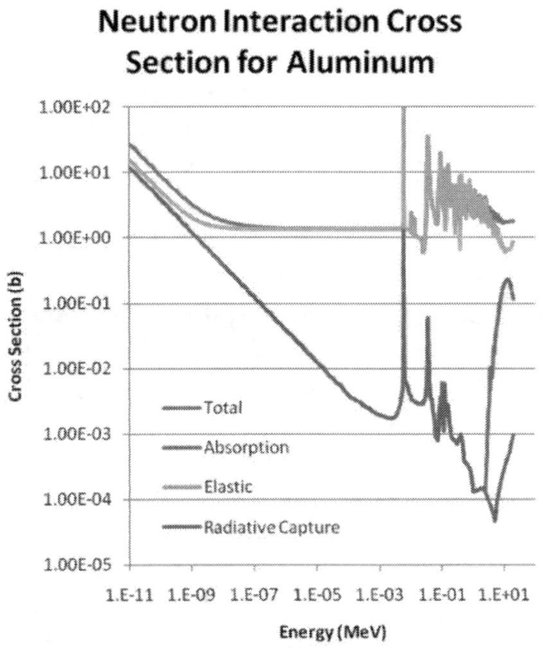

Figure 9. Neutron cross sections for aluminum (KAERI)

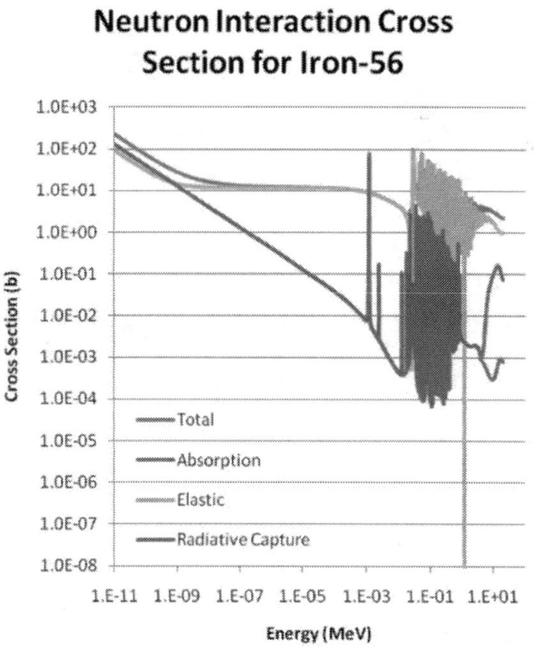

Figure 10. Neutron cross sections for iron-56 (KAERI)

element except oxygen, but resonant peaks in the radiative capture cross section plot between 1 keV and 1 MeV may also rival the elastic scattering cross sections for silicon, aluminum, and iron. Generally, absorption becomes more common at energies greater than 1 MeV.

GEANT4 Model Description

After describing the model inputs, the focus now turns to building the GEANT4 model from these inputs. This model uses a real-world geometric environment as the basis for the simulation, an intense gamma ray and neutron source with user-variable spectral energy distributions as the prompt radiation source, and a detection system which can produce a simulated radiation detector measurement at any point in the world environment. The following section describes each portion of the GEANT4 code which was modified for this forensic analysis case study.

Material Definitions & World and Building Geometries

The DetectorConstruction file in GEANT4 stores the world geometry. This includes the world extents and the "targets" and defines the target material from a list of materials. To construct the three composite materials (A36 steel, stone, and red brick) used in this model, composition data in Table 1 were directly input into the DetectorConstruction file as mass fractions of the composite material.

The world geometry is also defined within the DetectorConstruction file, where the world defines the physical limits of the model space. The "world length" in this model is approximately 600 meters, or about two city blocks in Salt Lake City, producing a 600 meter cube. The world environment is further defined to be composed of air, with its center located at 250 South Main Street, as shown in Figure 4. The world center is assigned the (0, 0, 0) coordinate, which acts as the location reference for all other object geometries.

Lastly, 14 buildings were added as "targets" to the DetectorConstruction file, each with its own shape, dimensions, position, and material. While several object geometries are available within GEANT4, only rectangle objects were used to describe the building shapes. Each rectangle requires dimensions provided as the half-length, which is described in the first three columns of Table 2. The last three columns from Table 2 were used to describe each object's center point, where each target is given a position vector to its center in reference to the world center. These values were all calculated from the geometries stored in the GIS. Each target was randomly assigned a material from the four available composite materials. The results of this process produced the world environment shown in Figures 11 & 12.

Gamma Ray and Neutron Source Description

In order to incorporate a user-defined spectral energy distribution at the radiation source, this model used a GeneralPurposeSource (GPS) to replace the more common ParticleGun. The GPS provides greater flexibility when defining the source characteristics and allows the user to define the characteristics of the source in a macro file accompanying the executable model. In the macro file for this model, the gamma ray and neutron sources were split into two separate isotropic sources with separate spectral energy distributions. Each energy distribution was defined by a histogram using percentages occurring between energies, using the distributions in Figures 2 & 3 as guides.

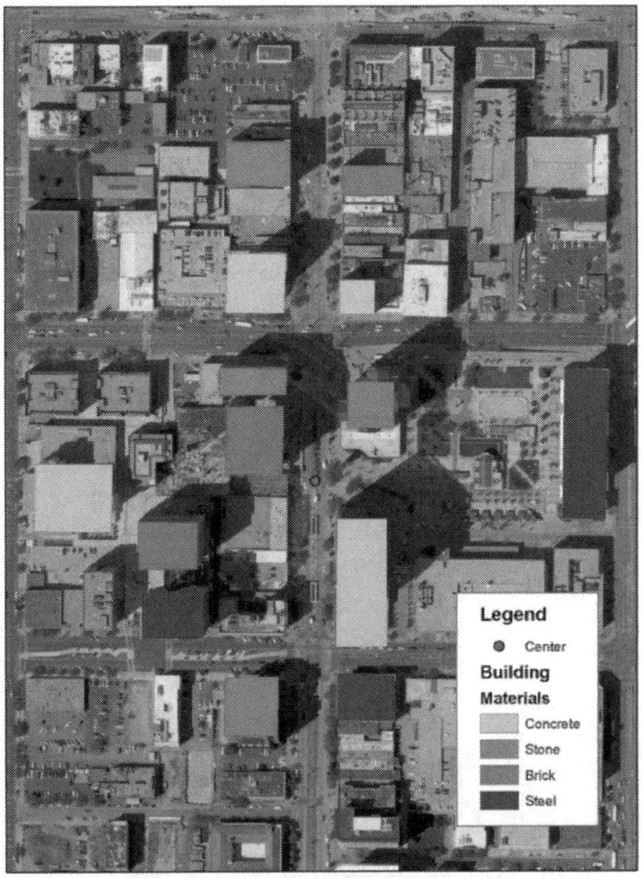

Figure 11. Finalized world environment with color-coded building attribute descriptions

Figure 12. 3D wireframe representation for the final GEANT4 model

Detection System

The last component to be defined in the model is the detection system. Two options were considered: (1) create a sensitive detector which provides detected gamma ray and neutron spectral energy distributions at any location specified by the user, or (2) generate a scoring mesh to produce large-scale data collection capabilities. The scoring mesh divides the environment into pre-defined cells (in this case, boxes) and counts the particles passing through that cell. The scoring mesh was used in this model at the present time due to time limitations and the ability to produce output visualizations from the environment-wide simulation data.

Defining the scoring mesh takes place in the macro file. A boxMesh is first created to define the scoring mesh. To measure both the neutrons and gamma rays in the world environment, a separate mesh is created for counting neutrons and gamma rays. Mesh commands in the /score directory are then used to define the size of each box and the number of boxes distributed in each direction. Once the mesh is defined, multiple scorers are defined with a filter associated with each scorer. The scorer is a detector, where this detector is defined to count the number of particles passing through it (nOfStep), and the filter specifically limits the scorer to counting specific particles with certain properties. In this case, the filters are used to count either neutrons or gamma rays of different energies, where each energy range is defined in the /score/filter command so that the measurement is divided into a histogram-like distribution. This data can then be output into a text file (CSV) for further analysis.

GEANT4 Model Outputs

Data outputs from GEANT4 can be targeted for multiple uses with multiple programs, but this model specifically uses HepRep and Matlab for output data analysis and visualization. The macro files accompanying the executable file initiate trajectory recording for visualization in HepRep. These visualizations show the path each particle takes from the source and includes information about particle type, energy, and interactions which take place in the world environment for each particle trajectory. Visualizations such as those shown in Figure 13 present some preliminary results from the model.

First, the visualized results show many scattering interactions, matching our expectation after examining the attenuation coefficients and cross section. Ionization taking place at the building surfaces, which is best observed in the left half of Figure 13, likely indicates pair production interactions from higher energy gamma rays. Being able to visualize these interactions is important when considering which areas may require greater attention post-detonation for clean-up and finding good locations for shelters based on the amount and types of radiation present in different areas in the urban environment.

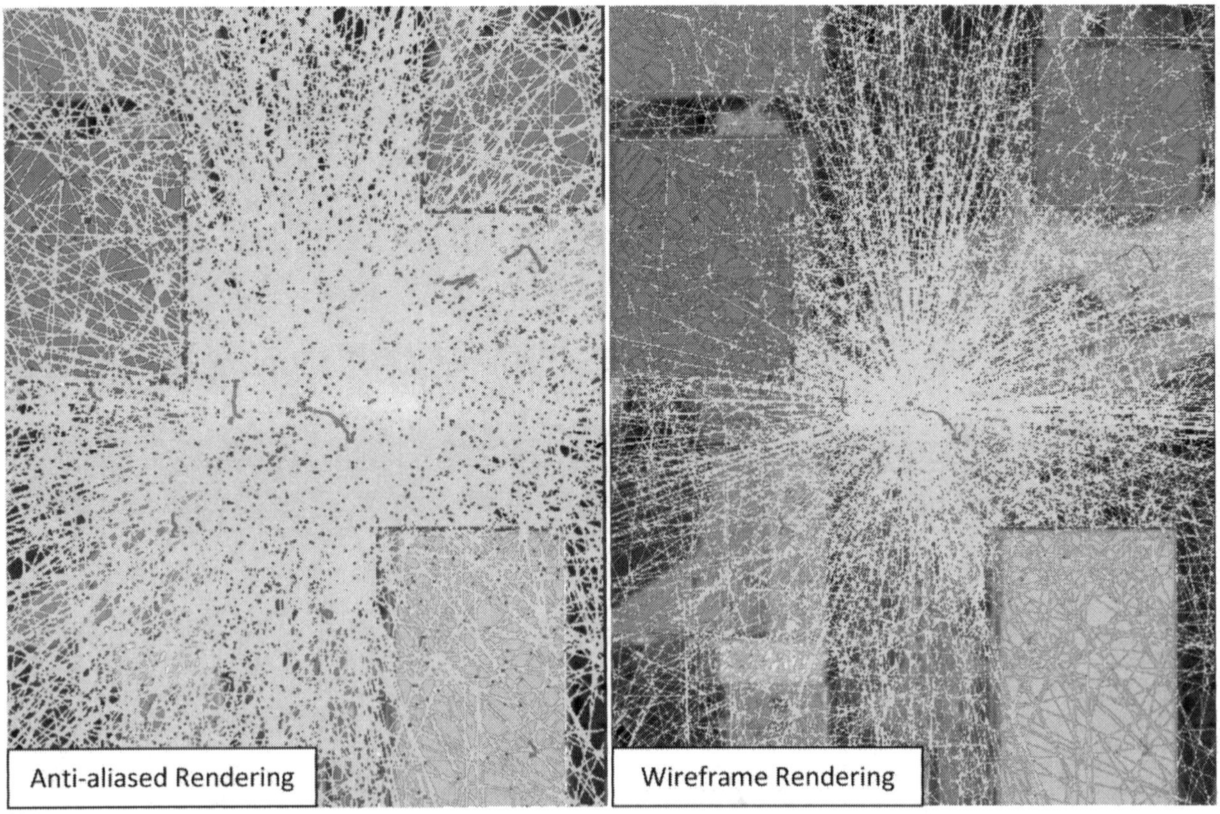

Figure 13. HepRep visualization results overlaid on background aerial photography, demonstrating interactions (left) and attenuation/shielding (right) at building surfaces

From the scoring mesh, Comma-Separated Value (CSV) tables, stored as text files, record the number of neutrons and gamma rays passing through each cell in the mesh. When the data format is converted into matrix form, MATLAB can be used to visualize the number of particles passing through different regions of the entire world environment, as shown in Figure 14. The scoring mesh can be defined with large cell sizes (low resolution results) to reduce run times and minimize data file sizes, but the amount of useful information is quite limited (as shown in Figure 14). However, with higher resolution outputs and more particles running in the simulation, maps such as those shown in Figure 15 could be used to show the effects of building locations on particle movement to a greater degree.

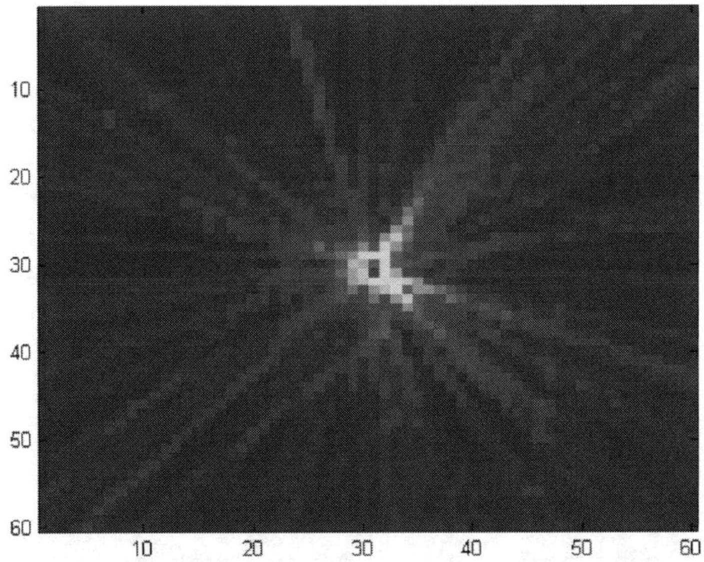

Figure 14. Neutron particle density for the world environment from scoring mesh

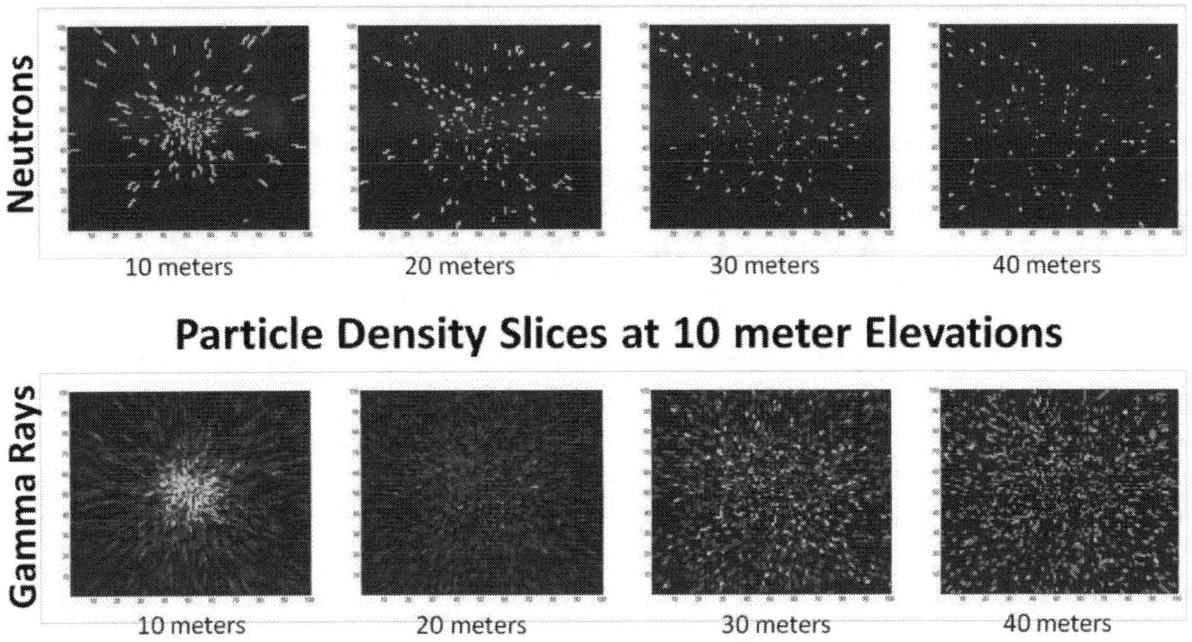

Figure 15. High resolution particle density maps at 10 meter elevation intervals

Figure 15 shows the particle density maps for a very high resolution scoring mesh over the entire simulation environment with a large number of source particles. The neutron density maps appear to show some locations at low elevations with low particle densities, with more random densities observed at higher elevations. This coincides with expected elastic interactions with the building materials, reducing the particle densities in building regions. However, we observe interesting results in the gamma ray density maps. The lowest elevation shows very high densities at the source, and the higher elevations appear to be more random, but the 20 meter elevation map shows an odd reduction in overall density. This is likely due to a coding error or an inconsistent visualization property (i.e. inconsistent color scale) since the effect is not reproduced at higher elevations, where a trend would be observed.

Conclusion

The GEANT4 simulation presented here is a very simplified version of a more capable system that could be implemented in post-detonation nuclear forensics practice. By comparing a measured spectral energy distribution with those estimated by the model with consideration for geographical location and proximity to the source, practitioners may develop better estimates of the source material composition. Also, visualization tools accompanying GEANT4 in HepRep, and the ability to output files which can be used in Matlab for visualization, bring useful tools for examining the effects of different radiation sources in urban environments and the effects of the nearby building environment on exposure just after detonation.

Future Work

Generally, this model is very limited in its current state due to its size, its less accurate building geometry, and its scoring system. As a 600 meter cube, the simulation produces a lot of data because it tracks so many particles in such a large space. By adjusting the limits of the world volume in the detector construction file, the executable could be made to be more efficient by no longer calculating particles below the ground elevation in the area. Similarly, changes could be made in the detector construction file to better define building geometries based upon interior dimensions, or simply using building shells rather than solid objects. Since almost no buildings are completely solid, the current model overestimates the effects the buildings have on shielding and other interactions. Additionally, there is no ground layer in this model, so ground reflection effects are not captured as well. Lastly, the scoring mesh currently employed in the model provides a very simple way of estimating the spectral energy distribution at any point in the world environment, but its low resolution and creation of large data output files hinders analysis. When the resolution is increased to the point where the data is more useful, such large amounts of data are created that it significantly increases run times and produces large CSV files which must still be reformatted for analysis in another application. Using a custom scorer would likely be the best approach because it could be tailored for providing higher-resolution measurements, albeit at specific locations rather than for the entire environment, and better-formatted output files for simpler analysis and visualization.

Acknowledgements

This research was funded by grants from the Department of Energy and Department of Homeland Security as part of the developing nuclear forensics program at the University of Utah.

References

1. American Physical Society and American Association for the Advancement of Science (2008), "Nuclear Forensics: Role, State of the Art, and Program Needs." Web. <http://iis-db.stanford.edu>
2. Glasstone, S. and Dolan, P. J. (1977), "The Effects of Nuclear Weapons, Third Edition." U.S. Government Printing Office, Washington, D.C. Web. <http://www.dtic.mil>
3. *Geant4 User's Guide for Application Developers, GEANT4 Materials Database, Appendix Section 9*. European Organization for Nuclear Research (CERN). December 2010, Web. <http://geant4.cern.ch>
4. Blatt, Harvey and Tracy, Robert J. (1997). "Petrology (2nd ed.)." New York: Freeman. p. 66. ISBN 0-7167-2438-3.
5. Punmia, B.C., Jain, Ashok Kumar and Jain, Arun Kr., "Basic Civil Engineering." Firewall Media, 2003, p. 33. Web. <http://books.google.co.in/books?id=sWZxu_muxyIC>
6. *ASTM A36 Steel plate*. MatWeb: Material Property Data. Web. <http://www.matweb.com/search/datasheet.aspx?matguid=afc003f4fb40465fa3df05129f0e88e6&>
7. Berger, M.J., Hubbell, J.H., Seltzer, S.M., Chang, J., Coursey, J.S., Sukumar, R., Zucker, D.S., and Olsen, K. National Institute of Standards and Technology, Gaithersburg, MD. *XCOM: Photon Cross Section Database (version 1.5)*. 2010. Web. <http://physics.nist.gov/xcom>
8. *ENDFPLOT-2.0*. Korea Atomic Energy Research Institute (KAERI) Nuclear Data Evaluation Lab. Web. <http://atom.kaeri.re.kr/ >
9. *Utah GIS Portal*. Utah Automated Geographic Reference Center. Web. <http://gis.utah.gov/>
10. *Buildings of Salt Lake City*. Emporis.com. Web. <http://www.emporis.com/application/?nav=worldmap_ci_bu&id=102671&lng=3>

New Nuclear Forensics Education Program and Research At the University of Utah

Tatjana Jevremovic, Ph.D.
Nuclear Engineering Program, The University of Utah,
Salt Lake City, UT

Abstract

To enhance sustainable viability for nuclear forensics related disciplines, such as analytical, geological and radio chemistry, nuclear physics and engineering, the University of Utah Nuclear Engineering Program in collaboration with the Departments of Chemical Engineering, Chemistry and Civil and Environmental Engineering, is establishing an interdisciplinary academic program to support and advance education in the field of nuclear forensics science, technologies and methodologies. Specifically, a new nuclear forensics graduate program track will be offered at the University of Utah. New courses will be developed as a synergy between various disciplines and related to the existing or planned research pertaining to nuclear forensics. As such, this new graduate program will represent the first academic nuclear forensics education in State of Utah. Development of the education program in nuclear forensics is directly connected to new research trusts pertaining to: computational nuclear forensics, radio-analytical chemistry, illicit trafficking of nuclear materials, nuclear fuel cycle and nuclear research reactors data base development & analysis for incidents involving nuclear materials and nuclear fuel trafficking.

Rationale for, and Description of, New Graduate Program in Nuclear Forensics at the University of Utah Nuclear Engineering Program

With the grant received from DOE/DHS in fall of 2010 the Utah Nuclear Engineering Program is building a new graduate program in nuclear forensics. The program at the MS and PhD level has been established and approved. The course work has been defined and will be offered starting fall 2012. The four core (required) courses are described as follows.

As described in GAO-09-527R April 30, 2009, the technical nuclear forensics represents "the analysis of nuclear or radiological materials that are intercepted or the radioactive debris and prompt output signals (such as gamma rays) produced by a nuclear event that can contribute to the identification of the sources of these materials and the processes used to create them." The analytical techniques would then determine the nature of nuclear events in case of nuclear device detonation or release of radiological materials by analyzing the radioactive debris samples (*postdetonation nuclear forensics*). Nuclear forensic techniques also could potentially be used to determine the origin of nuclear or radiological materials or devices seized prior to their use in a weapon (*predetonation nuclear forensics*). The U.S. government's predetonation nuclear forensics capabilities have been demonstrated in investigations on seized nuclear material from illicit smuggling operations (GAO-09-527R, 2009).

Predetonation and postdetonation nuclear forensics combines a wide variety of academic disciplines, such as radiochemistry, nuclear engineering and physics, isotope geochemistry, materials sciences and analytical chemistry. Radiochemistry represents the basis for many of the techniques needed in analyzing the radioactive debris from nuclear events. There are less than 300 experts with nuclear forensics skills and expertise in the country directly involved in nuclear forensics analysis; none of these experts basically work full time in nuclear forensics according to the NTNFC survey conducted in 2008. In addition, the majority of these scientists is close to retirement age. Due to decline in academic programs pertaining to the core disciplines of nuclear forensics, it is difficult to replace the retired force in time yet preserve the hands-on experience and knowledge gained from the past nuclear test programs.

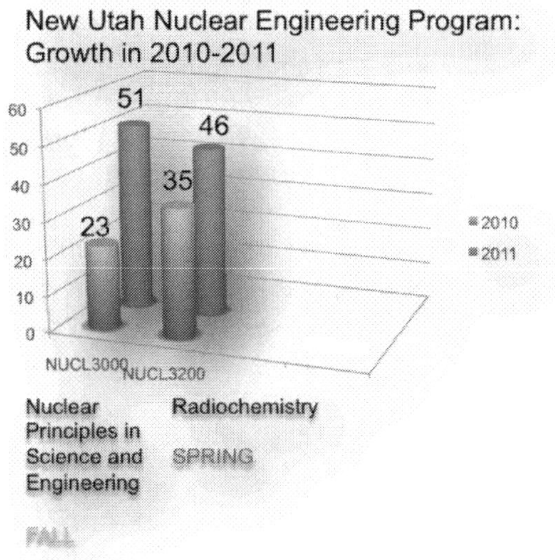

In the spring of 2010 the Utah Nuclear Engineering Program in collaboration with Chemistry Department started the course, "*Radiochemistry with Laboratory I*". The course attracted 35 students from various engineering and science disciplines. In spring of 2011 we had 46 students. It is worth mentioning that "*Radiochemistry with Laboratory I*" represents one of the four core courses for newly established Minor in Nuclear Engineering at the University of Utah. This course represents the base to expand into a new graduate level course, "*Analytical Nuclear Forensics*" that will be required as a prerequisite for "*Nuclear Forensics Technologies, Techniques and Methodologies*" course. This course will be focused at potential sources of forensics in fuel cycle. The steps from uranium ore mining to yellow cake production, UF_6 isotope separation, fuel pellet fabrication and fuel element fabrication will be discussed to identify the various points where terrorists are interested in nuclear materials along this processing route for both dirty bombs and nuclear weapons of mass destruction. In addition, chemical and nuclear processing methods used to produce radioisotopes will be discussed to identify the various points where terrorists are interested in these nuclear materials. Special attention will be given to the analytical chemical methods that can be used to measure the composition of nuclear decay products that would be present after a nuclear terrorism incident to determine the ultimate source of the nuclear materials that were used as well as tract the plume of radiation and fall out in the geographical region.

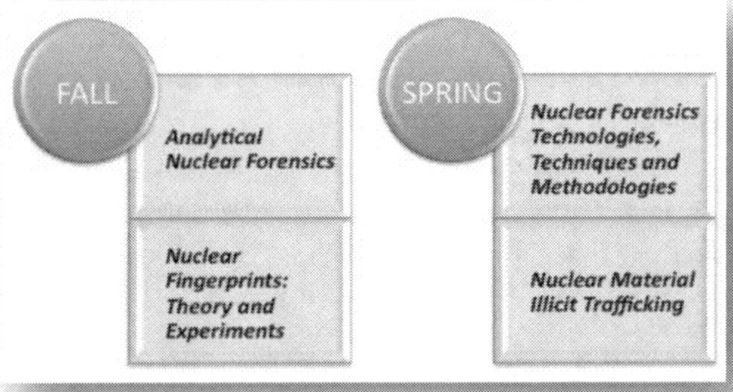

Between 1993 and 2007, there were 1,340 confirmed incidents of illicit trafficking and unauthorized activities involving nuclear and radiological materials worldwide (according to IAEA). Eighteen of these reported incidents involved nuclear material that could be used to produce a nuclear weapon. The worldwide illicit trafficking pathways employ civil infrastructure and particularly the global and domestic transportation networks in different modes such as air cargo, container ships, and freight trucks. The smugglers therefore study such structures in the countries of US and West Europe, for a potentially successful smuggling of nuclear and weapon-grade materials. The transportation sensor network design problem has striking similarities to the predetonation nuclear forensics. The significant similarities include: (1) both systems are complex dynamic spatial systems organized around a hierarchical network structure; (2) both carry flows that vary dynamically, with varying degrees of predictability from origins to destinations; and (3) both have origin-to-destination pairs and the traffic movement can be detected with point or point-to-point detectors. The main differences between the two networks lie in the levels of flow and the consequence of non-detection. Transportation networks see high volumes of flow and errors in demand detection perpetuated by sensors have minimal consequences. Therefore, based on the ongoing research and existing course, both on advancing sensor and traffic network design, we are developing a new graduate course on "*Nuclear Material Illicit Trafficking.*"

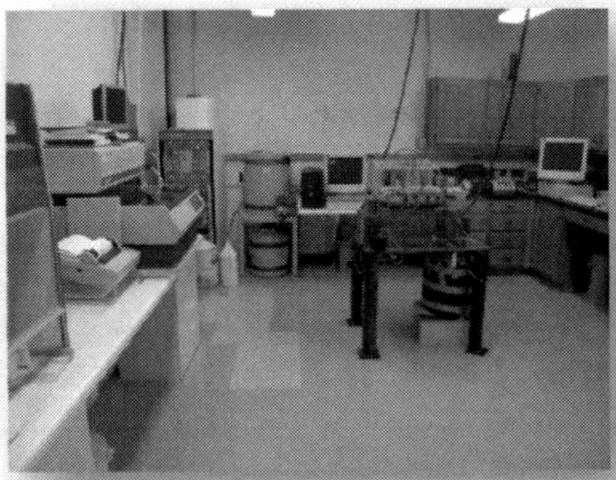

The necessity of detecting nuclear materials convertible to weapons (such as U^{233}, U^{235} and Pu^{239}) was recognized as early as the Manhattan Project. When a Congressional Committee asked Dr. J. Robert Oppenheimer how to detect nuclear materials hidden in baggage, he had a simple answer "*use a screwdriver.*" Oppenheimer referred to the fact that well-shielded fissionable materials are almost impossible to detect due to their weak nuclear signatures.

More recently, neutron and gamma interrogation techniques and nuclear fluorescence have provided very exciting opportunities for reliable, non-invasive detection. But, although nondestructive techniques have been developed to detect the fissionable materials, currently installed detection systems do not meet expectations for practical fast and accurate detection. Hence growing efforts are devoted to developing new sensors and sensing systems. Fast and accurate detection of illegally transported nuclear material is crucial in preventing potential terrorist attacks against the US; strengthen the global system of trade; enable new technology and possibly international standards for the global nuclear fuel cycle. Kilograms of fissile plutonium or enriched U^{233} or U^{235} are needed to construct a nuclear bomb. Thus we may assume two general scenarios; one in which small amounts of fissile materials are transferred over a long period of time (in personal baggage) or a larger amount is transported in trucks or containers. Despite growing technological developments, the underlying measurement / detection philosophy remains unchanged representing significant impediment to accurate and fast detection of smuggled nuclear materials. Our current research is focused at the development of fundamentally new approach in detecting weak signals; funded by the NSF/DNDO agency, our computational system development produced a new framework suitable for nuclear detection simulation and fast analysis of complex signals. This project's outcomes are used as a base to develop a new graduate course, *"Nuclear Fingerprints: Theory and Experiments"*. The experiments will be developed to emphasize some of the aspects of nuclear signals, i.e. nuclear fingerprints. The course will include computational simulations and the use experiments to benchmark such simulation models.

Establishing New Research in Nuclear Forensics at the University of Utah Nuclear Engineering Program

Although started in late fall of 2010, the team of four professors and a few students by the end of summer of 2011, have been able to produce a number of new results and publish numerous papers (see the references). The following is a brief overview of the new research areas developed and under further development at the Utah Nuclear Engineering Program:

Development of a co-simulation tool for nuclear fuel processing is underway to perform radiochemical plant material and energy balance calculations while at the same time simulating the nuclear criticality. As chemical engineers and nuclear scientists, we recognize that a tool of this type is needed as the nuclear industry moves forward, to avoid accidents, and to save the process engineer time and to save the industry money. With the resurgence of the nuclear industry in the USA there is a need to modernize the tools used for nuclear fuel processing and reprocessing. Today's modern tools for process design include dynamic and steady state mass and energy balance programs like ASPEN Plus and ProSim that make mass and energy balances for a series of unit operations, e.g. reactors, separators of various types that make up a nuclear fuel processing plant. In the design mode this software develops a detailed design of each unit operation giving specific dimensions for the equipment. Some programs, e.g. ASPEN Plus, account for populations of particles as well. The problems with the utilization of these tools for nuclear material processes are that:

1) They lack a physical property data base populated with actinide and lathanide materials properties, however, that can be remedied by purchasing proprietary data bases and tailoring these with experimental data.

2) They are more often than not aqueous based processes while the programs were developed for organic chemical systems. This is not an insurmountable problem but may require proprietary data bases for aqueous species, ions and complexes in solution, like the OLI data base in Aspen for example.

3) They lack physical property data based on isotopes making calculation of unit operations based upon isotope separations impossible.

4) They lack simultaneous calculations for nuclear criticality and radiation dose calculations.

Currently there is not a straight-forward way to analyze the criticality of radiochemical processes in one software package or one that is linked to a mass and energy balance software package. In order to design a chemical and nuclear process accurately, an integrated process simulation tool that has the capability to handle material and energy balances, isotopes, *and* nuclear interactions is needed which is the focus of this is proposed research. The first step is to collect the physical properties of the actinides and lanthanides to populate a proprietary data-base that will allow a process simulator program to perform calculations on actinide processes. The process simulator is ASPEN Plus because it has a dynamic simulator that is needed for ion exchange processes and start up and shut down where most nuclear accidents occur. To perform isotope separations the process simulator will need to be augmented for time dependent physical separation methods. By far the most important improvement in process simulator needed for the nuclear industry is the addition of radiation dose and criticality calculations for all unit operations in a nuclear chemical plant. It must be the focus of future work so that nuclear processes do not have accidents. The parameters needed from the process simulation software to perform criticality calculations are mass, concentration, geometry of the system. The parameters needed for criticality calculations that are not presently in the process simulator property data-base are neutron cross sections as a function of neutron energy and nuclear interactions. The compilation of nuclear cross sections for scattering, absorption, and fission as well as transition probabilities and the energies for the alpha, beta and gamma rays that decay off needs to be created to allow the software to perform criticality calculations and dose prediction. Criticality and dose prediction allows safety analysis to be carried out at the same time as process design to avoid an inadvertent nuclear accident. This can be done by adding this capability to the process simulation software as an additional linked calculation after the mass and energy balance calculations have converged. Criticality calculations involve the solution to the neutron transport equation specifically the effects of fissile material density in the solution on the thermal utilization factor, material and geometric buckling. The nuclear cross sections for scattering, absorption, and fission as well as transition probabilities must be compiled in the physical property data-base as a function of energy. Gamma radiation dose calculations are a direct result of the neutron fluxes that result at the boundary of the equipment. Various numerical approaches to this problem including finite difference, finite element and Monte Carlo calculations have been used in the past. We propose to use Quadrature Method of Moments (QMOM) for this problem as it is an efficient way to solve population balance equations within computation fluid dynamics methods and we have played an important role in their development, validation, verification and uncertainty quantification for particulate processes. This effort will use this understanding of QMOM and tailor it for the solution to neutron populations as predicted by the neutron transport equation. These solutions can also determine the nuclear energy produced that should be coupled back into the process energy balance.

Development of models, algorithms and tools for detecting and interdicting source-to-target patterns of nuclear material illicit trafficking. In order to ship the nuclear materials from a source location with special nuclear material to a target city, the smugglers have to employ the global and domestic transportation network in different modes such as container ships and freight trucks. We will study the radiation sensor network design problem based on an interdisciplinary approach from multiple fields including: nuclear physics, nuclear engineering, transportation engineering, optimization theory, and information theory. An information-theoretic framework will be developed to assist the decision-maker in optimizing the network-wide information gain, system reliability, and total operating cost, by deploying and networking fixed and mobile radiation sensors at different critical locations. The open-source software tools is under development to provide other teachers and researchers with free simulation packages that they can further develop and improve to better fit their needs. In particular, the following critical and theoretically challenging issues will be addressed:

1) How to locate additional mobile or re-locatable radiation detection sensors to improve system redundancy;

2) How to quantify the system-wide economic and risk impacts of miss rates and false alarm rates from different types of existing and new sensors;

3) How to design mobile sensor monitoring routes inside a container terminal to enable collaborative detection and spatial coverage of the sensor nodes.

The integrated simulation framework under development will be able to estimate the error range for each source-to-target flow pair, and their possible error correlation, through a large number of complex but realistic simulated instances under a given sensor network design plan. Different predicted smuggling flow patterns will be considered to ensure the robustness of the suggested sensor location plan. The simulation model is envisioned to be built on a high-performance computational environment to facilitate the most efficient use of emerging parallel hardware.

Based on the known **iMASS** (intelligent Model-Assisted Sensing System) project funded by the NSF/DNDO from 2007 to 2010we are developing an innovative methodology for fast and accurate interrogation of nuclear materials. The novelty of the **iMASS** approach originates from the synergism of smart and customizable Monte-Carlo simulation, advanced signal processing and intelligent information management.[1] The **iMASS** project was executed with great successes. Major goals set in the original scope of work have been accomplished and resulted in some very valuable findings, the most significant one being the ultra upgradability and extensibility of **iMASS** approach. The **iMASS** emerged as an innovative algorithm based on an intelligent information management engine that interactively and adaptively synchronizes the sensing elements with signal processing units to achieve fast and accurate detection of nuclear materials. After three years of research&development, **iMASS** project culminated in reaching its original goal of demonstrating the capabilities using Nuclear Resonance Fluorescence (NRF) as a test bed set of signals. In essence, **iMASS** integrates the following three components into a powerful, accurate yet timely efficient detection system:

- ***Advanced signal processing-*** **iMASS** consists of a set of advanced signal processing algorithms based on statistical and heuristic methods for detection with quantifiable probability and confidence level[2] [3].
- ***Smart Monte-Carlo simulation-*** **iMASS** integral part is a smart Monte-Carlo simulation package based on GEANT4 capable of reconfiguring itself in response to user's demands[4] [5].
- ***Intelligent information management*-** **iMASS** is supported by a highly sophisticated intelligent information management engine, which collects and integrates measurement data from detectors and performs an anticipatory decision making[6].

The expanded research includes two themes: upgrading **iMASS** for various applications, and extending the scope of **iMASS** by applying it to material tracking. **iMASS** represents a novelty concept that emphasizes and coordinates collaborative efforts among components to maximize performance at system level. Based on this philosophy, **iMASS** is designed as an open architecture in

---

[1] Perry, J., Xiao, S., & Jevremovic, T. (2009). iMASS: Evolved NRF Simulations for More Accurate Detection of Nuclear Threats. In *Proceedings of the 17th International Conference on Nuclear Engineering*. Brussels, Belgium.

[2] P. Forsberg, V. Agarwal, J. Perry, R. Gao, L. H. Tsoukalas and T. Jevremovic. PEAKSEEK: A Statistical Processing

[2] P. Forsberg, V. Agarwal, J. Perry, R. Gao, L. H. Tsoukalas and T. Jevremovic. PEAKSEEK: A Statistical Processing Algorithm for Radiation Spectrum Peak Identification. *21st Int. Conf. Tools of Artificial Intelligence, New Jersey, 2009.*

[3] A. Pantelopoulos, M. Alamaniotis and T. Jevremovic, S. Chung, S. Park, N. Bourbakis. LG-Graph based Detection of NRF Spectrum Signatures: Initial Results and Comparison. *21st Int. Conf. Tools of Artificial Intelligence, New Jersey, 2009.*

[4] Perry, J., Xiao, S., & Jevremovic, T. (2008). iMASS: Computational NRF Spectra Signal from Geant4. In *2008 20th IEEE International Conference on Tools with Artificial Intelligence* (pp. 541-546). Dayton, OH: IEEE. doi: 10.1109/ICTAI.2008.94

[5] Young, J., Perry, J., & Jevremovic, T. (2010). LightGrid - An Agile Distributed Computing Architecture for Geant4. *Nuclear Instruments and Methods in Physics Research Section A: Accelerators, Spectrometers, Detectors and Associated Equipment*, 614(1), 154-158.

[6] Alamaniotis, M., Young, J., Perry, J., Xiao, S., Agarwal, V., Forsberg, P., et al. (2009). Engineering Solution to Nuclear Material Detection at Ports: Introducing the Novel iMASS Paradigm. In *Proceedings of 21st IEEE International Conference on Tools with Artificial Intelligence*. Newark, NJ.

which every component has ample room for further improvements. When a facility owns any type of nuclear material, it is important to be able to track these materials in the interest of safety and security for the facility and the surrounding area. It is crucial to make sure that a facility can track these materials both in transport between facilities (as the NRC tracks) and to track within the facility itself. Part of this would be tracking how moving these materials could affect the people in the area during and after the transport of the material. The effects of the movements/tracking/trafficking of such materials is now analyzed in terms of radiation level change in predefined environment. We consider certain amounts of typical nuclear materials that move (scenario of insiders intending to take these materials out from the predefined environment) and show the spatial change of the radiation levels changing in time in regard to the known background.

References

1. Przybyla, J., Taylor, J., and Zhou, X. (2010) Locating Sensors for Detecting Source -to-Target Patterns of Special Nuclear Material Smuggling: A Spatial Information Theoretic Approach. Sensors. 10(9), 8070- 8091. published, 08/27/2010. http://www.mdpi.com/1424-8220/10/9/8070/
2. Schwerdt, I., Levinthal, J., Dances, C., Choe, D., Yang, H., and Jevremovic, T. (2011) Developing Experimental and Computational Nuclear Forensics Signals of Fast Neutron Irradiation Facility At TRIGA Reactor, in Proceedings of the 2011 AIChE Annual Conference, American Institute of Chemical Engineers, http://www.aiche.org
3. Voyles, A., Yang, H., and Jevremovic, T. (2011) GEANT4 Simulation of Irradiation Facilities and Neutron Sources At University of Utah TRIGA for Nuclear Forensics and Detection, in Proceedings of the 2011 AIChE Annual Conference, American Institute of Chemical Engineers, http://www.aiche.org
4. Taylor, J., Yang, H., Zhou, X., and Jevremovic, T. (2011) GEANT4 Model for Nuclear Forensic Analysis of Nuclear Detonation In Urban Environments, in Proceedings of the 2011 AIChE Annual Conference, American Institute of Chemical Engineers, http://www.aiche.org
5. Sherman, T., Yang, H., Choe, D., and Jevremovic, T. (2011) Monte Carlo Simulation of Tritium Generation Signals In the Utah TRIGA Reactor for Nuclear Forensics, in Proceedings of the 2011 AIChE Annual Conference, American Institute of Chemical Engineers, http://www.aiche.org
6. Andrew Voyles, Haori Yang and Tatjana Jevremovic, GEANT4 – Based Modeling of Research Reactor Irradiation Facilities for Nuclear Forensics, Presented at the 2nd National Conference in Advancing Tools and Solutions

for Nuclear Material Detection, Salt Lake City, Utah, May 2-3, 2011.

7. Todd Sherman and Tatjana Jevremovic, Role of Neutron Activation Analysis in Nuclear Forensics Program at the University of Utah, Presented at the 2nd National Conference in Advancing Tools and Solutions for Nuclear Material Detection, Salt Lake City, Utah, May 2-3, 2011.
8. Greg Moffitt, Charles Grissom and Tatjana Jevremovic, Radiochemistry for Nuclear Forensics at the University of Utah, Presented at the 2nd National Conference in Advancing Tools and Solutions for Nuclear Material Detection, Salt Lake City, Utah, May 2-3, 2011.
9. Jeffrey Taylor, Xuesong Zhou, Haori Yang and Tatjana Jevremovic, Nuclear Forensics in Urban Settings with the use of GEANT4, Presented at the 2nd National Conference in Advancing Tools and Solutions for Nuclear Material Detection, Salt Lake City, Utah, May 2-3, 2011
10. Ian Schwerdt, Joseph Levinthal, Chris Dances, Dong-OK Choe, Haori Yang and Tatjana Jevremovic Experimental Measurements and GEANT4 Modeling of fast Neutron Irradiation Facility at TRIGA for Nuclear Forensics, Presented at the 2nd National Conference in Advancing Tools and Solutions for Nuclear Material Detection, Salt Lake City, Utah, May 2-3, 2011.
11. Taylor, J. and Przybyla, J. (2011) Spatial-Information Approach to Analyzing and Planning Distributed Transportation Security Systems, National Security Innovation Competition Proceedings Report, National Homeland Defense Foundation, published, 05/19/2011. http://www.nhdf.org

(554a) Integration of a Membrane Based Extraction Process Into a Lignocellulosic Biorefinery

David Grzenia, Chemical and Biological Engineering, Colorado State University, Fort Collins, CO, Dan Schell, National Renewable Energy Laboratory, Golden, CO and **S. Ranil Wickramasinghe**, Ralph E Martin Department of Chemical Engineering, University of Arkansas, Fayetteville, AR

In the future, it is anticipated that multi-product biorefineries will be used to convert lignocellulosic biomass to fuels and chemicals, in which, it may be possible to use membrane-based separations to recover fuels and various products and to improve energy efficiency. To biochemically convert biomass a pretreatment step is required to increase the susceptibility of the biomass to enzymatic cellulose hydrolysis. However, many pretreatment processes release not only fermentable sugars but other compounds that are toxic to the fermenting microorganism. In this study, we investigated the feasibility of using membrane extraction for removal of acetic acid, sulfuric acid, furfural, HMF and other toxic compounds from dilute acid pretreated corn stover to reduce its toxicity. Acetic acid in its protonated form was extracted from the liquor portion of a corn stover hydrolysate across a membrane into an organic phase consisting of oleyl alcohol and Alamine 336, the latter a tertiary amine containing aliphatic chains of 8-10 carbon atoms. Other toxic compounds such as HMF and furfural are also co-extracted into the organic phase.

The sugars in hydrolysates that were detoxified using our membrane extraction process as well as conventional ammonium hydroxide detoxification, were fermented to ethanol by glucose-xylose fermenting *Zymomonas mobilis* 8b. The best ethanol yields and rates were achieved for hydrolysates using membrane extraction with oleyl alcohol as the organic phase followed by ammonium hydroxide conditioned hydrolysates. Hydrolysates that were detoxified using membrane extraction with octanol as the organic phase were unfermentable because highly toxic octanol was found in the hydrolysate. Our results indicate that membrane extraction may be an effective membrane detoxification technology.

See more of this Session: Separation Processes In Biorefineries I: Membrane and Chromatographic Separations
See more of this Group/Topical: International Congress on Energy 2011

Title: Modified Anopore^{TM} membranes for the microfiltration of cellulose hydrolysis mixtures

Yong-Hong Zhao and Daniel F. Shantz

Artie McFerrin Department of Chemical Engineering, Texas A&M University, College Station, TX 77843-3122, United States

ABSTRACT:

Three classes of nonfouling substances, i.e., poly[poly(ethylene glycol) methyl ether methacrylate] (POEM), poly(sulfobetaine methacrylate) (PSBMA), and perfluoroalkyl tagged poly(ethylene glycol) surfactants (fPEG), were covalently tethered onto the surface of Anopore^{TM} membranes to form a series of organic-inorganic hybrid membranes. These membranes were then used for the microfiltration of cellulose hydrolysis mixtures. It was found that the fluxes of the three synthesized hybrid membranes could be largely recovered by washing with methanol and water after the microfiltration of cellulose hydrolysis mixtures, showing much better antifouling properties than the bare Anopore^{TM} membrane. Furthermore, the stability of the membranes in acidic environment was enhanced significantly as compared to the unmodified membranes when the membrane surface was grafted with a thick enough polymer layer. The synthesized membranes were also characterized by scanning electron microscopy (SEM), Fourier transform infrared spectroscopy (FTIR) and X-ray photoelectron spectroscopy (XPS), which were used to understand the morphology and chemical compositions of the membrane surfaces, and the results of these methods correlate with the microfiltration data presented. Mathematical modeling of the microfiltration data indicates that the pore constriction model best describes the fouling behavior of the bare Anopore^{TM} membrane, while the cake filtration model fits the data most satisfactorily for the three groups of hybrid membranes. The current work presents an alternative approach to membrane design for biomass related separations.

Keywords: hybrid membranes; biomass separation; antifouling; surface modification

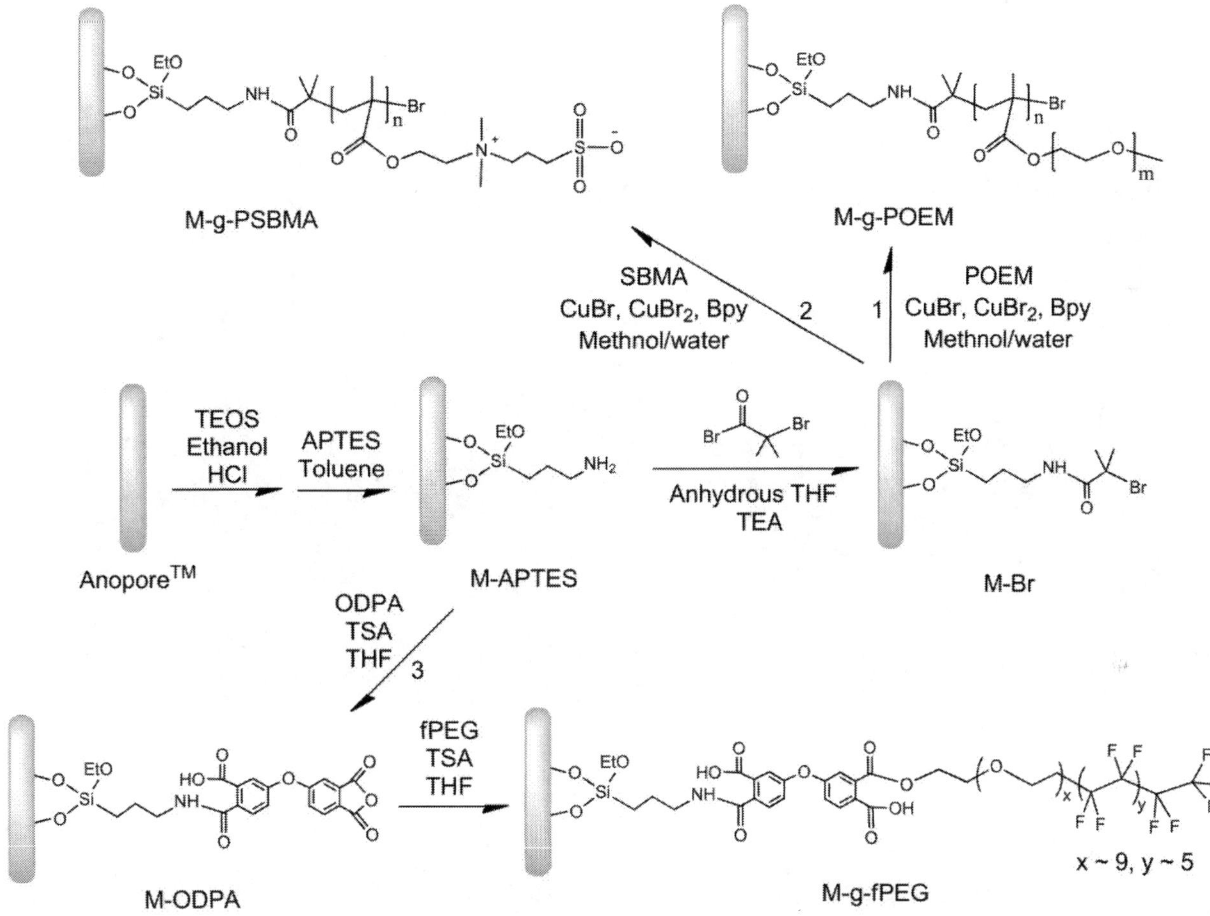

Surface modification protocols for forming hybrid membranes

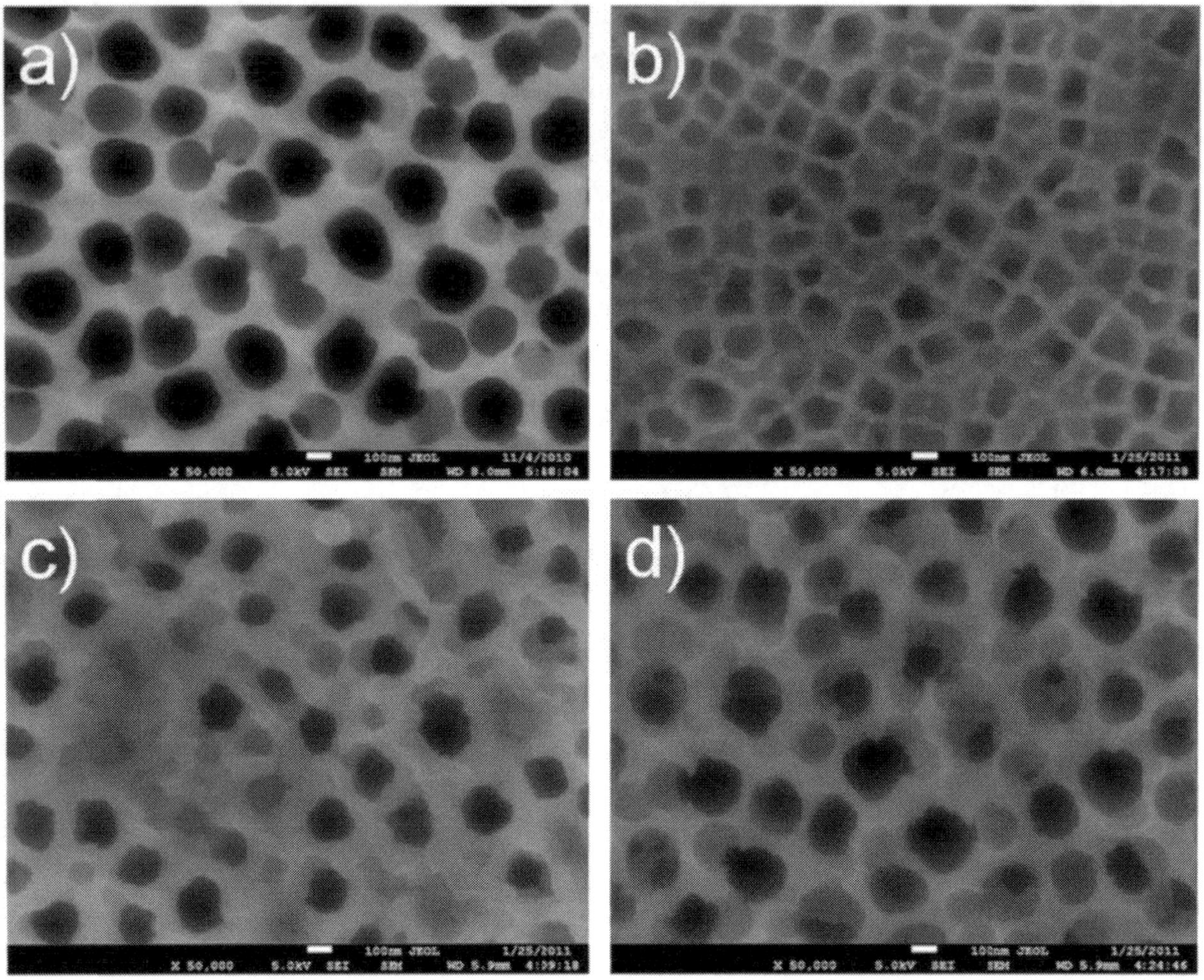

SEM images for a) bare Anopore™ membrane; b) M-g-POEM-24h;

c) M-g-PSBMA-24h and d) M-g-fPEG.

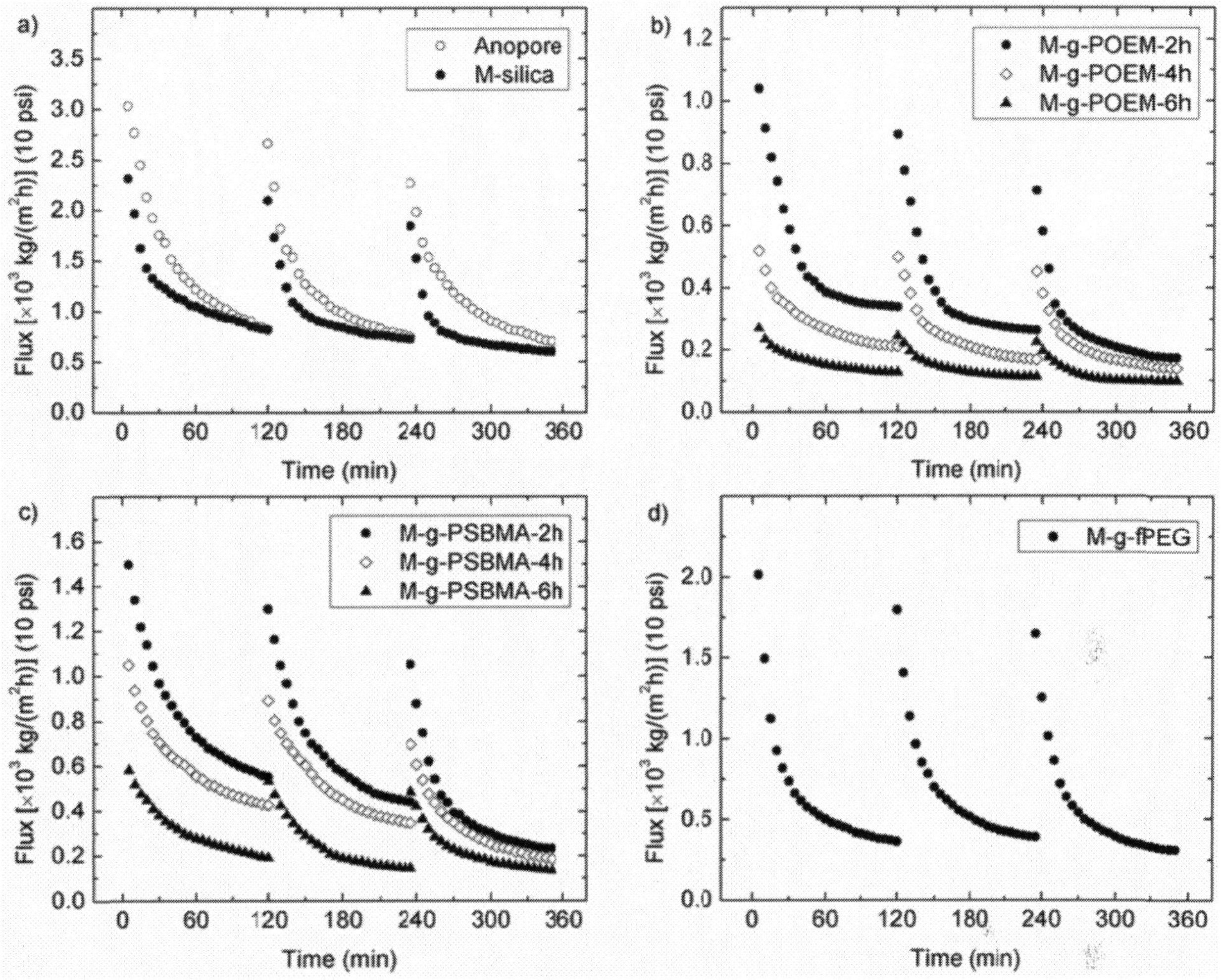

Flux measurement for a) Anopore™ and M-silica; b) M-g-POEM; c) M-g-PSBMA and d) M-g-fPEG. The solution is cellulose hydrolysis mixtures at pH 7.

(554c) Polymer Enhanced Microfiltration of Lignocellulosic Hydrolyzates

Rakesh Yasarla, State University of New York, Syracuse, NY and **Bandaru V. Ramarao**, Paper and Bioprocess Engineering, Empire State Paper Research Institute, Syracuse, NY

Lignocellulosic hydrolyzates produced by hydrothermal treatment of wood chips contain lignin and carbohydrates in colloidal forms that can interfere with downstream processing. Separation of such materials is necessary to improve the fermentability of sugars into biofuels and other bioproducts. Flocculation of such hydrolyzates prior to microfiltration improves their filterability. The effect of prior flocculation on filtration resistance, steady state fluxes and the influence of solid concentration and temperature are discussed.

See more of this Session: Separation Processes In Biorefineries I: Membrane and Chromatographic Separations
See more of this Group/Topical: International Congress on Energy 2011

(554d) Acetic Acid Removal From the Hydrolysate Using Reactive Extraction In a Membrane Extractor/Reactor

Rahul Patil[1], Glenn Lipscomb[1] and Frederick F. Stewart[2], (1)Department of Chemical and Environmental Engineering, University of Toledo, Toledo, OH, (2)Interfacial Chemistry, Idaho National Laboratory, Idaho Falls, ID

Production of bioethanol from lignocellulosic biomass requires pretreatment of the biomass to hydrolyze hemicellulose, separate lignin and reduce cellulose crystallinity. During the pretreatment by dilute acid hydrolysis, several grams of acetic acid per liter of hydrolysate is generated. Acetic acid needs to be removed from the hydrolysate since it can dramatically reduce the efficiency of the subsequent fermentation step by inhibiting enzyme activity. Lime treatment, the current method of hydrolysate detoxification, consumes significant quantities of calcium hydroxide, involves solids handling and does not produce any valuable products.

We propose to remove the acetic acid with an alternative process whereby acetic acid is extracted into 1-octanol in a membrane extractor and reacted with 1-octanol *in situ* using a membrane containing an embedded catalyst to yield 1-octyl acetate. Acetic acid concentration in the hydrolysate will be reduced from 12.5 g/L to < 1 g/L. The product ester will be removed periodically from the organic phase. The product has very good solvent properties and can be sold to improve process economics.

In our study, we prepare membranes with a suitable catalyst (initial efforts utilize strong acid ion exchange resins) embedded in the membranes to achieve extraction and reaction in a single step. The effects of catalyst concentration, operating temperature and flow rates are investigated to optimize the performance of the process. Thus, the membrane extractor/reactor removes acetic acid from the hydrolysate and produces a valuable by-product in an environmentally friendly process.

See more of this Session: Separation Processes In Biorefineries I: Membrane and Chromatographic Separations
See more of this Group/Topical: International Congress on Energy 2011

(554e) Evaluation of Membrane Separations Opportunities within Lignocellulosic Biorefining Processes: Tailoring Membrane Properties

Jennifer M. Leberknight[1], Amitkumar Gautam[2] and Todd J. Menkhaus[2], (1)Chemical and Biological Engineering, South Dakota School of Mines and Technology, Rapid CIty, SD, (2)Chemical and Biological Engineering, South Dakota School of Mines and Technology, Rapid City, SD

Evaluation of Membrane Separations Opportunitieswithin Lignocellulosic BiorefiningProcesses: Tailoring Membrane Properties

Jennifer Leberknight, Amitkumar Gautam, and Todd J. Menkhaus

South Dakota School of Mines and Technology

Department of Chemical and Biological Engineering

501 East Saint Joseph Street

Rapid City, SD 57701

Tel: (605)-394-2422

Email: Todd.Menkhaus@sdsmt.edu

Membraneseparations can be integrated into a biorefinery at many locations and offertremendous potential to reduce water usage and energy consumption, decreaseproduction costs, simplify purification operations, and offer opportunities forrecovery of valuable co-products. Here weexplore five locations for integration of membrane technology into the basic biochemicalproduction process for fuels and/or chemicals.The biomass is initially pretreated and enzymatically hydrolyzed toconvert the cellulose and hemicellulose into fermentable sugars, after which asolid-liquid clarification operation to remove solids (primarily lignin) may beemployed (Membrane Separation 1). Theliquid stream can then be further separated to recycle the hydrolysis enzymes (MembraneSeparation 2) and/or to concentrate the sugars by reverse osmosis (Membrane Separation3). Increased fermentation efficiency couldbe achieved by integrating a membrane step after fermentation to separate andrecycle the fermentation organism (Membrane Separation 4), while membraneextraction or pervaporation can be used for *in-situ*product recovery (Membrane Separation 5).Within the various membrane processes, fouling is a major challenge toovercome. We have correlated membranefouling tendencies within the different membrane separation locations tomeasureable properties of the membrane and foulants such as ionic charge,hydrophilicity, size, and surface roughness to determine why membranesfoul. Through a combination of analytical andmodeling approaches we postulate fouling mechanisms and identify methods tomitigate fouling. Finally, simple empiricalmodels using variable concentrations of fouling components and quantitativemembrane properties have been developed to help establish predictive performanceabilities. As detailed examples, bothmicrofiltration (Separations 1 and 4) along with pervaporation of ethanol froma fermentation broth (Separation 5), were evaluated. Both selectivity and fluxmust be considered in new fouling-resistant membrane designs. The unique soluble compounds (phenolics,furans, and organic acids) in the lignocellulosic derived feeds greatly complicatethe fouling and separation, as will be discussed.

See more of this Session: Separation Processes In Biorefineries I: Membrane and Chromatographic Separations
See more of this Group/Topical: International Congress on Energy 2011

(554f) Membrane Filtration of Lignocellulosic Hydrolyzates: Comparison of Crossflow and Dead-End Microfiltration

Bhavin Bhayani, Empire State Paper Research Institute, Syracuse, NY and Bandaru V. Ramarao, Paper and Bioprocess Engineering, Empire State Paper Research Institute, Syracuse, NY

Depleting oil reserves have encouraged the development of technologies to harness the natural resources to sustain the energy needs. Renewable lignocellulosic biomasses have been focus of many studies but advances in the purification and recovering of the products from the product medium has been very limited. Amongst many separation processes membrane separation offers better prospects to simplify purification operations, offers valuable by-products, decrease energy consumption and reduce water requirements. A membrane separation system can be either crossflow or dead end and both have unique advantages and disadvantages. While Cross flow filtration offers higher feed velocity i.e. cross flow velocity and better control over cake layer development, dead end filtration can result in energy savings. The focus of this study is to characterize the flow and the effect of cake layer development under different pressures and flow rate. A detailed comparison of flow characteristics has been studied. It is also found that the temperature of the hydrolyzate influences separation in a very significant manner.

See more of this Session: Separation Processes In Biorefineries I: Membrane and Chromatographic Separations
See more of this Group/Topical: International Congress on Energy 2011

(554g) Effect of Column Temperature On HPLC Separation of Acetic Acid and Levulinic Acid

Rui Xie and Maobing Tu, Auburn University, Auburn, AL

High performance liquid chromatography (HPLC) is a routine method in the analysis of degradation products in the prehydrolysate or the hydrolysis liquor of lignocellulosic biomass. Cation exchange column chromatography with a metal counter ion has been used for the quantitative analysis of carbohydrates with water as the eluent. A cation exchange column with exchangeable hydrogen ion (H^+) has been employed in quantification of organic acids and furan derivatives with acidic water as the mobile phase. In particular, the Aminex cation-exchange HPX-87H column has been developed for determining organic acids, furan derivatives and alcohols. Although this column has been broadly used in biomass analysis, potential issues with co-elution of target analytes have been reported by several groups. Scarlata and Hyman (2010) suggested that Aminex HPX-87H column was not suitable for biomass sugar analysis because xylose, galactose and fructose are co-eluted together, and glucose and mannose are co-eluted together. The co-elution of acetic acid and levulinic acid has not been addressed indicated the value of acetic acid in biomass hydrolysate was not reliable at high severity due to continuous degradation of furfural and HMF to levulinic acid.

As one of the principle components in hemicelluloses, acetic acid always has a noticeable amount in the hydrolysate of lignocellulosic biomass. Characterization of this organic acid becomes even more important during fermentation due to its inhibition to the microorganisms. Levulinic acid, one of the normal degradation compounds of HMF, is present in the hydrolysate at low concentrations. The separation of these two acids could be easily ignored due to their co-elution in the hydrolysate liquor. Moreover, the separation performance of HPLC column gradually deteriorates during its lifetime worsening co-elution of these two acids. Solutions to improve the separation of these organic acids included the use of a dual column system, employment of a capillary electrophoresis to separate and quantify the concentration of acetic acid and levulinic acid from softwood hydrolysate. However, no method has been developed to resolve co-eluting acetic acid and levulinic acid by optimizing the HPLC chromatographic conditions. In the present study, the improvement of HPLC separation of acetic acid and levulinic acid on the Aminex HPX-87H column was investigated by varying column temperature, flow rate, and concentration of sulfuric acid in the mobile phase. Resolution and capacity factors were used as indicators for measuring the effectiveness of separation using pure standard compounds as well as real biomass hydrolysate from woody biomass.

See more of this Session: Separation Processes In Biorefineries I: Membrane and Chromatographic Separations
See more of this Group/Topical: International Congress on Energy 2011

(555a) Imidazole and Imidazole-Amine Hybrid Solvents for CO2 Capture

Jason E. Bara and Matthew S. Shannon, Chemical & Biological Engineering, University of Alabama, Tuscaloosa, AL

Imidazoles and imidazole-amine hybrids are unique classes of organic solvents that feature properties similar to aqueous amine solvents and ionic liquids (ILs), both of which have received a great deal of attention for CO2 capture applications. However, to date, the utility of the imidazole platform has not been considered, despite its potential advantages over both ILs and aqueous amine solvents. Imidazole-based solvents possess many desirable features for CO2 capture processes including low viscosities (< 5 cP), very low vapor pressures (<< 1 mm Hg at ambient temperature), high CO2 capacities (> 100 g/L), and a much lower cost relative to ILs. When used in combination with an amine, the inherent basicity (H + accepting capability) of the imidazole platform allows the entire solvent volume to be reactive toward CO2, producing a highly effective solvent is obtained that can be used under the low partial pressures found in post-combustion flue gas (2 psia). Imidazole-amine hybrids are able to exceed the 1:2 (mol: mol) stochiometric limitations of CO2-amine chemistry in a non-aqueous solvent, and are found to absorb up to 1:1 (mol:mol) CO2 per amine, while maintaining viscosities as low as 10 cP in a highly CO2-rich state. Because these solvents exhibit very low volatility under both CO2-lean and CO2-rich conditions, solvent regeneration energy can be minimized (i.e. minimal losses due to latent heat).

At the benchscale, we have demonstrated the viability of imidazoles and imidazole-amine hybrids for CO2 capture. By systematically varying the structures of the imidazole and/or amine components, we have characterized the key physical (viscosity, density), thermodynamic (vapor-liquid equilibrium and heat of reaction) and chemical properties needed to model solvent performance within CO2 capture processes, and have developed structure-property relationships for selecting imidazoles and imidazole-amine hybrid solvents likely to have the greatest utility.

The discussion will focus on our work relating to solvent chemistry, characterization of physical and thermodynamic properties, structure-property relationships, and our associated modeling efforts. Consideration will be given to how these solvents might be scaled from the lab to the field with an emphasis on potential impacts of water, SO2 and other flue gas contaminants as well as process improvements that might be achieved through modification of the conventional absorber-stripper process design for a non-aqueous solvent.

See more of this Session: Solvent-Based Carbon Capture Processes
See more of this Group/Topical: Sustainable Engineering Forum

(555b) Towards Industrially Viable Non-Aqueous Amine Solvent Systems

Jackson Switzer[1], Kyle Flack[2], Amy Rohan[1], Swetha Sivaswamy[1], Elizabeth Biddinger[1], Manish Talreja[1], Manjusha Verma[2], Pamela Pollet[2], Charles Eckert[1] and Charles Liotta[2], (1)School of Chemical and Biomolecular Engineering, Georgia Institute of Technology, Atlanta, GA, (2)School of Chemistry and Biochemistry, Georgia Institute of Technology, Atlanta, GA

Ionic liquids representattractive solvents for CO_2 capture due to their tunable propertiesand low vapor pressures. We have developed a class of amine solvents for use innon-aqueous conditions that react reversibly with CO_2 to form ionicliquids. We have also looked at several polyamine-containing compounds toincrease the number of reactive amine sites for added CO_2 capacity. Weare optimizing the solvents for CO_2 capture from coal-fired powerplants with a water-saturated flue gas stream. The solubility of water in boththe amine precursor and in the ionic liquid formed will influence the systemdynamics. The addition of water into the solvent system changes both capacityand viscosity. Under the right conditions, the addition of water can boostcapacity by promoting bicarbonate formation. Water can lower viscosity bydisrupting packing of the ionic liquid. Since bicarbonate has a relatively lowtemperature of reversal, the addition of water also affects the temperature atwhich CO_2 is released from the solvent system. The meritof these advantages will be compared against the parasitic energy penalty forheating water during regeneration. We will additionally present opportunities forfurther research and optimization of the solvents.

See more of this Session: Solvent-Based Carbon Capture Processes
See more of this Group/Topical: Sustainable Engineering Forum

(555d) CO2 Separation From Coal-Fired Power Plants by Regenerable Mg(OH)2 Solutions

Hari Krishna Bharadwaj[1], Joo-Youp Lee[1], Xiang Zhou[2], Lei Cheng[2] and Tim.C Keener[2], (1) Chemical Engineering Program, University of Cincinnati, Cincinnati, OH, (2)Environmental Engineering Program, University of Cincinnati, Cincinnati, OH

AICHE ABSTRACT

CO_2 Separation from Coal-Fired Power Plants by Regenerable $Mg(OH)_2$ Solutions

Hari K. Bharadwaj†, Joo-Youp Lee† *, Xiang Zhou‡, Lei Cheng‡, Tim C. Keener‡

†Chemical Engineering Program, ‡ Environmental Engineering Program, School of Energy, Environmental, Biological, and Medical Engineering, University of Cincinnati, Cincinnati, OH 45221-0012

World energy demand is expected to increase by 57% from 2004 to 2030 which will undoubtedly produce a concurrent increase in atmospheric GHG levels unless something is done to mitigate the release of CO_2 to the atmosphere. In response to this, a method is proposed of effectively, efficiently and reliably separating CO_2 from flue gases for potential sequestration purposes by means of the absorption of CO_2 in slurry containing magnesium compounds conducted at 52 °C, followed by a regeneration step conducted at 65 °C. $Mg(OH)_2$ is readily available commercially, and/or may be easily and economically reclaimed from coal-fired power plants using lime-based flue gas desulphurization systems (FGD). The CO_2 scrubbing system will be operated as a regenerable system where a concentrated CO_2 gas stream is created for further processing, and the magnesium reclaimed for further absorption of CO_2. The process has many advantages including proven absorption technology, no modification of the existing boiler heat-transfer configuration, low temperature CO_2 regeneration, and the use of magnesium by-products for further SO_3 control. This presentation will demonstrate our current research efforts in CO_2 absorption and regeneration including magnesium chemistry, lab-scale absorber operation, magnesium carbonate formation, and vapor-liquid-solid equilibrium for CO_2 stripper design and operation.

Keywords: CO_2 separation, Coal-fired power plants, magnesium hydroxide

See more of this Session: Solvent-Based Carbon Capture Processes
See more of this Group/Topical: Sustainable Engineering Forum

(555f) Amino Acid Solvents for CO2 Absorption

Le Li, Chemical and Engineering, University of Texas at Austin, Austin, TX and Gary Rochelle, Chemical Engineering, University of Texas at Austin, Austin, TX

Amino acid solvents were tested for CO_2 capture performance at optimized absorber conditions. The solvents are: 6 m potassium glycine (GlyK), 6.5 m potassium β-alanine (β-AlaK), 3 m / 5 m potassium taurine / homotaurine (TauK/HtauK), 6 m potassium sarcosine (SarK), and 4.5 m sodium sarcosine (SarNa). A Wetted Wall Column (WWC) was used to measured the absorption/desorption rates and CO_2 solubility of each solvent at variable CO_2 loadings and temperatures (40C, 60C, 80C, 100C). Solvents are analyzed at coal fired power plant flue gas conditions and gas turbine combined cycle (GTCC) plant conditions. The operation lean/rich CO_2 loading is assumed to correspond to CO_2 equilibrium partial pressures of 500 Pa/ 5000 Pa for coal, and 100 Pa / 1000 Pa for GTCC. The absorption/desorption rates, cyclic capacity, and heat of CO_2 absorption are reported for each solvent at both conditions and compared against 7 m monoethanolamine (MEA). All amino acid solvents have low capacities at 0.2-0.3 mol CO_2/mol alk, which is 50% of 7 m MEA. The absorption rate of 6 m SarK is competitive against 7 m MEA. 3 m / 5 m TauK /HtauK has an attractive high heat of absorption at 80 kJ/mol.

Key words: Amino acid; Solvent screening; Natural gas; Absorption/desorption rates; Cyclic capacity; Heat of CO_2 absorption

1. Introduction

Amino acid solvents are attractive for post combustion CO_2 absorption because of their low environmental impact, with characteristics such as zero volatility, low ecotoxicity, and high biodegradability [1]. To absorb CO_2, amino acids must be activated in water with the addition of an equi-molar amount of base. In the presence of added base, the amino group on the amino acid reacts with CO_2 like amines [2,3]. Potassium (K^+) was used as the base in four tested solvents: 6 m SarK, 6.5 β-AlaK, 6 m GlyK, and 3 m/5 m TauK/HtauK; sodium (Na^+) was used for 4.5 m SarNa. Many amino acid solvents precipitate with CO_2 loadings [4,5]. As aqueous solvents, this physical property limits solvent cyclic capacity and the potential for flexible operation at rich loadings.

Typical coal fired power plants generate flue gas with 12% CO_2 and 5-8% O_2. GTCC plants with similar power capacity generate flue gas with 3% CO_2, 15% O_2, and twice the molar flow rate of coal flue gas. These differences in flue gas properties results in changes in solvent performance. When used for GTCC, desirable solvent properties include: stability towards oxidative degradation, good absorption performance at lean CO_2 loadings, and low volatility.

2. Experimental Method and Data Analysis

$$P^*_{CO_2} = a + b/T + c\cdot\alpha + d\cdot\alpha/T + e\cdot\alpha^2$$

Experimental data were collected using a WWC, the same as the apparatus and method used by Chen [6] and Dugas [7]. The absorption/desorption rates are reported using liquid film mass transfer coefficients (k_g'). A semi-empirical VLE model (Equation 1) is used to model experimental data and represent solvent CO_2 solubility (Figure 1).

(1)

This model is used to calculate solvent capacity and heat of CO_2 absorption (Figure 2).

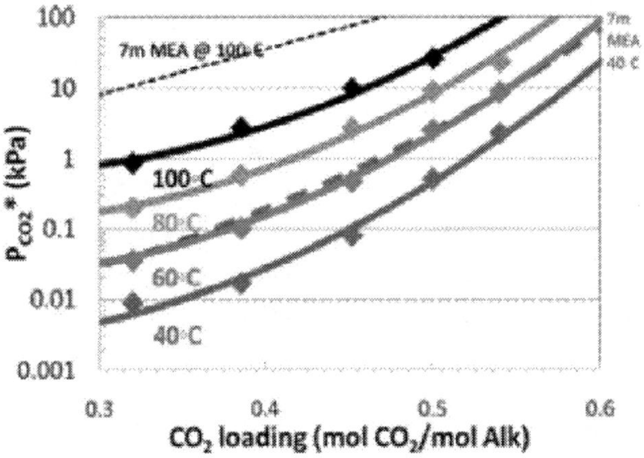

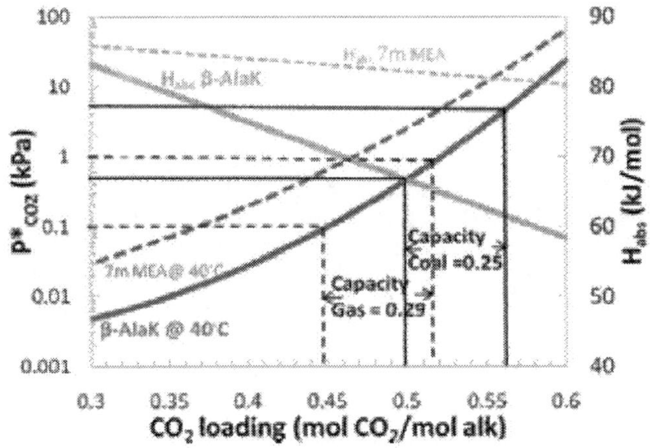

Figure 1: CO_2 solubility in 6.5 m β-alaK. Filled points: measured data. Solid lines: model prediction (Eq.1). Dashed lines: 7 m MEA model, Ref [6]

Figure 2: Cyclic capacity and heat of CO_2 absorption anlaysis for 6.5 m β-alaK

3. Results

Table 1: Summary of absorption properties of tested amino acid solvents, compared against 7 m MEA [7].

Amino acid (m)	**CO_2Capacity** (mol CO_2/kg Solution)		**$k_g'_{avg}$ (@40°C)** (x 10^{-7} mol CO_2/s Pa m^2)		**Mid Dabs (kJ/mol)** P*$_{CO2}$ =1.5kPa	P*$_{CO2}$ =0.5kPa
	Coal	Gas	Coal	Gas	Coal	Gas
GlyK (3.55)	0.25	0.25	3	10.2	64	69
GlyK (6)	0.35*	0.35	0.2*	3.2	64*	
SarK (6)	0.22	0.236	5	18.9	56.5	64
Tau/Htau (3/5)	0.195*	0.23	2.2*	10.3	74.5*	80
β AlaK (6.5)	0.25*	0.29	2*	7.4	64*	67
MEA (7)	0.47	0.55	4.3	11.7	82	83

4. References

[1] Eide-Haugmo I, et al. Environmental impact of amines. Proceedings of the 9th International Conference on Greenhouse Gas Control Technologies.1620 Nov 2008, Washington DC, USA

[2] Hook RJ. An investigation of some sterically hindered amines as potential carbon dioxide scrubbing compounds. Industrial & Engineering Chemistry Research. 1997;36:17791790.

[3] Kumar PS, Hogendoorn JA, Feron PHM, Versteeg GF. Kinetics of the reaction of CO_2 with aqueous potassium salt of taurine and glycine. American Institute of Chemical Engineers Journal. 2006;49(1):203-213

[4] Kumar PS, Hogendoorn JA, Feron PHM, Versteeg GF. Equilibrium solubility of CO_2 in aqueous potassium taurate solutions: Part 1. Crystallization in carbon dioxide loaded aqueous salt solutions of amino acids. Industrial & Engineering Chemistry Research. 2003;42:28322840.

[5] Majchrowicz ME, Brilman DWF, Groeneveld MJ. Precipitation regime for selected amino acid salts for CO_2 capture from flue gases. Energy Procedia. 2009;1:979984.

[6] Chen X, Closmann F, Rochelle GT. Accurate screening of amines by the wetted wall column. Energy Procedia 2010.

[7] Dugas R, Rochelle G. Absorption and desorption rates of carbon dioxide with monoethanolamine and piperazine. Energy Procedia 2009; 1(1): 1163-1169.

See more of this Session: Solvent-Based Carbon Capture Processes
See more of this Group/Topical: Sustainable Engineering Forum

(560a) Evaluation of Modified Feedstocks for Deconstrctionability

Seema Singh, Joint BioEnergy Institute, Emeryville, CA

Today, carbon-rich fossil fuels, primarily oil, coal and natural gas, provide 85 percent of the energy consumed in the United States. Fossil fuel use increases CO2 emissions, increasing the concentration of greenhouse gases and raising the risk of global warming. The high energy content of liquid hydrocarbon fuels makes them the preferred energy source for all modes of transportation. In the US alone, transportation consumes around 13.8 million barrels of oil per day and generates over 0.5 gigatons of carbon per year. This has spurred research into alternative, non-fossil energy sources. Among the options (nuclear, concentrated solar thermal, geothermal, hydroelectric, wind, solar and biomass), only biomass has the potential to provide a high-energy-content transportation fuel. Biomass is renewable resource that is carbon-neutral.

Currently, biofuels such as ethanol are produced largely from grains, but there is a large, untapped resource (estimated at more than a billion tons per year) of plant biomass that could be utilized as a renewable, domestic source of liquid fuels. Well-established processes convert the starch content of the grain into sugars that can be fermented to ethanol. Plant-derived biomass contains cellulose, which is more difficult to convert to sugars. The development of cost-effective and energy-efficient processes to transform cellulose in biomass into fuels is hampered by significant roadblocks, including the lack of specifically developed energy crops, the difficulty in separating biomass components, low activity of enzymes used to deconstruct biomass, and the inhibitory effect of fuels and processing byproducts on organisms responsible for producing fuels from biomass monomers.

The Joint BioEnergy Institute (JBEI) is one of three US Department of Energy Bioenergy Research Centers that will address these roadblocks in biofuels production. JBEI draws on the expertise and capabilities of three national laboratories (Lawrence Berkeley National Laboratory (LBNL), Sandia National Laboratories (SNL), and Lawrence Livermore National Laboratory (LLNL)), two leading US universities (University of California campuses at Berkeley (UCB) and Davis (UCD)), and a foundation (Carnegie Institute of Washington at Stanford University) to develop the scientific and technological base needed to convert the energy stored in cellulose into transportation fuels and commodity chemicals. Established scientists from the participating organizations are leading teams of researchers to solve the key scientific problems and develop the tools and infrastructure that will enable other researchers and companies to rapidly develop new biofuels and scale production to meet US transportation needs, and to develop and rapidly transition new technologies to the commercial sector. This talk will present a summary of the organization and efforts at JBEI, highlight the efforts on the evaluation of modified feedstock for deconstructionability and development of novel biomass pretreatment methods that enable the efficient conversion of biomass into next-generation biofuels.

(560b) Discovery and Characterization of Cellulolytic Enzymes From the Wood Wasp Symbiont Streptomyces Sp. ActE

Taichi E. Takasuka, Adam J Brook, Cameron R. Currie and Brian G. Fox, Great Lakes Bioenergy Center, University of Wisconsin, Madison, WI

One of the major hurdles for producing cost effective biofuels is the enzymatic conversion of lignocellulosic biomass into fermentable sugars. The US Department of Energy has issued the challenge of discovering and designing enzymes with novel biomass-deconstruction capabilities. This study focuses on the discovery of better enzymes and accessory proteins for biomass deconstruction from a novel bacterial species. *Streptomyces* sp. ActE, an isolate from wood-boring wasps of the genus *Sirex*. Its genomic sequence has been determined and initial physiological studies suggested high cellulolytic potential. We analyzed the genome-wide gene expression of ActE grown on insoluble substrates such as cellulose, xylan, chitin and others as the sole carbon source in the culture medium, and compared these expression profiles with glucose-grown control ActE cells. The secretomes obtained from growth on different substrates were analyzed by shotgun LC-MS/MS and enzymatic assays. From the combination of the genomic, transcriptomic, and secretomic analyses, we discovered ~200 genes that have carbon source-specific regulation of gene expression or protein secretion. This work provides the first extensive systems biology characterization of the capabilities of the genus *Streptomyces* for wide variety of polysaccharide utilization.

This work was funded by the DOE Great Lakes Bioenergy Research Center (DOE BER Office of Science DE-FC02-07ER64494).

See more of this Session: Advances In Biofuels: DOE Bioenergy Research Centers II
See more of this Group/Topical: Sustainable Engineering Forum

(560c) Dissecting the Cellulosome Enzyme Complex: Assembly and Function

Yannick J. Bomble[1], Mark Nimlos[2], Michael E. Himmel[1] and Michael F. Crowley[1], (1)Biosciences Center, National Renewable Energy Laboratory, Golden, CO, (2)National Bioenergy Center, National Renewable Energy Laboratory, Golden, CO

Most bacteria and fungi use free enzymes to degrade plant cell walls. However, a small number of bacteria have developed a fundamentally different approach, where enzymes are tethered to a large protein scaffold forming a multi-domain complex known as, the cellulosome. The study of these large protein assemblies is an ongoing research topic that has already yielded numerous breakthroughs. However, the mechanism of assembly of the enzyme subunits onto the natural scaffoldin; as well as the modes of action of the cellulosome and its enzymatic components, are currently not well understood.

In this study, we focused on the cellulosome-integrating protein (CipA) of C. thermocellum; as well as cellulosomal enzymes from families 5, 9 and 48. These three enzymes are representative of the diversity of enzymes secreted by C. thermocellum. This work not only aims at understanding the mechanisms involved in the sequential binding of the cellulosomal enzymes to the CipA scaffold of C. thermocellum, but also the binding of CipA to secondary scaffolds, such as OlpB. We focused on the physical properties of the binding of component enzymes to the scaffoldin. The modularity of the enzymes was found to be one of the main influences on the cellulosome assembly process.

Additionally, we examined several modules from the multi-modular enzyme complex, CbhA. Several of these modules have undefined functions, such as the X1 modules found closely linked to the catalytic domain. We hypothesize that the X modules could serve as tethers between several modules in CbhA to fine-tune the positioning of this enzyme during cellulose deconstruction. We conducted steered molecular dynamics simulations (directly comparable to AFM pulling experiments) to evaluate the energy and the profile for unfolding these modules. We found that the energy required to partially unfold this domain is accessible by the cellulosome and the unfolding pathways are similar in the multiple simulations conducted.

See more of this Session: Advances In Biofuels: DOE Bioenergy Research Centers II
See more of this Group/Topical: Sustainable Engineering Forum

(560d) Construction of Efficient Gram-Negative Biocatalysts for Lignocellulose Conversion

David Keating, Michael Schwalbach, Jeffrey Gardner, Mary Tremaine, Patricia Kiley and Robert Landick, Great Lakes Bioenergy Center, University of Wisconsin, Madison, WI

Research within the Great Lakes Bioenergy Research Center aims to better understand the bottlenecks associated with bioconversion of lignocellulose to liquid transportation fuels, knowledge which will then be used to improve the efficiency of this process. Much of these studies are currently focused on the Gram-negative bacterium *Escherichia coli,* due to its sophisticated genetics, well-understood physiology, and widespread industrial use.

To better understand the limitations associated with microbial biocatalysis of lignocellulosic biofuels, we have engineered *E. coli* strains capable of high-level ethanol production and tolerance. To develop efficient *E. coli* ethanologens, we have constructed mutations that reduce alternative pathways of election flow, and introduced the pyruvate decarboxylase and alcohol dehydrogenase genes from *Zymomonas mobilis*. In addition, we have carried out extensive directed evolution of *E. coli*, which has resulted in strains capable of growth in the presence of 10% ethanol. These strains are currently being subjected to a systems-level microbial analysis during lignocellulosic bioconversion. In collaboration with the Dale laboratory, we have generated enzymatic hydrolysates of AFEX-pretreated corn stover at high solid loadings. Furthermore, we have developed methodology to measure extracellular amino acids, metabolic endproducts, intracellular metabolites, and global transcription. We have used these techniques to characterize the physiological state of our ethanologenic strains during fermentation in lignocellulosic hydrolysates. Our initial analysis has already generated several surprising findings. First, lignotoxins within the hydrolysate appear to strongly limit the production of biomass, but have a more modest effect the conversion of glucose to ethanol. However, these inhibitors appear to strongly limit the conversion of xylose to ethanol. Second, microbial growth in lignocellulosic hydrolysates is a dynamic process, in which global transcription and overall physiology change dramatically during the course of fermentation. Our systems-level analysis of these strains has identified transcriptional and metabolic changes that provide insight into the mechanism of action of the lignotoxins. We are using this information, as well as data from similar experiments with lignotoxin-tolerant mutants, to generate new ethanologens with improved productivities under these conditions. Furthermore, we expect these studies will produce an improved molecular understanding of the current limitations of ethanologenesis in these strains, which we expect will be applicable to diverse ethanologenic microorganisms.

See more of this Session: Advances In Biofuels: DOE Bioenergy Research Centers II
See more of this Group/Topical: Sustainable Engineering Forum

(560e) One-Step Production of Lactate and Isobutanol From Pretreated Biomass by Recombinant Cellulolytic Bacillus Subtilis

Xiaozhou Zhang, Biological Systems Engineering, Virginia Polytechnic Institute and State University, Blacksburg, VA, Chun You, Biological Systems Eng, Virginia Polytechnic Institute and State University, Blacksburg, VA and **Y.-H. Percival Zhang**, Biological Systems Engineering Department, Virginia Polytechnic Institute and State University, Blacksburg, VA

Consolidated bioprocessing (CBP) is low-cost cellulosic biomass bioconversion by integrating cellulase production, cellulose hydrolysis, and sugar fermentation into a single step. Although significant efforts had been made to create recombinant cellulolytic microorganisms, real recombinant cellulolytic microorganism that can produce sufficient secretory active cellulases, hydrolyze cellulose, and utilize soluble sugars for supporting cell growth and cellulase synthesis without any other organic nutrient is not available yet. We demonstrated that over-expression of *Bacillus subtilis* endoglucanase BsCel5 by more than 30-fold enables non-cellulose-utilizing *Bacillus subtilis* to grow on amorphous cellulose and pretreated lignocellulosic biomass as the sole carbon source without addition of costly organic nutrients. After directed evolution and screening on amorphous cellulose, both the expression/secretion level and specific activity of Cel5 and Cel9 had been increased successfully. An advanced recombinant cellulolytic *Bacillus subtilis* strain has been constructed with engineered cellulase and the genes knock out/knock in. Besides these, we also achieved the co-expression and secretion of family 5, 9 and 48 cellulases in a single *B. subtilis* strain. The one-step production of lactate and isobutanol from cellulose and pretreated biomass in the minimal medium without adding any organic nutrient had been demonstrated. We are reconstituting *in vivo* mini-cellulosome on the cell surface of *B. subtilis.* The recombinant cellulolytic B. subtilis would be an ultra-low-cost platform for producing biofuels and other value-added products from non-food biomass.

See more of this Session: Advances In Biofuels: DOE Bioenergy Research Centers II
See more of this Group/Topical: Sustainable Engineering Forum

(560f) Development of Alkene Biofuels

Harry R. Beller, Joint BioEnergy Institute (JBEI), Emeryville, CA and Ee-Been Goh, Fuels Synthesis, Joint BioEnergy Institute (JBEI), Emeryville, CA

Aliphatic hydrocarbons are appealingtargets for advanced cellulosic biofuels, as they arepredominant components of petroleum-based gasoline and diesel fuels and thuswould be compatible with existing engines and fuel distribution systems. Long-chainalkenes are useful as feedstocks that can be crackedto form diesel- or gasoline-range hydrocarbons. We have studied alkenebiosynthesis in the bacterium *Micrococcus luteus*,a close relative of which was reported in the 1960s to biosynthesize *iso-* and *anteiso-* branched,long-chain alkenes. The underlyingbiochemistry and genetics of alkene biosynthesis werenot elucidated in those studies. In fact, until very recently, there were no known sequences forprokaryotic enzymes catalyzing the synthesis of non-isoprenoidaliphatic hydrocarbons (> C_2). We have shown that heterologousexpression of a three-gene cluster from *M. luteus*(Mlut_13230-13250) in a fatty-acid overproducing *E. coli* strain resultedin production of long-chain alkenes, predominantly 27:3 and 29:3 (no. carbon atoms: no. C=C bonds). Heterologous expression ofMlut_13230 (*oleA*) alone produced no long-chainalkenes but unsaturated aliphatic monoketones,predominantly 27:2, and *in vitro* studies with the purified Mlut_13230protein and tetradecanoyl-CoA produced the same C_{27}monoketone. Gas chromatography-time of flight (GC-TOF) mass spectrometry confirmedthe elemental composition of all detected long-chain alkenes and monoketones (putative intermediates of alkenebiosynthesis). Negative controlsdemonstrated that the *M. luteus* genes wereresponsible for production of these metabolites. We have proposed a metabolic pathway for alkene biosynthesisstarting with acyl-CoA (or -ACP) thioestersand involving decarboxylative Claisencondensation as a key step, which we believe is catalyzed by OleA. Suchactivity is consistent with our data and with the homology of Mlut_13230 (OleA) to FabH, which catalyzes decarboxylative Claisencondensation during fatty acid biosynthesis. We will discuss how understanding of the alkenebiosynthetic pathway can lead to enhanced production.

See more of this Session: Advances In Biofuels: DOE Bioenergy Research Centers II
See more of this Group/Topical: Sustainable Engineering Forum

(579a) Understanding the Evolution of by-Product Synergy Networks by Network Analysis

Shweta Singh and Bhavik R. Bakshi, William G. Lowrie Department of Chemical and Biomolecular Engineering, The Ohio State University, Columbus, OH

By-Product synergy (BPS) networks aim to mimic nature by using waste from one process as a resource in another. They are increasingly popular due to their ability to reduce environmental impact while saving money. Exploiting this win-win opportunity is a primary driver in the development of BPS networks, as evidenced by the many successes of the networks developed by the US Business Council for Sustainable Development and similar efforts elsewhere in the world. Development of a BPS network enhances the economic and environmental efficiency, but like most complex networks, it is likely that BPS networks also exhibit a trade-off between efficiency and robustness. If such a trade-off exists then it would make sense to rely on criteria other than just maximizing efficiency for the design of BPS networks. This may reduce the chance of developing fragile networks that are vulnerable to collapse.

This work aims to understand the trade-off between efficiency and resilience in BPS networks and use this insight to design more robust networks. . The field of network analysis provides different studies that look at resilience of networks such as service oriented networks , supply chain logistics for military purposes and many others. Resilience is an essential property for the sustainability of networks, and indicates the ability of a network to maintain their function in case of perturbation. Several metrics like average path length, network disconnection, and others have been used to understand network structure. Such metrics will be used in this work to understand the evolution of BPS networks. Examples from real networks in the US and toy networks will be used to illustrate the approach. Besides BPS networks, this work will also consider the behavior of techno-ecological networks. These are networks that include ecological systems such as forests and wetlands that provide essential goods and services such as oxygen, water and carbon sequestration for the sustenance of the technological networks. It is also desired that BPS networks must be such that the functioning of industries in the network is not severely affected by the collapse of one or two industries. The insight from such work be explored for guiding the development of current and future BPS and techno-ecological networks.

See more of this Session: Complexity, Networks, Dynamics and Sustainability
See more of this Group/Topical: Sustainable Engineering Forum

(579b) Optimal Rerouting of Traffic Flows for Resilient Management of Recharging Station Networks

Kris Villez[1], Craig Rieger[2] and Venkat Venkatasubramanian[1], (1)School of Chemical Engineering, Purdue University, West Lafayette, IN, (2)Idaho National Laboratory, Idaho Falls, ID

One of the most important challenges to be met inthe transition towards an ecologically friendly society is theimplementation of better ways of transportation, which reduce theamounts of energy used and which permit the switch to "greener"sources of energy. One part of the solution is likely to be met bymeans of electrical vehicles. Because electrical vehicles are limitedin range and their "refueling" (i.e. recharging) takestime, a system in which batteries can be replaced quickly has beenproposed [1]. Still, for such a system to work, it is of criticalimportance that electric vehicles can efficiently reach a servicestation. We have studied this problem in the past by optimizing thelocation of service stations for a given traffic flow in the form ofan Integer Linear Program (ILP). Originally, this was solved for afailure-free setting [2]. More recently, we have extended this forscenarios in which parts of the infrastructure (roads, road junctionsand service stations) can fail [3].

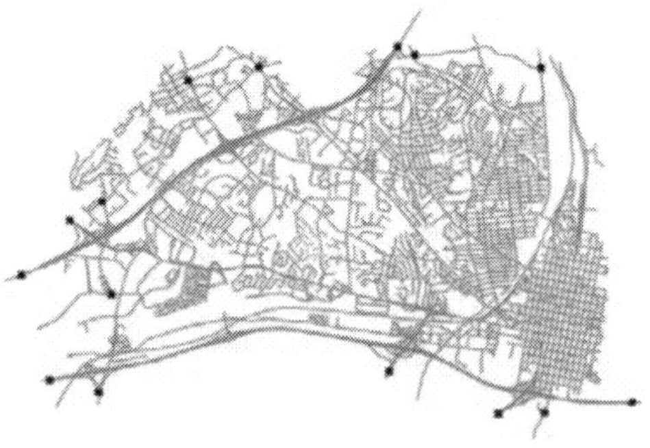

In ourvision, the previously adopted strategy to optimize the location ofservice stations is only one necessary element. Next to this, oneshould account for the fact that the route of electrical vehicles canbe modified. Indeed, the routes of electrical vehicles can be guidedand manipulated so to make sure that they can reach service stationsefficiently under varying conditions. We explore this option in thiswork by evaluation of the following algorithm.

Considerfirst that the set of service stations is fixed and the normaltraffic flow of conventional cars is known in the form of routes(paths) over a city network. In such case, we reroute the cars in agreedy fashion as follows:

The paths which contain a service station remain the same

The paths which do not contain a service station are modified as follows:

Select the most busy path (highest car flows)

For the selected path, find the shortest path from start location to destination which contain one of the service stations. Do this for all available service stations. To this end, we apply Dijkstra's algorithm to find (1) the shortest path between the start location and the selected service station and (2) the shortest path between the selected service station and the destination location.

If no paths could be found which include a service station, exclude the traffic flow associated with this path in subsequent steps and go to 7. Otherwise, continue to 4.

Following the above enumeration of shortest paths, sort the obtained paths from shortest to longest.

Now the flow in the original path is distributed over the paths starting with the shortest paths while not exceeding the capacity of any of the roads. If all computed paths are used to their maximum capacity, retain the excessive amount of cars on the original path.

If any of the roads in the city network become saturated, i.e., reach their maximum capacity, exclude these roads as available roads in the network for repeated optimization steps.

Stop the algorithm if any of the following conditions is met:

All edges in the network are saturated

All car flows have been excluded from optimization (step 3) or have been rerouted successfully (step 5).

We apply the above algorithm to the case studyused before, which is that of the city of Alexandria, VA. To computethe normal flow of conventional cars, we use the model as implementedin TRANSIMS.

REFERENCES

K. Galbraith, "Better Place unveils battery swap station," http://green.blogs.nytimes.com/2009/05/13/better-place-unveils-battery-swap-station/ , 2009

A. Shukla, J. Pekny, and V. Venkatasubramanian, "An optimization framework for cost effective design of refueling station infrastructure for alternative fuel vehicles," Computers & Chemical Engineering, In Press, Available online, 2011.

Kris Villez, Akshya Gupta, Venkat Venkatasubramanian, Craig Rieger. Resilient design of recharging station networks for electric transportation vehicles. Submitted to the 4th International Symposium on Resilient Control Systems (ISRCS2011).

(579c) Scheduling of Multiple Chemical Plant Start-Ups for Minimizing Regional Air Quality Impacts

Tianxing Cai, Dan F Smith Department of Chemical Engineering, Lamar University, Beaumont, TX and Qiang Xu, Dan F. Smith Department of Chemical Engineering, Lamar University, Beaumont, TX

Chemical industrial regions concentrate many chemical plants, whose emissions may case highly localized and transient air pollution events violating national ambient air quality standards (NAAQS). Flaring emissions, especially the intensive start-up flaring, emits huge amounts of pollutants, which has potentially significant impacts on local air quality. Give the situations that many chemical plants are spatially located in an industrial region and their startup flare emissions exhibit significant temporal variability. Such variability may cause an unexpected coincidence of flare emissions casing local air quality problems. Therefore, from the emission control point of view, the best practices for improving regional air quality should integrate both efforts on plant-wide emission minimization at every industrial point source, and regional-wide multi-plant emission variability control.

In this paper, a scheduling methodology for coordinating multiple chemical plant start-ups is developed to minimize regional air quality impacts. This methodology includes two stages of work. In the first stage, geometric locations of plants and air quality monitor stations are both taken into account. Each plant proposes its startup time window, emission profile, and expected start-up time. In the second stage, a scheduling model for regional chemical plants' startup is developed. The objective of the model is to reschedule multi-plant start-up sequence to achieve the minimum sum of delay time compared to the expected start-up time of each plant. Certainly, the major constraints include the atmosphere dispersion model constraints and predicted air quality indexes from air quality monitor stations satisfying NAAQS. The development provide quantitative decision support for multiple stake holders, including government environment agencies, chemical industry, and local communities.

See more of this Session: Complexity, Networks, Dynamics and Sustainability
See more of this Group/Topical: Sustainable Engineering Forum

(579d) Assessing Risks Due to Loss of Natural Capital: The Case of Pollination Services

Vikas Khanna, Department of Civil and Environmental Engineering, University of Pittsburgh, Pittsburgh, PA and Bhavik R. Bakshi, William G. Lowrie Department of Chemical and Biomolecular Engineering, The Ohio State University, Columbus, OH

Several recent events in the U.S. have highlighted the criticality and vulnerability of systems to sudden shocks such as natural disasters and deliberate attacks. Proper understanding of such disruptive scenarios and their impact using holistic and integrated systems modeling techniques is crucial for effective resource allocation and disaster management. Attempts to study the economic-environmental impact of sudden shocks or disruptive scenarios pose both methodological and empirical challenges for engineers and policy makers alike. This is because of the complexity and the highly interconnected nature of the industrial-ecological networks and the associated systems. Such complex systems are observed to have high efficiency, performance, and robustness, but are often fragile and vulnerable to unanticipated perturbations. Proper understanding of such disruptive scenarios and their impact is pivotal for guiding design of complex systems and their long-term sustainability.

The present work focuses on exploring the utility of Input-Output (IO) models for studying the effect of sudden shocks and quantifying the associated risks to complex industrial systems [1]. We are using the IO model to understand the impact of changes in the availability of natural resources including natural capital. This includes understanding the potential impact of loss of services such as pollination, water scarcities, and soil fertility or loss of certain key resources such as fossil fuels. Such information is used to determine sectors that are likely to face maximum risk due to sudden disruptions. Since the EIO model is based on empirically available data and considers a static and linear state of the economy, it is not able to simulate long-term effects of such disruptions. However, it is appropriate for gaining insight into the short-term effects of disruptions before any adaptation due to market forces or policies. Such simulation is relevant to understanding the effect of environmental changes as well as human-induced changes such as terrorism and natural disasters.

We illustrate the utility of our framework by studying the effect of loss of pollination services attributable to managed honeybees on the various industrial sectors. Natural systems provide a number of ecosystem goods and services to mankind. One such ecosystem service is the pollination provided by wild insects and managed honeybees. We restrict our focus to the critical ecological pollination service provided by domesticated or managed honeybees thus excluding the role of wild insects. Concerns have been expressed over the disappearance of 50-90 percent of the honeybee colonies across the U.S. as a result of the colony collapse disorder. Honeybees are responsible for pollinating more than 14 billion worth of crops in the United States [2]. There have been several attempts to quantify the direct economic value of pollination services provided by wild insects and honeybees. However, no studies exist in the literature to the best of our knowledge that try to study and quantify the effect of the loss of critical pollination services on complex industrial systems in a systematic manner. This is the first comprehensive attempt to quantify the direct and indirect effects of the loss of pollination services provided by domesticated honeybees on the complex industrial systems using an IO model. Besides quantifying the economic impact, the methodology can identify industrial sectors with greatest sensitivity under a given perturbation. It is observed that sectors that suffer the highest economic loss under agiven perturbation are not always the ones with the highest production degradation. In addition, the IO model is also able to identify sectors experiencing the inoperability and the economic losses as a result of mostly the indirect effect of the perturbation. The IO model is also attractive in this case as it is based on readily available empirical data for a specified time period and can simulate the short term effect of shocks before any adaptation due to market forces or policies. The implications of the results for guiding conservation policies will also be discussed.

References

1. R. E. Miller and P. D. Blair. Input-Output analysis. Foundations and extensions. Prentice-Hall, Inc., Englewood Cliffs, New Jersey, 1985.
2. M.E. Watanabe. Pollination worries rise as honey bees decline. Science, 265(5176):1170–1170, 1994.

See more of this Session: Complexity, Networks, Dynamics and Sustainability
See more of this Group/Topical: Sustainable Engineering Forum

(579e) Supply Chain Planning In Oil & Gas Industry: Energy and Clean Technologies Under Uncertainty

Ivan Ordoez Sr., **Ariel Uribe** and Andrs Joaqun Calderon, Colombian Petroleum Institute, ECOPETROL S.A., Piedecuesta, Colombia

Conventional planning strategies are complicated to develop because the future events are uncertain. For a supply chain of oil & gas industry is difficult to planning new production and transport networks because oil & gas reserves, energy generation technologies and availability of electric public net is very uncertain at long term. This work shows then behavior future of decision process in investments of new supplier energy systems when three elements act: 1) expansion in oil crude and gas production 2) dynamic of growing in electric public network 3) imperative use of clean technologies to generation of energy. In this work the topological connectivity among generation and consume point is a manipulated variable and this effect is evaluated on taking decisions process. For this propose a representative chain supply of oil & gas industry composed by around 300 oil fields and 33 transport terminals is represented by a MILP model in GAMS (General Algebraic Modeling System). This model is used to an optimization of several future scenarios. The analysis of results shows that decisions in new investments of clean technologies are too sensible with dynamics of expansion in oil crude and gas production and growing in electric public network. Several solutions can obtain according to supply chain connectivity where clean technologies to energy generation is economically viable if the supply chain will have generation and consumes points poorly connected.

See more of this Session: Complexity, Networks, Dynamics and Sustainability
See more of this Group/Topical: Sustainable Engineering Forum

(583a) Production of Concentrated Monosaccharride Solutions From Biomass Using CO2-H2O Mixtures At Varying Temperatures

Jeremy S. Luterbacher[1], Jefferson W. Tester[1] and Larry P. Walker[2], (1)Department of Chemical and Biomolecular Engineering, Cornell University, Ithaca, NY, (2)Department of Biological and Environmental Engineering, Cornell University, Ithaca, NY

Sustainably producing concentrated solutions of monosaccharrides from biomass is a key bottleneck in the conversion of lignocellulosic biomass to biofuels or bioproducts. Most pretreatment and enzymatic hydrolysis processes are run at low-solid concentration (<10 wt%) and use chemical catalysts, while high-solids enzymatic hydrolysis experiments are almost always performed with air-dried pretreatment mixtures.

This study uses a stirred 1 L reactor to study high solids (40 wt%) and high-pressure (200 bar) H_2O-CO_2 pretreatment of mixed hardwood and switchgrass. In addition, two-temperature stage pretreatment was introduced and optimized to take advantage of the temperature dependent biomass depolymerization reaction sequence (a short high-temperature stage at 200-210°C for 0 to 24 min was followed by a long low-temperature stage at 160-170°C for 60 min). Optimal glucan to glucose yields around 80% were obtained after 72 hrs of dilute hydrolysis (1 wt% glucan) for both substrates at similar pretreatment conditions (210°C, 16 min for hardwood and 210°C, 1 min for switchgrass, both followed by 60 min at 160°C). These optimally pretreated substrates were used as feedstock in high-solids enzymatic hydrolysis reactions, which gave glucose yields above 80% for both hardwood and switchgrass after 48 hrs of hydrolysis. Therefore, without additional chemical catalysts or any drying, two-temperature stage H_2O-CO_2 pretreatment coupled with high-solids enzymatic hydrolysis can produce monosaccharide solutions of 185 gr/L (170 gr/L of glucose and cellobiose 15gr/L of hemicellulose sugars) for mixed hardwood and 148 gr/L (115 gr/L of glucose and cellobiose 33gr/L of hemicellulose sugars) for switchgrass. This suggests that H_2O-CO_2 pretreatment is an attractive alternative to chemically catalyzed processes such as dilute acid pretreatment.

See more of this Session: Developments In the Pretreatment of Lignocellulosics for Bioconversion III
See more of this Group/Topical: Sustainable Engineering Forum

(583b) Structural Characterization of AHP-Pretreated Biomass

Muyang Li, Chemical Engineering and Material Science, Michigan State University, East Lansing, MI and David Hodge, Department of Chemical Engineering and Material Science, Michigan State University, East Lansing, MI

Alkaline hydrogen peroxide (AHP) pretreatment is exceptionally well-suited to grasses, yielding high digestibilities at low enzyme loadings, while generating relatively few fermentation innhibitors. For AHP pretreatment, the question of how structural and chemical compositional changes within the plant cell wall correlate to pretreatment effectiveness has not been effectively resolved, while knowledge of how selective modification of lignin during pretreatment may improve digestibility may yield insights into both improving pretreatments and tailoring plant cell walls for deconstruction. This study presents a comprehensive chemical and structural characterization of the changes in plant cell walls associated with switchgrass (*Panicum virgatum* cv. Cave-In-Rock), corn stovers (a commercial hybrid and inbred brown midrib lines bm1 and bm3), sugar maple (*Acer saccharum*) and hybrid aspen (*Populus tremula x tremuloides*) that were AHP pretreated at varied severities. Both the remaining solids and solubilized biomass hydrolyzates were subjected to a number of characterizations including total polysaccharide composition (monosaccharides, acetate, uronic acids), lignin content as both Klason and acetyl bromide lignin, and the ratio of H/G/S monolignols as determined by thioacidolysis GC/MS. Solid-state ^{13}C CP/MAS NMR and HSQC 2D NMR were applied to determine changes associated with intra-lignin, carbohydrate-carbohydrate, carbohydrate-lignin linkages and functional groups. Apparent molecular weight distributions of soluble polymeric lignins and hemicellulose aggregates were determined by HP-SEC. Py-GC/MS was utilized to characterize changes in the volatilized compounds before and after pretreatment implying pretreatment-induced alterations in pyrolysis-labile monolignol linkages.

See more of this Session: Developments In the Pretreatment of Lignocellulosics for Bioconversion III
See more of this Group/Topical: Sustainable Engineering Forum

(583c) Pretreatment of Corn Stover for Bioethanol Production Using Low-Moisture Anhydrous Ammonia (LMAA); Ammoniation and Pretreatment

Chang Geun Yoo, Agricultural and Biosystems Engineering, Iowa State University, Ames, IA, Nhuan-John Nghiem, Crop Conversion Science and Engineering, USDA-ARS-ERRC, Wyndmoor, PA, Kevin B. Hicks, Eastern Regional Research Center, Agricultural Research Service, Wyndmoor, PA and **Tae Hyun Kim**, Department of Agricultural and Biosystems Engineering, Iowa State University, Ames, IA

In the past several decades, various pretreatment methods have been suggested to enhance the enzymatic digestibility and fermentability of lignocellulosic biomass. Although a few of them may be effective, several cost barriers which prohibit scale-up exist including high chemical input and excessive water use. In order to solve these problems, a simpler pretreatment method using anhydrous ammonia was developed. In this method, corn stover with 30~70% moisture was used, which was contacted with anhydrous ammonia in a packed bed reactor. After the ammoniation step, biomass was treated at different reaction temperatures (40-120°C) and reaction times (24-96 h). Pretreated biomass was directly saccharified and fermented "with no additional washing step." Therefore, it is anticipated that this method can reduce energy cost substantially due to the extremely low liquid loading and moderate conditions applied. With the treated solids, we have obtained significantly improved ethanol fermentability (>85%) using *E.coli* KO11 strain.

In this investigation, it is speculated that ammonia swells cellulose structure effectively and ammoniation prevents methoxyl groups on lignin from absorbing cellulase enzymes, thus it enhances the enzyme hydrolysis rate and ethanol fermentation yield of lignocellulosic biomass. Ammoniation pretreatment can also supply assimilable nitrogen (up to 1.25 wt.% of the weight of dry biomass) for microorganism growth in the fermentor.

In this paper, various pretreatment conditions such as moisture levels, residual ammonia concentrations, reaction time, and reaction temperature were explored to reduce the pretreatment cost and its pertinent costs. Enzyme digestibility and ethanol fermentation test results will also be presented.

See more of this Session: Developments In the Pretreatment of Lignocellulosics for Bioconversion III
See more of this Group/Topical: Sustainable Engineering Forum

(583d) Feedstock Mixture Effects On Sugar Monomer Recovery Following Dilute Acid Pretreatment and Enzymatic Hydrolysis

Michael J. Brodeur-Campbell, Jordan Klinger and David Shonnard, Chemical Engineering, Michigan Technological University, Houghton, MI

Feedstock MixtureEffects on Sugar Monomer Recovery Following Dilute Acid Pretreatment andEnzymatic Hydrolysis

MichaelBrodeur-Campbell[1], Jordan Klinger[1], and DavidShonnard[1,2]

[1] Department of Chemical Engineering,Michigan Technological University

[2] Director Sustainable Futures Institute,Michigan Technological University

Increaseduse of lignocellulosic biomass for bioenergyproduction holds promise for improved domestic energy security, reducedgreenhouse gas (GHG) emissions, rural economic development, and developing amore sustainable energy infrastructure [Chum and Overend2001; Perlack, Wright, et al. 2005; U.S. DOE 2006]. Transportation fuels are especially difficultto replace as their replacement has to have a high energy density, be compatiblewith existing fuel distribution infrastructure, and be compatible with existingautomobile technology. In 2009, 97% oftransportation fuels were derived from fossil sources [EIA 2008], andtransportation accounted for one-third of carbon dioxide emissions in 2008 [EIA2009]. Liquid fuels derived from biomassare particularly attractive as a replacement for fossil-derived transportationfuels. Cellulose is the most abundantbiopolymer on Earth, and lignocellulosic feedstocks represent one of the fewnatural resources available in sufficient quantity to replace a significantfraction of liquid transportation fuel.

In acommercial biorefinery feedstocks are likely to come from a variety of sourcesand include many different materials, such as softwoods, hardwoods, agriculturalresidues, and herbaceous or woody energy crops.Much work has been done to understand and optimize the pretreatment andenzymatic hydrolysis conditions for pure feedstocks, but little is knownabout mixture effects on enzymatic hydrolysis.A primary concern in processing woody biomass to produce a range ofbiofuels is to maximize the production of desired intermediates, in this case,determining the quantity of monomer sugars that are recoverable. The biochemical platform as described by theNational Renewable Energy Laboratory (NREL) was the process studied. Briefly, lignocellulosic biomass which iscomposed primarily of hemicellulose, cellulose, and lignin is treated withdilute sulfuric acid to hydrolyze the hemicellulose into soluble sugars. The solid residue remaining after dilute acidpretreatment is then treated with cellulose-degrading enzymes to hydrolyze thecellulose into soluble sugars. Theremaining lignin may be utilized as a high energy co-product stream to provideheat and power for the cellulosic ethanol conversion processes.

This studyseeks to clarify the effects of processing mixtures of feedstocks on overall sugarrecovery. We wish to test the hypothesisthat sugar yields from mixed feedstocks can be predicted by a simple linearinterpolation model based on the behavior of the pure species. Experiments were performed on pure species(aspen, balsam, and switchgrass), and on 50/50 blends of aspen/balsam andaspen/switchgrass. Aspen was chosen as amodel species due to its high suitability for processing in a commercialbiorefinery. Balsam was chosen to mixwith aspen to study the effects of high lignin content on mixed feedstockprocessing, and switchgrass was chosen to mix with aspen to investigate theeffects of a high ash content energy crop.

One-halfgram (0.5 g) of dry sample was weighed and placed into one of a set of 9stainless steel Swagelok cylindrical reactor tubes, and 4.5 ml of dilutesulfuric acid (0.5% w/v) was added to achieve 10% (wt) solids loading in thereactor. Reactors were rapidly heated to160°C in a silicon oil bath (Dow Corning 550 fluid), and three tubes wereremoved from the oil bath at previously determined time intervals. From previous research involving pure samplesof aspen, balsam, and switchgrass (Jensen et al. 2010), the time to achievemaximum yields of monomer and oligomer sugars from combined dilute acidhydrolysis and enzymatic hydrolysis was determined; the "optimum" time. Threereactors were removed at one-half the optimal time, three reactors were removedat the optimal time point, and the last three were removed at two times theoptimal time. Reactor tubes removed fromthe oil bath were immediately submerged in an ice-water bath to quench thereaction. One milliliter of liquid fromeach reactor was removed, filtered, neutralized, and analyzed on an Agilent1200 series HPLC for glucose and xylose concentration. One milliliter of liquid from each reactorwas filtered and then further acidified to 4% (w/v) sulfuric acid, and treatedat 121°C for one hour tohydrolyze any soluble oligomers into monomer form. After neutralization these samples were analyzedby HPLC as above. Results from these twoliquid samples were combined into total soluble (monomer plus oligomer) sugarrecovery from dilute acid pretreatment.Solids were washed with excess distilled water on a glass-fiber filter(VWR) and carried on to enzymatic hydrolysis.

Solidresidues containing cellulose and lignin were subjected to enzymatic hydrolysisusing Accelerase 1500 (Genencore). Each of the three reactors from the samepretreatment time point were loaded with a different enzyme concentration 0.1, 0.25, and 0.5 ml Accelerase 1500 per gram ofcarbohydrate in the feedstock in order to study the interaction betweenpretreatment severity and enzyme loading on sugar yields. Reaction conditions were 50mM citrate buffer,pH 4.8, 50°C,250 RPM orbital shaking. Liquid sampleswere removed at 0, 1, 2, and 3 days, filtered, and analyzed for sugar concentrationsby HPLC as above.

Maximumyields of xylose from pretreatment were seen to vary with both time andfeedstock composition in the expected manner. Maximum glucose yields following enzymatichydrolysis were seen to be strongly affected by both pretreatment severity andenzyme loading. Results are summarizedin Table 1 below.

Table 1: Experimental sugar recoveries for Dilute AcidHydrolysis (DAH), Enzymatic Hydrolysis (EH), and experimental and predictedrecoveries for Total Sugar (TS) yield as percent of theoretical.

	DAH		EH		TS		TS (Predicted)	
	Xylose	**Glucose**	**Xylose**	**Glucose**	**Xylose**	**Glucose**	**Xylose**	**Glucose**
Balsam	72%	7%	0%	2%	72%	9%		
Aspen/Balsam	82%	5%	8%	27%	90%	32%	85%[1]	29%
Aspen (1)	83%	3%	4%	47%	87%	50%		
Aspen/Switchgrass	79%	5%	4%	53%	83%	58%	86%	60%
Switchgrass	69%	12%	5%	59%	74%	71%		
Aspen (2) w/ β-Glucosidase	93%	4%	6%	61%	100%	65%		

[1]Predicted value based on Aspen (2) pretreatment results.

Xyloseyields from pretreatment varied from 69-93% of the theoretical maximum. Pure switchgrass gave the lowest yield andpure aspen the highest. Each purespecies and mixture showed a peak hydrolysis yield near the expected optimaltime. Earlier time points showedincomplete hydrolysis with greater quantities of soluble oligomericsugars in the hydrolyzate. Later timepoints yielded sugar decomposition products such as furfural and hydroxymethylfurfural (HMF) in the hydrolyzate. Longer pretreatment reaction times alwaysresulted in higher glucose yields, but due to the degradation of xylose atlater time points total sugar yields were generally lower for extendedpretreatment times.

Glucoseyields following enzymatic hydrolysis varied from 2-61% with pure balsam givingthe lowest yield and pure aspen with supplemental β-glucosidase giving thehighest. The first aspen experiment wasrepeated due to incomplete pretreatment (indicated by lower than expected sugarrecoveries), and subsequently treated with the highest Acceleraseconcentration plus supplemental β-glucosidase to study the effect ofadditional enzyme loading on yield.

Using asimple linear interpolation model to predict sugar recoveries for mixedfeedstock streams based on the known yields obtainable for pure species gives aresult that is accurate to within about 2% of the experimental data for bothglucose and xylose recovery following dilute acid pretreatment, and to within3% for glucose recovery following enzymatic hydrolysis. Xylose recovery following enzymatichydrolysis for blends containing balsam could not be predicted accurately fromthese data, but since that is only a very small amount of the total sugar yieldthe discrepancy did not affect the model results for total sugar yield. Total sugar yield was predicted to within 5%of experimental results.

References:

Chum, Helena L. and Ralph P Overend (2001). *Biomass and renewable fuels*, FuelProcessing Technology 71, 187-195.

Perlack,R. D., L. L. Wright, et al. (2005). Biomass as Feedstock for a Bioenergy and Bioproducts Industry: The Technical Feasibility of aBillion-Ton Annual Supply. U. S. D. o. A. (USDA) and U. S. D. o. E. (DOE).

U.S. DOE (2006). *Breakingthe Biological Barriers to Cellulosic Ethanol: A Joint Research Agenda*,DOE/SC-0095, U.S. Department of Energy Office of Science and Office of EnergyEfficiency and Renewable Energy (www.doegenomestolife. org/biofuels/).

EIA (2008). *Emissionsof Greenhouse Gases in the United States*. U.S. Energy Information Administration Office of Integrated Analysis andForecasting (http://www.eia.doe.gov/oiaf/1605/ggrpt/carbon.html).

EIA (2009). *Annual Energy Review 2009*. U.S.Energy Information Administration, (http://www.eia.doe.gov/aer/ pecss_diagram.html).

Jensen, Jill R., Juan E. Morinelly, Kelsey R. Gossen,Michael J. Brodeur-Campbell, and David R. Shonnard (2010). Effectsof dilute acid pretreatment conditions on enzymatic hydrolysis monomer andoligomer sugar yields for aspen, balsam, and switchgrass. BioresourceTechnology 101, 2317-2325.

See more of this Session: Developments In the Pretreatment of Lignocellulosics for Bioconversion III
See more of this Group/Topical: Sustainable Engineering Forum

(583e) Hydrolysate De-Acidification Using Resin-Wafer Electrodeionization

Yupo J. Lin, Saurav Datta, Michael Henry, Anthony Fracaro and Seth Snyder, Argonne National Laboratory, Argonne, IL

Dilute acid pretreatment of lignocellulosic feedstock offers an economic route for the extraction of fermentable sugars from non-food biomass for biofuel production by bioconversion. However, removal of acids from the hydrolysate is a critical step to ensure optimal conditions for the subsequent bioprocessing/conversion steps, i.e., enzyme hydrolysis of the cellulose to release C6 sugars and for the final fermentation to produce biofuel. The removal (conditioning step) of inorganic and organic acids (released together with C5 sugar molecules from the hemicellulose) after the acid treatment stage poses a great economic challenge to the overall biofuel production cost. To address this challenge, we used a proprietary resin wafer electrodeionization (RW-EDI) technology to remove both the inorganic and organic acids in a single step. Preliminary process economic analysis indicates that the easily scalable and small foot-print unit operation of RW-EDI can offer an economic and viable means of addressing the hydrolysate conditioning challenge. Selective removal of inorganic acids and organic acids was demonstrated when appropriate operation conditions were employed for the RW-EDI operation. This unique feature makes possible the separate recovery of both inorganic acid (e.g., sulfuric acid) and organic acids, which act as fermentation inhibitors, so that the inorganic acid can be recycled to the dilute acid pretreatment stage. The RW-EDI also provides an option to remove other fermentation inhibitor such the furfural and hydroxy methyl furfural (HMF) during the hydrolysate conditioning. In this presentation, process design and performance of hydrolysate de-acidification using RW-EDI will be discussed. Pilot-scale demonstration of technology viability using synthetic hydrolysate will be also presented along with preliminary assessments of economic and process benefits.

See more of this Session: Developments In the Pretreatment of Lignocellulosics for Bioconversion III
See more of this Group/Topical: Sustainable Engineering Forum

(583f) Biomimicry of the Fungal Consolidation of Biomass Pretreatment with Saccharification

Justin T. Kaffenberger and Jonathan S Schilling, Bioproducts and Biosystems Engineering, University of Minnesota, St Paul, MN

Efficient breakdown of ligno-cellulosic material prior to fermentation for bio-fuel production is a key step in optimizing bio-fuel production. Current industrial practice involves chemical pretreatment and acid or enzymatic saccharification to prepare ligno-cellulosic biomass for fermentation. The commercial viability of bio-fuels hinges in large part on production cost. One approach to reducing production cost is process consolidation. Consolidation not only eliminates the capital costs associated with the additional processing steps, but can also trim operational costs by reducing total reaction time and down time between steps.

In nature, the brown-rot fungus *Postia placenta* successfully performs both chemical pretreatment and enzymatic saccharification of biomass in a consolidated manner. As evidenced by its recently sequenced genome, *P. placenta* produces a limited suite of cellulases, lacking genes for exoglucanases. Despite this limitation, it is still capable of rapidly degrading wood. Brown-rot fungi quickly reduce the degree of polymerization of wood with little weight loss through a reduction-oxidation pathway that yields highly reactive hydroxyl radicals through a chelator-mediated fenton reaction. While hydroxyl radicals would readily damage its cellulases, the fungus manages to conduct both reaction types, enzymatic and oxidative, in the same general location. Examination of the spatial and temporal relationship of these two reaction systems as brown-rot fungal degradation of wood progresses will be presented.

See more of this Session: Developments In the Pretreatment of Lignocellulosics for Bioconversion III
See more of this Group/Topical: Sustainable Engineering Forum

(586a) Comparison of Photoreactor Designs for Oxidation of Dilute Aqueous Waste Contaminants

Amanda M. Grannas[1], **Dorothy W. Skaf**[2], Kevin C. Brodwater[2] and Montanna Herdemian[2], (1) Chemistry, Villanova University, Villanova, PA, (2)Chemical Engineering, Villanova University, Villanova, PA

The viability of photocatalytic reactions in advanced oxidation processes for removal of dilute aqueous contaminants was evaluated using a variety of reaction systems. Using acetaminophen as analyte, the oxidation reaction was studied using various operating conditions and reactor designs to better understand how the reaction conditions impact conversion and potentially, product selectivity. Catalyst concentration, reagent concentration, metal loading on the catalyst, oxygen purging, solution pH, solution temperature, addition of oxidant, batch versus continuous operation, light intensity and semiconductor stabilization were investigated. Literature results for reaction intermediates and reaction kinetics were used to interpret experimental results. A main focus was optimization of reaction conditions that can make photocatalytic oxidation of dilute aqueous contaminants more sustainable and economical. Catalyst stability, consumption of additional reagents, value and/or safety of the degradation products and energy consumption were considered for the various reaction options.

See more of this Session: Environmental Applications of Nanotechnology and Nanomaterials
See more of this Group/Topical: Topical C: Environmental Aspects, Applications, and Implications of Nanomaterials and Nanotechnology

Study of the Photodeposition of Noble Metals on BiOCl for the Photocatalytic Decomposition of Rhodamine B

Liang Kong, Zheng Jiang, Henry H.-C. Lai, Rebecca Nicholls, Tiancun Xiao, Peter P. Edwards**

Department of Chemistry, Inorganic Chemistry Laboratory, University of Oxford, U.K.

Introduction

Extensive efforts have been made to discover robust semiconductor photocatalysts for efficient utilization of the most sustainable solar energy in photocatalytic water splitting, CO_2 reduction and environment cleanup[1-4]. The p-type BiOCl has been recently emerged as an active photocatalysts only under the illustration of UV light which accounts for approximately 5% energy of sunlight spectrum[5]. This photocatalyst is still facing the similar major challenges encountered by most photocatalysts: catalyst stability and energy loss associated with the recombination of photogenerated charge carriers (electrons and holes). A variety of strategies, such as the formation of heterojunctions[6], impurity doping and surface metallisation, have been attempted to extend the harvesting of sunlight into visible light region[7]. Surface metallisation using noble metals is one of the most effective methodologies for trapping photogenerated electrons due to the high conductivity, excellent electron-trap capability and additional plasmonic effect of the nano-sized noble metals. To date, the semiconductor surface metallisation mainly concentrated on the n-type semiconductors[8] rather than p-type semiconductor photocatalysts. Most recently, Ag/BiOBr and Pt/BiOI have demonstrated enhanced photocatalytic performance[9,10], whereas, the effects of noble metals deposition on the photocatalytic performance BiOCl has not been explored.

In this work, we deposited a small amount of various noble metals on BiOCl *via* a simple photo-deposition method and found that the noble metals have different influences on the structure, properties and photocatalytic performance of BiOCl. The interaction between noble metal and BiOCl has been discussed to interpret the effects of surface-deposited noble metal.

Results

Crystallinity and morphology of the metal/BiOCl (here after NM/BOC)

The XRD patterns (Fig. 1A) of the NM/BOC samples are well consistent with the tetragonal BiOCl phase (JCPDS no. 01-082-0485[11]) without obvious change of the characteristic diffraction peaks, suggesting the surface deposition of noble metals does not affect the crystal structure of BiOCl. There is no diffraction peaks of noble metals observed, which can be attributed to the low loading (0.5 wt%) and the high dispersion of noble metals on the surface of BiOCl. Although the lattice parameters of the BiOCl are not affected significantly, as shown in Table 1, the deposited noble metal species drastically decrease the crystallite size in following order of BiOCl > Rh/BOC > Pt/BOC > Pd/BOC > Ag/BOC.

Table 1. Crystal parameters, band-gap and photodegradation rate constant of NM/BOC

Sample ID	Crystallite size (Å)	Lattice Parameters		k_{RhB} (min^{-1})
		a (Å)	*c* (Å)	
BiOCl	451	3.892	7.376	0.247
0.5 wt% Pt-BiOCl	355	3.889	7.361	0.169
0.5 wt% Pd-BiOCl	325	3.887	7.365	0.194
0.5 wt% Rh-BiOCl	384	3.893	7.383	0.140
0.5 wt% Ag-BiOCl	292	3.888	7.387	0.253

As shown in Fig.1(B), Ag/BiOCl sample remains the flake-like morphology as same as the pure BiOCl (SEM not shown here). The average diameter of the flakes is approximately 200 nm. The TEM image (Fig. 1C)of the Ag/BiOCl sample further evidences its flake-like morphology and the nano-sized metal Ag particles in diameter approximately 5nm were highly dispersed on the surface of the BiOCl flakes in agreement with the XRD results. Other NM/BiOCl samples have the similar flake morphologies.

Fig. 2 (A)XRD patterns of as-syntheisised undoped BiOCl and samples with 0.5 wt% photodeposited Pt, Pd, Ag and Rh, (B)SEM and (C)TEM image of 0.5 wt% Ag/BiOCl sample

XPS analysis

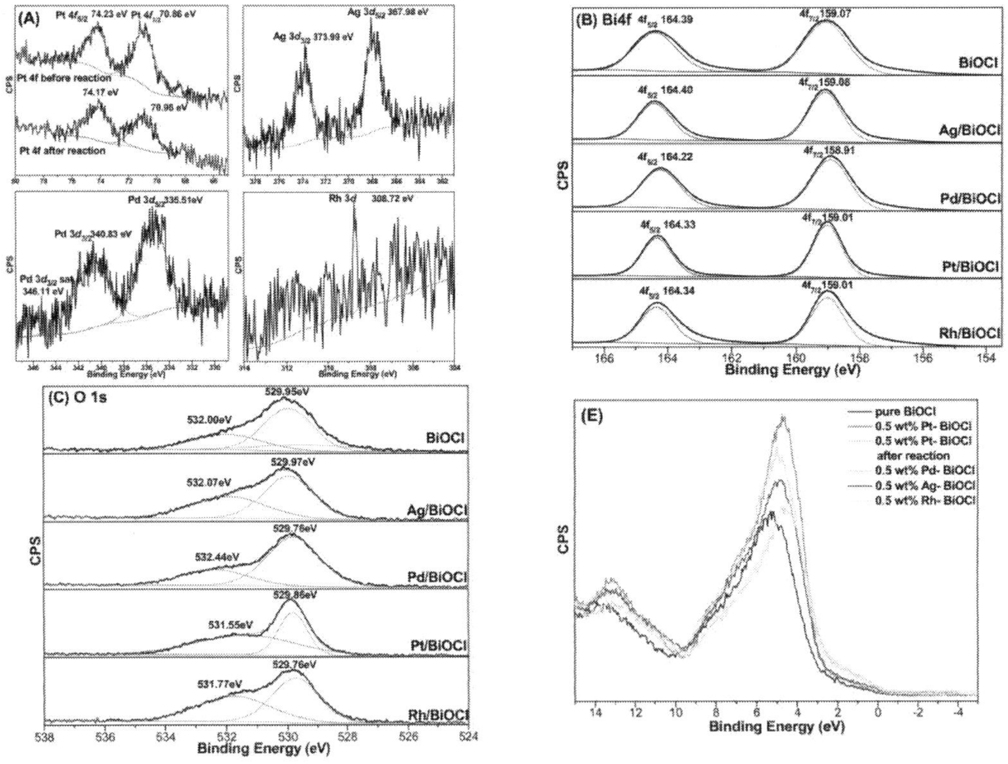

Fig.2 XPS spectra of (A) Bi 4*f*, (B) O 1*s*, (C) Pt 4*f* (before and after photocatalytic reaction), Ag 3*d*, Pd 3*d* and Rh 3*d*, and (E) VBM of all the NM/BiOCl samples

The characteristic binding energies (B.E.) of Bi, Cl, O and noble metals can be well identified from the XPS results of the corresponding samples. The XPS-derived surface atomic ratios of noble metal, Cl, O and Bi elements the samples are listed in Table 2. These results confirm the desired compositions of the BiOCl and NM/BOC samples. The noble metals in the study are all in metallic states as shown in Fig 2(A) without any signals associated with noble metal cations observed, suggesting the photoreduction is a successful strategy in deposition of noble metals on the BiOCl photocatalyst. Fig 2(B) shows the Bi $4f_{5/2}$ and Bi $4f_{7/2}$ XPS core lines with B.E located around 164.3 eV and 159.0 eV, revealing the Bi in BiOCl is Bi^{3+} [12]. The Bi 4*f* core lines shifted towards lower energies for the NM/BOC samples except for Ag/BOC whose Bi 4*f* core lines increased about 0.2 eV. The most significant red-shifts of Bi 4*f* B.E. are observed on the Pd/BOC sample. The detailed B.E. changes may be found in Table 3. There is no obvious changes can be observed from Cl $2p_{1/2}$ (199.4 eV) and Cl $2p_{3/2}$ (197.8 eV) peaks of Cl^- in the BiOCl and NM/BOC, suggesting the supported noble metal can not affect the inter-layered Cl^- [13,14]. There are two different oxygen species presented on the surface of the series samples (Fig. 2C). The O 1*s* B.E. around 532.0 eV is assigned to the Bi-O bond in BiOCl, while the O 1*s* peak at 529.9 eV due to the surface hydroxyl groups of the sample[15,16]. Moreover, the O 1s of Bi-O shift towards higher energy against the trend of Bi 4f on the NM/BOC samples except for that of Ag/BOC, which further confirms that surface metallisation affect the chemical environment of the composite. The valence band XPS of the samples are compared in Fig.2D, in which the shifts of VB edges (VBMs) towards lower energy on the NM/BOC than VB of BOC are clearly observed and the shifts significantly depend on the noble metal specials.

Table 2. Surface elemental composition (%) of the NM-BiOCl samples

Atomic Ratio	Bi	O	Cl	NM
BiOCl	39.2	38.2	22.6	-
Rh-BiOCl	36.3	52.4	10.9	0.49
Pd-BiOCl	37.4	36.1	25.9	0.56
Pt-BiOCl	38.3	34.2	27.3	0.18
Pt-BiOCl after reaction	33.4	43.8	22.7	0.19
Ag-BiOCl	36.8	41.9	20.9	0.46

Table 3. XPS characteristic peak positions of the NM-BiOCl samples

Sample ID	Peak position (eV)				
	Bi $4f_{5/2}$	**Bi $4f_{7/2}$**	**Cl $2p_{1/2}$**	**Cl $2p_{3/2}$**	**O 1*s***
BiOCl	164.4	159	199.5	197.8	530.1
Pt-BiOCl	164.2	158.9	199.3	197.7	529.8
Pt-BiOCl after reaction	164.3	159.0	199.4	197.8	529.9
Pd-BiOCl	164.2	158.9	199.3	197.7	529.7
Rh-BiOCl	164.3	159.0	199.3	197.7	529.8
Ag-BiOCl	164.4	159.1	199.6	197.9	530.1

Uv-Vis Spectra

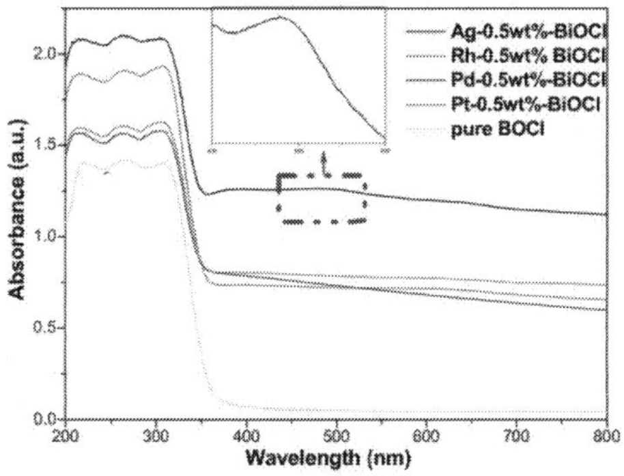

Fig. 3 UV-vis diffuse reflectance spectra of NM- BOC photocatalysts

Fig. 3 shows the UV/vis diffuse reflectance spectra of BiOCl and NM/BOC samples. T BiOCl can only absorb UV light with wavelength below 360 nm, while all the NM/BOC samples can harvest light through the whole UV-Vis region, suggesting even small amount of noble metals can greatly improve the light absorption in comparison to pure BiOCl. The enhanced absorption of NM/BiOCl samples both in UV and Visible regions may be attributed to the close field surface plasmonic effect (or the scattering effect)[9]. The Ag/BOC sample exhibits the strongest plasmonic effect among the NM/BOC samples. The indirect and direct band-gap energy of NM/BOC can thus be estimated from the data plot of $(\alpha \cdot hv)^{1/2}$ and $(\alpha \cdot hv)^2$ *versus* photo energy (hv) [5] and listed in Table 2, from which the introducing of noble metals decrease both the indirect band-gap and direct band-gap.

Photocatalytic activity

Photocatalytic decomposition of RhB, a chemically stable organic dye with a characteristic absorption peak at 554 nm, was chosen as a probe reaction to clarify the effects of deposited noble metals on the BiOCl. As shown in Fig. 4(A) and (B), all the samples can completely decompose Rh.B within 30 minutes UV light irradiation under the same reaction conditions. The photocatalytic activity highly dependes on the type of noble metal, with the activity changes in the order of Ag/BOC > pure BiOCl > Pd/BOC > Pt/BOC > Rh/BOC. The results suggest that the modification of noble metals cannot enhance the photocatalytic activity of the BiOCl samples even though they have enhanced absorption of irradiation.

Although most of the NM/BOC are less active than BOC except for Ag/BOC, the NM/BiOCl samples showed higher stability after UV-light irradiation. For example, the fresh and used 0.5 wt% Pt/BiOCl samples shows the very similar XRD patterns (Fig.4C) except for the stronger Bragg diffraction intentisty for the sample post photocatalysis. The enhanced intensity may be attributed to the growth of the crystallite during the photocatalytic reaction. The XPS spectra of fresh and used Pt/BiOCl samples were also compared, as shown in Table 2, indicating there is no obvious difference between the samples before and after the photocatalytic reaction. Both the XRD pattern and XPS spectra of the samples after reaction confirmed that the metal-modified samples are considerably stable under UV light irradiation.

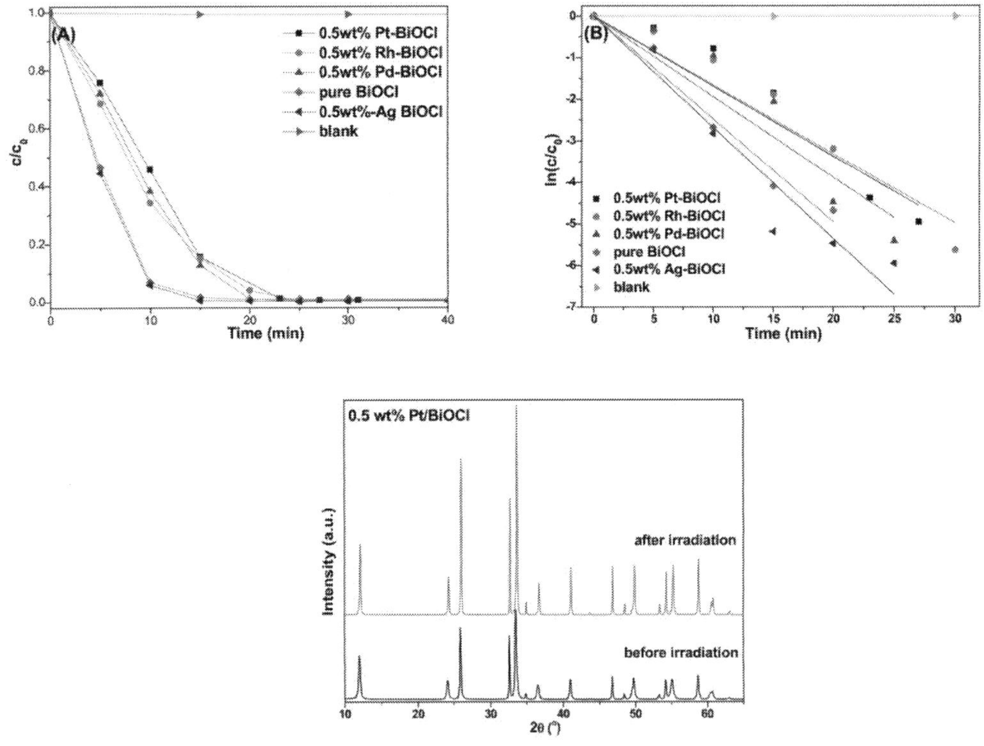

Fig. 4 (A) Temporal course of the degradation of RhB in the presence of different noble metal- BiOCl; (B) Kinetic linear curves of RhB phpotocatalytic degrdation over noble metal- BiOCl and (C) XRD patterns of 0.5 wt% Pt- BiOCl sample before and after UV-light irradiation

Working mechanism

Understanding the reason of the superior photocatalytic activity and the working mechanism of the Nm/BiOCl composite catalysts is essential for fundamental investigation of semiconductor photocatalysts. The formation of Schottky barrier between metals and semiconductor could enhance the photocatalytic activity due to their different Fermi levels, characterised by the work function of the metals and the band structure of the semiconductors[8]. Photoexcited electrons in the conduction band of the semiconductor will raise its Fermi level to more negative values[17], usually leading to enhanced photo-activity of most photocatlysts, such as TiO_2, while our experimental results differ from previous reports that the modification of noble metals decrease the activity of photocatalysts. According to the different work functions of the noble metals, we obtain the height of the noble metals in the reducing order of Pd > Pt > Ag > Rh. This trend is in accordance with the photocatalytic activity consequence of NM/BiOCl samples, i.e., the noble metal with higher Fermi level shows better photocatlaytic activity. However, the work functions of the NM are more positive than the CB of BiOCl, which leads to photogenerated electrons from the VB easily flow into noble metal rather than the CB of BOC and decreased photocatalytic activity of the NM/BOC due to the less opportunity to generate reactive O_2^{-}•. At the same time, the chemical potential of the photogenerated holes are less oxidative due to the shift of VB towards to more negative positions, which means the decreased oxidative capability of VB holes to oxidize the dye molecules. The overall results are decreased activity of the NM/BOC in comparison with pure BiOCl.

Conclusion

Noble metal-modified BiOCl photocatalysts have been synthesised *via* a two-step precipitation-photodeposition method. The metallic state modification on the surface of BiOCl significantly enhances the absorbance in the visible-light region, *i.e.* decreases the band-gap to about 2.7 eV, however, the modification cannot effectively enhance their photocatalytic activity with respect to the pure BiOCl in photodecomposion of RhB. The decreased photocatalytic activity of the NM/BOC can be well interpret by the work function difference between the NMs and BiOCl.

Acknowledgement

This work was financially supported by the principal's major fund at Jesus College, University of Oxford, Oxford Challenge Seed Fund, UK Photocatalysis Network, UKSHEC (EPSRC). Dr Jiang appreciates John Houghton Fellowship at Jesus College (Oxford) and the grants of Royal Society for international conference (TG092414) and international collaboration (TG101750).

Reference

(1)Ohtani, B. *Journal of Photochemistry and Photobiology C: Photochemistry Reviews* **2010**, *11*, 157-178.

(2)Jiang, Z.; Yang, F.; Yang, G.; Kong, L.; Jones, M. O.; Xiao, T.; Edwards, P. P. *Journal of Photochemistry and Photobiology A: Chemistry* **2010**, *212*, 8-13.

(3)Jiang, Z.; Al-Shahrani, F.; Lin, T.-W.; Cui, Y.; Xiao, T. In *Studies in Surface Science and Catalysis*; Dongyuan Zhao, S. Q. Y. T., Chengzhong, Y., Eds.; Elsevier, 2007; Vol. Volume 165; pp 355-359.

(4)Jiang, Z.; Xiao, T.; Kuznetsov, V. L.; Edwards, P. P. *Philosophical Transactions of the Royal Society A: Mathematical, Physical and Engineering Sciences* **2010**, *368*, 3343-3364.

(5)Zhang, K.-L.; Liu, C.-M.; Huang, F.-Q.; Zheng, C.; Wang, W.-D. *Applied Catalysis B: Environmental* **2006**, *68*, 125-129.

(6)Shamaila, S.; Sajjad, A. K. L.; Chen, F.; Zhang, J. *Journal of Colloid and Interface Science* **2011**, *356*, 465-472.

(7)Kong, L.; Jiang, Z.; Xiao, T.; Lu, L.; Jones, M. O.; Edwards, P. P. *Chemical Communications* **2011**, *47*, 5512-5514.

(8)Ohtani, B.; Kakimoto, M.; Nishimoto, S.; Kagiya, T. *Journal of Photochemistry and Photobiology A: Chemistry* **1993**, *70*, 265-272.

(9)Yu, C.; Fan, C.; Meng, X.; Yang, K.; Cao, F.; Li, X. *Reaction Kinetics, Mechanisms and Catalysis* **2011**, *103*, 141-151.

(10)Yu, C.; Yu, J. C.; Fan, C.; Wen, H.; Hu, S. *Materials Science and Engineering: B* **2010**, *166*, 213-219.

(11)Keramidas, K. G.; Voutsas, G. P.; Rentzeperis, P. I. *Zeitschrift fur Kristallographie* **1993**, *205*, 35-40.

(12)Xiao, Q.; Zhang, J.; Xiao, C.; Tan, X. *Catalysis Communications* **2008**, *9*, 1247-1253.

(13)Ge, L. *Journal of Molecular Catalysis A: Chemical* **2008**, *282*, 62-66.

(14)Drelinkiewicz, A.; Sobczak, J. W.; Sobczak, E.; Krawczyk, M.; Zieba, A.; Waksmundzka-Góra, A. *Materials Chemistry and Physics* **2009**, *114*, 763-773.

(15)V.I., N.; D., G.; Dzhurinskii, B. F.; N.P., S.; Y.V., S. *Zh. Neorg. Khimii* **1975**, *20*, 2307.

(16)Martensson, N.; Malmquist, P. A.; Svensson, S.; Basilier, E.; Pireaux, J. J.; Gelius, U.; Siegbahn, K. *Nouveau, J. Chim.* **1977**, *1*, 191

(17)Subramanian, V.; Wolf, E. E.; Kamat, P. V. *Journal of the American Chemical Society* **2004**, *126*, 4943-4950.

(586c) Synthesis and Characterization of Magnetic Nanoparticles for Enhanced Gas-Liquid Mass Transfer

Alexander P. Mathews, Department of Civil Engineering, Kansas State University, Manhattan, KS, Dambar B. Hamal, Department of Chemistry, Kansas State University, Manhattan, KS, Paul Owings, Civil Engineering, Kansas State University, Manhattan, KS and Ken J. Klabunde, Chemistry, Kansas State University, Manhattan, KY

Abstract

Magnetite particles weresynthesized by the coprecipitation of iron (II) and iron (III) chlorides in thepresence of a surfactant and ammine base. These particles had relatively lowinternal surface areas, but high external surface due to the small particlediameter. Silica coated magnetite particles were synthesized by the hydrolysisof tetraorthosilicate (TEOS) and the coprecipitation of iron chlorides in thepresence of different surfactants and an ammine base. Commercial magnetitenanoparticles were also coated with silica using TEOS and a base in thepresence or absence of a dispersant. All samples were characterized using XRD,SEM, TEM, and nitrogen adsorption to obtain diffraction patterns, particlemorphology data, and BET surface areas. Adsorption isotherm studies wereconducted for the adsorption of trichloroethylene from aqueous samples on tolow surface area magnetite nanoparticles and high surface area silica coatedmagnetite nanoparticles. Studies were also conducted on the effect of nanoparticleson the effect of high and low surface area nanoparticles on gas-liquid masstransfer in aqueous systems. This paper will present data on the synthesis andcharacterization of these nanoparticles and their potential to enhance masstransfer in gas-liquid systems.

See more of this Session: Environmental Applications of Nanotechnology and Nanomaterials
See more of this Group/Topical: Topical C: Environmental Aspects, Applications, and Implications of Nanomaterials and Nanotechnology

(586d) Visible-Light-Driven Photodegradation of Contaminants In Water Over Surface-Engineered BiOBr Semiconductor Micro/Nano-Structures

Zheng Jiang[1], Liang Kong[2], Tiancun Xiao[2] and Peter P. Edwards[2], (1)Jesus College, University of Oxford, Oxford, United Kingdom, (2)Department of Chemistry, Inorganic Chemistry Laboratory, University of Oxford, Oxford, United Kingdom

Visible-Light-Driven Photodegradation of Contaminants in Water over Surface-Engineered BiOBr Semiconductor Micro/Nano-Structures

Zheng Jiang[1,2], Liang Kong[1], Tiancun Xiao[1], Peter P. Edwards[1]

(1)Jesus College, University of Oxford, Oxford, OX1 3DW, U.K.; (2)Department of Chemistry, Inorganic Chemistry Laboratory, University of Oxford, Oxford, OX1 3QR, U.K.

Visible-Light-Driven Photocatalytic degradation of contaminants (PDC) over semiconductor photocatalysts represents an advanced, cost-efficient and promising water purification technology because it can utilize the large portion of energy in the free sun light[1,2]. However, the implementation of PDC technique is is still facing a great challenge over the effective and stable visible-light-responsive photocatalysts with desired structure and surface functionality. Extensive efforts have been made to enhance the adsorption of sunlight and depress the recombination of the photogenerated charges for maximum photocatalytic efficiency. Semiconductor BiOBr is an attractive photocatalyst with excellent Uv-Vis light illumination despite its adsorption of sunlight is limited by its relatively large bandgap[3,4]. This talk will focus on the strategies we developed to resolve the crucial issues over the BiOBr semiconductor by semiconductor surface-engineering. An efficient and controllable synthesis methodology has been developed to prepare BiOBr mico-/nano-structures with desired visible-light adsorption.

First, a rapid synthesis route was developed to prepare BiOBr nanoflowers with exposed highly reactive surfaces and it was further adopted to prepare heterojucntions composed of BiOBr and second semiconductors. The results reveal the surface-oriented BiOBr and its heterojunctions possess hierarchical architectures and ultrahigh activity under visible-light irradiation in photodegradation of several typical refractory model organic pollutants. The synergism effects of the enhanced surface absorption, enhanced light adsorption, mitigated light-scattering, and special hierarchical structures were unearthed through key characterizations. Second, visible-light-induced photodeposition was adopted to prepare the surface supported noble-metal/BiOBr photocatalysts and the effects of the NM on the BiOBr surface were explored. It was found that the surface NMs deteriorated the photocatalytic performance of the BiOBr photocatalysts, although the surface NMs significantly improve the stability of the BiOBr under long-term light exposure. The mechanism was proposed on basis of the surface engineering, and band structure of the multi-structures.

The correlation of the photocatalytic activity to its structure and surface will be discussed on basis of the detailed physicochemical characterization and theoretical analyses.

References:

1. Z Jiang, T Xiao,V Kuznetzov, P.P. Edwards, ***Phil. Trans. R. Soc. A***; 2010, 368; 3343-3364;

2. Z. Jiang, F. Yangb, N. Luo, B. T. T. Chu, D. Sun, H. Shia, T. Xiao, P. P. Edwards, ***Chem. Commun.***, 2008, 6372-6374

3. L. Kong, Z. Jiang, T. Xiao, L. Lu, M. O. Jones and P. P. Edwards, ***Chem. Commun.***, 2011, 47, 5512-5514

4. Z. Jiang, F. Yang, G. Yang, L. Kong, M. O. Jones, T. Xiao, P. P. Edwards, ***J PHOTOCHEM PHOTOBIOL A-CHEM***, 2010,212(1): 8-13.

See more of this Session: Environmental Applications of Nanotechnology and Nanomaterials
See more of this Group/Topical: Topical C: Environmental Aspects, Applications, and Implications of Nanomaterials and Nanotechnology

(586e) Magnetic Catch & Release: Reversible Organic Contaminant Adsorption and Enrichment From Water

Roland Fuhrer, Inge K. Herrmann, Evagelos K. Athanassiou, Robert N. Grass and Wendelin J. Stark, Institute for Chemical- and Bioengineering, ETH Zurich, 8093 Zurich, Switzerland

There is an increasingneed for fast and efficient extraction methods to remove low concentrated (ppblevels) organic compounds from water with high adsorbent recyclability. Surface-modified magnetic nanoparticles can be used in extraction processesas they readily disperse in common solvents and combine high saturationmagnetization with excellent accessibility. Reversible and recyclableadsorption and desorption through solvent changes and magnetic separationprovide technically attractive alternatives to classical solvent extraction. Carbon-coatedcobalt nanoparticles with a thin polymer layer were tagged withβ-cyclodextrin. The resulting material reversibly adsorbed organiccontaminants in water within minutes. Isolation of the immobilized inclusioncomplex was easily carried out within seconds by magnetic separation due to thestrong magnetization of the nanomagnets (metal core). The trapped moleculeswere fully and rapidly recovered by filling the cyclodextrin cavity with amicrobiologically well accepted substitute, e.g. benzyl alcohol.Phenolphthalein was used as a model compound for organic contaminants such aspolychlorinated dibenzodioxins (PCDDs) or bisphenol A (BPA). The nanomagnetscan be regenerated under mild conditions and reused in the next cycle at fullefficiency (16 recycles tested). Tagging cyclodextrins with magnetic, stable nanoparticles(carbon shell, polymer layer) makes them magnetoresponsive and may lead to anew generation of adsorbents in separation, contaminant enrichment or drugdelivery. Experiments at ultralow concentrations (160 ppb) underline the highpotential of cyclodextrin modified nanomagnets as a fast, recyclable extractionmethod for organic contaminants in large water streams or as an enrichment toolfor analytics.

Reference:

1) R. Fuhrer, I.K. Herrmann, E.K.Athanassiou, R.N. Grass, W.J. Stark, **Langmuir**,2011, 27(5), 1924-9

Figure:

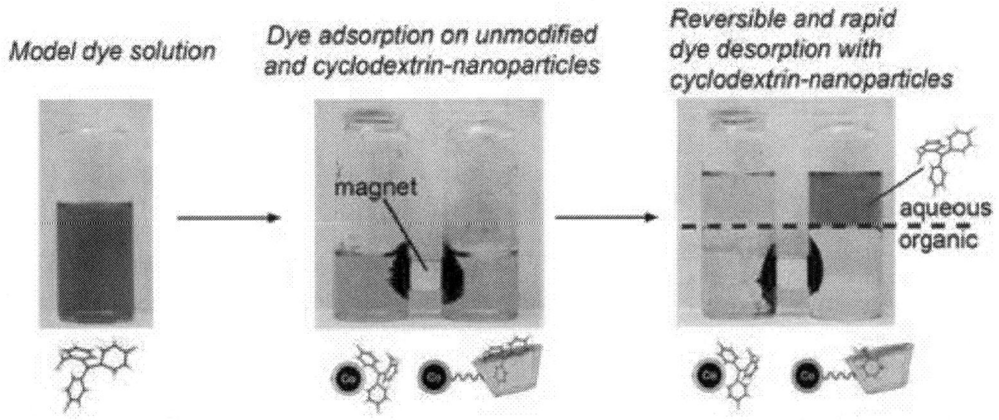

See more of this Session: Environmental Applications of Nanotechnology and Nanomaterials
See more of this Group/Topical: Topical C: Environmental Aspects, Applications, and Implications of Nanomaterials and Nanotechnology

(586f) Improved Photocatalytic Activity Under Visible Light Irradiation of Nanosized-TiO2 Co-Doped with Vanadium and Nitrogen

Renuka Jaiswal[1], Rupali Dholam[1], Nainesh Patel[2], Antonio Miotello[2] and **Dushyant C. Kothari**[1], (1) Centre for Nanosciences & Nanotechnology, University of Mumbai,, Mumabi, India, (2)Physics, Universita degli Studi di Trento, Povo (Trento), Italy

Heterogeneous nanocrystalline titania (TiO_2) has been extensively investigated due to its notable photocatalytic capability for the industrial waste water treatment. However, an obvious problem hindering the application of TiO_2 is its poor response to visible light due to its wide bandgap. Herein, TiO_2 photo-catalyst is synthesized by co-doping vanadium and nitrogen by sol-gel method to sensitize TiO_2 in visible light. The concentrations of V and N were varied and optimized to enhance photo-catalytic activity. In order to increase the active surface area, the crystalline powder was ball milled to nanosize. The photo-catalysts were characterized by XRD, FT-IR, SEM, EDS, BET surface area, XPS, and UV–vis diffuse reflectance spectra. The co-doped TiO_2 nanocrystal has a narrower band gap than singly V-doped TiO_2 and singly N-doped TiO_2. The XPS detected vanadium (V^{4+} and V^{5+}) and nitrogen ions on the surface of photo-catalysts, but no diffractive peak of vanadium oxide or titanium nitride was observed in XRD spectra. High energy ball-milling was able to increase the surface area due to reduction in particle size below 50 nm as confirmed by SEM. The co-doped TiO_2 photo-catalyst showed significantly better photocatalytic activity for the degradation of organic dyes such as Rhodamine B and methylene blue under visible light irradiation ($\lambda > 420$ nm) as compared with V-doped TiO_2 and N-doped TiO_2. The incorporation of N and V in TiO_2 lattice induces the appearance of passivated shallow acceptor and donor levels near the valence and conduction band, respectively, causing narrowing of the band gap. In addition, these levels can also act as traps for photoexcited holes or electrons to reduce the recombination between photo-generated charges thus contributing to enhance the photocatalytic activity.

See more of this Session: Environmental Applications of Nanotechnology and Nanomaterials
See more of this Group/Topical: Topical C: Environmental Aspects, Applications, and Implications of Nanomaterials and Nanotechnology

(588a) Flexible Energetically Self-Sufficient Co-Production of Hydrogen and Formic Acid

Jorge Pena Lopez, Chemical Engineering Department, University of California, Los Angeles, Los Angeles, CA and Vasilios Manousiouthakis, Chemical Engineering Department, University of California at Los Angeles, Los Angeles, CA

The use of fossil fuels (gasoline, Diesel, etc.) for light vehicle based transportation has resulted in two major environmental quandaries: poor air quality in largely populated cities and increased carbon dioxide emissions into the Earth's atmosphere. The use of hydrogen fueled vehicles has been proposed as an avenue to address the aforementioned dilemmas. However, hydrogen production technologies have yet to achieve a generation system competitive with, for example, the production of gasoline. Several options exist for hydrogen production; one of the most explored has been the reforming of natural gas. However, conventional natural gas reformers release carbon dioxide into the atmosphere. A solution to this issue is redirecting carbon dioxide into formic acid, thus preventing its emission into the atmosphere.

In this work we propose a novel method to synthesize energetically self-sufficient and environmentally friendly process flowsheets, and we show hydrogen production as a case study. To this end we make use of commercially available technologies for the production of formic acid [1]. The resulting process flowsheet goes through heat and power integration techniques that allow the maximum utilization of energy in favor of hydrogen production, maintaining a self-powered plant and the carbon dioxide emissions at zero. Using the simulation tool UniSim Design, we show that there exists an optimal themodynamical and operational point for the generated process flowsheet.

[1] Reutemann W, Kieczka H. Ullmann's Encyclopedia of Industrial Chemistry, "Formic Acid". John Wiley & Sons, Ltd. 2000

See more of this Session: Fundamentals of Hydrogen Production and Utilization
See more of this Group/Topical: Environmental Division

The Attractiveness of Ammonia Borane to Generate Hydrogen for PEM Fuel Cell Vehicles

Ahmad Al-Kukhun, Hyun Tae Hwang, and Arvind Varma

School of Chemical Engineering, Purdue University, 480 Stadium Mall Drive, West Lafayette, IN 47907

Current promising methods to release hydrogen from Ammonia Borane (NH_3BH_3, AB; 19.6 wt% H_2) including neat thermolysis, thermolysis in ionic liquid bmimCl with or without proton sponge, thermolysis with nano-BN and hydrothermolysis, were investigated for hydrogen yield and ammonia formation. It was found that even trace moisture influences AB dehydrogenation significantly. The hydrothermolysis (13.5 wt% H_2, 1 mol% NH_3) at 85°C and thermolysis in bmimCl with 3 wt% moisture at 110°C (13 wt% H_2, 0.2 mol% NH_3) methods were found to be the most promising. This study demonstrates that AB dehydrogenation is an attractive approach to generate hydrogen for PEM fuel cell vehicle applications.

(588c) Solar Thermochemical Hydrolysis of Metal Nitrides for H2 Production and Integrated Storage In Form of Ammonia

Ronald Michalsky and Peter Pfromm, Department of Chemical Engineering, Kansas State University, Manhattan, KS

Solar thermochemical hydrolysis of metal nitrides forH_2 production and integrated storage in form of ammonia

RonaldMichalsky and Peter H. Pfromm

Department of Chemical Engineering, Kansas State University, Manhattan, Kansas, USA

Conversion ofabundant but intermittent solar energy at high temperatures to high valuechemical energy stored in the products of a solar thermochemical reaction cycle,such as H_2 or syngas has been studied widely. Ammonia (NH_3)has been proposed more recently as hydrogen carrier for sustainabletransportation fuel. Easily liquefied and transported, NH_3 exceedsthe H_2 storage requirements of the Department of Energy. Modifieddiesel engines can combust NH_3 releasing mainly H_2O and N_2. H_2 can also be recovered catalytically on board a vehicle from $NH_3$for subsequent combustion.

Solarthermochemical NH_3 synthesis from steam and nitrogen produces asustainable solar fuel and avoids the storage problem for H_2.Production of metallic nitrides by reduction of their metal oxides is a high-temperatureand energy-intensive process that may take advantage of concentrated solarradiation as inexpensive and sustainable source for process heat. This workpresents a solar thermochemical NH_3 synthesis process sequence ofnitride hydrolysis splitting H_2O and absorbing protons released inthe formation of NH_3 at ambient pressure, and endothermic metaloxide reduction and nitridation driven by concentrated solar energy.

A thermodynamicrationale is presented. Various characteristic metals are selected andexperimentally explored studying yield and kinetics of breaking the N_2triple bond during nitridation and liberating NH_3 when hydrolyzingthe nitride. Experimental data will focus transition metal interactions on N_2uptake from the gas phase and nitrogen distribution in the solid bulk materialstudied using various solid-state analytical techniques. Protonation of thenitrogen liberated during steam hydrolysis of various ionic, covalent,intermediate and interstitial nitrides is examined establishing a mass balanceon the nitrogen atom. The yield of NH_3 will be correlated with thenitride ionicity and with the activation energy of nitride hydrolysis to thusdiscuss the trade-off between desirable high yields of metal nitridation andnitride hydrolysis on one side and undesirable strong bonds between the metalliccomponent and oxygen formed during hydrolysis.

See more of this Session: Fundamentals of Hydrogen Production and Utilization
See more of this Group/Topical: Environmental Division

(588d) Quantification of [Fe-Fe]-Hydrogenase, MrcA and FTHFS Gene In Mixed Anaerobic Cultures During Dark Hydrogen Fermentation

Subba Rao Chaganti[1], Daniel D. Heath[1] and Jerald A. Lalman[2], (1)Great Lakes Institute for Environmental Research, University of Windsor, Windsor, ON, Canada, (2)Department of Civil and Environmental Engineering, University of Windsor, Windsor, ON, Canada

Energy security, climatic change and decreasing fossil fuel inventories are major issues driving researchers to develop technologies to produce cheap and dependable energy supplies. Hydrogen (H2) is a promising potential energy carrier for meeting future energy needs because it is eco-friendly, it has a high heating value (2.75 times higher energy content than hydrocarbon fuels) and it can be converted into electricity via fuel cells (Benemann, 1996; Momirlan and Vezrioglu, 2002). Currently, H2 is mainly produced from fossil fuel-based feedstock (Oil/naptha reforming, coal gasfication, methane steam reforming etc.,) chemicals because many commercial petrochemical processes are cost effective (Ewan and Allen, 2005). However, the dark fermentation process is emerging as a practical route for producing H2 because of fast reaction rates, technical simplicity and less area (1-5%) requirements compared to the direct photolysis or indirect photolysis or photo fermentation (Levin et al., 2004; Li and Fang 2007).

The successful operation of the dark fermentation process requires a good understanding of how mixed microbial cultures behave under a variety of operating conditions. Identifying microbial communities can provide descriptions of the interaction between different groups; however, this approach may not provide sufficient information about the function of the communities. Noticeable changes in the microbial community composition within the H2 fermenting reactor is expected during operation. Quantifying the gene expression and monitoring fluctuations within the community structure will provide an understanding of the changes linked to the operation of bioreactors. This approach can possibly serve as a rapid tool for bioprocess monitoring.

In the present study, genes responsible for H2 production and H2 consumption were monitored as a means of establishing changes within the microbial community structure. The Fe-Fe hydrogenase gene is monitored because it is related to H2 production. Methanogens and acetogens are major H2 consumers and monitoring includes examination of the McrA gene and formyltetrahydrofolate synthetase (FTHFS) gene, respectively. A monitoring protocol for three different sources of cultures (A, B and C) at pH 5.5 and 37oC was developed using qRT-PCR. Terminal Restriction Fragment Length Polymorphism (T-RFLP) analysis was also performed to characterize changes in the community. The results indicated that the expression levels of [Fe-Fe]-hydrogenase, MrcA and FTHFS gene varied with the culture source. The [Fe-Fe]-hydrogenase gene expression was shown to correlate with H2 production. Increasing MrcA expression as well as increasing FTHFS and [Fe-Fe]-hydrogenase activity was correlated with low H2 yields of approximately 1.65±0.25 mol.mol-1 glucose in culture B. In comparison, cultures A and C with lower FTFHS and higher [Fe-Fe]-hydrogenase activities had higher H2 yields of 2.4±0.12 and 2.5±0.23 mol.mol-1 glucose, respectively. This data clearly suggest that monitoring the [Fe-Fe]-hydrogenase gene alone is insufficient to gain an understanding of how mixed cultures function and genes related to H2-consumers (MrcA and FTHFS) should be monitored and down regulated to achieve higher H2 yields.

References

Bennemann, J. 1996. Hydrogen biotechnology: progress and prospects. Nat. Biotechnol. 14, 1101-1103.

Ewan, B.C.R., Allen, R.W.K. 2005. A figure of merit assessment of the routes to hydrogen. Int. J. Hydrogen Energ. 30 (8), 809-819.

Levin, D.B., Pitt, L., Love, M. 2004. Biohydrogen production: prospects and limitations to practical application. Int. J. Hydrogen Energ. 29 (2), 173-185.

Li. C., Fang, H.H.P. 2007. Fermentative hydrogen production from wastewater and solid wastes by mixed cultures. Crit. Rev. Env. Sci. Tec. 37 (1), 1-39.

Momirlan, M., Veziroglu, T.N. 2002. Current status of hydrogen energy. Renew. Sust. Energ. Rev. 6 (1-2), 141-179.

See more of this Session: Fundamentals of Hydrogen Production and Utilization
See more of this Group/Topical: Environmental Division

(588e) Hydrogen Car Fill-up: Thermodynamic Modeling Using the Generic Cubic Equation of State

Fernando Olmos, Chemical and Biomolecular Engineering, UCLA, Los Angeles, CA and Vasilios Manousiouthakis, Chemical Engineering Department, University of California at Los Angeles, Los Angeles, CA

In order to provide hydrogen fueling stations with required information about hydrogen's pressure, temperature, and mass flowrate, hydrogen was modeled as a non-ideal gas employing the generic cubic equation of state (GC-EOS), along with its special four cases (VdW, RK, SRK, PR). With the purpose of validating the GC-EOS hydrogen thermodynamic property predictions, comparisons were carried out with NIST hydrogen's data, real-life hydrogen fueling station data, and with a mass balance equation in integrable form. Both the results of the GC-EOS and the mass balance model matched the NIST data and the hydrogen fueling station data describing the time evolution of the molar volume inside of the hydrogen vehicle tank. In addition, based on the concept of residual properties, an explicit expression for constant-pressure and constant-volume heat capacity was derived based on the GC-EOS as a function of molar volume and temperature. The results from the heat capacities, based on GC-EOS, matched NIST hydrogen's data at low pressures, and at pressures ranging from 100 to 1000 bar there is deviation less than 2%. Finally, the fueling station was modeled employing a total mass and energy balance, along with the GC-EOS, establishing that, in the absence of heat transfer, the maximum temperature achieved by hydrogen inside of the vehicle's tank does not depend on how high or low the mass flowrate is, but rather it is a function of the amount of mass accumulated in the tank.

See more of this Session: Fundamentals of Hydrogen Production and Utilization
See more of this Group/Topical: Environmental Division

(606a) Zero Liquid Discharge (ZLD) System for Flue-Gas Derived Water From Oxy-Combustion Process

Sivaram Harendra, Danylo Oryshchyn, Thomas Ochs, Stephen J. Gerdemann and John Clark, National Energy Technology Laboratory, U.S. Department of Energy, Albany, OR

Researchers at the National Energy Technology Laboratory (NETL) located in Albany, Oregon, have patented a process – Integrated Pollutant Removal (IPR) that uses off-the-shelf technology to produce a sequestration ready CO_2 stream from an oxy-combustion power plant. Capturing CO_2 from fossil-fuel combustion generates a significant water product which can be tapped for use in the power plant and its peripherals. Water condensed in the IPR® process may contain fly ash particles, sodium (from pH control), and sulfur species, as well as heavy metals, cations and anions. NETL is developing a treatment approach for zero liquid discharge while maximizing available heat from IPR. Current treatment-process steps being studied are flocculation/coagulation, for removal of cations and fine particles, and reverse osmosis, for anion removal as well as for scavenging the remaining cations. After reverse osmosis process steps, thermal evaporation and crystallization steps will be carried out in order to build the whole zero liquid discharge (ZLD) system for flue-gas condensed wastewater. Gypsum is the major product from crystallization process. Fast, in-line treatment of water for re-use in IPR seems to be one practical step for minimizing water treatment requirements for CO_2 capture. The results obtained from above experiments are being used to build water treatment models.

See more of this Session: Pretreatment/Treatment of Industrial Wastewater for Reuse
See more of this Group/Topical: Environmental Division

(606b) Sorptive Removal of Copper and Lead From Aqueous Solution by An Fe-Mn Binary Oxide

Lei Wang, Department of Civil Environemtnal Engineering, National University of Singapore, Singapore, Singapore, Gaosheng Zhang, Yantai Institute of Coastal Zone Research, Chinese Academy of Sciences, Yantai, China and **J. Paul Chen**, Department of Civil & Environmental Engineering, National University of Singapore, Singapore, Singapore

Abstract: Heavy metals sorption onto the surfaces of Fe-Mn binary oxides may play an important role in the mobilization and transformation of heavy metal because of the common occurrence of these oxides in the environment. However, the study of sorption behaviors of heavy metals on Fe-Mn binary oxides is less reported. Therefore, this study investigated systematically the sorption kinetics, edges and isotherms of Cu^{2+} and Pb^{2+} on a synthesized Fe-Mn binary oxide using a simultaneous oxidation and coprecipitation method. Additionally, the influences of coexisting Ca^{2+}, Mg^{2+} and humic acid on their sorption were also examined. The sorption of Cu^{2+} and Pb^{2+} on the Fe-Mn binary oxide was fast and was almost completed within two hours. The prepared Fe-Mn binary oxide had high sorption capacity towards Cu^{2+} and Pb^{2+}, especially the latter. The maximal sorption capacities are 0.96 and 1.42 mmol/g, respectively. The sorption of Cu^{2+} and Pb^{2+} was obviously pH-dependent and high pH value favored their sorption. The presence of Ca^{2+}, Mg^{2+} and humic acid had no great effect on Cu^{2+} and Pb^{2+} sorption. The Fe-Mn binary oxide has also the potentials to be used as a scavenger for heavy metals from water, due to its high sorption capacity and low cost.

See more of this Session: Pretreatment/Treatment of Industrial Wastewater for Reuse
See more of this Group/Topical: Environmental Division

(606c) Supported-Gas-Membrane Process for Removal and Recovery of Amines From Water Streams

Pengfei Shan[1], Liqiang Liu[2], **Yingjie Qin**[3] and Jie He[1], (1)the School of Chemical Engineering & Technology, Tianjin University, Tianjin, China, (2)Chembrane Engineering & Technology, Inc., Tianjin, China, (3)Chembrane Research & Engineering,Inc., Newark, NJ

Amines are widely used in a range of industries. For example, methylamine is mainly used for medicine (hormone or caffeine), agricultural chemicals (carbaryl, rogor, and so on) and explosive; dimethylamine is used in dimethylformamide (DMF) synthesis, agricultural chemicals, pharmacy, tannery and organic chemical industries; trimethylamine is used in the field of feed additive, weedicide, ion exchange resins, and so on ; ethylamine is used in the field of synthetizing dyes, extraction, emulsifier, rubber vulcanizing, medicine and also in petroleum refining; diethylamine is used for organic synthesis and curing agent for epoxy; triethylamine is mainly used as organic solvent, synthetizing dyes, inhibitor, catalyst, preservative, and so on; butylamine can be used as rubber accelerator, pesticides and so on. On the other hand, diethylamine-containing wastewater is produced as a by-product from DMF hydrolysis in the application of DMF in artificial leather industry. With the extensive application of amines and amine-derivatives, environmental impact resulting from amine emission can not be ignored, when more and more amine-containing wastewaters from chemical plants, refinery plants and pharmaceutical plants are discharged into rivers, seas and also influence greatly on the environment of urban community especially in China.

Amines are volatile, more basic than ammonia, more toxic than ammonia, and less biodegradable than ammonia. There are several conventional methods which have been applied to the treatment of amine-containing wastewater, such as chemical precipitation, biodegradation, ion exchange, air stripping, and so on. In these methods, air stripping is the most widely used in chemical industry, however with problems of high cost of equipment, high electric power or steam consumption, and secondary pollution of air around.

Nowadays, research on the applications of membrane-based absorption, and supported-gas-membrane (trans-membrane striping/absorption) is quite extensive for removing volatile and reactive components such as O_2, CO_2, NH_3, H_2S, SO_2, HCN, NO, Cl_2 Br_2, and I_2 from gaseous or water streams to another reactive aqueous solution. But until now, there is no report for application of gas-membrane stripping or supported gas membrane process to remove amine from its solution. Compared to conventional methods to remove amine from aqueous solution described above, supported gas membrane process has several prominent features and benefits. Firstly, the stripping of amine from the aqueous solution and the absorption of amine in another aqueous solution which is usually an acidic absorbent solution occur simultaneously in a membrane contactor by replacing a stripping (desorption) column and a absorption column, which provides a maximum driving force for ammonia removal. Secondly, the feed and stripping streams are located in each side of the microporous hydrophobic membrane and cannot influence each other, so the flexibility is greatly improved as to conventional column. Thirdly, the membrane contactor has high packing density and the hollow fiber membrane has small diameter, therefore the membrane contactor can supply much higher surface area per volume compared to conventional columns. Fourthly, compared to conventional separation contactors, the membrane contactor (hollow fiber modules) encounters no back-mixing, wall flow, by-pass flow, dead angle, flooding, foaming entrainment etc, so the overall transfer coefficient of this membrane-based separation process is much higher. And finally, when compared to air stripping or steam stripping, the operational cost of supported gas membrane separation process, almost eliminates the consumption of steam and only consume much lees amount of electricity to drive the wastewater to pass through the membrane module.

Aqueous solutions containing 500 – 10000 mg/L ammine was tested for supported-gas-membrane stripping by using aqueous solution of 5 – 20% sulfuric acid as absorbing solution. The amines tested included methylamine, dimethylamine, trimethylamine, ethylamine, diethylamine, triethylamine and tert-butylamine. Aqueous solution of ammonia was also tested as a reference feed. Experimental data demonstrated that supported–gas-membrane stripping is a highly efficient separation process to remove, to recover, to concentrate and to purify amine from it aqueous solution even containing salt and other volatile organics. It was found that the overall mass transfer coefficient is in such an order: trimethylamine > ammonia > triethylamine > diethylamine > ethylamine > dimethylamine > methylamine. The overall mass transfer coefficient is a function of the apolarity and volatility of amine (indicated by Henry's law coefficient), and of its molecular weight (indicated by diffusion coefficient of amine in aqueous phase and in gaseous phase).

A pilot plant-scale demonstration unit was operated for removal/recovery of diethylamine from its aqueous solution containing ca. 2000 mg/L diethylamine, which was distillate drained from a typical artificial leather plant. Techno-economic anylysis demonstrated supported gas membrane process is a highly efficient method to remove amine from wastewaters, even though the ideal by-product is highly purified amine in the concentrated solution.

See more of this Session: Pretreatment/Treatment of Industrial Wastewater for Reuse
See more of this Group/Topical: Environmental Division

RECYCLING BIOBUTANOL FERMENTATION SPENT BROTH BY ULTRAFILTRATION USING HOLLOW FIBER MEMBRANE

Michelle C. Almendrala, School of Chemical Engineering and Chemistry, Mapua Institute of Technology, Manila, Philippines. Email: michael.almendrala@yahoo.com;
Shang Tian-Yang, Chemical and Biomolecular Engineering, The Ohio State University, Columbus, Ohio, USA. Email: yang.15@osu.edu

1. Introduction

Butanol has been considered as a better alternative energy for fossil fuels than ethanol to solve the growing energy demands and rising oil prices. It is considered as the next generation biofuel which is more advanced and have properties very similar to gasoline. However, all fermentation processes like biobutanol production is a water-intensive process, thus, producing large quantity of fermentation spent broth that requires treatment. To minimize the discharge of effluent and prevent water pollution and groundwater contamination, recycling and reuse of wastetwater can be a sustainable solution. Innovative technologies are now being considered in water treatment and reclamation such as membrane filtration. Previous study of recycling spent ethanol broth for the subsequent fermentation was able to replace as much as 75% of the water and salts for subsequent batches. A 64% volume reduction of wastewater was achieved in this manner [1]. In a novel integrated fermentation for lactic acid fermentation, about 75% of spent broth was reused for three consecutive biomass recycle fermentations with no significant decrease in fermentation performance [2]. In a batch ethanol fermentation process, distillery waste was recycled using ceramic ultrafiltration membrane. Then, permeate was reused in the saccharification step and almost the same ethanol yield was obtained up to eight recycles and a slight increase in the fermentation by-products was observed [3]. Recent work on recycling ethanol fermentation broth was able to reuse permeate in the next cooking cycle using ceramic microfiltration membrane. Similar ethanol yield in up to thirteen recycling steps with only a slight increase in fermentation by-products was observed [4]. Another similar work using a two-stage anaerobic treatment process was used to totally recycle the wastewater from the ethanol fermentation step over 13 recycles [5].

In this study fermentation spent broth from butanol production will be recycled using ultrafiltration polysulfone hollow fiber membrane. Hollow fiber membrane can process viscous fluids and yield high surface membrane area-to-volume ratio which permits a high permeate flux. The operating differential pressure or known in membrane separation as the transmembrane pressure can affect the permeation flux and energy consumption of the membrane filtration process. Thus, permeate flux during ultrafiltration will be studied at various operating temperatures and transmembrane pressures. A mathematical equation will be developed to demonstate permeate flux decline. Ultrafiltration data will be used in the future optimization and scale up of the process. Since butanol will be the next generation biofuel, the data that will be gathered from this work will be used as an initial study in the full permeate recycling and reuse of biobutanol spent broth. Performance of fermentation using the recycled broth was investigated and compared with a control batch using fresh P2-medium. Analysis of the by-products of fermentation and their accumulation during the recycling will not be included in this work.

2. Methodology

2.1 *Organism*

A mutant strain of bacteria *Clostridium beijerinckii ATCC* 5505 (Bioprocessing Laboratory, Chemical and Biomolecular Engineering, Ohio State University, Ohio, USA) capable of producing a mixture of solvents consisting of acetone, *n*-butanol, and ethanol was used. The inoculum was prepared by injecting 3 ml of the spore suspension into 500 ml of the medium, incubated for 12 hours at 37°C.

2.2 *Bacterial Culture and Medium*

Clostridium beijerinckii ATCC 5505 was cultured in 500 clostridial medium which has the following typical formula: 30 g/L glucose, 3.0 g/L of yeast extract, 1 g/L of peptone, minerals and vitamins. For the batch fermentation, one liter medium mixture composition was 80 to 100 g of glucose, 1 to 3 g yeast extract, buffer: 0.75 g K_2HPO_4 and 0.5 g KH_2PO_4, minerals: 0.2 g $MgSO_4 \cdot 7H_2O$, 0.01 g $MnSO_4 \cdot H_2O$, 0.01 g $FeSO_4 \cdot 7H_2O$, 0.01 NaCl, and vitamins. Distilled water was used to dissolve the chemicals and permeate from ultrafiltered spent broth. The dissolved materials in distilled water was autoclaved at 15 psig and 121°C for 30 min and cooled. The sterilized medium concentrate was then mixed aseptically with recycled broth in a 5-L fermenter. The medium was flushed continuously with oxygen-free nitrogen gas for one hour to maintain anaerobic conditions in the vessel.

2.3 *Recycling Fermentation Spent Broth*

Figure 1 shows the bench-scale butanol fermentation process and recycling scheme. Fibrous-bed bioreactor FBB system with medium recirculation used in the fermentation process to produce n-butanol from pure glucose (P2-medium) using *Clostridium bjeirilenckii* ATCC 5505. The fibrous bed bioreactor was made of a glass column packed with spiral wound cotton towel. The column reactor was aseptically connected to a sterile 5-L stirred-tank fermentor (Marubishi MD-300) containing ~ 1 to2 L medium. The medium pH was adjusted to 5.1- 5.4 using pure NH_4OH, and the reactor temperature was controlled at 37ºC. The medium was continuously circulated at 20–50 mL/min via a peristaltic pump and flushed with filtered nitrogen to ensure anaerobiosis. The reactor was then inoculated with the cells by injecting 500 mL of a stock culture, and approximately 2 to 3 days were allowed for cell growth and immobilization in the fibrous matrix. After immobilization, a new batch of fermentation medium was inoculated and allowed for fermentation for 3 to 4 days. Samples were taken at regular intervals for the analysis of cell, substrate and product concentrations. The fibrous bed bioreactor was used for about 3 months and replaced once the fermentation results showed a large concentration of lactate. The butanol and other solvent products were separated from the broth by gas stripping. The spent broth was clarified by ultrafiltration hollow fiber membrane then completely reused. The permeate from the ultrafiltration was recycled to the cooking step for the next fermentation. The membrane permeate was completely recycled and mixed with tap water.

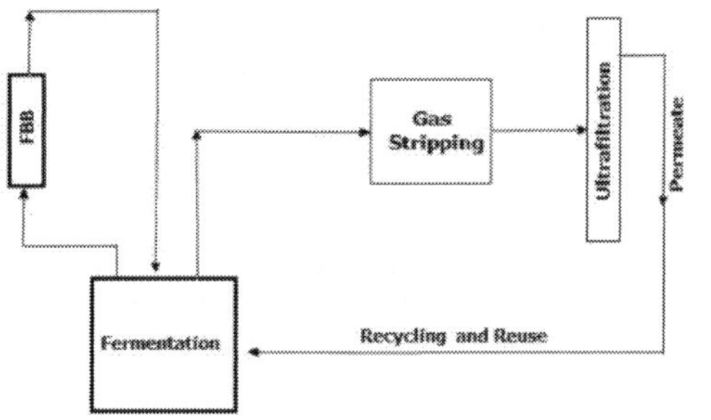

Figure 1. Recycling and reuse of fermentation spent broth.

2.4 *Ultrafiltration*

The recycling process for the fermentation of biobutanol using ultrafiltration hollow fiber membrane cartridge is illustrated in Figure 2. The butanol and other solvents produced from the batch fermentation was removed by gas stripping at 50ºC. The polysulfone hollow fiber membrane (Romicon HF 0.8-0.75 PM500, Koch Membrane Systems, Inc., Wilmington, MA) with a molecular weight cut-off of 500 kDalton was used to separate suspended solids and high molecular weight organics from the fermentation spent broth. The permeate was completely recycled and mixed with fresh medium for reuse in the next fermentation.

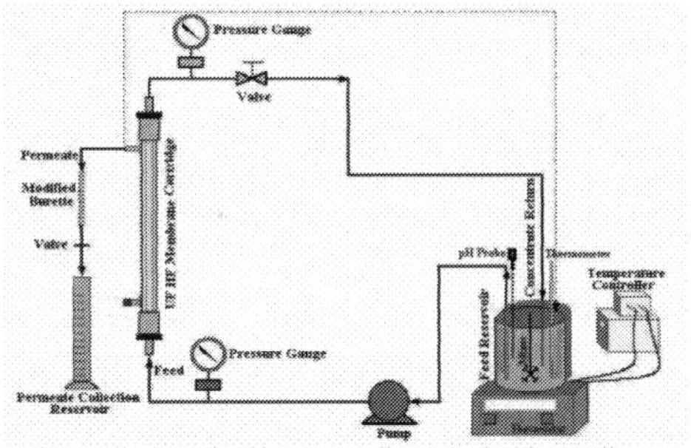

Figure 2. Ultrafiltration set-up for spent broth recycling.

2.4 *Ultrafiltration Flux*

Flux decline during crossflow UF of the fermentation broths of was studied, in which polysulfone membrane with a molecular weight cut-off of 500 kDa was used. All experiments were performed at a stirring speed of 10rpm but at different transmembrane pressures and temperatures. A mathematical model will be developed to analyze flux behavior, which involves permeation flux variation with time.

2.5 *Analytical Methods*

The cell density was measured by optical density (OD) at a 600-nm wavelength with a spectrophotometer (Model 3400; Sequoia-Turner, Mountain View, CA). The pH was measured with a pH/ORP Controller (model 5652-10; Cole Parmer, Vernon Hills, IL). After removing the bacterial cells by centrifuging at 16,100 *g* for 5 min, the clear fermentation broth was subjected to analysis for residual glucose and product concentrations. Glucose concentration was measured using a YSI model 2700 Select Biochemistry Analyzer (Yellow Springs, OH). Acetone, butanol, ethanol, acetic acid, and butyric acid were determined using a gas chromatograph (Varian 3400 GC) equipped with a flame ionization detector and an SP4270 integrator. The glass column (Supelco, Bellefonte, PA) was packed with 80/120 Carbopack BAW/6.6% Carbowax 20M. The oven temperature was programmed from 100ºC to 185ºC at a rate of 15ºC/min after an initial holding time of 1 min. The injector and detector temperatures were set at 200 and 225ºC, respectively. Nitrogen was the carrier gas, set at a flow rate of 20 mL/min. The fermentation kinetics and performance using recycled broth were evaluated.

3. Results and Discussion

3.1 Ultrafiltration of Spent Broth

3.1.1 Effect of Transmembrane Pressure on Permeation Flux

Figure 3 depicts the relationship between transmembrane pressure and flux for biobutanol fermentation broth. When the ultrafiltration process was started at a high pressure of 25 psi, much lower flux was observed and longer time of filtration was required in recycling the spent broth. At 19 psi higher permeate flux resulted in shorter ultrafiltration time. This behavior of flux can be explained by higher fouling at higher transmembrane pressure. The deposition of particles from fermentation which includes cells, organic by-products and other particles of fermentation broth must have formed a gel polarized layer on the membrane wall which added resistance to permeate flux [6, 7].

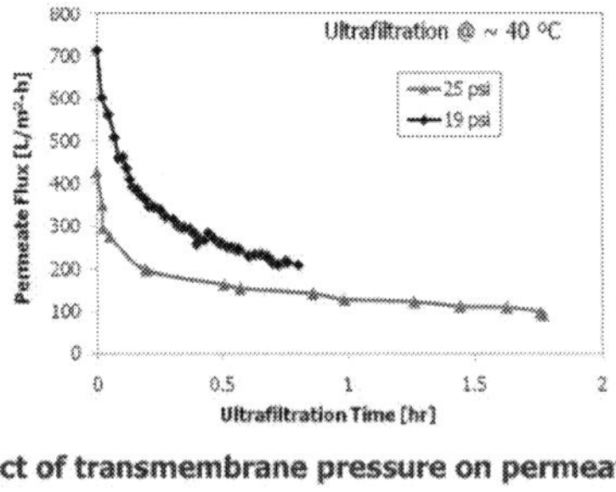

Figure 3. Effect of transmembrane pressure on permeate flux at 40°C.

3.1.2 Effect of Operating Temperature on Permeation Flux

Figure 4 shows the effect of temperature in permeate flux. Higher temperature resulted in higher permeation rate. Ultrafiltration at high temperature of 45°C gave the highest rate of permeation. The permeate flux decrease at a faster rate at high temperature but at 20 and 15ºC the flux remains almost constant but lower values. At relatively low temperature fouling is not significant in ultrafiltration of biobutanol spent broth. Deposition of the particles on membrane surface was slow, thus, polarization layer was not developed. At higher temperature the cells in the broth must have precipitated near the membrane surface thus causing gel polarization.

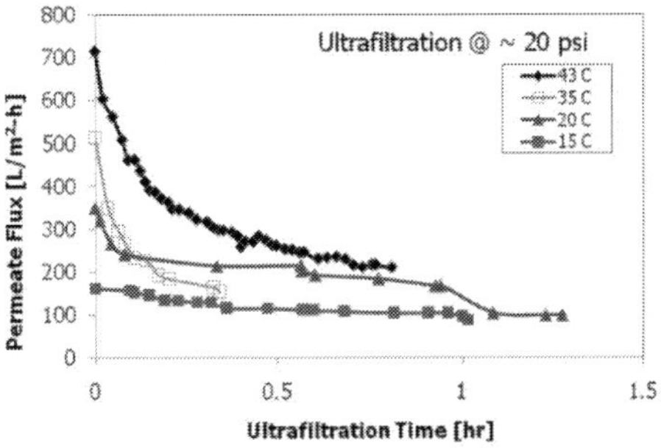

Figure 4. Effect of temperaure in permeate flux at 20 psi.

3.1.3 Permeate flux decline

An analysis of permeate flux decline encountered during ultrafiltration (UF) of biobutanol spent broth is presented by using a predictive model developed at constant temperature and transmembrane pressure. The model predictions match closely with the experimental flux behavior. The flux and time relationship developed is given by the equation:

$$J = \frac{220}{0.364 + t}$$

where J = permeate flux, L/m^2-h and t = u;trafiltration time, [s]. The complete form of permeate flux decline with time was obtained by solving for the two constants (a = 220 and b = 0.364) by least square method. Comparison of the calculated and measured time change of the flux is shown in Fig. 5. A good agreement between the experimental and calculated data was obtained. The standard deviation (S.D.) is 0.4 % under the ranges studied.

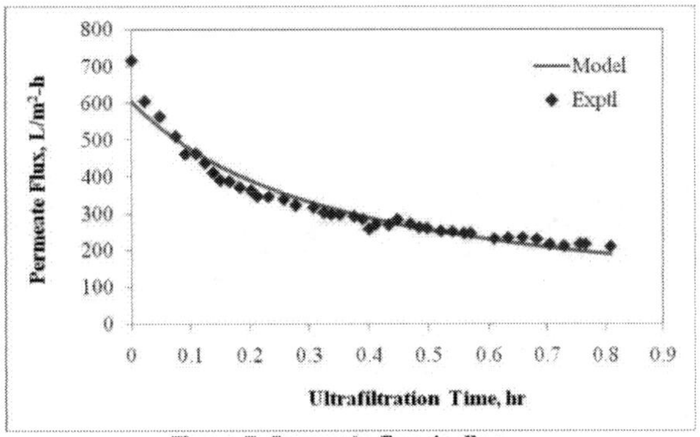

Figure 5. Permeate flux decline

3.2 Fermentation Medium: Effect of Increased Salt Concentration

The recycled medium used in all batch fermentation was approximately 80% of the actual medium as shown in Table 1. For the recycled medium the buffer was not added to be able to reduce the effect of excess ions in the broth which may inhibit the fermentation process. As seen in the result obtained in 2nd Recycle Run2 in Table 1, when buffer was added in the second recycle the butanol titer obtained was the lowest. In addition, butanol titer and productivity also decreased in the last batch of fermentation as shown in Table 2. Therefore, the accumulation of salt in the fermentation broth had negative effect on cell growth and rate of fermentation [9]. The addition of buffer solution and minerals in each recycle favors the accumulation of metal ions in the fermenter that under a certain critical concentration of these ions the cell growth inhibition will occur[3].

Table 1. Fermentation medium composition.

Medium Composition	Control Batch		1st Recycle		2nd Recycle	
	Concn	*Unit*	*Run 1*	*Run 2*	*Run 1*	*Run 2*
Glucose	90 - 100	(g/L)	88.20	102.50	121.78	122.79
Yeast Extract	2	(g/L)	1.25	2.25	1.16	2.00
KH_2PO_4	0.75	(g/L)	0.00	0.00	0.00	0.50
K_2HPO_4	0.75	(ml/L)	0.00	0.00	0.00	0.75
*Vitamins	1	(ml/L)	0.83	0.93	1.05	1.00
*Minerals	5	(ml/L)	4.17	4.66	4.36	6.99
DI Water	Vol	(ml)	200.00	209.00	320.00	274.80
Recycled Broth	Vol	(ml)	1000	1400	1400	1128
% Recycled Broth			83.33	87.01	81.40	80.41

*Prepared in the Bioprocessing Laboratory, Chemical and Biomolecular Engineering, Ohio State University

3.4 Effect of Recycle in ABE Fermentation Products Concentration

The ratio of acetone-butanol-ethanol in fermentation broth using fresh medium is 3:6:1 as in Control Batch while the ABE ratio using the recycled medium is 2:7:1 based on data in Table 2. The variation in the ratio of the two types of fermentation process may be attributed to the presence of inhibitors, that may affect the formation of ABE products. It was observed that there was a two-fold increase in acetic acid concentration in the first recycle and three-fold increase in the second recycle. There was a six-fold increase in butyric acid concentration in the first recycle and nine-fold increase in the second recycle. This can be attributed to the high concentration of glucose in the recycled medium which favors the formation of the acids.

Table 2. ABE Fermentation Performance

Fermentation Performance	Control Batch	1st Recycle		2nd Recycle	
		Run 1	*Run 2*	*Run 1*	*Run 2*
Fermentation Time, hr	94.72	81.58	70.18	90.30	88.95
Acetone, g/L	9.83	5.00	3.95	3.80	2.58
Ethanol, g/L	1.43	1.60	2.59	2.88	4.32
Butanol Titer, g/L	19.39	14.85	14.43	13.04	11.20
Butanol Productivity, g/L-hr	0.21	0.18	0.21	0.14	0.13
Butanol Yield, %	21.07	19.82	37.97	13.81	21.21
Total Acids	6.62	19.46	19.31	27.14	35.95
Total ABE	30.64	21.44	20.97	19.72	18.09
% Glucose consumption	99.58	85.02	46.34	60.98	42.43

3.5 Effect of recycling in fermentation performance

In this work, the fermentation broth was recycled up to two recycles. Several runs were conducted for the first recycle and second recycle. According to Figure 6 and the results obtained in Table 2, a conclusion was reached that the fermentation using recycled broth is feasible since there was a significant amount of butanol produced during the fermentation. The percentage yield of butanol using the fresh and recycled medium was maintained at approximately 21% and the time of fermentation was not prolonged. Although titer is lower in the recycled medium the productivity of butanol in the firat recycle was almost comparable to that in the fresh medium. The productivity in the second recycle as shown in Table 2 was reduced to 14 g/L-hr. The accumulation of the low molecular weight organics and salts that can pass through the ultrafiltration membrane Therefore, some organic compounds and salts could not be separated by ultrafiultration and accumulation is likely to happen along with the consecutive recycle steps. The increase in organic compound in the recycled broth was observed as the color of the permeate becomes darker. The quantitative analysis of the organic compounds in the permeate as well as the cocentration of the ions was not included in this study. The glucose utilization was reduced in the second recycle which was indicated by the increase in residual total glucose in the fermentation broth.

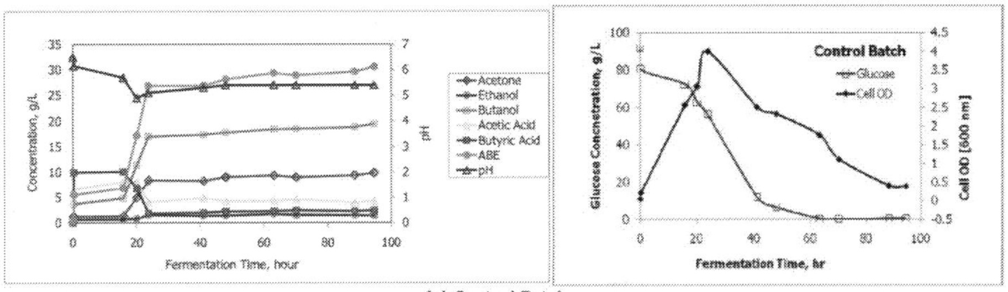

(a) Control Batch

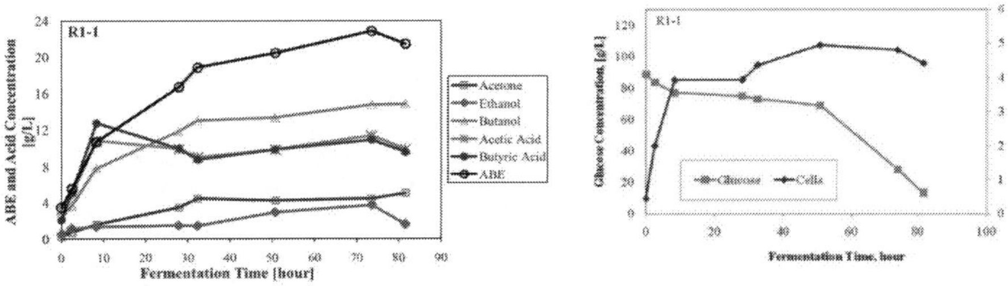

(b) Recycle 1- Run1

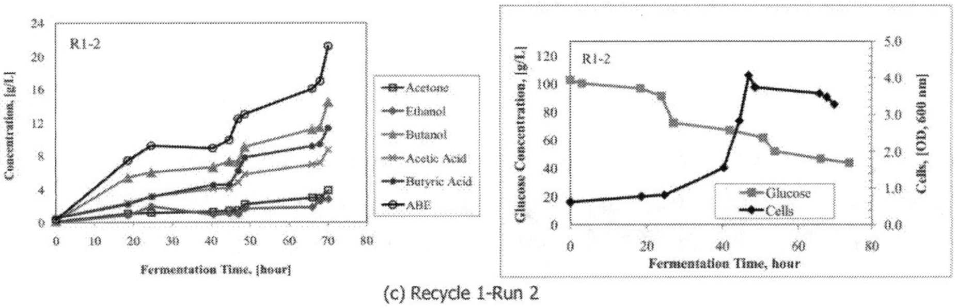

(c) Recycle 1-Run 2

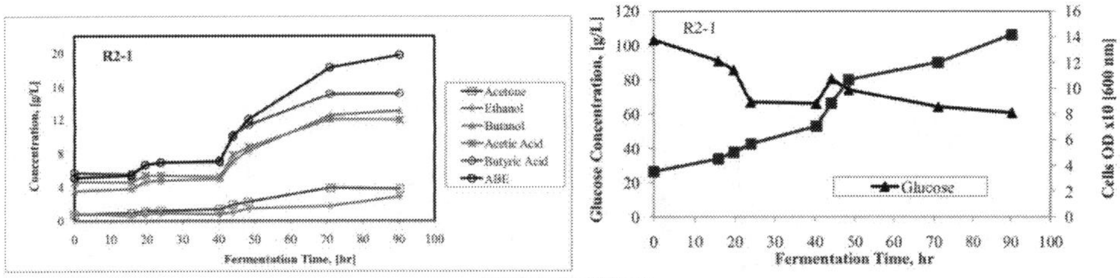

(d) Recycle 2-Run1

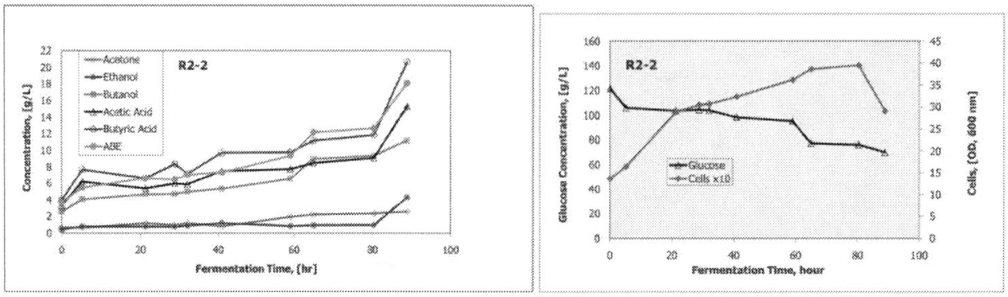

(e) Recycle 2-Run2

Figure 6. Fermentation kinetics of biobutanol

In Figure 7, the effect of the number of recycles in butanol fermentation was demonstrated. The concentration of butanol versus the time of filtration showed that there was decrease in butanol production as the number of recycles increase. This effect can be explained by the accumulation of metal ions and low molecular weight organic compounds in the fermentation broth. The ultrafiltration process was not able to remove the excess ions and small organic molecules in the fermentation broth. The accumulation of these particles as the number of recycles increase causes the inhibition to bacterial growth, thus, reducing the performance of fermentation.

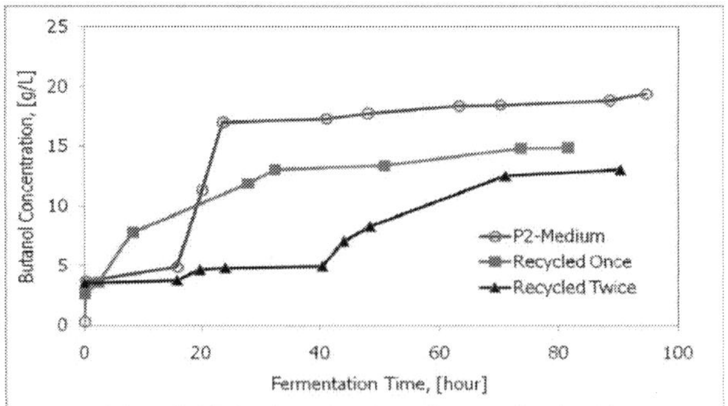

Figure 7. Effect of number of recycles to butanol yield.

4. Conclusion

A mathematical model was developed to show the relationship of flux decline with time. Permeation rate of spent broth was observed to increase at higher values of temperature and pressure. Performance of fermentation using the recycled broth showed that the biobutanol titer was reduced from ~21 g/L to ~14 g/L and ~12 g/L in the first and second recycle, respectively. After two consecutive recycle of the broth, glucose consumption continuously decreased which resulted in lower productivity of butanol. The fermentation time, on the other hand, was not prolonged but productivity was reduced from 0.22g butanol/L-hr to approximately 0.17 productivity in the first recycle and 0.14 in the second recycle.The reduction in yield, however, was observed to be less significant from 23% to 21%. The decrease in productivity can be attributed to the accumulation of low molecular weight organics, and salts in the recycled permeate which inhibit fermentation. Based on the results ultrafiltration can be considered as a cost effective method of recycling fermentation broth up to second recycle in this study.

The recovery of clarified fermentation broth for reuse was possible up to 80 to 90% v/v of the spent broth. The ultrafiltration process can be operated at low temperature and pressure since all data obtained for the rate of permeate flux were significantly high and no severe fouling was observed during the operation. In addition, data indicated that full filtrate recycling during butanol production is feasible since the yield of butanol in fermentation with recycled medium did not change significantly. The fermentation time in all recycles was not prolonged and the productivity of butanol in the first recycle was comparable to the results obtained using the fresh medium. The accumulation of metal ions and the low molecular weight organic compounds in the broth at some crtitical concentration inhibit the cell growth and therefore, reduce the fermentation rate, and decrease butanol yield and productivity. To improve the process there is a need for an additional separation process to reduce the concentration of the metal ions and low molecular weight organic compounds in the recycled broth. Another component of the broth that must be reduced is the amount of total residual glucose in the recycled broth. The glucose utilization was reduced as the number of recycles increase and this increased glucose concentration will cause the increased production of acetic and butyric acid which may inhibit the cell growth for the production of butanol. There are several ways in which the amount of inhibitors can be reduced in the process. An integrated ultrafiltration and nanofiltration membrane separation processes will be able to reduce the concentration of the accumulated metal ions and the low molecular weight organic compounds. In order to reduce the residual sugar and other organic inhibitors in the spent broth, separation of the concentrated solution of sugar and organic waste in the broth by freeze concentration is possible. Therefore, integrating one of these separation process in recycling the broth will make the process efficient and therefore, will result in less pollution with waste water. Another objective that will be met by the process is high and efficient butanol production with a low cost in raw materials and energy consumption. Further research on the determination of the organic compounds in the spent broth and concentration of metal ions in each recycle must be conducted to be able to determine the amount of metal ions that can be reused in the fermentation, thus, reducing the amount of material needed in the fermentation and reduce the cost of fermentation process.

Acknowledgment

This research was supported by grants from the Philippine-American Educational Foundation-Fulbright Program (PAEF-Fulbright Philippines), the Department of Chemical and Biomolecular Engineering of the Ohio State University, and by the Bioprocessing research group Dr. Shang-Tian Yang, Ohio, USA.

References

1. Hsiao, T.Y., Glatz, C.E., Glatz, B.A.(1994), Broth recycle in a yeast fermentation. Biotechnol Bioeng. Vol. 44(10):1228 34. Retrieved from http://www.ncbi.nlm.nih.gov/ pubmed/18618549 on July 8, 2008. Hsiao. "Broth recycle in a yeast fermentation." *Biotechnology and Bioengineering*, v. 44 issue 10, 2004, p. 1228.
2. Ye, K., Jin, S., Shimizu, K. (1996), Cell recycle and broth reuse fermentation with cross-flow filtration and ion exchange resin. Dept. of Biochemical Engineering and Science, Kyushu Institute of Technology, Japan. 08
3. Kim, JS., Kim, BG., and Lee, CH., (1999), Distillery waste recycle through membrane filtration in batch alcohol fermentation. Biotechnology Letters, Springer Netherlands, Vol 21, No. 5, 401-405
4. Ding, Z., L. Zhang, Y. Fang, L. Xu, K. Zhang, G.Y. Shi, (2009). Application of full permeate recycling to very high gravity ethanol fermentation from corn, Korean J. Chem. Eng., 26(3), 719-723.
5. Zhang, Q.H., X. Luc, L. Tanga, Z.G. Maoa, J.H. Zhanga. (2010). A novel full recycling process through two-stage anaerobic treatment of distillery wastewater for bioethanol production from cassava. Journal of Hazardous Materials 179, 635–641.
6. Fane, A.G. Ultrafiltration: :Factors influencing flux and rejection, in R. J. Wakeman (ed.) Progress in Filtration and Separation, Elsevier (1986) 101 – 179.
7. Wang, L. and Song, L. Flux decline in crossflow microfiltration and ultrafiltration: Experimental verification of fouling dynamics. J. Memb. Sci., 160 (1999) 41-50.
8. Cheryan, M. Ultrafiltration Handbook. Technomic Publishing Co. Inc., Lancaster, PA. 1986.
9. Nanba A, Nishizawa Y, Tsuchiya Y, Nagai S (1987) J. Ferment. Technol. 65: 277–283.

(606e) An Investigation Into the Use of Fluorinated Hydrating Agents In the Desalination of Industrial Wastewater

C N. Petticrew[1], D. Ramjugernath[1], C. Buckley[1], Amir H. Mohammadi[2] and P. Naidoo[1], (1)Chemical Engineering, University of KwaZulu-Natal, Durban, South Africa, (2)CEP-Tep, MINES ParisTech, Fontainebleau, France

Salts are removed from industrialwastewater due to their fouling and equipment corrosion effects. Currentdesalination technologies include reverse osmosis, multi stage flashdistillation, vapour compression, electrodialysis and thermal distillation. Thecurrent desalination technologies are sufficient in retrieving pure water froma low solute concentrated feed, however such processes are energy intensive, withscaling of saturated sulphates and membrane damage due to the presence ofchloride (Khawaji et al., 2008).Thereis therefore a need for recovering water from the concentrated brine solutionat ambient temperatures and pressures. Membrane distillation and gas hydratetechnology are currently being investigated (Osman et al., 2010) as suitablefeasible alternative technologies. During the process of membranedistillation, crystal disengagement and a rapid decline in the distillate flux mayoccur, resulting in the flux tending to zero (Mariahet al., 2006). As a result, desalination using gas hydrate technology, inparticular using hydrofluorocarbons as formers is currently being investigated. Hydrates are composed of guest molecules, hydrofluororcarbons, and hostmolecules, water. When hydrates form, normally under high pressures and lowtemperatures, the host molecules form a cage, enclosing the guest molecule.This guest molecule may be in the form of a gas or a liquid. Desalination usinggas hydrate technology includes the formation of hydrates by mixing industrialwastewater and the former at the hydrate forming temperature and pressure.Hydrate crystals will form leaving a concentrated salt solution. The solutionwill be filtered or decanted and the hydrate crystals will be washed to removeexcess salts attached to the crystals. The former will then be separated fromthe fresh water and re-injected into the feed, while the fresh water can bereused and the salts recovered (McCormack and Anderson,1995). The use ofhydrofluorocarbons as hydrate formers reduces the hydrate dissociation pressurewhile increasing temperature. Criteria for choosing the best fluorine formersrequire the former to be environmentally acceptable; non-toxic; non-flammable; chemicallystable; a class 2 hydrate; available in commercial quantities; low cost;compatible with standard materials; as well as contain a high critical point (McCormackand Anderson, 1995). Thus R22, and R134a were chosen as possible formers andwill each be investigated in the presence of various concentrations of NaCl and$CaCl_2$. According to Javanmardi and Moshfeghian (2003), to enable gashydrate technology to be used as a competing desalination technology, energycosts must be reduced. A gas hydrate promoter, cyclopentane, which will allowhydrates to form at ambient temperatures and moderate pressures will beinvestigated in the presence of the fluorinated refrigerants and various salts.

These systems will be measured using a static high pressure equilibrium cell (Tshibangu,2010) and the results will be modeled. The hydrate phase will be modeled usingthe method developed by Eslamimanesh et al. (2010), while the liquid phase willbe modeled using the Aasberg Peterson method (Aasberg Peterson et al., 1991).

References

1. Aasberg-Petersen.K, Stenby.E, Fredenslund.A, (1991), Prediction ofHigh-Pressure Gas Solubilities in Aqueous Mixtures of Electrolytes, Ind. Eng.Chem. Res., (30), 2180-2185

2. Eslamimanesh.A, Mohammadi.A, Richon.D, (2010), Thermodynamic Model forPredicting Phase Equilibria of Simple Clathrate Hydrates of Refrigerants,Fontainebleau

3. Javanmardi J, Moshfeghian, (2003), Energy consumption and economicevaluation of water desalination by hydrate phenomenon, Applied Thermal Engineering,23, 845-857.

4. Khawaji.A,Kutubkhanah.I, Wie.J, (2008), Advances in seawater desalination technologies, *DesalinationJournal* (221), 47-69

5. Mariah.L, Buckley.C, Naidoo.P, Brouckaert.C, Ramjugernath.D, Curcio.E,Driolo.E and Jaganyi.D, (2006), Thermodynamic Modelling of Vapour Pressuresduring the membrane distillation and crystallization of concentrated mixedbrines, MSc Thesis, Chemical Engineering, University of KwaZulu-Natal

6. McCormack.A, Anderson.R, (1995), Clathrate Desalination PlantPreliminary Research Study, Water Treatment Technology Program Report No. 5, ThermalEnergy Storage Inc, California

7. Osman.M, Schoeman.J and Baratta.L, (2010), Desalination / concentrationof reverse osmosis and electrodialysis brines with membrane distillation, *Desalinationand water Treatment,* (24),293-301

8. Tshibangu, M. M. (2010).Measurements of HPVLE data for fluorinated systems, MSc Thesis, University ofKwa-Zulu Natal

See more of this Session: Pretreatment/Treatment of Industrial Wastewater for Reuse
See more of this Group/Topical: Environmental Division

(606f) Titanium-Loaded Photocatalytsts for Removal of Organic Dyes From Wastewater Using Visible Light Irradiation

Ahmed Arafat, Muhammad A. Daous and Essam Y. Al-Shareef, Chemical & Materials Engineering Department, King Abdulaziz University, Jeddah, Saudi Arabia

Heterogeneous photocatalytic oxidation (HPO) processes employing catalystssuch as TiO_2, ZnO, etc., using UV lighthas shown promising results for the degradation of persistent organicpollutants, and their conversion to more biologically degradable and less toxicsubstances. However, a large number of organic contaminants cannot be fullydegraded by such methods. In order to meet increasingly more stringentenvironmental regulations and laws, more efficient processes for the removal oforganic pollutants are needed. Among the most promising and advanced methodsbeing investigated, photocatalysis at semiconductorsand at TiO_2, in particular, receives a considerable attention asreflected by the recent large body of research published in this field.Titanium dioxide, however, can be easily degraded when subjected to therelatively harsh UV irradiation. Moreover, use of UV irradiation is notattractive in a large industrial scale water treatment processes because of itspotential harmful effects on humans. Both of the problems cited above could be resolvedif UV irradiation could be replaced by a harmless visible light irradiation.

Titanium dioxidenanoparticles successfully anchored into 3-D mesoporous material (TUD-1) usinga sol-gel preparation strategy were prepared and tested as photocatalysts,employing only visible light ($\lambda = 447$ nm), for the treatment ofwastewaters containing azo dyes. Samples of differentSi/Ti ratio, ∞, 100, 75, 50 and 25 and 10 *w/w* were prepared andcharacterized by X-ray diffraction (XRD), transmission electron microscopy(TEM), ultraviolet spectroscopy (UV), and nitrogen adsorption at 77K. The newly developed Ti-TUD-1 photocatalysts showed excellent activity towards theremoval of the tested Azo dyes solutions. Up to 98%of azodyes were successfully removed from thecontaminated wastewater. Moreover, the catalysts showed remarkable stabilityunder the relatively mild irradiation conditions used. Repeated photodegradation tests demonstrated remarkablereproducibility and showed no decrease in photocatalyticactivity after 5 cycles.

See more of this Session: Pretreatment/Treatment of Industrial Wastewater for Reuse
See more of this Group/Topical: Environmental Division

(613a) Comparison of Amine-Based Absorbents for Post-Combustion Carbon Capture

Sumedh S. Warudkar[1], Kenneth R. Cox[1], Michael S. Wong[2] and George J. Hirasaki[1], (1)Chemical and Biomolecular Engineering, Rice University, Houston, TX, (2)Chemistry Department, Rice University, Houston, TX

Asa part of this work, we have performed a comparison of 4 amine-based absorbentsusing the process simulation software ProMax with TSWEET.Absorbents were compared at pulverized coal fired power plant flue gas conditions.The criteria for comparison of absorbents are the energy requirement andabsorber and stripper dimensions for 90% CO_2 capture (separation +compression) from a 500 MW pulverized coal fired power plant which are looselyrepresentative of the operating and capital cost respectively. Due to the largeflow-rate (~1100 MMSCFD) of flue gas that must be processed by the separationsystem, use of 3 absorber/stripper trains is proposed. Separated $CO_2$is compressed to a pipeline pressure of 8MPa using a 4 stage compressor train.The amine absorbents considered for comparison are monoethanolamine (MEA),diethanolamine (DEA), diglycolamine (DGA) and 2-amino-2-methyl-1-propanol(AMP). The parameters explored as a part of this study includes absorbentconcentration, lean absorbent entry temperature into the absorber column,stripper pressure. On the basis of our findings, we report that diglycolamine(DGA) and diethanolamine (DEA) are superior absorbents than monoethanolamine(MEA) and 2-amino-2-methyl-1-propanol (AMP) for CO_2 capture. Ahigher stripper pressure reduces the overall energy requirement for CO_2capture by decreasing the stripping steam requirement as well as reducing thecompression costs. Figure 1 and Figure 2 are plots of change in stripperdiameter and overall energy duty for CO_2 separation with change inthe stripper pressure.

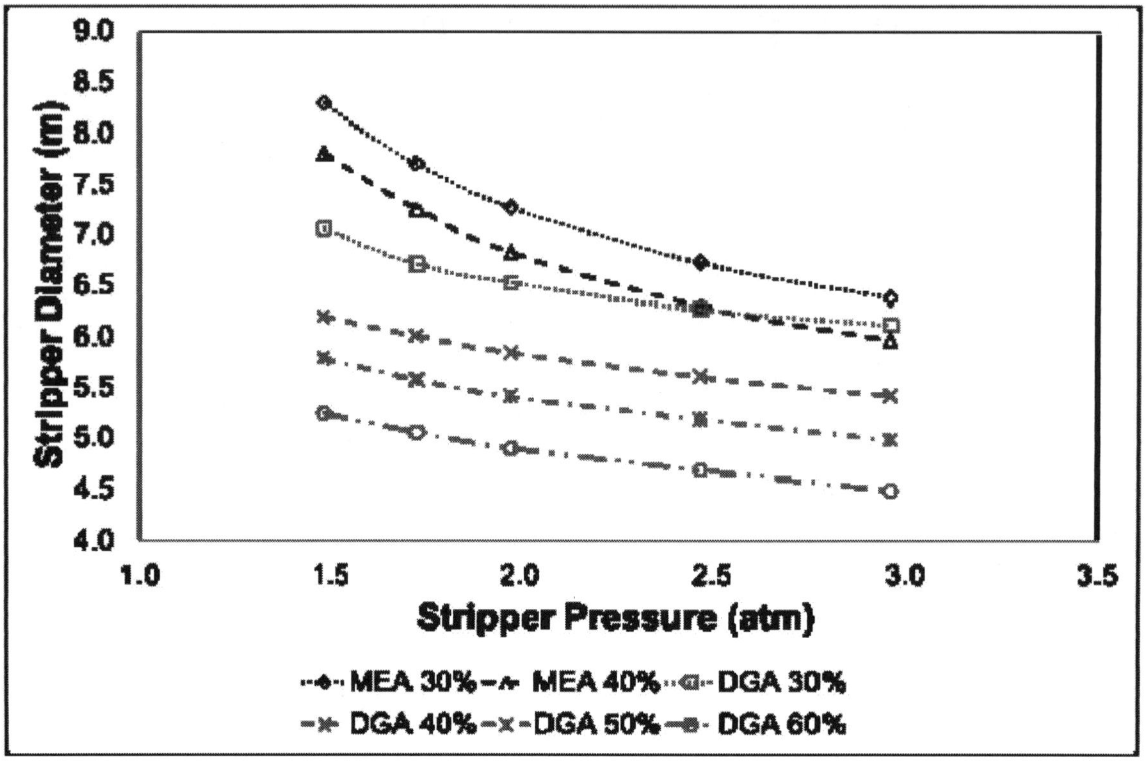

Figure1: Stripper Diameter Vs. Stripper Pressure

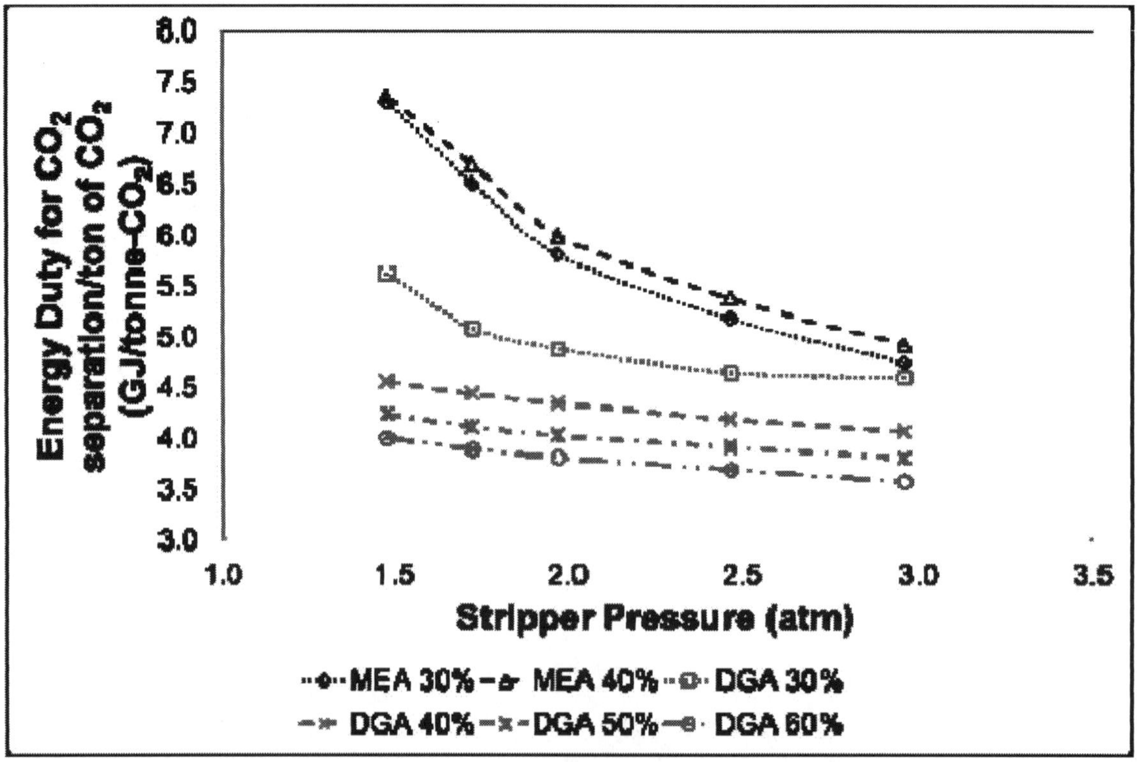

Figure2: Energy Duty for CO_2 separation Vs. Stripper Pressure

See more of this Session: Sorbent Processes for Carbon Capture
See more of this Group/Topical: Sustainable Engineering Forum

(613b) Study of CaO Carbonation for CO2 Capture Process

Lydie Rouchon, **Loc Favergeon** and Michle Pijolat, Centre SPIN, Ecole Nationale Suprieure des Mines, Saint-Etienne, France

Anthropogenic carbon dioxide (CO_2) emissions, major contributors to the greenhouse effect, are considered as the main cause of global warming. So, decrease of CO_2 emitted by large industrial combustion sources or power plants, is an important scientific goal. One of the approaches is based on CO_2 separation and capture from flue gas, followed by sequestration in a wide range of geological formations. In this aim, CO_2 is captured by sorbents like calcium oxide (CaO) in multi-cycle process of carbonation/decarbonation. However, it was shown that the most important limitations of such process are related to the reversibility of reaction. CaO rapidly loses activity towards CO_2, so the maximum extent of carbonation decreases as long as the number of cycles increases [1].

In order to well understand the processes and parameters influencing the capture capacity of CaO-based sorbents, it appears important to understand the mechanism of calcium oxide carbonation and to get details on the kinetic law governing the reaction, which has not been really studied up to now. To investigate this mechanism, CaO carbonation kinetics was followed by means of thermogravimetric analysis (TGA) on divided materials. Special care was given to the validation of the usual kinetic assumptions such as steady state and rate-determining step asumptions [2-3].

The aim was to obtain a complete model describing the reaction in order to explain the influence of intensive variables such as carbonation temperature and CO_2 partial pressure. TGA curves obtained under isothermal and isobaric conditions showed a strong slowing down of the reaction speed once a given fractional conversion was reached. The slowing down and its corresponding fractional conversion was observed to depend on carbonation temperature and CO_2 partial pressure. To explain these results, the evolution of texture and microstructure of the solid during processing was regarded as essential. Reaction at the grain scale induces a volume increase from CaO to $CaCO_3$ which causes a change in the porosity characteristics at the aggregates scale, which could block the access of the gas to the core of aggregates. So carbonation reaction can be described by a two scales model: at a nonporous grain scale for the chemical reaction and at the aggregate scale, for the CO_2 intergranular diffusion. The kinetic modeling is able to couple both modeling scales in order to explain the kinetic slowing down and the influence of temperature and CO_2 partial pressure on the reaction rate.

The coupling is realized by the combination of the thermohydraulic modeling (heat transfer, mass transfer and hydrodynamic) at the reactor scale based on the finite element method, with the kinetics of heterogeneous grains population in aggregates structured: at reactor scale, for heat, hydrodynamic and mass transport problems, a finite element formulation was built. A classical way is followed to obtain weak formulations for each state variable: temperature, charge and partial pressures. At microscopic scale, the reaction fractional conversion α is followed for a representative population of grains. By finite difference it is possible to evaluate the reaction rate $d\alpha/dt$ for a given temperature and given partial pressures. This rate allows to evaluate heat and mass sources produced by the reaction. So, using these sources terms, microscopic reactions have an impact on the spatial and temporal evolution of the thermodynamic processes at the reactor scale. Inversely since it modifies the areic frequency of nucleation and the areic reactivity of growth, thermodynamic influences fractional conversion and reaction rate of the microscopic reaction.

[1] J.C. Abanades, D. Alvarez. E*nergy & Fuels,* **2003**, *vol. 17*, pp. 308-315.

[2] M. Pijolat, F. Valdivieso, M. Soustelle. *Thermochimica Acta,* **2005**, *vol. 439 (1-2)*, p 86.

[3] M.Pijolat, M. Soustelle. *Thermochimica Acta*, **2008**, *vol. 478 (1-2)*, pp. 34-40.

See more of this Session: Sorbent Processes for Carbon Capture
See more of this Group/Topical: Sustainable Engineering Forum

(613c) Multiphase Porous Medium Adsorption/ Desorption Model for CO2 Capture Applications

Denis Lukanin[1], Amit Halder[2] and Ameya Joshi[2], (1)Corning SNG, St. Petersburg, Russia, (2) Modeling and Simulation, Corning Incorporated, Corning, NY

Multiphase porous medium adsorption/desorption model for CO2 capture applications

Denis Lukanin[1*], Amit Halder[2], and Ameya Joshi[2]

[1]Corning Science Center, Corning SNG, St.Petersburg (Russia)

[2]Modeling and Simulation Department, Corning Inc., Corning,NY (USA)

*lukanindp@corning.com

Introduction

During the past decade, there has been much interest in CO2 capture and sequestration technologies due to emerging need for green-house gas mitigation. Adsorption of CO2 from gas mixtures by solid adsorbents gives one possible solution to the problem. Most widely used solid adsorbent reactors are conventionally based on packed beds loaded with various types of zeolite, activated carbon, or polymer adsorbent materials [1]. A number of adsorption-regeneration strategies are used in industry, such as the Pressure Swing Adsorption (PSA), Vacuum Swing Adsorption (VSA), Temperature Swing Adsorption (TSA), carrier gas purge, and their combinations [3].

In this work, we present a comprehensive bed-scale numerical tool allowing for the simulation of packed-bed adsorption reactor operation. The model is developed on the basis of FLUENT® commercial fluid dynamics solver and involves computation of non-isothermal gas mixture flow through a packed bed, accompanied with the mechanism of CO2 extraction from feed mixture and its accumulation in the solid phase. We have also incorporated a thermal effect of adsorption in the model and developed an algorithm for separate computation of fluid- and solid-phase heat transfer, which was not readily available in the software. Various adsorbent regeneration (desorption) regimes can also be modeled with the developed tool.

For the verification of the model, we use experimental data on VSA CO2 adsorption cycle in 13X Zeolite-based packed beds borrowed from the papers by Xiao et al (2008) [2] and Zhang et al (2008) [3].

Model description

The model is based on numerical solution of Navier-Stokes equations describing mixture gas flow dynamics in the porous medium that represents the packed bed. The set of equations is accompanied with a transport equation for CO2 mass fraction. A separate CO2 accumulation equation in the solid phase is incorporated in the model.

Non-isothermal nature of flow in the packed bed is modeled with two separate energy equations for solid and gas phases. We account for the heat released during adsorption by incorporating the respective heat source term depending on the amount of adsorbed CO2 in the solid phase energy equaiton.

Results and Discussion

Verification of the model has been done using experimental data on CO2 adsorption taken from the paper [3] by Zhang et al. This reference contains time-evolution curves of outlet CO2 concentration for two adsorption experiments at different inlet flow rates: Q=115LPM and Q=66LPM. We used the first experiment to adjust unknown free parameters of the model. With these adjusted parameters we ran the simulation for the low flow rate and obtained pretty reasonable agreement between computed and experimental results.

References

1. Wei Liu et. al *JOM* Vol. 61 No. 4 36-44 (2009)

2. Penny Xiao et. al *Adsorption* 14 575–582 (2008)

3. Jun Zhang, Paul A. Webley, Penny Xiao, *Energy Conversion and Management* 49 346–356 (2008)

See more of this Session: Sorbent Processes for Carbon Capture
See more of this Group/Topical: Sustainable Engineering Forum

(613d) Engineering Amine-Based Solid Adsorbents for Carbon Capture From Dilute CO2 Streams

Sunho Choi[1], McMahan L. Gray[2], Taku Watanabe[3], David Sholl[3] and Christopher W. Jones[1], (1) School of Chemical & Biomolecular Engineering, Georgia Institute of Technology, Atlanta, GA, (2)U.S. Department of Energy, National Energy Technology Laboratory, Pittsburgh, PA, (3)School of Chemical and Biomolecular Engineering, Georgia Institute of Technology, Atlanta, GA

A prominent methodology to tune the adsorption properties of solids for applications in CO2 capture is to functionalize them with amines sites, as the amine-CO2 interaction is strong under a variety of conditions [1]. Specifically, amine-modified solids such as porous silicates and MOFs are now well-studied under conditions that simulate post-combustion CO2 capture. In addition, we have recently shown that supported amine materials are promising solids for removal of CO2 from ultra-dilute gas sources such as the ambient air [2-4].

This talk will focus on the engineering of inorganic-organic hybrids amine adsorbents to enhance their CO_2 adsorption performance, with a specific focus on CO_2 adsorption capacity, adsorption kinetics and stability to practical desorption conditions. The structures of various hybrid silica-based and MOF-based adsorbents were modified by tuning the both the host substrate and the surface amine functionalities. Structure-property relationships for these materials are under development, and based on these studies, we will discuss the potential to design inorganic-organic hybrid adsorbents with desired CO_2 adsorption characteristics, such as the ability to extract CO_2 efficiently from ambient air.

[1] S. Choi, J. H. Drese, C. W. Jones, ChemSusChem 2009, 2, 796-854.

[2] S. Choi, J. H. Drese, P. M. Eisenberger, C. W. Jones, AIChE Annual Meeting, Nashville, Tennessee, 2009.

[3] S. Choi, J. H. Drese, P. M. Eisenberger, C. W. Jones, Env. Sci. Technol., 2011, 45, 2420-2427.

[4] S. Choi, M. L. Gray, C. W. Jones, ChemSusChem, 2011, 10.1002/cssc.201000355

See more of this Session: Sorbent Processes for Carbon Capture
See more of this Group/Topical: Sustainable Engineering Forum

(613e) Development and Evaluation of a Novel Hot Carbonate Absorption Process with Crystallization-Enabled High Pressure Stripping for Post-Combustion CO2 Capture

Yongqi Lu, Illinois State Geologic Survey, University of Illinois at Urbana-Champaign, Champaign, IL, Xinhuai Ye, Illinois State Geological Survey, University of Illinois at Urbana-Champaign, Champaign, IL, Massoud Rostam-Abadi, Department of Environmental Engineering, University of Illinois at Urbana-Champaign, Urbana, IL and Andrew Jones, Department of Energy, National Energy Technology Laboratory, Pittsburgh, PA

A novel Hot Carbonate Absorption Process with Crystallization-Enabled High Pressure Stripping (Hot-CAP) is being developed at the University of Illinois at Urbana-Champaign to overcome the energy use disadvantage of the monoethanolamine (MEA)-based processes. The Hot-CAP uses a carbonate salt, such as potassium carbonate (K_2CO_3, PC) or sodium carbonate (Na_2CO_3, SC), as a solvent. The process integrates four major unit operations including CO_2 absorption at elevated temperature (60-80° C), crystallization of bicarbonate from CO_2-rich solution to recover the carbonate, CO_2 stripping of bicarbonate slurry at high-pressure (up to 40 bar), and reclamation of a sulfate salt as the byproduct of sulfur dioxide (SO_2) removal.

Process feasibility studies were performed for these four major unit operations based on vapor-liquid equilibrium, solid solubility, and kinetic data that are available in the literature. Preliminary results show that to achieve 90% CO_2 removal, the energy use of the Hot-CAP is about 40-50% less than that of its MEA counterpart due to a low heat of absorption (7-17 kcal/mol CO_2 with the heat of bicarbonate crystallization included), a higher CO_2 working capacity (1-1.5 times higher than MEA), and reduced compression work (>50% lower than the MEA processes). Process engineering data for these unit operations specific to the Hot-CAP condition will be experimentally obtained, and current efforts are focused on measuring the kinetic data of CO_2 absorption into the carbonate solutions at elevated concentrations and temperatures, while also investigating the role of promoters/catalysts in promoting the CO_2 absorption rate. This presentation will provide a summary of the results from the process feasibility analysis and the absorption kinetics study, which are the initial tasks in a three-year, DOE-funded research project focused on the development of an advanced solvent-based CO_2 capture process for coal-based power plants.

Kinetics of CO_2 Absorption into Aqueous Potassium Carbonate Solutions Promoted by Carbonic Anhydrase Enzyme

Xinhuai Ye[1], Yongqi Lu[1], Massoud Rostam-Abadi[1,2], Andrew Jones[3]
[1]Illinois State Geological Survey, University of Illinois at Urbana-Champaign;
[2] Department of Environmental Engineering, University of Illinois at Urbana-Champaign;
[3] U.S. Department of Energy, National Energy Technology Laboratory, Pittsburgh, PA

Abstract

A novel Integrated Vacuum Carbonate Absorption Process (IVCAP) for post-combustion CO_2 capture is being developed at the University of Illinois at Urbana-Champaign. A key technical aspect of this process is to employ a biocatalyst, *carbonic anhydrase* (CA) enzyme, to enhance the rates of CO_2 absorption into the potassium carbonate (PC) solutions. This work investigates the kinetics, catalytic activity, and chemical and thermal stability of the CA enzyme to promote the CO_2 absorption into PC solutions.

The rates of CO_2 absorption into 20 wt% PC solutions containing various amounts of the CA enzyme have been measured in a stirred cell reactor under the pseudo-first-order conditions, and the values of the first-order reaction rate constant of the CA enzyme, k_e, for enhancing the CO_2 hydration were determined using classical theory of gas absorption. The results showed that the values of k_e were between 9.0 and 15.0 $L/mg{\cdot}s$ in the 20% PC solutions with carbonate conversion rate ranged between 10 and 50% at a temperature range of 25-50°C. The rates of CO_2 absorption into 20 wt% PC solutions were promoted by 3~6 times using 300 mg/L CA enzyme at a temperature between 40 and 50°C.

The CA enzyme demonstrated good chemical stability to several flue gas impurities as evidenced by the fact that the increase of ionic strength with anions of Cl^-, NO_3^-, and SO_4^{2-} in the solution did not significantly impact the k_e values. The long-term thermal stability study revealed that the CA enzyme maintained about 90% activity at 40°C over the 6 month period, whereas about 50% of the activity was lost at 50°C within two months.

(644a) Environmental Risk-Relevant Molecular Modeling of Adsorbed NDMA In MFI Zeolites

Elaheh Kamaloo[1], N. A. Deskins[2], Nikolas Kazantzis[3] and Robert W. Thompson[1], (1)Chemical Engineering, Worcester Polytechnic Institute, Worcester, MA, (2)Department of Chemical Engineering, Worcester Polytechnic Institute, Worcester, MA, (3)Chemical Engineering, WPI, Worcester, MA

Protecting the water supply is a top environmental concern and of great importance. Zeolites have been widely used to separate harmful molecules from water streams and ensure a safe, clean environment. N-Nitrosodimethylamine (NDMA) is a suspected carcinogenic and toxic chemical. NDMA can be found in water resources as a result of industrial facilities or as a by-product of water or wastewater treatment. Therefore removal of NDMA from drinking water is an important human safety concern. Studying of the adsorption and stability of NDMA inside a zeolite provides useful information about the feasibility and operating conditions of NDMA removal. Computer modeling can play a significant role in the study of NDMA removal (and other toxics), particularly since experimental studies with NDMA are undesirable. Current work presents quantum mechanical modeling (density functional theory) of NDMA adsorption in purely siliceous MFI, MFI with extra-framework Na and H, and MFI with silanol nests. Stability of NDMA inside the zeolite is investigated by calculating binding energies at various sites and in various configurations. Preliminary results provide a promising prospect on the development of scientifically sound guidelines on the utility of high-silica zeolites for NDMA removal as well as transparent criteria for the determination of key zeolite features (e. g. composition of the zeolite) that need to be incorporated into the synthesis stage.

See more of this Session: Contaminants In Water: Genesis, Rapid Detection and Sustainable Removal Processes
See more of this Group/Topical: Environmental Division

(644b) Detection of Disinfection by-Products In Drinking Water Systems Via Biomimetic Metabolism

Alvaro A. Rodriguez and Chelsea N. Monty, Chemical and Biomolecular Engineering, The University of Akron, Akron, OH

Disinfection by-products (DBPs) are generated by the reaction of disinfectants (e.g. chlorine) with natural organic matter present during drinking water treatment. DBPs metabolize in the body to form reactive intermediates which can be carcinogenic, genotoxic and hepatotoxic. The United States Environmental Protection Agency (US EPA) has established regulations for this type of compounds including trihalomethanes (THMs), haloacetic acids (HAAs), bromate and chlorite. However, emerging DBPs such as haloacetonitriles (HAN) have been found to be more toxic than their counterparts are. Real-time detection systems are essential for water distribution facilities to maintain public health and safety of potable water. While analytical techniques are frequently used offering selectivity and sensitivity, long processing times are seen as a drawback. On the other hand, biomimetic models offer long shelf-life and easy to use methods that have been developed in the prediction of drug metabolism. The purpose of this work is to develop an integrated biomimetic system for Phase I and II of metabolism monitoring glutathione (GSH) depletion. Assays of this kind can be utilized to screen for toxicity of samples by employing metalloporphyrins as structures that mimic the active site of cytochrome P450 offering the ability of detection of DBPs and anthropogenic compounds in drinking water systems. The aim of this presentation is to show the ability of biomimetic chemistry to generate in vivo metabolites of chloroform by optimizing variables such as pH, catalyst concentration, substrate concentration and oxidant while keeping temperature constant. Fourier Transform Infrared spectroscopy (FTIR) is used in Phase I of metabolism for the biomimetic reaction to monitor changes in absorbance for radical groups. Phase II of metabolism is carried out by GSH conjugation of metabolites generated in Phase I in the presence of cetyltrimethylammonium bromide (CTAB), catalyst that mimics the role of glutathione transferase (GST) in the detoxification step. CTAB is a cationic surfactant that reduces the surface tension of water inducing the formation of micelles. Critical micelle concentration determination ensures that the right concentration is in use in the optimization of Phase II of metabolism. Future work includes the application of electrochemical methods to determine GSH depletion in the integrated biomimetic chemistry Phase I and II.

See more of this Session: Contaminants In Water: Genesis, Rapid Detection and Sustainable Removal Processes
See more of this Group/Topical: Environmental Division

(644c) Electrolysis for Bromide Removal and Reduction In Formation Potential of Brominated Disinfection Byproducts From Drinking Water

Sirikarn Surawanvijit, Chemical&Biomolecular Engineering, University of California, Los Angeles, Los Angeles, CA and Yoram Cohen, Chemical & Biomolecular Engineering, University of California, Los Angeles, Los Angeles, CA

Meeting the disinfectant and disinfection by-product rules is a balancing act for water utilities as they are required to decrease health risks due to microorganisms with disinfectants, while simultaneously reducing potential the health risks attributed to disinfection byproducts (DBPs), such as trihalomethanes (THMs) and haloacetic acids (HAAs). A recent concern is that water from sources impacted by salt water intrusion (e.g., in coastal regions) contain a significant concentration of bromides (210-500 ug/L), which typically leads to the formation of brominated disinfection byproducts in chlorinated and ozonated water. Studies have suggested that brominated DBPs are more toxic than their chlorinated analogs. Accordingly, bromide removal from source water prior to disinfection is essential. In the present study, the effectiveness of electrolysis for bromide removal from drinking water, as well as the formation potential of brominated DBPs in the electrolyzed water, were investigated experimentally in a specially constructed electrolysis reactor. In principle, as water is allowed to pass through electrode pairs in the reactor, negatively charged bromide ions are attracted to the anode and undergo oxidation at the anode surfaces. As a result, the bromide ion is converted into bromine which can then volatilize from the treated water source. The design and operational criteria of the proposed process were optimized by testing electrolytic reactors of different designs under various hydraulic retention times and electric current settings. The results illustrated significant reduction in bromide concentration (44%) and formation potential of brominated THMs (up to 85%) and bromated (up to 73%) at an applied current as low as 1 A. Subsequently, the proposed process was scaled up to a 2.5-gpm reactor in which a similar bromide reduction level was achieved.

See more of this Session: Contaminants In Water: Genesis, Rapid Detection and Sustainable Removal Processes
See more of this Group/Topical: Environmental Division

(644d) A Simple Immuno-Assay to Detect Drugs of Abuse In Wastewater

Rachel Anderson and Otakuye Conroy-Ben, Civil and Environmental Engineering, University of Utah, Salt Lake City, UT

The consumption of illicit drugs worldwide has resulted in these compounds entering wastewater treatment plants and the environment. Previous studies in Europe and the United States have shown that drug use can be detected in raw sewage and wastewater. European metropolitan cities displayed patterns of recreational drug abuse peaking on weekends, and use is most prevalent in large cities. In the U.S., methamphetamine is the most frequently detected drug of abuse in wastewater treatment facilities. By analyzing controlled drugs in wastewater, it is possible to estimate the drug consumption of the community. The cost and difficulty of methods used to detect these compounds have impeded their widespread application.

The objective of this study is to detect, quantify, and monitor drugs of abuse trends through wastewater analysis using a simple and affordable method. Neogen Forensic plates were used to test wastewater and river water for drugs of abuse. Each well contains pre-plated antibodies which coordinate to the enzyme-substrate complex. After the designated reaction time is reached, the well is washed of any unreacted enzyme-substrate. A substrate specific to free antibody (K-Blue) is then added. The amount of K-Blue bound to free antibody is inversely related to the amount of drug present, and is measured using a UV-VIS spectrophotometer (l = 540 nm).

Wastewater and wastewater impacted water was tested for the presence of illicit drugs, and quantified. Compounds analyzed in this study include caffeine, oxycodone, cotinine, MDMA, cocaine, THC, and RHC. The Snyderville Water Reclamation Plant (4 MGD) utilizes an oxidation ditch followed by media filtration and UV disinfection prior to discharge into a local freshwater stream. The Central Valley Water Reclamation Facility (50 MGD) uses a trickling filter for biological treatment, followed by chlorine disinfection prior to discharge in the Jordan River. Drug activity was also measured along the length of the Jordan River, which receives wastewater effluent from three treatment plants.

All of the controlled substances analyzed in this study could be detected during wastewater treatment and in river water using the enzyme immuno-assay. The target drugs were more prevalent in wastewater treatment plants that received influent from urban areas. Drug activity decreased after secondary treatment in the Snyderville and Central Valley facilities. A smaller decrease was observed at the Wellsville and Plain City lagoons. Overall drug activity increased downstream in the Jordan River.

See more of this Session: Contaminants In Water: Genesis, Rapid Detection and Sustainable Removal Processes

Novel Microbial Fuel cell Based Biosensor for Continuous Measurement of BOD (Biochemical Oxygen Demand) in Wastewater

Shailesh Kharkwal[a, b,c], Min Lu[a], How Yong Ng[b,c]
and Sam Fong Yau Li[a,c]

[a] *Department of Chemistry, Faculty of Science, National University of Singapore, 3 Science Drive 3, Singapore 117543;*
[b]*Centre for Water Research, Department of Civil and Environmental Engineering, National University of Singapore, 9 Engineering Drive 1, Singapore 117576;*
[c]*NUS Graduate School for Integrative Sciences and Engineering, Centre of Life Science, 28 Medical Drive Singapore 117456*

In recent years microbial biosensors have demonstrated great potential as an alternative to the conventional analytical method for BOD measurement in wastewater. These Biosensors have certain advantages over the conventional 5-day BOD analysis, such as rapid response, portability of sensing device, high accuracy, and less rigorous and time consuming. Various BOD sensors have been reported in literature. They can be generally classified as respirometer type and bioluminescence based biosensors. Respirometer type biosensors (Hikuma et al., 1979; Sangeetha et al., 1996; Yang et al., 1997), mainly made by immobilizing yeast, *Bascillus subtilis*, and *Serratia marcescens* cells, take advantages of the high reaction rates of microorganisms interfaced to electrodes to measure the dissolved oxygen (DO) consumption (Riedel et al., 1988; Kim and kwon, 1999). While, the luminous bacteria, such as *Photobacterium phosphoreum,* isolated from marine sources, emits light at a constant rate in proportion to the amount of assimilable organic compounds in wastewater. However, the main disadvantage of respirometer and bioluminescence based biosensors is that both of them are not able to cope with a broad range of organic substrates, e.g., some complex compounds in wastewater may interfere with the performance of sensor. In this regard, microbial fuel cell (MFC) based BOD sensors have been developed and tested. MFC-type BOD sensors have a number of advantages over different kind of biosensors such as broad substrate versatility, long operational stability, high result reproducibility, etc.

This work aims to develop an efficient and stable MFC-type BOD sensor system which is affordable. Keeping the economy in mind, low cost manganese dioxide materials, α-MnO2, β-MnO2 and γ-MnO2 were tested as alternative cathode catalysts for oxygen reduction reaction

(ORR) in air-cathode MFC. Three MnO_2 species were further used as cathode catalysts in air-cathode cubic MFCs, in which mixed culture was inoculated as biocatalysts and domestic wastewater was used as the substrate in the anode chamber. It was found that β-MnO_2 yielded the best performance with a power density of 97.8 mW·m-2 that is quite comparable to Pt (Min Lu et al; 2011). This study demonstrated that using β-MnO_2 on CNT support instead of Pt could potentially improve the feasibility of scaling up MFC for practical applications such as BOD sensing device by lowering production cost. The performance of β-MnO_2 based MFC is further measured as BOD sensor using artificial wastewater (AWW). The performance of sensor is evaluated in terms of its COD range, response time, reproducibility and operational solubility. The sensor output, generated current or voltage, have a linear relationship with BOD up to 500 ppm, and a stability of more than 7 months.

Results

The first step towards using MFCs as BOD sensor was to optimize the feed flow rate with respect to the maximum steady state voltage and lowest hydraulic retention time (HRT). Different ranges of flow rates from 0.6 ml/min to 1.1 ml/min were tested. The optimal flow rate which provided a maximum steady state voltage of 23.1 mV and lowest HRT of 28.3 min was estimated at 1 ml/min (Table 1). No further increase in voltage was observed at flow rates above 1 ml/min.

Table1: Optimization of flow rate with voltage generation and HRT.

Flow Rate (ml/min)	Voltage (mV)	HRT (min)
0.6	18.6	47.16
0.7	20.7	40.42
0.8	21.6	35.37
0.9	22.0	31.44
0.93	22.1	30.43
0.95	22.5	29.78
0.98	22.8	28.87
1.0	**23.1**	**28.30**
1.1	22.6	25.72

The response time of the MFC may be defined as the time required to reach 95 per cent of the new steady-state voltage (Kim et al., 2004). To achieve the fastest response time of the BOD sensor, a calibration of steady-state current versus COD (50–500 ppm) was conducted with an external load of 5 Ω. The fastest response time of the sensor was estimated to be between 2-3 h for the same COD concentration. All the experiments were done in duplicates. Linear relationship between voltage and COD was observed up to a COD concentration of 500 ppm. Regression coefficient of r^2 = 0.985 is the best among all previously reported values, while the linear relationship with r^2 = 0.952 was observed for COD of up to 500 ppm (Fig 1). Experiment run was repeated to check the reproducibility of the results. The corresponding voltage value for specific COD concentration can be seen in Fig 2. Repetition of experiment showed high result reproducibility.

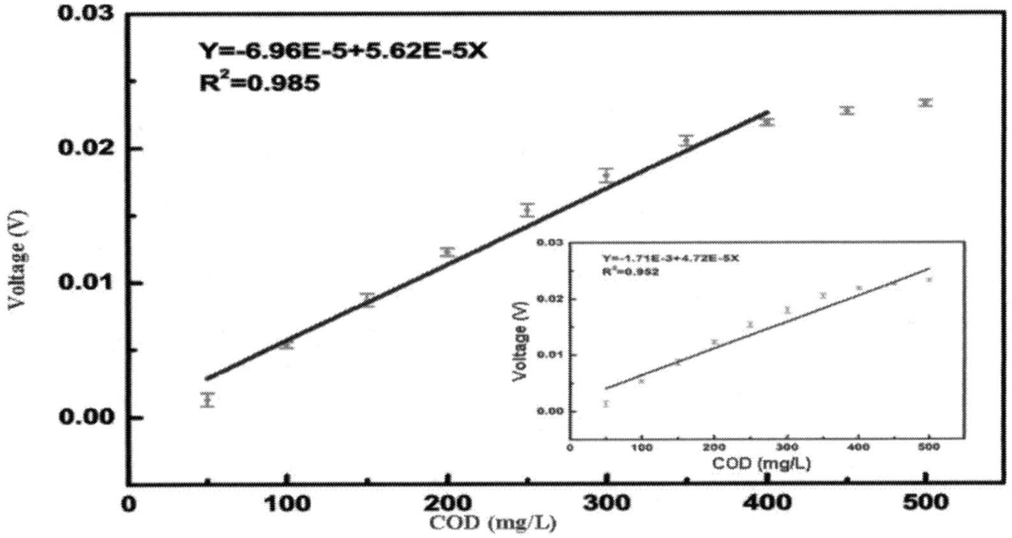

Fig 1: Steady state voltage as a function of COD concentration of synthetic wastewater feed. Voltage values were average of duplicate MFCs.

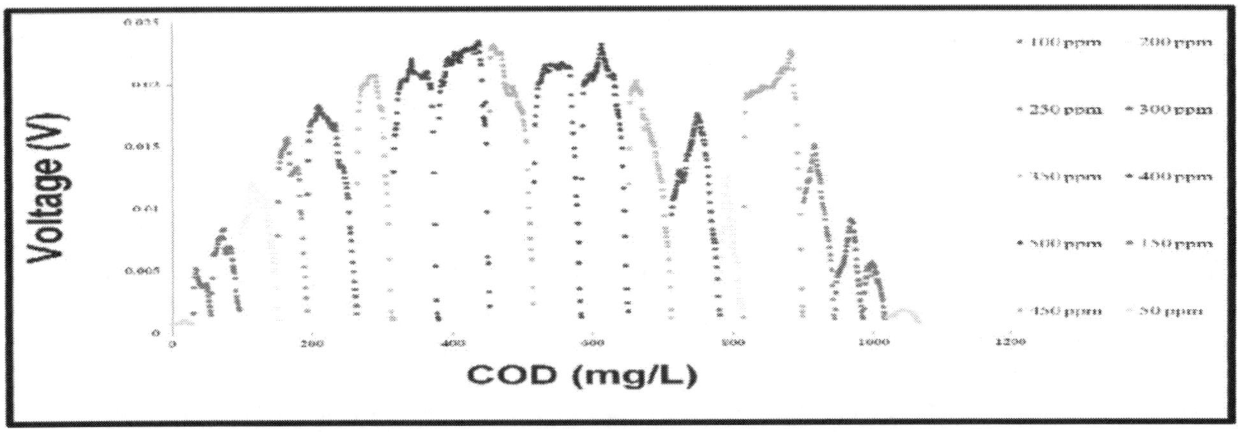

Fig 2: Pattern of voltage generation under different COD concentrations.

Conclusion

This study is first to use cheaper MnO_2 catalyst (in place of expensive Pt catalyst) based MFC as BOD sensor. Higher range of linear relationship between MFC sensor output and COD concentration (50 to 500 ppm), fast response time of the MFC sensor (2-3 h), more than 7 months of operational stability, and good reproducibility show the potential of MFC as a BOD sensor. Future work comprises calibration of MFC based BOD sensor with domestic wastewater and its use for measuring lower concentration of labile organics.

References:

1. Gil, G.C., Chang, I.S., Kim, B.H., Kim, M., Jang, J.K., Park, H.S., Kim,H.J., 2003. Operational parameters affecting the performance of a mediator-less microbial fuel. Biosens. Bioelectron. 18, 327–334.

2. Hikuma, M., Suzuki, H., Yasuda, T., Karube, I., Suzuki, S., 1979. Amperometric estimation of BOD by using living immobilized yeast. Eur. J. Appl. Microbiol. Biotechnol. 8, 289–297.

3. Hyun, C.K., Tamiya, E., Takeuchi, T., Karube, I., Inoue, N., 1993. Novel BOD sensor based on bacterial luminescence. Biotechnol. Bioeng. 41, 1107–1111.

4. Kim, M.-N., Kwon, H.-S., 1999. Biochemical oxygen demand sensor using Serratia marcescens LSY 4. Biosensors & Bioelectronics 14, 1–7.

5. Kim, B.H., Park, H.S., Kim, H.J., Kim, G.T., Chang, I.S., Lee, J., Phung, N.T., 2004. Enrichment of microbial community generating electricity using a fuel-cell-type electrochemical cell. Appl. Microbiol. Biotechnol. 63, 672–681.

6. Lu M., Kharkwal S., Ng H. Y., Li S. F. Y., 2011. Carbon nanotube supported MnO2 catalysts for oxygen reduction reaction and their applications in microbial fuel cells. Biosens. Bioelectron. 26 (12), 4728-4732.

7. S, Sangeetha., G, Sugandhi., M, Murugesan., Mudhav, M., V., et al., 1996. Torulopsis candida based sensor for the estimation of biochemical oxygen demand and its evaluation. Electroanalysis. 8, 698–707.

8. Riedel, K., Renneberg, R., Kühn, M., Scheller, F., 1988. A fast estimation of biochemical oxygen demand using microbial sensors. Appl. Microbiol. Biotechnol. 28, 316–318.

Ultrafiltration/Nanofiltration Based Technique for the Removal of Endocrine Disruptors from Surface Water

Abstract ID# 222410

Ana Gil, Georgina Sarmento, Miguel Minhalma, Elizabeth Duarte and Maria Norberta De Pinho

Introduction

Membrane technologies and more specifically pressure-driven membrane processes like microfiltration (MF), ultrafiltration (UF), nanofiltration (NF) and reverse osmosis (RO) (1 - 9) are playing a crucial role on the production of drinking water of high quality. This is attributed to the fact that these technologies address in a very specific way: 1) The efficient removal of residual organic matter (NOM) that acts as a precursor of the formation of disinfection by-products that are health hazards; 2) The removal of conventional micropollutants such as pesticides, alkylphthalates and endocrine disruptors chemicals (EDs). The EDs and namely the 4-Nonylphenil (NP) can be integrated in surface waters through diverse mechanisms, like direct discharge of industrial and domestic wastes, the discharge of effluents from biological treatment plants and drains of agriculture into rivers and soils.

The present work addresses both the removal of NOM by Ultrafiltration and the removal of NP by Nanofiltration. Two sets of experiments were carried out in concentration mode: i) Ultrafiltration of surface water from Tagus River (Valadas, Portugal); ii) Nanofiltration of surface water from Sado River (Setúbal, Portugal).

Results

Ultrafiltration of Tagus River Water (10)

The variation of the permeation fluxes with the water recovery rates is displayed in Fig. 1.

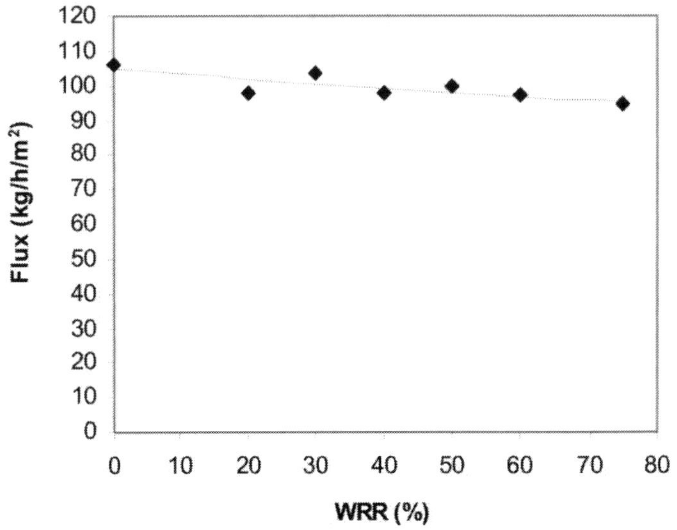

Figure 1. Permeation fluxes of Tagus River water as a function of water recovery rates.

Membrane: ETNA10PP; Operating Conditions: Transmembrane Pressure = 4 bar; Feed flow rate = 554 L/h.

The rejection coefficients to the total organic carbon (TOC) and to the dissolved humic substances (UVA^*_{254nm}) are displayed in Fig. 2 in a wide range of water recovery rates.

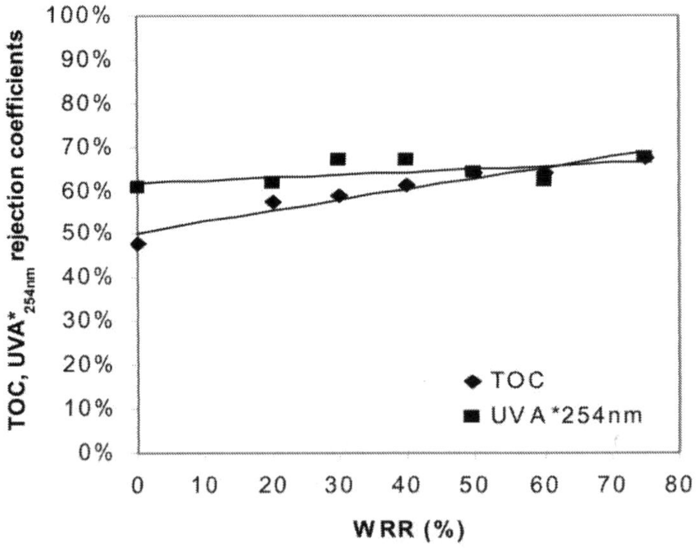

Figure 2. Organic matter rejection coefficients as a function of water recovery rates.

Membrane: ETNA10PP; Operating conditions: Transmembrane Pressure = 4 bar; Feed flow rate = 554 L/h.

Nanofiltration of Sado River Water (11)

The decline of the permeation fluxes with the increase of the water recovery rates is displayed in Figures 3 and 4 for surface water without and with inoculation of 4-Nonylphenol, respectively.

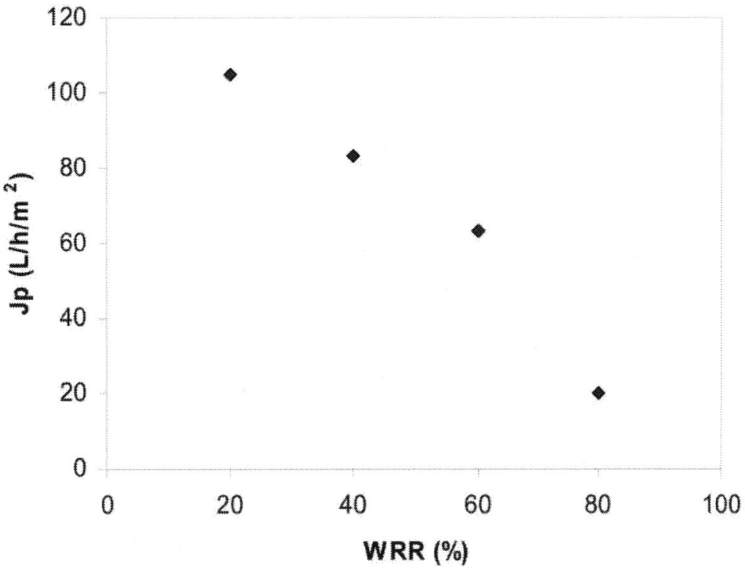

Figure 3. NF permeate fluxes as a function of water recovery rates for the Raw Sado River water. Membrane: NF 270; Transmembrane pressure = 30 bar.

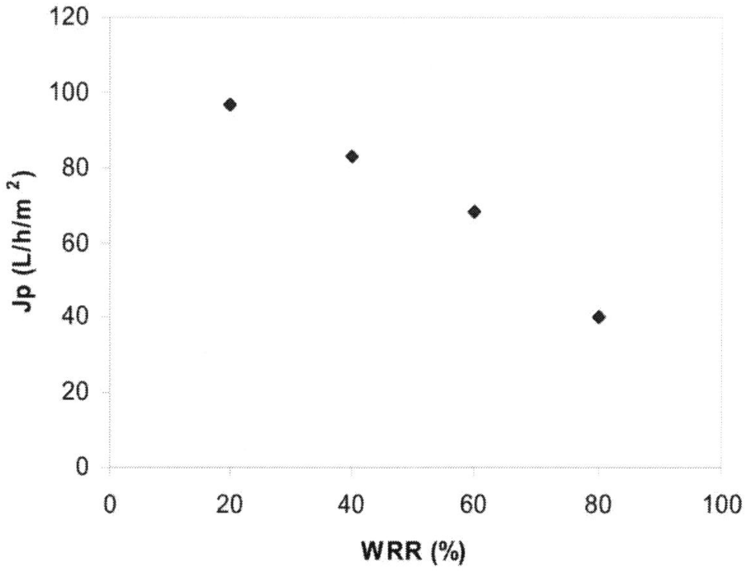

Figure 4. NF permeate fluxes as a function of water recovery rates for the Sado River water inoculated with 4-Nonylphenol.

Membrane: NF 270; Transmembrane pressure = 30 bar.

In the nanofiltration of Sado River water with and without inoculation of NP the rejection coefficients to NOM are 77 % and to NP are 100 %.

Conclusions

The decline of the surface water permeation fluxes with the water recovery rates is more pronounced in the operation of nanofiltration than in the operation of ultrafiltration.

In nanofiltration the flux decline is less pronounced when the water is inoculated with NP. At the water recovery rate of 80 % the permeate flux is 40 L/h/m^2 for water inoculated with NP and 20 L/h/m^2 for the raw water.

The average apparent rejection coefficient obtained for NOM is 77 % and for the endocrine disruptor, NP, is 100%, for the surface water with and without inoculation of the endocrine disruptor. The higher rejection coefficients to endocrine disruptors when compared to the ones of natural organic matter give evidence that there is a NOM fraction of high molecular weight retaining the endocrine disruptor in the concentrate stream and a permeate stream with NOM of lower molecular weight that is free of endocrine disruptor.

References

(1) Hong, S. and Elimelech, M., Chemical and physical aspects of natural organic matter (NOM) fouling of nanofiltration membranes, J. Membr. Sci., 132 (1997) 159-181.

(2) Cho, J.W., Amy, G. and Pellegrino, J., Membrane filtration of natural organic matter: initial comparison of rejection and flux decline characteristics with ultrafiltration and nanofiltration membranes, Water Res., 33 (11) (1999) 2517-2526.

(3) Kiso,Y., Sugiura, Y., Kitao, T. and Nishimura, K., Effects of hydrophobicity and molecular size on rejection of aromatic pesticides with nanofiltration membranes, J. Membr. Sci., 192 (1-2) (2001) 1-10.

(4) Kiso, Y., Mizuno, A., Othman, R., Jung, Y.J., Kumano, A. and Ariji, A., Rejection properties of pesticides with a hollow fiber NF membrane (HNF-1), Desalination, 143 (2) (2002) 147-157.

(5) Costa, A.R., de Pinho, M.N., The role of membrane morphology on ultrafiltration for natural organic matter removal, Desalination, 145 (1-3) (2002) 299 - 304.

(6) Costa, A.R., de Pinho, M.N., Coagulation/flocculation/ultrafiltration for natural organic matter removal in drinking water production, Water Science and Technology: Water Supply, 4-5 (2004) 215-222.

(7) Costa, A.R., de Pinho, M.N., Performance and cost estimation of nanofiltration for surface water treatment in drinking water production, Desalination, 196 (1-3) (2006) 55-65.

(8) Costa, A.R., de Pinho, M.N., Elimelech, M., Mechanisms of colloidal natural organic matter fouling in ultrafiltration, J. Membr. Sci., 281 (2006) 716–725.

(9) Jung, Y.J., Kiso, Y., Park, H.J., Nishioka, K. and Min, K.S., Rejection properties of NF membranes for alkylphenols, Desalination, 202 (1-3) (2007) 278-285.

(10) Costa, A. R., Mechanisms of membrane fouling by colloidal organic matter and optimization of membrane processes for surface water treatment, PhD Thesis (2007), Instituto Superior Técnico, Technical University of Lisbon, Portugal.

(11) Salvaterra, A. F., Sarmento, G., Minhalma, M., and De Pinho, M. N., Nanofiltration of surface water for the removal of endocrine disruptors, Desalination and Water Treatment (2011), In press.

(658b) The Importance of a Well-Designed Cross-Flow System In the Removal Study of Adsorbing Estrogens by Nanofiltration Membranes

Andrea JC Semiao, School of Engineering, The University of Edinburgh, Edinburgh, United Kingdom

Life on Earth, including the survival of humans and other living species, is strongly dependent on the production of clean water for potable usage. Endocrine disruptors, such as hormones, have a potential harmful effect on human health and should therefore be removed from potable water resources. However, so far, such disruptors have not been completely removed by conventional wastewater treatment processes and are consequently discharged into potable water resources, contaminating them. On the other hand, nanofiltration (NF) is widely used in water treatment as is the case of the Mery-sur-Oise plant in Paris that treats the water from the Seine River to recover it as potable water, through the removal of pesticides and other contaminants. Both the difficulty in removing the endocrine disruptors that exhibit a potential health risk and the broad use of NF for waste water treatment make the understanding of the removal mechanisms of the above-mentioned pollutants a crucial issue.

At laboratory scale, cross-flow systems with a slit channel have been widely used in the research of membrane water treatment, particularly when the focus lies on removal of trace contaminants. Slits constitute a laboratory model of spiral-wound modules allowing for the study of the fundamentals of mass and momentum transport mechanisms in the feed channel by controlling its hydrodynamics. Spiral-wound modules are, in turn, the most common membrane geometry used in water treatment, due to their compact geometry with high surface/volume ratio.

This work aims at demonstrating the importance of having a well-designed cross-flow system and without oil leakages from the pumping system in the removal of endocrine disruptors by showing how both an ill-designed cross-flow system and the presence of oil in the aqueous solution impact on the results of their removal by NF membranes. For that, a comparison between hormone filtration results yielded by both type of systems is performed. Additionally, further experiments performed with the well-designed system aim at defining, understanding and quantifying the physical mechanisms of momentum and mass transfer controlling the adsorption of hormones onto polymeric NF membranes, viewing the development of a predictive model.

In an ill-designed system, pump delivering flow rates to the cross-flow membrane cell above a limit value (Reynolds number based on the hydraulic slit diameter, Re_{dh}, *above 10,000) causes the membrane active layer to be ripped off from the support layer, and yields a E2 retention below 15% whereas that obtained for a* Re_{dh} *number of 2800 in a well-designed cross-flow system is higher than 80%. On the other hand, the presence in the aqueous solution of oil contamination originated from the pump of the ill-*

designed system caused the E2 retention to be much lower than expected: 25% in the presence of oil against 65% in the absence of oil obtained in the well-designed system. These and other results led to the abandonment of the ill-designed system since, in opposition to it, the well-designed one allowed to mimic spiral-wound hydrodynamic conditions and carry out the work aiming at understanding the fundamentals of the removal mechanisms of hormones by NF membranes. Special care was taken in the new system to re-design the pump head of a commonly used diaphragm pump in cross-flow systems to make sure that there was no oil leakage. These findings are very important for pump manufacturers and users.

Once these changes were made, it was possible to proceed with the experiments viewing the definition of a predictive model of adsorption of hormones onto polymeric NF membranes. First, knowledge from experiments was acquired on how hydrodynamics and operating conditions - pressure, circulating Reynolds numbers (Re_h) and feed concentration - affect hormone adsorption: adsorption is governed by the initial concentration at the membrane surface that, in turn, depends on the hormone feed concentration, operating Re_h and pressure. Membrane retention, however, depends on the initial polarisation modulus, given by the fraction of the initial concentration at the membrane surface divided by the initial feed concentration. Then, based on these results, a sorption model was developed, predicting very well both feed and permeate transient concentrations for two different hormones in the normal range of operating pressures and Re_h of spiral-wound membrane modules. These findings are an important advancement in determining which membrane would be more suitable to effectively remove hormones since the model predicts well the feed and permeate concentrations with a substantial reduction of experimental work.

See more of this Session: Micropollutants, Disinfection, and Other Considerations in Water Treatment
See more of this Group/Topical: Topical 1: Water Technology for Developed and Developing Countries (see also Separations Division)

(658c) SW-RO Membranes for the Removal of Ammonium Thiocyanate From Coking Wastewaters In Steel Manufacturing

Susana Luque[1], **Jose R. lvarez**[1], M. El-Mansour[1], F.E. Antn[1], B. Gonzlez[2] and B. P. Vivas[2], (1) Department of Chemical and Environmental Engineering, University of Oviedo, Oviedo, Spain, (2) ArcelorMittal Research & Development, Global R&D Asturias, Avils, Spain

The liquid effluents from a coke plant in an integrated steel manufacturing facility usually contain ammonium cation, thiocyanate and cyanide anions, phenols and suspended solids. Cyanides, cyanates and thiocyanates are hazardous substances with a negative effect on living species and the environment. The generation of these pollutants is inherent to the manufacturing process, and the wastewaters containing them cannot be discharged without treatment, according to the current regulations. Chemical treatment, incineration and biological digestion are the most common treatment methods. However, when the thiocyanate ion concentration exceeds 400 ppm, the microbial activity is hindered, and thus, an additional chemical pre-treatment is needed. Incineration can also be used to treat concentrated thiocyanate effluents, but at a high cost.

Reverse osmosis (RO) has proved to be a feasible technology in the recovery of thiocyanate from the aforementioned condensates. Several experiments have been carried out at laboratory scale to choose a suitable membrane and to gather experimental data which would allow developing a model to design and scale up the filtration process. Commercially available RO spiral wound membranes were selected. The membrane showing the best behaviour was the so-called SW30-HR membrane from DOW FILMTECTM as it gave adequate values of rejection and permeability. The effect of process parameters has also been addressed and recommendations for plant operation have been proposed.

Onsite pilot plant runs have also been conducted in order to evaluate RO performance when operating longer term with the actual industrial stream. A pilot unit equipped with two 4" spiral wound modules were used. The results have been in close agreement with those obtained at laboratory scale.

A mathematical model which represents the process has then been proposed, validated and used to design a full scale plant which accomplishes the desired separation. Coking wastewaters effluents usually have thiocyanate concentrations between 1000 and 3000 ppm, but eventually, it is possible to have condensates that reach 9000 ppm. Therefore, several case studies and sensitivity analysis have been carried out based on both these typical and extreme values. An economic evaluation is also included.

ACKNOWLEDGMENTS

Financial support by PCTI (RIA Project, PEST08, Principality of Asturias, Spain)

(658d) Environmental and Economic Evaluation of Wastewater and Water Disinfection Technologies

Tapas K. Das, School of Engineering, Saint Martin's University, Lacey, WA

This paper describes some of the common use of disinfectants and disinfection practices, along with most modern and emerging technologies that are proving to be more efficient in disinfecting varieties of microorganisms present in wastewater and drinking water. In rural areas of developing nations waterborne disease such as typhoid, cholera, hepatitis, and gastroenteritis, infect and kill many infants and children each day, according to the World Health Organization (WHO). With increasing emphasis on promoting a sustainable water ecological future and concern over introducing a toxic chemical in the water, the design of the disinfection process is increasingly leaning toward technologies that destroy the pathogens while balancing the effect of this disinfected wastewater on the population of aquatic biota or a drinking water supply. Disinfection is the most important water treatment process for preventing the spread of infectious diseases. For the past several decades, chlorine has been the disinfectant of choice. The discovery and subsequent regulation of some chemical by-products of chlorine (e.g., trihalomethanes) has increased the popularity of other disinfectants such as ozone, hydrogen peroxide, UV radiation, sonication, etc. In this presentation, UV radiation will be used as the primary disinfection method, and will be tested in conjunction with chlorine , chlorine dioxide, sodium hypochlorite Fenton reagent and O_3 to enhance the germicidal effectiveness of protozoa, *cryptosporidium parvum*, *giardia duodenalis cysts*, and other microorganisms. The effectiveness of removing pathogens, odorous compounds/color and the environmental and economic impacts of wastewater and drinking water disinfection technologies will be assessed.

See more of this Session: Micropollutants, Disinfection, and Other Considerations in Water Treatment
See more of this Group/Topical: Topical 1: Water Technology for Developed and Developing Countries (see also Separations Division)

(658e) Chemical Properties and Degradation Efficiency of Radio-Frequency Plasmas In Water

Selma Mededovic, Dept. of Chemical and Biomolecular Engineering, Clarkson University, Potsdam, NY

To treat contaminated water, advanced oxidationtechnologies (AOT) utilize various precursors and processes to produce hydroxylradicals (OH); species which are, because of their highoxidation potential, capable of degrading virtually any organic compound in theaqueous solution. However, in some of the AOTs, the production of OHcanbe quite limited depending on, for example, the typeof the OHprecursor (H_2O vs. H_2O_2)and the turbidity of the solution. Thus, there is a requirement for atechnology that will overcome the drawbacks associated with the current AOTsand the preliminary results from this study indicate that radio-frequency (RF)plasma in water, a new "type" of plasma formed by low-voltage high-frequency(13.56 MHz) electrical discharge, could be that technology. Compared to theconventional high-voltage streamer-like discharges, the volume of the RF plasmais several times larger and the intensity of the plasma-generated UV light appearsto be higher. Thus, the goal of this study is to chemically and spectroscopicallycharacterize RF plasma and asses its efficiency for the degradation ofRhodamine B dye.

See more of this Session: Micropollutants, Disinfection, and Other Considerations in Water Treatment
See more of this Group/Topical: Topical 1: Water Technology for Developed and Developing Countries (see also Separations Division)

Determination of Bisphenols Using Ultraviolet Spectroscopy

Alfred R. Conklin, Jessica Veite, Jennifer Wilson

Chemistry Department, Wilmington College Wilmington, Ohio

Background

Bisphenols are an environmental concern because of both their wide use and evidence connecting them to biological effects especially relating to their estrogenic activity (1, 2). There are a family of bisphenols which are commonly used as monomers in common plastics, including drink containers and liners for drink and food cans. Bisphenols can leach out of plastic containers into their contents. When consumed the consumers are exposed to bisphenols.

Exposure to bisphenols has been related to a change in the size of reproductive organs, sperm production, and the onset of puberty. This has lead to concern over exposure to bisphenols particularly in young children (3,4,5).

The most commonly cited bisphenol is Bisphenol A, see Figure 1. There is however a family of bisphenols derived by making substitution on the methylene group connecting the two phenol moieties. As is seen in the Figure, bisphenols contain two phenol moieties.

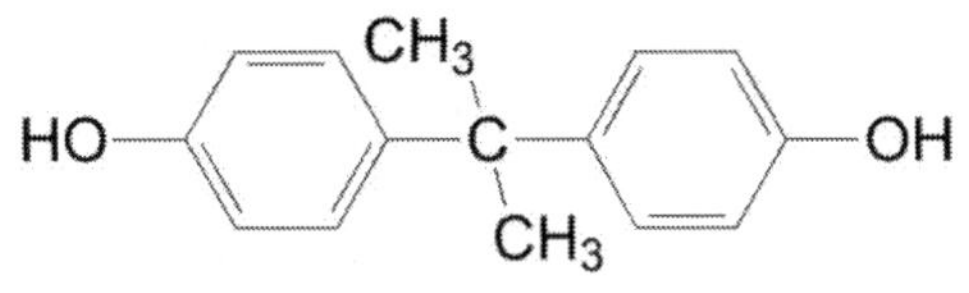

Figure 1. Structure of Bisphenol A.

Gas chromatography-mass spectroscopy (GC-MS) has been used to determine the bisphenol content of foods, including water (6). Solutions containing bisphenol can be injected directly into the GC-MS and analyzed. If the concentration is low BPA can be absorbed on an appropriate absorbent and concentrated before analysis (7, 8). It is also possible to dervatize the bisphenol before analysis (9). High performance liquid chromatography HPLC has been used to isolate and detect bishenols. (10,11). Molecularly imprinted polymers along with GC and HPLC and stable isotopes have also been used in bisphenol analysis (12, 13, 14).

Because phenol has an ultraviolet spectrum it is to be expected that bisphenol would also have an ultraviolet spectrum and that the absorptivity would be high. This is indeed the case. It is thus possible to determine the amount of bisphenols in both aqueous and ethanolic solutions using ultraviolet spectroscopy.

In many cases plastics are used in the bottled water industry so using ultraviolet spectroscpy to determine the bisphenols in aqueous solutions should be possible. This type of analysis is simple and requires few resources. If additional confirmation of the existence or quantity of bisphenol is needed a GC-MS analysis can be carried out.

Experimental.

Bisphenol A [2,2-Bis(4-hydroxyphenyl)ethane] 97% was obtained from Sigma Aldrich as small spheres. Spheres were ground to a fine powder, using a mortar and pestle before use.

Deionized Distilled water was prepared and stored in glass until used (DID water).

Absolute Ethanol 90.25 % containing 4.75% methanol and 5.00% isopropanol as denaturants was used in all experiments involving ethanol.

Syringe filters, Nylon, cellulose acetate, PVDF, PTFE, PSE with 0.45 and 0.2 μm, pores were obtained from Fisher Scientific.

Ultraviolet spectroscopy was carried out using Quartz, 10mm path length, 0.7 ml cells. Ultraviolet spectra were obtained using a Shimadzu UV-1650PC spectrophotometer. GC-MS analysis was carried using a Thermo ISQ with a Focus GC and a TG-5MS 30m x 0.25mm x 0.25μm capillary column. The GC inlet temperature was set at 220 °C and the transfer line at 250 °C with a helium carrier gas flow rate of 1.2 ml per minute. The oven was set at 125 °C for 0.5 minutes then increased to 250 °C at 20 °C/minutes and held at this temperature for 2 minutes. 1 μl injections were used for all samples.

Procedure

Solutions containing varying amounts of BPA but typically in the range of 2 to 200 ppm bisphenol A in both DID and ethanol were prepared. Aqueous solutions were analyzed using ultraviolet spectroscopy. Ethanolic solutions were analyzed using both ultraviolet spectroscopy and GC-MS. In both cases standard curves were prepared using Excel and the linear fit for the lines and their r^2 calculated.

BPA standards were filtered through various syringe filters and the ultraviolet absorption before and after filtration measured. Each standard was filtered three different times, except as noted, and the standard deviation determined.

Results

Both aqueous and ethanolic BPA solutions analyzed by ultraviolet spectroscopy and GC-MS produced standard curves with $r^2 = 0.99$, see Figure 2. Small additions of either water to ethanolic, or ethanol to aqueous solution resulted in small but significant changes in absorption.

Absorption of BPA from the standard solution by syringe filters varied between 96% to 4 % of the standard BPA solution contained a calculated 175 mg BPA per liter solution. This

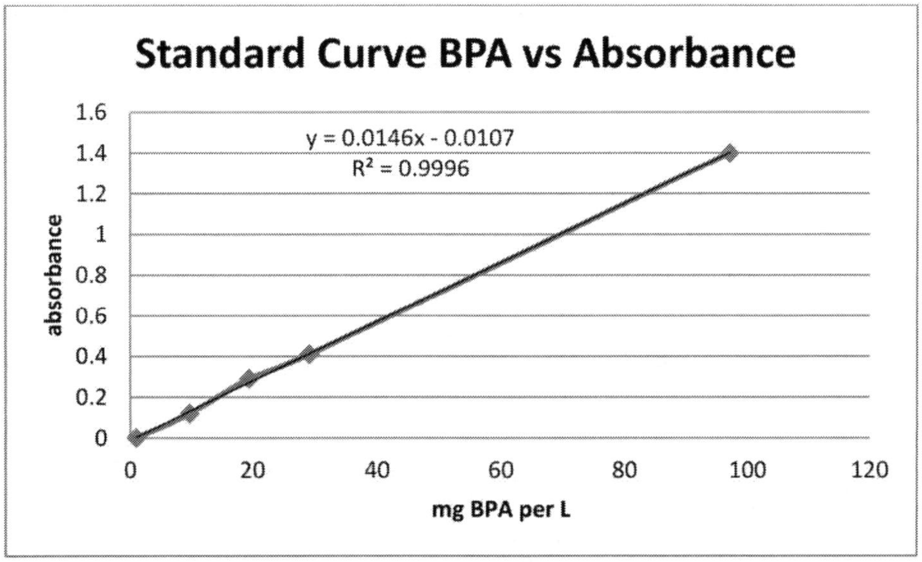

Figure 1. An example of a graph of BPA vs absorption at 277 nm. Note that although the abosrpance for 100 mg/l is outside the normally used range of ultraviolet spectroscopy the r^2 is still 0.999.

Table 1. Adsorbance of BPA by various syringe filters containing different filter material

Filter Brand	Material	Pore Size	BPA Adsorbance after filtration	Standard Deviation
None Standard BPA Solution(1)*			2.51 (2)*	±0.03
Whatman	Nylon	0.45	1.51	±0.2
Fisher	Nylon	0.45	0.11	±0.06
Gelman	Nylon	0.45	0.78	±0.3 (3)*
Whatman	Cellulose Acetate	0.2	0.64	±0.02
Whatman	Cellulose Acetate with GMF	0.2	0.26	±0.1
Whatman	PVDF	0.2	1.51	±0.3
Whatman	PTFE	0.2	2.11	±0.1
Millex	PVDF	0.22	0.74	±0.1
Millex	PTFE	0.45	2.44	±0.06
Corning	PSE	0.2	0.94	±0.7(3)*

Notes: (1)* an aqueous solution
(2)* standard solution was not filtered.
(3)* initial measurements showed large variation so 6 replicates were determined.

solution was chosen so that the syringe filters absorbing the most BPA would still provide results which would be on scale and could be graphed. Solutions filtered through Fisher Nylon had the lowest adsorbance (lowest amount of BPA) after filtration, and Millex PTEF had the highest absorption after filtration (highest amount of BPA) but was not significantly different from the control. Filtering with Whitman PTEF and Nylon also produced a filtrate with high adsorbance. All other results were significantly lower (statistically) than the control although not always significantly different from each other. Variations in pore size did not seem to affect adsorbance although source of filter material did. Table 1 gives complete data for various filters tested. All measurements were replicated 3 times except as noted.

Conclusion

Ultraviolet spectroscopy can be used to determine the occurrence and concentration of Bisphenols in simple solutions. Caution must be used in using ultraviolet spectroscopy when the extractant may be contaminated with another solvent or extractant. Also it is possible that some components of both drinks and foods may interfere with either the λ_{max} or absorption. However with relativly pure water samples, ultraviolet spectroscopy works well. In cases where there is some doubt GC-MS can be used to confirm the results.

Syringe filters adsorb Bisphenol, with PTFE generally adsorbing the least and Nylon adsorbing the most. Syringe filter filtration should be avoided when determining the concentration of BPA in solutions. If syringe filtration must be done the results must be corrected for the adsorption of BPA by the filters.

References

1. Vom Saal, Frederick S., Paul S. Cooke, David L. Buchanan, Paola Palanza, Kristina A. Thayer, Susan C. Nagel, Stefano Parmigiani, Wade V. Welshons (1998), "A Physiologically Based Approach To the Study of Bisphenol a and Other Estrogenic Chemicals On the Size of Reproductive Organs, Daily Sperm Production, and Behavior," Toxicology and Industrial Health, 14:239-260.

2. Knaak James B. and Lloyd J. Sullivan (1966), "Metabolism of bisphenol A in the rat," Toxicology and Applied Pharmacology 8:175-184.

3. Howdeshell, Kembra L, Andrew K. Hotchkiss, Kristina A. Thayer, John G. Vandenbergh and Frederick S. vom Saal (1999), "Environmental toxins: Exposure to bisphenol A advances puberty," Nature 401:763-764.

4. Laws, Susan C., Stephan A. Carey, Janet M. Ferrell, Gerald J. Bodman, Ralph L. Cooper (2000), "Estrogenic Activity of Octylphenol, Nonylphenol, Bisphenol A and Methoxychlor in Rats," Toxicol. Sci. 54:154-167.

5. Staples Charles A., Philip B. Dome, Gary M. Klecka, Sondra T. Oblock and Lynne R. Harris (1998), "A review of the environmental fate, effects, and exposures of bisphenol A", Chemosphere 36:2149-2173.

6. Basheer Chanbasha and Hian Kee Lee (2004), "Analysis of endocrine disrupting alkylphenols, chlorophenols and bisphenol-A using hollow fiber-protected liquid-phase microextraction coupled with injection port-derivatization gas chromatography–mass spectrometry," J. Chromatography A, 1057:163-169.

7. Braun P., M. Moeder, St. Schrader, P. Popp, P. Kuschk and W. Engewald (2003), "Trace analysis of technical nonylphenol, bisphenol A and 17α-ethinylestradiol in wastewater using solid-phase microextraction and gas chromatography–mass spectrometry," J. Chromatography A, 988:41-51.

8. Chang Chia-Min, Chi-Chi Chou and Maw-Rong Lee (2005), "Determining leaching of bisphenol A from plastic containers by solid-phase microextraction and gas chromatography–mass spectrometry," Analytica Chimica Acta, 539:41-47.

9. Salafranca Jesús, Ramón Batlle and Cristina Nerín (1999), "Use of solid-phase microextraction for the analysis of bisphenol A and bisphenol A diglycidyl ether in food stimulants," J. Chromatography A, 864:137-144.

10. González-Casado A., N. Navas, M. del Olmo, J.L.Vílchez (1998), "Determination of Bisphenol A in Water by Micro Liquid-Liquid Extraction Followed by Silylation and Gas Chromatography-Mass Spectrometry Analysis," J. Chromatographic Science, 36:565-570.

11. Kataoka Hiroyuki, Megumi Ise, Shizuo Narimatsu (2002), "Automated on-line in-tube solid-phase microextraction coupled with high performance liquid chromatography for the analysis of bisphenol A, alkylphenols, and phthalate esters in foods contacted with plastics," J. of Separation Science, 25:77–85.

12. Szap Peter, Ivar Kesse and Jack Klapp (1978), "The Analysis of Bisphenol a by High Performance Liquid Chromatography," J. Liquid Chromatography 1:89-96.

13. Kawaguchi Migaku, Yoshio Hayatsu, Hisao Nakata, Yumiko Ishii, Rie Ito, Koichi Saito and Hiroyuki Nakazawa (2005), "Molecularly imprinted solid phase extraction using stable isotope labeled compounds as template and liquid chromatography–mass spectrometry for trace analysis of bisphenol A in water sample," Analytica Chimica Acta, 539 83-89.

14. Navarro-Villoslada F., Blanca San, Vicente and María C. Moreno-Bondi (2004), "Application of multivariate analysis to the screening of molecularly imprinted polymers for bisphenol A,"Analytica Chimica Acta, 504:149-162.

(690a) Development of New Redox Flow Battery Systems

Liyu Li[1], Wei Wang[2], Soowhan Kim[2], Vijayakumar Murugesan[2], Zimin Nie[2], Baowei Chen[2], Qingtao Luo[2], Feng Chen[2], Jun Liu[2], Jianzhi Hu[2] and Gary Yang[2], (1)Hydrocarbon Processing Group, Pacific Northwest National Laboratory, Richland, WA, (2)Pacific Northwest National Laboratory, Richland, WA

Concerns over the environmental consequences of burning fossil fuels and their remaining reserves have led to an increasing use of renewable energy generated from sources such as solar and wind. However, the intermittent, varied nature of the renewable resources makes it difficult to integrate these valuable energies into the electrical grids. A solution to this problem is to employ large-scale electrical energy storage (EES), which is widely considered as an effective approach to improve the reliability, power quality, and economy of these renewable energies. Among the most promising large-scale EES technologies are the redox flow batteries (RFBs). RFBs are special electrochemical systems that can repeatedly convert and store multi-megawatt-hours (MWhs) of electricity energy to chemical energy and convert the chemical energy back to electricity energy when needed. The advantages of RFBs include short response time, low self discharge, long life-time, and independent tunable power and storage capacity, which make it possible to independently scale up the electricity storage capacity and power generation capacity. With the support of US Department of Energy's Office of Electricity, PNNL has carried out extensive efforts in the development of advanced redox flow batteries. Substantial progress has been achieved in the past two years, including two new redox chemistries (V-Fe system and all vanadium mixed acid system), which will be reported in this presentation

See more of this Session: Energy Storage and Electrochemical Processes I
See more of this Group/Topical: International Congress on Energy 2011

Study on preparation and characterization of the composite bipolar plate for Vanadium Redox Flow Battery

XIE Xiao-Feng[1], WANG Wen-Pin[2] , WANG Shu-Bo[1], WANG Jin-Hai[1],

LV Ya-Fei[2], MATHUR V. K[3]

1 Institute of Nuclear and New Energy Technology, Tsinghua University, Beijing 100084, China

2 College of Material Science And Engineering, Beijing University of Chemical Technology, Beijing 100029,China

3.Department of Chemical Engineering, University of New Hampshire, NH 03824, USA

Bipolar plate, as one of the main components of all-vanadium redox flow battery (VRB), can collect the current generated by electrochemical reaction, and can separate positive and negative electrolyte. The electrolyte is a sulfuric acid solution containing vanadium ions (V(IV/V,II/III)/ V(V)). Among these ions, the V(II) has the most strong oxidation ability. Therefore, the bipolar plate not only should have good conductivity, but also have high resistance to chemical and electrochemical corrosion.

Bipolar plate is a carbon composite material, which is a mixture of carbon materials and a polymer. The polymers used are polyethylene, nylon, polypropylene (PP), rubber modified polypropylene, polytetrafluoroethylene and polyvinylidene fluoride, etc. Polymer is helpful to form the conductive network, and improve the corrosion resistance of bipolar plate.

Maleic anhydride grafted polypropylene(g-PP) is a non-polar molecules in the main chain with a strong polar side groups maleic anhydride. It can increase the bonding and the compatibility between polar materials and non-polar materials for enhancing the dispersion between the conductive filler and the polymer matrix, and to improve conductive property and corrosion resistance of the bipolar plate.

In this work, composite bipolar plates, by polypropylene (PP), polypropylene-maleic anhydride graft copolymer (g-PP) as polymer matrix, and flake graphite(GC) as conductive filler, were prepared using a solution intercalation method. Four probes method, potentiodynamic experiments, potentiostatic and cyclic voltammetry(CV) were used to investigate the conductivity, corrosion resistance and electrochemical performance of bipolar plates. The results show that g-PP made the graphite uniformly dispersed in the polymer, which significantly improved compatibility between the conductive filler and the polymer matrix. The conductivity of the composite bipolar plate was about 77.8 $S\cdot cm^{-1}$. When the voltage was 1.2 V, the corrosion current was about $5.74\times10^{-2} A\cdot cm^{-2}$. The prepared bipolar plate could meet the requirements of VRB.

EXPERIMENTALS

The conductivity of samples (5 cm×1.5 cm) was measured by a four probe resistance tester. Other tests carried out were: Cyclic voltammetry: -1.5~2.0V, scanning speed

10mv·s-1; potentiodynamic polarization curves: -1.5~2.0V, scanning speed 5v·s-1; potentiostatic test (1.2 V, scan time 3600 s). Electrochemical measurements used three-electrode cell with the sample as the working electrode. A Pt electrode as the counter electrode and a saturated calomel electrode as reference electrode (SCE) were used. A homemade VRB electrolyte was used for the experiments.

A certain amount of GC, PP, g-PP and xylene were placed in three-necked bottle, stirred, heated to reflow temperature and kept under reflux for a period of time. The product was precipitated by a solvent, leached, dried, and hot-pressed to form the composite conductive plate.

RESULTS AND DISCUSSION

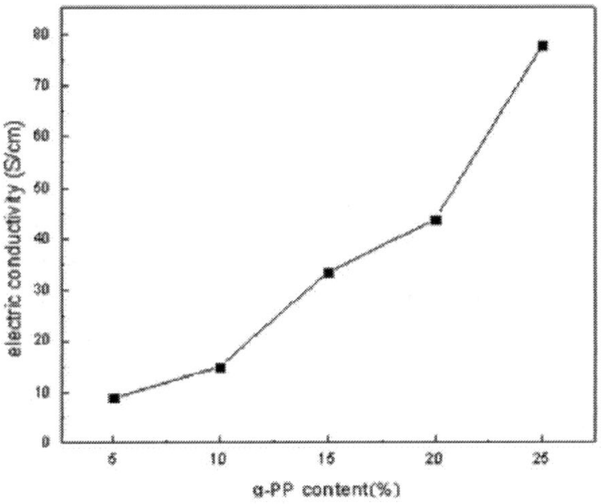

Fig.1 Effect of g-PP contents on conductivity

GC is the most familiar of the natural graphite materials. It has good conductivity. For a GC content of 70wt%, bipolar plates were prepared by using the g-PP content 10wt%, 15wt%, 20wt% or 25wt%. The effect of g-PP contents on conductivity of bipolar plates are shown in Fig. 1. It can be seen that with an increase in g-PP content, the conductivity of bipolar plate increases steeply. It means that g-PP addition contributes to the uniform distribution of the GC in the matrix, resulting in the conductive network structure formation and thus conductive property improvements. Polyacrylonitrile graphite felt (PAN-based) as electrode materials is widely used in VRB. Its conductivity commonly is about 50 $S \cdot cm^{-1}$. It is preferred to have the g-PP content greater than or equal to 20 wt% for collecting current.

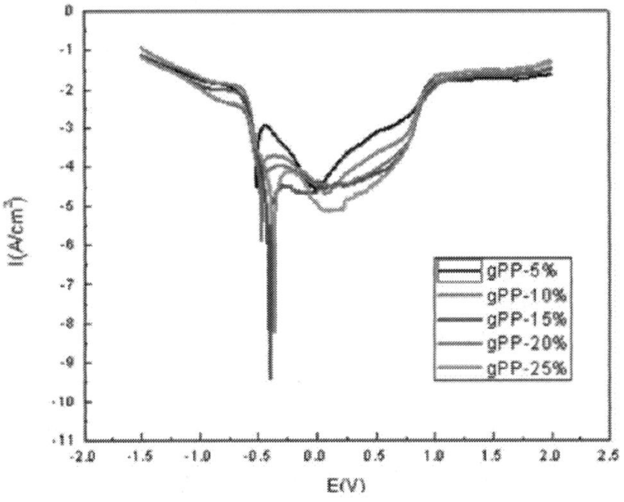

Fig.2 Potentiodynamic polarization curves

Fig.2 shows the potentiodynamic polarization curves of bipolar plates. It can be seen that as the voltage increases, anode passivation of bipolar plate occurs. That means the corrosion of bipolar plate is mainly due to oxygen corrosion. As the g-PP content increases, the corrosion current of bipolar plate decreases first and then increases. When the g-PP content of bipolar plate is 20 wt%, the corrosion current is the lowest.

The effect of g-PP contents on corrosion current are showed in Fig.3. It can be seen that with the increase in g-PP content, the electrochemical corrosion current of bipolar plate decreases rapidly, and then runs through the transition state into the steady-state. When g-PP contents is about 5%, the corrosion current density is 1.8×10^{-2} A/cm^2. With the g-PP content increasing, the corrosion current density reduces to 9.1×10^{-3} A/cm^2. Moreover, once corrosion occurs, it will cause intergranular corrosion or pitting corrosion, and then the corrosion rate further increases. When the g-PP content is 20wt%, the corrosion current of bipolar plate is close to the minimum, which is approximately 8.9×10^{-3} A·cm^{-2}.

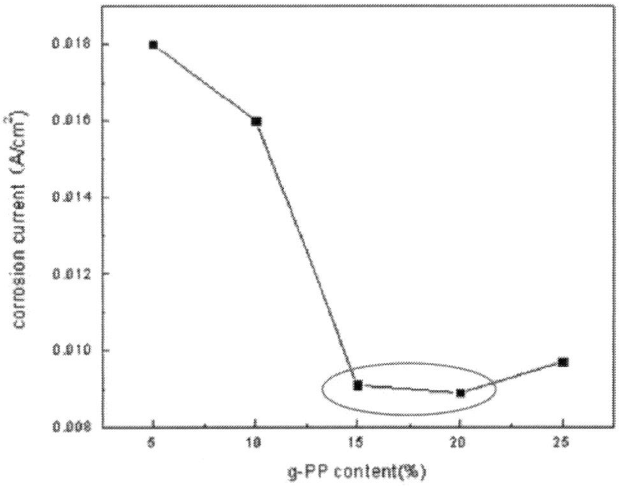

Fig.3 Effect on g-PP contents on corrosion current

Fig.4 shows cyclic voltammetry of bipolar plates under various g-PP contents. With the g-PP content increasing, more V(II) and V(III), V(IV) and V(V) transformation peaks were observed with reversibility. When g-PP is about 10%, it has the high activity. If it was used as bipolar plate in battery, the bipolar plate was easily corroded. For oxygen evolution peak of bipolar plate at anode, the higher the oxygen evolution potential, the more stable the battery is. When the g-PP content is 20wt%, there is low activity and high oxygen evolution potential of the bipolar plate providing the best the corrosion resistance of the bipolar plate.

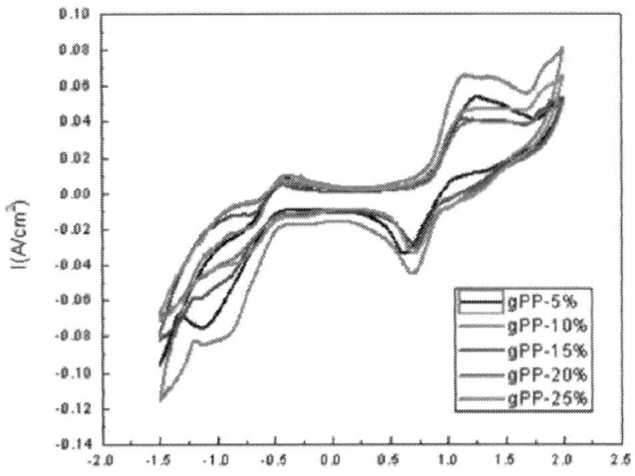

Fig.4 Cyclic voltammetry of bipolar plates

With the increase of g-PP content, the dispersion of GC in the polymer matrix is more uniform, the holes in the bipolar plate surface are reduced, and vanadium ions have difficulty in attaching to the surface of bipolar plate. As the g-PP content increases to 25wt%, the functional group COO^- content of the bipolar plate surface increases. COO^- can contribute to vanadium ion adsorption from solution to the surface. With the reaction of vanadium ion adsorption and desorption, the charge transfer should obviously change. It makes the electrochemical reaction of the vanadium ions on the surface of the bipolar plate to easily happen. These results are in agreement with the previous investigation.

CONCLUSIONS

Analysis of potentiodynamic, potentiostatic experiments and cyclic voltammetry shows when the GC content is constant, within a certain range, conductivity of bipolar plate is proportional to the g-PP content. The g-PP is added to make the dispersion between GC and the polymer matrix and improve the compatibility between the conductive filler and the polymer matrix significantly. Addition of g-PP leads to improving the corrosion resistance of bipolar plate. Conductivity and corrosion resistance of composite bipolar plate can meet the requirements of VRB.

ACKNOWLEDGEMENTS

The work was funded by Tsinghua University Initiative Scientific Research Program (201010800623), State Key Laboratory for Automotive Safety and Energy(KF10132), State Key Lab of Power Systems(SKLD10KZ03).

REFERENCES

[1] WANG Wenpin , WANG Jinhai, WANG Shubo, XIE Xiaofeng , LV Yafei. Properties of Conductive Bipolar Plate for All-Vanadium Redox Flow Battery [J]. The Chemical Industry and Engineering Society of China, 2011, 62(SI): 203-207

[2] **http://www.asbury.com/Natural-Flake-Graphite.html**

[3] Xia Liping, Lin Changwu, Luo Dongmei, Zhang Aimin. Study on properties of fluoro rubber composite current collector for vanadium redox battery [J]. Power Sources, 2010, 134(7): 672-675

[4] Chen maobin, Zhang Shengtao, Li Xiaobing, Liu Lian, Liu Xiaojiang, Meng Fanming. Preparation and performance of polytetrafluoroethylene conducting plastic on vanadium redox flow battery collector [J]. Polymer Materials Science And Engineering, 2009, 25(6): 121-124

[5] Chen Maobin, Lan Wei, Zhang Shengtao, Cui Yanhua, Meng Fanming, Liu Xiaojiang. Performance of Conductive Polyethylene Composite [J]. Applied Chemistry, 2009, 26(5): 619-621

Investigation on electrode reaction mechanism for vanadium redox flow battery

XIE Xiaofeng[1], LI Xiaoshan[1], YANG Chun[1], WANG Jinhai[1], WANG Shubo[1], Shang Yuming[1], MATHUR V. K[2]

1. Institute ofNuclear andNew EnergyTechnology, Tsinghua University, Beijing100084, China

2.Department of Chemical Engineering, University of New Hampshire, NH 03824, USA

Abstract:

The effect of glycerine addition to the anodic electrolyte, on the anodic reaction, was investigated. The results show that 1% glycerine added to the electrolyte had positive effect on the performance of electrolyte reaction on the surface electrode. The solution resistance, capacitance value and polarization resistance also improved. The reaction of VO_2^+ to VO^{2+} was helped as a result of lower polarization resistance.

Key words: Vanadium Redox Flow Battery ; Glycerine ; Electrochemical impedance spectroscopy; Electrolyte, Cyclic voltammetry; Potentiodynamic

1. Intruduction

Vanadium redox flow battery (VRB) employs vanadium ions of different valences as liquid electrolytes. It is independent of storage capacity, output power, normal atmospheric temperature operation, fast response, and long operating life. The VRB system can widely be used for load leveling, distributing power, standby power supply, and municipal transport. Cyclic voltammetry and Electrochemical impedance spectroscopy (EIS) methods are useful in characterizing the polarization resistance corrosion rate and electrochemical mechanism.

When the positive electrolyte vanadium concentration is up to 2M or the temperature is elevated above 310K, the fully charged electrolyte has poor stability. The precipitate of V(V) would block the electrode and prevent the electrolyte from flowing through the pipe. In order to keep the electrolyte from precipitating, and improve the energy storage density of electrolyte used in VRB, one method is to increase the concentration of vanadium ions. Many workers have chosen inorganic and organic additives to be added to the electrolyte for solving this problem. Some additives added to electrolyte can broaden the electrochemical window and increase electrochemical activity. The dissolved V(V) in concentrated sulfuric acid solution becomes more stable and effective at high concentration of vanadium. Glycerine has been used to enhance the conductivity and stability of anodic electrolyte for improving the performance of VRB. In this work, the effect of glycerine addition, to the anodic electrolyte, on the mechanism has been studied.

2 EXPERIMENTS

2.1 Preparation of anodic electrolyte

$VOSO_4 \cdot 3H_2O$ was dissolved in the deionized water, and H_2SO_4 was added to form $VOSO_4$+3 M solution. The solution was added to anodic electrolytic cell, and H2SO4 of the same concentration was added to cathodic electrolytic cell. The electrolytic cell was separated by an ion exchange membrane (Nafion 117, DuPont), and the lead electrodes were used for anode and cathode electrolysis. The prepared electrolyte was 1.8 mol/L, with the active area of $16cm^2$ (4cm × 4cm) and electrolytic current density of 20 ~ $200mA/cm^2$.

2.2 Electrochemical tests

A three-electrode electrochemical cell was used for electrochemical tests. The graphite electrode (with 5mm diameter) was a working electrode. The reference electrode was SCE, and a platinum mesh was used as auxiliary electrode. Different concentration (v/v) of glycerine was added in the anodic electrolyte.

The electrode characterization included the determination of the polarization resistance and electrochemical mechanism. A Zanher IM6ex instrument was used to conduct the potentiodynamic, cyclic voltammetry and EIS measurements. Potentiodynamic tests were used to measure the corrosion resistance. The initial potentials were -0.15 V, and the final potentials were 2.0 V, the scan rate was 1 mVs-1. Cyclic voltammetry was used to measure the electrochemical activities. The scan range was -0.8 ~ 1.6 V. The EIS measurements were performed at different potential and the perturbation amplitude was 5 mV. The applied frequencies ranged from 100K to 10^{-1} Hz.

3. RESULTS AND DISCUSSION

A potentiodynamic test is commonly employed to measure a material's resistance to corrosion. Potentiodynamic tests were conducted to estimate the corrosion resistances. The "corrosion resistances" here was considered as the electrolyte's ionization after the glycerine was added. Fig. 1 presents the potentiodynamic polarization scans at different glycerine contents. From the Tafel curves in Fig. 1, the related polarization kinetic parameters were analyzed by potentiodynamic polarization model in Zanher IM6ex electrochemical workstation. From the potentiodynamic polarization results and using Eq. (1), polarization resistance, Rp was estimated. Table 1 presents the polarization parameters of anodic electrolyte. The polarization resistances for 0.1% and 10% glycerine were 44744 Ω·cm2 and 13027 Ω·cm2, respectively.

$$R_P = \frac{\beta_a \beta_c}{2.3 i_{corr} (\beta_a + \beta_c)} \qquad (1)$$

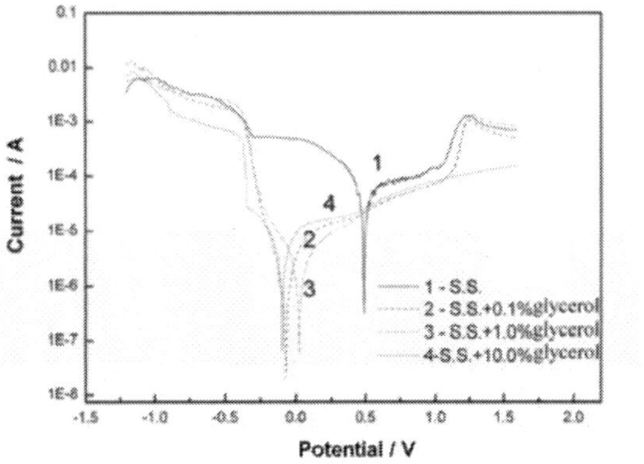

Fig. 1 Potentiodynamic polarization scans of anodic electrolyte

Table 1 Polarization parameters of anodic electrolyte

	βa/mV	-βc/mV	Rp/Ω	Icorr/μA
Non glycerin	86.2	79.9	1654	10.9
0.1% glycerin	64.1	62.3	44744	0.307
1.0% glycerin	54.0	52.9	28617	0.406
10% glycerin	57.1	60.7	[illegible]	[illegible]

Cyclic voltammetry or CV is a type of potentiodynamic electrochemical measurement. Cyclic voltammetry was used to study the electrochemical properties of the glycerine in solution. The results are shown in Fig. 2. When glycerine was added to the anodic electrolyte, the quantity and status of vanadium is changing between the electrolyte and electrode. With the glycerine content increased from 0.1% to 10%, the oxidation peak current decreased. The hydroxy groups on glycerine are bound with the vanadium ions, resulting in the decrease of surface electrode charge, transferring more vanadium ions to the bulk solution.

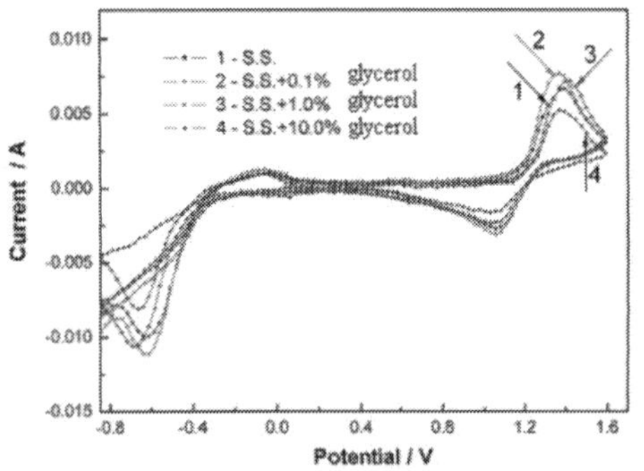

Fig. 2 Cyclic voltammetry scans of anodic electrolyte

EIS reveals information about the reaction mechanism of an electrochemical process: different reaction steps will dominate at certain frequencies, and the frequency response shown by EIS can help identify the rate limiting step. The EIS was employed to investigate the electrode reaction mechanism in anodic electrolyte containing glycerine as additive. The spectrums were simulated and analyzed through the equivalent circuit of $R_\Omega(C_d(R_pW))$. The reaction mechanism of glycerine added to electrolyte is shown in Fig 3.

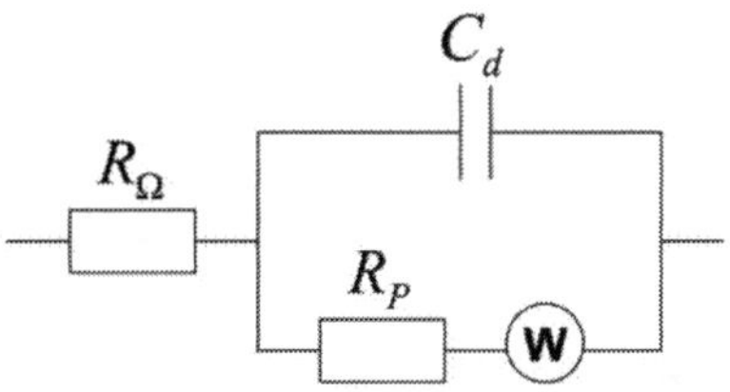

Fig 3. Equivalent circuit using RΩ(Cd(RpW))

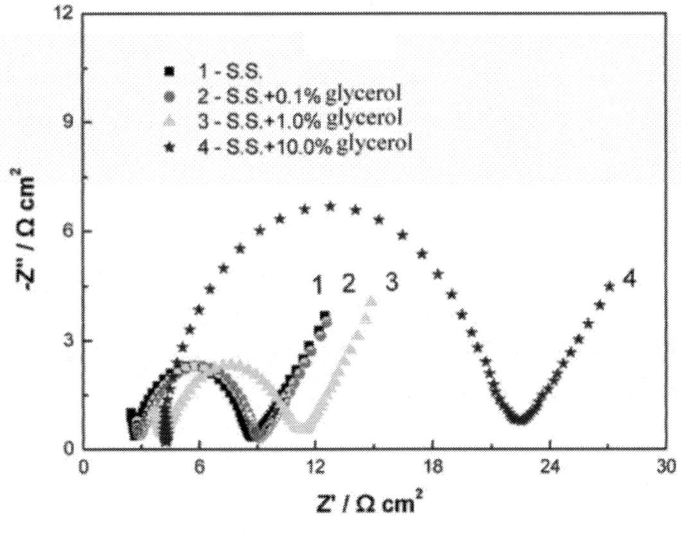

Fig.4 Impedance spectra for anodic electrolytes

The results of EIS are shown in Fig 4. From the impedance curve, with the glycerine concentration increasing, the impedance seems to increase. It means there is an optimal amount for glycerine that can be added to the electrolyte. Therefore, we should conduct the simulation of equivalent circuit to explore the mechanism of reaction with different concentration of glycerine added in. The results show the solution resistance increases by the addition of glycerine. It is difficult for glycerine to ionize in water. When the non-conducting glycerine is added to the electrolyte, the solution resistance should increase. With the increase in glycerine concentration, the resistance of electrolyte increases from 2.795Ω·cm^2 to 4.16 Ω·cm^2. In fact, the molecular structure of glycerine has three hydroxyls,

Table 2 Parameters of the equivalent circuit

	RΩ/(Ω·cm^2)	Cd/(S·cm^2)	Rp/(Ω·cm^2)	W/(S·cm^2)
Non glycerin	2.572	7.735	6.106	0.2478
0.1% glycerin	2.795	8.809	6.25	0.2676
1.0% glycerin	3.855	4.04	7.413	0.245

which are prone to bind with vanadium ions. With the glycerine content increasing from 0% to 10%, the polarization resistance increased. It means the glycerine added to electrolyte is unfavorable for reaction VO_2^+ to VO^{2+}. When the concentration exceeds 1%, the unnecessary glycerine might occupy the reaction site which can catalyze the redox reaction of vanadium ions. Therefore, when more glycerine is added to electrolyte, the reaction impedance increases. It means less glycerine is helpful for VO_2^+ to react to VO^{2+}. The results are in agreement with potentiodynamic test and Cyclic voltammetry. It indicates that glycerine contents when increased beyond certain limits will decrease the reaction rate.

4. CONCLUSIONS

The results show that glycerine added to anodic electrolyte greatly improved oxidation peak current, reaching a maximum when glycerine was about 1%(v/v). Meanwhile, with the increase in glycerine content, the solution resistance ($R_Ω$), the double electric layer capacitor (C_d) and the polarization resistance (R_p) reached their minimum. However the concentration polarization reached its maximum when glycerine added was about 1%(v/v).

The addition of glycerine increased solution resistance ($R_Ω$) which is disadvantageous to the vanadium ion movement in electrolyte solution. Moreover, the glycerine could greatly decrease the corrosion of electrolyte materials. With the increase in additive, the anti-corrosion tends to decrease. When the addivties was up to 0.1%(v/v), the anti-"corrosive ability" was the best.

ACKNOWLEDGEMENTS

The work was funded by Tsinghua University Initiative Scientific Research Program (201010800623), State Key Laboratory for Automotive Safety and Energy(KF10132).

REFERENCES

1) Bard, Allen J.; Larry R. Faulkner (2000-12-18). Electrochemical Methods: Fundamentals and Applications (2 ed.). Wiley. ISBN 0471043729.

2) http://en.wikipedia.org/wiki/Electrochemical_impedance_spectroscopy

(690d) Active Species Concentration Effects On Performance for a Non-Aqueous All-Vanadium Redox Flow Battery

Aaron A. Shinkle, Alice E. S. Sleightholme, Levi T. Thompson and Charles W. Monroe, Department of Chemical Engineering, University of Michigan, Ann Arbor, MI

Redox flow batteries(RFBs) are being developed for large-scale energy storage, uninterruptablegrid-independent power supplies and load-leveling systems for solar or windpower. Aqueous chemistries are used for all current commercial RFBs, therefore, cell potentials are limited to thestability range of water (01.23 V vs. SHE). Since a battery's cellpotential determines its energy and power density, aqueous chemistries limitthe possibilities for the performance of RFBs.

Several non-aqueous systems have beenreported recently, including systems based on ruthenium, uranium, chromium,manganese, and vanadium coordination complexes [1-5]. In the case ofvanadium acetylacetonate, the use of a non-aqueous electrolyte results in a stablecell potential window of 2.2 V [5], more than 50% wider than the aqueoussingle-metal vanadium chemistry (1.26 V [6]). However, very few resultsfrom battery charge/discharge experiments have been shown for this non-aqueousvanadium system.

This talk will examine thecharge/discharge profile for the vanadium acetylacetonate ($V(acac)_3$)RFB system in acetonitrile, with tetraethylammonium tetrafluoroborate ($TEABF_4$)support. The effects of active-species concentration on batteryperformance were examined. Linear sweep voltammetry will be used todetermine the maximum charge/discharge rate of the battery cell as a functionof electrolyte composition. Figure 1 shows the first five cycles of acharge/discharge result for 0.4 M $V(acac)_3$,0.5 M $TEABF_4$ in acetonitrile. Coulombic, energy, and powerefficiency for charge/discharge experiments were examined as a function ofconcentration to illustrate the effects of active species crossover. Efficiency and energy density were balanced to determine an optimal activespecies concentration range for operation of the battery. Lastly, thenon-aqueous system was compared to the optimized aqueous all-vanadium system toidentify areas of future work.

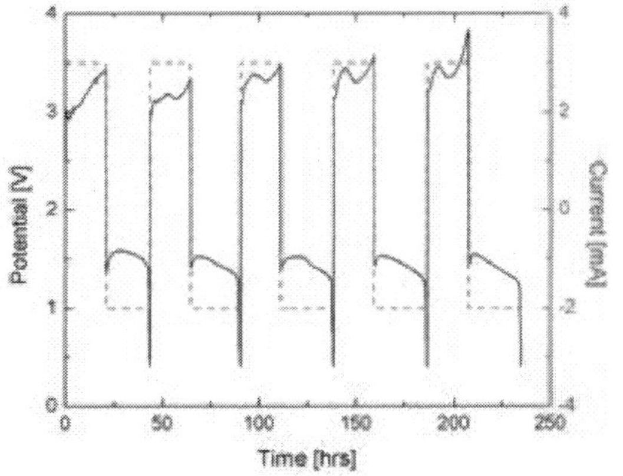

Figure1: First five cycles of a charge/discharge experiment for an all-vanadiumnon-aqueous redox flow battery chemistry. The system was charged at 3 mA until 50% theoretical stateof charge and discharged at 2 mA until 0.4 V. The system shown is 0.4 M vanadium acetylacetonate supported by 0.5 Mtetraethylammonium tetrafluoroborate in acetonitrile solvent. An H-cellconfiguration was used with graphite plate electrodes and a NeoseptaAHA anion exchange membrane.

REFERENCES

1. Matsuda Y, Takasu Y, Morita M, Tanaka K, Okada M, Matsumura-Inoue T(1985) Denki Kagaku 53:632

2. YamamuraT, Shiokawa Y, Yamana H,Moriyama H (2002) Electrochim Acta48:43

3. LiuQ, Shinkle AA, Li Y, Monroe CW, Thompson LT, Sleightholme AES (2010) Electrochem Comm 12:1634

4. SleightholmeAES, Shinkle AA, Liu Q, Li Y, Monroe CW, Thompson LT (2011) J Power Sources196:5742-5745

5. LiuQ, Sleightholme AES, Shinkle AA, Li Y, Thompson LT (2009) ElectrochemComm 11:2312

6. Poncede Leon C, Frias-Ferrer A, Gonzalez-Garcia J, Szanto DA, Walsh FC (2006) J Power Sources 160:716-732

See more of this Session: Energy Storage and Electrochemical Processes I
See more of this Group/Topical: International Congress on Energy 2011

(690e) Energy Storage and Generation From Thermopower Waves: Covalently Functionalized Thermal and Electrical Conduits

Joel T. Abrahamson[1], Changsik Song[2], Jennifer H. Hu[1], Jared M. Forman[1], Sayalee G. Mahajan[1], Nitish Nair[1], Wonjoon Choi[3], Eun-Ji Lee[2] and Michael S. Strano[1], (1)Department of Chemical Engineering, Massachusetts Institute of Technology, Cambridge, MA, (2)Department of Chemistry, Sungkyunkwan University, Suwon, South Korea, (3)Mechanical Engineering, Massachusetts Institute of Technology, Cambridge, MA

We have demonstrated through simulation and experiment that the nonlinear coupling between an exothermic chemical reaction and a nanowire or nanotube with large axial heat conduction accelerates the thermal reaction wave along the nano-conduit [1-3]. The thermal conduit rapidly transports energy to unreacted fuel regions, and the reaction wave induces a concomitant thermopower wave of high power density, producing electrical current in the same direction. At up to 7 W/g, this can be substantially larger than the power density offered by current micro-scale power sources (e.g. fuel cells, batteries) and even about seven times greater than that of commercial Li-ion batteries. MEMS devices and wireless sensor networks would benefit from such high power density sources to enable functions such as communications and acceleration hampered by present power sources.

Recently, we have covalently attached energetic molecules of mono-, di-, and trinitrobenzenes to single-walled carbon nanotubes via diazonium chemistry. Although covalent functionalization introduces defects that can scatter electrons and phonons, thermopower waves can still rapidly propagate and produce electricity on these decorated nanotubes. Differential scanning calorimetry confirms that the energetic molecules release heat, but thermopower waves require additional fuel adsorbed on the nanotubes to propagate.

[1] W. Choi, S. Hong, J. T. Abrahamson, *et al. Nature Materials*, 9, (2010), 423.

[2] W. Choi, J. T. Abrahamson, J. M. Strano, M. S. Strano. *Materials Today,* 13, (2010), 22.

[3] J. T. Abrahamson, *et al. ACS Nano* 5, (2011), 367.

See more of this Session: Energy Storage and Electrochemical Processes I
See more of this Group/Topical: International Congress on Energy 2011

(691a) Fouling and Cleaning Behavior of Forward Osmosis Membrane (keynote)

Baoxia Mi, Civil and Environmental Engineering Department, The George Washington University, Washington, DC

Forward osmosis (FO) is considered a promising process in sea/brackish water desalination and wastewater reuse. FO membrane process uses natural osmotic pressure generated by a concentrated draw solution to pull water across a semi-permeable membrane from the feed solution. The diluted draw solution is then reconcentrated to recycle the draw solution as well as to produce purified water. Compared to conventional pressure-driven membrane processes (i.e. reverse osmosis/nanofiltration), FO process has the potential to reduce energy consumption and generate less brine discharge. In addition, our study found that the lack of hydraulic pressure in FO membrane operation is advantageous in terms of minimizing irreversible membrane fouling. Many types of membrane fouling (e.g., alginate fouling, gypsum scaling) in FO processes can be fully reversed by rinsing with pure water without using chemical cleaning reagents. This talk will focus on the fouling and cleaning behavior of single foulants (alginate, BSA, gypsum) as well as that of combined foulants in FO membrane processes. A microsope-equipped direct observation setup is used to monitor fouling layer formation on membrane surfaces through a glass window to characterize the kinetics of membrane fouling and cleaning. The adhesion force between foulant and membrane surface is determined by atomic force microscopy (AFM) to better understand the fouling mechanisms.

See more of this Session: Forward Osmosis for Water Treatment
See more of this Group/Topical: Topical 1: Water Technology for Developed and Developing Countries (see also Separations Division)

(691b) Permeation of Interacting Solutes In Forward Osmosis

William Phillip[1], Jui Shan Yong[2] and Menachem Elimelech[2], (1)Department of Chemical and Biomolecular Engineering, University of Notre Dame, Notre Dame, IN, (2)Chemical & Environmental Engineering, Yale University, New Haven, CT

Osmotically driven membrane processes are an emerging set of technologies that show promise in water and wastewater treatment, desalination, and power generation. The majority of research in the field has focused on predicting and enhancing water permeation through membranes, however, the effective operation of these systems requires that the reverse flux of draw solute from the draw solution into the feed solution be minimized. In this study, a model is developed that describes the reverse permeation of draw solution across an asymmetric membrane in forward osmosis operation. This model accounts for solute-solute interactions that can arise between solutes within both the active layer and support layer of the asymmetric membrane. Experiments were carried out to validate the model predictions. Using independently determined membrane transport coefficients, strong agreement between the model predictions and experimental results was observed.

See more of this Session: Forward Osmosis for Water Treatment
See more of this Group/Topical: Topical 1: Water Technology for Developed and Developing Countries (see also Separations Division)

(691c) Network-Based Finite Element Analysis for Studying the Effect of Support Structure On Internal Concentration Polarization During Forward Osmosis

Weiyi Li, Yiben Gao and Chuyang Y. Tang, Singapore Membrane Technology Centre and School of Civil and Environmental Engineering, Nanyang Technological University, Singapore, Singapore

Forward osmosis is one of the most promising technologies for water treatment with low energy requirement and energy generation (pressure retarded osmosis). A typical forward osmosis membrane is composed of a very thin active layer and a thick porous support layer. In contrast to the reverse osmosis, the forward osmosis process is mainly driven by the osmotic pressure difference between the draw solution and the feed solution. Previous studies found that severe flux decline could be caused due to the loss of effective driving force, which is extremely sensitive to the concentration profile developed within the porous support layer, namely the internal concentration polarization.

However, most of classical models accounting for the internal concentration polarization during forward osmosis are based on a linear approximation of the support structure while the complex geometry is neglected or lumped into the macroscopic phenomenological coefficients. Little work has been done to systematically study the underlying mechanisms accounting for the effects of the porous substructures on the coupled transport phenomena.

In the current work, the support structure was approximated by a three-dimensional network with stochastic disconnectivity, which was mathematically characterized by a blockage probability function. Then, a variety of membrane morphologies were simulated by varying the blockage probability in different coordinate directions. Both the bulk flow and the solute flow in the networks were numerically evaluated by using the finite element analysis, and therefore visualized to provide deeper insights into the interplay between the elaborate porous structures and the osmotically-driven convective diffusion. The presented modeling results also quantitatively revealed the profound correlations between the macroscopic transport parameters and the characteristics of the support substructures, thereby offering a very useful tool to optimize the support structure for controlling the internal concentration polarization during forward osmosis.

See more of this Session: Forward Osmosis for Water Treatment
See more of this Group/Topical: Topical 1: Water Technology for Developed and Developing Countries (see also Separations Division)

(691d) New Approaches to Forward Osmosis Membrane Design

Ngoc Bui[1], Jason Arena[1] and **Jeffrey R. McCutcheon**[2], (1)Chemical, Materials & Biomolecular Engineering, University of Connecticut, Storrs, CT, (2)Department of Chemical, Materials and Biomolecular Engineering, University of Connecticut, Storrs, CT

Forward osmosis (FO) and pressure retarded osmosis (PRO) rely on salinity gradients to drive water across a membrane. A major limitation to the widespread development and use of these technologies is the lack of an appropriately designed membrane. Today's thin film composite (TFC) membrane design consists of a three tiered structure consisting of a polyester (PET) nonwoven, an phase inversion precipitated polyethersulfone (PSu), and a polyamide selective layer. While this design has yielded highly permselective membranes for hydraulically driven separations, they exhibit poor performance during osmotically driven processes due to the low porosity and intrinsic hydrophobicity of the PSu layer that limits pore wetting. If the pore structure does not completely wet, the effective porosity of the support is reduced and internal concentration polarization is enhanced through a reduced solute diffusivity.

This paper will present two approaches to membrane design that aim to mitigate the impact of internal concentration polarization. First, we are able to modify commercial TFC membrane support layers using polydopamine which chemically alter the hydrophobic support layers for increased hydrophilicity. We are also building radically new TFC membrane support structures using electrospun nanofibers. Modification of TFC membranes has yielded a 5-12 fold increase in flux performance over the native membranes while our new membranes have exhibited a 2-5 fold improvement in flux over the best commercially available FO membranes.

See more of this Session: Forward Osmosis for Water Treatment
See more of this Group/Topical: Topical 1: Water Technology for Developed and Developing Countries (see also Separations Division)

(691e) Development of a Pilot-Scale Forward-Osmosis Desalination Process

Young Kim, Dong-Wook Oh, Yu-Chang Kim, Kong Hoon Lee and Jungho Lee, Energy Plant Research Division, Korea Institute of Machinery and Materials, Daejeon, South Korea

A pilot-scale forward-osmosis (FO) desalination system is designed to produce 3 ton per day of fresh water. In the front part of the system, fresh water is permeated from the feed, or seawater, to the draw solution in membrane modules. Subsequently, the draw solute is separated from water to produce the potable water in a membrane, or a column distillation system. The concentration of draw solution is designed with considering the capability of separation systems. The capacity of the separation systems is determined to enable treating the amount of draw solution required for target production of fresh water. The developed system is examined in performance tests. The results contribute to assessing the feasibility of forward-osmosis desalination systems.

See more of this Session: Forward Osmosis for Water Treatment
See more of this Group/Topical: Topical 1: Water Technology for Developed and Developing Countries (see also Separations Division)

(691f) Zero-Energy Osmotic Water Purifier for Ethiopia: Providing Clean Drink to Vulnerable Populations

Ethan Butler, Chemical, Materials & Biomolecular Engineering Dept., University of Connecticut, Storrs, CT and **Jeffrey R. McCutcheon**, Department of Chemical, Materials and Biomolecular Engineering, University of Connecticut, Storrs, CT

Globally, 894 million people do not have access to the basic amount of freshwater needed for drinking, cleaning and cooking. This is a catastrophe because water directly impacts every measure of development used by the Human Development Index (HDI): healthcare, education, and the economy. Many of the afflicted populations are also victims of natural disaster or migration. Refugees in particular are vulnerable populations that desperately need clean, sterile fluids for hydration and nutrition. Many water purification technologies are commercially available, but few are suitable for implementation in the rural of developing countries. Traditional methods of distillation, filtration, and disinfection require electrical inputs or chemical feed stocks that are difficult, if not impossible, to supply to rural areas. Some low tech options, such as sand filtration, offer an inexpensive solution to filtering large contaminants from water but do very little to remove harmful heavy metals and other dissolved contaminants.

This study investigates Ethiopia as a possible site for implementing a commercial forward osmosis process for producing a clean, nutritious drink from any fresh or brackish water surface water. *Forward osmosis (FO)* is an emerging technology that could be used to remove biological, organic, and inorganic contaminants from water. It works by utilizing an *osmotic pressure* gradient created by a "draw solution" to pull water via osmosis across a selectively permeable membrane while rejecting contaminants. For the specific process to be evaluated here, a sugar-nutrient-electrolyte solution mixture is used as the draw solution. This solution is diluted during osmosis, thus producing a clean, lightly sweetened, safe and nutritious drink. The concentrated contaminated water is returned to the source. Since this process occurs naturally, no energy input is required and since there is no exhaustive media (such as with adsorption), there is no hazardous waste associated with maintenance. The water is taken *directly* from a pond, stream or river with the concentrate being returned and diluted into the source.

The objective of this project was to assess the efficacy of a commercial forward osmosis system intended for use in rural Ethiopia and Ethiopian refugee camps. The particular system used during this study is the Hydration Technology Innovation's (HTI) Village System for its claimed field durability and capacity (200 liters of nutritious drink per day). The system employs a membrane capable of removing dissolved solids (organic and inorganic) and biological organisms (bacteria and viruses), though these performance metrics have yet to be tested and widely disseminated. In fact, no peer reviewed study has ever evaluated this system when functioning under real conditions. UCONN

Engineers without Borders has teamed up with HTI to evaluate their system performance under real conditions (in this case, purifying water from an on-campus lake). This study will show our preliminary pilot scale studies of using the HTI system hybridized with other low-tech options for pretreatment in order to preserve membrane life and performance. We will also present our findings from an assessment trip to Ethiopia to better understand how such a system can be implemented.

See more of this Session: Forward Osmosis for Water Treatment
See more of this Group/Topical: Topical 1: Water Technology for Developed and Developing Countries (see also Separations Division)

(714a) The Role of Cellulose Accessibility On Enzymatic Saccharification of Lignocelluloses

Junyong Zhu, USDA Forest Products Laboratory, Madison, WI, Zhiguang Zhu, Biological Systems Engineering, Virginia Polytechnic Institute and State University (Virginia Tech), Blacksburg, VA, Y.-H. Percival Zhang, Biological Systems Engineering Department, Virginia Polytechnic Institute and State University, Blacksburg, VA, Yonghao Ni, 3Limerick Pulp and paper Center, University of New Brunswick, New Brunswick, Canada, ZiBin He, Limerick Pulp and Paper Centre, Virginia Polytechnic Institute and State University (Virginia Tech), Blacksburg, VA and Xiaolin Luo, South China University of Technology, Guangzhou, China

So much research has been dedicated to understand the effects of various factors on enzymatic hydrolysis of lignocelluloses. It has been understood that removal of hemicelluloses is very important to improve lignocellulosic substrate digestibility. It has also been identified that improving enzyme accessibility to cellulose is more important than delignification to enzymatic saccharification. This study is attempted to develop a comprehensive understanding of various factors on enzymatic hydrolysis of lignocelluloses by taking a progressive approach. In the first phase, a set of lignocellulosic substrates were produced by drying to different degrees from the same never dried sample. Therefore, these dried substrates have exactly the same chemical composition and structure. However, drying induced fiber hornification produces irreversible change of cell wall structure when rewetting, i.e., fibrils are tightly packed together through hydrogen bonding to produce permanent fiber pore shrinkage and loss of pores in cell wall. As a result, the enzyme accessibilities of these samples were significantly reduced comparing with the never- dried sample and varied with the degree of drying. This set of samples is an excellent candidate to verify the effect of enzyme accessibility on enzymatic hydrolysis. In the second phase, two sets of samples, one produced from fractionation of the same sample, and another from disk milling of the same chemically pretreated biomass to different degrees, were used to determine the effect of substrate size/specific surface or degree of size reduction on enzymatic hydrolysis. Finally, different pretreatment methods were used to explain the effect of chemical composition on enzymatic hydrolysis. Both enzymatic hydrolyses and measurements of cellulase accessiblity to cellulose (CAC) were carried out to quantify the contributions of various factors to enzymatic hydrolysis of lignocelluloses.

See more of this Session: Recalcitrance of Woody Biomass
See more of this Group/Topical: International Congress on Energy 2011

(714b) Structural Studies On Recrystallized Ionic-Liquid (IL) Treated Lignocellulosic Biomass

Indira Samayam[1], B. Leif Hanson[2], Paul Langan[3], Christopher J. Barr[1] and **Constance Schall**[1], (1) Chemical and Environmental Engineering, University of Toledo, Toledo, OH, (2)Instrumentation Center, The University of Toledo, Toledo, OH, (3)Biosciences Division - Los Alamos National Laboratory, Los Alamos, NM

Terrestrial lignocellulosic biomass is composed of three major components; cellulose, a linear polymer of glucose (a hexose sugar); hemicellulose, a complex branched polymer of xylose (a pentose sugar) and other sugar derivatives; and lignin, a polyphenyl propanoid macromolecular assembly that is covalently cross-linked to hemicellulose. These components can serve as a source of carbon based feedstock for fuel and chemical production in much the same way that crude oil serves as the carbon feedstock in petrochemical refineries.

The deconstruction of lignocellulosic biomass into simple sugars constitutes a core barrier for producing value added products from the sugar platform. The cellulosic component has a crystalline and oriented fibrillar structure in biomass, whereas the hemicellulosic and lignin components have less ordered structures that encrust the cellulose fibers. The crystalline structure of native cellulose fibers, called cellulose I, can be a major impediment to its hydrolysis to monomeric sugars. Pretreatment can disorder the crystalline cellulosic component and result in structural changes in the cellulose fibers. These structural changes in cellulose fiber can be achieved by modifying the pretreatment conditions. Depending upon the type of pretreatment, the conversion of cellulose fibers to other crystal forms, such as cellulose II, cellulose III_{II} and cellulose III_{I}, or amorphous forms, can greatly improve their susceptibility to hydrolysis. To investigate the effects of varying the parameters of IL pretreatment on the structure of their cellulosic components, X-ray powder diffraction (XRD) and X-ray fiber diffraction from biomass samples was used with unoriented and partially (fiber) orientated cellulosic components . A comparative study on enzymatic hydrolysis and sugar analysis related to the observed changes in cellulose structure to biomass digestibility was also performed. The improvement in hydrolysis is hypothesized to be primarily due to structural changes in the cellulose constituent.

See more of this Session: Recalcitrance of Woody Biomass
See more of this Group/Topical: International Congress on Energy 2011

(714c) Kinetics and Mechanism of Cellulase Hydrolysis of Nanocrystalline Cellulose

Xiao Zhang, Chemical Engineering and Bioengineering, Washington State University, Richland, WA and Xiaohui Ju, Washington State University, Richland, WA

Kinetics and Mechanism of Cellulase Hydrolysis of Nanocrystalline Cellulose

Xiaohui Ju, Xiao Zhang

School of Chemical Engineering and Bioengineering, Centre for Bioproducts and Bioenergy, Washington State University,

Abstract

Decrystallization of cellulose is a major bottleneck in enzymatic hydrolysis of lignocellulosic biomass. The kinetic and mechanism of enzyme degradation of nanocrystalline cellulose prepared from bleached kraft pulp were investigated. The efficiency of enzymatic degradation of nanocrystalline cellulose by three commercial enzymes was determined. All these cellulase preparations demonstrated the ability to degrade nanocrystalline cellulose. The mechanism of enzyme decrystallization of cellulose was investigated by XRD and NMR. This presentation will discuss some of the new findings in cellulose crystal structural changes during enzymatic hydrolysis.

See more of this Session: Recalcitrance of Woody Biomass
See more of this Group/Topical: International Congress on Energy 2011

(714d) Incremental Disassembling of Woody Biomass and Production of Biochemicals

Shijie Liu, Paper and Bioprocess Engineering, SUNY College of Environmental Science and Forestry, Syracuse, NY

Woody biomass is a reliable renewable source of raw materials for energy and chemicals. Woody biomass is structurally composed of cellulose, lignin, hemicelluloses and other minor components. The compositions vary from species to species. The complexity / heterogeneity of woody biomass presents challenges to conversion technologies, which is perceived as recalcitrance. Pretreatment and detoxification have been two unfavorable but widely discussed processes todate. At SUNY ESF, a strategy of incremental disassembling of woody biomass is adopted in which hemicelluloses and extractives are separated first through hot-water extraction, cellulose and lignin are next. Hot-water extraction uses water as solvent without any other added chemicals. Acetyl groups in the woody biomass dissolve into the liquor and serves as catalyst. In each step of the separation, high value applications are examined. For example, hemicelluloses are hydrolyzed and fractionated via membrane for xylo-oligomers, xylose, acetic acid, methanol, and aromatics, etc. The residual solid mass containing mostly cellulose and lignin has wide applications: fiber board, wood fuel pellet, pulp and paper, and sugars / aromatics. Incremental disassembling is emphasized for value preservation of various components. A clever integration of the processes makes the conventional pretreatment and detoxification processes obsolete. The sugar stream produced at SUNY ESF has been successfully fermented to ethanol, butanol, lactic acid, and PHA. There is no deliberate detoxification that generates waste byproducts.

See more of this Session: Recalcitrance of Woody Biomass
See more of this Group/Topical: International Congress on Energy 2011

(714e) Effect of Steam Explosion On Degradability and Accessiblity of Loblolly Pine In Bioethanol Applications

Yuzhi Kang, Sang Beom Kim, Prabuddha Bansal, Dr Matthew Realff and Andreas S. Bommarius, School of Chemical and Biomolecular Engineering, Georgia Institute of Technology, Atlanta, GA

Abstract

Among the various biomass pretreatment methods, steam explosion is one of the most widely used due to its cost-effective disruption of the plant structure and hence enhanced enzymatic accessibility. The operational parameters including temperature, residence time and acid concentration have pronounced effect on crystallinity (CrI) of the feedstock loblolly pine. CrI was determined from X-ray diffraction pattern with the method of least squares and it was found to decrease as pretreatment severity increase.

Three different substrates, Avicel, phosphoric acid swollen cellulose (PASC), and SO_2 steam exploded Loblolly pine (SELP), were subjected to enzymatic hydrolysis and subsequent adsorption study. Both SELP samples (2.9-419-5 and 5-417-5) were found more readily degraded by the CelluclastÒ enzyme system than Avicel. Despite the complexity of both substrates and enzyme mixtures, the SELP adsorption isotherm can be fitted to a Langmuir isotherm with $R^2 > 0.99$. The saturation point was reached at ~6.24 mg/ml total enzyme loading. Adsorption data was correlated with initial rates and a linear relationship between initial rate and adsorption amount was confirmed.

See more of this Session: Recalcitrance of Woody Biomass
See more of this Group/Topical: International Congress on Energy 2011

CHANGE OF STRUCTURE OF CORNCOB WITH DILUTE ACID AND LIME PRETREATMENT AND A POTENTIAL PROCESS FOR XYLOSE CO-PRODUCTION OF CELLULOSIC ETHANOL

Ji-hong Li[1], Shi-Zhong Li[1], Zhi-pei Yan[1], Qing Li[2], Liang-cai Peng[3]

[1]Institute of New Energy Technology, Tsinghua University, Beijing, China

[2]Colledge of Science, Huazhong Agricultural University, Wuhan, China

[3]Colledge of Plant Science & Technology, Huazhong Agricultural University, Wuhan, China

Abstract

Degree of polymerization (DP) of cellulose is considered an important factor to effect on the conversion of Lignocellulose. The higher DP of cellulose, the more difficult conversion of lignocellulose is. In this study, corncobs were pretreated with dilute acid process for production of xylose, and then the residues were pretreated with lime. The enzymatic hydrolysis and the structures involved the composition, DP and crystallinity index of these two types of feedstock were determined. Comparison of structures resulted from these two pretreatment revealed that DP was not the crucial factor to determined efficiency of enzymatic hydrolysis. Lime pretreatment could improve the conversion of cellulose although the residues with lime pretreatment had higher DP and percentage of crystallized cellulose. Moreover, the results also confirmed lignin was not the key factors for enzymatic hydrolysis of lignocellulose because the residues with dilute acid pretreatment had the similar percentages of lignin to that with lime pretreatment. The crucial effect of lime pretreatment was to remove the inhibitor for cellulase and increase the percentages of cellulose. With obtained xylose, the residues with further lime pretreatment had expected efficient enzymatic hydrolysis of cellulose. So this process could be a potential process for xylose co-production of cellulosic ethanol to decrease the cost of cellulosic ethanol.

(714g) Green and High Efficient Enzyme Hydrolysis Process for Sweet Sorghum Bagasse

Xinshu Zhuang, Wen Wang, Qiang Yu, Wei Qi, Qiong Wang, Xuesong Tan, Yu Zhang and Zhenhong Yuan, Guangzhou Institute of Energy Conversion,CAS, Guangzhou, China

In order to improve enzymatic hydrolysis efficiency, several approaches including liquid hot water (LHW) pretreatment, adding surfactant, magnetic stirring and fed-batch hydrolysis were investigated, in which Sweet sorghum bagasse (SSB) was chosen as the feedstock. The experiment results showed that with the LHW pretreatment, the SSB conversion ratio reached 35.91% , nearly 75% xylan and 33% lignin were removed. Using the LHW-pretreated SSB, glycan conversion ratio raised from 16.6% to 37.9% by adding surfactant Tween80 with the optimum dosage of 0.175ml/g DM. The virulence test showed that Tween80 nearly had no influence to the growing of ethanol fermentation yeast-*Saccharomyces cerevisiae*. Comparing stirring and shaking mode for the enzyme hydrolysis, the glycan conversion rate was 57.3% in stirring mode, higher than 44.4% obtained from shaking mode. The image of Scanning Electron Microscope (SEM) showed that LHW pretreatment could destroy the tight structure of SSB and the comminution of pretreated material could smaller the particles to enhance the hydrolysis process. Finally the LHW-pretreated SSB, Tween80 and stirring mode were applied to fed-batch process, the concentration of glucose, cellobiose and xylose reached 88.8g/l, 30.7g/l and 18.9g/l respectively, and the glycan conversion ratio was 89.3% under the optimum condition. These research works lay the foundation for the construction of a green and high efficient sugar platform to facilitate the cellulosic ethanol development.

These research work are financially supported by the National Key Technology R&D Program (2011BAD22B01), National High Technology Research and Development Program of China (2010AA101606) and the Knowledge Innovation Project of the Chinese Academy of Sciences (KSCX1-YW-11-A3, KSCX2-YW-G-063)

[1]Corresponding author. yuanzh@ms.giec.ac.cn (Z. Yuan)

See more of this Session: Recalcitrance of Woody Biomass
See more of this Group/Topical: International Congress on Energy 2011

Membrane Technology after Four Decades: Well Established but Still Growing in Scope

David J Paulson of Water Think Tank, LLC, Minnetonka, Minnesota, USA

Abstract

Membrane technology is a broad category of media-based separations that can include such hybrid forms as embedded catalyst and ion exchange as gas and vapor phase separations. This paper, however, is focused on the most commercially prevalent use of membranes, namely pressure driven liquid separations in the membrane classes of reverse osmosis, nanofiltration, ultrafiltration, and microfiltration (RO/NF and UF/MF). It tracks their technical and commercial development history, their current status and near- to mid-term future developments. Due the success in the three areas of product development, scope of applications and commercial acceptance, there is much ground to cover. Some of this history is poorly understood today but very relevant to future growth of the technology. Some of it contradicts the mythology that has grown over the 40-plus years of commercial success and intense academic study of membrane technology. This paper looks at the last 40 years of the technology and the growth and evolution of the membrane industry.

The relative order of development of the four classes and their overlapping roles and influences on each other are described. The paper also addresses some of the key personnel, business entities and happenstances (planned and accidental) that pushed the technology ahead. Some examples of what is covered include the fact that RO was not the first modern commercial membrane class, how the term nanofiltration came to be coined, and how one of the first NF applications commercialized led to a U.S. Supreme Court decision on fundamental patent law. Why nanofiltration is purposely confused with the current nano-materials technology buzz, and why the definition of UF still has no accepted formal standards, or even an informal standard accepted in the industry, is also explained.

Despite a drumbeat of marketing claims to the contrary, there are only three major advances in product technology that have occurred among these pressure-driven processes since reverse osmosis was commercially proven in the early 1970s. A fourth could be included, depending on how low the commercial bar is lowered to define success. These are all described, as is the next major advance likely to be realized. This paper also addresses the value of each of these advances in commercial terms as well the increased scope of applications they have allowed. Membrane applications occur in very many market segments but for each they can be placed in three main categories; water purification, waste treatment and industrial processes. The growing potential for use in resource extraction industries presents an interesting subcategory of industrial processes that overlaps significantly on both other categories.

Water purification is the most common category, and aside from the health-critical applications, is fairly mundane from a technical perspective. Here the topic of "emerging contaminants" is frequently in the news, so the potential treatment of potable water with RO/NF, and why they are usually a poor choice, is described.

Membrane technology is suitable for many applications of waste treatment, so was applied early as a simple step to mitigate pollution of the environment. But an economic benefit drove technologists to apply it for a more useful dual purpose: Concentrating waste for disposal was soon only a partial step toward dewatering, purifying and recovering valuable materials out of waste. Only a few membrane companies made a strategic decision to push into waste and process applications, but they did look upstream in industrial processes to find more useful places to recover materials such as food sugars, catalysts, fine chemicals and plating metal salts. This drove the notion of "fractionation" or separating various sized solutes to allow economic harvest of valuable materials. Solutes in size ranges of 200 to 1000 Daltons drove the UF class concept of molecular weight cut-off (MWCO) into that of "loose RO" membranes - precursors to the NF class. This evolution improved the economics of installing membranes greatly, and the now common applications purifying amino acids, dyes, herbicides and more, were developed. Today, this concept is accepted as common chemical engineering wisdom.

Wisdom was and still is frequently missing among novices is the tendency to view of membrane pore structure as having discrete and well defined pores, and the even more inaccurate belief that only the membrane pores cause separation. The unfortunate idea that a pore size rating is a sufficient indicator of the separation that will occur online has led many to reject membranes as unworkable for their use, when only a more sophisticated understanding of the dynamics of the solute/membrane interaction is required. This is most important in the UF class and both the cause and how to how to overcome this problem of engineering attitude is described. Examples of successful, as well as failed, UF class separations are presented.

The first case of energy savings by RO membrane application probably occurred in the woods of southern Quebec in the early 1970s. The purpose of employing membrane for energy recovery was largely evaluated by comparing the thermal evaporation to pressurized membrane costs for simple water removal. The idea of recovering heat in industrial liquids is still rarely exploited, but advances in temperature resistance of membrane and element construction is incrementally expanding potential here. This paper identifies several examples of the intersection of new membrane materials and improved energy recovery, addressing new materials, lower pump energy costs and new applications currently under review.

The paper concludes with example applications that have increased economic potential due to the incremental improvements made steadily in membrane technology, probable step-change improvements and the one or two genuine breakthroughs that are on the horizon. These are presented by showing the progression of both the technology and improvements in understanding how to use it to project how the expansion of membranes continues.

(715b) Zeolite-Polyamide Nanocomposite Membranes for Sea-Water Desalination

Mary Laura Lind, School for Engineering of Matter, Transport, and Energy, Arizona State University, Tempe, AZ

Many exciting and new advanced materials concepts have been proposed for preparing better performing desalination membranes including aligned-carbon nanotubes, protein-polymer hybrids, zeolite-coated ceramics, chlorine-tolerant sulfonated and fluorinated polymers, and inorganic-organic nanocomposites. Following a brief introduction to the state of the art in nanocomposite desalination membranes, this presentation will focus on recent developments related to "thin film nanocomposite" (TFN) reverse osmosis (RO) membrane materials. Previously, it has been hypothesized that zeolite addition increased membrane permeability through molecular sieving action. In this talk, we will discuss the impacts of zeolite molecular sieve pore size and hydration state on the performance and physical-chemical properties of TFN membranes designed for use in seawater desalination. We will present extensive characterization results that suggest molecular sieving only partly explains the TFN membranes' higher water permeability and suggest that polyamide film structure (i.e., crosslinking) is different for zeolite containing polyamide films. From these results we hypothesize new insights into the mechanisms by which zeolite nanoparticles influence the observed separation performance of zeolite-polyamide TFN membranes.

See more of this Session: Reverse Osmosis Processes and Membrane Technologies
See more of this Group/Topical: Environmental Division

(715c) Analysis of Zeolite Contribution to Thin Film Nanocomposite Membrane Reverse Osmosis Performance

Andrew Lundfelt, Arizona State University, Tempe, AZ and Mary Laura Lind, Semte, Arizona State University, Tempe, AZ

Hydrophilic zeolite nanoparticles have been incorporated into polyamides to form thin film nanocomposite (TFN) membranes, which studies show impact their performance and transport properties for reverse osmosis applications[1]. This work has focused on the use of Linde type A zeolite nanocrystals because of their hydrophilicity and ideal pore size (ca 4 angstroms), which is appropriate to allow water molecules but not hydrated ions to pass. Recent studies of these thin film nanocomposite membranes have shown an increase in permeability while maintaining similar solute rejection levels to their corresponding thin film composites[1-4]. Various hypotheses have been proposed about the mechanism by which zeolites impact TFN membrane performance[1-4]; these include that zeolites may act as preferential flow channels for water, zeolites may create defects through the thin film, and that zeolites may alter the structure of the polyamide.

In this study, we systematically evaluate zeolite loading of TFN membranes. In particular, we present both TFN seawater performance along with rigorous characterization and quantification of the thin film surface chemistry and structure. Precise quantification of the zeolite loading of the thin films has been performed with thermal and imaging techniques. We then compare the results of the TFN membranes with quantified nanoparticle loading to that of the individual components and what is expected theoretically.

References

1. Jeong, B.H., Hoek, E.M., Yan, Y.S., Subramani, A., Huang, X., Hurwitz, G., Ghosh,

A.K., and Jawor, A. (2007). Interfacial polymerization of thin film

nanocomposites: a new concept for reverse osmosis membranes. J. Memb. Sci, *294 (1)*.

2. Lind, M.L., Ghosh, A.K., Jawor, A., Huang, X., Hou, W., Yang, Y., and Hoek, E.M.

(2009). Influence of zeolite crystal size on zeolite-polyamide thin film

nanocomposite membranes. Langmuir, *25 (17)*, 10139-10145.

3. Lind, M.L., Jeong, B., Subramani, A., Huang, X., and Hoek, E.M. (2009). Effect of mobile cation on zeolite-polyamide thin film nanocomposite membranes. J. Mater. Res., *24 (5)*, 1624-1631.

4. Lind, M.L., Suk, D.E., Nguyen, T.V. et al. (2010). Tailoring the Structure of Thin Film Nanocomposite Membranes to Achieve Seawater RO Membrane Performance. Env. Sci. Tech., *44 (21)*, 8230-8235.

See more of this Session: Reverse Osmosis Processes and Membrane Technologies
See more of this Group/Topical: Environmental Division

(715d) Anti-Adhesive Properties of Polysulfone Membranes Modified with PAA/PAH Polyelectrolyte Multilayers

Li Tang and Kai Loon Chen, Department of Geography and Environmental Engineering, The Johns Hopkins University, Baltimore, MD

Microfiltration (MF) membranes are increasingly used in drinking water treatment and wastewater reuse because of their effectiveness in removing waterborne pathogens (e.g., bacteria, viruses, and protozoa cysts). However, due to the ubiquity of microorganisms in influent waters, MF processes are often hindered by biofouling which reduces clean water production, shortens membrane lifespan, and increases energy demands. To overcome these limitations, we investigate the surface modification of polymeric MF membranes with the use of polyeletrolyte multilayers (PEMs) in order to enhance their resistance to biofouling. Specifically, commercially available polysulfone (PS) membranes are modified with poly(allylamine hydrochloride) (PAH) / polyacrylic acid (PAA) multilayers in order to enhance the membranes' resistance to bacterial adhesion (*E. coli* MG1655). The properties of the modified membrane surface are characterized by using ATR-FTIR, XPS, AFM, and SEM. Crossflow membrane filtration experiments are conducted with *E. coli* suspensions prepared at pH 7.1 and 10 mM NaCl. By using a fluorescent microscope, the deposition of the fluorescent *E. coli* cells can be observed through a window on top of the membrane housing in real time. The images of deposited bacteria on the membrane surface are taken using a digital camera at equal time intervals throughout the entire deposition process. The initial bacterial deposition rate (first 30 min) is obtained by enumerating the deposited bacteria as a function of time. The reversibility of bacterial attachment is also investigated by flushing the deposited bacteria using the same background solution at an elevated crossflow velocity and zero transmembrane pressure. The effects of PEMs modification are determined by comparing both the bacterial deposition kinetics and the degree of bacterial release from the modified membranes with those of the original membrane. In addition, the influence of the number of bilayers within the PEMs on the membrane hydraulic resistance and anti-adhesive properties is investigated.

See more of this Session: Reverse Osmosis Processes and Membrane Technologies
See more of this Group/Topical: Environmental Division

(715e) Integrated Membrane Bioreactor Process for Treatment of Reclaimed Water for Groundwater Recharge

Woonhoe Kim[1], **Varadarajan Ravindran**[2] and Massoud Pirbazari[1], (1)Sonny Astani Department of Civil and Environmental Engineering, University of Southern California, Los Angeles, CA, (2) University of Southern California, Los Angeles, CA

Membrane separations promise to yield substantial economic, energy, and environmental benefits leading to enhanced global competitiveness by significantly reducing energy consumption, increasing capital productivity, and lowering waste generation. Many challenges must be overcome before membrane technologies can be widely used in wastewater treatment, water reclamation and water reuse. The technological barriers include membrane fouling, low permeate flux, low separation factors or selectivity, and poor rejection characteristics. The membrane bioreactor (MBR) process has shown excellent potential for water reclamation and groundwater recharge. In order to make the process more efficient, economical and versatile for water reclamation and reuse applications, superior membranes must be developed with better aqueous transport and anti-fouling characteristics to resist permeate flux decline attributed to organic and biological fouling attributable to natural organic matter (NOM), exopolymeric substances (EPS), and other macromolecules. This would significantly reduce energy costs which represent a substantial portion of total operation costs associated with membrane systems.

In the realm of water reclamation and reuse, the MBR process with ultrafiltration can be made more energy efficient, economical and versatile than reverse osmosis or nanofiltration by using exhibiting better transport properties, higher permeate flux, and greater fouling resistance. The MBR process is specially designed the control of membrane fouling and permeate flux decline by a combination of powder activated carbon (PAC) sorption and fluid management. Therefore, the present research was directed at overcoming these problems to develop superior foulant-resistant membranes that can be used in MBR process for treatment of reclaimed water to be subsequently used in groundwater recharge. The batch-mode flat-sheet membrane tests represent an economic means for evaluating the potential for membrane fouling and rejection characteristics. These test were conducted in the Osmonics Sepa CF Type I plate-and-frame test cells, generally recommended by the U.S. Environmental Protection Agency for performance membrane evaluations in water treatment and water reclamation systems.

Preliminary flat-sheet membrane tests were performed as a prelude to evaluate the nature of fouling and the fouling potential of polymeric membranes, and to obtain the best membrane material for the fabrication of superior foulant-resistant hollow-fiber membranes for use in MBR systems. These membranes could involve the combination of polymeric materials impregnated with nano-materials. In these tests, the permeate flux and the effluent total organic carbon (TOC) were determined to evaluate permeate flux decline patterns. The reclaimed water was obtained from a water reclamation facility in

Southern California. In these tests the conditions in a membrane bioreactor system were simulated using PAC, acclimated microorganisms, and the results provided insight into membrane fouling, permeate flux deterioration, and TOC rejection in a system containing a typical MBR feed. Additionally. surface characterization techniques were employed for evaluating fouling potential and morphological details of membrane surfaces include atomic force microscopy (AFM), Fourier transform infrared spectroscopy (FTIR), and x-ray photoelectron spectroscopy (XPS). These studies provided useful information on several aspects including the nature and extent of membrane fouling (and permeate flux recovery), and alteration to membrane material caused by fouling.

See more of this Session: Reverse Osmosis Processes and Membrane Technologies
See more of this Group/Topical: Environmental Division

EFFICIENCIES OF ANHYDRIDE BASED POLYMERS AS ANTISCALANTS IN WATER DESALINATION PLANTS
AIChE 2011 ANNUAL PROCEEDINGS - Paper # 715f

Y. A. Roomi and M. R. Riazi, Kuwait University, Kuwait*
**Corresponding author contact: mrriazi@gmail.com*

Introduction

The presence of a safe and reliable source of water is an essential prerequisite for the establishment of a stable community. The number of countries classified as water scarce (i.e. suffering for shortage of fresh water resources below 1000 m^3 / cap / y) is increasing every year [1]. They are mainly located in the Middle East and North Africa regions, even though some of them can be found in Europe, Asia and the Caribbean Area.

Hence it is of vital importance to overcome the potable water deficiency by means of sea water desalination [2]. For theses reasons, modern desalination technologies were mainly developed in Gulf countries, where the first desalination plants were built in the 1960s.

Still today, most of the huge desalination projects are located in this area, as confirmed by the geographical distribution of the market demand, mainly located in the Middle East regions [3,4]. In brief this real need of fresh water for different purposes (potable water, agriculture and industrial end-users), forced additional efforts to improve desalination techniques.

In these water treatment and water desalination processes scale formation has been recognized as a widespread problem [5]. Seawater and brackish water always have the tendency for scale formation and fouling problems due to dissolved salts and finely suspended solids [6]. The precipitation of scaling salts such as $CaSO_4$ and $CaCO_3$ on membranes leads to significant water flux decline [7-9]. Scale formation of soluble salts is one of the major factors limiting the application of reverse osmosis (RO) membranes [10-16].

Commonly, scale consists of calcium and magnesium salts of carbonates and sulphates, hydroxides and phosphates. The scale forming characteristics of water are dependent on physical, chemical and bacteriological factors such as pH, temperature, pressure, dissolved minerals and carbon dioxide, bicarbonate, dissolved gases, microorganisms, algae etc.

To date, control of super saturation and scaling is dominated by experimental expertise and lacks the theoretical understanding of the complex process of crystallization. The strong economical impact of scaling forces engineers to cope with emerging molecular views on crystallization because molecular simulation has already revealed some of the controlling mechanism and relations. But till now it is only partly understood in which way antiscalants interact with crystallization steps resulting in retardation or complete inhibition.

Growth inhibiting antiscalants like polyacrylate, adsorb to the crystal with their negatively charged functional groups. A further growth of the affected crystal surface is inhibited as long as the antiscalant is adsorbed to the surface and is not overgrown. Nucleation inhibiting antiscalants like sodium- hexametaphosphate (Calgon) adsorb to a nucleus in an early stage and are poisoning it or more likely form poisoned crystal nuclei around them. They bind metal ions permanently and form a nucleus, which cannot rearrange its structure to a regular nucleus and hence cannot grow beyond the critical size to a stable crystal.

It is often stated that antiscalants adsorb onto formed crystals or associate/complex with incipient nuclei (or crystals) and that these phenomena govern the inhibition of scale formation [17]. The precise mechanism of scale inhibition is not clearly understood at this time. However, it is known that in supersaturated solutions of sparingly soluble salts, a significant delay in crystal nucleation and subsequent growth is observed in response to AS treatment. The scale inhibition capability of antiscalants is related to chemical structure, molecular weight, active functional groups and solution pH –parameters.

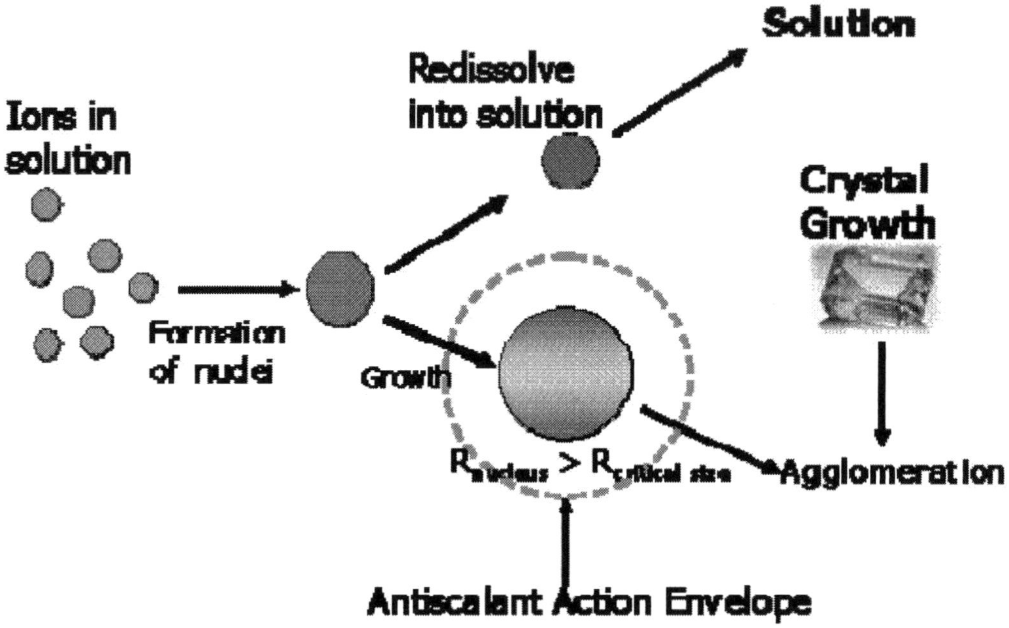

Figure 1. Schematic illustration of the action of antiscalants

Inhibition of scale in MSF plants is achieved by one of the following treatment methods:

- Use of acid to cause depletion of carbonate present in seawater. This is known as the acid treatment method.
- Use of commercially available scale inhibitors, known as the chemical additive treatment method.
- Combined use of the above two, i.e., use of scale inhibitors at a reduced dosing rate with partial depletion of carbonate by acid. This is known as the hybrid treatment method.

Acid treatment is an old method that has been used in the early MSF desalination plants. In this method acid, mostly H_2SO_4, reacts with the carbonate present in seawater. This reaction yields H_2O and CO_2, hence, preventing formation of $CaCO_3$. Precise control of pH is very crucial for both the inhibition of scale and prevention of corrosion.

In the second and third methods certain chemical additives, when dosed at the rate of few mg/L into seawater, are capable of suppressing or delaying the initiation of scale formation. In addition to this, such chemical additives have another effect which causes precipitated particles to remain dispersed and suspended in the flowing water by preventing adhesion between individual particles and between particles and internal tube surfaces.

Many different synthetic water-treatment polymers have been employed in a wide range of water treatment application as dispersants for particulate matter, inhibitors of mineral scale formation and deposition, and corrosion inhibitors [18-20]. Polymers containing carboxylic acid and/or sulphonic acid functionality have been found to be particularly useful.

Maleic acid based polymers are particularly effective in the desalination of seawater by the multi-stage flash evaporation (MSF). Their action is enhanced when maleic acid is co-polymerized with a small proportion of other monomers, such as acrylates. Water treatment polymers are added to aqueous systems in a predetermined concentration, which is effective to inhibit the formation and deposition of mineral scale and to inhibit corrosion. Once the water treatment polymers have been introduced into the aqueous systems, their concentration must be monitored by some means in order to maintain the polymer concentration at a predetermined level.

In seawater desalination, the primary reaction that triggers the alkaline scale formation is the thermal decomposition of bicarbonate ion at moderate temperatures (see reaction-1 below) [21]. The CO_3^{2-} ion precipitates as $CaCO_3$ when the solubility limits are exceeded (reaction 2). At a slightly higher temperature, the CO_3^{2-} hydrolyses to OH and precipitates $Mg(OH)_2$ (reactions 3 to 5).

$$2HCO_3 \rightarrow CO_3^{2-} + CO_2 + H_2O \qquad (1)$$

$$Ca^{2+} + CO_3^{2-} \rightarrow CaCO_3\downarrow \qquad (2)$$

$$CO_3^{2-} + H_2O \rightarrow OH^- + HCO_3\downarrow \qquad (3)$$

$$CO_3^{2-} + H_2O \rightarrow CO_2 + 2OH^- \qquad (4)$$

$$2OH^- + Mg^{2+} \rightarrow Mg(OH)_2\downarrow \qquad (5)$$

Calcium carbonate is partially soluble and can form hard scale. In a system without scaling inhibition an acid feed, HCO^{-3}, goes to CO^{-3} and then to OH^-, as shown in the principle scale-forming reaction above in which the bicarbonate ion decomposes above 50°C to form calcium carbonate. Scales formed by these thermal decomposition's are known as alkaline and include calcium carbonate ($CaCO_3$) and magnesium hydroxide $Mg(OH)_2$.

Calcium sulfate is a third type of scale, which is the most difficult to remove using chemicals. At high temperatures (about 120°C) in high concentrated brines, calcium sulfate scale will form. All three types of scale formed impair the heat transfer across heat exchangers and condensers. This constitutes a major cost factor because of the need for frequent, expensive maintenance, especially in countries with extensive desalination processes.

This study is a continuation of our earlier works in which we synthesized homopolymers and copolymers based on maleic-anhydride, discussed the reaction mechanism and kinetic models [22,23]. This study will test the effects of the water-soluble maleic-anhydride based polymers as scale control additives and will present their inhibition efficiency. In our previous study we also showed application of such polymers as anti scale controls in oil production facilities [24].

Synthesis and Experiments

Maleic anhydride (MA) is a widely available local material which is cost effective in comparison with other materials. Production of copolymer from MA occurs at two steps. In the first step n-octyl-acrylamide (AAO) monomer is produced and copolymer is produced in the second step as shown in Figure 2. Series of Low molecular weight maleic –anhydride homopolymers and copolymers have been synthesized in our laboratory as explained in our previous presentation [24].

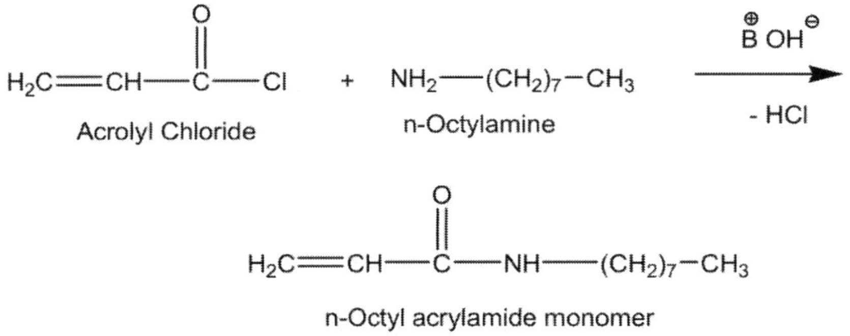

STEP- 1 : Reaction Scheme for preparation of n-Octyl acrylamide monomer (AAO)

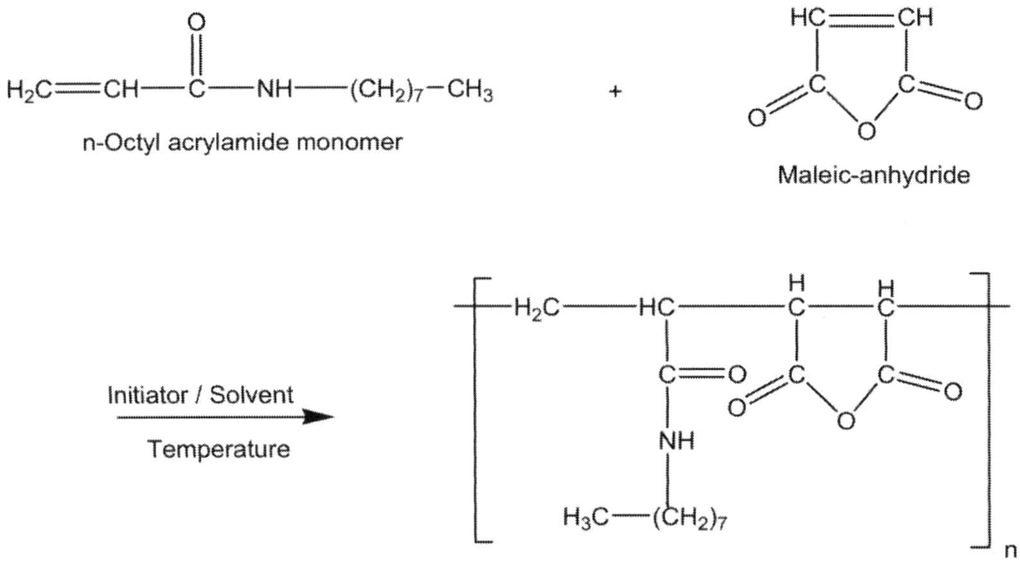

STEP- 2 : Reaction Scheme for preparation of Maleic-anhydride- n-Octyl acrylamide copolymer

Figure 2. Copolymer production in two steps

Copolymerization of MA with AAO was conducted with the objective of characterizing the reaction and to generate a series of low molecular weight co-polymers of varying composition. The reaction was conducted at four different concentrations. The samples were labeled as: MA / 30 AAO, MA / 50 AAO, MA / 60 AAO and MA / 40 AAO indicating 70:30, 50:50, 40:60 and 60:40 mol % of MA:AAO, respectively.

A typical polymerization was carried out as follows. MA dissolved in 200ml of o-xylene was stirred and heated to reflux under a N_2 atmosphere until the temperature reached 70^0C. AAO and initiator dissolved in 50 ml of solvent were added over a period of five minutes. Refluxing was continued for a further 9 hours and the polymer was allowed to separate from the solution. Excess of solvent was distilled off and the oily layer so obtained was recrystallized in benzene to give the copolymer product. The series of n-octylacrylamide-maleic anhydride copolymers prepared by the free radical mechanism were analyzed by ^1H-NMR and ^{13}C-NMR spectroscopy to determine the copolymer composition.

Inhibition Efficiency of Synthesized Polymers

Synthesized polymers are all highly effective inhibitors of $CaCO_3$ and $CaSO_4$ precipitation. Precipitation tests for $CaCO_3$ were made with 0.1, 0.3 and 0.5 ppm while for $CaSO_4$ tests were conducted at 0.25, 0.5, 1, 2, 3, 5 and 7 ppm polymer concentrations. Test methods used for inhibition efficiency of polymers were according to ASTM D 1126-02 and NACE TM 0374-95 [25,26]. To calculate the percentage inhibition of synthesized polymers following equation was used:

$$\%\ \text{inhibition}=\frac{[Ca^{2+}]_{antiscalant}-[Ca^{2+}]_{blank}}{[Ca^{2+}]_{original}-[Ca^{2+}]_{blank}}\text{x}100$$

where:

$[Ca^{2+}]_{antiscalant}$ = concentration of soluble Ca^{2+} in an antiscalant containing solution (mg/L)
$[Ca^{2+}]_{original}$ = concentration of soluble Ca^{2+} in original solution (mg/L)
$[Ca^{2+}]_{blank}$ = concentration of soluble Ca^{2+} in a blank solution (mg/L)

For Magnesium scale inhibitor test, a similar method was used as the calcium scale inhibition test (ASTM D 1126-02 & NACE TM 0374-95). The concentration of total hardness and calcium was determined by titration with EDTA. The soluble magnesium was calculated as per the standard ASTM D 1126-02. To calculate the percentage inhibition of synthesized polymers following equation was used:

$$\%\ \text{inhibition} = \frac{[Mg^{2+}]_{antiscalant}-[Mg^{2+}]_{blank}}{[Mg^{2+}]_{original}-[Mg^{2+}]_{blank}}x100$$

Figure 3 shows $CaSO_4$ and $Mg(OH)_2$ anti-precipitation activity for copolymers obtained from (MA / AAO, 5:5) monomer ratio and varying BPO concentration as the initiator [24]. The efficiencies of various synthesized polymers for scale removal of $CaSO_4$, $Mg(OH)_2$ and $CaCO_3$ for both homopolymers and copolymers are presented in Figures 4, 5 and 6

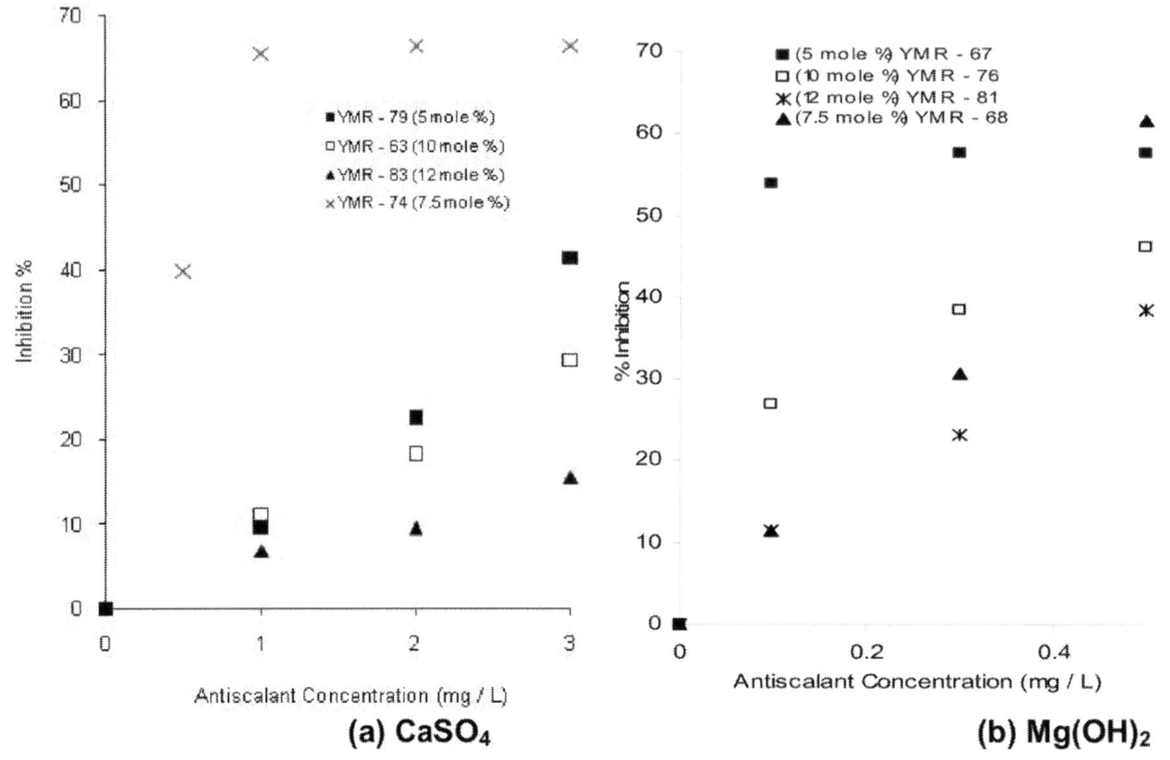

Figure 3: (a) $CaSO_4$ and (b) $Mg(OH)_2$ anti-precipitation activity for copolymers obtained from (MA / AAO, 5:5) monomer ratio and varying BPO concentration

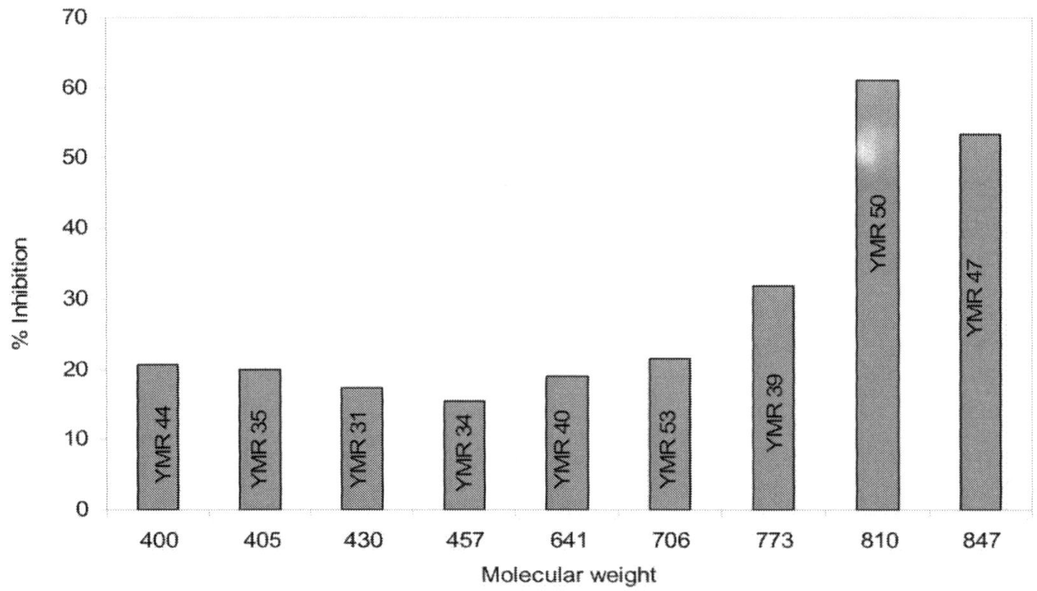

Figure 4: Inhibition % for $CaSO_4$ scales with molecular weight of MA homopolymers

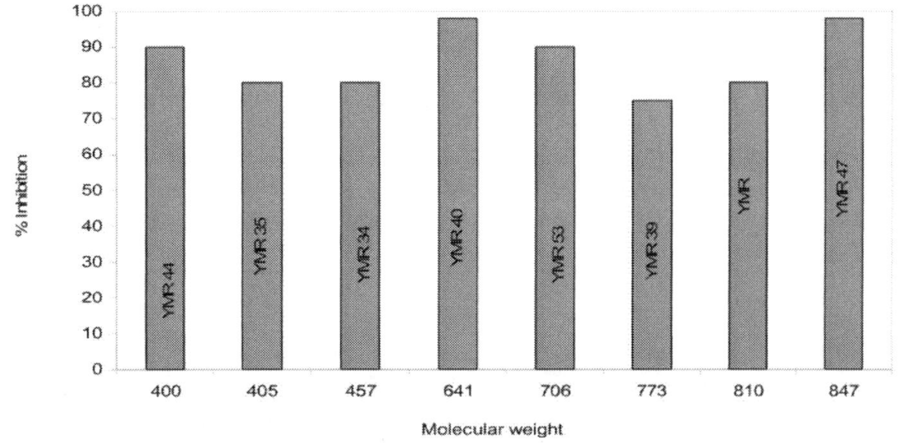

Figure 5: Inhibition % for $Mg(OH)_2$ scales with molecular weight of MA homopolymers

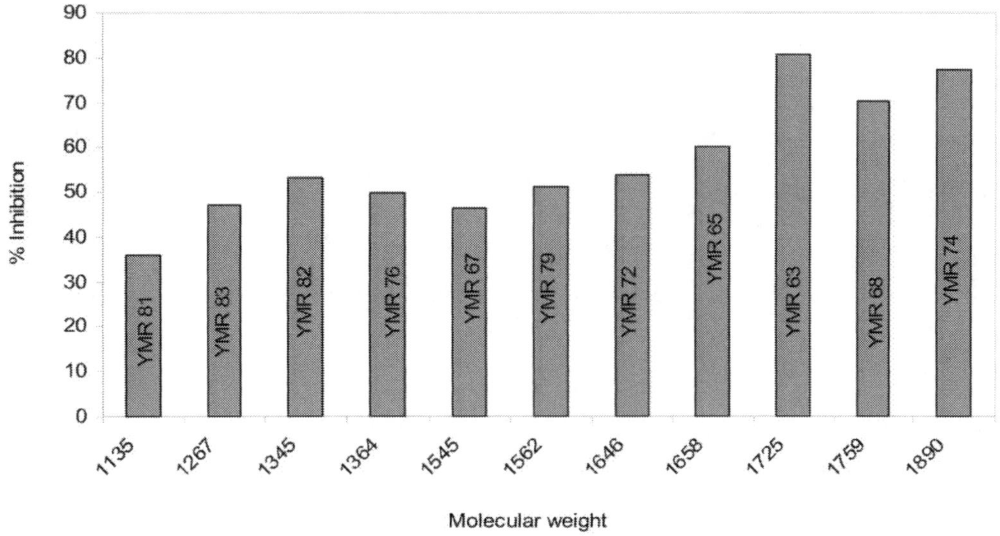

Figure 6: Inhibition % for $CaCO_3$ scales with molecular weight of MA copolymers

Scale Inhibitors and Inhibition Constant

The general reaction predicting the scale formation step and inhibition by anti-scalant is shown in Fig. 7. Inhibitor constant K_i which actually is a dissociation constant is given by the following equation and is determined by a Lineweaver-Burk plot or Dixon's Plot [27] as the ratio of $K_i=k_{-1}/k_1$. In Dixon's plot, the scale inhibition constant (K_i) between a scale inhibitor and an inorganic crystal is estimated from the reciprocal of the rate of scale formation (1/v) which is plotted against an increasing initial concentration of scale inhibitor employed as shown in Fig. 8.

The effects of inhibition constant determined from above method on efficiency of scale prevention are shown in Figure 9. From these figures it is observed that the value of inhibition constant decreases with increase in inhibition % and this is in confirmation with the theory (the first equation in Figure 7) that the reaction between the antiscalant and the inorganic crystal is a reversible reaction with non-covalent interactions such as hydrogen bonds, hydrophobic interactions or ionic bonds. Multiple weak bonds between the inhibitor and the active site combine to produce strong and specific binding. Therefore in order for the reaction to proceed in forward direction and produce more of unproductive complex the value of K_i should decrease.

The change of crystal habit, size and modifications brought about by synthesized polymer addition were examined through scanning electron microscope. The SEM photographs for the $CaCO_3$ with and without the presence of the polymer are presented in Figure 10a (without polymer) and in Figures 10 b with a synthesized polymer. Results presented in this paper indicate that synthesized polymers have potential applications in water desalination plants as antiscalant agents.

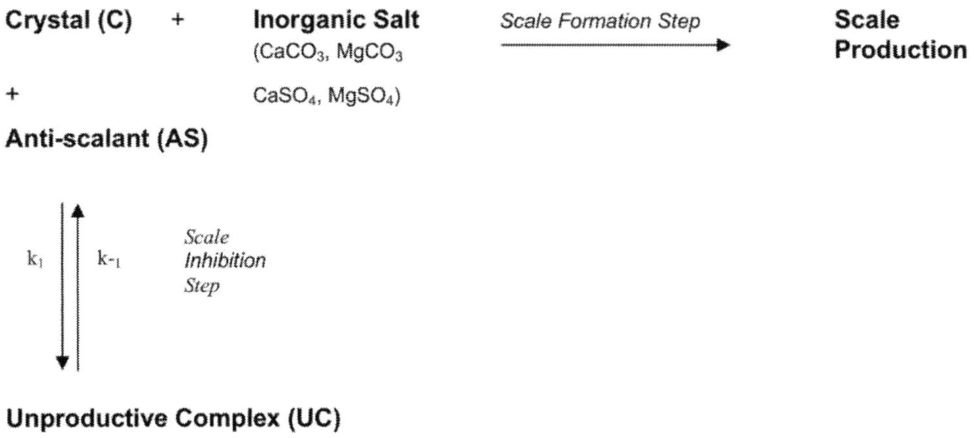

Figure 7: Scale formation and Inhibition Step

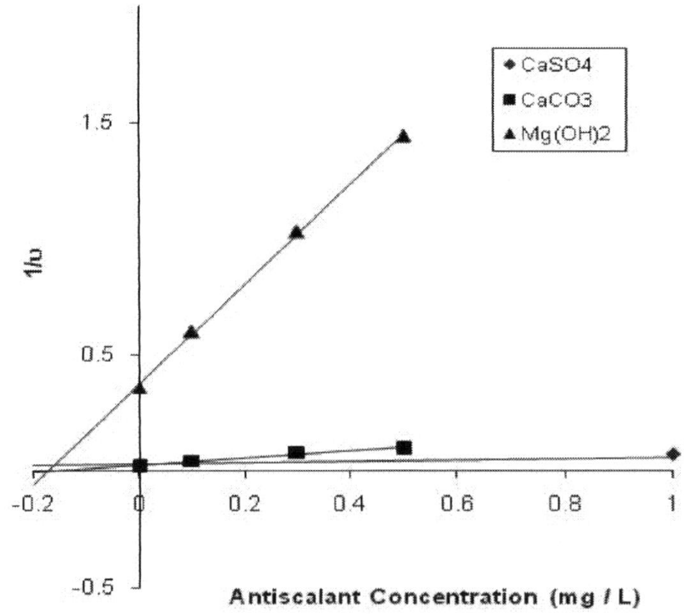

Figure 8: Scale inhibition constant estimation for a synthesized polymer

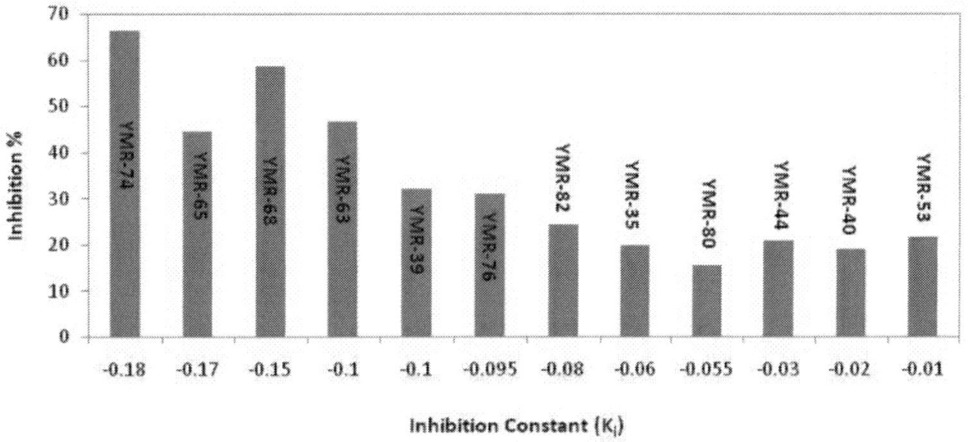

Figure 9: Inhibition Constant versus Inhibition % for $CaSO_4$ scales

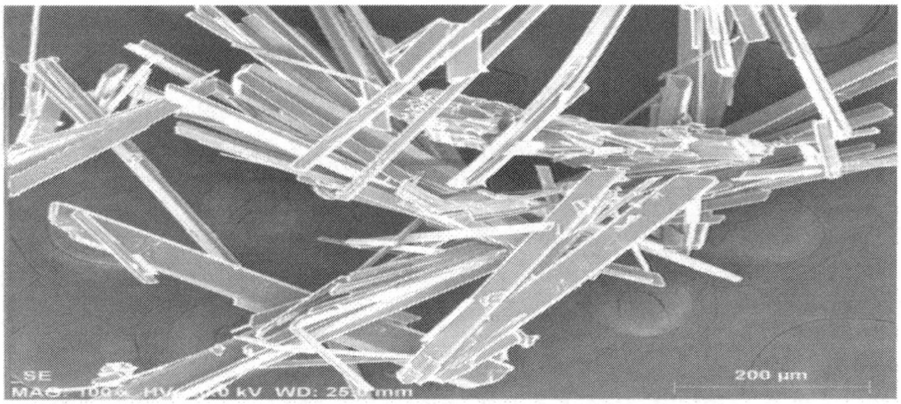

Fig (10a)

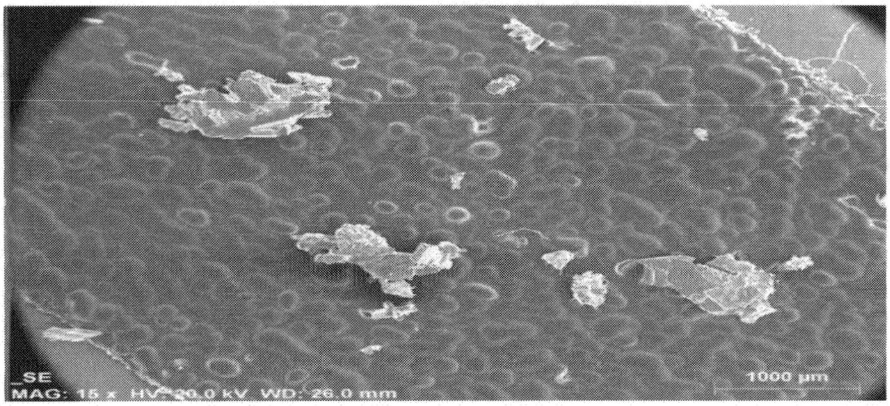

Fig (10b)

Figure 10: SEM photograph for the $CaSO_4$ in the (a) absence of scale inhibitor (BLANK) (b) presence of synthesized a synthesized scale inhibitors YMR-50 (b)

Acknowledgment

The funding for this project was provided by the Research Administration of Kuwait University under Project No. RE 05/06 which is appreciated.

References

1. I. Bremere, M. Kennedy. A. Stikker and J. Schippers (2001), Desalination, 138, 7-15.
2. G. Gusmano, G. Montesperelli, G. Forte, E. Olzi and A. Benedetti (2005), Desalination, 183, 187-194.
3. K. Genthner (2001), Desalination, 141, 99-100.
4. Middle East regional study on water study and demand development: Concluding report prepared by Israeli, Jordanian and Palestinian study teams February 1998 (2001), Desalination, 136, 105-108.
5. Konstantinos D. Demadis, Eleftheria Mavredaki, Aggeliki Stathoulopoulou, Eleftheria Neofotistou, Chris Mantzaridis, (2007), Desalination, v 213, n 1-3, 38-46.
6. M. Al-Ahmad and F. A. Aleem (1994), Desalination, 96, 409.
7. Oleg D. Linnikov (2000), Desalination, Volume 128, Issue 1, Pages 47-55.
8. J. Borden, J. Gilron and D. Hasson (1987), Desalination, 66, 257.
9. Hai-Yan Li, Wei Ma, Lu Wang, Ru Liu, Lin-Sen Wei and Qiang Wang, (2006), Desalination, v 196, n 1-3, 237-247.
10. Stephen P. Chesters (2009), Desalination 238, 22–29.
11. S. Lee and R. M. Lueptow (2003), Desalination, 155, 131.
12. C. Tzotzi, T. Pahiadaki, S.G.Yiantsios, A.J. Karabelas, N. Andritsos, (2007), Journal of Membrane Science, v 296, n 1-2, 171-184.
13. P. Shakkthivel, R. Sathiyamoorthi and T. Vasudevan (2004), Desalination, 164, 111-123.
14. P. Shakkthivel and T. Vasudevan (2006), Desalination, 197, 179-189.
15. Yang, Qingfeng (2006), Chinese Journal of Chemical Engineering, v14, n 2, 178-183.
16. Q. Zhao, Y. Liu and S. Wang (2005), Desalination, 180, 133-138.
17. Wen-Yi Shih, K. Albrecht, J. Glater and Y. Cohen (2004), Desalination, 169 (3), 213-221.
18. Al-Shammari, M. Safar and Al-Dawas, (2000), Desalination, 128, 1-16.
19. S. Ghizellaouia, M. Euvrardb, J. Ledionc and A. Chibania (2007), Desalination, 206, 185–197.
20. Yunxia Zhang, W. U. Jihuai, H.A.O. Sancun, L.I.U. Minghua (2007), Chinese Journal of Chemical Engineering, Volume 15, Issue, Pages 600-605.
21. Speigler, K.S. and Laird, A.D.K., (1980), *Principles of Desalination*, Part-B, Academic Press, New York.
22. Al-Roomi, Y. M. and K. F. Hussian (2005) Polymer Bulletin, Vol 55, 81-93.
23. Al-Roomi, Y. M. and K. F. Hussian (2006), Journal of Applied Polymer Science, Vol 102, Issue 4, 3404-3412.
24. Al-Roomi, Y. M., K. F. Hussian, M. R. Riazi (2010), AIChE Annual Meeting, Salt Lake City, UT., November 7-12.
25. ASTM International, West Conshohocken, PA, USA. http://www.astm.org/
26. NACE International (The *National Association of Corrosion Engineers*) Standards, Houston, TX, USA. http://www.nace.org/
27. Dixon, M.,1953. Biochem. J. **55**, 170–171.

(733a) Renewable Fuel From Activated Sludge Using Fluidized-Bed Catalytic Cracking (FCC) Process

Emmanuel D. Revellame, William Holmes, Rafael Hernandez and William T. French, Dave C. Swalm School of Chemical Engineering, Mississippi State University, Mississippi State, MS

Annual production of activated sludges in the United States ranges from 7 – 8 million tons. Currently, disposal technologies of these sludges include incineration and landfilling, both of which are heavily regulated and could create adverse environmental impacts. A fraction of the activated sludges can be used as fertilizer once it complies with regulations for Class A biosolids. Activated sludges contain microbial consortia known to produce lipidic materials and many other organic compounds, which could be used to produce biofuels and specialty chemicals. Several studies showed that reasonable yields (2 – 6% weight) of biodiesel could be produced from activated sludge. At these yields however, activated sludge biodiesel is less economical than petroleum diesel. This study describes the production of renewable fuel from activated sludge using Fluidized Catalytic Cracking (FCC). This process could convert most of the compounds present in sludge into fuel, and thus, increase fuel yields and reduce cost.

Several catalysts, sludge to catalyst ratios, sludge moisture content, fluid flow rates and temperatures were tested. Initial results using ZSM-5 at 500°C with air showed considerable production of aromatic compounds, from benzene to methylnaphthalene.

See more of this Session: Conversion of Solid Wastes to Energy and/or Product
See more of this Group/Topical: International Congress on Energy 2011

(733b) Simultaneous Extracellular Polymeric Substance (EPS) and Lipid Production by Activated Sludge Via Fermentation of Glucose

Patrisha J. Pham, Rafael Hernandez, Hien Nguyen, Andro H. Mondala and W. Todd French, Dave C. Swalm School of Chemical Engineering, Mississippi State University, Mississippi State, MS

The use of activated sludge microflora to convert sugars, derived from the hydrolysis of lignocellulose biomass, into lipids for biofuel production is of current research interest in our group. Lipid accumulation in this system has been enhanced by manipulating the carbon-nitrogen (C:N) ratio and initial sugar loading on activated sludge bioreactors with glucose as the main carbon source. Activated sludge is a flocculated suspension composed of microorganisms that can also produce extracellular polymeric substances (EPS) *in situ* during fermentation. Microbial EPS which are secreted largely by bacteria in harsh environments are a matrix rich in polymers including polysaccharides, glycoproteins, proteins, nucleic acids and phospholipids. Thus, studying EPS production along with lipid formation is essential for understanding the technical and economic value of the whole process. In this study, samples were obtained from aerobic batch fermentation inoculated with 20 % (v/v) activated sludge. The total incubation time for the batch fermentation experiment was seven days. EPS from the collected samples were extracted using a Dowex cation exchange resin in phosphate buffered saline solution and precipitated with 95% v/v cold ethanol and dialyzed. Fourier transform-infrared spectroscopy of freeze-dried EPS revealed characteristic IR peaks for polysaccharides and proteins. Total polysaccharide and protein concentrations were estimated in each EPS sample using phenol-sulfuric and modified Lowry methods, respectively. An increase in lipid % (g/g cell dry weight) and a decrease in specific EPS % (g/g cell dry weight) was observed from day 0 to 7 in all treatments. Increasing the carbon loading while maintaining the same C:N ratio increased lipid % and decreased specific EPS %. However for a low C:N ratio, lipid % remained constant while specific EPS % decreased. These results will be useful for manipulation of EPS and lipid production and consequently enable development of valuable strategies for improving the process.

See more of this Session: Conversion of Solid Wastes to Energy and/or Product
See more of this Group/Topical: International Congress on Energy 2011

(733c) The Hydrolysis of Nylon6 In Ionic Liquid to Recycle Nylon Waste

Xiao-Li Xia and Hai-Jun Wang, Jiangnan University, Wuxi, China

To date, several processes for Nylon6 depolymerization have been put forward. Herein, we first examined several ionic liquids for the hydrolysis of Nylon6. These ionic liquids are Brønsted acidic and they have been used both as a catalyst and environmental solvent for the hydrolytic reaction of Nylon6. The final result showed that using 1-butyl-3-methylimidazolium chloride ([Bmim]Cl) as solvent and catalyst for hydrolysis of Nylon6 ,the high conversion of Nylon6 and yield of Aminocaproic acid were obtained. In order to confirm the product, FT-IR and ^{1}H NMR were performed. After the experimental, theoretically, we also used high level ab initio methods with the 6-31G(d) basis sets to calculate the interaction between [Bmim]Cl and Nylon6. Based on our calculated interaction, we conclude that hydrogen bonds exist between [Bmim]Cl and Nylon6 which explain that Nylon6 dissolve in [Bmim]Cl.

See more of this Session: Conversion of Solid Wastes to Energy and/or Product
See more of this Group/Topical: International Congress on Energy 2011

(733d) Process Development for Efficient Utilization of Waste Plastic to Liquid Product for Rural Development In Nigeria: Gasification Option

Bamikole Amigun, Environmental Biotechnology and Conservation Department, National Biotechnology Development Agency (NABDA), Abuja, Nigeria, Christie Onyia, Environmental Biotechnology and Conservation Department, National Biotechnology Development Agency, Abuja, Nigeria and Bamidele Ogbe Solomon, Director General's Office and consultancy, National Biotechnology Development Agency (NABDA), Abuja, Nigeria

Waste production is a very important problem in all countries especially in developed countries and Nigeria is no exception. Waste plastics create a very serious environmental challenge because of their huge quantities and their disposal problems, due to their non-biodegradable attributes (or the biodegradation process is very slow). Plastic waste poses a serious environmental problem that is not addressed by recourse to landfill. Open dumping has been the predominant solid waste disposal option in Nigeria. Their destruction by incineration poses serious air pollution problems due to the release of airborne particles and carbon dioxide into the atmosphere. Only a limited amount of plastics can be recycled, while most plastics could be used, for example in energy production. The management of plastic waste using gasification in fluidized bed technology is a technically feasible option. Apart from eliminating the waste, a chemical recycling is produced during this process, obtaining a gas (syngas) which can be used in different applications such as ethanol through bacteria conversion of the syngas intermediate. Ethanol can be further converted to ethanol gel fuel to substitute the use of paraffin and traditional firewood in rural communities. This is doubly environmentally friendly since it will reduce the volume of plastic waste being disposed of in landfill whilst producing green fuel without generating green house gases. This application could give some prospect of self-reliant energy supplies at local levels, with potential economic, ecological, social, and security benefits; thereby improving the socio-economic and political wellbeing of the rural people.

See more of this Session: Conversion of Solid Wastes to Energy and/or Product
See more of this Group/Topical: International Congress on Energy 2011

(733e) An Anaerobic Bacterial Treatment of Corn Stover to Prepare a Biorefining Feedstock

James MacLellan, Biosystems and Agricultural Engineering, Michigan State University, East Lansing, MI

Anaerobic Bacterial Treatment of Corn Stover to Prepare a Biorefining Feedstock

James MacLellan, Zhengbo Yue, Robert Kraemer, Wei Liao*

Department of Biosystems and Agricultural Engineering, Michigan State University

East Lansing, Michigan 48823

Abstract

The objective of this study is to apply an anaerobic bacterial consortium to uptake the hemicellulose in corn stover to prepare a relatively cellulose-enriched feedstock for biofuel production. Previous research demonstrated that submerged anaerobic bacterial culture is capable of converting hemicellulose in raw lignocelluloses to methane gas. The in-depth study is needed to optimize the culture, enhance the hemicellulose utilization efficiency, and improve the quality of treated cellulose as a biorefining feedstock. An anaerobic bacterial consortium from animal manure will be screened and enriched as the seed for this study. Two feedstocks of raw corn stover and steam-treated corn stover will be used to compare the effects of structure on the anaerobic culture of hemicellulose uptake. A completely Randomized Design (CRD) on seed-to-feedstock ratio, culture temperature, and culture time will be carried out. Fiber composition, methane production, and microbial consortia will also be analyzed. An optimal condition can then be established based on the quality and quantity of cellulose in the treated corn stover. The ethanol production on the treated corn stover will be fulfilled in order to elucidate the effects of anaerobic bacterial treatment on biofuels production.

*: Corresponding author

Dr. Wei Liao

Department of Biosystems & Agricultural Engineering

Michigan State University

202 A.W. Farrall Hall

East Lansing, MI 48824-1323

Tel: 517-432-7205

Fax: 517-432-2892

E-mail: liaow@msu.edu

See more of this Session: Conversion of Solid Wastes to Energy and/or Product
See more of this Group/Topical: International Congress on Energy 2011

(733f) Towards Developing Anaerobic Digestion Based Biorefinery - Research On Food Wastes As Feedstock

Shulin Chen[1], Liang Yu[2], Jingwei Ma[2] and Craig Frear[2], (1)Department of Biological Systems Engineering, Washington State University, Pullman, WA, (2)Washington State University, Pullman, WA

Anaerobic digestion (AD) is a proven technology for waste treatment. However, as the organic material becomes more valuable and being treated as resources, traditional AD process is no longer adequate. The Bioprocessing and Bioproduct Engineering Laboratory at Washington State University has been devoting its effort in developing the second generation of AD technology with the concept of biorefinery, featured by producing various products from improved AD processes. Key to an AD biorefinary is innovative design of the digestors. This presentation reports the recent results from a study aimed at evaluating a two stage-digester designed for treating food wastes operated within a solid concentration range of 15-20%. The digester had a unique two-stage configuration that resulted in high biogas production rate than the existing technologies based on the laboratory data. The experimental observations also led to a simple mixing design that has low capital and operational costs. The information provides some new insights in AD research and development related to municipal solid waste applications.

See more of this Session: Conversion of Solid Wastes to Energy and/or Product
See more of this Group/Topical: International Congress on Energy 2011

(737a) Development and Testing of An Asymmetric Supercapacitor Based On a Carbon Foam Electrode Impregnated with Nickel Hydroxide

Wen Nee Yeo[1], Tony N. Rogers[1] and Bahne C. Cornilsen[2], (1)Department of Chemical Engineering, Michigan Technological University, Houghton, MI, (2)Department of Chemistry, Michigan Technological University, Houghton, MI

Supercapacitors have attracted significant attention in recent years due to their potential to power hybrid vehicles, portable audio equipment, and memory back-up devices. The trend in supercapacitors is to combine a double-layer electrode with a pseudo-capacitance electrode in an asymmetric capacitor configuration.[1] The double-layer electrode is usually an activated carbon (AC) since it has high surface area, good conductivity, and relatively low cost.[2, 3] The pseudo-capacitance electrode usually consists of RuO_2, MnO_2, $Ni(OH)_2$, etc., with the most commonly studied material being $Ni(OH)_2$.[4-7]

In this work, a graphitic carbon foam was impregnated with nickel hydroxide active mass using electrochemical deposition to make a nickel-carbon foam electrode. It served as the pseudo-capacitance electrode in an asymmetric supercapacitor. Carbon foam was used as a conductive holder of the active mass because of its electrical conductivity, pore size, and pore connectivity. The objective was to reduce the cost and weight of the supercapacitor while maintaining or increasing its capacitance and gravimetric energy storage density. The specific capacitance of the nickel-carbon foam electrode at different current densities was measured in constant-current flooded cell cycling tests. The measured specific capacitance of 2641 F/g at 5 mA/cm^2 was higher than a reported value of 2080 F/g for an Al-substituted α-$Ni(OH)_2$ electrode.[6] The nickel-carbon foam electrode was then combined with an AC at different ratios of the negative and positive electrode masses to study asymmetric cell performance. The mass ratio for optimum cell discharge capacity was found to be ~3.7.

References

[1] A.I. Belyakov, 3rd European Symposium on Supercapacitors and Applications (ESSCAP) Roma, Italy, 2008.

[2] K. Babel, K. Jurewicz, *Journal of Physics and Chemistry of Solids* 65 (2004) 275-280.

[3] P.-L. Taberna, Geoffroy Chevallier, P. Simon, D. Plee, T. Aubert, *Materials Research Bulletin* 41 (2006) 478-484.

[4] Jong Hyeok Park, Sungwook Kim, O Ok Park, J.M. Ko, *Applied Physics A* 82 (2006) 593-597.

[5] D.-D. Zhao, S.-J. Bao, W.-J. Zhou, H.-L. Li, *Electrochemistry Communications* 9 (2007) 869-874.

[6] J.-W. Lang, L.-B. Kong, M. Liu, Y.-C. Luo, L. Kang, *Journal of Solid State Electrochemistry* 14 (2010) 1533-1539.

[7] Q. Huang, X. Wang, J. Li, C. Dai, S. Gamboa, P.J. Sebastian, *Journal of Power Sources* 164 (2007).

See more of this Session: Energy Storage and Electrochemical Processes II
See more of this Group/Topical: International Congress on Energy 2011

(737b) Effects of Annealing On $Zr_8Ni_{19}X_2$ (X = Ni, Mg, Al, Sc, V, Mn, Co, Sn, La, Hf) for Ni/MH Battery Operation

Jean Nei[1], Kwo Young[2], Simon K. Y. Ng[1] and Steven O. Salley[1], (1)Chemical Engineering and Materials Science, Wayne State University, Detroit, MI, (2)Energy Conversion Devices Inc./Ovonic Battery Company, Rochester Hills, MI

Zr, Ti-based AB_2 metal hydride (MH) alloys are considered to be one of the most promising candidates to replace the misch metal based AB_5 alloys presently used as the negative electrode in nickel/metal hydride (Ni/MH) batteries. Since most of the studies on AB_2 MH alloys focus on multi-element and highly disordered families, the understanding of basic Ti-Ni and Zr-Ni intermetallic alloys is crucial for fully utilizing the synergetic effects between the main AB_2 Laves phases and the non-Laves secondary phases.

For the best balance between large storage capacity (lower hydride heat of formation) and high discharge capability (higher hydride heat of formation), a stoichiometry of A:B close to 1:2 is desirable. However, there is no AB_2 intermetallic alloy for either Zr-Ni or Ti-Ni binary systems. The Zr-Ni phase diagram shows several intermetallic alloys, and the closest intermetallic compounds to AB_2 are Zr_7Ni_{10} and Zr_8Ni_{21}. Zr_7Ni_{10} has slightly stronger metal-hydrogen bond strength and solidifies congruently from liquid. Zr was replaced partially by Ti to reduce the metal-hydrogen bond strength. Other substituting elements were also studied in order to improve the electrochemical properties. Zr_8Ni_{21} was chosen as the focus of this report. Zr_8Ni_{21} does not solidify directly from the liquid as in the case of Zr_7Ni_{10}. Instead, the Zr_2Ni_7 phase is first solidified from the liquid with a Zr-to-Ni ratio of 8:21 and later reacts with the remaining liquid to form Zr_8Ni_{21} alloy peritectically. If the condition of equilibrium is not reached during cooling, one or more secondary phases will precipitate out. Therefore, an annealing process was adopted for this study.

The effects of annealing at 960 °C for 8 hours on the phase structures of a series of partially substituted Zr_8Ni_{21} alloys were studied. The substituting elements included Mg, Al, Sc, V, Mn, Co, Sn, La, and Hf. In general, the abundance of dominant phase increased and the number of smaller secondary phases diminished after annealing. Similar to the choice of C14/C15 main phase in AB_2 alloys, the number of outer-shell electrons plays an important role in determining the main phase of the

A_8B_{21}samples. As number of outer shell electrons of the substituting element increases, the main phase evolves from tetragonal Zr_7Ni_{10}, to orthogonal Zr_7Ni_{10}, Zr_2Ni_7 and then Zr_8Ni_{21}.

The annealing effects on the gas phase and electrochemical hydrogen storage properties were studied. It's observed that as the Zr-content in the main phase of the alloy increases (orthogonal Zr_7Ni_{10} > tetragonal Zr_7Ni_{10} > Zr_8Ni_{21} > Zr_2Ni_7), the gas phase storage capacity increases. After annealing, due to the increase in the abundance of main phase, the capacities of alloys with higher capacities before annealing increase while others decrease. The sample with Hf-supplement shows the highest electrochemical discharge capacity of 200 mAh/g. After annealing, all samples with the same main phase as the as-cast ones show degradation in electrochemical capacity due to the reduction in number and abundance of secondary phases. All supplements show improvement in surface exchange current from the base alloy. Except for La and Hf supplements, annealing reduces the exchange current density. The bulk hydrogen diffusion constant decreases for most of the supplements except V and Sn. All supplements, excluding Sn, show improvement in bulk diffusion after annealing. The gas phase full capacity has strong correlation to the full electrochemical discharge capacity. Among all samples, as-cast La-supplement shows the best overall electrochemical properties.

See more of this Session: Energy Storage and Electrochemical Processes II
See more of this Group/Topical: International Congress on Energy 2011

Optimal Charge Rates for a Lithium Ion Cell

Saeed Khaleghi Rahimian, Sean Rayman, Ralph E. White

Department of Chemical Engineering, University of South Carolina, Columbia, SC, USA

Extended Abstract

The optimum charge rate for a lithium ion cell at each cycle is determined to maximize the useful life of the cell that is defined as the number of cycles before the end of discharge voltage (EODV) drops below 3.0 V or the cell discharge capacity becomes less than 20% of the original discharge capacity. First, we showed that by applying a dynamic optimization routine the number of cycles can be increased by approximately 29.4 % with respect to the case with one optimal charge current. The dynamic optimization results indicated that the optimum charge rates are the minimum currents at which the constraints for the useful life are satisfied. This is due to the minimum charge rate resulting in minimum side reaction rate and capacity fade. Then, a new method has been developed to find the optimal charge without using optimization algorithms. The cell capacity and the EODV are both a monotonic ascending function of the charge current; as a result the optimal charge current can be obtained by some iterations of the charge –discharge processes for each cycle. The new approach improves the useful cell life by approximately 41.6 % with respect to using only one optimal charge current in a few minutes while the previous optimization algorithm takes at least one day. Thus, the new technique enables us to find the best charge profile in online application in reasonable computation time by using any rigorous model for the cell simulation.

Introduction

The capacity of the Li-ion batteries decreases with the number of charge-discharge cycles due to undesirable side reactions. Since these batteries are used for tens to thousands cycles, depending on application, there is a demand to seek the best operating conditions to improve the cycling performance to minimize the capacity fade, and thereby increasing cell life. Only a few studies have attempted to optimize the charge rates for battery longevity and energy efficiency using optimization algorithms. S. Bashash et. al. applied multiobjective genetic algorithm to obtain the charge pattern of a plug-in hybrid electric vehicle to simultaneously minimize the total cost of fuel and electricity and the battery life degradation [1]. In previous work [2] we showed that the lithium ion cell life can be improved by ~30% with respect to the case with single optimal charge rate used by optimization of charge currents, as a function of cycle number during cycling. Choosing twenty rates, the optimization needs to evaluate the objective function (number of cycles) more than 80000 times and if the objective function evaluation takes only 1 sec, at the best condition, it will take about one day to find the optimal profile. In this work, a new approach is proposed to find the best charge rate at each cycle that improves the useful life 10% more than the dynamic optimization in much less computation time.

Lithium-ion cell Simulation

A single particle (SP) physics based model, which includes capacity fade, was applied to simulate the cell performance under low-earth-orbit (LEO) cycling conditions [2]. The main assumptions associated with the SP model are as follows:

- The concentration of the electrolyte and the potential in the solution phase are constant and uniform for all time across the cell sandwich (cathode, separator, anode).

- Positive and negative electrode potentials depend on time only.
- The model considers the capacity fade by a continuous occurrence of a side reaction (reduction of ethylene carbonate) only during the charge near the surface of the negative electrode. Moreover, the state of the charge (SOC) of the positive electrode is diminished during cycling.
- The concentration of Li ions in the spherical particle is estimated using a two term polynomial approximation.

A volume average technique is applied to predict the diffusion of the lithium ions for the cathode ($LiCoO_2$ with no Ni) and the carbon anode [4]:

$$\frac{dx_{p,avg}}{dt}=\frac{-3J_p}{FR_p c_{p,\max}} \qquad x_{p,surf}-x_{p,avg}=\frac{-J_p R_p}{5FD_{s,p}c_{p,\max}} \qquad J_p=\frac{I_{app}}{S_p}$$

$$\frac{dx_{n,avg}}{dt}=\frac{-3J_n}{FR_n c_{n,\max}} \qquad x_{n,surf}-x_{n,avg}=\frac{-J_n R_n}{5FD_{s,n}c_{n,\max}} \qquad J_n=\frac{-I_{app}}{S_n}-J_s$$

where $x_{i,avg}$ is the ratio of the solid average concentration to the maximum solid concentration for each electrode ($c_{i,max}$) , $x_{i,surf}$ is the ratio of the solid surface concentration to the maximum solid. J_s, the side reaction rate expression, is calculated by using Tafel kinetics:

$$\begin{cases} J_s=-i_{0f}\exp\left(\frac{-\alpha_{c,f}F}{RT}\eta_s\right)[charge] \\ J_s=0\,[discharge] \end{cases}$$

$$\eta_s=\phi_n-U_{ref,f}-\frac{I_{app}}{S_n}R_{film}$$

R_{film} is defined as:

$$R_{film}=R_{SEI}+\frac{\delta_{film}}{k_f}$$

and the rate at which the film thickness increases is calculated by:

$$\frac{d\delta_{film}}{dt} = \frac{-J_s M_f}{\rho_f F}$$

The SOC of the positive electrode updated at the end of each charging process as follows [4]:

$$\theta_{p,N} = \theta_{p,N-1} - \theta_{1,N-1}$$

where θ_l is the loss of SOC obtained by dividing the capacity loss to the maximum capacity of the cell (subscript N-1 indicates the previous cycle):

$$\theta_1 = \frac{Q_S}{Q_{\max}}$$

$$Q_s = -\int_0^{t=chsrge\ time} (J_s S_n) dt$$

$$Q_{\max} = \int_0^{t=chsrge\ time} I_{app} dt \qquad (first\ cycle)$$

LEO cycling [3], which contains the following steps, was applied for the cell simulation after the cell is completely charged initially:

1) Constant current discharge (0.6857 C-rate (40 % DOD)) for 35 min discharge time. Unless the voltage drops below 3.0 V or the cell capacity reaches the 20% of the total discharge capacity, go to step 2.
2) Constant current charging (up to 1 C rate) for 61 min charge time. If the voltage reaches the EOCV (4.05 V), go to step 3, if not go to step 1.
3) Constant voltage (4.05 V) charging for the remaining charge time, go to step 1.

Dynamic Optimization Results

To compare the results of the dynamic optimization and the new method, the best profile obtained by the optimization using single charge rate and twenty charge rates are presented in this work as well as [2]. Figures 1 and 2 show the variation of the cell capacity and the end of

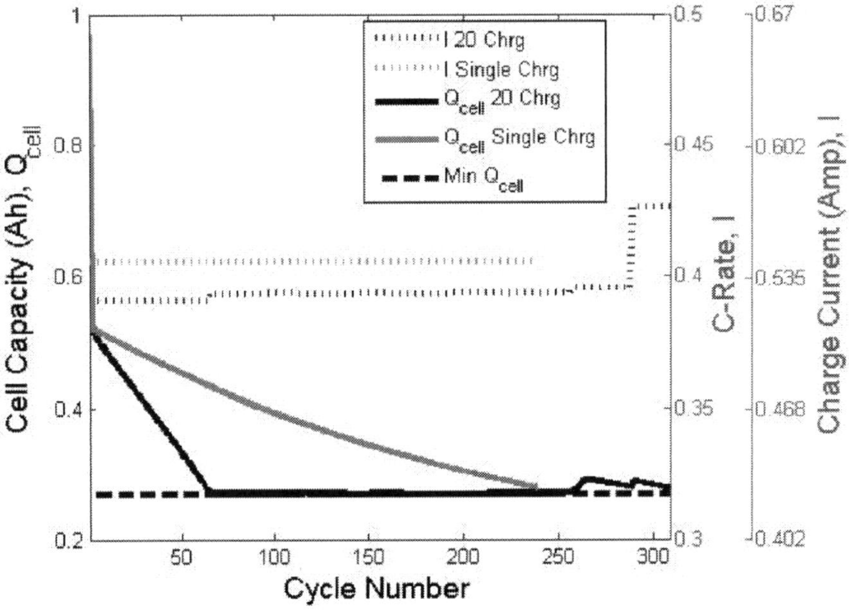

Figure 1: Remaining cell capacity at the end of discharge versus cycle number for single and 20 charge currents [2]

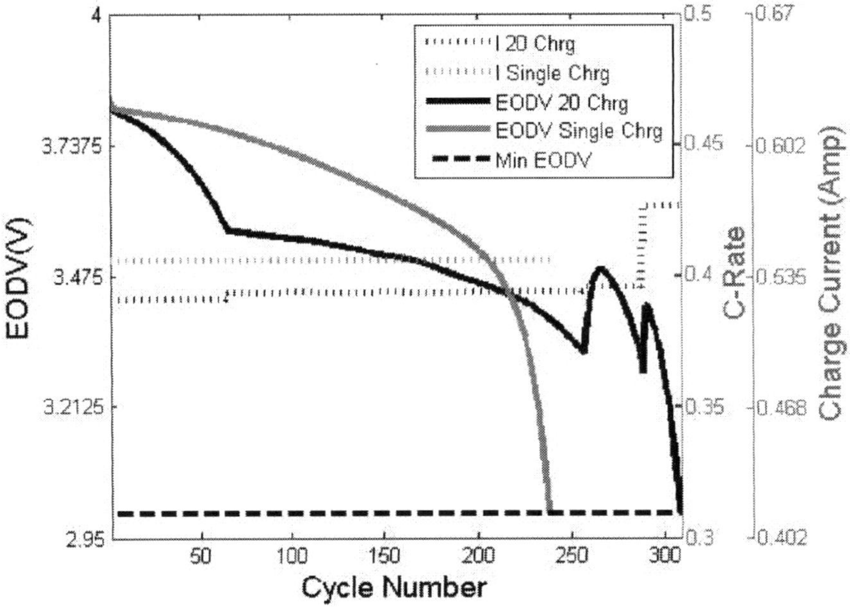

Figure 2: EODV versus cycle number for single charge and 20 charge currents [2]

voltage (EODV) respectively as a function of cycle number for single and twenty charge rates. Using twenty charge currents increases the number of cycles to 309, 29.4 % increase with respect to one optimum current.

New Approach

The optimization results indicate that the best charge current is the minimum rate at which, the cell capacity constraint and the EODV constraint are not violated. As a result we tried to develop a new method to find the optimal rate at each cycle as shown in Figure 3. By using this algorithm, any other constraints (e.g. end of charge voltage (EOCV) at 4.0 V) can be included without any difficulties.

Results and Discussion

The optimal charge rate and the cell capacity as a function of the cycle number obtained by the new approach that improves the number of cycles to 337, 41.6 % increase with respect to one optimal charge rate, shown in Figure 4. Variation of the EODV and the EOCV with the cycle number is presented in Figure 5. Adding the constraint that keeps the EOCV greater than 4.0 V (*constraint in Figure 3), decreases the number of cycles to 317. This is due to increase in charge rate in some cycles to not allow the EOCV dropping below the constraint. The computation time for the new approach is 228 sec on a Dell Precision T7500, with 2 Quad Core 2.53 GHz Zenon Processors CPUs and 12.285 GB of RAM.

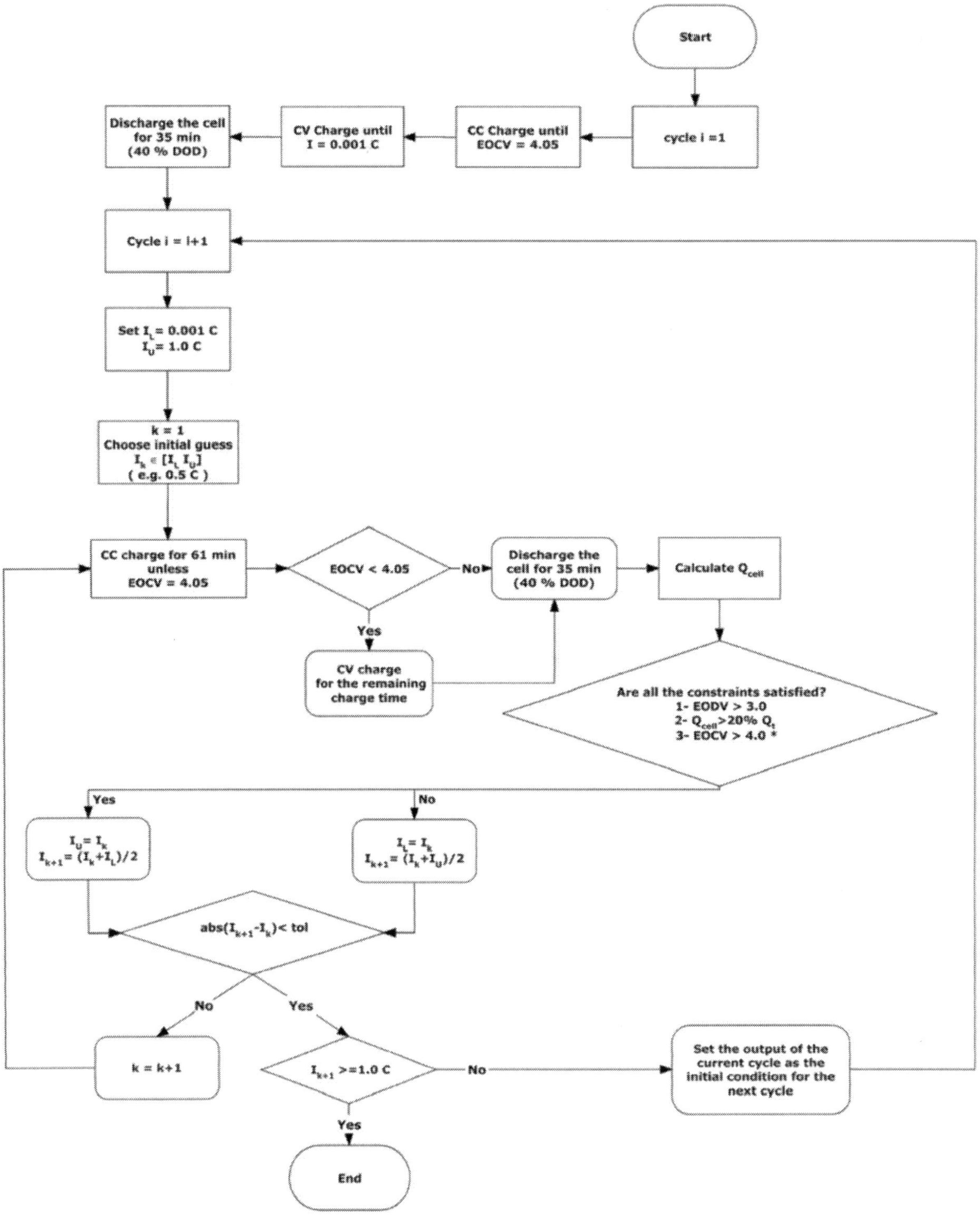

Figure 3: The new approach flowchart

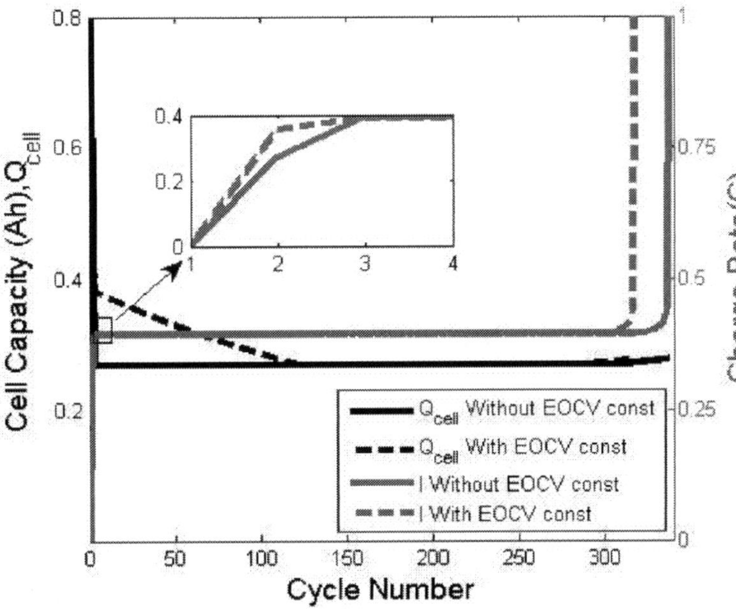

Figure 4: Remaining cell capacity at the end of discharge and Charge rate versus cycle number obtained by new approach with and without EOCV constraint

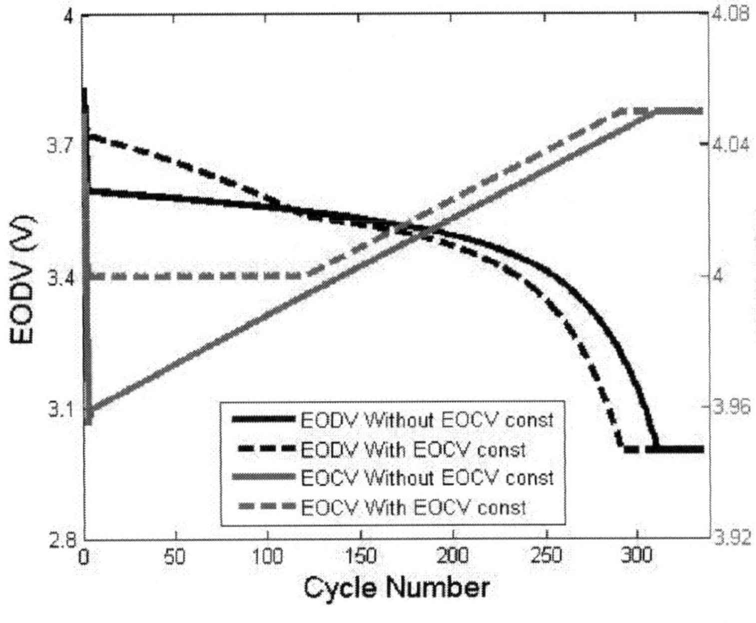

Figure 5: EODV (black) and EOCV (gray) versus cycle number obtained by new approach with and without EOCV constraint

Conclusion

A new approach was developed to provide the optimal charge rates at each cycle to maximize the useful life of a Lithium-ion cell. The computation time is greatly reduced by this approach with respect to the optimization method. For online application, the new method only requires the prediction of the next cycle by using prior cycling data while the entire cell life needs to be predicted for the optimization. Moreover, the new approach allows one to use a more rigorous model (i.e. P2D) to find the best charge rates without prohibitively increasing the computational time.

References

1. S. Bashash, S. J. Moura, J. C. Forman, H. K. Fathy, *J. Power Sources*, **196**, 541 (2011).
2. S. Khaleghi Rahimian, S.C. Rayman, and R.E. White, Maximizing the life of a lithium-ion cell by optimization of charging rates, *J. Electrochem. Soc.* **152** (12), A1302–A1308 (2010).
3. J. Lee, Y. K. Anguchamy, B. N. Popov, J. Power Sources 162 (2006) 1395.
4. P. Ramadass, B. Haran, P. M. Gomadam, R. E. White, B. N. Popov, J. Electrochem. Soc. 151(2) (2004) A196.

Basic model of secondary zinc air batteries to quantify environmental impacts

Daniel Schröder[a], Ulrike Krewer[a,b]

[a]Portable Energy Systems, Chair for Process Systems Engineering, Otto-von-Guericke University Magdeburg, Universitätsplatz 2, 39106 Magdeburg, Germany
[b]Otto Hahn Group/Portable Energy Systems, Max Planck Institute for Dynamics of Complex Technical Systems, Magdeburg, Sandtorstrasse 1, 39106 Magdeburg, Germany

---

Zinc air batteries (ZABs) offer a high theoretical specific energy density (1353 $\mathrm{Wh/kg}$[1]) and therefore appeal for application in portable and mobile devices. The availability of bi-functional air electrode catalysts, such as non precious perovskite structured metal, makes ZABs electrically rechargeable and even more attractive.
Figure 1 illustrates the reaction and transport processes occurring in a ZAB during discharge mode. In general, the battery consists of the zinc electrode, the air electrode and the separator. Equations (1)–(3) state the occurring reactions in caustic electrolyte.

$$\mathrm{Zn} + 4\,\mathrm{OH}^- \overset{r_1}{\rightleftharpoons} \mathrm{Zn(OH)}_4^{2-} + 2\,\mathrm{e}^- \tag{1}$$

$$\mathrm{Zn(OH)}_4^{2-} \overset{r_2}{\rightleftharpoons} \mathrm{ZnO} + 2\,\mathrm{OH}^- + \mathrm{H_2O} \tag{2}$$

$$\mathrm{H_2O} + \frac{1}{2}\mathrm{O_2} + 2\,\mathrm{e}^- \overset{r_3}{\rightleftharpoons} 2\,\mathrm{OH}^- \tag{3}$$

---

[1]Linden, E., 2004. Handbook of Batteries, 3rd ed., McGraw-Hill, p.1.13

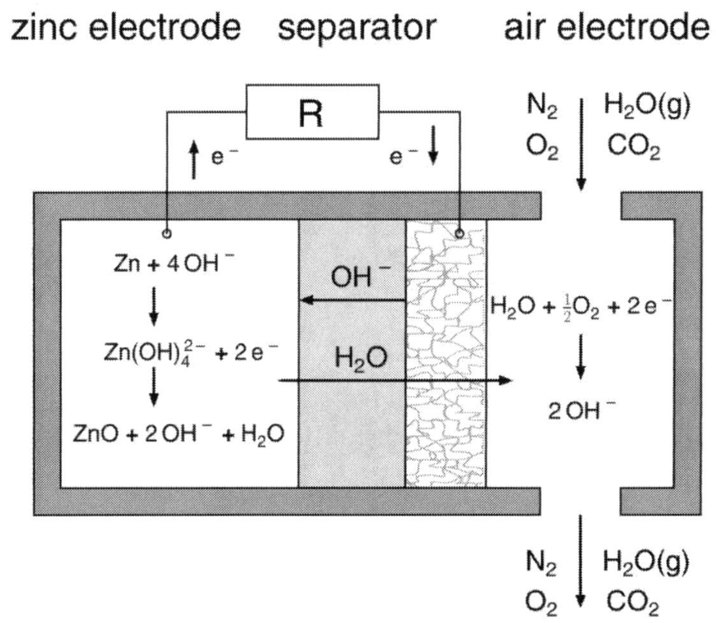

Figure 1: Schematic view of a zinc air battery showing the occurring reaction and transport processes during discharge. The charge process is then vice versa.

It is essential that OH^- and H_2O are transported through the separator during ZAB operation. As well, the ZAB is open to the environment which implies that uptake or loss of gaseous water, O_2 and even CO_2, might influence the operation behavior of the ZAB. For example, carbonates can be formed due to the reaction of the caustic electrolyte with CO_2.
Research on ZABs is almost exclusively experimental and focuses on the characterization of materials for zinc electrode, air electrode or separator, as well as on investigating the performance of entire ZABs.
It is suggested that ZABs are affected by the surrounding humidity; this holds especially for secondary batteries due to the longer operation time[2]. Models can help to analyze the influence of humidity on ZAB performance and get insight into the water balance of a ZAB.
Relatively few models exist to describe ZABs. They are mainly based on the macroscopic approach for porous battery electrodes by Newman and Tobias[3]. These models apply a set of coupled discretized non-linear dif-

[2]Haas, O., Müller, S. and Wiesener, K., 1996. Chem. Ing. Tech., 68(5), pp.524-542.
[3]Newman, J. and Tobias, C.W., 1962. J. Electrochem. Soc., 109, pp.1183-1191.

ferential algebraic equations (DAEs) to describe the transport and reaction phenomena occurring in the ZAB. They are used to analyze and optimize the ZABs design dimensions such as separator thickness and zinc electrode thickness. So far the existing models are not used to investigate the environmental impact on ZAB operation.

Our aim is to investigate the impact of the environmental gaseous water amount, O_2 amount and CO_2 amount on ZAB operation. For this, a flexible and expandable model is designed. In a first step, we introduce an isothermal basic model to describe the reaction and transport processes in secondary ZABs.

Our approach is as follows: The zinc electrode and the air electrode are described as an ideal continuously stirred tank reactor (CSTR), respectively. Each CSTR electrode consists of a fixed metal phase and a liquid phase containing the electrolyte and the dissolved and ionic species. The metal phase at the zinc electrode consists of Zn and ZnO. The metal phase at the air electrode is of inert catalyst structure. Both metal phases do not change their structure with time, however the zinc electrode metal phase ideally replaces its elements Zn and ZnO. The liquid volume in each CSTR electrode is varying with time and can be increased for example by uptake of gaseous water into the air electrode.

The here considered basic model consists of five molar balances for the ionic and solid species in the zinc electrode, three molar balances for the ionic and dissolved species at the air electrode, two charge balances for each electrode and two differential equations for the varying liquid electrolyte volume in the electrodes. These equations are a system of ordinary differential equations (ODEs). In our work, we solve the resulting ODE system with appropriate initial conditions using Matlab®.

The respective electrode over-potentials at the zinc electrode and the air electrode, originating from charge balances, and the ohmic drop through the separator, depending on the conductivity of the electrolyte, enable to calculate the total battery voltage.

The electrochemical reactions (1) and (3), are described by a Butler-Volmer approach. The chemical reaction (2) is applied as precipitation/saturation mechanism in the liquid electrolyte KOH.

Diffusion of OH^- and H_2O and migration of OH^- via the separator are included into the molar balances as exchange terms between the two electrodes. Both electrodes are apart by a certain distance, here the separator thickness in the range of μm. Due to the presence of a counter-ion (K^+) in the liquid electrolyte, electro-neutrality holds. Migration of $Zn(OH)_4^{2-}$ through the separator is omitted.

At this point we investigate three modifications of the basic model:

- Modification (a) is a reference, where gaseous water is neither entering nor leaving the ZAB. Here the O_2 amount in the air electrode is set constant and does not change with time.
- Modification (b) accounts for varying O_2 amount in the air electrode, whereas uptake or loss of water is neglected.
- Modification (c) allows gaseous water to enter or leave while the ZAB is operated with constant O_2 amount in the air electrode.

From transient simulations with galvanostatic charge and discharge cycles we obtain three results for the respective modification:

- Modification (a) shows that OH^- and H_2O transport is sufficient to maintain battery operation.
- Modification (b) shows how cycling may affect the reaction rates at the air electrode and the related ZAB performance
- Modification (c) shows how cycling may affect the water content and the related ZAB performance.

Further investigations target to include the carbonation at the air electrode.

(737e) Novel O2/H2O-Selective Membrane for Metal/Air Batteries

Wei Liu, Pacific Northwest National Laboratory, Richland, WA

This presentation will describe new membrane materials and structures that can be used to encapsulate the cathode of a metal/air battery for its operation under ambient air conditions. The membrane barrier is needed to allow transport of oxygen from air into the cathode for the electrochemical reaction and to block permeation of water vapor that can degrade the battery. In this work, two kinds of membranes are found effective for selective permeation of oxygen over water vapor: immobilized oil membrane, supported dense polymer thin-film membrane. The immobilized oil membrane is prepared by loading silicone oil into the pores of a thin (~50μm), porous support sheet of suitable pore sizes and surface property for retaining of the oil by capillary force. The supported thin-film membrane was prepared by coating a dense layer of PTFE (polytetrafluoroethylene) material (<10mm thick) on a thin porous metal sheet support (~50μm). Both membranes enable stable operation of a lithium/air battery in ambient air (~20% RH) in a much longer time than the conventional porous diffusion membrane. With attachment of a quality membrane on the cathode, the battery operated in ambient air even shows better performance than the same battery housed in a dry box. Systematic membrane preparation, characterization, and battery testing results will be presented. The membrane has large potential for other metal/air battery systems.

See more of this Session: Energy Storage and Electrochemical Processes II
See more of this Group/Topical: International Congress on Energy 2011

(762a) Desalination of the Colorado River Water: A Hybrid Approach

Anand Venkatesan and Phillip C. Wankat, School of Chemical Engineering, Purdue University, West Lafayette, IN

Reverse Osmosis (RO) is the most important technology for water desalination. The 96 million US gallons per day (gpd) Yuma desalination plant for desalination of irrigation return flows (3200 mg/L total dissolved salts (TDS)) in Colorado River (CR) is designed to operate at about 70% water recovery. Gabelich et al [1], studied RO desalination of CR water (887-1184 mg/L) in a pilot plant and demonstrated high recovery (≈95%). In this paper we design a hybrid system for desalination of CR water at the same capacity as the Yuma plant, for a feed with TDS = 941 mg/L.

The hybrid process comprised of Ion exchange (IEX) water softening and pH adjustment, integrated in between two RO steps. IEX treatment and pH adjustment of the concentrate from the Primary RO (PRO) is done before it is sent to the Secondary RO (SRO). IEX treatment of PRO concentrate helped remove Ca^{2+}, Mg^{2+} and Ba^{2+} which are the principal scale forming precursors and pH adjustment increased the solubility of Mg^{2+} salts. Use of concentrated retentate from SRO for regeneration of the IEX column eliminates the need for regeneration chemicals [2,3]. The hybrid process was evaluated for achieving high water recovery (≥95%).

The PRO and the SRO were modeled as cascades and optimal configurations were developed. Recovery in the PRO without the addition of antiscalants was 75%. The IEX was simulated with multiple columns feeding simultaneously and a single column regenerating. IEX treatment and pH adjustment of PRO concentrate allowed operation of SRO at 70% recovery and if antiscalants are added to control silica scaling SRO can be operated at recoveries as high as 95%. The overall recoveries were 92% without antiscalant addition and 98% with antiscalant addition in SRO. IEX/RO operation at higher recoveries reduces the amount of brine to be disposed. Economic analysis is done to study the impact of reduced brine on the cost of product water.

1. C. J. Gabelich et al., Journal of Membrane Science 301 (2007) 131-141
2. T. Vermeulen et al., Desalination, 47 (1983) 149-159
3. A. Venkatesan and P. C. Wankat, Desalination, 271 (2011) 122-131

See more of this Session: Water Reuse and Hybrid Water Treatment Processes
See more of this Group/Topical: Topical 1: Water Technology for Developed and Developing Countries (see also Separations Division)

TOWARD SUSTAINABILITY AND LOW COST IN ELECTRODIALYSIS REVERSAL DESALINATION

Maung Thein Myint, New Mexico State University, Las Cruces, NM 88003, mmyint@nmsu.edu, Abbas Ghassemi, New Mexico State University, Las Cruces, NM 88003, aghassem@nmsu.edu,

Summary

The improvements of electro-dialysis reversal (EDR) are discussed from the literature review. To be sustainable and low cost desalination, the areas of design and operation have to be *further* improved, and brine has to be reused in bio-energy production are suggested in details. Tertiary treated municipal wastewater (TTMWW) is proposed to use as feed in concentrate stream and electrode rinsing water in EDR to recover almost all of the fresh brackish groundwater (seawater) pumped for completely sustainable.
Key words: acid, chemical, electrode design, hydraulic leak, reusing concentrate from desalination; reusing tertiary treated MWW as concentrate in EDR.

Biography

Maung Thein Myint's Education

Yangon Technology University	*Civil Engineering*	*BE*	*1987*
NewMexico State University	*Environmental Engineering*	*MS*	*2001*
New Mexico State University	*Engineering*	*PhD*	*2005*
New Mexico State University	*Institute for Energy & the Environment; Post Doc*		*2009-present*

Abbas Ghassemi,

Education

University of Oklahoma	*Chemical Engineering*	*BS*	*1979*
New Mexico State University	*Chemical Engineering*	*MS*	*1989*
New Mexico State University	*Chemical Engineering*	*PhD*	*1990*

Appointments

Director, Institute for Energy & the Environment	*New Mexico State University*	*2006-present*
Professor, Chemical Engineering	*New Mexico State University*	*1999-present*
Executive Director, WERC	*New Mexico State University*	*1999-present*
Director, Institute for Energy & the Environment	*New Mexico State University*	*2008-present*

Introduction

Electro-dialysis reversal (EDR) desalination is known for its *excellent methods* to clean membrane. Membranes are cleaned by *reversing* the polarity between positive and negative polarity and by *switching* the hydraulic flow between concentrate and dilute streams in every fixed circle of polar reversal interval. Due to this excellence, researchers are trying to find ways to improve EDR in spacers, membrane's ability to withstand calcium sulfate, relaxing the fouling layer in membrane, and perm-selectivity of membrane. The spacer model was improved from Mark III to Mark IV to promote the turbulence and to increase the utilization area from 64 to 74%; to decrease power consumption from 0.14 to 0.10 kWh/m^3; and to increase the conductivity reduction from 27 to 36% (Grebenyuk and Grebenyuk, 2002). Acrylic based anion- and cation- exchange membranes that are highly resistance to irreversible resin fouling by organics, were introduced to use in 1981 and 1991 (Allison, 1995). The aliphatic anion selective membrane (AR 204 SXZL) ability to withstand sulfate fouling was improved to a saturation level of 440% $CaSO_4$ to gain a calculated water recovery rate (R_c) of 93.5% in a high level (42%) of SO_4^{2-} in feed water without pretreatment but with acid and anti-scalant additions in concentrate stream (Elyanow et al., 1981).

The costs of HCl and SHMP dosages in Dirab and Labakha-hawaita, Saudi Arabia are 15 and 83 times higher than power cost (Valcour, 1985). To avoid these high cost and un-sustainability with the chemical, EDR was successfully operated with low mean ion resident time in concentrate stream (*MIRTc*) by maintaining the similar LSI and $CaSO_4$ saturation level in the concentrate stream. For examples, EDR was successfully operated at LSI 2.29 and 358.9% $CaSO_4$ saturation level with the higher *R* 79.1% without adding any anti-scalants and without any acids in Turek et al., 2009 lab by *slowing down the velocity* in concentrate steam to gain the lower *MIRTc* and lower *dose* in single pass without any recirculation of concentrate from effluent into feed of concentrate stream. Moreover, Wisniewski et al., 2001 also demonstrated to operate ED without any recirculation to gain higher R 89.7% by *decreasing the volume* of concentrate for low dose and low *MIRTc*. The advantages of operating EDR in lower dose and lower *MIRTc* are to reduce the contact time between foulants and the surface of membrane in the concentrate steam, to reduce the dose to which is not high enough to be toxic to membrane, to increase the life time of membrane, and to eliminate chemical usage.

In the *classical* EDR, dimensions, flow, and velocity of dilute and concentrate are equal; LSI and $CaSO_4$ saturation level are used to control the scaling and fouling processes in concentrate (AWWA, 1995) as such LSI<+2.16 for preventing $CaCO_3$ from fouling and $CaSO_4$ saturation level<200 for averting $CaSO_4$ from precipitation. If LSI is more than allowable limit, acid is added in concentrate to keep $CaCO_3$ continuing dissolving (Katz, 1979); if $CaSO_4$ saturation level in concentrate is more than the allowable limit, antiscalant such as sodium hexametaphosphate (SHMP) is added in concentrate stream to maintain $CaSO_4$ enduring dissolving (Valcour, 1985). EDR however, was successfully *modernized* to operate with the higher water recovery rate (R) *without any anti-scalant and without acid*; this new EDR operated LSI at 2.29 and $CaSO_4$ saturation level 358.9% at *lower dose and lower* $MIRT_c$ (Turek et al., 2009). Dose and $MIRT_c$ are proposed to address the controlling process. By lowering *R* and polar reversal interval, EDR can be operated at MIRTc<130 min; at MIRTc<130min, desalting cost/energy can be minimized by eliminating chemicals requirement (Myint et al., 2010a, 2011a, and 2011b).

Brine disposal cost could range from 5 to 33% (Khordagui, 1997; Mohamed et al., 2005). In the case of inland sites, this minimum cost increases to the order of 15% of the costs of desalination (Glueckstern and Priel, 1996; Oren, et al., 2010). The cost for concentrate disposal by evaporation pond is 1.18-10.04 $\$/m^3$ (Greenlee et al., 2009). At present, all the best available disposal methods are highly questionable to the environmental concerns, and therefore everywhere in literature (i.e., Myint et al., 2010b) called for the recovery the salt or reusing of brine concentrate wasted from desalination in bioenergy and microalgae biodiesel production to be *sustainable* and to *dramatically reduce* the cost.

Suggestion

Water leaked significantly from the membranes, due to sheet membranes are stacked together, in higher R and higher demineralization rate as shown in Elyanow et al., 1981 that the metered water recovery rate (R_m) 86.0% was different from calculated R_c (93.5%). Although membrane has the ability to withstand the saturation level of $CaSO_4$ to 440%, the set objective was not achieved due to the hydraulic leak. The Research in reverse osmosis stated that acid and anti-scalant addition hinders the permeability rate (Hasson et al., 2007). To gain the higher R_c (equal R_m) with lower desalination cost, researches in hydraulic leaks and acids and antiscalant additions have to be revised in EDR.

Moreover, ED/EDR is generally assembled with 100 cell pairs for low salinity feed water to 700 cell pair for the high salinity brackish water (Jain and Reed, 1985) in one electrical stage due to the current transferring area available in electrodes. The positive and negative electrodes that have the same sizes of the membrane are usually placed in the top and bottom of the membrane cell pairs. The feed water is supplied into influent of the cell pair and collected from the effluent of the cell pairs. The TDS concentration gradually decreases from influent to effluent due to the voltage (direct current) supplied in the electrodes. The constant voltage is typically supplied to the electrodes uniformly along the surface. The current is carried by ionic molar flux (Kim et al., 2011); the current near the effluent is small due to the less TDS which leads to larger resistance ($R = V/I$) in the constant voltage potential application case. The electrode design has to be updated with the higher current intensity in influent and lower current intensity in effluent gradually along with the TDS concentration profile.

One way to be sustainable is to reuse tertiary treated municipal wastewater (TTMWW) as feed in concentrate stream and rinsing water in electrode channels. TTMWW has been successfully used as makeup in cooling system of cooling tower of coal-fired power plant such as Redhawk Power Plant in Arizona which uses TTMWW for over 90% of the cooling system makeup (Li et al., 2011). The trivial cost for tertiary treatment is much lower than the cost paid for dam water or desalinated water (Papaiacovou, 2001). *EDR membrane* has demonstrated to handle continuous microfiltration pretreated MWW effluent from the EDAR del Sureste to produce the high quality of the treated water pilot plant studies (Pino and Durham, 1999) and field scale in Canary Islands (Veza, 2001). Other successful story of field scale EDR membrane desalting TTMWW are 300,000 GPD in Moody Gardens in Galveston, Texas and two of 1 MGD in Middle East without organic fouling and biogrowth, however, with colloidal fouling in half of the plants (Allison, 1995).

The ultrafiltration (UF) with module ZW-10 has capability to in the removal of *E. coli* from secondary treated MWW (Table 1). In the case of *E. coli*, in particular, UF removed 100% of *E. coli* (Lubello et al., 2003). The barriers also proved capable, although not absolute, in reducing TC as was expected. UF with module ZW-10 reduces turbidity (94.5%), TSS (98.7%), and COD (around 35%) from secondary treated MWW (Table 2) and produces permeate with low SDI values (usually $< 3\%$) (Lubello et al., 2003). By using UF treated TTMWW as feed into concentrate stream and electrode rinsing water, therefore, all of the freshwater may be desalted into a drinking water, and all the TDS will be collected in concentrate stream in TTMWW. This idea will result in overall higher fresh water recovery rate.

UF with module ZW-10 removes 100% of the *E. coli*; dilute water rinses out all the concentrate (TTMWW) which attaches in surface of membrane from previous cycle for 30 seconds before product water produce in EDR (AWWA, 1995); dilute product water does not directly contact with TTMWW. Although, reusing TTMWW as feed in concentrate and electrode rinsing waters in EDR which produces drinking after the *E. coli* treatment in dilute product water, is technically feasible and completely sustainable, perception of the public may be problem.

Conclusion

With the updating design in electrodes; improving in hydraulic leaks; reducing in chemical usage in concentrate stream; reusing TTMWW as feed in concentrate steam and electrode rinsing water; and reusing brine in bio-energy production; EDR will have a capability to produce almost 100% of water recovery rate from fresh brackish groundwater (seawater) pumped with the low desalination cost.

Acknowledgement

We thank and appreciate the funding agency the Office of Naval Research (Contract # N00014-08-1-0304) from the USA and Tier I: Reclamation Cooperative Agreement Program.

Reference

1. Allison, R. P. (1995) Electrodialysis reversal in water reuse applications. *Desalination* **103**, 11-18.
2. AWWA (1995) Manual of Water Supply Practices- M38, First Edition: Electro-dialysis and Electro-dialysis Reversal. American Water Works Association, Denver, CO.,.
3. Elyanow, D., Sieveka, E., and Mahoney, J. (1981) The determination of super-saturation limits in an EDR unit with aliphatic anion membranes. Ninth Annual Conference and International Trade Fair of the National Water Supply Improvement Association, Volume II. May 3-June 4.
4. Glueckstern, P. and Priel, M. (1996) *Desalination* **108**, 19.
5. Grebenyuk, V. D. and Grebenyuk, O. V. (2002) Electrodialysis: From an Idea to Realization. *Russian Journal of Electrocchemistry* **38**(8) 806-809.
6. Greenlee, L. F., Lawler, D. F., Freeman, B. D., Marrot, B., Moulin, P. (2009) Reverse osmosis desalination: water sources, technology, and today's challenges. *Water Research* **43**, 2317-2348.
7. Hasson, D., Drak, A., Komlos, C., Yang, Q., Semiat, R. (2007) Detection of fouling on RO modules by residence time distribution analyses. *Desalination* **204**, 132–144.

8. Jain, S. M. and Reed, P. B. Electro-dialysis. In *Comprehensive Biotechnology: The Principles, Applications, and Regulations of Biotechnology in Industry, Agriculture and Medicine, Volume 2: The principles of Biotechnology: Engineering Considerations*; Moo-Young, M., Cooney, C. L., Humphrey, A. E., Eds.; Pergamon Press: Oxford 1985; pp 575-590.
9. Katz, W.E. (1979) The electro-dialysis reversal (EDR) process. *Desalination* **28**, 31-40.
10. Khordagui, H. (1997) Environmental Aspects of Brine Reject from Desalination Industry in ESCWA Region. ESCWA, Beirut, Lebanon.
11. Kim, Y., Walker, W. S., and Lawler, D. F. (2011) Electrodialysis with spacers: effects of variation and correlation of boundary layer thickness. *Desalination* **274**, 54-63.
12. Li, H., Chien, S-H., Hsieh, M-K., Dzombak, D. A., and Vidic, R. D. (2011) Escalating Water Demands for Water Production and the Potential for Reuse of Treated Municipal Wastewater. *Environ. Sci. Technol.*, **45**, 4195–4200.
13. Lubello, C., Gori, R., de Bernardinis, A. M., and Simonelli, G. (2003) Ultrafiltration as tertiary treatment for industrial reuse. *Water Science and Technology: Water Supply* **3**(4) 161-168.
14. Mohamed, A. M. O., Maraqa, M. and Handhaly, J. A. (2005) Impact of land disposal of reject brine from desalination plants on soil and groundwater. *Desalination* **182**, 411-433.
15. Myint, M. T., Ghassemi, A. and Nirmalakhanda N. (2010a) Design of An ILEDR for Brackish Groundwater: A Literature Review Approach. *Desalination and Water Treatment* **24,** 150–175.
16. Myint, M. T., Ghassemi, A. and Nirmalakhanda N. (2010b) Review of Nondestructed Anaerobic Digested Sludge to Reinstate Concentrate from Electrodialysis Reversal Desalination in Bioenergy Production. *Environmental Engineering Science* **27**(10) 811-823.
17. Myint, M. T., Ghassemi, A. and Nirmalakhanda N. (2011a) Low Energy Desalination: Low Dose & Low Mean Ion Resident Time in Concentrate of Electrodialysis Reversal. 7th IWA leading-edge conference on water and wastewater technologies 2010Phoenix, Arizona, USA. *Water Science Technology* **63**(9) 1855-1863.
18. Myint, M. T., Ghassemi, A. and Nirmalakhanda N. (2011b) Modeling in Desalination: Electro-dialysis Reversal. *Desalination and Water Treatment* **27**, 255-267.
19. Oren, Y., Korngold, E., Daltrophe, N., Messalem, R., Volkman, Y., Aronov, L., Weismann, M., Bouriakov, N., Glueckstern, P., Gilron, J. (2010) Pilot studies on high recovery BWRO-EDR for near zero liquid discharge approach. *Desalination* **261**, 321–330.
20. Papaiacovou, I. (2001) Case study — wastewater reuse in Limassol as an alternative water source. *Desalination* **138**(1-3) 55-59.
21. Pino and Durham, 1999 Wastewater reuse through dual-membrane processes: opportunities for sustainable water resources. *Desalination* **124**(1-3) 271-277.
22. Turek, M., Was, J., and Dydo, P. (2009) Brackish water desalination in RO- single pass EDR system. *Desalination and WaterTreatment* **7**: 263-266.
23. Veza, J. M. (2001) Desalination in the Canary Islands: an update. *Desalination* **133,** *259-270.*
24. Wisniewski, C., Persin, F., Cherif, T., Sandeaux, R., Grasmick, A., and Gavach, C. (2001)

De-nitrification of drinking water by the association of an electrodialysis process and a membrane bioreactor: feasible and application. *Desalination* **139**, 199-205.

25. Valcour Jr, H.C. (1985) Recent applications of EDR. Conference Proceedings: Second World Congress on Desalination and Water Re-use. Bermuda, Nov 17-22.

Table **1**. Results of microbiological analysis treated by UF with ZW-10

n.	TC (MPN/100 ml)		E. Coli (UFC/100 ml)	
	UF concentrate	UF permeate	UF concentrate	UF permeate
1	380,000	1,950	29,000	0
2	93,000	11,000	1,900	0
3	93,000	11,000	1,000	0

Lubello et al., 2003.

Table **2**. Quality of feed, permeate and concentrate treated by UF with ZW-10

Parameters	Feed (TTMWW)			Permeate from UF			Concentrate from UF		
	Mean	Range	STD	Mean	Range	STD	Mean	Range	STD
COD (mg/L)	41.4	27–68	10.2	27.4	17–53	8.2	77.6	34–116	22.7
Turbidity (NTU)	7.35	2.2–32	6.9	0.4	0.18–0.70	0.18	29.9	7.7–70	16.2
TSS (mg/L)	23.2	3.7–71.1	15.9	0.32	0.71	0.18	45.5	15–74.8	16.7

Lubello et al., 2003

"Municipal wastewater treatment using electrocoagulation and adsorption for reuse in industrial processes and green areas"

Hector Moreno, Alejandro Romero, Jose R. Parga, Jewel Andrew Gomes, Eric Peterson.

One of the biggest challenges of the 21st century is to provide water, primarily potable, to the majority of the world population. This is a critical issue, especially in developing countries. One way to augment that effort is to reuse water. Currently, some emerging wastewater treatment technologies based on electrochemistry are available that are competitive and advantageous over conventional technologies (e.g., Electrocoagulation (EC), electro-flotation, electro-decantation and electro-oxidation). Although EC has been known for more than a century, it has not been comprehensively studied. EC has been commercialized for the removal of specific contaminants and wastewater treatment. This work demonstrates how EC has been studied experimentally and is proposed as an alternative method for the biological treatment of municipal wastewater. Experiments were conducted to determine the optimum operational conditions. Results show a remarkable removal efficiency for: chemical oxygen demand (COD) 77 to 94%; for coliforms 80%, and; colony forming units (CFU) 99.98%; within 30 seconds residence time.

Key words: Electrocoagulation, COD, CFU, municipal wastewater.

Introduction

Wastewater is the liquid end-product (by-product), of municipal, agricultural, and industrial activity. The chemical composition of wastewater reflects the origin from which it came. The term wastewater implies it is a waste product to be discarded. However the water we use (e.g. drinking, washing, bathing, etc.) ultimately ends up back in the stream, river, lake, or groundwater to be withdrawn, treated, and used again. So wastewater must be treated and/or reused in an environmentally sound manner before it is discarded [1].

Water pollutants consist of floating materials, settable solids, organic and inorganic suspended solids, organic and inorganic dissolved solids and liquids, oil, fats and grease, dissolved gases and microorganisms. All water pollutants have different characteristics, poseing separate threats for human health and their removal.
Wastewater treatment plants are devoted to remove most of the pollutants while delivering water with little or no risk back in the stream, river, lake, groundwater or industrial reuse.

Conventional wastewater treatment include: screening, coagulation – flocculation, sedimentation, biological treatment, filtration and disinfection.

Table 1 show typical concentrations of some key parameters that determine the efficiency of wastewater treatment and what will be its usage after the treatment, table 1 shows the typical concentrations. Fecal coliforms and CFU are especially important in removing because communicable diseases can be transmitted by the pathogenic organisms in wastewater.

Table 1. Typical contaminant concentration of untreated domestic wastewater (After Metcalf and Eddy (1991).

		Concentration		
Contaminants	**Unit**	**Weak**	**Medium**	**Strong**
BOD_5 at 20° C	mg L^{-1}	110	220	400
Chemical oxygen demand (COD)	mg L^{-1}	250	500	1000
Grease	mg L^{-1}	50	100	150
Total coliform	CFU 100 mL^{-1}	10^6-10^7	10^7-10^8	10^8-10^9
Total coliform	number per milliliter	10^4		10^5

Biological Oxygen Demand (BOD), is defined as the amount of oxygen required by bacteria to decompose organic matter for a specified period of time (usually 5 days), under aerobic conditions. The amount of oxygen reported with this method represents only the carbonaceous oxygen demand (CBOD) or the easily decomposed organic matter. BOD is commonly used to measure natural organic pollution. Its analysis technique uses a titrametric determination of O_2 consumption due to the biological oxidation of sample.

An alternative to the BOD test for determining the oxygen consuming potential of a wastewater sample is the chemical oxygen demand (COD) test. As the name implies, carbonaceous oxygen demand is oxidized chemically in the COD test. Like BOD, the units for COD are in milligrams of oxygen per liter (mg L^{-1}).

Chemical Oxygen Demand (COD), is the amount of oxygen required to oxidize organic and oxidizable inorganic matter present in water or wastewater. Its analysis technique determines oxidant consumed during a chromic acid digestion of the sample in standard conditions.

The advantage of this test is that it is quick and reproducible. Analysis of COD is much more practical than analyses for specific pollutants.

A disadvantage is that it measures everything, including non-biodegradable organic matter, that can be oxidized by potassium dichromate. There are also some incompatibilities with the COD test. For example, low molecular weight fatty acids and aromatic hydrocarbons may not be well oxidized during the test while inorganic ions (e.g. chloride and nitrite for instance) may be oxidized. The addition of certain catalysts during the test can eliminate most of these interferences [2].

EC Removal efficiency

There is an extremely high variability in organic compound removal efficiencies for organic compounds, BOD and COD using EC. This is one reason why EC is not universally accepted as a conventional technology. Removal efficiency for COD might show negative, medium and surprising high values. It all depends on the compounds that contribute to COD and their reactivity with iron. For example suspended solids in small quantities are not a problem for EC, organic compounds reacting with iron (II) and iron (III) to form insoluble compounds are also removed. Organic compounds reacting with iron (II) to form insoluble compounds and with iron (III) to form soluble compounds or inversely will be partially removed. Organic compounds that do not react with iron (II) or iron (III) will not be affected by the EC process [3].

Current theory of EC involve several successive stages:

1) ***Generation of metal ions.***

Anode:

$$Fe \rightarrow Fe^{+2} + 2e^- \quad (1)$$

$$Fe \rightarrow Fe^{+3} + 3e^- \quad (2)$$

2) ***Hydrolysis of metal ions and generation of metal hydroxides***

$$Fe + 6H_2O \rightarrow Fe(H_2O)_4(OH)_{2\,(aq)} + 2H^{+1} + 2e^{-1} \quad (3)$$

$$Fe + 6H_2O \rightarrow Fe(H_2O)_3(OH)_{3\,(aq)} + 3H^{+1} + 3e^{-1} \quad (4)$$

3) ***With iron electrodes, Hydrogen evolution at the cathode.***

Cathode:

$$2H^+ + 2\,e^- \rightarrow H_{2(g)}\uparrow \quad (5)$$

Comments:

Electrochemistry depends on thermodynamics and kinetics. The rate of reaction depends on the dissolution of iron and the removal of $[H^+]$ via H_2 evolution. This reaction occurs fast for low pH values (strong acids) and for a weak acid the rate will depend on the pKa of the acid.

4) ***Destabilization of the contaminants, particulate suspension, breaking of emulsions, and aggregation of the destabilized phases to form flocs.***

This part relative to colloids and suspended matter is accepted since in small quantities they are not a problem for EC.

5) ***Theory of EC consider that chemical reactions and precipitation can occur during the EC process or that other cation and/or the hydroxyl ion (OH) form a precipitate with the pollutant.***

EC can be considered an accelerated corrosion process with Green Rust (GR) recognized as an important intermediate phase in corrosion of Fe (0).

Green rust (GR). GRs are layered Fe(II)-Fe(III) hydroxides having a pyroaurite-type structure consisting of alternating positively charged hydroxide and hydrated anion layers. Some Fe (II) of the octahedral sheets of $Fe(OH)_2$ is replaced by Fe(III). This results in positive layers of charge that are balanced by the inclusion of anions between the layers. Its exact nature depends on the interlayer anion. Various forms have been synthesized and studied. Results of many studies show GR conforming to a general chemical composition and stoichiometry, that can be represented with the general formula: $[Fe^{II}_{(6-x)}\ Fe^{III}_{\ x}(OH)_{12}]^{x+}[(A)_{x/n}\cdot yH_2O]^{x-}$, where x ranges from 0.9 to 4.2, A is an n-valent anion (typically CO_3^{2-}, Cl^- or SO_4^{2-} and y denotes the varying amounts of interlayer water (typically y ranges from 2 to 4 for most GRs) [4].

GR's, unlike most iron oxides and their form, have great specific internal surface areas and consequently great reactivities. They are reactive ion exchangers and sorbents (Hansen et al., 2001). The amphoteric surface hydroxyl groups lead to both sorption of heavy metals (cations) and organic anions, e.g. linear alkyl benzene sulfonates (LAS), the major synthetic surfactant in laundry detergents and cleaning products, and inorganic anions, e.g. silicate, arsenate and selenate. Polar non-charged compounds are sorbed into the interlayer. Cations (e.g. Cu(II), Ni(II), Zn(II), Cd(II), Co(II) and Mg(II)) may isomorphically substitute for Fe(II) during GR formation (co precipitation). For instance, nickelous-ferric GR, end products obtained by substitution of Fe(II) ions by Ni(II) ions in GR (Refait & Génin (1993, 1997) and Refait et al. (1994, 1998)). This incorporation of cationic inorganic contaminants into the structures of GR may provide an effective means of sequestering contaminants in the subsurface.

The reactivity of GR is responsible of the removal of cationic and anionic pollutants, by substituting atoms of Fe (II) or (III) or OH radicals.

Overall reaction. In general, hydrogen gas and GR are formed at the cathode as shown in equation (6) [5]:

$$6Fe + (12+x)\, H_2O \rightarrow \tfrac{1}{2}\,(12+x)H_{2(g)}\uparrow + x\; Fe(OH)_3 * (6-X)\; Fe(OH)_{2\,(s)} \qquad (6)$$

6) ***Formation of rust (dehydrated hydroxides) occurs as shown in the following:***

$$2Fe(OH)_3 \leftrightarrow Fe_2O_3 + 3H_2O \qquad \text{(hematite, maghemite)} \qquad (7)$$

$$Fe(OH)_2 \leftrightarrow FeO + H_2O \qquad (8)$$

$$2Fe(OH)_3 + Fe(OH)_2 \leftrightarrow Fe_3O_4 + 4H_2O \qquad \text{(magnetite)} \qquad (9)$$

$$Fe(OH)_3 \leftrightarrow FeO(OH) + H_2O \qquad \text{(goethite, lepidocrocite)} \qquad (10)$$

Hematite, maghemite, rust, magnetite, lepidocrocite and goethite have been identified as EC by-products.

Experimental

The corresponding experiments, used the following: Lab materials, beakers of different capacities, iron electrodes of 59x32x2 mm placed 6 mm apart, Steren MUL-285 multimeters, a model ELI-055 Steren voltage converter 110 ac V – 15 dc V, an Arrow 82608, 0.56 KW four poles interruptor to reverse current, a LabTech Hot plate stirrer; Model LMS-1003, for filtration, 150 mm Whatman # 1 filter papers, an HI/839800 Hanna Digestor, COD Hanna vials and an HI83099 Hanna Photometer for COD determination.

A picture of the experimental setup is shown in figure 1. The wastewater sample is in the beaker size rector; as electrodes, iron plates are used and current flow is reversed every minute if required with a four pole interrupter, to avoid electrode polarization and uneven plate consumption. Plates are connected to an amperimeter and a voltmeter to measure current and voltage.

Figure 1. Experimental setup

All samples were taken from the "Las Fuentes" collector, next to the El Campestre wastewater treatment plant in Torreon; Coahuila, Mexico, and were run depicted in the process flow diagram shown as Figure 2. A sample was also treated with an adsorbent agent (WAA). Three replicates were made with the medium value reported.

THE GRAPHIC ON THIS PAGE WAS NOT AVAILABLE ON THE E-MEDIA VERSION OF THIS PAPER.

Figure 2. Process flow diagram of the experimental tests.

MPN and CFU were determined using EPA standard methods in the Biology Faculty of the Coahuila Autonomus University, Campus Torreon.

RESULTS AND DISCUSSION

Experiment 1: run to assess the quality of the method, on sample 1, for 5 minutes. Parameters and results are shown in Table1.

Table 1. Parameters and results for experiment 1.

	V	A	Conductivity (μS)		pH		Initial COD (mg/lt)	Final COD (mg/lt)	% COD Removal
			Initial	Final	Initial	Final			
	12.58	0.54	923	764	9.07	9.39	245	122	50.24
WAA	-12.96	-0.49		902		9.47		81	66.93

Experiment 2: five minutes residence time showed good results, with experiment run for ten minutes to test the effect of residence time. Parameters and results for this test are presented in Table 2.

Table 2. Parameters and results for experiment 2.

	V	A	Final Conductivity (μS)	Final pH	Initial COD (mg/lt)	Final COD (mg/lt)	COD Removal (%)
			889	9.59	245	123	49.79
WAA	13.4	0.47	842	9.56		74	69.79

Results are similar to those for experiment 1, so it is advisable to run EC for five minutes, but stopping the process every minute and determining COD removal efficiency.

Experiment 3: to determine the effect of residence time on COD removal efficiency. Results can be seen in Table 3 and Figure 3.

The best result, adsorbent agent added, was attained within one minute of residence time, with little difference between trials.

Some samples turned yellowish red, which indicating the presence of excess iron and its oxidation.

Table 3. Determination of the effect of residence time on COD removal (Exp 3)

Residence time t_r (min)	V	A		Conductivity (µS)	pH	Initial COD (mg/lt)	Final COD (mg/lt)	COD Removal (%)
1	13.01	0.53		859	9.2	245	144	41.22
			WAA	917	9.55		82	66.53
2	15.03	0.46		850	9.21		141	42.44
			WAA	908	9.75		84	65.71
3	14	0.46		926	9.21		121	50.61
			WAA	834	9.77		98	60
4	13.24	0.48		908	9.27		136	44.49
			WAA	804	9.91		104	57.55
5	13.23	0.47		912	9.3		127	48.16
			WAA	833	9.68		86	64.89

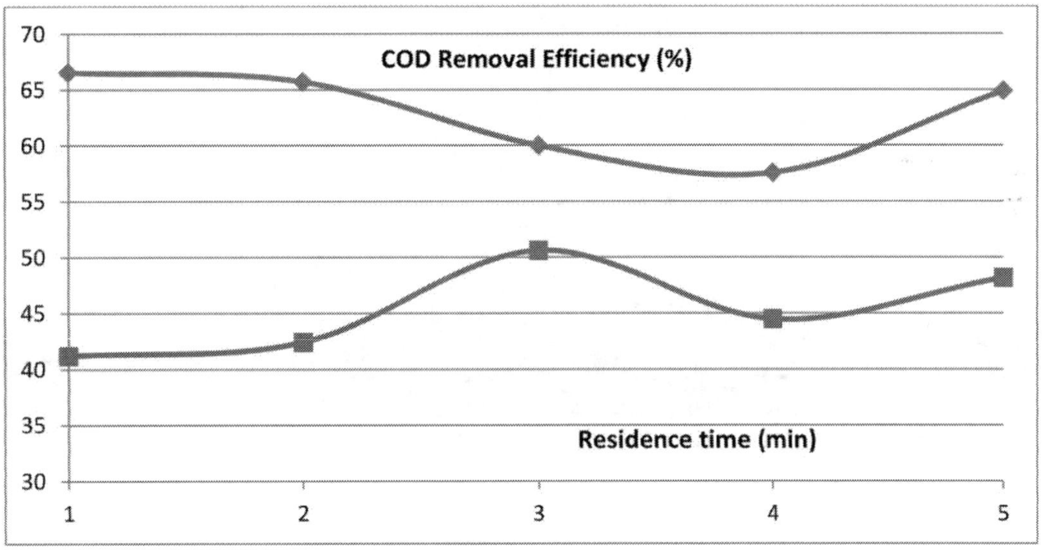

Figure 3. Graphic of COD removal efficiency vs. Residence time for experiment 3.

Experiment 4: To continue residence time on COD removal efficiency, samples were run at 15, 30 and 45 seconds of residence time. Results are seen in Table 4, and Figures 4 - 7 show the process. In figure 4, green rust is seen after EC. Figure 5, shows that after filtration, some color and odor are still present. Figure 6, reveals that after adding adsorbent agent and filtration, color and odor are gone.

Table 4. Determination of the effect of residence time on COD removal (Exp 4)

Residence time t_r (sec)	**V**	**A**	**Watts**		**Conductivity (µS)**	**pH**	**Initial COD mg/lt)**	**Final COD (mg/lt)**	**COD Removal (%)**
15	11.22	0.62	6.95		926	8.56		69	71.83
				WAA	1057	9.59		25	89.79
30	12.08	0.57	6.88		910	8.67		61	75.10
				WAA	1003	10.02	245	14	94.28
45	11.02	0.6	6.61		923	8.75		65	73.46
				WAA	984	10.04		46	81.22

Figure 4. Treated wastewater after EC at 15, 30 and 45 sec (Experiment 4)

Figure 5. Treated wastewater after filtration, from left to right 15, 30 and 45 sec

Figure 6. Treated wastewater after addition of adsorbent agent and filtration , from left to right 15, 30 and 45 sec

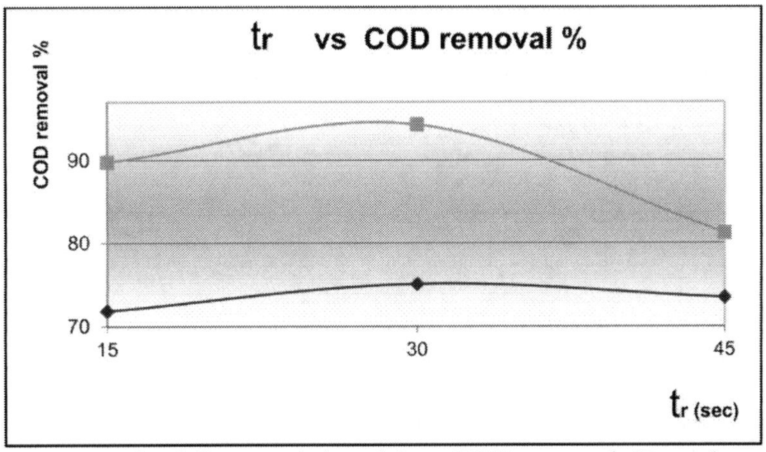

Figure 7. Graph of Removal time (t_r) vs COD removal, (Experiment 4)

Less than a minute was needed to completely treat wastewater. A COD removal efficiency of 94.28 % was achieved within 30 seconds using the adsorbent agent. With these parameters the energy consumption is just 7 watts.

Experiment 5: Three additional samples run at a residence time of 30 seconds with results presented in tables 5 - 7.

Table 5. COD removal with 30 seconds of residence time (Exp 5.1).

Residence time t_r (sec)	V	A		Conductivity (μS)	pH	Initial COD (mg/lt)	Final COD (mg/lt)	COD Removal (%)
30	11.18	0.58		950	8.01	352	187	46.87
			WAA	1000	7.63		80	77.27
30	10.2	0.59		940	8.5	282	122	56.73
			WAA	1000	9.0		65	76.95
30	10.2	0.59		900	7.3	342	92	73.09
			WAA	860	7.27		51	85.08
Average								58.89
			WAA					79.77

Results are consistent with previous ones presented results and with those obtained for biological wastewater treatment plants.

Experiment 6: is designed to determine the removal efficiency of MPN of coliforms in wastewater with EC. A sample was run using the residence time determined in previous experiments. Results are presented in table 6.

Table 6. EC parameters, Initial and final MPN. (Experiment 6)

Residence time t_r (min)	V	A		Conductivity (µS)	pH	Coliforms MPN		Coliforms Removal (%)
						Initial	Final	
30	11.64	0.56		1006	7.42	>2400	1100	54.16
			WAA	985	7.28		480	80

Experiment 7: examines the EC process for CFU, again run for 30 seconds with results shown in Table 7.

Table 7. Parámetros iniciales y finales de la EC de la (Prueba 6)

Residence time t_r (min)		Conductivity (µS)	pH	CFU/ml		CFU Removal (%)
				Initial	Final	
30		900	7.30	2,130,000	21,000	99.01
	WAA	860	7.27		400	99.98

The attained results are excellent as the removal efficiency for CFU is almost 100%.

CONCLUSIONS

It has is demonstrated that using EC in municipal wastewater treatment is efficient and effective for removing COD, Coliforms and CFU. Residence times less than only a minute are required to complete the wastewater treatment, with 30 sec established as an optimal residence time. COD removal is good and comparable to oxidation pond biological treatment where residence times range from 3 to 6 days. Conventional methods, (e.g. biological) also generate odors and a lot of sludges. EC generates just the exact required amount of coagulant. EC wastewater treatment might substitute primary and secondary treatment in conventional treatment plants due versatility and substantially lower residence times.

According to the data seen in table 8, for a residence times between 1 to 10 minutes, COD removal is almost constant (64 % WAA). With residence times of 15 to 45 seconds the mean value for COD removal is 84% with a variance of 48 which makes the results consistent.

For all residence times COD removal is higher WAA.

Table 8 . Comparison of COD removal for various residence times.

Experiment	**Residence time t_r (min)**	**COD removal (%)**		**Experiment**	**Residence time t_r (sec)**	**COD removal (%)**	
			WAA				**WAA**
1	5	50.24	66.93	4	15	71.83	89.79
2	10	49.79	68.79		30	75.1	94.28
3	1	41.22	66.53		45	73.46	81.22
	2	42.44	65.71	5.1	30	73.09	85.08
	3	50.61	60.0	5.2	30	46.87	77.27
	4	44.49	57.55	5.3	30	56.73	76.95
	5	48.16	64.89				
Mean (μ)		46.71	64.06	Mean (μ)		66.18	84.10
Variance (σ^2)		15.43	14.01	Variance (σ^2)		134.88	48.61

The maximum permissible values of pollutants allowed for the official Mexican norm NOM-003-ECOL-1997, for residual treated wastewaters for reuse in services to the public are presented in table 9

Table 9. NOM-003-ECOL-1997

Type of reuse	Monthly average				
	Fecales Coliforms MPN/100 ml	Helminto eggs (e/l)	Oil & grease mg/lt	BOD_5	TSS mg/lt
Service to the public with direct contact.	240	1	15	20	20
Service to the public with occasional or indirect contact.	1000	5	15	30	30

Comparison with an original simple coliform removal efficiency of 80%, (e.g. > 2400 MPN of coliformes) a minimum 480 MPN of coliforms is attained while the norm indicates a maximum value of 1000 for services with indirect or occasional contact to the public. Results comply with the regulations at less than half the maximum value allowed, and treated wastewater may be used for reuse in green areas and industrial process as well.

Objective are accomplished, investigate the application of EC combined with the use of an adsorbent agent for wastewater treatment and show that it is possible to determine parameters for optimum operation, and foundation for industrial application in a continuous process were established.

Acknowledgements

The authors wish to acknowledge the support for this project to the National Council of Science and Technology (CONACYT) and to the Dirección General de Educación Superior Tecnológica (DGEST) from Mexico.

[1] WASTEWATER GENERATION & TREATMENT: DOMESTIC SEWAGE , Published By : Dr. B. Sengupta, Member Secretary, Central Pollution Control Board, Delhi – 32, at ENVIS Centre – 01, Text : Dr. R.C. Trivedi, Sh. R.M. Bhardwaj, Dr. S. Agrawal, Sh. Nazimuddin, Ms. Kokil Malhotra and Sh. Mahendra Pandey]

[2] Environmental and Ecological Chemistry, Edited by : Aleksandar Sabljic, Rudjer Boskovic Institute, Croatia, Part 4. CHEMISTRY OF WASTEWATER, Timothy G. Ellis, Department of Civil, Construction and Environmental Engineering, Iowa State University, Ames, Iowa, USA 2004.

[3] Hector A. Moreno-Casillas , David L. Cocke , Jewel A.G. Gomes, Paul Morkovsky , J.R. Parga, Eric Peterson ; **Electrocoagulation mechanism for COD removal**; Separation and Purification Technology 56 **(2007)** 204–211

[4] G. Chavez Lucia H. "The role of green rust in the environment: a review" Revista Brasileira de Engenharia Agricola e Ambiental, v9, n2, 284-288, 2005.

[5] H. A. Moreno C, D. L. Cocke, J. A. G. Gomes, P. Morkovsky, J. R. Parga, E. Peterson and C. Garcia; Electrochemical Generation of Green Rust using Electrocoagulation; ECS Transactions, 3 (18) 67-78 (2007 10.1149/1.2753225, The Electrochemical Society

(762d) Transport and Remediation of Chromium (VI) In Groundwater Using An Integrated Electrokinetic and Nanoscale Particle Technology

Charlotte Chan, **Ryan Thacher**, Lewis Hsu and Massoud Pirbazari, Sonny Astani Department of Civil and Environmental Engineering, University of Southern California, Los Angeles, CA

Hexavalent chromium [Cr(VI)] is one of the EPA's target contaminants due its prevalence in natural waters in concentrations above levels regarded as safe. When present in soils of low hydraulic conductivity such as clay, conventional treatment technologies become ineffective due to the difficulty and energy consumption in pumping through tightly packed soil. An emerging novel technology, electrokinetic remediation, uses electric potential to transport ionic contaminants through saturated soils and groundwater, eliminating problematic issues associated with state-of-the-art remediation technologies. Strategic placement of electrodes surrounding a contaminated zone with low voltage application facilitates transport of ions through aquifers.

Hexavalent chromium in water is found as the anions chromate, dichromate, or chromic acid, and has been shown to migrate by electrokinetic phenomena towards the anode. At the anode, acidic conditions generated by electrolysis of water and reductive electrochemical conditions promote Cr(VI) reduction to Cr(III), which is a more stable oxidation state with lower solubility and negligible toxicity. Nonetheless, to enhance the remediation process, it is proposed to inject zero-valent iron nanoparticles (nZVI) directly into the subsurface. nZVI have tremendous potential in remediation work due to their high surface area to volume ratio, high reaction rate, and the ability to travel through soil pore space. Furthermore, as compared to macroscale iron particles, nZVI are less expensive and have the added flexibility of being able to inject them into the subsurface at multiple locations and at great depth.

A concern regarding the use of nZVI in natural systems is their tendency to agglomerate and form larger particles, or adsorb to soil particles. This reduces their reactivity and mobility in the subsurface, rendering them potentially inactive. Many conditions affect the degree of agglomeration, such as pH, the presence of multi- or mono-valent cations, and natural organic matter. The likelihood of agglomeration can be determined by the zeta-potential of individual nanoparticles. This is a measure of electrokinetic potential that exists across the interface of all solids and liquids, and as the absolute value of zeta-potential increases, the suspension becomes more stable.

This study explores the use of nZVI in electrokinetic systems for the treatment of Cr(VI) in clay soils. Zeta-potential analysis was performed to determine the stability of surface-stabilized nZVI under experimental conditions. Results indicated a significant negative surface charge on the nZVI, which is increased by the presence of both Cr(VI), and natural organic matter found in all soils and groundwater. The clay soil was also shown to be highly stable under neutral to basic conditions due to a negative surface charge, thereby repelling nZVI in solution.

Integration of nZVI with an electrokinetic soil-column apparatus was conducted to simulate the effectiveness of these two technologies in a natural system. A 20cm column was loaded with a clay soil spiked with Cr(VI), and electric potential was continuously supplied to electrodes housed in reservoirs on either end of the soil column. The reduction and transport of Cr(VI) was investigated with two different methods of nZVI application; as a permeable reactive barrier (PRB) and direct injection. Use of nZVI as PRB effectively reduced Cr(VI) on the cathode-side of the barrier, but had no impact on Cr(VI) on the anode side. Direct injection of nZVI into the electrokinetic cell near the cathode increased reduction rates within the first 72 hours of treatment.

It is found that nZVI can remain stable and dispersed upon injection into a simulated groundwater environment, and move through interstitial space under an electric field to enhance remediation of groundwater and soils contaminated with Cr(VI).

See more of this Session: Water Reuse and Hybrid Water Treatment Processes
See more of this Group/Topical: Topical 1: Water Technology for Developed and Developing Countries (see also Separations Division)

(762e) Starvation of a Mixed Culture of Chlorella Vulgaris and Scenedesmus Sp. for Enhanced Removal of Ammonia From Wastewater Effluent and Its Impact On Lipids Formation

Bala Padmini Lingaraju[1], **Jinsoo Kim**[2], Joo-Youp Lee[1], Kaniz F. Siddiqui[3] and Zhouyang Liu[4], (1) Department of Chemical and Materials Engineering, University of Cincinnati, Cincinnati, OH, (2) Department of Chemical & Materials Engineering, University of Cincinnati, Cincinnati, OH, (3)The Metropolitan Sewer District of Greater Cincinnati, Cincinnati, OH, (4)Environmental Engineering Program, University of Cincinnati, Cincinnati, OH

Starvation of a Mixed Culture of *Chlorella vulgaris* and *Scenedesmus sp*. for Enhanced Removal of Ammonia from Wastewater Effluent and Its Impact on Lipids Formation

Bala P. Lingaraju[1], Jinsoo Kim[1], Joo-Youp Lee[1], Zhouyang Liu[2], Kaniz F. Siddiqui[3]

[1] Chemical Engineering Program, University of Cincinnati, OH 45221-0012

[2] Environmental Engineering Program, University of Cincinnati, OH 45221-0012

[3] The Metropolitan Sewer District of Greater Cincinnati, Cincinnati, OH 45204-2002

Abstract

Microalgae have proven to have the capability to remove nitrogen and phosphorus from wastewater. Investigation of a possible synergy between wastewater treatment and renewable energy production was also considered. However, one of the major barriers against the proposed idea is the slow uptake kinetics. In our preliminary study, a mixed culture of *Chlorella vulgaris and Scenedesmus sp.* starved of nitrogen has shown a faster ammonia uptake rate from secondary wastewater effluent. High density culture and suspended algae has shown faster uptake rate compared to settled algae. In this study, the residual ammonia uptake kinetics has been investigated in a batch system in terms of initial algal seed density using two different seeds grown under nitrogen-deficient and –sufficient conditions in order to determine an initial algal cell density that can achieve 90% ammonia removal within 48 hours. The change in lipids content and profile before and after the ammonia uptake has also been analyzed using solvent extraction followed by gas chromatography. TEM/confocal microscope image analysis and Nile red stain method were also used to validate the change in lipids content.

Keywords: *Chlorella vugaris*, Wastewater treatment, Ammonia removal, Lipids analysis.

See more of this Session: Water Reuse and Hybrid Water Treatment Processes
See more of this Group/Topical: Topical 1: Water Technology for Developed and Developing Countries (see also Separations Division)

(762f) Influent Solids Retained High Biomass Reactor Treating Dairy Manure Wastewater and Kinetics Study

Jingwei Ma, Department of Biological Systems Engineering, Washington State University, Pullman, WA, Liang Yu, Washington State University, Pullman, WA and Shulin Chen, Biological Systems Engineering, Washington State University, Pullman, WA

In large-scale dairies, flushing system is employed as a major means of manure collection for the sake of its efficiency and less labor requirements. The large amount of wastewater needs to be treated for reuse. Anaerobic digestion converts the organic carbon in dairy manure to biogas that can be used as renewable energy to reduce greenhouse and odor gas emission. Flushed dairy manure poses challenges to conventional anaerobic digestion technology. To address this issue, fixed-film digesters with submerged high specific surface area media were introduced to retain active biomass through bacterial attachment. Nevertheless, fixed-film was classified as an inappropriate process for dairy manure anaerobic digestion due to its vulnerability to media clogging. Apparently, a process with no need of exotic media addition would be preferred for anaerobic digestion of flushed dairy manure.

In this study, a strategy using fiber in the dairy manure as biomass carriers for anaerobic digestion of solid wastes was investigated by using anaerobic sequential batch reactor (ASBR) in lab scale, floor scale and pilot scale. Operating parameters, such as settling time, hydraulic retention time (HRT), and organic loading rate (OLR) were evaluated for process optimization. In lab scale experiments, it was revealed that settling time play a key role in biomass retention in form of biofilm. Short settling time favors biofilm formation due to high selection pressure. Bench scale results showed optimum HRT for volumetric methane production rate and methane productivity were 4d and 8 d, respectively. Pilot study confirmed these results and showed 6d HRT should be optimum for ASBR treating flushed dairy manure. A kinetic study was carried out for process control and design. Four basic models, i.e. first-order, Grau, Monod and Chen & Hashimoto were compared with the aim to choose appropriate models. Chen & Hashimoto model based model gave the best simulation with R^2 of 0.99. However, Monod based model showed a poor performance since influent concentration was not include in steady state effluent concentration. At optimized condition (6 d HRT), the volumetric methane production rate of 0.18L/L/d and VS removal of 40% were achieved. It was proved that ASBR using manure fiber as support media not only improved methane production but also reduced the necessary HRT to achieve a similar treating efficiency compared with other study.

See more of this Session: Water Reuse and Hybrid Water Treatment Processes
See more of this Group/Topical: Topical 1: Water Technology for Developed and Developing Countries (see also Separations Division)

AUTHOR INDEX

AUTHOR INDEX

AUTHOR INDEX

AUTHOR INDEX

AUTHOR INDEX

AUTHOR INDEX

AUTHOR INDEX

AUTHOR INDEX

AUTHOR INDEX

AUTHOR INDEX

AUTHOR INDEX

AUTHOR INDEX